# Frontiers in Earth Sciences

Series Editors: J. P. Brun, O. Oncken, H. Weissert, C. Dullo

Onno Oncken
Guillermo Chong
Gerhard Franz
Peter Giese
Hans-Jürgen Götze
Victor A. Ramos
Manfred R. Strecker
Peter Wigger
(Editors)

# The Andes
## Active Subduction Orogeny

With 287 Figures, 159 in color

**Editors**

**Onno Oncken**
Department 3 Geodynamik
GeoForschungsZentrum Potsdam
Telegrafenberg, 14473 Potsdam, Germany
oncken@gfz-potsdam.de

**Guillermo Chong**
Departamento de Geología
Universidad Católica del Norte
Avenida Angamos 610, Antofagasta, Chile
gchong@ucn.cl

**Gerhard Franz**
Petrologie, Institut für Angewandte Geowissenschaften
Technische Universität Berlin
Ernst-Reuter-Platz 1, 10587 Berlin, Germany
gerhard.franz@tu-berlin.de

**Peter Giese**

**Hans-Jürgen Götze**
Abteilung Geophysik, Institut für Geowissenschaften
Christian-Albrechts-Universität Kiel
Otto-Hahn-Platz 1, 24118 Kiel, Germany
hajo@geophysik.uni-kiel.de

**Victor A. Ramos**
Departamento Ciencias Geológicas
Universidad de Buenos Aires
Ciudad Universitaria, 1428 Buenos Aires, Argentina
andes@gl.fcen.uba.ar

**Manfred R. Strecker**
Institut für Geowissenschaften
Universität Potsdam
Karl-Liebknecht-Strasse 24, 14476 Potsdam, Germany
strecker@geo.uni-potsdam.de

**Peter Wigger**
Fachrichtung Geophysik, Institut für Geologische Wissenschaften
Freie Universität Berlin
Malteserstrasse 74–100, 12249 Berlin, Germany
wigger@geophysik.fu-berlin.de

Cover figure:
Bathymetry data: Smith WHF, Sandwell DT (1997) Global seafloor topography from satellite altimetry and ship depth soundings. Science 277:1957–1962. Topography data: SRTM30 NASA. Induction arrows from the South Chilean margin (see Fig. 8.8 of this volume). Seismogram from a North Chilean subduction zone earthquake (recorded 2005 by P. Wigger).

Library of Congress Control Number:   2006933828

Additional material to this book can be downloaded from http://extras.springer.com

| | | |
|---|---|---|
| ISSN | 1863-4621 | Springer Berlin Heidelberg New York |
| ISBN-10 | 3-540-24329-1 | Springer Berlin Heidelberg New York |
| ISBN-13 | 978-3-540-24329-8 | Springer Berlin Heidelberg New York |

This work is subject to copyright. All rights are reserved, whether the whole or part of the material is concerned, specifically the rights of translation, reprinting, reuse of illustrations, recitations, broadcasting, reproduction on microfilm or in any other way, and storage in data banks. Duplication of this publication or parts thereof is permitted only under the provisions of the German Copyright Law of September 9, 1965, in its current version, and permission for use must always be obtained from Springer.  Violations are liable to prosecution under the German Copyright Law.

**Springer is a part of Springer Science+Business Media**
springeronline.com
© Springer-Verlag Berlin Heidelberg 2006

The use of general descriptive names, registered names, trademarks, etc. in this publication does not imply, even in the absence of a specific statement, that such names are exempt from the relevant protective laws and regulations and therefore free for general use.

Cover design: Erich Kirchner, Heidelberg
Typesetting: Klaus Häringer, Stasch · Bayreuth (stasch@stasch.com)
Production: Almas Schimmel

Printed on acid-free paper    30/3141/as – 5 4 3 2 1 0

*In Memoriam*

*Peter Giese*
Prof. (em.) Dr. rer. nat.
1931–2005

On 25 May 2005, aged 73, the tireless and respected Peter Giese, Professor of Geophysics at the Free University of Berlin, passed away. This exceptional scientist initiated the Collaborative Research Centre (SFB) 267, "Deformation Processes in the Andes", and served as Speaker for the SFB from 1993 until his retirement in 1998. The project was successfully completed in 2005 and ranks among Prof. Giese's many outstanding achievements.

Peter Giese was born on 6 August 1931 in Berlin-Kreuzberg. He came to geophysics in a roundabout way, which proved to have considerable significance for his entire scientific life. He began his studies in geology and attained his first degree in 1954. Subsequently, he changed to the field of geophysics and in Munich wrote his Doctoral thesis on Rayleigh waves in the Winter semester of 1956/1957. This was followed by two years in the oil industry, which included work in Turkey.

In April 1959, Prof. Giese began work as an academic assistant at the Institute of Geophysics at the University of Munich. His preferred research area was the study of seismic refraction with emphasis on the European Alps. He held leading positions in several national and international research projects and set new standards in the evaluation and interpretation of measured data. His work in seismic refraction culminated in 1968 with his Habilitation thesis where he presented a model for the Earth's crust under the Alps. This model maintains its validity today.

Peter Giese was appointed full Professor in the Department of Geophysics of the Free University of Berlin in the Summer semester of 1970. Owing to his comprehensive geoscientific background, Prof. Giese was almost predestined to further the reintegration of the individual geoscientific fields, in both research and teaching, with a comprehensive curriculum reform. Teaching and the advancement of the next generation of scientists were always one of his primary concerns, and his lectures and classes were correspondingly highly valued.

Prof. Giese was also strongly active in the reunification of geoscience between East and West Germany through his work as Chairman of the founding committee for the GeoForschungsZentrum (German Research Centre for Earth Sciences) based in Potsdam. In addition, he was substantially involved in two of the largest and most significant geoscientific research programs in Germany: DEKORP (German Continental Reflection Seismology Program) and KTB (Continental Deep Drilling Program).

As a mark of appreciation by the whole geoscientific community, Prof. Giese was awarded the Gustav Steinmann Medal in 1997 by the Geological Association of Germany. In 2004, the German Geophysical Society named him an Honorable Member during their annual conference in Berlin.

After his retirement, Prof. Giese continued to participate in academic life and the work of the institute as one of the last full professors ("Ordinarius"). Only a few weeks before his death in April 2005, the SFB 267 held its International Final Symposium in Potsdam and, despite his illness, Prof. Giese attend the closure event of the SFB 267 for a short time.

With the death of Prof. Peter Giese, we have lost an active, exceptional scientist, a committed academic teacher, a remarkable personality and a great person whom many of us lovingly and respectfully called "Master".

We will always remember him with honor.

# Preface

Convergent plate margins and subduction zones are first-order features shaping the Earth. Convergent continental margins combine the majority of processes that affect the internal architecture thermal and geochemical character of continental lithosphere. In addition, the close relationships between active deformation and uplift, magmatism and associated crustal growth, ore formation, the release of more than 90% of global seismic energy at convergent margins, make these plate boundaries important natural laboratories where mass and energy flux rates can be studied at various scales. Since the advent of plate tectonic theory, it has been recognized that all of these phenomena are intimately related and often governed by feedback mechanisms. Accordingly, subduction orogeny has become an international, high-priority theme in process-oriented, earth-system analysis. In this context, Dewey and Bird (1970) have defined the Andes as the type representative for orogeny and associated processes at convergent margins in their benchmark paper. The Andes, therefore, provide an excellent natural laboratory for studying the above processes.

This rationale has guided the Earth Science departments at the Free University of Berlin, the Technical University Berlin, Potsdam University and the GeoForschungsZentrum Potsdam (GFZ; the German Research Centre for Earth Sciences) in shaping the 12-year collaborative research program (SFB 267) – 'Deformation processes in the Andes'. Since 1993, members of the SFB 267, under the leadership of the late Prof. Peter Giese until 1998, have been actively working on establishing a broad scientific basis for covering a variety of features in the central Andes. The research has encompassed all earth science methodologies, from geophysical imaging to space geodetic monitoring, and from geochemical analysis to geological observations.

Our goal was to develop a quantitative understanding of the mutual relationship between internal and external mechanisms in Cenozoic deformation and plateau-building at the Andean convergent margin. To this end, we analyzed two key areas: the central Andes, which is governed by plateau-style deformation, and the North Patagonian Andes, which is of an entirely different style despite the same plate kinematic conditions. Owing to the differences in architecture and evolution, we consider both areas an ideal setting for testing models and assumptions on subduction orogeny.

The results of this research program have been published (and submitted) in over 200 papers to peer review journals. The present volume summarizes these results and adds more recent insights aimed at contributing to the international discussion on Andean dynamics. There are four main sections, complemented by a collection of materials on DVD. According to the rationale of the program, these sections include: several papers that deliver various large-picture aspects of the Andes; papers on the elements of a subduction system; a series of papers that analyze in more detail the contribution of various 'agents of change' that control subduction orogeny; and, lastly, several papers that provide images of the deep subsurface and numerical models aimed at unraveling the role of various key mechanisms contributing to Andean orogeny.

**Acknowledgments**

The results obtained during this study would never have been possible without the support of a large number of people and institutions. We would, therefore, like to take this opportunity to acknowledge the substantial assistance from which our program has benefited.

First and foremost, this includes our partners in the South American countries of Chile, Bolivia and Argentina, who have been instrumental to the success of the SFB 267. For outstanding logistical help we particularly thank the Universidad Católica del Norte, Antofagasta, and the Universidad de Concepción (both in Chile), the Universidad Autonoma Tómas Frías, Potosi (Bolivia) and the Universidad Nacionál de Salta (Argentina). Much of our work, especially the field operations, have also benefited from the direct or indirect support of a number of institutions in our host countries. These include the governmental institutions of SERNAGEOMIN (Chile), SERGEOMIN (Bolivia), IGM (Chile), and SEGEMAR (Argentina), and also a wide spectrum of industry partners including ENAP, CODELCO, Repsol-YPF, and YPFB.

The generous financial support for this program over its 12 years was provided by the German Science foundation (DFG) and it is acknowledged with particular gratitude. We thank all colleagues from the various review committees for accompanying our program with unceasing readiness to invest considerable amounts of time and for their indulgence, constructive criticism and moral support. SFB 267 project groups at the Earth Sciences institute of the Free University of Berlin also benefited from substantial additional grants provided by the Free University's "Research Commission". Likewise, several groups at the GFZ received additional financial support from that institution. Our thanks also go to a number of funding agencies that have added to the DFG's support by providing both grants for South American students and young scientists from other countries (the German Academic Exchange Service (DAAD) and the Humboldt Foundation) and funds for piggyback projects (the Federal Ministry of Education and Research (BMBF) and the Volkswagen Foundation).

Technical and editorial assistance throughout the production of this volume, a heroic effort that we would particularly like to acknowledge, was provided by Beate Latif, Allison Britt, and Petra Paschke. Last but not least, we are indebted to the large number of scientists who reviewed the papers in this volume. We are grateful for their valuable time and their constructive criticism, which has helped to make the volume what it is. They are:

| | | |
|---|---|---|
| Rick Allmendinger | Muriel Gerbault | Claudio Rosenberg |
| Ronald Amundson | Taras Gerya | Fritz Schlunegger |
| Chris Beaumont | Ingo Grevemeyer | Hans-Ulrich Schmincke |
| Michael Bevis | Robert Hack | Dave Scholl |
| Ulrich Bleil | Adrian Hartley | Hans Seck |
| Christoph Breitkreuz | Francisco Hervé | David Snyder |
| Andrew Calvert | Gregory Hoke | Rubén Somoza |
| José Cembrano | Teresa Jordan | Manuel Suarez |
| Umberto Cordani | Jonas Kley | Stuart Thomson |
| Claire Currie | Heidrun Kopp | Paola Vanucchi |
| John Dehls | Anne Krabbenhöft | Kelin Wang |
| Louis Dorbath | Nina Kukowski | |
| Susan Ellis | Rolf Meißner | and four reviewers who prefer to remain anonymous. |
| Steve Foley | Euan Nisbet | |
| Andrés Folguera | Javier Quinteros | |

The Editors
Berlin and Potsdam 2006

# Contents

**Part I**
**The Big Picture** .................... 1

O. Oncken · D. Hindle · J. Kley · K. Elger · P. Victor · K. Schemmann
**1** Deformation of the Central Andean Upper Plate System –
Facts, Fiction, and Constraints for Plateau Models .................... 3

R. B. Trumbull · U. Riller · O. Oncken · E. Scheuber · K. Munier · F. Hongn
**2** The Time-Space Distribution of Cenozoic Volcanism
in the South-Central Andes: a New Data Compilation
and Some Tectonic Implications .................... 29

G. Franz · F. Lucassen · W. Kramer · R. B. Trumbull · R. L. Romer · H.-G. Wilke
J. G. Viramonte · R. Becchio · W. Siebel
**3** Crustal Evolution at the Central Andean Continental Margin:
a Geochemical Record of Crustal Growth, Recycling and Destruction .................... 45

J. Klotz · A. Abolghasem · G. Khazaradze · B. Heinze · T. Vietor · R. Hackney · K. Bataille
R. Maturana · J. Viramonte · R. Perdomo
**4** Long-Term Signals in the Present-Day Deformation Field of the Central and
Southern Andes and Constraints on the Viscosity of the Earth's Upper Mantle .. 65

C. R. Ranero · R. von Huene · W. Weinrebe · C. Reichert
**5** Tectonic Processes along the Chile Convergent Margin .................... 91

**Part II**
**Elements of the Subduction System** .................... 123

A. Hoffmann-Rothe · N. Kukowski · G. Dresen · H. Echtler · O. Oncken · J. Klotz
E. Scheuber · A. Kellner
**6** Oblique Convergence along the Chilean Margin: Partitioning,
Margin-Parallel Faulting and Force Interaction at the Plate Interface .................... 125

C. Sick · M.-K. Yoon · K. Rauch · S. Buske · S. Lüth · M. Araneda · K. Bataille · G. Chong
P. Giese† · C. Krawczyk · J. Mechie · H. Meyer · O. Oncken · C. Reichert · M. Schmitz
S. Shapiro · M. Stiller · P. Wigger
**7** Seismic Images of Accretive and Erosive Subduction Zones
from the Chilean Margin .................... 147

C. M. Krawczyk · J. Mechie · S. Lüth · Z. Tašárová · P. Wigger · M. Stiller · H. Brasse
H. P. Echtler · M. Araneda · K. Bataille
**8** Geophysical Signatures and Active Tectonics
at the South-Central Chilean Margin .................... 171

D. Völker · M. Wiedicke · S. Ladage · C. Gaedicke · C. Reichert · K. Rauch · W. Kramer
C. Heubeck

**9** Latitudinal Variation in Sedimentary Processes in the Peru-Chile Trench off Central Chile ............................................................. 193

N. Kukowski · O. Oncken

**10** Subduction Erosion – the "Normal" Mode of Fore-Arc Material Transfer along the Chilean Margin? ............................................................. 217

J. Lohrmann · N. Kukowski · C. M. Krawczyk · O. Oncken · C. Sick · M. Sobiesiak · A. Rietbrock

**11** Subduction Channel Evolution in Brittle Fore-Arc Wedges – a Combined Study with Scaled Sandbox Experiments, Seismological and Reflection Seismic Data and Geological Field Evidence ............. 237

## Part III
**Tectonics and Surface Processes – Responses to Change** ................ 263

R. N. Alonso · B. Bookhagen · B. Carrapa · I. Coutand · M. Haschke · G. E. Hilley
L. Schoenbohm · E. R. Sobel · M. R. Strecker · M. H. Trauth · A. Villanueva

**12** Tectonics, Climate, and Landscape Evolution of the Southern Central Andes: the Argentine Puna Plateau and Adjacent Regions between 22 and 30° S ........ 265

E. Scheuber · D. Mertmann · H. Ege · P. Silva-González · C. Heubeck · K.-J. Reutter
V. Jacobshagen

**13** Exhumation and Basin Development Related to Formation of the Central Andean Plateau, 21° S ............................................. 285

K.-J. Reutter · R. Charrier · H.-J. Götze · B. Schurr · P. Wigger · E. Scheuber · P. Giese†
C.-D. Reuther · S. Schmidt · A. Rietbrock · G. Chong · A. Belmonte-Pool

**14** The Salar de Atacama Basin: a Subsiding Block within the Western Edge of the Altiplano-Puna Plateau ............................................. 303

U. Riller · H.-J. Götze · S. Schmidt · R. B. Trumbull · F. Hongn · I. A. Petrinovic

**15** Upper-Crustal Structure of the Central Andes Inferred from Dip Curvature Analysis of Isostatic Residual Gravity ......... 327

M. Haschke · A. Günther · D. Melnick · H. Echtler · K.-J. Reutter · E. Scheuber · O. Oncken

**16** Central and Southern Andean Tectonic Evolution Inferred from Arc Magmatism ............................................. 337

R. I. Hackney · H. P. Echtler · G. Franz · H.-J. Götze · F. Lucassen · D. Marchenko
D. Melnick · U. Meyer · S. Schmidt · Z. Tašárová · A. Tassara · S. Wienecke

**17** The Segmented Overriding Plate and Coupling at the South-Central Chilean Margin (36–42° S) ............................................. 355

T. Vietor · H. Echtler

**18** Episodic Neogene Southward Growth of the Andean Subduction Orogen between 30° S and 40° S – Plate Motions, Mantle Flow, Climate, and Upper-Plate Structure ............................................. 375

J. Glodny · H. Echtler · O. Figueroa · G. Franz · K. Gräfe · H. Kemnitz · W. Kramer
C. Krawczyk · J. Lohrmann · F. Lucassen · D. Melnick · M. Rosenau · W. Seifert

**19** Long-Term Geological Evolution and Mass-Flow Balance of the South-Central Andes ............................................. 401

P. M. Blisniuk · L. A. Stern · C. P. Chamberlain · P. K. Zeitler · V. A. Ramos · E. R. Sobel
M. Haschke · M. R. Strecker · F. Warkus

**20** Links between Mountain Uplift, Climate, and Surface Processes
in the Southern Patagonian Andes ........................ 429

## Part IV
**The System at Depth: Images and Models** .................... 441

G. Asch · B. Schurr · M. Bohm · X. Yuan · C. Haberland · B. Heit · R. Kind · I. Woelbern
K. Bataille · D. Comte · M. Pardo · J. Viramonte · A. Rietbrock · P. Giese†

**21** Seismological Studies of the Central and Southern Andes ........ 443

F. R. Schilling · R. B. Trumbull · H. Brasse · C. Haberland · G. Asch · D. Bruhn · K. Mai
V. Haak · P. Giese† · M. Muñoz · J. Ramelow · A. Rietbrock · E. Ricaldi · T. Vietor

**22** Partial Melting in the Central Andean Crust: a Review of Geophysical,
Petrophysical, and Petrologic Evidence .................... 459

S. Medvedev · Y. Podladchikov · M. R. Handy · E. Scheuber

**23** Controls on the Deformation of the Central and Southern Andes (10–35° S):
Insight from Thin-Sheet Numerical Modeling ................ 475

A. Y. Babeyko · S. V. Sobolev · T. Vietor · O. Oncken · R. B. Trumbull

**24** Numerical Study of Weakening Processes in the Central Andean Back-Arc . 495

S. V. Sobolev · A. Y. Babeyko · I. Koulakov · O. Oncken

**25** Mechanism of the Andean Orogeny: Insight from Numerical Modeling ..... 513

## Part V
**The Andean Information System: Data, Maps and Movies** ........ 537

H.-J. Götze · M. Alten · H. Burger · P. Goni · D. Melnick · S. Mohr · K. Munier · N. Ott
K. Reutter · S. Schmidt

**26** Data Management of the SFB 267 for the Andes –
from Ink and Paper to Digital Databases ................... 539

K.-J. Reutter · K. Munier

**27** Digital Geological Map of the Central Andes between 20° S and 26° S ........ 557

S. Schmidt · H.-J. Götze

**28** Bouguer and Isostatic Maps of the Central Andes ............... 559

W. B. W. Schnurr · A. Risse · R. B. Trumbull · K. Munier

**29** Digital Geological Map of the Southern and Central Puna Plateau,
NW Argentina ........................................... 563

D. Melnick · H. P. Echtler

**30** Morphotectonic and Geologic Digital Map Compilations
of the South-Central Andes (36°–42° S) ..................... 565

K. Munier · J. Levenhagen · H. Burger

**31** Introduction to the Attached DVD ......................... 569

# List of Contributors

*Amir Abolghasem*
Institut für Geologie, Universität Hannover
Appelstrasse 11A, 30167 Hannover, Germany
*abolghasem@geowi.uni-hannover.de*

*Ricardo N. Alonso*
Universidad Nacional de Salta
Buenos Aires 177, 4400 Salta, Argentina
*smresalta@arnet.com*

*Michael Alten*
Institut für Geologische Wissenschaften, Freie Universität Berlin
Malteserstrasse 74–100, 12249 Berlin, Germany
*Jessica-Aileen.Alten@gga-hannover.de*

*Manuel Araneda*
Servicios Geofísicos En Ingeniería y Minería – SEGMI
San Sebastian 2750, Las Condes, Santiago, Chile
*segmi@netexpress.cl*

*Günter Asch*
GeoForschungsZentrum Potsdam
Telegrafenberg, 14473 Potsdam, Germany
*asch@gfz-potsdam.de*

*Andrey Y. Babeyko*
GeoForschungsZentrum Potsdam
Telegrafenberg, 14473 Potsdam, Germany
*babeyko@gfz-potsdam.de*

*Klaus Bataille*
Departamento de Ciencias de la Tierra, Universidad de Concepción
Concepción, Chile
*bataille@udec.cl*

*Raúl Becchio*
Instituto de Geología del Noroeste Argentino, Universidad Nacional de Salta
Buenos Aires 177, 4400 Salta, Argentina
*tato@unsa.edu.ar*

*Arturo Belmonte-Pool*
Santiago, Chile
*belmonte_arturo@yahoo.com*

*Peter M. Blisniuk*
Institut für Geowissenschaften, Universität Potsdam
Karl-Liebknecht-Strasse 24, 14476 Potsdam, Germany
*blisniuk@rz.uni-potsdam.de*

*Mirjam Bohm*
GeoForschungsZentrum Potsdam
Telegrafenberg, 14473 Potsdam, Germany
*mirjam@gfz-potsdam.de*

*Bodo Bookhagen*
School of Earth Sciences, Stanford University
94301 Stanford CA, United States of America
*bodo@pangea.Stanford.EDU*

*Heinrich Brasse*
Fachrichtung Geophysik, Institut für Geologische Wissenschaften, Freie Universität Berlin
Malteserstrasse 74–100, 12249 Berlin, Germany
*h.brasse@geophysik.fu-berlin.de*

*David Bruhn*
GeoForschungsZentrum Potsdam
Telegrafenberg, 14473 Potsdam, Germany
*dbruhn@gfz-potsdam.de*

*Heinz Burger*
Institut für Geologische Wissenschaften, Freie Universität Berlin
Malteserstrasse 74–100, 12249 Berlin, Germany
*hburger@zedat.fu-berlin.de*

*Stefan Buske*
Fachrichtung Geophysik, Institut für Geologische Wissenschaften, Freie Universität Berlin
Malteserstrasse 74–100, 12249 Berlin, Germany
*buske@geophysik.fu-berlin.de*

*Barbara Carrapa*
Institut für Geowissenschaften, Universität Potsdam
Karl-Liebknecht-Strasse 24, 14476 Potsdam, Germany
*carrapa@geo.uni-potsdam.de*

*C. Page Chamberlain*
Department of Geological and Environmental Sciences, Stanford University
94305 Stanford CA, United States of America
*chamb@pangea.stanford.edu*

*Reynaldo Charrier*
Departamento de Geología, Universidad de Chile
Casilla 13518, Correo 21, Santiago, Chile
*rcharrie@cec.uchile.cl*

*Guillermo Chong*
Departamento de Geología, Universidad Católica del Norte
Avenida Angamos 610, Antofagasta, Chile
*gchong@ucn.cl*

*Diana Comte*
Departamento de Geofisica, Universidad de Chile
Av. Blanco Encalada 2085, Santiago, Chile
*dcomte@dgf.uchile.cl*

*Isabelle Coutand*
Université de Lille
Lille, France
*Isabelle.Coutand@univ-lille1.fr*

*Georg Dresen*
Department 3 Geodynamik, GeoForschungsZentrum Potsdam
Telegrafenberg, 14473 Potsdam, Germany
*dre@gfz-potsdam.de*

*Helmut P. Echtler*
Department 3 Geodynamik, GeoForschungsZentrum Potsdam
Telegrafenberg, 14473 Potsdam, Germany
*helle@gfz-potsdam.de*

*Harald Ege*
Institut für Geologische Wissenschaften, Freie Universität Berlin
Malteserstrasse 74–100, 12249 Berlin, Germany

*Kirsten Elger*
GeoForschungsZentrum Potsdam
Telegrafenberg, 14473 Potsdam, Germany

*Oscar Figueroa*
Departamento de Ciencias de la Tierra, Universidad de Concepción
Casilla 160 C, Concepción, Chile
Ofigueroa@udec.cl

*Gerhard Franz*
Petrologie, Institut für Angewandte Geowissenschaften, Technische Universität Berlin
Ernst-Reuter-Platz 1, 10587 Berlin, Germany
gerhard.franz@tu-berlin.de

*Christoph Gaedicke*
Bundesanstalt für Geowissenschaften und Rohstoffe (BGR)
Stilleweg 2, 30655 Hannover, Germany
Gaedicke@bgr.de

*Peter Giese †*

*Johannes Glodny*
GeoForschungsZentrum Potsdam
Telegrafenberg, 14473 Potsdam, Germany
glodnyj@gfz-potsdam.de

*Patrick Goni*
Institut für Geologische Wissenschaften, Freie Universität Berlin
Malteserstrasse 74–100, 12249 Berlin, Germany
patrick_goni@web.de

*Hans-Jürgen Götze*
Abteilung Geophysik, Institut für Geowissenschaften, Christian-Albrechts-Universität Kiel
Otto-Hahn-Platz 1, 24118 Kiel, Germany
hajo@geophysik.uni-kiel.de

*Kirsten Gräfe*
GeoForschungsZentrum Potsdam
Telegrafenberg, 14473 Potsdam, Germany

*Andreas Günther*
Bundesanstalt für Geowissenschaften und Rohstoffe (BGR)
Stilleweg 2, 30655 Hannover, Germany
andreas.guenther@bgr.de

*Volker Haak*
GeoForschungsZentrum Potsdam
Telegrafenberg, 14473 Potsdam, Germany

*Christian Haberland*
Institut für Geowissenschaften, Universität Potsdam
Karl-Liebknecht-Strasse 24, 14476 Potsdam, Germany
haber@geo.uni-potsdam.de

*Ron Hackney*
Abteilung Geophysik, Institut für Geowissenschaften, Christian-Albrechts-Universität Kiel
Otto-Hahn-Platz 1, 24118 Kiel, Germany
rhackney@geophysik.uni-kiel.de

*Mark R. Handy*
Fachrichtung Geologie, Institut für Geologische Wissenschaften
Freie Universität Berlin
Malteserstrasse 74–100, 12249 Berlin, Germany
mhandy@mail.zedat.fu-berlin.de

*Michael Haschke*
University School of Earth, Ocean and Planetary Sciences, Cardiff, UK
HaschkeM@cardiff.ac.uk

*Bertram Heinze*
Helmholtz Association
German-Russian House Moscow
Malaya Pirogovskaya 5, 119435 Moscow, Russian Federation
*bertram.heinze@helmholtz.de*

*Benjamin Heit*
GeoForschungsZentrum Potsdam
Telegrafenberg, 14473 Potsdam, Germany
*heit@gfz-potsdam.de*

*Christoph Heubeck*
Institut für Geologische Wissenschaften
Freie Universität Berlin
Malteserstrasse 74–100, 12249 Berlin, Germany
*cheubeck@zedat.fu-berlin.de*

*George E. Hilley*
Department of Geological and Environmental Sciences
Stanford University
94301 Stanford CA, United States of America
*hilley@pangea.stanford.edu*

*David Hindle*
IfM-GEOMAR
Wischhofstrasse 1–2, 24148 Kiel, Germany
*dhindle@ifm-geomar.de*

*Arne Hoffmann-Rothe*
Bundesanstalt für Geowissenschaften und Rohstoffe (BGR)
Stilleweg 2, 30655 Hannover, Germany
*a.hoffmann-rothe@bgr.de*

*Fernando Hongn*
Instituto de Bio y Geociencias, Universidad Nacional de Salta
Buenos Aires 177, 4400 Salta, Argentina
*hongn@unsa.edu.ar*

*Volker Jacobshagen*
Institut für Geologische Wissenschaften, Freie Universität Berlin
Malteserstrasse 74–100, 12249 Berlin, Germany
*vojac@zedat.fu-berlin.de*

*Antje Kellner*
Department 3 Geodynamik, GeoForschungsZentrum Potsdam
Telegrafenberg, 14473 Potsdam, Germany
*akellner@gfz-potsdam.de*

*Helga Kemnitz*
GeoForschungsZentrum Potsdam
Telegrafenberg, 14473 Potsdam, Germany
*heke@gfz-potsdam.de*

*Giorgi Khazaradze*
Universidad de Barcelona
08003 Barcelona, Spain
*gkhazar@ub.edu*

*Rainer Kind*
GeoForschungsZentrum Potsdam
Telegrafenberg, 14473 Potsdam, Germany
*kind@gfz-potsdam.de*

*Jonas Kley*
Institut für Geowissenschaften
Friedrich-Schiller-Universität Jena
Burgweg 11, 07749 Jena, Germany
*jonas.kley@uni-jena.de*

*Jürgen Klotz*
Department 1 Geodäsie und Fernerkundung
GeoForschungsZentrum Potsdam
Telegrafenberg, 14473 Potsdam, Germany
*klotz@gfz-potsdam.de*

*Ivan Koulakov*
GeoForschungsZentrum Potsdam
Telegrafenberg, 14473 Potsdam, Germany

*Wolfgang Kramer*
Department 3 Geodynamik, GeoForschungsZentrum Potsdam
Telegrafenberg, 14473 Potsdam, Germany

*Charlotte M. Krawczyk*
Department 3 Geodynamik, GeoForschungsZentrum Potsdam
Telegrafenberg, 14473 Potsdam, Germany
*lotte@gfz-potsdam.de*

*Nina Kukowski*
Department 3 Geodynamik, GeoForschungsZentrum Potsdam
Telegrafenberg, 14473 Potsdam, Germany
*nina@gfz-potsdam.de*

*Stefan Ladage*
Bundesanstalt für Geowissenschaften und Rohstoffe (BGR)
Stilleweg 2, 30655 Hannover, Germany
*s.ladage@bgr.de*

*Jörn Levenhagen*
Freie Universität Berlin
Malteserstrasse 74–100, 12249 Berlin, Germany
*leven@zedat.fu-berlin.de*

*Jo Lohrmann*
Department 3 Geodynamik, GeoForschungsZentrum Potsdam
Telegrafenberg, 14473 Potsdam, Germany
*jo@gfz-potsdam.de*

*Friedrich Lucassen*
Department 4 Chemie der Erde, GeoForschungsZentrum Potsdam
Telegrafenberg, 14473 Potsdam, Germany
*lucassen@gfz-potsdam.de*

*Stefan Lüth*
Fachrichtung Geophysik, Institut für Geologische Wissenschaften
Freie Universität Berlin
Malteserstrasse 74–100, 12249 Berlin, Germany
*stefan@geophysik.fu-berlin.de*

*Katrin Mai*
GeoForschungsZentrum Potsdam
Telegrafenberg, 14473 Potsdam, Germany

*Dimitri Marchenko*
GeoForschungsZentrum Potsdam
Telegrafenberg, 14473 Potsdam, Germany
*dm@gfz-potsdam.de*

*Rodrigo Maturana*
Instituto Geográfico Militar
Nueva Santa Isabel 1640, Santiago, Chile
*rmaturana@igm.cl*

*James Mechie*
Department 2 Physik der Erde, GeoForschungsZentrum Potsdam
Telegrafenberg, 14473 Potsdam, Germany
*jimmy@gfz-potsdam.de*

*Sergej Medvedev*
Physics of Geological Processes, University of Oslo
PO Box 1048 Blindern, 0316 Oslo, Norway
*sergeim@fys.uio.no*

*Daniel Melnick*
Department 3 Geodynamik, GeoForschungsZentrum Potsdam
Telegrafenberg, 14473 Potsdam, Germany
*melnick@gfz-potsdam.de*

*Dorothee Mertmann*
Institut für Geologische Wissenschaften, Freie Universität Berlin
Malteserstrasse 74–100, 12249 Berlin, Germany
*mertmann@zedat.fu-berlin.de*

*Heinrich Meyer*
Bundesanstalt für Geowissenschaften und Rohstoffe (BGR)
Stilleweg 2, 30655 Hannover, Germany
*heinrich.meyer@bgr.de*

*Uwe Meyer*
Bundesanstalt für Geowissenschaften und Rohstoffe (BGR)
Stilleweg 2, 30655 Hannover, Germany
*U.Meyer@bgr.de*

*Sabine Mohr*
Fachrichtung Geophysik, Institut für Geologische Wissenschaften
Freie Universität Berlin
Malteserstrasse 74–100, 12249 Berlin, Germany

*Kerstin Munier*
Fachrichtung Geologie, Institut für Geologische Wissenschaften
Freie Universität Berlin
Malteserstrasse 74–100, 12249 Berlin, Germany

*Miguel Muñoz*
Av. Jorge Matte 2005, Santiago, Chile
*mmhgeo@hotmail.com*

*Onno Oncken*
Department 3 Geodynamik, GeoForschungsZentrum Potsdam
Telegrafenberg, 14473 Potsdam, Germany
*oncken@gfz-potsdam.de*

*Norbert Ott*
Fachrichtung Geophysik, Institut für Geologische Wissenschaften, Freie Universität Berlin
Malteserstrasse 74–100, 12249 Berlin, Germany
*drno@zedat.fu-berlin.de*

*Mario Pardo*
Departamento de Geofísica, Universidad de Chile
Av. Blanco Encalada 2086, Santiago, Chile
*mpardo@dgf.uchile.cl*

*Raúl Perdomo*
Universidad Nacional de La Plata
La Plata, Argentina
*perdomo@presi.unlp.edu.ar*

*Ivan Alejandro Petrinovic*
IBIGEO-CONICET and Universidad Nacional de Salta
Mendoza N° 2, 4400 Salta, Argentina
*petrino@unsa.edu.ar*

*Yuri Podlachikov*
Physics of Geological Processes, University of Oslo
PO Box 1048 Blindern, 0316 Oslo, Norway
*y.y.podladchikov@fys.uio.no*

*Juliane Ramelow*
Department 4 Chemie der Erde, GeoForschungsZentrum Potsdam
Telegrafenberg, 14473 Potsdam, Germany
jule@gfz-potsdam.de

*Victor A. Ramos*
Departamento Ciencias Geológicas, Universidad de Buenos Aires
Ciudad Universitaria, 1428 Buenos Aires, Argentina
andes@gl.fcen.uba.ar

*César Rodríguez Ranero*
Instituto de Ciencias del Mar
Paseo Marítimo de la Barceloneta 37–49, 08003 Barcelona, Spain
cranero@icm.csic.es

*Klaus Rauch*
Fachrichtung Geophysik, Institut für Geologische Wissenschaften
Freie Universität Berlin
Malteserstrasse 74–100, 12249 Berlin, Germany
klausheinz.rauch@web.de

*Christian Reichert*
Bundesanstalt für Geowissenschaften und Rohstoffe (BGR)
Stilleweg 2, 30655 Hannover, Germany
christian.reichert@bgr.de

*Claus-Dieter Reuther*
Geologisch-Paläontologisches Institut
Universität Hamburg
Bundesstrasse 55, 20146 Hamburg
reuther@geowiss.uni-hamburg.de

*Klaus-Joachim Reutter*
Institut für Geologische Wissenschaften
Freie Universität Berlin
Malteserstrasse 74–100, 12249 Berlin, Germany
kjreutt@zedat.fu-berlin.de

*Edgar Ricaldi*
Departamento de Física, Universidad Mayor de San Andrés
La Paz, Bolivia
ericaldi@fiumsa.edu.bo

*Andreas Rietbrock*
Department of Earth and Ocean Sciences
University of Liverpool
L693BX Liverpool, United Kingdom
ariet@liverpool.ac.uk

*Ulrich Riller*
Museum für Naturkunde, Humboldt-Universität Berlin
Invalidenstrasse 43, 10115 Berlin, Germany
Ulrich.riller@museum.hu-berlin.de

*Andreas Risse*
Department 4 Chemie der Erde, GeoForschungsZentrum Potsdam
Telegrafenberg, 14473 Potsdam, Germany
risse@gfz-potsdam.de

*Rolf L. Romer*
Department 4 Chemie der Erde, GeoForschungsZentrum Potsdam
Telegrafenberg, 14473 Potsdam, Germany
romer@gfz-potsdam.de

*Matthias Rosenau*
Department 3 Geodynamik, GeoForschungsZentrum Potsdam
Telegrafenberg, 14473 Potsdam, Germany
rosen@gfz-potsdam.de

*Kerstin Schemmann*
Department 3 Geodynamik, GeoForschungsZentrum Potsdam
Telegrafenberg, 14473 Potsdam, Germany
*kschem@gfz-potsdam.de*

*Ekkehard Scheuber*
Fachrichtung Geologie, Institut für Geologische Wissenschaften, Freie Universität Berlin
Malteserstrasse 74–100, 12249 Berlin, Germany
*scheuber@zedat.fu-berlin.de*

*Frank R. Schilling*
Department 4 Chemie der Erde, GeoForschungsZentrum Potsdam
Telegrafenberg, 14473 Potsdam, Germany
*fsch@gfz-potsdam.de*

*Sabine Schmidt*
Abteilung Geophysik, Institut für Geowissenschaften, Christian-Abrechts-Universität Kiel
Otto-Hahn-Platz 1, 24118 Kiel, Germany
*sabine@geophysik.uni-kiel.de*

*Michael Schmitz*
Fundación Venezolana de Investigaciones Sismológicas FUNVISIS
Caracas, Venezuela
*mschmitz@funvisis.org.ve*

*Wolfgang B. Schnurr*
Fachrichtung Geologie, Institut für Geologische Wissenschaften, Freie Universität Berlin
Malteserstrasse 74–100, 12249 Berlin, Germany

*Lindsay Schoenbohm*
Department of Geological Sciences, Ohio State University
Columbus OH 43210, United States of America
*schoenbohm.1@osu.edu*

*Bernd Schurr*
IDC / Scientific Methods, CTBTO
Wagramerstrasse 5, 1400 Vienna, Austria
*bernd.schurr@ctbto.org*

*Wolfgang Seifert*
Department 3 Geodynamik, GeoForschungsZentrum Potsdam
Telegrafenberg, 14473 Potsdam, Germany
*ws@gfz-potsdam.de*

*Serge Shapiro*
Fachrichtung Geophysik, Institut für Geologische Wissenschaften, Freie Universität Berlin
Malteserstrasse 74–100, 12249 Berlin, Germany
*shapiro@geophysik.fu-berlin.de*

*Christof Sick*
Fachrichtung Geophysik, Institut für Geologische Wissenschaften, Freie Universität Berlin
Malteserstrasse 74–100, 12249 Berlin, Germany
*c.sick@gazdefrance-peg.com*

*Wolfgang Siebel*
Institut für Geowissenschaften, Universität Tübingen
Wilhelmstrasse 56, 72074 Tübingen, Germany
*wolfgang.siebel@uni-tuebingen.de*

*Patricio Silva González*
Institut für Geologische Wissenschaften, Freie Universität Berlin
Malteserstrasse 74–100, 12249 Berlin, Germany
*patriciosilva@gmx.net*

*Edward R. Sobel*
Institut für Geowissenschaften, Universität Potsdam
Karl-Liebknecht-Strasse 24, 14476 Potsdam, Germany
*sobel@rz.uni-potsdam.de*

*Monika Sobiesiak*
Department 2 Physik der Erde, GeoForschungsZentrum Potsdam
Telegrafenberg, 14473 Potsdam, Germany
polar@gfz-potsdam.de

*Stephan Sobolev*
Department 2 Physik der Erde, GeoForschungsZentrum Potsdam
Telegrafenberg, 14473 Potsdam, Germany
stephan@gfz-potsdam.de

*Libby A. Stern*
Department of Geological Sciences, University of Texas
Austin TX 78712, United States of America

*Manfred Stiller*
Department 2 Physik der Erde, GeoForschungsZentrum Potsdam
Telegrafenberg, 14473 Potsdam, Germany
manfred@gfz-potsdam.de

*Manfred R. Strecker*
Institut für Geowissenschaften, Universität Potsdam
Karl-Liebknecht-Strasse 24, 14476 Potsdam, Germany
strecker@geo.uni-potsdam.de

*Zuzana Tašárová*
Abteilung Geophysik, Institut für Geowissenschaften, Christian-Albrechts-Universität Kiel
Otto-Hahn-Platz 1, 24118 Kiel, Germany
zuzana@geophysik.uni-kiel.de

*Andrés Tassara*
Departamento de Geofísica, Universidad de Chile
Av. Blanco Encalada 2002, Santiago, Chile

*Martin H. Trauth*
Institut für Geowissenschaften, Universität Potsdam
Karl-Liebknecht-Strasse 24, 14476 Potsdam, Germany
trauth@geo.uni-potsdam.de

*Robert B. Trumbull*
Department 4 Chemie der Erde, GeoForschungsZentrum Potsdam
Telegrafenberg, 14473 Potsdam, Germany
bobby@gfz-potsdam.de

*Pia Victor*
Department 3 Geodynamik, GeoForschungsZentrum Potsdam
Telegrafenberg, 14473 Potsdam, Germany
pvictor@gfz-potsdam.de

*Tim Vietor*
GeoForschungsZentrum Potsdam
Telegrafenberg, 14473 Potsdam, Germany
tvietor@gfz-potsdam.de

*Arturo Villanueva*
Facultad de Ciencias Naturales
Universidad Nacional de Tucumán
Av. Miguel Lillo 205, 4000 Tucumán, Argentina
avillanueva_pe@yahoo.com

*José G. Viramonte*
Instituto de Geología del Noroeste Argentino, Universidad Nacional de Salta
Av. Buenos Aires 177, 4400 Salta, Argentina
viramont@unsa.edu.ar

*David Völker*
Institut für Geologische Wissenschaften, Freie Universität Berlin
Malteserstrasse 74–100, 12249 Berlin, Germany
voelker@zedat.fu-berlin.de

*Roland von Huene*
IfM-GEOMAR
Wischhofstrasse 1–2, 24148 Kiel, Germany

*Frank Warkus*
Institut für Geowissenschaften, Universität Potsdam
Karl-Liebknecht-Strasse 24, 14476 Potsdam, Germany
warkus@geo.uni-potsdam.de

*Wilhelm Weinrebe*
IfM-GEOMAR
Wischhofstrasse 1–2, 24148 Kiel, Germany
wweinrebe@geomar.de

*Michael Wiedicke*
Bundesanstalt für Geowissenschaften und Rohstoffe (BGR)
Stilleweg 2, 30655 Hannover, Germany
Michael.Wiedicke-Hombach@bgr.de

*Susann Wienecke*
Geological Survey of Norway
Leiv Eirikssons vei 39, 7491 Trondheim, Norway
susann.wienecke@ngu.no

*Peter Wigger*
Fachrichtung Geophysik, Institut für Geologische Wissenschaften, Freie Universität Berlin
Malteserstrasse 74–100, 12249 Berlin, Germany
wigger@geophysik.fu-berlin.de

*Hans-Gerhard Wilke*
Departamento de Ciencias Geológicas, Universidad Católica del Norte
Casilla 1280, Antofagasta, Chile
hwilke@ucn.cl

*Ingo Woelbern*
Institut für Geophysik, Johann-Wolfgang-Goethe-Universität Frankfurt am Main
Feldbergstrasse 47, 60323 Frankfurt am Main, Germany
woelbern@geophysik.uni-frankfurt.de

*Mi-Kyung Yoon*
Institut für Geophysik, Universität Hamburg
Bundesstraße 55, 20146 Hamburg, Germany
mi-kyung.yoon @zmaw.de

*Xiaohui Yuan*
Department 2 Physik der Erde, GeoForschungsZentrum Potsdam
Telegrafenberg, 14473 Potsdam, Germany
yuan@gfz-potsdam.de

*Peter K. Zeitler*
Department of Earth and Environmental Sciences, Lehigh University
Bethlehem PA 18015, United States of America
peter.zeitler@lehigh.edu

# Part I
# The Big Picture

**Chapter 1**
Deformation of the Central Andean Upper Plate System – Facts, Fiction, and Constraints for Plateau Models

**Chapter 2**
The Time-Space Distribution of Cenozoic Volcanism in the South-Central Andes: a New Data Compilation and Some Tectonic Implications

**Chapter 3**
Crustal Evolution at the Central Andean Continental Margin: a Geochemical Record of Crustal Growth, Recycling and Destruction

**Chapter 4**
Long-Term Signals in the Present-day Deformation Field of the Central and Southern Andes and Constraints on the Viscosity of the Earth's Upper Mantle

**Chapter 5**
Tectonic Processes along the Chile Convergent Margin

The introductory part comprises five chapters which deal with several of the large-scale and first-order features that characterize the present architecture and the past evolution of the Central and Southern Andes. In the first chapter, O. Oncken et al. present a first complete quantitative assessment of the spatial and temporal pattern of deformation accumulation in the Central Andes. By means of time-series analysis they evaluate the alternative mechanisms that are considered to be most important for driving Andean deformation. They conclude that two factors: the rate of advance of the upper plate, and the rheologic properties of the plate interface, played the key roles in plateau building at the convergent plate margin.

R. Trumbull et al. (Chap. 2) explore the time-space evolution of Cenozoic magmatism in the Central Andes, a key feature of the convergent plate margin, using a new compilation of volcanism and analysis by various statistical methods. They assess regional patterns of volcanic intensity with time, and compare these with the deformation record and with key structural-geometric features of the subduction system. The analysis reveals complex relationships of magmatism with processes operating at both regional and local scales, but there is clearly no direct relationship with upper-plate deformation.

The compositional evolution of the Andean crust since the late Proterozoic is analyzed by G. Franz et al. in Chap. 3, with an emphasis on the issue of crustal growth vs. destruction at the convergent margin. Although convergent margins are commonly controlled by the addition of juvenile crust, the Andean case is shown not to be one of major continental growth, but of long term steady-state, where processes of growth, recycling, and destruction balance out.

J. Klotz et al. (Chap. 4) provide a complete image of the ongoing deformation of the Central to South American Plate margin employing GPS measurements. Their results clearly indicate that elastic deformation associated with the seismic cycle governs the present-day situation in the fore-arc, whereas in the back-arc, the GPS record coincides with the signal of long-term geological deformation, documenting a progressive southward decrease of shortening velocity.

C. Ranero et al. conclude this part from the marine perspective, showing the extreme diversity of the offshore fore-arc architecture from Northern to Southern Chile based mainly on high-resolution geophysical data. Subduction erosion is shown to dominate in the north, whereas the fore-arc processes gradually change to allow build-up of a small accretionary wedge in Southern Chile. These variations result in very different global mass flux budgets and tectonic styles of the fore-arc along the convergent margin.

# Chapter 1

# Deformation of the Central Andean Upper Plate System – Facts, Fiction, and Constraints for Plateau Models

Onno Oncken · David Hindle · Jonas Kley · Kirsten Elger · Pia Victor · Kerstin Schemmann

**Abstract.** We quantitatively analyse the spatial pattern of deformation partitioning and of temporal accumulation of deformation in the Central Andes (15–26° S) with the aim of identifying those mechanisms responsible for initiating and controlling Cenozoic plateau evolution in this region. Our results show that the differential velocity between upper plate velocity and oceanic plate slab rollback velocity is crucial for determining the amount and rate of shortening, as well as their lateral variability at the leading edge of the upper plate. This primary control is modulated by factors affecting the strength balance between the upper plate lithosphere and the Nazca/South American Plate interface. These factors particularly include a stage of reduced slab dip (33 to 20 Ma) that accelerated shortening and an earlier phase (45 to 33 Ma) of higher trenchward sediment flux that reduced coupling at the plate interface, resulting in slowed shortening and enhanced slab rollback. Because high sediment flux and transfer of convergence into upper plate shortening constitute a negative feedback, we suggest that interruption of this feedback is critical for sustaining high shortening transfer, as observed for the Andes. Although we show that climate trends have no influence on the evolution of the Central Andes, the position of this region in the global arid belt in a low erosion regime is the key that provides this interruption; it inhibits high sediment flux into the trench despite the formation of relief from ongoing shortening. Along-strike variations in Andean shortening are clearly related to changes of the above factors. The spatial pattern of distribution of deformation in the Central Andes, as well as the synchronization of fault systems and the total magnitude of shortening, was mainly controlled by large-scale, inherited upper plate features that constitute zones of weakness in the upper plate leading edge. In summary, only a very particular combination of parameters appears to be able to trigger plateau-style deformation at a convergent continental margin. The combination of these parameters (in particular, differential trench-upper plate velocity evolution, high plate interface coupling from low trench infill, and the lateral distribution of weak zones in the upper plate leading edge) was highly uncommon during the Phanerozoic. This led to very few plateau-style orogens at convergent margins like the Cenozoic Central Andes in South America or, possibly, the Laramide North American Cordillera.

## 1.1 Introduction and Geodynamic Framework

To this date, the question of why and how a plateau-type orogen formed with crustal thickening at the leading edge of western South America remains one of the hotly debated issues in geodynamics. During the Cenozoic, the Altiplano and Puna Plateaux of the Central Andes (average elevation some 4 km, with an extent of 400 × 2 000 km) developed during continuous subduction of the oceanic Nazca Plate in a convergent continental margin setting – a situation that is unique along the 60 000 km of convergent margins around the globe. The key challenge is to understand why mechanical failure of the later plateau extent developed along the central portion of the leading edge of South America only, as well as why and how this feature developed only during the Cenozoic, although the cycle of Andean subduction has been ongoing since at least the Jurassic.

Since the 1980s, a plethora of models has been published that attempt to find a solution to this 'geodynamic paradox' (Allmendinger et al. 1997). These fall into distinct classes with respect to the key mechanisms, although some of the models involve a combination of several processes. Suggested mechanisms include

- upper plate deformation related to changes in plate convergence parameters and plate reorganization (e.g. Pardo-Casas and Molnar 1987; Coney and Evenchick 1994; Scheuber et al. 1994; Silver et al. 1998; Somoza 1998);
- changes in the geometry and properties of the downgoing slab, including topographic features of the oceanic plate (ridges, oceanic plateaus), with both changes affecting the mechanics of the upper plate system (e.g. Gephart 1994; Giese et al. 1999; Gutscher et al. 2000a, 2000b; Yañez et al. 2001);
- plate-scale patterns of sub-lithospheric mantle flux related to upper plate motion and deformation (e.g. Russo and Silver 1996; Silver et al. 1998; Marrett and Strecker 2000; Heuret and Lallemand 2005);
- mantle-driven thermal processes affecting the upper plate lithosphere (e.g. Isacks 1988; Wdowinski and Bock 1994; Allmendinger et al. 1997; Mahlburg Kay et al. 1999; ANCORP-Working Group 2003; Garzione et al. 2006);
- spatial and temporal variations in the strength and properties of the upper plate foreland and the plate interface (e.g. Allmendinger and Gubbels 1996; Hindle et al. 2002; Lamb and Davis 2003); and
- climate-related variations in surface erosion affecting deformation through surface material flux (e.g. Masek et al. 1994; Horton 1999; Montgomery et al. 2001; Lamb and Davis 2003; Sobel et al. 2003).

The only accessible information that is available for evaluating these partly competing models is the response of the upper plate to these geodynamic processes through time. Among other aspects, virtually all of the above models rely on the observed quantity, distribution, and timing of crustal deformation and thickening of the Central Andes as a key element. Yet, although a wealth of related information on the Andes is available, its nature is highly heterogeneous and has never been summarized systematically, with the exception of the distribution of shortening (Kley and Monaldi 1998). Only for selected transects have the magnitude of shortening, structures, and timing of deformation been jointly evaluated with higher resolution (Jordan et al. 1993; Echavarria et al. 2003; Victor et al. 2004; Elger et al. 2005).

These latter studies provide an important advance in understanding accumulation of deformation, but have not resolved the above controversy. Hence, a more precise spatial and temporal resolution of deformation periods and strain partitioning patterns across the entire plateau, and their effect on crustal thickening, is still required. The Central Andes lend themselves to such an analysis because of the unique preservation of syn- to post-tectonic deposits in local- to large-scale basins throughout the entire orogen. These deposits often have well established ages (through dating of volcanic deposits) and cross-cutting relationships with neighbouring or underlying structures. This situation permits an assessment of the accumulation of deformation which is unparalleled in its detail when compared to nearly all other orogenic belts.

In this paper, we attempt to systematize published data on shortening and timing of deformation between 15 and 26° S (Fig. 1.1) with the aim of identifying spatial and temporal patterns of deformation partitioning and shortening rates along the entire plateau. The focus of our discussion is on the best-studied area, the Altiplano domain between 18 and 22° S, with only some general inferences for the less well-analyzed domains north and south. We use these data to quantitatively constrain crustal deformation in three dimensions over several time steps. We also derive time series of deformation accumulation in cross section. We evaluate the consequences for the pattern of crustal thickening, formation of potential topography, and sediment dispersal and deposition, using the geological record as an additional constraint. Finally, we use all these quantitative data to investigate their relationships with various other time series data such as plate kinematic parameters, magmatism, and climate. This might aid the assessment of conflicting theories and unveil key mechanisms, and their roles, in the formation of the Andean Plateau.

## 1.2 Distribution of Deformation

The Central Andes between 10 and 30° S consist of major structural units which approximately parallel the trend of the mountain range. In the center, where the Andes attain their maximum width, they comprise, from west to east, (*1*) the offshore and onshore fore-arc region, (*2*) the Western Cordillera that marks the location of the presently active magmatic arc, (*3*) the Altiplano-Puna high plateau, and (*4*) an eastern belt of fold and thrust structures comprising the Eastern Cordillera, the Sub-Andean Ranges and the Sierras Pampeanas (Fig. 1.1). Except for the Altiplano-Puna Plateau that begins near 14° S in southern Peru and terminates near 27° S in northern Argentina as well as the Sierras Pampeanas (including the Santa Barbara System; Fig. 1.1) that do not continue north of 23° S, these units can be traced with somewhat varying character along this entire segment of the Andean chain.

The geology of the fore-arc region is complicated by earlier magmatic arcs and associated sedimentary basins of Jurassic to Paleogene age. It records at least two pre-Neogene shortening events, the Cretaceous Peruvian stage and, most notably, the late Eocene Incaic phase. The structural style of these early contractional phases varies from major low-angle overthrusts (Mégard 1984; Carlotto et al. 1999) to thick-skinned inversion structures with an important strike slip component (Reutter et al. 1996; Haschke and Guenther 2003).

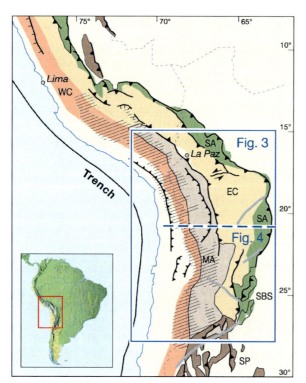

**Fig. 1.1.** Geological sketch map showing the main units building the Central Andes. *WC:* Western Cordillera; *EC:* Eastern Cordilleras; *SBS:* Santa Barbara System; *SP:* Sierras Pampeanas; *SA:* Subandean Ranges; *MA:* present magmatic arc (*hatched area*). *Inset* shows location of Central Andes

The Neogene tectonism of the fore-arc area varies in space and time. Late Neogene to Recent tectonic erosion has induced subsidence and dominantly normal faulting in the offshore fore-arc (Clift et al. 2003; von Huene and Ranero 2003). The onshore fore-arc in Chile also shows normal faulting with uplift in the west (Coastal Cordillera) and thrust or reverse faults of limited throw in the east (Hartley et al. 2000; Kuhn 2002; Victor et al. 2004). This eastern onshore fore-arc region underwent a relatively uniform oceanward tilt in the Neogene, forming the "western monocline" or Altiplano west flank (Isacks 1988; Victor et al. 2004).

Young volcanics cover almost all the Western Cordillera, which apparently experienced little Neogene deformation (Scheuber and Reutter 1992). Both the Altiplano-Puna and the Eastern Cordillera are underlain by late Proterozoic to Paleozoic strata that were affected by one or more folding events of regionally differing ages before the Cretaceous (Kennan et al. 1995; Jacobshagen et al. 2002). Cenozoic, west- and east-vergent thrusting and folding was coeval with the uplift of the Eastern Cordillera and the deposition of thick syntectonic strata on the Altiplano-Puna. The basal detachment to this deformation may be thermally or stratigraphically controlled.

The Sub-Andean Ranges form a thin-skinned foreland fold-and-thrust belt from Peru to northernmost Argentina. Its stratigraphically-controlled detachment horizons vary, but are mostly located in Paleozoic shales near the base of a thick, conformable sedimentary cover. From northern Argentina southward, the structural style of foreland thrusting changes first to thick-skinned inversion structures (Cristallini et al. 1997; Kley and Monaldi 2002) and then to large-scale basement thrusts in the Sierras Pampeanas and along the margin of the Puna. Similar foreland basement thrusts of smaller magnitude also occur in Peru, e.g. the Shira uplift.

The Central Andes now have a relatively dense network of balanced cross sections enabling the estimation of the areal distribution of regional strain in the mountain range. A considerable number of new, cross section-based shortening estimates have been published since an earlier compilation by Kley and Monaldi (1998), particularly for the Altiplano and Eastern Cordillera. Even so, most cross sections cover the Sub-Andean foreland belt and transects over the entire Andes are rare.

A problem with any compilation of balanced section estimates is the inherent subjectivity of the techniques used by different authors. In order to get conservative estimates, we checked many published cross sections and subtracted any shortening attributed to large-scale doubling of stratigraphic packages or large buried duplexes, except where well documented. We also used strain compatibility along strike as an important additional criterion when assessing the plausibility of estimates. This practically rules out the extremely high estimates of bulk contraction recently proposed (McQuarrie 2002), and also some very low estimates of shortening in the Altiplano-Puna and Eastern Cordillera. All shortening estimates are summarized in Table 1.1 and shown in plan view in Fig. 1.2. Total post-Eocene shortening in the widest part of the Central Andes is about 250–275 km and appears to vary little between 17° and 21° S. To the northwest, and to the south of this central segment, shortening decreases rapidly to a poorly constrained value of some 120 km.

Contraction by folding and thrusting is the dominant mode of Cenozoic deformation in the Central Andes. However, a Paleogene to early Neogene phase of normal faulting and transtension has been described from the Altiplano and its western border. The extension produced by this event probably did not exceed a few kilometers (Jordan and Alonso 1987; Elger et al. 2005) and is therefore not included in our quantitative assessments. During the main late Oligocene to Recent contractional phase, the shortening gradients on both limbs of the Andean arc were, in part, achieved by large, oblique strike slip or transfer zones (de Urreiztieta et al. 1996; Müller et al. 2002; Riller and Oncken 2003). The most clear-cut of these structures is the sinistral Cochabamba fault zone (Kennan et al. 1995; Sheffels 1995), while the others are broad bands of distributed shear (de Urreiztieta et al. 1996; Kley 1999).

Strike-slip motion on these fault zones is estimated at as much as several tens of kilometers. They have also accommodated some orogen-parallel extension caused by

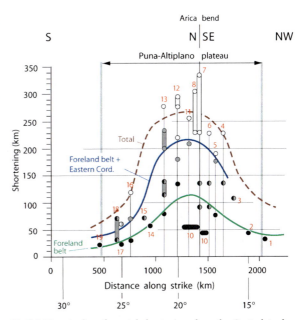

**Fig. 1.2.** Magnitudes of crustal shortening along the Central Andes according to various published estimates. *Thin vertical lines* indicate transects where total shortening estimates have been made or can be compiled. *Numbers* of data points/transects are keyed to Table 1.1. *Thick curves* are our smoothed average estimates for the magnitudes and variations of shortening in the foreland belt, foreland belt plus Eastern Cordillera, and entire orogen

**Table 1.1.** Distribution of shortening between 14 and 27° S

| No. | °S lat. | Pacific piedmont | Altiplano-Puna | Eastern Cordillera | Foreland belt | Total | Total, preferred | Type of data | Reference |
|---|---|---|---|---|---|---|---|---|---|
| 1 | 14 | | | | 31[a] | | | Balanced cross section | Gil et al. 1999 |
| 2 | 15 | | | 33[a] | 42[a] | >75 | | Balanced cross section | Carlotto 1998; Gil et al. 1999 |
| 3 | 16 | | | | 105 | | | Balanced cross section | Roeder and Chamberlain 1995 |
| 4 | 16.5 | | | | 90[a] – 135 | 195[a] – 230 | 210 | Balanced cross section | Baby et al. 1989; Roeder 1988 |
| 5 | 17 | | 14 | 103 | 74 | 191 | | Balanced cross section | Baby et al. 1997 |
| 6 | 17.5 | | | | 90[a] – 135 | 195[a] – 230 | | Balanced cross section | Roeder 1988 |
| 7 | 18 | | 9 | >132 | 90[a] – 135 | >231 – >276 | 255 – 275 | Balanced cross section | McQuarrie and DeCelles 2001; Roeder and Chamberlain 1995 |
| 8 | 18.5 | | 47 | >142, c. 157 | 40 – 60 | >229, c. 254 | | Balanced cross section | Baby et al. 1993; McQuarrie and DeCelles 2001; Sheffels 1990 |
| 9 | 18 – 18.5 | | >35 | >35 | 40 – 60 | >75 | | Structural cross section | Baby et al. 1993; Lamb and Hoke 1997 |
| 10 | 17.5 – 19.5 | | | | 40 – 60 | | | Balanced map | Baby et al. 1993 |
| 11 | 19 | | (42 – 47) | 210 | 210 | 252 – 257 | | Balanced cross section | Sheffels 1990 |
| 12 | 20 | | 42 – 60 | >43 | 136 | >221 | | Balanced cross section | Hérail et al. 1990; McQuarrie and DeCelles 2001; Elger et al. 2005 |
| 13 | 21 | 3 | 60 | 35 – 95 | 115 – 140 | 191 – 276 | | Balanced cross sections; reflection seismics | Baby et al. 1992; Dunn et al. 1995; Elger et al. 2005; Kley 1996; Müller et al. 2002; Victor et al. 2004 |
| 14 | 22.5 | | (38 – 50) | (40 – 95) | 55 – 60 | 120 – 200? | 160 | Reflection seismic; balanced cross section | Allmendinger and Zapata 2000; Echavarria et al. 2003; Kley and Monaldi 1999 |
| 15 | 23 | | (38 – 50) | 70 | 70 | | 120 | Balanced cross section | Kley (unpublished data) |
| 16 | 24 | | 50 | 40? | 30 | 120? | | Balanced cross sections, structural cross section | Cladouhos et al. 1994; Kley and Monaldi 2002 |
| 17 | 25 | | 37 | 37 | 21 | >58 | | Balanced cross sections, structural cross section | Coutand et al. 1999; Kley and Monaldi 2002 |
| 18 | 25.5 | | c. 30 | 30 – 70 | 30 – 70 | 60 – 100? | 80 | Balanced cross sections | Cristallini et al. 1997; Grier et al. 1991 |
| 19 | 27 | | (c.30) | | 20 | >20 | 50 | Fault length/ displacement ratios | Allmendinger 1986; Jordan and Allmendinger 1986 |

( ): Interpolated/extrapolated from adjacent transects.
[a] Measured from cross-section (no value given in ref.) or alternative estimate (this chapter).

the slightly divergent motion of the thrust belts on both limbs of the Andean arc. Minor strike slip and normal faults have affected the northern and southernmost limbs of the plateau from the late Miocene to Pliocene (Mercier et al. 1992; Cladouhos et al. 1994; Marrett et al. 1994; Allmendinger and Gubbels 1996; Marrett and Strecker 2000). Large-scale, active, orogen-parallel normal faults with throws of up to several kilometers occur only in the high Cordilleras of Peru (Dalmayrac and Molnar 1981; Mercier et al. 1992; McNulty and Farber 2002) and peter out along strike towards the Altiplano Plateau.

## 1.3 Timing of Deformation

Deformation in the units described above has accumulated during various periods of complex shortening and plateau formation. Figure 1.3 summarizes our current knowledge of the deformation periods in the Andes between approximately 15 and 27° S in a series of maps showing time windows that were found to be appropriate for imaging the orogen-scale partitioning mode. We compiled sources that described in detail the age relationships

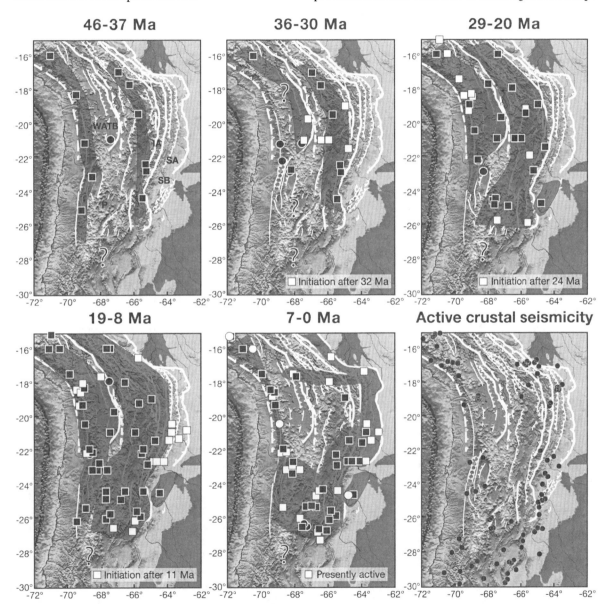

**Fig. 1.3.** Temporal pattern of deformation in Central Andes. *Squares* show locations with documented shortening deformation ages, *circles* indicate oblique extensional to transtensional kinematics (see electronic supplements for complete databank and sources). *White squares and circles* in diagram 7–0 Ma mark areas that are presently active, but might have started earlier than recently. *Shaded areas* are interpreted zones of contemporaneous deformation based on structural and topographic continuity. Crustal seismicity (hypocenters < 50 km, excluding plate interface) with magnitudes $M_w$ > 4.5 from database by Engdahl et al. (1998)

between growth deposits and structures based on stratigraphic and structural analysis, isotopic age dating, fission track analysis, and estimates of the duration of shortening based thereon (see electronic supplement for database in file 'Deformation Data Bank').

The maps show a distinct evolution with an initial stage (46–37 Ma) involving shortening only in the Precordillera/Western Cordillera and along the axis of the Eastern Cordillera. This well established Incaic event has low shortening magnitudes with significant oblique slip in the west (Reutter et al. 1996; Müller et al. 2002; Haschke and Guenther 2003; Ege 2004; Horton 2005). Also, it started earlier in the west (46 Ma) than in the east (40 Ma). Subsequently (36–30 Ma), shortening only affected the axis of the Eastern Cordillera. During the final stage of this period, two domains of shortening in the west and east were again established.

The first main stage of Andean shortening (29–20 Ma) followed the above stage ('Andean crisis' of Sempere et al. 1990). This affected the entire plateau from its western flank (Muñoz and Charrier 1996; Charrier et al. 2002; Victor et al. 2004; Farías et al. 2005), across the central plateau (Baby et al. 1997; Lamb and Hoke 1997; Rochat et al. 1999; Elger et al. 2005), to the entire Eastern Cordillera (Sempere et al. 1990; Müller et al. 2002; Horton 2005). During this period, deformation also expanded southwards into the Puna domain (e.g. Coutand et al. 2001). In the period between 19 and 8 Ma, this pattern continued with increasing deformation in the Puna while that in the Altiplano slowed and motion on the east flank of the plateau began. In the final period (7–0 Ma), deformation shifted to the flanks of the Altiplano Plateau – with high shortening rates only at the eastern flank and in the still active Puna domain. Current deformation with ongoing seismicity is entirely focussed on the margins and limbs of the Central Andes and is consistent regionally with domains of young geological deformation.

These images highlight several fundamental consequences, all of them of key relevance for dynamic models and some of them in contrast to conventional wisdom. These are:

1. the present shape of the plateau (as defined by the 3 500 m contour) was already defined by initial deformation in the Paleogene though plateau elevations were not attained until much later;
2. along strike, the extent of domains deforming coevally (data sampled in time windows larger than 5 Ma) is in the order of magnitude of the plateau itself and indicates apparent along-strike synchronization between at least 15 and 23° S;
3. deformation started earlier than often stated in many parts of the system; and
4. deformation did not spread regularly across strike from any part of the system. Instead, the pattern indicates several subsystems extended along strike that accumulated shortening partly independently from each other with across strike spreading and coalescence only in the final stages.

This low resolution image of partitioning is complemented by a detailed analysis of shortening and shortening rates across the plateau at 21° S ± 1° (Fig. 1.4). Using the data on horizontal shortening in this area, as established by Müller et al. (2002), Kley and Monaldi (2002), Echavarria et al. (2003), Victor et al. (2004), and Elger et al. (2005), as well as recently reported ages from synkinematic growth deposits and unconformities (see above authors and Ege 2004; Horton 2005), we analyzed the evolution of shortening in more detail slightly modifying the strategy originally defined by Jordan et al. (1993; for details see Victor et al. 2004, and Elger et al. 2005). Applying this strategy, we also recalculated the data of Echavarria et al. (2003) measured for the Subandean belt at 22° 30' S where shortening (i.e. 60 km) is less than the 90 km of shortening reported at 21° S (Kley 1996; Baby et al. 1997); recalculation involved data projection along strike to the section at 21° S based on the structural continuity of the faults and folds analysed and correction for larger shortening values. Using sedimentation rates from growth strata thicknesses and available isotopic ages, we estimated onset and end of activity for each structure neighbouring the deposits. Along with the related shortening, we calculated average shortening rates for every structure by evenly distributing the shortening measured for each structure over its period of maximum and minimum activity. To estimate the maximum or minimum activity, we added, respectively subtracted, the errors in the isotopic age to the ages at the bases and tops of syntectonic deposits as estimated from sedimentation rates. In addition, we estimated the potential error in determining shortening at ±10% (+20% in less well constrained sections, i.e. the Eastern Cordillera). Summing these numbers for all kinematically-linked faults provides minimum shortening rates (minimum shortening and maximum period of fault activity) and maximum shortening rates (maximum shortening and minimum period of fault activity) for a thrust system.

These data were collected separately for all individual structures building the major structural units of the Southern Central Andes: the Chilean Precordillera, Western Altiplano, Central Altiplano, Eastern Altiplano, Eastern Cordillera and Interandean zone, and Subandean ranges. Figure 1.4 shows the summed results for each structural unit and the estimated periods of activity for each major fault, from west to east. In addition to the above results in plan view (Fig. 1.3), two main stages of shortening can be identified across the Altiplano up to the western border of the Eastern Cordillera. Following a stage of Eocene shortening, confined to the Precordillera and Western Cordillera, each with local shortening rates

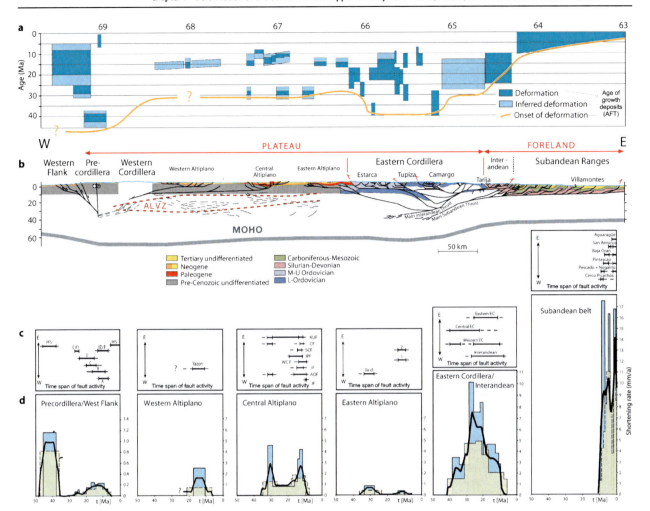

**Fig. 1.4.** Distribution of deformation ages across the Southern Central Andes (21° S) based on published and own data (modified from Elger et al. 2005). **a** Compilation of deformation ages: Western Flank (Victor et al. 2004), Precordillera (Haschke and Günther 2003), Altiplano (Elger et al. 2005; Ege 2004; Silva-González 2004), Eastern Cordillera (Gubbels et al. 1993; Müller et al. 2002), Interandean (Kley 1996; Ege 2004), and Subandean (Kley 1996). **b** Balanced cross section at 21° S compiled from Victor et al. (2004; Altiplano West Flank), Elger et al. (2005; Altiplano), and Müller et al. (2002, Eastern Cordillera and Subandean), Moho and Andean Low Velocity Zone (ALVZ) from receiver function data (Yuan et al. 2000). Line drawing in the middle crust indicates locations of strong reflectivity in the ANCORP seismic line (cf. ANCORP working group 2003). **c** Estimates of maximum and minimum periods of active faulting and related folding in the units building the Altiplano. **d** Cumulative shortening rates of the thrust systems building the plateau calculated from fault activity periods and heave; bold line delineates average shortening rate based on sliding average of three million year sampling width

around 1 mm yr$^{-1}$, the entire western and central Altiplano experienced shortening in the Oligocene (35–25 Ma) with local shortening rates of 0.1 to 3.0 mm yr$^{-1}$ (not resolved in Western Altiplano for lack of exposure). This stage was followed by an early Miocene lull in deformation, but shortening resumed with distinct local acceleration to some 1.7–3.0 mm yr$^{-1}$ (west and center) in the middle to late Miocene (20–10 Ma). In most of the Altiplano, rates were higher during the later stage while the Eastern Altiplano exhibits the opposite relationship. At 7–8 Ma, deformation ceased nearly everywhere except for the continuation of very slow dextral slip on the Precordilleran Fault System.

In contrast to this evolution, the Eastern Cordillera deformed over a long time span starting some 40 million years ago (cf. Müller et al. 2002; Ege 2004; Horton 2005 for relevant data). Deformation spread from its center to the east and west with a maximum shortening rate of 6–9 mm yr$^{-1}$ between 30 and 17 Ma. Deformation practically ceased between 12 and 8 Ma, as evidenced by the formation of the San Juan del Oro surface (Gubbels et al. 1993; Kennan et al. 1995). During this period, the Subandean fold and thrust belt was initiated. It propagated eastwards somewhat discontinuously, as recently established by Echavarria et al. (2003), with shortening rates at some 8–14 mm yr$^{-1}$ and with maximum rates occurring around 7 and 2 Ma. The current shortening rate at this latitude, established by GPS measurements, is 9 ± 1.5 mm yr$^{-1}$ (Bevis et al. 2001; Klotz et al. 2001) and is focused nearly entirely at the present deformation front. Since elastic strain accumulation from interseismic locking of the plate boundary is negligible in this back-arc area this value is close to the present shortening rate.

Superposing and adding the results from the individual thrust systems shows the temporal partitioning of shortening rates across the central Andes, as well as the bulk evolution of the shortening rate. We recognize five stages, each of which is distinct for its main locus of deformation (Fig. 1.5). In addition, Elger et al. (2005) suggested that this procedure and balancing results indicate that the western and central parts of the system are synchronized and kinematically coupled, probably by a joint detachment system. Moreover, these parts are not in phase with the Eastern Cordillera and Subandean belt (Fig. 1.5a), but rather seem to have operated alternately. From this distinct autonomy of the two belts, each built from a system of kinematically-linked and simultaneously active faults, and from their separation by an undeformed basin system, Elger et al. (2005) concluded that there are two, largely independent, thrust belt systems. This interpretation is substantiated by recent balanced sections assembled across the Andes at this latitude and shown in Fig. 1.4b (cf. also the two belts of deformation apparent in Fig. 1.3).

The first stage (approximately 46 to 30 Ma) of bulk shortening in the Southern Central Andes (Fig. 1.5b) was at a rate of 0–8 mm yr$^{-1}$, followed by a stage between 30 and 10 Ma with more rapid and strongly fluctuating rates that ranged from 5–10 mm yr$^{-1}$. From 10 Ma, the shortening rate was 6–14 mm yr$^{-1}$ until the present value of 9 mm yr$^{-1}$ was attained. Since the map pattern (Fig. 1.3) indicates significant synchronization of domains along the plateau (sampling over time scales of 5–10 Ma), we consider the above results from 21° S as representative for the entire Altiplano between 15 and 23° S, with along-strike rate changes probably being related to equivalent minor changes in shortening magnitude (see above).

This argument, however, is probably only valid for fluctuations over long time spans (> 5–10 Myr) as smaller fluctuations are more prone to error or, if real, may reflect local variations in the activity of individual structures along strike. However, sedimentary cycles in the nearby Chaco foreland basin clearly indicate stages of enhanced erosion around 9–7 Ma, and again since 2–3 Ma (Echavarria et al. 2003). Similar local fluctuations in sedimentary cycles, typically encompassing 1–3 million years, were also found by Elger et al. (2005) in the syntectonic deposits of the southern Altiplano. The number of such cycles varies among neighbouring basins clearly indicating that they do not correlate across the plateau. Hence, all observations support the existence of short-term fluctuations in the bulk shortening rate that may only partly be synchronized, superposed on long wavelength fluctuations and the general trend that exhibits long-term synchronicity along strike the primary geological units.

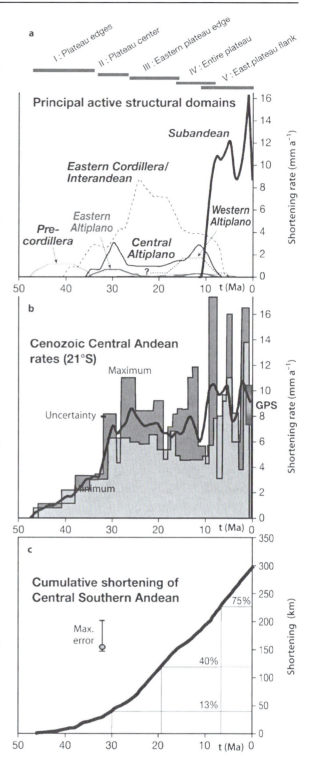

**Fig. 1.5.** Cumulative shortening and shortening rate evolution of Southern Central Andes (21° S). **a** Stacked average shortening rates of individual thrust systems building the Southern Altiplano (from Fig. 1.4d) and stages of principal deformation; **b** summed minimum and maximum shortening rates; bold line depicts average based on sliding three million year average, **c** cumulative shortening

## 1.4 Crustal Thickening and Elevation

Based on the pattern of upper crustal shortening and its accumulation history, we next explore the associated pattern of lower crustal deformation, erosion and sedimentation, and total crustal thickening. As additional constraints, we employ the distribution of present-day crustal thickness (Yuan et al. 2002), topography, exhumation pattern as deduced from fission track data (Benjamin et al. 1987; Ege 2004), and the location of sedimentary basins along with sediment dispersal directions (Horton et al. 2001).

The large deficit in shortening required to produce the present-day topography and geophysically-observed range in crustal thickness has long been recognized (Isacks 1988; Allmendinger et al. 1997; Kley and Monaldi 1998; Yuan et al. 2002). Explanations for removing this deficit have tended to be based on finding additional shortening, most recently in an Eastern Cordillera backthrust belt (McQuarrie and DeCelles 2001), and in a more speculative Western Cordillera thrust belt of Late Cretaceous-Paleocene age, for which no direct structural evidence exists (DeCelles and Horton 2003; not to be confused with the Western Altiplano thrust belt of Neogene age, described above).

The more detailed history of tectonic activity across the Central Andes presented here still shows that most bulk shortening occurred in the post-30 Ma period of Eastern Cordillera and Subandean activity. Moreover, it assumes values of shortening close to those which Kley (1999) compiled into a Central Andean block model, yielding a plan view displacement field for cumulative shortening with a conservative estimate of ~275 km in the center of the belt.

The trends of the displacement vectors, implying some along-strike (north-south) contraction and material flowing into the axis of the bend, have recently been offered as an alternative way of filling, at least some of, the shortening deficit (Hindle et al. 2005). A substantial amount of material, which would not be taken into account in two-dimensional cross sections, is shown to contribute to crustal thickening in this way. Hindle et al. (2005) modeled this process with the shortening data in Table 1.1, assuming a two-layer, "simple shear" type Central Andes – basically identical to the kinematic assumption of McQuarrie and DeCelles (2001).

This is equivalent to the bulk shortening in the Eastern Cordillera and Subandean thrust belts, compensated by distributed, ductile shortening beneath the plateau region. It has produced new estimates for the predicted distribution of crustal thickness and total volume, which show a smaller deficit than previously thought (~-10% to +2%). However, this distribution of material does not match that seen today, suggesting redistribution by erosion and, possibly, lower crustal ductile flow driven by differential topography (e.g. Husson and Sempere 2003).

Another possible distribution of shortening between the upper and lower crust was suggested by Isacks (1988) and Allmendinger and Gubbels (1996). Instead of an entirely "simple shear" configuration, relaying all shortening from the fold thrust belts to a ductile lower crust beneath the Altiplano, they suggested that shortening in the plateau and Eastern Cordillera was mostly "pure shear", i.e. lower crustal shortening was not detached from the upper crust, and Subandean shortening was subsequently in simple shear mode, relaying down to the Altiplano lower crust.

One way to evaluate which of these modes of shortening is more plausible is by evaluating the potential topography developed from either model. We here define potential topography as the value resulting from crustal deformation under the kinematic conditions stated (e.g. for simple shear, see geological section in Fig. 1.3a; and Hindle et al. 2005) and from observing isostacy.

The "simple shear" evolution of topography with time is shown in Fig. 1.7. It assumes an initial 40 km thick, pre-shortened crust with a shortening history from 30 to 10 Ma in the Eastern Cordillera and from 10 Ma to the Present in the Subandean, whilst thickening in the lower crust of the Altiplano runs continuously from 30–0 Ma, and oth-

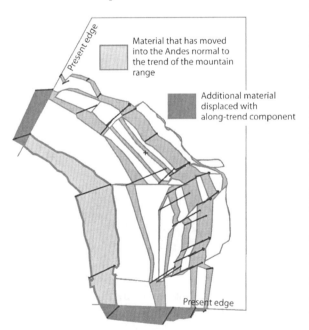

Fig. 1.6. Block model of Andean shortening showing the motion of a large amount of material in a convergent flow towards the axis of the bend. Model shows crustal blocks in their retro-deformed position according to the shortening estimates presented in Kley (1999; see also Table 1.1) and the displacement vectors which they follow to reach their present-day positions

**Fig. 1.7.**
Predicted evolution of topography in the Central Andes based on the displacement field of Kley (1999) and a two-layer distribution of crustal shortening. Snapshots show development of topography at (**a**) 20 and 0 Ma and (**b**) 25, 20, 10 and 0 Ma. **a** Kinematic assumption is simple shear thickening from underthrusting of crust below basal detachment to the west, by the amount of upper plate shortening. **b** Pure shear assumption of plateau deformation until 10 Ma followed by simple shear underthrusting of deep crust underlying the Subandean belt to the west (cf. Allmendinger and Gubbels 1996). Vectors are gradient vectors at each time interval. For kinematic constraints used, refer to Hindle et al. (2005)

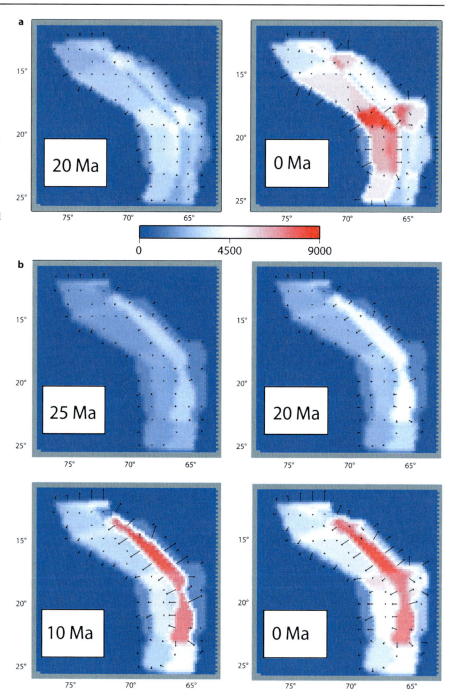

erwise follows the same boundary conditions as in Hindle et al. (2005). This model does not contain any effects from the redistributive mechanisms of crustal material (erosion and lower crustal flow). It also uses a very simple isostatic relationship between crustal thickness and topography, which is probably incompatible with both the Subandean region, where flexural support is proven to occur (Whitman 1994), and the Puna region, where delamination of the lithospheric mantle influences topography (Yuan et al. 2002).

Also shown in Fig. 1.7 are gradient vectors of topography. These should indicate the direction in which either erosion/redeposition or intracrustal flow must have operated or is still active. Topography is built continuously in the Altiplano, resulting in maximum predicted elevations of 7.5 km (crustal thickness ~85 km) along the axis of the Altiplano. The gradient vectors in the Altiplano region remain relatively constant in direction and increase in magnitude. Significantly, they suggest a constant movement of material parallel to the Andes along the Central

Altiplano. The vectors pointing west at the western edge are due to an unrealistic representation of the Western Cordillera in the model and do not reflect any real trend – with exception of the here observed surface mass transport to the Chilean Longitudinal Valley (e.g. Victor et al. 2004). Between the Altiplano and the Eastern Cordillera, an eastward transport of material is suggested throughout Andean history.

Comparing the predicted patterns for a simple shear evolution of the Andes with those identified from provenance analysis at surface (e.g. Horton et al. 2001, 2002), predicted topography and major basins show general inconsistency with both provenance observations and the large, internally-drained basin that constitute major parts of the Altiplano. In contrast, Fig. 1.7b shows the model based on a mixed pure and simple shear history of shortening. This model shows an excellent match between elevation distribution and sediment dispersal patterns (cf. Horton et al. 2001), taking into account that the western part of the system (Western Cordillera) is not modeled. The model also reproduces the development of internal drainage in the Central Altiplano, with infill mainly from the Eastern Cordillera during this time span (e.g. Horton et al. 2001). Hence, the Eastern Cordillera is shortened entirely by pure shear (30–10 Ma) but Subandean shortening is confined to the upper crust and compensated by assumed "ductile shortening" below the Altiplano.

This model shows strong redistribution from the west flank of the Eastern Cordillera to the Altiplano and from its eastern flank to the early Andean foreland. Some small redistributive flow parallel to the orogen within the Altiplano is also suggested. The onset of Subandean shortening results in only small changes to the vector trends. However, again, we must bear in mind that the effect of erosion would reduce topography in the Eastern Cordillera and create transient effects in the gradient that are not modeled here. Nevertheless, a simplified, pure shear configuration has much more success in matching the known provenances of sediments than a simple shear-only model. This strongly suggests that a simple shear-only concept of Andean shortening is, in fact, not possible.

## 1.5 Parameter Correlation Analysis

### 1.5.1 Key Observations

Linking the various quantitative results, we conclude several deformation-related aspects to be key elements for understanding the evolution of the Central Andes. These, in particular, include the following:

- Between approximately 17 and 21° S, there are no significant variations in either the timing or magnitude of shortening – the system is well synchronized. Further north and south there are only gradual changes. Both of these observations match the later extent of the plateau and thereby also constrain the spatial scaling properties of localization causes for plateau formation;
- The early stage of pure shear affecting the entire modern plateau and its lateral extent defines a mechanical condition of similar spatial scale in the lithosphere that is succeeded by failure of the boundary of the Brazilian shield at approximately 10 Ma.
- Plateau growth occurred in distinct stages and originated from two belts that expanded laterally and coalesced – thereby controlling the redistribution of surface and intracrustal material – with alternate peaks in activity prior to synchronized shutdown.
- With respect to shortening in the Altiplano domain, the Puna appears to have been delayed and may be more complex. The observational data base for the northern flank of the Altiplano is even more incomplete, but the current extensional kinematics in the Peruvian Plateau domain suggests no symmetry to the Puna.
- The bulk Central Andean shortening rate evolved in three stages (45–30 Ma, 30–10 Ma, and 10–0 Ma) – independent of partitioning mode – and shows increasing fluctuations.

These observations obviously relate to the mechanical properties of the upper plate system, to the kinematic boundary conditions under which it evolved, and to time-dependent changes of one or several of the mechanisms suggested as driving factors for plateau formation. Hence, we subsequently analyse the consequences of length scaling and of temporal evolution in more detail, exploring the associated properties of some of the key mechanisms listed in the introduction. This analysis is necessarily incomplete as some processes are as yet not directly identifiable in proxies available as time series data to form a basis for correlation analysis. This is particularly true for mantle-related processes involving delamination of mantle lithosphere, asthenospheric wedge circulation, basal mantle torque, etc., all of which are more amenable through numerical modeling.

### 1.5.2 Length Scales of Deformation Mechanisms

The large-scale, lateral continuity of structural units of more than 800 km, and their synchronicity at the 5–10 Ma time scale, largely precludes minor features of the oceanic plate from having controlled plateau evolution. This includes collision with oceanic ridges or minor oceanic plateaus, as hypothesized for the passage of the Juan Fernandez ridge during the Miocene (Yañez et al. 2001), or related plateaux (Giese et al. 1999; Gutscher et al. 2000b). All of these topographic features on the downgoing plate in the central Andean segment tend to lack the lateral extent, as well as longevity, to have been able to cause the large-scale, synchronized deformation pattern in the arc/back-arc domain. A more detailed inspection of the deformation time series (Fig. 1.5) clearly shows that the suggested time of passage (22–16 Ma) of the Juan Fernandez Ridge at the latitude analysed (Yañez et al. 2001) has no expression in our data.

In contrast to the restricted influence of topographic features on the oceanic plate, segments of the slab dipping at different angles – such as shallow dip and "flat slab" – tend to exhibit the same spatial extent as the upper plate orogenic architecture. These slab segments have been repeatedly hypothesized on various grounds (Isacks 1988; Allmendinger et al. 1997; Mahlburg Kay et al. 1999, see below) and are expected to particularly influence the thermal, i.e. mechanical state of the overlying mantle from a sequence of hydration ('flat slab stage') and subsequent melting ('slab steepening'). Also, Gephart (1994) has remarked on the intriguing symmetry between the Andean surface topography and the dip of the present-day slab segments – implying a close connection between upper plate deformation and slab geometry. Accordingly, slab geometry should translate into variations in the strength of the collision zone, either via lateral changes in plate coupling area or through modification of the geothermal gradient, i.e. the mechanical properties of the intervening upper plate mantle. Yet, this symmetry is a present-day feature, the past evolution of which is not constrained. The only temporal constraint proposed (Allmendinger et al. 1997; Mahlburg Kay et al. 1999) is based on the spatial and geochemical patterns of volcanism and suggests a flat-slab stage for the Central Andes at around 33–26 Ma, when a magmatic lull is identified for this region. In Figs. 1.3 and 1.5, this period can be seen to be related to a shift of active shortening towards the back-arc (see below for further discussion of the time signal).

Geophysical properties of the plateau crust may seem to provide additional support. The temperature field – albeit only represented by few data (Springer and Förster 1998) – and the existence of a crustal s-wave, low velocity anomaly from receiver function data, attributed to fluids or melts (Yuan et al. 2000), all share a significant lateral extent that seems to match the extent of the plateau, at least across strike. Along with the lateral match in the occurrence of Neogene surface magmatism, approximately confined by the 3 500 m elevation line (Allmendinger et al. 1997), these observations have been interpreted by the above authors to reflect a perturbed thermal state of the plateau crust and the underlying mantle (cf. Isacks 1988; Wdowinski and Bock 1994) for most of the Neogene. In addition, resolving lithosphere thickness from various geophysical data by Tassara et al. (2006) clearly shows that the lateral extent of the plateau is outlined by an identically shaped domain of anomalously thin mantle lithosphere. Yet, for the latter it is not clear whether it is a consequence or cause of a mechanically weak part of the leading edge of the upper plate.

While the geophysical observations are snapshots of the present-day distribution of properties, the spatial correlation with more long-lived magmatism may indicate the potential for an equally long-lived thermal state with this lateral extent (see below). This view may be supported by contrasting observations in the Southern Andes (36–40° S), where the Nazca Plate is subducting at a similar rate and dip, whereas no significant shortening and crustal thickening has occurred in the back-arc domain: Here, no series of geophysical anomalies of the Central Andean size indicating hot, weak crust has been detected (see Sobolev et al. 2006, Chap. 25 of this volume).

Another argument pertaining to length scale has been forwarded by Allmendinger and Gubbels (1996; see also Allmendinger et al. 1983) and suggests a link with foreland strength. The occurrence of a thick Lower Paleozoic basin in front of the Altiplano that thins substantially north of the Andean bend and south of the Altiplano (at some 24° S), perfectly correlates with the lateral extent of the thin-skinned Subandean belt and the maximum in crustal shortening (cf. Figs. 1.8 and 1.2). Towards the south, where deformation at the front changes to thick-skinned, basement-involving style, shortening decreases, implying that lateral variations in crustal strength indeed control the spatial scaling of plateau formation. However, as seen from Fig. 1.8, it is mainly the shortening style in the upper crust that is affected by sediment thickness distribution. The lateral extent of the plateau itself, as indicated by the location of the western and eastern limits of the plateau (3 500 m elevation contour) largely matches first order discontinuities affecting the entire crust.

In the west, the position and geometric outline of the Paleogene volcanic arc is coincident with the Eocene shortening domain and obviously controlled the western boundary of shortening throughout plateau evolution. Sempere et al. (2002) suggested that the eastern limit coincides with the site of a narrow Permian to Jurassic rift system. The southern continuation can be related to the trace of Cretaceous extension and crustal thinning (Salta Rift and Potosi Basin). Again, both eastern rift systems trace the present-day eastern plateau edge, which is indicated by the 3 500 m contour. Both plateau limits also constrain the lateral extent of deep crustal, pure shear thickening, and localize later uplift and maximum crustal thickening, as shown from topography modeling at these boundaries (cf. 3 500 m outline to Fig. 1.7b). Comparing Figs. 1.8 and 1.3, it becomes obvious that the above features strongly controlled the sites where deformation was initiated and the subsequent synchronization of active fault systems along their trend.

In conclusion, as originally conjectured by Allmendinger et al. (1983) the location and spatial scaling properties of deformation partitioning would seem to be pre-

**Fig. 1.8.**
Changes in structural style of foreland thrusting and correlation with structural and stratigraphic, inherited discontinuities. Data from Baby et al. (1995), Mathalone and Montoya (1995), Pankhurst et al. (1998); Rapela et al. (1998), Salfity and Marquillas (1994), Sempere et al. (2002) and other sources. Basement thrusts in the south are interpreted to reactivate metamorphic fabrics (trend of foliation shown schematically; e.g. Schmidt et al. 1995). *Thin dashed line* shows the plateau outline as reflected by elevations beyond 3 500 m

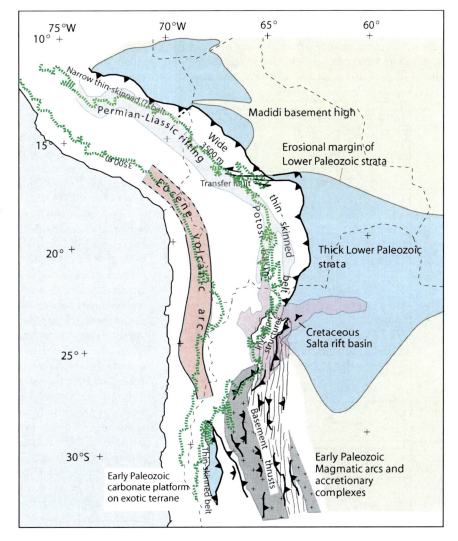

conditioned by the primary features that build the crust in this part of South America and constitute the main weak zones. These particularly include those mechanical heterogeneities that are due to Paleogene magmatic arcs, Mesozoic rifts and Paleozoic sedimentary basins. Conversely, none of the individual topographic or structural features of the Nazca Plate can be seen to have had a significant role. In addition, the spatial extent of crustal pure shear shortening in the earlier stages is laterally confined by the deep-reaching properties of the above heterogeneities and, probably, by the thermal state of the mantle, as influenced by changes in slab geometry.

### 1.5.3 Deformation Time Series and Plate Kinematics

All of the above arguments lack the temporal aspect, and explain neither the onset of shortening nor changes in the rate of shortening, all of which we consider vital when probing for controlling mechanisms and their evolution through time. Therefore, we next perform a time series analysis of various features where equivalent data are available. Our first step is to correlate the plate kinematic parameters with the evolution of shortening in the upper plate; this is one of the classical parameters repeatedly suggested to exert a primary influence on the deformation of the leading edge of the upper plate. From inspection of Fig. 1.9a and 1.9b, it becomes immediately obvious that neither the convergence rate between the Nazca and South American Plates, nor convergence obliquity (see: Pardo-Casas and Molnar 1987; Somoza 1998) seem to relate to the evolution of the shortening rate in the Central Andes. At best, a weak correlation between an increased shortening rate and enhanced convergence rate may be indicated for the initial stage. For all of the Neogene, the upper plate shortening rate appears to be anti-correlated with the plate convergence rate, discounting any significant influence of the latter on upper plate deformation.

Convergence may be decomposed to yield the absolute drift of South America. Acceleration of this drift since the Miocene was suggested by Silver et al. (1998) to have driven plateau formation. They noted that plate kinematic changes, resulting from slowing of the motion of Africa and the continued opening of the South Atlantic, were responsible for the westward acceleration of South America. Earlier, Coney and Evenchick (1994) had suggested, on a less quantitative basis, that increased westward motion of South America following South Atlantic opening had already started during the Cretaceous. More recently, Heuret and Lallemand (2005) have argued that upper plate velocity with respect to slab rollback is critical in determining upper plate deformation mode based on a global analysis of subduction systems. And our data do, in fact, exhibit a general match between the shortening rate and the evolution of the South American westward drift through the Cenozoic – starting even earlier than the Miocene as suggested by Silver et al. (1998).

Upper plate shortening commenced when absolute westward drift exceeded about 1.7 cm yr$^{-1}$ at the latitudinal range of 18–22° S. This value is equivalent to the rate of motion of the trench or subduction hinge owing to rollback of the subducting Nazca Plate. If this value has remained stable through time, integration of the differential velocity between the South American drift rate and the slab rollback rate ($\delta v$, shaded area in Fig. 1.9c) should approximately correspond to the total shortening of the upper plate. We obtain 280 km, matching within error bounds the value given in the shortening database (see Table 1.1, 21° S). Moreover, the present difference between slab rollback and upper plate velocity, i.e. 1.1 cm yr$^{-1}$, should approximately correspond to the current shortening rate of 0.9–1.2 cm yr$^{-1}$ as obtained from GPS studies along the Central Andes in the region of active foreland shortening (Bevis et al. 2001; Klotz et al. 2001).

This result implies that the Cenozoic slab rollback rate next to the Central Southern Andes has, at best, slightly fluctuated through time around the above value of approximately 1.7 cm yr$^{-1}$ (Fig. 1.9d). This fluctuation can be assessed because the rate of slab rollback or subduction hinge migration $v_{sh}$ is kinematically determined by the difference between the upper plate velocity $v_{up}$, the deformation rate of its leading edge $v_{dr}$, and the mass removal rate from subduction erosion $v_{ser}$:

$$v_{sh} = v_{up} - v_{dr} - v_{ser} \qquad (1.1)$$

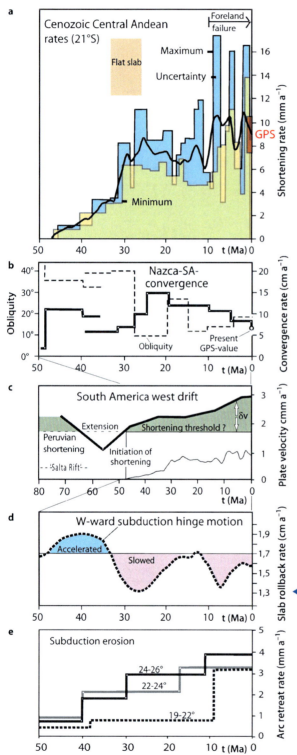

◀ **Fig. 1.9.** Kinematic factors controlling deformation. **a** Bulk minimum and maximum shortening rate and sliding average from Fig. 1.5b; **b** convergence rate and obliquity between Nazca plate and Central South America from Pardo-Casas and Molnar (1987, before 35 Ma) and Somoza (1998, after 35 Ma); present GPS-velocity from Klotz et al. (2001); **c** West drift of South America from Silver et al. (1998) for the past 80 million years, including curve of average shortening rate from Fig. 1.5b (for explanation of kinematic regimes, see text); **d** variation of westward velocity of the subduction hinge (slab rollback) as calculated from upper plate velocity minus shortening rate and subduction erosion rate using sliding five million year average; and **e** arc retreat rate from mapping position and migration of volcanic front (based on volcano data base by Trumbull et al. 2006, Chap. 2 of this volume)

We estimate the subduction erosion rate by assuming a constant trench to arc distance, i.e. constant slab dip to a depth of approximately 100 km, and derive mass flux rate at the plate tip by measuring the change in the position of the volcanic front between 19 and 22° S through time (data base shown in Trumbull et al. 2006, Chap. 2 of this volume; see also Kukowski and Oncken 2006, Chap. 10 of this volume, for more details on subduction erosion evolution). The assumption of a constant arc-trench gap may induce some error in the result. However, from the very continuous value of this parameter along the Andean margin between 8 and 36° S, independent of deeper variations in slab dip ('flat slabs' at 10 and 32° S), we estimate this feature to also have been quite stable over time. The subduction erosion rate shows a continuous increase over time during the Cenozoic at several latitudes, with similar results. This trend is coincident with the increase in the westward drift of South America and the shortening rate of the upper plate (Fig. 1.9e).

Since the input data for calculating the variations in subduction hinge migration are subject to some error, we have attempted a first order estimate of the effect of propagating the errors of the individual components. Subduction erosion rate has no relevant role here, as the value is one to two orders of magnitude smaller than the other input data. Also, the impact of using the here reported maximum or minimum shortening rates instead of the averaged shortening rate is small with exception of larger fluctuations for the past 10 Ma (maximum deviations from the result shown in Fig. 1.9d may then reach twice the deviation from the average 1.7 cm yr$^{-1}$ shown). Both input data do not affect the general trend of the curve in Fig. 1.9d. The single most relevant source of error is the westward drift of South America, i.e. its fluctuations that may affect and modify the general shape of the curve. Silver et al. (1998) based their calculation of the absolute South America motion on the motion of Africa with respect to Atlantic Basin hot spots as well as on the South Atlantic fracture zones and sea floor magnetic anomalies. They calculated velocities from stage poles averaged over time windows of ca. 10 Ma decomposing angular velocity vectors into the plate motion parallel (ca. E-W rate) and orthogonal (ca. N-S velocity component) without providing error estimates. An error of 20–30% in these data would be able to significantly modify the curve or even to erase any trend. Since temporal resolution of subduction erosion velocity and west drift velocity is low, we smoothed the resulting time series by averaging over a 5 million year sliding time window. The result shows that fluctuations in the hinge rollback velocity about the above background value have remained below some 20% (Fig. 1.10d).

The robustness of this emerging relationship is emphasised by analysing the preceding, late Cretaceous to Eocene, evolution of the Central South American margin. Extension of South America's leading edge, as recorded by various authors for this time span (Rochat et al. 1999; Jordan et al. 2001; Elger et al. 2005), is coincident with a period of very slow westward drift (1.0–1.5 cm yr$^{-1}$) for South America (differential velocity $\delta v$ between 0 and $-0.7$ cm yr$^{-1}$; Fig. 1.9c). Moreover, the preceding middle-late Cretaceous stage of enhanced westward drift, at some 2.2 cm yr$^{-1}$, perfectly matches the well established stage of Peruvian phase shortening found in the Cretaceous arc area along large parts of the Andes (e.g. Mégard 1984). However, simultaneous development of the Salta Rift over most of the Cretaceous in the back-arc domain (e.g. Viramonte et al. 1999) indicates additional complexity during this period. Initiation of the South Atlantic opening with widespread rifting at this stage is clearly contemporaneous with westward acceleration, allowing shortening at the leading plate edge during concomitant extension of the central and eastern parts of South America.

Clearly, subduction hinge motion at the rate of absolute upper plate motion minus deformation of its leading edge is a kinematic necessity. Yet, the relative stability of the rate of rollback through time (fluctuating less than 20%) at this latitude, in spite of varying upper plate velocities (variation of 300%), indicates that the potential of the Nazca Plate to accommodate velocity changes in the upper plate by changes in the rate of slab rollback is restricted. Hence, the differential velocity between the upper plate and slab rollback of the oceanic plate is indeed critical for determining kinematics, amount, and the rate of shortening at the leading edge of the South American upper plate, as suggested by Coney and Evenchick (1994), Silver et al. (1998), and Heuret and Lallemand (2005).

Obviously, at first sight the quantities in this argument are only valid for the portion of the Andes south of the Arica inflection, where the trend of the plate boundary (7°) is nearly perpendicular to the current azimuth of plate acceleration (275°; cf. Silver et al. 1998). With increasing curvature of the Bolivian orocline resulting from the shortening gradient depicted in Fig. 1.2, the northern flank of the plateau (north of 17° S) acquires increasing obliquity to this absolute plate motion vector (present obliquity is 66°). However, calculation of the motion of the oceanic plate hinge with the above routine considering the amount of shortening at 15–16° S (only some 150 km versus 275 km at 19° S) and the vector of absolute plate motion, we arrive at the same hinge rollback velocity for the northern flank in the reference frame of the plate motion direction as with the southern flank. But: in the reference frame of the plate margin, the plate-margin normal component of hinge rollback in the North obviously is smaller with some 0.9 to 1 cm yr$^{-1}$ because of the above obliquity. This observation would seem to support arguments that hinge rollback of the Nazca Plate is not entirely independent from upper plate velocity. Rather, the plate-margin normal value observed at the southern plateau branch (1.7 cm yr$^{-1}$ ± 0.3) may reflect a maximum value achievable under the circumstances responsible.

## 1.5.4 Deformation Time Series and Mechanics: Temperature

Slab rollback velocity is generally interpreted to be limited by various parameters (see Mantovani et al. 2001, and Heuret and Lallemand 2005, and references therein). These include variations in slab pull force, slab anchoring, mantle viscosity, mantle flow patterns, slab length (distance to a slab window that enables the flow of mantle from below to above the subducting plate, as required by rollback), and the strength of the plate interface (see also Hassani et al. 1997). Conversely, changes in the shortening rate of the upper plate may be seen to reflect changes in the integral strength of the overriding lithosphere. To a first approximation, therefore, comparison of both parameters may help to identify changes in the balance of bulk mechanical strength for the two-plate system, i.e. changes in the balance between (*a*) the strength of the plate interface, (*b*) the strength of the upper plate lithosphere, (*c*) the strength of the boundary to the eastern foreland and shield, and (*d*) the resistance of the lower plate to slab rollback. Relative weakening of the plate interface would support overriding and hinge rollback; weakening of the upper plate should entail enhanced shortening. The shortening rate, as well as the rate of slab rollback, may hence be matched with a time series that may be considered a proxy for any one of the above key mechanical components.

We therefore, next, explore the variations in the rate of slab hinge rollback in more detail. Inspection of the time series in Fig. 1.9d shows several departures from a constant rollback rate. The first stage of slow shortening is accompanied by a stage of enhanced rollback between 47 and 33 Ma. Subsequent acceleration of shortening between 33 and 20 Ma occurred when rollback slowed. Final acceleration of the shortening rate at about 10 Ma coincides with a final short pulse of slowed hinge motion. From the addition of uncertainty in the data involved, this final pulse might not be interpretable. The earlier two features, in contrast, are more robust and independent of the time filter used to smooth the data and of the higher uncertainty involved with the various input data. As a first conclusion, this development of rollback rate reflects either (*a*) a changing ability of the upper plate to override the lower plate, thereby enhancing slab retreat or (*b*) a higher absorption of deformation in the upper plate from increasing resistance of the lower plate to retreat.

We first test evidence indicative of upper plate strength. As mentioned above, the thermal evolution of the upper plate, and therefore its mechanical properties, may have evolved through time. This seems to be obvious from the match between the lateral extent of the plateau and the surface magmatism (Allmendinger et al. 1997). In addition, delamination of lithospheric mantle at various scales, as suggested by Kay and Mahburg-Kay (1993) for the Puna and Beck and Zandt (2002) as well as Garzione et al. (2006) for the Altiplano may be an important heat source affecting the thermal state of the lithosphere. Magmatism and unusually fast surface uplift rates are the only available indirect proxies to reflect the impact of this process. Recently, Garzione et al. (2006; see also companion paper by Ghosh et al. 2006) concluded that the rapid late Miocene rise of the Bolivian Altiplano by 2.5–3.5 km (between 10.3 and 6.8 Ma) based on oxygen isotope data from fossil soils indicates wholesale mantle lithosphere delamination as cause since the overall Andean shortening rate observed only explains some 30% of this value for the given time frame. We note, however, that all their data were derived from a single locality within the eastern flank of the Corque syncline, one of many structures building the plateau. This structure experienced uplift in the hangingwall position of a major west-dipping fault during exactly the above time frame (see sections and age dating by Hérail et al. 1993; Lamb and Hoke 1997; McQuarrie and DeCelles 2001). With respect to the elevation of the regional of the stratigraphic level hosting the fossil soils, the uplift of the sample sites from this local structure amounts to 1.3–2.5 km as seen from the nearby sections from Lamb and Hoke (1997) or 4–7 km in the sections slightly further south by Hérail et al. (1993) and McQuarrie and DeCelles (2001). Hence, we note that these data may not be interpreted to indicate general high uplift rate caused by delamination for the entire Altiplano. They rather reflect local structurally controlled uplift supported by syntectonic sedimentation superseding widespread uplift from overall shortening and crustal thickening. Yet, from geophysical data indicating very little mantle lithosphere material left underneath major parts of the Altiplano (e.g. Beck and Zandt 2002) lithospheric material has probably been lost in the past. We therefore conclude that the removal of lithosphere may as well have occurred by the piecemeal removal model by Beck and Zandt (2002) rather than in a single major act covering the entire Altiplano.

Magmatism to the surface may be slower to react, because of the slow diffusion of a thermal signal in the lithosphere, delays in melt segregation and transport to the surface, the respective influence of the kinematic regime on near surface melt transport and eruption, and other parameters that may each contribute to changes in volcanic flux that are not directly related to the deformation rate nor to the thermal state of the mantle. However, magmatism may provide a longer history over a broader region than the above sampled soils and may yield less ambiguous clues to the thermal evolution of the underlying mantle. To minimize bias resulting from the use of multiply-dated, individual volcanic layers we only make use of the number of individual volcanoes (no tuffs, ashes, ignimbrites, etc.; see Trumbull et al. 2006, Chap. 2 of this volume, for data base) collected with a 3 million year sliding

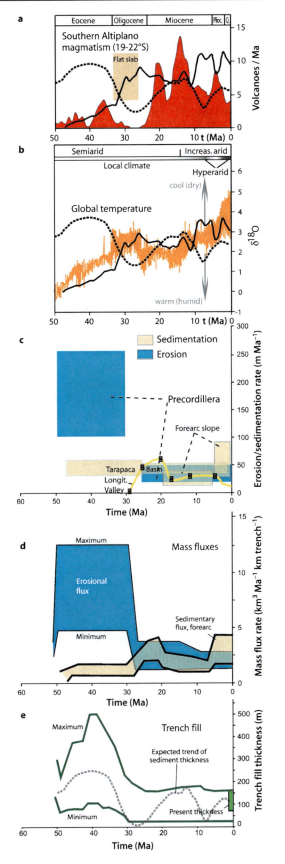

window, thus accounting for isotopic error and the typical volcano lifetime (Fig. 1.10a). Also, mantle-derived magmatism as opposed to crustal contributions probably has lower residence times in the crust providing a more instantaneous indicator of changes in the thermal regime.

Even so, factors such as the incomplete sampling of volcanoes and decreasing preservation with age, as well as accessibility of sediment-covered older volcanoes, are contained in the data and require consideration. Therefore, the lack of an obvious correlation between the shortening rate and the volcanic flux in Fig. 1.10a is not surprising. In contrast, we tend to observe an anti-correlation one of which is known to be quite robust: the magmatic lull between 33–26 Ma is a well established feature; and the significant spatial expansion of active magmatism into the back-arc after 26 Ma is just as robust. Surprisingly, however, the stage between 33–26 Ma that has been related to a time of shallow subduction (the 'flat slab' of Isacks 1988; Allmendinger et al. 1997; Mahlburg Kay et al. 1999) is exactly coincident with a period of reduced rate of slab rollback and higher efficiency of transferring convergence into shortening of the upper plate. We therefore conclude that variations in the thermal state of the mantle only become relevant for crustal shortening rates when flat slab subduction provides enhanced coupling over an enlarged area between both plates in a cooled, and mechanically stronger, environment. Interestingly, modeling by Sobolev et al. 2006 (Chap. 25 of this volume) also indicates temporal anti-correlation of magmatism and upper plate shortening, possibly related to stages of delamination following earlier lithospheric thickening. Delamination, from the above magmatic data and the model inferences, may then have started as early as 25 Ma.

◀ **Fig. 1.10.** Mechanical factors from thermal weakening or climatic evolution 18–22° S. **a** Number of isotopically dated volcanoes ($N = 190$), collected in three million year sliding windows, between 19 and 22° S (see Trumbull et al. 2006, Chap. 2 of this volume, for database) superposed with bulk shortening rate from Fig. 1.9a and rollback fluctuation from Fig. 1.9d as well as with period suggested for flat slab subduction (Mahlburg Kay et al. 1999). **b** Global climatic trend from $\delta^{18}O$ data (Zachos et al. 2001) superposed with bulk shortening rate and slab rollback fluctuation from Fig. 1.9a,d; *top of figure* includes local climate evolution in northern Chile from Alpers and Brimhall (1988), Gaupp et al. (1999), and Hartley and Chong (2002). **c** Erosion and sedimentation rates in the Chilean Precordillera and Longitudinal Valley; Precordillera erosion rates from Maksaev and Zentilli (1999) and Kober et al. (2002); fore-arc sedimentation rates from Victor et al. (2004), González (1989) and Kudrass et al. (1998). **d** Estimated range of integrated erosional and sedimentary mass flux for North Chilean Precordillera and fore-arc basins. **e** Range of trench fill thickness as calculated from Eq. 1.2 including present-day average thickness (Kudrass et al. 1998; von Huene and Ranero 2003); superposed is the trend of expected thickness from the assumption that plate interface strength shows inverse correlation with the amount of incoming sediment

## 1.5.5 Deformation Time Series and Mechanics: Climate and Plate Interface

Climate has been repeatedly suggested to have had an impact on the formation of the high Andes Plateau. Since the work by Masek et al. (1994), standard models have related precipitation, surface erosion and mass redistribution in the Andes with the local response of orogenic wedges. It has been suggested that this response is responsible for focussing deformation on the more humid eastern flank of the plateau or for sustaining vertical growth in the more arid parts of the system, within the framework of critical wedge theory (e.g. Horton 1999; Montgomery et al. 2001; Sobel et al. 2003). Several authors (e.g. Pope and Willett 1998; Montgomery et al. 2001; Sobel et al. 2003) even argue that the position of the Central Andes in the arid belt of the southern hemisphere is a key element for enabling the lateral growth of a mechanically-limited plateau, with tectonics substantially more efficient than erosion.

With exception of the analysis by Sobel et al. (2003), most of the analyses focus on the present (to late Quaternary) climatic pattern and erosion regime, with little evidence to demonstrate the past evolution and coupling of both climate and tectonics. In essence, this work seems to show that it is mainly the distribution or focussing of deformation at the eastern orogenic front and the wedge taper which are affected by precipitation and erosion patterns (cf. the broad high-taper zone of active Subandean shortening in the Beni region at 16–18° S (Horton 1999), versus the well focussed, active seismicity at the orogenic tip of a low-taper wedge south of 21° S, Fig. 1.3). Also, the low erosion regime is critical for supporting surface uplift via the development of internal drainage with the storage of sediment within the orogen (Sobel et al. 2003; Elger et al. 2005). In contrast to the localization effect of erosion on deformation and the shape of the thrust wedge, we find that both the shortening rate derived from GPS measurements along the eastern Andean front between 15 and 25° S (0.9–1.2 cm yr$^{-1}$; Bevis et al. 2001; Klotz et al. 2001) and the cumulative geological shortening over the past 10 million years (see Table 1.1) are quite independent of the substantial precipitation gradient in the same area (1 200 to 650 mm yr$^{-1}$ from north to south; New et al. 2001).

More recently, Lamb and Davis (2003) have argued that global cooling during the Cenozoic, rather than local climate-tectonics coupling, has been the key mechanism driving the deformation history of the upper plate system. Their argument is based on two key components:

a Global cooling since the Eocene, which has accelerated since the middle Miocene, has increased aridity and decreased the sediment input into the trench.

b Fluid input into the subduction system is reduced as total pore volume is lower with less trench sediment, thus increasing the effective frictional strength of the plate boundary through time and resulting in increased coupling and accelerated shortening in the upper plate.

They base their argument on the global sea surface temperature (SST) curve by Zachos et al. (2001). As discussed above, the mechanical relevance of changes in interplate frictional strength that affect the state of stress and the kinematic regime of the upper plate has been demonstrated from modeling studies (Hassani et al. 1997).

A first test of these arguments is again provided by checking the shortening time series against the global SST curve from $\delta^{18}$O data (Zachos et al. 2001; Fig. 1.10b). The shortening rate increases over time and is loosely correlated with global cooling. A correlation with the evolution of slab retreat is, at best, present during the Paleogene (humid climate = higher surface mass flux and faster retreat). In both cases, a time lag of at least one to three million years should be apparent in the tectonic response of the upper plate as this much time, at least, is required to redistribute the sedimentary layer along the plate boundary (based on the convergence rate). While these correlations would seem to provide some support for the arguments put forward by Lamb and Davis (2003), the mechanism of climate impact is not immediately evident since it is unclear how the global climate signal has been converted into local temporally-varying plate interface lubrication from changes in trench sedimentation. Sediment flux into the trench is governed by non-linear relations and feedbacks between climate, precipitation, and erosion, all of which are influenced by factors such as deformation-controlled topographic build-up, base level changes, and rock erodibility.

Rather than focussing on the reconstruction of climate history, we attempt a reconstruction of the changes in mass flux of the Chilean fore-arc system for the entire Cenozoic. Since the data situation is heterogeneous, with little information from the fore-arc slope, we only assess the minimum and maximum flux consistent with the data available. Trench sediment thickness $T_{ts}$ is derived from the following equation:

$$T_{ts} = \left( \frac{(\dot{V}_{vf} + A\dot{V}_{er})(1-CD)}{(1-0.01\phi)} + \dot{V}_{ps} + \dot{V}_{mw} - \dot{V}_{fs} \right) \frac{1}{PC} \quad (1.2)$$

where $\dot{V}_{er}$ is eroded rock volume from the arc and fore-arc; $\dot{V}_{vf}$ is the volcanic volume eroded from the volcanic edifices (this material will not usually be 'felt' by AFT data, because of the short life time and high erodibility of volcanoes); $\dot{V}_{ps}$ is the volume of pelagic sediment on the incoming oceanic plate; $\dot{V}_{mv}$ is the sediment volume derived from mass wasting at the plate tip; $\dot{V}_{fs}$ is sediment volume stored in the fore-arc (all volume fluxes in km$^3$ Myr$^{-1}$ km$^{-1}$ trench length); $\dot{PC}$ is plate convergence rate; CD is chemical denudation; and $\varphi$ is porosity in % from converting solid rock into unconsolidated porous sediment.

The pelagic sediment thickness on the Nazca Plate is estimated to be between 30 and 150 m (Kudrass et al. 1998). Eroded volume from the Precordillera and western Altiplano is derived from apatite fission track data (Maksaev and Zentilli 1999) and from estimates based on cosmogenic nuclide dating (Kober et al. 2002). Both data sets are calculated over the area of the eroded Altiplano margin drained toward the fore-arc. The calculation of fore-arc sediment volume is based on stratigraphic and thickness data from the Longitudinal Valley (based on data from Victor et al. 2004) as well as sediment thicknesses observed in the offshore Tarapacá Basin (González 1989) and on the fore-arc slope (Kudrass et al. 1998), in all cases integrated over the depositional area.

The area analysed is constrained between 19 and 22° S by the plateau section with the shortening rate data shown here and the existence of a fluvial link between source and sink. From this database (Fig. 1.10c), we calculate total erosional and sedimentary fluxes though time (Fig. 1.10d). Assuming no chemical denudation in this arid environment and an average porosity of 30% for the clastic sediments, the difference between these two fluxes plus the incoming sediment thickness forms the basis for calculating trench thicknesses through time from Eq. 1.2 (Fig. 1.10e). It must be noted, however, that the results may be affected by the loss or addition of mass from trench-parallel sediment transport. After the hypothetic passage of the Juan Fernandez Ridge (ca. 18 Ma, Yañez et al. 2001), no sediment is transported into the system from the south. Earlier thicknesses may therefore represent minimum values.

As becomes obvious in Fig. 1.10e, the possible range of trench fill thicknesses has always remained well below some 500 m in spite of the large uncertainty. The data provide weak indications that a stage of maximum trench fill thickness (150–450 m) occurred approximately between 45 and 30 Ma and has since remained stable below the present-day maximum of 160 m. Comparison with Fig. 1.10b shows that this trend neither mimics global cooling nor the local climate for the western plateau flank, as established by Alpers and Brimhall (1988), Gaupp et al. (1999), and Hartley and Chong (2002). Hence, the response of the upper plate system in terms of flux variability is not related to the Cenozoic climate variations in this area (cf. Fig. 1.10b and e). Rather, the formation of local topography at the site of the modern Precordillera (i.e. the Early Tertiary arc) during the Eocene period of high shortening rate at the western plateau rim (Incaic phase; cf. Precordillera in Fig. 1.4d) and the convergence rate have had the key roles in controlling trench fill thickness during this time.

If the tectonic response of the upper plate were strongly influenced by the mechanism suggested by the model of Lamb and Davis (2003), changes in trench fill thickness should correlate with the propensity of the subduction hinge and slab to retreat or to resist overriding, as identified above (fast shortening = high interplate friction and slow retreat from low sediment input; slow shortening = low friction and high retreat rate from high sediment input). Possibly, such a case can be made for the period between 45 and 30 Ma (Fig. 1.10e) when potentially higher sediment flux indeed correlates with higher retreat rate. For the subsequent period of slowed retreat, ending about 20 Ma, the relationship is not convincing, probably because it is superseded by another effect (see above for the role of flat slab subduction). In the Late Neogene (< 10 Ma), slowed slab retreat during increasing upper plate velocity appears to enforce increased shortening and subduction erosion rates, pointing to a strengthening of the plate interface with respect to the upper plate. While this last-named feature is coincident with foreland failure at the eastern flank of the plateau (see above) and development of a thrust belt above a weak detachment (supporting the strength balance argument), we noted earlier that this feature of the hinge motion evolution is not robust enough for interpretation. Altogether, these relationships indeed suggest that interplate coupling has been high throughout the Cenozoic – possibly as high as suggested by Bevis et al. (2001) for the present – and that it is a requirement for the efficient transfer of convergence into shortening.

Independent support for the trench fill argument is derived from the fact that the efficiency of subduction erosion increased with time. Lallemand et al. (1994) and Clift and Vannucchi (2004) observed that trench fill is a key parameter affecting the subduction erosion of the upper plate. On average, circum-Pacific convergent margins switch from subduction erosion to accretion, with development of an accretionary wedge, when trench fill thickness exceeds ca. 1 000 m. Decreasing sediment thickness should therefore promote subduction erosion. The Central Andean margin, the archetype margin controlled by long-term subduction erosion (e.g. Rutland 1971; von Huene and Ranero 2003), has apparently remained below this threshold throughout the Cenozoic. Also, we note that the average surface roughness of the Nazca Plate, some 500 m, is higher than the maximum reconstructed trench sediment thickness (Fig. 1.10e). This would prevent deposition of a continuous fluid-rich layer weakening the plate interface that could suppress the abra-sive removal of material from the tip and base of the upper plate. The topographic lows of the downgoing plate are then filled either through mass wasting of the lower slope or abrasion of the wedge base (von Huene and Ranero 2003). Lamb and Davis (2003) expect this material to have a lower average porosity and to thus contribute less to the lowering of the effective frictional strength.

In summary, all the detected responses of the tectonic system to sediment flux are constrained to the Eocene period of potentially higher trench infill. Possible enhanced 'lubrication' between 45 and 30 Ma is nearly entirely controlled by deformation-dependent, enhanced topography and higher erosional flux from the Chilean Precordillera owing to the Incaic shortening stage. In conclusion, therefore, neither the global climate trend nor its local consequence has had a significant impact on the Central Andean shortening evolution. We expect a significant impact of global or local climate change on trench processes, with consequences for the mode of upper plate deformation, only if it induces sedimentary flux into the trench well above the amplitude of surface roughness of the oceanic plate and, more particularly, when it exceeds the critical threshold imposed by the transport capacity of the subduction channel (i.e. about 1 km, Clift and Vannucchi 2004).

## 1.6 Discussion and Conclusions

We observe that the bulk shortening rate has evolved in three stages (45–30 Ma: an increase from 0 to 8 mm yr$^{-1}$; 30–10 Ma: a fluctuation between 6 and 10 mm yr$^{-1}$; and 10–0 Ma: a rise to 8–14 mm yr$^{-1}$ until the attainment of the modern value of 9 mm yr$^{-1}$). Analysing various time series data, we identify the evolution of the shortening rate of the Central Andean Plateau to have been dominated by the increase in the rate of westward drift by the South American Plate (cf. Coney and Evenchick 1994; Silver et al. 1998). Apparently, at this latitude, the upper plate needed to exceed a critical velocity threshold of the plate-margin normal component of approximately 1.7 cm yr$^{-1}$ of lateral motion to enable it to shorten. Lower rates of drift relate to the extension of the leading edge of Central South America. This value defines a slab rollback rate that fluctuates only slightly (±20%) throughout the Cenozoic as the upper plate drift accelerated from 1.0 to 2.9 cm yr$^{-1}$ to the west. In support of the suggestion by Russo and Silver (1996), we identify the plate-margin normal velocity difference between the hinge/slab rollback and the drift of the upper plate as the essential parameter determining the shortening rate of the leading edge of the South American upper plate.

This relationship is superseded by minor variations that can be shown to be linked to two main aspects: a stage of flat slab subduction and a stage of enhanced erosion and trench infill. A change in subduction geometry involving subduction at low angle with slowed retreat rate resulted in a period of enhanced transfer of shortening into the upper plate (33–20 Ma). This 'flat slab' stage, suggested by Isacks (1988; see also Allmendinger et al. 1997; Mahlburg Kay et al. 1999) on the grounds of volcanic evolution, can be seen to have increased the shortening rate and to have shifted the site of active shortening towards the continent. Both these effects were probably controlled by higher interplate coupling across a larger area.

Prior to this stage, slower upper plate shortening, related to a stage of higher trench fill thickness resulting from enhanced erosion during the Eocene, may be caused by weakening of the plate interface with faster slab retreat. However, in contrast to the suggestion by Lamb and Davis (2003), this effect is not driven by global cooling or the local climatic evolution in the Cenozoic. It is entirely owing to the rapid formation of relief during Eocene shortening in the arc domain and its equally rapid erosion. Trench fill can be shown to have continuously remained well below the capacity of the subduction channel and also below the amplitude of topographic relief of the fractured Nazca Plate. Effects from Cenozoic climate changes can, therefore, be shown to have remained below the thresholds expected to significantly influence the shortening rate of the upper plate.

In contrast to the situation in the Central Andes, slab rollback at the southern margin of the Andes (40° S) – with > 2 km of trench fill since the Pliocene (see Kukowski and Oncken 2006, Chap. 10 of this volume) – must have occurred at a similar rate as the westward drift of the upper plate. We conclude this from the southward decrease of shortening in the upper plate to nearly absent shortening at that latitude, particularly after the onset of glacially-derived sediment infill (see Vietor and Echtler, Chap. 18 of this volume) and subsequent increase of trench fill thickness from below 1 km to more than 2 km (Kukowski and Oncken, Chap. 10 of this volume). Consequently, the slab rollback rate in the south must have experienced an increase over time, as opposed to the nearly stable value in the bend area of the Central Andes. This latitude-dependence of rollback velocity along the Andean margin may have been controlled by one, or several, of the mechanisms usually invoked to limit rollback (see Mantovani et al. 2001, and references therein). These mechanisms also include variations in slab pull force, mantle viscosity, mantle flow patterns, and slab length (distance to a slab window enabling the escape of lower plate mantle into the upper plate domain, as required by rollback).

However, from the correlation of several features, such as trench fill, differential trench-upper plate velocity, and slab geometry, with the Central Andean deformation record, we suspect an intricate relation between mass flux into the trench, plate coupling, slab rollback, and upper plate deformation that hints at feedback relationships in a system of coupled processes. For example, modeling by Hassani et al. (1997, comp. also Sobolev et al. 2006, Chap. 25 of this volume) suggests that slab rollback is strongly influenced by the frictional strength of the plate interface. Increased strength minimizes rollback and inhibits upper plate extension because of higher coupling, and vice versa. This relationship, in consequence, will cause a negative feedback from the following cycle: Shortening in the upper plate increases surface relief and subsequently also increases erosion, resulting in a higher trench-ward mass flux. This entails a decrease in coupling, an enhanced rollback rate and reduced efficiency of shortening. Conversely, extension of the upper plate creates accommodation space for trapping sediment, thereby reducing the mass flux towards the trench and increasing the interplate frictional strength, hence reducing rollback and supporting shortening.

This negative feedback between upper plate deformation and the potential of surface mass flux to affect the strength of the plate boundary tends, therefore, to stabilize the trench-upper plate system. This leads to a steady state of the latter with minimized rate of upper plate deformation. Overcoming this negative feedback to maintain higher shortening rates thus requires additional mechanisms that interrupt or disturb the feedback cycle. Arid conditions and slow shortening rates, for example, have the potential to prevent significant sediment flux into the trench, thereby maintaining high frictional strength at the plate interface.

Hence, upper plate shortening of the South American margin may not have been governed solely by the differential velocity between trench and upper plate. The inability of the central parts of the orogenic system (15–30°) to generate significant sediment flux in a climate belt with little erosion potential and high convergence rate tended to minimize the thickness of the trench fill, also affecting shortening of the upper plate in this region. Moreover, it appears that the slab rollback velocity of the Nazca Plate is also, at least, partly limited by the sediment flux into the trench. This assessment builds on the observation that the sediment flux is controlled by the latitude-dependent, climatic regime and deformation-induced topography, with a systematic trend of decreasing shortening along the Central and Southern Andean margin towards the south. In conclusion, therefore, the position of the Central Andean Plateau in the southern hemisphere, global arid belt is not a coincidence but a requirement for substantial upper plate shortening through the maintenance of low trench-ward sediment flux.

The spatial distribution of deformation, and its magnitude, in the Central Andes shows a correlated length of linked synchronized fault systems at time windows greater than five million years that match the later extent of the plateau between 17–22° S. The deformation of the limbs of the plateau beyond these limits, the Puna and its equivalent in Peru, appears to be delayed with respect to that in the Altiplano domain. From comparison with the scaling properties of various causes suggested, we conclude that the localization of deformation was mainly controlled by: (a) a thermally perturbed domain in the upper plate mantle possibly related to partial removal of the mantle lithosphere (cf. Isacks 1988; Wdowinski and Bock 1994; Allmendinger et al. 1997; Mahlburg Kay et al. 1999) that distributed upper plate deformation in a wide zone of diffuse pure shear contraction; (b) major mechanical heterogeneities that localize the initial crustal deformation and laterally delimit the plateau as well as the domain of pure shear crustal thickening (Eocene arc in the west versus Mesozoic rifts in the east); and (c) the large-scale, thick Paleozoic basin in the back-arc that controlled foreland failure and allowed the formation of a thin-skinned thrust belt with maximum shortening in bulk simple shear style. All three effects determine the primary mechanical instability of the Central South American leading edge, causing strain localization, synchronization of faults, and the amount of shortening accumulation. In contrast, the impact of oceanic plate properties – convergence rate, obliquity, or topographic features (ridges, plateaux, ocean islands) – appears to have been negligible, with exception of its effect on the lateral extent of the thermal perturbation in the upper mantle.

In summary, only a very particular combination of parameters appears to be able to trigger plateau-style deformation at a convergent continental margin. We suggest that this combination mainly involves the evolution of the differential trench-upper plate velocity, strong coupling at the plate interface owing to low trench infill associated with a low erosion regime, and the lateral distribution of weak zones in the leading edge of the upper plate. Some of these mechanisms or conditions are partly related via feedback cycles. This unique combination has only very rarely persisted during Earth history. As noted by Russo and Silver (1996, see also Jordan and Allmendinger 1986, and Coney and Evenchick 1994), cordillera formation at the eastern margins of the Pacific, including both the Cenozoic Andes and, possibly also, the Laramide/Sevier North American Cordillera, are among the few identified representatives of plateau-style deformation that are related to rapid upper plate displacements. Yet, while plate-scale mantle flux is probably responsible for the large-scale kinematics of the plate system and cordillera formation at the leading edge of the upper plate, as suggested by these authors (see also Silver et al. 1998), the particular style and distribution pattern of shortening in the Central Andes, as opposed to the much lower deformation in the Northern and Southern Andes, requires the additional establishment of the above conditions.

## Acknowledgments

Research for the above project was funded by the German Science Foundation (DFG) under the auspices of the Collaborative Research Center 267 'Deformation Processes in the Andes' and the Leibniz-Program (On7/10-1). This support is gratefully acknowledged. A great number of people have helped through discussion of our results as well as in finalizing the manuscript. OO, KE, and PV wish to thank our Chilean colleagues, Guillermo Chong, Hans Wilke, and Gabriel González (Universidad Catolica del Norte, Antofagasta, Chile) for the introduction to the field and many stimulating discussions. Also, we gratefully acknowledge support by the former National Bolivian Oil Industry (YPFB) which provided the seismic sections in the context of the ANCORP cooperation, and by the Chilean Oil Industry (ENAP). We extend our particular thanks to David Tufiño and Ramiro Suárez (YPFB) as well as Oscar Aranibar and Eloy Martínez (Chaco SA). JK would like to thank his colleagues in Salta, especially C.R. Monaldi for years of continuous support and collaboration. Thanks also to V. Carlotto (Lima) for providing data and discussion on shortening in Peru. B. Trumbull provided the data on volcanoes; M. Dziggel has drawn some of the figures; a large number of colleagues have contributed to the development of the ideas in this paper, most notably T. Vietor, S. Sobolev, and N. Kukowski, although we claim complete responsibility for any erroneous interpretations and biased views. Ed Sobel and Allison Britt have taken great efforts in reviewing an earlier version of this manuscript and in improving its language. Last, but not least, Victor Ramos, Rick Allmendinger and Rubén Somoza provided thoughtful reviews that helped to clarify a number of issues. We thank all these colleagues for their collaboration and help.

## References

Allmendinger RW (1986) Tectonic development, southeastern border of the Puna Plateau, northwestern Argentine Andes. Geol Soc Amer Bull 97:1070–1082

Allmendinger RW, Gubbels T (1996) Pure and simple shear plateau uplift, Altiplano-Puna, Argentina and Bolivia. Tectonophysics 259(1–3):1–13

Allmendinger RW, Zapata TR (2000) The footwall ramp of the Subandean decollement, northernmost Argentina, from extended correlation of seismic reflection data. Tectonophysics 321(1):37–55

Allmendinger RW, Ramos VA, Jordan TE, Palma M, Isacks BL (1983) Paleogeography and Andean structural geometry, Northwest Argentina. Tectonics 2:1–16

Allmendinger RW, Jordan TE, Kay SM, Isacks BL (1997) The evolution of the Altiplano-Puna Plateau of the Central Andes. Ann Rev Earth Planet Sci 25:139–174

Alpers CN, Brimhall GH (1988) Middle Miocene climatic change in the Atacama Desert, northern Chile: evidence from supergene mineralization at La Escondida. Geol Soc Amer Bull 100(10):1640–1656

ANCORP-Working Group (2003) Seismic imaging of an active continental margin – the Central Andes (ANCORP '96). J Geophys Res 108: doi 10.1029/2002JB001771

Babeyko AY, Sobolev SV, Trumbull RB, Oncken O, Lavier LL (2002) Numerical models of crustal scale convection and partial melting beneath the Altiplano-Puna Plateau. Earth Planet Sci Lett 199(3–4):373–388

Baby P, Hérail G, Lopez JM, Oller J, Pareja J, Sempere T, Tufiño D (1989) Structure de la Zone subandine de Bolivie: influence de la géométrie des séries sédimentaires antéorogéniques sur la propagation des chevauchements. Comptes Rendus, Série 2, 309:1717–1722

Baby P, Hérail G, Salinas R, Sempere T (1992) Geometry and kinematic evolution of passive roof duplexes deduced from cross section balancing: example from the foreland thrust system of the southern Bolivian Subandean zone. Tectonics 11(3):523–536

Baby P, Guillier B, Oller J, Montemurro G (1993) Modèle cinématique de la Zone subandine du coude de Santa Cruz (entre 16° S et 19° S, Bolivie) déduit de la construction de cartes equilibrées. Comptes Rendus, Série 2, 317:1477–1483

Baby P, Moretti I, Guillier B, Limachi R, Mendez E, Oller J, Specht M (1995) Petroleum system of the northern and central Bolivian Sub-Andean Zone. In: Tankard AJ, Suárez RS, Welsink HJ (eds) Petroleum basins of South America. AAPG Mem, pp 445–458

Baby P, Rochat P, Mascle G, Hérail G (1997) Neogene shortening contribution to crustal thickening in the back arc of the Central Andes. Geology 25(10):883–886

Beck SL, Zandt G (2002) The nature of orogenic crust in the central Andes. J Geophys Res 107: doi 10.1029/2000JB000124

Benjamin MT, Johnson NM, Naeser CW (1987) Recent rapid uplift in the Bolivian Andes: evidence from fission-track dating. Geology 15(7):680–683

Bevis M, Kendrick E, Smalley R Jr, Brooks B, Allmendinger R, Isacks B (2001) On the strength of interplate coupling and the rate of back arc convergence in the central Andes: an analysis of the interseismic velocity field. Geochem Geophys Geosyst 2: doi 2001GC000198

Carlotto V (1998) Evolution andine et raccourcissement au niveau de Cusco (13°–16° S, Pérou). PhD thesis, Université de Grenoble

Carlotto V, Carlier G, Jaillard E, Sempere T, Mascle G (1999) Sedimentary and structural evolution of the Eocene-Oligocene Capas Rojas Basin: evidence for a late Eocene lithospheric delamination event in the southern Peruvian Altiplano. 4th ISAG, Göttingen, pp 141–145

Charrier R, Baeza O, Elgueta S, Flynn JJ, Gans P, Kay SM, Muñoz N, Wyss AR, Zurita E (2002) Evidence for Cenozoic extensional basin development and tectonic inversion south of the flat-slab segment, southern Central Andes, Chile (33°–36° S.L.). J S Amer Earth Sci 15(1):117–139

Cladouhos TT, Allmendinger RW, Coira B, Farrar E (1994) Late Cenozoic deformation in the central Andes: fault kinematics from the northern Puna, northwestern Argentina and southwestern Bolivia. J S Amer Earth Sci 7(2):209–228

Clift PD, Vannucchi P (2004) Controls on tectonic accretion versus erosion in subduction zones: implications for the origin and recycling of the continental crust. Rev Geophys 42: doi 10.1029/2003RG000127

Clift PD, Pecher I, Kukowski N, Hampel A (2003) Tectonic erosion of the Peruvian forearc, Lima Basin, by subduction and Nazca Ridge collision. Tectonics 22(3):1023 doi 10.1029/2002TC001386

Coney PJ, Evenchick CA (1994) Consolidation of the American cordilleras. J S Amer Earth Sci 7:241–262

Coutand I, Chauvin A, Cobbold PR, Gautier P, Roperch P (1999) Vertical axis rotations across the Puna Plateau (northwestern Argentina) from paleomagnetic analysis of Cretaceous and Cenozoic rocks. J Geophys Res 104(10):22965–22984

Coutand I, Gautier P, Cobbold P, De Urreiztieta M, Chauvin A, Gapais D, Rossello E, Lopez-Gammundi O (2001) Style and history of Andean deformation, Puna Plateau, northwestern Argentina. Tectonics 20:210–234

Cristallini E, Cominguez AH, Ramos VA (1997) Deep structure of the Metan-Guachipas region: tectonic inversion in northwestern Argentina. J S Amer Earth Sci 10(5-6):403–421

Dalmayrac B, Molnar P (1981) Parallel thrust and normal faulting in Peru and constraints on the state of stress. Earth Planet Sci Lett 55(3):473–481

De Urreiztieta M, Gapais D, Le CC, Cobbold PR, Rossello E (1996) Cenozoic dextral transpression and basin development at the southern edge of the Puna Plateau, northwestern Argentina. Tectonophysics 254(1-2):17–39

DeCelles PG, Horton BK (2003) Early to middle Tertiary foreland basin development and the history of Andean crustal shortening in Bolivia. Geol Soc Amer Bull 115(1):58–77

Dunn JF, Hartshorn KG, Hartshorn PW (1995) Structural styles and hydrocarbon potential of the Subandean thrust belt of southern Bolivia. In: Tankard AJ, Suárez RS, Welsink HJ (eds) Petroleum basins of South America. AAPG Mem 62:523–543

Echavarria L, Hernandez R, Allmendinger R, Reynolds J (2003) Subandean thrust and fold belt of northwestern Argentina: geometry and timing of the Andean evolution. AAPG Bull 87(6):965–985

Ege H (2004) Exhumations- und Hebungsgeschichte der zentralen Anden in Südbolivien (21° S) durch Spaltspur-Thermochronologie an Apatit. PhD thesis, Freie Universität Berlin, http://www.diss.fu-berlin.de/2004/64/index.html

Elger K, Oncken O, Glodny J (2005) Plateau-style accumulation of deformation – the Southern Altiplano. Tectonics 24: doi 10.1029/2004TC001675

Engdahl ER, Van der Hilst RD, Buland RP (1998) Global teleseismic earthquake relocation with improved travel times and procedures for depth determination. Bull Seismo Soc Amer 88(3):722–743

Farías M, Charrier R, Comte D, Martinod J, Hérail G (2005). Late Cenozoic uplift of the western flank of the Altiplano: evidence from the depositional, tectonic, and geomorphologic evolution and shallow seismic activity (northern Chile at 19°30' S). Tectonics 24(4): doi 10.1029/2004TC001667

Garzione CN, Molnar P, Libarkin JC, MacFadden BJ (2006) Rapid late Miocene rise of the Bolivian Altiplano: evidence for removal of mantle lithosphere. EPSL 241, doi 10.1016/j.epsl.2005.11.026

Gaupp R, Kött A, Wörner G (1999) Palaeoclimatic implications of Mio-Pliocene sedimentation in the high-altitude intra-arc Lauca Basin of northern Chile. Palaeogeogr Palaeoclimat Palaeoecol 151:79–100

Gephart JW (1994) Topography and subduction geometry in the Central Andes: clues to the mechanics of a noncollisional orogen. J Geophys Res 99(6):12279–12288

Ghosh P, Garzione CN, Eiler JM (2006) Rapid uplift of the Altiplano revealed through $^{13}C$-$^{18}O$ bonds in paleosol carbonates. Science 311: doi 10.1126/science.1119365

Gil W, Baby P, Marocco R, Ballard JF (1999) North-south structural evolution of the Peruvian Subandean Zone. 5$^{th}$ ISAG, pp 278–282

Giese P, Scheuber E, Schilling F, Schmitz M, Wigger P (1999) Crustal thickening processes in the Central Andes and the different natures of the Moho-discontinuity. J S Amer Earth Sci 12:201–220

González E (1989) Hydrocarbon resources in the coastal zone of Chile. In: Ericksen GE, Cañas Pinochet MT, Reinemund JA (eds) Geology of the Andes and its relation to hydrocarbon and mineral resources. Circum-Pacific Council for Energy and Mineral Resources Earth Science Series, Houston, Texas, pp 383–404

Grier ME, Salfity JA, Allmendinger RW (1991) Andean reactivation of the Cretaceous Salta Rift, northwestern Argentina. J S Amer Earth Sci 4(4):351–372

Gubbels TL, Isacks BL, Farrar E (1993) High-level surfaces, plateau uplift, and foreland development, Bolivian Central Andes. Geology 21:695–698

Gutscher MA, Maury R, Eissen JP, Bourdon E (2000a) Can slab melting be caused by flat subduction? Geology 28(6):535–538

Gutscher MA, Spakman W, Bijwaard H, Engdahl ER (2000b) Geodynamics of flat subduction: seismicity and tomographic constraints from the Andean margin. Tectonics 19(5):814–833

Hartley AJ, Chong G (2002) Late Pliocene age for Atacama Desert: implications for the desertification of western South America. Geol Soc Amer Bull 30(1):43–46

Hartley AJ, May G, Chong G, Turner P, Kape SJ, Jolley EJ (2000) Development of a continental forearc: a Cenozoic example from the Central Andes, northern Chile. Geology 28(4):331–334

Haschke M, Guenther A (2003) Balancing crustal thickening in arcs by tectonic vs. magmatic means. Geology 31(11):933–936

Hassani R, Jongmans D, Chéry J (1997) Study of plate deformation and stress in subduction processes using two-dimensional numerical models. J Geophys Res 102:17951–17965

Heuret A, Lallemand S (2005) Plate motions, slab dynamics and back-arc deformation. Phys Earth Planet Interior 149:31–51

Hérail G, Baby P, López M, Oller J, López O, Salinas R, Sempéré T, Beccar G, Toledo H (1990) Structure and kinematic evolution of Subandean thrust system, Bolivia. ISAG, Grenoble, pp 179–182.

Hérail G, Rochat P, Baby P, Aranibar O, Lavenue A, Mascle G (1993) El Altiplano Norte de Bolivia: Evolucion Geologica Terciaria. Actas del II Simposio de Estudios Altiplánicos, 19–21 Octubre 1993, Santiago de Chile, pp 33–44

Hindle D, Kley J, Klosko E, Stein S, Dixon T, Norabuena E (2002) Consistency of geologic and geodetic displacements during Andean orogenesis. Geophys Res Lett 29(8): doi 10.1029/2001GL013757

Hindle D, Kley J, Oncken O, Sobolev SV (2005) Crustal flux and crustal balance from shortening in the Central Andes. Earth Planet Sci Lett 230:113–124

Horton BK (1999) Erosional control on the geometry and kinematics of thrust belt development in the Central Andes. Tectonics 18(6):1292–1304

Horton BK (2005) Revised deformation history of the central Andes: inferences from Cenozoic foredeep and intramontane basins of the Eastern Cordillera, Bolivia. Tectonics 24: doi 10.1029/2003TC001619

Horton BK, Hampton BA, Waanders GL (2001) Paleogene synorogenic sedimentation in the Altiplano Plateau and implications for initial mountain building in the Central Andes. Geol Soc Amer Bull 113(11):1387–1400

Horton BK, Hampton BA, Lareau BN, Baldellon E (2002) Tertiary provenance history of the northern and central Altiplano (Central Andes, Bolivia): a detrital record of plateau-margin tectonics. J Sed Res 72(5):711–726

Husson L, Sempere T (2003) Thickening the Altiplano crust by gravity-driven crustal channel flow. Geophys Res Lett 30(5): doi 10.1029/2002GL016877

Isacks BL (1988) Uplift of the Central Andean Plateau and bending of the Bolivian Orocline. J Geophys Res 93(B4):3211–3231

Jacobshagen V, Müller J, Wemmer K, Ahrendt H, Manutsoglu E (2002) Hercynian deformation and metamorphism in the Cordillera Oriental of Southern Bolivia, Central Andes. Tectonophysics 345(1-4):119–130

Jordan TE, Allmendinger RW (1986) The Sierras Pampeanas of Argentina: a modern analogue of Rocky Mountain foreland deformation. Amer J Sci 286:737–764

Jordan TE, Alonso RN (1987) Cenozoic stratigraphy and basin tectonics of the Andes Mountains, 20°–28° South latitude. AAPG Bull 71:49–64

Jordan TE, Allmendinger RW, Damati JF, Drake RE (1993) Chronology of motion in a complete thrust belt: the Precordillera, 30–31 degrees S, Andes Mountains. J Geol 101:135–156

Jordan TE, Burns WM, Veiga R, Pangaro F, Copeland P, Kelley S, Mpodozis C (2001) Extension and basin formation in the Southern Andes caused by increased convergence rate: a mid-Cenozoic trigger for the Andes. Tectonics 20(3):308–324

Kay RW, Kay SM (1993) Delamination and delamination magmatism. Tectonophysics 219:177–189

Kennan L, Lamb S, Rundle C (1995) K-Ar dates from the Altiplano and Cordillera Oriental of Bolivia: implications for Cenozoic stratigraphy and tectonics. J S Amer Earth Sci 8(2):163–186

Kley J (1996) Transition from basement-involved to thin-skinned thrusting in the Cordillera Oriental of southern Bolivia. Tectonics 15(4):763–775

Kley J (1999) Geologic and geometric constraints on a kinematic model of the Bolivian orocline. J S Amer Earth Sci 12(2):221–235

Kley J, Monaldi CR (1998) Tectonic shortening and crustal thickness in the Central Andes: how good is the correlation? Geology 26(8):723–726

Kley J, Monaldi CR (1999) Estructura de las Sierras Subandinas y del Sistema de Santa Bárbara. In: González BG, Omarini RH, Viramonte JG, Bossi GE, Coira B, Sureda RJ (eds) Geología del Noroeste Argentino, Tomo I. Relatorio-Congreso Geologica Argentino, Salta 14(1):415–425

Kley J, Monaldi CR (2002) Tectonic inversion in the Santa Barbara system of the central Andean foreland thrust belt, northwestern Argentina. Tectonics 21(6): doi 10.1029/2002TC902003

Klotz J, Khazaradze G, Angermann D, Reigber C, Perdomo R, Cifuentes O (2001) Earthquake cycle dominates contemporary crustal deformation in Central and Southern Andes. Earth Planet Sci Lett 93:437–446

Kober F, Schlunegger F, Ivy OS, Wieler R (2002) The dependency of cosmogenic nuclides to climate and surface uplift in transient landscapes, Abstracts 12th annual VM Goldschmidt conference. Pergamon Oxford International, p 408

Kudrass HR, Von Rad U, Seyfried H, Andruleit H, Hinz K, Reichert C (1998) Age and facies of sediments of the northern Chilean continental slope – evidence for intense vertical movements. 03G0104A, BGR Hannover

Kuhn D (2002) Fold-and-thrust belt structures and strike-slip faulting at the SE margin of the Salar de Atacama Basin, Chilean Andes. Tectonics 21(4): doi 10.1029/2001TC901042

Kukowski N, Oncken O (2006) Subduction erosion at the Chile-Peru margin. In: Oncken O, Chong G, Franz G, Giese P, Götze H-J, Ramos VA, Strecker MR, Wigger P (eds) The Andes – active subduction orogeny. Frontiers in Earth Science Series, Vol 1. Springer-Verlag, Berlin Heidelberg New York, pp 217–236, this volume

Lallemand SE, Schnuerle P, Malavieille J (1994) Coulomb theory applied to accretionary and nonaccretionary wedges: possible causes for tectonic erosion and/ or frontal accretion. J Geophys Res 99(6):12033–12055

Lamb S, Davis P (2003) Cenozoic climate change as a possible cause for the rise of the Andes. Nature 425:792–797

Lamb S, Hoke L (1997) Origin of the high plateau in the Central Andes, Bolivia, South America. Tectonics 16(4):623–649

Mahlburg Kay S, Mpodozis C, Coira B (1999) Neogene magmatism, tectonism, and mineral deposits of the Central Andes (22° to 33° S Latitude). Soc Econom Geol Spec Publ 7:27–59

Maksaev V, Zentilli M (1999) Fission track thermochronology of the Domeyko Cordillera, northern Chile: implications for Andean tectonics and porphyry copper metallogenesis, Latin American mineral deposits. Can Inst Mining Metall Petrol, Montreal, Canada, pp 65–89

Mantovani E, Viti M, Babbucci D, Tamburelli C, Albarello D (2001) Back arc extension: which driving mechanism? In: Jessell MW (ed) General Contributions. J Virt Explor 3:17–45

Marrett RA, Strecker MR (2000) Response of intracontinental deformation in the Central Andes to late Cenozoic reorganization of South American Plate motions. Tectonics 19(3):452–67

Marrett RA, Allmendinger RW, Alonso RN, Drake RE (1994) Late Cenozoic tectonic evolution of the Puna Plateau and adjacent foreland, northwestern Argentine Andes. J S Amer Earth Sci 7(2):179–207

Masek JG, Isacks BL, Gubbels TL, Fielding EJ (1994) Erosion and tectonics at the margins of continental plateaus. J Geophys Res 99(7):13941–13956

Mathalone JMP, Montoya M (1995) Petroleum Geology of the Sub-Andean Basins of Peru. In: Tankard AJ, Suárez SR, Welsink HJ (eds) Petroleum Basins of South America. AAPG Mem 62:423–444

McNulty B, Farber D (2002) Active detachment faulting above the Peruvian flat slab. Geology 30(6):567–570

McQuarrie N (2002) Initial plate geometry, shortening variations, and evolution of the Bolivian Orocline. Geology 30(10):867–870

McQuarrie N, DeCelles P (2001) Geometry and structural evolution of the Central Andean backthrust belt, Bolivia. Tectonics 20(5):669–692

Mégard F (1984) The Andean orogenic period and its major structures in central and northern Peru. J Geol Soc Lond 141:893–900

Mercier JL, Sébrier M, Lavenu A, Cabrera J, Bellier O, Dumont JF, Machare J (1992) Changes in the tectonic regime above a subduction zone of Andean type: the Andes of Peru and Bolivia during the Pliocene-Pleistocene. J Geophys Res 97(8):11945–11982

Montgomery DR, Balco G, Willett SD (2001) Climate, tectonics, and the morphology of the Andes. Geology 29(7):579–582

Müller JP, Kley J, Jacobshagen V (2002) Structure and Cenozoic kinematics of the Eastern Cordillera, southern Bolivia (21° S). Tectonics 21(5): doi 10.1029/2001TC001340

Muñoz N, Charrier R (1996) Uplift of the western border of the Altiplano on a westvergent thrust system, northern Chile. J S Amer Earth Sci 9(3-4):171–181

New M, Lister D, Hulme M, Makin I (2002) A high-resolution data set of surface climate over global land areas. Climate Res 21:1–25

Pankhurst RJ, Rapela CW, Saavedra J, Baldo E, Dahlquist J, Pascua I, Fanning CM (1998) The Famatinian magmatic arc in the central Sierras Pampeanas: an Early to Mid-Ordovician continental arc on the Gondwana margin. In: Pankhurst RJ, Rapela CW (eds) The Proto-Andean margin of Gondwana. Geol Soc Lond Spec Publ 142:343–367

Pardo Casas F, Molnar P (1987) Relative motion of the Nazca (Farallon) and South American Plates since late Cretaceous time. Tectonics 6(3):233–248

Pope DC, Willett SD (1998) Thermal-mechanical model for crustal thickening in the Central Andes driven by ablative subduction. Geology 26(6):511–514

Rapela CW, Pankhurst RJ, Casquet C, Baldo E, Saavedra J, Galindo C, Fanning CM (1998) The Pampean orogeny of the southern Proto-Andes: Cambrian continental collision in the Sierra de Córdoba. In: Pankhurst RJ, Rapela CW (eds) The Proto-Andean margin of Gondwana. Geol Soc Lond Spec Publ 142:181–217

Reutter K-J, Scheuber E, Chong G (1996) The Precordilleran fault system of Chuquicamata, northern Chile: evidence for tectonic inversion along arc-parallel strike slip faults. Tectonophysics 259:213–228

Riller U, Oncken O (2003) Growth of the central Andean Plateau by tectonic segmentation is controlled by the gradient in crustal shortening. J Geol 111:367–384

Rochat P, Hérail G, Baby P, Mascle G (1999) Bilan crustal et contrôle de la dynamique érosive et sédimentaire sur les mécanismes de formation de l'Altiplano. Comptes Rendus, Série 2, 328(3):189–195

Roeder D (1988) Andean-age structure of Eastern Cordillera (Province of La Paz, Bolivia). Tectonics 7(1):23–39

Roeder D, Chamberlain RL (1995) Structural Geology of Sub-Andean Fold and Thrust Belt in Northwestern Bolivia. In: Tankard AJ, Suárez RS, Welsink HJ (eds) Petroleum Basins of South America. AAPG Mem 62:459–479

Russo RM, Silver PG (1996) Cordillera formation, mantle dynamics, and the Wilson cycle. Geology 24(6):511–514

Rutland RWR (1971) Andean orogeny and ocean floor spreading. Nature 233:252–255

Salfity JA, Marquillas RA (1994) Tectonic and sedimentary evolution of the Cretaceous-Eocene Salta Group basin, Argentina. In: Salfity JA (eds) Cretaceous tectonics of the Andes. Vieweg, Braunschweig, pp 266–315

Scheuber E, Reutter KJ (1992) Magmatic arc tectonics in the Central Andes between 21 and 25° S. Tectonophysics 205:127–140

Scheuber E, Bogdanic T, Jensen A, Reutter KJ (1994) Tectonic development of the north Chilean Andes in relation to plate convergence and magmatism since the Jurassic. In: Reutter KJ, Scheuber E, Wigger PJ (eds) Tectonics of the southern Central Andes, structure and evolution of an active continental margin. Springer, Berlin Heidelberg New York, pp 121–139

Schmidt CJ, Astini RA, Costa CH, Gardini CE, Kraemer PE (1995) Cretaceous Rifting, Alluvial Fan Sedimentation, and Neogene Inversion, Southern Sierras Pampeanas, Argentina. In: Tankard AJ, Suárez SR, Welsink HJ (eds) Petroleum Basins of South America. AAPG Mem 62:341–358

Sempere T, Hérail G, Oller J, Bonhomme MG (1990) Late Oligocene-Early Miocene major tectonic crisis and related basins in Bolivia. Geology 18(10):946–949

Sempere T, Carlier G, Soler P, Fornari M, Carlotto V, Jacay J, Arispe O, Neraudeau D, Cardenas J, Rosas S, Jimenez N (2002) Late Permian-Middle Jurassic lithospheric thinning in Peru and Bolivia, and its bearing on Andean-age tectonics. Tectonophysics 345:153–181

Sheffels BM (1990) Lower bound on the amount of crustal shortening in the central Bolivian Andes. Geology 18:812–815

Sheffels BM (1995) Is the bend in the Bolivian Andes an orocline? Petroleum basins of South America. In: Tankard AJ, Soruco RS, Welsink HJ (eds) Petroleum basins of South America. AAPG Mem 62:511–522

Silva González P (2004) Der südliche Altiplano im Tertiär: Sedimentäre Entwicklung und tektonische Implikationen. PhD thesis, Freie Universität Berlin, http://www.diss.fuberlin. de/2004/125/index.html

Silver PG, Russo RM, Lithgow BC (1998) Coupling of South American and African Plate motion and plate deformation. Science 279:60–63

Sobel ER, Hilley GE, Strecker MR (2003) Formation of internally drained contractional basins by aridity-limited bedrock incision. J Geophys Res 108(7): doi 10.1029/2002JB001883

Sobolev SV, Babeyko AY, Koulakov I, Oncken O, Vietor T (2006) Mechanism of the Andean orogeny: insight from the numerical modeling. In: Oncken O, Chong G, Franz G, Giese P, Götze H-J, Ramos VA, Strecker MR, Wigger P (eds) The Andes – active subduction orogeny. Frontiers in Earth Science Series, Vol 1. Springer-Verlag, Berlin Heidelberg New York, pp 513–536, this volume

Somoza R (1998) Updated Nazca (Farallon)-South America relative motions during the last 40 My: Implications for mountain building in the central Andean region. J S Amer Earth Sci 11(3):211–215

Springer M, Förster A (1998) Heat flow density across the Central Andean subduction zone. Tectonophysics 291:123–139

Tassara A, Götze HJ, Schmidt S, Hackney R (2006) Three-dimensional density model of the Nazca Plate and Andean continental margin. Submitted to J Geophys Res

Trumbull RB, Riller U, Oncken O, Scheuber E, Munier, K, Hongn F (2006) The time-space distribution of Cenozoic arc volcanism in the South-Central Andes: a new data compilation and some tectonic implications. In: Oncken O, Chong G, Franz G, Giese P, Götze H-J, Ramos VA, Strecker MR, Wigger P (eds) The Andes – active subduction orogeny. Frontiers in Earth Science Series, Vol 1. Springer-Verlag, Berlin Heidelberg New York, pp 29–44, this volume

Victor P, Oncken O, Glodny J (2004) Uplift of the western Altiplano Plateau: evidence from the Precordillera between 20 degrees S and 21 degrees S: northern Chile. Tectonics 23: doi 10.1029/2003TC001519

Vietor T, Echtler H (2006) Episodic Neogene southward growth of the Andean subduction orogen between 30° S and 40° S – plate motions, mantle flow, climate, and upper-plate structure. In: Oncken O, Chong G, Franz G, Giese P, Götze H-J, Ramos VA, Strecker MR, Wigger P (eds) The Andes – active subduction orogeny. Frontiers in Earth Science Series, Vol 1. Springer-Verlag, Berlin Heidelberg New York, pp 375–400, this volume

Viramonte J, Kay SM, Becchio R, Escayola M, Novitski I (1999) Cretaceous rift related magamatism in central-western South America. J S Amer Earth Sci 12(2):109–121

von Huene R, Ranero CR (2003) Subduction erosion and basal friction along the sediment-starved convergent margin off Antofagasta, Chile. J Geophys Res 108(2): doi 10.1029/2001JB001569

Wdowinski S, Bock Y (1994) The evolution of deformation and topography of high elevated plateaus. 2. Application to the Central Andes. J Geophys Res 99(4):7121–7130

Whitman D (1994) Moho geometry beneath the eastern margin of the Andes, northwest Argentina, and its implications to the effective elastic thickness of the Andean foreland. J Geophys Res 99(8):15227–15289

Yañez GA, Ranero CR, von Huene R, Diaz J (2001) Magnetic anomaly interpretation across the southern Central Andes (32°–34° S): the role of the Juan Fernandez Ridge in the late Tertiary evolution of the margin. J Geophys Res 106(4):6325–6345

Yuan X, Sobolev SV, Kind R, Oncken O, Bock G, Asch G, Schurr B, Graeber F, Rudloff A, Hanka W, Wylegalla K, Tibi R, Haberland C, Rietbrock A, Giese P, Wigger, P, Roewer P, Zandt G, Beck S, Wallace T, Pardo M, Comte D (2000) Subduction and collision processes in the Central Andes constrained by converted seismic phases. Nature 408:958–961

Yuan X, Sobolev SV, Kind R (2002) Moho topography in the Central Andes and its geodynamic implications. Earth Planet Sci Lett 199(3–4):389–402

Zachos J, Pagani M, Sloan L, Thomas E, Billups K (2001) Trends, rhythms, and aberrations in global climate 65 Ma to present. Science 292(5517):686–693

## Electronic Supplement

Databank containing all available data with information on timing and location of deformation is contained in the DVD accompanying this volume under the following filename: Deformation Data Bank.

# The Time-Space Distribution of Cenozoic Volcanism in the South-Central Andes: a New Data Compilation and Some Tectonic Implications

Robert B. Trumbull · Ulrich Riller · Onno Oncken · Ekkehard Scheuber · Kerstin Munier · Fernando Hongn

**Abstract.** The coincidence of late Paleogene to Neogene shortening and crustal thickening with vigorous volcanic activity in the central Andes has long invited speculation about a causal relationship between magmatism and deformation. In aid of understanding this and related issues, we present here a new compilation of radiometric ages, geographic location and dominant rock type for about 1 450 Cenozoic volcanic and subvolcanic centers in the south-central Andes (14–28° S). This paper describes variations in the time-space distribution of volcanism from 65 to 0 Ma, with emphasis on the post-30 Ma period where Andean-style shortening deformation and volcanism were most intense. The central Andes are unusual for the abundance of felsic ignimbrites and their distribution is shown separately from the intermediate to mafic volcanic centers which are here termed the "arc association". Overall, the time-space patterns of volcanic activity for the ignimbrite and the arc association are similar but ignimbrite distribution is more patchy and more closely associated spatially with the plateau region.

The distribution of volcanic activity as a function of longitude and age, as well as cumulative frequency curves of volcanic centers as a function of age reveal major differences in arc productivity, i.e., number vs. age of volcanic centers, from north to south along the arc. Eocene and early-mid Oligocene activity was confined to a narrow belt in the Precordillera. Post-30 Ma activity was shifted to the east and spread over a much broader area than earlier arcs, probably due to a shallower subduction angle. This phase of volcanism began at about the same time from north to south (ca. 25 Ma) but the peak activity shifted progressively southward with time. Cumulative frequency curves demonstrate that 50% of volcanic output accumulated north of 20° S by 16 Ma, whereas this level was reached for 20–23°, 23–26° and 26–28° S segments at about 12, 10 and 8 Ma, respectively. Plate reconstructions place the subducted part of the Juan Fernández Ridge beneath the arc at these latitudes between about 25 and 5 Ma, but age-frequency diagrams of volcanism show no evidence that ridge subduction influenced arc productivity.

The spatial distribution of volcanism shows some influence by crustal structures on a local scale (tens of km), most notably the preferential clustering of volcanic centers at intersections of the frontal arc with NW-SE-trending lineament zones. However, on a regional scale the time-space distribution of volcanic centers and the distribution of active shortening domains in the central Andes varied independently. The evidence does not support the concept that Andean crustal thickening and plateau formation were preconditioned by thermal weakening of the crust. A comparison of volcanic output vs. shortening rates for latitude 19–22° S confirms that the onset of intense deformation in the Oligocene preceded that of volcanism by about 10 Ma, and the increase in volcanic activity at about 20–16 Ma has no expression in shortening rates. After plateau formation, however, beginning at about 10 Ma, both shortening rate and volcanic output increased together and reached their highest levels. This period experienced extremely large-volume ignimbrite eruptions from the Altiplano-Puna volcanic complex. The ignimbrite magmas represent an episode of widespread crustal melting, and it is likely that the rise in shortening rates reflects melt-enhanced weakening of the crust.

Variations in the location and width of the CVZ arc respond to changes in slab dip, but the complex distribution of volcanic centers in time and space shown in this study belies a simple relationship. A condition that must be met for correlation between surface volcanism and slab dip is a near-vertical ascent of magmas, both in the mantle wedge and through the crust. We conclude that in the Neogene arc of the central Andes, vertical ascent of magmas is disturbed by effects of crustal heterogeneity, intense deformation, lithospheric thickening and partial delamination, and crustal melting.

## 2.1 Introduction

The Neogene history of the central Andes records a time of intense volcanism and extreme horizontal crustal shortening and vertical thickening which produced the world's second highest continental plateau. The plateau is host to an extensive volcanic province including the highest stratovolcanoes on Earth and one of the largest concentrations of felsic ignimbrites (see Coira et al. 1993 for a review). The overall spatial coincidence of intense Neogene volcanism and the central Andean Plateau has long invited speculation on the relationship between magmatism and deformation, especially as related to plateau formation. Whereas magmatic addition to the crust was once thought to be important (James 1971a; Thorpe and Francis 1979), it is now generally accepted that magmatism has made only a minor volumetric contribution to crustal thickening (e.g., Francis and Hawkesworth 1994; Allmendinger et al. 1997). However, the thermal effect of magmatism may well be significant and Isacks (1988) championed the popular idea that a spreading of magmatism over a broad area, induced by relatively flat subduction in the late Oligocene, might have preconditioned the upper plate for plateau formation by thermal weakening. James and Sacks (1999) proposed a variant of this idea, suggesting that the important process for weakening the upper plate was not heating but hydration from dewatering of a shallow slab.

The conceptually simple test of the thermal weakening concept – the relative timing of magmatic and deformation events – is difficult in practice, not least because of incomplete age information for deformation and volcanism. Differences in the scale of observation can lead to different conclusions depending on the scale at which the relevant factors operate. At the regional or lithospheric scale, these may involve the lower and upper crust as well as the mantle wedge. At the local scale, factors affecting magma storage, ascent and eruption within the brittle upper crust may be more relevant. The few studies that have addressed the relationship between volcanism and deformation came up with ambiguous or conflicting conclusions. Wörner et al. (2000) pointed out that the first sedimentological evidence for surface uplift of the Western Cordillera in northern Chile precedes the first major magmatic event. Similarly, Elger et al. (2005) concluded that the onset of shortening in the southern Altiplano Plateau of Bolivia preceded volcanism. On the other hand, Victor et al. (2004) found that phases of active shortening in the western flank of the Altiplano Plateau in northern Chile show a fair correlation with age ranges of large ignimbrite eruptions on the plateau but do not correlate with the activity of the arc volcanoes. Ramos et al. (2002) studied the eastward migration of volcanic and deformation fronts near 33° S and made the important observation that magmatism lagged behind deformation in the cordillera, whereas the opposite was true for the shift into the foreland, i.e., in the Sierras Pampeanas. They attributed this difference in relative timing to the variable strength of the crust in the two regions.

The basic question remains unanswered: whether there has been a causal relationship between magmatism and deformation in the Central Andes and if so, which of the two processes played the determining role and at what scale(s) did they interact? This paper revisits the issue using a new compilation of age and spatial data for magmatism in the south-central Andes. We incorporate recent progress in quantifying the space-time distribution and rates of crustal shortening (Oncken et al. 2006, Chap. 1 of this volume and references therein), and we also examine the possible influence of Juan Fernández Ridge subduction on volcano distribution based on the projected track of the subducted ridge from Yañez et al. (2001).

An important aspect of volcanism in the CVZ that needs to be considered in any regional survey is its compositional diversity. In particular, there is an abundance of felsic, caldera-sourced ignimbrites whose volumes locally far exceed the products of the mafic to intermediate stratovolcanoes and other types of volcanic features (cones, domes, intrusive dykes, etc.; see Table 2.1). The latter are lumped together in this paper as the "arc association" in distinction to the "ignimbrite association". It was considered important to show and discuss the distribution of these two associations separately because, as described below, the origin of the magmas involved and thus their geodynamic significance can differ.

The basis for the compilation are data assembled by various groups and individuals within the 12-year SFB-267 program, augmented by a recently-published compilation of volcanism in Peru by Rosenbaum et al. (2005). For this paper the data have been combined, screened for duplicates and brought into a consistent format and scope for all entries. The Excel spreadsheet appended contains age, location and rock type data for about 1 450 volcanoes and related dykes, stocks and ignimbrites, located between 14° S and 28° S latitude and 64° W to 74° W longitude. The northern and southern limits of the compilation are bound approximately by the Peruvian and Chilean flat-slab segments of the CVZ (Fig. 2.1). The compilation begins at about 65 Ma and carries on to the Holocene, but the focus is on the period post-40 Ma, during which most of the shortening and magmatism in the plateau region and its margins took place. Readers can refer to Coira et al. (1982), Scheuber et al. (1994), Petersen (1999) and Haschke et al. (2002) for discussions of the earlier arcs.

**Table 2.1.** List of attributes given in the compilation and key to numerical values

| | Description/explanation | | |
|---|---|---|---|
| **Volcano type** | | | |
| 1 | Intrusions (stocks, dykes) | | |
| 2 | Lava (domes, cones, flows) | | |
| 3 | Pyroclastic deposits (ash, tuff, ignimbrite) | | |
| 4 | Caldera | | |
| 5 | Stratovolcano, composite volcano | | |
| 6 | Monogenetic centre (mafic cone) | | |
| **Rock type** | | | |
| 1 | Basalt, basaltic andesite | | |
| 2 | Andesite | | |
| 3 | Andesite-dacite mix | | |
| 4 | Dacite | | |
| 5 | Dacite-rhyolite mix, rhyolite | | |
| **Age class** | | | |
| 1 | Pre-Oligocene | | > 36 Ma |
| 2 | Lower Oligocene | 30 | – 35.9 Ma |
| 3 | Upper Oligocene | 24 | – 29.9 Ma |
| 4 | Lower Miocene | 16 | – 23.9 Ma |
| 5 | Middle Miocene | 11 | – 15.9 Ma |
| 6 | Upper Miocene | 6 | – 10.9 Ma |
| 7 | Lower Pliocene | 3 | – 5.9 Ma |
| 8 | Upper Pliocene | 1.6 | – 2.9 Ma |
| 9 | Pleistocene | 0.01 | – 1.5 Ma |
| 10 | Holocene or active | | |

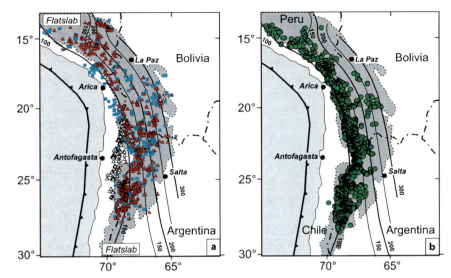

**Fig. 2.1.**
Location maps of the central Andes showing the area above 3 000 m in *dark shading* and the slab depth contours from Cahill and Isacks (1992). The latter are the basis for defining flat-slab regions in Peru and Chile as shown; **a** gives the location of dated volcanic and subvolcanic centers compiled for this study and listed in the data appendix. Symbols: *black triangles* for pre-30 Ma centers, *red triangles* for the post-30 Ma arc association, blue squares for post-30 Ma ignimbrites (see text for explanation); **b** shows volcanic features compiled from Landsat images by de Silva and Francis (1991)

## 2.2 Data Compilation and Reliability

Setting up an internally consistent and workable data set from diverse sources required compromises between data quality and completeness. Our philosophy was to include a maximum number of observations so as to achieve the best possible coverage and to accept the limitations of individual precision in age and location. To achieve a consistent and useful data structure, we imposed a simple, standard set of attributes for volcanic types, rock types and ages. These attributes are assigned numerical values for convenience in graphics applications (symbol codes, data sorting, etc.), as explained in Table 2.1. A problem was the incomplete or imprecise rock descriptions and/or inconsistent nomenclature given in the literature. This made it impossible to simply take the given description for each entry and still achieve a workable structure in the dataset. Our attributes for "volcano types" comprise six possibilities although descriptions in the data sources are obviously more variable. Only one type was assigned for each entry, so for example, if the source describes a dated lava or pyroclastic flow from a particular stratovolcano, the entry is listed under type 5 (stratovolcano) and not lava, whereas if samples are described only as "lava", "cone" or "dome" the entry is assigned to type 2. Descriptions of "pumice" or "tuff" are entered as type 3.

The key variables in the compilation are age and location of volcanic centers. The names of individual volcanoes or volcanic fields are given if they were given in the original reference but this is often not the case, or the description was imprecise (e.g., lava flow east of volcano xx). All locations are given in decimal degrees. The vast majority of entries include the geochronology method, material analyzed and uncertainty estimates, but we elected to also compile ages for which some of this information was lacking in the referenced source. A common problem encountered in this work was that more than one age was cited in the literature for a particular volcanic center, that conflicting locations were given in different sources, or that duplicate entries for a single center resulted from compilation and merging of existing datasets into one. Obviously, a volcano can have a history of activity that is not well represented by a single age, and it is common that application of more than one dating method results in significantly different dates. Duplicate entries for the same volcano were generally easy to recognize from identical geographic coordinates and age values. We removed as duplicates all entries whose geographic coordinates agreed within 0.01° and which had identical ages within the stated uncertainty. In cases where significantly different ages for a given locality resulted from different authors or by different dating methods, we included both ages assuming the range is real. Despite best efforts to avoid mistakes and repetition in the compilation, we cannot rule out that some duplicates and errors in location or other attributes persist. However, we suggest that they are insignificant given the large number of entries and the wide age and regional coverage represented in the compilation. One potential bias that cannot be avoided is that the very low erosion rates in the central Andes favors burial of older volcanic deposits under more recent ones. This means that the relative proportion of older volcanism will be underestimated somewhat. The extent of this problem is difficult to assess but it should be kept in mind. The bias will obviously be greater for the flat-lying and relatively thin ignimbrite deposits than for the volcanic domes and stratovolcanoes. Finally, there are certain to be omissions of some dated volcanoes and ignimbrites, and we make no claim of completeness. In particular, the national geological surveys have active dating programs related to their regional mapping and dozens of new age data can be expected each year. However, the data structure of the compilation makes additions and modifications simple.

The location of all dated samples in our compilation is shown in Fig. 2.1 and compared with the distribution of volcanic centers in the same area identified from Landsat images by de Silva and Francis (1991). This is an important comparison because the satellite compilation is essentially free of any sampling bias whereas field studies are inherently limited or at least influenced by aspects like access, national borders and financial constraints. The similarity in distribution of volcanic centers in the two maps is quite good, which gives us confidence that the data compilation provides a fairly complete regional coverage of volcanism in this part of the CVZ, without major systematic gaps. The only systematic difference between our data coverage and that of de Silva and Francis (1991) is that our pre-30 Ma group (black triangles) are lacking in the Landsat interpretation, and this is clearly because the older centers are eroded and not morphologically prominent. The ignimbrite and tuff entries present a special problem in terms of location because the erupted sources are commonly unknown and the given location reflects only the sampling site. Many of the Neogene ignimbrite deposits in northern Chile, Argentina and southern Bolivia represent extremely large eruptions from caldera complexes that themselves are several tens of km across, so the location of dated samples can only roughly represent the source location. An illustration of this problem is shown by the occurrence of a few ignimbrite points in Figs. 2.1 and 2.4 that are far to the west of the arc front, some extending almost to the coastline near 18° S and 20° S. These points represent samples collected from distal pyroclastic flows that travelled down the steep western topographic margin of the plateau. These examples are probably extreme cases of the scale of dispersion of ignimbrite deposits relative to their source but the problem is a general one and should be kept in mind. An alternative would be to plot only the known ignimbrite calderas with their ages but this would reduce the number of ignimbrite points from about 400 to only about 20. Our approach has been to include all entries that were described as ignimbrites or tuffs with the given location, and to create a separate category for calderas (4) so that the known source calderas can be plotted separately (see Fig. 2.4f). For the case of long-lived resurgent caldera systems, the assigned age represents the eruption of the main caldera-forming ignimbrite, for example the 4.1 Ma Atana ignimbrite for La Pacana (Gardeweg and Ramirez 1987) or the 2.2 Ma Galán ignimbrite for Cerro Galán (Francis et al. 1989).

## 2.3 Results: the Time-Space Variations of Volcanism

To reiterate, the data presentations and discussions in this paper distinguish the separate contributions of the ignimbrite association on the one hand (volcano types 3 and 4, see Table 2.1) and the "arc association" on the other (volcano types 1, 2, 5 and 6). It was decided not to attempt to separate the arc and back-arc volcanism because the E-W changes in arc location and width over time make this distinction uncertain without other information.

### 2.3.1 Changes in Arc Width and Position

A fundamental characteristic of magmatism in the central Andes which has long been recognized is the progressive eastward migration of the Mesozoic and early Paleogene magmatic arcs (e.g., James 1971b; Rutland 1971; Coira et al. 1982; Mpodozis and Ramos 1989; Scheuber and Reutter 1992). This migration is attributed to subduction erosion of the continental margin (see Kukowski et al. 2006, Chap. 10 of this volume, and references therein). The Jurassic to Eocene arcs maintained a fairly constant width of 50–100 km as they shifted eastwards but the late Paleogene (Oligocene-Miocene) arc, at first, was much wider and extended across much of the present plateau region. Changes in the position and width of the arc over time has commonly been shown by diagrams of age versus longitude (Scheuber et al. 1994; Allmendinger et al. 1997; Wörner et al. 2000). An example given in Fig. 2.2 from Haschke et al. (2002) emphasizes the successive eastward shifts of arc activity from about 200 Ma to 30 Ma, and the extreme broadening of the area of volcanic activity after about 30 Ma.

The age data from our compilation are plotted against longitude for five N-S segments from 14° S to 28° S in Fig. 2.3. Note that the apparent westward shift of the arc front and the greater E-W extent of volcanism in segments north of 20° S compared with those further south is due

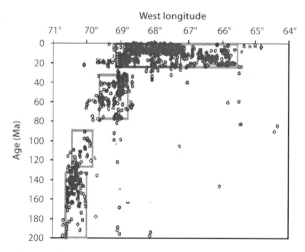

**Fig. 2.2.** Relation between age and longitude for Mesozoic and Cenozoic magmatism in the Central Andes (19–28° S) from Haschke et al. (2002). The *gray rectangles* emphasize the episodic eastward shifts of the magmatic arc since the Jurassic and the extreme widening of the magmatic arc after about 30 Ma

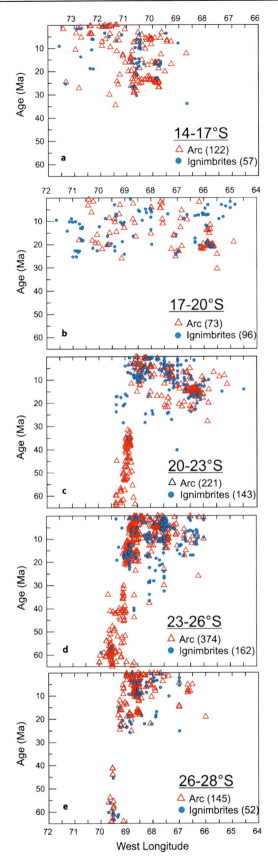

to the westward bend of the Andes orogen (see Fig. 2.1). Relative to the trench, the arc width and the position of the arc front are actually similar in all segments. The diagrams distinguish ignimbrite and arc associations by blue dots and red triangles, respectively. For the most part, the distribution patterns of the two associations are similar, but there are some important differences in detail. One is the broader E-W extent of ignimbrite deposits relative to volcanic centers. This may be in part related to the greater dispersion of ignimbrites from their sources, and it can also be due to differences in magma origin of the two associations (crustal melting vs. slab-fluxed melting in the mantle wedge – see discussion below). Another point is that ignimbrites older than 30 Ma appear to be almost completely lacking. It is not clear to what extent this simply reflects preferential erosion of tuff and ignimbrite deposits and is not a true feature of the Paleogene arc. Erosion likely is an important factor since many of the pre-30 Ma entries are subvolcanic stocks and dykes and some of these have felsic compositions.

The single most important point to note in Fig. 2.3 is that there are strong variations in the intensity and distribution of volcanism in the CVZ along the arc. This has been often stated in the past (e.g., Coira et al. 1982; 1993; Allmendinger et al. 1997; Kay et al. 1999). Volcanism in general appears most highly concentrated in the two central segments (20–23° S and 23–26° S) and it drops off to the north and the south. Furthermore, there is important volcanic activity before 30 Ma in the three segments south of 20° S (Fig. 2.3c–e) but this is lacking farther north (as Petersen 1999, shows, Eocene activity is present in the arc north of 12° S). The diagrams in Fig. 2.3 also emphasize the narrow E-W extent of pre-30 Ma volcanism and the moderate, systematic eastward migration of the early arc with time. This eastward migration corresponds to about 1.5 km Myr$^{-1}$ which agrees with independent rates of subduction erosion summarized by Kukowski et al. (2006, Chap. 10 of this volume).

Near the end of the Paleogene at about 30 Ma there is an age gap in volcanism on the order of 10 Ma in segments south of about 20° S, before re-establishment of a much broader and more complex Miocene and younger activity (< 25 Ma). There is some isolated volcanism within this age gap (see Fig. 2.3c,d) and it is interesting that these exceptions mostly involve ignimbrites. Allmendinger et al. (1997) described a broadening of the late

◂ **Fig. 2.3.** Relation between age and longitude for volcanic activity from our data compilation in the interval 65 Ma to 0 Ma. Separate diagrams (**a, b, c, d, e**) were generated for 5 segments of latitude to illustrate differences in volcanic distribution along the arc. The arc and ignimbrite associations are discriminated by *blue dots* and *red triangles*, respectively, and the number of data points for each is given in parentheses

Oligocene/early Miocene arc in the CVZ and its subsequent narrowing toward the west, which they attributed to slab steepening. Our data compilation confirms this trend for the 20–23° S segment (Fig. 2.3c). However, in the two segments south of 23° S, post-30Ma volcanism remained focused at about the same E-W position as before (69–68.5° W) and the arc front maintained its regular eastward migration at about 1–1.5 km Myr$^{-1}$ as before the volcanic gap. There was an important broadening of post-30 Ma volcanism in these segments, too, but it was less extreme than at 20–23° S, it began later (ca. 15 Ma in the 23–26° S segment and after 10 Ma at 26–28° S), and magmatism did not subsequently narrow again to the same extent. The segments north 17–20° S are different again. From 20–17 °S there is a very broad and diffuse spread of volcanism and no regular changes with age. The pattern at 14–17 °S is less spread out as in the adjacent segment to the south and shows a slight westward shift of the magmatic front with time (exaggerated on the figure due to the curvature of the arc).

### 2.3.2 Distribution Maps of Volcanic Activity and Deformation

The time-space distribution of volcanic activity in the study area is illustrated on a series of maps in Fig. 2.4 and compared with domains of shortening deformation taken from Oncken et al. 2006 (Chap. 1 of this volume). The latter are shown as yellow patches, with question marks denoting the southern limit of age information. It was difficult to choose optimum age intervals for such maps that neither impose artificial divisions nor ignore natural ones suggested by geological information. The most objective way to depict changes of volcanism with time is to use the smallest intervals allowed by the age uncertainties. This is impractical for print but can be shown with a computer animation of the distribution maps. An animation file is given in the electronic appendix that uses one million-year-intervals for the range 0 Ma to 24 Ma and two-million-year intervals for 25 Ma to 65 Ma, accounting for the lower age precision for the older centers. The ignimbrite and arc associations are distinguished by blue squares and red triangles, respectively, on the animation and on Fig. 2.4. We leave it to interested readers to examine these maps in detail and illustrate in Fig. 2.4a–e exemplary "snapshots" of volcanic distribution in the age intervals: 0 to 8, 8 to 20, 20 to 30, 30 to 37 and 37 to 46 Ma. Choice of these particular intervals was made to allow combining the patterns of volcanism with maps showing the active domains of shortening.

The shifting patterns of volcanic activity and the domains of active shortening on Fig. 2.4 show some overlap but there are enough differences to negate a causal relationship between magmatism and deformation. For the two earliest intervals of 46 Ma to 30 Ma (Fig. 2.4a,b) deformation affected both the western and eastern plateau margins, whereas magmatism was confined to the western margin and was at a very low level of activity. Volcanic activity increased after 30 Ma, and in the age intervals 30 Ma – 20 Ma and 20 Ma – 8 Ma volcanism and deformation covered similar broad areas across much of the region (Fig. 2.4c,d). The distribution of the two phenomena diverge again in the 8 Ma – 0 Ma interval (Fig. 2.4e). Whereas volcanism concentrated again in the Western Cordillera (except the ignimbrites, see below) the zone of active shortening moved to the Eastern Cordillera and Subandean Ranges. The spread of volcanism across the plateau in the late Oligocene-Miocene involved both the ignimbrite and arc associations (Fig. 2.4c,d), but the distribution of ignimbrites tends to be more patchy and irregular than that of the arc association. Moreover, important concentrations of ignimbrites remained active in the plateau region during the late Miocene and Pliocene at around 19° S (Frailes and Morococola fields, see Fig. 2.4f) and 23° S (the Altiplano-Puna Volcanic Complex, APVC) while the arc association retreated to the western Cordillera during that time.

### 2.3.3 Volcanic Clustering and the Influence of Crustal Lineaments

The maps on Fig. 2.4 show strong local clustering of volcanism in the post-20 Ma age intervals, and these clusters have been are evaluated in more detail by calculating point densities from the volcanic distribution maps and adding color contour plots of these alongside the maps in Fig. 2.5a–c. For the point density calculations, the area was divided into a grid of 20 km mesh which was chosen to exceed twice the typical stratovolcano diameter. Each cell in the grid was assigned a number giving the total of volcanic centers within a 100 km search radius, and these grid values were then color-contoured. A potential problem interpreting clusters or gaps in any compilation based on field sampling is incomplete or uneven coverage. For the CVZ, uneven coverage can arise owing to variable degrees of access or scientific motivation, and also to the fact that younger deposits, particularly regional ignimbrite sheets, tend to cover older ones. To help judge the problem of sampling bias, we add an independent map of volcanic distribution and its point density plot from the de Silva and Francis (1991) Landsat compilation. Comparison with the Landsat compilation suggests that a negative bias (gaps in coverage) is not a serious problem but there is one example of positive sampling bias, in the region of the Maricunga mineral belt of Chile at 26–28° S, 68–69° W. Here, the motivation of economic interest re-

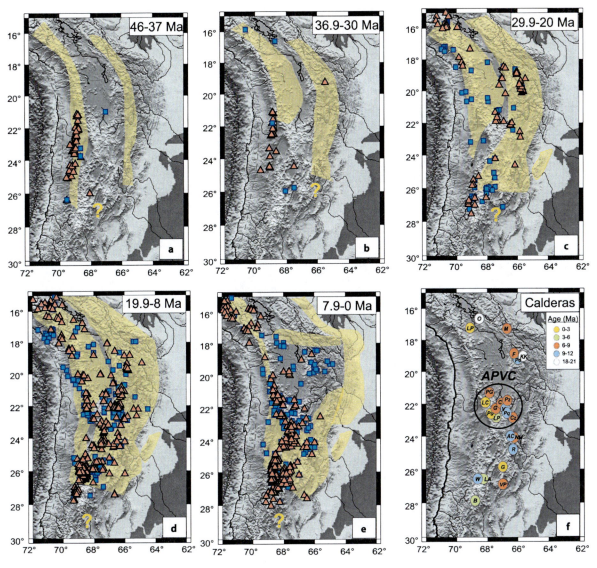

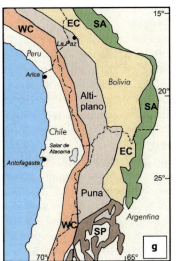

**Fig. 2.4.** *Maps* **a–e** show the time-space distribution of volcanic activity in the central Andes along with domains of active shortening deformation (*yellow shading*, from Oncken et al. 2006, Chap. 1 of this volume; question marks indicate southern limit of age information). The arc and ignimbrite associations are depicted by *red triangles* and *blue squares*, respectively. Map **f** shows the location and ages of major source calderas for felsic ignimbrites. Several of these calderas were long-lived and the assigned age represents the caldera-forming event. The Altiplano-Puna Volcanic Complex of de Silva (1989) is labeled (*APVC*). Caldera names: *O*: Oxaya (inferred location from Schröder 1999), *LP*: Lauca-Perez, *M*: Morococola, *F*: Frailes, *KK*: Kari-Kari, *Po*: Porcera, *PG*: Pastos Grandes, *Pz*: Panizos, *C*: Corutu, *LC*: Laguna Colorado, *G*: Guacha, *V*: Villama, *Pq*: Pairique, *Pu*: Purico, *LP*: LaPacana, *Cz*: Coranzuli, *AC*: Aguas Calientes, *NM*: Negra Muerta, *R*: Ramadas, *G*: Galán, *LA*: Laguna Amarga, *W*: Wheelwright, *VP*: Vicuna Pampa, *B*: Bonete (Incapillo). Map **g** shows approximate limits of the major morpho-tectonic units of the central Andes mentioned in the text: Altiplano and Puna Plateaus, Western Cordillera (*WC*), Eastern Cordillera (*EC*), Subandean Ranges (*SA*), Sierras Pampeanas (*SP*)

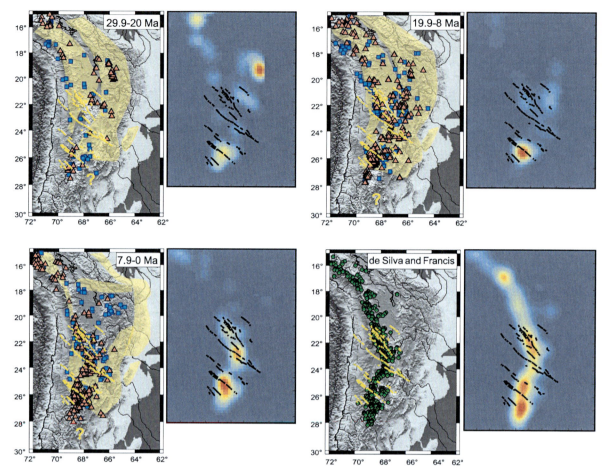

**Fig. 2.5.** Maps of volcanic distribution in the age intervals: 29–20 Ma, 19.9–8 Ma and 7.9–0 Ma from Fig. 2.4 and for comparison, the volcano distribution from Landsat images (de Silva and Francis 1991 – displayed are the two morphologically youngest categories in the compilation). The emphasis is on clustering of volcanism and the influence of major NW-SE-trending lineament zones, the latter shown as *dashed lines* as described by Salfity (1985). Point density plots are added to each map as color panels to better show the relative intensity of volcanic clusters. Color coding for point density is relative to the total number of points in each segment, with *red* for highest and *blue* for lowest density

sulted in detailed mapping and performance of over 200 age determinations for volcanic rocks (Kay et al. 1994a; McKee et al. 1995). The Maricunga area shows up as the single most prominent cluster in all age intervals shown. Proof that this cluster is an artifact of the intense work done in that area is given by comparing the cluster map made from volcanic distributions in the Landsat compilation, which is free of sampling bias.

A useful application of the point density analysis is to assess the influence of major NW-SE-striking lineament zones on the localization of volcanism. These lineament zones were compiled by Salfity (1985). Their exact geologic nature, age of formation and duration of activity are not well known, but it has long been noted that the transverse lineaments play host to several of the prominent outliers of the Neogene arc into the Puna Plateau (e.g., Coira et al. 1993). The lineament zones have attracted special attention recently because of their potential role in localizing hydrothermal systems and ore deposits, including the Chuquicamata, La Escondida, and El Salvador Cu porphyry deposits in Chile (Chernikoff et al. 2002; Matteini et al. 2002; Richards and Villenueve 2002), as well for their possible influence on formation of "trapdoor" calderas (Riller et al. 2001; Petrinovic et al. 2005, Ramelow et al. 2005). Comparison of the lineament zones with maps of volcanic activity on Fig. 2.5 shows a poor correspondence in the early and mid Miocene intervals. The influence of transverse lineaments appears to be stronger for the late Miocene and younger volcanic activity (7–0 Ma). Not only are there linear outliers of volcanic activity along lineaments east of the main arc, but also the strongest clusters of volcanoes along the arc correspond to regions intersected by the lineament zones (Fig. 2.5). This correspondence also shows up in the Landsat compilation of volcanoes of de Silva and Francis (1991) and is therefore not affected by any sampling bias. The preferential occurrence of volcanism at the lineaments may be related to a change in the deformation regime from dominantly

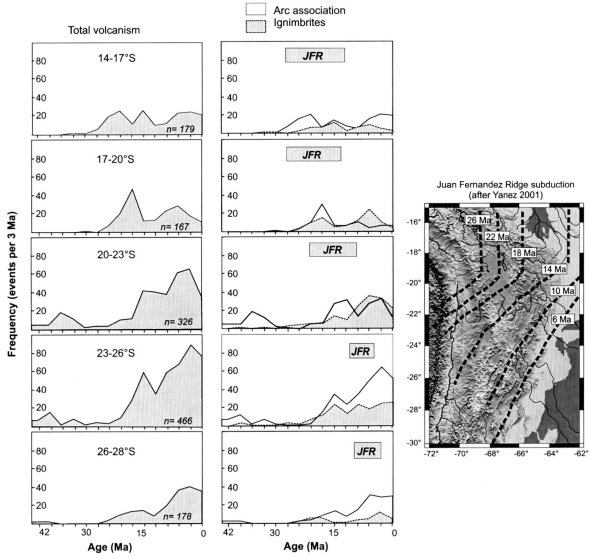

**Fig. 2.6.** Frequency-age distribution of volcanic activity from 45 to 0 Ma, displayed separately for the 3° latitudinal sections as in Fig. 2.3. The frequency axis gives total number of events in 3-Myr intervals, with points plotted at the start of each interval (at 0 for 0 to 3 Ma etc.). Separate diagrams are shown for total volcanism (*left panels*) and for arc versus ignimbrite associations (*right panels*). Gray bars in the right panels represent the time during which the Juan Fernández Ridge (*JFR*) was beneath the arc at the respective latitude, based on reconstructions of the JFR position from Yañez et al. (2001), see *inset*

vertical crustal thickening to orogen-parallel stretching accomplished chiefly by dilation on the lineaments (Riller et al. 2001).

### 2.3.4 Arc Productivity and Influence of the Juan Fernández Ridge

Changes in arc productivity with time and location along the arc are examined in frequency histograms of volcanism plotted against age in Fig. 2.6. The bin width for each histogram was set at 3 million years to accommodate typical age uncertainties of 0.5 to 2 Ma. The frequency value for each interval was plotted at the starting age of that interval, thus the value at 0 Ma represents the summed events for 0 to 3 Ma and so on. Diagrams are shown separately for consecutive 3-degree segments of latitude as in Fig. 2.3, and separate curves are shown for total volcanism as well as for the ignimbrite and arc associations alone. Readers should note that some of the smaller peaks contain less than 10 events and may partly reflect the vagaries of sampling instead of true episodicity of volcanism. Also, given the overall low degree of erosion in the central Andes, the frequency of older volcanism shown on the diagrams should be taken as minimum estimates because of the greater likelihood of younger volcanic cover.

We urge caution in interpreting individual peaks and valleys on these frequency distribution plots, but nevertheless consider several first-order characteristics to be important and robust:

1. There is generally a very low level of activity in all segments before 30 Ma as noted above, and significant pre-30 Ma activity appears to be limited to south of 20° S (compare Fig. 2.3).
2. Magmatic activity begins to increase in the 24 Ma to 27 Ma age interval all along the arc. The timing of this onset is roughly the same for both the arc and ignimbrite associations but thereafter, variations in activity of these two associations appear to be independent of each other.
3. Volcanic productivity overall is higher in segments to the south of 20° S than those farther north, and this southward increase in arc output is pronounced in the post-10 Ma time interval.

Several authors have suggested that the past history of Juan Fernández Ridge subduction influenced the arc activity in the CVZ (e.g., Gutscher et al. 2000; Yañez et al. 2001; Kay and Mpodozis 2002). If this is true, it should be evident in the frequency – age curves in Fig. 2.6. Yañez et al. (2001) reconstructed the past trajectory of the ridge based on plate motion vectors and a sketch of its positions within the study area is shown on Fig. 2.6. Using that reconstruction, we can estimate the time interval during which the subducted ridge was located beneath the volcanic arc in each of the five arc segments and those intervals are marked on the right-hand panels of Fig. 2.6 by gray bars labeled JFR. Specifically, the time intervals were derived by determining when the ridge positions from Yañez et al. (2001) crossed mid-latitude of each 3° segment for an E-W extent of 5° longitude centered on the arc, i.e., 70–65° W for the three segments south of 20° S and 72–67° W for the northern two segments. The E-W extent for each latitude is suggested as a conservative estimate of the zone where ridge subduction should affect magma production within the mantle (see also Kay and Mpodozis 2002). As an example, the midpoint of arc segment 20–23° S is 21.5° S and for that latitude, the ridge axis passed beneath longitude 70–65° W between about 24 Ma and 14 Ma. From the lack of any systematic coincidence of the JFR positions in Fig. 2.6 with highs or lows of the frequency curves, we conclude that southward passage of the subducted Juan Fernández Ridge beneath the CVZ at 14 Ma to 28 Ma had no inhibiting effect on volcanic productivity. If anything, the opposite is the case as both ridge subduction and the peak in arc activity migrate southward through the area with time. By no means was ridge subduction associated with a gap in volcanism as is the case with the Chilean and Peruvian flat-slab segments of the modern arc (Fig. 2.1).

The histograms in Fig. 2.6 suggest that the peak phase of volcanic output in the CVZ becomes progressively younger to the south, and the total arc output also appears to increase in this direction. This southward trend in arc activity is underscored in Fig. 2.7 in a series of cumulative frequency curves which reflect the relative rates at which magmatic activity in each segment built up since 40 Ma. A useful reference line on these diagrams is the age by which 50% of total activity in each segment had accumulated. In terms of total volcanism, both segments north of 20° S reached the 50% level by about 16 Ma, whereas the respective ages for the three segments to the south are 12, 10 and 8 Ma. The diagram for arc association alone shows the same south-stepping trend, but it is difficult to compare the cumulative frequency curve for 20–23° S (yellow) with the others because it includes much more pre-30 Ma activity. There is an interesting parallel between our evidence for a southward younging of the peak of arc volcanism in the CVZ and seismic tomography studies showing present-day variations in velocity and velocity attenuation in the upper mantle beneath the same region. The tomography results are summarized by Schurr et al. (2006) and Sobolev et al. 2006 (Chap. 25 of this volume). They show that the upper mantle beneath the arc (85 km depth layer) has lower P- and S-wave velocities and higher wave attenuation values south of about 23° S than it does further north, suggesting that magmatism is stronger in the south.

Note that the north-south variations in accumulation rates for the ignimbrite association are smaller than those of the arc association and they appear not to vary systematically with latitude. The age for 50% accumulation of ignimbrites is about 17 Ma in the 14–17° S segment and between 12 Ma and 9 Ma in the others. This is in keeping with the more patchy distribution of ignimbrite activity mentioned above, and may again reflect the differences in magma sources for the two volcanic associations.

## 2.4 Discussion

### 2.4.1 Magmatism and Deformation

The maps in Fig. 2.4 demonstrate that, on a regional scale, the domains of active magmatism and shortening deformation in the CVZ migrated with time largely independently of each other. Neither the onset of "Andean" deformation in the late Eocene, nor the increase in the intensity and areal extent of deformation at around 30–20 Ma were preceded by any peak in magmatic activity. In fact, the opposite is true. A similar conclusion that magmatism followed deformation was reached for the northern CVZ by Sébrier and Soler (1991) and by McQuarrie et al. (2005). In addition to the age disparity, there are differences in the regions affected by magmatism and shorten-

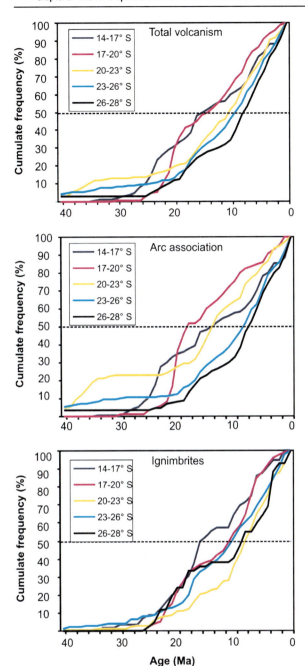

**Fig. 2.7.** Cumulate frequency against age for total volcanism (*top*) and for arc versus ignimbrite associations, individually. Separate curves are shown for the 3° latitude segments. The *horizontal line* at 50% is shown as a measure of differences in the rate of activity north to south. In terms of total volcanism for example (*top diagram*), 50% of all activity north of 20° S was achieved by 16 Ma, whereas the same level of activity was reached later in successive segments to the south

ing. Shortening deformation began at both plateau margins in the late Eocene/early Oligocene, then intensified and spread across much of the plateau in the late Oligocene and Miocene, and finally shifted to the eastern margin after about 10 Ma. Volcanism in the plateau region began later, but once started it also spread across much of the plateau in early and mid-Miocene times, but then shifted back to the western plateau margin in the late Miocene while deformation concentrated in the east. Thus, the evidence weighs against a causal relationship between magmatism and crustal deformation on the orogenic scale and specifically, there is no support for the concept that thermal weakening of the crust was a precondition for crustal shortening and thickening that led to formation of the central Andes Plateau. Oncken et al. (2006, Chap. 1 of this volume) and Sobolev et al. (2006, Chap. 25 of this volume) discuss alternative mechanisms of plateau formation in detail and conclude that crustal strength *per se* was not a controlling factor but rather the strength of the interface between the upper and the lower plates, which is largely a function of trench fill. They also attribute a major role to the overriding velocity of the South American Plate. Those factors are only weakly linked to the processes that govern melt generation in the mantle wedge (e.g., subduction geometry and rate, slab and mantle temperature), which we believe explains the lack of correlation between patterns of volcanic and tectonic activity at the regional scale.

However, a different relationship between magmatism and deformation becomes evident from observations at a local scale (a few km to tens of km), where structures promoting or impeding magma ascent in the upper crust may be as important, or more so, than the factors that govern bulk crustal deformation and mantle melting. A comparison of magmatism and deformation patterns at the local scale is beyond our scope but two examples that demonstrate a structural control on the spatial distribution of the frontal arc volcanoes are: (*1*) the eastward deflection of volcanism around the Salar de Atacama, which is underlain by a mechanically rigid crustal block (Schurr and Rietbrock 2004; Riller et al. 2006, Chap. 15 of this volume); and (*2*) the occurrence of eastward outliers of the arc along the NW-SE lineament zones. The point density maps for post-20 Ma arc activity in Fig. 2.5 demonstrated a clustering of volcanism where these lineament zones (dashed lines) intersect the frontal arc, and Riller et al. (2001) suggested that the trans-arc lineaments may have controlled the location of ignimbrite calderas. Thus, at the local scale structures and discrete zones of active deformation probably did influence the spatial patterns of arc activity by controlling magma ascent in the upper crust.

Unfortunately there is far less quantitative information on the age-space distribution of deformation in the study area than for volcanic activity, so a detailed comparison of the two is rarely possible. An exception is the region of latitude 19–22° S, where E-W shortening rates since the Eocene have been estimated from age and displacement data of thrust fault systems (Victor et al. 2004; Elger et al. 2005; Oncken et al. 2006, Chap. 1 of this volume). The cumulative shortening rates for 40 Ma to present are

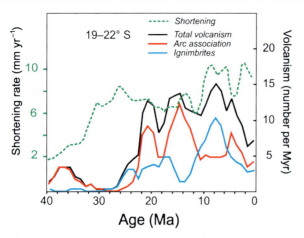

**Fig. 2.8.** Synoptic diagram for latitude 19–22° S, showing the variations in cumulative shortening rates (3 Myr sliding average) across the central Andes from Oncken et al. (2006, Chap. 1 of this volume), along with the corresponding variations in volcanic output (total, arc association and ignimbrites also plotted as 3 Myr sliding average)

shown as a dashed green curve on Fig. 2.8 and compared with volcanic productivity curves for the corresponding latitude from our compilation. All curves on the diagram represent sliding average values calculated for 3 Myr age intervals. The same caveats apply to interpretation of the minor peaks and valleys in this diagram as discussed for Fig. 2.6 (see also Oncken et al. 2006, Chap. 1 of this volume), but we believe the following features are significant. The strong rise in shortening rates during the early Oligocene (35 Ma to 25 Ma) corresponds with a period where volcanic activity was particularly low, and not just in this arc segment but overall (see Fig. 2.6). Between about 30 Ma and 10 Ma, the average shortening rates remained rather constant at 6–8 mm yr$^{-1}$ and they show no response to the large increases in volcanic output at about 22 Ma and 15 Ma; if anything, rates decrease slightly during this interval. However, at about 10 Ma both the shortening rate and total volcanism curves increase to their highest levels. It is important that the rise in volcanicity at this time is due to ignimbrite eruptions (blue curve), not the arc association. In the period of about 12 to 6 Ma, vast felsic ignimbrite deposits formed from caldera complexes in the Altiplano-Puna volcanic complex (APVC, de Silva 1989; de Silva and Francis 1991), which is one of the largest concentrations of ignimbrites worldwide. It is well established that the APVC ignimbrite magmas originate from crustal melting (de Silva 1989; Coira et al. 1993; Ort et al. 1996; Kay et al. 1999; Lindsay et al. 2001; Schmitt et al. 2001). Establishment of a partially molten zone in the crust could well explain the rise in shortening rates observed from 10 Ma onward. The volcanicity curves drop off again after about 6 Ma whereas the shortening rates remain high, with fluctuation. This does not contradict the melt-weakening interpretation because there is abundant geophysical evidence (seismic and electrical conductivity anomalies) that a widespread partial melt zone persists at about 20 km depth today (Brasse et al. 2002; Chmielowski et al. 1999; Schilling et al. 2006, Chap. 22 of this volume). In this context, we point out that by no means all ignimbrites in the CVZ formed by crustal melting. For example, Siebel et al. (2001) and Schnurr et al. (in press) studied over 20 ignimbrite units of mid-Miocene to Pliocene age in eastern Chile and western Argentina at 25–27° S and concluded that almost all of them formed by differentiation of arc magmas and not by crustal melting. The reliable way to distinguish between ignimbrites derived from crustal melting and those representing differentiated arc magmas is with radiogenic isotope data (Siebel et al. 2001), and these are available for only a fraction of the 400 ignimbrite entries in our compilation. It is not recommended to use the distribution of ignimbrites as a proxy for the thermal state of the crust without good information on their composition.

### 2.4.2 Volcanic Distribution and Subduction Geometry

A postulate in many interpretations of the Neogene history of the CVZ is that past changes in slab dip can be deduced from variations in the location, width and composition of the magmatic arc (e.g., Kay et al. 1999; Coira et al. 1993; Kay and Mpodozis 2002; Ramos et al. 2002). An empirical dependence of arc-trench distance and arc width on subduction angle has been demonstrated for many subduction zones (Tatsumi and Eggins 1995). However, subduction angle is clearly not the only factor influencing volcanic distribution even in comparatively simple island arcs, and even less so in a continental setting such as the CVZ. The complex variations of volcanism shown in this study defy explanation by subduction geometry alone. A good illustration of this is the age versus longitude plot of arc activity (Fig. 2.3). The classic scenario for the CVZ invokes shallow subduction in the Oligocene that shuts off arc activity, followed by a wide band of volcanism as the slab re-steepens and the asthenospheric wedge is "refilled" from the east, and finally by a phase of arc narrowing and shifting trenchward as slab dip continues to increase (Coira et al. 1993; Allmendinger et al. 1997; Kay et al. 1999). The patterns of arc (and ignimbrite) volcanism shown in Fig. 2.3c (20–23° S) fit this scenario quite well but there are problems applying it to adjacent segments north and south. In the 17–20° S segment (Fig. 2.3b), the volcanic distribution has a broad E-W extent and it stays that way from 25 Ma onward, yet local seismicity in the subduction zone indicates that slab dip in this segment is about the same as to the south. In the segment of 23–26° S (Fig. 2.3d), volcanism starts up after the Oligocene gap at the same longitude as before and remains concentrated in the western Cordillera thereafter, although there is a broadening of the arc at about 15 Ma. The same observation holds for the next segment to the south (26–28° S). Either past changes in subduction geometry were strongly disparate from north to south – yet ending up with a similar slab dip throughout the region today – or other factors affecting arc distribution complicated or overrode the effect of slab dip. We favor the latter explanation because of the particular geodynamic features of subduction orogeny in the CVZ. These include extreme horizontal shortening and thickening, which must have affected the mantle lithosphere as well as the crust; the possible underthrusting of Brazilian shield lithosphere towards the west; and the detachment of high-density, lower crust and upper-mantle material as a consequence of crustal thickening (Kay and Kay 1993; Kay et al. 1994b). Consider that the simple relationship between subduction angle and volcanic distribution requires near-vertical ascent of fluids and magma in the mantle wedge and through the crust. Evidence for non-vertical distribution of melts and fluid in the mantle wedge today is shown by seismic tomography studies (Schurr et al. 2003; 2006). Added to this complexity is the effect of long ascent paths for magmas through the overthickened and tectonically complex continental crust, with the consequence that patterns of surface volcanism are skewed in a temporal and/or spatial sense with respect to melt generation in the mantle. A prominent example is the eastward bend of the volcanic arc around the Salar de Atacama, which may well be controlled by upper-crustal discontinuities (Riller et al. 2006, Chap. 15 of this volume). And finally, the effect of a mid-crustal melt zone in the CVZ on the rheologic and density structure of the crust can also be expected to compromise a vertical ascent of arc magmas from mantle source to surface. These factors above are more than enough to complicate the underlying relationship between arc location and slab dip, and to cause the muddle in observed variations of volcanic distribution with age along the arc.

## 2.5 Conclusions

The new compilation of nearly 1 450 entries for late Cenozoic volcanism in the south-central Andes (14–28° S) allows examination of the time-space distribution patterns of volcanic activity separately for felsic ignimbrites and for intermediate to mafic arc-related volcanism (arc association). The variations in those patterns since about 40 Ma, and their comparison with time-space domains of active deformation in the same region lead to the following main conclusions:

1. The onset of deformation preceded that of magmatism by as much as 10 million years. The distribution patterns of volcanic activity and domains of shortening deformation vary independently at the regional scale and show no convincing evidence of a causal relationship. Spatially, both phenomena affected large parts of the plateau region in the Miocene but deformation then concentrated at the eastern flank of the plateau whereas arc volcanism retreated to the western flank.
2. The evidence refutes the hypothesis that thermal weakening of the crust was a precondition for the extreme Neogene horizontal shortening and crustal thickening that produced the central Andean Plateau. However, comparison of shortening rate estimates with volcanicity for 19–22° S suggests that melt-enhanced crustal weakening may have been important in the late Miocene. A sharp rise in shortening rates beginning at about 10 Ma corresponds to the most intense period of ignimbrite-forming eruptions in the central Andes (Altiplano-Puna volcanic complex at 22–24° S). The ignimbrite magmas derived from crustal melting, and geophysical anomalies suggest that extensive zones of partial melting persist in the middle crust below this area today.
3. On the local scale, the distribution of volcanism is influenced by upper-crustal structures. Examples are the eastward deflection of the arc around the Salar de Atacama, the occurrence of volcanic outliers east of the arc front along NW-SE-trending lineament zones, and the clustering of volcanism where the lineament zones intersect the frontal arc. At this scale of observation, patterns of surface volcanism are determined as much, or more, by factors affecting magma storage and ascent in the upper crust as by the regional-scale factors that govern bulk deformation and the generation of melts in the mantle wedge.
4. Statistical analysis of arc productivity with time for 3° segments of latitude shows that the onset of volcanism after the Oligocene gap was coeval at all latitudes (c. 25 Ma), but the main activity shifted southwards with time. Whereas 50% of arc activity north of 20° S had accumulated by 16 Ma, the three southern segments, 20–23° S and 23–26° S and 26–28° S achieved this level at about 12 Ma, 10 Ma and 8 Ma, respectively.
5. Comparison between the N-S variations in arc productivity over time and the reconstructed position of the subducted Juan Fernández Ridge under the study area shows that ridge subduction did not strongly affect arc productivity.
6. Changes in the position and width of the Neogene arc in the CVZ are commonly used to infer past changes in subduction geometry (slab dip). However the complexity of arc distribution patterns with time revealed in this study and especially the major changes in patterns along the arc cannot be explained by subduction geometry alone. The simple relationship between distribution of volcanism and slab dip is distorted by other factors related to the geodynamic setting of the central Andes. Such factors include those that influence

the thermal structure of the mantle wedge (upper-plate shortening and thickening; lithospheric delamination; westward flow of Brazilian continental lithosphere), and others that influence the ascent paths of magmas through the thick and actively deforming crust.

## Acknowledgments

This work was supported by the collaborative research program SFB-267 "Deformation Processes in the Andes". For discussions and help with the statistical analysis and data management for the appendix we owe our thanks to Heinz Burger (Berlin). Kerstin Schemmann and Ulrich Micksch (Potsdam) helped with the final animation of volcano distributions. Numerous graduate students and student assistants from Berlin and Potsdam contributed to the compilation efforts over the years. In particular Michael Haschke and Kerstin Elger are acknowledged for sharing their compilations with us. Special thanks go to Silvina Guzmán and Carolina Montero López (Salta, Argentina) for help during a last phase of intense work extending data coverage to the Puna region. Tim Vietor, Nina Kukowski and Stephan Sobolev contributed stimulating discussions and sage comments on earlier versions of the manuscript. The paper benefited from constructive reviews by Hans-Ulrich Schminke and Christoph Breitkreuz, and editorial assistance of Gerhard Franz.

## References

Allmendinger RW, Jordan TE, Kay SM, Isacks BL (1997) The evolution of the Altiplano-Puna plateau of the Central Andes. Ann Rev Earth Planet Sci 25:139–174

Brasse H, Lezaeta P, Rath V, Schwalenberg K, Soyer W, Haak V (2002) The Bolivian Altiplano conductivity anomaly. J Geophys Res 107: doi 10.1029/2001JB000391

Cahill T, Isacks BL (1992) Seismicity and shape of the subducted Nazca plate. J Geophys Res 97:17503–17529

Chernikoff CJ, Richards JP, Zappettini EO (2002) Crustal lineament control on magmatism and mineralization in northwestern Argentina: geological, geophysical and remote sensing evidence. Ore Geol Rev 21:127–155

Chmielowski J, Zandt G, Haberland S (1999) The central Andean Altiplano-Puna magma body. Geophys Res Lett 26:783–786

Coira B, Davidson J, Mpodozis C, Ramos V (1982) Tectonic and magmatic evolution of the Andes of Northern Argentina and Chile. Earth Sci Rev 18:302–332

Coira B, Kay SM, Viramonte J (1993) Upper Cenozoic magmatic evolution of the Argentine Puna – a model for changing subduction geometry. Int Geol Rev 35:677–720

De Silva SL (1989) Altiplano-Puna volcanic complex of the central Andes. Geology 17:1102–1106

De Silva SL, Francis P (1991) Volcanoes of the Central Andes. Springer, New York

Elger K, Oncken O, Glodny, J (2005) Plateau-style accumulation of deformation: Southern Altiplano. Tectonics 24: doi 10.1029/2004TC001675

Francis PW, Hawkesworth CJ (1994) Late Cenozoic rates of magmatic activity in the Central Andes and their relationships to continental crust formation and thickening. J Geol Soc London 151:845–854

Francis PW, Sparks RSJ, Hawkesworth CJ, Thorpe RS, Pyle DM, Tait SR, Mantovani MS, McDermott F (1989) Petrology and geochemistry of volcanic rocks of the Cerro Galán caldera, northwest Argentina. Geol Mag 126:515–547

Gardeweg M, Ramirez CF (1987) La Pacana caldera and the Atana Ignimbrite. A major ash-flow and resurgent caldera complex in the Andes of Northern Chile. Bull volcanology 49:547–566

Gutscher MA, Spakman W, Bkjwaard H, Engdahl ER (2000) Geodynamics of flat subduction: seismicity and topographic constraints from the Andean margin. Tectonophys 19:814–833

Haschke M, Siebel W, Günther A, Scheuber E (2002) Repeated crustal thickening and recycling during the Andean orogeny in North Chile (21°–26° S). J Geophys Res 107: doi 10.1029/2001JB000328

Isacks BL (1988) Uplift of the Central Andean Plateau and bending of the Bolivian Orocline. J Geophys Res 93:3211–3231

James DE (1971a) Andean crustal and upper mantle structure. J Geophys Res 76:3246–3271

James DE (1971b) Plate tectonic model for the evolution of the central Andes. Geol Soc Am Bull 71:3325–3346

James DE, Sacks IS (1999) Cenozoic formation of the central Andes: a geophysical perspective. In: Skinner B (ed) Geology and ore deposits of the Central Andes. Soc Econ Geol Spec Publ 7:1–26

Kay RW, Kay SM (1993) Delamination and delamination magmatism. Tectonophysics 219:177–189

Kay SM, Mpodozis C (2002) Magmatism as a probe to the Neogene shallowing of the Nazca plate beneath the modern Chilean flatslabs. J South Am Earth Sci 15:39–57

Kay SM, Mpodizis C, Tittler A, Cornejo P (1994a) Tertiary magmatic evolution of the Maricunga mineral belt in Chile. Int Geol Rev 36:1079–1112

Kay SM, Coira B, Viramonte J (1994b) Young mafic back arc volcanic rocks as indicators of continental lithospheric delamination beneath the Argentine Puna plateau, Central Andes. J Geophys Res 99:24323–24339

Kay SM, Mpodozis C, Coira B (1999) Neogene magmatism, tectonism, and mineral deposits of the Central Andes (22 degrees to 33 degrees S latitude). In: Skinner B (ed) Geology and ore deposits of the Central Andes. Soc Econ Geol Spec Publ 7:27–59

Kukowski N, Oncken O (2006) Subduction erosion – the "normal" mode of fore-arc material transfer along the Chilean margin? In: Oncken O, Chong G, Franz G, Giese P, Götze H-J, Ramos VA, Strecker MR, Wigger P (eds) The Andes – active subduction orogeny. Frontiers in Earth Science Series, Vol 1. Springer-Verlag, Berlin Heidelberg New York, pp 217–236, this volume

Lindsay JM, Schmitt AK, Trumbull RB, De Silva SL, Siebel W, Emmermann R (2001) Magmatic evolution of the La Pacana Caldera system, central Andes, Chile: compositional variation of two cogenetic, large-volume felsic ignimbrites and implications for contrasting eruption mechanisms. J Petrol 42:459–486

Matteini M, Mazzuoli R, Omarini R, Cas R, Maas R (2002) The geochemical variations of the upper Cenozoic volcanism along the Calama-Olacapato-El Toro transversal fault system in central Andes (24° S): petrogenetic and geodynamic implications. Tectonophysics 345:211–227

McKee EH, Robinson AC, Rybuta JJ, Cuitino L, Moscoso RD (1994) Age and Sr isotopic composition of volcanic rocks in the Maricunga Belt, Chile: implications for magma sources. J S Am Earth Sci 7:167–177

McQuarrie N, Horton BK, Zandt G, Beck S, DeCelles PG (2005) Lithospheric evolution of the Andean fold–thrust belt, Bolivia, and the origin of the central Andean plateau. Tectonophysics 399: 15–37

Mpodozis C, Ramos V (1989) The Andes of Chile and Argentina. In: Ericksen GE, Canas Pinochet MT, Reinemund JA (eds) Geology of the Andes and its relation to hydrocarbon and mineral resources. Circum-Pacific Council for Energy and Mineral Resources, Earth Science Series 11:59–89.

Oncken O, Hindle D, Kley J, Elger K, Victor P, Schemmann K (2006) Deformation of the central Andean upper plate system – facts, fiction, and constraints for plateau models. In: Oncken O, Chong G, Franz G, Giese P, Götze H-J, Ramos VA, Strecker MR, Wigger P (eds) The Andes – active subduction orogeny. Frontiers in Earth Science Series, Vol 1. Springer-Verlag, Berlin Heidelberg New York, pp 3–28, this volume

Ort M, Coira BL, Mazzoni MM (1996) Generation of a crust-mantle magma mixture: magma sources and contamination at Cerro Panizos, central Andes. Contrib Min Petrol 123:308–322

Petersen U (1999) Magmatic and metallogenic evolution of the Central Andes. In: Skinner B (ed) Geology and ore deposits of the Central Andes. Soc Econ Geol Sp Publ 7:109–153

Petrinovic IA, Riller U, Brod JA (2005) The Negra Muerta Volcanic Complex, southern Central Andes: geochemical characteristics and magmatic evolution of an episodically active volcano-tectonic complex. J Volcanol Geoth Res 140:295–320

Ramos VA, Cristallini EO, Pérez DJ (2002) The Pampean flat-slab of the Central Andes. J S Am Earth Sci 15:59–78

Ramelow J, Riller U, Romer R, Oncken O (2005) Kinematic link between episodic trap door collapse of the Negra Muerta Caldera and motion on the Olacapato-El Toro Fault Zone, southern central Andes. Int J Earth Sci: doi 10.1007/s00531-005-0042-x

Richards JP, Villeneuve M (2002) Characteristics of the late Cenozoic volcanism along the Archibarca lineament from Cerro Llullaillaco to Corrida de Cori, northwest Argentina. J Volcanol Geoth Res 116:161–200

Riller U, Petrinovic I, Ramelow J, Strecker M, Oncken O (2001) Late Cenozoic tectonism, caldera and plateau formation in the central Andes. Earth Planet Sci Lett 188:299–311

Riller U, Götze H-J, Schmidt S, Trumbull RB, Hongn F, Petrinovic IA (2006) Upper-crustal structure of the Central Andes inferred from dip curvature analysis of isostatic residual gravity. In: Oncken O, Chong G, Franz G, Giese P, Götze H-J, Ramos VA, Strecker MR, Wigger P (eds) The Andes – active subduction orogeny. Frontiers in Earth Science Series, Vol 1. Springer-Verlag, Berlin Heidelberg New York, pp 327–336, this volume

Rosenbaum G, Giles D, Saxon M, Betts PG, Weinberg RF, Duboz C (2005) Subduction of the Nazca Ridge and the Inca Plateau: insights into the formation of ore deposits in Peru. Earth Planet Sci Lett 239:18–52

Rutland RWR (1971) Andean orogeny and sea floor spreading. Nature 233:252–255

Salfity JA (1985) Lineamentos transversales al rumbo andino en el Noroeste Argentino. Actas 4 Congreso Geologico Chileno 2: A119–A127

Scheuber E, Bogdanic T, Jensen A, Reutter KJ (1994) Tectonic development of the north Chilean Andes in relation to plate convergence and magmatism since the Jurassic. In: Reutter KJ, Scheuber E, Wigger P (eds) Tectonics of the Southern Central Andes, Springer, New York, pp 121–139

Scheuber E, Reutter KJ (1992) Magmatic arc tectonics in the Central Andes between 21° and 25° S. Tectonophysics 205:127–140

Schilling FR, Trumbull RB, Brasse H, Haberland C, Asch G, Bruhn D, Mai K, Haak V, Giese P, Muñoz M, Ramelow J, Rietbrock A, Ricaldi E, Vietor T (2006) Partial melting in the Central Andean crust: a review of geophysical, petrophysical, and petrologic evidence. In: Oncken O, Chong G, Franz G, Giese P, Götze H-J, Ramos VA, Strecker MR, Wigger P (eds) The Andes – active subduction orogeny. Frontiers in Earth Science Series, Vol 1. Springer-Verlag, Berlin Heidelberg New York, pp 459–474, this volume

Schmitt AK, De Silva SL, Trumbull RB, Emmermann R (2001) Magma evolution in the Purico ignimbrite complex, northern Chile: evidence for zoning of a dacitic magma by injection of rhyolitic melts following mafic recharge. Contrib Mineral Petrol 140:680–700

Schnurr WBW, Trumbull RB, Clavero J, Hahne K, Siebel W, Gardeweg M (in press) Twenty million years of felsic magmatism in the southern Central Volcanic Zone of the Andes: geochemistry and magma genesis of ignimbrites from 25–27° S, 67–72°W. J Volc Geoth Res

Schröder W (1999) Geochemisch-petrographische Korrelation känozoischer Ignimbrite in den nördlichen Zentralanden, im Dreiländereck N-Chile, W-Bolivien und S-Peru. PhD thesis, Univ. of Potsdam

Schurr B, Rietbrock A (2004) Deep seismic structure of the Atacama basin, northern Chile. Geophys Res Lett 31: doi 10.1029/2004GL019796

Schurr B, Asch G, Rietbrock A, Trumbull R, Haberland C (2003) Complex patterns of fluid and melt transport in the central Andean subduction zone revealed by attenuation tomography. Earth Planet Sci Lett 215:105–119

Schurr B, Rietbrock A, Asch G, Kind R, Oncken O (2006) Evidence for lithospheric detachment in the central Andes from local earthquake tomography. Tectonophysics 415: 203–223

Sébrier M, Soler P (1991) Tectonics and Magmatism in the Peruvian Andes from late Oligocene Time to the Present. In: Harmon RS, Rapela CW (eds) Andean magmatism and its tectonic setting. Geol Soc Am Sp Paper 265:259–278

Siebel W, Schnurr WBW, Hahne K, Kraemer B, Trumbull RB, Van den Bogaard P, Emmermann R (2001) Geochemistry and isotope systematics of small- to medium-volume Neogene-Quaternary ignimbrites in the southern central Andes: evidence for derivation from andesitic magma sources. Chem Geol 171:213–237

Sobolev SV, Babeyko AY, Koulakov I, Oncken O (2006) Mechanism of the Andean orogeny: insight from numerical modeling. In: Oncken O, Chong G, Franz G, Giese P, Götze H-J, Ramos VA, Strecker MR, Wigger P (eds) The Andes – active subduction orogeny. Frontiers in Earth Science Series, Vol 1. Springer-Verlag, Berlin Heidelberg New York, pp 513–536, this volume

Tatsumi Y, Eggins S (1995) Subduction Zone Magmatism. Frontiers in Earth Science Blackwell, Cambridge

Thorpe RS, Francis PW (1979) Variations in Andean andesite compositions and their petrogenetic significance. Tectonophysics 57:53–70

Victor P, Oncken O, Glodny J (2004) Uplift of the western Altiplano plateau: evidence from the Precordillera between 20° and 21°S (northern Chile). Tectonics 23: doi 10.1029/2003TC001519

Wörner G, Hammerschmidt K, Henjes-Kunst F, Lezaun J, Wilke H (2000) Geochronology (40Ar/39Ar, K–Ar and He-exposure ages) of Cenozoic magmatic rocks from Northern Chile (18–22° S): implications for magmatism and tectonic evolution of the central Andes. Rev Geol Chile 27:205–240

Yañez G, Ranero CR, von Huene R, Diaz J (2001) Magnetic anomaly interpretation across a segment of the Southern Central Andes (32–34° S): implications on the role of the Juan Fernández Ridge in the tectonic evolution of the margin during the upper Tertiary. J Geophys Res 106:6324–6345

## Electronic Appendices

- *Appendix 1*: (Excel file) Compilation of age and location of Cenozoic volcanoes and related features in the central Andes.
- *Appendix 2*: (Word file) List of source references for data compilation
- *Appendix 3*: (Quicktime file) Animation of volcanic distribution maps from 65 Ma to 0 Ma.

# Chapter 3

# Crustal Evolution at the Central Andean Continental Margin: a Geochemical Record of Crustal Growth, Recycling and Destruction

Gerhard Franz · Friederich Lucassen · Wolfgang Kramer · Robert B. Trumbull · Rolf L. Romer · Hans-Gerhard Wilke
José G. Viramonte · Raul Becchio · Wolfgang Siebel

**Abstract.** Active continental margins are considered as the principal site for growth of the continental crust. However, they are also sites of recycling and destruction of continental crust. The Andean continental margin has been periodically active at least since the early Paleozoic and allows the evaluation of the long-term relevance of these processes. The early Paleozoic orogeny at ca. 0.5 Ga recycled and homogenized the ~2 Ga old early Proterozoic crust of the Brazilian Shield, which was previously orogenized at ca. 1 Ga, consistent with global models of prominent crustal growth at 2 Ga and a near constant mass of continental crust in the Phanerozoic. The metamorphic and magmatic evolution and the isotopic signatures of the early Paleozoic rocks do not indicate significant crustal growth, either by accretion of exotic terranes and island arcs or by juvenile additions from a mantle source. The dominant inferred mode of crustal evolution in the Paleozoic was recycling of older crust. Destruction of continental crust by subduction erosion is prominent in sections of the present active margin and is also likely to have occurred in the past orogens. Voluminous juvenile magmatism is only observed in the Jurassic – lower Cretaceous extensional magmatic arc. Compositions of mantle-derived magmas from the early Paleozoic to the Cainozoic, as well as late Cretaceous mantle xenoliths, indicate that depleted mantle was already present beneath the early Paleozoic orogen. The old subcontinental, enriched mantle related to the Brazilian shield and bordering Proterozoic mobile belts was modified by asthenospheric mantle in the younger subduction systems. In summary, this transect of the Andes is not a site of major continental growth, but a site where long-term processes of growth, recycling and destruction balance out.

## 3.1 Introduction

Growth of the continental crust is closely linked to the phenomenon of subduction, and active continental margins are generally considered to be the principal sites for the formation of continental crust. Ernst (2000) commented that 'the subduction process ... appears to be responsible for the formation and growth of continents. ... This constitutes the geologic regime where new continental crust is forming.' Similarly, Middlemost (1997)

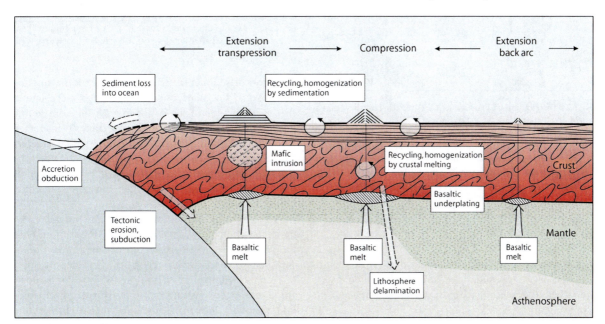

**Fig. 3.1.** Schematic compilation of the processes of crustal growth, recycling and destruction at an active continental margin; *circles* indicate recycling, *solid arrows* growth and *dashed arrows* destruction of continental crustal material. The transpressional regime is situated in the fore-arc, shown here together with the extensional regime for the sake of simplicity. Tectonic shortening, which leads to thickening of the crust and area reduction and may eventually lead to lithospheric delamination, is not shown

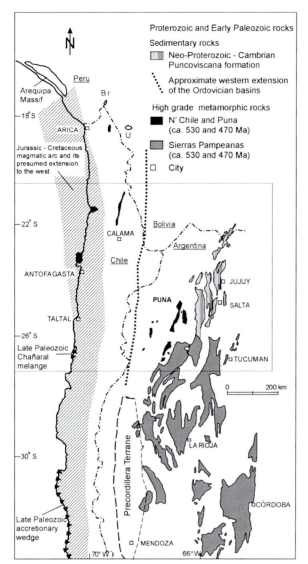

**Fig. 3.2a.** Distribution of Proterozoic (Arequipa Massif, Br = Berenguela, U = Uyarani; ca. 1.2–1.0 Ga) and early Paleozoic rocks between ca. ~16 and 33° S and the position of the Precordillera Terrane (modified from Lucassen et al. 2000); northern extension of rocks from the late Paleozoic accretionary wedge of southern and central Chile (Hervé 1988; Willner et al. 2004; this volume) and the late Paleozoic Chañeral Melange (Bell 1987); extension of the early Jurassic – Cretaceous magmatic arc (presumed extension to the west: Von Hillebrandt et al. 2000). The *box* shows the approximate area covered by Fig. 3.2b and by the detailed geological map (see Reutter and Munier 2006, Chap. 27 of this volume; Schnurr et al. 2006, Chap. 29 of this volume)

states for continental margins 'It is also where new continental crust evolves and the process of subduction inserts new chemical heterogeneities and complexities into the upper mantle.'

In a subduction environment, possible processes of continental crustal growth include the addition of magma from the mantle to the crust in volcanic arcs, the tectonic

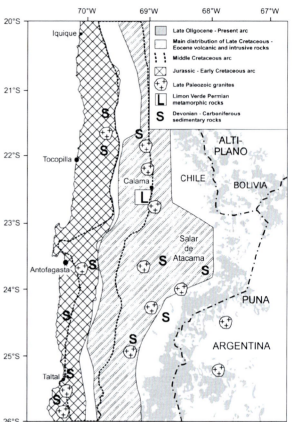

**Fig. 3.2b.** Late Paleozoic – Cainozoic magmatic rocks from the various magmatic arcs intruded or overlay early Paleozoic crust. Devonian – Carboniferous sedimentary rocks frequently form the host rocks of the late Paleozoic intrusions. The Permian high-grade metamorphic rocks of Limon Verde south of Calama are unique in the area. Map modified after Reutter et al. 1994; E. Scheuber, Freie Universität Berlin, pers. comm.

accretion of sedimentary material from the subducting to the overriding plate, obduction of oceanic crust from the lower plate, and the collision with the continental margin of continental or island arc terranes transported on the subducting plate (Fig. 3.1). From a global perspective, only the addition of mantle-derived magma and the accretion of oceanic crust can be strictly considered as net continental growth, because the other processes only redistribute continental material.

Active continental margins are also potential sites for the destruction of continental crust and lithosphere, both on a local and a global scale. Surface area reduction of continental crust (which we consider as the local perspective) results when the crust is thickened by tectonic shortening (homogeneous shortening, overthrusting and folding). Volume reduction of crust and associated lithosphere (the global perspective) can occur because of density-driven delamination of mafic material from the base of the crust as a result of shortening (Fig. 3.1). The latter

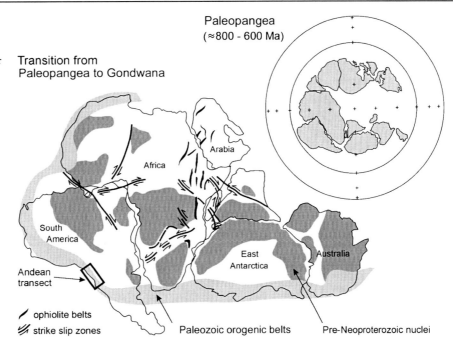

Fig. 3.3.
Paleogeographic reconstructions for Paleopangea and Gondwana (modified from Piper 2000), showing the position of the Andean transect as part of the Paleozoic orogenic belts at the continental margin

process recycles crust (together with mantle lithosphere) into the deeper mantle.

Finally, the destruction of continental crust at active margins can also occur by the subduction of crustal material into the mantle, either as continental sediments deposited in the trench and/or as tectonically eroded crustal rocks from the lip of the overriding plate. Clift and Vanucchi (2004), among others, pointed out that tectonic erosion at continental margins is a major factor in the destruction of continental crust. These processes will leave traces in both the crust and the mantle lithosphere. Most evidence is preserved in the continental crust and revealed through the composition, ages and conditions of formation of metamorphic, magmatic and sedimentary rocks. The evolution of the upper mantle is less tractable from geologic studies, but is an important factor for the formation and destruction of continental material. In this paper we review the available evidence for crustal and mantle evolution in the Central Andes.

Since the first plate tectonic models were developed, the Central Andes have been considered as the type example of an active continental margin (e.g. Mitchell and Reading 1969), and its over-thickened crust and high plateau are the typical example for 'Cordilleran-type orogeny' (Fig. 3.2). The Andean orogeny dominates the present form of this continental margin and its most prominent features (volcanic arc and plateau), but, in fact, the Central Andean crust records at least 600 Ma of history as the leading edge of the South American continent (Table 3.1; Fig. 3.3). It also preserves relics from the 2 Ga history of the western part of the South American craton. This long history is documented by the ages of magmatism and metamorphism, by the depositional ages of Phanerozoic sedimentary rocks (Table 3.1), and by the chemical and isotopic composition of the diverse crustal rocks (Figs. 3.4 to 3.7). The early Paleozoic and Late Paleozoic to Cainozoic history was dominated by prolonged periods of subduction. The mantle evolution of the margin is less well constrained due to the discontinuous record of mantle-derived magmatic rocks, and the fact that mantle xenoliths are essentially absent in arc magmas. Nevertheless, there are enough data to provide some useful constraints on mantle composition.

The purpose of this paper is to summarize and discuss crustal and mantle features of the Central Andes between 21° S and 27° S (as the core area, but extending to 18° S and 32° S) as a case study of the development of an active continental margin. We evaluate how representative the presently active state of the system is regarding long-term evolution, particularly with respect to crustal growth or destruction. The discussion is based on information about the geological history of the area as well as on a large volume of chemical and isotopic data from the continental crust and mantle-derived rocks. Much of the data reviewed in this paper were acquired between 1992 and 2004 in the SFB 267 research program 'Deformation processes in the Andes' and between 1988 and 1991 in an earlier program 'Mobility of active continental margins'. Our work focused on the Phanerozoic crystalline basement and on the chemical-petrological characteristics of Cainozoic magmas. A further aspect was the nature of the upper mantle and its changes with the evolution of the Central Andes. As a large data set has been built up in these years, the present paper can only give a brief over-

**Table 3.1.** Evolution of the continental crust

| Time (Ma) | Evolution of the continental crust |
|---|---|
| 30–Recent | Area of the Early Paleozoic Mobile Belt: Formation of the Puna–Altiplano Plateau. Cenozoic Andean andesite and ignimbrite volcanism sample the Early Paleozoic crust. Strong arc normal shortening and tectonic thickening crust (for references see the text). |
| <120 | Area of the Early Paleozoic Mobile Belt: Eastward movement of the magmatic arcs (1, 2). Possible subduction erosion of the Late Paleozoic, Triassic and Jurassic magmatic arc rocks and Early Paleozoic crust (2 and text). Mainly transpression – transtension (1, 2). |
| 200–120 Start of the 'Andean cycle' | Area of the Early Paleozoic Mobile Belt: Formation of juvenile 'magmatic' crust in the Jurassic–Lower Cretaceous arc. Direct evidence for crustal growth (for references see text). Phases of arc normal extension and transtension and transpression (2). |
| 330–200 | Area of the Early Paleozoic Mobile Belt: Widespread granitoid intrusions mainly in N Chile (ca. 300–220 Ma; 4, 5, 6) and an extended Permian rhyolite province (7, 8) recycle Early Paleozoic crust; Late Triassic transition to mantle derived melts in the arc (9, 10) high grade metamorphic rocks of Sierra de Limón Verde (270 Ma; 11). Extension to transpression. |
| 400–330 | Early Paleozoic Mobile Belt: Passive margin evolution (12, 13) with sedimentary rocks recycling Early Paleozoic crust (14). Exhumation of the metamorphic basement largely finished. K-Ar cooling ages of metamorphic rocks from micas and first erosional unconformities (15, 16). |
| 500–400 Local name 'Famatinian cycle' | Early Paleozoic Mobile Belt: Formation of the mobile belt; high T/moderate P metamorphism (470–400 Ma; 16, 17, 18, 19, 20, 21, 22) and intense crustal derived magmatism mainly recycles Proterozoic crust (490–400 Ma; 17, 18, 23, 24, 25, 30); oldest K-Ar cooling ages from hornblende in metamorphic rocks (15, 16). Ordovician sediments recycle the Early Paleozoic and Proterozoic crust (14, 26, 27). |
| ca. (900) 600–500 (400) Panafrican or Brasiliano cycle (Local name: Pampean cycle) | Early Paleozoic Mobile Belt: Formation of the mobile belt; Neoproterozoic–Eocambrian Sediments (28, 29) recycle Proterozoic basement (14, 25, 29) and form part of the protoliths of the Early Paleozoic basement. High T/moderate P metamorphism (530–500 Ma; 16, 17, 18, 19, 20, 21, 22) and crustal derived magmatism recycles Proterozoic crust (17, 18, 23, 24, 25, 30); South American craton: Formation of mobile belts between 700–500 Ma; high T metamorphism and associated granitoid magmatism (31, 32, 33). |
| ca. 1 300–900 Sunsás cycle (Contemporaneous with the Grenvillian cycle of North America) | South American craton: Major metamorphic–magmatic cycle with mobile belts in-between older cratonic regions (31, 33, 34); granulite facies metamorphism and granitoid magmatism in the Arequipa Massif (Peru) and Bolivia (19, 36, 37, 38). Early Paleozoic Mobile Belt: Upper intercept ages of inherited zircons in the Early Paleozoic crust (17, 18, 19; 38). |
| ca. 2 000 Early Proterozoic | South American craton: First order crust-formation period at ca. 2 Ga around and west of the Paleoproterozoic to Archean parts of South America; (33, 39, 40). Early Paleozoic Mobile Belt: Nd model ages (6, 14, 16, 23, 25, 26) and upper intercept U-Pb ages of inherited zircons ca. 1.6–2 Ga (8, 19;29, 38). |

(1) Haschke et al. 2002; (2) Scheuber et al. 1994; (3) Scheuber and Gonzales 1999; (4) Berg and Baumann 1985; (5) Brown 1991 (6) Lucassen et al. 1999a (7) Kay et al. 1989; (8) Breitkreuz and van Schmus 1996; (9) Morata et al. 2000; (10) Bartsch 2004; (11) Lucassen et al. 1999b; (12) Bahlburg and Hervé 1997; (13) Augustsson and Bahlburg 2003; (14) Bock et al. 2000; (15) Becchio et al. 1999; (16) compilation in Lucassen et al. 2000; (17) Damm et al. 1990; (18) Damm et al. 1994; (19) Wörner et al. 2000; (20) Lucassen and Becchio 2003; (21) Höckenreiner et al. 2003; (22) Büttner et al. 2005; (23) in: Pankhurst and Rapela (eds) 1998; (24) Pankhurst et al. 2000; (25) Lucassen et al. 2001; (26) Egenhoff and Lucassen 2003; (27) Zimmermann and Bahlburg 2003; (28) Aceñolaza et al. 1999; (29) Schwartz and Gromet 2004; (30) Lucassen et al. 2002a; (31) De Brito Neves and Cordani 1991; (32) Trompette 1997; (33) in: Cordani et al. 2000a; (34) Teixeira et al. 1989; (35) Litherland et al. 1989; (36) Wasteneys et al. 1995; (37) Martignole and Martelat 2003 (38) Loewy et al. 2004; (39) Sato and Siga 2002; (40) Rino et al. 2004.

view of the most important features. The interested reader is referred to the previous papers where details from individual studies are published.

The main part of this paper, therefore, summarizes the key geologic and petrologic features of the crust and upper mantle for this transect of the Central Andes, from the earliest recorded period to the present Andean cycle. We place strong emphasis on the isotope systems Rb-Sr, Sm-Nd and U-Th-Pb because they give the most direct information on mantle versus crustal provenance of the rocks concerned. We will show that the history of the Andean active margin, although it has certainly been a major site of geological activity, did not result in significant net crustal growth. Our perspective is local in a sense, but, because the Central Andes region has been an active continental margin through most of the Phanerozoic, we believe that the long history of crustal development recorded provides useful insights for the global perspective as well.

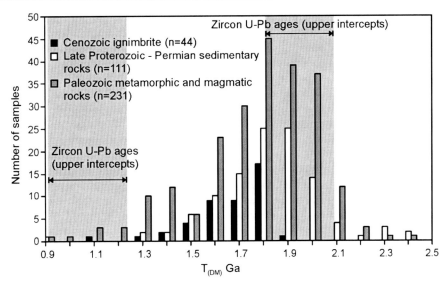

**Fig. 3.4.** Sm-Nd model ages of metamorphic, magmatic and sedimentary rocks of the Central Andes between 21° S and 32° S (single stage evolution model; Goldstein et al. 1984), compared to the range of U-Pb ages from zircon. The Sm-Nd data ($n = 342$; 4 samples outside the range) indicate the major crustal growth period in the early to middle Proterozoic (see text); the zircon ages indicate two distinct ages of crystallization. The Nd signature is also typical for Paleozoic rocks of the Coastal and Main Cordillera between 36 and 41° S (Lucassen et al. 2004). The Nd isotope signature of the large-scale Cainozoic ignimbrite resembles the isotope signature of the older crust. Figure modified from Lucassen et al. (2001); data south of 27° S from Pankhurst and Rapela (1998) and unpublished data of F. Lucassen and R. Becchio. The range of zircon ages includes the data of Damm et al. (1990, 1994), Wörner et al. (2000), Loewy et al. (2004, and references therein)

## 3.2 Proterozoic Development of the Central Andes

There is no record of Archean crust in the Central Andes and the earliest information comes from the distribution of Nd model ages of Phanerozoic igneous and metamorphic rocks ($T_{DM}$; Fig. 3.4). This shows a well-defined peak at 1.8–2.0 Ga, and a shoulder towards ages of 1.3 Ga. Model ages older and younger than this range are scarce. The material reworked in the Phanerozoic orogenies formed during a major period of crustal growth between 1.6 and 2.0 Ga, which is evident in many areas of the South American craton and common west of the Early Proterozoic and Archean cores (e.g., Goldstein et al. 1997; Cordani et al. 2000b; Sato and Siga 2002). Similar age ranges (2.2–1.9 Ga and 1.5–1.0 Ga) were also identified from analysis of detrital zircon populations from the Amazon (Rino et al. 2004) and Orinoco (Goldstein et al. 1997) river sediments, which – in contrast to the Phanerozoic rocks at the western edge of the continent – show the whole spectrum of ages including the Earliest Proterozoic and Archean ages (peak at 2.6 to 2.8 Ga and tail to Archean ages > 3 Ga) of the South American craton.

The major orogenic cycle during the Mesoproterozoic in the western section of the continent was the Sunsás orogeny at 0.9–1.3 Ga. Within the transect described in this paper (Fig. 3.2), metamorphic or magmatic crystallization ages corresponding to the Sunsás orogeny are absent, but such rocks do occur immediately to the north in the Arequipa Massif (southern Peru) and in a few small isolated outcrops in Bolivia (Table 3.1; Fig. 3.2a). Their ages are similar to those in the Sunsás Belt, which is located east of the Bolivian Andes (Litherland et al. 1989) and in the Andes of Colombia (Restrepo-Pace et al. 1997). The peak metamorphism is of high to ultra-high temperature type, the main lithologies described are prevailingly felsic, and protolith ages are mainly Paleoproterozoic to Mesoproterozoic (e.g. Wasteneys et al. 1995; Restrepo-Pace et al. 1997; Wörner et al. 2000; Martignole and Martelat 2003). The distribution of $T_{DM}$ model ages indicates that juvenile additions to the crust in this period are of minor importance compared with the older episodes (Fig. 3.4) and we conclude, that if the Sunsas orogeny was important in the transect, it did mainly rework ca. 2.0 Ga old basement.

Evidence for inherited material of ca. 1–2 Ga of age is common in the Paleozoic and younger magmatic, metamorphic and sedimentary rocks in the Central Andes. The inherited material is evident in the upper intercepts of zircon U-Pb discordia arrays, as cores of single zircons revealed by in-situ dating methods, or as Sm-Nd average crustal residence ages described above (see also Fig. 3.4; Table 3.1). Existing models for Paleo- to Mesoproterozoic supercycles of crustal evolution in South America (Wasteneys et al. 1995; Goldstein et al. 1997; Restrepo-Pace et al. 1997; Wörner et al. 2000; Martignole and Martelat 2003; Condie 2002; Rino et al. 2004) are in good agreement with these Proterozoic ages for crustal formation and metamorphism in the Central Andes transect. Their similarity with those of the North American Grenvillian Belt, and the fact that their occurrence extends considerably N-S along the Andes, led to speculation of a possible collision of North and South America in Grenvillian/Sunsás time (e.g. Restrepo-Pace et al. 1997, and references therein). We do not consider the Proterozoic evolution further in this paper since key areas for Proterozoic crust occur outside the transect and the basement rocks within the transect have been completely reworked in the Paleozoic.

Little is known about the crustal evolution of our transect area for the time between the Sunsás orogeny and the early Paleozoic, which is described in the following section. The oldest sedimentary rocks in this age range are Neoproterozoic to early Cambrian (e.g. Puncoviscana Formation; Aceñolaza et al. 1999). Paleogeographic reconstructions (e.g. Unrug et al. 1996; Piper 2000) and geotectonic interpretations (e.g. Astini et al. 1995; Aceñolaza et al. 2002) suggest that the section of the margin considered here was located at the western edge of the South American Plate from the late Proterozoic onwards (Fig. 3.3).

Evidence for the presence – at least of remnants – of Proterozoic mantle lithosphere in the region is found in some alkaline rocks along the Cretaceous rift (see below) in the present back-arc region south of 27° S and north of 21° S (Lucassen et al. 2002b, 2005). Their isotopic characteristics are typical of old enriched mantle lithosphere, with $^{87}Sr/^{86}Sr$ initial ratios between 0.7040 and 0.7065, $\varepsilon_{Nd}$ values between –1 and –9, and relatively unradiogenic Pb-isotope ratios. Similar isotope compositions of alkaline basaltic magmas are known from the cratonic parts of South America, e.g. the Paraná province of the Brazilian shield (e.g. Carlson et al. 1996; Gibson et al. 1996, 1999). In the region of 21 to 27° S, however, there is no evidence for such an enriched lithospheric mantle in the rift magmas or their xenoliths. If enriched mantle was once present in this region, it has apparently been replaced or overprinted by the depleted mantle that now dominates the region (see below).

## 3.3 The Early Paleozoic Orogeny

Early Paleozoic rocks, both sedimentary and metamorphic, cover large areas of southern Bolivia and northwestern Argentina but are scarce in northern Chile (Fig. 3.2). The Neoproterozoic – early Cambrian sedimentary rocks partly form the protoliths of the high-temperature metamorphic rocks in northwestern Argentina (Willner et al. 1985; Lucassen et al. 2001; Schwartz and Gromet 2004). Studies of these rocks over an area between ~18 and 32° S ($5 \times 10^5$ km$^2$) show that the crust is surprisingly homogeneous in lithology and composition (with some very local exceptions). The dominant rock types in the high-grade basement of northwestern Argentina and northern Chile are felsic gneisses and migmatites of upper amphibolite facies, locally transitional to granulite facies.

Uniform pressure estimates between 5 and 7 kbar in the region indicate uniform mid-crustal exposure levels. High-pressure (high-P) rocks indicating exhumation from the root of the orogen, or those typical of accretionary complexes, seem to be absent. The uniformity of high-temperature (high-T) metamorphic conditions and absence of high-P rocks indicate large-scale uniform crustal thickening and possible plateau formation similar to the present Andean orogen (Lucassen and Franz 2005b).

Ages of metamorphic crystallization (Table 3.1) cluster in two groups, 530–500 Ma (Pampean) and ~470–400 Ma (Famatinian). Granitoid magmatism is widespread and voluminous, especially in the early Ordovician (560–400 Ma, see Table 3.1) whereas mafic igneous rocks are rare. The apparent grouping of metamorphic crystallization ages could be affected by the small size of the database. A continuous high temperature regime in the crust is indicated by voluminous and widespread granitoid magmatism (560–400 Ma, see Table 3.1), especially in the Early Ordovician (500–470 Ma).

Magmatic rocks occur in the same areas as the high-T metamorphism (Fig. 3.2). Most magmatic rocks are granitoid intrusions, and their chemical composition indicates various depths of melt generation and hybridization in the lower to mid crust (e.g. Damm et al. 1994; Coira et al. 1999; Pankhurst and Rapela 1998; Pankhurst et al. 2000). All melts show considerable contribution of crustal material (see below) or are crustal melts compositionally related to the metasediments. The composition of scarce early Paleozoic mafic intrusions in the plateau area, and west of it, resembles rocks from arc magmatism (Damm et al. 1990; Coira et al. 1999; Kleine et al. 2004; Zimmermann and Bahlburg 2003).

Detritus from the evolving early Paleozoic orogen is partly preserved in extensive Ordovician siliciclastic sediments of southern Bolivia (e.g. Egenhoff and Lucassen 2003, and references therein), which extend into northwestern Argentina. The Argentine Ordovician sequences contain some volcaniclastic rocks (e.g. Bock et al. 2000; Zimmermann and Bahlburg 2003, and references therein), but they are volumetrically unimportant and, in any case, do not represent significant mafic, potentially juvenile magmatism during this time. In northern Chile and the northern section of the Sierras Pampeanas, uplift and erosion of the orogen was completed in the Devonian (Lucassen et al. 2000, and references therein). In the southern section of the Sierras Pampeanas erosional unconformities occur in the Carboniferous (e.g. Ramos 2000).

The chemical (Fig. 3.5) and Nd-Sr isotope compositions (Fig. 3.6a) of various early Paleozoic sedimentary, felsic igneous and metamorphic rocks show similar compositional features for the different lithological or age groups in the region of 21 to 27° S. Although the metamorphic conditions indicate a deep middle crust, the rock

**Fig. 3.5.**
Average major and trace element composition of rock units of the Central Andes from 21–27° S (normalized to average upper crust; Taylor and McLennan 1995). **a** The early Paleozoic crust has a uniform felsic composition similar to average upper-continental crust, except for Ca, Na and Sr. Early Paleozoic gneisses from northern Chile show a different chemical composition compared with the early Paleozoic rocks from Argentina and Bolivia, indicating a slightly less evolved protolith for the Chilean rocks. **b** Late Paleozoic granitoid and Mesozoic – Recent rocks of the various magmatic arcs and, for comparison, the gneisses of northern Chile and the paragneisses of Argentina. The Jurassic rocks, which are mantle derived, differ strongly from average upper crust, late Paleozoic granitoids, and typical rocks of the Cainozoic arc. Large-volume Cainozoic ignimbrites are compositionally similar to the Argentine early Paleozoic crust. Data Sources: Early and late Paleozoic rocks (Lucassen et al. 1999a; Lucassen et al. 2001; Egenhoff and Lucassen 2003), Jurassic magmatic arc (Lucassen et al. 2002b, Kramer et al. 2004), Cainozoic large-volume ignimbrite (Lindsay et al. 2001), average Central Volcanic Zone (CVZ) andesite (GEOROC database)

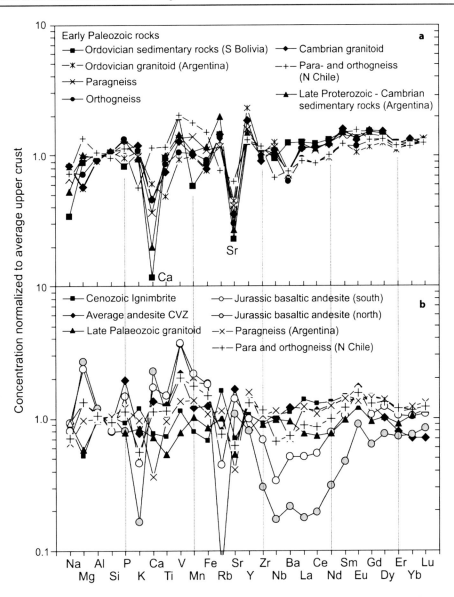

compositions are close to average upper-crustal values of Taylor and McLennan (1995). The main compositional differences from average upper crust are the lower Na, Ca and Sr contents in the rocks from the Central Andes transect, which may indicate a lower content of plagioclase (Eu is also slightly depleted compared to upper crust). The depletions in Sr, Eu and Ca relative to average upper crust, along with the slightly higher Rb and rare earth element (REE) contents suggest that the Central Andean Paleozoic crust was already chemically evolved. As discussed below, the isotopic and trace element data generally show compositional features typical for an old continental crust (Bock et al. 2000; Lucassen et al. 2001, 2002a; Egenhoff and Lucassen 2003).

The inherited age pattern of this crust indicates several metamorphic-magmatic-sedimentary cycles. The result of these processes of homogenization is well displayed by the uranogenic Pb isotope compositions of the igneous rocks. The Pb isotope ratios indicate a progressive homogenization over time of the initially varied Pb isotope compositions during the early Paleozoic, leading finally to a rather uniform Pb isotope composition of the crust (Fig. 3.7a). Compositional variations in the uranogenic Pb isotopes can be explained by long-term separation of crustal sections which formed earlier, e.g. in the Sunsás or older metamorphic-magmatic events.

In summary, crustal evolution during the early Paleozoic orogeny appears to have involved primarily reworking or recycling of pre-existing, ~2 Ga old felsic crust. Juvenile additions of magma from the mantle in this time are small. Extensive production of crustal melts during high-T regional metamorphism contributed to compositional homogenization on a regional scale, especially for Pb isotopes but probably also for the Sr isotope systems

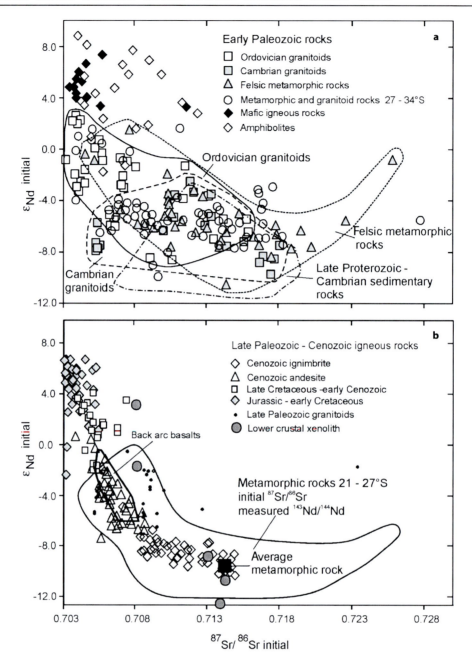

**Fig. 3.6.** Initial Nd-Sr isotope composition of major rock units of the Central Andes from 21 to 27° S. **a** Isotope ratios of rock groups of different age from Bolivia, Chile and Argentina of the early Paleozoic felsic magmatic and metamorphic basement, compared with the basement between 27–34° S (shown as a single group), and mafic magmatic rocks and amphibolite. The two regional groups show a very similar compositional range. The magmatic rocks are either crustal melts or a hybrid containing depleted mantle and felsic crust. The mafic rocks represent the composition of the mantle below the early Paleozoic arc, which is of the depleted mantle type. The Sr isotope ratios of many amphibolites and some mafic igneous rocks scatter to values of the felsic crust caused by Rb and Sr mobility during metamorphism, but the Sm-Nd composition of most rocks does not seem to be affected. **b** Late Paleozoic – Recent rocks show similar compositional trends between a depleted sub-arc mantle (represented by the Jurassic – early Cretaceous igneous rocks) and the early Paleozoic felsic crust. The possible isotopic composition of the Paleozoic lower crust is indicated by lower-crustal xenoliths, which plot into the field of the initial isotope ratios of the early Paleozoic rocks. The isotopic composition of the back-arc basalts is similar to the composition of the andesites. Data sources: (**a**) modified from Lucassen et al. (2001), additional data from Pankhurst and Rapela (1998), Egenhoff and Lucassen (2003), Kleine et al. (2004), and F. Lucassen and R. Becchio (unpublished data). (**b**) Rogers and Hawkesworth (1989), Kay et al. (1994), Trumbull et al. (1999), Lindsay et al. (2000, and references therein), Lucassen et al. (1999c, 2001, 2002b, and references therein), Haschke et al. (2002), Kramer et al. (2004) and Bartsch (2004)

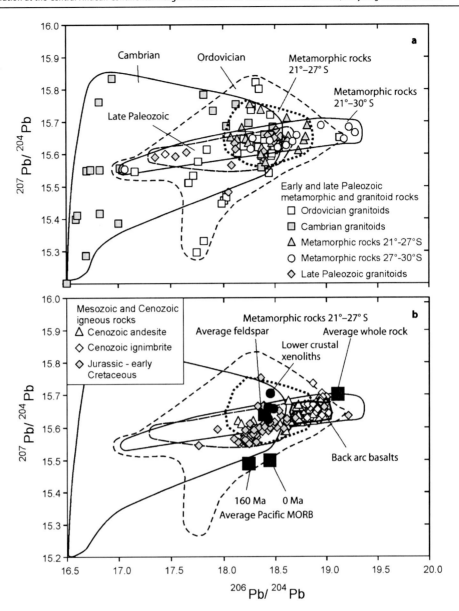

**Fig. 3.7.**
Uranogenic Pb initial isotope composition (**a**) Paleozoic rocks and (**b**) Mesozoic – Recent rocks. **a** All data shown are from feldspar. The Pb isotopic heterogeneities seen in the Cambrian and Ordovician rocks were homogenized during Paleozoic high-grade metamorphism and crustal melting. **b** The present-day Pb isotope composition of the crust is bracketed by the average feldspar composition of the metamorphic rocks and the lower-crust xenoliths, both representing the U-depleted lower crust and the average whole-rock composition representing the undepleted upper crust. The isotopic composition of andesite and ignimbrite represents this variation between these lower- and upper-crustal sources. Jurassic rocks show some affinity to the average Pacific MORB, which is the best available proxy for the Pb isotope composition of the sub-arc mantle. Modified from Lucassen et al. (2002a). Other data sources (see Fig. 3.6b). Average Pacific MORB from the data compilation in Lucassen et al. (2002b)

(Figs. 3.6a and 3.7a). Crystallization ages of the high-grade metamorphic and magmatic rocks indicate a long-standing regime (> 100 Ma) of high temperature (> 600 °C) in the mid-crust over a very large area. With the present data set, no distinctive thermal peaks or spatial patterns in metamorphic conditions can be distinguished. Thermal anomalies of this size and duration (i.e. equal to the lithospheric time constant) indicate a long-lasting and deep-seated origin, probably a shallow asthenosphere related to the active mantle wedge of a subduction zone.

Information about the mantle component involved in the early Paleozoic orogeny comes from meta-igneous mafic rocks, which are small in volume but not uncommon. Compositionally, the rocks are subalkaline basalt to basaltic andesite (Lucassen et al. 2001, and references therein).

Ages of this magmatism are unknown but many of the metabasic rocks represent former dikes and their structural position with respect to the regional metamorphic fabric indicates an early Paleozoic emplacement. The principal mantle component in these rocks was compositionally very different from the enriched subcontinental mantle of the Proterozoic. Instead, most mafic magmas originated from depleted mantle according to their Sr and Nd isotopic compositions (Figs. 3.6a and 3.8a) and their source(s) are interpreted as the mantle wedge active in the various periods of early Paleozoic subduction. The array of the early Paleozoic magmatic rocks in the Sr–Nd isotope diagram (Fig. 3.6a) has the same form as that for the Andean Mesozoic – Cainozoic arc magmatism (Fig. 3.6b), which supports the inference of arc activity in the early Paleozoic.

## 3.4 Late Paleozoic – Triassic

The uplift and erosion of the early Paleozoic orogen was followed in the Devonian to lower Carboniferous by an episode of passive margin-type sedimentation and only minor magmatism (Bahlburg and Hervé 1997). The Devonian passive margin extends as far south as 48° S (Augustsson and Bahlburg 2003) and a largely uniform evolution of this large section of the margin is assumed at least until the Jurassic (Glodny et al. 2006, Chap. 19 of this volume). A renewed onset of magmatism began contemporaneously along the margin at ca. 300 Ma, between at least 21 and 40° S, and continued into the Permian and Triassic (Table 3.1; Berg and Baumann 1985; Parada 1990; Brown 1991; Lucassen et al. 1999a, 2004; Glodny et al. 2006, Chap. 19 of this volume).

Isolated late Paleozoic intrusions in the uppermost levels of the crust, mainly granite with minor diorite (Fig. 3.2), have similar compositions to the early Paleozoic granites (Fig. 3.5b). The isotopic signatures of these granites indicate considerable proportions of recycled early Paleozoic continental crust (Figs. 3.6b and 3.7b). The same is true for the silicic volcanic rocks from the extensive Permian silicic igneous province of central Chile (there known as the Choiyoi group; Kay et al. 1989; Llambias et al. 2003), the northern part of which extends into the plateau region (Breitkreuz and Van Schmus 1996). Triassic magmatism is fairly common but of minor volume and it has been poorly investigated. Available evidence suggests that recycling of older crust dominated (Berg and Baumann 1985; 32° S, Morata et al. 2000), and only for the latest Triassic have mantle-derived volcanic rocks been described (Morata et al. 2000; Bartsch 2004).

Large-scale crustal thickening after exhumation and erosion of the early Paleozoic orogen, and before the Cainozoic plateau formation, is ruled out by the sedimentary record for the late Paleozoic (e.g. Bahlburg and Breitkreuz 1991) and Mesozoic (Prinz et al. 1994; Salfity and Marquillas 1994). The record indicates low average-sedimentation rates in continental platform or shallow marine environments. The dominant tectonic style in the late Paleozoic to Mesozoic was transtension to transpression, and there are no indications for large-scale compression (Scheuber et al. 1994; Ramos 2000).

Metamorphism accompanied by deformation in the late Paleozoic–Mesozoic sedimentary rocks is absent or of very low grade. An exception is the Sierra de Limón Verde (Fig. 3.2b), where Permian high-grade metamorphism (ca. 700 °C, 12 kbar; Lucassen et al. 1999b) is unique in the region. The tectonic mechanism for the exhumation of the Limón Verde rocks is speculative, but could well be related to the Permian transpressional to transtensional regime. The local character of the exhumation of lowermost crust is underlined by the contemporaneous sedimentation of detritus from Limón Verde rocks during final uplift in the Triassic (Lucassen et al. 1999b).

Good constraints on the nature and evolution of the subcontinental mantle during this time are unavailable. We can assume from the overall geologic quiescence that the mantle composition did not change significantly from its state in the early Paleozoic, i.e., a depleted mantle persisted.

The magmatic record of the late Paleozoic into the Triassic indicates crustal recycling as an important process and the variable but generally moderate contents of mantle derived material in the granites (e.g. Lucassen et al. 1999a, 2004) resulted only in minor juvenile addition to the crust, because the magmatism is not voluminous. There is also no apparent continental growth by accretion in this time period. A late Paleozoic accretionary wedge is well known in the southern Central Andes (e.g. Hervé 1988; Glodny et al. 2006, Chap. 19 of this volume; Fig. 3.2a), but this was either not formed, or subsequently destroyed, in our area. Only the late Paleozoic Chañaral Melange, exposed south of 27° S, is a possible remnant of an accretionary wedge that once extended farther to the north (Bell 1987; Fig. 3.2a).

## 3.5 Jurassic – Lower Tertiary

A dramatic change in the nature of arc-related magmas took place in late Triassic – early Jurassic and this is considered as the onset of the Andean Cycle, although subduction had already commenced in the late Carboniferous after the passive margin configuration. Large volumes of mantle-derived magmas were extruded and intruded during the Jurassic to lower Cretaceous, mainly in the Coastal Cordillera (Fig. 3.1; e.g. Palacios 1978; Buchelt and Tellez 1988; Rogers and Hawkesworth 1989; Pichowiak 1994; Lucassen and Franz 1994; Lucassen et al. 2002b; Kramer et al. 2005). They form the extensive La Negra volcanic province and the related coastal batholith.

This magmatic belt is also prominent in the southern Central Andes (e.g. Vergara et al. 1995). The time span of the La Negra volcanism (early to middle Jurassic, ca. 180–150 Ma) is constrained mainly by stratigraphic evidence from intercalated marine sediments (Table 3.1; von Hillebrandt et al. 2000), by the formation ages of the batholith (ca. 190–140 Ma; Pichowiak et al. 1994; Dallmeyer et al. 1996) and by the early Cretaceous ages of dikes that represent the last magmatic phase (Lucassen and Franz 1994).

The La Negra magmas extruded mainly from fissure eruptions close to sea level (Bartsch 2004), and lava sequences up to several thousand meters thick accumulated during subsidence in a generally extensional tectonic regime (Scheuber et al. 1994). The time-space evolution of this magmatism shows a westward shift of activity in the mid-Jurassic in the northern part of the transect at Iquique (Kossler 1998; Kramer et al. 2005), but this migration is not obvious further south. The principle host rocks are early Paleozoic metamorphic rocks, and late Paleozoic sediments and intrusions, indicating that magmas intruded or were deposited on old continental crust.

The La Negra rocks comprise mainly basaltic andesite, with subordinate amounts of both more basic and more evolved compositions that locally can be volumetrically important. Local occurrences of alkaline volcanic rocks, commonly rich in sodium, can be interpreted as either melts with an important slab component (Kramer et al. 2005) or as magmas formed by local remelting of underplated material (see Petford and Gallagher 2001) from the Jurassic – lower Cretaceous magmatism. The coastal batholith is dominantly of basaltic andesite to andesite composition, but it also contains layered (ultra-)mafic intrusions and locally important bodies of granodiorite to granite.

Negative Nb-Ta anomalies in trace element distribution-patterns and a relatively low La/Yb ratio for the subalkaline volcanic and intrusive rocks are indicative of a magmatic arc setting. The low total REE abundances and flat REE patterns (Fig. 3.5b) argue for a high degree of melting (e.g. Lucassen and Franz 1994; Kramer et al. 2005; Bartsch 2004). Despite differences in detail among Jurassic igneous rocks, in terms of composition, age and spatial distribution (described for the area N of 20° S by Kramer et al. 2005; and at ca. 25° S by Bartsch 2004), they show considerable similarities along strike and constitute a compositionally distinct group, different from both the older (late Paleozoic) and younger (Cainozoic) magmatic suites in the region (Fig. 3.5b).

Initial Nd-Sr isotopic compositions of Jurassic – Lower Cretaceous magmatic rocks fall within in the range of the depleted mantle (Fig. 3.6b). Some samples show an elevated Sr isotope ratio, attributed to seawater alteration at the time of extrusion (Kramer et al. 2005). Initial Nd isotopic compositions for most samples show little variation ($+3.5 < \varepsilon_{Nd} < +7.5$) considering the long time span and large area represented by the dataset (Fig. 3.6b). The high values for Nd isotope ratios indicate important amounts of juvenile additions to the crust. Crustal contributions are considered to be small because the Pb isotope ratios of these mantle-derived magmas (Fig. 3.7b), which are very sensitive to additions of crustal material, are less radiogenic than those of the surrounding Paleozoic crust or average subducted sediment, which we take as a proxy for the upper crust.

Rocks from deeper sections of the coastal batholith (ca. 10 to 15 km depth) are low-P-granulite facies orthogneiss of Jurassic age (170–140 Ma; Lucassen and Thirlwall 1998), partially retrogressed to amphibolite (Lucassen and Franz 1996; Scheuber and Gonzales 1999). The metamorphism occurred in an extensional regime during long-standing magmatic activity, deformation and heating related to periodic additions to the batholith (Lucassen et al. 1996). The granulite and amphibolite are compositionally indistinguishable from their magmatic protoliths (Lucassen and Franz 1994; Lucassen et al. 2002b).

The late Cretaceous – lower Tertiary magmatic arc migrated east with time (e.g. Rogers and Hawkesworth 1989; Döbel et al. 1992; Scheuber et al. 1994; Haschke et al. 2002) and the magmas produced show an increasing contribution of Paleozoic crustal material in their isotope signatures (Fig. 3.6b). The variable composition of these magmas, and the fact that their occurrence is dispersed over a larger area, and their minor volume compared with the Jurassic – lower Cretaceous magmatism, makes an estimate of their volumetric importance to crustal growth difficult. In difference to the Jurassic – lower Cretaceous arc- there are no voluminous additions of mafic material seen in the geophysical image of the crust or in the composition of respective magmas (references or see below). The eastward shift of the magmatic arc with time is generally attributed to subduction erosion at the continental leading edge, but it is uncertain exactly how many kilometers of continental crust have been eroded since the Late Triassic. The earliest volcanic arc of the Andean Cycle must have been, in part, west of the present coastline (Fig. 3.2a; von Hillebrandt et al. 2000; Bartsch 2004), and it can be presumed that of the order of 200 km of continental crust, including parts of the Jurassic juvenile additions, have been removed by subduction erosion.

An important feature of the Cretaceous evolution, with potentially crucial consequences for understanding the Andean orogeny in the Cainozoic, is the development of a N-S-trending rift system quite far inland from the arc. The rift is now preserved as inverted basins in the eastern part of the cordillera, i.e., the Salta Rift in northwestern Argentina (Galliski and Viramonte 1988; Viramonte et al. 1999), but the rift was once more extensive, reaching as far west as the Chilean Precordillera (Salfity and Marquillas 1994). Late Cretaceous, alkaline, basaltic rocks erupted along this rift, and the upper-mantle/lower-crustal xenoliths they contain provide good constraints on the composition of the mantle and on the thermal structure of the lithosphere (Lucassen et al. 1999c, 2002b, 2005, submitted). Thermobarometry studies and isotopic dating of the xenoliths indicate high temperatures in the lower crust (ca. 900 °C) and upper mantle (> 1 000 °C) during the late Cretaceous (ca. 100 Ma), meaning that the Cainozoic orogeny was preceded by a N-S-oriented thermal anomaly in an area that now forms the eastern edge of the Andean Plateau. The basanite host rocks have depleted mantle signatures (Fig. 3.8) and the isotopic compositions of most mantle xenoliths also indicate a depleted mantle source, with some variations in isotopic composition. The variations in the U-Th-Pb systems indicate the addition of crustal material to the source and subsequent radiogenic growth probably since the early Paleozoic (Fig. 3.8).

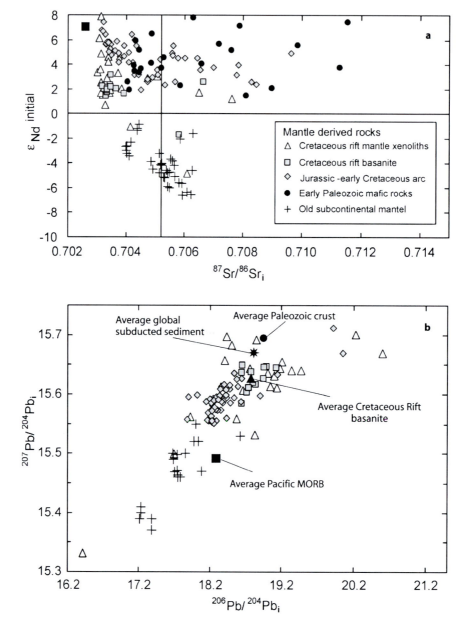

**Fig. 3.8.** Nd-Sr isotope (**a**) and uranogenic Pb isotope (**b**) composition of mantle-derived rocks. **a** The Jurassic – lower Cretaceous igneous rocks show compositional affinities to average Pacific MORB; many Sr isotope ratios reflect seawater alteration. The Cretaceous upper-mantle xenoliths from the rift have a broad range of Nd isotope compositions, but most samples resemble that of contemporaneous basanite from the rift. The mantle below the rift is also depleted, but slightly less radiogenic in its Nd isotope composition than the sub-arc mantle. The early Paleozoic mafic rocks are shown for comparison. There is no evidence for substantial contribution of old subcontinental mantle between 21–27° S. **b** Considering a MORB type source for the Jurassic – lower Cretaceous igneous rocks, the contributions of the crustal sources to these rocks are minor because the Pb isotope composition of many of these rocks is not dominated by crust. Pb isotope compositions, especially the $^{206}Pb/^{204}Pb$ ratios of many mantle xenoliths and basanite samples, are more radiogenic than those of the Jurassic to lower Cretaceous rocks. They are also different from the local crustal sources. The old subcontinental mantle is less radiogenic in uranogenic Pb, but shows much higher $^{208}Pb/^{204}Pb$ ratios at given $^{206}Pb/^{204}Pb$ than our samples (not shown; Lucassen et al. 2002b) and a substantial contribution of this mantle to our samples can be excluded. Data sources: Carlson et al. (1996), Gibson et al. (1996, 1999), Lucassen et al. (2001, 2002b), Kramer et al. (2004) and Bartsch (2004); for some Jurassic rocks, F. Lucassen and W. Kramer (unpublished data); average global subducted sediment, Plank and Langmuir (1998); Jurassic – lower Cretaceous igneous rocks and MORB are corrected for in situ decay to 160 Ma, all other Cretaceous rocks to 100 Ma, the early Paleozoic mafic rocks to 450 Ma

The upper mantle sampled by Late Cretaceous rift magmas must comprise at least part of the lithosphere that was later tectonically shortened in the Cainozoic and is now located under the Central Andean Plateau. From the evidence summarized above, we can conclude that the upper mantle from the Paleozoic to the Cretaceous (and most probably till the present) from 21 to 27° S has a depleted signature. The range of isotopic compositions of basic igneous rocks from the Jurassic to lower Cretaceous magmatic arc of the Coastal Cordillera, with $\varepsilon_{Nd}$ of +3.5 to +7.5 (Figs. 3.6a and 3.8a), indicate a mantle composition significantly less depleted than the contemporaneous mid-ocean-ridge-basalt (MORB) source (Fig. 3.8) and, also, the composition is irregularly variable in time and space. This heterogeneity is especially well displayed in the trace element and isotope composition of clinopyroxene from mantle xenoliths (Lucassen et al. 2005a). We argue that these variations are due to incompletely homogenized mixtures of components in the lithosphere that reflect its geologic history, perhaps even including scarce remnants of the Proterozoic lithosphere.

For the main issue of crustal growth versus destruction, the rift-related mafic rocks represent negligible net addition to the crust because their volumes are small and occurrences restricted to small plugs and dikes. Even if one assumes an order of magnitude larger amount of intruded or underplated material at the base of the crust, the early Cretaceous rift magmatism is volumetrically small. However, the rift magmatism and its underlying thermal anomaly are highly significant in terms of their weakening effect on the lithosphere and on localizing deformation in the subsequent Cainozoic orogeny that, as is discussed below, may eventually have triggered lithospheric delamination.

## 3.6 The Cainozoic Orogeny

During the Cainozoic, arc magmatism continued to migrate eastward, albeit with important variations in intensity. Paleocene and Eocene magmatism was located in the present Precordillera of Chile. The composition and isotope ratios of these early Tertiary magmatic belts are similar to those of the Neogene arc (dominantly andesite) and suggest that the arc magmas were strongly affected by contamination from the crust. Juvenile (basaltic) magmas in these belts are rare, although it is reasonable to assume that the andesites are associated with more mafic intrusive and/or cumulate material in the lower crust (Haschke et al. 2002). Nevertheless, there is no magmatism in this period to compare with the juvenile Jurassic crust in the present fore-arc, and seismic velocity data from the Precordillera indicate that high-density material does not contribute much to the total crustal thickness. Oligocene and younger volcanic rocks are located in the West Cordillera and the Altiplano-Puna Plateau region to the east. There are variations in the distribution of volcanism across this region that suggest a broad Oligocene – early Miocene arc, which subsequently narrowed and became focused in the area of the present volcanic chain (Coira et al. 1982, 1993). This variation in arc location and width was interpreted to reflect a change in slab dip from shallow in the Oligocene to steeper today.

The intense tectonic shortening that led to the near doubling of crustal thickness in the Central Andes, and the development of the world's second-largest continental plateau, began around 30 Ma (e.g. Almendinger et al. 1997). The contribution of Cainozoic juvenile magmas to the crustal thickness is considered to be negligible (e.g. Francis and Hawkesworth 1994). The erupted magmas from the Tertiary and Quaternary arcs are dominantly andesite to silicic andesite, and these magmas have incorporated 20–30% of continental crust (e.g. Harmon et al. 1984; Hildreth and Moorbath 1988; Wörner et al. 1994; Trumbull et al. 1999; Siebel et al. 2001). The volume contribution to crustal growth by the remaining 70–80% of magma, which was derived from the mantle wedge, is impossible to evaluate with any certainty, but geophysical evidence from both seismic and gravity studies (Beck and Zandt 2002; Yuan et al. 2000) demonstrate that only the lowermost part of the 60–80 km-thick crust can be mafic.

Two other volcanic associations are found in the Central Andes in addition to the andesite volcanoes of the frontal arc, but neither has a significant impact on crustal growth, albeit for different reasons. The first of these comprises the regionally extensive silicic ignimbrites that erupted from large resurgent caldera complexes primarily in the late Miocene to Pliocene (Coira et al. 1993; de Silva 1989; Francis et al. 1989; Ort et al. 1996; Lindsay et al. 2001). The ignimbrite magmas are dominated by crustal sources (> 70%; e.g. Francis et al. 1989) and therefore do not contribute greatly to crustal growth despite their large volumes. The second volcanic association comprises basaltic magmas that erupted primarily behind the arc. There are isolated examples of basaltic volcanism in the late Oligocene to early Miocene from Chile (Segerstrom basalts; Kay et al. 1994) and Bolivia (Chiar Khollu; Davidson and de Silva 1995) but the greatest concentration is in the late Pliocene and Quaternary in the back-arc region of northwestern Argentina (Kay et al. 1994). These basalts represent juvenile magmas, but their volumes are too small to contribute to crustal growth. On the contrary, these back-arc basalts have been suggested to indicate an episode of delamination of the lower crust and lithospheric mantle beneath the Central Andean Plateau (Kay et al. 1994).

The Nd, Sr and uranogenic Pb isotope ratios of the late Cainozoic andesite-dacite and ignimbrite associations show a rather restricted range of compositions, suggesting a mixing trend between an arc-source and the early Paleozoic crust (Figs. 3.6b and 3.7b). These diagrams illustrate the problem that estimating the degree of crustal involvement in a given suite of rocks depends on assumptions about the end-members. For example, there may be contributions to the crustal 'component' from lower crustal rocks that were depleted in Rb and U during the early Paleozoic orogeny. These will have less radiogenic Sr and Pb isotope ratios (Fig. 3.6a,b) than their counterparts in the mid-crust.

The composition of the sub-arc mantle wedge is poorly constrained because all of the arc magmas are differentiated and have been contaminated by crustal material either in the source via subduction processes, or during their ascent through the crust. Nevertheless, Nd and Sr isotope signatures of andesites from the main arc, the ignimbrites and most basalts from the back-arc region follow the same mixing trajectory between Paleozoic crust and a depleted mantle. Helium isotope signatures indicate an asthenospheric source for the primary mantle magma of the hybrid back-arc basalts (Pilz et al. 2001).

## 3.7 Discussion

The long record of compositional variations, metamorphism and tectonics preserved at the Andean continental margin (Table 3.1) is the basis for identifying the processes contributing to growth, destruction or modification of the continental crust and upper mantle, and for assessing their long-term importance. This discussion is focussed on the processes of continental growth and destruction at active margins as outlined in the introduction.

### 3.7.1 Crustal Growth: Juvenile Magmas and Tectonic Accretion

The data on age structure, metamorphic grade and isotopic-geochemical characteristics of the early Paleozoic metamorphic, magmatic and sedimentary rocks in the Central Andes (Table 3.1) support the interpretation of a coherent, autochthonous early Paleozoic mobile belt at the western leading edge of South America (Fig. 3.2; Aceñolaza and Miller 1982; Damm et al. 1994). No early Paleozoic exotic terranes have been recognized between 18 and 27° S. The geotectonic setting at this time was probably a non-collisional active continental margin with an orogen comparable to the Cainozoic orogen (Lucassen and Franz 2005b).

This interpretation contrasts with suggestions that the Paleozoic basement is a mosaic of accreted terranes (e.g. Ramos et al. 1986; Ramos 2000). The arguments against late Proterozoic to Phanerozoic terrane accretion models for the areas between 18 and 27° S have been discussed in detail before (Becchio et al. 1999; Bock et al. 2000; Lucassen et al. 2000; Franz and Lucassen 2001) and rely mainly on the similar ages of metamorphic and magmatic crystallization, the age pattern of the inherited material, chemical and isotopic composition of sedimentary, magmatic and metamorphic rocks, and the type and distribution of early Paleozoic metamorphism, e.g. the absence of high pressure rocks typical of suture zones.

South of 28° S, the Argentine Precordillera is widely believed to be a Laurentia-derived exotic terrane based on its stratigraphic and faunal record (see reviews in Ramos and Keppie 1999). If the Argentine Precordillera accreted in the Ordovician as proposed, this would constitute a significant contribution to growth of the local crust. Our data are not inconsistent with terrane accretion south of 28° S, but, if it took place, we can conclude that the process is not discernable from the early Paleozoic pressure-temperature-time record of the adjacent high-grade metamorphic basement (Fig. 3.2; Lucassen and Becchio 2003).

Continental growth by accretion of oceanic crust and continental sediments from the lower plate is considered to be globally minor, based on observations of presently active margins (Clift and Vanucchi 2004, and references therein), and the Central Andean record is no exception. The existence and preservation of accretion complexes could be an important indicator of long-term stability of the trench-margin relationships. The development of accretionary wedges depends mainly on convergence geometry and rates, morphology of the lower plate and sediment thickness in the trench, which in turn may be linked to climate. These factors are unlikely to remain constant over the time span considered here, so it is probable that accretionary and non-accretionary periods alternated during the history of the active margin, as observed in the fore-arc of the southern Central Andes (Glodny et al. 2006, Chap. 19 of this volume). However, Phanerozoic accretionary units are generally absent in the transect area, except for the local Chañaral melange.

Early Paleozoic magmatism was extensive and has been well studied, but the addition of mantle magmas was apparently volumetrically unimportant to the growth of the continental crust throughout this period. The magmas produced were largely granitic and their isotopic composition clearly shows that recycling of Proterozoic crust was the dominant source. The same is true for late Paleozoic igneous rocks, which were derived from early Paleozoic and Proterozoic crustal sources. There could, potentially, have been substantial intrusions of juvenile magmas in the lower crust that are not now exposed. However, the excellent geophysical coverage of the Central Andes shows that dense mafic material is a very minor component of the crust today (Asch et al. 2006, Chap. 21 of this volume), and there are no indications in the composition of Mesozoic magmas for a thick Paleozoic mafic lower crust. Igneous rocks derived from partial melting of mafic crust are compositionally distinct (e.g. in the coastal batholith of Peru: Atherton and Petford 1993; Petford and Atherton 1996; and tonalite-trondhjemite-granodiorite rocks in the early Paleozoic orogen: Pankhurst et al. 2000) and such rocks are rare or absent among the Paleozoic intrusions as well as among the Cainozoic volcanic rocks.

The occurrence of magmas derived from mafic (lower) crust in the Central Andes seems restricted to the Mesozoic province of the Coastal Cordillera and Precordillera, where a mafic to intermediate crust formed in the Jurassic and Lower Cretaceous (Lucassen et al. 1994, 2002b; Haschke et al. 2002). The tectonic regime operating at this period was extensional to transpressional and caused by strongly oblique subduction (e.g. Scheuber et al. 1994). This mafic crust in the Coastal Cordillera has a clear geophysical expression (seismic velocity and gravity), which increases our confidence in concluding that thick mafic lower crust is absent in the Paleozoic belt east of the Coastal Cordillera (see other contributions of this volume). Indeed, the Jurassic – lower Cretaceous magmatism in the Coastal Cordillera is the only important episode of crustal growth from mantle magmas in this section of the Central Andes. Some reworking of mafic crust is found in the Paleocene and Eocene magmatic arcs of the Precordillera (Haschke et al. 2002).

### 3.7.2 Crustal Destruction: Subduction Erosion and Lithospheric Delamination

The process of subduction erosion in the Central Andes today is related to morphological roughness of the oceanic plate and the climate-related lack of sediment infill of the trench. It is evidenced by the young tectonic features of the fore-arc and continental slope (e.g. von Huene and Ranero 2001; Kukowski and Oncken 2006, Chap. 10 of this volume). The climate, which plays a key role in the sediment supply from the continent, was likely dominated by arid conditions from the Jurassic onwards (Hartley et al. 2005 and references therein) and adds a long-term component to the subduction erosion model, which is based mainly on observations of active features. Presently, in northern Chile the loss of crust by subduction erosion is greater than the replenishment by magmatic additions, and the result produces a trench retreat of ~3 km $Ma^{-1}$, or 40 to 45 $km^3$ per Ma and km at 22° S during the last 8 Ma (Kukowski and Oncken 2006, Chap. 10 of this volume). Clift and Vanucchi (2004) show that arc-retreat rates of this order can be maintained for the short-term (tens of Ma), but for the long-term (200 Ma) history relevant to the Central Andes, an average retreat rate of 1 km $Ma^{-1}$ seems realistic.

The overall volumetric importance of subduction erosion in the Central Andes since the early Paleozoic is difficult to assess because it depends on knowing the initial configuration of the continental margin and convergence parameters, as well as the assumption of a constant distance between trench and magmatic front of the eastward-migrating Mesozoic magmatic arcs (e.g. Scheuber et al. 1994). A rough estimate based on the latter assumption, is that up to ~200 km of crust have been lost since the late Mesozoic (Scheuber et al. 1994, and references therein; Kramer et al. 2005; Kukowski and Oncken 2006, Chap. 10 of this volume). This eroded crust comprised parts of the Jurassic magmatic arc, as well as the original late Paleozoic fore-arc, based on the assumption of a northern continuation of the Chañaral Melange. This estimate implies that at least a part of the subducted continental crust was juvenile crust formed in the Mesozoic.

A second potential mechanism for destruction of continental crust at active margins is delamination of the roots of overthickened crust into the mantle (e.g. Kay and Kay 1993; Meissner and Mooney 1998, and references therein). This mechanism is density-driven and therefore restricted to mafic rocks that develop sufficient high density by high-P-phase transitions (basalt-eclogite) to sink into the upper mantle. Delamination is suggested to have taken place under the ultra-thick Central Andean Plateau in the late Pliocene (Kay et al. 1994). The arguments for this proposal come from the occurrence and composition of back-arc basalts, from seismic evidence of low-velocity upper mantle with high S-wave attenuation, and from the coincidence of high plateau elevation with only moderate tectonic shortening. Sobolev et al. (2006, Chap. 25 of this volume) can show by numerical modeling that lithospheric delamination is in fact a realistic scenario for the Central Andes.

Delamination may also have been an important process for removing mafic crust in the early Paleozoic orogen. In a period of more than 100 Ma the Paleozoic belt that extended over thousands of kilometers along the continental margin (Fig. 3.3) must have formed an impressive orogen and it is very likely that in such a subduction system lithosphere was delaminated and destroyed parts of the lower crust. There is no direct petrological evidence for delamination, but, on the other hand, geophysical evidence suggests a dominantly felsic crust under the now-exposed Paleozoic belt, and there are no younger igneous rocks with compositions consistent with a mafic crustal source. By contrast, the Jurassic–early Cretaceous crust of the Coastal Cordillera in the Central Andes and the Cainozoic crust in the Southern Andes have the mafic to intermediate compositions that would be expected for active margins where subduction magmatism continuously delivers mafic magma to the base of the crust. A subduction setting is certain for the Cainozoic orogen in the Central Andes and was also most likely the case in the early Paleozoic, so the apparent lack of mafic crust in these two examples probably means that it has been lost. In summary, even if there are good indications for mafic additions to the crust by andesite magmatism in the Cainozoic, this is likely to be balanced out by lithospheric delamination.

Loss of crustal material into the mantle requires some return flow. Both aspects are relevant for the long-term evolution of the crustal and mantle reservoirs in general (e.g. Jacobsen 1988; Kramers and Tolstikhin 1997), but in this paper we restrict ourselves to the specific aspects of arc-evolution. As described above, the composition of the Andean upper mantle differs from that of the major MORB-source reservoir despite its generally depleted character. The compositional differences could reflect input of crustal material into the mantle since the early Paleozoic, either by subduction or by delamination. Long-term separation from the convective mantle could have contributed to the specific evolution of the Sr, Pb, Nd radiogenic isotope systems of the lithospheric mantle, as indicated by the trace element contents and isotope signatures of mantle xenoliths (Lucassen et al. 2002b, 2005, submitted). The pre-Paleozoic lithospheric mantle was possibly similar to that beneath other sections of the Brazilian shield and, if so, a nearly complete exchange with depleted sub-arc mantle took place starting in the early Paleozoic orogeny.

The present data set and isotopic data from mantle-derived igneous rocks in the southern Central Andes (Lucassen et al. 2004, and references therein) allow us to speculate that the convective asthenospheric mantle beneath the Andean continental margin has developed a uniform composition, over a large space and time domain, which is distinct from a common Pacific MORB source. The Andean margin has been active for a very long time in a more or less constant position at the edge of old subcontinental lithosphere. The uniform composition of the mantle-derived magmas can be created by mixing material from a MORB source with old continental crust or with isotopically similar continental sediments deposited on the oceanic plate. Alternatively, the rocks could represent magmas derived from a uniform, compositionally modified MORB-like source.

During the Phanerozoic evolution of the Andean active margin, processes of crustal recycling dominate, but processes of crustal growth and destruction are occasionally important. Global estimates of crustal growth assume a nearly constant volume or a slight volume increase of the crust since the Precambrian, with 80 to 90 vol% of the present crust having formed by the beginning of the Phanerozoic (e.g. Jacobsen 1988; Kramer and Tolstikhin 1997; Collerson and Kamber 1999; Condie 2002). The global system of mid-ocean ridges and subduction zones has been the principle setting for mass flux between the mantle and crust, at least for the Phanerozoic. Provided that subduction erosion is not a short-lived transitional state of subduction zones, but is characteristic for their long-term evolution, the condition of constant crustal volume would require considerable addition of juvenile material from the mantle to balance the loss. The place for compensating crustal losses must be the continental and intra-oceanic arcs.

In the Central Andes between 21 and 27° S, the isotope and trace element compositions of all major Paleozoic magmatic, metamorphic and sedimentary units indicate recycling of a Proterozoic felsic crust formed ca. 2 Ga ago (Figs. 3.4, 3.6 and 3.7; and reworked in the Mesoproterozoic, in accordance with major cycles of crustal growth and orogenic reworking of South America (Cordani et al. 2000b; Sato and Siga 2002) The dominant processes in the early Paleozoic imply conservation of the existing felsic crustal material. The same holds true for the Cainozoic orogen, where juvenile additions to the crust by arc magmas are of minor importance volumetrically.

Regionally, significant crustal growth was restricted to the extensional-transpressional Jurassic to lower Cretaceous arc. This may imply that large-scale crustal growth by juvenile magmatic additions in continental arcs requires certain geotectonic situations. An oblique-subduction setting with extensional tectonics may be necessary to allow efficient mantle melting and intrusion of magmas into the crust. Conversely, orthogonal subduction may favor destruction of the crust by subduction erosion under conditions of arid climate and lack of trench sediments. The strong influence of the long-term prevailing climate conditions is indicated by the increasing preservation of a late Paleozoic – early Mesozoic accretionary wedge towards the south. A near steady state situation of the active margin since Mesozoic is described south of 38° S (Glodny et al. 2006, Chap. 19 of this volume, and references therein). Where, as in the Central Andes, subduction orogeny leads to major crustal shortening and thickening, further loss of the crust can occur by lithospheric delamination. This delaminated lower crust is prevailing juvenile, mafic and relatively young because it was probably replenished between or during the respective orogenies by magmatic underplating (c. 1200–900 Ma, 560–400 Ma, 30 Ma) and only the more differentiated juvenile rocks contributed to the crust. The statement that active continental margins and the continental subduction environment are the major site for growth of the continents and the continental crust is, in this simplified form, definitely not correct.

## Acknowledgments

The authors thank all members of the Special Research Project for sharing ideas and for many discussions, especially P. Giese †, O. Oncken and P. Wigger for leading and organizing the project. The DFG and the DAAD is thanked for generous financial support during the last years. J. Glodny's critical reading of an earlier version and reviews by U. Cordani and E. Nisbet are gratefully acknowledged.

# References

Aceñolaza FG, Miller H (1982) Early Paleozoic orogeny in southern South America. Precambrian Research 17:133–146

Aceñolaza FG, Aceñolaza GF, Estaban S (1999) Biostratigrafia de la formación Puncoviscana y unidades equivalentes en el NOA. In: Bonorino GG, Omarini R, Viramonte J (eds) Geologia del noroeste Argentino. XIV Congreso Geológico Argentino, Salta, Argentina, Relatorio, pp 91–114

Allmendinger R, Jordan TE, Kay SM, Isacks BL (1997) The evolution of the Altiplano-Puna plateau of the Central Andes. Annual Review Earth Planetary Sciences 25:139–174

Asch G, Schurr B, Bohm M, Yuan X, Haberland C, Heit B, Kind R, Woelbern I, Bataille K, Comte D, Pardo M, Viramonte J, Rietbrock A, Giese P (2006) Seismological studies of the Central and Southern Andes. In: Oncken O, Chong G, Franz G, Giese P, Götze H-J, Ramos VA, Strecker MR, Wigger P (eds) The Andes – active subduction orogeny. Frontiers in Earth Science Series, Vol 1. Springer-Verlag, Berlin Heidelberg New York, pp 443–458, this volume

Astini RA, Benedetto JL, Vaccari NE (1995) The Early Paleozoic evolution of the Argentine Precordillera as a Laurentian rifted, drifted, and collided terrane: a geodynamic model. Geol Soc Am Bull 107:253–273

Atherton MP, Petford N (1993) Generation of sodium-rich magmas from newly underplated basaltic crust. Nature 362:144–146

Augustsson C, Bahlburg H (2003) Active or passive continental margin? Geochemical and Nd isotope constraints of metasediments in the backstop of a Pre-Andean accretionary wedge in southernmost Chile (46°30'- 48°30' S). Geol Soc London Spec Pub 208:253–268

Babeyko AY, Sobolev SV, Trumbull RB, Oncken O, Lavier LL (2002) Numerical models of crustal scale convection and partial melting beneath the Altiplano-Puna plateau. Earth Planet Sci Lett 199:373–388

Bahlburg H, Hervé F (1997) Geodynamic evolution and tectonostratigraphic terranes of northwestern Argentina and northern Chile. Geol Soc America Bull 109:869–884

Bartsch V (2004) Magmengenese der obertriasssischen bis unterkretazischen Vulkanite in der Küstenkordillere von Nord-Chile zwischen 24° und 27° S. PhD thesis, Technische Universität Berlin

Becchio R, Lucassen F, Franz G, Viramonte J, Wemmer K (1999) El basamento paleozoico inferior del noroeste de Argentina (23–27° S) – metamorfismo y geocronologia. In: Bonorino GG, Omarini R, Viramonte J (eds) Geologia del noroeste Argentino. XIV Congreso Geológico Argentino, Salta, Argentina, Relatorio, pp 58–72

Bell CM (1987) The origin of the Upper Paleozoic Chañaral mélange of N Chile. J Geol Soc 144:599–610

Berg K, Baumann A (1985) Plutonic and metasedimentary rocks from the Coastal range of northern Chile: Rb-Sr and U-Pb isotopic systematics. Earth Planet Sci Lett 75:101–115

Bock B, Bahlburg H, Wörner G, Zimmermann U (2000) Tracing crustal evolution in the southern Central Andes from Late Precambrian to Permian using Nd and Pb isotopes. J Geol 108:515–535

Breitkreuz C, Van Schmus WR (1996) U/Pb geochronology and significance of Late Permian ignimbrites in Northern Chile. J S Am Earth Sci 9:281–293

Brown M (1991) Comparative geochemical interpretation of Permian-Triassic plutonic complexes of the Coastal Range and Altiplano (23°30' to 26°30' S), northern Chile. Geol Soc Am Special P 265:157–171

Buchelt M, Tellez C (1988) The Jurassic La Negra Formation in the area of Antofagasta, northern Chile (lithology, petrography, geochemistry). In: Bahlburg H, Breitkreuz C, Giese P (eds) The Southern Central Andes. Springer, Heidelberg, pp 172–182

Carlson RW, Esperanca S, Svisero DP (1996) Chemical and Os isotopic study of Cretaceous potassic rocks from southern Brazil. Contrib Mineral Petrol 125:393–405

Clift P, Vannucchi P (2004) Controls on tectonic accretion versus erosion in subduction zones: implications for the origin and recycling of continental crust. Rev Geophys 42: doi 10.1029/2003RG000127

Coira B, Davidson J, Mpodozis C, Ramos V (1982) Tectonic and magmatic evolution of the Andes of northern Argentina and Chile. Earth Sci Rev 18:303–332

Coira B, Kay SM, Viramonte J (1993) Upper Cenozoic magmatic evolution of the Argentine Puna – a model for changing subduction geometry. Int Geol Rev 35:677–720

Coira BL, Mahlburg Kay S, Peréz B, Woll B, Hanning M, Flores P (1999) Magmatic sources and tectonic setting of Gondwana margin Ordovician magmas, northern Puna of Argentina and Chile. In: Ramos V, Keppie D (eds) Laurentia-Gondwana Connection before Pangea. Geol Soc Am Spec P 336:145–170

Collerson KD, Kamber BS (1999) Evolution of the continents and atmosphere inferred from Th-U-Nb systematics of the depleted mantle. Science 283:1519–1522

Condie KC (2002) Continental growth during 1.9-Ga superplume event. J Geodyn 34:249–264

Cordani UG, Milani EJ, Thomaz Fihlo A, Campos DA (2000a) Tectonic evolution of South America. Rio De Janeiro, p 856

Cordani UG, Sato K, Teixeira W, Tassinari CCG, Basei MAS (2000b) Crustal evolution of the South American Platform. In: Cordani UG, Milani EJ, Thomaz Fihlo A, Campos DA (eds) Tectonic evolution of South America. Rio De Janeiro, pp 19–40

Dallmeyer RD, Brown M, Grocott J, Taylor GK, Treloar PJ (1996) Mesozoic magmatic and tectonic events within the Andean plate boundary zone, 26°–27°30' S, North Chile: constraints from $^{40}Ar/^{39}Ar$ mineral ages. J Geol 104:19–40

Damm KW, Pichowiak S, Harmon RS, Todt W, Kelley S, Omarini R, Niemeyer H (1990) Pre-Mesozoic evolution of the central Andes: the basement revisited. Geol Soc Am Spec P 241:101–126

Damm KW, Harmon RS, Kelley S (1994) Some isotope and geochemical constraints on the origin and evolution of the Central Andean basement (19°–24° S). In: Reutter KJ, Scheuber E, Wigger PJ (eds) Tectonics of the Southern Central Andes. Springer, Heidelberg, pp 263–275

Davidson JP, De Silva SL (1995) Late Cenozoic magmatism of the Bolivian Altiplano. Contrib Mineral Petrol 119:387–408

De Brito Neves BB, Cordani UG (1991) Tectonic evolution of South America during the Late Proterozoic. Precambrian Res 53:23–40

De Silva SL (1989) Altiplano-Puna volcanic complex of the Central Andes. Geology 17:1102–1106

Döbel R, Hammerschmidt K, Friedrichsen H (1992) Implications of $^{40}Ar-^{39}Ar$ dating of early Tertiary volcanic rocks from the North Chilean Precordillera. Tectonophysics 202:55–82

Egenhoff SO, Lucassen F (2003) Chemical and isotopic composition of Lower to Upper Ordovician sedimentary rocks (Central Andes/ South Bolivia): implications for their source. J Geol 111:487–497

Ernst WG (2000) Earth systems: processes and issues. Cambridge University Press

Francis PW, Hawkesworth CJ (1994) Late Cainozoic rates of magmatic activity in the Central Andes and their relationship to continental crust formation and thickening. J Geol Soc London 151:845–854

Francis PW, Sparks RSJ, Hawkesworth CJ, Thorpe RS, Pyle DM, Tait SR, Montovani MS, McDermott F (1989) Petrology and geochemistry of volcanic rocks of the Cerro Galan caldera, northwest Argentina. Geol Mag 126:515–547

Franz G, Lucassen F (2001) Comment on the paper "Puncoviscana folded belt in northwestern Argentina: testimony of Late Proterozoic Rodinia fragmentation and pre-Gondwana collisional episodes" by Omarini et al. (2000) Int J Earth Sci (Geologische Rundschau) 90:890–893

Galliski MA, Viramonte JG (1988) The Cretaceous paleorift in northwestern Argentina: a petrological approach. J S Am Earth Sci 1: 329–343

Gibson SA, Thompson RN, Dickin AP, Leonardos OH (1996) Erratum to 'High-Ti and low-Ti mafic potassic magmas: key to plume-lithosphere interaction and continental flood-basalt genesis'. Earth Planet Sci Lett 141:325–341

Gibson SA, Thompson RN, Leonardos OH, Dickin AP, Mitchell JG (1999) The limited extent of plume-lithosphere interactions during continental flood-basalt genesis: geochemical evidence from Cretaceous magmatism in southern Brazil. Contrib Mineral Petrol 137:147–169

Glodny J, Echtler H, Figueroa O, Franz G, Gräfe K, Kemnitz H, Kramer W, Krawczyk C, Lohrmann J, Lucassen F, Melnick D, Rosenau M, Seifert W (2006) Long-term geological evolution and mass-flow balance of the South-Central Andes. In: Oncken O, Chong G, Franz G, Giese P, Götze H-J, Ramos VA, Strecker MR, Wigger P (eds) The Andes – active subduction orogeny. Frontiers in Earth Science Series, Vol 1. Springer-Verlag, Berlin Heidelberg New York, pp 401–428, this volume

Goldstein SL, ONions RK, Hamilton PJ (1984) A Sm-Nd study of atmospheric dust and particulates from major river systems. Earth Planet Sci Lett 70:221–236

Goldstein SL, Arndt NT, Stallard RF (1997) The history of continent from U-Pb ages of zircons from Orinoco river sand and Sm-Nd isotopes in Orinoco basin river sediments. Chem Geol 139:271–286

Harmon RS, Barreiro BA, Moorbath S, Hoefs J, Francis PW, Thorpe RS, Deruelle B, McHugh J, Viglino JA (1984) Regional O-, Sr-, and Pb-isotope relationships in late Cenozoic calc-alkaline lavas of the Andean Cordillera. J Geol Soc London 141:803–822

Haschke M Siebel W Günther A Scheuber E (2002) Repeated crustal thickening and recycling during the Andean orogeny in north Chile (21°–26° S). J Geophys Res Solid Earth 107: doi 10.1029/2001JB000328

Hervé F (1988) Late Paleozoic subduction and accretion in southern Chile. Episodes 11:183–188

Hildreth W, Moorbath S (1988) Crustal contributions to arc magmatism in the Andes of Central Chile. Contrib Mineral Petrol 98:455–489

Höckenreiner M, Söllner F, Miller H (2003) Dating the TIPA shear zone: an early Devonian terrane boundary between the Famatinian and Pampean systems (NW Argentina). J S Am Earth Sci 16:45–66

Jacobsen SB (1988) Isotopic constraints on crustal growth and recycling. Earth Planet Sci Lett 90:315–329

Kay SM, Ramos VA, Mpodozis C, Sruoga P (1989) Late Paleozoic to Jurassic silicic magmatism at the Gondwana margin: analogy to the middle Proterozoic in North America? Geology 17:324–328

Kay SM, Coira B, Viramonte J (1994a) Young mafic back arc volcanic rocks as indicators of continental lithospheric delamination beneath the Argentine Puna plateau, central Andes. J Geophys Res 99:24323–24339

Kay SM, Mpodozis C, Tittler A, Cornejo P (1994b) Tertiary magmatic evolution of the Maricunga mineral belt in Chile. Int Geol Rev 36:1079–1112

Kleine Th, Mezger K, Zimmermann U, Münker C, Bahlburg H (2004) Crustal evolution of the early Ordovician Proto-Andean margin of Gondwana: trace element and isotope evidence from Complejo Igneo Pocitos (Northwest Argentina). J Geol 112:503–520

Kossler A (1998) Der Jura in der Küstenkordillere von Iquique (Nordchile)- Paläontologie, Lithologie, Stratigraphie, Paläogeographie. Berliner geowiss Abh (A)197, Berlin, pp 226

Kramer W, Siebel W, Romer RL, Haase G, Zimmer M, Ehrlichmann R (2004) Geochemical and isotopic characteristics and evolution of the Jurassic volcanic arc between Arica (18°30' S) and Tocopilla (22° S), North Chilean Coastal Cordillera. Chemie der Erde

Kramers JD, Tolstikhin IN (1997) Two terrestrial lead isotope paradoxes, forward transport modeling, core formation and the history of the continental crust. Chem Geol 139:75–100

Kukowski N, Oncken O (2006) Subduction erosion – the "normal" mode of fore-arc material transfer along the Chilean margin? In: Oncken O, Chong G, Franz G, Giese P, Götze H-J, Ramos VA, Strecker MR, Wigger P (eds) The Andes – active subduction orogeny. Frontiers in Earth Science Series, Vol 1. Springer-Verlag, Berlin Heidelberg New York, pp 217–236, this volume

Lindsay JM, Schmitt AK, Trumbull RB, De Silva SL, Siebel W, Emmermann R (2001) Magmatic evolution of the La Pacana Caldera system, central Andes, Chile: compositional variation of two cogenetic, large-volume felsic ignimbrites and implications for contrasting eruption mechanisms. J Petrol 42:459–486

Litherland M, Annells RN, Darbyshire DPF, Fletcher CJN, Hawkins MP, Klinck BA, Mitchell WI, OConnor EA, Pitfield PEJ, Power G, Webb BC (1989) The Proterozoic of Eastern Bolivia and its relationship to the Andean Mobile Belt. Precambrian Res 43:157–174

Llambias EJ, Quenardelle S, Montenegro T (2003) The Choiyoi group from central Argentina: a subalkaline transitional to alkaline association in the craton adjacent to the active margin of the Gondwana continent. J S Am Earth Sci 16:243–257

Loewy SL, Connelly JN, Dalziel IWD (2004) An orphaned basement block: the Arequipa-Antofalla basement of the central Andean margin of South America. Geol Soc Am Bull 116:171–187

Lucassen F, Franz G (1994) Arc related Jurassic igneous and metaigneous rocks in the Coastal Cordillera of northern Chile/Region Antofagasta. Lithos 32:273–298

Lucassen F, Franz G (1996) Magmatic arc metamorphism: Petrology and temperature history of metabasic rocks in the Coastal Cordillera of Northern Chile/Region Antofagasta. J Metamorphic Geol 14:249–265

Lucassen F, Franz G (2005) The early Paleozoic orogen in the Central Andes: a non-collisional orogen comparable to the Cainozoic high plateau? Geol Soc London Spec Pub 246:257–273

Lucassen F, Thirwall MF (1998) Sm-Nd formation ages and mineral ages in metabasites from the Coastal Cordillera, northern Chile. Geologische Rundschau 86:767–774

Lucassen F, Fowler CMR, Franz G (1996) Formation of magmatic crust at the Andean continental margin during early Mesozoic: a geological and thermal model of the North Chilean Coast Range. Tectonophysics 262:263–279

Lucassen F, Franz G, Thirlwall MF, Mezger K (1999a) Crustal recycling of metamorphic basement: Late Paleozoic granites of the Chilean Coast Range and Precordillera at ~22°S. J Petrol 40:1527–1551

Lucassen F, Franz G, Laber A (1999b) Permian high pressure rocks – the basement of Sierra de Limón Verde in N-Chile. J S Am Earth Sci 12:183–199

Lucassen F, Lewerenz S, Franz G, Viramonte J, Mezger K (1999c) Metamorphism, isotopic ages and composition of lower crustal granulite xenoliths from the Cretaceous Salta Rift, Argentina. Contrib Mineral Petrol 134:325–341

Lucassen F, Becchio R, Wilke HG, Thirlwall MF, Viramonte J, Franz G, Wemmer K (2000) Proterozoic-Paleozoic development of the basement of the Central Andes (18°–26°) – a mobile belt of the South American craton. J S Am Earth Sci 13:697–715

Lucassen F, Becchio R, Harmon R, Kasemann S, Franz G, Trumbull R, Wilke HG, Romer RL, Dulski P (2001) Composition and density model of the continental crust at an active continental margin – the Central Andes between 21° and 27°S. Tectonophysics 341:195–223

Lucassen F, Harmon, R, Franz G, Romer RL, Becchio R, Siebel W (2002a) Lead evolution of the Pre–mesozoic crust in the Central Andes (18°–27°): Progressive homogenisation of Pb. Chem Geol 186:183–197

Lucassen F, Escayola M, Franz G, Romer RL, Koch K (2002b) Isotopic composition of Late Mesozoic basic and ultrabasic rocks from the Andes (23–32°S) – implications for the Andean mantle. Contrib Mineral Petrol 143:336–349

Lucassen F, Becchio R (2003) Timing of high-grade metamorphism: Early Paleozoic U-Pb formation ages of titanite indicate long-standing high-T conditions at the western margin of Gondwana (Argentina, 26–29°S). J Metamorphic Geol 21:649–662

Lucassen F, Trumbull R, Franz G, Creixell C, Vasquez P, Romer RL, Figueroa O (2004) Distinguishing crustal recycling and juvenile additions at active continental margins: the Paleozoic to Recent compositional evolution of the Chilean Pacific margin (36 – 41°S). J S Am Earth Sci 17:103–119

Lucassen F, Franz G, Viramonte J, Romer RL, Dulski P, Lang A (2005) The late Cretaceous lithospheric mantle beneath the Central Andes: evidence from phase equilibria and composition of mantle xenoliths. Lithos 82:379–406

Martignole J, Martelat JE (2003) Regional-scale Grenvillian-age UHT metamorphism in the Mollendo-Cama block (basement of the Peruvian Andes). J Metamorphic Geol 21:99–120

Meissner R, Mooney W (1998) Weakness of the lower continental crust: a condition for delamination, uplift and escape. Tectonophysics 296:47–60

Middlemost E (1997) Magmas, rocks and planetary development. Longman, pp 299

Mitchell AH, Reading HG (1969) Continental margins, geosynclines and ocean floor spreading. J Geol 77:629–646

Morata D, Aguirre L, Oyarzún M, Vergara M (2000) Crustal contribution in the genesis of the bimodal Triassic volcanism from the Coastal Range, central Chile. Revista Geológica de Chile 27:83–98

Ort MH, Coira BL, Mazoni MM (1996) Generation of a crust-mantle magma mixture: magma sources and contamination at Cerro Panizoz, central Andes. Contrib Mineral Petrol 123:308–322

Palacios MC (1978) The Jurassic Paleovolcanism in Northern Chile. PhD thesis, Eberhard-Karls-Universität Tübingen

Pankhurst RJ, Rapela CW (1998) The Proto-Andean margin of Gondwana. Geol Soc London Spec Pub 142:383

Pankhurst RJ, Rapela CW, Fanning CM (2000) Age and origin of coeval TTG, I- and S-type granites in the Famatinian Belt of the NW Argentina. In: Barbarin B, Stephens WE, Bonin B, Bouchez JL, Clarke D, Cuney M, Martin H (eds) Fourth Hutton symposium on the origin of granites and related rocks. Trans Royal Soc Edinburgh Earth Sci 91:151–168

Parada M (1990) Granitoid plutonism in central Chile and its geodynamic implications: a review. In: Kay SM, Rapela CW (eds) Plutonism from Antarctica to Alaska. Geol Soc Am Spec P 24:51–66

Petford N, Atherton MP (1996) Na-rich partial melts from newly underplated basaltic crust: the Cordillera Blanca Batholith, Peru. J Petrol 37:1491–1521

Petford N, Gallagher K (2001) Partial melting of mafic (amphibolitic) lower crust by periodic influx of basaltic magma. Earth Planet Sci Lett 190:483–499

Pichowiak S (1994) Early Jurassic to Early Cretaceous magmatism in the Coastal Cordillera and the Central Depression of north Chile. In: Reutter KJ, Scheuber E, Wigger PJ (eds) Tectonics of the Southern Central Andes. Springer, Heidelberg, pp 203–218

Pilz P, Hammerschmidt K, Niedermann S (2001) Noble gas investigations of volcanic rocks and geothermal emanations in the back-arc of the Central Volcanic Zone (CVZ). III Simposio Sudamericano de geologia isotopica, Pucon, Chile, Conference Abstract

Piper JDA (2000) The Neoproterozoic Supercontinent: Rodinia or Paleopangaea? Earth Planet Sci Lett 176:131–146

Plank T, Langmuir CH (1998) The geochemical composition of subducting sediment and its consequences for the crust and mantle. Chem Geol 145:325–94

Prinz P, Wilke HG, Hillebrandt A (1994) Sediment accumulation and subsidence history in the Mesozoic marginal basin of northern Chile. In: Reutter KJ, Scheuber E, Wigger PJ (eds) Tectonics of the Southern Central Andes. Springer Verlag, Heidelberg, pp 219–232

Ramos VA (2000) The Southern Central Andes. In: Cordani UG, Milani EJ, Thomaz Fihlo A, Campos DA (eds) Tectonic evolution of South America. Rio De Janeiro, pp 561–604

Ramos V, Keppie D (1999) Laurentia-Gondwana connection before Pangea. Geol Soc Am Spec P 336:276

Ramos VA, Jordan TE, Allmendinger RW, Mpodozis C, Kay SM, Cortés JM, Palma MA (1986) Paleozoic terranes of the central Argentine Chilean Andes. Tectonics 5:855–880

Restrepo Pace PA, Ruiz J, Gehrels G, Cosca M (1997) Geochronology and Nd isotopic data of Grenville-age rocks in the Colombian Andes: new constraints for Late Proterozoic-Early Paleozoic paleocontinental reconstructions of the Americas. Earth Planet Sci Lett 150:427–441

Reutter K-J, Munier K (2006) Digital geological map of the Central Andes between 20°S and 26°S. In: Oncken O, Chong G, Franz G, Giese P, Götze H-J, Ramos VA, Strecker MR, Wigger P (eds) The Andes – active subduction orogeny. Frontiers in Earth Science Series, Vol 1. Springer-Verlag, Berlin Heidelberg New York, pp 557–558, this volume

Rino S, Komiy, T, Windley BF, Katayama I, Motoki A, Hirata T (2004) Major episodic increases of continental crustal growth determined from zircon age of river sands: implications for mantle overturns in the Early Precambrian. Phys Earth Planet Ints 146:369–394

Rogers G, Hawkesworth CJ (1989) A geochemical traverse across the North Chilean Andes: evidence for crust generation from the mantle wedge. Earth Planet Sci Lett 91:271–285

Salfity JA Marquillas RA (1994) Tectonic and sedimentary evolution of the Cretaceous-Eocene Salta Group basin, Argentina. In: Salfity JA (ed) Cretaceous tectonics of the Andes. Vieweg, pp 266–315

Sato K, Siga O Jr (2002) Rapid growth of continental crust between 2.2 and 1.1 Ga in the South American Platform: integrated Australian, European, North American and SW USA crustal evolution study. Gondwana Res 5:165–173

Scheuber E, Gonzales G (1999) Tectonics of the Jurassic-Early Cretaceous magmatic arc of the north Chilean Coastal Cordillera (22–26° S): a story of crustal deformation along a convergent plate boundary. Tectonics 18:895–910

Scheuber E, Bogdanic T, Jensen A, Reutter KJ (1994) Tectonic development of the north Chilean Andes in relation to plate convergence and magmatism since the Jurassic. In: Reutter KJ, Scheuber E, Wigger PJ (eds) Tectonics of the Southern Central Andes. Springer, Heidelberg, pp 7–22

Schnurr WBW, Risse A, Trumbull RB, Munier K (2006) Digital geological map of the Southern and Central Puna Plateau, NW Argentina. In: Oncken O, Chong G, Franz G, Giese P, Götze H-J, Ramos VA, Strecker MR, Wigger P (eds) The Andes – active subduction orogeny. Frontiers in Earth Science Series, Vol 1. Springer-Verlag, Berlin Heidelberg New York, pp 563–564, this volume

Schwartz JJ, Gromet LP (2004) Provenance of a late Proterozoic to early Cambrian basin, Sierras de Córdoba, Argentina. Precambrian Res 129:1–21

Siebel W, Schnurr WBW, Hahne K, Kraemer B, Trumbull RB, Van den Bogaard P, Emmermann R (2001) Geochemistry and isotope systematics of small- to medium-volume Neogene-Quarternary ignimbrites in the southern central Andes: evidence for derivation from andesitic magma sources. Chem Geol 171:213–237

Sobolev SV, Babeyko AY, Koulakov I, Oncken O (2006) Mechanism of the Andean orogeny: insight from numerical modeling. In: Oncken O, Chong G, Franz G, Giese P, Götze H-J, Ramos VA, Strecker MR, Wigger P (eds) The Andes – active subduction orogeny. Frontiers in Earth Science Series, Vol 1. Springer-Verlag, Berlin Heidelberg New York, pp 513–536, this volume

Teixeira W, Tassinari CCG, Cordani UG, Kawashita K (1989) A review of the geochronology of the Amazonian Craton: tectonic implications. Precambrian Res 42:213–227

Trompette R (1997) Neoproterozoic (~600Ma) aggregation of Western Gondwana: a tentative scenario. Precambrian Res 82:101–112

Trumbull RB, Wittenbrink R, Hahne K, Emmermann R, Büsch W, Gerstenberger H, Siebel W (1999) Evidence for Late Miocene to Recent contamination of arc andesites by crustal melts in the Chilean Andes (25°–26°S) and its geodynamic implications. J S Am Earth Sci 12:135–155

Unrug R (1996) The assembly of Gondwanaland. Episodes 19:11–20

Viramonte JG, Kay SM, Becchio R, Escayola M, Novitski I (1999) Cretaceous rift related magmatism in central-western South America. J S Am Earth Sci 12:109–121

Vergara M, Vi B, Nystrom JO, Cancino A (1995) Jurassic and early Cretaceous island arc volcanism, extension, and subsidence in the Coast Range of central Chile. Geol Soc Am Bull 107:1427–1440

Von Hillebrandt A, Bartsch V, Bebiolka A, Kossler A, Kramer W, Wilke HG, Wittmann S (2000) The paleogeographic evolution in a volcanic-arc/back-arc setting during the Mesozoic in northern Chile. Zeitschrift für angewandte Geologie SH1 2000. Hannover, pp 87–93

von Huene R, Ranero R (2003) Subduction erosion and basal friction along the sediment-starved continental margin off Antofagasta, Chile. J Geophys Res 108: doi 1029/2001JB0001569

Wasteneys HA, Clark AH, Farrar E, Lagridge RJ (1995) Grenvillian granulite-facies metamorphism in the Arequipa Massif, Peru: a Laurentia-Gondwana link. Earth Planet Sci Lett 132:63–73

Willner AP, Miller H, Jezek P (1985) Geochemical features of an Upper Precambrian-Lower Cambrian greywacke/pelite sequence (Puncoviscana trough) from the basement of the NW-Argentine Andes. Neues Jahrbuch für Geologie und Paläontologie, Monatshefte, pp 498–512

Willner AP, Glodny J, Gerya TV, Godoy E, Massonne H-J (2004) A counterclockwise PTt-path of high pressure-low temperature rocks from the Coastal cordillera accretionary complex of South Central Chile: constraints for the earliest stage of subduction mass flow. Lithos 75:283–310

Windley, BF (1986) The evolving continents. 2nd edition, reprinted with corrections. Wiley and Sons, Chichester, pp 399

Wörner G, Moorbath S, Horn S, Entenmann J, Harmon RS, Davidson JP, Lopez Escobar L (1994) Large- and fine-scale variations along the Andean arc of northern Chile (17.5°–22°S). In: Reutter KJ, Scheuber E, Wigger PJ (eds) Tectonics of the Southern Central Andes. Springer, Heidelberg, pp 77–92

Wörner G, Lezaun J, Beck A, Heber V, Lucassen F, Zinngrebe E, Rößling R, Wilke HG (2000) Geochronology, metamorphic petrology and geochemistry of basement rocks from Belén (N. Chile) and C. Uyarani (W. Bolivian Altiplano): implications for the evolution of Andean basement. J S Am Earth Sci 13: 717–737

Zimmermann U, Bahlburg H (2003) Provenance analysis and tectonic setting of the Ordovician clastic deposits in the southern Puna Basin, NW Argentina. Sedimentology 50: 1079–1104

# Chapter 4

# Long-Term Signals in the Present-Day Deformation Field of the Central and Southern Andes and Constraints on the Viscosity of the Earth's Upper Mantle

Jürgen Klotz · Amir Abolghasem · Giorgi Khazaradze · Bertram Heinze · Tim Vietor · Ron Hackney · Klaus Bataille
Rodrigo Maturana · Jose Viramonte · Raul Perdomo

**Abstract.** As part of the South American Geodynamic Activities project we observed the present day deformation field in the territories of Chile and Argentina using the Global Positioning System. The results clearly show that the earthquake cycle dominates the contemporary surface deformation of the central and southern Andes. Compared to geological timescales, the transient elastic deformation related to subduction earthquakes presents a short-term signal which can be explained by interseismic, coseismic, and postseismic phases of interplate thrust earthquakes. We constructed the Andean Elastic Dislocation Model (AEDM) in order to subtract the interseismic loading from the observed velocities. The estimated parameters of the AEDM, and the amount and depth of coupling between the subducting Nazca and overriding South American Plates, represent long-term features and show that the seismogenic interface between both plates is fully locked and that the depth of coupling increases from north to south.

The prominent signals in the residual velocity field (i.e. observed velocities minus AEDM) are obviously due to postseismic relaxation processes; they are visible in the area of the 1995 $M_w$ 8.0 Antofagasta earthquake and in the area of the 1960 $M_w$ 9.5 Valdivia earthquake. Although postseismic deformations, compared to geologic timescales, are short-term signals, those signals are valuable constraints on important long-term features of Andean evolution, i.e., the viscosity of the upper mantle and lower crust. The observed surface data are best fitted with a three-dimensional finite element model in which we incorporate a mantle viscosity of $4 \times 10^{19}$ Pa s.

The most obvious long-term deformation signal is manifested in the back-arc of the subduction zone where the Brazilian Shield thrusts beneath the Subandean zone. The style and amount of back-arc shortening changes along strike of the orogen, increasing from zero in the south (latitude < –38° S) to values in the order of 10 mm yr$^{-1}$ close to the Bolivian Orocline. In the fore-arc, whilst we see indications for long-term E-W extension, we did not find any apparent slip partitioning. In addition to this long-term signal, we suggest that the asymmetry of interseismic and coseismic deformation may lead to tectonic structures in the fore-arc. If the coseismic deformation does not release all of the accumulated deformation, then, over many earthquake cycles, part of the interseismic deformation may be transformed into permanent long-term plastic deformation.

## 4.1 Objectives

The primary objective of our investigation was to observe and understand the contemporary surface deformation, including its temporal and spatial variation along the Andean subduction zone (ASZ). In detail, we aimed to study:

1. The kinematics and dynamics of present-day deformation processes along the central and southern Andes, by means of campaign-style GPS measurements.
2. The relationship between the different phases of the seismic cycle and the observed crustal deformations through three-dimensional (3D) elastic dislocation modeling.
3. The relationship between the short-term (i.e. earthquake related deformations) and long-term deformation processes (i.e. mountain building processes, crustal shortening) by comparing geodetic results with seismic studies and with results obtained through neotectonic field investigations.
4. The viscosity of the upper mantle and lower crust using the observed postseismic surface deformation.

The tectonic setting is described in different contributions of this volume – e.g., see Oncken et al. (Chap. 1), Sobolev et al. (Chap. 25), Haschke et al. (Chap. 16), Ranero et al. (Chap. 5). The present-day deformation of the Andes is dominated by transient effects related to the subduction earthquake cycle (Klotz et al. 2001). The earthquake-related elastic deformation presents a relatively short-term signal compared to geological timescales. Conversely, the permanent plastic deformation that contributes to the formation of tectonic features, such as mountains, is a long-term signal. The separation of short-term effects from the much smaller long-term effects is a major challenge.

## 4.2 Data

### 4.2.1 GPS Data

In order to observe the present-day deformation of the Earth's surface in the area of the Central and Southern Andes, we initiated the South American Geodynamic Activities (SAGA) project as a cooperative effort including numerous organizations and institutions in the host countries. The deformation data was derived by repeated observations of a high-precision Global Positioning System (GPS) network which we have established in Chile and

**Fig. 4.1.**
**a** Velocity vectors of the SAGA GPS network based on the 1994, 1995 and 1996 occupations. The epicenters, and the coseismic rupture areas of the 1995 $M_w$ 8.0 Antofagasta and 1960 $M_w$ 9.5 Chile earthquakes are shown with *white stars* and *thick dashed lines*, respectively. The Nazca/South American plate convergence vector is based on the estimate of Angermann et al. (1999). Shading represents the ETOPO5 topographic data (NOAA 1998). **b** Velocity vectors of the SAGA-North GPS network based on the 1996 and 1997 occupations. The epicenter of the 1995 $M_w$ 8.0 Antofagasta earthquake is shown as a *star*; *thick dashed lines* outline the approximate extent of the fault rupture area; the focal mechanism is based on the CMT solution

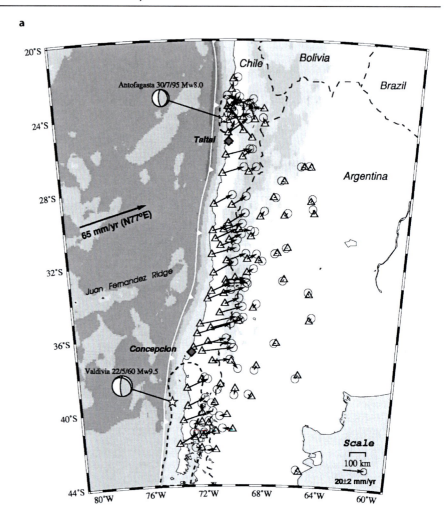

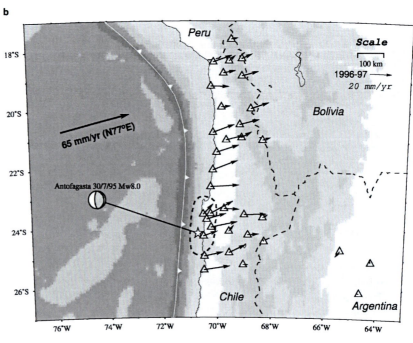

the western part of Argentina. Currently this geodetic network consists of 230 sites. The realized GPS observations come from a repeated field survey of parts of the SAGA network and 8 permanent GPS sites, including stations of the International GPS Service (IGS).

Since the initiation of the SAGA project in 1993, there have been seven GPS campaigns spanning various parts of Chile and western Argentina. These measurements enabled us to obtain the first data for the contemporary crustal deformation field along the Central and Southern Andes between latitudes 17° S and 43° S (Fig. 4.1; Klotz et al. 1999; Klotz et al. 2001; Khazaradze and Klotz 2003). In addition, we captured the deformations associated with the $M_w$ 8.0 Antofagasta earthquake of 30 July 1995, which were crucial in understanding the processes related to the earthquake deformation cycle (Klotz et al. 1999). Following this earthquake, a dense part of the network in northern Chile and northwestern Argentina, between latitudes 22 and 26° S, comprising ~70 sites (hereafter referred to as SAGA-North) was re-observed. Seventeen selected sites from the SAGA-North network were re-observed in 1996, and in 1997 we re-observed the entire SAGA-North network. We surveyed the southern section of the SAGA network (between latitudes 37 and 43° S) in 1994 and 1996, but unlike the northern section, no large earthquake ($M_w > 7$) occurred within this area during this time.

Every site was occupied for at least three consecutive days with daily observations exceeding 20 hours. All campaign data, together with data from selected IGS sites, were processed with the Earth Parameter and Orbit System

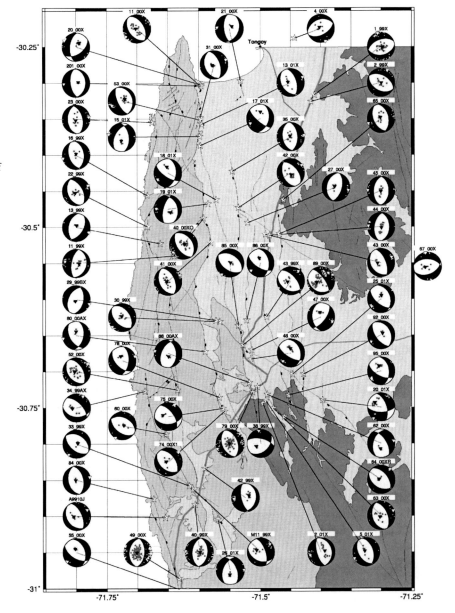

**Fig. 4.2a.**
Fault-plane solutions inferred from 1513 neotectonic fault-slip data sets presented as focal mechanisms at individual outcrops. **a** Extensional faulting; **b** Strike-slip faulting; **c** Reverse faulting; *White sectors*: extensive quadrants; *Black sectors*: compressive quadrants; Borderlines between quadrants show dip and azimuth of fault planes and auxiliary planes in the lower half sphere. *Filled circles* and *white crosses* show orientation of individual P- and T-axes (as derived from fault-slip data analysis at individual outcrops), respectively. *Stars*: location of outcrops. *Thick gray line*: Panamerican Highway. *CA:* Cretaceous Arc; *CB:* Cenozoic Basin; *CC:* Coastal Cordillera

**Fig. 4.2b.**
Strike-slip faulting

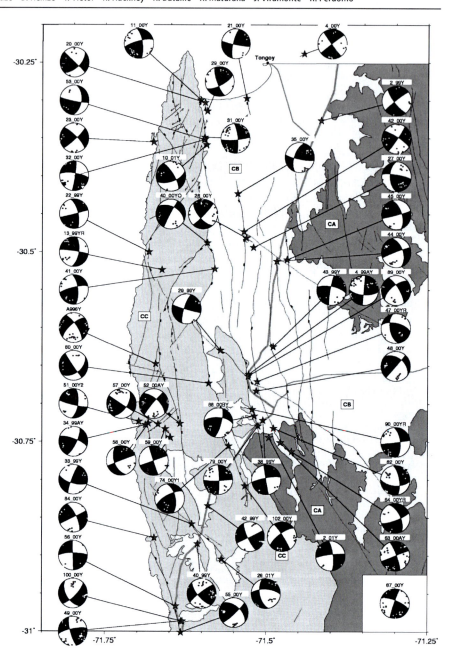

(EPOS) software from the GeoForschungsZentrum (GFZ) (Angermann et al. 1997), using the combined IGS satellite orbits and the Earth orientation parameters. No other constraints were imposed. The fiducial-free solution was transformed to the ITRF97 reference frame applying a Helmert transformation constrained by the IGS stations. Detailed descriptions of our data processing procedure and a table of GPS velocities can be found in Klotz et al. (2001). The velocity vectors presented in Fig. 4.1a,b are given in a reference frame fixed to a nominal 'stable' South America (SA) that is defined using the IGS stations Kourou (KOUR), Fortaleza (FORT), and La Plata (LPGS). The mean value of the IGS station position residuals, reflecting the achieved network precision, ranges between 2–4 mm and 5–7 mm for horizontal and vertical components, respectively. Since the expected rates of surface uplift in the Andean subduction zone do not exceed 2 mm yr$^{-1}$ (Jordan et al. 1997), 2–4 years of observations with the above-stated vertical precision are not sufficient to resolve these motions with an adequate confidence. For this reason we concentrate further discussion solely on the horizontal deformation rates. All GPS-derived vectors are available at *http://sfb267.geoinf.fu-berlin.de/web/en/more.htm*.

**Fig. 4.2c.**
Reverse faulting

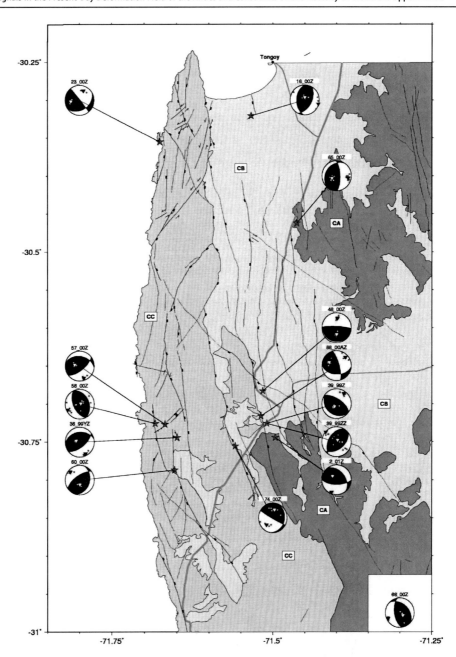

### 4.2.2 Neotectonic and Seismicity Data

In order to investigate the rates of seismic loading, as well as the earthquake characteristics along the Chilean margin, we compared results from the GPS investigations with findings from both geological field investigations and seismicity studies. Detailed neotectonic and paleoseismological field studies in the fore-arc region south of La Serena, between 30 and 31° S, had the goals of better understanding the recent deformation in an area with abundant late Tertiary to Quaternary surface deformation and where the GPS residual motions suggest significant intraplate deformation. For our field studies, we chose the semi-arid area between latitudes 30° S (La Serena) and 31° S (Punta Talca), in the northern central part of Chile (Fig. 4.2). This area was chosen for its reasonable outcrop conditions; the numerous geologic, structural, and morphologic evidence of active faulting; a well-developed river system (Rio Limari, Estero Punitaqui); and apparent sequences of marine terraces (Radtke 1987). The sequences of river and marine terraces may aid in constraining recent deformation rates (Paskoff 1970; Veit 1993; Lavenu and Cembrano 1997). In addition, significant changes in trench azimuths at 30° S and 33° S constrain an apparent 'geometrical trench segment', that might be worth testing for its relation to interplate and intraplate seismicity. Details of the method to derive the fault plane solutions and all data are described in (Heinze 2003).

## 4.3 Methods

### 4.3.1 Elastic Dislocation Model

In order to subtract the dominating interseismic loading signal from the observed velocities, we constructed a 3D model of the seismogenic interface between the subducting Nazca and the overriding South American Plates and used elastic dislocation theory (Okada 1985) to predict the amount of interseismic deformation on a given set of observation points (Savage 1983; Wang et al. 2001).

The main assumptions of the elastic dislocation modeling can be summarized as follows:

1. Earthquake cycles can be approximated by the elastic behavior of the crust. Although the real earth is not elastic, and its viscoelastic properties should also be taken into account for more realistic modeling of surface deformation throughout the earthquake cycle, it has been shown that for modeling deformation rates at the later stages of the interseismic period, as is the case for the central section of ASZ investigated in the given study elastic dislocation models are adequate (e.g. Dragert et al. 1994; Wang et al. 2001).

2. Surface deformation depends directly on fault slip-rates which are equal to the plate convergence velocity at the seaward and landward free-slip zones.
3. Fault slip-rates are zero at the locked zone (i.e. plates are fully coupled) and increase landward from zero to the full slip rate (i.e. plate convergence rate) within the transition zone (Fig. 4.3).
4. Surface deformation rates remain constant throughout the interseismic period.
5. Widths of the locked and transition zones remain constant throughout the interseismic period.
6. Net permanent deformation throughout the entire earthquake cycle is equal to zero. This implies that all of the strain accumulated throughout the interseismic period is released during an abrupt coseismic rupture on the fault, followed by rapid postseismic relaxation. This assumption is discussed in Sect. 4.5.4.

Our preferred model, called the Andean Elastic Dislocation Model (AEDM), has been selected after experimenting with various input parameters in more than twenty other models. To calculate deformation rates predicted by the AEDM at a given set of GPS stations, we used a modified version of the program DISL3D, by Kelin Wang of the Pacific Geoscience Center, Canada, (Wang et al. 2001), with the initial development of the program attributed to Flück (1996) and Flück et al. (1997). The main principle of the model (and the code) is based on Okada's (1985, 1992) formulation for dislocations due to a point-source force applied to an infinitesimal surface area. The final goal of estimating deformation rates on

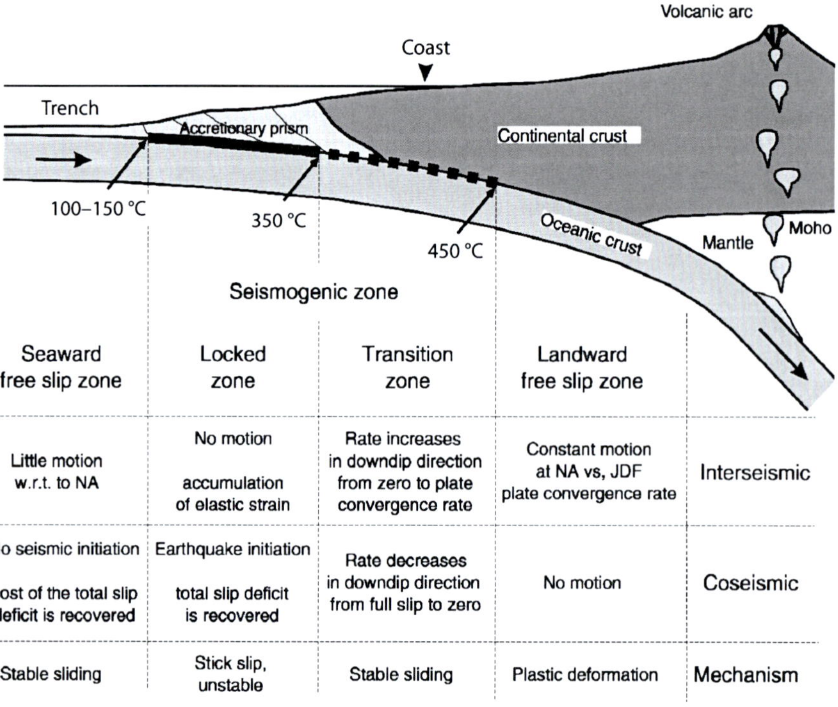

**Fig. 4.3.**
Schematic representation of subduction as used in an elastic dislocation model. The thrust fault is divided into four major zones. The main behavioral characteristics of each zone associated with different phases of earthquake cycle are summarized in the underlying table. The figure is adopted from Flück (1997) based on Hyndman and Wang (1993)

| Seaward free slip zone | Locked zone | Transition zone | Landward free slip zone | |
|---|---|---|---|---|
| Little motion w.r.t. to NA | No motion accumulation of elastic strain | Rate increases in downdip direction from zero to plate convergence rate | Constant motion at NA vs, JDF plate convergence rate | Interseismic |
| No seismic initiation most of the total slip deficit is recovered | Earthquake initiation total slip deficit is recovered | Rate decreases in downdip direction from full slip to zero | No motion | Coseismic |
| Stable sliding | Stick slip, unstable | Stable sliding | Plastic deformation | Mechanism |

the surface of the Earth is achieved by dividing the fault surface along the thrust interface into numerous small triangular elements and performing summation for all of the triangles. Such an approach enables us to account for non-planar fault surfaces, as is the case within our study area where the subducting Nazca Plate is characterized by along-strike variation in dip (e.g. Cahill and Isacks 1992) and coupling depth (Tichelaar and Ruff 1991, 1993).

### 4.3.2 Calculation of Strain Rates

We studied the distribution of the strain field using geodetic data, geological fault-slip data (infinitesimal approach) measured in the fore-arc, and earthquake focal mechanisms based on Harvard CMT solutions. Geodetic strain rates were approximated using two different methods.

The first method includes conversion of site positions into a geocentric coordinate system assuming a spherical Earth and a consequent inversion for principal strains, translation, and rigid body rotation components for the center of a network. Because this approach can only be justified if the region considered deforms homogeneously (bulk strain), we divided the study area into nine subregions that were assumed to represent units of homogeneous deformation (Fig. 4.4). A detailed description of this method can be found in Khazaradze (1999).

The second method is based on linearly interpolated site motions. In this case, the inversion of principal directions and rates was performed using the full-velocity

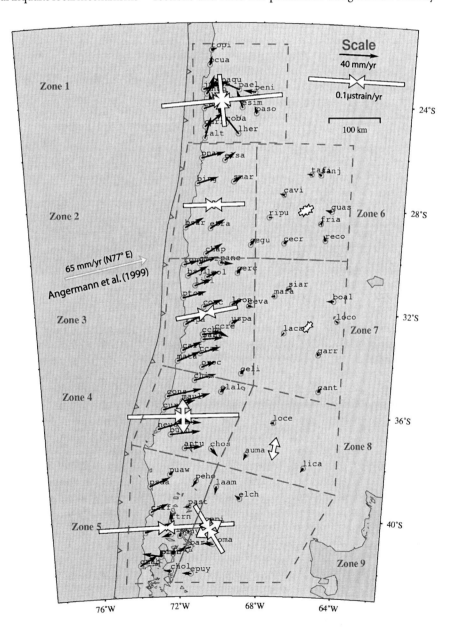

**Fig. 4.4.**
Horizontal strain rates for the SAGA network based on GPS velocities estimated from the 1994, 1995 and 1996 GPS campaigns. Subsets of stations delimited with dashed rectangles were chosen to separate stations located in the fore-arc and back-arc, as well as introduce lateral segmentation. Strain crosses are located at the center of mass of an individual subregion. Numerical results are given in Table 4.1

gradient tensors determined in a 1.5° grid that is oriented parallel to the surface trace of the trench. Because the focus of the second approach was on interplate loading, we concentrated on site velocities that are not expected to reflect coseismic or post-earthquake transient deformation. Therefore, the applied data sets exclude the area affected by post-earthquake activities following the 1995 Antofagasta earthquake but include data from the northernmost part of the study area (Fig. 4.5).

Geological field data were taken at locations that are assumed to reflect the regional trend, i.e., only data from those outcrops that are related to active faulting (preferably in recent stratigraphic units) and where structural and/or rheological inhomogeneities suggest deviation from an overall trend were used. Because this procedure is subjective, related measurements were made independently by different specialists at the different outcrops. Large deviations from an average solution were accepted and an approach described by Michel (1994) was utilized to deduce principal strain directions from fault-slip data. Composite CMT solutions for selected earthquakes were approximated by summing up seismic moment tensors and by approximating composite directions without weighting the results using the earthquake moments. The selected seismic data were assumed to dominantly reflect interplate deformation. The resolved shortening directions from this, and the other approaches introduced above, were compared to the plate convergence direction between the Nazca and South American Plates. The angular deviations between the directions of shortening, earthquake slip directions and plate convergence were taken as a proxy for slip partitioning. Additionally, supposing that interplate loading is co-linear with interplate deformation release, we checked the angular deviation between the best-fit fault slip directions used in the AEDM and the direction of plate convergence, with the assumption that this might further elucidate partitioning of oblique convergence into trench-normal and trench-parallel components.

### 4.3.3 Intraplate Earthquake Faulting from Geologic Field Investigations

Neotectonic field investigations and interpretations of satellite images (e.g. Landsat TM) were used to constrain active faults, quantify principal strain directions, fault dimensions and recent surface displacements, and study their implications for active intraplate earthquake deformation. To estimate principal strain directions, populations of fault-slip data were separated according to their relative and/or absolute ages and then investigated using approaches described by Meschede (1994) and Doblas (1998). Besides contributing to an unproved assessment

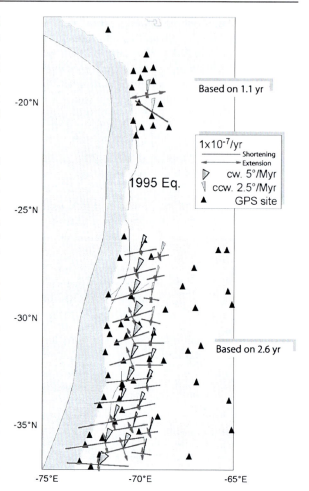

**Fig. 4.5.** Principal strain rates along the fore-arc using interpolated GPS velocities

of the intraplate earthquake hazard, one major objective was to interactively test results from the static modeling approach described in Heinze (2003) and its reliability for quantifying local intraplate deformation.

Additional modeling investigated possible interactions of interplate earthquakes and intraplate fore-arc deformation between 30 and 34° S (Fig. 4.6). For this, scaling laws (Scholz et al. 1986; Wells and Coppersmith 1994) and the approximate interface geometry, as inferred from the AEDM, were applied to locate large to major historical earthquakes (NEIC 2001) along the plate interface. Three-dimensional dislocation modeling in an elastic half-space was applied using information from all historical earthquakes with moments larger than $4 \times 10^{19}$ Nm, detected in the catalogues, that appear to be related to the plate interface along this part of the study area. The redistribution of stresses/strains (Das and Scholz 1981; King et al. 1994; Deng and Sykes 1997) was approximated for certain grids and for straight faults approximating the traces of active faults studied in the field.

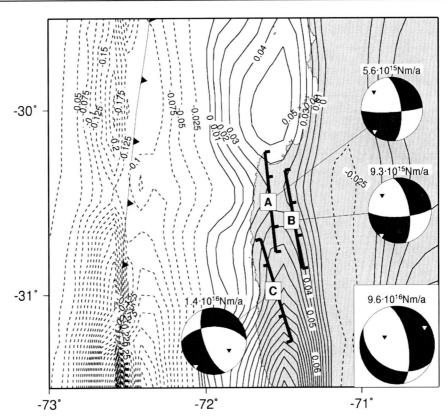

**Fig. 4.6.**
Strain rate inferred from dislocation modeling of 339 interplate events of magnitude $M_w$ 5 or higher since 1647. *Solid contour* and *dashed lines* respectively show positive (extensional) and negative (contractional) strain rates in μstrain yr$^{-1}$. Lines A, B and C represent model intraplate faults of 60 km length, 11 km width and 65° dip. Focal mechanisms depict movement along fault and moment rate, inferred from strain redistribution due to successive rupture of the subduction zone. Rate and slip were calculated from strain components parallel to the strike and dip of the individual fault planes. The *inset* shows the focal mechanism and moment rate inferred from GPS-residuals (see Fig. 4.8)

### 4.3.4 Three-Dimensional Viscoelastic Finite Element Model

We used numerical modeling for studying postseismic relaxation processes in the areas of the 1995 Antofagasta and 1960 Valdivia earthquakes. Since modeling of a subducting slab is beyond the limitations of existing analytical dislocation models, we simulated the observed postseismic deformations of SAGA-South and SAGA-North with 3D finite element models. The finite element method allows complicated geometries and is flexible enough to account for internal discontinuities in laterally heterogeneous models.

Our models consist of a 30 km thick elastic oceanic lithosphere which subducts beneath a 40 km thick elastic continental plate (Fig. 4.7). A viscoelastic mantle underlies the elastic lithospheric layers. The crust is assumed to behave with linear elasticity, while Maxwellian viscoelasticity is applied to shear deformation in the mantle. Mantle volumetric deformation is assumed to be elastic.

We specified Poisson solids with rigidities of 28 GPa and 40 GPa for the continental and oceanic plates, respectively. The mantle the rigidity was set to 70 GPa and the Poisson ratio to 0.25. Mantle viscosity varies in different models.

To model dislocation, the seismogenic zone was specified by pairs of nodes on the interface of the two crustal plates. Each node of a pair is located on one crustal plate (Fig. 4.7). Initially they occupy identical locations:

$$x_k^i = x_k^j; \quad k = 1, 2, 3$$

where indices $i$ and $j$ are node numbers in a node pair. The rupture is modeled by constraining the coordinate differences between nodes of each pair:

$$u_k^i - u_k^j = s_k; \quad k = 1, 2, 3 \qquad (4.1)$$

where the $u$ components are unknown and $s_k$ are the slip components. In this way we model the fault slip by given values of $s_k$. The relative positions of node pairs remain constant during the stress relaxation period as a consequence of Eq. 4.1.

Boundary conditions are applied so that material cannot flow out of the model:

$$u_x(x_{\min}, t) = u_x(x_{\max}, t) = u_y(y_{\min}, t)$$
$$= u_y(y_{\max}, t) = u_z(z_{\min}, t)$$

where $x = x_{\min}$ and $x = x_{\max}$ are the western and eastern borders of the model, and $y = y_{\min}$ and $y = y_{\max}$ are the southern and northern boundaries, and $z = z_{\min}$ is the bottom of the model. The fixed four bottom corners of the model define the reference frame of the deformation

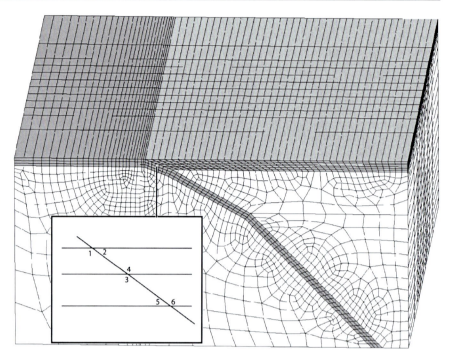

**Fig. 4.7.**
Three-dimensional finite element model of the subduction zone with an elastic continental plate (*light gray*) and an elastic oceanic plate and slab (*dark gray*) overlaying a viscoelastic mantle (*white*). The *inset* depicts node pairs on the seismogenic zone: odd numbered nodes are located on the oceanic plate, while even numbered nodes are attributed to the continental plate. Corresponding nodes occupy identical positions before the rupture. Their relative positions after the rupture are locked

analysis. In order to simulate two slipping phases of rupture and after-slip, we applied constrain equations equal to Eq. 4.1 in two time steps: Step 1 is the rupture phase when a slip of 20 m is posed in a few seconds on the rupture area; Step 2 constrains after-slip as a slow slide in the down dip extension of the rupture area over a time interval of a few years. The maximum after-slip occurs on the upper end of the after-slip area at the beginning of this step and decreases linearly with time and distance

Node pairs are locked in the third analysis step. Rupture and after-slip change the stress state within the mantle as well as in the crust and the new stress field relaxes due to mantle viscosity. For the model of the great Valdivia earthquake, we applied a rupture of 17 m on a 850 km along strike segment (Barrientos and Ward 1990). The surface deformation field is calculated for $y = 35$ years.

A flexible script allowed us to easily change the following parameters:

- model dimensions,
- element size,
- crustal material properties, i.e., elastic parameters,
- mantle material properties, i.e., elastic parameters and viscosity,
- coseimic slip,
- thickness of the continental crust,
- thickness of the oceanic crust,
- slab geometry,
- rupture depth,
- after-slip size,
- after-slip area, i.e., the depth extent of the after-slip,
- time period of the after-slip.

We analyzed the sensitivity of those parameters on the surface velocity field during extensive tests and summarized the results (Sect. 4.4).

## 4.4 Results

### 4.4.1 Andean Elastic Dislocation Model (AEDM)

The parameter that we estimated from our elastic dislocation model is the width of the seismogenic zone along the subduction thrust interface, which is defined by the up-dip and down-dip limits of the coupled zone, and the amount of coupling along the interface. Studies of great subduction earthquakes and their aftershocks from around the world have shown that ruptures associated with these events are limited to a depth of 50 km (Pacheco et al. 1993; Tichelaar and Ruff 1993). The up-dip limit of the seismogenic zone does not extend to the surface (i.e. the trench), but the seaward portion of the thrust interface also deforms (Byrne et al. 1988). Most commonly, this observation is attributed to the presence of unconsolidated and semi-consolidated sediments on the incoming oceanic plate (Hyndman and Wang 1993; Tichelaar and Ruff 1993). Hyndman and Wang (1993) suggest that this mineralogic transformation and the consequent locking of the subduction thrust interface takes place at a depth where temperatures reach 100–150 °C (Fig. 4.3). For the purpose of our studies the exact location of the up-dip limit of the seismogenic zone is not crucial, since the deformation rates observed onshore are not sensitive to this parameter. However, the location of the up-dip limit is important in modeling coseismic displacements and associated tsunamis (e.g. Flück et al. 1997).

In order to minimize the number of unknown input parameters, we attempted to incorporate as many independent geophysical observations as possible in our modeling efforts. The main independent constraints for the construction of an initial model of the subducting Nazca Plate are the shape of the upper surface of the subducting Nazca Plate (i.e. slab geometry) and the convergence direction and rate between the Nazca and South American Plates. Although we have constrained the width of the seismogenic zone (its up-dip and down-dip limits) along the thrust interface directly from the GPS data, we have also adhered to the constraints imposed by the maximum extent of interplate seismicity (e.g. Tichelaar and Ruff 1991, 1993), as well as the results obtained through thermal modeling (e.g. Oleskevich et al. 1999).

The geometry (i.e. the dip in vertical cross sections) of the subducting Nazca Plate can be divided into *shallow*, *intermediate* and *deep* parts. However, for this study the deep structure of the Nazca Plate is not essential, since the seismogenic zone of the thrust interface does not extend below 60 km depth and, therefore, it is unlikely that the observed surface deformation is significantly influenced by the geometry of the slab below 100 km depth. The *shallow* part of the Nazca Plate extends from the deformation front to a depth of about 30 km. We have estimated the location of the deformation front (i.e. the trench) from the ocean floor bathymetry (Smith and Sandwell 1997). The geometry of the slab is determined from offshore and onshore seismic reflection/refraction studies (e.g. Wigger et al. 1994; Schmitz et al. 1999). The *intermediate* part of the slab, from 30 km to 100 km depth, is mainly constrained by locations of intra-slab earthquakes recorded by global and local seismic arrays (e.g. Graeber and Asch 1999; Husen et al. 1999).

Additional constraints on slab geometry come from teleseismic receiver function studies (Yuan et al. 2000). The intermediate structure of the slab (down to 50 km depth) has also been imaged by wide-angle seismic profiling (e.g. Wigger et al. 1994; Flueh et al. 1998; Patzwahl et al. 1999; The ANCORP Working Group 1999). The deep part of the Nazca Plate (below −100 km) has been imaged either directly by studying the hypocentral location of earthquakes occurring within the subducting slab or with the aid of seismic tomography (e.g. Engdahl et al. 1995; Masson and Delouis 1997).

The direction and rate of convergence between the Nazca and South American Plates used in our modeling is based on the present-day velocity (Angermann et al. 1999). The entire length of the SAGA network has been subdivided into three overlapping sections (for the purpose of clarity, these sections are shown without an overlap in Fig. 4.8), which we have modeled separately. In the following discussion, these sections will be referred to as

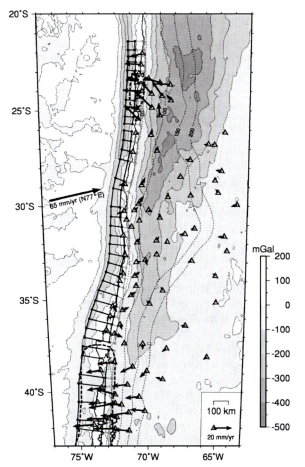

**Fig. 4.8.** Gravity anomalies and the seismogenic zone along the Nazca/South America thrust interface. The gravity anomalies are overlain by residual velocities obtained by subtracting the interseismic signal predicted by the AEDM from the observed velocities shown in Fig. 4.1. The grid with nodal points depicts the outline of the 100% locked (*black outline*) and the transition (*gray outline*) zones. *Dashed contour lines* represent the depth of the subducting Nazca slab from a revised model of its geometry (Tassara et al., in prep.). *Thick dashed lines* mark rupture zones from the 1995 Antofagasta and 1960 Valdivia earthquakes. *Stars* mark the hypocenters of these earthquakes (from the National Earthquake Information Center for the 1995 earthquake; and as suggested by Krawczyk (2003) for the 1960 earthquake). Gravity data are free-air anomalies offshore (Andersen and Knudsen 1998) and Bouguer anomalies onshore (Blitzkow, pers. comm., 2003; Götze et al. 1990, 1994)

the Northern, Central and Southern sections. In the initial model, the up-dip and down-dip limits of the locked and transition zones were initially constrained to 30 and 45 km depth, respectively. Later, the width was varied to achieve the best fit on the coastal stations, where the observed velocities are at their maximum. Besides the visual examination of the predicted and observed velocities, the L1 norm minimization method has also been used to quantitatively evaluate the goodness of fit. This was done by subtracting the model predicted velocities from the observed velocities (component-wise) and then summing up the differences for all the stations. The seismogenic zone along the Nazca/South America thrust interface and the residual velocities obtained by subtracting the interseismic signal predicted by the AEDM from the observed velocities are shown in Fig. 4.8.

### 4.4.2 Strain Rates and Slip Partitioning

The principal strain directions shown in Fig. 4.4 represent results from the regional inversion of velocity gradients from the first strain-rate inversion method discussed in Sect. 4.3.2. Related numerical results representing the principal strain rates, rigid body rotations and translations with standard errors are listed in Table 4.1.

Principal strain rates derived by using the interpolated site velocities are shown in Fig. 4.5. In addition, the vorticity is depicted by the relative sizes of associated segments. Results suggest that shortening rates along the fore-arc area south of 21.5° S reach 0.1–0.2 µ strain yr$^{-1}$. Shortening in the direction of ENE-WSW to E-W clearly dominates over WNW-ESE to N-S extension, indicating that the dominant extension might be oriented vertically and that deformation is mostly accommodated by thrust faulting. An increase in shortening rates to the south of latitude 30° S coincides with the apparent N-S deepening of the locked part of the interface as suggested from the AEDM (Fig. 4.8). High strain rates and almost uniform principal strain directions, as indicated by the data to the south of 21.5° S, are opposed to the lower average strain rates with changed attitudes found within the area to the north of 21.5° S. This suggests that straining conditions within this northern area are different from the rest of the investigated area.

The raw observations shown in Fig. 4.1 suggest that the oblique convergence vector is not partitioned into components parallel and perpendicular to the trench along trench-parallel strike-slip faults (see also Hoffmann-Rothe et al. 2006, Chap. 6 of this volume). Also, the best-fitting elastic model (AEDM) requires a fault slip direction that is parallel to the plate convergence. All results indicate to us that no obvious slip partitioning due to the oblique convergence of the Nazca and South American Plates is occurring within our study area. Geodetic studies from Brooks et al. (2003) between 26° S and 36° S, as well as findings from Ruegg et al. (2002) between 35° S and 37° S, confirm this result.

### 4.4.3 Fore-arc Extension Derived from Geologic Field Investigations

Results from the geologic field investigations suggest that E-W extension along N-S to NNW-SSE striking faults currently dominates over strike-slip and reverse faulting within the coastal fore-arc area south of La Serena (Fig. 4.2). Recent minor N-S directed transcurrent faulting and E-W directed thrust faulting were, however, detected at different locations, but these features are not discussed further. Fig. 4.2 shows focal mechanism diagrams as computed using the fault-slip data taken from mesoscale extensional faults from different investigated outcrops.

**Table 4.1.** Regional strain, rotation and translation from the 1994, 1995 and 1996 GPS campaigns

| Zone | Principal Strain Rates (µ strain/yr) | | | Rotation | Translation | |
|---|---|---|---|---|---|---|
| | $\varepsilon_1$[a] | $\varepsilon_2$[b] | | (deg Ma$^{-1}$) | (mm yr$^{-1}$) | (deg) |
| | $\varepsilon_1 \pm 1\sigma$ | $\varepsilon_2 \pm 1\sigma$ | $\theta^f \pm 1\sigma$ | $\Omega^d \pm 1\sigma$ | $C^e \pm 1\sigma$ | $C_a^e \pm 1\sigma$ |
| 1 | −0.051 ±0.039 | −0.124 ±0.042 | 84 ±4 | +0.1 ±0.3 | 10.3 ±0.6 | 2 ±4 |
| 2 | −0.006 ±0.044 | −0.062 ±0.050 | 89 ±6 | +0.5 ±0.3 | 15.5 ±1.2 | 63 ±4 |
| 3 | −0.008 ±0.030 | −0.015 ±0.031 | 62 ±48 | −0.6 ±0.4 | 1.9 ±1.2 | −34 ±37 |
| 4 | +0.002 ±0.025 | −0.063 ±0.028 | 79 ±3 | +0.5 ±0.2 | 14.8 ±0.7 | 73 ±3 |
| 5 | −0.003 ±0.013 | −0.011 ±0.015 | 42 ±42 | −0.5 ±0.3 | 1.6 ±1.2 | −70 ±43 |
| 6 | +0.032 ±0.049 | −0.114 ±0.051 | 90 ±2 | +1.8 ±0.4 | 22.6 ±0.9 | 81 ±2 |
| 7 | +0.020 ±0.093 | −0.002 ±0.087 | −80 ±30 | −0.2 ±0.7 | 3.2 ±2.0 | −139 ±36 |
| 8 | +0.004 ±0.044 | −0.139 ±0.050 | 88 ±2 | +2.8 ±0.3 | 4.8 ±0.9 | 42 ±11 |
| 9 | +0.029 ±0.054 | −0.055 ±0.051 | −32 ±9 | −1.0 ±0.7 | 6.8 ±1.2 | −100 ±10 |

[a] Minimum principal strain rate in µ strain/yr. Positive value is extension represented by outward pointing arrows.
[b] Maximum principal strain rate in µ strain/yr. Negative value is compression represented by inward pointing arrows.
[c] Azimuth of $\varepsilon_2$ in degrees measured clockwise from North with $1\sigma$ errors.
[d] Rigid body translation with $1\sigma$ errors.
[e] Azimuth of the rigid body translation azimuth in degrees, measured clockwise from North with $1\sigma$ errors.

Figure 2a shows extensional faulting, Fig. 4.2b shows strike-slip faulting and Fig. 4.2c depicts reverse faulting. Fault-plane solutions are inferred from 1 513 neotectonic fault-slip data sets. With a few exceptions along some E-W striking faults, resolved extensional directions trend E-W to ENE-WSW.

An example of recent faulting in late Tertiary to Quaternary sediments is shown in Fig. 4.9. Abundant evidence of rock crushing suggests instantaneous movements and local fault-related colluvial deposits in the form of 'wedges' propose intraplate surface-rupturing events. Most of the extensive (lengths > 50 km) active faults investigated, trend NS to NNW-SSE and exhibit some sinistral strike-slip motion in addition to normal faulting. Individual events could not be correlated (and dated) between outcrops on the same faults. Offsets of several decimeters to meters, that are apparently event-related, and long straight fault segments (> 50 km), suggest that large normal faulting events may have occurred in the area frequently and recently.

The focal mechanism and moment rate shown in the inset of Fig. 4.6 represent differential residual motions of the AEDM with respect to the frontal coast area. It was derived using residual velocities along stations in the forearc, located in the area east and west of the studied active faults (Fig. 4.2) between 71.25 and 71.75° W and 30.25 and 31° S. The focal mechanism diagram of Fig. 4.6 assumes that the differential motion is accommodated by faulting along the studied intraplate faults. The results from geology and geodesy coincide, taking into account the 95% confidence limits between 5° and more than 20° for the principal directions. Apparent rates derived using the residuals to the AEDM suggest $9 \pm 4$ mm yr$^{-1}$ of horizontal, $4.5 \pm 2.2$ mm yr$^{-1}$ of dip-slip, and $2.9 \pm 1.4$ mm yr$^{-1}$ of strike-slip motion along 60° dipping and –10° striking (azimuth) faults (larger section). Values roughly one-third smaller were derived from the smaller section.

### 4.4.4 Constraints on the Viscosity of Upper Mantle and Lower Crust

In order to reproduce the observed seaward motion of the postseismic deformation field in the area of the great 1960 Valdivia earthquake, we constructed the 3D finite element model discussed in section 3.4. Firstly, the present-day surface deformation was corrected using the effect of the current interseismic strain accumulation, due to the locking of Nazca and South America plates according to the AEDM. Thus, the finite element model gives solely the delayed response of the earthquake rupture. As discussed in section 3.4, we first analyzed the sensitivity of different model parameters to the observable horizontal velocity field on the continental crust. The following parameters were selected as reference model:

- Model dimensions: $x_{left} = -500$ km, $x_{right} = 1\,000$ km, $z_{depth} = 700$ km
- Element side length:
  - in the crust, in the vicinity of the rupture area: 10 km
  - in the crust, further from the rupture area: 20 km
  - in the mantle, close to the crust and slab: 20 km
  - in the mantle, next to the model boundaries: 300 km
- Viscosity: $4 \times 10^{19}$ Pa s
- Effective elastic thickness of the continental lithosphere: 40 km

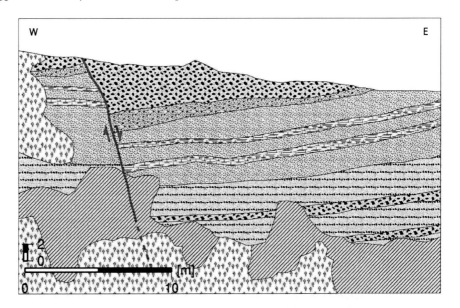

**Fig. 4.9.**
Cross section of the Puerto Aldea fault 5 km S of El Rincón (Site 13/01). Plio-Pleistocene Limarí Formation offset by 4.0 to 4.7 m. *1* – Clayey silt with soil formation in the capping beds. *2* – Clayey silt with caliche formation (white bands). *3* – Silty sand with up to 10% components ranging from mm to cm in size (moderately rounded). *4* – Sand (fine-coarse grained, gray-green). *5* – Alluvial layer (poorly rounded cobbles). *6* – Fluvial layer with crushed cobbles at the intersection with the fault plane (> 75% well rounded cobbles). *Thick line (continuous/dashed)* – known/presumed fault plane

- Effective elastic thickness of the oceanic lithosphere: 30 km
- Rupture depth: 40 km
- Coseismic slip: 17 m
- Maximum after-slip size: 17 m
- After-slip depth range: 40–70 km
- After-slip time interval: 10 years

Material is not allowed to leave the model. The reference frame is defined by four bottom corners, therefore the location of the deformation with respect to the boundary may affect the model results. Our model fixes perpendicular motions on the model boundaries, unless they occur on the upper free boundary. The coseismic slip on the rupture zone is 20 m. Model displacements are computed with respect to the model boundaries, causing the results to be sensitive to the model dimensions. Fig. 4.10a shows that a model extending from 500 km west to 1 000 km east of the trench with a depth of 700 km is large enough not to affect the derived velocity field. There is no need to have a homogeneous mesh over the entire model, therefore the mesh is dense only near the rupture zone with a maximum element side of 10 km. Within the crust and further from the rupture zone, the maximum element side length is increased up to 20 km. The mantle is meshed so that the elements close to the crust and slab are of the same size as those of the neighbouring areas, while the side lengths on the boundaries can be as large as 300 km. In order to check the quality of this configuration, we halved the above values resulting in an effect on the surface velocity that was less than 0.1 mm yr$^{-1}$. A strong dependency exists between the surface velocity field and the mantle viscosity, and a possible range of mantle viscosities, together with the observed velocities, is shown in Fig. 4.10b. A lower viscosity corresponds to faster relaxation. The surface velocity of a model with a viscosity of $3 \times 10^{19}$ Pa s is up to 1.5 cm yr$^{-1}$ faster than a model with a viscosity of $6 \times 10^{19}$ Pa s. The observed data are best explained when we incorporate a mantle viscosity of $4 \times 10^{19}$ Pa s.

Further sensitivity studies include, among others, the effective elastic thickness of both continental and oceanic lithosphere; the depth of the coseismic rupture; and the effect of the after-slip size. The thickness of the elastic continental lithosphere changes the overall shape of the velocity curve. A thicker continental lithosphere pushes the velocities of the coastal area more inland and slows the seaward motion (Fig. 4.10c). Similarly, altering the thickness of the elastic oceanic lithosphere also changes the shape of the surface velocity curve (Fig. 4.10d): a thick oceanic lithosphere pushes the coastal area velocities more seaward. If the dip of the seismogenic zone is given, then a greater rupture depth (i.e. the downward end of the rupture zone is deeper) extends the rupture further inland. Therefore, a deeper rupture, and the corresponding after-slip, moves the minimum point of the velocity curve – which is the maximum absolute velocity – landward, so that the minimum point remains above the after-slip area (Fig. 4.10e). The depth range, the amount, and the time period of after-slip do not show considerable impact on the observed velocities. Figure 4.10f shows the results of varying the maximal after-slip values at the beginning of the after-slip: a difference of 10 m in the after-slip size does not affect the results within the given accuracy of the surface observations. All in all, judging from our model we can state that postseismic after-slip does not have a significant effect four decades after the Valdivia earthquake.

This finding is in contrast to the results of the Antofagasta earthquake. Here, the observed postseismic displacements one and two years after the event cannot be explained by viscoelastic stress relaxation. Figure 4.11 depicts the contribution of a hypothetical postseismic viscoelastic relaxation together with GPS observations made 15 months (Fig. 4.11a) and 27 months (Fig. 4.11b) after the 1995 Antofagasta earthquake. The GPS velocities shown are the residual vectors resulting after the subtraction of the interseismic AEDM values from the observed displacements. Obviously, viscoelastic processes cannot be the driving mechanism of short-term postseismic motions in this area since the model does not fit the data. In contrast, the surface velocities 40 years after the 1960 Valdivia earthquake is well explained by our viscoelastic finite element model (Fig. 4.10b) and the good fit confirms the assumption that viscoelastic processes are the main driving mechanisms decades after great earthquakes. Our observations are best fitted when we incorporate a viscosity of $4 \times 10^{19}$ Pa s.

## 4.5 Discussion and Conclusions

The GPS-results of Klotz et al. (2001) and Khazaradze and Klotz (2003) show for the first time the present-day deformation field at the southern Andean subduction zone in Chile and the back-arc in Argentina. The changing spatial-temporal pattern of geodetically-measured crustal deformation rates in the Andes is governed by different phases of the earthquake deformation cycle. Therefore, the present-day patterns of deformation are significantly different from those in the geological record and do not reflect the recent evolution of the Andes. The transient elastic deformation related to subduction earthquakes represents a short-term signal compared to geological timescales. Conversely, the permanent plastic deformation that contributes to the formation of tectonic features can be characterized as a long-term signal. Most of the transient deformation can be explained by interseismic, coseismic, and postseismic phases of interplate thrust earthquakes. The possible occurrence of slow

**Fig. 4.10.**
Sensitivity of different model parameters to the surface velocity in a cross section.
**a** Effect of model dimensions on the velocity field after 40 years. Enlarging the model has no considerable impact on the results, while the computation time increases rapidly.
**b** Sensitivity of the surface velocity field to the mantle viscosity. Increasing the mantle viscosity, e.g. from $4 \times 10^{19}$ Pa s to $6 \times 10^{19}$ Pa s, results in modeled surface velocities which are outside the error bars of the observed velocities.
**c** Effect of the effective elastic thickness of the continental lithosphere. **d** Effect of the effective elastic thickness of the oceanic lithosphere. The thickness of the reference model is 40 km for the continental lithosphere and 30 km for the oceanic lithosphere. **e** An extension of the rupture depth moves the maximum velocity further inland. **f** The after-slip mechanism – i.e. size, time interval and depth range of after-slip – has no significant effect on the surface velocity field after 40 years. Changing the initial after-slip value does not affect the model results within the given accuracy of the observed vectors

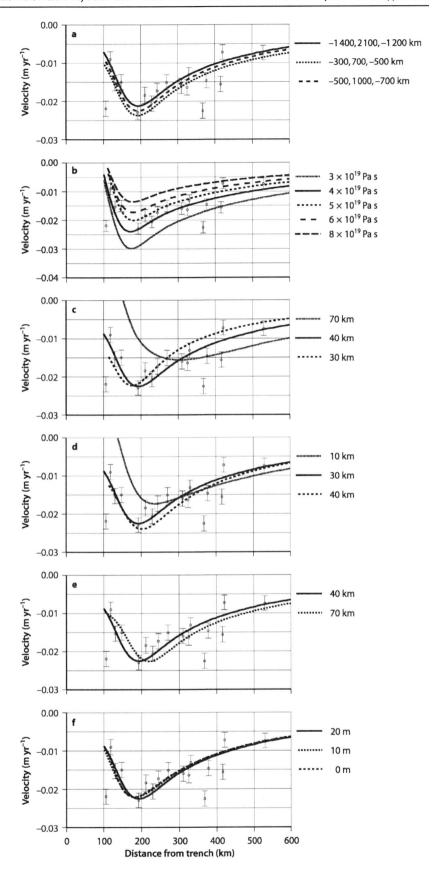

**Fig. 4.11.**
Contribution of hypothetical postseismic viscoelastic relaxation of the Antofagasta earthquake versus: **a** observed displacements 15 months after the event; **b** observed displacements 27 months after the event. *Triangles* are the residual velocities of the AEDM. The viscoelastic model cannot fit the observations in the Antofagasta area

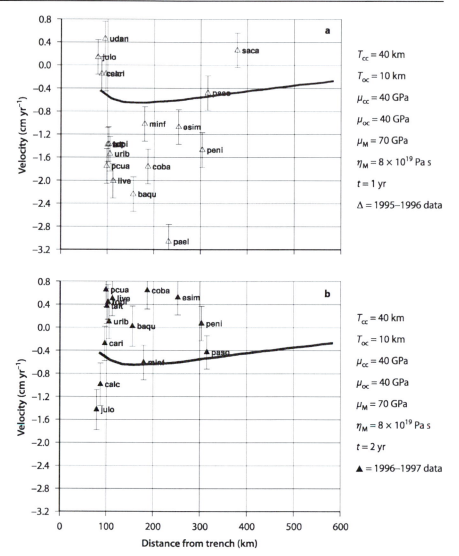

earthquakes along the plate interface, as detected along the Cascadia subduction zone (Dragert et al. 2001), can also be characterized as a transient signal. Intraplate earthquakes occurring within the crust may also contribute to the observed vectors, though it is more difficult to detect the interseismic strain accumulation related to these events.

If the short-term deformation field predicted by the elastic dislocation models is correct, then the residual velocity field, obtained by subtracting the predicted velocities from observations, should represent long-term plastic deformation. In principle, the long-term deformation field consists of components that include crustal shortening, block rotations, translation of a fore-arc sliver and plastic deformation associated with slowly creeping faults and minor earthquakes. A small part of the derived velocity vectors is due to observational and processing errors, including local site effects. Along the Andes, the most obvious long-term deformation signal is visible in the back-arc of the subduction where the Brazilian Shield underthrusts the Subandean zone. Also, the derived depth of the seismogenic zone is a long-term feature. In the following sections we discuss the significance of our results in terms of the long-term features of the Andean subduction zone.

## 4.5.1 Coupling of Plates

The term 'coupling of plates' often causes confusion since it is not a clear and unique expression to describe a mechanical interaction between two plates. If we investigate fault processes using geodetic observations, the term coupling describes the state of current slip between rocks on each side of a fault, e.g. the interface between the subducting Nazca Plate and the overriding South American Plate. A fully coupled fault may also be described as fully locked, having a coupling ratio of '1', or being 100%

coupled. A decoupled fault is described as free slipping, having a coupling ratio of '0', or being 0% coupled. In this case, the slip at the interface will be at plate convergence rate. A partially coupled fault (coupling ratio between 0 and 1) implies slip slower than the plate convergence rate. Since this type of coupling is valid for the short time period of geodetic observations, we may call this interaction 'geodetic coupling'. It solely describes the kinematics regardless of fault properties or stress. A good clarification of the semantics of coupling is given by Wang and Dixon (2004).

The term 'seismic coupling' is also a kinematic concept and, likewise it does not refer to stress conditions or fault properties. The seismic coupling factor is the relation of seismic slip to the total plate convergence. The factor is '1' if all of the convergence takes place as earthquake slip, but in general it is lower. Seismic coupling describes the relation between seismic and aseismic slip and is only valid for the relatively short time-period of seismic observations. Seismic coupling cannot be higher than the geodetic coupling.

Another coupling-term may be called 'geologic coupling' which describes the interaction between two plates over geological timescales. This term is not a kinematic description but refers to long-term permanent signals and mountain-building processes. In principle, the long-term 'geologic coupling' is independent from earthquake-related deformation whereas the geodetic coupling always refers to the interseismic period. During an earthquake, the geodetic coupling is very low or zero, but may be high during the much longer interseismic period.

Clearly, the degree of coupling is a long-term feature of the Andean subduction zone. Also, the depth of coupling, or the width of the seimogenic zone, is most likely a long-term feature. Our 3D forward modeling results suggest that the observed velocities in the central section represent interseismic strain accumulation due to a fully locked thrust interface between the subducting Nazca and the overriding South American Plates. The geodetically estimated depth of coupling (that equals the width of the locked zone plus half the width of the transition zone) increases from ~33 km depth in the north to ~50 km depth in the south (Fig. 4.8). This depth range agrees well with the maximum depth of seismic coupling deduced from the distribution of interplate earthquakes along the ASZ (Tichelaar and Ruff 1991) and with the depth of the seismogenic zone estimated from thermal modeling (Oleskevich et al. 1999).

There is also a correlation between the depth-extent of coupling and the characteristics of the fore-arc gravity field (Fig. 4.8). This correlation suggests that coupling is related to the mass distribution associated with the fore-arc. Where the locking depth is greatest, positive anomalies at the coastline are broadest. In the north, the shallower locking depth is associated with a more sharply defined positive coastal anomaly. While the negative anomalies in the trench reflect the depth of the trench and the degree of sediment fill, the positive anomalies centered on the coastline are strongly influenced by the depth of the slab below the fore-arc (e.g. Hackney et al. 2006, Chap. 17 of this volume). Where positive anomalies are broadest, the slab is shallowest, and a larger part of the slab is in contact with the overriding continental crust. In contrast, narrower anomalies reflect a deeper slab beneath the fore-arc and a situation in which less of the slab is in contact with the continental crust.

If the gravity anomalies are an indication of the down-dip extent of locking, then they are also an indicator of the long-term nature of this locking. Song and Simons (2003) have pointed out that the amplitude of anomalies is much greater than the anomaly changes that would be induced by deformation during the seismic cycle. Therefore, the gravity anomalies reflect the long-term summation of deformation from individual earthquake cycles. To cause these anomalies, the particular features of the subduction zone that lead to a particular locking situation must be stable over long time periods.

The fact that the transition from the shallow to the deep coupling depth in our model (between latitudes 30–35° S) coincides with the area where the slab becomes steeper, suggests also that the deeper portions of the slab might affect the depth of coupling. Earlier studies have uncovered the intriguing coincidence between the changing dip of the slab at 33° S and the location where the Juan Fernández Ridge enters the Chile trench (Jordan et al. 1983). In addition, at approximately the same latitude, two interesting features are observed: an abrupt termination of back-arc seismicity (Chinn and Isacks 1983); and a drastic increase in the amount of the incoming sediment caused by a change in the climatic environment (Bangs and Cande 1997).

The down-dip limit of the seismogenic zone of the thrust interface can be described by two depth limits, as shown in the schematic diagram of subduction (Fig. 4.3). The first limit is the maximum depth of the fully locked portion of the seismogenic zone, which is characterized by stick-slip behavior (i.e. earthquakes nucleate in this zone). This down-dip limit of the locked zone (i.e. depth of seismic coupling) is most commonly attributed to thermal effects taking place at a temperature of 350°C (Hyndman and Wang 1993; Tichelaar and Ruff 1993), but the geometrical effects suggested by gravity anomalies may also be important. At this critical temperature, crustal rocks are known to undergo transformation from velocity weakening to velocity strengthening (i.e. seismic to aseismic transition) (Scholz 1990). The second limit, the down-dip limit of the transition zone, corresponds to a temperature of approximately 450 °C. The zone between the down-dip limits corresponding to 350 °C and 450 °C is often referred to as a transition zone. Although earthquakes probably nucleate within the locked zone, the rupture may propagate deep into the transition zone (see Oleskevich et al. 1999). In the case of the subduction of old (i.e. cold) oceanic crust, a temperature of 350 °C on the thrust interface is often reached below 50–60 km depth, which is too deep for elastic coupling since thrust-type earthquakes are not observed at those depths. In this case, as suggested by Tichelaar and Ruff (1991, 1993) and Oleskevich et al. (1999), the governing factor for the extent of the seismic coupling is the depth where the downgoing slab meets the mantle wedge. This intersection is commonly encountered at depths of 35 to 50 km, the average depth of the fore-arc continental Moho. In this case, geometrical effects related to slab dip may become important because a steeper slab would encounter the mantle sooner than a shallower slab.

### 4.5.2 Variation of Back-arc Shortening

The most obvious long-term deformation signal in the present-day deformation field is visible in the back-arc where the Brazilian Shield underthrusts the Subandean zone. Current deformation due to ongoing Andean compression is apparently concentrated in the thrust belt at the eastern margin of the Altiplano/Puna regions (Allmendinger et al. 1997). The contemporary rate of shortening has previously been estimated from seismic and geodetic investigations. The seismically derived deformation rate in the Sub-Andes is about 2 mm yr$^{-1}$ (Suarez et al. 1983), but the period of seismic recording may not be long enough to cover the entire seismic cycle and part of the deformation may be released aseismically (Wang 2004). Therefore, a rate of 2 mm yr$^{-1}$ provides a lower boundary for the shortening rate. The reported long-term average shortening of the southern Central Andes, 210–370 km shortening in 27 million years (Schmitz 1994), corresponds to 8–14 mm yr$^{-1}$.

All geodetic data, including the results of Norabuena et al. (1998), Bevis et al. (1999), Klotz et al. (1999) and Bevis et al. (2001), show localized shortening in the fold and thrust belt north of 25° S. Norabuena et al. (1998) reported a rate between 10–12 mm yr$^{-1}$, but Bevis et al. (1999) reprocessed those GPS data and combined them with new observations to derive a present day shortening rate of 8.9 ± 1.6 mm yr$^{-1}$. Bevis et al. (2001) interpret the back-arc convergence between 12 and 22° S as a convergence between two rigid plates, the South American Plate and the Andean mountain belt as a rigid microplate. They concluded that 8.5% of the Nazca/South American Plate convergence (5–6 mm yr$^{-1}$) is accommodated in the back-arc. Bevis et al. (2001) also concluded that the best-constrained GPS velocity estimates imply that the rate of shortening in the southern Subandean zone decreases from 9.5 mm yr$^{-1}$ near 18.5° S to about 6.5 mm yr$^{-1}$ near 21.5° S.

Klotz et al. (1999) reported a slower rate of 3–4 mm yr$^{-1}$ further south at 24° S. The strong decrease of shortening rates from north to south do not agree with a rigid microplate motion around an Euler pole coaxial with the pole of the Nazca-South American plate convergence. Although the three-plate model reduces the number of parameters to describe the back-arc convergence for an extended area to only three values describing the Euler vector, it does not give further insight into the back-arc shortening process.

The central part of our study area that is not affected by postseismic relaxation processes shows a significant deformation signal only in the E-W components of zones 3 and 7 between 30 and 34° S (Fig. 4.12a). The presence of a long-term deformation signal within the back-arc is significant: the negative slope of 0.011 ± 0.003 μstrain yr$^{-1}$ corresponds to ~6 mm yr$^{-1}$ E-W shortening over a distance of 550 km. This result falls within the range of the 2–7 mm yr$^{-1}$ shortening rate estimated from geological observations spanning the last 25 Ma with preference to the higher end of the range (Kley and Monaldi 1998). South of 34° S, we did not detect a significant shortening signal in the back-arc.

Brooks et al. (2003) interpret the back-arc convergence between 26 and 36° S as a convergence between the rigid Andean plate and the rigid South American Plate and fit their data with localized deformation along the Argentinean Precordillera fold and thrust belt. This view is consistent with the narrow band of shallow crustal earthquakes from the nearly complete catalogue of earthquakes of $M_w$ > 5.2 (Engdahl et al. 1998) and also with geological reconstructions showing eastward migration of the Argentinean Precordillera through active deformation near its tip (Zapata and Allmendinger 1996). Brooks et al. (2003) suggest that the contemporary back-arc boundary creeps continuously at ~4.5 mm yr$^{-1}$. This shortening rate does not contradict our findings.

Overall, we conclude that the rate and style of present-day shortening varies with latitude: in the north it is best described by localized deformation at higher rates (i.e. up to 9.5 mm yr$^{-1}$), further south the signal is smaller and may also be described by distributed deformation. Stronger constraints for geodynamic modeling of mountain building processes requires a higher spatial resolution of the velocity field over longer time spans.

### 4.5.3 Long-Term Signals in the Fore-arc

In the fore-arc, the GPS measurements mainly reflect short-term earthquake related deformations. Therefore, long-term tectonic signals are hard to detect. Nevertheless, we see an indication of trench-perpendicular (E-W) extension within the fore-arc of the central section of our study

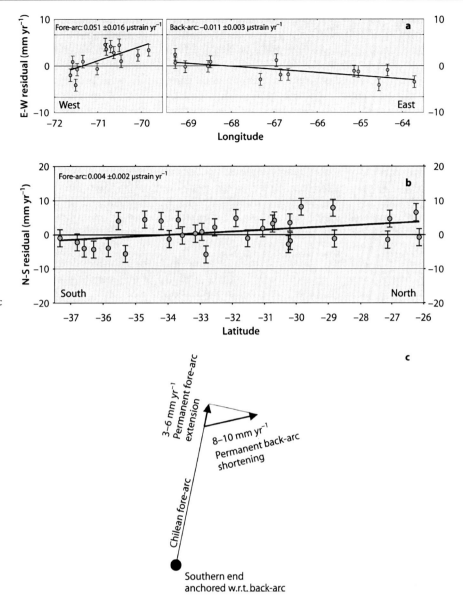

**Fig. 4.12.**
Long-term deformation signal in the fore-arc (*zones 2–4*) and back-arc (*zones 6–8*) based on the residual velocities of the AEDM and obtained by subtracting the elastic dislocation model predictions from the observed velocities. **a** East-West component for the stations located in the fore-arc and the back-arc. **b** North-South component for the stations located within the fore-arc. Uncertainties are given at 1σ level. **c** Relation between back-arc shortening and along strike fore-arc extension: stronger shortening in the northern part of 8–10 mm yr$^{-1}$ and slow or no shortening in its southern part requires a strike-parallel extension of the fore-arc of 3–6 mm yr$^{-1}$

area (see Sect. 4.3 and 4.2). The strain rate derived from the least squares fit to the E-W residual velocity component is 0.051 ± 0.016 μstrain yr$^{-1}$ (Fig. 4.12a). Neotectonic field studies from the same area, between latitudes 30° S and 31° S, reveal N-S striking Quaternary normal faulting, which is also an indication for E-W extension (Heinze 2003). The amount of E-W extension inferred from neotectonic analysis and crustal balancing of the 25 km wide fore-arc zone is about 0.6–3.3% for the last 1.2–6.5 Ma. The resolved strain rates range between 0.001–0.028 μstrain yr$^{-1}$. The GPS derived residual velocities reveal almost identical extension directions (ENE-WSW). In contrast to the relatively high rate of fore-arc strain in the central section, the northernmost section of our study area (21.5–18° S) shows less evidence of intraplate strain build-up.

The N-S component of the observed and the residual vectors in the fore-arc shows an indication of trench-parallel extension over the entire lateral extent of our study area. Figure 4.12b illustrates the least squares fit to the N-S component. The extension equals 0.004 ± 0.002 μstrain yr$^{-1}$ which corresponds to a rate of 4 km Myr$^{-1}$ over a distance of 1 000 km. This result is in agreement with latitudinal differences in the back-arc shortening process as shown in Fig. 4.12c, which illustrates that stronger shortening (8–10 mm yr$^{-1}$) in the northern part of the study area of and slow or no shortening in the southern part (south of 34° S) requires a strike-parallel extension of the fore-arc of approximately 3 to 6 km Myr$^{-1}$. Hence the geodetically observed trench-parallel stretching is most likely a long-term signal of the upper-plate deformation field.

Another potential long-term signal in the fore-arc is the partitioning of the trench-parallel component of oblique convergence along a vertical strike-slip fault by the development of a fore-arc sliver. However, our data do not show evidence for this kind of strain partitioning in the upper plate along the entire study area, including in the central section of our study area where the signal is not contaminated by postseismic effects (see also Hoffmann-Rothe et al. 2006, Chap. 6 of this volume). This finding is in agreement with the results of McCaffrey (1994), who examined the deviations of slip vectors of interplate thrust earthquakes from expected plate convergence directions and concluded that the effective rheology of the fore-arc in South America is elastic, and thus less prone to internal deformations and more susceptible to the occurrence of great subduction earthquakes.

A different phenomenon of partitioning associated with oblique plate convergence is called 'strain partitioning' by Bevis and Martel (2001). They showed that above the locked portion of the plate interface, surface velocity is more oblique than the plate convergence vector and less oblique further inland. This effect is purely elastic and will be recovered during coseismic strain release. So far, other authors did not consider long-term signals in the present-day deformation field of the fore-arc. Ruegg et al. (2002) observed velocities at 13 GPS sites in the fore-arc between 35° S and 37° S and interpreted those vectors only in terms of interseismic strain accumulation. Also, Brooks et al. (2003) did not consider long-term signals of their derived velocity field in the fore-arc. As mentioned above, the analysis of the residual velocity field and strain rates within the Chilean fore-arc is somewhat speculative as the residual velocities include possible modeling uncertainties of large interseismic vectors in the fore-arc.

### 4.5.4 Structures in the Fore-arc as a Consequence of Earthquake Deformation

Part of the short-term elastic deformation may contribute to the long-term deformation field, if the coseismic strain release does not match the deformation accumulated during the interseismic phase. Over many earthquake cycles, even a very small portion of the stored elastic strain not released in following earthquakes would lead to significant permanent tectonic structures.

Figure 4.13 shows the three components of the interseismic and coseismic displacements as a function of the distance to the trench. The coseismic strain release does not match the interseismic accumulation of strain for all three components. The landward limit of the locked zone coincides approximately with the coastline which has a distance to the trench of about 100 km. The elastic interseismic accumulation of strain produces subsidence seawards of the coastline and uplift landwards. Thus, the coastline coincides with the change from subsidence to uplift. King et al. (1988) and Stein et al. (1988) noted the qualitative agreement between the elastic vertical deformation and the long-term plastic deformation. Due to viscoelasticity and sediment erosion and deposition the vertical component of the interseismic and coseismic deformation cannot be symmetric.

The situation is more interesting when looking at the whole 3D picture instead of the vertical component alone. Figure 4.14a–c depicts the horizontal components of interseismic, coseismic and postseismic surface deformation. The coseismic strain release (Fig. 4.14b) shows a concentric fault-slip distribution. It is reasonable to assume that earthquakes recur with similar characteristics and thus produce a similar concentric surface deformation. This pattern is generally attributed to the presence of localized asperities along the thrust interface, which cause uneven strain accumulation during the interseismic period. However, our geodetic observations and the GPS results from Brooks et al. (2003) suggest uniform interseismic strain build up that reflects a spatially homogeneous coupling along the seismogenic interface (Fig. 4.14a). If we assume that the coseismic slip distribution of further hypothetical interplate earthquakes in the study area is similar to the slip observed during the Antofagasta earthquake, than the strain release would not be spatially homogeneous. This implies that the homogeneous interseismic deformation field cannot be released by the coseismic strain pattern. Also, postseismic processes such as after-slip or viscoelastic relaxation do not change the pattern since those processes only add to the coseismic deformation pattern (Fig. 4.14c). This leads to the conclusion that a significant part of the strain applied to the deformation zone by elastic loading and unloading may remain as plastic deformation. An example of a possible remaining deformation field is shown in Fig. 4.14d.

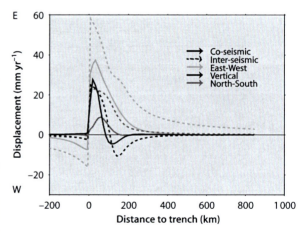

**Fig. 4.13.** East-West, North-South, and vertical components of the modeled interseismic and coseismic displacements. For all three components, the coseismic strain release does not match the interseismic strain accumulation

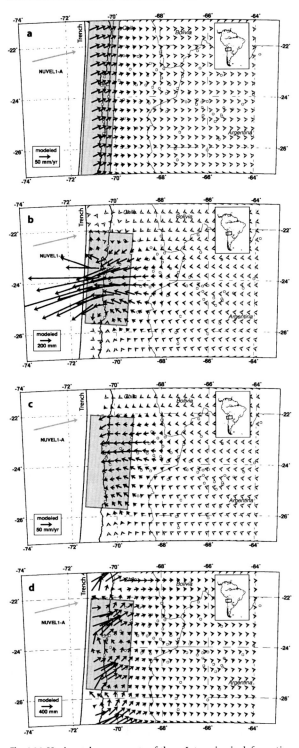

Fig. 4.14. Horizontal components of the: a Interseismic deformation for a fully coupled seimogenic zone; b coseismic deformation of the 1995 Antofagasta earthquake; c postseismic deformation field one year after the 1995 Antofagasta earthquake, and; d interseismic deformation field after 60 years minus three hypothetical earthquakes similar to the Antofagasta earthquake. The remaining deformation field shows northward displacements in the fore-arc and localized strain in East-West direction

The E-W component of the discrepancies shown in Fig. 4.14d might be released by different short-term processes. These are, among others, aseismic strain release or temporal variations in strain build-up. However, if other processes are invoked to explain the remaining E-W component, a significant northward shift of the fore-arc will remain (Fig. 4.14d). The remaining N-S component is not equivalent to the model of slip partitioning along trench-parallel strike-slip faults since it does not involve concentration of strain along faults. This suggests that an assumed elastic deformation of the upper plate may cause significant long-term parallel motion of the fore-arc in the order of mm yr$^{-1}$. Further continuous, long-term monitoring of contemporary surface deformation will be needed in order to reliably distinguish between alternative processes.

### 4.5.5 Constraints on the Viscosity of Upper Mantle and Lower Crust

The prominent signals of the residual velocity field shown in Fig. 4.8 are the seaward motion of inland sites between 23° S to 25° S and 38° S to 43° S, and are obviously due to postseismic relaxation processes. Vectors in the northern part show the integrated displacements 3 months and 15 months after the 1995 $M_w$ 8.0 Antofagasta earthquake, whereas vectors in the southern part show the velocities 35 years after the great 1960 $M_w$ 9.5 Valdivia earthquake. In contrast to strong postseismic signals detected one year after the occurrence of the 1995 Antofagasta earthquake, the deformation field observed two years later is already dominated by the interseismic strain build-up (Fig. 4.1b). Postseismic displacements can be explained either by a continuing after-slip along the deeper sections of the coseismic rupture surface and/or by viscoelastic relaxation processes occurring within the lower crust and the upper mantle (Hearn 2003; Wang 2004). The deformation in the northern section is most likely due to a continuing afterslip occurring a year after the earthquake, while the viscoelastic relaxation effect is most likely responsible for the movements observed in the south. Therefore, we were able to put some constraint on the viscosity of the upper mantle with respect to the lower crust by modeling the observed postseismic deformation from the area of the 1960 Valdivia earthquake. The rheological behavior is a fundamental constraint in understanding the long-term deformation and the recent evolution of the Andes. The postseismic velocities in the area of the 1960 Valdivia earthquake are best fitted to our 3D finite element model when we incorporate a mantle viscosity of $4 \times 10^{19}$ Pa s. The good fit of the model values to the GPS-vectors is shown in Fig. 4.15. The constraint on the viscosity of the Earth's mantle corresponds with the reported values derived in the same area or its vicinity as discussed below.

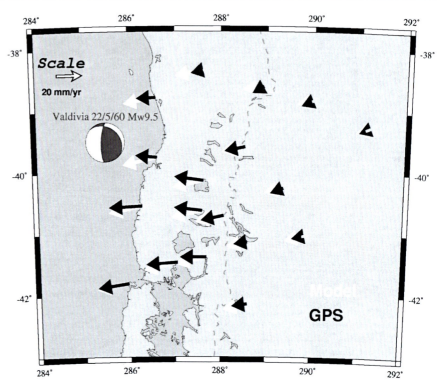

**Fig. 4.15.**
Plan view of the area of the 1960 Valdivia earthquake. Modeled velocities (*white*) fit the observed velocities (*black*) reasonably well

Barrientos et al. (1992) studied post-seismic uplift data of tide gauge records in Puerto Montt at 41.2° S and explained it by a propagating creep on the down-dip extension of slip in a purely elastic model. Nelson and Manley (1992) found 1.4 m of uplift at Isla Mocha (38.4° S) during the 1960–1965 period and 1.76 m of uplift during the following 24 years, and interpret the uplift as aseismic after-slip. The application of an elastic dislocation model in the depth range of 50–90 km is questionable since high temperatures will cause ductile behavior at the plate interface. Piersanti (1999) modeled the uplift at Puerto Montt (Barrientos et al. 1992) and Isla Mocha (Nelson and Manley 1992) using a spherical, self-gravitating, viscoelastic model. This model led to a viscosity range of $8 \times 10^{19}$ Pa s to $1 \times 10^{20}$ Pa s. Ivins and James (1999) investigated the sensitivity of a modeled surface uplift to the history of ice loading in the past 5 000 years. They proposed two models with weak and strong structures. The weak model with $5 \times 10^{19}$ Pa s and a relatively thin elastic crust shows considerable sensitivity to the load history. The strong regime, in contrast, shows insignificant sensitivity to the details of the load history. It consists of a thicker elastic crust underlain by a mantle of $\eta \geq 6.5 \times 10^{20}$ Pa s viscosity. Finally, they speculated that according to different evidence, the lower viscosity regime is very likely to exist in Patagonia and gave a lower viscosity value of $5 \times 10^{19}$ Pa s, which is slightly above our estimate. The discrepancy could be even greater considering that post-glacial rebound models give viscosities of interior continental regions.

Khazaradze et al. (2002) presented results of a 3D viscoelastic finite element model and found that the GPS data are consistent with a viscosity of $3 \times 10^{19}$ Pa s. Hu et al. (2004) extended this model considerably and found the best fit by incorporating a continental mantle viscosity of $2.5 \times 10^{19}$ Pa s. The model also explains the postseismic uplift observations of Barrientos et al. (1992) by introducing a oceanic mantle viscosity of $10^{20}$ Pa s. Our result is slightly higher than the estimations of Khazaradze et al. (2002) and Hu et al. (2004) as we could fit the same data with a lower viscosity value of $3 \times 10^{19}$ Pa s, for example, by increasing the thickness of the continental crust from 40 km to 60 km or decreasing the thickness of the oceanic crust from 30 km to 10 km, or a combination of both. Increasing the coseismic slip vector or decreasing the rupture depth resulted in lower viscosities as well. Also, the introduction of a slightly different slip geometry affected the result in the order of $\pm 1 \times 10^{19}$ Pa s, with a shallower slab leading to lower continental mantle viscosity. The discrepancies with other published results, including results from other subduction zones (Wang 2004), indicate that the upper continental mantle viscosity in the area of the great 1960 Valdivia earthquake is constrained within $\pm 1$–$2 \times 10^{19}$ Pa s.

## Acknowledgments

This work was supported by the collaborative research program SFB 267 'Deformation Processes in the Andes' of the

'Deutsche Forschungsgemeinschaft' and by the GeoForschungsZentrum Potsdam. We are grateful to many organizations and individuals who contributed to the acquisition of the GPS data. We thank Mike Bevis and Heidrun Kopp for their careful reviews of the manuscript. The maps in this paper were generated using the public domain Generic Mapping Tools (GMT) software (Wessel and Smith 1995).

## References

Allmendinger RW, IsacksBL, Jordan TE, Kay SM (1997) The evolution of the Altiplano-Puna plateau of the Central Andes. Ann Rev Earth Sci 25:139–174

Andersen OB, Knudsen P (1998) Global marine gravity field from the ERS-1 and Geosat geodetic mission altimetry. J Geophys Res 103:8129–8137

Angermann D, Baustert G, Galas R, Zhu SY (1997) EPOS.P.V3 (Earth Parameter & Orbit System): software user manual for GPS data processing. In: Scientific Technical Report 97/14. GeoForschungsZentrum, Potsdam, Germany, pp 52

Angermann D, Klotz J, Reigber C (1999) Space-geodetic estimation of the Nazca-South America Euler vector, Earth Planet Sci Lett 171(3):329–334

Bangs NL, Cande SC (1997) Episodic development of a convergent margin inferred from structures and processes along the southern Chile margin. Tectonics 16(3):489–503

Barrientos SE, Ward SN (1990) The 1960 Chile earthquake: inversion for slip distribution from surface deformation. Geophys J Int 103:589–598

Barrientos SE, Plafker G, Lorca E (1992) Postseismic coastal uplift in southern Chile. Geophys Res Lett 19:701–704

Bevis M, Martel S (2001) Oblique plate convergence and interseismic strain accumulation. Geochem Geophys Geosyst 2: doi 2000GC000125

Bevis M, Kendrick EC, Smalley R Jr, Herring T, Godoy J, Galban F (1999) Crustal motion north and south of the Arica deflection: comparing recent geodetic results from the Central Andes. Geochem Geophys Geosyst 1:1–12

Bevis M, Kendrick E, Smalley R Jr, Brooks BA, Allmendinger RW, Isacks BL (2001) On the strength of interplate coupling and the rate of back arc convergence in the central Andes: an analysis of the interseismic velocity field. Geochem Geophys Geosyst 3:doi 10.129/2001GC000198

Brooks BA, Bevis M, Smalley R Jr, Kendrick E, Manceda R, Laurý´a E, Maturana R, Araujo M (2003) Crustal motion in the Southern Andes (26°–36°S): do the Andes behave like a microplate? Geochem Geophys Geosyst 4(10): doi 10.1029/2003GC000505

Byrne DE, Sykes LR, Davis DM (1988) Estimating seismic potential of subduction zones using the seismic front and forearc morphology. US Geol Surv Open File Rep

Cahill T, Isacks BL (1992) Seismicity and shape of the subducted Nazca Plate. J Geophys Res 97(12):17503–17529

Chinn DS, Isacks BL (1983) Accurate source depths and focal mechanisms of shallow earthquakes in western South America and in the New Hebrides island arc. Tectonics 2(6):529–563

Creager KC, Ling Yun C, Winchester JP, Engdahl ER (1995) Membrane strain rates in the subducting plate beneath South America. Geophys Res Lett 22(16)2321–2324

Das S, Scholz CH (1981) Off-fault aftershock clusters caused by shear stress increase? Bull Seismol Soc Am 71(5):1669–1675

Deng J, Sykes LR (1997) Stress evolution in Southern California and triggering of moderate-, small-, and micro-size earthquakes. J Geophys Res 102(11):24411–24435

Doblas M (1998) Slickenside kinematic indicators. Tectonophysics 295(1–2):187–197

Dragert H, Hyndman RD, C. RG, Wang K (1994) Current deformation and the width of the seismogenic zone of the northern Cascadia subduction thrust. J Geophys Res 99(B1):653–668

Dragert H, Wang K, James T (2001) A silent slip event on the deeper Cascadia subduction interface. Science 292:1525–1528

Engdahl ER, Van der Hilst RD, Berrocal J (1995) Imaging of subducted lithosphere beneath South America. Geophys Res Lett 22(16):2317–2320

Engdahl ER, Van der Hilst RD, Buland RP (1998) Global teleseismic earthquake relocation with improved travel times and procedures for depth determination. Bull Seismol Soc Am 88: 722–743

Flück P (1996) 3-D dislocation model for great earthquakes of the Cascadia subduction zone. Diploma thesis, Swiss Federal Institute of Technology (ETH), Zürich, Switzerland

Flück P, Hyndman RD, Wang K (1997) Three-dimensional dislocation model for great earthquakes of the Cascadia subduction zone. J Geophys Res 102(9):20539–20550

Flueh ER, Vidal N, Ranero CR, Von Hojka AHR, Bialas J, Hinz K, Cordoba D, Danobeitia JJ, Zelt C (1998) Seismic investigation of the continental margin off- and onshore Valparaiso, Chile. Tectonophysics 288:251–263

Götze H-J, Lahmeyer B, Schmidt S, Strunk S, Araneda M (1990) Central Andes Gravity Data Base. EOS 71(16):401, 406-7

Götze H-J, Lahmeyer B, Schmidt S, Strunk S (1994) The lithospheric structure of the Central Andes (20°–26°S) as inferred from quantitative interpretation of regional gravity. In: Reutter, Scheuber, Wigger (eds) Tectonics of the Southern Central Andes. Springer, Heidelberg, pp 7–21

Graeber F, Asch G (1999) Three dimensional models of P-wave velocity and P-to-S velocity ratio in the southern central Andes by simultaneous inversion of local earthquake data. J Geophys Res 104(9):237–220

Hackney RI, Echtler HP, Franz G, Götze H-J, Lucassen F, Marchenko D, Melnick D, Meyer U, Schmidt S, Tašárová Z, Tassara A, Wienecke S (2006) The segmented overriding plate and coupling at the south-central Chilean margin (36–42° S). In: Oncken O, Chong G, Franz G, Giese P, Götze H-J, Ramos VA, Strecker MR, Wigger P (eds) The Andes – active subduction orogeny. Frontiers in Earth Science Series, Vol 1. Springer-Verlag, Berlin Heidelberg New York, pp 355–374, this volume

Hearn EH (2003) What can GPS data tell us about the dynamics of post-seismic deformation? Geophys J Int 155:753–777

Heinze B (2003) Active intraplate faulting in the forearc of North Central Chile (30°–31° S): implications from neotectonic field studies, GPS data, and elastic dislocation modeling, Scientific Technical Report 03/07. GeoForschungsZentrum, Potsdam, Germany

Hoffmann-Rothe A, Kukowski N, Dresen G, Echtler H, Oncken O, Klotz J, Scheuber E, Kellner A (2006) Oblique convergence along the Chilean margin: partitioning, margin-parallel faulting and force interaction at the plate interface. In: Oncken O, Chong G, Franz G, Giese P, Götze H-J, Ramos VA, Strecker MR, Wigger P (eds) The Andes – active subduction orogeny. Frontiers in Earth Science Series, Vol 1. Springer-Verlag, Berlin Heidelberg New York, pp 125–146, this volume

Hu Y, Wang K, He J, Klotz J, Khazaradze G (2004) Three-dimensional viscoelastic finite element model for post-seismic deformation of the great 1960 Chile earthquake. J Geophys Res (in press)

Husen S, Kissling E, Flueh E, Asch G (1999) Accurate hypocentre determination in the seismogenic zone of the subducting Nazca Plate in northern Chile using a combined on-/offshore network. Geophys J Int 138(3):687–701

Hyndman RD, Wang K (1993) Thermal constraints on the zone of major thrust earthquake failure: the Cascadia subduction zone. J Geophys Res 98(2):2039–2060

Ivins ER, James TS (1999) Simple models for late Holocene and presentday Patagonian glacier fluctuations and predictions of a geodetically detectable isostatic response. Geophys J Int 138:601–624

Jordan TE, Isacks BL, Allmendinger RW, Brewer JA, Ramos VA, Ando CJ (1983) Andean tectonics related to geometry of subducted Nazca Plate. Geol Soc Am Bull 94(3):341–361

Jordan TE, Reynolds JH III, and Erikson JP (1997) Variability in age of initial shortening and uplift in the Central Andes. In: Ruddiman WF (ed) Tectonic uplift and climate change. Plenum Press, New York, pp 41–61

Khazaradze G (1999) Tectonic deformation in western Washington State from Global Positioning System measurements. PhD thesis, Univ of Washington, Seattle

Khazaradze G, Klotz J (2003) Short and long-term effects of GPS measured crustal deformation rates along the South-Central Andes. J Geophys Res 108(B6): doi 10.1029/2002JB001879

Khazaradze G, Qamar A, Dragert H (1999) Tectonic deformation in western Washington from continuous GPS measurements. Geophys Res Lett 26:3153–3156

Khazaradze G, Wang K, Klotz J, Hu Y, He J (2002) Prolonged postseismic deformation of the 1960 great Chile earthquake and implications for mantle rheology. Geophys Res Lett 29(22):2050

King GCP, Stein RS, Rundle J (1988) The growth of geological structures by repeated earthquakes, 2: Field examples of continental dip-slip faults. J Geophys Res 93:13319–13331

King GCP, Stein RS, Lin J (1994) Static stress changes and the triggering of earthquakes. Bull Seismol Soc Am 84(3):935–953

Kley J, Monaldi CR (1998) Tectonic shortening and crustal thickness in the Central Andes: how good is the correlation? Geology 26(8):723–726

Klotz J, Angermann D, Michel GW, Porth R, Reigber C, Reinking J, Viramonte J, Perdomo R, Rios VH, Barrientos S, Barriga R, Cifuentes O (1999) GPS-derived deformation of the Central Andes including the 1995 Antofagasta M (sub w) = 8.0 earthquake. Pure Appl Geophys 154:3709–3730

Klotz J, Khazaradze G, Angermann D, Reigber C, Perdomo R, Cifuentes O (2001) Earthquake cycle dominates contemporary crustal deformation in Central and Southern Andes. Earth Planet Sci Lett 193:437–446

Krawczyk C, SPOC Team (2003) Amphibious seismic survey images plate interface at 1960 Chile earthquake. EOS 84(32):301–312

Lavenu A, Cembrano J (1997) Quaternary state of stress in southern Chilean Andes between 32°–45° South latitude. Abs Prog Geol Soc Am 29(6):443

Liu M, Yang Y, Stein S, Zhu Y, Engeln J (2000) Crustal shortening in the Andes: why do GPS rates differ from geological rates? Geophys Res Lett 27:3005–3008

Masson F, Delouis B (1997) Local earthquake tomography in northern Chile using finite-difference calculations of P-travel times. Phys Earth Planet Int 104(4):295–305

McCaffrey R (1994) Global variability in subduction thrust zone-forearc systems. Pure Appl Geophys 142(1):173–224

Meschede M (1994) Methoden der Strukturgeologie. F. Enke, Stuttgart

Michel G (1994) Neo-kinematics along the North-Anatolian Fault (Turkey). Universität Tübingen

NEIC (2001) http://neic.usgs/gov/neis/bulletin/010213142208.html

Nelson AR, Manley WF (eds) (1992) Holocene coseismic and aseismic uplift of Isla Mocha, south-central Chile. Quaternary International 15-16, Pergamon, Oxford UK

NOAA (1998) Digital relief of the surface of the Earth. National Geophysical Data Center, Boulder, Colorado

Norabuena E, Leffler-Griffin L, Mao A, Dixon T, Stein S, Sacks SI, Ocola L, Ellis M (1998) Space geodetic observations of Nazca-South America convergence across the central Andes. Science 279:358–362

Okada Y (1985) Surface deformation due to shear and tensile faults in a half-space. Bull Seismol Soc Am 75:1135–1154

Okada Y (1992) Internal deformation due to shear and tensile faults in a half-space. Bull Seismol Soc Am 82(2):1018–1040

Oleskevich DA, Hyndman RD, Wang K (1999) The updip and downdip limits to great subduction earthquakes: thermal and structural models of Cascadia, south Alaska, SW Japan, and Chile. J Geophys Res 104(B7):14965–14992

Pacheco JF, Sykes LR, Scholz CH (1993) Nature of seismic coupling along simple plate boundaries of the subduction type. J Geophys Res 98(8):133–159

Paskoff R (1970) Recherches geomorphologiques dans le Chili semiaride. Biscaye Freres

Patzwahl RJ, Mechie J, Schulze A, Giese P (1999) Two-dimensional velocity models of the Nazca plate subduction zone between 19.5°S and 25°S from wide-angle seismic measurements during the CINCA95 project. J Geophys Res 104(4):7293–7317

Piersanti A (1999) Postseismic deformation in Chile: constraints on the asthenospheric viscosity. Geophys Res Lett 26:3157–3160

Plafker G (1972) Alaskan Earthquake of 1964 and Chilean Earthquake of 1960: implications for Arc Tectonics. J Geophys Res 77:901–925

Plafker G, Savage JC (1970) Mechanism of the Chilean earthquake of May 21 and 22, 1960. Geol Soc Am Bull 81:1001–1030

Radtke U (1987) Marine terraces in Chile (22°–32°S): geomorphology, chronostratigraphy and neotectonics: preliminary results II. Quaternary of South America and Antarctic Peninsula 5:239–256

Ruegg JC, Campos J, Madariaga R, Kausel E, De Chabalier JB, Armijo R, Dimitriv D, Georgiev I, Barrientos S (2002) Interseismic strain accumulation in south central Chile from GPS measurements, 1996-1999. Geophys Res Lett 29:1517

Savage JC (1983) A dislocation model of strain accumulation and release at a subduction zone. J Geophys Res 88(6):4984–4996

Schmitz M (1994) A balanced model of the southern central Andes. Tectonics 13:484–492

Schmitz M, Lessel K, Giese P, Wigger P, Araneda M, Bribach J, Graeber F, Grunewald S, Haberland C, Luth S, Rower P, Ryberg T, Schulze A (1999) The crustal structure beneath the Central Andean forearc and magmatic arc as derived from seismic studies – the PISCO 94 experiment in northern Chile (21°–23°S) J S Am Earth Sci 12(3):237–260

Scholz CH (1990) The mechanics of earthquakes and faulting. Cambridge Univ Press, Cambridge UK

Scholz CH, Aviles, CA, Wesnousky SG (1986) Scaling differences between large interplate and intraplate earthquakes. Bull Seismol Soc Am 76(1):65–70

Smith WHF, Sandwell DT (1997) Global sea floor topography from satellite altimetry and ship depth soundings. Science 277(5334):1956–1962

Song TA, Simons M (2003) Large trench-parallel gravity variations predict seismogenic behavior in subduction zones. Science 301:630–633

Stein RS, King GCP, Rundle J (1988) The growth of geological structures by repeated earthquakes. 1: Conceptual framework. J Geophys Res 93:13307–13318

Suarez G, Molnar, P, Burchfield BC (1983) Seismicity, fault plane solutions, depths of faulting and active tectonics of the Andes of Peru, Ecuador and southern Colombia. J Geophys Res 88:10403–10428

Tassara A, Schmidt S, Götze H-J (in prep) A three-dimensional density model of the Andean margin

The ANCORP Working Group (1999) Seismic reflection image revealing offset of Andean subduction-zone earthquake locations into oceanic mantle. Nature 397(6717):341–344

Tichelaar BW, Ruff LJ (1991) Seismic coupling along the Chilean subduction zone. J Geophys Res 96(7):11997–12022

Tichelaar BW, Ruff LJ (1993) Depth of seismic coupling along subduction zones. J Geophys Res 98(2)2017–2037

Veit H (1993) Upper Quaternary landscape and climate evolution in the Norte Chico (northern Chile): an overview. Mountain Res Develop 13(2):139–144

Wang K (2004) Elastic and viscoelastic models for subduction earthquake cycles. In: Dixon T (ed) Seismogenic zone of subduction thrusts

WWang K, Dixon T (2004) "Coupling" semantics and science in earthquake research. EOS 85(18)

ang K, He J, Dragert H, James T (2001) Three-dimensional viscoelastic interseismic deformation model for the Cascadia subduction zone. Earth Planet Space 53:295–306

Wells DL, Coppersmith KJ (1994) New empirical relationships among magnitude, rupture length, rupture width, rupture area, and surface displacements. Bull Seismol Soc Am 84(4):974–1002

Wessel P, Smith WHF (1995) New version of the Generic Mapping Tools released. EOS 76(33):329

Wigger PJ, Schmitz M, Araneda M, Asch G, Baldzuhn S, Giese P, Heinsohn WD, Martinez E, Ricaldi E, Roewer P, Viramonte J (1994) Variation in the crustal structure of the southern Central Andes deduced from seismic refraction investigations In: Reutter KJ, Wigger PJ (eds) Tectonics of the southern Central Andes: structure and evolution of an active continental margin. Springer, Berlin, pp 23–48

Yuan X, Sobolev SV, Kind R, Oncken O, Andes Seismology Group (2000) Subduction and collision processes in the Central Andes constrained by converted seismic phases. Nature 408(6815): 958–961

Zapata TR, Allmendinger RW (1996) Growth stratal records of instantaneous and progressive limb rotation in the Precordillera thrust belt and Bermejo basin, Argentina. Tectonics 15: 1065–1083

# Chapter 5

# Tectonic Processes along the Chile Convergent Margin

César R. Ranero · Roland von Huene · Wilhelm Weinrebe · Christian Reichert

**Abstract.** The Chile subduction zone, spanning more than 3 500 km, provides a unique setting for studying, along a single plate boundary, the factors that govern tectonic processes at convergent margins. At large scale, the Chile trench is segmented by the subduction of the Chile Rise, an active spreading center, and by the Juan Fernández hot spot ridge. In addition, the extreme climatic change from the Atacama Desert in the north to the glacially influenced southern latitudes produces a dramatic variability in the volume of sediment supplied to the trench. The distribution of sediment along the trench is further influenced by the high relief gradients of the segmented oceanic lithosphere.

We interpret new and reprocessed multichannel seismic reflection profiles, and multibeam bathymetric data, to study the variability in tectonic processes along the entire convergent margin. In central and south Chile, where the trench contains thick turbidite infill, accretionary prisms, some 50–60 km wide, have developed. These prisms, however, are ephemeral and can be rapidly removed by high-relief, morphological features on the incoming oceanic plate. Where topographic barriers inhibit the transport of turbidites along the trench, sediment infill abruptly decreases to less than 1 km thick and is confined to a narrow zone at the trench axis. There, all sediment is subducted; the margin is extending by normal faulting and collapsing due to basal tectonic erosion. The transition from accretion to tectonic erosion occurs over short distances (a few tens of km) along the trench.

In the turbidite-starved northern Chile trench, ~1 km of slope debris reaches the trench and is subsequently subducted. There, tectonic erosion is causing pronounced steepening of the margin, associated pervasive extension across the slope and into the emerged coastal area, and consequent collapse of the overriding plate. The volume of subducting material varies little along much of the margin. However, the composition of the material varies from slope debris of upper-plate fragments and material removed from the upper plate by basal erosion, to turbidites derived from the Andes.

## 5.1 Introduction

The continental margin of Chile is more than 3 500 km long and spans a broad range of tectonic conditions. The two principal factors that control tectonic processes and influence the evolution of the arc/fore-arc system are the character of the incoming oceanic plate and the variability in sediment supply to the trench. This latter factor is caused by contrasting climate along the Pacific side of the Andes (Fisher and Raitt 1962; Scholl et al. 1970; Schweller et al. 1981; Bourgois et al. 2000; Yañez et al. 2002).

The most prominent feature of the downgoing oceanic lithosphere is the Chile Rise, an active spreading center that marks the boundary between the Nazca Plate to the north and the Antarctic plate to the south (Fig. 5.1). The Nazca Plate currently moves at ~65 km Myr$^{-1}$ (Angermann et al. 1999; Kendrick et al. 2003), but probably moved at ~85 km Myr$^{-1}$ for the past several million years, in a direction almost perpendicular to the trench. The Antarctic plate moves at ~18 km Myr$^{-1}$ and is fairly perpendicular to the trench between ~46 and 53° S, but south of this, it is very oblique to the margin (DeMets et al. 1990). The spreading center trends roughly NW and approaches the trench at a low angle; its axis currently collides with the continent at ~46° S. This broad topographic swell of young oceanic lithosphere extends north to about 40° S, where the system of Valdivia fracture zones offsets the spreading center some 500 km to the west. North of the Valdivia fracture zones, the lithosphere approaching the trench is older and deeper (Tebbens et al. 1997). The Chile Rise first collided with the continent south of 48° S, in the Tierra del Fuego, at ~14 Ma and the triple junction has been migrating northwards ever since (Cande and Leslie 1986).

Another prominent feature on the Nazca Plate is the Juan Férnandez hot spot chain, albeit a smaller topographic feature than the spreading center. The Juan Férnandez Ridge is a gentle topographic swell crested by a series of disconnected, large seamounts that first collided with the Chile margin in the north at about 22 Ma. The ridge has moved progressively southwards to the current collision point located roughly 32–33° S (Yañez et al. 2001). These two large topographic features on the oceanic plate do not only influence tectonics in the areas where they collide with the continent, but they also compartmentalize the trench into segments. Each of these segments is characterized by different tectonic structures and there is considerable variation in the thickness of the trench turbidites.

The second factor that seems to indirectly exert a major influence on the Chilean margin tectonics is the variable climate along the continent in space and time. Climatic conditions govern to a large degree the erosional rates of the arc-fore-arc and, therefore, the amount of sedi-

**Fig. 5.1.**
Shaded relief map of the Nazca and Antarctic plates and South America continent (Smith and Sandwell 1994). Convergence vectors are after DeMets et al., (1990). *Black lines* are tracks of most of the seismic reflection data collected along the continental margin in the past ~15 years. *Bold tracks* are multichannel seismic reflection (MCS) lines, and thin track, high-resolution MCS lines. Sonne 104 tracks are offshore from Antofagasta. Sonne 101 (*thin*) and Sonne 104 and Sonne 161 (*bold*) tracks are offshore and north of Valparaiso. Offshore from Valparaiso to Valdivia are Vidal Gormaz (VG02, *thin*) and Conrad and Sonne 161 (*bold*) tracks. Near the triple junction are Conrad RC2901 tracks. Along southernmost Chile are RC2902 and OGS-Explora IT95 & IT97 tracks

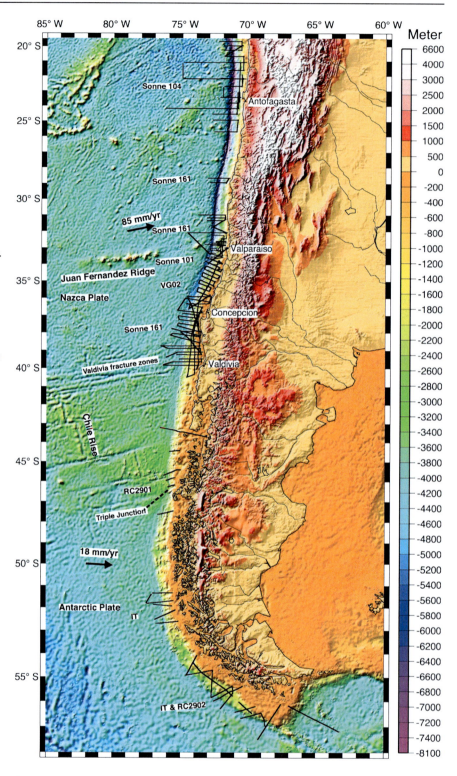

ment supplied to the trench. Sedimentation varies greatly along the arc/fore-arc region. North of ~28° S, the Atacama Desert has received little precipitation roughly since the early Miocene (Scholl et al. 1970) and the relatively small amount of erosion has been trapped in fore-arc basins. Few rivers traverse the area (Fig. 5.1) and the trench axis is largely empty of turbidites. In mid-latitude temperate climate, precipitation is moderate and there is a well developed river drainage system that supplies turbidites to the trench. At latitudes southern than ~40° S,

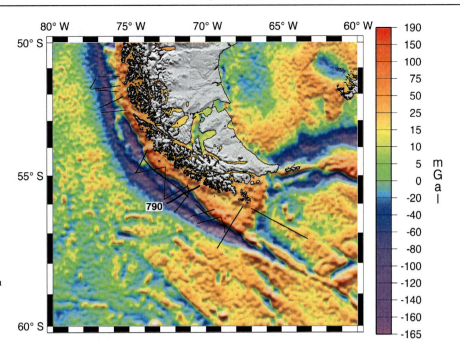

**Fig. 5.2.**
Shaded-relief gravity anomaly map from satellite altimetry offshore (Sandwell and Smith 1997) and topography onshore from southern South America. *Black lines* are tracks of seismic data from RV R. Conrad cruise RC2902 and RV OGS-Explora cruises IT95 & IT97. *Bold line* is RC2902 line 790 shown in Fig. 5.3. Note the steep gradient change in the gravity anomaly from positive to negative marking a roughly NW-SE-oriented lineament delineating the shelf break. The elongated body with negative gravity values seawards of the shelf marks the location of the accretionary prism and fore-arc basin

glacial – interglacial periods might have controlled the amount of sediment supplied to the trench. It has been proposed that this variability modulates the balance between accretionary and erosional processes along the mid and southern latitudes of the margin (Bangs and Cande 1997, Bourgois et al. 2000).

Still debated, however, is the effect of varied sediment supply during glacial – interglacial periods. Sediment might have been dominantly supplied during rapid glacial denudation of the Andes (Scholl et al. 1970; Bangs and Cande 1997), as has similarly been proposed for other glacially influenced and heavily sedimented areas like the Aleutian trench (von Huene and Scholl 1991). Alternatively, most ice-rafted sediment might have been deposited on the continental shelf during glaciations with relatively little sediment reaching the trench. During interglacial periods, the shelf is above sea level and, as the glaciers retreat, sediment is discharged at the shelf edge. Thus, abundant turbidites can reach the trench (Bourgois et al. 2000).

In this paper, we review studies of the Chile margin (Fig. 5.1) with a special focus on three tectonic settings: (*1*) the effect of spreading center subduction and subsequent margin evolution, (*2*) the collision of the Juan Férnandez Ridge, and (*3*) the contrasting tectonic processes between heavily sedimented and starved segments of the trench. We have reprocessed several multichannel seismic profiles, included new seismic lines, and merged bathymetric data from the triple junction. Additionally, we present new bathymetric data from the collision zone of the Juan Férnandez Ridge, and summarize data from the sediment-starved, northern margin offshore from Antofagasta.

## 5.2 The Tectonic Signature of Spreading Center Subduction and Margin Healing

To evaluate the effect of spreading center subduction and the subsequent evolution of margin structure, we reprocessed three seismic lines: line RC2902-790 from southernmost Chile (Figs. 5.1, 5.2 and 5.3), located where the Chile Rise collided with the margin at ~14 Ma; line RC2901-769, located only ~100 km south of the current triple junction (Figs. 5.4, 5.5 and 5.6) where the Chile Rise collided between 6 and 3 Ma (Cande and Leslie 1986); and line RC2901-745, located where the spreading center is currently subducting (Figs. 5.4, 5.5 and 5.7). We also merged multibeam bathymetry data around the triple junction (Figs. 5.4 and 5.5). Most bathymetric data were collected with Research Vessel (RV) L'Atalante (Bourgois et al. 2000) and we filled gaps in coverage around the Taitao Ridge with older data collected with RV R. Conrad (Bangs et al. 1992).

### 5.2.1 Southernmost Chile

Line RC2902-790 crosses the trench, the continental slope and the shelf (Fig. 5.2). The gravity map shows that along the margin the continental shelf break is marked by a sharp NW-SE-trending boundary with a steep gradient from positive to negative gravity values (Fig. 5.2). Seawards of the shelf, an elongated body characterized by negative gravity values extends parallel to the margin from about 56.5 to 52° S. Regional bathymetry (Fig. 5.1) and seismic images (Fig. 5.3) do not show a sudden deepening of the seafloor,

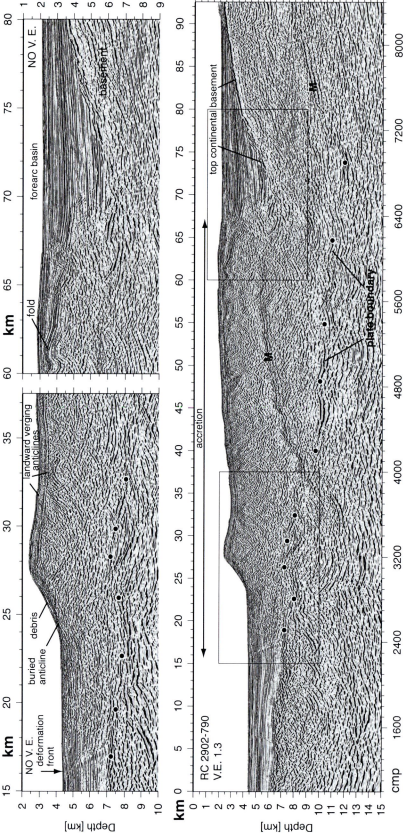

**Fig. 5.3.** Pre-stack depth migration of RC2902 line 790. The *lower panel* shows the profile with a 1.3 vertical exaggeration (V.E.) and the two close ups with no vertical exaggeration. The reflections marking the plate boundary are delineated by the *black-filled circles*. Note the shallow angle of the subducting plate and the ~40 km wide, 5–8 km thick, accretionary prism. Accretion involves an area ~10 km wide across the trench, that is filled with ~2.5 km of sediment. The upper strata of the fore-arc basin onlap the accretionary prism and are folded, indicating recent deformation. The continental basement has a smooth top surface onlapped by undeformed sediment of the fore-arc basin. *M* stands for multiple of the water layer; *cmp*: common mid point gather

indicating that the elongated body is probably of relatively low density. Gravity modeling along line 790 indicates that this body is formed by an accretionary prism and a deepwater fore-arc basin (Rubio et al. 2000). The accretionary prism and fore-arc basin have also been imaged in other seismic lines across the low-density body (Fig. 5.1; e.g. Polonia et al. 1999; Polonia et al. 2001). We concentrate on the structure of the deformation front, accretionary prism, and fore-arc basin and its infill. Rubio et al. (2000) presented the entire seismic line RC2902-790, but here we have reprocessed and pre-stack depth-migrated line 790 with an updated velocity model from the previous version.

From the deformation front to the fore-arc basin, the accretionary prism is about 50 km wide (~ km 16–66, Fig. 5.3) and slopes gently towards the trench. The area of accretion has similar dimensions in other seismic profiles across the lower density body (Polonia et al. 2001). In the trench, strata thicken slightly landwards indicating progressive tectonic thickening and, perhaps, faulting below the resolution of the data. The most seaward, major, seaward-vergent, thrust fault occurs at km 16–17 and seems to cut across the entire ~2.5 km thick sediment section, forming a gentle anticline (~ km 16–20, Fig. 5.3). The next thrust towards land is not well imaged, but another seaward-tilted, open anticline is centered at about km 25 and partially covered by slumped debris from the steep front of a prominent high.

The seaward limit of this older, accreting, sediment package is marked by an abrupt change from incoherent strata, indicating strong deformation in the core of the anticline, to little deformed, subhorizontal strata under the flat seafloor of the trench. This older anticline has thickened the sediment section to ~3.5 km. Even though the current frontal deformation indicates thickening of the prism over a short distance, this mechanism may not explain the 6 to 8 km thickness of most of the older part of the accretionary prism. At ~ km 23, the prism abruptly rises to form relief some 8 km wide and more than 1 500 m high. This high has been created by uplift over local relief in the subducting oceanic plate (Fig. 5.3) and is not a feature typical of the accretionary process.

The internal structure of the prism is generally not well imaged deeper than about 1 km below the seafloor, but a series of deeper discrete reflections indicate a broad tectonic structure. Folded strata are imaged across the approximate upper 1 km of the prism, and the flanks of some of these anticlines can be locally followed for several kilometers depth into the prism. Interestingly, several thrust anticlines have a landward vergence that is particularly clear in the frontal part of the prism (~ km 30–45, Fig. 5.3). Several landward-dipping reflections seem to cut across the prism and could represent out-of-sequence thrusts that thicken the prism, maintaining critical taper, leading to the final 6–8 km total thickness across much of the accretionary prism.

The fore-arc basin is some 30 km wide with deformed strata abutting and overlying the accretionary prism, and with undeformed strata onlapping the margin basement (~ km 60–90, Fig. 5.3). A similar fore-arc basin and basement configuration has been imaged along much of the margin (Polonia et al. 1999, 2001). The deformed, young, fore-arc sediment (e.g. km 60–65) indicates that thickening of the accretionary prism has continued until recently. In contrast, the subhorizontal and undeformed strata onlapping the margin basement indicate that the basement structure formed prior to the fore-arc basin and, thus, prior to the accretionary prism.

The basement has a characteristic seaward-dipping top surface defined by a very smooth reflection to ~4 km depth. There, it is faulted in a series of segments, several kilometers long, stepping down to about 7 km depth, where it is no longer imaged (Fig. 5.3). The basement probably consists of granitoids from the Patagonian batholith and remnants of a Paleozoic-Mesozoic subduction complex forming a seaward-dipping backstop to the accretionary prism (Polonia et al. 1999; Rubio et al. 2000). The remarkably smooth top of the basement suggests that it has been formed by erosion, either by glaciers or, perhaps, in a surf zone, and has subsequently subsided to current depths. The accretionary prism and fore-arc basin probably formed after the subduction of the spreading center in this area at ~14 Ma (Polonia et al. 1999).

The subduction of the Chile Rise has been widely associated with enhanced tectonic erosion and the destruction of the frontal part of the margin in the collision area (Cande et al. 1987; Bangs et al. 1992; Behrmann et al. 1994; Bourgois et al. 1996, 2000; Bangs and Cande 1997; Behrmann and Kopf 2001; Lagabrielle et al. 2000). The top of the basement was possibly smoothed by sub-aerial erosion, as mentioned above, during uplift of the margin over the subducting spreading center with subsequent collapse. The total subsidence displayed by the erosional unconformity is probably the result of both northward migration of the buoyant lithosphere of the spreading center and tectonic erosion at the base of the overriding plate.

### 5.2.2 Structure near the Triple Junction

Line RC2901-769, located about 100 km south of the triple junction, shows the structure of the margin where the Chile Rise collided with the continent at ~3–6 Ma (Fig. 5.4). It can be compared with the structure of line 790 to further understand the effect of spreading center subduction and the subsequent evolution of the margin.

Multibeam bathymetry shows the pronounced difference in margin structure north and south of the triple junction (Figs. 5.4 and 5.5; Bourgois et al. 2000). Turbidite sedimentation is diverted north and south from the collision point owing to topographic gradients. North of

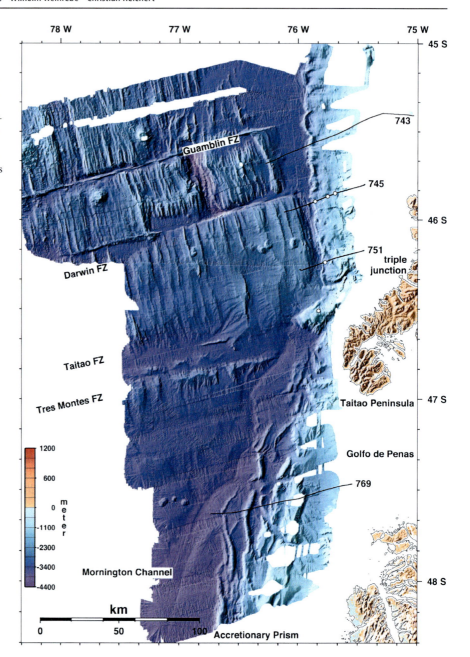

Fig. 5.4.
Shaded relief map from a compilation of multibeam bathymetry of the area around the triple junction. Most of the data were collected with RV Atlalante using the Simrad EM12 system (Bourgois et al. 2000). Some coverage gaps have been filled with data collected with a Seabeam system by RV R. Conrad (Bangs et al. 1992). *Black lines* are tracks of seismic reflection profiles collected during cruise RC2902. *Filled circles* are ODP-141 sites (Behrmann et al. 1992). *FZ*: fracture zone

the triple junction, the margin is characterized by a small, Plio-Quaternary, accretionary prism with much of the older material removed by tectonic erosion (Bangs et al. 1992; Behrmann et al. 1992, 1994; Bourgois et al. 2000). South of the triple junction, the trench is flooded with turbidites where the Golfo de Penas accretionary prism is developing. The accretionary prism widens abruptly to the south as the deformation front migrates seawards and a broader zone of trench turbidites is involved in the deformation (Cande and Leslie 1986; Bourgois et al. 2000).

Line 769 runs from the trench axis to the edge of the continental shelf, across an area where the deformation zone broadens, and it offers a picture of the ongoing rebuilding of the accretionary prism after the subduction of the Chile Rise (Fig. 5.6). This line images a transect of the margin some 92 km long, the same length as line 790 in southernmost Chile. Even though the same tectonic units are present in both line 769 and line 790, the topography of the seafloor is very different and structure indicates a different stage of margin development.

The area of accretion from the landward-most accreted sediment to the current deformation front is some 57 km wide (~ km 30–87, Fig. 5.6). Two apparent accretionary ridges are currently developing at the trench with the thrust anticlines centered approximately at km 70 and km 86, and with little intervening vertical deformation (1 and 2 in

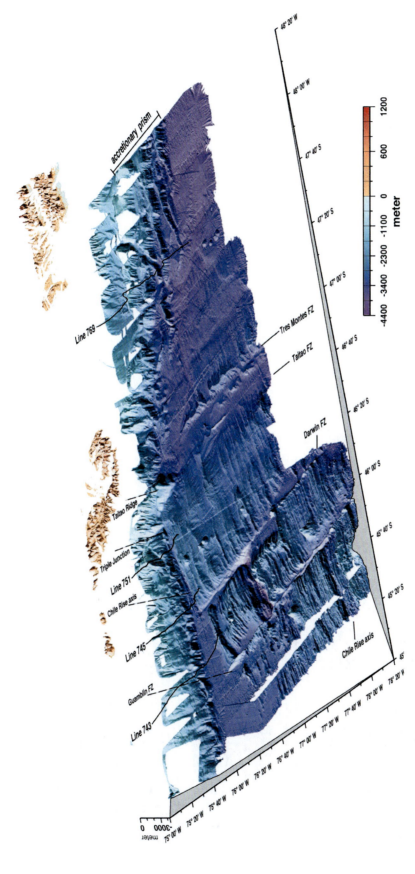

**Fig. 5.5.** Perspective view of shaded-relief bathymetry of the area around the triple junction, also shown in Fig. 5.4. Note the shallow bathymetry where the axis of the Chile Rise is colliding with the continent. Turbidites are diverted north and south. Towards the south is the area where the Chile Rise has already subducted, and the Golfo de Penas accretionary prism is developing, indicated by anticline ridges across a trench filled with turbidites (Bourgois et al. 2000). The trench north of the collision zone of the Chile Rise axis also contains a wide area with turbidites, but no large accretionary prism is developing. *FZ*: fracture zone

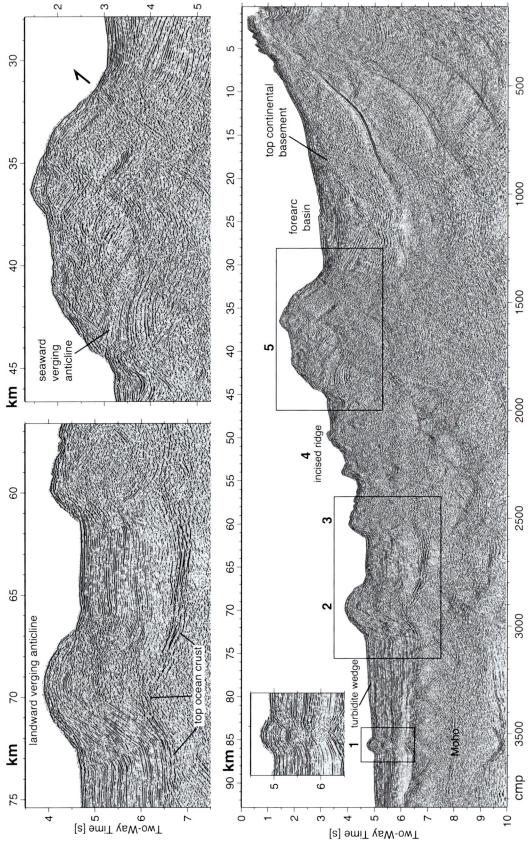

Fig. 5.6. Post-stack time migration of RC2902 line 769 across the Golfo de Penas accretionary prism (see Fig. 5.4 for location). The area of accretion is ~60 km wide. Note the general landward increase in dimension of the accretionary ridges. *Ridges 1 and 2* display a clear, landward vergence. *Ridge 5* seems to have overthrusted the older sediment of the fore-arc basin and then backthrust by out-of-sequence faulting, thus thickening the prism. The fore-arc basin has a configuration similar to that imaged for southernmost Chile in Fig. 5.3. *M* stands for multiple of the water layer; *cmp*: common mid point gather

Fig. 5.6). The first small ridge at km 86 is probably young and thrusting is incipient, but it seems to involve nearly the entire turbidite section. The anticline in the seismic image is towards the northern end of a bathymetric ridge that progressively decreases in size from south to north (Figs. 5.4 and 5.5). Thus, the image may show the area where thrusting is propagating northwards. Even though little overthrusting of strata has yet occurred, the thrust anticline has begun to verge landwards (inset of ridge 1 in Fig. 5.6).

The second ridge, centered at about km 70, displays three to four kilometers of shortening (measured on the folded turbidite strata that were horizontally deposited), and clear landward vergence, similar to some of the anticlines in line 790. The sediment section thickens from about 1.5 km at ridge 1 to some 2.5 km at ridge 2. Subhorizontal strata in the trench also thicken gradually landwards, probably by a combination of progressive horizontal shortening by small-scale deformation and recent turbidite fan deposition that thins seawards. One such deep-sea fan deposit appears at the top of the sediment sequence as a wedge-shaped unit that thins from ridge 2 to ridge 1.

Landwards of the two ridges are intervening areas of subhorizontal strata in a zone of increasing relief and complex deformation (ridges 3 to 5, ~ km 30–62 in Fig. 5.6). The bathymetry shows that these structures correspond to three increasingly higher and broader, margin-parallel ridges that are cut by large canyons. The flanks of ridges 3 to 5 are incised by gullies (Figs. 5.4 and 5.5), indicating that these ridges were formed during an older period of deformation than ridges 1 and 2. The internal structure of these older ridges is poorly imaged compared to the younger ridges, and this is perhaps owing to greater deformation. Both the width and height of the older ridges progressively increase landwards, a pattern observed across the entire accretionary prism (Figs. 5.4, 5.5 and 5.6). We interpret this gradual thickening to be a result of accretion and postulate that it may eventually lead to the formation of structure similar to that observed in southernmost Chile along line 790.

The fifth ridge is the largest and it exhibits a structure that indicates the mechanism of progressive prism thickening. A seaward-dipping reflection on the landward side of the ridge may correspond to overthrusting of the fore-arc basin, where older strata are tilted. An anticlinal structure is imaged in the seaward part of the ridge (km 40–45 at 2.5–4.5 s, Fig. 5.6), and the small basin between the fourth and fifth ridge (km 45–49, Fig. 5.6) shows synsedimentary deformation and shortening; both observations indicating seaward thrusting of the ridge. These structures indicate that the ridge has been uplifted and the prism thickened by both landward-verging thrusts, similar to the first and second ridges, and the subsequent, out-of-sequence, seaward-verging thrusts. This style of deformation is reminiscent of the thickening mechanism discussed above for the accretionary prism of southernmost Chile along line 790.

The fore-arc basin contains two main sedimentary units that are separated by an angular unconformity overlying a seaward-sloping, basement wedge. The oldest unit is tilted landwards and the younger upper unit gently onlaps the basement and ridge 5, indicating that deformation in this area is currently minor and has migrated seawards. The basin floor is a smooth surface that progressively deepens from some 400 m at the shelf break to roughly 4.5 km depth at the deepest part of the basin, where it is not further imaged. The fore-arc basin floor is very similar in character to the basin floor of southernmost Chile (cf. Figs. 5.3 and 5.6). Here, we also interpret that the smooth top of the basement is an erosional unconformity that formed at shallow depth when the area was uplifted as the buoyant lithosphere of the Chile Rise subducted beneath it. We also think that the current seaward-sloping attitude is due to both the northward migration of the shallow spreading center and the tectonic erosion of the base of the overriding continental plate.

The processes active during spreading center subduction can be studied in the segment north of the triple junction. Here, the high-relief flank of the ridge is subducting under the margin, and the rift valley of the spreading center forms currently the trench axis (Figs. 5.4 and 5.5). In this sector the continental margin shares the structural character of the segments to south where the spreading center collided in the past. However, the structural elements appear to be in an earlier phase of development caused by the ongoing collision with the spreading center.

Line RC2901-745 is located about 40 km north of the triple junction (Fig. 5.4), imaging the structures resulting from the ongoing collision of the spreading center and the continental margin (Bangs et al. 1992). The dimension and structural character of the ~20 km wide, seaward-sloping, top of continental basement imaged on line 745 (Fig. 5.7) are similar to the basement imaged under the continental slope south of the triple junction (Figs. 5.3 and 5.6). The trench contains turbidites some 100 m thick and the accretionary prism is roughly 12 km wide and formed during the Pliocene, probably in a single short episode (Behrmann et al. 1994).

It has been proposed that this segment of the margin is currently in a non-accretionary or erosional stage (Bangs and Cande 1997). We suggest that the subduction of one flank of the spreading center in this area has not only removed part of the pre-existing accretionary prism (Behrmann et al. 1994, Bourgois et al. 2000) but is also actively eroding the base of the continental plate across an area at least 20 km wide. Once the spreading center has migrated northward, the eroded basement may become the backstop to a large accretionary prism and the floor of a fore-arc basin similar to the structure imaged south of the triple junction.

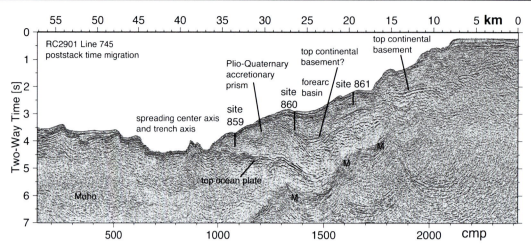

**Fig. 5.7.** Post-stack time migration of RC2902 line 745 where the Chile Rise flank is subducting (see Fig. 5.4 for location). The recent accretionary prism is ~20 km wide. The top of the basement is a smooth reflection similar to line 769 (Fig. 5.6) and 790 (Fig. 5.3). *M* stands for multiple of the water layer; *cmp*: common mid point gather

In summary, along most of the margin and seawards of the continental shelf, the three main tectonic units are: a continental basement with a seaward-dipping, smooth, upper surface; the fore-arc basin; and the zone of accretion. The structural configuration observed in the two seismic lines south of the triple junction share similar structural characteristics (Figs. 5.3 and 5.6). The width of the continental basement's upper surface typically ranges between 20 and 35 km in the two seismic profiles we analyzed and in five other lines in the area (Fig. 5.1; Polonia et al. 2001). The accretionary prism is 50–60 km wide along most of the margin. And an intervening fore-arc basin is present above the contact of the accretionary prism and continental basement. We interpret the configuration of the margin basement as caused by tectonic erosion during spreading center subduction and the accretionary prism created during subsequent margin rebuilding. The structural configuration of the margin shaped by ongoing subduction of the spreading center (Fig. 5.7) supports this interpretation.

A somewhat different structural configuration occurs along the margin between 51° S and 53° S, where the strike-slip Magellan fault system extends offshore across the shelf, and margin tectonics may be influenced by slip along those large faults (Polonia et al. 2001).

## 5.3 The Tectonic Signature of the Subduction of the Juan Fernández Ridge

Another large subducting structure that has fundamentally influenced margin tectonics for at least the past 20 million years is the Juan Fernández Ridge (Yañez et al. 2001, 2002). The Juan Fernández Ridge is a hot spot chain formed by intraplate volcanism ~900 km west of the trench (Fig. 5.1). The chain of aligned seamounts trends ~85° E and a gap between volcanic edifices, some 250 km wide, separates the younger seamounts from seamounts near the trench, where the ridge changes strike to a NE direction (Figs. 5.1 and 5.8).

The Juan Fernández Ridge influences margin tectonics in two main ways. Firstly, the ridge produces extensive deformation where it collides with the continent (von Huene et al. 1997). Secondly, the ridge is a topographic barrier to the transport of sediment along the trench axis. The ridge separates a heavily sedimented trench to the south that contains more than 2.5 kilometers thick turbidites, and extends almost to the triple junction, from a trench north of 32.5° S with turbidite infill of less than 1 km thickness (Schweller et al. 1981). The volume of trench infill away from ridges seems to fundamentally govern tectonic style along the margin.

Several surveys of multibeam bathymetry across the incoming plate and continental margin have resulted in a high-resolution map of morphotectonic and sedimentary structure showing different tectonic domains resulting from ridge subduction. The bathymetric data in Figs. 5.8 and 5.9 include data collected with the Hydrosweep system during legs 1, 2 and 3 of cruise Sonne 101 (von Huene et al. 1997) as well as legs 2 and 3 of cruises Sonne 103 and Sonne 104, and data collected with the Simrad EM12 system during Sonne 161. The previous data gaps between 32 and 34° S were smoothed through interpolation (von Huene et al. 1997; Laursen et al. 2002). The new bathymetry includes new data collected over more than three months, which, after data cleaning with Mbsystem (Caress and Chase 1996), permits gridding at 90 m rather than the previous 220 × 300 m. This has improved resolution and has covered a larger area. Four seismic profiles (Figs. 5.10–5.13) show the structure beneath the continental slope and trench axis in areas that were affected during different stages by the southward migration of the colliding Juan Fernández Ridge.

## 5.3.1 The Oceanic Plate

Seawards of the trench axis, the oceanic plate is thinly covered by ~100 m of pelagic sediment. Consequently, the multibeam bathymetry shows well the morphology of the igneous oceanic crust (Figs. 5.8 and 5.9). The topography of the oceanic plate in the vicinity of the seamount chain is distinctly different from the surrounding plate (Figs. 5.8 and 5.9). The oceanic plate that is not influenced by hot spot magmatism in the north displays a NW-SE-oriented, seafloor-spreading fabric of narrow ridges. These ridges are cut along faults, that roughly parallel the trench, forming during bending prior to subduction. In contrast, the area affected by hot spot magmatism almost completely lacks the NW-SE-oriented seafloor-spreading fabric, but exhibits numerous small volcanoes in addition to the two large edifices of the O'Higgins Seamounts and a NW-SE-elongated ridge closer to the trench. Seismic images show thicker, upper crustal, volcanic layering at the crest of the hotspot ridge (Fig. 5.13). All these features indicate that hot spot volcanism has obliterated most pre-existing seafloor-spreading fabric.

The hot spot-modified region also displays different bending behavior at the trench. There, bending-related faulting mostly strikes obliquely to the trench and paral-

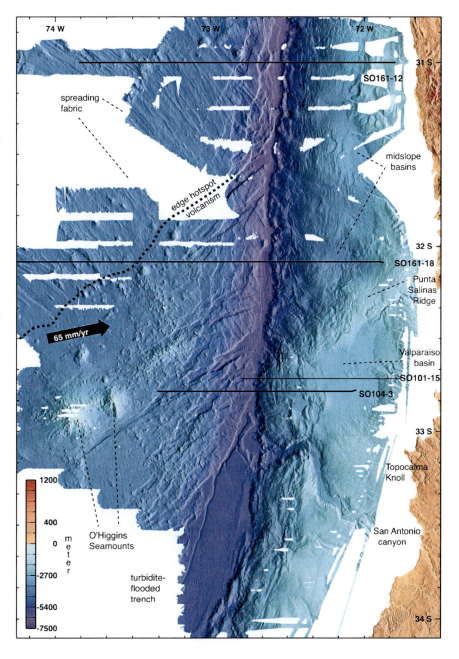

**Fig. 5.8.**
Shaded-relief bathymetry from central Chile illuminated from the east. *Black lines* are seismic tracks. The Juan Fernández Ridge is a wide (~150 km), gentle swell crested by the O'Higgins Seamounts and spotted by many small volcanoes. As it bends, it breaks along NE-trending faults. The plate unaffected by the hot spot has a seafloor-spreading fabric of narrow NW-SE ridges and breaks along trench-parallel faulting. The trench south of the ridge is flooded with turbidites and starved northwards. The flooded and starved trenches face, respectively, an accretionary prism and a depressed continental slope cut by normal faults. The Valparaiso Basin is bounded seawards by an extending, inactive, accretionary prism. The Punta Salinas Ridge is formed by uplift over large subducting seamounts. The ridge trends ~20° to the convergence vector (*black arrow*, DeMets et al. 1990) and is slowly migrating southwards

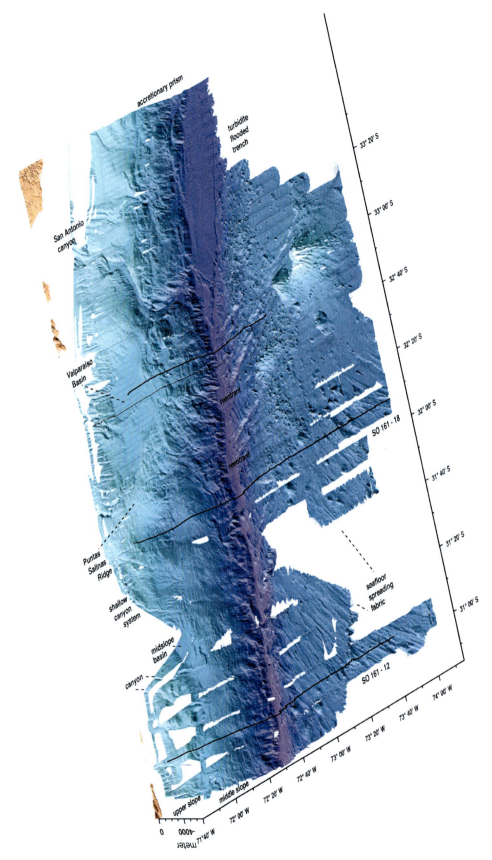

Fig. 5.9. Perspective view from 290° azimuth and 38° elevation of shaded-relief bathymetry from central Chile illuminated from 45° N. *Black lines* are seismic tracks. Note the change in slope depth and morphology along the margin. The crest of the hot spot ridge is located in front of the Valparaiso Basin. There, the front of the margin has been breached by seamounts and displays a fractured morphology. The seamounts are currently uplifting the Punta Salinas Ridge. North of the ridge crest, the trench is narrow with little sediment infill and the facing continental slope is depressed and extended by normal faulting. A flooded trench lies at the foot of an accretionary prism

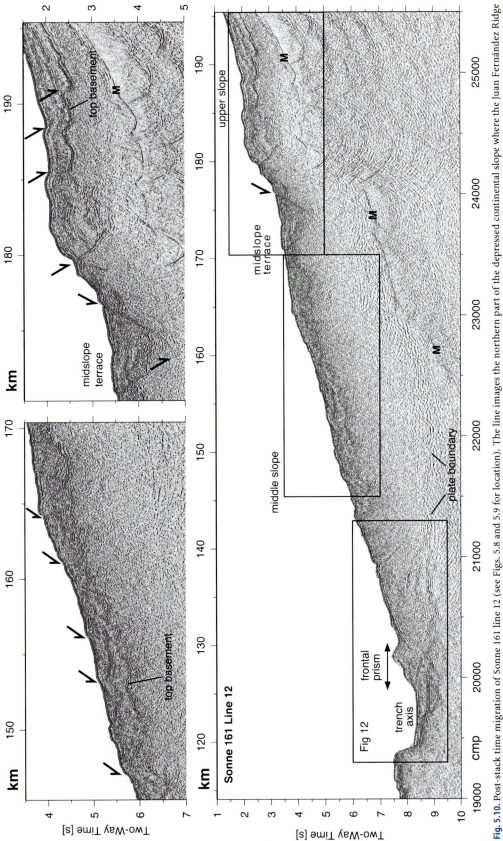

**Fig. 5.10.** Post-stack time migration of Sonne 161 line 12 (see Figs. 5.8 and 5.9 for location). The line images the northern part of the depressed continental slope where the Juan Fernández Ridge collided ~0.5 Ma, but where normal oceanic lithosphere currently subducts. The margin is fronted by a sediment prism ~5 km wide. The middle slope is covered by ~500 m of deformed sediment cut by numerous seaward-dipping faults. The change from middle to upper slope occurs abruptly across a scarp of a large seaward-dipping fault forming the mid-slope terrace. The upper slope is a less deformed terrain cut by landward-dipping faults. *M* stands for multiple of the water layer; *cmp*: common mid point gather

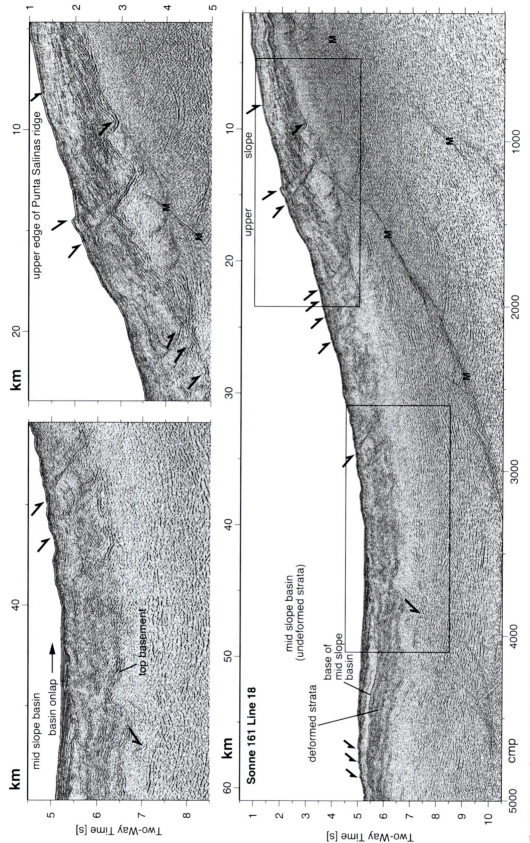

**Fig. 5.11.** Post-stack time migration of Sonne 161 line 18 (see Figs. 5.8, 5.9 and 5.14 for location). The image shows the structure of the middle and upper slope north of the Punta Salinas Ridge. The upper slope is cut by landward-dipping faults. The transition from upper to middle slope is gradual and partially covered by young sediment of a shallow mid-slope basin filling a terrace formed by the change from landward- to seaward-dipping faults. M stands for multiple of the water layer; *cmp*: common mid point gather

lel to the eastern trend of the hot spot chain, although minor trench-parallel faulting is locally observed (Fig. 5.8). Bathymetry shows that in the segment affected by hot spot magmatism, the bending-related faults, the alignment of large volcanic edifices, and the Punta Salinas Ridge crossing the continental slope, all strike obliquely to the trench and to the convergence vector (Fig. 5.8). Thus, the ridge is migrating southwards. The hot spot magmatism has influenced a swath of lithosphere, some 150 km wide, that subducts at ~55° to the trench axis from about 31.5 to 33.5° S. With the ~20° angle to the convergence vector (DeMets et al. 1990), it has migrated southwards some 375 km from 11 Ma to the present at a rate of 35 km Myr$^{-1}$ (Yañez et al. 2001).

### 5.3.2 Continental Margin Segmentation

The ridge has long been a barrier to the axial transport of turbidites along the trench (Fig. 5.8). The combined effect of trench-sediment distribution, shallow subduction angle, ridge relief on the oceanic plate, and southward ridge migration fundamentally controls margin tectonics. Clearly, the character of the incoming plate and trench infill are paralleled by the pattern of tectonic structures along the continental margin. We analyzed the tectonic structure and evolution of the margin using a new compilation of multibeam bathymetry (Figs. 5.8 and 5.9) and three seismic sections that image different stages of deformation related to ridge collision.

The continental slope bathymetry shows five areas with different morphotectonic structure: (*1*) the depressed continental slope north of the Punta Salinas Ridge; (*2*) tectonic uplift over the ridge; (*3*) the deep Valparaiso fore-arc basin; (*4*) the lower slope seawards of the Valparaiso Basin; and (*5*) the elevated margin south of the Topocalma knoll.

Seismic line SO 161-12 crosses the depressed slope to the north of the Punta Salinas Ridge, a prominent uplift across the continental slope formed by the underthrusting Juan Fernández Ridge, and shows the structure of the margin after ridge collision at ~3–4 Ma (Figs. 5.8 and 5.9). It shows a narrow trench axis, turbidite infill some 800 m thick, and margin structures that formed during and after ridge collision (Fig. 5.10). Seismic line SO 161-18 runs just north of the Punta Salinas Ridge and the landward-most segment of this line possibly images part of the uplifted area (Figs. 5.8 and 5.9). Line 18 images margin structure in an area where hot spot-influenced lithosphere is presently underthrusting, but high seamounts are subducting a short distance south under the Punta Salinas Ridge (e.g. Papudo Seamount). Thus, line 18 shows the frontal structure of the margin in a stage where ridge-related deformation is active, and in the recent past was affected by seamount subduction. Any subducted seamount would currently be un-

der the shelf break at the NE extension of the Punta Salinas Ridge (Figs. 5.8 and 5.11).

The third transect is formed by two seismic lines that image the structure from the upper slope to the ocean plate (Figs. 5.8 and 5.9). SO 101-15 is a high resolution, but low penetration, image showing structure across the shallow part of the Valparaiso fore-arc basin, but it images little structure under the outer slope. SO 104-3 images the structure of the seaward segment of the basin, under the lower continental slope and the trench axis (Fig. 5.12).

### 5.3.3 The Depressed Continental Slope

The area north of the Punta Salinas Ridge has a narrow trench axis and a steep continental slope divided into three margin-parallel domains with distinct morphological character: The base of the lower slope is formed by narrow ridges; the middle slope is steeper but contains a terrace in the shallower portion; and the upper slope is more coherently structured (Figs. 5.8 and 5.9).

One or two narrow, base-of-slope ridges meander along the trench and are only disrupted locally by subducting, high-relief features on the oceanic plate. The 5–10 km wide swath of narrow-ridge morphology terminates either abruptly upslope at a sharp, trench-parallel fault or less abruptly at smoother, steep terrain with numerous faults (Figs. 5.8 and 5.9). Seismic lines SO161-12 and -18 show a ocean plate that plunges more steeply than in the south at line SO104-3, and a narrow trench axis filled with turbidites some 500–800 m thick underthrust beneath the ridges at the base of slope (Fig. 5.13).

A locally bright, décollement reflection occurring beneath the ridges at the base of slope continues updip to the trench seafloor, indicating that little, if any, turbidite is frontally accreted (Fig. 5.13). Downdip, the image is less clear; the turbidites seem to rapidly change thickness. In line 12, the interpretation is obscured by changes in turbidite thickness over the subducting-plate topography. In line 18, smoother ocean-plate topography reveals a thick package of underthrust turbidites, indicating both little frontal accretion or underplating. At the transition from the ridges to the backstop, possible mud diapirs stand over the surrounding seafloor (Fig. 5.13). Compaction-driven, fluid flow sourced from the pore fluid of underthrusted strata may form these diapirs as it escapes along the contact between the frontal prism and the backstop.

We interpret the narrow ridges as having formed either by limited accretion or by the reworking of slope debris. Similar structures at the toe of the slope have been imaged in other thinly sedimented trenches, such as the Middle America Trench (Ranero et al. 2000; Ranero and von Huene 2000; von Huene et al. 2000; Ranero et al., in press.), where all incoming sediment seems to be subducted. Drilling results from offshore Costa Rica docu-

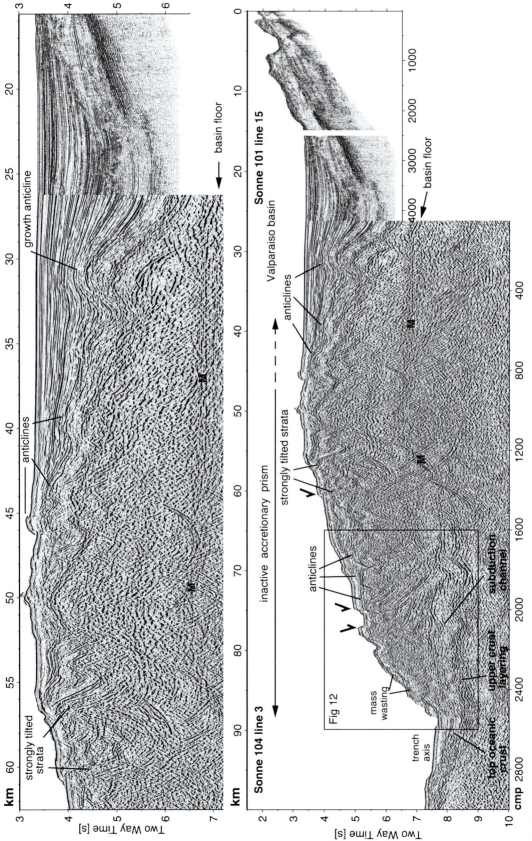

**Fig. 5.12.** Post-stack time migration of Sonne 101 line 15 and Sonne 104 line 3 (see Figs. 5.8, 5.9 and 5.14 for location). The image shows the structure across the Valparaiso Basin and the inactive accretionary prism. The basin contains a large, fault-propagating anticline that documents, along with several smaller anticlines and a syncline forming the depocenter, shortening and the long-term uplift of the seaward flank of the basin. The frontal part of the margin shows accreted sediment in thrust slices. However, the seaward-sloping prism, cut by normal faults, indicates that accretion is not active and the prism is extending and collapsing owing to basal erosion that supplies material to a thick subduction channel. *M* stands for multiple of the water layer; *cmp*: common mid point gather

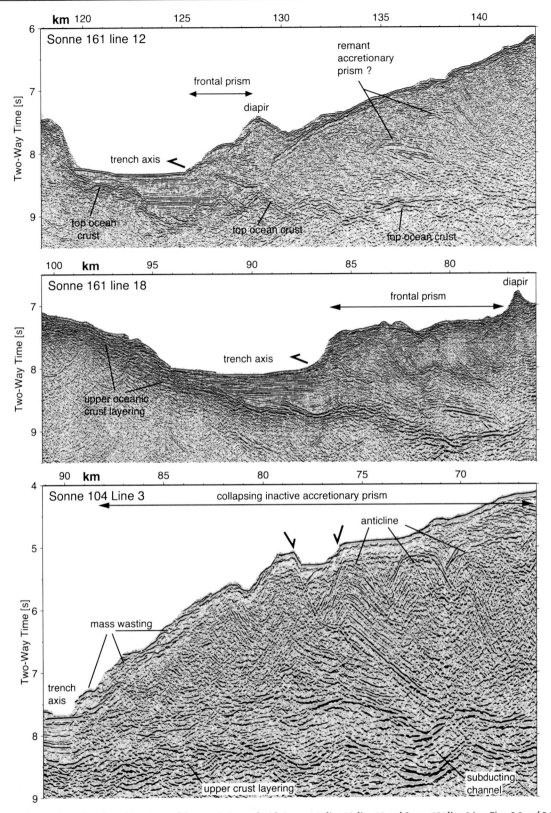

**Fig. 5.13.** Comparison of the frontal structure of the margin imaged with Sonne 161 line 12, line 18 and Sonne 104 line 3 (see Figs. 5.8 and 5.9 for location). Note the difference between the depressed slope structure (lines 12 and 18) and the margin front seawards of the Valparaiso Basin (line 3), where the ridge has more recently begun to underthrust. *cmp*: common mid point gather

ment that the small, frontal, sediment prism there is formed by reworked slope material (Kimura et al. 1997). Similar ridges are also observed along the northern Chile margin (see below), which has a turbidite-starved trench (von Huene and Ranero 2003).

The arcuate reflections characterizing the rock body beneath the lower slope in line 12 (~ km 133–142, Fig. 5.13) may represent strata in thrust slices, and that the backstop rock to the frontal prism is a currently inactive, accretionary prism partially eroded by the subducting ridge. A lack of trench-subparallel ridges overlying the reflections, (Figs. 5.10 and 5.13) and widespread, normal faulting shows that accretion is not active here. The character of the frontal prism changes little in the ~180 km from the northern SO161-12 line to just north of the Punta Salinas Ridge where the Juan Fernández Ridge currently subducts. This observation further supports the assertion of little, if any, accretion along this segment.

Upslope of the frontal prism, the middle slope is characterized by comparatively smooth topography disrupted only by pervasive, small-offset, normal fault scarps cutting the seafloor (Figs. 5.9 and 5.10). Faulting is generally margin parallel and the bathymetry illuminated from the east shows that scarps face dominantly seawards (Fig. 5.8). The transition from the middle to the upper slope occurs across a gentle terrace, where the mid-slope dip gradually shallows (Figs. 5.9 and 5.10). The mid-slope terrace changes in width across the area: In the north it terminates abruptly upslope at one or two steep fault scarps, whereas in the south it steadily increases dip towards the flank of the uplifted Punta Salinas Ridge (Figs. 5.9 and 5.11). The mid-slope terrace locally contains small basins characterized by smooth, unfaulted seafloor (Figs. 5.8 and 5.9) filled with a few hundred meters of landward-tilted sediment that unconformably overlie a package of deformed strata more than 1 km thick (Fig. 5.11).

The terrain of the upper slope in the northern area is cut by canyons, with the largest reaching to the mid-slope terrace (Fig. 5.9). This terrain is also cut by landward-dipping faults that have offsets up to several hundred meters, much larger than the seaward-dipping faults in the mid slope (Fig. 5.10). Along line SO161-12, these scarps of landward-dipping faults increase offset progressively downslope, cutting the top of the basement and the seafloor (~ km 180–195, Fig. 5.10). Those faults in the uppermost slope offset the top of the basement by less than 100 m and have no clear expression on the seafloor. An abrupt topographic transition from the upper to the middle slope occurs across a scarp of some 700 m, formed by a large, seaward-dipping fault that marks a sudden change in the dominant fault dip (~ km 177, Fig. 5.10). Although the upper to middle slope transition is only partially mapped in this area, the abrupt seafloor scarps indicate an ubiquitous, fault-controlled transition (Figs. 5.8 and 5.9).

The image from line SO161-18 is unlike that of line 12 and might represent a different stage of margin evolution in the wake of the migrating Punta Salinas Ridge. The perspective view of the bathymetry shows shallow canyons starting to develop some 30 km north of line 18, whereas along line 18 they do not exist (Fig. 5.9). Line 18 also shows that the upper slope is cut by landward-dipping faults, and the largest of these form scarps at the seafloor (Fig. 5.11). The upper slope shows a generally steeper dip compared to the northern area, but the transition to the middle slope is not a sharp fault scarp (Fig. 5.9). Seismic images show that fault dip changes from a landward to a seaward direction beneath the mid-slope terrace, but this change is partially masked in the bathymetry because a few hundred meters of largely undisturbed Recent sediment covers the older and faulted strata (~ km 40–50, Fig. 5.11). Recent sediment structure shows units separated by unconformities with depocenters shifting upslope, indicating that slope subsidence is progressively migrating landwards.

We interpret the structure of the northern margin segment, characterized by a tectonically extending upper plate, as both the result of the progressive, southward migration of the Juan Fernández Ridge and subduction erosion where the thickness of turbidite infill is reduced to less than 1 000 m and confined to a narrow axial zone. The observations described above indicate that subduction erosion affects the entire continental slope. We now discuss ridge subduction and margin evolution along a trench axis that is both thinly and heavily sedimented.

### 5.3.4 Seamount Subduction and Uplift of the Punta Salinas Ridge

The Punta Salinas Ridge is the most obvious morphological expression of the crest of the subducted Juan Fernández Ridge beneath the continental slope (Figs. 5.8 and 5.9). Several observations support the interpretation that the Punta Salinas Ridge has formed by uplift over the crest of the hot spot ridge (von Huene et al. 1997). A circular, magnetic anomaly corresponding to the subducted Papudo Seamount, of similar size to the O'Higgins edifices, is oriented slightly en échelon to the landward crest of the Juan Fernández Ridge (Yañez et al. 2001). The scar marking where the Papudo Seamount collided with the margin is visible as a reentrant in the lower slope (von Huene et al. 1997; Fig. 5.14). The subducted seamount uplifts the seaward end of the Punta Salinas Ridge, which continues across the entire slope at about 32° 10', where the continental shelf seems to extend seawards and a canyon system is lacking (Figs. 5.8 and 5.9).

Uplift of the uppermost slope is indicated by a large canyon, at the northern end (~33°20') of the Valparaiso Basin, that has been diverted southwards (Figs. 5.8 and 5.14). Subducting seamounts can leave a prominent furrow

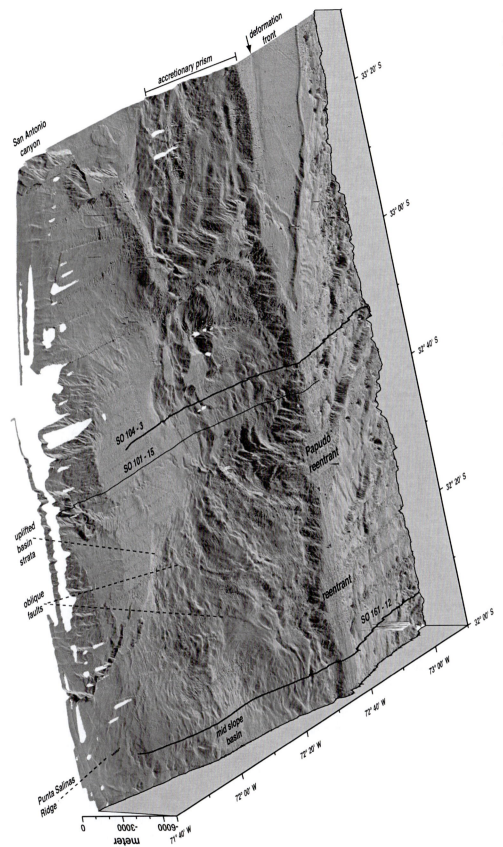

**Fig. 5.14.** Perspective view from 290° azimuth and 38° elevation of shaded-relief bathymetry illuminated from 45° N. The data shows the area where the crest of the Juan Fernández Ridge has recently begun subducting. Note the deformed accretionary prism seawards of the Valparaíso Basin and the undisturbed accretionary prism south of the San Antonio Canyon, where the trench is flooded with turbidites

in their wake, as observed across the upper plate offshore from Costa Rica. At Costa Rica, the overriding plate is made of igneous rock that has collapsed behind a subducting seamount along faults parallel to the subduction path. The large grooves, so formed, indicate tectonic removal of material from the base of the upper plate (Ranero and von Huene 2000; von Huene et al. 2000). However, offshore Chile, the frontal part of the margin is made of accreted sediments that deform plastically around the subducting seamounts and have not formed pronounced grooves.

The Papudo Seamount is currently under the Punta Salinas Ridge along the flank of the Valparaiso Basin, but the scar from its subduction is only visible in the lower slope. Another structure, possibly related to a subducting seamount, is observed about 40 km north of the entry point of the Papudo Seamount. There, the toe of the margin slope is indented and the upper plate shows two parallel faults separated by ~15 km and striking at high angle to the trench (Figs. 5.8 and 5.14). These oblique faults are unique along the margin and they extend, albeit discontinuously, from near the trench axis to the summit of the Punta Salinas Ridge. The faults may represent deformation related to a subducting seamount currently uplifting the Punta Salinas Ridge. Landwards of the faults, young sediments of the Valparaiso Basin are uplifted against the Punta Salinas Ridge, indicating deformation of the upper plate related to ridge migration.

In summary, the morphology of the Punta Salinas Ridge indicates that ridge-crest subduction affects the entire width of the continental slope. Deformation of the margin produced by relief on the subducting plate may be increased by the growth of the bending-related faulting in the oceanic plate after subduction (von Huene et al. 1977). Faulting of the ocean plate parallel to the Juan Férnandez Ridge starts some 100 km away from the trench with fault offset increasing trenchwards to > 1 km (Figs. 5.8, 5.9 and 5.14). Bend faulting probably remains active underneath the continental plate, leading to increased relief in the subducting plate and enhanced deformation of the overriding plate.

### 5.3.5 Valparaiso Basin, the Collapsing Lower Continental Slope and the Change from Thin to Thick Trench Infill

South of the Punta Salinas Ridge, the Valparaiso Basin, a morphologically distinct lower continental slope, and the narrowest trench axis differentiate the area of the collision of the crest of the Juan Fernández Ridge (Fig. 5.8). The Valparaiso Basin is a deep-water, fore-arc basin flanked to the north by the Punta Salinas Ridge and to the south by the Topocalma knoll, a basement high covered by thin sediment. Seawards, the basin is bounded by accreted thrust slices of turbidites. The effect of ridge subduction on the margin, and its relationship to basin formation, are controversial. The Juan Fernández Ridge might have recently migrated into this area, deforming the basin along the southern flank of the Punta Salinas Ridge (von Huene et al. 1997; Yañez et al. 2001). Alternatively, a quasi-stationary collision zone for the hot spot ridge over the past ~10 million years has deformed the margin and formed the Valparaiso Basin (Laursen et al. 2002).

The lower slope fronting the Valparaiso Basin rises above the segment of depressed eroded slope to the north. The structure in the seismic images is notably different from that to the north, and the frontal ~25 km shows thrust anticlines and the folded strata of an accretionary prism (Fig. 5.13). However, the structure and morphology seem to indicate that thrusting is waning (von Huene et al. 1997).

The steep, frontal scarp along the base of slope (Fig. 5.14) has formed by mass wasting, and is imaged as detached, seaward-slipping, sediment packages (Fig. 5.13). Margin-parallel lineaments across the lower slope, that might have been accretionary ridges as in the south, are currently cut by local, margin-perpendicular lineaments. Some fault scarps indicate extensional faulting (Figs. 5.8, 5.13 and 5.14).

The subduction channel under the prism is thicker than in the images from the north, despite thinner trench infill, which may indicate basal erosion of the prism. The steep topography of the accretionary prism (Fig. 5.14) and the observations discussed above indicate an extending and collapsing margin front, probably owing to basal tectonic erosion above the extension of the Juan Fernández Ridge crest. The ocean plate south of the ridge crest is covered by thick sediment at the trench and the accretionary prism is intact. This contrast indicates that tectonic erosion of the accretionary prism in this segment is probably a recent process associated with the progressive, southward migration of the ridge and that the change from accretion to subduction erosion along the trench occurs abruptly near the ridge crest.

The width of the accretionary prism and its relationship to the formation of the Valparaiso Basin has been debated. Using high-resolution seismic reflection data, Laursen et al. (2002) interpreted the lower slope as deformed continental basement and postulated that any pre-existing accretionary prism has been removed by tectonic erosion. In contrast, modeling of a two-dimensional wide-angle profile coincident with line Sonne 104-3 found velocities consistent with accreted material extending some 40 km landwards of the trench (Flueh et al. 1998). Further, modeling of a three-dimensional wide-angle data set extends the accretionary prism to ~50–60 km from the trench, beneath the seaward flank of the Valparaiso Basin (Zelt et al. 1999). Alternatively, low velocities could represent continental rock that is strongly deformed and invaded by fluids, but fast horizontal velocity gradients from < 4.5 km s$^{-1}$ to > 5.5 km s$^{-1}$ support a change in rock type under the seaward flank of the basin (Flueh et al. 1998; Zelt et al. 1999).

Magnetic data similarly indicates a rock-type boundary underneath the seaward part of the Valparaiso Basin (Yañez et al. 2001). Seismic reflection profile Sonne 104-3 only shows clear accretionary structures in the frontal ~25 km of the margin; farther landwards, only the upper 1–2 km beneath the slope is well imaged (Figs. 5.12 and 5.13). The lack of accretionary structures in the seismic record might be an imaging problem caused by the steepening and deformation of accreted slices by out-of-sequence thrusting during prism thickening, similar to the structures observed in southernmost Chile (Figs. 5.3 and 5.6). Several packages of steeply tilted strata, roughly occurring in the upper 1.0 to 1.5 km, overlaid by less deformed sediment possibly indicate steep thrusting and prism thickening (~ km 55–60, Fig. 5.12).

Seismic line Sonne 104-3 images the seaward flank of the Valparaiso Basin and high-resolution line Sonne 101-15 shows the landward flank (Figs. 5.8 and 5.12). The basin depocenter is a narrow syncline, about 10 km wide, centered 65 km from the trench axis. The basin infill structure is asymmetric; the landward basin flank is covered by seaward-dipping, undeformed strata and the seaward flank is formed by landward-dipping, folded strata. Deformation of the strata indicates that thrusting and folding progressed in time from the center to the seaward edge of the basin. Near the basin depocenter, the oldest strata display growth folding related to deep thrusting, but the youngest strata show little deformation (km 25–30, Fig. 5.12).

Basin infill thins seawards with strata gently dipping landwards, indicating progressive uplift in this area (~ km 35–55, Fig. 5.12). Basin formation has been attributed to tectonic erosion focused under the middle slope, about 60 km from the trench (Laursen et al. 2002). However, the northern, depressed slope segment, where tectonic erosion is active, shows pervasive, extensional faulting cutting the seafloor and little evidence for contractional structures, as observed in the Valparaiso Basin where no extensional faults are observed. Tectonic erosion elsewhere produces upper plate thinning and subsidence at the frontal part of the margin with subsidence progressing inboard (Ranero and von Huene 2000; Vannucchi et al. 2003, 2004; von Huene and Ranero 2003; von Huene et al. 2004). The Punta Salinas Ridge indicates that seamount subduction causes uplift across the entire slope that has not deformed the Valparaiso Basin.

Preserved accretionary structures seawards of the Valparaiso Basin (Fig. 5.13) indicate that tectonic erosion has recently started thinning the lower slope. Therefore, the younger sediment in the Valparaiso Basin might have been deposited at the back of a growing accretionary prism against a continental backstop, similar to the basins south of the triple junction. Progressive shortening and tilting of basin strata, and seaward migration of the deformation, may be explained by a growing and thickening accretionary prism. In contrast, mid-slope basins occurring at the tectonically eroded, depressed slope segment to the north are small and related to extensional faulting. Their subsidence is limited to a few hundreds of meters and they do not show contractional structures (Fig. 5.11).

### 5.3.6 The Accretionary Segment

South of the mouth of the San Antonio Canyon, the trench axis rapidly widens to ~40 km and is flooded by turbidite up to 2.5 km thick, owing to the oceanic plate bending deeper into the trench axis than at the Juan Fernández Ridge(Figs. 5.8, 5.9 and 5.14). The change in plate dip along this segment indicates that the oceanic lithosphere buried under thick trench sediment has not been affected by the hot spot magmatism. The morphology of the continental slope contrasts with the northern segment. The frontal ~40 km of the slope are formed by a series of margin-parallel, en échelon ridges of variable dimensions but increasing height upslope (von Huene et al. 1997).

Active frontal accretion is inferred from the bathymetry, which shows a series of low ridges deforming the trench turbidites near the toe of the slope (Figs. 5.8 and 5.9). These ridges are also imaged in seismic records (von Huene et al. 1997; Laursen et al. 2002). A prism some 10 km wide has been imaged with high-resolution seismic data (Laursen et al. 2002). Deeper penetration, seismic reflection data show accretionary structures in the frontal 20 to 30 km (von Huene et al. 1997; Flueh et al. 1998) and a relatively low velocity body extends ~40 km landwards of the trench, with indications of sediment underplating further landwards (Flueh et al. 1998).

The ridges across the frontal 40 km of the margin probably represent thrust anticlines that have progressively thickened the margin landwards. The accretionary ridges are well displayed across the southern wall of the cross-cutting San Antonio Canyon. Some trends seem to extend across the canyon into the slope fronting the Valparaiso Basin, where basal tectonic erosion is currently thinning that inactive part of the accretionary prism (Fig. 5.14). Upslope from the en échelon ridges, the slope is covered by a blanket of sediment that smoothes the topography and is only disturbed by shallow canyons and shallow slide scars (Figs. 5.8 and 5.9).

### 5.3.7 A Margin Evolution Model for Central Chile

The contrasting seafloor morphology and internal structure of the different segments of the continental margin can be interpreted in terms of the southward migration of the collision zone of the Juan Férnandez Ridge and its control on the distribution of trench turbidites. This distribution, in turn, governs the dominant tectonic process of accretion or subduction erosion along a particular segment of the margin (Figs. 5.8 and 5.9).

The accretionary segment in the south exemplifies margin structure unaffected by the subduction of the ridge. It shows an undisturbed upper slope and an accretionary prism that has formed where a normal oceanic plate plunges at a trench buried under thick sediments. The edge of the hot spot ridge is marked by a rapid shallowing of the trench axis and an abrupt decrease in the sediment infill. The edge of the obliquely converging ridge is underthrusting the frontal part of the margin facing the Valparaiso Basin, where an inactive accretionary prism is thinned by basal erosion accompanied by upper plate extension and collapse. The most spectacular deformation produced by the ridge is where the ridge crest and seamounts tunnel under the continent and uplift the slope. The Punta Salinas Ridge results from subducted relief, documenting that seamounts cresting the hot spot ridge deform the entire continental slope.

North of the Punta Salinas Ridge, is the depressed slope segment with a distinct three-fold structure. We interpret extension of the middle and lower slope to indicate tectonic erosion. Some thinning may have resulted from the past subduction of seamounts that removed upper plate material during tunneling, as observed elsewhere (Ballance et al. 1989; Ranero and von Huene 2000).

The trench contains a slightly thicker sediment infill than observed at the crest of the ridge, indicating that some trench turbidites are transported over the ridge. However, the turbidites in the trench axis, at less than 1 000 m thick, are not sufficiently thick to reestablish accretion, and the small frontal sediment prism has a constant width along the 200 km of surveyed segment. Most, or all, of the trench turbidites are currently being subducted.

The middle continental slope is pervasively faulted by dominantly seaward-dipping faults, indicating upper plate extension that is caused, most likely, by basal tectonic erosion. The upper slope displays a more stable morphology with canyons that end at the mid-slope terrace, marking the transition between the middle and upper slope. No canyon system crosses the middle slope, where tectonism and mass wasting may be too rapid to allow the development of a drainage system.

The upper slope is characterized by landward-dipping faults, also indicating progressive collapse of this part of the margin. Normal faulting suddenly changes orientation from mainly dipping landwards across the upper slope to dipping seawards across the middle slope. The abrupt change occurs across a large, seaward-dipping, normal fault that forms a scarp bounding the mid-slope terrace. The seaward-dipping faults probably represent a second generation of faults cutting a terrain, previously extended by landward-dipping faults, as upper plate thinning progresses and the continental slope collapses because of basal tectonic erosion. In summary, the subduction of the hot spot chain caused pronounced deformation at the collision zone and the amount of trench turbidite infill dramatically changes the tectonic style along the margin as the ridge migrates southwards.

## 5.4 Tectonics of the Sediment-Starved, Northern Chile Margin

### 5.4.1 Updating the Classic End-Member, North Chile, Convergent Margin Model

Based only on seismicity and singlebeam echosounder bathymetry, the north Chile convergent margin was once the Chilean subduction zone end-member model (Uyeda and Kanamori 1979; Jarrard 1986). It was diagrammed with a growing accretionary prism despite reports of margin retreat indicated by an unroofed Mesozoic volcanic arc along the coast (cf. Miller 1970, Rutland 1971, Scholl et al. 1980). Offshore, refraction seismic data required the slope to have a thin sediment cover over the basement (cf. Fisher and Rait 1962), and a few cores had recovered Miocene sediment near the seafloor (Bandy and Rudolfo 1964). However, many authors favored an idealized end-member model because, at the time, continental growth along all convergent margins was commonly assumed.

Not until the 1990s, was the quality of new data sufficiently irrefutable to make a convincing counter argument. Coastal basement rock was dredged offshore (Kudrass et al. 1998) and the overlying Miocene to Holocene sediment was cored (Hartley and Jolly 1995). The evidence now overwhelmingly favors truncation of the Mesozoic continental crust by subduction erosion at the trench, but reversing years of citations to the contrary is slow.

In popular models of subduction erosion, a direct contact between the upper and lower plate basement along the mega-shear of the subduction zone is assumed (cf. Hilde 1983). The sediment layer in the subduction channel that reduces friction between the plates was inferred to be largely absent along northern Chile because, as discussed in previous sections, the trench axis is essentially without sediment. During the past 20 million years, sediment from the Andes was trapped in basins on land (Hartley and Jolley 1995) and during some 10 million years of aridity the supply has been minimal. In addition, sediment transport along the trench axis from the south was blocked north of Valparaiso by the subducting Juan Fernández Ridge (Yañez et al. 2001).

Hence, the northern Chile convergent continental margin, despite its proximity to a high mountain range, is sediment starved and, thus, an excellent place to test the theoretical high-friction mechanism of subduction erosion. Here we summarize investigations of the past decade along northern Chile and present an updated kinematic model of erosion and seismogenesis resulting from the 1990s data acquisition and synthesis (cf. Sallares and Ranero 2005, von Huene and Ranero 2003, Husen et al. 2000, Delouis et al. 1998, Hartley and Jolly 1995).

## 5.4.2 Data near Antofagasta

Definitive data across the northern Chile margin acquired in the late 1990s includes multibeam bathymetry (Fig. 5.15), seismic reflection and refraction data (Fig. 5.16) and aftershocks of the $M_w$ 8.0 Antofagasta 1995 earthquake recorded on a land-sea seismic network. Multibeam bathymetry (Fig. 5.15) in the vicinity of Antofagasta images basement deformation very effectively because of thin slope sediment.

Extension by landward-dipping faulting occurs onshore across the Coastal Cordillera fore-arc (Delouis et al. 1998; Loveless et al. 2005). This extensional faulting style continues across the continental shelf (von Huene and Ranero 2003; Sallares and Ranero 2005). As seen in the bathymetry from some 600 km to the south (Fig. 5.8), canyons beginning on the shelf deepen across the upper slope and are cut by trench-parallel, landward-dipping, normal faults. This morphology merges into a smooth middle slope dominated by mass wasting where local, lens-shaped bodies, 20–35 km wide, extend downslope and are dissected by seaward-dipping, normal faults (von Huene and Ranero 2003).

The morphology of the lower slope is characterized by low ridges that are continuous with lower plate horsts subducting beneath the thin leading apex of the upper plate. Seawards of the trench axis, the horsts and grabens are very well developed and cut across a pervasive, ocean crust fabric that parallels magnetic anomalies. This fabric is reactivated as the crust bends downward into the trench axis. This morphological pattern is not only common to the margin north of Juan Fernández Ridge (Fig. 5.8), but is also commonly observed in high-resolution, multi-beam bathymetry across other erosional convergent margins.

The seismic image of structure in the area is revealed with a depth-processed record and a joint analysis of seismic reflection and refraction data (Fig. 5.16). Coherent reflections from 4 km below the seafloor are sufficiently strong to give reliable rock velocities during depth processing (von Huene and Ranero 2003). Deeper velocities are constrained with sea and land, wide-angle seismic records that are modeled with joint reflection and refraction travel time tomography (Sallares and Ranero 2005).

The depth-migrated reflection line ends 25 km from the coast. However, 1960s sparker data across the shelf, combined with subsequent mapping on the Mejillones Peninsula, reveals landward-dipping, normal faults with vertical displacements of some 250 m. The faults bound blocks that are capped by seaward-tilted strata (von Huene and Ranero 2003). In the middle slope, a different structure is imaged; it consists of five large, landward-rotated, fault blocks (Fig. 5.16). Seaward-dipping faults bounding the tilted blocks are defined by fault-plane reflections that extend 4 km into basement, where they merge with a near-horizontal zone of continuous reflectivity. The reflective zone forms a detachment beneath the sequence of rotated blocks and corresponds to the top of a profound velocity inversion (Sallares and Ranero 2005).

This upper plate zone divides a tectonically extending upper zone from a lower zone, but the lower zone tectonic structure is unresolved. Analysis of the detachment fault geometry using methods applied to slide bodies (Watts and Grilli 2002) shows only 3.5 MPa of frictional strength, or about half the average shear strength of consolidated marine sediment, despite its path through basement. Restoration on the basement-sediment unconformity in the rotated blocks indicates ~4.5 km of downslope extension and an original slope angle of about 16°. If the unconformity offshore is a continuation of the unconformity on land, margin subsidence of up to 6 km is indicated. Apparently, subsidence steepens the slope to a point where gravitational force exceeds material strength, resulting in extensional tectonism in the middle slope.

Lower slope structure and morphology are typical of a well-developed, frontal prism despite little terrigeneous sediment in the trench axis, so it must be composed of debris from mass wasting upslope. The prism's 3.5 km s$^{-1}$ seismic velocity indicates fluid saturated, erosional debris and disaggregated basement rock. A 2.4 g cc$^{-1}$ density and porosity >15%, obtained from modeling wide-angle seismic and gravity data (Sallares and Ranero 2005) and geometric analysis of conjugate fault dips (Davis and von Huene 1987), indicate a strength the same as accretionary prisms composed of trench sediment.

Thus, the upper plate's strong rock, with velocities of 5 km s$^{-1}$ along the coast and continental shelf, becomes a weakened rock body, with ~4.5 km s$^{-1}$ velocity, in the mid slope and disintegrates progressively downslope to a very weak material with relatively low (3.5 km s$^{-1}$) velocity and high porosity adjacent to the trench axis. This weak lower-slope material is readily deformed by subducting lower plate relief. These observations are consistent with previously discussed observations made north of the Juan Fernández Ridge and with shipboard-processed records north and south of Antofagasta (Block 1998).

Seismic images of the subducted lower plate beneath the frontal prism show a horst-and-graben structure with twice the relief of the unsubducted horsts that lie seawards of the trench axis. A continuation of horst-and-graben growth is consistent with continued bending of the subducted ocean plate, which is greater than the bending seawards of the trench axis. It is speculated that the growth of these subducting horsts is helped by rising fluid pressure in their bounding normal faults, which is caused by the increasing load of the landward-thickening upper plate. This subducting relief is overlain by low-velocity material in the subduction channel (Sallares and Ranero 2005). Only 1.5 to 2 km of structureless reflectivity separates the subducting horsts from the overlying, normal,

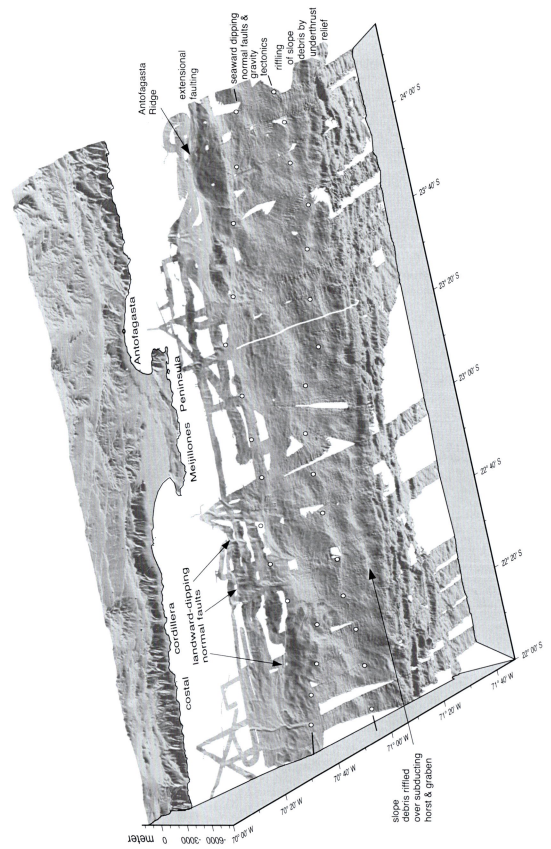

**Fig. 5.15.** Perspective view of shaded-relief bathymetry and topography from the sediment-starved trench of the northern Chile margin offshore from Antofagasta. *White line* is Sonne 104 line 13 shown in Fig. 5.16. *White-filled diamonds* bound the upper, middle and lower slopes with distinct morphotectonic character. Note the similarity of the submerged Antofagasta Ridge and Mejillones Peninsula, both formed as footwalls to large landward-dipping fault systems

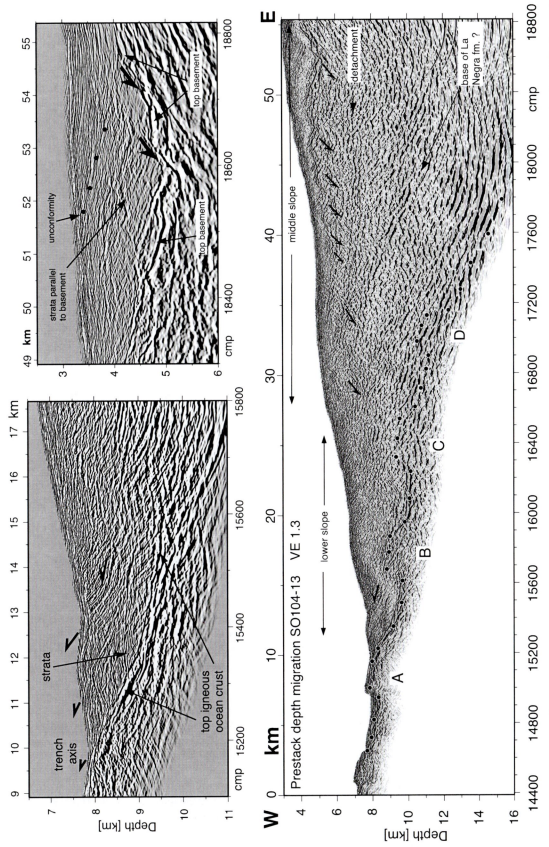

**Fig. 5.16.** Pre-stack depth migration of Sonne 104 line 13 offshore from Antofagasta (see Fig. 5.15 for location). Vertical exaggeration is 1.3. Note landward-dipping faults cutting the middle slope and soling out at a subhorizontal detachment at ~7 km depth. *Black-filled circles* mark the top of the subducting plate and letters indicate subducting horsts. *Left close up* shows the well-stratified debris filling an incoming graben and forming the frontal prism. *Right close up* shows the top of basement and overlying sediment cut and extended by normal faults. *cmp*: common mid point gather

fault plane reflections. This gap, separating contractional deformation from extensional deformation beneath the middle slope, is no more than 1.5 km wide, consistent with the absence of interplate earthquakes from which we conclude that interplate friction is low.

A true configuration of the plate interface and the beginning of the seismogenic zone is well constrained with controlled-source, wide-angle, seismic data, as well as relocated aftershocks from the $M_w$ 8.0 Antofagasta 1995 earthquake. The latter were recorded on an extensive, land-sea seismological array (Husen et al. 1999). Wide-angle data constrain the configuration of the plate interface to the depth where seismic rupture began and the Antofagasta 1995 earthquake aftershocks are relocated with 2 km precision. Velocities indicate that aftershocks occurred where the subduction channel had lost most of its fluid content and the porosity of the overriding plate had decreased to < 5% (Sallares and Ranero 2005). Velocity thus indicates a transition in physical properties along the subduction zone near the up-dip limit of seismogenic behavior. We speculate that increased strength and reduced fluid down dip increase friction and coupling to the level required for seismogenesis.

### 5.4.3 An Updated Kinematic Model

In our revised model of the northern Chile margin, we focus on the down-slope changes from the shelf to the trench axis. Specifically, we consider the following three themes: (*1*) the weakening of the margin wedge towards its apex; which is associated with (*2*) fluid migration and drainage of the subduction channel and the underlying, faulted, lower plate; and (*3*) removal of continental crustal mass at the trench and vertical unloading of the margin wedge owing to crustal thinning.

Subsidence of the outer shelf also steepens the slope and moderate bending stress is relieved by breaking along landward-dipping, "book-case" faults. The core of the margin beneath the shelf and upper slope is strong and not significantly invaded by rising fluid because the lower-plate fluid source, some 20 km deep, may be largely depleted. With a steepening slope, the down-slope force of gravity increases until at the middle slope it exceeds the shear strength of upper crustal layers. Gravity tectonics progressively breaks up the framework of the margin wedge and upwardly migrating fluid hydrofractures the underside of the wedge, promoting erosion and accelerated crustal thinning. Porosity averages > 10% and fluid from the subduction channel has reached the upper levels of the margin wedge.

Progressive weakening of the upper plate, an increase in gravity tectonics, and subducting plate relief continue to raise and lower the margin. Eventually, it disaggregates into fragments that are too small to be imaged with seismic reflection systems operating at the sea surface. The disaggregated, thin apex of the margin forms a frontal prism with the strength of accreting sedimentary prisms.

The prism elevates pore fluid pressure in subducting material and reduces interplate friction. As the subduction channel material loses pore fluid and bound water, it strengthens down dip and fractures brittlely, thus enhancing fracture permeability. This enables fluid from the subducting igneous ocean crust to migrate more freely into the plate interface and upper plate. Both the pore water and the bound water released by mineral transformations in the subduction channel are drained. After the plate bends downward, it unbends beneath the upper slope which exerts compressional stress in the upper igneous ocean crust. This collapses fracture permeability and impedes further upward fluid migration. Material changes and fluid depletion along the plate interface are inferred to significantly influence the beginning of seismogenic behavior.

It appears that any mechanical model of the northern Chile margin must be integrated with a fluid-driven tectonic system. Pore fluid and released bound water continuously drain down dip and lessen lower plate porosity. In turn, this fluid migration weakens the upper plate and the loss of strength increases towards the apex of the margin wedge. That weakened material in the frontal prism elevates pore fluid pressure to levels where all material in the trench axis subducts. Understanding the tectonics of an erosional margin requires an understanding of its hydrology.

## 5.5 Summary and Concluding Remarks

The > 3 500 km long Chile subduction zone provides a unique opportunity to study along a single subduction system the factors governing tectonic processes at convergent plate boundaries. Trench sediment supply, turbidite transport along the trench axis, and deformation produced by spreading center and hot spot ridge subduction are fundamental controls on the dominant tectonic processes along convergent margins (Fig. 5.17).

South of the triple junction, the Antarctic plate converges with the continental plate at some 18 cm yr$^{-1}$, with the vector ranging from margin-perpendicular in the north to strongly oblique in the south (Fig. 5.1). In this area, the trench is flooded with turbidites and convergence is relatively slow. The Chile Rise collided with the southernmost margin at ~14 Ma, leaving a truncated continental margin that can be recognized in gravity maps (Fig. 5.2). After the spreading center subducted, the edge of the continental margin subsided to some 4–5 km depth and an accretionary wedge, up to 8 km thick and about 60 km wide, developed against the truncated margin (Fig. 5.3). Here, most of the ~2.5 km of trench turbidite are currently accreted.

**Fig. 5.17.**
Shaded relief map of the Nazca and Antarctic plates and South America continent (Smith and Sandwell 1994). Convergence vectors are after DeMets et al. (1990). Lines along the trench denote the dominant tectonic process during the last few millions of years (< 5 Myr). North of ~33° S 0.5–1.0 km of sediments subducts and tectonic erosion along the base of the overriding plate may contribute several hundreds of meter of debris to the subduction channel. The subduction channel of the accretionary segment north of the triple junction similarly contains ~1.0–1.5 km of sediment. Along most of the segment south of the triple junction the majority of the sediment at the trench appears to be accreted

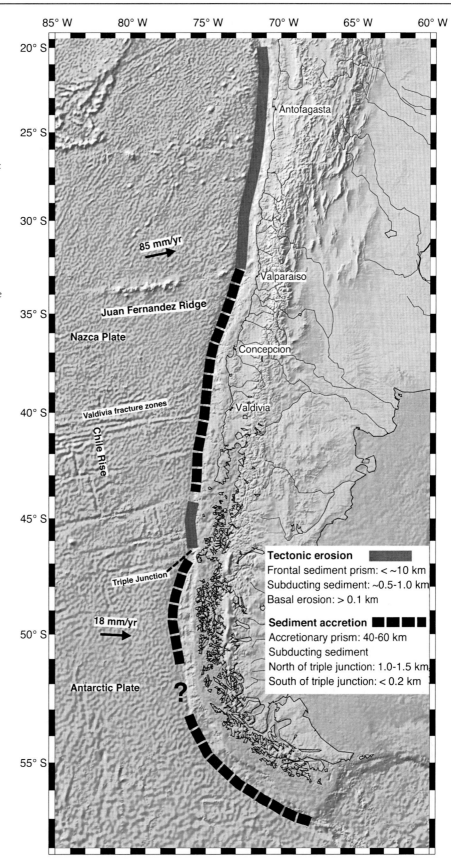

Further north, the Chile Rise subducted around 3–6 Ma, about 100 km south of the triple junction, leaving a similar, deep and truncated edge of the continental margin at the Golfo de Penas, where an accretionary prism has subsequently developed (Figs. 5.4 and 5.5). This accretionary structure provides a snapshot of the early phases of prism development (Figs. 5.4 to 5.6). Undeformed trench turbidites are ~1.5 km thick and the area of sediment accretion is some 60 km wide, which is, surprisingly, almost as broad as the entire prism in southernmost Chile. However, here the accretionary prism contains a smaller volume of material.

Along the Golfo de Penas accretionary prism, several discrete thrust anticlines indicate the mechanism of prism thickening. These anticlines increase in width and height landwards, and landward- and seaward-verging structures indicate that they have progressively thickened the prism by thrusting and back-thrusting (Fig. 5.6). The bathymetry shows that the accretionary ridges coalesce some 50 km to the south where accretion has been active longer. Thus, even though trench infill in this area is about 1.5 km thick and convergence is trench normal, the structure indicates that, if conditions are maintained, an accretionary prism as broad as that in southernmost Chile could develop.

The effects that glacial and interglacial periods have on trench turbidite infill and, thus, on tectonics and mass balance budgets (Bangs and Cande 1997; Bourgois et al. 2000, Behrmann and Kopf 2001) could be studied by drilling the trench sediments. This would unravel the deposition history of approximately the past two million years. Similarly, drilling the fore-arc basin sediment, deposited after spreading center subduction, could potentially provide the sedimentary record for the past ~14 million years in the south and ~3–6 million years near the Golfo de Penas.

Where the axis of the spreading center is currently colliding with the continental margin, a small sediment prism (~5 km wide) occurs. It fronts a seaward-dipping continental basement, indicating margin erosion (Fig. 5.7). North of the Darwin fracture zone, the Chile Rise is offset westwards by some 50 km and a deeper trench is filled with turbidites. Here, however, the frontal prism is formed by a ridge, only ~10 km wide, at the toe of the slope. The morphology and dimensions of this slope-toe ridge resemble the ridges at the slope apex north of the Juan Fernández Ridge collision zone, along the depressed slope segment where tectonic erosion is active (Figs. 5.8 and 5.9). However, the dimensions of the continental slope are dramatically different: the slope near the triple junction is ~30 km wide, whereas north of the Juan Fernández Ridge, the slope is ~80 km wide. This observation might hint that the subduction of the spreading center is more erosive than the subduction of the hot spot ridge. Only one of the flanks of the spreading center has so far been subducted and thus further tectonic erosion is expected to occur in this sector for the next several millions of years.

The collision of the Juan Fernández Ridge with the continental margin has two distinct effects. Firstly, the crest of the ridge rapidly erodes the margin, removing most of the pre-existing accretionary structure, and local uplift above large subducting seamounts fractures the entire continental slope (Figs. 5.8, 5.9 and 5.13). Secondly, the shallow ridge topography restrains the transport of turbidites along the trench, leading to a permanent modification of the dominant tectonic process along the margin, i.e. the change from frontal sediment accretion to tectonic erosion.

Where the trench contains a turbidite thickness of less than 1 km, the continental slope is fronted by a 5–10 km wide sediment prism that probably helps to increase fluid overpressures in underthrusting sediment and may, in fact, facilitate sediment subduction, as proposed for other thinly sedimented trenches (e.g. von Huene et al. 2000). Fluids from the underthrusting sediment escape at the contact between the frontal prism and the backstop, forming some 100 meter high mud diapirs. There is no increase in width of the frontal sediment prism for 200 km along the trench after ridge subduction, indicating that accretion is not active. Thus, north of the crest of the hot spot ridge, the slightly less than 1 km of sediment at the trench is largely, or completely, subducted.

This is in great contrast to the Golfo de Penas area where ~1.5 km thick sediment fills the trench and covers a wider area, forming a large accretionary prism. The initiation of tectonic erosion rapidly brings a fundamental change in tectonics across the entire continental slope. The structure of the depressed slope segment indicates thinning of the overriding plate by basal erosion and the concomitant tectonic extension and collapse of the margin rock framework.

South of the collision zone of the Juan Fernández Ridge, the trench is filled with more than 2 km of turbidites, and a broad accretionary wedge and an undisturbed morphology upslope of the prism indicate a more stable configuration (Fig. 5.8). Frontal sediment accretion dominates here, but 1.0–1.5 km of trench sediment has underthrust past the accretionary prism and has been subducted (von Huene et al. 1997; Diaz 1999). The area where frontal accretion currently dominates extends southwards until the region where trench sediment infill and tectonics are modified by the subduction of the Chile Rise (Bangs and Cande 1997).

The margin north of ~28° S exemplifies continental margin tectonics along a turbidite-starved trench. The margin shares a basic structural configuration with the depressed slope segment north of the crest of the hot spot ridge. A narrow, frontal, sediment prism abuts a truncated, extending and subsiding continental basement cut by normal faults (Figs. 5.15 and 5.16). However, tectonic erosion has been active for a longer period here and some structures of the margin indicate a more advanced stage of deformation. A steeper continental slope leads to widespread mass wasting that, just before subduction, fills ocean plate grabens with well-stratified debris ~1 km thick (Fig. 5.16). Thus, even though turbidites produced in southern latitudes do not reach here, a thick, fluid-rich layer of debris is being subducted.

In addition to surficial mass wasting, the middle continental slope contains large, lens-shaped blocks that are detaching along seaward-dipping faults. The upper slope displays a less disaggregated terrain cut by large landward-dipping faults of an extensional system that continues onshore (von Huene and Ranero 2003). The ~1 km thick subducting fluid-rich debris seems to play a fundamental role in the erosional tectonic processes that dominate the margin. Overpressured fluids expelled along the subducting channel probably hydrofracture the base of the overriding plate, thus promoting basal tectonic erosion and triggering extensional tectonics across the slope and even into the emerged fore-arc (von Huene and Ranero 2003; Sallares and Ranero 2005). The tectonic role of fluids expelled from the subduction channel of the accretionary segments of the margin is an important topic for future research.

The broad scale distribution of dominant processes that emerges from the observations gives some general information on material budget (Fig. 5.17). South of the triple junction most trench sediment is accreted along the majority of that segment. However, north of the triple junction a thick layer of fluid-rich terrigenous material subducts both at erosional and accretionary segments. The origin of the material transported in the subduction channel may be complex: It may range from trench turbidites coming from the subaerial erosion of the arc-fore-arc, to turbidites from slope mass wasting processes and also includes a poorly constrained volume of debris tectonically eroded from the underside of the overriding plate. Furthermore, tectonically eroded material may consist of volcanic rocks from an old arc, or consolidated sediment from a fossil accretionary prism. The contribution of each of those different terrigeneous materials clearly changes from the accretionary to the erosional segments. However, their relative volume, as well as the volume of pore and chemically-bound fluids, in the subduction channel might change gradually from north to south, as the progressive increase in trench turbidites and decreasing intensity in erosional processes may indicate.

Finally, further research should be devoted to understand how relative changes in material properties and fluid content in the subduction channel might influence shear coupling along and across the plate boundary and whether erosional plate boundaries are characterized by relatively higher friction (Lamb and Davis 2003; Sobolev and Babeyko 2005) or comparatively similar to accretionary plate boundaries (von Huene and Ranero 2003).

## Acknowledgments

RV Sonne cruises were funded by the German Ministry of Research and Education (BMBF). Jaques Bourgois and François Michaud kindly provided the digital multibeam bathymetry data from the Chile Triple Junction area. We thank M Scherwath for merging the bathymetric data around the triple junction. J. Alsop from Lamont Doherty Earth Observatory kindly provided the stacks of RV R. Conrad lines 745 and 769. J. Laursen and D. Klaeschen processed Sonne 101-15. All figures in this paper have been produced with the GMT package (Wessel and Smith 1998). Reviews by R. Allmendinger, J. Cembrano, A. Hartley and U. Riller are gratefully acknowledged. This is SFB574 contribution 73.

## References

Angermann D, Klotz J, Reigber C (1999) Space-geodetic estimation of the Nazca-South America Euler Vector. Earth Planet Sci Lett 171:329–334

Ballance et al. (1989) Subduction of a large Cretaceous seamount of the Louisville ridge at the Tonga Trench: a model of normal and accelerated tectonic erosion. Tectonics 8:953–962

Bandy OL, Rudolfo KS (1964) Distribution of foraminifera and sediments, Peru-Chile Trench area. Deep Sea Res 11:817–837

Bangs NL, Cande SC (1997) The episodic development of a convergent margin inferred from structures and processes along the southern Chile margin. Tectonics 16(3):489–505

Bangs NL, Cande SC, Lewis SD, Miller JJ (1992) Structural framework of the Chile margin at the Chile Ridge collision zone. Proc Ocean Drill Prog Initial Rep 141:11–21

Behrmann JH, Kopf A (2001) Balance or tectonically accreted and subducted sediment at the Chile Triple Junction. Int J Earth Sci 90:753–768

Behrmann JH et al. (1992) Proceedings of the Ocean Drilling Program, Initial Reports. Volume 141. Ocean Drill. Program, College Station TX

Behrmann JH, Lewis SD, Cande SC, ODP Leg 141 Scientific Party (1994) Tectonics and geology of spreading ridge subduction at the Chile Triple Junction: a synthesis of results from Leg 141 of the Ocean Drilling Program. Geol Rundsch 83:832–852

Block M (1998) Interpretations of MCS data. In: Hinz K et al. (eds) Crustal investigations off- and onshore Nazca/Central Andes (CINCA), BGR Report No. 117.613. Bundesanstalt für Geowissenschaften und Rohstoffe, Hannover, pp 69–102

Bourgois J, Martin H, Lagabrielle Y, Le Moigne J, Frutos Jara J (1996) Subduction-erosion related to spreading-ridge subduction: Taitao Peninsula (Chile margin triple junction area). Geology 24:723–726

Bourgois J, Guivel C, Lagabrielle Y, Calmus T, Boulegue J, Daux V (2000) Glacial-interglacial trench supply variation, spreading-ridge subduction, and feedback controls on the Andean margin development at the Chile triple junction area (45–48° S). J Geophys Res 105:8355–8386

Cande SC, Leslie RB (1986) Late Cenozoic tectonics of the southern Chile trench. J Geophys Res 91:471–496

Cande SC, Leslie RB, Parra JC, Hobart M (1987) Interaction between the Chile ridge and the Chile trench: geophysical and geothermal evidence. J Geophys Res 92:495–520

Caress DW, Chase DN (1996) Improved processing of Hydrosweep DS multibeam data on the RV Maurice Ewing. Marine Geophys Res 18:631–650

Davis DM, von Huene R (1987) Inferences on sediment strength and fault friction from structures of the Aleutian Trench. Geology 15:517–522

Delouis BH, Philip H, Dorbath L, Cisternas A (1998) Recent crustal deformation in the Antofagasta region (northern Chile) and the subduction process. Geophys J Int 132:302–338

DeMets C, Gordon RG, Argus DF, Stein S (1990) Current plate motions. Geophys J Int 101:425–478

Diaz JL (1999) Sediment subduction and accretion at the Chilean convergent margin between 35° and 40°S. PhD thesis, Christian-Albrechts-Universität zu Kiel

Fisher RL, Raitt RW (1962) Topography and structure of the Peru-Chile trench. Deep Sea Res 9:423–443

Flueh ER, Vidal N, Ranero CR, Hojka A, von Huene R, Bialas J, Hinz K, Cordoba D, Dañobeitia JJ, Zelt C (1998) Seismic investigation of the continental margin off- and onshore Valparaiso, Chile. Tectonophysics 288:251–263

Hartley AJ, Jolley EJ (1995) Tectonic implications of Late Cenozoic sedimentation from the Coastal Cordillera of northern Chile (22–24°S). J Geol Soc London 152:51–63

Hilde TWC (1983) Sediment subduction vs. accretion around the Pacific. Tectonophysics 99:381–397

Husen S, Kissling E, Flueh E, Asch G (1999) Accurate hypocenter determination in the seimogenic zone of the subducting Nazca plate in north Chile using a combined on-/offshore network. Geophys J Int 138:687–701

Husen S, Kissling E, Flueh ER (2000) Local earthquake tomography of shallow subduction in north Chile: a combined onshore and offshore study. J Geophys Res 105:28183–28198

Jarrard RD (1986) Relations among subduction parameters. Rev Geophys 24(2):217–284

Kendrick E, Bevis M, Smalley R Jr, Brooks B, Vargas RB, Lauría E, Fortes LPS (2003) The Nazca - South America Euler vector and its rate of change. J S Am Earth Sci 16:125–131

Kimura G, et al. (1997) Proceedings of the Ocean Drilling Program, Initial Reports, Volume 170. Ocean Drill Prog, College Station, TX

Kudrass HR, Von Rad U, Seyfied H, Andruleit H, Hinz K, Reichert C (1998) Age and facies of sediments of the northern Chilean continental slope – evidence for intense vertical movements, in Crustal investigations off- and onshore Nazca/Central Andes (CINCA). In: Hinz K et al. (eds) BGR Report No. 117.613, Bundesanstalt für Geowissenschaften und Rohstoffe, Hannover, pp 170–196

Lagabrielle Y, Guivel C, Maury R, Bourgois J, Fourcade S, Martin H (2000) Magmatic-tectonic effects of high thermal regime at the site of active ridge subduction: the Chile triple junction model. Tectonophysics 326:255–268

Lamb S, Davis P (2003) Cenozoic climate change as a possible cause for the rise of the Andes. Nature 425:792–797

Laursen J, Scholl D, von Huene R (2002) Neotectonic deformation of the central Chile margin: deepwater forearc basin formation in response to hot spot ridge and seamount subduction. Tectonics 21: doi 10,1029/2001TC901023

Loveless JP, Hoke GD, Allmendinger RW, González G, Isacks BL, Carrizo DA (2005) Pervasive cracking of the northern Chilean Coastal Cordillera: new evidence of forearc extension. Geology 33:973–976

Miller H (1970) Das Problem des hypothetischen "Pazifischen Kontinentes" gesehen von der chilenischen Pazifikküste. Geol Rundsch 59:927–938

Polonia A, Brancolini G, Torelli L, Vera E (1999) Structural variability at the active continental margin off southernmost Chile. J Geodyn 27:289–307

Polonia A, Brancolini G, Loreto MF, Torelli L (2001) The accretionary complex of southernmost Chile from the analysis of multi-channel seismic data. Terra Antartica 8:87–98

Ranero CR, von Huene R (2000) Subduction erosion along the Middle America convergent margin. Nature 404:748–752

Ranero CR, von Huene R, Flueh E, Duarte M, Baca D, McIntosh K (2000) A cross-section of the convergent Pacific margin of Nicaragua. Tectonics 19:335–357

Ranero CR, von Huene R, Weinrebe W, Barckhausen U (in press) Convergent margin tectonics of Middle America: a marine perspective. In: Alvarado G (ed) Central America, Geology, Hazards and Resources. AA Balkema Publisher

Rubio E, Torné M, Vera E, Diaz A (2000) Crustal structure of the southernmost Chilean margin from seismic and gravity data. Tectonophysics 323:39–60

Rutland RWR (1971) Andean orogeny and ocean floor spreading. Nature 233:252–255

Sallares V, Banero CR (2005) Structure of the North Chile erosional convergent margin off Antofagasta (23°30' S). J Geophys Res 110: doi 10.1029/2004JB003418

Sandwell DT, Smith WHF (1997) Marine gravity anomaly from Geosat and ERS-1 satellite altimetry. J Geophys Res 102:10039–10050

Scholl DW, Christensen MN, von Huene R, Marlow MS (1970) Peru-Chile trench sediments and sea-floor spreading. Geol Soc Am Bull 81:1339–1360

Scholl DW, von Huene R, Vallier TL, Howell DG (1980) Sedimentary masses and concepts about tectonic processes at underthrust ocean margins. Geology 8:564–568

Schweller WJ, Kulm LD, Prince RA (1981) Tectonics structure, and sedimentary framework of the Perú-Chile Trench. In: Kulm LD, et al. (eds) Nazca Plate: Crustal formation and Andean convergence. Mem Geol Soc Am 154:323–349

Smith WHF, Sandwell DT (1994) Bathymetric predictions from dense altimetry and sparse shipboard bathymetry. J Geophys Res 99:21803–21824

Sobolev SV, Babeyko AY (2005) What drives orogeny in the Andes? Geology 33:617–62

Tebbens SF, Cande SC, Kovacs L, Parra JC, LaBreque JL, Vergara H (1997) The Chile ridge: a tectonic framework. J Geophys Res 102:2035–2059

Uyeda S, Kanamori H (1979) Back-arc opening and the mode of subduction. J Geophys Res 84:1049–1061

Vannucchi P, Ranero CR, Galeotti S, Straub SM, Scholl DW, McDougall Ried K (2003) Fast rates of subduction erosion along the Costa Rica Pacific margin: implications for non-steady rates of crustal recycling at subduction zones. J Geophys Res 108: doi 10.1029/2002JB002207

Vannucchi P, Galeotti S, Clift PD, Ranero CR, von Huene R (2004) Long term subduction erosion along the Middle America Trench offshore Guatemala. Geology

von Huene R, Ranero CR (2003) Subduction erosion and basal friction along the sediment starved convergent margin off Antofagasta Chile. J Geophys Res 108: doi 10.1029/2001JB001569

von Huene R, Scholl D (1991) Observations at convergent margins concerning sediment subduction, subduction erosion, and the growth of continental crust. Rev Geophys 29:279–316

von Huene R, Corvalan J, Flueh ER, Hinz K, Korstgard J, Ranero CR, Weinrebe W, CONDOR Scientists (1997) Tectonic control of the subducting Juan Fernández Ridge on the Andean margin near Valparaiso, Chile. Tectonics 16:474–488

von Huene R, Ranero CR, Weinrebe W, Hinz K (2000) Quaternary convergent margin tectonics of Costa Rica, segmentation of the Cocos Plate, and Central American volcanism. Tectonics 19:314–334

von Huene R, Ranero CR, Vannucchi P (2004) A model for subduction erosion. Geology 32:913–916

Watts P, Grilli SR (2002) Tsunami generation by submarine mass failure, I: Wavemaker modes. J Wtrwy Port Coast Oc Engrg

Wessel P, Smith WHF (1998) New improved version of generic mapping tools released. EOS 79(47):579

Yañez GA, Ranero CR, von Huene R, Díaz J (2001) A tectonic interpretation of magnetic anomalies across a segment of the convergent margin of the Southern Central Andes (32°–34°S). J Geophys Res 106:6325–6345

Yañez GA, Cembrano J, Pardo M, Ranero CR, Selles D (2002) The Challenger-Juan Fernández-Maipo major tectonic transition of the Nazca-Andean subduction system at 33°–34°S: geodynamic evidences and implications. J S Am Earth Sci 15:23–38

Zelt CA, Hojka AM, Flueh ER, McIntosh KD (1999) 3D simultaneous seismic refraction and reflection tomography of wide-angle data from the central Chilean margin. Geophys Res Lett 26: 2577–2580

# Part II
# Elements of the Subduction System

**Chapter 6**
Oblique Convergence along the Chilean Margin: Partitioning, Margin-Parallel Faulting and Force Interaction at the Plate Interface

**Chapter 7**
Seismic Images of Accretive and Erosive Subduction Zones from the Chilean Margin

**Chapter 8**
Geophysical Signatures and Active Tectonics at the South-Central Chilean Margin

**Chapter 9**
Latitudinal Variation in Sedimentary Processes in the Peru-Chile Trench off Central Chile

**Chapter 10**
Subduction Erosion – the "Normal" Mode of Fore-Arc Material Transfer along the Chilean Margin?

**Chapter 11**
Subduction Channel Evolution in Brittle Fore-Arc Wedges – a Combined Study with Scaled Sandbox Experiments, Seismological and Reflection Seismic Data and Geological Field Evidence

This part explores the key processes and features affecting the fore-arc of a plate boundary system in more detail. In the opening chapter, A. Hoffmann-Rothe et al. (Chap. 6) analyze the patterns of deformation partitioning of the fore-arc from Northern to Southern Chile and relate them to a variety of parameters that show significant changes along strike. They find that mechanical weakening of the upper plate and the evolving fore-arc geometry have been the overwhelming factors influencing the patterns observed.

C. Sick et al. (Chap. 7) compare the deep seismic images of the Northern and Southern Chilean fore-arcs and are able to demonstrate that the tectonic erosion mode of subduction (in the north) and the accretionary mode in the south have very different reflection seismic responses even down to greater depths.

C. Krawczyk et al. (Chap. 8) complement the deep seismic image for the fore-arc in Southern Chile with a series of other geophysical data and by including the geological record. The role of mass transfer in the subduction channel stands out clearly in their analysis as a key element to explain both the present geophysical image and the rock record.

Again focusing in the same area, D. Völker et al. (Chap. 9) study the sedimentary record in the trench using seismic data and dredging. They demonstrate that large-scale sediment redistributions along the trench, and the Quaternary climate cycles, have controlled the filling history and the basin architecture.

On a more general level and regional scope, N. Kukowski et al. (Chap. 10) use a compilation of sediment flux and coastal uplift data to show that the subduction erosion mode appears to be the normal mode of mass transfer along the Andean margin. And therefore, climate-driven infilling of the trench is instrumental in shifting the mass flux mode from erosional to accretionary, with build-up of small wedges.

Finally, J. Lohrmann et al. (Chap. 11) employ analog sandbox simulations of subduction channel processes to understand the geophysical and geological record of the Chilean margin. Their experiments reveal intriguing kinematic complexity active over various time scales, which help to explain the nature of the seismological observations.

# Oblique Convergence along the Chilean Margin: Partitioning, Margin-Parallel Faulting and Force Interaction at the Plate Interface

Arne Hoffmann-Rothe · Nina Kukowski · Georg Dresen · Helmut Echtler · Onno Oncken · Jürgen Klotz · Ekkehard Scheuber · Antje Kellner

**Abstract.** The Chilean fore-arc exhibits margin-parallel strike-slip faulting and associated fore-arc sliver formation. Comparison of the long-term (geologic timescale) and short-term (human timescale) record of margin-parallel faulting along the oblique Chilean subduction margin between 15° S and 46° S reveals significant spatio-temporal heterogeneity. We have reviewed newly compiled data on the geometric, kinematic and mechanical properties and their variation along-strike of the Chilean margin and evaluated their competing influence on fore-arc deformation. Among the parameters considered are the plate kinematics (e.g., convergence obliquity and rate), overriding plate heterogeneities that affect its capability for localizing horizontal shear (e.g., thermal and structural weaknesses) or resistance to block motion (e.g., plate margin curvature) as well as properties governing or indicating force interaction at the plate interface (e.g., trench sediment-fill, geodetic and seismic coupling depth). Most remarkably, the short-term GPS-derived fore-arc velocity field, dominated by elastic loading processes, shows little variation along-strike of the margin, despite the significantly changing conditions (e.g., trench sediment-fill, mass transfer mode at the tip of the overriding plate, plateau or no plateau, and slab-dip variations). Variations in recent and past strike-slip motion do not appear to depend on the rate or obliquity of convergence, nor on the mode of mass transfer at the subduction front. The frictionally coupled area on the plate contact increases southwards and a decreasing taper along the Chilean margin can be reconciled in the framework of taper theory by a southward decrease of the effective coefficient of friction on the plate interface. The development of fore-arc slivers seems to be primarily controlled by mechanisms that cause effective rheological weakening of parts of the upper plate and/or by geometries that hamper margin-parallel sliver motion. While the seaward concave-shaped margin in North Chile hinders the margin-parallel motion of a fore-arc sliver, the present strike-slip activity of the Liquiñe-Ofqui Fault Zone in southern Chile is likely facilitated by the superposition of two conditions: a shallowly dipping slab and an exceptionally small arc to trench distance.

## 6.1 Introduction

Subduction zones with relative plate motion oblique to the trench frequently exhibit margin-parallel strike-slip fault systems in the overriding plate, thus bearing witness to the phenomenon of strain partitioning (Jarrard 1986a,b). In cases where the partitioning is complete, the plate interface accommodates pure thrust-faulting and all of the trench-parallel component of plate convergence is taken up within the upper plate (Fitch 1972). Shear focussed within a margin-parallel strike-slip fault system may eventually cause the formation of a fore-arc sliver that moves independently of the upper plate.

In the plate kinematic framework, convergence along the Peru-Chile trench has been almost always oblique since at least the middle Jurassic, though the sense of obliquity reversed once in the late Cretaceous (Figs. 6.1 and 6.2). The geological record of past and recent margin-parallel strike-slip faulting along the Chilean margin exemplifies that strain partitioning has played an important role in the overriding South American Plate (e.g. Scheuber and González 1999; Reutter et al. 1996; Cembrano et al. 2000). However, despite the relatively constant plate kinematic boundary conditions, none of the strike-slip systems that have been active during different stages of orogenic evolution covered the entire 3 500 km length of the Chilean Andes south of the orogenic bend at 20° S (Fig. 6.3). On the contrary, the fore-arc deformation pattern shows prominent spatio-temporal variations in margin-parallel fault activity.

The data base for the Chilean margin has grown considerably over the last decade and integrates extended knowledge of upper plate deformation fields as well as geophysical constraints, such as derived from GPS-studies and seismic and electromagnetic imaging (Khazaradze and Klotz 2003; ANCORP Working Group 2003; Brasse et al. 2002). Mass transfer modes at the tip of the subduction system (erosive, non-erosive, accretive), the morphologic expression of the Andes and the climatic boundary conditions, among other parameters, exhibit extreme variations along strike. The active Chilean margin between 15° S and 48° S constitutes an ideal system for examining how the overriding plate responds to oblique convergence, and for discussing the competing influences of both spatial and temporal (geological versus human timescale) controlling factors with respect to the along-strike variability of boundary conditions.

The factors addressed in this paper are: the overall framework of plate kinematics (convergence rate and obliquity; e.g., Jarrard 1986a; McCaffrey 1992; Beck 1991; Burbridge and Braun 1998); the properties controlling the force interaction ('coupling') of the two plates (trench

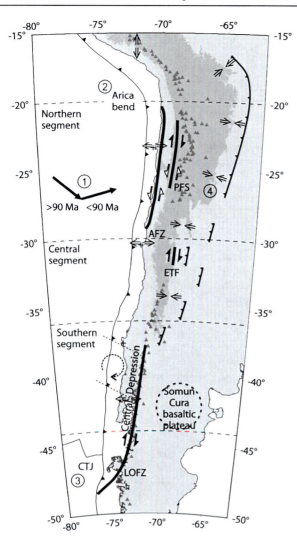

**Fig. 6.1.** Cartoon of the kinematic framework of the Chilean margin. Circled numbers refer to the major kinematic constraints described at the beginning of Sect. 6.1.2. The segmentation of the margin used in the paper is indicated. *Dark shading* marks topography of ≥ 2000 m elevation (Lindquist et al. 2004). *Pairs of solid arrows* denote the modern strike-slip fault direction, *hollow arrows* refer to past senses of slip. The schematic representation of compressional and extensional tectonic regimes is based on the World Stress Map database (Reinecker et al. 2004). *Plus symbols* indicate strong Coastal Cordilleran uplift (Heinze 2003). The *dashed line* and *circular arrow* indicate Neogene block rotations (Vietor et al. 2004). *Triangles* denote Holocene volcanoes. *CTJ*: Chile triple junction; *LOFZ*: Liquiñe-Ofqui Fault Zone; *AFZ*: Atacama Fault Zone; *PFS*: Precordilleran Fault System; *ETF*: El Tigre Fault

sediments, the age and temperature of the subducting plate, the roughness of the interface; e.g.; Chemenda et al. 2000; Yáñez and Cembrano 2004; McCaffrey 2002); the properties of the upper plate (zones of weakness, inherited structure, bulk rheological strength; e.g., Chemenda et al. 2000; McCaffrey et al. 2000); and the constraints that hamper margin-parallel fore-arc motion, often referred to as 'buttressing' (e.g., Beck et al. 1993; Wang 1996).

### 6.1.1 Characteristics of Oblique Subduction

In an oblique subduction setting, strain partitioning is evident if the particle displacement field in the fore-arc is not oriented parallel to the relative plate convergence vector driving the deformation: The ratio of margin-normal to margin-parallel motion at every point in the fore-arc is not equivalent to the ratio of the respective components of the convergence vector.

Short-term deformation over the human timescale, as constrained by geodetic observations, may be dominantly elastic. Results from elastic half-space dislocation modeling suggest that strain-partitioning in the upper plate is required when two obliquely converging plates cannot slip freely past each other, i.e. the plates are 'coupled' over some width at their interface (Fig. 6.3a; McCaffrey 2002; Bevis and Martel 2001; Savage 1983; see Sect. 6.4.3 for a definition of 'coupling'). Above the coupled zone, the subducting plate 'drags' the fore-arc wedge in the direction of convergence and the fore-arc becomes elastically strained.

Adopting a reference frame with a fixed upper plate, as in Fig. 6.3a, the vectors of horizontal surface velocity rotate towards trench-normal (decreasing obliquity) beyond the downdip end of coupling (EOC). This is because the transition from a coupled to an uncoupled interface imposes a more rapid landward decay of the trench-parallel component in the upper plate when compared to the trench-normal component (Bevis and Martel 2001; McCaffrey 2002).

We will use the term 'slip-partitioning' to indicate that at least some of the trench-parallel slip component is (permanently) accommodated by faults within the upper plate, either distributed over many structures or localized along a single, trench-parallel, strike-slip fault (Fig. 6.3b). In this case, the resulting pattern of the short-term, horizontal surface velocity field may strongly depend on the degree of slip-partitioning and the relative position of the EOC with respect to the strike-slip fault (McCaffrey 2002). Surface vectors laterally located between the EOC and the strike-slip fault system rotate towards the trench-parallel (increasing obliquity) because the trench-normal component decays landward from the EOC, while the strike-slip component is sustained due to the trench-parallel motion of the rigid fore-arc sliver. In this case GPS derived velocity vectors would also indicate slip-partitioning. As McCaffrey (2002) has shown, these model predictions can be successfully tested against natural case studies, such as the Sumatra subduction system.

Long-term geological records provide evidence for cumulative permanent deformation of the overriding plate. They allow us to infer average slip rates at margin-parallel strike-slip faults over the geological time scale (e.g., Sumatra Fault: Bellier and Sébrier 1995; Philippine Fault: Barrier et al. 1991; Liquiñe-Ofqui Fault Zone: Rosenau 2004).

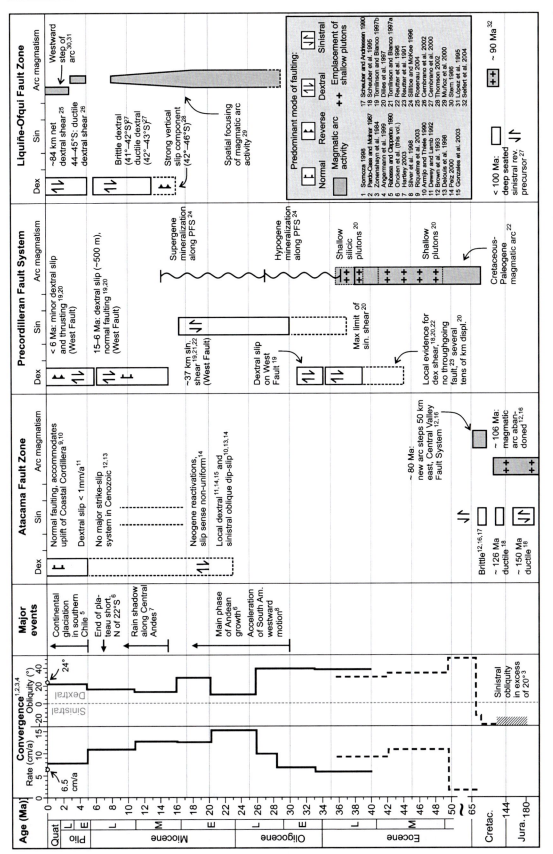

Fig. 6.2. Compilation of the main modes of faulting along the major, margin-parallel strike-slip faults of the Chilean margin in conjunction with magmatic arc activity. Convergence rate and obliquity are according to Somoza (1998) and Pardo-Casa and Molnar (1987). Plate obliquity for the Cretaceous is schematically represented after Zonenshayn et al. (1984). Present day convergence parameters (*open circles*) are averaged between 20 and 45° S based on Angermann et al. (1999)

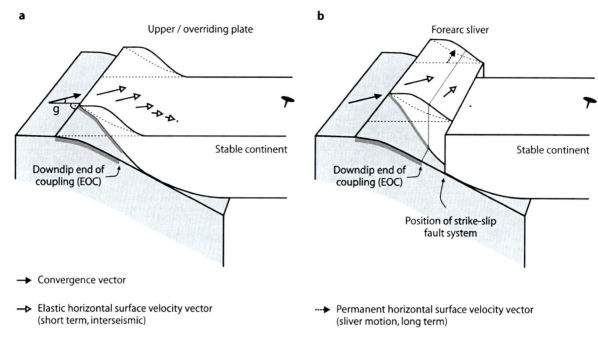

**Fig. 6.3.** Cartoon showing short-term (inter-seismic) and long-term (geologic) expected horizontal surface motion for an obliquely convergent subduction setting (modified after Bevis and Martel 2001). **a** In a purely elastically strained fore-arc without slip-partitioning the leading edge of the overriding plate is dragged with the subducting plate along the coupled zone of the plate interface. The horizontal surface velocity vectors will rotate towards trench-normal (decreasing obliquity) beyond the downdip end of coupling (EOC). **b** If the fore-arc permanently accommodates a part of the trench-parallel component of relative plate convergence in a localized strike-slip fault system, long-term (geologic) and short-term (inter-seismic) detected surface motions will differ. Surface vectors located between the EOC and the strike-slip fault system rotate towards trench-parallel (obliquity increases). Note that 'fore-arc sliver' in this context refers to the block between the trench and the strike-slip fault system

### 6.1.2 Kinematic Framework of the Chilean Margin

In general, the modern Chilean margin is framed by the following major kinematic constraints (Fig. 6.1):

1. Uniform convergence to the ENE between the Nazca and South American Plates since the late Cretaceous (Somoza 1998; Pardo-Casas and Molnar 1987), preceded by oblique subduction to the SE (Zonenshayn et al. 1984);
2. A concave seaward plate boundary at ~20° S that is paralleled by the 'Bolivian orocline' (e.g., McQuarrie 2002; Kley 1999);
3. Subduction of the Chile Ridge at ~46° S and northward movement of the triple junction since its impingement with South America in the middle Miocene (Forsythe and Nelson 1985); and
4. Back-arc shortening that governs the orogen from the bend (20° S) to about 38° S (Rosenau 2004; Melnick et al. in press) with contraction more or less in the direction of the topographic gradient (Reinecker et al. 2004).

We have subdivided the Chilean margin into three segments using margin-parallel strike-slip faults (Figs. 6.1 and 6.2). The Atacama Fault Zone (AFZ) and the Precordilleran Fault System (PFS) are located in the fore-arc of northern Chile along the western edge of the broad Andean Plateau. Both fault systems have been relatively inactive recently and relate predominantly to the long-term evolution of the Andes since the Jurassic. However, the AFZ still separates the uplifting Coastal Cordillera from the compressional inner fore-arc (e.g., Adam and Reuther 2000), and seismicity indicates right-lateral shear along the northern extension of the PFS (Comte et al. 2003; Fig. 6.4). The central segment (Fig. 6.1) is devoid of large-scale, trench-parallel faulting, although a short strike-slip fault system, the El Tigre fault, is located further landward (Siame et al. 1997). The southern segment features the intra-arc Liquiñe-Ofqui Fault Zone (LOFZ) in the low-relief and narrow southern Patagonian Andes of Chile (36–50° S). The LOFZ has exhibited dextral strike-slip motion since the Pliocene (Rosenau 2004).

Slip-partitioning along obliquely convergent margins can also manifest in rotating fore-arc blocks (Wallace et al. 2004). Large-scale Paleogene block rotations covering the entire fore-arc have been frequently inferred for northern Chile (Abels and Bischoff 1999; Arriagada et al. 2003; Beck 1998). These clockwise block rotations are accom-

**Fig. 6.4.**
Map of the Chilean margin showing the present deformation field in the form of GPS-derived velocity vectors. Campaigns included: SAGA – SAGA94-96 (Klotz et al. 2001), without co- and post-seismic vectors of the 1995 $M_w$ 8.1 Antofagasta earthquake, and SAGA96-97 (Khazaradze and Klotz 2003) campaigns; KEN01 – the CAP and SNAPP networks north of 23° S, as displayed in Kendrick et al. (2001); RUE02 – sites between 35 and 37° S (Ruegg et al. 2002); BRO03 – sites between 26 and 36° S (Brooks et al. 2003); CHL04 – sites between 18 and 24° S (Chlieh et al. 2004). All velocities are relative to a fixed South America reference frame. *Thick solid lines* indicate major fault systems (see Fig. 6.1 for abbreviations). The *short-dashed line* marks the surface projection of the position of the geodetically-estimated locking depth (Khazaradze and Klotz 2003). The Nazca-South American plate convergence vector is based on the estimate of Angermann et al. (1999). Age contours of the ocean floor are after Müller (1997). *Triangles* denote volcanoes with known eruptions during the Holocene (Siebert and Simkin 2002). *Gray dots* mark interplate thrust earthquakes during 1977–2004 (magnitude 1–10, Harvard CMT catalogue at http://www.eismology.harvard.edu/CMTsearch.html; thrust criteria: tension axis plunge between 45° to 90°, intermediate axis plunge between 0° to 45°). Focal mechanisms of shallow crustal, right-lateral, strike-slip earthquakes along the LOFZ are taken from Barrientos and Acevedo-Aránguiz (1992): Lonquimay, 24 Feb 1989, $M_w$ ~5.3; Chinn and Isacks (1983): Hudson, 28 Nov 1965, $M_w$ ~6.2; and Comte et al. (2003): Aroma, 24 July 2001, $M_w$ 6.3. *Dark quadrants* indicate compression at the source. *Hatched areas* schematically indicate aftershock distribution areas (rupture area) of the Antofagasta, 30 July 1995, $M_w$ 8.1 earthquake (Sobesiak 2004; Ruegg et al. 1996) and the Valdivia (Great Chile) earthquake, 22 May 1960, $M_w$ 9.5 (Barrientos 1990; Krawczyk and SPOC Team 2003). *CTJ*: Chile triple junction. The *inset* shows bathymetry and topography (Lindquist et al. 2004) and depth contours of the subducting slab according to the slab model by Tassara et al. (2006). *Blue lines* mark the main geophysical transects mentioned in Sect. 6.2.4. Cities: *An* – Antofagasta; *Sa* – Santiago de Chile; *Co* – Concepción; *Va* – Valdivia

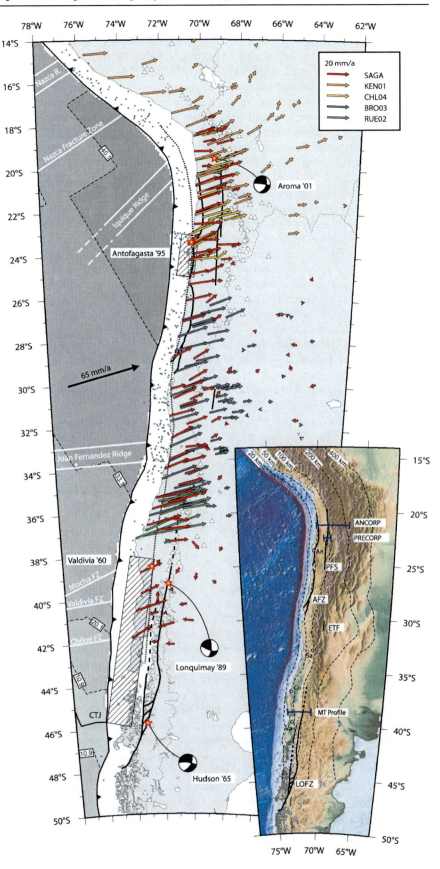

modated by NW-striking, sinistral, strike-slip faults. In contrast, there is no evidence for such rotation in northern Chile since the late Neogene (~11 Ma, Somoza et al. 1999, Arriagada et al. 2003). In southern Chile, the onshore/offshore fore-arc also exhibits segmentation into blocks along NW-striking faults (Echtler et al. 2003; Reuther et al. 2003; Kus et al. 2006). Paleomagnetic data show large clockwise block rotation since the late Miocene (Vietor et al. 2004) accompanied by left-lateral faulting, in accordance with dextral oblique convergence. With the reactivation of the LOFZ in the Pliocene, these faults have been predominantly reverse (Vietor et al. 2004).

In the following section, we briefly review (long-term) geological records of strike-slip faulting along the Chilean margin in concert with constraints derived from geophysical imaging of the fault zones. In Sect. 6.3, we examine the short-term, GPS-derived, horizontal surface displacement field for indications of present-day fore-arc sliver motion (see also Klotz et al. 2006, Chap. 4 of this volume).

## 6.2 Long-Term Records: Geological Evolution of Trench-Linked Strike-Slip Faults

### 6.2.1 Atacama Fault Zone (AFZ)

The Atacama Fault Zone (AFZ; Fig. 6.2) is the most prominent trench-parallel structure of the northern Chilean Coastal Cordillera, extending over more than 1000 km between 20 and 30° S (Cembrano et al. 2005; Gonzales et al. 2003; Riquelme et al. 2003). It formed during mid- to late Jurassic as a left-lateral, intra-arc fault of the Jurassic – early Cretaceous magmatic arc (Scheuber and González 1999), in accordance with sinistral oblique plate convergence of that time (Scheuber and Andriessen 1990; Fig. 6.2).

Two peaks of fault activity are dated at ~150 and ~126 Ma using mylonites from this fault zone (Scheuber et al. 1995). Both peaks followed the intrusion of large shallow plutons, suggesting that increased magmatic heat weakened the crust and triggered localized deformation. After magmatic activity migrated further to the east to the present-day Longitudinal Valley (Scheuber et al. 1994), ductile deformation turned to brittle sinistral slip at around 106 Ma (Taylor et al. 1998; Brown et al. 1993).

The left-lateral Central Valley Fault Zone possibly succeeded the AFZ (Taylor et al. 1998). Parts of the AFZ have been reactivated in Neogene to Quaternary times in response to the uplift of the Coastal Cordillera. Neotectonic kinematics are dominated by normal faulting and minor dextral strike-slip movements (González et al. 2003; Riquelme et al. 2003; Pelz 2000), though left-lateral slip is also observed (Delouis et al. 1998). However, it is likely that the AFZ has not been active as a major strike-slip system since the Miocene and, indeed, Dewey and Lamb (1992) estimated less than 1 mm yr$^{-1}$ of dextral strike-slip since the Pliocene.

### 6.2.2 Precordilleran Fault System (PFS)

The Precordilleran Fault System (PFS, also referred to as the Domeyko fault system, Fig. 6.2) is composed of several anastomosing fault systems, many of which can be traced for more than one hundred kilometers (e.g., Sierra Moreno Fault, West Fault, Limon Verde). These faults are of great economic importance, as they structurally controlled the formation of copper ore deposits (Ossandón et al. 2001). The PFS originated in the late Eocene and formed as an intra-arc, trench-parallel structure in the eastward-migrating magmatic arc of the Paleogene (Reutter et al. 1991). Syn-arc displacements were dextral in accordance with the dextral obliquity of plate convergence. After volcanism ceased in the Precordillera (~38 Ma), the PFS was reactivated several times as a fore-arc structure with mainly sinistral strike-slip displacements (Reutter et al. 1996). The West Fault (or Falla Oeste), being the main strand of the PFS, accommodated about 36 km of sinistral slip (Tomlinson and Blanco 1997b). Contemporaneously with the uplift of the Precordillera in the Neogene, the PFS once again switched its sense of displacement and predominantly accommodated normal and dextral slip of the order of several hundred meters (Tomlinson and Blanco 1997b; Dilles et al. 1997). Although the Precordillera is subject to crustal seismicity (see below), a clear correlation of earthquakes between 22° S and 24° S with the PFS has not been established (Belmonte 2002).

### 6.2.3 Liquiñe-Ofqui Fault Zone (LOFZ)

The Liquiñe-Ofqui Fault Zone (LOFZ, Fig. 6.2) is a fault system, up to 100 m wide, that parallels the Southern Volcanic Zone of the Andes for about 1100 km between 47.5° S, where the Chile Ridge is currently subducted, and 37.5° S, where it joins the southernmost segment of the Central Andean foreland fold-and-thrust belt in the back-arc (Hervé 1976; Cembrano et al. 1996, 2000; see also Vietor and Echtler 2006, Chap. 18 of this volume, and references therein). Cembrano et al. (2000) found evidence for a Mesozoic sinistral-reverse predecessor of the LOFZ. Its initiation as a localized shear zone at the end of the Miocene (~6 Ma) correlates with the impingement of the Chile Ridge with South America and subsequent northward migration of the triple junction (Nelson et al. 1994; Cembrano et al. 2002).

The active LOFZ can be traced as a set of NNE-trending lineaments with which Pliocene to present-day active volcanoes are associated, suggesting a lithospheric dimension of the deformation zone. Hervé (1977) inferred a late Cenozoic dextral sense of shear for the LOFZ and established its basic geodynamic role in accommodating margin-parallel shear imposed by oblique subduction. More recent structural, geochronological (Arancibia et al. 1999; Lavenu and Cembrano 1999; Cembrano et al. 2002), and thermochronological (Thomson 2002) studies demonstrate its Neogene dextral transpressive kinematics. Moreover, the southern segment is extensively exhumed while the northern segment (37.5–39° S) exhibits no exhumation (Rosenau 2004). Rosenau's (2004) kinematic model suggests that the LOFZ system accommodated 84 km (+66/–28 km) of northward translation of the fore-arc sliver, resulting in a Pliocene to Recent mean slip rate of about 17 mm yr$^{-1}$. This accounts for more than half of the margin-parallel component of oblique convergence into the intra-arc zone.

The northern termination of the LOFZ shows en-echelon dextral faults and grabens that have been interpreted as a crustal scale, extensional relay accommodating part of the slip along the LOFZ (Reuther et al. 2003; Melnick et al. 2004). The February 1989, $M_s$ 5.3 Lonquimay (Barrientos and Acevedo-Aránguiz 1992) and November 1965, $M_s$ 5.3 Hudson (Chinn and Isacks 1983) shallow crustal earthquakes demonstrate present-day dextral strike-slip motion along the LOFZ (Fig. 6.4).

### 6.2.4 Deep Geometry and Geophysical Properties

Seismic reflection and electromagnetic imaging provide constraints on the deep geometry of the fault zones. The location of the survey profiles addressed below are shown in the inset of Fig. 6.4.

Along the ANCORP'96 onshore deep seismic reflection profile across the Chilean margin at about 21° S, the surface traces of both the AFZ and the PFS coincide with lateral changes in the reflectivity pattern at depth (ANCORP Working Group 2003). The AFZ forms the boundary between a reflective lower crust in the west, extending from 35 to the subducting slab at 55 km depth, from a less reflective lower crust to the east. Moreover, aftershock seismicity of the 1995 Antofagasta $M_w$ 8.1 earthquake is confined to the west of the AFZ (Sobesiak 2004). A deep-reaching structural anomaly is further indicated by a zone of increased seismicity above the slab in a downward continuation from the AFZ surface expression (Fig. 7b of ANCORP Working Group 2003).

The West Fault branch of the PFS constitutes the western boundary of the highly reflective Quebrada Blanca Bright Spot at a depth of about 35 km. The latter is probably fed by fluids rising through deep reaching faults (ANCORP Working Group 2003). The image from the PRECORP seismic profile at 22.5° S shows another region of increased reflectivity at 15–25 km depth, the eastern edge of which again coincides with the West Fault (Yoon et al. 2003). Two crustal earthquakes beneath the western edge of the Precordillera probably indicate a northward prolongation of dextral shear beyond the PFS: the 2001 $M_w$ 6.3 and 2002 $M_w$ 5.6 Aroma events (Comte et al. 2003; see Fig. 6.4). Aftershocks were distributed subvertically between the surface and the Benioff zone at 90 km depth, indicating lithospheric-scale faulting (Comte et al. 2003).

Long-period electromagnetic investigations reveal well-conducting subsurface domains clearly associated with major faults in the fore-arc of both the central and southern Andes. All the anomalies with resistivities in the range of a few $\Omega$m reach depths of at least 15–20 km and constitute major zones of crustal weakness. At 21° S the ANCORP magnetotelluric profile (Brasse et al. 2002) traverses the AFZ close to the coast in the Coastal Cordillera (see Fig. 22.1 of Schilling et al. 2006, Chap. 22 of this volume, for cross-sectional magnetotelluric profiles). Induction vectors at long periods – derived from the ratio of vertical to horizontal magnetic field variations – were expected to point strictly from west to east due to the strong coast effect, but the oblique strike of the Atacama Fault with respect to the coast line leads to a strong deflection of as much as 90° (Lezaeta 2001).

Similar effects occur in the Precordillera: In all profiles at 22° S, 21° S and 20.3° S a good conductor is modeled at depths of 15–20 km, directly below the trace of the West Fault. It has to be noted that this conductive structure is *not* compatible with the aforementioned deeper bright spots discovered in reflection seismic studies (see Schilling et al. 2006, Chap. 22 of this volume). A near-surface magnetotelluric study of a segment of the West Fault, north of the Chuquicamata copper mine (~22° S), revealed a conductive anomaly resulting from present-day, upper crustal, fluid penetration into the fault zone (Hoffmann-Rothe et al. 2004a). Similar results were obtained from less densely investigated transects across other branches of the PFS (Janssen et al. 2002). Whether fracture permeability and presence of fluids also cause the mid-crustal conductors mentioned above remains speculative.

Using data derived from profiles at 38.9° S and 39.3° S, the middle crust below the volcanic arc in South Chile was modeled as a highly-conductive zone (Brasse and Soyer 2001). However, it is difficult to state if this zone of high conductivity has a causal relation to the Liquiñe-Ofqui Fault Zone.

### 6.3 Short-Term Records: Current Fore-arc Deformation (GPS)

On first inspection, the GPS velocity vectors compiled from various campaigns (listed in the caption of Fig. 6.4) indicate neither localized slip-partitioning on a strike-slip fault system nor northward translation of a fore-arc sliver at present (Klotz et al. 2006, Chap. 4 of this volume; Khazaradze and Klotz 2003): The velocity vectors point roughly parallel to the relative convergence direction of the Nazca and South America plates and their rates gradually decrease away from the trench (Fig. 6.4, cf. Fig. 6.3a). The absence of slip-partitioning is also indicated by the pattern of coseismic deformation resulting from the 1995 $M_w$ 8.0 Antofagasta earthquake (Klotz et al. 1999), as well as by geodetic results from the Arica Bend region, where convergence obliquity varies greatly (Bevis et al. 1999). To test whether the GPS data comprise a long-term permanent deformation signal one usually calculates the residual velocity field by subtracting a short-term deformation field from the observed velocities. For this region, the short-term deformation field has been obtained with elastic dislocation modeling (see Klotz et al. 2006, Chap. 4 of this volume). Here, we want to follow a different approach by having a closer look at published *raw* GPS velocities in order to be independent of uncertainties in the model.

Figure 6.5a shows margin-normal profiles of the GPS velocity vector obliquities for the three segments delineated in Sect. 6.1.2 (Fig. 6.1). Those GPS vectors located trenchward of the strike-slip faults may indicate northward translating fore-arc slivers if the vectors deviate to a more margin-parallel orientation, i.e. increased obliquity beyond the convergence obliquity (cf. Fig. 6.3b). Though there are some higher than average obliquities between the trench and the faults, a systematic deviation is not evident in the diagrams. In the case of an elastically strained fore-arc, one expects rotation of the velocity vectors to a more trench-normal orientation (decreasing obliquity) with increasing distance from the trench (cf. Fig. 6.3a). This is well depicted by Brooks et al. (2003) for the data set between 26° S and 36° S (Fig. 6.5a, central segment, Fig. 6.4).

Where the profile in Fig. 6.5b crosses a major fore-arc-sliver-bounding fault, the trench-parallel component of the velocity vectors is expected to exhibit a step towards lower values, and this may be evident across the AFZ (a step of ~5 mm yr$^{-1}$). Relatively high trench-parallel velocities are also present in the fore-arc bounded by the LOFZ, but a sharp drop to smaller values is not resolved. Note that the GPS vectors south of 37° S have to be interpreted with care since they appear to still be influenced by the postseismic effects of the 1960 $M_w$ 9.5 great Chilean (Valdivia) earthquake (Klotz et al. 2001; Hu et al. 2004). Due to stress relaxation after coseismic elastic rebound (vectors pointing towards the trench), the velocities will likely be biased toward smaller values.

The trench-parallel velocity components of the central segment exhibit a sharp decrease between 250–300 km away from the trench (Fig. 6.5b, middle). Interestingly, almost all vectors with velocities above ~10 mm yr$^{-1}$ within the central segment fall between 33° S and 36° S (data from this latitudinal range are shown with inversely colored symbols in Fig. 6.5b, middle), coinciding with the presence of volcanoes (Figs. 6.4 and 6.6d). This decrease in velocity across the volcanic axis may indicate decoupling of two domains across a weakened segment of crust.

The trench-normal velocity components show less complexity (Fig. 6.5c). The highest velocities range between 25 to 30 mm yr$^{-1}$ for all segments, and each individual data set reveals a rather continuous decline of displacement rates towards the continent.

In conclusion, the horizontal surface velocity field and its landward decay is in accordance with model predictions for an elastically strained fore-arc without the effect of fore-arc sliver motion. Closer inspection of the compiled GPS data may indicate some slip-partitioning, however, this has to be regarded with care due to the superposition of postseismic effects and the lack of data with respect to southern Chile. The GPS evidence for strain-partitioning along the Chilean margin is not comparable to results obtained for the Sumatra oblique subduction zone, for example, where fore-arc motion is clearly resolved (McCaffrey 2002; McCaffrey et al. 2000). In Chile, a contradiction remains between the short-term GPS record of strain-partitioning and the long-term record that provides clear evidence of margin-parallel strike-slip faulting, particularly in the southern segment. This may indicate that either: (*1*) slip-partitioning does not presently occur or (*2*) the lateral distance of the downdip EOC to the position of the strike-slip fault is too short to produce a noticeable effect on the obliquity of the horizontal surface velocity vectors (see Fig. 6.3b, McCaffrey 2002).

1. Fore-arc slip-partitioning can also be inferred by analyzing how the orientation of interplate thrust earthquake slip-vectors deviates from the direction of convergence (e.g., Jarrard 1986a; McCaffrey 1992, 2002; Yu et al. 1993). If the fore-arc accommodates margin-parallel strain plastically at the faults, the slip vectors of interplate thrust earthquakes are less oblique than the direction of convergence. McCaffrey (1996) derived slip-vector obliquities for the Chilean margin from 23° S to 38° S (Fig. 6.6j); there are no interpretable earthquakes beyond this latitude (Fig. 6.6g). Between 23° S, i.e. just south of the Arica-Bend, and 38° S the slip-vectors are rotated towards trench-normal and, on average, deviate from the convergence vector by 6°. This may indicate slip-partitioning to a minor degree, however, taking into account the error estimates, its significance has to be doubted.

This still leaves the possibility that the slip rates at the faults are too small to be detected over the human timescale. McCaffrey (1994) has pointed out that the frequent appearance of margin-parallel strike-slip faults in continental fore-arcs may simply be due to the prolonged existence of these settings providing sufficient time for large total slip. This is certainly conceivable for recent slip rate estimates as low as those for the AFZ.

However, yet another reason for the lack of evidence for present-day strike-slip motion may emerge from

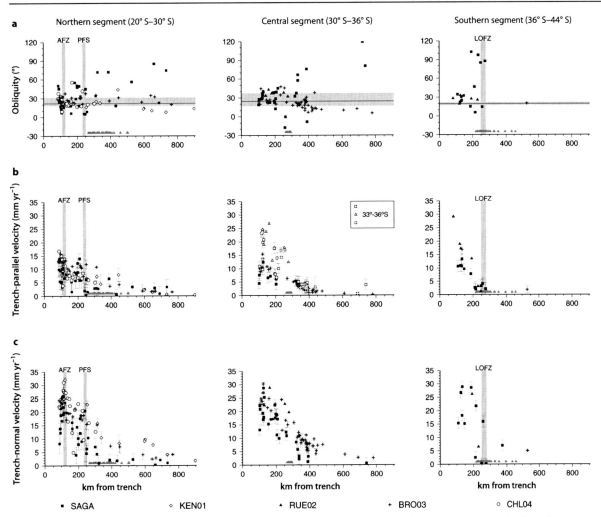

**Fig. 6.5.** Margin-normal profiles of the GPS data shown in Fig. 6.4 for three segments along the Chilean margin. *Gray triangles* at the bottom give the projected positions of volcanoes active in the Holocene. **a** Horizontal GPS surface vector obliquity, defined as the angle the vector makes with the trench-normal azimuth. The *gray horizontal bar* indicates the range of convergence vector obliquities for a given segment and the *thin line* denotes the average value. *Gray vertical bars* indicate positions of margin-parallel strike-slip fault systems. **b** Trench-parallel displacement component. Data in the central segment that fall in the latitudinal range between 33 and 36° S, i.e. south of the volcanic gap, are specially emphasized (*inset*). **c** Trench-normal displacement component

results obtained from analogue simulations of obliquely convergent accretionary settings (Hoffmann-Rothe et al. 2004b): The pattern of deformation in the fore-arc wedge exhibits a cyclicity that corresponds to the episodic accretionary imbrication processes at the deformation front (Gutscher et al. 1998). Monitoring of the experimental surface displacement field for such a setting with oblique subduction reveals that (Hoffmann-Rothe et al. 2004b): (*a*) Trench-normal shortening remains focused at the toe of the fore-arc wedge, as long as a new frontal thrust-imbricate is initiated, while margin-parallel shear reaches far into the fore-arc, and (*b*) spatial strain-partitioning is less pronounced in the subsequent underthrusting phase. An accretionary thrust-cycle, comprising the relatively short thrust-initiation phase (*a*) and the following underthrusting phase (*b*), may last some hundred thousand years (Adam et al. 2004; Kukowski et al. 2000; Kukowski and Oncken 2006, Chap. 10 of this volume). The recurrence interval of new thrust-imbricates may define internal clocking for periods of strong strain partitioning. Considerable slip at margin-parallel faults may thus only be expected over the relatively short time frames of the thrust-initiation phase, possibly not sampled in southern Chile at the present.

2. The geographical distance between the downdip EOC and the strike-slip fault depends on the depth of interplate coupling and on the dip of the shallow subducting slab. The downdip EOC determined from elastic dislocation modeling (Khazaradze and Klotz 2003), shows significant variation from north to south (Fig. 6.6g). In the north, the AFZ sits almost above the

EOC, while the PFS is laterally separated from the EOC by more than 150 km (Fig. 6.4). In the southern segment (south of 35° S), the coupling depth reaches to ~50 km (Fig. 6.6g) and the lateral distance of the LOFZ to the EOC also decreases (Fig. 6.4). South of 42° S, the geodetically estimated depth of coupling occurs laterally within 50 km of the volcanic arc (Fig. 6.6d). This may be the reason why sliver motion is not apparent in the southern Chile GPS vectors despite the geologically derived LOFZ slip rates of 17 mm yr$^{-1}$ since the Pliocene. However, additional data are needed in southern Chile to constrain fore-arc sliver motion more precisely.

## 6.4 Discussion of Factors Controlling Fore-arc Deformation

The potential factors controlling fore-arc deformation and its partitioning are subdivided into three categories: (1) the plate kinematic boundary conditions, primarily the obliquity (Sect. 6.4.1), (2) the upper plate heterogeneities affecting its rheology (capability for localizing horizontal shear) or resistance to block motion ('buttressing', Sect. 6.4.2), and (3) the plate coupling properties that govern the force interaction between the subducting and overriding plates at the plate interface (Sect. 6.4.3). Note that this classification is artificial as many parameters are strongly linked to each other by feedback processes. Figure 6.6 shows a compilation of the spatial variability of parameters along the margin between 14 and 48° S (references are listed in the figure caption). The following is a brief summary of the main trends.

Convergence obliquity is almost uniform between 21 and 48° S, with exceptions at 28° S and 34° S, where seamount chains impinge on the trench line (Fig. 6.6a). The age of the ocean floor (Fig. 6.6b) decreases almost linearly from the Arica bend (20° S) to 39° S, with two little steps across the Challenger Fracture zone (32° S) and the Mocha Fracture zone (38° S); the age gradient increases south of the Mocha Fracture zone towards the Chile triple junction.

The lateral distance of the trench to the volcanic front (Fig. 6.6d) is greatest in the region of the Atacama block at 23° S (cf. Reutter et al. 2006, Chap. 14 of this volume) and narrows by about 100 km south of 40° S. Active volcanoes are absent above the flat-slab segment (Fig. 6.6e) where the subducting slab trajectory beneath South America (at ~100 km depth) is almost horizontal. At shallow depths, the dip angle of the slab becomes smaller from 30° S towards 40° S (Fig. 6.6e). South of 40° S, the model is only weakly constrained, but it is likely that this tendency is sustained in the far south (Tassara et al. 2006).

However, there is a positive gravity anomaly at the coast that broadens towards the south, suggesting a decreasing slab dip at depths less than 50 km (Fig. 6.6k; Hackney et al. 2006, Chap. 17 of this volume). At 39° S, this gravity anomaly is interrupted by a region of negative gravity, which is interpreted by Hackney et al. as reflecting the deepening of the slab beneath the fore-arc.

Precipitation across the fore-arc varies from dry in the north to wet in the south (Fig. 6.6l), and is inversely correlated to the height of the Andean mountain chain and plateau. Changing climatic conditions along strike also influence the thickness of sediments in the trench (Bangs and Cande 1997; Lamb and Davis 2003): the trench is almost devoid of sediments north of the Juan Fernandez Ridge (33° S), while south of it, sediment of more than 2 km thickness fills the trench (Fig. 6.6c).

Sediment supply to the trench is reflected in the predominant tectonic mass transfer modes, with the northern half of the margin being tectonically erosive, while the southern half is accretive. However, the amount of material accreted relative to the subducted sediments varies (e.g., Diaz-Naveas 1999; Sick et al. 2006, Chap. 7 of this volume; Kukowski and Oncken 2006, Chap. 10 of this volume). The averaged margin slope angle ($\alpha$) also changes from north to south (Fig. 6.6f); it is relatively constant along the erosive northern margin, but decreases south of 30° S.

Since 1977, there has been more interplate thrust seismicity north of 35° S compared to the southern segment (Fig. 6.6g), in the latter of which lies the site of the great $M_w$ 9.5 great Chilean earthquake of 1960 (Barrientos and Ward 1990). Likewise, the interplate and crustal seismic energy release is generally lower in the southern segment (Fig. 6.6i). Note, however, that the data cover only part of the seismic cycle.

Tichelaar and Ruff (1991, 1993) suggested that the maximum depth of the interplate seismogenic zone changes at about 28° S from 36–41 km in the north to 48–53 km in the south, based on the distribution of interplate thrust-earthquakes ($M > 6$; Fig. 6.6g). However, with respect to other published estimates of the depth of the seismogenic zone, a rather consistent depth of approximately 50 km emerges for the entire length of the seismically active margin (e.g., Stauder 1973; Tichelaar and Ruff 1991; Suarez and Comte 1993; Comte et al. 1994; Comte et al. 1999; Delouis et al. 1996; Pacheco et al. 1993; Patzwahl et al. 1999; Pardo et al. 2002). The geodetically determined depth-limit of the currently locked segment, estimated using GPS surface velocities, varies from about 33 km depth in the northern segment to about 52 km depth in the south (Fig. 6.6g).

### 6.4.1 Convergence Obliquity and Sense of Slip

The sense of slip at the time of formation of all three major margin-parallel strike-slip systems is consistent with the obliquity at the respective times (Fig. 6.2): the AFZ initiated with sinistral slip as the Aluk (Phoenix) oceanic plate

**Fig. 6.6a–f.**
Compilation of along-strike variations in parameters defining the Nazca-South America subduction margin between 14 and 48° S.
**a** Convergence vector obliquity, the angle the convergence vector azimuth makes with the trench normal, and convergence rate based on estimates of Angermann et al. (1999). The trench line used for calculation is taken from Bird (2003). **b** Ocean floor age of the trench (Müller 1997) and corresponding heat flow estimation after Stein (2003). Positions of major morphologic features of the oceanic plate are indicated. **c** Thickness of trench fill compiled from Bangs and Cande (1997); Hampel et al. (2004). Also shown is the maximum depth of the trench based on Lindquist et al. (2004).
**d** Margin-normal, lateral distance of the trench to the coast (*circles*), the volcanic front (*triangles*) and the position of the downdip end of geodetic coupling as shown in Fig. 6.4.
**e** Averaged dip of the subducting slab for various depth spans on trench-normal transects of the slab model by Tassara et al. (2006). The *gray line* (trench – 50 km depth contour) is used as $\beta$ for the taper model (Sect. 6.4.4).
**f** Averaged margin slope angle along trench-normal transects of the bathymetry and topography grid by Lindquist et al. (2004). The *dashed line* shows the taper angle $(\alpha + \beta)$

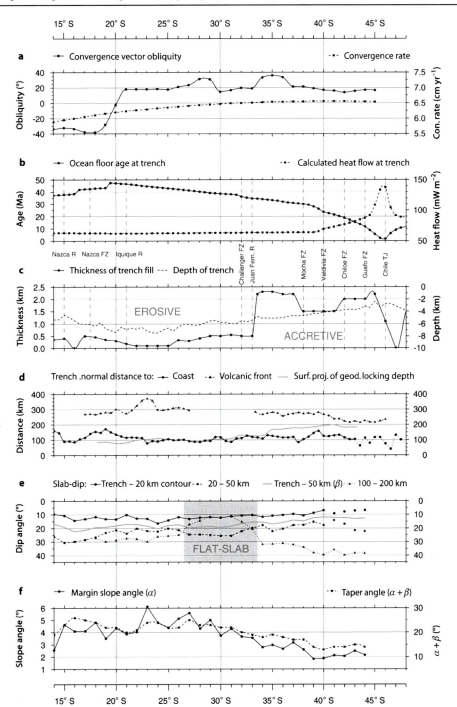

subducted to the SE (Fig. 6.1; Zonenshayn et al. 1984; cf. Scheuber and Gonzales 1999); the PFS and the LOFZ formed under the dextral obliquity of the present plate system, while convergence obliquities were about 35° (Pardo-Casas and Molnar 1987) and 20° (Somoza 1988), respectively.

The degree of obliquity is irrelevant for inducing strain-partitioning in the fore-arc as long as there is a gradient in the margin-parallel shear along the plate interface, the strongest constraint being an abrupt downdip EOC (Fig. 6.3a). However, if slip is to be accommodated permanently in the upper plate, the yield strength and the orientation of the stress tensor, with respect to potential faults, will be of importance. This is the basis for the depiction of oblique subduction systems that comprise a rigid fore-arc bound by the inclined subduction thrust fault and a strike-slip fault ('three plate approximations', e.g., Fitch 1972; Beck 1991; McCaffrey 1992; Platt 1993).

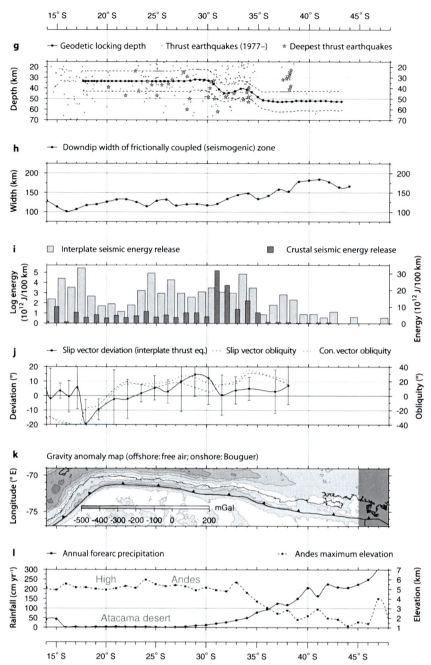

Fig. 6.6g–l. Compilation of along-strike variations in parameters defining the Nazca-South America subduction margin between 14 and 48° S. g Coupling depth estimation based on elastic dislocation modeling of GPS data (from Khazaradze and Klotz 2003). The *dashed lines* mark the downdip end of the fully coupled zone and the downdip end of the subsequent transition zone, respectively. *Gray dots* mark the depth of interplate thrust earthquakes since during 1977–2004 (magnitude 1–10, Harvard CMT catalogue at *http://www.seismology.harvard.edu/CMTsearch.html*; thrust criteria: tension axis plunge between 45° to 90°, intermediate axis plunge between 0° to 45°). *Stars* denote precisely located underthrusting events as determined from teleseismic data and local networks, compiled from Tichelaar and Ruff (1991); Suarez and Comte (1993); Comte et al. (1994); Pardo et al. (2002); Krawcyzk et al. (2003). h Calculated downdip width of the seimogenic zone (see Sect. 6.4.5 and Fig. 6.7). i Interplate seismic energy release (1977 to present) calculated from Harvard CMT catalogue (*light gray columns*). Upper crustal (<70 km) seismic energy release 250–800 km away from the trench (*dark gray columns*, 1900–1995; Gutscher et al. 2000). j Deviation of interplate, thrust earthquake, slip vectors from the convergence vector taken from McCaffrey (1996). Also shown is the slip vector obliquity (*dashed*) and the convergence vector obliquity (*dotted*) used in McCaffrey's (1996) analysis. k Gravity anomaly map. Free-air anomalies shown offshore (KMS2001 satellite altimetry model, Anderson and Knudsen 1998), Bouguer anomalies onshore (combined dataset from SFB267 – Götze et al. (1994), Tašárová (2004) and data from Denizar Blitzkow, Sao Paolo). l Average annual precipitation along the Chilean fore-arc (2° east of the trenchline) sampled from the 10 minute climatology grid of New et al. (2002) and profile of maximum elevation of the Andes extracted from topography (Lindquist et al. 2004)

These simple models require an angle of minimum obliquity depending on the relative strengths of the subduction thrust and the (pre-existing) strike-slip fault that has to be surpassed for the initiation of strike-slip motion on the margin-parallel fault. With respect to the geological record of strike-slip faulting, such conditions must have been met along the Chilean margin. However, as pointed out by McCaffrey (1992), in making a reliable estimation of the actual shear stress ratio at the two faults, one usually lacks sufficient control on the depth of the faults, nor does one have an independent measure of the minimum obliquity.

Three-dimensional analogue sandbox models of oblique subduction provide insights about the deformation patterns of fore-arcs beyond the above simple model with a rigid fore-arc (e.g. Martinez et al. 2002; Hoffmann-Rothe et al. 2004b; McClay et al. 2004). In these models, dry sand simulates brittle fore-arc rheologies, satisfying the Mohr-Coulomb failure criterion (Lohrmann et al. 2003). The surface pattern of fore-arc deformation in such models exhibits slip-partitioning, though not necessarily concentrated at a single margin-parallel strike-slip fault (Martinez et al. 2002; Hoffmann-Rothe et al. 2004b). Martinez et al. (2002) observed partitioning for small angles of obliquity and increasing margin-parallel slip rates with increasing obliquity. Simulations of doubly-vergent orogenic wedges with oblique collision show that for low obliquity (< 45°) margin-parallel slip can be effectively accommodated within individual oblique-slip thrust or strike-slip faults, whereas formation of a linked margin-parallel and sub-vertical strike-slip fault requires obliquities of at least 70° (McClay et al. 2004). These high values could imply that the criteria for the formation of such faults at low convergence obliquity should be sought among processes that effectively weaken parts of the upper plate (see Sect. 6.4.2).

The shear sense of margin-parallel strike-slip faults in Chile does not always correlate with the sense of convergence obliquity, as is evident from the sinistral slip phase of the PFS and the Neogene reactivations along the AFZ (Fig. 6.2). Such slip reversals are not obviously triggered by changes in convergence obliquity or rate. Co- and post-seismic antithetic shearing (with respect to the convergence obliquity) is conceivable if the 'rebound' of the elastically strained fore-arc is partly accommodated permanently at fore-arc faults. Heinze (2003) used this process to model the Plio-Pleistocene left-lateral component of deformation along N-S striking faults in the Altos de Talinay outer fore-arc region (30.5° S). However, this model is unsuitable for explaining large-scale, margin-parallel strike-slip offsets in conflict with the sense of convergence obliquity (as shown in Fig. 6.2). It is apparent that forces other than those controlled by oblique convergence dominate the dynamics of fore-arc deformation in these cases. To cite some examples:

- The Neogene uplift of the Coastal Cordillera, originating from either basal accretion (Adam and Reuther 2000) or flexural isostatic rebound due to the loss of continental material by tectonic erosion (Pelz 2000), is most likely accommodated by normal faulting along the pre-existing AFZ fault system.
- A regime of E-W extension is in agreement with left-lateral offsets observed at some faults (Delouis et al. 1998) and could also be sustained by the gravitational trenchward stretching effect of margin topography (Wang and He 1999).
- Sinistral strike-slip displacement on branches of the PFS is frequently explained with models of clockwise rotation of crustal blocks. Relative movement of these blocks is accommodated on N-S trending sinistral faults. Such rotational patterns could be a consequence of a dextral, E-W-directed shear couple resulting from the margin-parallel shortening gradient in the Central Andes (Tomlinson and Blanco 1997b). Rotations could also be driven by northward shear traction at the base of the brittle crust (Arriagada et al. 2000) or result from oroclinal bending (Kley 1999).

In conclusion, oblique plate convergence is a necessary prerequisite for strain-partitioning at margin-parallel strike-slip faults. However, temporal variations in the obliquity or rate are not the primary controlling factor for switching these fault systems 'on' and 'off'. As the angle of obliquity is almost uniform along the margin, along-strike variations in obliquity cannot cause the differences seen in contemporary strike-slip faulting activity between North and South Chile.

### 6.4.2 Localization of Horizontal Shear and Resistance to Margin-Parallel Fore-arc Motion

Many of the modern, trench-linked, strike-slip faults in continental overriding plates are located close to or in the magmatic arc (Jarrard 1986b; e.g., Great Sumatra Fault, Philippine Fault) implying that zones of shear weakness, caused by thermal softening, control their position (Beck 1983; Saint Blanquat et al. 1998). The history of the AFZ, PFS and LOFZ clearly support this model as all three fault zones were initiated as intra-arc faults (Fig. 6.2). The emplacement of shallow plutons appears to have preceded the formation of the AFZ and PFS, with younger intrusions and hydrothermal activity along the PFS locally controlled by the fault system (Ossandón et al. 2001). The AFZ was ductile during the early Cretaceous, but brittle faulting occurred after the magmatic arc migrated further to the east (Fig. 6.2), exemplifying how a structurally weakened zone can facilitate the localization of deformation. Dextral slip occurs along the modern LOFZ in southern Chile and is coincident with the resumption of magmatic arc activity in both space and time (see Fig. 6.2).

Concurrent with the flattening of the deep slab between 27–34° S (Fig. 6.6e), magmatic activity migrated to the east and ceased by about 5 Ma (Kay et al. 1999; Fig. 6.6d). The surface expression of a major strike-slip fault, occurring at a similar distance from the trench as the LOFZ does further south, is missing and shortening prevails instead (Vietor and Echtler 2006, Chap. 18 of this volume). Gutscher (2001) suggested that flat-slab subduction, in combination with cooler lithospheric temperatures and the resulting stronger rheology of the overriding plate, facilitates the transfer of deformation further landward and, with oblique convergence, the activation of margin-parallel strike-slip activity far from the trench. This model may also apply to the presently-active, 120 km long, dextral El Tigre Fault system located at 31° S in the San Juan Province of Argentina (Siame et al. 1997; Fig. 6.4). Active volcanism is also present south of ~33.5° S, where GPS vectors indicate a possible zone of decoupling along the volcanic axis between 34 and 36° S, despite lacking a major fault at the surface (see Sect. 6.3, Fig. 6.5b).

Based on elastic dislocation models, Savage (1983) proposed that with oblique subduction the strain field above the downdip end of coupling (EOC) on the plate interface will be dominated by margin-parallel shear. McCaffrey (2000) reasoned, with regard to the Sumatra Fault, that a magmatic arc that is in close lateral vicinity to the EOC (less than 100 km) may facilitate the localization of shear strain at a strike-slip fault. Following this rationale, and assuming a downdip end of the seismogenic zone at 50 km depth, one would not expect the present volcanic arc in northern Chile to localize margin-parallel shear, given its lateral distance of almost 200 km from the EOC (Fig. 6.6d). Conversely, tectonic erosion of the North Chile fore-arc has shifted the AFZ to an optimal position relative to the EOC (Fig. 6.4) so as to effectively focus transverse shear at this pre-existing zone of weakness. However, considerable strike-slip is not evident from the present deformation record, possibly indicating that zones of structural weakness are far more inefficient for focusing shear than thermally weakened zones.

The translation of a fore-arc sliver can also be hampered by an obstacle that resists sliver displacement at its leading edge (Beck 1983, Beck et al. 1993). In the area around the Arica bend (20° S), oblique convergence gradually changes from dextral in the south to sinistral in the north along the seaward concave margin (Figs. 6.4 and 6.6a), providing a geometrical 'buttress' as defined by Beck et al. (1993). Estimates of arc-parallel strain rates from slip vector deflection of interplate thrust earthquakes indicate arc-parallel compression in northern Chile (17–31° S; McCaffrey 1996). This is typically expected at the leading edge of a buttressed sliver (Jarrard 1986a; Beck et al. 1993;

Wang 1996). Moreover, the orogenic edifice itself may also prevent northward sliver translation along the northern segment (Beck et al. 1993). This concurs with the cessation of dextral slip at the PFS (West Fault segment) in the late Oligocene, when the main phase of Andean orogenic growth was triggered by the increase of westward motion of the overriding plate (Fig. 6.2; Silver et al. 1998; Oncken et al. 2006, Chap. 1 of this volume).

South of 35° S, the oceanic slab enters the subduction zone at an increasingly shallower angle (Fig. 6.6e) and the geodetically-estimated coupling depth increases to 50 km (Fig. 6.6g), bringing the EOC closer to the magmatic arc (Fig. 6.6d). In addition, the magmatic arc is shifted closer to the trench (Figs. 6.4 and 6.6d), possibly indicating higher temperatures at the top of the slab or in the mantle wedge (England et al. 2004). The increase in temperature could result from the subduction of the Chile Ridge and the associated passage of asthenospheric 'slab windows' (Gorring et al. 1997) and/or an asthenospheric thermal anomaly that manifests in the production of plateau basalts (e.g., Somun Cura basaltic plateau; De Ignacio et al. 2001). This combination of factors, which brings a thermally weakened zone close to the EOC, could be responsible for the modern strike-slip activity of the LOFZ at its southern termination.

The northern termination of the LOFZ, at about 38° S, also suggests the involvement of a buttress. Lavenu and Cembrano (1999) proposed a buttressing effect related to the seaward-concave curvature of the margin around the impinging Juan Fernandez Ridge (33° S). The authors further speculated that the northward transition from steep to flat slab subduction (beyond a depth of 100 km; Fig. 6.6e) at about 34° S may additionally hamper northward sliver translation. Neotectonic field data show N-S-directed compression in the fore-arc beyond the northern end of the LOFZ (e.g., Reuther et al. 2003; Lavenu and Cembrano 1999). The analysis of earthquake slip data between 39 and 31° S, however, shows less significant arc-parallel compression than in northern Chile (McCaffrey 1996).

To conclude, the most plausible control factor for the localization of horizontal shear on large-scale, margin-parallel strike-slip faults is a thermally weakened zone within the upper plate. The seemingly optimal condition of a pre-existing fault (AFZ) in close vicinity to the downdip EOC on the plate interface is not sufficient in the presence of a geometrical buttress such as the Arica Bend.

### 6.4.3 Force Interaction at the Plate Interface

The degree of force interaction between the converging plates of a subduction system is often evoked for controlling fore-arc deformation and slip-partitioning (e.g., see Chemenda et al. 2000, and references therein). In the following discussion, the term 'plate coupling' refers to the 'coupling force' established at the plate interface. This coupling force is a function of the shear stresses on the interface that need to be overcome and, most importantly, of the frictionally coupled area and its position along the slab (Wang and He 1999; see also Wang and Dixon 2004; Klotz et al. 2006, Chap. 4 of this volume). The wider (in down-dip view) and the deeper the frictionally coupled area, the greater the plate coupling force and the further the arcward reach of compression (Wang and He 1999). A plate interface, with a smaller integrated force that resists shearing, will exhibit less upper plate deformation and more relative slip between the plates then will be a 'stronger' interface. It is important to note that neither the magnitude of subduction earthquakes, which depends on the rupture area involved, nor 'seismic coupling', being the long-term ratio of seismic to aseismic slip, is a measure for the plate coupling force or strength of the interface (e.g., Wang and He 1999).

Is there evidence for present-day changes in interplate coupling forces along the Chilean margin that may contribute to the variability of strike-slip activity? A number of published and new parameters are reviewed regarding their implications for interface friction and shear stresses, as well as extent and position of the coupled area:

1. *Distribution of seismicity.* On the basis of a statistical analysis of upper plate seismic energy release that shows relatively high values between 31 and 34° S (Fig. 6.6i, right), Gutscher et al. (2000) inferred an increased coupling force along the flat slab segment due to the larger (viscous) plate contact area. However, long-term coupling characteristics should be evaluated from many seismic cycles along the entire length of the margin, to avoid misinterpretation due to short-term variations in seismicity. For example, the Harvard CMT seismic record shows a distinctive short-term increase of seismicity in the above mentioned latitudinal range since 1992. Similarly, the region of relative crustal and interplate seismic quiescence (Fig. 6.6g) and reduced seismic energy release south of 35° S (Fig. 6.6i) roughly coincides with the rupture zone of the 1960 $M_w$ 9.5 great Chilean earthquake (Barrientos and Ward 1990). The only exception to this coincidence is in the vicinity of the Arauco Peninsula at about 37.5° S, where seismic energy release is locally higher (Bohm et al. 2002). The very effective release of the accumulated elastic strain during this large earthquake can be attributed to a smooth plate interface that can rupture over an enormous area, which is probably caused by the subduction of a coherent sediment layer (Tichelaar and Ruff 1991). In summary, however, the distribution of seismicity does not constrain the along-strike variation in plate coupling force.

2. *Trench sediments.* The thick sediment fill in the trench south of the Juan Fernandez Ridge (Fig. 6.6c) could reduce the friction along the plate interface because subducted sediments carry large amounts of water. This may weaken the subduction fault by causing high fluid pressures, as water is released, while permeability is low (Wang et al. 1995; Lamb and Davis 2003). Also, the well stratified, wet and poorly consolidated sediments themselves could constitute a layer of reduced rheological strength or contain potential detachment layers that weaken the plate interface (Shreve and Cloos 1986; Lamb and Davis 2003, and references therein). Downward transport of sediments is further facilitated where sediment-filled fracture zones subduct, as is the case south of 37° S, and this can also lower the melting temperatures at the mantle wedge (Singer et al. 1996). Lamb and Davis (2003) suggested that interface friction is effectively reduced if the trench fill thickness reaches ~500 m. von Huene and Ranero (2003) estimated that about 1.6 km of the 2.5 km thick sediment pile is subducted just south of the Juan Fernandez Ridge.

3. *Mode of mass transfer.* High roughness of the subducting seafloor, as seen north of 33° S, where the sedimentary cover is thin, likely facilitates erosion of the continental margin (Lallemand et al. 1994). The process of subduction erosion, however, does not necessarily imply high basal friction (von Huene and Ranero 2003) and increased plate coupling forces. From a global perspective, a controlling influence of either erosive or accretive mass transfer along subduction margins for the formation of margin-parallel strike-slip faults has to be questioned, as the formation of fore-arc slivers is observed in both of these settings (e.g., erosive: Peru, Kukowski et al. (2004); southern Kurils, Chemenda et al. (2000); accretive: Sumatra, McCaffrey et al. (2000); Cascadia, Wang (2000), Adam et al. (2004)).

4. *Age of oceanic plate and convergence rate.* Ruff and Kanamori (1983) and Jarrard (1986a) proposed that interplate shear stresses are a function of the age of the subducted oceanic plate and the convergence rate. Though their original rational of correlating the prevalence of large earthquakes to strong plate coupling forces is outdated (see above), the theoretical reasoning is still common: Young and buoyant oceanic crust, as well as fast convergence, increase plate coupling due to increased normal stresses acting on the plate interface. The Chilean margin contradicts this correlation, according to a model calculated by Yáñez and Cembrano (2004) in which shear forces on a viscous slip zone between the oceanic and continental plates balance the buoyancy forces associated with the trench and margin topography. Their results indicate decreasing interface strength from the Arica bend region towards the Chile Ridge, where oceanic crust is youngest.

5. *Downdip width of frictionally coupled zone.* The downdip limit of the seismogenic zone can be regarded as a proxy for the depth extent of the frictionally coupled plate interface (e.g., Wang and Suyehiro 1999). The downdip transition from unstable to stable aseismic sliding has been proposed to approximately coincide with the 350 °C isotherm or, where fore-arc crust is thin, with the intersection between the upper-plate Moho and the plate interface (e.g., Oleskevich et al. 1999). Several studies estimated the depth extent of the seismogenic zone along the Chilean margin from inter-plate thrust earthquakes, yielding a depth of about 50 km with no substantial along-strike variation (see Sect. 6.4, Fig. 6.6g). This consistent depth suggests that any variation in the thermal regime of the subduction zone, either resulting from the decreasing age of the subducting oceanic plate toward the south (Fig. 6.6b) or from the thicker insulating sedimentary cover in southern Chile (Fig. 6.6c), does not effectively influence the depth extent of the seismogenic zone (e.g., Wang 2000). A worldwide study further supports the proposition that the width of the seismogenic zone is not correlated to the age of the oceanic plate nor to its convergence rate (Pacheco et al. 1993).

   Using the slab geometry above the 50 km depth contour (split into dip values for slab segments from the trench to 20 km depth, and from 20 to 50 km depth, Fig. 6.6e), and taking into account an average 10 km depth for the updip limit of the seismogenic zone (Oleskevich et al. 1999), we calculate an estimate of the downdip width of the frictional plate contact along the margin (Fig. 6.6h): It evolves from around 90 km width along the northern segment, to 97 km along the central segment, increasing to 128 km along the southern segment. This would imply an increased coupling force in southern Chile if the effective coefficient of basal friction remains unchanged.

6. *Fore-arc wedge geometry.* We have attempted a qualitative estimate of the variation in the effective coefficient of interface friction along strike, using Dahlen's (1990) solution of the critical taper problem of a dry cohesionless wedge (his equation 74 and following, Fig. 6.7). It is assumed that the geometry of the fore-arc wedge above the frictionally-coupled plate interface is an expression of the ratio of basal and internal strength of the fore-arc. It is important to keep in mind that this approach goes beyond the theory's original formulation for accretionary wedges or foreland fold and thrust-belts, and requires the fore-arc wedge to be in a critical state. This is presumed as: (*i*) the converging system has been unchanged for a very long time, enabling the margin to attain a critical taper, especially along the northern segment; (*ii*) the taper varies smoothly along the trench (Fig. 6.6f); and (*iii*) the fore-arc is seismically active along its entire length (on the verge of Coulomb failure) and features (normal) faults cutting through the crust to the interface (see also Adam and Reuther 2000; Melnick et al. 2004). The taper geometry of the fore-arc wedge between the trench line and the 50 km depth contour of the slab is estimated from the slab dip ($\beta$, Fig. 6.6e, gray curve) and the approximated slope angle ($\alpha$, Fig. 6.6f). With the assumption that the fore-arc crust has constant internal strength ($\Phi = 30°$, Byerlee 1978), the decreasing taper angle ($\alpha + \beta$) towards the south reflects a decreasing effective coefficient of interface friction by 16%

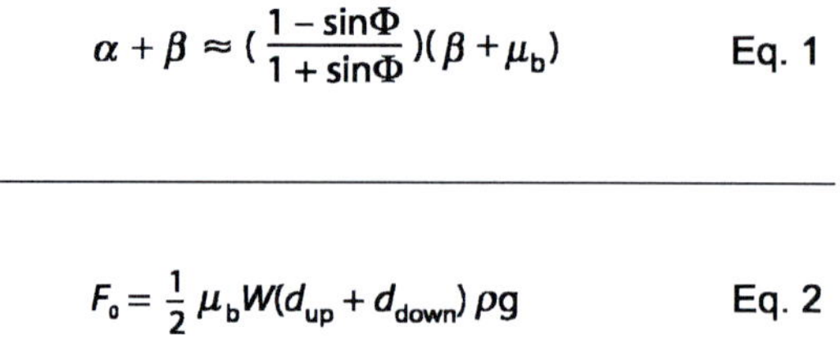

|  | Northern segment (20°–30° S) | | Southern segment (36°–44° S) | |
|---|---|---|---|---|
| $\Phi$ (°) | 30 | | 30 | |
| $\alpha$ (°) | 4.7 | | 2.3 | |
| $\beta$ (°) | 18.4 | | 13.4 | $\alpha + \beta \approx (\frac{1 - \sin\Phi}{1 + \sin\Phi})(\beta + \mu_b)$    Eq. 1 |
| $\mu_b$ | 0.56 (0.89) | | 0.47 (0.59) | |
| $\mu_b(N) / \mu_b(S)$ | | 1.20 | | |
| $W$ (m) | 124 000 | | 167 000 | |
| $d_{up}$ (m) | 3600 | | 6712 | $F_0 = \frac{1}{2} \mu_b W(d_{up} + d_{down}) \rho g$    Eq. 2 |
| $d_{down}$ (m) | 50 933 | | 50 110 | |
| $\mu_b(N) / \mu_b(S)$ | | 1.40 | | |

$\Phi$: angle of internal friction; $\alpha$: averaged slope angle; $\beta$: averaged slab dip angle (note: $\alpha$ and $\beta$ are in radians for calculation); $\mu_b$: calculated basal friction coefficient; $W$: mean downdip width of frictionally coupled zone for given segments; $d_{up}$: updip depth of $W$; $d_{down}$: downdip depth of $W$; Eq. 1 (approximat solution) from Dahlen (1990), note that the effective friction coefficients have been calculated using Dahlen's accurate solution, approximations given in parentheses; Eq. 2 from Wang and He (1999).

**Fig. 6.7.** Evaluation of interface frictional properties along the northern and southern Chilean margin

(Fig. 6.7). This concurs with the reasoning that sediment supply lubricates the interface.

7. *Interface shear force estimation.* The ratio of effective basal friction coefficients between the northern and southern segment can also be evaluated using a different set of empirical input data related to the width and position of the plate contact. We use Wang and He's (1999) analytical approximation of the shear force as integrated effect of shear stress on the plate contact interface (their equation 7, see Fig. 6.7). Under the assumption of constant coupling forces along the trench an effective coefficient of basal friction along the northern segment, about 1.4 times higher (29%) than the southern segment, would be required to compensate for the increased downdip width of the frictionally coupled plate contact towards the south, as estimated above (Fig. 6.7). This is higher than the friction ratio obtained by taper theory (16%, item 6 above). The downdip depth of 50 km probably over-estimates the extent of the seismogenic zone in southern Chile, given that the Moho can be imaged at a depth of about 35–40 km (Bohm et al. 2002; Yuan et al. in press) and that the great Chile $M_w$ 9.5 earthquake nucleated at similar depth (Krawcyzk et al. 2003). In the case of a reduced downdip depth of e.g. 45 km, the friction ratio obtained would be similar to the value resulting from taper theory. Alternatively, this could be achieved by assuming a slightly higher net coupling force in southern Chile, showing that it is difficult to obtain quantitative results as long the plate contact geometry is not precisely known. However, as Wang and He's approach ignores the contribution of tectonic forces, which are assumed to be constant along the trench, and particularly makes no assumptions about a critical taper geometry, the similarity of the two ratios obtained may support the proposition that the fore-arc wedge geometry in fact reflects the prevailing ratio of internal to basal friction of the fore-arc (item 6).

In conclusion, establishing the current state of plate coupling in southern Chile compared to northern Chile remains ambiguous as, to date, it is not known how to balance the various processes that might affect force interaction at the plate boundary. A prevalence of arguments suggest reduced interface friction along the southern Chile segment (items 2, 4, 6, 7). This is probably compensated by an increasing downdip width of the frictionally coupled interface toward the south (item 5), which could imply that the plate coupling forces along the Chilean margin remain more or less constant on a regional scale. With respect to the geodetic locking depth, which deepens from ~35 km beneath the northern segment to ~50 km beneath the southern segment (Fig. 6.6g; Khazaradze and Klotz 2003; Klotz et al. 2006, Chap. 4 of this volume), it could be argued that modeling of the observed fore-arc GPS velocities requires a larger downdip width of the geodetically locked plate interface at which slip is imposed at full convergence rate in order to balance a reduced effective coefficient of friction.

The above findings make it difficult to give a definite answer to the question posed at the beginning of Sect. 6.4.3. However, a direct positive relationship between the degree of force interaction at the plate interface and fore-arc sliver motion, as proposed by Jarrard 1986a, cannot be inferred. On the contrary, presuming the plate contact for the southern segment is actually weakened by the large quantity of sediments that have filled the trench since the Pliocene (Bangs and Cande 1997), then the temporal correlation with the onset of strike-slip on the LOFZ would actually suggest an inverse relationship.

## 6.5 Conclusions

The long and complex history of margin-parallel strike-slip faulting and associated fore-arc sliver formation is an important component of the evolving Chilean fore-arc. By comparing the long-term geological, as well as the short-term seismological and geodetical records of margin-parallel faulting with the pattern and variability of various parameters along the trench, we have attempted to evaluate the potential factors controlling the formation of fore-arc slivers. A single dominating parameter controlling the mode of fore-arc deformation does not emerge on either timescale. On the contrary, it is most remarkable that, despite along-strike changes of conditions, such as trench-fill, mass transfer mode, plateau or no plateau, slab-dip variations, etc., the GPS derived fore-arc velocity field that is dominated by elastic loading effects (human timescale) appears to be fairly homogenous. The spatial and temporal heterogeneity of fore-arc strike-slip faulting along the Chilean margin that even includes slip-sense reversals, exemplifies that, on a geological timescale, the forces imposed on the margin by the plate kinematics can be 'overruled' by more short-lived regional forces (e.g., in response to underplating, gravitational collapse, oroclinal bending). The identification of a single predominant controlling parameter may thus even be impossible.

Variations in recent strike-slip motion do not appear to depend on convergence obliquity nor rate, nor on the mode of mass transfer at the subduction front. Estimates of the variation in interplate coupling force along the trench remain ambiguous. A southward increase of the frictionally coupled area at the plate contact, and a decreasing taper angle of the fore-arc wedge, can be reconciled in the framework of taper theory by a southward decrease of the effective coefficient of friction at the plate interface. With respect to the modern activity of the Liquiñe-Ofqui Fault Zone, a positive correlation of increased coupling force and the appearance of margin-parallel strike-slip faulting is not evident.

The development of for-earc slivers seems to be primarily controlled by mechanisms that cause effective rheological weakening of parts of the upper plate and by geometries that hamper margin-parallel sliver motion, such as the seaward concave-shaped margin in northernmost Chile. The Pliocene activation of the Liquiñe-Ofqui Fault Zone was likely facilitated by the exceptionally short distance of the volcanic arc from the trench, combined with a shallowly dipping slab in the upper 50 km that transferred stresses at the base of the overriding plate further towards the arc. Thermal weakening of the upper plate could be a consequence of the passage of the Chile Ridge system beneath the overriding plate (slab windows), and the close vicinity of a mantle anomaly at the southern end of the Nazca slab is most likely responsible for the Somun Cura basaltic plateau and associated magmatism.

## Acknowledgments

This work was funded by the Deutsche Forschungsgemeinschaft within the framework of the collaborative research project SFB267 'Deformation processes in the Andes'. We wish to thank Andres Tassara for providing the slab-model from which the slab-dip data were taken, as well as Ron Hackney and Denizar Blitzkow for providing the fore-arc gravity map. This work benefited from discussions with Heinrich Brasse, Daniel Melnick, Tim Vietor and Sergei Medvedev. Our special thanks to Susan Ellis and Kelin Wang for thoughtful reviews and suggestions that undoubtedly helped to improve the manuscript. We gratefully acknowledge the help of Henning Zirbes with compiling the data presented in Figs. 6.4 and 6.6. Figures 6.3 to 6.5 were generated using the public domain software GMT (Wessel and Smith 1998). Thanks to Allison Britt for improving the language of our manuscript.

## References

Abels A, Bischoff L (1999) Clockwise block rotations in northern Chile: indications for a large-scale domino mechanism during the middle-late Eocene. Geology 27(8):751–754

Adam J, Reuther CD (2000) Crustal dynamics and active fault mechanics during subduction erosion. Application of frictional wedge analysis on to the North Chilean Forearc. Tectonophysics 321:297–325

Adam J, Klaeschen D, Kukowski N, Flueh E (2004) Upward delamination of Cascadia Basin sediment infill with landward frontal accretion thrusting caused by rapid glacial age material flux. Tectonics 23(3): doi 10.1029/2002TC001475

ANCORP Working Group (2003) Seismic imaging of a convergent continental margin and plateau in the Central Andes (Andean Continental Research Project 1996 (ANCORP'96)). J Geophys Res 108(B7): doi 10.1029/2002JB001771

Anderson OB, Knudsen P (1998) Global marine gravity field from ERS-1 and Geosat geodetic mission altimetry. J Geophys Res 103: 8129–8137

Angermann D, Klotz J, Reigber C (1999) Space-geodetic estimation of the Nazca-South America Euler vector. Earth Planet Sci Lett 171:329–334

Arancibia G, Cembrano J, Lavenu A (1999) Dextral transpression and deformation partitioning in the Liquine-Ofqui Fault Zone, Aisen, Chile (44–45°S). Rev Geologica Chile 21(1):3–22

Armijo R, Thiele R (1990) Active faulting in northern Chile: ramp stacking and lateral decoupling along a subduction plate boundary. Earth Planet Sci Lett 98:40–61

Arriagada C, Roperch P, Mpodozis C (2000) Clockwise block rotations along the eastern border of the Cordillera de Domeyko, Northern Chile (22°45'–23°30'S). Tectonophysics 326:153–171

Arriagada C, Roperch P, Mpodozis C, Dupont Nivet G, Cobbold PR, Chauvin A, Cortés J (2003) Paleogene clockwise tectonic rotations in the forearc of central Andes, Antofagasta region, northern Chile. J Geophys Res 108(B1): doi 10.1029/2001JB001598

Bangs NL, Cande SC (1997) Episodic development of a convergent margin inferred from structures and processes along the southern Chile margin. Tectonics 16(3):489–503

Barrientos SE, Ward SN (1990) The 1960 Chile earthquake: inversion for slip distribution from surface deformation. Geophys J Int 103:589–598

Barrientos SE, Acevedo Aránguiz PS (1992) Seismological aspects of the 1988-1989 Lonquimay (Chile) volcanic eruption. J Volcanol Geotherm Res 53:73–87

Barrier E, Huchon P, Aurelio MA (1991) Philippine Fault: a key to Philippine kinematics. Geology 19:32–35

Beck ME Jr (1983) On the mechanism of tectonic transport in zones of oblique subduction. Tectonophysics 93:1–11

Beck ME Jr (1991) Coastwise transport reconsidered: lateral displacements in oblique subduction zones, and tectonic consequences. Phys Earth Planetary Interiors 68:1–8

Beck ME Jr (1998) On the mechanism of crustal block rotations in the central Andes. Tectonophysics 299: 75–92

Beck ME Jr (1999) On the mechanism of crustal block rotations in the central Andes – Reply. Tectonophysics 213:467–469

Beck ME Jr, Rojas C, Cembrano J (1993) On the nature of buttressing in margin-parallel strike-slip fault systems. Geology 21: 755–758

Bellier O, Sébrier M (1995) Is the slip rate variation on the Great Sumatran fault accommodated by forearc stretching? Geophys Res Lett 22:1969–1972

Belmonte A (2002) Crustal seismicity, structure and rheology of the upper plate between the Precordillere and the volcanic arc in northern Chile (22°S–24°S). PhD thesis, Freie Universität Berlin, http://www.diss.fu-berlin.de/2002/202/

Bevis M, Martel SJ (2001) Oblique plate convergence and interseismic strain accumulation. Geochem Geophys Geosyst 2: doi 2000GC000125

Bevis M, Smalley R Jr, Herring T, Godoy J, Galban F (1999) Crustal motion north and south of the Arica deflection: Comparing recent geodetic results from the Central Andes. Geochem Geophys Geosyst 1: doi 1999GC000011

Bird P (2003) An updated digital model of plate boundaries. Geochem Geophys Geosyst 4(3): doi 10.1029/2001GC000252

Bohm M, Lüth S, Echtler H, Asch G, Bataille K, Bruhn C, Rietbrock A, Wigger P (2002) The Southern Andes between 36° and 40°S latitude: seismicity and average seismic velocities. Tectonophysics 356:275–289

Brasse H, Soyer W (2001) A magnetotelluric study in the Southern Chilean Andes. Geophys Res Lett 28(19):3757–3760

Brasse H, Lezaeta P, Rath V, Schwalenberg K, Soyer W, Haak V (2002) The Bolivian Altiplano conductivity anomaly. J Geophys Res 107(B5): doi 10.1029/2001JB000391

Brooks BA, Bevis M, Smalley R Jr, Kendrick E, Manceda R, Lauría E, Maturana R, Araujo M (2003) Crustal motion in the Southern Andes (26°–36°S): do the Andes behave like a microplate? Geochem Geophys Geosyst 4(10): doi 10.1029/2003GC0005054

Brown M, Diáz F, Grocott J (1993) Displacement history of the Atacama fault system 25°00′S–27°00′S, northern Chile. Geol Soc America Bull 105:1165–1174

Burbridge DR, Braun J (1998) Analogue models of obliquely convergent continental plate boundaries. J Geophys Res 103(B7): 15221–15237

Byerlee J (1978) Friction of rocks. Pure Applied Geophys 116:615–626

Cembrano J, Hervé F, Lavenu A (1996) The Liquine Ofqui fault zone: a long-lived intra-arc fault system in southern Chile. In: Geodynamics of the Andes, Vol. 259(1–3):55–66, Elsevier, Amsterdam

Cembrano J, Schermer E, Lavenu A, Sanhueza A (2000) Contrasting nature of deformation along an intra-arc shear zone, the Liquiñe-Ofqui fault zone, southern Chilean Andes. Tectonophysics 319: 129–149

Cembrano J, Lavenu A, Reynolds P, Arancibia G, López G, Sanhueza A (2002) Late Cenozoic transpressional ductile deformation north of the Nazca-South America-Antarctica triple junction. Tectonophysics 354:289–314

Cembrano J, González G, Arancibia G, Ahumada I, Olivares V, Herrera V (2005) Fault zone development and strain partitioning in an extensional strike-slip duplex: a case study from the Mesozoic Atacama fault system, Northern Chile. Tectonophyscis 400:105–125

Chemenda A, Lallemand S, Bokun A (2000) Strain partitioning and interplate friction in oblique subduction zones: constraints provided by experimental modeling. J Geophys Res 105(B3):5567–5581

Chinn DS, Isacks BL (1983) Accurate source depths and focal mechanisms of shallow earthquakes in western South America and in the New Hebrides island arc. Tectonics 2(6):529–563

Chlieh M, De Chalabier JB, Ruegg JC, Armijo R, Dmowska R, Campos J, Feigl KL (2004) Crustal deformation and fault slip during the seismic cycle in the North Chile subduction zone, from GPS and InSAR observations. Geophys J Int 158:695–711

Comte D, Pardo M, Dorbath L, Dorbath C, Haessler H, Rivera L, Cisternas A, Ponce L (1994) Determination of seismogenic interplate contact zone and crustal seismicity around Antofagasta, northern Chile using local data. Geophys J Int 116:553–561

Comte D, Dorbath L, Pardo M, Monfret T, Haessler H, Rivers L, Frogneux M, Glass B, Meneses C (1999) A double-layered seismic zone in Arica, northern Chile. Geophys Res Let 26(13):1965–1968

Comte D, Dorbath C, Dorbath L, Farías M, David C, Haessler H, Glass B, Correa E, Balmaceda I, Cruz A, Ruz L (2003) Distribución temporal y en profundidad de las réplicas del sismo superficial de Aroma, norte de Chile del 24 de Julio de (2001) X Congreso Geológico Chileno 2003, Universidad de Concepción, Chile

Dahlen FA (1990) Critical taper model of fold-and-thrust belts and accretionary wedges. Annu Rev Earth Planet Sci 18:55–99

De Ignacio C, López I, Oyarzun R, Márquez A (2001) The northern Patagonian Somuncura plateau basalts: a product of salb-induced, shallow asthenospheric upwelling? Terra Nova 13(2):117–121

Delouis B, Cisternas A, Dorbath L, Rivera L, Kausel E (1996) The Andean subduction zone between 22 and 25°S (northern Chile): Precise geometry and state of stress. Tectonophysics 259:81–100

Delouis B, Philip H, Dorbath L, Cisternas A (1998) Recent crustal deformation in the Antofagasta region (northern Chile) and the subduction process. Geophys J Int 132:302–338

Dewey JF, Lamb S H (1992) Active tectonics of the Andes. Tectonophysics 205:79–95

Díaz Naveas JL (1999) Sediment subduction and accretion at the Chilean Convergent Margin between 35°S and 40°S. PhD thesis, University of Kiel

Dilles J, Tomlinson AJ, Martin M, Blanco N (1997) The El Abra and Fortuna complexes: a porphyry copper batholith sinistrally displaced by the Falla Oeste. In: VIII Congreso Geológico Chileno, ACTAS Vol III – Nuevos Antecedentes de la Geologí a del Distrio de Chuquicamata, Periodo 1994–1995, Sessión 1: Geología Regional: pp 1878–1882, Universidad Catolica del Norte, Chile

Echtler H, Glodny J, Gräfe K, Rosenau M, Melnick D, Seifert W, Vietor T (2003): Active Tectonics controlled by inherited structures in the long-term stationary and non-plateau South-Central Andes. EGS-AGU-EUG Joint Assembly, Nice, France, Abstract EAE03-A-10902

England P, Engdahl R, Thatcher W (2004) Systematic variation in the depth of slabs beneath arc volcanoes. Geophys J Int 156:377–408

Fitch TJ (1972) Plate convergence, transcurrent faults, and internal deformation adjacent to Southeast Asia and the western Pacific. J Geophys Res 77:4432–4460

Forsythe RD, Nelson E (1985) Geological manifestation of ridge collision: evidence for the Golfo de Penas, Taito basin, southern Chile. Tectonics 4:477–495

González G, Cembrano J, Carrizo D, Macci A, Schneider H (2003) Link between forearc tectonics and Pliocene-Quaternary deformation of the Coastal Cordillera, Northern Chile. J S Am Earth Sci 16:321–342

Gorring ML, Kay SM, Zeitler PK, Ramos VA, Rubiolo D, Fernandez MI, Panza JL (1997) Neogene Patagonian plateau lavas: continental magmas associated with ridge collision at the Chile triple junction. Tectonics 16(1):1–17

Götze H-J, Lahmeyer B, Schmidt S, Strunk S (1994) The lithospheric structure of the Central Andes (20°–26°S) as inferred from quantitative interpretation of regional gravity. In: Reutter KJ, Scheuber E, Wigger P (eds) Tectonics of the Southern Central Andes. Springer, Heidelberg, pp 7–21

Gutscher MA (2001) An Andean model of interplate coupling and strain partitioning applied to the flat subduction zone of WS Japan (Nankai Trough). Tectonophysics 333:95–109

Gutscher MA, Kukowski N, Malavielle J, Lallemand S (1998) Episodic imbricate thrusting and underthrusting: Analog experiments and mechanical analysis applied to the Alaskan Accretionary Wedge. J Geophys Res 103:10161–10176

Gutscher MA, Spakman W, Bijwaard H, Engdahl ER (2000) Geodynamics of flat subduction: seismicity and tomographic constraints from the Andean margin. Tectonics 19(5):814–833

Hackney RI, Echtler HP, Franz G, Götze H-J, Lucassen F, Marchenko D, Melnick D, Meyer U, Schmidt S, Tašárová Z, Tassara A, Wienecke S (2006) The segmented overriding plate and coupling at the south-central Chilean margin (36–42°S). In: Oncken O, Chong G, Franz G, Giese P, Götze H-J, Ramos VA, Strecker MR, Wigger P (eds) The Andes – active subduction orogeny. Frontiers in Earth Science Series, Vol 1. Springer-Verlag, Berlin Heidelberg New York, pp 355–374, this volume

Hampel A, Kukowski N, Bialas J (2004) Ridge subduction at an erosive margin: the collision zone of the Nazca Ridge in southern Peru. J Geophys Res 109: doi 10.1029/2003JB002593

Hartley AJ (2003) Andean uplift and climate change. J Geol Soc London 160:7–10

Heinze B (2003) Active intraplate faulting in the forearc of North Central Chile (30°–31° S), Scientific Technical Report STR03/07. PhD thesis, GeoForschungsZentrum Potsdam, http://www.gfz-potsdam.de/bib/zbstr.htm

Hervé M (1976) Estudio geológico de la falla Liquiñe-Reloncaví en la área de Liquiñe: antecedentes de un movimiento transcurrente (Provincia de Valdivia). Actas I Congreso Geológico Chileno B, pp 39–56

Hervé F (1977) Petrology of the crystalline basement of the Nahuelbuta Mountains, southcentral Chile. In: Ishikawa T, Aguirre L (eds) Comparative Studies on the Geology of the Circum-Pacific Orogenic Belt in Japan and Chile, 1st Report. Japan Society Prom. Science Tokyo, pp 1–51

Hoffmann-Rothe A, Ritter O, Janssen C (2004a) Correlation of electrical conductivity and structural damage at a major strike-slip fault in Northern Chile. J Geophys Res 109: doi 10.1029/2004JB003030

Hoffmann-Rothe A, Kukowski N, Oncken O (2004b) Phase dependent strain partitioning in obliquely convergent settings. Bollettino di Geofisica 45:93–97

Hu Y, Wang K, He J, Klotz J, Khazaradze G (2004) Three-dimensional viscoelastic finite element model for postseismic deformation of the great 1960 Chile earthquake. J Geophys Res 109: doi 10.1029/2004JB003163

Janssen C, Hoffmann-Rothe A, Tauber S, Wilke H (2002) Internal structure of the Precordilleran fault system (Chile) – insights from structural and geophysical observations. J Struc Geol 24:123–143

Jarrard RD (1986a) Relations among subduction parameters. Reviews of Geophysics 24(2):217–284

Jarrard RD (1986b) Terrane motion by strike-slip faulting of forearc slivers. Geology 14:780–783

Kay SM, Mpodozis C, Coira B (1999) Neogene magmatism, tectonism, and mineral deposits of the Central Andes (22° to 33° S Latitude). In: Skinner BJ (ed) Geology and ore deposits of the Central Andes. Soc Economic Geol Spec Publ 7:27–59

Kendrick E, Bevis M, Smalley R Jr, Brooks BA (2001) An integrated crustal velocity field for the central Andes. Geochem Geophys Geosyst 2: doi 10.1029/2001GC000191

Khazaradze G, Klotz J (2003) Short- and long-term effects of GPS measured crustal deformation rates along the south central Andes. J Geophys Res 108: doi 10.1029/2002JB001879

Kley J (1999) Geologic and geometric constraints on a kinematic model of the Bolivian orocline. J South American Earth Sci 12:221–235

Klotz J, Angermann D, Michel G W, Porth R, Reigber C, Reinking J, Viramonte J, Perdomo R, Rios VH, Barrientos S, Barriga R, Cifuentes O (1999) GPS-derived deformation of the Central Andes including the 1995 Antofagasta Mw = 8.0 earthquake. Pure Appl Geophys 154:3709–3730

Klotz J, Khazaradze G, Angermann D, Reigber C, Perdomo R, Cifuentes O (2001) Earthquake cycle dominates contemporary crustal deformation in Central and Southern Andes. Earth Planetary Science Letters 193:437–446

Klotz J, Abolghasem A, Khazaradze G, Heinze B, Vietor T, Hackney R, Bataille K, Maturana R, Viramonte J, Perdomo R (2006) Long-term signals in the present-day deformation field of the Central and Southern Andes and constraints on the viscosity of the Earth's upper mantle. In: Oncken O, Chong G, Franz G, Giese P, Götze H-J, Ramos VA, Strecker MR, Wigger P (eds) The Andes – active subduction orogeny. Frontiers in Earth Science Series, Vol 1. Springer-Verlag, Berlin Heidelberg New York, pp 65–90, this volume

Krawczyk C, SPOC Team (2003) Amphibious seismic survey images plate interface at 1960 Chile earthquake. EOS 84(32):301, 304–305

Kukowski N, Oncken O (2006) Subduction erosion – the "normal" mode of fore-arc material transfer along the Chilean margin? In: Oncken O, Chong G, Franz G, Giese P, Götze H-J, Ramos VA, Strecker MR, Wigger P (eds) The Andes – active subduction orogeny. Frontiers in Earth Science Series, Vol 1. Springer-Verlag, Berlin Heidelberg New York, pp 217–236, this volume

Kukowski N, Schillhorn T, Flueh ER, Huhn K (2000) A newly identified strike slip plate boundary in the north-east Arabian Sea. Geology 28:355–358

Kukowski N, Hampel A, Krabbenhöft A, Bialas J (2004) Varying rates and modes of subduction erosion along the Peruvian margin. AGU Fall Meeting, San Francisco

Kus J, Block M, Reichert C, Diaz Naveas J, Ladage S, Urbina O, SO161-cruise participants (2006) NW-SE structural segmentation of the south-central Chilean offshore forearc area between 35°S and 40°S. Marine Geology (submitted)

Lallemand SE, Schnürle P, Malavielle J (1994) Coulomb theory applied to accretionary and nonaccretionary wedges: possible causes for tectonic erosion and/or frontal accretion. J Geophys Res 99(B6):12033–12055

Lamb S, Davis P (2003) Cenozoic climate change as possible cause for the rise of the Andes. Nature 425:792–797

Lavenu A, Cembrano J (1999) Compressional and tranpressional stress pattern for Pliocene and Quaternary brittle deformation in fore arc and intra arc zones (Andes of Central and Southern Chile). J Struc Geol 21:1669–1691

Lezaeta P (2001): Distortion analysis and 3D modeling of magnetotelluric data in the Southern Central Andes. PhD thesis, Free University Berlin, http://www.diss.fu-berlin.de/2001/108/

Lindquist KG, Engle K, Stahlke D, Price E (2004) Global topography and bathymetry grid improves research efforts. EOS 85(19):186

Lohrmann J, Kukowski N, Adam J, Oncken O (2003) The impact of analogue material properties on the geometry, kinematics, and dynamics of convergent sand wedges. J Struc Geol 25:1691–1711

López Escobar L, Cembrano J, Moreno H (1995) Geochemistry and tectonics of the Chilean Southern Andes Quaternary volcanism (37°–46°S). Rev Geológica Chile 22(2):219–234

Martinez A, Malavielle J, Lallemand S, Collot JY (2002) Strain partitioning in an accretionary wedge, in oblique convergence: analogue modelling. Bull Soc Géol France 173(1):17–24

McCaffrey R (1992) Oblique plate convergence, slip vectors, and forearc deformation. J Geophys Res 97(B6):8905–8915

McCaffrey R (1994) Global variability in subduction thrust zone-forearc systems. PAGEOPH 142(1):173–224

McCaffrey R (1996) Estimates of modern arc-parallel strain rates in fore arcs. Geology 24(1):27–30

McCaffrey R (2002) Crustal block rotations and plate coupling. In: Stein S, Freymueller JT (eds) Plate boundary zones. Geodynamics Series 30:101–122

McCaffrey R, Zwick PC, Bock Y, Prawirodirdjo L, Genrich J, Stevens C W, Puntodewo SSO, Subarya C (2000) Strain partitioning during oblique convergence in northern Sumatra: geodetic and seismologic constraints and numerical modeling. J Geophys Res 105(B12):28363–28376

McClay KR, Whitehouse PS, Dooley T, Richards M (2004) 3D evolution of fold and thrust belts formed by oblique convergence. Mar Petr Geol 21:857–877

McQuarry N (2002) Initial plate geometry, shortening variations, and evolution of the Bolivian orocline. Geology 30(10):867–870

Melnick D, Rosenau M, Folguera A, Echtler H (2006) Late Cenozoic tectonic evolution, western flank of the Neuquén Andes between 37 and 39° south latitude. In: Kay SM, Ramos VA (eds) Late Cretaceous to recent magmatism and tectonism of the Southern Andean Margin at the latitude of the Neuquén Basin (36–39° S). Geol Soc Am Spec P

Müller RD, Roest WR, Royer JY, Gahagan LM, Sclater JG (1997) Digital isochrons of the world's ocean floor. J Geophys Res 102(B2):3211–3214

Muñoz J, Troncoso R, Duhart P, Crignola P, Farmer L, Stern CR (2000) The relation of the mid-Tertiary coastal magmatic belt in south-central Chile to the late Oligocene increase in plate convergence rate. Rev Geológica Chile 27:177–203

Nelson E, Forsythe R, Arit I (1994) Ridge collision tectonics in terrane development. J South American Earth Sci 7(3–4):271–278

New M, Lister D, Hulme M, Makin I (2002) A high-resolution data set of surface climate over global land areas. Climate Res 21:1–25

Oleskevich DA, Hyndman RD, Wang K (1999) The updip and downdip limits to great subduction earthquakes: thermal and structural models of Cascadia, south Alaska, SW Japan, and Chile. J Geophys Res 104(B7):14965–14991

Oncken O, Hindle D, Kley J, Elger K, Victor P, Schemmann K (2006) Deformation of the central Andean upper plate system – facts, fiction, and constraints for plateau models. In: Oncken O, Chong G, Franz G, Giese P, Götze H-J, Ramos VA, Strecker MR, Wigger P (eds) The Andes – active subduction orogeny. Frontiers in Earth Science Series, Vol 1. Springer-Verlag, Berlin Heidelberg New York, pp 3–28, this volume

Ossandón G, Fréraut R, Gustafson LB, Lindsay DD, Zentilli M (2001) Geology of the Chuquicamata mine: a progress report. Econ Geology 96:249–270

Pacheco JF, Sykes LR, Scholz CH (1993) Nature of seismic coupling along simple plate boundaries of the subduction type. J Geophys Res 98(B8):14133–14159

Pardo Casas F, Molnar P (1987) Relative motion of the Nazca (Farallon) and South American plates since late Cretaceous time. Tectonics 6:233–248

Pardo M, Comte D, Monfret T (2002) Seismotectonics and stress distribution in the central Chile subduction zone. J South American Earth Sci 15:11–22

Patzwahl R, Mechie J, Schulze A, Giese P (1999) Two-dimensional velocity models of the Nazca plate subduction zone between 19.5°S and 25°S from wide-angle seismic measurements during CINCA95 project. J Geophys Res 104(B4):7293–7317

Pelz K (2000) Tectonic erosion along the central Andean forearc (20–24°S). PhD thesis, GeoForschungsZentrum Potsdam, Scientific Technical Report STR00/20

Platt JP (1993) Mechanics of oblique convergence. J Geophys Res 98(B9):16239–16256

Rabassa J, Clapperton CM (1990) Quaternary glaciations of the Southern Andes. Quaternary Sci Rev 9:153–174

Reinecker J, Heidbach O, Tingay M, Connolly P, Müller B (2004) The 2004 release of the World Stress Map. http://www.world-stress-map.org

Reuther CD, Potent S, Bonilla R (2003) Crustal stress history and geodynamic processes of a segmented active plate margin: South-Central Chile: the Arauco Bío-Bío trench arc system. X Congreso Geológico Chileno 2003, Universidad de Concepción, Chile

Reutter KJ, Scheuber E, Helmcke D (1991) Structural evidence of orogen-parallel strike slip displacements in the Precordillera of northern Chile. Geologische Rundschau (Int J Earth Sci) 80:135–153

Reutter K-J, Scheuber E, Chong G (1996) The Precordilleran fault system of Chuquicamata, Northern Chile: evidence for reversals along arc-parallel strike-slip faults. Tectonophysics 259:213–228

Reutter K-J, Charrier R, Götze H-J, Schurr B, Wigger P, Scheuber E, Giese P, Reuther C-D, Schmidt S, Rietbrock A, Chong G, Belmonte-Pool A (2006) The Salar de Atacama Basin: a subsiding block within the western edge of the Altiplano-Puna Plateau. In: Oncken O, Chong G, Franz G, Giese P, Götze H-J, Ramos VA, Strecker MR, Wigger P (eds) The Andes – active subduction orogeny. Frontiers in Earth Science Series, Vol 1. Springer-Verlag, Berlin Heidelberg New York, pp 303–326, this volume

Riquelme R, Martinod J, Hérail G, Darrozes J, Charrier R (2003) A geomorphological approach to determining the Neogene to recent tectonic deformation in the coastal Cordillera of northern Chile (Atacama). Tectonophysics 361:255–275

Rosenau M (2004) Tectonis of the Southern Andean intra-arc zone (38°–42°S), PhD thesis, Freie Universität Berlin

Ruegg JC, Campos J, Armijo R, Barrientos S, Briole P, Thiele R, Arancibia M, Cañuta J, Duquesnoy T, Chang M, Lazo D, Lyon-Caen H, Ortlieb L, Rossignol JC, Serrurier L (1996) The Mw = 8.1 Antofagasta (North Chile) Earthquake of July 30 (1995) first results from teleseimic and geodetic data. Geophys Res Lett 23(9):917–920

Ruegg JC, Campos J, Madariaga R, Kausel E, De Chabalier J B, Armijo R, Dimitrov D, Georgiev I, Barrientos S (2002) Interseismic strain accumulation in south central Chile from GPS measurements, 1996–1999. Geophys Res Lett 29(11): doi 1517, 10.1029/2001GL013438

Ruff L, Kanamori H (1983) Seismic coupling and uncoupling at subduction zones. Tectonophysics 99:99–117

Saint Blanquat M, Tikoff B, Teyssier C, Vigneresse JL (1998) Transpressional kinematics and magmatic arcs. In: Holdsworth RE, Strachan R, Dewey JF (eds) Continental transpressional and transtensional tectonics. Geol Soc London Spec Publ 135:327–340

Savage JC (1983) A dislocation model of strain accumulation and release at a subduction zone. J Geophys Res 88(B6):4984–4996.

Scheuber E, Andriessen PAM (1990) The kinematic and geodynamic significance of the Atacama Fault Zone, Northern Chile. J Structural Geology 12(2):243–257

Scheuber E, González G (1999) Tectonics of the Jurassic-early Cretaceous magmatic arc of the north Chilean Coastal Cordillera (22°–26°S): a story of coupling and decoupling in the subduction zone. Tectonics 18(5):895–910

Scheuber E, Bogdanic T, Jensen A, Reutter K-J (1994) Tectonic development of the North Chilean Ades in relation to plate convergence and magmatism since the Jurassic. In: Reutter K-J, Scheuber E, Wigger P (1994) Tectonics of the Southern Central Andes. Springer-Verlag, Berlin Heidelberg New York, pp 121–139

Scheuber E, Hammerschmidt K, Friedrichsen H (1995) $^{40}Ar/^{39}Ar$ and Rb-Sr analyses from ductile shear zones from the Atacama Fault Zone, Northern Chile: the age of deformation. Tectonophysics 250:61–87

Schilling FR, Trumbull RB, Brasse H, Haberland C, Asch G, Bruhn D, Mai K, Haak V, Giese P, Muñoz M, Ramelow J, Rietbrock A, Ricaldi E, Vietor T (2006) Partial melting in the Central Andean crust: a review of geophysical, petrophysical, and petrologic evidence. In: Oncken O, Chong G, Franz G, Giese P, Götze H-J, Ramos VA, Strecker MR, Wigger P (eds) The Andes – active subduction orogeny. Frontiers in Earth Science Series, Vol 1. Springer-Verlag, Berlin Heidelberg New York, pp 459–474, this volume

Seifert W, Rosenau M, Echtler H (2005) The evolution of the South Central Chile magmatic arcs: Crystallization depths of granitoids estimated by hornblende geothermobarometry – implications for mass transfer processes along the active continental margin. N Jb Geol Paläont 236:115–127

Shreve RL, Cloos M (1986) Dynamics of sediment subduction, melange formation, and prism accretion. J Geophys Res 91(B10):10229–10245

Siame LL, Bourlès DL, Sébrier M, Bellier O, Castano JC, Araujo M, Perez M, Raisbeck GM, Yiou F (1997) Cosmogenic dating ranging from 20 to 700 ka od a series of alluvial fan surfaces affected by the El Tigre fault, Argentina. Geology 25(11):975–978

Siebert L, Simkin T (2002) Volcanoes of the world: an illustrated catalog of holocene volcanoes and their eruptions. Smithsonian Institution, Global Volcanism Program Digital Information Series, GVP-3, http://www.volcano.si.edu/world/

Sick C, Yoon M-K, Rauch K, Buske S, Lüth S, Araneda M, Bataille K, Chong G, Giese P, Krawczyk C, Mechie J, Meyer H, Oncken O, Reichert C, Schmitz M, Shapiro S, Stiller M, Wigger P (2006) Seismic images of accretive and erosive subduction zones from the Chilean margin. In: Oncken O, Chong G, Franz G, Giese P, Götze H-J, Ramos VA, Strecker MR, Wigger P (eds) The Andes – active subduction orogeny. Frontiers in Earth Science Series, Vol 1. Springer-Verlag, Berlin Heidelberg New York, pp 147–170, this volume

Sillitoe RH, McKee EH (1996) Age of supergene oxidation and enrichment in the Chilean porphyry copper province. Economic Geol 91:164–179

Silver PG, Russo RM, Lithgow-Bertelloni C (1998) Coupling of South America and African plate motion and plate deformation. Science 279:60–63

Singer BS, Leeman WP, Thirlwall MF, Rogers NW (1996) Does fracture zone subduction increase sediment flux and mantle melting in subduction zones? Trace element evidence from Aleutian arc basalts. In: Bebot GE, Scholl DW, Kirby SH, Platt JP (eds) Subduction: top to bottom. AGU, Geophysical Monograph 96:285–291

Sobesiak M (2004) Fault plane structure of the 1995 Antofagasta earthquake (Chile) derived from local seismological parameters. PhD thesis, University of Potsdam

Somoza R (1998) Updated Nazca (Farallon)-South America relative motions during the last 40 My: implications for mountain building in the central Andean region. J South American Earth Sci 11(3):211–215

Somoza R, Singer S, Tomlinson A (1999) Paleaomagnetic study of upper Miocene rocks from northern Chile: implications for the origin of late Miocene–Recent tectonic rotations in the southern Central Andes. J Geophys Res 104(B10):22923–22936

Stauder W (1973) Mechanism and spatial distribution of Chilean earthquakes with relation to subduction of the oceanic plate. J Geophys Res 7(23):5033–5061

Stein CA (2003) Heat flow and flexure at subduction zones. Geophys Res Lett 30(23): doi 10.1029/2003GL018478

Stern C (1989) Pliocene to present migration of the volcanic front, Andean Southern Volcanic Front. Rev Geológica Chile 16(2):145–162

Suarez G, Comte D (1993) Comment on 'Seismic coupling along the chilean subduction zone' by B. W. Tichelaar and L. R. Ruff. J Geophys Res 98(B9):15825–15828

Tassara A, Schmidt S, Götze H-J (2006) Three-dimensional density model of the oceanic Nazca plate and the Andean continental margin. J Geophys Res

Tašárová, Z (2004) Gravity data analysis and interdisciplinary 3D modelling of a convergent plate margin (Chile, 36–42 S). PhD thesis, Free University Berlin, http://www.diss.fu-berlin.de/2005/19

Taylor GK, Grocott J, Pope A, Randall DE (1998) Mesozoic fault systems, deformation and fault block rotation in the Andean forearc: a crustal scale strike-slip duplex in the Coastal Cordillera of northern Chile. Tectonophysics 299:93–109

Tichelaar BW, Ruff LJ (1991) Seismic coupling along the Chilean subduction zone. J Geophys Res 96:11997–12022

Tichelaar BW, Ruff LJ (1993) Depth of seismic coupling along subduction zones. J Geophys Res 98:2017–2037

Thomson S (2002) Late Cenozoic geomorphic and tectonic evolution of the Patagonian Andes between latitudes 42°S and 46°S: an appraisal based on fission-track results from the transpressional intra-arc Liquiñe–Ofqui fault zone. Geol Soc Am Bulletin 114(9):1159–1173

Tomlinson AJ, Blanco N (1997a) Structural evolution and displacement history of the West Fault system, Precordillera, Chile: part I, synmineral history. In: VIII Congresso Geológico Chileno, ACTAS Vol III – Nuevos Antecedentes de la Geologí a del Distrio de Chuquicamata, Periodo 1994–1995, Sessión 1: Geología Regional, Universidad Catolica del Norte, pp 1873–1877

Tomlinson AJ, Blanco N (1997b) Structural evolution and displacement history of the West Fault system, Precordillera, Chile: part II, postmineral history. In: VIII Congresso Geológico Chileno, ACTAS Vol III – Nuevos Antecedentes de la Geologí a del Distrio de Chuquicamata, Periodo 1994–1995, Sessión 1: Geología Regional, Universidad Catolica del Norte, pp 1878–1882

Vietor T, Echtler H (2006) Episodic Neogene southward growth of the Andean subduction orogen between 30° S and 40° S – plate motions, mantle flow, climate, and upper-plate structure. In: Oncken O, Chong G, Franz G, Giese P, Götze H-J, Ramos VA, Strecker MR, Wigger P (eds) The Andes – active subduction orogeny. Frontiers in Earth Science Series, Vol 1. Springer-Verlag, Berlin Heidelberg New York, pp 375–400, this volume

Vietor T, Echtler H, Müller H, Oncken O, Ladage S (2004) Late Miocene block rotations in the South Chilean offshore forearc (38.5°S) linked to backarc shortening: paleomagnetic and bathymetric evidence. EOS 85(47) Fall Meeting Supplements (Abstract)

von Huene R, Ranero C R (2003) Subduction erosion and basal friction along the sediment-starved convergent margin of Antofagasta, Chile. J Geophys Res 108: doi 10.1029/2001JB001569

Wallace LM, Beavan J, McCaffrey R, Darby D (2004) Subduction zone coupling and tectonic block rotations in the north Island, New Zealand. J Geophys Res 109: doi 10.1029/2004JB00324

Wang K (1996) Simplified Analysis of horizontal stresses in a buttressed forearc sliver at an oblique subduction zone. Geophys Res Letters 23(16):2021–2024

Wang K (2000) Stress-strain 'paradox', plate coupling, and forearc seismicity at the Cascadia and Nankai subduction zones. Tectonophysics 319:321–338

Wang K, Dixon T (2004) 'Coupling' semantics and science in earthquake research. EOS 85(15):180

Wang K, He J (1999) Mechanics of low-stress forearcs: Nankai and Cascadia. J Geophys Res 104(B7):15191–15205

Wang K, Suyehiro K (1999) How does plate coupling affect crustal stresses in Northeast and Southwest Japan? Geophys Res Lett 26(15):2307–2310

Wang K, Mulder T, Rogers GC, Hyndman RD (1995) Case for low coupling stress on the Cascadia subduction fault. J Geophys Res 100(B7):12907–12918

Wessel P, Smith WHF (1998) New, improved version of the Generic Mapping Tools released. EOS 79(47):579

Yáñez G, Cembrano J (2004) Role of viscous plate coupling in the late Tertiary Andean tectonics. J Geophys Res 109: doi 10.1029/2003JB002494

Yoon M, Buske S, Lüth S, Schulze A, Shapiro SA, Stiller M, Wigger P (2003) Along-strike variations of crustal reflectivity related to the Andean subduction process. Geophys Res Lett 30(4): doi 10.1029/2002GL015848

Yu G, Wesnousky SG, Ekström G (1993) Slip Partitioning along major convergent plate boundaries. Pure Appl Geophysics 140(2):183–210

Yuan X, Asch G, Bataille K, Bock G, Bohm M, Echtler H, Kind R, Oncken O, Wölbern I (2006) Deep seismic images of the Southern Andes. In: Kay SM, Ramos VA (eds) Late Cretaceous to recent magmatism and tectonism of the Southern Andean margin at the latitude of the Neuquén Basin (36–39°S). Geol Soc Am Spec P

Zonenshayn LP, Savostin L, Sedov A (1984) Global paleogeodynamic reconstruction for the last 160 million years. Geotectonics 18:181–195

# Seismic Images of Accretive and Erosive Subduction Zones from the Chilean Margin

Christof Sick · Mi-Kyung Yoon · Klaus Rauch · Stefan Buske · Stefan Lüth · Manuel Araneda · Klaus Bataille
Guillermo Chong · Peter Giese† · Charlotte Krawczyk · James Mechie · Heinrich Meyer · Onno Oncken
Christian Reichert · Michael Schmitz · Serge Shapiro · Manfred Stiller · Peter Wigger

**Abstract.** Modern seismic imaging methods were used to study the subduction processes of the South American convergent margin. The data came from reflection and from wide-angle/refraction experiments acquired within the framework of the Collaborative Research Center SFB267 'Deformation Processes in the Andes'. Two areas of differing character and subduction type were investigated: an erosive margin to the north (19–26° S) and an accretionary margin to the south (36–40° S). Results from different seismic models yield three main transects that give an overall impression about the internal structure below the Chilean margin. At the erosive margin, we find that the upper part of the subducting oceanic lithosphere is characterized by a horst-and-graben structure that coincides with the coupling zone between the plates. Strong coupling between oceanic crust and fore-arc in the case of a horst-continent collision is also indicated by plate-parallel faults beneath the lower continental slope, which we interpret as the upper parts of the subduction channel. In this context, the subduction channel represents the downgoing Nazca Plate as well as those portions of the continental crust which moved landward. Low seismic velocities below the coastline also represent parts of the subduction channel and of the hydrofractured base of the upper crust near the plate interface. Between 45 and 60 km depth, a double reflection zone marks the upper and lower boundary of the subducted oceanic crust. Off southern Chile, the ocean bottom is characterized by relatively smooth morphology. In contrast, in the south, the trench is filled with sediments and contains an axial channel (Figs. 7.16 to 7.18) extending in N-S direction along the trench axis within the investigation area. The periodicity of the reflected seismic signal within these sediments correlates with the main glacial cycle during the Quaternary. The recent accretionary wedge is built up from strongly heterogeneous unconsolidated sediments. Frontal accretion takes place within the southern working area except for the region around the Arauco Peninsula, which shows uplift due to basal accretion and antiformal stacking. Below the Coastal Cordillera, the heterogeneity of the modern accretionary wedge and the antiformal stack structure of the Permo-Triassic accretionary wedge complicate imaging at depths greater than about 30 km. Thus, we obtain an image of the top of the subduction channel as a thin reflector segment only to about 25 km depth.

## 7.1 Introduction

The South American convergent margin exhibits significant diversity of key convergence parameters such as mass transfer, pattern of seismicity, upper plate deformation and kinematics (e.g. von Huene and Scholl 1991; Isacks 1988). These processes, in particular the role of the plate interface (i.e. its geometry and properties) for the evolution of the convergence system, have been the subject of some debate (e.g. Jordan et al. 1983; Jensen et al. 1984; Tebbens and Cande 1997; Yanez et al. 2002). The mode of mass transfer (tectonically erosive versus accretive) and the seismogenic plate interface, especially, are poorly understood, mainly due to a lack of high resolution images encompassing the entire plate boundary system, from the downgoing plate across the offshore and onshore fore-arc system and arc into the continental back-arc. The subduction of the Nazca Plate below the South American Plate shows different characteristics along the Chilean continental margin. While in northern Chile subduction erosion is active (von Huene et al. 1999), accretionary wedge growth occurs south of 28° S (e.g. Bangs and Cande 1997). The deeper geometry of the material flux, however, is unknown. In addition, beneath the north Chilean margin, especially below the Precordillera and the Western Cordillera, earlier investigations found a vertical offset between seismic reflections attributed to the downgoing Nazca and the subduction zone seismicity (e.g. ANCORP Working Group 1999). Thus, the internal structure of the Wadati-Benioff-Zone, and its relation to the plate interface, may exhibit more complexity than usually assumed and requires the inclusion of more recent seismological information. Other reflection structures within the upper plate, such as the Quebrada Blanca Bright Spot, have been observed (e.g. ANCORP Working Group 1999, 2003), but whether they are the result of crustal melts, fluids, or lithological layering is still controversial.

In the framework of the Collaborative Research Center (SFB 267), these and other seismic investigations in Chile were carried out onshore and offshore in the last decade to provide clearer images of the above aspects (Fig. 7.1). During this period, four land experiments took place, PISCO (Proyecto de Investigación Sismológica de la Cordillera Occidental) in 1994 (Lessel 1997; Schmitz et al. 1999), PRECORP (Precordilleran Research Program) in 1995 (Yoon et al. 2003), ANCORP (Andean Continental Research Program) in 1996 (ANCORP Working Group 1999, 2003; Buske et al. 2002) and the reconnaissance profile ISSA (Integrated Seismological experiment in the Southern Andes) in 2000 (Bohm et al. 2002; Lüth et al. 2003). In cooperation with the Federal Institute for Geosciences and Natural Resources (Bundesanstalt für Geowissenschaften und Roh-

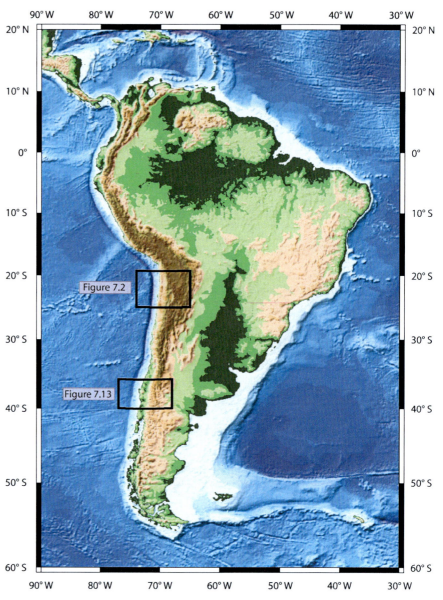

**Fig. 7.1.**
Topographic map of South America (Database: *http://topex.ucsd.edu/*). The *black rectangles* mark the investigation areas, where the northern one corresponds to the CINCA area, including the ANCORP and PRECORP profiles and the southern one to the SPOC area and the ISSA profile. For details see Figs. 7.2 and 7.13, respectively

stoffe (BGR)), the research vessel "SONNE" was used to accomplish the CINCA (Crustal Investigations off- and onshore Nazca/Central Andes; von Huene et al. 1999) and SPOC (Subduction Processes Off Chile; Reichert and SPOC Scientific Shipboard Party 2002; Krawczyk et al. 2003) marine seismic reflection experiments in 1995 and 2001, respectively. At the same time, a number of onshore stations recorded the airgun shots during these offshore measurements (Patzwahl et al. 1999; Lüth et al. 2003). Besides these onshore wide-angle/refraction recordings during the CINCA and SPOC experiments, several offshore wide-angle lines were acquired. This paper, presents, discusses and builds a common framework of the representative results from seismic imaging, and the corresponding interpretations for each of the different seismic experiments, to create a picture of the deformation processes in the Andes.

## 7.2 The Erosive Margin

### 7.2.1 Working Area

Most of the data sets used were acquired in Chile, except for parts of the ANCORP profile and the refraction line ISSA, which were partly acquired in Bolivia and Argentina, respectively (Fig. 7.2). The CINCA study area extends laterally between 19° S and 26° S with the seaward part extending to approximately 74° W longitude. During the experiment, about 4 500 km of marine reflection, 1 300 km of wide-angle/refraction and 460 km of onshore wide-angle seismic data were acquired. In addition to the streamer lines, the airgun shots were recorded by several onshore stations. The wide-angle profiles of the PISCO experiment were also included

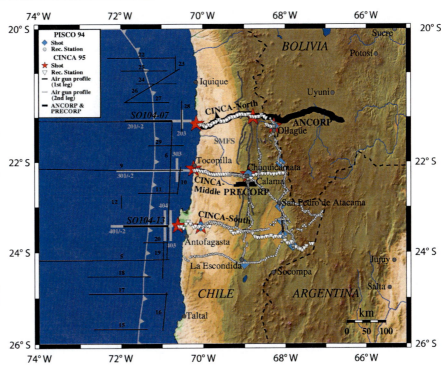

**Fig. 7.2.**
Overview of the CINCA area. Offshore: The *solid black lines* denote the airgun profiles acquired during the CINCA experiment, whereas the *thicker solid black lines* mark profiles SO104-07 and SO104-13. The latter are oriented nearly perpendicular to the trench (shown by the *gray line with triangles*). The *solid gray lines* mark the offshore wide-angle profiles (e.g. Flüh et al. 1997). Onshore: The *thick gray lines* show the locations of the deep reflection experiments ANCORP and PRECORP. Shot locations of the CINCA wide-angle profiles are depicted by *black stars*, and of PISCO by *gray diamonds*. The corresponding receiver locations are shown as *white triangles* and *gray circles*, respectively

(Fig. 7.2; gray diamonds and circles) and detailed information about PISCO can be found in Schmitz et al. (1999).

In 1996, the ANCORP line was acquired at 21° S along a 385 km east-west profile (maximal N-S extension of 50 km) as a combined refraction and reflection experiment. It started at the Chilean coast and traveled eastwards ending in Bolivia (Fig. 7.2). The 1995 deep seismic reflection profile PRECORP, located at 22.5° S near Calama, was about 50 km long. Some images of both experiments have already been published: PRECORP by Yoon et al. (2003); and ANCORP by the ANCORP Working Group (1999, 2003), Buske et al. (2002) and Yoon et al. (2003).

### 7.2.2 Wide-Angle and Refraction Seismics

In 1994, the PISCO experiment was performed between 21° S and 24.5° S (Fig. 7.2) with wide-angle/refraction seismic profiling comprising two N-S striking profiles at 68° W (320 km long) and 69° W (70 km long) (Lessel 1997; Schmitz et al. 1999). A total of 33 shots were recorded, each by 70 stations. The combined offshore and onshore wide-angle experiment of the 1995 CINCA project was done in two stages. In the first stage, airgun shots were deployed at five positions and recorded by an array at the coast consisting of 12 seismic stations spaced 1.85 km apart. In the second stage, three airgun lines were repeated for inline recordings on ocean bottom hydrophones (OBH) and onshore seismic stations. The onshore stations also recorded chemical onshore and offshore explosions at both ends of the profiles (Fig. 7.2; Hinz et al. 1995).

The wide-angle experiment along the ANCORP transect, carried out as a roll-along survey with a 25 km receiver spread, was realized by repeated shots at larger offsets resulting in 10 shot gathers with a maximum offset of approximately 230 km (ANCORP Working Group 1999). The shot points were located 30–60 km distant to each other and the receiver spacing was approximately 90–100 m. Most of the shot gathers had a maximum shot-receiver distance of less than 160 km resulting in high resolution to approx. 40 km depth but only providing restricted information from greater depths. The structure of the deeper continental crust was investigated using seismic refraction profiles measured in earlier experiments (Wigger et al. 1994; Schmitz et al. 1999). The data, processing and interpretation of the CINCA part are described in detail by Patzwahl (1998) and Patzwahl et al. (1999) and ANCORP by the ANCORP Working Group (1999, 2003) and by Lüth (2000).

The velocity-depth models resulting from the PISCO data (Fig. 7.3; Lessel 1997) show different characteristics. In the Precordillera (Fig. 7.3b) beneath 20 km, a zone of low velocity (about 5.9–6.3 km s$^{-1}$) is located which is more pronounced in the northern part of the model. Furthermore, some discontinuities are visible between 50–70 km depth where the velocity at the lowermost discontinuity increases to about 8.0 km s$^{-1}$. At 68° W in the Western Cordillera (Fig. 7.3a), the model shows lower velocities near the surface. In the region of Ollagüe, a low velocity zone (LVZ) is located within the upper continental crust

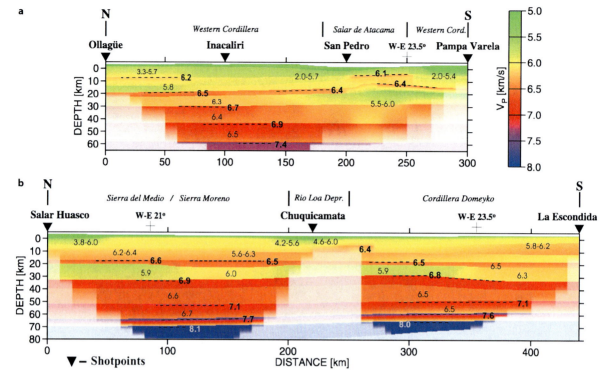

**Fig. 7.3.** 2D velocity models from (**a**) 68° W and (**b**) 69° W resulting from the PISCO data (see Lessel (1997) for details). The model at 69° W was determined by combining the PISCO data with data from earlier investigations (e.g. Wigger et al. 1994). The *light zone* in (**b**) marks an area where no data were recorded

between 10 km and 20 km depth (also seen in Fig. 7.5). The boundary between upper and lower continental crust has been observed at about 20 km depth (Schmitz et al. 1999) in both models, but differences in depth on each profile indicate a slight eastward dip-component. South of San Pedro, this boundary rises to approx. 15 km depth and is overlain by a zone of high velocity (5–15 km depth). The continental Moho in the Precordillera has been observed at approximately 65–70 km depth, whereas in the Western Cordillera it has not been detected. For more details about the PISCO experiment and results refer to Lessel (1997) and Schmitz et al. (1999). Images of the continental Moho estimated with the help of the receiver function method are presented and discussed by Asch et al. in this volume (Chap. 21).

Modeling of the CINCA data resulted in nine E-W models, three of which were derived from a combination of onshore and offshore shots and recordings. A simplified cross section along the transect at 23.25° S is shown in Fig. 7.4. In this velocity model, the oceanic crust and the upper and lower continental crust are identified. The results of the CINCA wide-angle and refraction measurements extend the structural model derived from earlier campaigns to the coastal region and the offshore fore-arc. We summarize the most important aspects (Patzwahl 1998):

- In the outer fore-arc, the subduction angle of the Nazca Plate (approx. 10°) does not vary along strike to 40 km depth.
- Within the limit of resolution, the boundary between oceanic and continental plate, derived from wide-angle observations, coincide with the hypocenters of the aftershock series of the Antofagasta 1995 earthquake (Husen et al. 1999; Patzig 2000).

The central result from the evaluation of the ANCORP wide-angle data is a detailed and well resolved P-wave velocity model extending to an intermediate depth of about 40 km (Fig. 7.5). The velocity model basically confirms results from earlier studies (Wigger et al. 1994; Patzwahl 1998), but it is better resolved along the transect. The P-wave velocity in the upper continental crust (to a depth of about 30 km) decreases from the fore-arc towards the magmatic arc. Under the magmatic arc and the Bolivian Altiplano (at distance between 150 km and 300 km from the coast, see Fig. 7.5) a low velocity layer ($\approx 5.0$–$5.5$ km s$^{-1}$) is modeled between 15 and 20 km depth. Beneath the Altiplano, to maximum depth of 10 km, low P-wave velocities (less than 4.5 km s$^{-1}$) indicate thick sedimentary layers. The structure of the lower continental crust and the oceanic crust is constrained only under the Coastal Cordillera (0–80 km from the coast). A geophysical Moho has been found which rep-

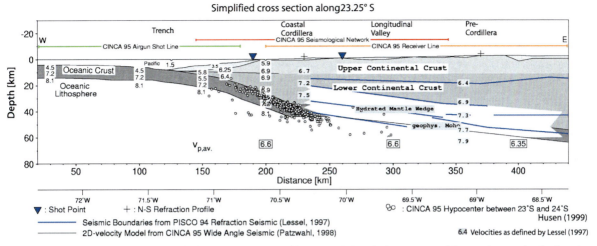

**Fig. 7.4.** Results from wide-angle seismics for line SO104-13 (Patzwahl 1998) including the hypocenters of the Antofagasta aftershock series recorded by the CINCA network (Husen et al. 1999). The *numbers* denote P-wave velocities directly above and below the seismic boundaries. *Gray shaded values* depict velocities estimated by Lessel (1997)

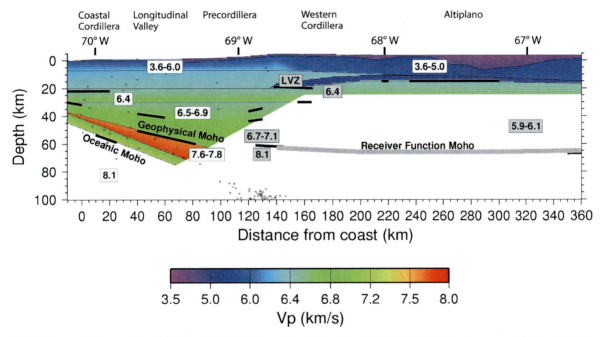

**Fig. 7.5.** P-wave velocity model resulting from ANCORP wide-angle and refraction data (Lüth 2000). The *circles* represent the local seismicity recorded by the ANCORP network (Haberland and Rietbrock 2001). Velocity values and ranges indicate the velocity variation within respective layers. *Gray boxes* highlight velocity estimates from earlier studies (e.g. Patzwahl 1998)

resents the apparent Moho under the fore-arc above the oceanic crust (e.g. Scheuber and Giese 1999). This geophysical Moho is characterized by Moho-PmP-like phases in the wide-angle seismograms. It is located at 55 km depth (Fig. 7.5) and indicates the transition from the lower continental crust to a possible hydrated mantle wedge with P-wave velocities slightly below normal upper mantle velocities. The oceanic Moho lies some 55–60 km depth under the Coastal Cordillera. This velocity model provides the basis for the depth conversion of the near-vertical reflection data discussed below.

Neither the ANCORP wide-angle results (Fig. 7.5) nor the cross section of line 13 (Fig. 7.4) show the LVZ which is visible within the Precordillera velocity model (Fig. 7.3b). The absence of this zone within the E-W sections indicates that it is limited to the Main Cordillera, and thus not detectable due to the resolution within the traverse of line 7 and line 13, respectively.

**Table 7.1.** Acquisition parameters of CINCA lines SO104-07 and SO104-13

| Profile | Location (deg) | Profile length (km) | Max. offset (km) | Shot spacing (m) | Receiver spacing (m) | No. of channels | Recording time (s) | Sampling rate (ms) |
|---|---|---|---|---|---|---|---|---|
| SO104-07 | 21 | 120 | 3 | 50 | 25 | 108 | 14 | 4 |
| SO104-13 | 23.25 | 110 | 3 | 50 | 25 | 108 | 14 | 4 |

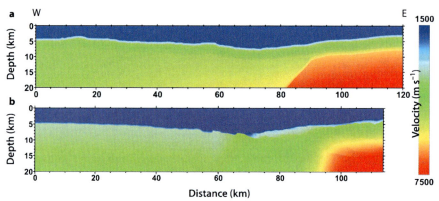

**Fig. 7.6.**
2D velocity models for (**a**) line SO104-07 and (**b**) SO104-13. The eastern part of the lower model was kindly provided by César Ranero (Geomar, Kiel). Both models were used to calculate the travel-time tables required for the prestack Kirchhoff depth migration

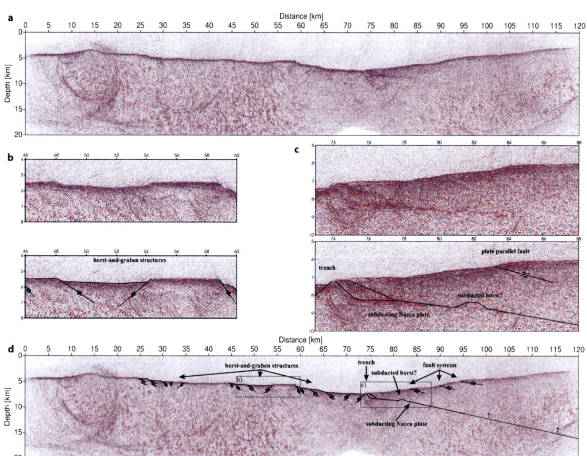

**Fig. 7.7.** Depth migrated section of CINCA line SO104-07. **a** Without interpretation; **b** enlargement of the horst-and-graben structures west of the trench; **c** enlargement of the subducting Nazca Plate with one of the plate parallel faults; **d** interpreted depth migrated section of CINCA line SO104-07. The *gray rectangles* mark the locations of the enlargements. The interpreted sections are also presented on the DVD

## 7.2.3 The Marine Reflection Experiments

Within the CINCA experiment, 22 seismic reflection profiles were acquired. Here, we present some of the results of two profiles, line SO104-07 and line SO104-13 (Fig. 7.2). Both are oriented approximately perpendicular to the trench. The acquisition parameters of both profiles are shown in Table 7.1.

To improve the images of oceanic crustal structures and of those structures at and below the continental slope, especially of the downgoing Nazca Plate, we used Kirchhoff prestack depth migration. This technique requires reliable velocity information of the subsurface. A detailed velocity model for the eastern part of line SO104-13 (Caesar R. Ranero, IFM-Geomar, Kiel, pers. comm.), and an additional velocity analysis of the data, enabled us to build extended models for both areas of interest (Fig. 7.6). The description of the data preprocessing and an extended discussion of the velocity models can be found in Sick (2005).

Figure 7.7 shows the non-interpreted (top) and interpreted (bottom) depth migrated section of profile SO104-07. To the west of the trench, at about 15 km, we observe a small seamount with a height between 600 and 700 m. At 25 km, the first horst-and-graben structures can be found which continue to the trench at about 74 km (Fig. 7.7b). Below the ocean bottom, east of the trench, the subducting plate can be observed down to a depth of approx. 12 km (Fig. 7.7d). Directly at the trench, the geometry of the slab is ambiguous and the bump in the slab at 82 km is likely a subducted horst. Above the subducted oceanic crust, several slightly eastward-dipping parallel fault systems are visible. Their outcrops correlate with escarpments at the continental slope.

The horst-and-graben structures in the southern area (line SO104-13, Fig. 7.8) are more pronounced than in the north (Fig. 7.7) with the horst flanks up to 1 000 m high (Fig. 7.8b). Below the seafloor, within the oceanic crust, strong reflectors occur at 7 km depth between 15 km and 40 km along the profile, indicating the boundary between upper and lower crust. Also, the oceanic Moho can be observed between 12 km and 15 km depth (Fig. 7.8c). To the east of the trench, the plate boundary can be observed to a depth of 15 km. While in the northern section only one possible subducted horst is present, we find at least

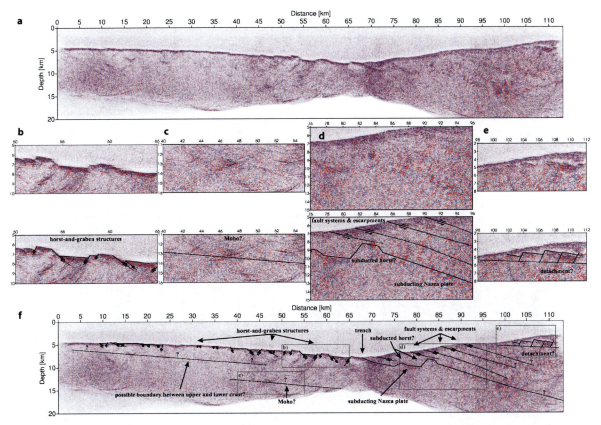

**Fig. 7.8.** Image of line SO104-13 at 23.25° S. **a** Non-interpreted depth section; **b** enlargement of the horst-and-graben structures; **c** enlargement of the oceanic Moho; **d** enlargement of the subducting Nazca Plate and plate parallel faults; **e** enlargement of the normal faults; **f** interpreted section of line SO104-13. The *gray rectangles* mark the locations of the enlargements. The interpreted sections are also presented on the DVD

two in this area (Fig. 7.8d). Within the continental crust, a series of reflectors parallel to the subducted oceanic plate is present and, in some cases, these reflectors extend through the continental crust to a depth of 10 km. Similar features are also visible in the depth image of line 7 (Fig. 7.7). As in the north, these faults surface at the location of trenchward facing escarpments visible in the continental slope to as much as 8 km above the plate interface. In the upper part of the continental slope, between 100 km and 110 km, a normal fault system transects the upper continental crust and merges into a slightly landward-dipping detachment (Fig. 7.8e).

Both depth migrated sections show new interesting details in the images of the subsurface and of the subducting Nazca Plate, especially in the continental crust below the continental slope. Comparing the images, the subduction angle is slightly steeper in the southern area (Fig. 7.8). Both the horst-and-graben structures at the ocean bottom to the west of the trench and the subducted horst-like structures in the subsurface east of the trench are clearly visible in both migrated sections, but they are more pronounced at 23° S. However, at 21° S, we observe neither normal faults nor a detachment as compared to the south (Fig. 7.8), but the Nazca Plate can be observed in both images down to a depth of about 15 km.

The continental slope faults paralleling the subducting plate are also more pronounced in the southern area. The outcrops of these structures, and the associated escarpments at the continental slope, correlate well with the bathymetric data (Fig. 7.9; von Huene et al. 1999). From the fault geometry and the shape of the escarpments, we conclude that they are active thrusts parallel to the plate boundary operating at low strain rates. The surface is offset, but there is no conspicuous evidence for large displacements from cross-cutting structures or overthrust sediment. These features were not previously identified (e.g. von Huene and Ranero 2003), and are here interpreted to mark the upper part of the subduction channel where material is initially underthrust as large fault-bounded slabs.

**Table 7.2.** Acquisition parameters of ANCORP and PRECORP

| Experiment | Location (deg) | Profile length (km) | Max. offset (km) | No. of shots/ receivers | Shot spacing (km) | Receiver spacing (m) | Recording time (s) |
|---|---|---|---|---|---|---|---|
| PRECORP | 22.5 | 120 | 14.3 | 38/144 | 2.4 | 100 | 60 |
| ANCORP | 21 | 385 | 25.1 | 131/252 | 6.25 | 100 | 60 |

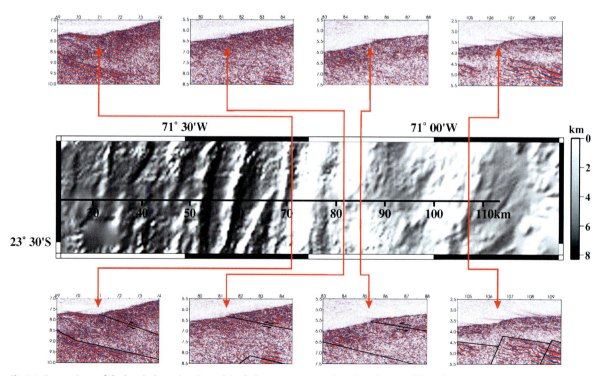

**Fig. 7.9.** Comparison of the break-through points of the faults at 70, 82, 85 and 107 km along profile with the bathymetry (von Huene et al. 1999), respectively. The bathymetry is illuminated by a light source located N-W of the image

## 7.2.4 The Onshore Profiles PRECORP and ANCORP

The PRECORP and ANCORP experiments were carried out to gain reflections from depths to 100 km, the required maximum target depth needed for obtaining images of the subducting Nazca Plate. The acquisition parameters for both experiments are shown in Table 7.2.

The preprocessed data sets were imaged using a three-dimensional (3D) Kirchhoff prestack depth migration scheme implemented from topography. The true shot and receiver locations had been taken into account during the travel time calculation and the computation of the migrated depth sections. The required macro velocity model was obtained by extending the two-dimensional (2D) model obtained from refraction seismic results (see Fig. 7.5) to 3D. The migration scheme was as follows: firstly, each common shot gather was migrated separately; secondly, envelope stacking of the migrated sections yielded a 3D image with a total length of 370 km (E-W), a width of 50 km (N-S) and a depth of 100 km; and finally, the third imaging step was a selective stacking procedure of E-W oriented slices located near the profile line, disregarding those parts far away. This so called "offline-stacking" increased the signal to noise ratio further and provided a 2D depth image. A detailed description of the processing as well as the results can be found in Yoon (2001) and Yoon et al (2003).

The ANCORP depth section (Fig. 7.10) shows two parallel, clearly distinguishable reflectors at the western end of the profile (0–15 km). The upper reflector, the Nazca reflector, can be followed to a depth of approx. 80 km. It reflects strongly over most of its extent, except for small gaps in the area around 20–50 km along profile. At depths greater than 80 km, the Nazca reflector becomes disrupted and weaker, probably because of the existence and the strong reflectivity of the Quebrada Blanca Bright Spot (QBBS) occurring above it, between 115–160 km along profile. The lower parallel reflector, about 7 km below the Nazca reflector, probably marks the oceanic Moho. The QBBS occurs between 15–40 km depth and has a maximum vertical extent of about 15 km. Below the Altiplano Plateau, about 300 km from the coast, strong reflections are visible between 15 km and 35 km depth (Altiplano reflections, Fig. 7.11), but due to the strong scattering caused by this heterogeneous zone, no further reflection events are found deeper. This zone of high reflectivity corresponds with the Altiplano conductor, as determined with magnetotelluric methods (Brasse et al. 2002).

The compilation of the 2D section with earthquake data shows a good agreement between the hypocenter locations and the position of the Nazca reflector at the western end of the ANCORP profile (0–20 km, Fig. 7.10a): the top of the hypocenters follow the upper boundary of the oceanic crust. Further to the east, the offset between the hypocenter locations and the Nazca reflector becomes apparently larger with depth. The hypocenter locations were calculated using a velocity model derived from seismic tomography (Graeber and Asch 1999). This model contains relatively higher velocities than the model from Lüth (2000) used for migration of the reflection data, possibly causing the apparent offset. However, a comparison of depth sections calculated with different velocity models indicated that the observed offset can not necessarily be explained by the difference of the velocity models (Yoon 2005). An additional study of depth migrated reflection data using different frequency bandwidths indicated that the Nazca reflector ($x = 70–110$ km, Fig. 7.2a) itself represents the hydrated zone in the continental mantle, whereas the lower edge of the reflector marks the oceanic crust (Yoon 2005). The latter is in good agreement with the observation that the hypocenters are mainly located within the oceanic crust and in the oceanic mantle (Rietbrock and Waldhauser 2004) and explains the apparent offset.

Compared to the 3D processing of the ANCORP data, the migration of the PRECORP data set was done in 2D consisting of two major steps: firstly, the migration was performed for early arrivals (TWT: 0–15 s) without any amplitude balancing; and secondly, later arrivals (TWT: 15–40 s) were migrated after correction for geometrical spreading. Finally, both depth sections were stacked together yielding a 100 km depth section. The final depth sections are shown in compilation with the results of local earthquake analysis (Graeber and Asch 1999; Haberland and Rietbrock 2001).

The 2D depth image of the PRECORP data set shows a reflector segment at 65 km depth between 50 and 60 km along the profile (Fig. 7.10b). This reflector segment is associated with the Nazca reflector. Poor data quality weakens the reflections between 60 and 95 km along profile, but between 95 and 125 km the Nazca reflector is clear again at 80–85 km depth. Above this reflector, at depths of 15–25 km, the so-called Calama Bright Spot (CBS) is visible. The CBS appears with a lateral extension of approximately 15 km and a maximum vertical extension of about 5–10 km. Due to the limited recording geometry it is not clear, whether the lateral extension of the CBS is real. The offset between the hypocenter locations and the Nazca reflector remains, but its extent is less compared to that observed in the ANCORP data.

The application of Kirchhoff prestack depth migration to the ANCORP data set yielded images of the subduction zone with some interesting details. The 2D section shows an oceanic crust which is about 7 km thick, corresponding to the results derived from wide-angle refraction measurements in this region (Patzwahl et al. 1999). Compared to previous poststack images (Stiller, pers. comm.), the reflections from the Nazca reflector do not break down completely at 80 km depth, but continue deeper. Furthermore, the nature of reflectivity of the Nazca reflector in the ANCORP section seems to

**Fig. 7.10.**
**a** 2D ANCORP depth section; **b** PRECORP depth section. The hypocenter locations are indicated by *black dots* (Graeber and Asch 1999; Haberland and Rietbrock 2001). *WFS*: West Fissure Fault System; *SMFS*: Sierra-de-Moreno Fault System; *QBBS*: Quebrada Blanca Bright Spot; *CBS*: Calama Bright Spot. Note that the CBS is located to the east of the SMFS (see also Yoon et al. 2003)

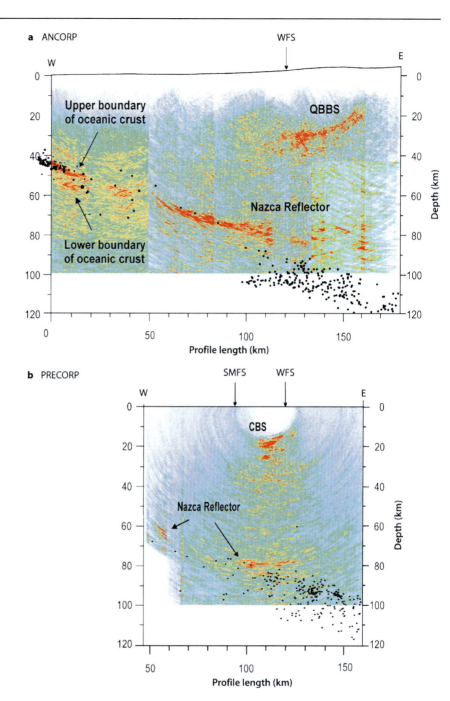

change with depth; the reflections appear very sharp at depths between 40 km and 60 km, but become disrupted and blurred at depths greater than 70 km. We assume that this phenomenon is probably due to both the influence of the heterogeneous overburden on the seismic signal and the complexity of the reflector itself at greater depths.

Each of the depth sections exhibits a bright spot: the Quebrada Blanca Bright Spot (QBBS) in the ANCORP section (Fig. 7.10a) and the Calama Bright Spot (CBS) in the PRECORP section (Fig. 7.10b). The QBBS, located at depths of 15–40 km, dips to the west and also has a northward-dipping component, as determined from the 3D section. The CBS occurs nearly 160 km further to the south and 10–15 km closer to the surface. It also has a westward-dipping component, and is nearly 15 km further west with respect to the trend of structures and topography at surface than the QBBS. Interestingly, the western edge of the QBBS correlates with the north-south-striking Precordilleran Fault System (PFS). Towards the south, the PFS (Scheuber and Reutter 1992) locally splits into the West Fissure (WFS; Figs. 7.2 and 7.10) and the Sierra-

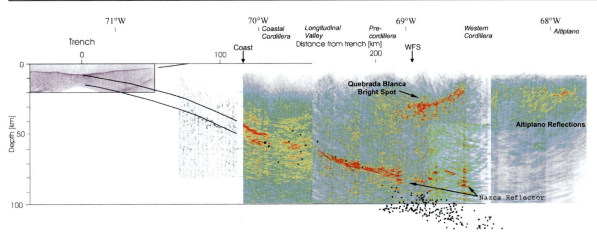

**Fig. 7.11.** Combination of the results at 21° S. From *left* to *right*: Image of line SO104-07 (see Fig. 7.7); boundaries and reflectors resulting from the wide-angle seismics (Patzwahl et al. 1999). Depth section of the ANCORP line

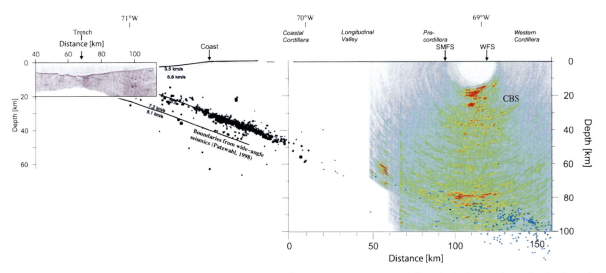

**Fig. 7.12.** Combined results at 23.25° S. From left to right: Image of line SO104-13 (see Fig. 7.8); boundaries from wide-angle seismics and hypocenters of the Antofagasta aftershock series (Patzig et al. 2002); depth section of the PRECORP line

de-Moreno Fault System (SMFS; Figs. 7.2 and 7.10b; Günther et al. 1997). Here, the CBS is located 15 km further to the east of the SMFS and west of the WFS. Thus, the Precordilleran Fault System always delimits the bright spots, but with changes in spatial relationship. The north dipping component of the QBBS, and the upward shift of the CBS's location, possibly indicates that both bright spots are somehow connected, or, at least, are caused by the same geological structure or processes. Nevertheless, the origin and the geological nature of these bright spots are not yet fully understood.

### 7.2.5 The Erosive Margin: Conclusions and Discussion

The investigations presented above yield new interesting details about the structural geometry of the subsurface.

In the offshore images, the seafloor west of the trench is dominated by horst-and-graben structures (Figs. 7.7 and 7.8). Along the profiles, the frequency of these structures increases laterally towards the trench and are more pronounced in the southern profile at 23.25° S. Other ocean bottom structures have been observed, e.g. a seamount in the north (Fig. 7.7) and escarpments east of the trench. In both profiles, the trench is nearly devoid of sediments. Below the seafloor, the boundary between the upper and lower oceanic crust has been identified on line SO104-13. The wide-angle results also image the boundary between the upper and lower continental crust of the overriding South American Plate (Fig. 7.4).

A combination of the images from 21° S and 23.25° S is shown in Fig. 7.11 and Fig. 7.12, respectively. Figure 7.11 illustrates a combination of the ANCORP line and the offshore line SO104-07 with a depth converted stacked section of the CINCA wide-angle data, including the resulting boundaries and reflectors from processing and modeling (Patzwahl et al. 1999). Figure 7.12 displays line SO104-13, the boundaries from wide angle seismic data and the PRECORP line. Both images allow continuous observation of the Nazca reflector down to approx. 80 km depth. From the continuous link between the Nazca reflector and the modeled top of the Nazca Plate near the trench from wide angle seismic results, the ANCORP Working Group (1999, 2003) concluded that this reflection is related to the top of the downgoing slab. The depth migrated parts from ANCORP and PRECORP image the top of the plate below the continent, but only in the western part of the ANCORP section can the oceanic Moho be seen (about 7 km below the Nazca reflector). In deeper regions of the ANCORP section, the reflector becomes disrupted and weaker probably due to the strong reflectivity of the overlying QBBS. Unlike the northern region, the Nazca reflector within the PRECORP image is visible below the CBS (Fig. 7.10b), and, as discussed previously, these bright spots might be connected. In contrast to the near-vertical seismic sections, the results from wide-angle seismic data (Fig. 7.11) display only the oceanic Moho as the velocity gradient between continental and oceanic crust is too small to obtain the critical angle.

All sections provide sharp imagery of the subsurface geometry and the position of the subducting slab. Comparing the images at 21° S with those at 23.25° S, we can conclude that to a depth of 40 km the variation of subduction angle is negligible from north to south within the working area. A low velocity zone (LVZ) is located where the subduction angle increases, approximately below the coastline and at the down-dip end of the seismogenic coupling zone as observed from aftershock seismicity of the 1995 Antofagasta earthquake (cf. Figs. 7.10a and 7.11). Hence, we consider this LVZ to represent parts of the subduction channel and of the hydraulically fractured base of the upper crust in the vicinity of the seismogenic plate interface.

In the models resulting from the wide-angle seismics at 23.25° S, the hypocenters of the 1995 Antofagasta aftershock series correlate well with the upper boundary of the subducted plate (Figs. 7.4 and 7.12). Also, at the western end of the ANCORP section, the hypocenters of the local earthquakes correlate with the top of the oceanic crust. Below the seismogenic coupling zone this correlation is increasingly offset as the hypocenters become more widely distributed. These results possibly indicate that the focal mechanisms of the earthquakes are changing with depth. Another possible explanation for this offset is that the velocity models used for imaging the local events and the reflection data differ, thus biasing the reflector depths. First comparisons of travel times calculated for a new velocity model derived from Rietbrock and Haberland (2001) (which differs only slightly from the Graeber and Asch (1999) model) and Lüth (2000) showed that the average velocity of the Rietbrock model is about 0.5 km s$^{-1}$ faster than the model after Lüth (2000) (Fig. 7.5). However, the location error of the hypocenters is about 5 km. The impact these differences in the models make to the processing of the ANCORP data, and the contribution of new relocalized hypocenters, will be part of further investigations.

A similar, but smaller, offset occurs between the PISCO earthquake locations and the Nazca reflector on the PRECORP section. From the width of the intermediate depth seismicity (about 20 km), its depth interval (70–120 km), and the thickness of the oceanic crust (7 km), the ANCORP Working Group (1999) and Yuan et al. (2000) have suggested that this offset is related to both dehydration embrittlement at the level of progressive eclogitization of the oceanic crust, and serpentinite breakdown in the underlying mantle, with the fluids expelled and trapped in the overlying mantle wedge. The oceanic Moho was detected within the wide-angle experiment on both lines 7 and 13 and west of the trench on the depth image of CINCA line 13 (Fig. 7.8) at about 15 km depth. Earlier investigations of line 13, processed with the help of CRS stack, detected the oceanic Moho at the same depth (Buske et al. 2002). The modeling results from the wide-angle seismics indicate the oceanic Moho is located some 40 km beneath the Coastal Cordillera in the southern section (Fig. 7.4) and at approx. 60 km depth in the north (Fig. 7.5). Along the ANCORP profile, receiver functions have helped determine a continental Moho about 60–70 km beneath the Altiplano (e.g. Yuan et al. 2000).

Other subsurface structures observed in the offshore images, are the normal faults with a detachment at 105 km on line 13 and the faults parallel to the plate interface (Figs. 7.7 and 7.8). The latter mark the upper parts of the subduction channel and show the material transfer between oceanic crust and fore-arc. The strong reflectivity of these features might be a result of fluid infiltration (e.g. Bangs and Shipley 1999), a material contrast or cataclastic fabric. Similar structures can be observed in sandbox experiments (Lohrmann et al. 2006, Chap. 11 of this volume).

## 7.3 The Accretionary Margin

### 7.3.1 Working Area

In 2001, the SPOC project investigated an area in the southern Central Andes between 36° S and 40° S with the aim of studying the fore-arc and the seismogenic coupling zone (Fig. 7.13). The upper plate in this region is segmented by large and prominent NW-SE oriented fault zones, namely the Bio-Bio and the Gastre fault zones. In 1960, the largest instrumentally-recorded earthquake occurred north of Valdivia with a magnitude of 9.5 (Cifuentes 1989). The rupture caused a coseismic slip of up to

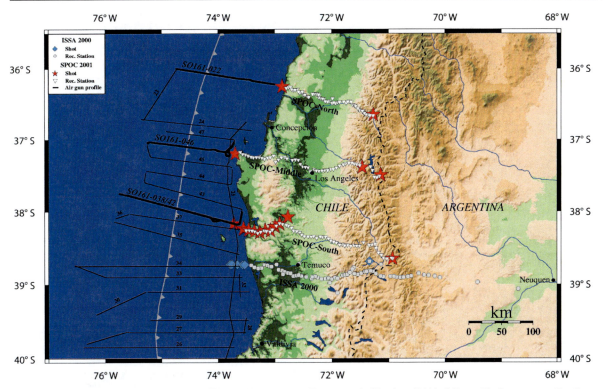

**Fig. 7.13.** Main area of the SPOC experiment. Offshore: The airgun profiles are marked by the *solid black lines* with the corresponding line numbers. Onshore: The shot locations of the SPOC experiment are depicted by *black stars* and the receiver locations are shown by *white triangles*. The near-vertical reflection experiment covered the offshore-onshore transition at 38.25° S. The southernmost profile, ISSA, is marked by the *gray circles*, whereas the shot locations are marked by *gray diamonds*

40 m, a vertical displacement up to 2 m, and a tsunami around 15 m high (Plafker and Savage 1970; Kanamori and Cipar 1974; Klotz et al. 2001).

During the experimental period, some 5 300 km were surveyed along 48 2D lines with multi-channel reflection seismics (MCS) off central Chile (Reichert et al. 2002). Three of the east-west traverses were used for combined onshore/offshore, wide-angle/refraction seismic observations (Fig. 7.13). Simultaneously, we recorded the near-vertical reflection experiment (SPOC) which covered the offshore-onshore part along the wide-angle line SPOC-South. In 2000, at about 39° S, we also shot the 320 km long reconnaissance seismic refraction profile ISSA consisting of 65 stations with an average spacing of 4–5 km (white circles in Fig. 7.13).

### 7.3.2 Wide-Angle and Refraction Seismics

The ISSA experiment comprised a seismic refraction profile along 39° S that recorded chemical shots fired in both the Pacific Ocean and the Main Cordillera, as well as a network recording local seismicity. The SPOC-Land experiment in 2001 (Krawczyk et al. 2003), an onshore extension of the SONNE Cruise 161 (Reichert et al. 2002), extended the seismic measurements. It comprised three offshore-onshore seismic wide angle profiles, a seismic network, and a seismic reflection profile. Acquisition parameters and first results are described by Krawczyk et al. (2003) and a detailed description of the data processing is given in Lüth et al. (2004).

The velocity model (Fig. 7.14) resulting from the ISSA wide-angle data yields three main features (cf. Lüth et al. 2003):

1. In the uppermost 20 km, the crust shows a lateral heterogeneity, where the average velocity in the fore-arc is approx. 6.1 km s$^{-1}$ and in the arc, 6.4 km s$^{-1}$. Beneath the magmatic arc, the upper crust is roughly 10 km thick and the lower crust is characterized by high velocities (about 6.8 km s$^{-1}$) at shallow depth. In contrast to this, the boundary between upper and lower continental crust below the Coastal Cordillera, located at 15 km depth, is characterized by velocities between 6.5–6.6 km s$^{-1}$.
2. The oceanic Moho was observed by the seismic refraction profile to approx. 55 km depth beneath the Coastal Cordillera.
3. Between the lower continental crust and the subducting oceanic crust, a wedge shaped structure occurs beneath the fore-arc. The upper eastern part of this layer can be attributed to the continental mantle, which is supposed

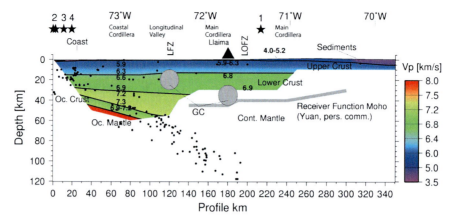

**Fig. 7.14.** P-wave velocity model for the ISSA profile (39°S) as derived from iterative travel time modeling. The velocity values ($v_p$) are given near the layer boundaries. *Black dots* indicate the earthquake hypocenters located from the ISSA passive seismological experiment (Bohm et al. 2002). The continental Moho, estimated with the help of the receiver function method (receiver function Moho), is positioned according to Yuan et al. (2004). The *gray circles* locate good electrical conductors (Brasse and Soyer 2001). *LFZ*: Lanalhue Fault Zone; *LOFZ*: Liquiñe-Ofqui Fault Zone

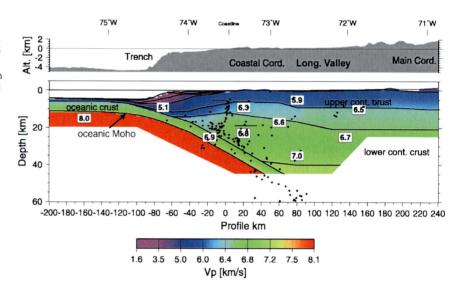

**Fig. 7.15.**
P-wave velocity model resulting from wide-angle data at 38.25° S along the SPOC south profile. Velocity values ($v_p$) are shown in km s$^{-1}$ (Krawczyk and the SPOC Team 2003). The earthquake hypocenters are from Bohm et al. (2002)

to be deeper than 40 km depth under the arc (Yuan et al. 2004), but this is not the case in the upper western part where the P-wave velocity is considerably lower than that usually recorded from typical mantle.

The most comprehensive data set from the SPOC-Land experiment comes from the southern profile, along 38.25° S (Fig. 7.13). Here, a wide-angle OBS/OBH (Ocean Bottom Seismic/Ocean Bottom Hydrophone) profile was acquired offshore. Simultaneously, the airgun pulses were recorded by an onshore profile, and the near-vertical reflection profile was acquired on the same transect. The transect from the Pacific Nazca Plate to the Chilean Main Cordillera yielded a 2D P-wave velocity model. This model comprises the oceanic crust, to approx. 45 km depth under the Coastal Cordillera, and the continental crust from the trench to the Main Cordillera. The continental crust is characterized by gradually increasing P-wave velocities from the trench to the Main Cordillera (see Fig. 7.15), whereas very low near-surface velocities (< 3.5 km s$^{-1}$) characterize the westernmost part between trench and coastline. As wide-angle measurements were also recorded on two other transects, along 37° S and 36.25° S (Fig. 7.13), the geometry of the subducting oceanic crust can be compared amongst the three transects (Lüth et al. 2004). Along the central transect, crossing the Arauco Peninsula, the subduction angle of the oceanic crust is shallowest (~12°) to an approximate depth of 45 km whereas the subduction angle is approximately 7° steeper in the other two transects (Lüth et al. 2004). The resulting velocity model is used for depth conversion of the offshore reflection data and the prestack migration of the near-vertical reflection data.

**Fig. 7.16.**
Prestack depth migration of line SO161-38/42 (converted to time so that t is easier to compare it with Figs. 7.17 and 18)

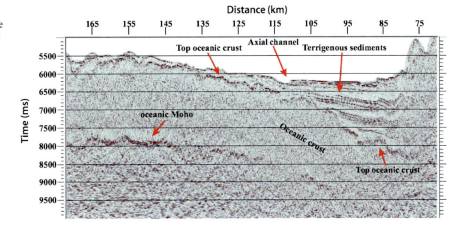

**Fig. 7.17.**
Terrigenous trench sediments (prestack time migrated data). **a** Line SO161-22; **b** line SO161-38/42; **c** line SO161-46

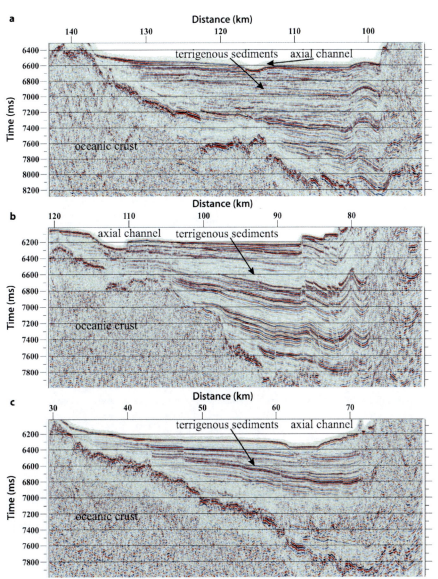

## 7.3.3 Offshore Reflection Seismics

The overall aim of the SPOC experiment was to investigate the different conditions that might influence the subduction process such as, the age of the oceanic crust, its structure and composition, its sedimentary cover, its thermal state, the subduction angle and obliquity, and the amount of terrigenous sediment from the continent.

Three selected 2D lines (SO161-22, SO161-38/42 and SO161-046; Fig. 7.13) from the offshore reflection seismic survey were processed and combined with the onshore seismic data to get a continuous image of the subsurface across the southern Chilean subduction system. The lines were recorded with a streamer length of 3 km using 132 channels with a shot point interval of 50 m and recording time of 14 s. Here, the objectives were firstly, to produce a high-resolution subsurface image of the section crossing the deep sea, the trench with sedimentary infill, the shelf slope and the shallow water area, and secondly, to derive the correct depth of the layers using structural inversion. For this purpose, the seismic data were processed in both time and depth domains. We applied both poststack Kirchhoff time migration and, to improve the resolution and the continuity of the horizons, prestack Kirchhoff time migration. The objective for the depth-domain processing was to determine the most accurate position of the reflectors. We depth migrated all three lines using an interval velocity-depth model derived from refraction seismics (Fig. 7.15; Lüth et al. 2003). A detailed description is given in Rauch (2004, 2005).

The images of the terrigenous sediments in the trench show remarkable deformation structures with clear faulting, as well as a significant periodicity of the reflected signals (Fig. 7.17). The structural interpretation of the processed lines gives a better understanding of the pre- and post-tectonic sedimentation, of the fault patterns, and of the coupling between the top of the oceanic crust and the trench sediments. Fault mapping (Fig. 7.18) clearly shows that the entire sedimentary infill of the trench is affected and is currently accreted to the tip of the upper plate.

In two of the three processed lines (SO161-022 and SO161-38/42) a significant periodicity of the reflections in the trench sediments is obvious, but this effect is miss-

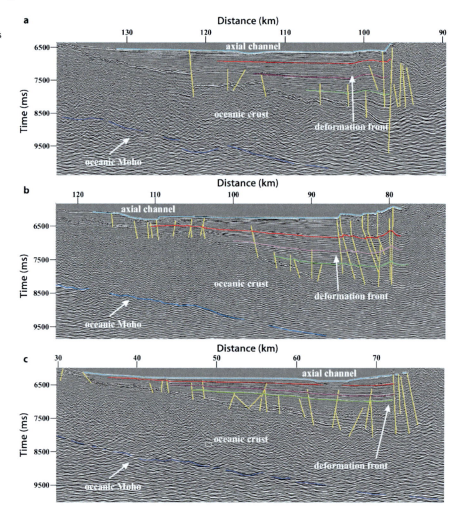

**Fig. 7.18.**
Prestack time migrated sections with interpreted faults and selected horizons. **a** Line SO161-22; **b** line SO161-38/42; **c** line SO161-46

ing in line SO161-046 (Fig. 7.17). A possible explanation might lie in the location of the lines relative to the canyon mouth. The terrigenous sediments, transported through the canyons into the deep water, are dislocated and deposited in the trench by turbidites in a general south to north direction, due to the northward-dipping ocean bottom. In lines SO161-22 and SO161-36/42 the sediments are transported over a short distance from the Bio Bio and Imperial/Tolten fans, respectively, while line SO161-46 is more than 100 km northwards. The structural interpretation also shows that there is no relative shift between the trench sediments and the moving Nazca Plate due to high friction. This is an important consideration for calculating the age of the terrigenous sediments using the convergence rate of the Nazca Plate (for details see Rauch 2004).

### 7.3.4 Offshore-Onshore Near-Vertical Reflection Experiment

The data set that provided the depth image shown in Fig. 7.19 was acquired during a combined offshore-onshore seismic campaign at 38.25° S. The near-vertical reflection experiment covered the offshore-onshore transition along a 54 km long E-W profile providing a preliminary examination of the structural and reflectivity potential along the seismogenic plate interface (Krawczyk et al. 2003). The line was set up in three spreads and recorded a series of two offshore and 12 onshore chemical shots fired at 10 different locations. This geometry provided a depth section with a total length of about 90 km. Again, the parallel wide-angle profile provided the velocity model from refraction data analysis (Fig. 7.15; Lüth et al. 2003). Krawczyk et al. (2003) presented a poststack migrated depth section of this profile (Fig. 7.20). Here, we focus on the prestack depth migration result.

The data preprocessing consisted mainly of bandpass filtering to suppress high frequency peaks (50 Hz signals) and low frequency noise, muting the first arrivals, and the removal of strong noise at later arrival times by applying a bottom mute. An automatic gain control was applied for amplitude equalization and better detection of reflection events, especially at low amplitude levels (see Yoon (2005) for details). The resulting depth migrated images reveal that cross-line stacking of the envelopes, i.e. stacking 2D slices from the 3D volume along strike, calculated from the depth section yielded the most promising final image. Neither amplitude normalization nor equalization has been applied.

Figure 7.19 shows a complex reflection pattern with several distinct reflectors. Twenty kilometers west of the coast, a thin, slightly east-dipping reflector is visible at 15 km depth ("A" in Fig. 7.19) and possibly indicates the upper boundary of the subduction channel (cf. Fig. 20). The reflector continues eastward to a depth of about 25 km. Between 0 km and 40 km along profile, the depth image shows several strong reflector segments with changing dips and a strong background reflectivity (B). In the east (40–70 km), near horizontal reflector segments are visible at depths of 12 km and 23 km (C). The image does not reveal structures below 35 km depth, such as the subducting oceanic crust in the easternmost part and the continental Moho.

### 7.3.5 The Accretionary Margin: Conclusions and Discussion

The data acquired in South Chile show very different results compared to those from North Chile. The trench shows substantial sediment infill onlapping an oceanic plate that is nearly devoid of sediment. The sediments reflect with significant periodicity (Figs. 7.16 to 7.18). Faulting of the entire sedimentary trench-fill indicates an almost total accretion with no underthrusting, in contrast to earlier suggestions for this latitude by Bangs and Cande (1997). Bangs and Cande (1997) noted from reprocessing industry reflection lines that the resulting accretionary wedges were rather small, yielding a maximum age of no more then Late Pliocene based on calculating the cross sectional area, trench-fill thickness and convergence velocity. They interpreted this young age as a reflection of enhanced trench sedimentation during glacial erosion of the Andean Cordillera to the east. Using the same procedure of age dating, we calculate an age of the current trench-fill of some 400 000–550 000 years. Hence the 4–5 seismic sequences would approximate a 100 000 year cycle. This corresponds well with the main glacial cycle during the Quaternary and corroborates the interpreted nature of the trench-fill.

The processing of the SPOC wide-angle/refraction data yielded a reliable 2D P-wave velocity model that includes the most important horizons shown in Fig. 7.15. We localized the boundary between upper and lower continental crust at about 6 and 10 km depth below the Coastal and Main Cordilleras, respectively. In contrast, the model from the ISSA profile at 39° S shows a thickness of the upper crust of 15 km. Comparison of both models indicates the upper/lower crustal boundary dips in N-S direction beneath the Coastal Cordillera (Figs. 7.14 and 7.15; Lüth et al. 2003). The top of the oceanic crust and the oceanic Moho have been clearly identified in the images from the offshore reflection seismics (Figs. 7.16 to 7.18) as well as in the wide-angle results. In the wide-angle results, the oceanic crust is continuously imaged to about 45 km depth under the Coastal Cordillera. Along the central transect crossing the Arauco Peninsula, the shallow dip-angle of the oceanic slab beneath the Coastal Cordillera indicates a possible relationship between the strong uplift of the Arauco Peninsula (compared to the areas

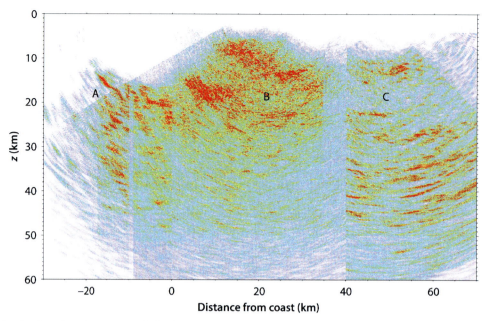

**Fig. 7.19.** SPOC depth section after 2D prestack depth migration (at 38.25° S). *A*: This thin eastward-dipping reflector corresponds to the upper boundary of the subduction channel (Fig. 7.20); *B*: region with strong background reflectivity; *C*: horizontal reflector segments. See text for details

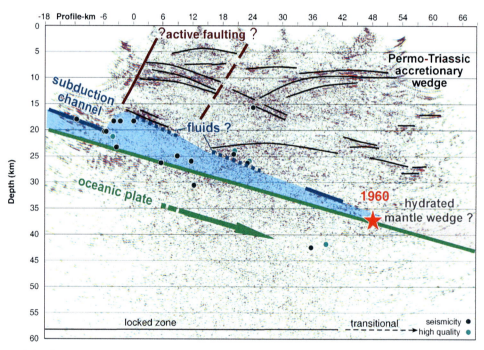

**Fig. 7.20.** Poststack depth migrated reflection seismic profile (SPOC) at 38.25° S. Interpretation from Krawczyk et al. (2004), seismicity data from Bohm et al. (pers. comm.). An extended discussion of this image can be found in Krawczyk et al. (2006, Chap. 8 of this volume)

north and south of it) and the varying geometry of the subducting oceanic crust (Bohm et al. 2002; Lüth et al. 2004). The very low P-wave velocities in the region to the east of the trench (see Fig. 7.15) suggest the presence of a recent accretionary wedge composed of unconsolidated and strongly heterogeneous sediments (Krawczyk and the SPOC Team 2003; Lüth et al. 2004). It is clear that this heterogeneous material strongly scatters the seismic energy and thus attenuates the reflected signals from the oceanic crust within the depth section shown in Fig. 7.19. Hence, the possible position of the Nazca reflector in the depth section is indicated only as a thin and disrupted reflector segment.

Neither in the near-vertical reflection data nor in the wide-angle data along the southern SPOC profile can the continental Moho be observed. Only the receiver function technique localized the continental Moho at a depth of about 40 km (Yuan et al. 2004), as shown in the ISSA model (Fig. 7.14). Beneath the fore-arc, at approx. 30 km depth (Fig. 7.14), the smooth velocity increase is due either to the presence of strongly hydrated and serpentinized mantle, or to subducted meta-sediments underplating the base of the continental crust. Bohm et al. (2002) observed uplift in parts of the coastal area which also might indicate relatively light (seismically slow) material being underplated. Based on sand box modeling, this area is the postulated center of basal accretion and antiformal stacking at depth (Lohrmann et al. 2000).

Beneath the Coastal Cordillera, an eastward-dipping reflector can be seen within the depth section above the oceanic crust. (Fig. 7.19). This reflector is visible to a depth of 25 km and corresponds to the upper part of the subduction channel pictured within the poststack depth section (Fig. 7.20). Other reflector segments observed between 0 and 40 km along profile within the prestack depth image, exactly underly subhorizontal to mildly folded large-scale antiformal stacks at the surface that have the same geometry. These stacks are built from exhumed trench sediment with intercalated oceanic rocks that were underplated and uplifted in Permo-Triassic times. Hence, Krawczyk et al. (2004) concluded that these features indicate the internal structure of the Permo-Triassic accretionary wedge (Fig. 7.20) continuing down to near the present plate interface.

## 7.4 Erosive Margin Versus Accretionary Margin

The highly different subduction conditions within the areas investigated enable us to compare our results with respect to the variations between the erosive margin, located in the north, and the accretionary margin in southern Chile. The ocean bottom in the southern region exhibits a thin cover (< 200 m) of pelagic sediments (Reichert et al. 2001; Rauch 2004). Nearly the same situation is found in the north, where the sediments do not exceed 300 m thickness (Hinz et al. 1995). In contrast to the south, where the morphology of the ocean bottom is more or less smooth (Reichert et al. 2003), the images from the CINCA area shows distinct horst-and-graben structures in the bending zone near the trench. Trench infill has only been observed where submarine canyons are present, e.g., south of 33° S, whereas canyons are sparse to absent in the north (Thornburg and Kulm 1987). Other reasons for the lack of trench sediments in Figs. 7.7 and 7.8 are the extremely arid climate in the Central Andes and the existence of inland basins that reduce the supply of sediments to the slope (Hartley and Chong 2002). The entire trench infill detected between 36° S and 40° S, is currently accreted to the tip of the upper plate with no underthrusting. The existence of a recent, steeply sloping accretionary wedge has been confirmed by the very low velocities found beneath the continental slope (Fig. 7.15). On the other hand, the trench region of the erosive margin comprises a frontal prism composed of more or less stratified detritus (e.g. von Huene et al. 1999) which can be seen in Fig. 7.7 at 75 km and Figure 8 at 70 km, respectively. An axial channel within the trench exists only in southern Chile (Figs. 7.16 to 7.18). In both areas, the images show a typical oceanic crustal thickness of about 7 km indicated, by the depth of the Moho.

The subduction angles, which are generally shallow along the Peru-Chile trench as the subducting lithosphere is young, thin and buoyant (e.g. Jarrard 1986; Stern 2002; Clift and Vannucchi 2004), are of the same magnitude in both areas, with an exception beneath the Coastal Cordillera at the Arauco Peninsula. Here, the subduction angle is about 7° less than in the other regions where it averages approx. 20° (Patzwahl 1998; Lüth et al. 2004). Bohm et al. (2002) and Lüth et al. (2004) interpreted this shallower subduction angle as a result of strong uplift in the Arauco region.

Ranero and von Huene (2000) pointed out that the geometry of erosive margins is strongly connected to the topography of the incoming oceanic crust. Surface structures like seamounts, aseismic ridges or horst-and-graben structures (Figs. 7.7 and 7.8) facilitate frontal erosion and indicate a strong coupling between oceanic crust and fore-arc (Ranero and von Huene 2000). Basal tectonic erosion also occurs beneath the coast and the frontally and basally eroded material is transported within the seismic coupling zone, defined by shallow earthquakes (< 50 km depth) (e.g. Fig. 7.10a; Scheuber and Giese 1999), and subducts further into the mantle wedge (e.g. Pelz 2000). The frontally eroded and subducted rocks may contain huge amounts of fluid especially when a graben structure transports part of the slope debris below the overriding plate. These fluids may then be discharged into the boundary between oceanic and continental crust (von Huene and Ranero 2003). One of the most important controlling parameters for the geometry of the accretionary margin, and especially for the steepness of the accretionary wedge, is the friction along the base of the wedge (e.g. Davis et al. 1983; Gutscher et al. 1996). High friction is suggested by both the structural interpretation of the offshore data (Fig. 7.18) and the steep slope angle of the accretionary wedge.

The results from the wide-angle seismics (ANCORP profile; Fig. 7.5) show that the crustal thickness in the Central Andean magmatic arc is up to 20 to 30 km greater than in the south (Fig. 7.14). Furthermore, the average velocities show other interesting differences between the Central and the Southern Andes. In the north, crustal velocities in the arc are less than those observed in the fore-arc, but the opposite trend is seen to south (Wigger et al. 1994). This difference is caused by the completely different development of the continental crust during the Andean cycle. In the Central Andes, frontal erosion at the trench since the Jurassic has resulted in an eastward migration of the volcanic front of about 200 km (Scheuber et al. 1994), whereas the Southern Andes has experienced no considerable migration of the magmatic arc (Hervé 1994; Jordan et al. 2001).

The present fore-arc crust in the Central Andes was the magmatic arc in Jurassic and early Cretaceous times and the original crust was replaced to a large extent by plutonic and volcanic rocks (Scheuber et al. 1994). These rocks are responsible for the high seismic velocities visible in the wide-angle results (Figs. 7.4 and 7.5). However, the active magmatism in the Central Andean arc is not only connected with high heat flow (Springer and Förster 1998), high seismic attenuation (Haberland and Rietbrock 2001) and electrical conductivity (Echternacht et al. 1997), but also with the presence of low velocity zones, resulting in low average velocity values. Contrary to this, the east-west variations of magmatic activity during the Andean cycle in the Southern Andes were caused by changing subduction rates and obliquity (Jordan et al. 2001). Here, the subduction geometry has barely changed following the build up of the late Paleozoic accretionary complex. The gradually decreasing crustal seismic velocities from east to west (Figs. 7.14 and 7.15) reflect both the decreasing age of the accretionary complex and the westwardly-increasing tectonic activity caused by the underplating of subducted sediments at the base of the continental crust (Fig. 7.14; Lüth et al. 2003).

## 7.5 Conclusions

The results from the seismic investigations of the erosive and accretionary margins in Chile, not only provided new information about the internal geometry of the subsurface and the kinematics of the subduction processes, but also confirmed results from earlier studies. We selected the sections and models presented above in such a way that the images resulting from the different seismic methods along the three main transects (21° S, 23.25° S and 38° S) give an overall impression about the internal structure below the Chilean margin within the working areas.

The topography of the seafloor off northern Chile between 19° S and 26° S is rough with a thin sedimentary cover. Subducted horst-like structures indicate a strong coupling between the oceanic crust and fore-arc, whereas the subducting grabens, filled with debris from the frontal prism, transport fluids beneath the fore-arc, increasing the pore fluid pressure and thus reducing the friction along the plate interface (e.g. von Huene and Ranero 2003). The strong coupling and the resulting compressional forces from horst-continent collision possibly caused parts of the overriding plate to move, resulting in plate-parallel faults (Figs. 7.7 and 7.8) observed beneath the lower continental slope and the escarpments at the seafloor. A connection between these faults and the plate interface, possibly explains their strong reflectivity due to the infiltration of subducted fluids. However, material contrast or a cataclastic fabric are also reasonable interpretations for these features.

Basal, as well as frontal, erosion takes place at and beneath the outer fore-arc. Besides influences of deeper processes as discussed e.g. by Isacks (1988), Jordan et al. (1983, 2001), frontal erosion is one of the most important process resulting in a progressive eastward migration of the volcanic front (e.g. Scheuber et al. 1994). The dimensions and the geometry of the subduction channel are strongly related to the plate parallel faults which define its upper part beneath the continental slope. Under the coastline at 21° S, parts of the subduction channel are represented by a LVZ (Patzwahl 1998). The lower boundary of the oceanic crust has been clearly identified in the wide-angle results and the depth sections from the ANCORP and PRECORP profiles. Above the Nazca Plate, at about 55 km depth, the transition from the lower continental crust to a possibly hydrated mantle wedge is indicated by a geophysical Moho (Fig. 7.5). This depth region corresponds to the observations by Schmitz et al. (1999) who interpreted the Moho boundary beneath the Precordillera (at 65–70 km depth in Fig. 7.3b) as being produced by the phase boundary between the serpentinite and amphibole-bearing peridotites in the hydrated mantle wedge.

Off southern Chile, the ocean bottom is characterized by smooth morphology (Reichert et al. 2003). Here, an obvious axial channel is present at the bottom of the trench which is filled with terrigenous sediments (Figs. 7.16 to 7.18). The reflected signals from these sediments show a periodicity that can be correlated with the main glacial cycle (e.g. Rauch 2004). A recent small accretionary wedge, consisting of strongly heterogeneous unconsolidated sediment complexes, forms the outer fore-arc. Within the working area, almost all of the trench sediments are accreted frontally to tip of the plate. However, the region around the Arauco Peninsula, delimited by the Bio-Bio and Gastre fault zones to the north and the south, respectively, shows strong uplift (several km). This uplift of the Peninsula since the Late Miocene with acceleration towards the Quaternary, might be due to basal accretion and antiformal stacking, as observed during sandbox experiments (Lohrmann et al. 2000; Bohm et al. 2002). Here, the lower part of the trench sediments underthrust the frontal accretionary wedge. The strong uplift marks the position of active underplating (Lohrmann et al. 2000).

The sandbox experiments by Lohrmann et al. (2000) also showed that a variation in sediment supply results in changing rates of underplating and uplift. Another interesting point is that the crust in this region is nearly devoid of seismicity compared to the areas north and south (e.g. Bohm et al. 2002). The uplift also shows a possible relationship to the subduction angle, which is about 7° shallower beneath the Arauco Peninsula than seen in the other two transects.

Unfortunately, the top of the subduction channel has only been observed to a depth of about 25 km beneath the Coastal Cordillera (Fig. 7.19), as the heterogeneous interior of the modern accretionary wedge and the antiformal stack-structure of the Permo-Triassic mantle wedge (Fig. 7.20) produced strong seismic scattering. The location of the suggested hydrated mantle wedge in Fig. 7.20 corresponds to the eastern part of the wedge-shaped structure within the ISSA profile (Fig. 7.14).

## Acknowledgments

The Collaborative Research Center 267 (SFB 267) was financed by the Deutsche Forschungsgemeinschaft (DFG), the Freie Universität Berlin (FU), the Technische Universität Berlin (TU), the GeoForschungsZentrum Potsdam (GFZ) and the Universität Potsdam. Additional funding came from the German Federal Ministry of Education and Research (BMBF) for the ANCORP, CINCA and the ISSA projects. The marine data sets were acquired and kindly provided by the Federal Institute for Geosciences and Natural Resources (BGR), Hannover, Germany and the Leibniz Institute of Marine Sciences (IFM-GEOMAR). In this context we are grateful for support by the crew of the RV SONNE during cruise SO104 (CINCA) and SO161 (SPOC). Moreover, support for the field work, organization and technical accomplishment was given by a number of partners from South America: Universidad Católica del Norte Antofagasta, Chile; Universidad de Chile, Santiago; Universidad de Concepción, Chile; SERNAGEOMIN, Santiago, Chile; ENAP-SIPETROL, Santiago, Chile; CODELCO, Chile; Compañia Minera Punta de Lobos S. A., Chile; Compañia Minera Doña Inés de Collahuasi, Chile; Compañia Quebrada Blanca, Chile; Universidad Mayor de San Andres, La Paz, Bolivia; SERGEOMIN, La Paz, Bolivia; and Universidad Nacional de Salta, Argentina. The instrumental equipment for the seismic experiments was provided by the GIPP of the GeoForschungsZentrum Potsdam (GFZ), Germany, the Freie Universität Berlin, Germany, the Federal Institute for Geosciences and Natural Resources (BGR), Hannover, Germany and the Leibniz Institute of Marine Sciences (IFM-GEOMAR). Dedicated field work was provided by numerous students and scientists from South America and Germany.

## References

ANCORP Working Group (1999) Seismic reflection image revealing offset of Andean subduction zone earthquake locations into oceanic mantle. Nature 397:341–344

ANCORP Working Group (2003) Seismic imaging of a convergent margin and plateau in the central Andes (Andean Continental Research Project 1996 (ANCORP'96). J Geophys Res 108(B7): doi 10.1029/2002JB001771

Asch G, Schurr B, Bohm M, Yuan X, Haberland C, Heit B, Kind R, Woelbern I, Bataille K, Comte D, Pardo M, Viramonte J, Rietbrock A, Giese P (2006) Seismological studies of the Central and Southern Andes. In: Oncken O, Chong G, Franz G, Giese P, Götze H-J, Ramos VA, Strecker MR, Wigger P (eds) The Andes – active subduction orogeny. Frontiers in Earth Science Series, Vol 1. Springer-Verlag, Berlin Heidelberg New York, pp 443–458, this volume

Bangs NLB, Cande SC (1997) Episodic development of a convergent margin inferred from structures and processes along the southern Chile margin. Tectonics 16(3):489–503

Bangs NLB, Shipley TH (1999) Fluid accumulation and channelling along the northern Barbados Ridge decollement thrust. J Geophys Res 104(B9):20399–20414

Bohm M, Lüth S, Echtler H, Asch G, Bataille K, Bruhn C, Rietbrock A, Wigger P (2002) The Southern Andes between 36° and 40° latitude: seismicity and average seismic velocities. Tectonophysics 356(4):275–289

Brasse H, Soyer W (2001) A magnetotelluric study in the Southern Chilean Andes. Geophys Res Lett 28(19):3757–3760

Brasse H, Lezaeta P, Rath V, Schwalenberg K, Soyer W, Haak V (2002) The Bolivian Altiplano conductivity anomaly. J Geophys Res 107(B5): doi 10.1029/2001JB000391

Buske S, Lüth S, Meyer H, Patzig R, Reichert C, Shapiro SA, Wigger P, Yoon M (2002) Broad depth range seismic imaging of the subducted Nazca Slab, North Chile. Tectonophysics 350(4):273–282

Cifuentes IL (1989) The 1960 Chilean Earthquakes. J Geophys Res 94(B1):665–680

Clift P, Vannucchi P (2004) Controls on tectonic accretion versus erosion in subduction zones: Implications for the origin and recycling of the continental crust. Rev Geophys 42(2): doi 10.1029/2003RG000127

Davis D, Suppe J, Dahlen FA (1983) Mechanics of fold-and-thrust belts and accretionary wedges. J Geophys Res 88(B2):1153–1172

Echternacht F, Tauber S, Eisel, M., Brasse H, Schwarz G, Haak V (1997) Electromagnetic study of the active continental margin in northern Chile. Phys Earth Planet Int 102:69–88

Flüh ER, Grotzki N, Husen S, Ranero CR, Reichert C, Vidal N (1997) Seismic refraction investigations of the deep crustal structure offshore Chile. VIII Congreso Geologico Chileno, Antofagasta, Chile

Graeber FM, Asch G (1999) Three-dimensional models of P-wave velocity and P-to S-velocity ratio in the southern Central Andes by simultaneous inversion of local earthquake data. J Geophys Res 104(B9):20237–20256

Günther A, Haschke M, Reutter K-J, Scheuber E (1997) Repeated reactivation of an ancient fault zone under changing kinematic conditions: the Sierra de Moreno fault system (SMFS) (North Chilean Precordillera). Actas VIII Congreso Geologico Chileno 1:85–89

Gutscher MA, Kukowski N, Malavieille J, Lallemand S (1996) Cyclical behavior of thrust wedges: insights from high basal friction sandbox experiments. Geology 24:135–138

Haberland C, Rietbrock A (2001) Attenuation tomography in the western Central Andes: a detailed insight into the structure of a magmatic arc. J Geophys Res 106(B6):11151–11167

Hartley AJ, Chong G (2002) Late Pliocene age for the Atacama desert: implications for the desertification of western South America. Geology 30:43–46

Hervé F (1994) The Southern Andes between 39° and 44°S latitude: the geological signature of a transpressive tectonic regime related to a magmatic arc. In: Reutter K-J, Scheuber E, Wigger P (eds) Tectonics of the Southern Central Andes. Springer Verlag, Berlin Heidelberg New York, pp 223–248

Hinz K, Adam J, Bargeloh HO, Block M, Damm V, Dohmann H, Fritsch J, Kewitsch P, Krieger K, Neben S, Puskeppeleit K, Reichert C, Schlumschinski B, Schrader U, Schreckenberger B, Sievers J, Jiminez HN (1995) Crustal investigations off- and onshore Nazca/Central Andes (CINCA), SONNE cruise 104, leg 1–3. Technical Report BGR 113.998, Bundesanstalt für Geowissenschaften und Rohstoffe, Hannover

Husen S, Kissling E, Flueh E, Asch G (1999) Accurate hypocentre determination in the seismogenic zone of the subducting Nazca Plate in Northern Chile using a combined on-/offshore network. Geophys J Int 138(3):687–701

Isacks BL (1988) Uplift of the Central Andean plateau and bending of the bolivian orocline. J Geophys Res 93(B4):3211–3231

Jarrard RD (1986) Relations among subduction parameters. Rev Geophys 24(2):217–283

Jensen OL, Jordan TE, Ramos VA, Allmendinger RW, Isacks B (1984) Andean tectonics related to geometry of subducted Nazca Plate: discussion and reply. Geol Soc Am Bull 95(7):877–880

Jordan TE, Isacks BL, Allmendinger RW, Brewer JA, Ramos VA, Ando C (1983) Andean tectonics related to geometry of subducted Nazca Plate. Geol Soc Am Bull 94(3):341–361

Jordan TE, Burns WM, Veiga R, Pángaro F, Copeland P, Kelley S, Mpodozis C (2001) Extension and basin formation in the southern Andes caused by increased convergence rate: a mid-Cenozoic trigger for the Andes. Tectonics 20(3):301–324

Kanamori H, Cipar JJ (1974) Focal process of the great Chilean earthquake May 22, 1960. Phys Earth Planet Int 9:128–136

Klotz J, Khazaradze G, Angermann D, Reigber C, Perdomo R, and Cifuentes O (2001) Earthquake cycle dominates contemporary crustal deformation in Central and Southern Andes. Earth Planet Sci Lett 193(3–4):437–446

Krawczyk CM, SPOC Team (2003) Amphibious seismic survey images plate interface at 1960 Chile earthquake. EOS 84(32):301, 304–305

Krawczyk CM, Stiller M, Lüth S, Mechie J, SPOC Research Group (2003) Image of the seismogenic coupling zone in Central Chile: the amphibious experiment SPOC (Subduction Processes Off Chile). EGS-AGU-EGU Joint Assembly, Nice, France, EAE03-A-01228

Krawczyk CM, Lohrmann J, Oncken O, Stiller M, Mechie J, Bataille K, SPOC Research Group (2004) Basal accretion as mechanism for crustal growth in the 1960 Chile earthquake area. Geophys Res Abs 6, EGU04-A-02306

Krawczyk CM, Mechie J, Lüth S, Tašárová Z, Wigger P, Stiller M, Brasse H, Echtler HP, Araneda M, Bataille K (2006) Geophysical signatures and active tectonics at the south-central Chilean margin. In: Oncken O, Chong G, Franz G, Giese P, Götze H-J, Ramos VA, Strecker MR, Wigger P (eds) The Andes – active subduction orogeny. Frontiers in Earth Science Series, Vol 1. Springer-Verlag, Berlin Heidelberg New York, pp 171–192, this volume

Lessel K (1997) Die Krustenstruktur der Zentralen Anden in Nordchile (21–24°S), abgeleitet aus 3D-Modellierungen refraktionsseismischer Daten. Berliner Geowiss. Abh., Reihe B, Band 31

Lohrmann J, Kukowsky N, Adam J, Echtler H, Oncken O (2000) Identification of controlling parameters of the accretionary southern Chilean margin (38–40°S) with analogue models. 17. Geowiss. Lateinamerika-Kolloquium, Stuttgart

Lohrmann J, Kukowski N, Krawczyk CM, Oncken O, Sick C, Sobiesiak M, Rietbrock A (2006) Subduction channel evolution in brittle fore-arc wedges – a combined study with scaled sandbox experiments, seismological and reflection seismic data and geological field evidence. In: Oncken O, Chong G, Franz G, Giese P, Götze H-J, Ramos VA, Strecker MR, Wigger P (eds) The Andes – active subduction orogeny. Frontiers in Earth Science Series, Vol 1. Springer-Verlag, Berlin Heidelberg New York, pp 237–262, this volume

Lüth S (2000) Ergebnisse weitwinkelseismischer Untersuchungen und die Struktur der Kruste auf einer Traverse über die Zentralen Anden bei 21°S. Berliner Geowiss. Abh., Reihe B, Band 37

Lüth S, SPOC Research Group (2003) Subduction Processes Off Chile (SPOC) – results from the amphibious wide-angle seismic experiment across the Chilean subduction zone. EGS-AGU-EGU Joint Assembly, Nice, France, EAE03-A-04129

Lüth S, Wigger P, Mechie J, Stiller M, Krawczyk C, Bataille K, Reichert C, Flueh E (2004) The crustal structure of the Chilean forearc between 36° and 40°S from combined offshore and onshore seismic wide-angle measurements – SPOC (2001) Bolletino di Geofisica Teorica ed Aplicata, Special Issue, GeoSur 2004

Lüth S, Wigger P, ISSA Research Group (2003) A crustal model along 39° from a seismic refraction profile ISSA2000. Rev Geol Chile, 30(1):83–101

Patzig R (2000) Lokalbeben-Tomographie der Umgebung von Antofagasta (Nordchile) sowie Betrachtungen der Magnituden-Häufigkeits-Parameter in dieser Region. PhD thesis, Freie Universität Berlin

Patzig R, Shapiro, S, Asch, G, Giese, P, Wigger, P (2002) Seismogenic plane of the northern Andean Subduction Zone from aftershocks of the Antofagasta (Chile) 1995 earthquake. Geoph Res Let 29: doi 10.1029/2001GL013244

Patzwahl R (1998) Plattengeometrie und Krustenstruktur am Kontinentalrand Nord-Chiles aus Weitwinkelseismischen Messungen. Berliner Geowiss. Abh., Reihe B, Band 3

Patzwahl R, Mechie J, Schulze A, Giese P (1999) 2-D-velocity models of the Nazca Plate subduction zone between 19.5°S and 25°S from wide-angle seismic measurements during the CINCA'95 project. J Geophys Res 104(B4):7293–7317

Pelz K (2000) Tektonische Erosion am zentralandinen Forearc (20°–24°S). PhD thesis, Freie Universität Berlin

Plafker G, Savage JC (1970) Mechanism of the Chilean earthquakes of May 21 and 22, 1960. Geol Soc Am Bull 81(4):1001–1030

Ranero CR, von Huene R (2000) Subduction erosion along the Middle America convergent margin. Nature 404:748–755

Rauch K (2004) Zyklische Reflexionen terrigener Sedimente im Peru-Chile-Graben. PhD thesis, Freie Universität Berlin

Rauch K (2005) Cyclicity of Peru-Chile trench sediments between 36 and 38 degrees S: A footprint of paleoclimatic variations? Geoph Res Let 32: doi:10.1029/2004GL022196

Reichert C, Schreckenberger B, Adam J, Barckhausen U, Bargeloh H-O, Behrens T, Block M, Bönnemann C, Canuta Canuta J, Celedon Canessa V, Contreras Gonzales S, Damaske D, Diaz Naveas J, Franke D, Gaedicke C, Giglio Munoz S, Hermosilla Jarpa A, Kallaus G, Kewitsch P, Krawczyk CM, Kus J, Ladage S, Lindemann FM, Ranero C, Schrader U, Sepulveda J, Sievers J, Surburg E, Urbina Arce O, Zeibig M (2001) Subduction processes off Chile (SPOC) SONNE cruise SO161 leg 2 & 3. Technical Report BMBF Forschungsvorhaben 03G0161A, Bundesanstalt für Geowissenschaften und Rohstoffe, Hannover

Reichert C, SPOC Scientific Shipboard Party (2002) Subduction processes off Chile: initial geophysical results of SONNE Cruise SO-161(2+3). Geophys Res Abs EGS02-A-05338

Reichert C, Schreckenberger B, Flueh ER, Krawczyk C, SO161-2/3-Bordwissenschaftler (2003) Subduktionsvariabilität am aktiven Kontinentrand Chiles (SO 161-2 und -3). In: Tanner B (ed) Statusseminar 2003, Meeresforschung mit FS SONNE. Forschungszentrum Jülich, Projektträger des BMBF und BMWi, Bereich MGS, Außenstelle Warnemünde, Hamburg, Germany, pp 39–42

Rietbrock A, Haberland C (2001) A tear in the subducting Nazca slab: Evidence from local earthquake tomography and high precision hypocenters. EOS 82(47) Fall Meeting Supplement T31A-0822

Rietbrock A, Waldhauser F (2004) A narrowly spaced double-seismic zone in the subducting Nazca plate. Geophys Res Lett 31(10): doi 10.1029/2004GL019610

Scheuber E, Giese P (1999) Architecture of the Central Andes – a compilation of geoscientific data along a transect at 21°S. J South Am Earth Sci 12:103–107

Scheuber E, Reutter K-J (1992) Magmatic arc tectonics in the Central Andes between 21° and 25°S. Tectonophysics 205:127–140

Scheuber E, Bogdanic T, Jensen A, Reutter K-J (1994) Tectonic development of the North Chilean Andes in relation to plate convergence since the Jurassic. In: Reutter K-J, Scheuber E, Wigger P (eds) Tectonics of the Southern Central Andes. Springer Verlag, Berlin Heidelberg New York, pp 121–140

Schmitz M, Lessel K, Giese P, Wigger P, Araneda M, Bribach J, Graeber FM, Grunewald S, Haberland C, Lüth S, Röwer P, Ryberg T, Schulze A (1999) The crustal structure beneath the Central Andean forearc and magmatic arc as derived from seismic studies – the PISCO 94 experiment in northern Chile (21°–23°S). J South Am Earth Sci 12:237–260

Sick CMA (2005) Structural investigations off Chile: Kirchhoff Prestack depth migration versus Fresnel volume migration. PhD thesis, Freie Universität Berlin

Springer MH, Förster A (1998) Heat flow density across the Central Andean subduction zone. Tectonophysics 291:123–139

Stern RJ (2002) Subduction zones. Rev Geophys 40(4): doi 10.1029/2001RG000108

Tebbens SF, Cande SC (1997) Southeast Pacific tectonic evolution from the early Oligocene to Present. J Geophys Res 102:12061–12084

Thornburg T, Kulm LD (1987) Sedimentation in the Chile trench: depositional morphologies, lithofacies and stratigraphy. Geol Soc Am Bull 98:33–52

von Huene R, Ranero CR (2003) Subduction erosion and basal friction along the sediment-starved convergent margin off Antofagasta, Chile. J Geophys Res 108(B2): doi 10.1029/2001JB001596

von Huene R, Scholl DW (1991) Observations at convergent margins concerning sediment subduction, subduction erosion, and the growth of continental crust. Rev Geophys 29:279–316

von Huene R, Weinrebe W, Heeren F (1999) Subduction erosion along the North Chile margin. J Geodyn 27:345–358

Wigger P, Baldzuhn S, Giese P, Heinsohn WD, Schmitz M, Araneda M, Martínez E, Ricaldi E, Viramonte A (1994) Variations of the crustal structure of the Southern Central Andes deduced from seismic refraction investigations. In: Reutter K-J, Scheuber E, Wigger P (eds) Tectonics of the Southern Central Andes. Springer Verlag, Berlin Heidelberg New York, pp 23–48

Yanez G, Cembrano J, Pardo M, Ranero C, Selles D (2002) The Challenger-Juan Fernandez-Maipo major tectonic transition of the Nazca-Andean subduction system at 33–34°S: geodynamic evidence and implications. J South Am Earth Sci 15(1):23–38

Yoon M (2001) Application of true amplitude prestack migration to a deep seismic dataset (ANCORP, Chile & Bolivia). Diploma Thesis, Freie Universität Berlin

Yoon M (2005) Deep seismic imaging in the presence of heterogeneous overburden: numerical modeling and case studies from the Central and Southern Andes. PhD thesis, Freie Universität Berlin

Yoon M, Buske S, Schulze R, Lüth S, Shapiro SA, Stiller M, Wigger P (2003) Alongstrike variations of crustal reflectivity related to the Andean subduction process. Geophys Res Lett, 30(4):1160, doi 10.1029/2002GL015848

Yuan X, Sobolev SV, Kind R, Oncken O, Bock G, Asch G, Schurr B, Graeber F, Rudloff A, Hanka W, Wylegalla K, Tibi R, Haberland C, Rietbrok A, Giese P, Wigger P, Röwer P, Zandt G, Beck S, Wallace T, Pardo MC (2000) Subduction and collision process in the Central Andes constrained by converted seismic phases. Nature 408: 958–961

Yuan X, Asch G, Bataille K, Bock G, Bohm M, Echtler H, Kind R, Oncken O, Wölbern I (2004) Deep seismic images of the Southern Andes. In: Kay SM, Ramos VA (eds) Late cretaceous to recent magmatism and tectonism of the Southern Andean Margin at the latitude of the Neuquen Basin (36–39°S) Geol Soc Am Spec P, submitted

# Chapter 8

# Geophysical Signatures and Active Tectonics at the South-Central Chilean Margin

Charlotte M. Krawczyk · James Mechie · Stefan Lüth · Zuzana Tašárová · Peter Wigger · Manfred Stiller · Heinrich Brasse
Helmut P. Echtler · Manuel Araneda · Klaus Bataille

**Abstract.** The ISSA 2000 (Integrated Seismological experiment in the Southern Andes) and SPOC 2001 (Subduction Processes Off Chile) onshore and offshore projects surveyed the Chilean margin between 36 and 40° S. This area includes the location of the 1960 earthquake ($M_w$ = 9.5) that ruptured the margin from ~38° S southwards for ~1 000 km. Together with gravity and magnetotelluric components, the active-passive seismic experiments between 36 and 40° S provide the first, complete, high-resolution coverage of the entire seismogenic plate interface.

The observed offshore mode of sediment subduction corresponds well with the landward extension of the reflection seismic profile at 38°S (westernmost portion of line SPOC-South), which shows material transported downwards in a subduction channel. From the slow uplift of the Coastal Cordillera, we conclude that basal accretion of parts of this material controls the seismic architecture and growth of the south Chilean crust. There is almost no seismicity observed along the entire, approximately 130 km wide, seismogenic coupling zone. Furthermore, the study area is characterized by a 25–35 km thick crust beneath the Longitudinal Valley, with high-conductivity zones at 20–40 km depth that correspond to large fault zones. Below the volcanic arc, the crust is generally 35–45 km thick, with a maximum thickness of 55 km at ~36° S.

The slab steepens southwards along the margin (13°–21°), and a wedge-shaped body at the plate interface can be either interpreted as hydrated mantle with 20–30% serpentinization or, when divided, as mafic crustal material in the upper part and serpentinized mantle in the lower part. The lower plate could suffer slab rollback while the upper-plate kinematic segmentation exhibits fore-arc extension, possibly combined with corner-flow and active lower-plate retreat. At the top of the active subduction channel, underplating, fore-arc uplift and serpentinization are key processes at the south-central Chilean margin.

## 8.1 Introduction

The Andean convergent margin varies significantly along its 8 000 km length. It has areas of both long-term erosion and accretion (Clift and Vanucchi 2003), as well as different morphological segments and segmented deformation (e.g. Gansser 1973; Mpodozis and Ramos 1989; Dewey and Lamb 1992; Kley et al. 1999; Lamb and Davis 2003). The geodynamic evolution of this system, and the relationship between deep and surface structures, are controversially debated and poorly understood. Images of the offshore fore-arc and its onshore continuation are complex. The segments vary in geophysical character, and there are several fault systems that have been active during the past 24 million years. The high and broad central Andes (3–36° S, mean elevation 4 km, up to 800 km wide) are separated from the low and narrow Patagonian Andes (38–46° S, mean elevation < 1 km, ~300 km wide) by the transitional segment of the south-central or northern Patagonian Andes (36–42° S; Hervé 1994). This transitional segment has some apparent peculiarities and is the focus of the investigations presented and discussed here (Fig. 8.1).

Driven by the convergence between the Nazca and South American Plates, compressional deformation occurs in the central Andes. Since the Miocene, shortening and crustal thickening has taken place north of 37° S (Isacks 1988; Allmendinger et al. 1997; Jordan et al. 1997; Giambiagi et al. 2003) and decreases southwards (Ramos 1989; Vietor and Echtler 2006, Chap. 18 of this volume) with non-partitioned deformation (Hoffmann-Rothe et al. 2006, Chap. 6 of this volume). In contrast, shortening has not occurred since the late Miocene south of 37° S and strong strain-partitioning is evident. Only the active fore-arc wedge shows high rates of uplift (~2–3 km Myr$^{-1}$; Melnick et al. 2006) and accretion (Bangs and Cande 1997). Since the Pliocene, active deformation has manifested itself at the Liquiñe-Ofqui fault zone (Fig. 8.1; Hervé 1976), a dextral, strike-slip system that accommodates oblique subduction and the ridge-push forces of the Chile Rise (Thomson 2002; Cembrano et al. 2002).

These geological characteristics of the transitional margin segment and its unknown structure and activity at depth, down to the plate interface and, especially, within the seismogenic coupling zone, motivated our investigations that began in 1999. These investigations included geological field work (e.g. Melnick et al. 2006; Vietor and Echtler 2006, Chap. 18 of this volume), active seismic experiments (SO-161 Shipboard Scientific Party 2002; Lüth et al. 2003a,b; Krawczyk and the SPOC Team 2003; Rauch 2005) and passive seismological observations and tomography (Yuan et al. 2000, 2006; Bohm et al. 2002; Bohm 2004), as well as three-dimensional (3-D) modeling of gravity (Tašárová 2004; Tašárová et al. 2006; Hackney et al. 2006, Chap. 17 of this volume) and magnetotelluric data (Brasse et al. 2006), all aimed at constraining the tectonophysical characteristics of this part of the Andean active margin. This paper attempts to summarize and integrate these geophysical and tectonic observations and to discuss the possible state of the system and ongoing processes.

## 8.2 Tectonic Setting

The oceanic crust that constitutes the Nazca Plate, which subducts under the south-central Andes, is 25–35 million years old (37–39° S; Tebbens and Cande 1997). GPS modeling shows that convergence rates are 66 mm yr$^{-1}$ (Angermann et al. 1999) or ~80 mm yr$^{-1}$ averaged over the last five million years (Somoza 1998).

Along strike, prominent morphotectonic segments and their differing tectonic evolution have been attributed to variations in the geometry and physical properties of the downgoing plate (e.g. Jordan et al. 1983; Yañez et al. 2002; Wagner et al. 2005). Major changes in the lower plate occur at the Valdivia fracture zone system that intersects the margin at ~40° S. This fracture zone separates oceanic crust produced at the Chile Rise to the south from the oceanic crust of the East Pacific Rise to the north. These oceanic crusts differ in age, thickness and number of fracture zones (Tebbens and Cande 1997).

Perpendicular to the margin, the western flank of the northern Patagonian Andes is generally subdivided from west to east into the following main morphotectonic units (Fig. 8.1): (1) the Coastal Cordillera, formed by a Permo-Triassic accretionary complex (Western Series) south of ~38.2° S and a late Paleozoic magmatic arc (Eastern Series; Hervé 1988) mainly north of the above latitude; (2) the Longitudinal Valley, a basin filled by Oligo-Miocene sedimentary and volcanic rocks covered by Plio-Quaternary sediments; (3) the Main Cordillera, formed by a Meso-Cenozoic magmatic arc and intra-arc volcano-sedimentary basins; and (4) the Mesozoic Neuquén Basin and the Cretaceous-Tertiary foreland basin to the east.

In the Patagonian segment of the Andes, any accretionary flux at the Chilean subduction zone is added to the Coastal Cordillera (Bangs and Cande 1997), and the Main Cordillera presumably receives all of its mass flux from below (magmatic underplating). Accretive mass flux from the east through an eastern thrust-front occurred

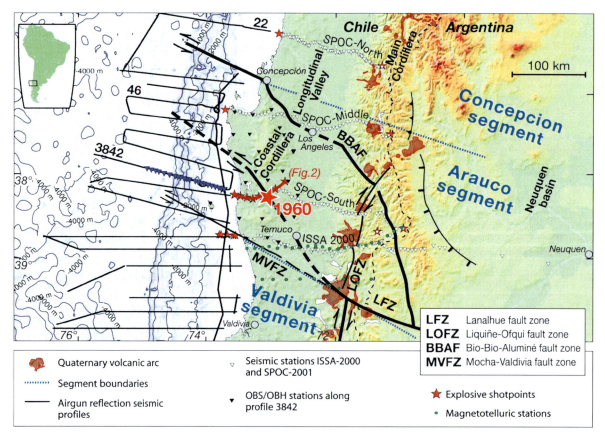

Fig. 8.1. Locations of the discussed geophysical surveys in south-central Chile, with main tectonic elements and fault zones (after Melnick and Echtler 2006, Chap. 30 of this volume). *LFZ:* Lanalhue fault zone; *LOFZ:* Liquiñe-Ofqui fault zone; *BBAF:* Bio-Bio-Aluminé fault zone; *MVFZ:* Mocha-Valdivia fault zone; *red areas:* Quaternary volcanic arc; *blue dotted lines* mark segment boundaries; *black lines:* airgun reflection seismic profiles; *yellow and black triangles:* seismic stations deployed along refraction and reflection seismic profiles during the ISSA-2000 and SPOC-2001 projects; *blue triangles:* OBS/OBH stations that recorded airgun shots along profile 3842; *red stars:* explosive shotpoints; *green dots:* magnetotelluric stations

mainly during the middle and late Miocene (Ramos 1989; Melnick et al. 2006; Vietor and Echtler 2006, Chap. 18 of this volume). Late Miocene, undeformed basalts overlying the deformed Neogene deposits indicate that frontal thrusting stopped at 9–6 Ma (Ramos 1989).

## 8.3 Geophysical Structure of the Andean Margin between 36–40° S

Since 1999, a variety of geological and geophysical surveys has been acquired at the Andean convergent margin from 36–40° S (Fig. 8.1). The tectonophysical results are summarized here.

### 8.3.1 Near-Vertical Reflection Seismic Data

The steep-angle component of the SPOC (Subduction Processes Off Chile) survey aimed to describe the seismogenic coupling zone and find the relationships between processes at depth and at surface. Therefore, an onshore, near-vertical incidence reflection seismic profile was shot in order to reveal the structural elements of the fore-arc in south-central Chile at 38.2° S along the westernmost portion of profile SPOC-South (see Krawczyk and the SPOC Team 2003 for technical details; Fig. 8.1 for location).

The receiver spread for the reflection line was 54 km long with three spread set-ups, each 18 km long. The set-

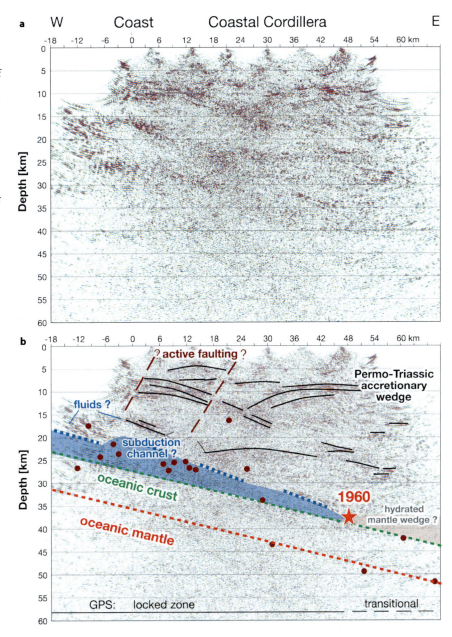

**Fig. 8.2.**
**a** SPOC near-vertical incidence, land seismic reflection profile after depth migration. The line covers the westernmost 70 km of the onshore part of the SPOC-South refraction profile. **b** In the interpreted section, the subducting plate is projected into the section from the wide-angle model. Immediately above, a possible subduction channel that transports material from W to E below the overriding plate and the presumed hypocenter of the 1960 Chile earthquake are marked. The upwardly convex reflective structures are interpreted to represent the Permo-Triassic accretionary wedge. Seismicity is sparse within a 25 km-wide clip area

up was twice shifted eastwards from the coast to the western margin of the Longitudinal Valley. With each set-up, 180 geophone-groups were deployed at 100 m spacing. Two explosive shots in the Pacific Ocean, ten borehole shots at six different locations within the onshore reflection line, and two shots east of the active spread were recorded.

This near-vertical reflection seismic land experiment thereby resulted in a 72 km-long, common-depth-point line, with twofold coverage in the innermost 45 km and singlefold coverage at the margins, extending the seismic imaging to the Pacific Ocean and to the Longitudinal Valley. Importantly, this profile crossed the presumed hypocenter of the 1960 Valdivia earthquake (located by Cifuentes 1989 and relocation after Krawczyk and the SPOC Team 2003; Engdahl and Villasenor 2002) and was augmented westwards across the offshore fore-arc during cruise SO-161, Leg 2 (SO-161 Shipboard Scientific Party 2002; Rauch 2005).

Offshore line 3842 (see Fig. 8.1 for location) shows the following main features: (a) thick successions of terrigenous trench sediments with a maximum thickness of ~3 km at the deformation front; (b) a thin (max. 150 m) pelagic cover over the faulted oceanic plate; and (c) the oceanic Moho occurring subhorizontally at about 10 km depth beneath the pelagic domain (Rauch 2005; Sick et al. 2006, Chap. 7 of this volume). The dip of the Moho towards the continent is not clearly imaged in the reflection seismic section and can only be estimated to amount to about 12–15°.

The onshore, depth-migrated section images different, strong and continuous, reflection bands in the upper and middle crust to at least 25 km depth (Fig. 8.2). Over the entire profile, two strong reflection bands, 2–3 km thick and slightly upwardly convex, are observed in the overriding plate between 7 and 18 km depth (Fig. 8.2a, profile-km 0–25 and 25–55). In the eastern half of the profile, a similar feature occurs between 20 and 30 km depth, some 10 km below the upper-crustal structures, but with weaker reflectivity and less continuity. In the western half of the profile, prominent reflections dip eastwards from about 15 km depth to ~38 km depth. These reflective bands are strongest below the coast, with reflectivity becoming less pronounced eastwards. Finally, in the central part of the seismic reflection profile, some relatively weaker reflections are seen in the oceanic plate between 30 and 45 km depth (Fig. 8.2).

The downgoing Nazca Plate is not structurally resolved and can only be deduced from the wide-angle data (see below). Further, signals describing a continental Moho towards the eastern end of the section lack clarity. In general, the regional seismicity is extremely sparse along the profile (clip area: 25 km) and observations are mainly restricted to the upper plate.

## 8.3.2 Refraction Seismic Data

As part of the SPOC project, offshore and onshore seismic refraction measurements were made in order to investigate the crustal structure in the area (Lüth et al. 2003b). Together with the results of the ISSA-2000 seismic profile (Lüth et al. 2003a), velocity models along four transects are now available between 36 and 39° S (see Fig. 8.1 for location).

The most comprehensive seismic data set was acquired along the SPOC-South profile (approximately along 38.2° S). Offshore, 22 OBS/OBH stations, provided and operated by

**Table 8.1.** Root-mean square (*RMS*) traveltime errors from the wide-angle velocity model along profile SPOC-South. The traveltime errors and numbers (*N*) of used traveltimes are indicated for each OBS/OBH (*OB*), for the two chemical shots (*S1* offshore, *S2* in the Main Cordillera) and for the land based receivers (*LR*) recording the airgun shots along the profile SO161-3842

| Station/shot | RMS error (s) | N |
|---|---|---|
| OB 46 | 0.099 | 393 |
| OB 45 | 0.058 | 368 |
| OB 44 | 0.079 | 328 |
| OB 43 | 0.125 | 407 |
| OB 42 | 0.114 | 420 |
| OB 41 | 0.131 | 315 |
| OB 40 | 0.243 | 581 |
| OB 39 | 0.089 | 14 |
| OB 38 | 0.153 | 438 |
| OB 37 | 0.167 | 581 |
| OB 36 | 0.157 | 129 |
| OB 35 | 0.112 | 214 |
| OB 33 | 0.203 | 107 |
| OB 32 | 0.129 | 284 |
| OB 31 | 0.118 | 184 |
| OB 30 | 0.090 | 232 |
| OB 29 | 0.221 | 408 |
| OB 28 | 0.132 | 458 |
| OB 27 | 0.159 | 327 |
| OB 26 | 0.144 | 296 |
| OB 25 | 0.107 | 364 |
| OB 24 | 0.133 | 275 |
| S1 | 0.134 | 42 |
| LR 1 | 0.052 | 213 |
| LR 2 | 0.101 | 597 |
| LR 3 | 0.299 | 582 |
| LR 4 | 0.193 | 1044 |
| LR 5 | 0.175 | 426 |
| LR 6 | 0.109 | 742 |
| LR 7 | 0.131 | 750 |
| LR 8 | 0.170 | 550 |
| LR 9 | 0.232 | 171 |
| LR 10 | 0.265 | 449 |
| LR 11 | 0.168 | 169 |
| LR 12 | 0.126 | 922 |
| LR 13 | 0.134 | 80 |
| S2 | 0.113 | 19 |
| All | 0.163 | 13879 |

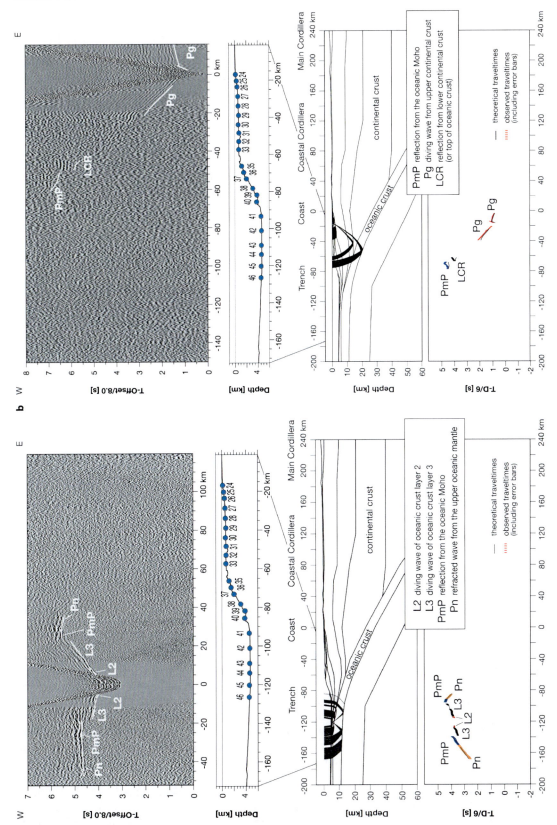

**Fig. 8.3.** Seismic data, topographic profile, ray paths and travel-time curves for offshore survey stations (**a**) OBH 45 and (**b**) OBH 24, from top to bottom. *White lines* in the seismic data indicate the modeled travel times from the picked phases. The phases are: *L2* – diving wave of oceanic crust layer 2; *L3* – diving wave of oceanic crust layer 3; *PmP* – reflection from the oceanic Moho; *Pn* – refracted wave from the upper oceanic mantle; *Pg* – diving wave from upper continental crust; *LCR* – reflection from lower continental crust (or top of oceanic crust)

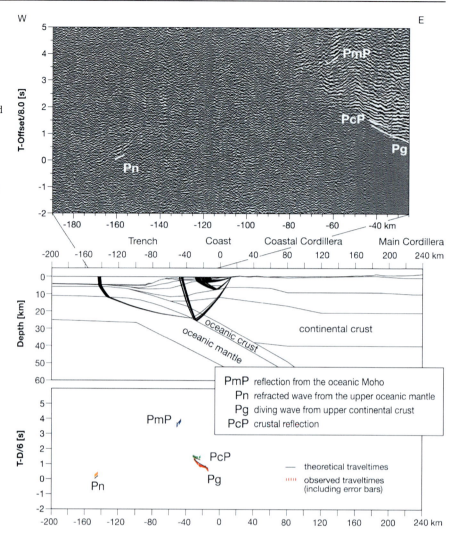

**Fig. 8.4.**
Seismic data, ray paths and travel-time curves for Station 3 of the amphibious survey (SPOC-South profile), from top to bottom. *White lines* in the seismic data indicate the modeled travel times from the picked phases. The phases are: *Pg* – diving wave through the upper continental crust; *PcP* – crustal reflection, possibly from the bottom of a velocity inversion; *PmP* – oceanic Moho reflection; *Pn* – refraction from the upper oceanic mantle

IfM-Geomar, were used for a wide-angle seismic survey along the airgun profile 3842 (offshore survey; e.g. Fig. 8.3). The airgun shots fired along the same profile were recorded by 26 three-component receivers onshore (amphibious survey; e.g. Fig. 8.4). Finally, chemical explosions were fired into an onshore receiver profile of 52 three-component receivers (onshore survey; e.g. Fig. 8.5). Along the SPOC-Middle and SPOC-North profiles, only an amphibious and an onshore survey were carried out; offshore measurements were not made along these transects.

The interpretation of the seismic wide-angle reflection and refraction data comprises forward modeling of travel-time curves that were picked from the receiver and shot gathers, respectively. The forward modeling was done with two-point ray-tracing (Zelt and Smith 1992). The forward modeling along the SPOC-South profile is based on 22 OBS/OBH stations, two chemical shots and 13 land based receivers recording the airgun shots along profile SO-161 3842 (Table 8.1). The overall root-mean square (RMS) traveltime error is 0.163 s, which is in the order of the picking error. The picking error depends strongly on the signal to noise ratio (S/N) of the data and is estimated to be between 50 ms (first breaks at small offsets) and 200 ms (Pn in noisy data), depending on the type of picked phases. The RMS errors for the single shot and receiver gathers, respectively, vary between 0.059 s and 0.299 s, due to the varying S/N ratios and due to the projection of all traveltime data on a 2-D line without taking into account possible 3-D effects.

Recordings from the offshore, seismic wide-angle survey (see Fig. 8.3 for data examples) constrain the seismic P-wave velocity distribution of the oceanic crust and upper mantle, as well as of the western tip of the continental crust. The oceanic crust consists of three layers and is 6–7 km thick. In the trench and on the lower continental slope, a layer of sediments can be identified owing to relatively low P-wave velocities (less than 2.5 km s$^{-1}$). This layer reaches a maximum thickness of about 2 km in the trench.

For the amphibious survey, onshore stations recorded airgun shots (e.g. see Fig. 8.4). Due to the offset range

**Fig. 8.5.**
Seismic data, ray paths and travel-time curves for the western shotpoint of the onshore survey (SPOC-South profile). *Gray lines* in the seismic data indicate the modeled travel times from the picked phases. The phases are: *Pg* – diving wave through the upper continental crust; *PcP(1)* – crustal reflection, possibly from the bottom of a velocity inversion; *Pc(1)* – upper crustal refraction; *PcP(2)* – crustal reflection from the top of the lower continental crust; *Pc(2)* – lower crustal refraction; *PmP(oc)* – oceanic Moho reflection

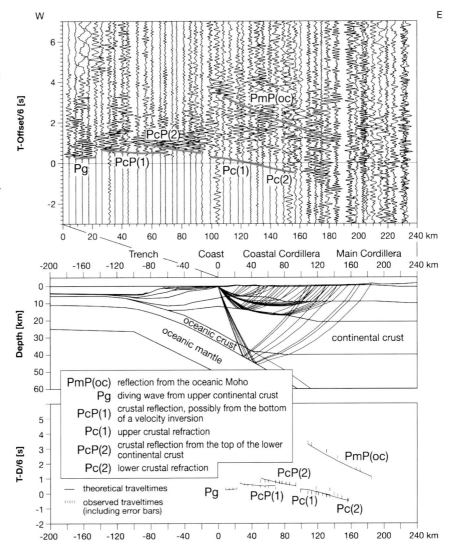

(min. 20 km, max. ~200 km), the focus of these recordings is on the geometry of the oceanic Moho beneath the coast. Additionally, the P-wave velocity distribution in the upper 20 km of the continental crust can be investigated using the first breaks between 20 and 80 km offset.

The recordings of the onshore survey (e.g. see Fig. 8.5) constrain the crustal structure of the continental crust between the Coastal Cordillera and the Main Cordillera. The deepest reflector observed by this survey is the oceanic Moho at approximately 40 km depth under the Coastal Cordillera.

After a velocity model was determined that explains the correlated phases observed along the SPOC-South transect, where all three experiment components are available, the offshore part of the SPOC-South model was projected onto the SPOC-Middle and SPOC-North transects. Then, phases were picked from the recordings of the amphibious and onshore surveys along these transects and, using the same forward-modeling method, velocity models were built.

The main features of these models are the oceanic crust and upper mantle as well as the continental crust between the offshore fore-arc and the Main Cordillera (Fig. 8.6, in which only areas of confidence are shown). The oceanic crust is approximately 6–7 km thick and the oceanic Moho is observed at about 40 km depth beneath the Coastal Cordillera. The continental crust is characterized by P-wave velocities that generally increase from west to east. The upper continental crust is approximately 10–15 km thick and the lower crust is ~25 km thick in the eastern part of the SPOC-North model. The thickness of the continental crust was only resolved along the SPOC-North profile. Along the other SPOC transects, no continental Moho was detected (Fig. 8.6).

The only other constraint available for indicating the continental Moho comes from receiver functions that imaged it along a profile at about 39° S (Yuan et al. 2006). These results show that the continental Moho lies at a depth of approximately 40 km below the Main Cordillera,

**Fig. 8.6.**
Refraction seismic profiles in south-central Chile from 36–39° S from the ISSA and SPOC projects (modified after Lüth et al. 2003a, 2003b; Krawczyk and the SPOC Team 2003), overlain by regional seismicity data (Bohm et al. 2002) projected from 25 km-wide clip areas onto the respective profiles. The P-wave velocities are color coded and quantified (*numbers*) in km s$^{-1}$. In the models only the areas of confidence are shown

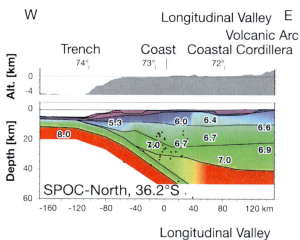

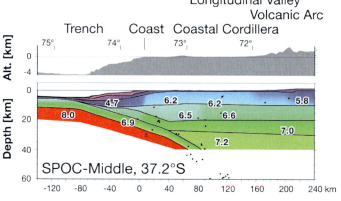

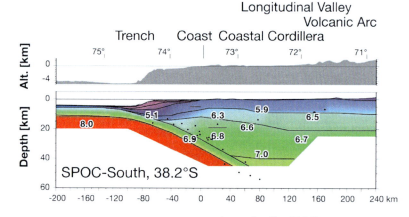

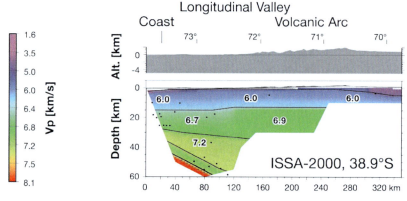

shallowing westwards to ~30 km depth below the Longitudinal Valley. Beneath the Lanalhue fault zone, the limit of the structural resolution of this method is reached, so the suggested Moho depth of 25 km at this location (Yuan et al. 2006) is not included here.

The topography and geometry of the oceanic Moho can be compared using the results of the four, wide-angle seismic transects (Fig. 8.7). For this purpose, the profiles are shifted laterally so that the eastern margin of the trench is at the same point for all transects. The topographical profiles along the three SPOC-transects almost coincide, except for the upper and lower continental slopes. The geometries of the oceanic Moho differ significantly along the SPOC- and ISSA-transects beneath the Coastal Cordillera. Subduction of the oceanic crust is shallower along the SPOC-Middle profile at 37.25° S than in the other transects, which show an increasing dip both towards the north and south (Fig. 8.7).

### 8.3.3 Electrical Resistivity Distribution from Magnetotelluric Soundings

At the end of 2000, long-period, magnetotelluric sounding data (MT, period range 10–20 000 s) were collected in the southern Andes along two profiles at 38.9° S and 39.3° S (Fig. 8.8), extending from the Pacific coast to the Chile/Argentina border. Two-dimensional (2-D) models (Brasse and Soyer 2001) were obtained by employing a smooth-inversion algorithm (Rodi and Mackie 2001).

The most prominent, low resistivity (or high conductivity) zones are found in the fore-arc (labelled A in Fig. 8.8a), below the volcanic arc (B; Fig. 8.8a) and further towards the Argentinian border (B'; Fig. 8.8a). Whereas anomaly A is spatially correlated with the surface trace of the Lanalhue fault, anomaly B maps the Liquiñe-Ofqui fault system at greater depth. However, anomaly B could also be, at least partly, caused by partial melt zones below the volcanic arc. Note that the magnitude of these conductivity anomalies is substantially less pronounced than below the Altiplano Plateau of the central Andes (Brasse et al. 2002).

This model was obtained by inverting MT impedances alone, with no regard for the transfer functions between the vertical and horizontal magnetic fields (tippers). If these tippers are plotted in the standard form for induction vectors, with their real parts pointing away from conductors in a 2-D environment, a dramatic, first-order effect is observed over the whole working area, extending from 38.9° S to 40.9° S (Fig. 8.8b). The vectors do not point from west to east, i.e. away from the coast (with the ocean as a large conductive body producing the so-called coast effect) as expected. Instead, they are systematically deflected to the NE. This behaviour cannot be explained by simple 2-D models nor by 3-D models with realistic geometries, even after extensive testing.

The only feasible way to account for an anomalous, vertical, magnetic field over a large area is to assume (macro-) anisotropic structures in the crust (for a comprehensive treatise on the effects of anisotropy on MT transfer functions see e.g. Pek and Verner 1997; Weidelt 1999). Under anisotropic conditions, the conductivity is expressed as a $(3 \times 3)$-tensor, or with symmetry arguments taken into account, as three conductivities in the spatial directions and three angles (strike, dip and slant). This leads to a sixfold-increased model space in the anisotropic domain (and subsequent, cumbersome modeling efforts) and it cannot be expected that all parameters can be equally resolved. We concentrate here on conductivity ratios and strike angle, $\alpha$, (measured anticlockwise from

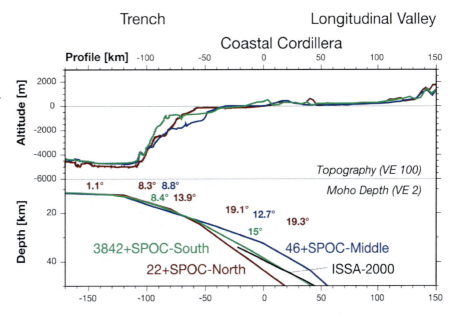

**Fig. 8.7.**
Comparison of the SPOC and ISSA transects. *Top*: Topographical profiles of the three SPOC transects. The common reference point of the profiles is the eastern margin of the trench. *Bottom*: Depth profiles of the oceanic Moho along the three SPOC transects (*red, green and blue lines*) and the ISSA-2000 profile (*black line*)

**Fig. 8.8.**
Results of a magnetotelluric survey in south-central Chile (modified after Brasse and Soyer 2001; Li et al., unpublished data). **a** The 2-D, isotropic, electrical resistivity (reciprocal of conductivity) model from inversion of magnetotelluric data along the ISSA-2000 profile at 38.9° S (see map in **b**) shows three, prominent, high-conductivity zones. The modeling of the corresponding induction vectors at a period of 3 300 s (**b**) results in (**c**) a 2-D resistivity model with an anisotropic crust that explains the observed induction vectors at long periods. The strike angle, α, of the anisotropic blocks is measured anticlockwise from geographic north

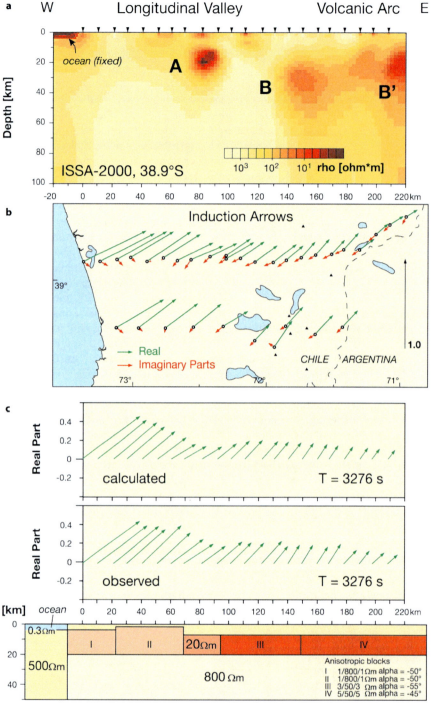

north), leaving dip and slant constant at 90° and 0°, respectively (Li et al., unpublished data).

Accordingly, a 2-D algorithm that includes anisotropy (Li 2000) was employed. Figure 8.8c explains the observed real part of the induction vectors, at least qualitatively. The model incorporates the ocean, with crude bathymetry, which is responsible for the large magnitude of induction vectors (approx. 0.8) in the Coastal Cordillera. The anisotropy of the crust causes the deflection. The model also incorporates an isotropic conductive zone that accounts for the slow decrease of vector magnitudes from west to east, and which is similar to zone A (Fig. 8.8a), found from 2-D isotropic inversion of impedances. The slightly higher conductive zone to the east is below the arc and back-arc volcanoes.

The mean strike angle is approximately –45 to –50°, i.e. the conductive direction is roughly NW-SE, which strongly concurs with the fault directions delineated by Melnick and Echtler (2006, Chap. 30 of this volume; see also Fig. 8.1). The MT modeling study indicates that these faults reach deep into the crust. The anisotropy pattern also overprints the major tectonic feature in the region, the Liquiñe-Ofqui fault (strike N 10° E), which is thought to largely control volcanism. Apparently, though, the NW-SE-striking faults have a more important influence on the conductivity structure (a prominent example is the Villarrica-Quetrupillan-Lanin volcanic lineament).

At present, we are unable to constrain the depth extent of the anisotropic layers, but the modeled lower boundary of 20 km appears to be a minimum estimate. The anisotropy is particularly pronounced below the Coastal Cordillera. Summarizing, this model indicates a deeply fractured fore-arc and arc crust, which facilitates the migration of fluids (or melts, if conditions are appropriate).

### 8.3.4 3-D Gravity Modeling

Geoid and gravity anomalies are thought to indicate local-scale segmentation of both the overriding continental and the subducting oceanic plates between 36 and 42° S along the Andean margin (Fig. 8.9a; Tašárová 2006; Hackney et al. 2006, Chap. 17 of this volume). A 3-D density model was constructed for the area from 36–42° S using the IGMAS (Interactive Gravity and Magnetics Application System) software (e.g. Götze 1978; Götze and Lahmeyer 1988).

In numerical simulations, subsurface structures are described as closed polyhedrons with constant density or susceptibility. With this indirect approach, using forward modeling, the computed gravity effect of a given structure is compared to the measured field, and input parameters (density and geometry) are changed until the best fit is reached (e.g. Fig. 8.9b). The modeled geological bodies are defined along vertical cross sections that are connected by triangulation, thereby creating a volume.

In total, more than 14 000 irregularly distributed land stations, together with marine gravity data from the SPOC experiment (Götze et al. 2002), were used for the modeling. The onshore gravity database is a compilation of several data sets (Araneda et al. 1999a, 1999b; Wienecke 2002) that were reprocessed after new control measurements.

The 2-D section through the 3-D model is shown along the SPOC-South profile at 38.2° S (Fig. 8.9). Obvious features include the pronounced gravity low (free air anomaly) of the trench ($-120 \times 10^{-5}$ m s$^{-2}$, profile-km 0–20) and the positive Bouguer anomalies in the fore-arc (centered above the coastline). These positive anomalies are caused by the shallow subducting plate together with the structures of the upper plate (accretionary wedge, high density structure at the wedge base, and crustal thinning underneath the Longitudinal Valley; $0–50 \times 10^{-5}$ m s$^{-2}$, profile-km 40–170).

Despite the crustal thinning beneath the Longitudinal Valley, the gravity values decrease to $-100$ to $-120 \times 10^{-5}$ m s$^{-2}$ towards the Longitudinal Valley and the main volcanic arc (profile-km 170–280), mainly because of the decreased mantle density beneath the arc, the crustal thickening towards the east, and the basin fill in the Longitudinal Valley. Taking into account factors such as field-measurement errors and topography correction, the accuracy of the computed Bouguer anomaly values for the entire database lies in the range of ±5 to $10 \times 10^{-5}$ m s$^{-2}$.

Within the 3-D gravity field between 36–42° S, three fore-arc/arc segments have been identified that are also evident in the geology (Melnick and Echtler 2006, Chap. 30 of this volume; cf. Fig. 8.1). The northern Arauco segment (36.5–39° S) shows a pronounced gravity high of 60 to $80 \times 10^{-5}$ m s$^{-2}$ in the fore-arc, whereas the middle Valdivia segment (39–40° S) is characterized by low fore-arc gravity ($-30$ to $0 \times 10^{-5}$ m s$^{-2}$) (Fig. 8.9a). The southernmost Bahía Mansa segment (40–42° S, not shown on the map) has a fore-arc gravity high of $60 \times 10^{-5}$ m s$^{-2}$ with a longitudinal offset, compared to the northern segment. Based on the results of forward-density modeling (Tašárová 2004; Tašárová 2006), the changes within the observed gravity field along the margin are dominantly caused by along-strike variations in the depth to the slab beneath the fore-arc region.

The position of the slab (calculated from the coastline to 20 km eastwards in the onshore fore-arc) is shallowest in the northern Arauco segment (22–35 km), where it is responsible, at long wavelengths, for the observed gravity high. The low gravity field of the middle Valdivia segment is reproduced in the density model by a deep oceanic plate beneath the fore-arc region (40–48 km). This middle segment is anomalous because it is the only part of the western continental margin of South America where a gravity high is not observed.

The gravity high in the southern segment, shifted landwards, can be explained by a thin crust (25 km), as also suggested by the geological data (Melnick and Echtler 2006, Chap. 30 of this volume), beneath a well-developed Longitudinal Valley with a sediment infill of some 3 km thickness. The depth to the slab beneath the fore-arc region in the southern segment is intermediate (32–38 km) at the coastline. Taken at a constant distance of 130 km east of the trench, the depth to the slab varies from 33–37 km in the northern segment (onshore fore-arc), through 39–42 km in the middle segment (offshore fore-arc), to 39–40 km beneath the southern segment (onshore fore-arc).

The position of the slab is dependent on the composition and, thus, on the density of the fore-arc accretionary wedge. Seismic data that constrain the density model are available in the northern segment at 36–39° S, whereas the results from the remaining segments are based on the interpretation of geology and gravity data only. Hence, both the position of the lower plate and the density of the

**Fig. 8.9.**

**a** Combined free-air (offshore) and Bouguer (onshore) anomaly map of south-central Chile. The onshore gravity database is a compilation of several data sets (Araneda et al. 1999a; 1999b; Wienecke 2002) that were reprocessed after new control measurements at more than 14 000 irregularly distributed land stations (*gray lines and patterns*; Tasarova 2006). The offshore data are a combination of the SPOC measurements (Götze et al. 2002; Schreckenberger et al. 2002) and the KMS global free air anomaly data (Andersen and Knudsen 1998). A typical gravity low, which increases southwards, is observed at the trench in the northern part of the study area. Two pronounced Bouguer gravity maxima dominate the fore-arc region. A negative Bouguer anomaly characterises the entire volcanic arc (*black triangles* mark volcanoes). The locations of the four geophysical cross sections (cf. Figs. 8.6 and 8.10) are indicated. **b** A cross section from the gravity model along the SPOC-South profile at 38.2° S. The computed gravity effect of a given structure (*black curve*) is compared with the measured anomaly (*red curve*, red stars mark positions of field stations) until the best fit is reached (*top*). The color-coded bodies (*numbers* indicate densities in g cm$^{-3}$) are overlain by interfaces from the seismic refraction model (*blue lines*) and by the seismicity (*black dots*), while the main volcanic arc is denoted by black triangles (*bottom*)

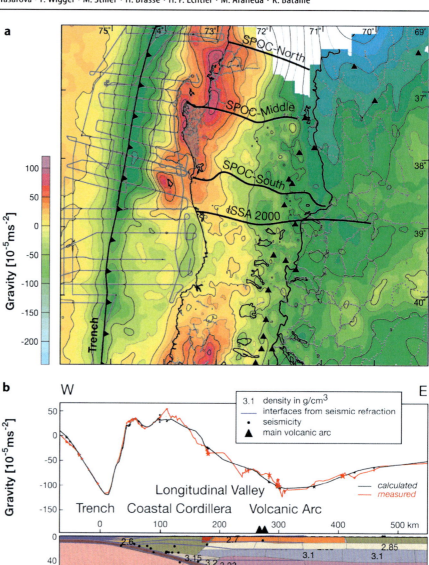

upper plate in the middle and southern segments of the study area are not well known and the nature of the fore-arc structures remains uncertain there.

### 8.3.5 Regional Seismology

The ISSA seismological network operated from January to April 2000 (36–40° S) and covered a region of 400 × 550 km. There were 62 seismological stations with an average station spacing of 50 km. A total of 333 local seismic events were gathered and, simultaneously, a high-quality data subset was inverted for determining one-dimensional velocity structure, hypocentral coordinates and station delays (Bohm et al. 2002). Thus, this experiment provided important teleseismic and seismological constraints in the Arauco segment (Bohm et al. 2002; Bohm 2004; Yuan et al. 2005).

The results show that crustal seismicity to 40 km depth is concentrated in the fore-arc region, with no earthquakes occurring beneath the Main Cordillera and along the Liquiñe-Ofqui fault zone. In a rather narrow zone below the coastal area and along the Bio-Bio-Aluminé and Lanalhue fault zones, maximum crustal seismicity at 20–30 km depth occurred at and above the plate interface (Bohm et al. 2002). Almost no crustal seismicity was observed between these two fault zones despite this area matching the region of highest uplift and topography in the forearc (Melnick and Echtler 2006, Chap. 30 of this volume). In the Arauco segment, focal mechanisms of selected earthquakes from the ISSA 2000 and CMT catalogues indicate an exclusively compressional character to 50 km depth, whereas deeper than this both compressional and extensional events are observed (Bruhn 2003; Bohm 2004).

Intermediate-depth seismicity, as revealed by the depth-frequency distribution of local earthquakes, reaches its first maximum at 20 km depth, reflecting the zone of seismic coupling. The second maximum at 60 km could be caused by dehydration embrittlement in the Nazca Plate (Bohm et al. 2002). Benioff seismicity is observed to 150 km depth resulting in the first accurate image of the Wadati-Benioff zone in this region. The Benioff zone defines an east-dipping plane with continuous seismicity to 120 km depth.

The observed Arauco cluster (Chinn and Isacks 1983; Bohm 2004) argues for relatively frequent seismic release and interface seismicity, normal Gutenberg-Richter behaviour, low coupling and, thus, less stress storage. In contrast, interseismic quiescence and postseismic relaxation is called for in the area around 38° S and southwards (Khazaradze and Klotz 2003), which could point to higher coupling and higher stress accumulation. Thus, from a seismological point of view, a possible segment boundary would be expected at around 38.2° S, where the coastal plain facies of the Arauco peninsula terminate, rather than at around 39° S.

## 8.4 Geodynamic Significance of the Tectonophysical Inventory

Gravity and MT-data modeling, regional seismicity, receiver functions, tomography and wide-angle data reveal spatial variations within the seismogenic zone between 36 and 39° S, in continental Moho depths (30–40 km) and in slab dip below the coastline (13–21°). With respect to the patterns of inherited and reactivated upper-plate structures and associated differences in the rheological behaviour of the overriding plate, we discuss the physical signatures observed across the 1960 Chile earthquake rupture plane.

By superposing the results of the different experiments, several combined east-west, geophysical cross sections through the south-central Chilean margin (36–39° S) have been compiled (Figs. 8.10 and 8.11). North-south variations are best illustrated by the results from those experiments that have greatest spatial coverage, e.g. gravity, local-earthquake, P-wave tomography, regional seismicity and seismic refraction (Fig. 8.10). In contrast, east-west variations are best shown by a cross section along the SPOC-South profile at 38.2° S in which the more detailed results from the deep, near-vertical seismic reflection experiments are highlighted (Fig. 8.11).

### 8.4.1 Moho Depth and Crustal Thickness

Where the seismic refraction results and/or the receiver functions provide good control on the continental Moho along the ISSA-2000 profile, the Moho depths derived by the two seismic methods and gravity all agree (Fig. 8.10). With respect to the oceanic Moho along the ISSA-2000 profile, the depths derived from the receiver functions are often shallower than those determined from the seismic refraction and gravity. In general, the study area is characterized by a crustal thickness in the range of 25–35 km in the fore-arc beneath the Longitudinal Valley, and 35–45 km below the volcanic arc, with a maximum under the arc of 55 km (Fig. 8.10). In the following paragraphs, we describe the available cross sections from north to south and address some specific features at the end of this section.

Along the SPOC-North profile where the seismic refraction method provides good control between 200 and 300 km distance, there is good agreement with the gravity results on the depths to the continental Moho. West of 200 km distance along the SPOC-North profile, the continental Moho depths determined by gravity rise to levels where seismic refraction velocities of about 7.0 km s$^{-1}$ are found.

Along the SPOC-Middle and SPOC-South profiles, no reflected or refracted phase associated with the continental Moho was found in the seismic refraction data and, so, the continental Moho has been drawn where velocities of about 7.2 km s$^{-1}$ are reached. This is based on the ISSA-2000 profile results where P-velocities for the uppermost mantle beneath the onshore fore-arc of about 7.2 km s$^{-1}$ were found (Fig. 8.6; Lüth et al. 2003b). Along the SPOC-Middle and SPOC-South profiles, the continental Moho depths derived by gravity and seismic refraction generally differ and, especially in the western parts of these profiles, the continental Moho depths derived by gravity rise to levels where velocities of no more than 7.0 km s$^{-1}$ are reached.

Beneath the ISSA-2000 profile, the continental crust in the vicinity of the volcanic arc has a thickness of around 40 km. Further north, below the SPOC-South profile, the continental crust is similarly thick (Fig. 8.10). Below the SPOC-Middle and SPOC-North profiles, north of the Bio-Bio-Aluminé fault (which marks the boundary between the central Andes and the south-central Andes), gravity indicates that the continental crust thickens northwards with a maximum value along the SPOC-North profile of 55–57 km east of the volcanic arc.

**Fig. 8.10.**
Geophysical cross sections of the three SPOC seismic refraction profiles and the ISSA-2000 profile (cf. Fig. 8.6). The cross sections show slices of the 3-D gravity and local earthquake tomography models, together with major boundaries and features from the receiver functions (*red lines* marking oceanic and continental Moho) and the seismic refraction (*thick black lines*) and magnetotelluric (patterns of *sloped black lines*) models. Densities from the gravity model are in g cm$^{-3}$. Regional seismicity data (Bohm et al. 2002) have also been projected from 20 km-wide clip areas onto the respective profiles. The region of possibly hydrated/serpentinized mantle is labelled *A*, "normal" mantle is labelled *B* and the 7 km s$^{-1}$ layer in the seismic refraction velocity models is labelled *C*. The continental Moho and regions of possibly serpentinized mantle from the gravity model are marked by *white lines*

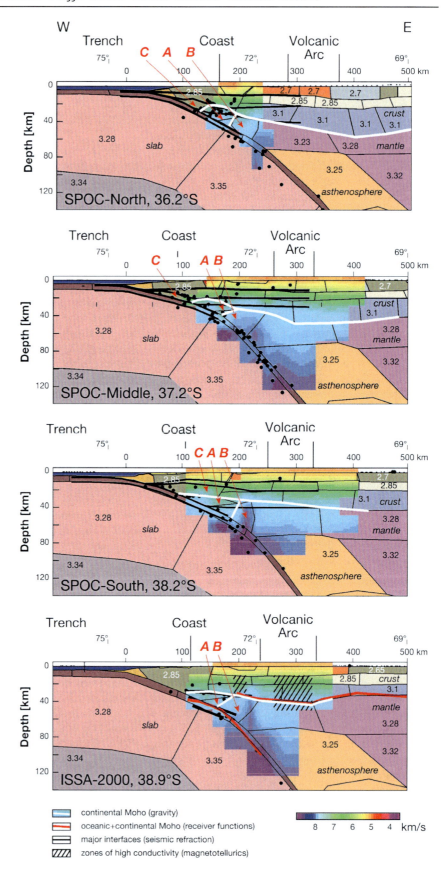

The continental crust thins towards the coast and the continental Moho meets the plate boundary at a depth of about 30 km. This is much shallower than the down-dip depth of the seismogenic coupling zone of some 40–50 km as derived from GPS dislocation modeling (Khazaradze and Klotz 2003) and the ISSA earthquake catalogue (Bruhn 2003). The exception is the SPOC-South profile where the junction may be at as much as 40 km depth if more weight is given to the seismic refraction than the gravity (Fig. 8.10).

In the Valdivia segment (39–40° S), the most striking feature of the fore-arc gravity field is the absence of a positive Bouguer gravity anomaly, which is in contrast to almost the entire South American margin (Fig. 8.9a; see also Hackney et al. 2006, Chap. 17 of this volume). Coupled negative-positive (trench – fore-arc) gravity anomalies are a characteristic feature of subduction-zone gravity fields globally and, thus, there has to be a regional feature causing this distinctly shaped anomaly in the study area.

Based on the results of the density modeling, there are two possible explanations for the observed gravity signal. Either the fore-arc continental crust has material of lower density or the slab position beneath the fore-arc region is deeper than to the north. In this respect, the geochemical data (Glodny et al. 2006, Chap. 19 of this volume) indicate that rocks composing the fore-arc paleo-accretionary complex between 39–40° S are denser than those north of 39° S. Therefore, differences observed in the gravity field cannot be explained by low-density crustal material. Hence, the observed gravity field was reproduced by a deeper position of the slab beneath this segment.

Regional seismicity is not only concentrated within the Wadati-Benioff zone but also within the continental crust, especially along the SPOC-North profile (Fig. 8.10). The high P-velocity anomaly derived from the local earthquake tomography west of the volcanic arc between 40 and 60 km depth can be seen to reside almost completely in the uppermost mantle.

Both of the regions of high electrical conductivity identified along the ISSA-2000 profile lie completely in the crust. The eastern high-conductivity zone can be easily attributed to melts below the volcanic arc (Fig. 8.10, bottom), but the western one might describe fractured crust within the onshore fore-arc at 39° S. Possibly, a deep-reaching connection exists between the fault structures mapped at surface (see Fig. 8.1) and deeper crustal levels, allowing fluids to ascend beneath the Longitudinal Valley.

### 8.4.2 Existence of a Serpentinized Mantle Wedge

In the area examined in southern Chile, a particular problem is posed by the body labelled A in the cross sections (Fig. 8.10). This body is, to a large extent, responsible for the mismatch between the gravity-derived and seismic-refraction-derived positions of the continental Moho. In the gravity model, this body is actually sub-divided into two parts, with the upper part having a density of 3.15 g cm$^{-3}$ and the lower part having a density of 3.2 g cm$^{-3}$.

Evidence for highly hydrated and serpentinized fore-arc regions that are affected by fluid devolatizing from the subducting slab is found world-wide, e.g. the mantle-wedges beneath northern Chile (ANCORP-Working Group 1999), Japan (Kamiya and Kobayashi 2000), the southern Cascadia margin (Rondenay et al. 2001; Bostock et al. 2002) and Kamchatka (Saha et al. 2005). If both gravity model layers are interpreted to be serpentinized mantle, then the continental Moho, as predicted from the density model, would meet the subducting plate some 10–15 km shallower than that based on the tomography model. Therefore, the unidentified body A can be interpreted as mafic crustal material in the upper part and serpentinized mantle in the lower part.

A mafic composition for the lower crustal rocks could be related to the intruded granitoids of the Nahuelbuta Mountains. These granitoids were formed in the late Paleozoic, represent the subduction-related magmatic arc, and a mafic lower crust at its base would be expected. Also, both the Nahuelbuta Mountains and the unidentified, high-density body occur only in the Arauco segment of the study area.

Further evidence that the upper part of the unidentified body could be crustal material is seen in the results from the SPOC wide-angle, reflection/refraction seismic profiles, where the intersection between the continental Moho and the subducting plate is at 30–40 km depth (Fig. 8.10). The upper limit of the unidentified body in the density model is often far shallower than the interpreted continental Moho from the seismic refraction models, and is situated in a layer with P-wave velocities of 7.0 km s$^{-1}$. The only profile where the upper boundary of the high density body coincides with the observed continental Moho from the seismic refraction models is the ISSA-2000 line, owing to the pronounced decrease of the observed Bouguer gravity anomaly values. Also, the late Paleozoic arc does not continue to this latitude and, so, the entire body here could represent serpentinized mantle only.

An argument against the unidentified body consisting of serpentinized mantle in its lower part is the depth of the seismic coupling zone, which, based on the ISSA earthquake catalogue (Bruhn 2003), reaches to 50 km depth. If the unidentified body is serpentinite, a reduction in seismicity could be expected. However, thrust earthquakes occur in the contact zone between the slab and the presumed serpentinite. This situation may be thought unlikely as it has been suggested that the down-dip limit of subduction thrust earthquakes is controlled by the formation of serpentinite in the fore-arc mantle (Peacock and Hyndman 1999, and references therein). However, numerical modeling studies (e.g. Sobolev and Babeyko 2005, and references therein) require friction coefficients significantly less than 0.1 along the plate interface. Such small friction coefficients are much lower than any from known minerals and, thus, it may indeed be possible to have thrust earthquakes where the slab is in contact with presumed serpentinite. Alternatively, the down-dip limit of subduction thrust-earthquakes could be related to the brittle-ductile transition.

Assuming that body A represents hydrated mantle, then approximately 20% serpentinized peridotite would have a density of 3.15–3.2 g cm$^{-3}$ and a P-velocity of 7.5–7.6 km s$^{-1}$ (Hacker et al., 2003; Hacker and Abers, 2004; Table 8.2). While the densities are in good agreement, the P-velocity of the 20% serpentinized peridotite is somewhat higher than the observed P-velocities of 7.0–7.3 km s$^{-1}$ in body A. However, the observed P-velocities could have errors of at least ±0.2 km s$^{-1}$, and if the densities in the gravity model were 0.05–0.10 g cm$^{-3}$ too high, then approximately 30% serpentinized peridotite could possibly explain body A. Conversely, a search of a databank containing velocity measurements for 416 rocks of many different types (Stadtlander et al. 1999 and references therein) reveals that approximately 20% serpentinized peridotite can have a density of about 3.15 g cm$^{-3}$ and a P-velocity of around 7.2 km s$^{-1}$. Thus, it is concluded that hydrated mantle with 20–30% serpentinization is one possible explanation for body A.

An alternative explanation for body A is that it is lower crustal material. In this case, utilizing the velocity-density relationship of Sobolev and Babeyko (1994), velocities of 7.0–7.3 km s$^{-1}$ do produce densities of 3.1–3.25 g cm$^{-3}$. Another alternative explanation is that body A comprises metamorphosed trench sediments derived from the Western Series. In the area under investigation, they consist of ~30–40% metabasites within metapelites, but only 100% metabasite would yield the required velocity and density values (Table 8.2). The calculation with ~30% metabasites and 70% metasediments gives values far below those observed, and is therefore unlikely.

It is generally possible to detect serpentinized peridotite seismologically by mapping Poisson's ratio, as has been demonstrated in Japan (Kamiya and Kobayashi 2000). In the area investigated here, we observe high Poisson's ratios (0.28–0.29; Bohm 2004), but because serpentinized mantle (0.28–0.30) and "normal" lower crust (0.27–0.28) both have high Poisson's ratios (Hacker et al. 2003; Hacker and Abers 2004; Sobolev and Babeyko 1994; Stadtlander et al. 1999) an unequivocal decision is precluded. The high Poisson's ratio of the body labelled A is associated with low seismicity and weak interplate coupling, which is, at least, consistent with the ductility of serpentinite.

The high P-velocities derived from the local earthquake tomography west of the volcanic arc between 40 and 60 km depth might constitute the remains of normal mantle material sandwiched between body A to the west and the hotter, somewhat slower and less dense mantle beneath the volcanic arc to the east (Fig. 8.10). Further east of the region resolved by the tomography, normal mantle material exists again as indicated by the regions in the cross sections with normal mantle density.

Along the ISSA-2000 profile, where it is best defined, the continental Moho shallows westwards from about

**Table 8.2.** Selected physical properties of different rock compositions chosen to explain the origin of the enigmatic body labelled A that is interpreted to lie between 25 and 40 km depth along the geophysical profiles (see Fig. 8.10 and text; unless indicated in the table, calculations use the formulas of Hacker and Abers 2004). Velocities and densities are calculated for temperatures of 400–500 °C and pressures of about 1 GPa, which are approximately the conditions where body A occurs in the south-central Chile subduction zone

|  | $v_P$ (km s$^{-1}$) | $v_S$ (km s$^{-1}$) | $v_P / v_S$ | Poisson's ratio | Density (g cm$^{-3}$) |
|---|---|---|---|---|---|
| Mafic rock, gabbro | 6.83 – 7.04 | 3.69 – 3.85 | 1.83 – 1.86 | 0.29 – 0.30 | 3.14 |
| Granulite facies | 6.99 | 3.84 | 1.82 | 0.28 | 3.2 |
| Garnet-amphibolite facies | 7.21 | 3.94 | 1.83 | 0.29 | 3.31 |
| Garnet-granulite facies | 7.57 | 4.19 | 1.8 | 0.28 | 3.37 |
| Normal mantle peridotites (lherzolite, harzburgite) | 8.07 – 8.20 | 4.65 – 4.75 | 1.73 | 0.25 | 3.31 – 3.32 |
| Serpentinite (96%) | 5.98 | 2.63 | 2.27 | 0.38 | 2.82 |
| 29% serpentinized lherzolite | 7.05 | 3.66 | 1.93 | 0.32 | 2.98 |
| 30% serpentinized peridotite (and Bohm 2004) | 7.34 | 3.84 | 1.91 | 0.31 | 3.14 |
| 20% serpentinized peridotite | 7.6 | 4.2 | 1.81 | 0.28 | 3.15 – 3.2 |
| Normal lower crust (and Sobolev and Babeyko 1994) | 7.0 – 7.3 | 3.9 – 4.1 | 1.77 – 1.81 | 0.27 – 0.28 | 3.1 – 3.2 |
| Metabasites occurring in southern Chile (100%) | 6.91 | 3.76 | 1.84 | 0.29 | 3.15 |
| Metamorphosed trench sediments southern Chile (30% metabasites) | 6.27 | 3.76 | 1.67 | 0.22 | 2.85 |
| Observed value of body A (this manuscript incl. Bohm 2004) | 6.8 – 7.5 | – | 1.80 – 1.84 | 0.28 – 0.29 | 3.15 – 3.2 |

40 km below the volcanic arc to about 35 km beneath the Longitudinal Valley to around 30 km depth where it meets the plate boundary (Fig. 8.10). At least some of this westward shallowing is probably due, in part, to thickening of the crust beneath the volcanic arc caused by the addition of magmatic material to the crust in this region, and probably due, in part, to the thinning of the crust below the Longitudinal Valley, which occurred when extensional processes formed the Longitudinal Valley basins. However, crustal thinning under the Longitudinal Valley between 36 and 39° S is likely to be very limited as the basins, comprising interbedded sedimentary and volcanic strata of late Oligocene to early Miocene age, are limited and have a maximum thickness of no more than 3–4 km (Martin et al. 1999; Jordan et al. 2001; Tašárová 2004). Utilizing a stretching model (McKenzie 1978; Le Pichon and Sibuet 1981) and applying the principles of Airy isostasy, it is estimated that the change in crustal thickness from the volcanic arc to the Longitudinal Valley, owing to magmatic thickening below the arc and extensional thinning below the Longitudinal Valley, is no more than 5 km. This would explain the observed change in crustal thickness, especially along the ISSA-2000 profile. Also, with this line of reasoning, the high P-velocities derived from the local earthquake tomography west of the volcanic arc between 40 and 60 km depth are unlikely to be mostly caused by Moho uplift beneath the Longitudinal Valley as a consequence of extension or magmatic thickening.

## 8.5 Subduction Tectonics

East-west variations in the structural image across the south-central Chilean subduction zone are shown in the cross section along the SPOC-South profile at 38.2° S (Fig. 8.11). It provides the first, complete, high-resolution coverage of the entire seismogenic plate interface. Here, both offshore (Rauch 2005) and onshore (Krawczyk and the SPOC Team 2003), deep, near-vertical, seismic reflection data complement the wide-angle (Lüth et al. 2003b) and tomographic (Bohm 2004) images.

In the onshore, depth-migrated section (Fig. 8.2), the upwardly convex, reflective structures occurring in the upper plate to approximately 30 km depth are interpreted as representing the Permo-Triassic accretionary wedge above the subducting Nazca Plate. The Permo-Triassic accretionary wedge is exposed at the surface also with upwardly convex structures. A major structural element is interpreted at the plate boundary between 18 and 38 km depth. It is marked as an approximately 3 km-thick subduction channel that is transporting sedimentary material from west to east below the overriding South American Plate (Fig. 8.2b). Other independent aspects also requiring the existence of a subduction channel are discussed below.

The offshore part of the reflection profile reveals that the up to 2–3 km-thick trench fill (Fig. 8.11) could serve as a feeder for the subduction channel, supplying the material necessary for crustal growth by basal accretion. The onshore structural image (Fig. 8.2; Krawczyk et al. 2004) suggests that this mechanism controls the seismic architecture and growth of the south Chilean crust.

Further, extensional tectonics and uplift of the Coastal Cordillera are observed at the surface (Melnick and Echtler 2006, Chap. 30 of this volume), but since the rates of uplift are low, the amount of basal accretion must also be low. This means that the bulk of the material has to be transported downwards otherwise an orogen with a very thick crust would be formed, which is not observed here. Thus, such a subduction channel, as described above, is also required from mass balance considerations.

Reflective elements that are thought to mark the upper boundary of the subduction channel are speculated to represent fluid traps. If the presence of a subduction channel is considered, then low to very low coupling in the frontal part of the plate interface is required, thus per-

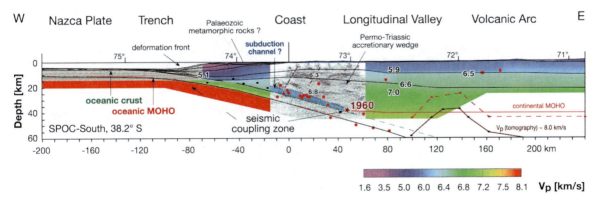

**Fig. 8.11.** Combined E-W seismic profile crossing marine and onshore domains (offshore and onshore fore-arc, Coastal Cordillera, Longitudinal Valley, volcanic arc) at about 38.2° S. The wide-angle velocity model is overlain by structural and tomographic information and interpretation to 60 km depth. Data are compiled according to Krawczyk and the SPOC Team 2003; Lüth et al. 2003b; Rauch 2005; Bohm et al. 2002; Bohm 2004

mitting wet sediments to be transported downwards. In relation to the uplift and extensional tectonics observed at the surface (Melnick and Echtler 2006, Chap. 30 of this volume), the marked reflectivity could correspond to a zone of dehydration, embrittlement and active basal accretion, as simulated in scaled analogue experiments (see Lohrmann et al. 2002, 2006, Chap. 11 of this volume).

The available, reflection seismic images and other deductions suggest the subduction channel to exist structurally between the deformation front and the 1960 earthquake hypocenter, thus covering a distance of at least 130 km to 35 km depth. How this would be represented deeper and eastwards cannot be determined at present. However, the wide-angle velocities do not contradict the subduction-channel interpretation and allow interpolation between the two reflection-seismic data sets and towards the Main Cordillera. Because the presence of a subduction channel requires low coupling in the frontal part of the plate interface, it might not be surprising that there is almost no seismicity observed along the entire, approximately 130 km-wide, seismogenic coupling zone at 38.2° S (Fig. 8.11).

These findings might be, in part, comparable to the observations made in northern Chile along the ANCORP profile (Fig. 8.12) and to related investigations around 21° S (ANCORP Working Group 1999; 2003; Buske et al. 2002; Patzig et al. 2002; Patzwahl et al. 1999; Sick et al. 2006, Chap. 7 of this volume; Yoon et al. 2003; Yuan et al. 2000). There, the zone of seismic coupling is well defined by seismicity data that indicates a ~60 km wide seismic coupling zone (Fig. 8.12), which is only half as wide as that determined for southern Chile (Fig. 8.11).

A subduction channel may also be present beneath northern Chile. Observations, such as a ~3 km thick zone of seismicity close to the coast, thrusts parallel to the plate margin, and tectonic erosion, imply the existence of a subduction channel. More proximate evidence is given by reflection and wide-angle seismic data. Patzwahl et al. (1999) observed a low-velocity zone between 25 and 40 km depth directly above the downgoing plate in the fore-arc region close to the coast. This wedge-shaped feature was interpreted as eroded and underplated material (Patzwahl 1998). Furthermore, reflectivity patterns paralleling this feature (due east of the coast in Fig. 8.12; cf. also detailed structural images of ANCORP-Working Group 2003) suggest that this structure is a possible subduction channel in northern Chile.

Looking from north to south, the data presented here and other geological evidence show a sudden decrease in seismicity south of 38° S and indicate a steepening of the slab from 37–42° S. The slab dip below the coastline is 19° in SPOC-North, 13° in SPOC-Middle, 15° in SPOC-South, 20° in ISSA-2000 (Figs. 8.10 and 8.11), and 19° at 41° S, as suggested by gravity (Tašárová 2004) and geological (Echtler et al. 2003) modeling.

At the same time, high tomographic velocities appear due west of the volcanic arc ($v_p$ ~7.5 and 8.0 km s$^{-1}$, Bohm 2004; Fig. 8.11). The subduction angle, as observed from wide-angle data at 37°, is shallower at this latitude than in all other lines (Fig. 8.6). This spatially coincides with the Bio-Bio-Aluminé fault zone which cuts through the area, or could imply an increased buoyancy of the oceanic crust under the Arauco Peninsula, contributing to the uplift of the peninsula.

The different findings along the investigated profiles can be explained if one assumes an active, lower-plate retreat in southern Chile. However, since the oceanic plate becomes younger southwards as the Chile Rise is approached, it is necessary to counteract the increasing buoyancy forces of the younger, warmer, and less dense material in the south, even allowing for a steepening of the slab and less pronounced fore-arc topography.

According to the laboratory modeling and terminology of Kincaid and Griffith (2003), this requires a horizontal return-flow around the subducting plate and a vertical corner-flow in the mantle wedge of the overriding plate. Both requirements are fulfilled in southern Chile. Vertical corner-flow is assured by asthenospheric upwelling from the

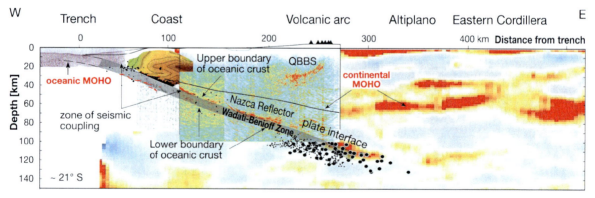

**Fig. 8.12.** Combined E-W seismic profile crossing marine and onshore domains (offshore and onshore fore-arc, Coastal Cordillera, Longitudinal Valley, volcanic arc, Altiplano, Eastern Cordillera) at about 21° S. Structural and tomographic data are shown to 150 km depth. *QBBS*: Quebrada Blanca Bright Spot. Data are compiled according to ANCORP Working Group 1999, 2003; Buske et al. 2002; Patzig et al. 2002; Patzwahl et al. 1999; Sick et al. 2006, Chap. 7 of this volume; Yoon et al. 2003; Yuan et al. 2000

Somuncura Hotspot, which has been active since the Oligocene (~25 Ma; de Ignacio et al. 2001), and horizontal corner-flow increases towards the Chile Rise where a slab window opened at ~12 Ma (Gorring et al. 1997).

Dating of the meta-sedimentary nappe piles exposed at the surface along the reflection seismic profile reveals that these rocks are Permo-Triassic (Glodny et al. 2006, Chap. 19 of this volume). They experienced their main deformation in the early Permian (~285 Ma). The basal-accretionary mass flow, responsible for their addition to the upper plate, ended at ~250 Ma and exhumation took place at ~200 Ma (Glodny et al., in press).

With a given thickness of some 20–30 km for the Paleozoic accretionary wedge in this area, and presuming that the position of the wedge complex remained stable, this would yield a value of about 0.6–1 km basal accretion per Myr. This type of calculation is not yet available at other convergent margin systems as there is a lack of exact age-dating combined with petrological and structural evidence for a complete mass-transfer path. Nonetheless, the net crustal growth rate in southern Chile is estimated to lie at the lower end of the global rates observed to date (cf. review of Clift and Vanucchi 2004).

## 8.6 Summary

The study area is generally characterized by a crustal thickness in the range of 25–35 km in the fore-arc beneath the Longitudinal Valley, and the presence of high-conductivity zones at 20–40 km depth that correspond to large fault and fracture zones. Magnetotelluric investigations confirm that these faults extend deep into the crust. They enhance the potential for fluid or melt migration and might even be responsible for the distribution of arc volcanoes. Below the volcanic arc, the crust is generally 35–45 km thick, with a maximum of 55 km at ~36° S. The subduction geometry along the margin between 36 and 39° S shows a discontinuous steepening of the slab (13–20°).

A wedge-shaped body lying at the tip of the fore-arc mantle can be interpreted as either mafic crustal material in the upper part and serpentinized mantle in the lower part, or hydrated mantle with 20–30% serpentinization, or normal lower crust.

The Permo-Triassic accretionary wedge above the subducting Nazca Plate and an approximately 3 km-thick subduction channel at 18–38 km depth show that basal accretion controls the seismic architecture and growth of the south Chilean crust at about 38.2° S. Here, there is almost no seismicity observed along the entire, approximately 130 km-wide, seismogenic coupling zone.

While the lower plate may suffer slab rollback, the upper plate shows seismic and kinematic segmentation, possibly combined with corner-flow, active lower-plate retreat and fore-arc extension. In addition, underplating, fore-arc uplift and dehydration/serpentinization are the main processes taking place at the tip of the active subduction channel at the south-central Chilean margin.

## Acknowledgments

This work was funded by the German Ministry for Education and Research and the German Research Council within the framework of various programmes (grants 03G0161, SFB 267), and by grants from the GFZ Potsdam and FU Berlin for seismic data acquisition. The seismic stations were made available by the Geophysical Instrument Pool Potsdam and the FU Berlin. We thank Johannes Glodny and Mirjam Bohm for discussion and Onno Oncken in addition for a first review of the manuscript. L. Dorbath, G. Franz and an anonymous reviewer provided helpful comments for improvement of the manuscript.

## References

Allmendinger R, Jordan T, Kay SM, Isacks B (1997) The evolution of the Altiplano-Puna Plateau of the Central Andes. Ann Rev Earth Planet Sci 25:139–174

ANCORP-Working Group (1999) Seismic reflection image revealing offset of Andean subduction-zone earthquake locations into oceanic mantle. Nature 397:341–344

ANCORP-Working Group (2003) Seismic imaging of a convergent continental margin and plateau in the central Andes (Andean Continental Research Project 1996 (ANCORP'96)). J Geophys Res 108(B7): doi 10.1029/2002JB001771

Andersen OB, Knudsen P (1998) Global marine gravity field from the ERS-1 and Geosat geodetic mission altimetry. J Geophys Res 103(C4):8129–8137

Angermann D, Klotz J, Reigber C (1999) Space-geodetic estimation of the Nazca-South America Euler vector. Earth Planet Sci Lett 171:329–334

Araneda M, Avendaño MS, Götze H-J, Schmidt S, Munoz J, Schmitz M (1999a) South central Andes gravity, new data base. 6th International Congres of the Brazilian Geophysical Society Abstracts

Araneda M, Avendaño MS, Schmidt S, Götze H-J, Muñoz J (1999b) Hoja Puerto Montt, Carta gravimétrica de Chile. SERNAGEOMIN de Chile, Subdirección de Geología, p. 18: Santiago, ISSN: 0717-2796

Bangs NL, Cande SC (1997) Episodic development of a convergent margin inferred from structures and processes along the southern Chile margin. Tectonics 16(3):489–503

Bohm M (2004) 3-D Lokalbebentomographie der südlichen Anden zwischen 36° und 40° S. Scientific Technical Report STR04/15, GeoForschungsZentrum Potsdam, http://www.gfz-potsdam.de/bib/pub/str0415

Bohm M, Lüth S, Echtler H, Asch G, Bataille K, Bruhn C, Rietbrock A, Wigger P (2002) The Southern Andes between 36–40°S latitude: seismicity and average seismic velocities. Tectonophysics 356:275–289

Bostock MG, Hyndman RD, Rondenay S, Peacock SM (2002) An inverted continental Moho and serpentinization of the forearc mantle. Nature 417:536–538

Brasse H, Soyer W (2001) A magnetotelluric study in the Southern Chilean Andes. Geophys Res Lett 28(19): 3757–3760

Brasse H, Lezaeta P, Rath V, Schwalenberg K, Soyer W, Haak V (2002) The Bolivian Altiplano conductivity anomaly. J Geophys Res 107(B5): doi 10.1029/ 2001JB000391

Brasse H, Li Y, Kapinos G, Eydam D, Mütschard L (2006) Uniform deflection of induction vectors at the South Chilean continental margin: a hint at electrical anisotropy in the crust. In: Ritter O, Brasse H (eds) Kolloquium Elektromagnetische Tiefenforschung Haus Wohldenberg, Holle, ISSN 0946-7467, pp 281-287

Bruhn C (2003) Momententensoren hochfrequenter Ereignisse in Südchile. PhD thesis, University of Potsdam, Germany

Buske S, Lüth S, Meyer H, Patzig R, Reichert C, Shapiro S, Wigger P, Yoon M (2002) Broad depth range seismic imaging of the subducted Nazca Slab, North Chile. Tectonophysics 350(4):273-282

Cembrano J, Lavenu A, Reynolds P, Arancibia G, Lopez G, Sanhueza A (2002) Late Cenozoic transpressional ductile deformation north of the Nazca-South America-Antarctica triple junction. Tectonophysics 354(3-4):289-314

Chinn DS, Isacks BL (1983) Accurate source depths and focal mechanisms of shallow earthquakes in western South America and in the New Hebrides island arc. Tectonics 2(6):529-563

Cifuentes IL (1989) The 1960 Chilean Earthquake. J Geophys Res 94 (B1):665-680

Clift P, Vannucchi PL (2004) Controls on tectonic accretion versus erosion in subduction zones: implications for the origin and recycling of the continental crust. Rev Geophys 42: doi 10.1029/2003RG000127

Dewey JF, Lamb SH (1992) Active tectonics of the Andes. Tectonophysics 205(1-3):79-95

Echtler, HP, Vietor T, Goetze H-J, Bohm M, Asch G, Lohrmann J, Melnick D, Tašárová Z (2003) Active tectonics controlled by inherited structures in South Central Chile (36°-42°S) - new tectonophysical insights. 10th Congreso Geológico Chileno, Concepción, Chile

Engdahl ER, Villasenor A (2002) Global seismicity: 1900-1999. In: Lee, Kanamori, Jennings, Kisslinger (2002) Int. handbook of earthquake and engineering seismology. pp 665-690

Gansser A (1973) Facts and theories on the Andes. J Geol Soc London 129(1):93-131

Giambiagi LB, Ramos VA, Godoy E, Alvarez P, Orts S (2003) Cenozoic deformation and tectonic style of the Andes, between 33° and 34° south latitude. Tectonics 22:1041-1059

Glodny J, Lohrmann J, Echtler H, Gräfe K, Seifert W, Collao S, Figueroa O (2005) Internal dynamics of a paleoaccretionary wedge: insights from combined isotope tectonochronology and sandbox modelling of the South-Central Chilean forearc. Earth Planet Sci Lett 231:23-39

Glodny J, Echtler H, Figueroa O, Franz G, Gräfe K, Kemnitz H, Kramer W, Krawczyk C, Lohrmann J, Lucassen F, Melnick D, Rosenau M, Seifert W (2006) Long-term geological evolution and mass-flow balance of the South-Central Andes. In: Oncken O, Chong G, Franz G, Giese P, Götze H-J, Ramos VA, Strecker MR, Wigger P (eds) The Andes – active subduction orogeny. Frontiers in Earth Science Series, Vol 1. Springer-Verlag, Berlin Heidelberg New York, pp 401-428, this volume

Glodny J, Gräfe K, Echtler H (2006) Mesozoic to Quaternary continental margin dynamics in South Central Chile (36°-42°S): the apatite and zircon fission track perspective. Int J Earth Sci (in press)

Gorring ML, Kay SM, Zeitler PK, Ramos VA, Rubiolo D Fernandez ML, Panza JL (1997) Neogene Patagonian plateau lavas: continental magmas associated with ridge collision at the Chile triple junction. Tectonics 16 (1), 1-17

Götze H-J (1978) Ein numerisches Verfahren zur Berechnung der gravimetrischen Feldgrößen drei-dimensionaler Modellkörper. Arch Met Geoph Biokl A25:195-215

Götze H-J, Lahmeyer B (1988) Application of three-dimensional interactive modeling in gravity and magnetics. Geophysics 53(8):1096-1108

Götze H-J, Schmidt S, Schreckenberger B (2002) Preliminary results: gravity. In: Flüh ER, Kopp H, Schreckenberger B (eds) Cruise Report SO161-1 and -4, Subduction Processes of Chile. GEOMAR Report 102:106-117

Hacker BR, Abers GA (2004) Subduction Factory 3: an Excel worksheet and macro for calculating the densities, seismic wave speeds, and $H_2O$ contents of minerals and rocks at pressure and temperature. Geochem Geophys Geosyst, 5, Q01005, doi 10.1029/2003GC000614

Hacker BR, Abers GA, Peacock SM (2003) Subduction factory 1. Theoretical mineralogy, densities, seismic wave speeds, and $H_2O$ content. J Geophys Res 108 (B1), doi 10.1029/2001JB001127

Hackney RI, Echtler HP, Franz G, Götze H-J, Lucassen F, Marchenko D, Melnick D, Meyer U, Schmidt S, Tašárová Z, Tassara A, Wienecke S (2006) The segmented overriding plate and coupling at the south-central Chilean margin (36-42° S). In: Oncken O, Chong G, Franz G, Giese P, Götze H-J, Ramos VA, Strecker MR, Wigger P (eds) The Andes – active subduction orogeny. Frontiers in Earth Science Series, Vol 1. Springer-Verlag, Berlin Heidelberg New York, pp 355-374, this volume

Hervé F (1976) Estudio geologico de la falla Liquine-Reloncavi en el area de Liquine: antecedentes de un movimiento transcurrente (Provincia de Valdivia). Actas Congr Geol Chil 1:B39-B56

Hervé F (1988) Late Paleozoic subduction and accretion in Southern Chile. Episodes 11:183-188

Hervé F (1994) The Southern Andes between 39 degrees and 44 degrees S latitude – the geological signature of a transpressive tectonic regime related to a magmatic arc. In: Reutter KJ, Scheuber E, Wigger PJ (eds) Tectonics of the southern Central Andes: structure and evolution of an active continental margin. pp 243-248

Hoffmann-Rothe A, Kukowski N, Dresen G, Echtler H, Oncken O, Klotz J, Scheuber E, Kellner A (2006) Oblique convergence along the Chilean margin: partitioning, margin-parallel faulting and force interaction at the plate interface. In: Oncken O, Chong G, Franz G, Giese P, Götze H-J, Ramos VA, Strecker MR, Wigger P (eds) The Andes – active subduction orogeny. Frontiers in Earth Science Series, Vol 1. Springer-Verlag, Berlin Heidelberg New York, pp 125-146, this volume

De Ignazio C, Lopez I, Oyarzun R, Marquez (2001) The northern Patagonia Somuncura Plateau basalts: a product of slab-induced, shallow asthenospheric upwelling? Terra Nova 13(2):117-121

Isacks BL (1988) Tectonics of the Central Andes. Adv Space Res 9(1): 79-84

Jordan TE, Isacks B, Allmendinger RW, Brewer JA, Ramos VA, Ando CJ (1983) Andean tectonics related to geometry of subducted Nazca Plate. Geol Soc Am Bull 94(3):341-361

Jordan TE, Reynolds JH, Ericsson JP (1997) Variability in age of initial shortening and uplift in the Central Andes. In: Ruddiman WF (ed) Tectonic uplift and climate change. pp 41-61

Jordan TE, Burns WM, Veiga R, Pángaro F, Copeland P, Kelley S, Mpodozis C (2001) Extension and basin formation in the southern Andes caused by increased convergence rates. A midcenozoic trigger for the Andes. Tectonics 20:308-324

Kamiya SI, Kobayashi Y (2000) Seismological evidence for the existence of serpentinized wedge mantle. Geophys Res Lett, 27(6): 819-822

Khazaradze G, Klotz J (2003) Short- and long-term effects of GPS measured crustal deformation rates along the south central Andes. J Geophys Res doi 10.1029/2002JB001879

Kincaid C, Griffith RW (2003) Laboratory models of the thermal evolution of the mantle during rollback subduction. Nature 425:58-62

Kley J, Monaldi CR, Salfity JA (1999) Along-strike segmentation of the Andean foreland: causes and consequences. Tectonophysics 301(1-2):75-94

Krawczyk CM, SPOC Team (2003) Amphibious seismic survey images plate interface at 1960 Chile earthquake. EOS 84(32):301, 304-305

Krawczyk CM, Lohrmann J, Oncken O, Stiller M, Mechie J, Bataille K, SPOC Research Group (2004) Basal accretion as mechanism for crustal growth in the 1960 Chile earthquake area. Geophys Res Abs 6 EGU04-A-02306

Lamb S, Davis P (2003) Cenozoic climate change as a possible cause for the rise of the Andes. Nature 425(6960):792–797

Le Pichon X, Sibuet JC (1981) Passive margins: a model of formation. J Geophys Res 86(B5):3708–3720

Li Y (2000) Numerische Modellierungen von elektromagnetischen Feldern in 2- und 3-dimensionalen anisotropen Leitfähigkeitsstrukturen der Erde nach der Methode der Finiten Elemente. PhD thesis, University of Göttingen

Lohrmann J (2002) Identification of parameters controlling the accretive and tectonically erosive mass-transfer mode at the South-Central and North Chilean forearc using scaled 2D sandbox experiments. Scientific Technical Report STR02/10, GeoForschungs-Zentrum Potsdam, http://www.gfz-potsdam.de/bib/zbstr.htm

Lohrmann J, Kukowski N, Krawczyk CM, Oncken O, Sick C, Sobiesiak M, Rietbrock A (2006) Subduction channel evolution in brittle forearc wedges – a combined study with scaled sandbox experiments, seismological and reflection seismic data and geological field evidence. In: Oncken O, Chong G, Franz G, Giese P, Götze H-J, Ramos VA, Strecker MR, Wigger P (eds) The Andes – active subduction orogeny. Frontiers in Earth Science Series, Vol 1. Springer-Verlag, Berlin Heidelberg New York, pp 237–262, this volume

Lüth S, Wigger P, Araneda M, Asch G, Bataille K, Bohm M, Bruhn C, Giese P, Quezada J, Rietbrock A (2003a) A crustal model along 39° S from a seismic refraction profile – ISSA 2000 Rev Geol Chile 30:83–101

Lüth S, Mechie J, Wigger P, Flueh ER, Krawczyk CM, Reichert C, Stiller M, Vera E, SPOC Research Group (2003b) Subduction Processes Off Chile (SPOC) – results from the amphibious wide-angle seismic experiment across the Chilean subduction zone. Geophys Res Abs 4

Martin MW, Rodriguez C, Godoy E, Duhart P, McDonough M, Campos A, Kato TT (1999) Evolution of the late Paleozoic accretionary complex and overlying forearc-magmatic arc, south central Chile (38°–41°S): constraints for the tectonic setting along the southwestern margin of Gondwana. Tectonics 18(4):582–605

McKenzie DP (1978) Some remarks on the development of sedimentary basins. Earth Planet Sci Lett 40:25–32

Melnick D, Echtler HP (2006) Morphotectonic and geologic digital map compilations of the South-Central Andes (36°–42° S). In: Oncken O, Chong G, Franz G, Giese P, Götze H-J, Ramos VA, Strecker MR, Wigger P (eds) The Andes – active subduction orogeny. Frontiers in Earth Science Series, Vol 1. Springer-Verlag, Berlin Heidelberg New York, pp 565–568, this volume

Melnick D, Rosenau M, Folguera A, Echtler HP (2006) Neogene tectonic evolution of the Neuquén Andes western flank (37–39°S). In: Kay SM, Ramos VA (eds) Evolution of an Andean Margin: a tectonic and magmatic view from the Andes to the Neuquén Basin (36–39°S lat). Geol Soc Am Spec P 407:73–95, doi 10.1130/2006.2407(04)

Mpodozis C, Ramos VA (1989) The Andes of Chile and Argentina. In: Ericksen GE, Pinochet MTC, Reinemund JA (eds): Geology of the Andes and its relation to hydrocarbon and mineral resources. Circum–Pacific Council for Energy and Mineral Resources, Earth Science Series, pp 59–90

Patzig R, Shapiro S, Asch G, Giese P, Wigger P (2002) Seismogenic plane of the northern Andean subduction zone from aftershocks of the Antofagsta (Chile) 1995 earthquake. Geophys Res Lett, 29(8): doi 10.1029/2001GL013244

Patzwahl R (1998) Plattengeometrie und Krustenstruktur am Kontinentalrand Nord–Chiles aus weitwinkelseismischen Messungen. Berliner Geowiss Abh B30

Patzwahl R, Mechie J, Schulze A, Giese P (1999) Two-dimensional velocity models of the Nazca plate subduction zone between 19.5°S and 25°S from wide-angle seismic measurements during the CINCA95 project. J Geophys Res 104(B4):7293–7317

Peacock SM, Hyndman RD (1999) Hydrous minerals in the mantle wedge and the maximum depth of subduction thrust earthquakes. Geophys Res Lett 26(16):2517–2520

Pek J, Verner T (1997) Finite difference modelling of magnetotelluric fields in 2-D anisotropic media. Geophys J Intern 128:505–521

Ramos VA (1989) Andean Foothills structures in northern Magallanes Basin, Argentina. AAPG Bull 73(7):887–903

Rauch K (2005) Cyclicity of Peru-Chile trench sediments between 36° and 38°S: a footprint of paleoclimatic variations? Geophys Res Lett 32: doi 10.1029/2004GL022196

Rodi W, Mackie RL (2001) Nonlinear conjugate gradients algorithm for 2-D magnetotelluric inversions. Geophysics 66:174–187

Rondenay S, Bostock MG, Shragge J (2001) Multiparameter two-dimensional inversion of scattered teleseismic body waves, 3. Application to the Cascadia data set. J Geophys Res 106:30795–30807

Saha A, Basu AR, Jacobsen SB, Poreda RJ, Yin QZ, Yogodzinski GM (2005) Slab devolatilization and Os and Pb mobility in the mantle wedge of the Kamchatka arc. Earth Planet Sci Lett 236(1–2):182–194

Schreckenberger B, Götze H-J, Schmidt S, Kewitsch P, Barckhausen U (2002) Gravity. In: Flüh ER, Kopp H, Schreckenberger B (eds): Cruise Report SO161-1&4, Subduction Processes of Chile. GEOMAR Report 102:81–94

Sick C, Yoon M-K, Rauch K, Buske S, Lüth S, Araneda M, Bataille K, Chong G, Giese P, Krawczyk C, Mechie J, Meyer H, Oncken O, Reichert C, Schmitz M, Shapiro S, Stiller M, Wigger P (2006) Seismic images of accretive and erosive subduction zones from the Chilean margin. In: Oncken O, Chong G, Franz G, Giese P, Götze H-J, Ramos VA, Strecker MR, Wigger P (eds) The Andes – active subduction orogeny. Frontiers in Earth Science Series, Vol 1. Springer-Verlag, Berlin Heidelberg New York, pp 147–170, this volume

SO-161 Shipboard Scientific Party (2002) Cruise Report SO-161, SPOC. BGR internal report, BGR, Hannover

Sobolev SV, Babeyko AY (1994) Modeling of mineralogical composition, density, and elastic wave velocities in anhydrous magmatic rocks. Survey in Geophysics 15:515–544

Sobolev SV, Babeyko AY (2005) What drives orogeny in the Andes? Geology 33:617–620

Somoza R (1998) Updated Nazca (Farallon)-South America relative motions during the last 40 My: implications for mountain building in the Central Andean region. J South Am Earth Sci 11(3):211–215

Stadtlander R, Mechie J, Schulze A (1999) Deep structure of the southern Ural mountains as derived from wide-angle seismic data. Geophys J Int 137:501–515

Tašárová Z (2004) Gravity data analysis and interdisciplinary 3D modelling of a convergent plate margin (Chile, 36–42°S). PhD thesis, Freie Universität Berlin, http://www.diss.fu-berlin.de/2005/19

Tašárová Z (2006) An improved gravity database and a three-dimensional density model of the Chilean convergent plate margin (36°–42°S). Geophys J Int, submitted

Tebbens SF, Cande SC (1997) Southeast Pacific tectonic evolution from early Oligocene to present. J Geophys Res 102(B6):12061–12084

Thomson SN (2002) Late Cenozoic geomorphic and tectonic evolution of the Patagonian Andes between latitudes 42° S and 46° S – an appraisal based on fission-track results from the transpressional intra-arc Liquine-Ofqui fault zone. Geol Soc Am Bull 114(9):1159–1173

Vietor T, Echtler H (2006) Episodic Neogene southward growth of the Andean subduction orogen between 30° S and 40° S – plate motions, mantle flow, climate, and upper-plate structure. In: Oncken O, Chong G, Franz G, Giese P, Götze H-J, Ramos VA, Strecker MR, Wigger P (eds) The Andes – active subduction orogeny. Frontiers in Earth Science Series, Vol 1. Springer-Verlag, Berlin Heidelberg New York, pp 375–400, this volume

Wagner LS, Beck S, Zandt G (2005) Upper mantle structure in the south central Chilean subduction zone (30° to 36° S). J Geophys Res 110: doi 10.1029/2004JB003238

Weidelt P (1999) 3-D conductivity models: implications of electrical anisotropy. In: Oristaglio M, Spies B (eds) Three-Dimensional Electromagnetics. Soc Expl Geophys, Tulsa, pp 119–137

Wienecke S (2002) Homogenisierung und Interpretation des Schwerefeldes entlang der SALT-Traverse zwischen 36°–42°S. Unpublished Diploma thesis, Freie Universität Berlin, Germany

Yañez G, Cembrano J, Pardo M, Ranero C, Selles D (2002) The Challenger-Juan Fernandez-Maipo major tectonic transition of the Nazca-Andean subduction system at 33–34 degrees S: geodynamic evidence and implications. J South Am Earth Sci 15(1):23–38

Yoon M, Buske S, Lueth S, Shapiro SA, Stiller M, Wigger P (2003) Along-strike variations of crustal reflectivity related to the Andean subduction process. Geophys Res Lett 30(4):9/1–9/4

Yuan X, Sobolev SV, Kind R, Oncken O, Bock G, Asch G, Graeber F, Hanka W, Wylegalla K, Tibi R, Haberland C, Rietbrock A, Giese P, Wigger P, Röwer P, Zandt G, Beck S, Wallace T, Pardo M, Comte D (2000) Subduction and collision processes in the Central Andes constrained by converted seismic phases. Nature 408(21):958–961

Yuan X, Asch G, Bataille K, Bock G, Bohm M, Echtler H, Kind R, Oncken O, Wölbern I (2006) Deep seismic images of the Southern Andes. In: Kay SM, Ramos VA (eds)Evolution of an Andean Margin: a tectonic and magmatic view from the Andes to the Neuquén Basin (36–39°S lat). Geol Soc Am Spec P 407:61–72, doi 10.1130/2006.2407(03)

Zelt CA, Smith RB (1992) Seismic traveltime inversion for 2-D crustal velocity structure. Geophys J Intern 108(1):16–34

# Latitudinal Variation in Sedimentary Processes in the Peru-Chile Trench off Central Chile

David Völker · Michael Wiedicke · Stefan Ladage · Christoph Gaedicke · Christian Reichert · Klaus Rauch
Wolfgang Kramer · Christoph Heubeck

**Abstract.** Four cruises of the German research vessel RV SONNE (cruises SO101, SO103, SO104 and SO161) surveyed the Chilean continental margin and oceanic plate using seismic measurements across the Peru-Chile Trench, swath-mapping bathymetry, sediment echosounding, dredges and gravity-core sampling. In this paper, we present data from cruise SO161 derived from the sediment-filled sector of the trench between 35 and 44° S. South of 33°10′ S, sediment fill in the trench ranges from 2 200 to 3 500 m thickness. The sediment volume decreases northwards, as the trench width narrows from 80 km at 41° S to 25 km at 33° S. Turbidity currents enter the trench mainly via nine canyon systems that are deeply incised into the continental slope. Reflection patterns from the trench fill exhibit a cyclicity that can be linked to Milankovic cycles. Turbiditic deposits at elevated positions within the trench indicate Pleistocene mass-wasting events that were able to overcome a height difference of some hundred meters. Within the trench, a fraction of the turbidity currents is channelled by a northward-dipping, axial channel. This axial channel has eroded up to 200 m into the trench fill and from 42° S, it extends northwards some 1 000 km, terminating at the foot of the Juan Fernandez Ridge. The channel has no continuous precursor and might have evolved its present-day form during the last glaciation.

## 9.1 Introduction

The continental margin of Chile is characterized by the oblique subduction of the Nazca Plate beneath the South American Plate. It is presently subducting at a rate of 6.6 cm yr$^{-1}$ with a convergence angle of 77° in relation to the trend of the trench (Angermann et al. 1999, Fig. 9.1). The subducted plate ages northwards: the crust is only 10 million years old at 44° S, but is 20 million years old at 41° S and is 33 million years old at 35° S (Müller et al. 1997, Fig. 9.1). The Nazca Plate subducts at angles of 9–25° (Patzwahl et al. 1999) and 15–20° (Lüth and Wigger 2003) at the subduction front. This produces the Peru-Chile trench which deepens from 4 100 m near southern Chile (44° S) to 6 400 m off the coast of central Chile, and deepens further to 8 200 m off northern Chile (Lindquist et al. 2004).

The Juan Fernandez Ridge (JFR) enters the subduction zone off the Valparaiso coast (33° S), whereas the Chile Ridge is subducting at 45° S (Fig. 9.1). The JFR forms a topographic high within the trench and separates two distinct sedimentary domains. North of the JFR, the trench forms a narrow depression with steep walls, ranges in depth from 6 100 m to more than 7 000 m and contains little sediment (Schweller et al. 1981; von Huene et al. 1997; Flueh et al. 1998; Laursen et al. 2002). South of the JFR, the trench is partly, or completely, filled with sediment. This article is about this sediment fill in the trench sector between the JFR and the more southerly located Chile Ridge. We discuss the processes of sediment input and redistribution within the trench.

While it is known that south of 45° S, accretion ceased with the subduction of the Chile Ridge (Herron et al. 1981; Cande et al. 1987, Behrmann et al. 1994), and that the northern Chilean margin is clearly an erosive margin (e.g. von Huene and Scholl 1991; Laursen et al. 2002; Adam and Reuther 2000), the nature of the transition between these two regimes is poorly understood. We consider knowledge about this sedimentary basin to be essential for the following reasons:

1. As large quantities of sediment are potentially subducted into the mantle, the flux rates and physical properties of the sediments delivered into the trench are likely to be major controlling parameters on the strength of the interface between the Nazca and the South American Plates (e.g. Lamb and Davies 2003).
2. Accordingly, variations in the sediment supply and, as a consequence, variations in the friction coeficcient of the plate interface ("lubrification") are used to explain spatial and lateral variations in sediment accretion, tectonic fore-arc erosion, and upper-plate shortening along the Andean active margin (Adam and Reuther 2000; Vietor and Echtler 2006, Chap. 18 of this volume).
3. Knowledge of sedimentation rates is basic for any attempt to calculate mass fluxes between the plates and for establishing the accretionary history of this sector of the continental margin.
4. There is strong potential for the sedimentary record in this area to bear evidence of mass-wasting events, such as submarine landslides, because of the high sedimentary input and the high seismicity at the plate contact. These conditions pose a potential tsunami risk for the greater region.

**Fig. 9.1.**
Shaded bathymetric map of the Chilean continental margin and trench using the global bathymetry data of Lindquist et al. (2004). The age of the subducting Nazca Plate, major fracture zones and the plate boundary, as provided by Müller et al. (1997), were added as lines. The outlined region between 36 and 40.5° S was covered by swath bathymetry during RV SONNE cruise SO161 (see Fig. 9.2). The Juan Fernandez Ridge enters the trench at 33° S, the Chile Ridge lies south of the mapped region. Transverse mercator projection

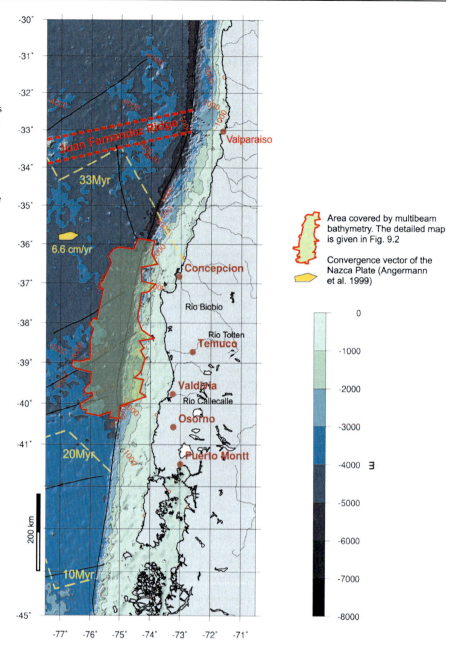

This article intends to address the following questions:

1. What are the pathways and transport modes for sediment entering the trench?
2. How is sediment redistributed within the trench?
3. What changes in sediment input and sedimentation rates are documented?

Examples of sediment-filled trenches and slopes at other convergent margins that have a similar tectonic situation, in terms of convergence rate and direction, vary greatly in geomorphology and sediment redistribution processes. The Cascadia margin, offshore Oregon, has sectors of steep, non-eroded lower slope in addition to well-preserved landslides (McAdoo et al. 1997, 2004). The Nankai trench off Japan accumulates turbidity-current-derived sediments via both submarine canyons and mass-wasting events. Within the trench, a trench-axial channel system distributes the sediment (Taira and Niitsuma 1986; Underwood et al. 1993; Bangs et al. 1999). The Makran margin off Pakistan exhibits a fold-and-thrust belt along the lower accretionary margin composed of steep, trench-parallel, sediment ridges that influence the course of channelled, downslope, sediment transport (Kukowski et al. 2001).

Each of the examples exhibits a specific interplay between seismicity, amount of sediment accretion, and downslope sediment transport in the form of both channelled turbidites and submarine landslides. McAdoo et al. (2004) claim that a relationship exists between the seismic character of a convergent margin, its likelihood for tsunamogenic earthquakes and its geomorphology. They conclude from their comparison of the Cascadia Margin with several other convergent margins that (*1*) slopes of convergent margins with a high frequency (every 100–300 years) of large earthquakes ($M > 8.0$) have smooth lower slopes, as shown by the bathymetry, and (*2*) that in contrast, margins that have produced historical tsunamogenic earthquakes tend to have well-preserved, non-eroded, steep slopes, well-preserved landslide deposits and are substantially rougher in terms of bathymetry. The authors explain these observations by the high frequency (in the first case) versus absence (in the latter case) of earthquake-triggered, erosive sediment flows that act as smoothing agents. One aim of our study is to add the Chile Margin to this discussion in a qualitative way.

For our study, we compiled results on the trench fill from several groups working on geophysical and geological data acquired in 2001 and 2002 on cruise SO161 of the German Research Vessel SONNE. Other aspects of the individual legs of cruise SO161 are summarized by Sick et al. (2006, Chap. 7 of this volume) and Ranero et al. (2006, Chap. 5 of this volume).

## 9.2 Characteristics of the Chilean Slope and Trench between 33° S and 41° S

Global seafloor topography (Smith and Sandwell 1997, Lindquist et al. 2004) and low-resolution, full-coverage bathymetry of the Chilean margin (Zapata 2001) show a shelf of intermediate width (20–30 km) between 33 and 44° S. The edge of this shelf lies at a water depth of some 150 to 250 m. The inclination of the continental slope generally ranges between 2.5 and 4.0°. The map of Zapata (2001), however, shows an irregular slope that is characterized by small plateaus, lineaments and escarpments with slopes up to 10°.

Six major, and some minor, submarine canyon systems can be distinguished. From south to north, these are the Bueno, Callecalle, Tolten, Imperial, and Biobio Canyons. North of the Biobio Canyon, three unnamed canyons and the large San Antonio Canyon off Valparaiso possibly form major sedimentary pathways to the trench (Thornburg et al. 1990, Laursen and Normark 2002). The mean distance between canyons along the coast is 70–100 km. The larger canyons commence on the shelf directly off large river mouths (Zapata 2001).

The role the canyon systems play in supplying sediment to the trench is evidenced by the submarine fan systems that exist at the exit of most of the larger canyons (Thornburg et al. 1990). These submarine fans are asymmetrical and their northern (downslope) morphology contrasts to their southern (upslope) morphology. They have depositional features (fan lobes) to the south and very coarse grained deposits (lag deposits) and erosional structures (furrows) to the north (Thornburg et al. 1990). At the exit of the San Antonio Canyon, sediments have ponded behind an accretionary ridge have been deposited as overbank deposits to the south of a distributary channel that breaches the ridge (Laursen and Normark 2002).

The trench in the area under discussion has a weak morphological expression owing to the sedimentary infill. Instead, a flat, broad plain, some 25–100 km across, is observed. It is inclined to the north and its average depth increases from 4 500 m at 42° S to 5 400 m at 36° S (Smith and Sandwell 1997, Lindquist et al. 2004).

In seismic sections across the trench, the sediment fill is visible as a seaward-thinning, wedge-shaped deposit formed by subhorizontal strata that onlap the sedimentary cover of the Nazca Plate (von Huene et al. 1997; Laursen and Normark 2002; also Figs. 9.9 and 9.10). This sedimentary sequence is approximately 2–3 km thick and predominantly composed of turbidites (Fig. 9.9; Bangs and Cande 1997; Mix et al. [online]; Laursen and Normark 2002; Sick et al. 2006, Chap. 7 of this volume).

Parts of the sediment volume are incorporated into the accretionary wedge, are underplated or are carried into the subduction zone (Bangs and Cande 1997; Ranero et al. 2006, Chap. 5 of this volume; Sick et al. 2006, Chap. 7 of this volume). The sediment column being carried into the subduction zone is divided by the decollement zone. This zone seperates a deeper stack that is being subducted without major deformation from an upper stack that is frontally or basally accreted. In some comparable cases, such as the Nankai subduction zone, the decollement zone is recognizable within the trench fill as a seismic reflector of particular character (Bangs et al. 1999). In the section of the Peru-Chile Trench currently examined, Bangs and Cande (1997) noted a large variability between regions where practically the entire trench sediment stack is accreted (at 38° 08' S) and regions where only the uppermost 350 m of sediment are incorporated into an accretionary wedge (at 39° 45' S). These authors also note that the relatively small width (20–30 km) of the accretionary wedge is not consistent with a continuous history of accretion. In the profiles of Bangs and Cande (1997), the decollement is not directly observable as a prominent reflector within the trench sediments.

The most prominent feature within the morphologically flat area of the trench sediment is formed by a S-N-trending channel, the Chile Trench Axial Channel (Thornburg and Kulm 1987, Laursen and Normark 2002). This feature cuts into the sediment fill and shows the same northward inclination as the surface of the trench fill. It is connected to the distributary channels coming from the sediment fans (Thornburg and Kulm 1987).

The sedimentary fill of the trench is sourced from the Andes and its western fore-arc. Erosion, and the thus generated sediment supply to the trench, is likely to be principally controlled by climate (e.g., Andean glaciation, fluvial runoff). In particular, as the area today represents the north to south transition from temperate to glacial climate zones, Pleistocene climatic changes have caused extensive fluctuations in glaciation (Rabassa and Clapperton 1990) that very likely affected sediment supply to the trench significantly (e.g. Bangs and Cande 1997).

The sediment input from land to the continental slope and trench is, at present, mainly fluvial (Lamy et al. 1998). The overall rate of sediment supply is related to the rate of precipitation at the western flank of the Andes, which shows a pronounced latitudinal gradient with high rainfall rates in southern Chile (200–800 cm yr$^{-1}$) and intermediate to low rainfall rates (100 cm yr$^{-1}$) northwards at Valparaiso (New et al. 1999). The larger rivers emerge from the Andes, cross the Central Valley of Chile, and then cut through the Coastal Cordillera, from which some minor rivers also emerge. To our knowledge, there are no published direct measurements of the sediment discharge from the major river systems.

Site 1232 of Ocean Drilling Program (ODP) leg 202 (Mix et al., online) was drilled into the sediment fill on the seaward side of the trench at 39°53.45' S, 75°54.08' W (Fig. 9.1). The upper 362 meters of the sediment sequence represent the Pleistocene. These sediments consist of gray, silty clay and clay, interbedded with numerous, graded layers of silty sand. They were interpreted as the basal parts of distal turbiditic deposits (Mix et al., online). Thickness of turbiditic beds range from a few centimeters to 108 cm. Both interbedded lithologies show a mineral assemblage that is consistent with andesitic sources from the southern Andes. The stratigraphy of the cored sequences indicates high overall sedimentation rates (> 450 m Myr$^{-1}$ from 0.46 Ma to the present), resulting in a frequency of turbidity currents in the range of one event per hundred years. The distance of the ODP site to the lowermost continental slope (85 km) and the thickness of individual turbidite layers (> 1 m) suggest large depositional events.

## 9.3 Data Acquisition and Processing

On cruise SO161-3, a set of almost E-W-running, seismic profiles was collected offshore from southern Chile. The data are digital, multi-channel recordings made in a frequency range between 20–150 Hz using an airgun array with a total chamber volume of 51.2 l and shot intervals of 18 ± 0.3 s. The source was generally fired at a depth of ~7 m below the sea surface. Seismic reflection data were collected by a digital streamer 3 000 m long. At an average ship speed of 5.4 knots, the distance between shot points was 50 ± 1 m. The ship's track was controlled by an integrated navigation system and dynamic positioning.

Three, selected, two-dimensional (2D) seismic lines (022, 046 and 3842, see Fig. 9.2) were processed in both time and depth domains to obtain the correct depths of the reflectors within the trench sediments (Rauch 2005).

Complementary to the multi-channel seismic data, a total of 3 900 km of bathymetric and sediment-echosounder profiles were acquired. Sediment-echosounding was performed with the hull-mounted, parametric PARASOUND-system (Krupp Atlas Elektronik). PARASOUND makes use of the acoustic, nonlinear properties of water to generate a signal in the range of 3.5 to 5.5 kHz with a narrow opening angle. This focussed signal results in a comparatively small signal footprint at the seafloor with a diameter of about 7% of the water depth. Thus, the lateral resolution is significantly improved in comparison to conventional, 3.5 kHz, echosounding systems. The major drawback of focussing acoustic energy is the signal loss where the inclination of the seafloor exceeds some degrees. Consequently, good profiles were obtained only on the shelf, in the sediment-filled trench and on the Nazca Plate. PARASOUND data were digitized and recorded with the data acquisition system PARADIGMA (Spiess 1993) for further processing and plotting.

Swath bathymetry was performed using Kongsberg's SIMRAD, EM 120, multibeam echosounder. The EM 120 system is capable of mapping the seafloor from shelf to deep sea with high resolution and accuracy and uses a signal frequency of 12 kHz. Signals are emitted as an array of 191 single beams, covering a swath angle of up to 150°. The opening angle of the single beams is as narrow as 1°. The sounding rate depends on water depth with a maximum rate of about 5 Hz in shallow water. The cruise tracks were designed to obtain sufficient coverage, which resulted in a bathymetric data set of good quality and extent. We integrated the cruise data into a bathymetric grid with a node density of 200 m using the MB-System software (Caress and Chayes 1995, 1996). We produced maps and three-dimensional (3D) figures with the Generic Mapping Tools (GMT) software of Wessel and Smith (1998). Additional 3D figures, depth profiles and calculation of areas were produced with the geographic information system GRASS (Neteler and Mitasova 2004).

The sampling tool for soft sediments was a standard, 1.5 t, gravity corer that retrieves sediment cores with a 10 cm diameter. The cores were routinely opened and described on-board. In particular, three gravity cores from

**Fig. 9.2.**
Perspective view of the Nazca Plate, sediment-filled trench and continental margin off southern Chile (35–41° S), produced using multibeam data. The Nazca Plate bears SW-NE-trending ridges and seamount chains that become progressively buried by sediments when entering the trench. The SE-NW trending spreading fabric of the oceanic crust is overprinted by horst-and-graben structures caused by plate bending. The sedimentary fill of the trench produces a low-gradient topography within the trench and in low-lying parts of the Nazca Plate. This trench narrows from south to north. Single, conical features within the trench represent seamounts that have been buried in the trench. Within the trench, the axial channel cuts into the flat-lying sediments. It is connected to major canyon systems by distributary channels. The canyon systems cut deep into the continental margin. At the Callecalle and Biobio Canyon exits, extensive fan lobes can be seen, whereas the Tolten Canyon has none apparent. The northern canyons (Biobio and Tolten) have eroded to the level of the trench. In contrast, the exit of the Callecalle Canyon lies well above the trench

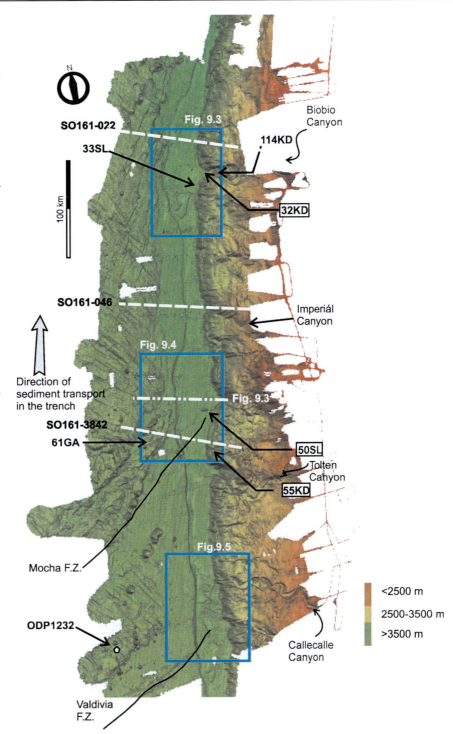

elevated positions within the trench were selected for a detailed study of grain size, mineral content and foraminifera.

Vitric glass fragments from explosive volcanic eruptions can be highly specific to their sources (Heiken 1972; Heiken and Wohletz 1985) and contain a wealth of diagnostic information about their modes of subaerial and aqueous transport. Sand samples from six stations (Table 9.2) were, therefore, investigated using optical and SEM microscopy. The provenance of the vitric components can be constrained with the petrographic, chemical and experimental techniques described by Heiken (1972) and Heiken and Wohletz (1985) (Table 9.1).

**Table 9.1.** Investigated samples and their vitric components

| Sample station, type and depth[a] (m) | Setting[a] | Sediment/rock[a] | Age[a] | Vitric volcanic ash particles | Petrochemical classification |
|---|---|---|---|---|---|
| 32KD 3595 | Accretionary front at Biobío Canyon | Fine- to medium-grained litharenite, weakly lithified | | Vesicular (bubble-wall shards, rounded pumice fragments with tubular vesicles) | Predominantly BA, some A |
| 33SL 4550 | Turbidites close to the mouth of Biobío Canyon | Turbiditic fine sand to silty clay | | Blocky fragments with low vesicularity | Predominantly A to D |
| 50SL 4380 | Seamount close to accretionary front, trench | Clayey to sandy turbidite | Lower Pleistocene to Holocene | Blocky fragments with low vesicularity (partly with rounded edges) | Predominantly D to R |
| 51GA 3820 | Seamount at western rim of the trench | Fine-sandy surface sediment | | Predominantly blocky fragments with low vesicularity, microfracture fans, attached spheres, subord. vesicular to reticulate | BA to A (with partly high Ti and P), subord. D |
| 55KD 3603 | Accretionary front at Toltén Canyon flank | Litharenite, sand to clay turbidite, weakly lithified | Miocene to Pliocene | Blocky, pyramidal, sometimes tubular-vesicular, spheres | BA to A |
| 114KD 2881 | South flank of the Biobío Canyon | Lithified turbidites of sandstone to mudstone | | Bubble-wall shards, blocky to reticulate fragments | BA to R |

[a] According to Cruise Report SO161-5 SPOK. *KD:* Box dredge; *SL:* gravity corer; *GA:* TV-Grab; *BA:* basaltic andesite; *A:* andesite; *D:* dacite; *R:* rhyolite.

## 9.4 Observations

### 9.4.1 Nature and Thickness of the Trench Fill

#### 9.4.1.1 The Submarine Canyons off Chile and Their Fan Systems

The submarine canyon systems off the coast of Chile are diverse (see Fig. 9.2). The Biobio and Tolten Canyons have incised up to 1000 m into the slope bedrock and their mouths have cut to the base of the slope. They do not meander; rather, their thalweg deflects around local topographic relief and generally follows pre-existing, NE-SW- and NW-SE-oriented, tectonic lineaments (Ladage et al. in prep; Sick et al. 2006, Chap. 7 of this volume). In contrast, the Callecalle and Bueno Canyons in the south of the study area are not as deeply incised (only up to 300 m) and have not fully eroded through the lower slope down to the trench depth (Fig. 9.6). The Callecalle Canyon meanders along the upper and middle slope, where the relief is low and its thalweg has low gradients of 0.8° and 1.8° across the upper and middle slopes, respectively. At the lower slope, the gradient of the thalweg increases to 5.8° and the canyon's sinuosity is low. The smaller Bueno Canyon does not meander, although it also has a low gradient thalweg across the middle slope (mean gradient 2.2°).

The Biobio Canyon has a v-shaped profile on the upper slope, whereas its floor becomes irregular to flat on the lower slope (Fig. 9.6). Tolten and Callecalle Canyons form steep walls and U-shaped valleys.

At the exit of the Biobio Canyon, we can distinguish (1) a relatively small (220 km$^2$), conical, submarine fan extending from the canyon exit and (2) elongate, lobate deposits that are intersected by distributary channels that cover an area of 620 km$^2$ south (upslope) of the canyon exit (Fig. 9.3). The latter deposits are easily recognizable because they are elevated (lying 4600–4650 m below sea level) with respect to the flat base of the trench fill south of the fan, which has a average depth of 4700 m. In contrast, north (downslope) of the canyon exit, the main channel of the canyon unites with a central axial channel to form a single depression over the entire width of the trench. Here, the surface of the trench fill lies 4900 m below sea level. This means that the entry of the Biobio Canyon into the trench causes a lowering of the trench fill base by 200 m over a distance of 70 km from south of the exit (36° 58′ S) to north of the exit (36° 26′ S).

At the exit of the Tolten Canyon, a single channel extends from the canyon mouth and connects directly to the trench axial channel (Fig. 9.4). The southern bank of this connecting channel is higher (>100 m) than the northern bank and resembles a levee. This levee is scarred by numerous gullies.

The Callecalle Canyon terminates at the lower slope, about 500 m above the level of the trench sediments (Fig. 9.5). The lobate submarine fan is sculpted by straight channels into a series of parallel ridges and basins. To the south of the southernmost basin, the seafloor is elevated and smooth. The slopes of the channels and basins, in contrast, contain numerous gullies.

**Fig. 9.3.**
Shaded map and perspective view of the Biobio Canyon exit and submarine fan. The Biobio fan consists of a small, conical fan and lobes that were exclusively deposited to the south of the canyon exit. The fan deposits shift the course of the axial channel seawards

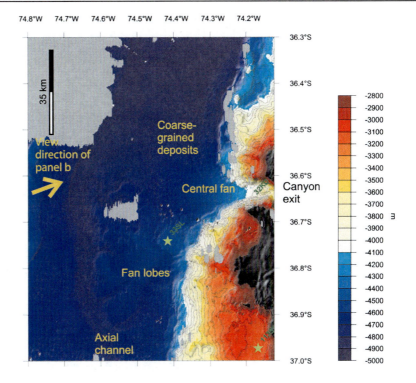

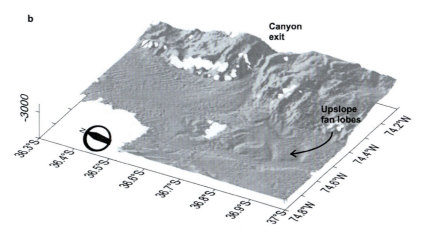

### 9.4.1.2 Morphology and Acoustic Character of the Recent Trench Fill

Between 33 and 42° S, the Peru-Chile Trench off Chile is morphologically defined as an elongate depression with a generally flat surface adjacent to the base of the continental slope (Fig. 9.2). The flat-lying trench fill extends onto the Nazca Plate and its E-W extension varies, in relation to the morphology of the oceanic plate, between 25 and 100 km. In general, the width of the flat depression decreases northwards. Also, its relative depth, in relation to the Nazca Plate, increases from 250 m at 39°30' S to 450 m at 37°30' S, and to 700 m at 36° S. The overall northward inclination of the base of the trench fill is 0.12°.

The seismic facies of the flat-lying areas within the trench, as shown by sediment echosounder (Parasound) profiles, are fairly uniform. Profiles show low signal penetration of 20 to 30 m and indistinct energy return. This acoustic character is produced by a large number of flat-lying, semi-parallel, densely spaced, internal reflectors that are continuous for several kilometers (Fig. 9.7a, echotype A, acoustic trench facies). We found no indication of sediment waves or drift sediments that could be interpreted in terms of bottom water currents in the trench.

The seaward limit of the flat trench fill is marked by the notably rough topography of the Nazca Plate (Fig. 9.2). Its

**Fig. 9.4.**
Shaded map and perspective view of the Tolten Canyon exit. A single distributary channel connects the canyon exit to the axial channel. South of the connecting channel, we can distinguish an elevated region that may represent former overbank deposits. These deposits are eroded at the southern slope of the connecting channel, which is discernible by the numerous gullies along the slope. A chain of seamounts is being transported towards the subduction front. Some of those seamounts are elevated 200–400 m above the flat trench floor

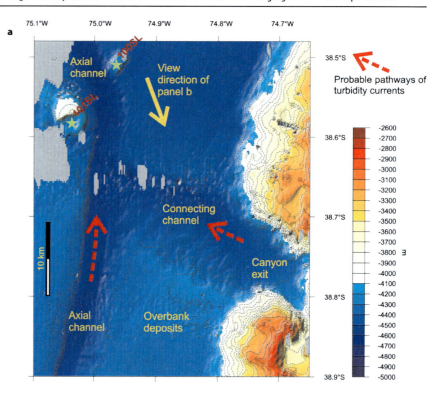

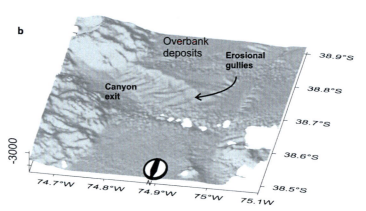

surface diplays a characteristic, linear, topographic pattern caused by a crustal horst-and-graben structure, masked by a thin sedimentary cover. Acoustically, this cover is characterized by evenly spaced, continuous, and very distinct reflectors (Fig. 9.7b, echotype B, acoustic Nazca drape facies). These reflectors parallel each other and follow the undulating seafloor surface. The signal penetration depth is significantly higher (> 50 m) than within the trench. The transition between the trench facies and the Nazca drape facies takes place at a well-defined height of 200 m above mean trench depth.

Some variations in echotype A are observed to the south of the Biobio and Tolten Canyon exits. In both cases, signal penetration is higher (up to 40 m) and reflectors are more continuous. North of the Biobio Canyon exit, signal penetration decreases and the bottom echo is obscured by small-scale hyperbolae (Fig. 9.7c).

Between 37.6° S and 38.2° S, two to three, trench-parallel, foothill chains run parallel to the base of the continental slope (Fig. 9.8). They show a relatively steep seaward-facing slope (~10°), while the landward slope is less inclined.

The Nazca Plate carries seamount chains, up to 2 km high, and elongated ridges aligned subparallel to the Valdivia and Mocha Fracture Zones (the two prominent fracture zones of the survey area; Fig. 9.2). As the edifices approach the subduction zone, and the underlying crust subsides, they are progressively buried by trench sediments. The tops of buried seamounts form isolated hills within the trench.

The overall morphology of the lowermost slope is complex with a large number of downslope gullies that are partly blocked by ridges (Fig. 9.8b). The morphology appears rough because the slope angle varies between 0 and 12° and several areas are inclined landwards with angles up to 10°.

**Fig. 9.5.**
Shaded map and perspective view of the Callecalle fan. The sediment fan forms a roughly delta-shaped structure that is sculpted by straight channels into a series of parallel ridges and basins. To the south of the southernmost basin, the seafloor may represent the original paleo- surface of a once-larger fan system

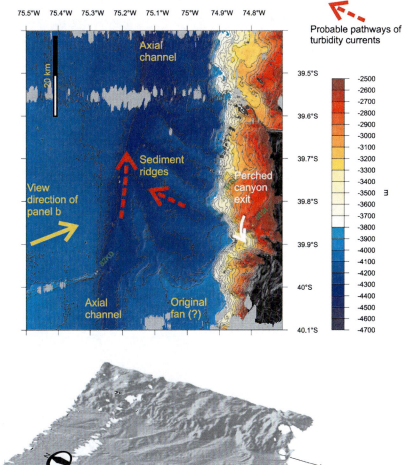

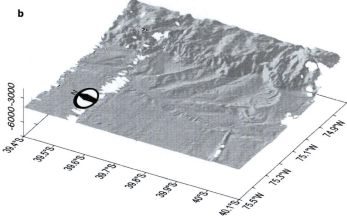

### 9.4.1.3 Thickness of the Trench Fill and the Reflection Pattern

The trench fill is wedge shaped and onlaps the pelagic/hemipelagic sedimentary cover of the oceanic crust (Fig. 9.9). South of the Mocha Fracture Zone, trench sediments prograde up to 100 km westwards onto the Nazca Plate. Seismic images show that the transition is time-transgressive. This is indicated by the progressive onlap of the horizontally stratified trench sequence over the drape of the incoming Nazca Plate (Fig. 9.9). Further north, the horst-and-graben structures of the Nazca Plate confine the trench sediments to a width of some 50 km at 36° S and 25 km at 33° S.

At the deformation front, the thickness of the undeformed trench fill sequence is commonly 1.7–1.8 s TWT, corresponding to about 1 800–2 000 m (assuming a constant seismic velocity of the trench sediments of 2200 m s$^{-1}$). Large thickness variations occur over the local relief of the Nazca Plate. A case in point is the buried extension of the Mocha Fracture Zone, where the thickness of the trench fill varies between 1.35–2.3 s TWT (corresponding to 1 500–2 500 m). However, a general change in its thickness along the trench axis between 40 and 36° S is not observed (Schlüter et al., submitted).

In most of the seismic profiles across the trench (e.g. Fig. 9.10), the seismic reflection pattern of the trench fill can be divided vertically into four self-similar, repetitive units (Schlüter et al., submitted; Rauch 2005). The units form a progression of subparallel internal reflectors with sequence boundaries onlapping the pelagic cover of the Nazca Plate. Each unit appears similar in terms of amplitude and

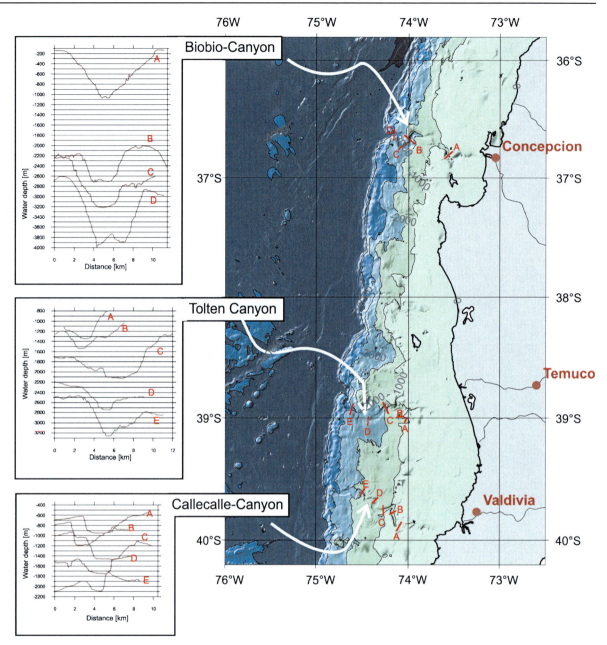

**Fig. 9.6.** Profiles across the major submarine canyon systems of the 36° S to 41° S sector of the Chilean continental margin. Vertical exaggeration of the profiles is 5, map scale 1:4450000

succession of reflectors in a way that they are interpreted as cyclic sequence stratigraphic units, formed by turbidites.

The similarity of the seismic units and thus cyclicity of the signals was confirmed by performing a cross-correlation analysis on the pre-stack, depth-migrated sections of lines 3842 and 022 (Rauch 2005). A wavelet, extracted from the seismic data between 4000–4500 m depth and averaged over 21 adjacent traces was correlated with 100 selected, migrated and stacked traces using the method of Sheriff (1991). A maximum of the resulting correlation function indicates a good fit or high similarity of both.

The result suggests significant periodicity and four reflection cycles (Figs. 9.10 and 9.11), which we interpret as strong indication of a periodicity in the sedimentation pattern of the trench.

In order to be able to relate this periodicity to possibly causal periodic global changes in climate, selected seismic reflectors were tentatively dated (Rauch 2005), using some assumptions. (*1*) The maxima of the correlation function correspond to geological layers which at a time in the past formed the seafloor of the trench. (*2*) The contacts of these reflectors with the pelagic units of the Nazca

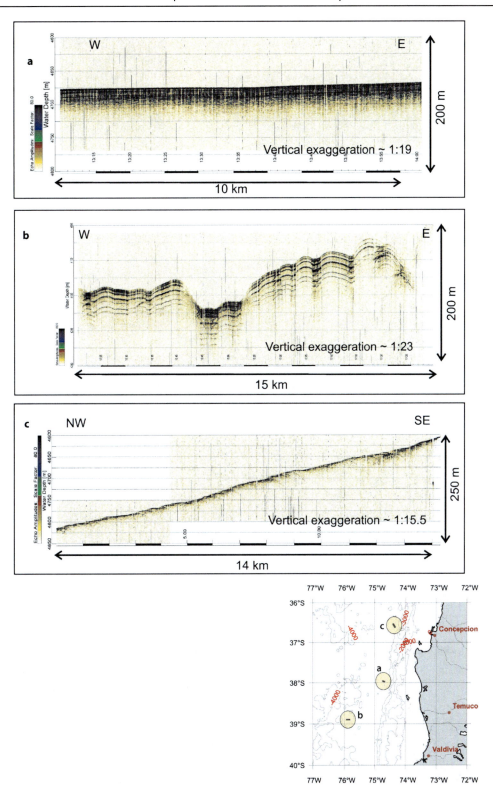

**Fig. 9.7.** Prevailing types of acoustic facies, derived from PARASOUND, sediment-echosounder data, off southern central Chile. **a** This echo type is dominant within the trench and represents the youngest turbiditic infill (acoustic trench facies). **b** Echotype found on the Nazca Plate. It represents the (hemi)-pelagic drape of sediments. The dividing line between the two follows a specific height contour above the trench floor and defines the seaward extent of the turbidites. **c** This echotype is seen exclusively to the north of the Biobio Canyon exit. It represents coarse-grained sediments, resulting from winnowing by the accelerated turbidity currents leaving the canyon

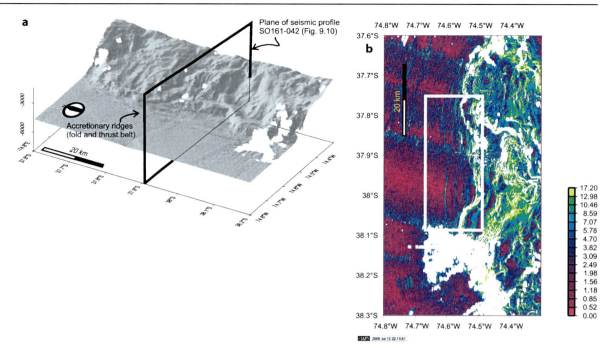

**Fig. 9.8.** Sedimentary accretionary ridges at the subduction front, between 37.6 and 38.3° S, and morphology of the lowermost continental slope. **a** A perspective view of the area looking from WSW. The three, parallel, sediment ridges represent a small fold-and thrust belt caused by compressive sediment deformation (reverse faults) at the subduction front. A seismic profile view of the fold-and thrust-belt is seen in lines 161-042 and 161-038 (Figs. 9.10 and 9.11). **b** Map view of the calculated slope angle (in degrees) along the lowermost slope of that area. The ridges exhibit a seawardly steep slope (angle ~10°) and a gentle slope landwards

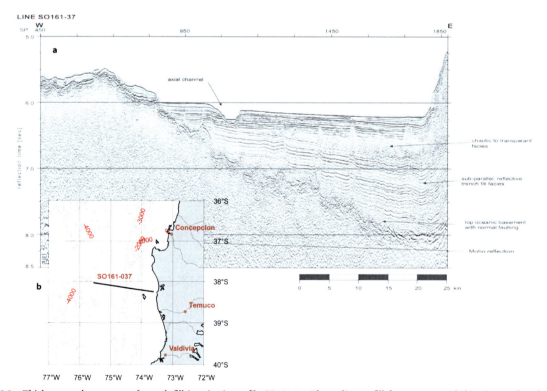

**Fig. 9.9. a** Thickness and geometry of trench fill in seismic profile SO161-37. The sediment fill forms a seaward-thinning, wedge-shaped deposit that is mainly built from subhorizontal strata that onlap the hemipelagic sediment cover of the Nazca Plate. Note the time-transgressive onlap of the basin fill on strata of the Nazca Plate. The horst-and-graben structure of the bent Nazca Plate produces faults that propagate through the sediment column. The ~ 5 km-wide axial channel appears to be partly filled by sediments. **b** Location of profile

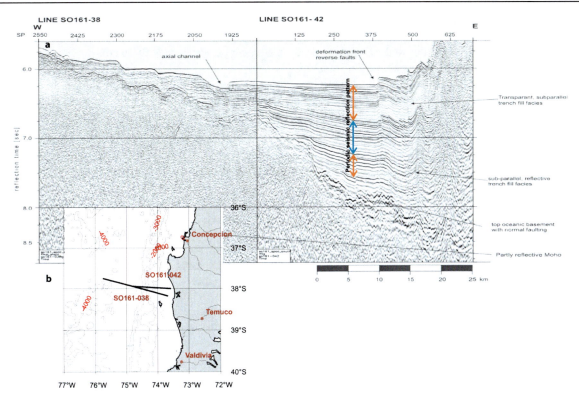

**Fig.9.10. a** Composite seismic profile across the trench, combined from the two seismic lines SO161-38 and SO161-42, showing the thickness and geometry of the trench fill. The sediment fill forms a seaward-thinning, wedge-shaped deposit that is mainly built from subhorizontal strata that onlap the hemipelagic sediment cover of the Nazca Plate. Within the subhorizontal reflectors, there is a periodic succession of strongly reflective (reflective trench-fill facies) and lesser reflective (transparent, subparallel trench-fill facies) sections. Four, self-similar, reflection cycles were seen by Rauch (2005). At the deformation front, reverse faulting produces sedimentary ridges, see Fig. 9.8. **b** Location of profile

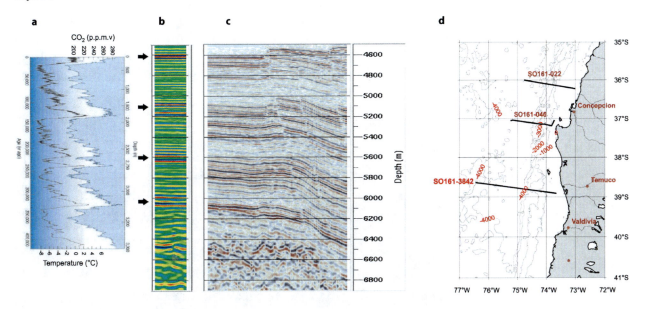

**Fig. 9.11.** Cyclicity in seismic reflection patterns of the trench fill and its relationship to global climatic proxydata. **a** Comparison between the Vostok curves of temperature and the atmospheric $CO_2$ content of Petit et al. (1999). **b** Graphical representation of the cross-correlation function for a section of seimic line 3842. **c** The seimic image of the cyclic reflection pattern in line 3842. The maxima of the correlation function are color-coded in *red* and marked by *arrows*. They correspond to temperature and $CO_2$ maxima of the Vostok curves (Rauch 2005). **d** Location of profile 3842, see also Fig. 9.2

Plate formed the onlap of the trench turbidites on the Nazca Plate at that time. (3) These contacts were slowly shifted eastwards (landwards) due to the convergence of the Nazca Plate while new sediment accumulated on top. (4) This eastward displacent in relation to the actual onlap of the present seafloor-forming turbidites on the pelagic cover of the Nazca Plate can be used to calculate the deposition age of the reflectors, if the changes in convergence rate with time are considered and a more or less costant width of the trench fill is assumed.

This periodicity of the correlation function was then compared with rhythmic patterns of global standard curves of temperature and $CO_2$ content (Vostok, Petit et al. 1999; Dome C, Epica 2004). The comparison is given in Fig. 11 and Table 9.2.

### 9.4.1.4 Sediment Composition of Trench Samples

Gravity cores retrieved from the trench fill, submarine fans and elevated regions within the trench retrieved revealed meters of multiply graded, turbiditic beds of fine sand to silty clay, interbedded with the silty clay that represents hemipelagic background sedimentation. Single, thick turbidite beds have a grain-size distribution in which about half of all grains measure > 100 μm, whereas the maximum interval of grain-size distribution for the intermittent pelagic beds is 10–20 μm.

### 9.4.1.5 Vitric Components of Volcaniclastic Sediments from the Trench and Canyons

Samples recovered from the trench and continental-slope canyons between 36.5 and 39° S (Table 9.1, Fig. 9.2) predominantly contain clayey to sandy, volcaniclastic sediments, or sedimentary rocks that include glass fragments. These fragments are lithic, crystalline, pale brown to black or, if secondary, colorless.

These vitric ash particles are derived from seamounts within or near the trench and have typically blocky or rounded shape (50-4 and 51GA, Fig. 9.12). Blocky particles might represent shallow or deep-sea, hyaloclastic ejecta (Bonatti 1970), whereas rounded particles (32-1, Fig. 9.12) are thought to owe their shape to transportation within rivers and submarine canyons. Volcanic ash samples recovered from the shelf and the continental slope predominantly contain particles rich in bubbles (bubble-wall shards, reticulite, 114-1-4, Fig. 9.12) as well as melt-derived glass spheres (55-1 and 51GA-1, Fig. 9.12) and

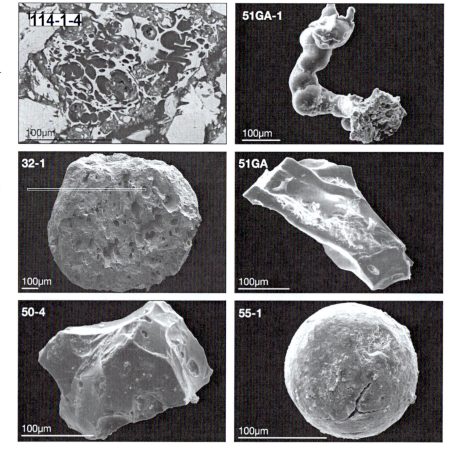

**Fig. 9.12.**
The morphology of sideromelan and tachylite particles in scanning electron photomicrographs (SEM images: U. Glenz, H. Kemnitz). *144-1-4*: vesicular to reticulate fragment showing lattice-like, framework, marginal, disintegrated to small, bubble-wall shards (backscattered electron image). *51GA-1*: vermiform bubble chain with highly vesicular attachment. *32-1*: pumice fragment with tubular vesicles, rounded by transport. *51GA* and *50-4*: blocky fragments with low vesicularity, *50-4*: with attached small spheres. *55-1*: glass sphere with microcracks (triple junction)

might have been transported by pumice rafts, ash clouds, ash falls, and turbidity currents. Those particles are also found at near-trench and seamount locations.

The chemistry of glass particles (symbols in Fig. 9.13) can be readily compared to bulk rock analyses from terrestrial lava flows between 38 and 39° S in the Lonquimay area (Kemnitz et al., in print). Both volcanic associations show a bimodal distribution. Referring to the predominantly mafic members of sample 51GA's glass association, an alkaline component was identified. Its high Ti and P content (Fig. 9.13) is characteristic of mafic seamount associations. However, the glassy ash composition of Miocene to Pliocene and younger sea sediment samples generally has a chemical spectrum similar to that of the arc volcanic rocks on land. Therefore, the bimodal volcaniclastic sediment input into the trench and nearby areas predominantly derives from the Andean arc volcanic rocks to a higher degree than was reported for the Chile Triple Junction at about 46° S (Strand 1995), where near-trench volcanism dominates.

### 9.4.1.6 Perched Turbidites within the Peru-Chile Trench

Three gravity cores were recovered from the summits and flanks of three unnamed seamounts that rise 200 m, 270 m and 600 m, respectively, above the surrounding abyssal plain that forms the surface of the trench fill (50SL at 38° 18.75' S, 74° 41.85' W; 100SL at 38° 30.60' S, 74° 57.90' W; and 101SL at 38° 35.00' S, 75° 02.27' W, Figs. 9.2, 9.4 and 9.7). Within these 3.2–7.3 m long cores, numerous turbiditic beds show medium-sand-sized, graded intervals, up to 12 cm thick, and sharp erosive contacts with the underlying sediment (Völker et al., submitted).

The mineralogy of the > 63 μm fraction of the 38–41 cm interval of core 50SL shows a marked similarity to volcaniclastic material of Andean provenance and compositionally resembles Recent sands from south-central Chilean shorefaces. Turbiditic beds of core 50SL contain few, but varied and diagnostic, benthic and planktonic fora-

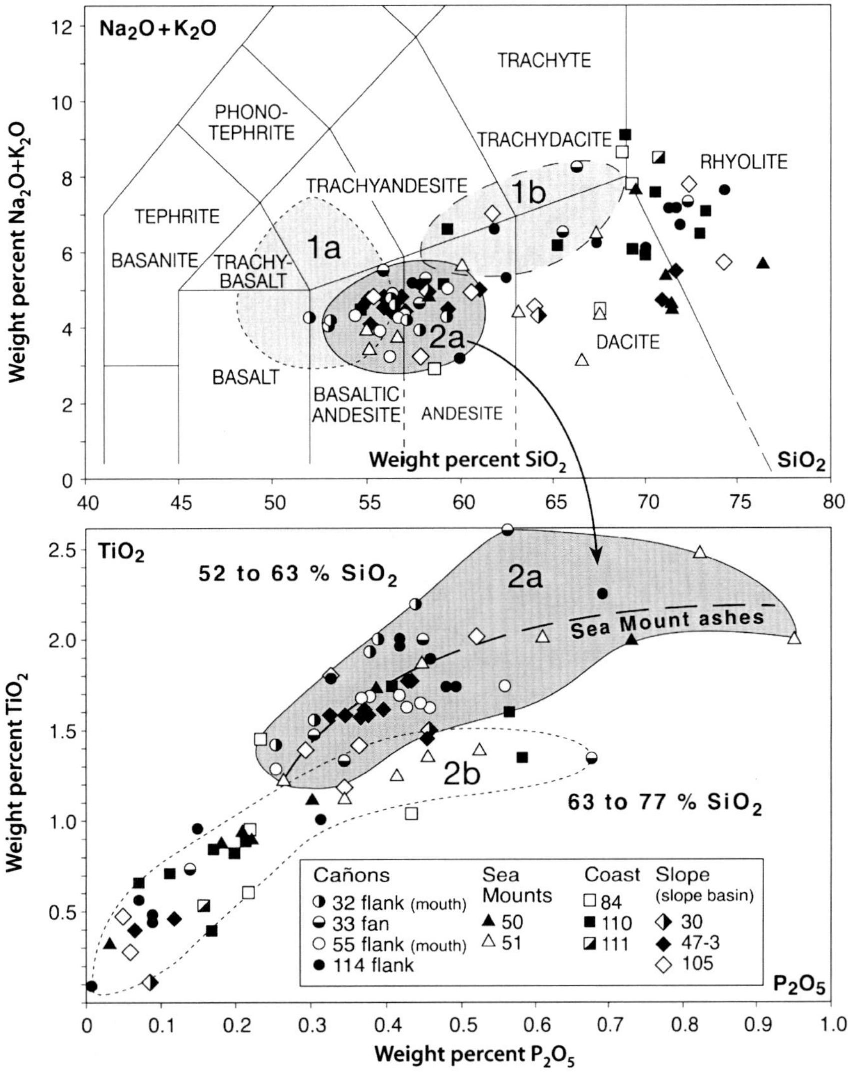

**Fig. 9.13.**
Petrochemical total alkali versus silica (TAS), more specifically $Na_2O + K_2O$ versus $SiO_2$ and $TiO_2$ versus $P_2O_5$ diagrams of offshore glass particles (electron microprobe analyses: W. Seifert). Mesozoic to Pliocene volcanic rocks from similar latitudes in the western Principal Cordillera are scattered within the pale gray fields of 1a and 1b (cf. Kemnitz et al. 2004). To differentiate between the basaltic-andesitic to andesitic glass samples that plot in field 2a of the TAS diagram, they are shown in a $TiO_2$ versus $P_2O_5$ diagram, where the glasses from seamounts display enlarged $TiO_2$ and $P_2O_5$ values

minifera (Völker et al., submitted), including representatives of: the outer-shelf biofacies; the shallow, oxygen-minimum fauna of the continental margin; and of the upper and lower middle bathyal biofacies. In summary, the benthic foraminifera in the samples represent a mixed fauna that cannot be related to an in-situ community at a water depth of ~4000 m.

The oxygen isotope pattern from the lower half (base to 1.5 m core depth) of core 50SL, exhibits a relatively uniform, "heavy" signal in the range of 2.5–3.5‰ (Fig. 9.14). This pattern can best be related to the last glacial period (MIS2). From 1.5 m core depth to its top, $\delta^{18}O$ values decrease sharply. This trend is interpreted as representing global atmospheric warming following the last glaciation and closely resembles standard, marine, oxygen isotope curves (Martinson et al. 1987). The topmost 30 cm section of core 50SL apparently represents Holocene sedimentation. The contrast between Pleistocene and Holocene thicknesses (> 5 m versus 0.3 m) suggests that Pleistocene sedimentation rates significantly exceeded Holocene rates. The two, thick, turbidite layers described above fall into the transition period between the last glacial period and the Holocene and are probably approximately 8000 to 13000 years old.

### 9.4.2 The Axial Channel

#### 9.4.2.1 Morphology of the Present Day Channel

Within the sediment-filled sector of the Peru-Chile Trench, a submarine channel of 3–5 km width and up to 200 m depth (the Chile axial channel) is observed (Figs. 9.2, 9.3, 9.4, 9.5, 9.8, 9.9 and 9.10). The beginnings of this channel lie north of the Strait of Ancud at 42° S, possibly at the exit of the Chacao Canyon (Thornburg et al. 1990, Laursen and Normark 2002). Between 40° S and 36 °S, for approximately 500 km, this channel is clearly continuous (Fig. 9.2). For lack of bathymetric data, we cannot follow the course of the axial channel between 36° S and 33° S. However, south of the Juan Fenandez Ridge, at about 33° S, a channel, some 200 m deep, is observed within the flat-lying sediments of the trench that terminates at the base of the ridge (see also Laursen and Normark 2002). We consider this feature to be the northern end of the same axial channel. If this is true, its total length exceeds 1000 km. The distributary channels of most submarine fans connect directly to the axial channel.

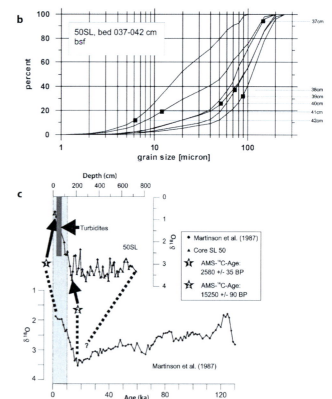

**Fig. 9.14.** Turbiditic bed in sediment core SL050 recovered from a small seamount within the Peru-Chile Trench. The seamount rises 180 m from the surrounding, flat seafloor. **a** Location of gravity core SL050, see also Fig. 9.2. **b** Cumulative grain-size distribution of samples within the 5 cm-thick, coarse-grained basal unit of a turbidite bed sampled 37–42 cm below the sea floor. **c** Oxygen isotope curves for core SL050. A correlation with the standard oxygen isotope curve of Martinson et al. (1987) gives a first age estimate

The axial channel has a flat base and well-defined lateral slopes that truncate adjacent, flat-lying, seismic reflectors (Fig. 9.15). In places, the slopes are deformed by gullies and slumps. Within the main channel bed, another, more confined, meandering internal channel forms bank deposits and occasionally undercuts the outer channel slopes.

There is no general distinction between sediment-echosounder patterns on either side of the channel. Where existent, channel levees are subdued, and opposing shoulders of the axial channel (across a distance of 3–5 km) often differ in height by some tens of meters in an unsystematic way.

The floor of the axial channel is inclined gently to the north. Its position is not always located in the deepest part of the trench, but varies between the western trench slope and the eastern deformation front, at the base of the continental slope. For instance, the upslope lobes of the Biobio and Callecalle Canyon systems displace the course of the axial channel oceanwards (Figs. 9.2 and 9.3). Regionally, the channel follows the graben structures of the bent Nazca Plate (Fig. 9.9; Sick et al. 2006, Chap. 7 of this volume).

### 9.4.2.2 Precursors of the Present Day Channel

There is no continuous precursor to the recent channel detectable in the seismic sections. Buried channels are clearly imaged in only two profiles of the SO161 data set. On a profile across the toe of the Callecalle deep-sea fan (SO161-44, Fig. 9.16), a 6 km-wide channel cuts into well-stratified turbidites and is buried by sediments approximately 500 ms [TWT] thick. The channel fill has stronger amplitude than the surrounding turbiditic succession.

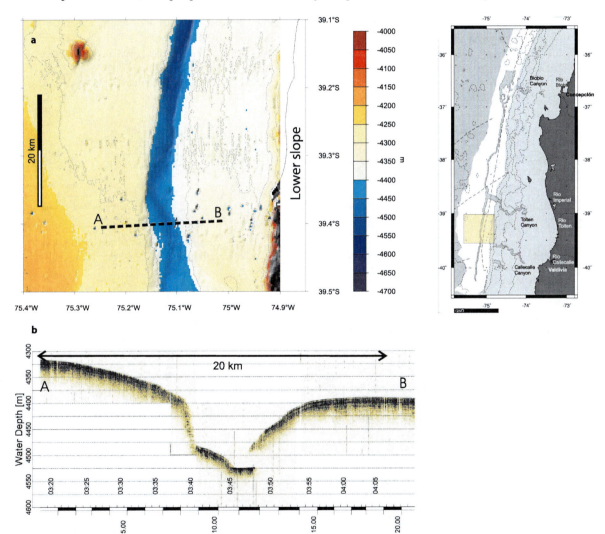

**Fig. 9.15.** Morphology of the Chile Trench Axial Channel in (**a**) map view and (**b**) as a PARASOUND cross section. The channel walls truncate subparallel strata of the trench fill. The steep, cutback sides are sometimes deformed by small gullies. The map view shows the meandering of the internal channels within the broader, outer channel bed. The profile shows the flat floor and the banks of the internal channel. Vertical exaggeration of profile: 1 : 28

It exhibits a discontinuous to hummocky reflection pattern and an erosional lower boundary over the older channel fill. The channel developed in two stages interrupted by a short time of turbiditic sedimentation, as evident by erosional truncations.

A second, buried channel is located in front of the Biobio Canyon, west of, and close to, the modern channel (profile SO161-26, Fig. 9.17). It is buried below well-stratified turbidites some 200 to 300 ms [TWT] thick. The ancient channel is narrower and shallower than its recent analogue. The thin sedimentary cover and its vicinity to the modern channel suggest a common origin

### 9.4.3 Sediment Deformation by Plate Convergence and Slab Bending

Normal faulting of the trench fill mainly occurs above the flexed, and eastwardly inclined, Nazca Plate. Inclination angles of the plate vary between 1.4 and 3.7°. The bending of the oceanic plate is accompanied by block faulting of the oceanic crust with prominent horst-and-graben structures. These normal faults tend to continue and penetrate into the sedimentary trench fill, and individual normal faults can reach offsets of up to 0.7 s [TWT] (line SO161-37, Fig. 9.9). Syn-sedimentary growth faulting is evident in the decreasing offsets of the pelagic unit towards surface. The sedimentary succession thickens while it migrates with the oceanic plate towards the toe of the accretionary wedge, where the sedimentation rate is highest. Nevertheless, some normal faults can be traced through the entire sedimentary trench fill and reach the seafloor, thus revealing recent fault activity.

Thrusting and reverse faulting in the trench fill occur near the base of the slope at the deformation front. In the central part of the study area, the deformation front shifts trenchwards, where a sequence of reverse faults documents an incipient stage of trench-fill deformation. At the base of the slope, folding of the sedimentary strata occurs (Sick et al. 2006, Chap. 7 of this volume). The surface expression of this process is revealed by the sedimentary ridges shown in Fig. 9.8.

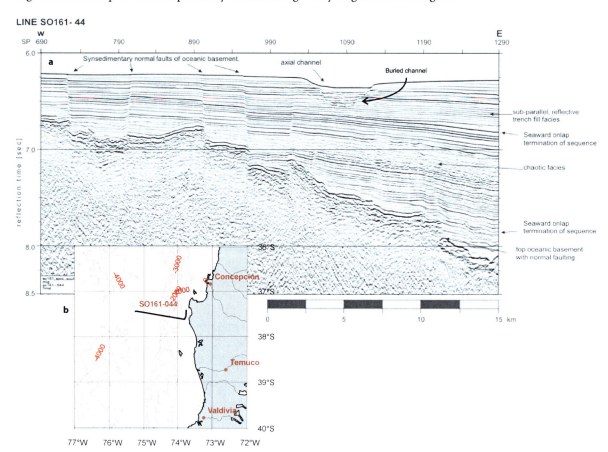

**Fig. 9.16. a** Precursor of the modern axial channel in seismic profile SO161-44. The subhorizontal strata of the trench fill onlaps the Nazca Plate. The horst-and-graben structure of the bent Nazca Plate produces faults that propagate through the sediment column. A 6 km-wide, buried channel cuts into well-stratified turbidites and is buried by sediments approximately 500 ms [TWT] thick. The channel fill has stronger amplitudes than the surrounding turbiditic succession. It exhibits a discontinuous to hummocky reflection pattern and an erosional boundary over the older channel fill. **b** Location of profile

## 9.5 Discussion

### 9.5.1 Transport of Sediment into the Trench

The bulk of coarse-grained sediment fill in the trench south of the Juan Fernandez Ridge is delivered by turbidity currents. The acoustic facies of the youngest trench fill, represented by echotype A (Fig. 9.7a), is typical for turbiditic sediments (e.g. Völker 1989; Kuhn and Weber 1993). In contrast, echotype B (Fig. 9.7b) represents the hemipelagic sediment cover of the Nazca Plate. The onlap of echotype A onto echotype B, and the observation that this transition is defined by a certain fixed height above the trench floor, indicates that this transition represents the onlap of turbiditic sediment over the hemipelagic drapes of the Nazca Plate. This time-transgressive onlap is observed in all seismic images. The transition documents the most distal, seaward limit of turbidite sedimentation. As it lies 200 m above the flat trench floor, turbidity currents evidently traverse the entire width of the trench (some 25 to 80 km), cross the axial channel and run up the western trench-fill slope to overcome this bathymetric difference.

As shown by the common characteristics of the widespread, volcanic, glassy ash particles from trench and canyon samples, transport by ash clouds, pumice rafts and submarine explosions superposes the turbidite sedimentation.

The submarine canyon systems provide the most probable pathways for turbidity currents from the shelf and upper slope as they form efficient traps for downslope transportation for the following reasons: (*1*) they are closely spaced along the continental margin; (*2*) they meander and branch and are, therefore, very likely to catch upper-slope-derived sediments, and (*3*) the majority cut into the shelf and, thus, trap longshore-transported sediment. The morphological differences in gradient, incision depth and depth levels of exits between the canyon systems indicate that the northern canyons in the study area have thalwegs closer to a state of equilibrium than the southerly ones.

The morphological development of submarine canyons is, among other processes, strongly controlled by the amount of sediment drained through the canyon (eg. Twidale 2003). Thus, the more mature canyons in the north probably transported more sediment than the southern ones and are, therefore, either older and/or had

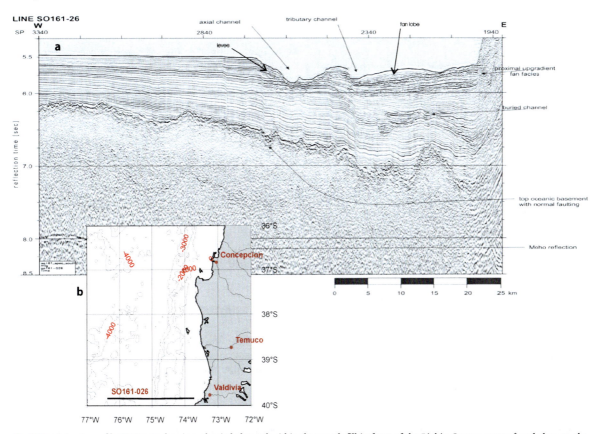

**Fig. 9.17. a** Seismic profile SO161-26 showing a buried channel within the trench fill in front of the Biobio Canyon, west of, and close to, the modern channel. It is buried beneath well stratified turbidites some 200 to 300 ms [TWT] thick. The ancient channel is narrower and shallower than its recent analogue, but the thin sedimentary cover and its vicinity to the recent channel suggest a common origin. **b** Location of profile

more sediment delivered. The southern shelf is broader and probably trapped large volumes of Pliocene and Neogene sediments in the Valdivia Basin (Mordojovich 1981; Reichert et al. 2002); only a minor proportion was delivered to the trench via the incipient Callecalle and Bueno Canyons, thus explaining their immature, "perched" morphology. Probably, the main phase of development for these canyons was during the Pleistocene, whereas the larger, northern canyons were likely active over a longer time span.

The turbidite beds recovered from abyssal hills within the trench appear to be products of individual, relatively powerful, turbidity currents. As the mineral composition and the foraminiferal faunal assemblage suggest upper-shelf to abyssal provenance, these turbidity currents might have resulted from large-scale slumping of the continental slope, rather than from confined turbidity currents within the canyons. The provenance of individual turbidite beds is a possible criterion for distinguishing earthquake-generated turbidites (variable, multi-sourced components) from storm-generated turbidites (Nakajima and Kanai 2000; Goldfinger et al. 2003). This would apply to the thickest turbidite bed in core SL050. As the discussion about the possible use of turbidites as indicators of paleo-seismicity is new, and the criteria for detecting earthquake-induced turbidites are still strongly debated (Goldfinger et al. 2003), we do not want to overstress this finding.

Oxygen isotope stratigraphy indicates that most of these turbiditic sequences were deposited during the last glacial, and that deposition ceased during the Holocene. Thick (> 1 m), turbiditic, silty sand beds recovered at ODP Site 1232, 85 km from the subduction front, might be product of the same events. The rough morphology of the lowermost slope, with structures that appear like gullies, may support the assumption of downslope, mass wasting apart from the larger canyon systems.

### 9.5.2 Sediment Redistribution within the Trench

The asymmetry of the Callecalle and Biobio fan systems – with deposition exclusively south of the canyon mouths and erosional features, as well as lag deposits, to the north – documents the preferred direction of turbidity currents entering the trench. The slight northward dip of the trench floor appears to have caused a difference in depositional conditions: Turbulent sediment clouds flowing southwards (upslope) quickly lost momentum and carrying capacity, whereas those flowing downslope, to the north, largely maintained these attributes. The variations in the acoustic facies within the trench support this assumption. The patches with higher signal penetration, south of the Biobio and Tolten Canyon mouths, are likely to represent finer-grained deposits, possibly levee/overbank sediments. In contrast, the areas of very low penetration and hyperbolic bottom echoes (Fig. 9.3c) represent lag deposits and a micro topographical pattern, such as longitudinal furrows. A similar sediment distribution was reported by Thornburg et al. (1990) for the same area.

The fact that the total thickness of the trench fill does not diminish towards the northern end of the sediment-filled sector of the Peru-Chile Trench, as well as the observation that just a few degrees further to the north the trench is sediment-starved, indicates a general south-to-north transport direction within the trench that is effectively dammed by the Juan Fernandez Ridge.

The axial channel, remarkable for its size and its lateral continuity, could be a primary pathway for the channelled flow of turbidity currents within the trench. The observation that the channel does not divide areas of different seismic facies or of different elevation can be interpreted in two ways:

1. The free flow of turbidity currents and channelled flow interplay. Most turbidity currents easily cross the channel and spread farther seawards, with only minor northward entry into the channel.
2. The axial channel is a relatively young feature and has cut into sediments that were deposited by sheet flows.

We are inclined to the latter option as the absence of ancient channels in most seismic sections lets us speculate that the present channel was not established before the Pleistocene. The limited distribution of ancient channels shows that they are local features linked to the deep-sea fans at the canyon exits. Their recent analogues are the tributaries on the deep-sea fan systems. Therefore, the modern, continuous, axial channel has no apparent precursors. The sedimentation regime might have changed significantly after the last glaciation, as supported by core-sample analyses (Völker et al., submitted). In any case, three phases of the build-up of the modern channel are documented: a primary incision (during the last glaciation), filling of the channel, and incision of the present day channel.

### 9.5.3 Changes in Sedimentation with Time

Numerous publications relate cyclic reflection patterns in sediments to variations in global orbital parameters (Milankovitch 1941; Vail et al. 1991). These variations are thought to exert a major influence on sedimentation mechanisms and rates via climatic changes. Recently, two Antarctic ice cores (Vostok, Petit et al. 1999; Dome C, Epica 2004) provided new and detailed temperature data spanning the last 420 000 years and 740 000 years, respectively.

Table 9.2 lists the age of the four maxima identified in the Vostok curves, as well as in the cross-correlation functions of the two seismic lines 3842 and 022 (Rauch 2005). The periodicity observed in the seismic data as calculated by dating selected reflectors lies in consistently the range of 100 000 years to 120 000 years for all of the cycles in the two seismic lines. This period is also seen in the maxima of the Vostok curves. We consider this a good match, if the basic and simple method used for the calculation of the horizon ages is taken into account.

In contrast to lines 3842 and 022, the reflections observable in line 046 exhibit no comparable periodicity and the cross-correlation function does not show any significant maximum. The cause of this difference may lie in the position of the line relative to the canyon mouths. Lines 022 and 3842 are located near the northern limit of the Tolten and Biobio fans, respectively (Fig. 9.1). Thus, the sedimentary material has been transported over a short distance. In contrast, line 046 is located near the southern border of the Biobio fan; the sediments were probably transported over a larger distance – more than 100 km. The transport distance seems to be crucial to the periodicity of the reflections, as the granularity, the pore volume and the pore fluid of the sediments could change substantially (Rauch 2005).

From the bathymetric appearance, the present-day fans of the Tolten and Callecalle Canyon systems (Figs. 9.4 and 9.5) seem to be relicts of much larger deposits. We interpret the modern bathymetry to have been formed by erosive dissection of the roughly delta-shaped fan deposits. The elevated regions that lie to the south of the canyon exits seem to be the last intact surfaces (paleoplains) of these deposits. It seems likely that this intensified erosion occurred during sea-level low stand at the last glacial maximum. We have no data to constrain this assumption, but it accords with investigations on the development of the San Antonio Canyon (at 33° S) by Laursen and Normark (2002). The authors state that little sediment is presently transported via the San Antonio Canyon, whereas they assume enhanced flushing of the canyon during the lowered sea level of the last glaciation.

We consider it a straightforward assumption to attribute the two phases in the construction of the axial channel, visible as a broader outer channel with a smaller channel meandering within, to the same timing. In this case, excavation of the larger channel was the product of intensified, turbiditic flows during the sea-level low stand of the last glacial maximum, when the hinterland of the southernmost extent of the axial channel was heavily glaciated. This structure is now being filled. Also, in this case, we cannot constrain this assumption.

## 9.6 Conclusions

The total volume of sediment stored in the Chilean sector of the trench diminishes from south to north. From 40° S to 33° S, this reduction is not manifest by a thinning of the sediment wedge but by a restriction of the width of the sediment fill from 85 km to 25 km. This latitudinal trend can be explained by the prominent gradient in precipitation, erosion, river runoff and sediment input from southern to central Chile. The present-day transition from extremely humid climate in the south to Mediterranean climate in central Chile probably had a counterpart in the Pleistocene, with extensive glaciation and associated sedimentation mainly in the southern sector. This latitudinal trend is, however, modified by the northward transport of sediment by turbidity currents within the trench. Evidence of this transport is seen in the northward-tilted axial transport channel, the asymmetry of the larger fan systems, and the clear distinction between a sediment-filled trench south of, and a sediment-free trench north of, the Juan Fernandez Ridge (which acts as a dam).

According to Bangs and Cande (1997), the volume of trench fill that is carried into the subduction zone varies over short latitudinal distances (e.g. 100 km) from the entire sediment fill (the maximum amount) to about 350 m of the uppermost sediment column. If such a configuration were stable over a period of the order of some 10 000 years, this should result in major latitudinal differences in the trench-fill volume and lower-slope morphology. The compressive ridges that are marked by thrusting and reverse faulting, observed in the sector between the Tolten and Biobio fans, may testify to enhanced accretion. However, this signal is not observable in terms of trench-fill thickness. Either the redistribution of sediment within the trench effectively levels such differences or the

**Table 9.2.**
Position of the maxima of the Vostok curve and the cross-correlation function on the timescale ($y$)

| Maxima of correlation | Vostok curve (temp.) ($y$) | Cross-correlation function ($y$) | |
|---|---|---|---|
| | | Line 3842 | Line 022 |
| 1 | 0 | 0 | 0 |
| 2 | 128 000 | 119 066 | 115 625 |
| 3 | 238 000 | 241 713 | 240 145 |
| 4 | 322 000 | 340 195 | 329 088 |

decollement depth at any given place is not continuous over larger time periods.

The bulk of the sediment fill is derived by turbidity currents delivered to the trench via the canyon systems. We assume that the majority of the fluvially-delivered sediment ends up in the trench because the shelf is kept free from young sediments by vigorous bottom currents (Huyer et al. 1991, Pizarro et al. 2002). Some slope basins may capture unchannelled, downslope transport. The turbidite-derived material is distributed within the trench by both sheet flows and channelled flows. The upper reach of turbiditic sedimentation on the seaward side of the trench lies 200 m above the trench floor and 50–80 km seawards of the lower continental foot. These deposits thus support the hypothesis that high energy turbidites have been frequent, at least, in the past, probably during the Pleistocene.

Seismic records of the trench fill show a pronounced cyclicity in the reflection intensity of flat-lying reflectors. These repetitive patterns can be correlated with Milankovic cycles, which leads to the speculation that the intensity of turbidite input can be linked to climatic signals. Sediment composition and the grain-size distribution of turbidite beds taken from elevated positions within the Peru-Chile Trench show that at least some of the turbidites are due to large-scale, slope-failure events that have affected the continental slope over a wide depth range. Exactly timing the layers is difficult because of the scarcity of planktonic foraminifera. For the same reason, we were not able to correlate single turbidite layers across the area. The question of whether the turbiditic record of the trench is dominantly influenced by climatic signals, or if it rather records seismogenic activity in this active region, cannot be definitely answered on the basis of our data. Nevertheless, we lean towards the idea of a complex, mixed signal, with climatic control of the sediment delivery to the shelf and slope basins and a triggering of large slump events by seismicity.

The combination of tectonic setting and sedimentary input is comparable to the Nankai trough (ODP leg 190, preliminary results), which similarly represents a slightly tilted, sediment-filled trench in a convergent margin setting. The trench axis of the Nankai trough is characterized by a 500 m-thick, seaward-thinning wedge of trench turbidites that overlie and diachronously onlap the pelagic Shikoku Basin sequence. As in Chile, the turbidite unit was supplied mostly via an axial transport system (Taira and Niitsuma 1986; Underwood et al. 1993). The geomorphology of the lower slope, conversely, is much rougher than that of the Nankai margin. Like the Cascadia margin, the slope shows sectors with steep gradients, but, unlike it, we observe numerous structures that can be attributed to erosive mass wasting. This mass wasting does not seem to result in a smooth, lower-continental slope but rather adds bathymetric roughness.

## Acknowledgments

This work was part of the sea-based SPOC project and is based on the data gathering of the RV SONNE crew. Additional data was provided by the bathymetric data center of the Bundesamt für Seeschifffahrt und Hydrographie Hamburg. We wish to express our gratitude to all the data hunters involved. First review and fruitful comments were given by Nina Kukowski.

## References

Adam J, Reuther C-D (2000) Crustal dynamics and active fault mechanics during subduction erosion. Application of frictional wedge analysis on to the North Chilean Forearc. Tectonophysics 321:297–325

Angermann D, Klotz J, Reigber C (1999) Space-geodetic estimation of the Nazca-South American Euler vector. Earth Planet Sci Lett 171: 329–334

Bangs NL, Cande SC (1997) Episodic development of a convergent margin inferred from structures and processes along the southern Chile margin. Tectonics 16:489–503

Bangs NL, Taira A, Kuramoto S, Shipley TH, Moore GF, Mochizuki K, Gulick SS, Zhao Z, Nakamura Y, Park J-O, Taylor BL, Morita S, Ito S, Hills DJ, Leslie SC, Alex CM, McCutcheon AJ, Ike T, Yagi H, Toyama G (1999) U.S.-Japan collaborative 3-D seismic investigation of the Nankai Trough plate-boundary interface and shallow-most seismogenic zone. EOS 80:F569

Behrmann JH, Lewis SD, Cande SC, ODP leg 141 scientific party (1994) Tectonics and geology of spreading ridge subduction at the Chile Triple Junction: a synthesis of results from leg 141 of the Ocean Drilling Program. Geol Rundsch 83: 832–852

Bonatti, E (1970) Deep sea volcanism. Naturwissenschaften 57: 379–384

Cande SC, Leslie RB, Parra JC, Hobart M (1987) Interaction between the Chile Ridge and Chile Trench: geophysical and geothermal evidence. J Geophys Res 92:495–520

Cruise Report SO161-5, SPOC (2002) Bundesanstalt für Geowissenschaften und Rohstoffe, Hannover

Diaz-Naveas JL (1999) Sediment subduction and accretion at the Chilean convergent margin between 35°S and 40°S. PhD thesis, Christian-Albrechts-Universität zu Kiel

Fekete BM, Vörösmarty CJ, Grabs W (2000) Global composite runoff fields based on observed river discharge and simulated water balances. Complex Systems Research Center, University of New Hampshire, UNH-GRDC Composite Runoff Fields v1.0, http://www.grdc.sr.unh.edu (version of 15.05.04)

Flueh ER, Vidal N, Ranero CR, Hojka A, von Huene R, Bialas J, Hinz K, Cordoba D, Danobeitia JJ, Zelt C (1998) Seismic investigation of the continental margin off- and onshore Valparaiso, Chile. Tectonophysics 288:251–263

Goldfinger C, Nelson CH, Johnson JE, Shipboard Party (2003) Holocene earthquake record from the Cascadia subduction zone and northern San Andreas Fault based on precise dating of offshore turbidites. Annu Rev Earth Planet Sci 31:555–577

Heiken G (1972) Morphology and petrography of volcanic ashes. Geol Soc Am Bull 83:1961–1988

Heiken G, Wohletz KH (1985) Volcanic ash. University of California Press, Berkeley

Herron EM (1981) Chile margin near latitude 38° S: evidence for a genetic relationship between continental and marine geologic features or a case of curious coincidences? Geol Soc Am Mem 154:755–760

Herron EM, Cande SC, Hall BR (1981) An active spreading center collides with a subduction zone: a geophysical survey of the Chile Margin triple junction. In: Nazca Plate: crustal formation and Andean convergence. Mem Geol Soc Am 154:683–702

Huyer A, Knoll M, Paluszkiewicz T, Smith R (1991) The Peru undercurrent: a study of variability. Deep Sea Res 38:1–38

Ingle JC, Keller G, Kolpack RL (1980) Benthic foraminiferal biofacies, sediments and water masses of the southern Peru-Chile Trench area, southeastern Pacific Ocean. Micropaleontology 26(2):113–150

Kemnitz H, Kramer W, Rosenau M (2004) Jurassic to Tertiary tectonic, volcanic, and sedimentary evolution of the Southern Andean intra-arc zone, Chile (38–39°S): a survey. N Jb Geol Paläont 236:19–24

Kuhn G, Weber M (1993) Acoustical characterisation of sediments by PARASOUND and 3.5kHz systems: related sedimentary processes on the Southeastern Weddelsea continental slope, Antarctica. Marine Geol 113:201–217

Kukowski N, Schillhorn T, Huhn K, Von Rad U, Husen S, Flueh E (2001) Morphotectonics and mechanics of the central Makran accretionary wedge off Pakistan. Marine Geol 173:1–19

Ladage S, Gaedicke C, Reichert C, Diaz-Naveas J (in prep) Bathymetry and morphotectonics of the Central Chilean margin (36–40°S)

Lamb S, Davis P (2003) Cenozoic climate change as a possible cause for the rise of the Andes. Nature 425:792–797

Lamy F, Hebbeln D, Wefer G (1998) Terrigenuous sediment supply along the Chilean continental margin: modern regional patterns of texture and composition. Geol Rundsch 87:477–494

Laursen J, Normark WR (2002) Late Quaternary Evolution of the San Antonio Submarine Canyon in the central Chile forearc (~33°S). Marine Geol 188:365–390

Laursen J, Scholl DW, von Huene R (2002) Evolution of the late cenocoic Valparaiso forearc basin in central Chile: Forearc basin response to ridge and seamount subduction. Tectonics 21:1–27

Lindquist K, Engle K, Stahlke D, Price E (2004) Global topography and bathymetry grid improves research efforts. EOS 85(19): doi 10.1029/2004EO190003

Lüth S, Wigger P (2003) A crustal model along 39°S from a seismic refraction profile – ISSA (2000) Rev Geol Chile, Vol 30(1):83–101

Martinson DG, Pisias NG, Hays JD, Imbrie J, Moore TC Jr, Shackleton NS (1987) Age dating and the orbital theory of the Ice Ages – development of a high-resolution 0 to 300,000-year chronostratigraphy. Quaternary Res 27(1):1–29

McAdoo BG, Orange DL, Screaton E, Lee H, Kayen R (1997) Slope basins, headless canyons and submarine paleoseismology of the Cascadia accretionary complex. Basin Res 9:313–324

McAdoo BG, Capone MK, Minder J (2004) Seafloor geomorphology of convergent margins: implications for Cascadia seismic hazard. Tectonics 23:1–15

Milankovitch M (1941), Kanon der Erdbestrahlung und seine Anwendung auf das Eiszeitproblem. Akad R Serbe 133

Mix AC, Tiedemann R, Blum P, et al (2003) Proc ODP Init Rep 202. http://www-odp.tamu.edu/publications/202_IR/202ir.htm (version of 2005-07-20)

Mordojovich C (1981) Sedimentary basins of Chilean Pacific offshore. In: Halbouty MT (ed) Energy resources of the Pacific region, Vol 12. American Association of Petroleum Geologists, Tulsa OK, United States, pp 63–82

Müller RD, Roest WR, Royer J-Y, Gahagan LM, Sclater JG (1997) Digital isochrons of the World's ocean floor. J Geophys Res 102(B2):3211–3214

Nakajima T, Kanai Y (2000) Sedimentary features of seismoturbidites triggered by the 1983 and older historical earthquakes in the eastern margin of the Japan Sea. Sedimentary Geol 135:1–19

Neteler M, Mitasova H (2004) Open Source GIS: a GRASS GIS Approach. The Kluwer international series in Engineering and Computer Science (SECS), Volume 773. Kluwer Academic Publishers, Boston

New MG, Hulme M, Jones PD (1999) Representing 20th century space-time climate variability. I: Development of a 1961–1990 mean monthly terrestrial climatology. J Climate 12:829–856

Patzwahl R, Mechie J, Schulze A, Giese P (1999) 2-D-velocity models of the Nazca Plate subduction zone between 19.5°S and 25°S from wide-angle seismic measurements during the CINCA 95 Project. J Geophys Res 104:7293–7317

Petit JR, Jouzel J, Raynaud D, Barkov NI, Barnola J-M, Basile I, Davis M, Delaygue G, Delmotte D, Kotlyakov VM, Legrand M, Lipenkov VY, Lorius C, Pépin L, Ritz C, Saltzmann E, Stievenard M (1999) Climate and atmospheric history of the past 420,000 years from the Vostok ice core, Antarctica. Nature 399:429–436

Pizarro O, Shaffer G, Dewitte B, Ramos M (2002) Dynamics of seasonal and interanual variability of the Peru-Chile Undercurrent. Geophys Res Lett 29(12),: doi 10.1029/ 2002GL 014790

Rabassa J, Clapperton CM (1990) Quaternary glaciations of the southern Andes. Quat Sci Rev 9:153–174

Ranero CR, von Huene R, Weinrebe W, Reichert C (2006) Tectonic processes along the Chile convergent margin. In: Oncken O, Chong G, Franz G, Giese P, Götze H-J, Ramos VA, Strecker MR, Wigger P (eds) The Andes – active subduction orogeny. Frontiers in Earth Science Series, Vol 1. Springer-Verlag, Berlin Heidelberg New York, pp 91–122, this volume

Rauch K (2005) Cyclicity of Peru-Chile trench sediments between 36° and 38°S: a footprint of paleoclimatic variations? Geophys Res Lett 32(8)L08302

Reichert C, SPOC Scientific Shipboard Party (2002) Subduction processes off Chile: initial geophysical results of SONNE Cruise SO-161 (2+3). Geophys Res Abs 4:EGS02-A-05338

Schlüter P, Ladage S, Gaedicke C (submitted) The central Chile trench fill: age estimation and distribution of trench sediments. Geo Marine Letters

Schweller WJ, Kulm LD, Prince RA (1981) Tectonics structure, and sedimentary framework of the Peru-Chile Trench. In: Kulm LD, Dymond J, Dasch EJ, Hussong DM (eds) Submarine fans and related turbidite systems. Springer-Verlag, New York, pp 23–28

Sick C, Yoon M-K, Rauch K, Buske S, Lüth S, Araneda M, Bataille K, Chong G, Giese P, Krawczyk C, Mechie J, Meyer H, Oncken O, Reichert C, Schmitz M, Shapiro S, Stiller M, Wigger P (2006) Seismic images of accretive and erosive subduction zones from the Chilean margin. In: Oncken O, Chong G, Franz G, Giese P, Götze H-J, Ramos VA, Strecker MR, Wigger P (eds) The Andes – active subduction orogeny. Frontiers in Earth Science Series, Vol 1. Springer-Verlag, Berlin Heidelberg New York, pp 147–170, this volume

Smith WHF, Sandwell DT (1997) Global sea floor topography from satellite altimetry and ship depth soundings. Science 277: 1956–1962

Spiess V (1993) Digitale Sedimentechographie – Neue Wege zu einer hochauflösenden Akustostratigraphie. Berichte des Fachbereichs Geowissenschaften der Universität Bremen 35

Strand K (1995) SEM microstructural analysis of a volcanogenic sediment component in a trench-slope basin of the Chile margin. In: Lewis SD, Behrmann JH, Musgrave RJ, Cande SC (eds) Proceedings of the Ocean Drilling Program. Scientific Results 141: 169–180

Taira A, Niitsuma N (1986) Turbidite sedimentation in the Nankai Trough as interpreted from magnetic fabric, grain size, and detrital modal analyses. In: Kagami H, Karig DE, Coulbourn WT et al. (eds) Initial Reptorts of the Deep Sea Drilling Program, vol 87. US Govt Printing Office, pp 611–632

Thornburg TM, Kulm LD (1987) Sedimentation in the Chile Trench: depositional morphologies, lithofacies, and stratigraphy. Geol Soc Am Bull 98:33–52

Thornburg TM, Kulm LD, Hussong DM (1990) Submarine-fan development in the southern Chile Trench: a dynamic interplay of tectonics and sedimentation. Geol Soc Am Bull 102:1658–1680

Twidale CR (2003) "Canyons" revisited and reviewed: Lester King's views of landscape evolution considered 50 years later. Geol Soc Am Bull 115(10):1155–1172

Underwood MB, Orr R, Pickering K, Taira A (1993) Provenance and dispersal patterns of sediments in the turbidite wedge of Nankai Trough. In: Hill IA, Taira A, Firth JV, et al (1993) Proceedings of the Ocean Drilling Program. Scientific Results 131:15–34

Vail PR, Audemard F, Bowman SA, Eisner PN, and Perez–Cruz C (1991) The stratigraphic signatures of tectonics, eustacy and sedimentology – an overview in cycles and events in stratigraphy. In: Einsele G, Ricken W, Seilacher A (eds) Springer-Verlag, Berlin Heidelberg New York

Vietor T, Echtler H (2006) Episodic Neogene southward growth of the Andean subduction orogen between 30° S and 40° S – plate motions, mantle flow, climate, and upper-plate structure. In: Oncken O, Chong G, Franz G, Giese P, Götze H-J, Ramos VA, Strecker MR, Wigger P (eds) The Andes – active subduction orogeny. Frontiers in Earth Science Series, Vol 1. Springer-Verlag, Berlin Heidelberg New York, pp 375–400, this volume

von Huene R, Scholl DW (1991) Observations at convergent margins concerning sediment subduction, subduction erosion, and the growth of continental crust. Rev Geophys 29:279–316

von Huene R, Corvalán J, Flueh ER, Hinz K, Korstgard J, Ranero CR, Weinrebe W, CONDOR scientists (1997) Tectonic control of the subducting Juan Fernández Ridge on the Andean margin near Valparaiso, Chile. Tectonics 16(3):474–488

Völker D (1989) Untersuchungen an strömungsbeeinflussten Sedimentationsmustern im Südozean. Interpretation sedimentechnographischer Daten und numerische Modellierung. Berichte des Fachbereich Geowissenschaften der Universität Bremen 115

Völker D, Reichel T, Wiedicke M, Heubeck C (submitted) Turbiditic cover of Southern Chilean seamounts: traces of "giant uphill" turbidity currents

Vorosmarty CJ, Fekete BM, Tucker BA (1996) Global River Discharge Database (RivDIS v1.0), Vol. 5: South America. International Hydrological Program, United Nations Educational, Scientific and Cultural Organization. Paris, France

Wessel P, Smith WHF (1998) New, improved version of the Generic Mapping Tools released. EOS 79:579

Zapata RA (2001) Estudio batimetrico del margen Chileno. In: Faculdad de Ciencias Fisicas y Matematicas, Departamento de Geofisica, vol. Universidad de Chile, Santiago de Chile

# Chapter 10

# Subduction Erosion – the "Normal" Mode of Fore-Arc Material Transfer along the Chilean Margin?

Nina Kukowski · Onno Oncken

**Abstract.** Subduction erosion shapes at least half of the world's convergent margins. However, its rates, and modes as well as spatial and temporal variation are poorly understood. Based on a compilation of published and newly derived estimates of subduction erosion along the Chilean part of the Andean margin, we discuss possible loci and modes of subduction erosion and also address the potential of subducting topographic highs for accelerating subduction erosion.

We also evaluate different approaches for estimating subduction erosion. Rates of subduction erosion computed from the offshore subsidence record and geometry of the margin are robust and, thus, reveal information on the regional variation in the efficiency of subduction erosion. Estimates of subduction erosion rates based on the migration of the volcanic arc front may be erroneous over short (neotectonic) timescales owing to the episodic nature of the volcanic-arc-front migration.

Information about the processes underlying subduction erosion comes from both natural observations and scaled physical experiments. Hydrofracturing at the base of the overriding fore-arc crust was identified as a process most convincingly explaining basal subduction erosion.

Subduction erosion off northern Chile is faster than that off Peru and Central America, and also faster than magmatic addition, thus making the central Andean subduction zone a site of net crustal loss. Consequently, the north Chilean part of the Andean margin is inferred to be a site of net destruction of continental crust.

From the geological record, we demonstrate that the south Chilean subduction zone, which has been in accretive mode since the Pliocene, has experienced subduction erosion since at least the middle Miocene at rates similar to the north. The change in mode is related to the doubling of sediment flux into the trench after the onset of continental glaciation of southern Chile at ~5 Ma. Hence, we identify sediment flux as the key variable controlling the mode of long-term material transfer, whereas ridge collision causes subduction erosion rates to exceed background values.

## 10.1 Introduction

With regard to their material-transfer modes, convergent margins are classified into erosive and accretive margins (e.g. von Huene and Scholl 1991 and references therein). At erosive margins, which make up at least half of the entire length of the world's convergent margins, trench-fill is thinner than a few hundred meters and no sediment is frontally or basally accreted. Instead, upper-plate material is removed, resulting in a high-taper fore-arc and the continental basement positioned close to the trench (Table 10.1, and references therein). At accretive margins, which equal about 30% of the world's convergent margins, a frontal, imbricate thrust wedge consisting of off-scraped and accreted sediments may be as wide as a few hundreds of kilometers. Whereas accretion of former oceanic plate material as well as magmatic addition e.g. contributes to the growth of continental crust, subduction erosion processes effectively destroys continental crust and returns it to the mantle. The relative balance of sediment accretion, subduction erosion and addition of mantle-derived material in the magmatic arc of a subduction zone determines if a convergent margin is a site of net crustal growth or destruction.

The structure of accretionary wedges is now well imaged in reflection seismic lines (e.g. Moore and Biju-Duval 1984; Fruehn et al. 1999; Gulick et al. 2004; Kopp et al. 2000; Kopp and Kukowski 2003), and many analogue and numerical simulations dedicated to convergent margin processes have focused on the growth and deformation of accretionary wedges (e.g. Davis et al. 1983; Gutscher et al. 1998a; Ellis et al. 2004). Based on such data, the processes and styles of sediment accretion are fairly well understood. Subduction erosion, however, is not so well understood despite being the more important material-transfer mode at convergent margins. A recent study on evaluating parameters that potentially control material transfer at convergent margins (Clift and Vannucchi 2004, and references therein) found that only high convergence rate and low trench-fill seem to favour subduction erosion.

Subduction erosion, as a process shaping some convergent margins, was recognized soon after the advent of plate tectonic theory (e.g. Scholl et al. 1970; Rutland 1971) and several concepts on possible mechanisms have been published since (e.g. Hilde 1983; Charlton 1988; von Huene and Culotta 1989; Lallemand et al. 1994; von Huene et al. 2004). Basically, subduction erosion acts in two different ways: Frontal erosion takes place at the tip of the fore-arc, i.e. at the trench, and redistributes material that has been transported downslope by small- and large-scale slumping and turbidity currents. Most of this material is of onshore or shelf origin and fills the space between topographic highs on the subducting plate. Basal subduction

Table 10.1. Features of an erosive margin

| Feature characteristic for subduction erosion | Related observations in northern Chile |
| --- | --- |
| Migration of the volcanic front | 0.95 km Myr$^{-1}$ on average since about 60 Ma at 20–25° S[a] |
| Continental margin retreat | 1.0–1.5 km Myr$^{-1}$ from 200 to 8 Ma at 20–26° S[b] |
| Large taper of the "margin wedge" | Nazca Plate dips between 7 and 11° between 22.5 and 23.5° S, lower slope may be as steep as 9°, resulting in a taper of about 14 to 20°[c] |
| High convergence rate | 6.8 cm yr$^{-1}$, was faster during the Miocene and Pliocene[d] |
| No or very minor sediment input | Sediment input less than 100 m thick throughout the trench off Chile between 19 and 25° S[e] |
| High roughness | Rough peak and valley topography of Nazca Plate[f] |
| Fore-arc extension | Major normal faults bordering lower, middle, and upper slope[g] |
| Fore-arc subsidence | May be as large as 3 000 m off Mejillones since the Miocene[h] |

[a] See compilation of magmatic data (Trumbull et al. this volume) and Fig. NK-DVD-1a/b.
[b] See Table 10.2, subduction-erosion and related references.
[c] Grotzki et al. (1998), Patzwahl et al. (1999), von Huene et al. (1999).
[d] Somoza (1998), Angermann et al. (1999), Norabuena et al. (1998).
[e] ANCORP working group (2003), Oncken et al. (this volume).
[f] Cf. von Huene et al. (1999), von Huene and Ranero (2003).
[g] Cf. Pelz (2000).
[h] Cf. Kudrass et al. (1998).

erosion, in contrast is a process which scrapes off material from the base of the overriding fore-arc.

A nearly total lack of sediment in the trench was identified from the first single-channel seismic data gathered across the offshore outer fore-arc (Scholl et al. 1970), leading to the early recognition of the Central Andean margin between 5 and 34° S as a prominent erosive margin (Rutland 1971; Kulm et al. 1977). In this region, subduction erosion has dominated mass transfer throughout the Cenozoic and possibly as long as since the Jurassic. In northern Chile, the availability of multi-disciplinary magmatic, geological, and geophysical data has enabled both the analysis of features affected by subduction erosion and the quantification of subduction erosion along several transects. In contrast, south of 34° S, the Andean margin has been accretive at high rates since the Pliocene (Bangs and Cande 1997). These opposing styles of mass transfer in two domains of the margin that share the same plate-kinematic boundary conditions, make the Andean margin a first-order locality for analysing and quantifying material transfer at convergent margins.

For this contribution, we have compiled published and new estimates for subduction erosion at the north Chilean margin and compared them with published rates of subduction erosion from Peru and Central America. We have addressed the processes contributing to subduction erosion by evaluating the geological record and by means of scaled sandbox experiments. We then used geophysical, geological, paleontological, and geochronological data to investigate the possible role and quantities of subduction erosion in south-central Chile.

## 10.2 Evidence for Long-Term Subduction Erosion at the North Chilean Margin between 18 and 34° S

The western edge of South America has been shaped by a subduction setting since at least the Jurassic. During the early Cenozoic, convergence took place between the Farallon plate and the South American Plate. Later, after the break-up of the Farallon plate at about 23 Ma (Lonsdale 2005), the Andean margin was shaped by subduction of the almost sediment-free, rough Nazca Plate beneath South America. Convergence was always oblique, with obliquity changing with time and latitude (e.g. Somoza 1998). Convergence velocity also fluctuated considerably and has been decreasing throughout the Neogene (Norabuena et al. 1999). However, it was always amongst the faster convergence rates observed on Earth. The presence of the cold, upwelling Humboldt Current offshore from central South America and the stability of South America's latitudinal position throughout the Cenozoic (and probably since the Jurassic) favoured long-term aridity (Hartley et al. 2005). Therefore, very minor amounts of terrigenous sediment have entered the trench at the north Chilean latitude; sediment thickness has remained below 500 m at most, generally below 200 m (Oncken et al. 2006, Chap. 1 of this volume).

Since the Jurassic, the magmatic arc front has migrated more than 200 km to the east (Rutland 1971; Scheuber et al. 1994), with the Jurassic arc now present along the Coastal Cordillera and on the slope of the fore-arc (Rötzler et al. 1998), only about 50 to 150 km east of the trench.

The Cretaceous to Eocene arc is now located in the Central Valley and in the Precordillera. Therefore, the present west to east segmentation of the fore-arc is largely built from a series of neighbouring magmatic arcs. Extensional tectonics are observed in the offshore fore-arc (Pelz 2000), the Coastal Cordillera, and the Central Valley (Hartley et al. 2000).

The geometry, velocity structure, and tectonic structure of the north Chilean fore-arc is well constrained by combined offshore-onshore, wide-angle seismic information (Grotzki et al. 1998; Patzwahl et al. 1999; ANCORP working Group 2003). It reveals a taper as large as 14 to 20°. The 15 to 20 km-wide, low-velocity prism present between the trench and the seaward edge of the pre-Tertiary, continental, crystalline crust is thought to consist of debris resulting from gravitational failure of the foremost lower slope (von Huene and Ranero 2003). Dredging undertaken during the 'CINCA' cruise recovered samples of volcanic rocks from water depths of about 4 000 m. Rötzler et al. (1998) suggested that these are equivalent to the Jurassic La Negra formation, indicating the proximity of the earlier volcanic arc to the trench.

## 10.3 Strategies for Quantifying Subduction Erosion

Because the various strategies employed to quantify subduction erosion depend on some fundamental assumptions about the location and mechanism of subduction erosion, we start our analysis by listing them.

1. Hilde (1983) considered that the peaks on a rough, down-going oceanic plate acted as a "chain-saw" and therefore rasped material from the overlying continental plate. However, more recent seismic images show that the valleys between the peaks on the oceanic plate are filled with loose debris during passage through the trench, which leads to a smoothened plate interface and positions the interplate thrust well above the top of the oceanic plate (von Huene and Culotta 1989). This calls the efficiency of the "chain-saw" into question.
2. Charlton (1988) proposed that subduction erosion is likely to occur if the dip of the downbent oceanic plate between the outer bulge and the trench is larger than the dip of the lower continental slope. However, his approach was entirely geometrical and did not consider the actual process of subduction erosion.
3. Subduction of oceanic crust with a very rough surface and subduction of topographic highs on the lower plate like seamounts or basement ridges were also recognized as favouring subduction erosion (Lallemand et al. 1994). The uplift and subsidence of the fore-arc related to the subduction of these features leads to enhanced rates of subduction erosion (e.g. Clift et al. 2003).
4. Fluid overpressure, finally, is thought to generate fractures in the basal region of the upper plate. This loosens fragments that then become incorporated into the subduction channel where they may undergo further downward transport (Lallemand et al. 1994; von Huene et al. 2004).

Although the proposed mechanisms are compatible with the geophysical and geological record of erosive margins, they do not provide a consistent process-oriented explanation for why and how subduction erosion occurs.

If the "chain-saw" (Hilde 1983) is regarded as the process governing subduction erosion, the volume eroded from the upper plate is set equal to the void volume, i.e. the volume between neighbouring peaks and valleys on the rough, down-going oceanic plate (Fig. 10.1). Accordingly, only frontal subduction erosion resulting from surficial mass wasting, as observed in reflection seismic data, is quantified in this way. Potential basal subduction erosion occurring arcwards of the trench is completely neglected by this approach, which may lead to a significant underestimation of subduction erosion rates at a specific margin.

If the geometry of the submarine fore-arc is known, trench migration over a given time period can be determined from the rate of subsidence, which is known, for example, from the paleo-bathymetric record of benthic foraminifera in drill holes (Vannucchi et al. 2004). If, in addition, the crustal thickness at the shelf-break is known, say from wide-angle seismic data along the same profile, this provides the total crustal volume loss in the given time interval. If drill-hole data or seismic reflection data allow for backstripping, then the isostatic effects resulting from the regional compensation of local subduction-erosion effects can also be taken into account (Clift et al. 2003). This approach accounts for all possible processes of subduction erosion and, therefore, provides the total volume of crustal material loss through subduction erosion from the trench to the coast.

However, inherent in this approach, as to the others so far described, is the assumption that the geometry of the fore-arc (i.e. slope angle and dip of the down-going plate) did not change over the time interval considered, such that the present-day geometry can be used (Clift et al. 2003; Vannucchi et al. 2004). Furthermore, with this approach, it is expected that subduction erosion only occurs in the offshore fore-arc. Consequently, values of subduction erosion derived in this way could underestimate subduction erosion if there is a contribution from the base of the onshore fore-arc.

Information on vertical movement and the paleo-geometry of a specific margin can be obtained from section balancing using swath bathymetry, depth-migrated seismic reflection profiles, structural analysis, the paleo-

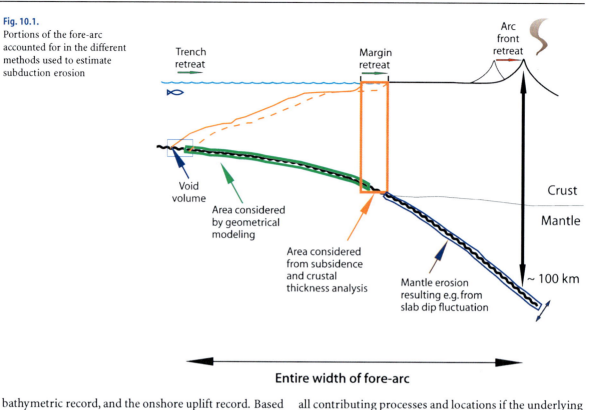

**Fig. 10.1.**
Portions of the fore-arc accounted for in the different methods used to estimate subduction erosion

bathymetric record, and the onshore uplift record. Based on this information, Pelz (2000) developed an incremental geometrical modeling approach for quantifying subduction erosion, including an estimate of the ratio of frontal versus basal subduction erosion. This concept is based on the observation that frontal erosion and basal subduction erosion are characterized by different material paths. Like the approach described before, this geometrical approach considers isostatic effects, and the rate of trench migration is obtained as a results of its application. Again, like the methods described above, the geometrical approach of Pelz (2000) provides an estimate for subduction erosion only in the offshore fore-arc (Fig. 10.1).

Migration of the magmatic arc front towards the backarc is thought to be caused by subduction erosion (Rutland 1971). Estimates of subduction erosion based on arc-front migration rates (Fig. 10.1) assume a stable geometry of the subduction zone over time with the result that trench migration equals arc migration. Assumptions about the average dip of the down-going slab (Jarrard 1986) and about the depth of magma generation (England et al. 2004) are inherent in this approach and are key sources of uncertainty. This approach, however, has the advantage of considering the entire fore-arc including the fore-arc mantle wedge, the base of which also may be eroded. Flattening of the slab has also been proposed to cause basal erosion along the fore-arc from the trench to the depth of magma generation (e.g. Kay et al. 2005 and references therein). In summary, arc front migration should provide a maximum estimate of subduction erosion containing all contributing processes and locations if the underlying assumptions can be shown to be valid.

Some of the parameters used to quantify subduction erosion, such as paleo-bathymetry or the depth of magma generation, yield a relatively large uncertainty and others, such as convergence rate or migration of the magmatic arc front, are averaged over long time intervals. Hence, the accuracy with which subduction erosion may be quantified is limited; estimates may be subject to about 30 to 50% error. However, within this range of error, results are robust and, as some of the approaches outlined above are based on different data, independent of each other.

From the above analysis and the involved mechanisms it is obvious that three independent processes contribute to the lithospheric volume lost through subduction erosion: (*i*) frontal erosion fills the void volumes between the topographic peaks on the lower plate and therefore contributes a volume $VSE_{void\_fill}$, (*ii*) basal subduction erosion scrapes off a volume $VSE_{basal\_eros\_crust}$ from the base of the fore-arc crust, and (*iii*) mantle erosion removes a volume $VSE_{mantle\_eros}$ from the continental fore-arc mantle wedge. All of these processes are fully independent of each other as they are controlled by different (and independent) parameters.

Subduction erosion leads to a migration of the magmatic arc front towards the hinterland. Therefore, a lithospheric estimate of the volume $VSE_{arc\_retr}$ lost by subduction erosion based on the rates of arc front migration should equal the sum of the rates of subduction erosion contributed by the three aforementioned processes. To estimate the net

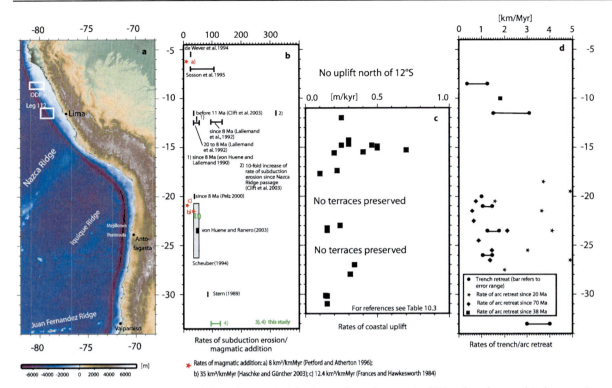

**Fig. 10.2. a** Topography of the central Andean fore-arc. **b** Rates of subduction erosion and magmatic addition along the central Andean margin (see Table 10.2 for additional information, e.g. subduction erosion type, crustal thickness). **c** Rates of coastal uplift. Time intervals and references are given in Table 10.3. **d** Rates of trench retreat and eastward migration of the magmatic arc front

material balance across the fore-arc, rates of magmatic addition need to be included in the budget equation.

Summarising, the contributors to subduction erosion are independent of each other, and add to a total value. Keeping in mind the large uncertainties associated with the determination of each of these rates, the above balance should enable us to extract the unknown contribution of one process if the others have been quantified. This may give some first order estimate for the mantle contribution to subduction erosion, which otherwise cannot be directly estimated from geophysical, geological, or paleontological data.

However, in the following analysis of subduction erosion along the Chilean continental margin we will restrict our quantification and quantitative comparisons to the loss of continental crust through subduction erosion, since suitable data from the South American and other subduction zones are available only for this component.

## 10.4 Rates of Crustal Loss through Subduction Erosion at the North Chilean Fore-Arc (18–34° S)

The multiformity of available data for the central Andean subduction zone (~7 to 34° S) has enabled us to apply all the methods outlined in the previous section to several transects across the fore-arc. We therefore compiled previously published rates of crustal subduction erosion, mainly from two corridors in the north Chilean fore-arc at 20 to 23° S (Mejillones) and 30 to 33° S (Juan Fernandez Ridge). We also present new estimates of crustal loss used for obtaining a solid base from which to discuss the potential spatial and temporal variability of subduction erosion and its causes along the central Andean margin (Fig. 10.2).

### 10.4.1 Subduction Erosion in the Mejillones Corridor

The episodic migration of the magmatic arc is best documented in the region between 21 and 26° S (Scheuber et al. 1994; Trumbull et al. 2006, Chap. 2 of this volume). The eastward shift of the magmatic arc front some 200 km since the Jurassic suggests an average migration rate of slightly more than 1 km Myr$^{-1}$. Assuming a magma generation depth of 100 km, an average dip of 20° for the subducting Nazca Plate (which yields a distance of 275 km between the trench and the magmatic arc), and a uniform geometry of the fore-arc throughout this period, Scheuber et al. (1994) equalled continental margin retreat rates with arc-migration rates and estimated the crustal contribution to subduction erosion to be between 37 and 54 cubic kilometers per million years per kilometer of trench (hereafter: km$^3$ Myr$^{-1}$ km$^{-1}$; Table 10.2). They did not quantify rates for a specific profile but regarded their estimate as valid for the whole region. Errors in their estimate could result

Table 10.2. Compilation of published and newly derived rates of subduction erosion along the Chilean margin

| Latitude (deg S) | Time interval (Ma) | Subsidence rates (m Myr$^{-1}$) | Trench retreat (km Myr$^{-1}$) | Erosion rates (km$^3$ Myr$^{-1}$ km$^{-1}$) | SE[a] type | Thickness of crust (km) | Reference |
|---|---|---|---|---|---|---|---|
| 20 | Since 8 | | 1 | 37 – 40 | I + II[b] | | Pelz 2000 |
| 22 | Since 8 | | | 40 – 45 | I + II | 32 – 34[c] | This study |
| 23.5 | Since 20 | 200 | 1.25 – 1.75 | 45 – 50 | I[d] | | von Huene and Ranero 2003 |
| 21 – 26 | Since 200 | | 1.05 – 1.45 | 37 – 50 | III | | Scheuber et al. 1994 |
| 30 | Since 15 | | | 84 | III[e] | | Stern 1991 |
| 33 | Since 10 | 300 – 500 | 3 – 4 | 96 – 128 | I + II | 32 – 36[f] | This study |
| 36 – 40 | 11 – 3 | 200 | | 25 – 35 | III | 27 – 30[g] | This study |
| 47 | 4.2 – 1.5 | | 8 | 231 – 443 | III | | Bourgois et al. 1996 |

[a] SE: Subduction erosion.
[b] Roman numbers indicate processes contributing to subduction erosion: I: void volumes, II: offshore basal erosion, III: total crustal loss (I + II = III).
[c] Crustal thickness from wide-angle data (Patzwahl et al. 1999).
[d] Porosity corrected volume.
[e] From crustal shortening.
[f] Subsidence rates from Laursen et al. (2002), crustal thickness from wide-angle data (Flueh et al. 1998).
[g] Crustal thickness from wide-angle data (Krawczyk et al. 2006, Chap. 8 of this volume).

from the assumption of fore-arc geometry being constant with time.

Material filling the volume between the topographic highs of the Nazca Plate (void volume) at the trench was estimated at more than 56 km$^3$ Myr$^{-1}$ km$^{-1}$ (von Huene et al. 1999), which would account for a high potential for frontal erosion off northern Chile. A later, corrected value for void volume yielded only 21 km$^3$ Myr$^{-1}$ km$^{-1}$ (Pelz 2000). This lower value is more compatible with the other estimates of subduction erosion at the north Chilean margin. Incremental geometrical modeling (Pelz 2000) yielded 37–40 km$^3$ Myr$^{-1}$ km$^{-1}$ for bulk subduction erosion since 8 Ma along a transect across the submarine fore-arc at about 20° S (Fig. 10.2, Table 10.2). von Huene and Ranero (2003) estimated the subduction erosion rate at 23.5° S to be 45–50 km$^3$ Myr$^{-1}$ km$^{-1}$ since 20 Ma, considering the void volumes on Nazca Plate before approaching the trench and assuming that they become significantly larger after the downgoing plate was bent at the outer bulge owing to slip along bending-related normal faults (Fig. 10.2, Table 10.2).

The reported estimates to date did not make use of the well-constrained thickness of the fore-arc crust at the shelf-break and the overall geometry of the submarine fore-arc, known from wide-angle seismic data (Grotzki et al. 1998; Patzwahl et al. 1999). Applying the approach outlined in Vannucchi et al. (2004) with the same estimate used by Pelz (2000) of about 1 800 m for subsidence (derived from deep-water sampling; Kudrass et al. 1998), we estimate bulk subduction erosion of the offshore fore-arc at a rate of 40–45 km$^3$ Myr$^{-1}$ km$^{-1}$ at 22° S during the past eight million years or so (Table 10.2).

### 10.4.2 Subduction Erosion in the Corridor Affected by the Juan Fernandez Ridge Subduction

At 30° S, the Andes underwent about 165 km of E-W shortening in the past 15 million years (Allmendinger et al. 1990). However, only about 80% of this observed shortening can be accounted for with section balancing, leaving an unexplained 33 km of lost crust. Subduction erosion was suggested as the most probable explanation for this crustal loss. Using a value of 38 km for the thickness of the crust, Stern (1991) estimated subduction erosion at a rate of 84 km$^3$ Myr$^{-1}$ km$^{-1}$. This approach for quantifying subduction erosion includes the entire fore-arc crust and, thus, is equivalent to net crustal loss.

Fast subsidence has taken place in the Valparaiso Basin at 33° S since about the late Miocene (Laursen et al. 2002). The crustal thickness is accurately known from wide-angle seismic data (Flueh et al. 1998). Using the strategy outlined in Vannucchi et al. (2004), we estimate rates of 96 to 128 km$^3$ Myr$^{-1}$ km$^{-1}$ for subduction erosion of the offshore fore-arc crust (Table 10.2, Fig. 10.2).

## 10.5 Loci and Modes of Subduction Erosion

### 10.5.1 Evidence from Field Observations

Off northern Chile, a low-velocity prism, up to 20 km wide next to the trench, does not reveal a reflection pattern typical for an accretionary wedge (von Huene and Ranero 2003; Sallares and Ranero 2005; Sick et al. 2006, Chap. 7 of this

**Table 10.3.** Compilation of published rates of coastal uplift along the central Andean margin

| Latitude (deg S) | Uplift rate (m kyr$^{-1}$) | Time interval (Ma) | Reference | Remark |
|---|---|---|---|---|
| 6 – 12 | | | | No uplift observed |
| 12 | 0.25 | 1.7 – 0 | le Roux et al. 2000 | |
| 14.3 – 14.9 | 0.3 | | Hsu 1992 | |
| 14.8 | 0.25 | 1.6 – 0.4 | Goy et al. 1992 | A bit speculative |
| 14.8 | 0.46 | 0.4 – 0 | Goy et al. 1992 | |
| 14.9 – 15.1 | 0.5 | | Hsu 1992 | |
| 15.3 – 15.6 | 0.2 | | Hsu 1992 | |
| 15.3 | 0.7 | 0.5 – 0 | Macharé and Ortlieb 1992 | Present collision of Nazca Ridge |
| 15.5 | 0.4 | 2 – 0 | Macharé and Ortlieb 1992 | |
| 17.4 | 0.22 | 0.3 – 0 | Zazo et al. 1994 | |
| 17.6 (Ilo) | 0.16 | Since 0.1 | Ortlieb et al. 1996b | |
| 17.7 | 0.1 | | Hsu et al. 1990 | |
| 18 – 20 | | | | No terraces preserved |
| 23 (Hornitos) | 0.24 | 0.3 – 0 | Ortlieb et al. 1996a | |
| 23 – 23.5 | 0.15 | 0.33 – 0 | Ortlieb et al. 1995 | |
| 23.5 | 0.1 – 0.2 | 2 – 0 | Ortlieb et al. 1996b | Mean rate for whole Quaternary |
| 24 – 26 | | | | No terraces preserved |
| 27 | 0.34 | 0.5 – 0 | Marquardt et al. 2004 | |
| 28 | 0.31 | 0.5 – 0 | Marquardt et al. 2004 | |
| 29 | 0.08 | Since 2.6 | le Roux et al. 2005 | Three terraces preserved, but the area seems to have been tectonically quiet |
| 30.15 | 0.14 | Holocene | Ota and Paskoff 1993 | |
| 30.2 – 31 | 0.1 – 0.2 | Late Quaternary | Ota et al. 1995 | Uplift also at the end of the Pliocene |

volume). This was interpreted to reflect a composition of slumped or reworked material. Extensional failure of the submarine fore-arc through large-scale normal faulting has been documented at numerous sites along the north Chilean fore-arc (von Huene and Ranero 2003; Sick et al. 2006, Chap. 7 of this volume). Considerable subsidence is documented from the middle slope (Kudrass et al. 1998) and the fore-arc basin (Laursen et al. 2002). These observations seem to rule out basal accretion of frontally eroded material beneath the offshore fore-arc. Instead, they indicate a significant contribution of basal erosion to bulk subduction erosion. Geometrical modeling indicates that basal erosion may be three times as efficient as frontal erosion and that it takes place at the base of the offshore fore-arc (Pelz 2000).

Suggested mechanisms responsible for the observed Pliocene to present-day coastal uplift along the central Andean margin south of 14° S are: (*1*) underplating of frontally eroded material from the tip of the fore-arc (Adam and Reuther 2000); (*2*) subduction of a basement high, the Iquique Ridge (Hartley and Jolley 1995); and (*3*) isostatic compensation of tectonic erosion (Pelz 2000). Quaternary rates of coastal uplift are not uniform along the margin (Table 10.3). Strongly enhanced uplift is observed from terraces at the present collision site between the Nazca Ridge and the South American margin. In contrast, subduction of the Iquique Ridge does not seem to have increased coastal uplift at its projected continuation. However, some of its branches may be responsible for uplifting Mejillones peninsula further south.

Based on coastal segments not affected by the subduction of ridges or seamounts, the background signal of coastal uplift is probably about 0.15 m kyr$^{-1}$ (Table 10.3). Enhanced coastal uplift between 27 and 28° S since about 0.5 Ma occurs where the trend of the coast changes from roughly NNE-SSW to N-S. To the south, coastal uplift again has occurred at very low rates since the Pliocene, motivating Le Roux et al. (2005) to interpret this area as tectonically quiescent.

The distribution and rates of coastal uplift indicate that uplift has not been modified by the subduction of the Juan

Fernandez Ridge. This ridge collided with the margin at about 20° S at 22 Ma and has since migrated southwards to its present position, initially at a rate of up to 20 cm yr$^{-1}$, and then slowing down at 11 Ma to 3.5 cm yr$^{-1}$ (Yañez et al. 2001). The Juan Fernandez Ridge shows only moderate crustal thickening (Kopp et al. 2004) and mainly consists of individual hot-spot-derived seamounts; therefore, it is a much smaller feature than the Nazca Ridge and does not cause significant buoyancy. The same is most probably true for the Iquique Ridge, but, to date, this ridge has not been subjected to geophysical imaging.

Coastal uplift is not enhanced at the current collision sites of the Iquique and Juan Fernandez Ridges with South America. Consequently, ridge subduction should be discarded as a possible mechanism for causing coastal uplift along the north Chilean margin. According to Victor et al. (2004), the Longitudinal Valley in the onshore fore-arc has remained in a vertically stable position since ~20 Ma, relative to sea level. This observation argues against basal accretion of formerly eroded material. Considering that all observed uplift of the Coastal Cordillera can be explained by isostatic rebound (Pelz 2000), then basal erosion most probably takes place along the whole submarine part of the fore-arc up to the coastline. An exception may be formed by individual peninsulas – e.g. Mejillones Peninsula, Caldera, etc. – which exhibit higher Pliocene to recent uplift rates. Here, some material may be underplating as suggested by Adam and Reuther (2000). Otherwise, the eroded material most likely seems to be transported within the subduction channel to at least as far as beneath the western edge of the Precordillera.

### 10.5.2 Evidence from Scaled Physical Experiments

Valuable information on how mass-transfer processes at convergent margins function and interact has come from scaled physical experiments (e.g. Malavieille 1984; Byrne et al. 1993; Gutscher et al. 1998a). In these experiments, granular materials such as sand, mortar and micro glass beads, which have been shown to approximate the deformation behaviour of sediments and upper-crustal rocks (Marone 1998; Lohrmann et al. 2003), are used to investigate the parameters controlling deformation and mass transfer under well-determined laboratory conditions. This approach follows the critical taper theory (Davis et al. 1983), which postulates self-similar wedge evolution. Consequently, the style of deformation and the geometry of the wedge are not depending on the velocity of convergence. Therefore, scaling between nature and experiment is achieved by scaling the physical properties of the natural and analogue materials, respectively. This leads to a consistent factor of about $10^{-5}$ for both, physical properties and length (Hubbert 1937).

Most previous analogue experimental studies on subduction-zone-material transfer have dealt with the formation of accretionary wedges, whereas only a few studies have addressed subduction erosion (Kukowski et al. 1994; Gutscher et al. 1998a; Kukowski et al. 1999; Adam et al. 2000; Lohrmann 2002). These studies showed that subduction erosion can occur in two different scenarios:

1. As soon as material is forced to leave the experimental box through a "subduction gate" at the lower edge of the rigid back-wall, the rear and basal part of the wedge undergo tectonic erosion (Kukowski et al. 1994; Gutscher et al. 1998a). This process is independent of any other boundary condition.
2. If no material is supplied at the tip of the laboratory wedge, erosion at the base of the wedge near its tip occurs in addition to frontal erosion from over-steepening and subsequent surficial mass wasting.

In the first, "open", experimental set-up, subduction erosion is forced by the continuous removal of material at the rear base of the apparatus, and the overall mass budget is a function of the relative thickness of the incoming material to the width of the "subduction gate" (Kukowski et al. 1994) which constrains the transport capacity of the subduction channel. In the second set-up, however, basal erosion is controlled by the contrast in strength between the wedge and the base along which it moves (Fig. 10.3a). This mode of basal erosion is restricted to the very front of the sand wedge, where the strength contrast between the wedge and the conveyor-belt (which was made of sand-paper) is close to negligible.

In comparison, further to the rear, the contrast in strength increases as the strength of the flow zone decreases (owing to, for example, dilatancy associated with particle flow during thrust motion). During subduction erosion, the front slope of the wedge attains its angle of repose and maintains it through dominantly small-scale, but frequent, slumping. Material at the tip of the wedge resulting from this gravitational mass wasting and material basally eroded is transported to the rear within the "subduction channel" (Lohrmann et al. 2006, Chap. 11 of this volume) and basally accreted as soon as the load of the overburden becomes too large to allow further subduction on top of the downgoing plate (cf. Gutscher et al. 1998b; Masek and Duncan 1998). In the experiment, the zone affected by basal erosion and material transport is about 8 to 10 mm wide. Increasing the width of the "subduction gate" at the rear of the experimental apparatus – i.e. the transport capacity of the channel – does not change the processes of subduction erosion, but leads to increased rates of wedge retreat, an increased ratio of basally versus frontally eroded material, and a decrease in, or even absence of, basal accretion.

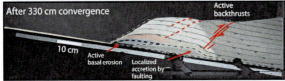

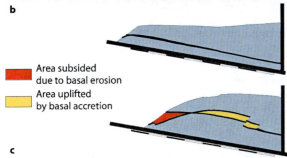

**Fig. 10.3.** a Initial and final stages of a scaled sandbox experiment, with no material input and output, revealing the fundamental processes of subduction erosion (after Kukowski et al. 1999). b Initial and final stages of a sandbox experiment in which thin, weak zones that separate lens-shaped areas are incorporated (after Lohrmann 2002). c Areas undergoing subsidence and uplift due to subduction erosion in the experiments shown in the preceding figures (after Kukowski et al. 2004)

It has been proposed that the heterogeneous strength of the fore-arc wedge (e.g. the "mega-lenses" introduced by Ranero and von Huene 2000) increases the potential of subduction erosion. Adopting this concept in sandbox experiments, by incorporating thin weak layers of micro glass beads into the initial sand wedge (Lohrmann 2002; Kukowski et al. 2004), indeed showed that the presence of the weak layers led to increased frontal and basal erosion, accompanied by considerable subsidence in the overburden and the formation of fault-bounded basins (Fig. 10.3b,c). Direct evidence of the efficiency of mass re-distribution is shown by the larger areas affected by basal erosion and basal accretion. In all experiments with boundary conditions independent of time and in which the initial wedge was made of only one material (sand or a mixture of sand and mortar, respectively), the rates of the different modes of material transfer became constant after an adjustment phase. This may indicate that subduction erosion processes are "self-controlled". The volumetric contribution of basal erosion to total erosion is about 10% in the closed set-up and 20% or more in the open set-up. This value is much lower than that seen in nature (see above), indicating that additional mechanisms need to be included (see below).

## 10.6 The Role of Seamounts and Basement Ridges in Enhancing Subduction Erosion

The collision of oceanic-plate topographic features such as basement ridges, spreading ridges and seamounts with the upper plate is a widespread phenomenon along the Earth's convergent margins. Swath bathymetry has revealed that the margin steepens around a subducting seamount, which leads to enhanced surficial mass wasting (e.g. Dominguez et al. 1998) and the development of a re-entrant in the wake of the seamount (Dominguez et al. 1998; Lallemand et al. 1989). In contrast, subducting ridges are not associated with local morphological features, but may lead to regional and strong coastal uplift (e.g. Hsu 1992). Whereas subducting seamounts affect a margin only locally, and probably for not longer than one to two million years, subduction of ridges oriented at a high angle to the plate boundary has regional effects and is a longer lasting process.

When a topographic high is subducting, the overlying continental crust is uplifted and undergoes extension (Fig. 10.4a). This may lead to faulting and weakening of the overburden. In the wake of a down-going seamount or ridge, the uplifted portion of the fore-arc undergoes fast and strong subsidence. The subsided fractured crust may then be subject to enhanced basal erosion (Fig. 10.4b). The weakening of the base of the overburden would reduce the contrast in strength between it and the subduction channel. A low strength contrast was identified as facilitating basal subduction erosion in physical experiments (cf. previous section).

Significantly enhanced subduction erosion has been quantified from the offshore record at several margins. Examples include the areas affected by the subduction of the actively spreading Chile Ridge, the Nazca and Louisville Ridges. Off Lima, Nazca Ridge subduction caused

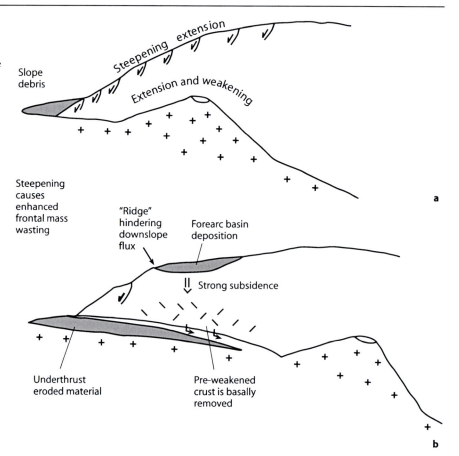

Fig. 10.4.
Sketch illustrating how subduction erosion is facilitated by seamount subduction, **a** at the extensional phase; **b** shows the enhanced basal erosional phase

erosion since 11 Ma at rates of about 320 km$^3$ Myr$^{-1}$ km$^{-1}$ which is up to 10 times higher than the long-term rates of crustal loss through subduction erosion observed at the same latitude at the Peruvian margin (Fig. 10.2; Clift et al. 2003). The subduction of the Louisville Ridge has contributed as much as an additional 40 km$^3$ Myr$^{-1}$ km$^{-1}$ to subduction erosion at the Tonga margin since 5 Ma, where the rate of long-term subduction erosion since 34 Ma has been 27–28 km$^3$ Myr$^{-1}$ km$^{-1}$ (Clift and McLeod 1999). Subduction erosion caused by the Chile Ridge in the area of the Taitao peninsula was estimated at 231–443 km$^3$ Myr$^{-1}$ km$^{-1}$ (Table 10.2; Bourgois et al. 1996) and, therefore, is as fast as that caused by the Nazca Ridge.

Hence, the contribution of ridge-enhanced subduction erosion is much less in the case of the Louisville Ridge compared to the Nazca Ridge. However, it still is considerably higher than average rates of long-term subduction erosion. Subsidence rates of the offshore Costa Rican forearc have at least doubled since the onset of subduction of the Cocos Ridge at 5 Ma. At the central Chilean margin (33 to 34° S), at the latitude where the Juan Fernandez Ridge is presently subducting, subduction erosion is more than twice as effective as that off Mejillones (Table 10.2, Fig. 10.2). We suggest that the most probable explanation for this observation is an acceleration of subduction erosion owing to the subduction of the Juan Fernandez Ridge.

However, due to its limited size, the increase of subduction erosion is not very pronounced.

We speculate that subduction of the Juan Fernandez Ridge may have caused a widening of the subduction channel, enabling it to efficiently remove sediments originating from glacial flux in south-central Chile (i.e. at ~42° S, the northernmost latitude at which southern-hemisphere, post-Miocene glaciation reached the coast). These sediments have been transported along the trench (Völker et al. 2006, Chap. 9 of this volume) as far north as the Juan Fernandez Ridge, which also serves as a barrier to further northward transport (cf. also Vietor and Echtler 2006, Chap. 18 of this volume). The enhanced rates of subduction erosion in the wake of the subduction of oceanic basement ridges confirm that a reduction in the strength contrast between the base of the fore-arc crust and the subduction channel accelerates basal subduction erosion.

## 10.7 Evidence for Subduction Erosion in South-Central Chile (35 to 45° S)

In contrast to the central Andean margin, the modes of material transfer are significantly more complex and variable at the south-central Chilean margin between about 35 and 45° S. At these latitudes subduction most prob-

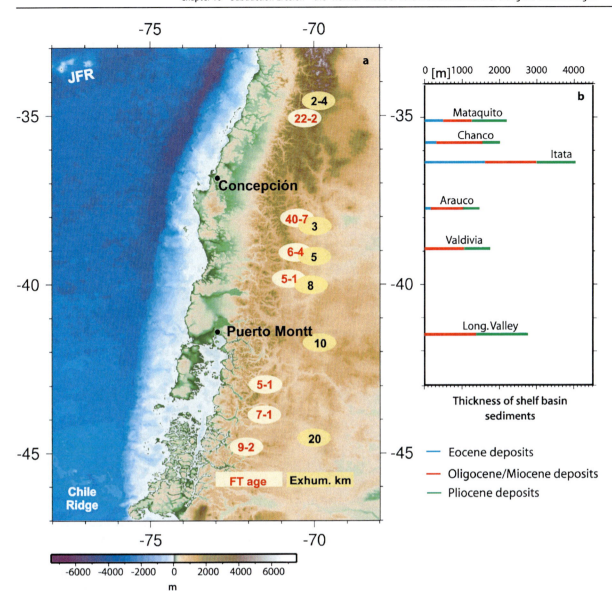

**Fig. 10.5. a** Topography of the south-central Chilean margin. Annotated are the fission track ages (*yellow*) and the exhumation in km (*white*). **b** Sediment record from the shelf and onshore basins

ably began in the late Carboniferous (Willner et al. 2004) and has persisted until the present day (Glodny et al. 2006, Chap. 19 of this volume and references therein).

Since the Pliocene, coincident with southern-hemisphere glaciation, an accretionary wedge has built up along the south-central Chilean margin (Bangs and Cande 1997). This wedge is juxtaposed against an accretionary complex of late Carboniferous to Triassic age. The lack of an accretionary wedge of Mesozoic and Tertiary age, as well as the Neogene eastward migration of the volcanic arc at 33–37° S (Kay et al. 2005) suggests that subduction erosion has been the dominant mode over most of this time. Yet, periods of westward advancement of the arc at 38–39° S (Stern 1989) indicate a non-uniform behaviour along strike and over time.

Bangs and Cande (1997) and Melnick and Echtler (2006) have argued that the Pliocene to Recent change from subduction erosion to accretion, as well as the related change in kinematics, observed in the main Andean Cordillera and the fore-arc basins was related to the glacially controlled increase of sediment flux to the trench. This increase modified the mode of tectonic mass flux and the mechanical coupling at the plate interface. The exhumation of Miocene granitoids (Seifert et al. 2005) and fission-track dating in the Main Cordillera (e.g. Thomson 2002; Adriasola et al. 2005; Glodny et al. 2006, Chap. 19 of this volume) show that denudation rates have increased substantially since about 5 Ma, which was approximately the time of the onset of continental glaciation in Chile (Rabassa and Clapperton 1990). Also, a significant southward gradient in glacial exhumation of the Main Cordillera is observed (see Seifert et al. 2005; Fig. 10.5), in line with the extent of glaciation. Although some eroded material has been stored in the basins of the Longitudinal Valley and the offshore fore-arc (Mordojovich 1981; Gonzáles 1989; Jordan et al. 2001), a significant amount of sediment must have reached the trench.

In order to explore the role of trench-fill, we have here attempted a more highly resolved reconstruction compared to the estimate by Bangs and Cande (1997; see also Melnick and Echtler 2006). Using the sediment mass-flux equation by Oncken et al 2006. (Chap. 1 of this volume) we have reconstructed the total erosional and sedimentary flux since the early Miocene.

The incoming pelagic sediment thickness on the Nazca Plate is estimated to be between 30 and 150 m (Kudrass et al. 1998). The volume of eroded material from the Andean Cordillera is derived from the literature as referenced throughout this section and, for the Coastal Cordillera, from Echtler (pers. comm.). From the observation of mass wasting at the tip of the steep offshore slope, we have added a minor contribution, estimated at some 10%, of material loss from the accretionary wedge to subduction erosion. Also, we have included additional mass flux from the erosion of volcanoes (~3 km$^3$ Myr$^{-1}$ km$^{-1}$ of arc length, i.e. 10% of the reported magmatic flux).

These data sets are calculated over the area of the eroded Andean margin between the Chile triple junction to the south and the Juan Fernandez Ridge to the north. Both ridges form the present-day limit of trench-parallel mass transport. We used reconstructions by Yanez et al. (2001) and Gorring et al. (1997) to restore their migration. We calculated fore-arc sediment volume using stratigraphic and thickness data from the Longitudinal Valley and offshore basins (Mordojovich 1981; Jordan et al. 2001), in all cases integrated over the depositional area. From this database (Fig. 10.5), we calculated total erosional and sedimentary fluxes though time (Fig. 10.6). For conversion of solid rock into sediment, we assumed minor chemical denudation in this semi-arid to humid environment (<10%) and an average porosity of 30% for the clastic sediments.

The results (Fig. 10.6) show a moderate trench-fill of between 500 and 1 000 m during the Miocene and a strong increase to ~2 km thickness since 5–6 Ma, which is the present-day value. Interestingly, it has been observed that this change relates to a change of deformational behaviour in the fore-arc (Melnick and Echtler 2006). Rapid surface subsidence and extension from the middle Miocene to the early Pliocene has since been succeeded by a switch to shortening and rapid uplift of the fore-arc basins. Melnick and Echtler (2006) interpreted this as a change from basal subduction erosion (in the Miocene) to tectonic underplating (since the Pliocene).

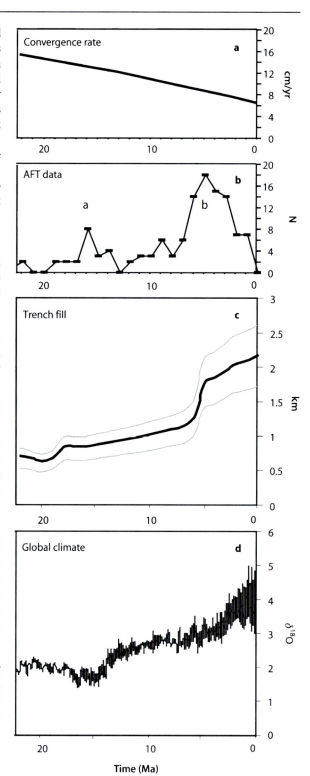

**Fig. 10.6. a** Convergence rate between the Nazca Plate and South America (Norabuena et al. 1999). **b** Frequency of fission-track data as a proxy for exhumation (*AFT* = apatite fission track). **c** Neogene variation in trench-fill (for data sources see text). **d** Trends of benthic oxygen isotope ratios (Zachos et al. 2001)

Surface subsidence of the offshore fore-arc, from nearshore conditions in the middle Miocene to bathyal depths (> 1500 m; Encinas et al. 2005), over ~8–10 Myr is of the same order of magnitude as that in northern Chile, yielding an estimated rate for crustal erosion of 25–35 km$^3$ Myr$^{-1}$ km$^{-1}$. Also, from the eastward migration of the volcanic arc during this period, by some 20–40 km at 36–37° S (Kay et al. 2005), the rate of total subduction erosion estimated from arc retreat of ~2–4 km Myr$^{-1}$ is similar to the recent north Chilean subduction erosion rate for this period (equivalent to a volume erosion rate of ~180–350 km$^3$ Myr$^{-1}$ km$^{-1}$).

The fast Pliocene build-up of an accretionary prism clearly was facilitated by the strong increase of glacial sediment flux. The prism shows, however, some particular features. Reflection seismic data reveal that 20 to 80% of the current trench-fill is being subducted (Diaz-Naveas 1999; Rauch 2005; Behrmann and Kopf 2001). The accreted portion decreases southward from about 36° S towards the latitude of the present-day Chile Ridge collision with South America and increases again further southwards (Table 10.4). From the prism volumes observed, it becomes evident that the existing wedges may have been built during the short time span of only a half to two million years throughout this portion of the Chile margin (Table 10.4). This is a considerably shorter amount of time than since the onset of glaciation with its associated high sediment flux. Hence, the portion of the trench-fill accreted to the tip of the plate must have grown over time to the present-day value, while the amount of subducted and underplated material must have decreased.

## 10.8 Discussion

Our new estimates for subduction erosion along the Chilean convergent margin, together with our compilation of published estimates, provide the basis for discussion of subduction erosion in space and time and, also, for comparison with other margins. Complemented by published rates of magmatic addition, these data enable a quantitative evaluation of crustal balance.

### 10.8.1 Comparison of Subduction Erosion Rates along the North Chilean Margin with Those along the Peruvian and Central American Margins

It has long been recognized that the Peruvian and Central American margins have been mainly shaped by subduction erosion (e.g. Moore et al. 1980; von Huene and Lallemand 1990). For both these margins, estimates of subduction erosion from the offshore record are available allowing a direct quantitative comparison on the base of equivalent data and methodologies, with the north Chilean margin.

At the Peruvian margin in the Lima region, rates of subduction erosion were first estimated at 60 to 90 km$^3$ Myr$^{-1}$ km$^{-1}$ through geometric reconstruction (von Huene and Lallemand 1990; Lallemand et al. 1992). However, the signal produced by Nazca Ridge subduction was not separated in these early studies. Later, Ocean Drilling Programme Leg 112

**Table 10.4.** Volume fluxes along the south-central Chilean margin

| Profile | Latitude (deg S) | $d_w$ at trench (m)[a] | Thickness of sediment (m) | %acc/sub[b] | Time[c] (Myr) | Accreted volume (km$^3$) | Average slope[d] (deg) |
|---|---|---|---|---|---|---|---|
| 728 | 35.95 | 4950 | 2400 | 69/31 | 0.69 | 92 | 2.45 |
| SO161-22 | 36.16 | 4900 | 2400 | | | | 2.44 |
| 727 | 36.33 | 4900 | 2000 | 50/50 | 0.64 | 51 | 2.33 |
| SO161-46 | 37.12 | 4690 | 2400 | | | | 2.32 |
| 730 | 37.92 | 4700 | 2100 | 40/60 | 1.27 | 87 | 2.91 |
| SO141-38 | 38.00 | 4620 | 2000 | | | | 2.92 |
| 732 | 39.25 | 4400 | 1600 | 19/81 | 3.25 | 79 | 1.81 |
| SO141-40 | 43.50 | 3800 | 3000 | | | | 2.69 |
| 734 | 44.50 | 3600 | 3400 | 20/80 | 4.01 | 219 | 2.47 |
| 769 | 47.50 | 3750 | 3300 | 60/40 | 2.35 | 374 | 2.59 |

[a] $d_w$: Water depth at the trench.
[b] Percentage of accreted and subducted material.
[c] Time to build up the accretionary wedge.
[d] Average slope from the trench to the coast.
Estimates based on information in Diaz-Naveas (1999), Behrmann and Kopf (2001), Rauch (2005) and GINA bathymetry-topography data (Lindquist 2004).

drill-hole records (Suess and von Huene 1988; Resig 1990) and the use of deep-sea submersibles (Sosson et al. 1994) have provided information on subsidence with time and wide-angle data (Krabbenhöft et al. 2004) provide the detailed geometry and velocity structure of the margin. Together, these data enabled to extract the long-term background rate of subduction erosion, which is as low as about 35 to 40 km$^3$ Myr$^{-1}$ km$^{-1}$, by subtracting the enhancement of subduction erosion caused by the passage of the Nazca Ridge from the bulk signal (Clift et al. 2003). Subduction erosion at the Costa Rica margin since 17–16 Ma was estimated at 34 to 36 km$^3$ Myr$^{-1}$ km$^{-1}$ (Vannucchi et al. 2001), whereas at the Guatemala margin, subduction erosion at rates of 11.3 to 13.1 km$^3$ Myr$^{-1}$ km$^{-1}$ since 19 Ma (Vannucchi et al. 2004) is considerably slower.

In conclusion, a comparison of these data with the published and newly derived data for the Chilean margin reveals that subduction erosion is more efficient along the north Chilean margin than at the Peruvian and Central America margins. It belongs to the highest rates of subduction erosion worldwide (Clift and Vannucchi 2004).

### 10.8.2 Variability of Subduction Erosion at the Chilean Margin and Net Crustal Balance

#### 10.8.2.1 Fate of the Material Lost through Subduction Erosion

Along the north Chilean margin, the eroded material mainly consists of detritus from Paleozoic basement and Mesozoic arc detritus, which is transported and deposited in the trench. Most of the probably high porosity is lost when this former trench-fill is compacted in the subduction channel. Material originating from the base of the fore-arc – including the mafic Jurassic arc basement – has more than 20 km of overburden and may be already characterized by very low porosity when it is incorporated into the subduction channel. High-pressure metamorphism of these materials leads to rocks that become denser than average during their passage through the subduction channel. Given that there is no compelling evidence for the accretion of subduction-channel material to the base of the fore-arc (e.g. Pelz 2000; Victor et al. 2004) – with exception of the above noted peninsulas –, subduction-erosion-derived material might well be returned to the mantle. Therefore, the loss of continental crust could equal the rates of subduction erosion at this latitude.

#### 10.8.2.2 Fluctuations of Subduction Erosion with Time

The available estimates suggest that subduction erosion not only varies along the central Andean margin, but may also fluctuate in time. At the Peruvian margin, subduction erosion seems to have accelerated since the middle Miocene at latitudes that have either not been affected by Nazca Ridge subduction, or where the enhancement of subduction erosion caused by ridge-margin collision has been separated and subtracted from the bulk rate.

At 20° S, subduction erosion might have accelerated since 6 Ma (Pelz 2000) compared to middle Miocene rates. Long-term (~70 Ma) rates of continental margin retreat and arc retreat seem to be about equal at 1 to 1.5 km Myr$^{-1}$ (Table 10.2). Extracting the longitudinal position of the volcanic arc front over time for 1° latitudinal corridors between 18 and 28° S (cf. Trumbull et al. 2006, Chap. 2 of this volume, and database on DVD) yields average rates of arc migration that are considerably higher for the Neogene than for the entire Cenozoic (Fig. NK-DVD-1a/b). Similar observations have been reported by Kay et al. (2005) for the northernmost Southern Volcanic Zone between 33 and 36° S.

In comparison, at the central Peruvian margin, 79 to 86 km of eastward migration of the magmatic arc since 60 Ma implies a rate of 1.3 to 1.4 km Myr$^{-1}$ (Cobbing 1999). As 70 km of this amount have only occurred since 38 Ma, this would indicate that the rate has accelerated to 1.8 km Myr$^{-1}$ since then. The available data also slightly imply a further increase in migration rates to about 2 km Myr$^{-1}$ since 11 Ma.

In summary, although our interpretation has to be looked at with caution due to the small number and low temporal resolution of the subduction-erosion estimates, the available data are compatible with an increase of subduction erosion and, therefore, the acceleration of arc migration in the Neogene. Convergence between the Nazca and South American Plates, however, decelerated during the Neogene (Norabuena et al. 1999). Consequently, we conclude that at the South American margin, high convergence rates do not correlate with faster subduction erosion, which is opposite to the observation of Clift and Vannucchi (2004). They, however, did not consider transient boundary conditions in their parameter analysis.

Periods of subduction erosion have also occurred at the south-central Chilean margin since the late Paleozoic. From arc-migration data and fore-arc basin subsidence, a Miocene to early Pliocene phase of subduction erosion is most conspicuous, with rates matching those observed in northern Chile. The change in fore-arc kinematics, surface motion and the switch from subduction erosion to accretion during the Pliocene are well correlated with the drastic increase in glacially controlled sediment influx. Hence, the south-central Chilean case provides a particularly illuminating example of the most relevant factors controlling subduction erosion and accretion.

While no threshold for sediment thickness can, as yet, be identified from our data, it is clear that doubling the sediment thickness from < 1 km to > 2 km within one to two million years entailed a fundamental change in both the mass-flux mode and fore-arc kinematics. Interestingly, there was a slightly delayed response from the upper plate. Redistributing the trench sediment over the entire seismogenic plate interface, thereby modifying its mechanical properties, requires some two million years at the present convergence rate of 65 mm yr$^{-1}$. While this may explain the delayed kinematic response of the fore-arc system, it does not explain the temporal increase of the portion of trench fill accreted at the tip of the plate and the formation of an accretionary wedge. It, therefore, would appear that changes between frontal and basal accretion are dependent on additional factors and not on sediment flux only.

### 10.8.3 The Relative Contribution of Frontal, Basal, and Mantle Erosion to Subduction Erosion

From our data, it becomes obvious that the estimate of subduction erosion from the migration of the magmatic arc front is higher than the estimates obtained from the offshore record for the entire Neogene period (Table 10.2). The two- to threefold higher rates of crustal subduction erosion calculated from the arc retreat (90–140 km$^3$ Myr$^{-1}$ km$^{-1}$), when compared to those from the offshore data (30–50 km$^3$ Myr$^{-1}$ km$^{-1}$) over the past ten million years, indicates that the assumptions underlying the arc-retreat approach may not always be substantiated. The only exception is the coinciding rates calculated for the collision site of the Juan Fernandez Ridge with South America (see below).

Between 20 and 24° S, the rates of crustal subduction erosion estimated with different approaches are quite similar to each other when compared over very long time periods. Whereas the offshore fore-arc contributes 37 to 50 km$^3$ Myr$^{-1}$ km$^{-1}$ of subducted volume, the average rates of crustal loss through subduction erosion estimated from the migration of the magmatic arc front since the Jurassic are 37 to 54 km$^3$ Myr$^{-1}$ km$^{-1}$ (we did not include the von Huene et al. (1999) estimate in this evaluation as it has been found to be erroneous (e.g. Pelz 2000)). However, the studies just mentioned used different time intervals in their estimates of subduction erosion.

During the Neogene, migration of the magmatic arc front became faster (= 3 km Myr$^{-1}$ versus an average of 1–1.5 km Myr$^{-1}$), implying that the estimate of subduction erosion from this migration could also be higher if only this time interval had been considered. Since the depth of the volcanic front to the present-day slab is around the global average of 110 km everywhere (England et al. 2004), and since there is no available evidence to show mantle cooling (required to accelerate eastward arc migration), we conclude that the slab geometry must have evolved to a slightly shallower geometry since the middle Miocene, thus increasing the fraction of fore-arc mantle removal. This however, is on the order of magnitude of between 0 and 17 km$^3$ Myr$^{-1}$ km$^{-1}$ and therefore considerably smaller than the amount of crust lost by subduction erosion.

Transects across the collision zone of the Juan Fernandez Ridge show some differences in the rates of subduction erosion. Because the distance between the sections is as large as 300 km, these differences could reflect local variations in the efficiency of subduction erosion. Compared to the north, subduction erosion is at least twice as fast at the site of Juan Fernandez Ridge collision site.

The seafloor seaward of the trench at these latitudes is considerably smoother than that further to the north. From this we infer that the potential importance of basal subduction erosion may well have increased by the subduction of the Juan Fernandez Ridge. Neogene rates of magmatic-arc-front migration are higher than in northern Chile (Kay et al. 2005), implying that, in addition, mantle erosion may be an important contributor to subduction erosion here. Compared to the areas north and south, the coincident results derived from all methods at this latitude with respect to the crustal component indicate a stable subduction geometry over the Neogene period.

### 10.8.4 Subduction Erosion Versus Magmatic Addition

The antagonist to subduction erosion is magmatic addition, which has been estimated to have added 20 to 40 km$^3$ Myr$^{-1}$ km$^{-1}$ of material to the Earth's convergent margins (Kay 1980, Reymer and Schubert 1984). Recent studies indicate that magmatic addition along some of the west Pacific margins (Dimalanta et al. 2002) and the Aleutian margin (Holbrook et al. 1999) is significantly faster, but magmatic addition at several latitudes along the central Andean margin is distinctly lower (Fig. 10.2).

For the Peruvian margin at about 9° S, Atherton and Petford (1996) estimated magmatic addition at 8 km$^3$ Myr$^{-1}$ km$^{-1}$, averaged since 100 Ma. Rates of magmatic addition for latitudes of 21 to 22°S have been estimated at 4 km$^3$ Myr$^{-1}$ km$^{-1}$ (Francis and Rundle 1976), 12.7 km$^3$ Myr$^{-1}$ km$^{-1}$ since 10 Ma (Francis and Hawkesworth 1994) and 35 km$^3$ Myr$^{-1}$ km$^{-1}$ for the Eocene arc (Haschke and Günther 2003). The last value may be at the high end, but for longer periods that also include episodes of magmatic quiescence average rates of magmatic addition could be lower. Given that the available data may be regarded as representative throughout the Cenozoic, magmatic addition would be significantly slower than subduction erosion. Therefore, we conclude that the central Andean margin is undergoing a net loss of continental crust at rates of 10 to 40 km$^3$ Myr$^{-1}$ km$^{-1}$.

### 10.8.5 Hydrofracturing As a Possible Cause of Basal Subduction Erosion

During transport within the subduction channel, fluids expelled due to compaction or released through metamorphic reactions can considerably elevate the fluid pressure, which then may also change the mechanics of a convergent margin (e.g. Davis et al. 1983; Mourgues and Cobbold 2006). Records from fossil subduction channels confirm that hydrofracturing is probably a very common and widespread process within subduction channels (e.g. Fisher 1996 and references therein). Building up of highly elevated fluid pressures leads to very low effective strength that may satisfy the Mohr-Coulomb criteria for fracturing. This finally results in a strongly increased potential for faulting and mechanical weakening.

Related mechanisms such as seismic pumping (Sibson et al. 1975), episodic high-pressure flow (Byerlee 1993) and pore-space reduction (Miller et al. 1996) have been proposed to help explain earthquake processes. Such processes could also explain basal subduction erosion as fluid pressure can significantly increase in the overburden when the interplate thrust is locked and probably acts as a permeability barrier (Byerlee 1993, Husen and Kissling 2001). This strongly weakens the fault and when hydraulic fracturing occurs near the base of the continental fore-arc crust, it decreases the strength contrast between the upper plate and subduction-channel material and may result in the efficient detachment of material (of variable size). Such fragments may be transported within the subduction channel and, therefore, possibly increase its width.

Using an average rate of 80 mm yr$^{-1}$ and subduction erosion rates for the entire offshore fore-arc of between 37 and 50 km$^3$ Myr$^{-1}$ km$^{-1}$, a layer with a minimum thickness of 460 to 625 m needs to be transported within the subduction channel for the material velocity to equal subduction velocity. A possible increase in the porosity of frontally eroded material may add some 100 m more to the subduction channel near the trench. Recent observations on the thickness of the zone of aftershock seismicity following subduction-related earthquakes at the Chilean margin, conversely indicate that the kinematically-defined subduction channel may be three and more kilometers wide. This coincides with reflection bands of equivalent thickness in deeper parts of the subduction channel beneath the onshore fore-arc (ANCORP Working Group 2003).

Sandbox experiments revealed that rates of transport within the subduction channel mostly do not equal lower-plate velocity, but are considerably lower (Lohrmann et al. 2006, Chap. 11 of this volume). As there is a continuous supply of material from the trench and the base of the fore-arc, the lower bulk velocity of the subduction-channel material requires that the thickness of the channel increases, especially in the absence of basal accretion of subduction-channel material. Hence, the thinner eroded layer and the thicker observed subduction channel are consistent with each other.

## 10.9 Conclusions

Our compilation of the published rates of subduction erosion at the Chilean convergent margin with the additional estimates from this study now enable us to derive a relatively detailed picture of how subduction erosion is affecting this margin. As subduction erosion is quantified by employing several different approaches, which themselves are based upon different types of data and over different periods of time, care needs to be taken when comparing rates of subduction erosion. The main conclusions of our analysis follow.

Estimates of crustal loss derived by estimating the rates of subduction erosion using the offshore subsidence record and margin geometry are robust. Therefore, these estimates enable us to also address the potential regional variability of subduction erosion.

In northern Chile, net crustal loss through subduction erosion is about 40–45 km$^3$ Myr$^{-1}$ km$^{-1}$ and is equally as fast as, or faster than, that at the Peruvian and Central American margins. As the rates of magmatic addition are distinctly lower, the central Andean margin is a site of net loss of continental crust.

In the region affected by the collision of the Juan Fernandez Ridge with the South American margin, subduction erosion since the Miocene has taken place at rates of about 100 km$^3$ Myr$^{-1}$ km$^{-1}$, which is at least twice as fast as that in northern Chile. The increase of subduction erosion is most probably caused by the Juan Fernandez Ridge, however, its effect is significantly smaller than those caused by the larger Nazca and Chile Ridges.

The south-central Chilean margin was erosive prior to about 3 Ma. At this time, dramatically increased, glacial, terrigenous flux led to the recent accretion of a frontal prism. We found subduction erosion to be as fast as about 30 km$^3$ Myr$^{-1}$ km$^{-1}$ between 11 and 3 Ma, which is about 75% of the rates in northern Chile.

The contribution of basal subduction erosion seems to be very important along the Chilean margin, with the subduction of topographic features enhancing it. Hydrofracturing at the base of the fore-arc crust is proposed as a mechanism of basal subduction erosion.

### Acknowledgments

This study has been funded by the DFG within the framework of SFB 267, TP F1. We thank many of our SFB colleagues, especially H. Echtler, T. Vietor, M. Haschke, and B. Trumbull for discussion and feedback on our hypotheses. The GINA data set (Lindquist 2004) has been used for the topographic maps. Figures 10.2, 10.4 and NK-DVDa/b were produced with the GMT software (Wessel and Smith 1998). We are grateful to A. Britt for improving the language of our manuscript. The manuscript benefitted from careful and insightful reviews by A. Krabbenhöft and K.-J. Reutter.

# References

Adam J, Reuther C-D (2000) Crustal dynamics and active fault mechanics during subduction erosion. Application of frictional wedge analysis to the north Chilean forearc. Tectonophysics 321: 297–325

Adam J, Kukowski N, Lohrmann J (2000) Mechanics and mass transfer patterns of tectonic erosive convergent margins: quantitative results from analogue sandbox models and their implications for natural forearc systems. AGU Fall Meeting, EOS 81/F

Adriasola AC, Thomson SN, Brix MR, Hervé F, Stöckhert B (2005) Postmagmatic cooling and late Cenozoic denudation of the North Patagonian in the Los Lagos region of Chile, 41°S to 42.15°S. Int J Earth Sci, doi 10.1007/s00531-005-0027-9

Allmendinger RW, Figueroa D, Snyder D, Beer J, Mpodozis C, Isacks BL (1990) Foreland shortening and crustal balancing in the Andes at 30°S latitude. Tectonics 9:789–809

ANCORP Working Group (2003) Seismic imaging of a convergent continental margin and plateau in the central Andes (Andean Continental Research Project 1996, ANCORP96). J Geophys Res 108(B7): doi 10.1029/2002JB001771

Angermann D, Klotz J, Reigber C (1999) Space-geodetic estimation of the Nazca-South America Euler vector. Earth Planet Sci Lett 171:329–334

Atherton MP, Petford N (1996) Plutonism and the growth of Andean crust at 9°S from 100 to 3 Ma. J S Am Earth Sci 9:1–9

Bangs NL, Cande SC (1997) Episodic development of a convergent margin inferred from structures and processes along the southern Chile margin. Tectonics 16:489–503

Behrmann JH, Kopf A (2001) Balance of tectonically accreted and subducted sediment at the Chile Triple Junction. Int J Earth Sci 90:753–768

Bourgois J, Martin H, Lagabrielle Y, Le Moigne J, Frutos Jara J (1996) Subduction erosion related to spreading-ridge subduction: Taitao peninsula (Chile margin triple junction area). Geology 24: 723–726

Byerlee JD (1993) Model for episodic flow of high-pressure water in fault zones before earthquakes. Geology 21:303–306

Byrne DE, Wang W-H, Davis DM (1993) Mechanical role of backstops in the growth of forearcs. Tectonics 12:123–144

Charlton TR (1988) Tectonic erosion and accretion in steady-state trenches. Tectonophysics 149:233–243

Clift PD, MacLeod CL (1999) Slow rates of subduction erosion estimated from subsidence and tilting of the Tonga forearc. Geology 27:411–414

Clift PD, Vannucchi P (2004) Controls on tectonic accretion versus erosion in subduction zones: implications for the origin and recycling of the continental crust. Rev Geophys 42: doi 10.1029/2003RG000127

Clift PD, Pecher IA, Kukowski N, Hampel A (2003) Tectonic erosion of the Peruvian forearc, Lima Basin, by subduction and Nazca Ridge collision. Tectonics 22(3) doi 10.1029/2002TC001386

Cobbing EJ (1999) The Coastal Batholith and other aspects of Andean magmatism in Peru. In: Castro A, Fernandez C, Vigneresse JL (eds) Understanding granites: integrating new and classical techniques. Geol Soc Lond Spec Pub 168:111–122

Davis D, Suppe J, Dahlen FA (1983) Mechanics of fold-and-thrust Belts and accretionary wedges. J Geophys Res 88:1153–1172

Diaz-Naveas JL (1999) Sediment subduction and accretion at the Chilean convergent margin between 35° and 40°S. PhD thesis, Christian-Albrechts-Universität Kiel

Dimalanta C, Taira A, Yumul GP, Tokuyama H, Mochizuki K (2002) New rates of western Pacific island arc magmatism from seismic and gravity data. Earth Planet Sci Lett 202:105–115

Dominguez S, Lallemand SE, Malavieille J, von Huene R (1998) Upper plate deformation associated with seamount subduction. Tectonophysics 293:207–224

Ellis S, Schreurs G, Panien M (2004) Comparisons between analogue and numerical models of thrust wedge development. J Struct Geol 26:1659–1675

Encinas A, Finger K, Nielsen S, Lavenu A, Buatois L, Peterson D (2005) Late Miocene coastal subsidence in Central Chile: tectonic implications. 6th International Symposium of Andean Geodynamics, Barcelona, IRD, pp 246–249

England P, Engdahl R, Thatcher W (2004) Systematic variation in the depths of slabs beneath arc volcanoes. Geophys J Int 156: 377–408

Fisher DM (1996) Fabrics and veins in the forearc: a record of cyclic fluid flow at depths of <15 km. In: Bebout GE, Scholl DW, Kirby SH, Platt JP (1996) Subduction – top to bottom. AGU Geophysical Monograph 96:75–90

Flueh ER, Vidal N, Ranero CR, Hojka A, von Huene R, Bialas J, Hinz K, Cordoba D, Dañobeitia, Zelt C (1998) Seismic investigation of the continental margin off- and onshore Valparaiso, Chile. Tectonophysics 288:251–263

Francis PW, Hawkesworth CJ (1994) Late Cenozoic rates of magmatic activity in the Central Andes and their relationships to continental crust formation and thickening. J Geol Soc Lond 151:845–854

Francis PW, Rundle CC (1976) Rates of production of the main Andean magma types. Geol Soc Am Bull 87:474–480

Fruehn J, von Huene R, Fisher MA (1999) Accretion in the wake of terrane collision: the Neogene accretionary wedge off Kenai Peninsula. Tectonics 18:263–277

Glodny J, Echtler H, Figueroa O, Franz G, Gräfe K, Kemnitz H, Kramer W, Krawczyk C, Lohrmann J, Lucassen F, Melnick D, Rosenau M, Seifert W (2006) Long-term geological evolution and mass-flow balance of the South-Central Andes. In: Oncken O, Chong G, Franz G, Giese P, Götze H-J, Ramos VA, Strecker MR, Wigger P (eds) The Andes – active subduction orogeny. Frontiers in Earth Science Series, Vol 1. Springer-Verlag, Berlin Heidelberg New York, pp 401–428, this volume

González E (1989) Hydrocarbon resources in the coastal zone of Chile. In: Ericksen GE, Cañas Pinochet MT, Reinemund JA (eds) Geology of the Andes and its relation to hydrocarbon and mineral resources. Circum-Pacific Council for Energy and Mineral Resources Earth Science Series 11:383–404

Gorring ML, Kay SM, Zeitler PK, Ramos VA, Rubiolo D, Fernandez ML, Panza JL (1997) Neogene Patagonian plateau lavas: continental magmas associated with ridge collision at the Chile triple junction. Tectonics 16:1–17

Goy JL, Macharé J, Ortlieb L, Zazo C (1992) Quaternary shorelines in southern Peru: a record of global sea-level fluctuations and tectonic uplift in Chala Bay. Quaternary International 15/16:99–112

Grotzki NR, Flueh ER, Reichert CJ, Patzwahl R, Mechie J, Giese P (1998) Modeling of seismic wide angle refraction/reflection data. In: Hinz K (ed) Crustal investigations off- and onshore Nazca/Central Andes (CINCA) Rep BGR 117.613:69–102

Gulick SPS, Bangs NLB, Shipley TH, Nakamura Y, Moore G, Kuramoto S (2004) Three-dimensional architecture of the Nankai accretionary prism's imbricate thrust zone off Cape Mutoro, Japan: prism reconstruction via en echelon thrust propagation. J Geophys Res 109: doi 10.1029/2003JB002654

Gutscher M-A, Kukowski N, Malavieille J, Lallemand SE (1998a) Material transfer in accretionary wedges from analysis of a systematic series of analog experiments. J Struct Geol 20:407–416

Gutscher M-A, Kukowski N, Malavieille J, Lallemand SE (1998b) Episodic imbricate thrusting and underthrusting: analogue experiments and mechanical analysis applied to the Alaskan accretionary wedge. J Geophys Res 103:10161–10176

Hartley AJ, Jolley EJ (1995) Tectonic implications of Late cenozoic sedimentation from the Coastal Cordillera of northern Chile (22–24°S). J Geol Soc Lond 152:51–63

Hartley AJ, May G, Chong G, Turner P, Kape SJ, Jolley EJ (2000) Development of a continental forearc: A Cenozoic example from the Central Andes, northern Chile. Geology 28:331–334

Hartley AJ, Chong G, Houston J, Mather AE (2005) 50 million years of climatic stability: evidence from the Atacama Desert, northern Chile. J Geol Soc London 162:421–424

Haschke M, Günther A (2003) Balancing crustal thickening in arc by tectonic vs. magmatic means. Geology 31:933–936

Hilde TWC (1983) Sediment subduction versus accretion around the Pacific. Tectonophysics 99:381–399

Holbrook WS, Lizzaralde D, McGeary S, Bangs N, Diebold J (1999) Structure and composition of the Aleutian island arc and implications for crustal growth. Geology 27:31–34

Hsu JT (1992) Quaternary uplift of the Peruvian coast related to the subduction of the Nazca Ridge: 13.5° S to 15.6° S. Quat Int 15–16:87–97

Hubbert MK (1937) Theory of scale models as applied to the study of geological structures. Geol Soc Am Bull 48:1459–1520

Husen S, Kissling E (2001) Postseismic fluid flow after the large subduction earthquake of Antofagasta, Chile. Geology 29(9):847–850

Jarrard RD (1986) Relations among subduction parameters. Rev Geophys 24:217–284

Jordan TE, Burns WM, Veiga R, Pángaro F, Copeland P, Kelley S, Mpodozis C (2001) Extension and basin formation in the southern Andes caused by increased convergence rate: a mid-Cenozoic trigger for the Andes. Tectonics 20:308–324

Kay RW (1980) Volcanic arc magmas: implications of a melting-mixing model for element recycling in the crust-upper mantle system. J Geol 88:497–522

Kay SM, Godoy E, Kurtz A (2005) Episodic arc migration, crustal thickening, subduction erosion, and magmatism in the south-central Andes. Bull Geol Soc Am 117(1/2):67–88

Kopp H, Kukowski N (2003) Backstop geometry and accretionary mechanics of the Sunda margin. Tectonics 22: doi 1029/2002TC001420

Kopp C, Fruehn J, Flueh ER, Reichert C, Kukowski N, Bialas J, Klaeschen D (2000) Structure of the Makran subduction zone from wide-angle and reflection data. Tectonophysics 329:171–191

Kopp H, Flueh ER, Papenberg C, Klaeschen D (2004) Seismic investigations of the O'Higgins Seamount Group and Juan Fernandez Ridge: aseismic ridge emplacement and lithosphere hydration. Tectonics 23: doi 10.1029/2003TC001590

Krabbenhöft A, Bialas J, Kopp H, Kukowski N, Hübscher C (2004) Crustal structure of the Peruvian continental margin from wide-angle seismic studies. Geophys J Int 159:749–964

Krawczyk CM, Mechie J, Lüth S, Tašárová Z, Wigger P, Stiller M, Brasse H, Echtler HP, Araneda M, Bataille K (2006) Geophysical signatures and active tectonics at the south-central Chilean margin. In: Oncken O, Chong G, Franz G, Giese P, Götze H-J, Ramos VA, Strecker MR, Wigger P (eds) The Andes – active subduction orogeny. Frontiers in Earth Science Series, Vol 1. Springer-Verlag, Berlin Heidelberg New York, pp 171–192, this volume

Kudrass HR, Von Rad U, Seyfried H, Andruleit H, Hinz K, Reichert C (1998) Age and facies of sediments of the northern Chilean continental slope – Evidence for intense vertical movements. In: Hinz K (ed) Crustal investigations off- and onshore Nazca/Central Andes (CINCA). Bundesanstalt für Geowissenschaften und Rohstoffe, Report 117.613:170–196

Kukowski N, von Huene R, Malavieille J, Lallemand SE (1994) Sediment accretion against a buttress beneath the Peruvian continental margin as simulated with sandbox modeling. Geol Rundschau 83:822–831

Kukowski N, Adam J, Lohrmann J (1999) Erosive mass transfer at convergent margins: constraints from analog models and application of Coulomb wedge analysis. ISAG 99:400–404

Kukowski N, Lohrmann J, Hampel A (2004a) Visage of a tectonically erosive margin: interplay of geophysical data and analogue modelling. Nordisk Geologisk Vinter Møte, Uppsala

Kukowski N, Hampel A, Bialas J, Huebscher C (2004b) Transtensional tectonics caused by subduction erosion at the offshore Peruvian margin. AGU Fall Meeting, San Francisco CA

Kulm LD, Schweller WJ, Masias A (1977) A preliminary analysis of the subduction process along the Andean continental margin near 6° to 45° S. In: Talwani, Pitman WC III (eds) Island arcs, deep-sea trenches, and back-arc Basins, AGU Maurice Ewing Series 1, pp 285–301

Lallemand SE, Culotta R, von Huene R (1989) Subduction of the Daiichi Kashima seamount in the Japan trench. Tectonophysics 160:231–247

Lallemand SE, Schnürle R, Manoussis S (1992) Reconstruction of subduction zone paleogemoetries and quantification of upper plate material losses caused by tectonic erosion. J Geophys Res 97:217–239

Lallemand SE, Schnürle P, Malavieille J (1994) Coulomb theory applied to accretionary and nonaccretionary wedges: possible causes for tectonic erosion and/or frontal accretion. J Geophys Res 99:12033–12055

Laursen J, Scholl DW, von Huene R (2002) Neotectonic deformation of the central Chile margin deepwater forearc basin formation in response to hot spot ridge and seamount subduction. Tectonics 21(3): doi 1029/2001TC901023

Lindquist KG (2004) Global topography and bathymetry grid improves research efforts. EOS 85(19):186–187

Lohrmann J (2002) Identification of parameters controlling the accretive and erosive mass transfer mode at the South-Central and North Chilean forearc using 2D scaled sandbox experiments. GFZ Scientific Technical Reports STR02/10

Lohrmann J, Kukowski N, Adam J, Oncken O (2003) The impact of analogue material properties on the geometry, kinematics, and dynamics of convergent sand wedges. J Struct Geol 25:1691–1711

Lohrmann J, Kukowski N, Krawczyk CM, Oncken O, Sick C, Sobiesiak M, Rietbrock A (2006) Subduction channel evolution in brittle forearc wedges – a combined study with scaled sandbox experiments, seismological and reflection seismic data and geological field evidence. In: Oncken O, Chong G, Franz G, Giese P, Götze H-J, Ramos VA, Strecker MR, Wigger P (eds) The Andes – active subduction orogeny. Frontiers in Earth Science Series, Vol 1. Springer-Verlag, Berlin Heidelberg New York, pp 237–262, this volume

Lonsdale P (2005) Creation of the Cocos and Nazca plates by fission of the Farallon plate. Tectonophysics 404:237–264

Maacharé J, Ortlieb L (1992) Plio-Quaternary vertical motions and the subduction of the Nazca Ridge, central coast of Peru. Tectonophysics 205:97–108

Malavieille J (1984) Modélisation expérimentale des chevauchements imbriqués: application aux chaines de montagnes. Bull Soc Géol France 7:129–138

Marone C (1998) Laboratory-derived friction laws and their application to seismic faulting. Ann Rev Earth Planet Sci 26:643–696

Marquardt A, Lavenu A, Ortlieb L, Godoy E, Comte D (2004) coastal neotectonics in southern Central Andes: uplift and deformation of marine terraces in Northern Chile (27°S). Tectonophysics 394:193–219

Masek JG, Duncan CC (1998) Minimum-work mountain building. J Geophys Res 103(B1):907–918

Melnick D, Echtler HP (2006) Inversion of forearc basins in south-central Chile caused by rapid glacial age trench fill. Geology, Sep. 2006 issue

Miller SA, Nur A, Olgaard DL (1996) Earthquakes as a coupled shear stress-high pore pressure dynamical system. Geophys Res Lett 23:197–200

Moore JC, Biju-Duval B (1984) Tectonic synthesis, Deep Sea Drilling Program Leg 78A: Structural evolution of offscraped and underthrust sediment, northern Barbados ridge complex. Deep Sea Drilling Program Initial Reports 78:601–621

Moore GF, Shipley TH, Lonsdale PT (1980) Subduction erosion versus sediment offscraping at the toe of the Middle America trench off Guatemala. Tectonics 5:513–531

Mordojovich C (1981) Sedimentary basins of Chilean Pacific offshore. In: Halbouty MT (ed) Energy resources of the Pacific region. AAPG Studies in Geology 12:63–82

Mourgues R, Cobbold PR (2006) Thrust wedges and fluid overpressures: sandbox models involving pore fluid, J Geophys Res 111: doi 10.1029/2004JB003441

Norabuena EO, Leffler-Griffin L, Mao A, Dixon T, Stein S, Sacks IS, Ocola L, Ellis M (1998) Space geodetic observations of Nazca-south America convergence across the Central Andes. Science 279:358-396

Norabuena EO, Dixon TH, Stein S, Harrison CGA (1999) Decelerating Nazca-South America and Nazca-Pacific plate motion. Geophys Res Lett 26(22):3405-3408

Oncken O, Hindle D, Kley J, Elger K, Victor P, Schemmann K (2006) Deformation of the central Andean upper plate system – facts, fiction, and constraints for plateau models. In: Oncken O, Chong G, Franz G, Giese P, Götze H-J, Ramos VA, Strecker MR, Wigger P (eds) The Andes – active subduction orogeny. Frontiers in Earth Science Series, Vol 1. Springer-Verlag, Berlin Heidelberg New York, pp 3–28, this volume

Ortlieb L, Ghaleb B, Hillaire-Marcel C, Goy JL, Machare J, Zazo C, Thiele R (1995) Quaternary vertical deformation along the southern Peru – northern Chile coast: marine terrace data. XIV INQUA Congress Berlin 1995, Terra Nostra 2(95):205

Ortlieb L, Zazo C, Goy JL, Hillaire-Marcel C, Ghaleb B, Cournoyer L (1996a) Coastal deformation and sea-level changes in the northern Chile subduction area (23°S) during the last 330 ky. Quat Sci Rev 15:819 – 831

Ortlieb L, Zazo C, Goy JL, Dabrio C, Macharé J (1996b) Pampa del Palo: an anomalous composite marine terrace on the uprising coast of southern Peru. J S Am Earth Sci 9:367–379

Ota Y, Paskoff R (1993) Holcene deposits on the coast of north-central Chile: Radiocarbon ages and implications for coastal changes. Rev Géologica de Chile 20:25–32

Ota Y, Miyauchi T, Paskoff R, Koba M (1995) Plio-Quaternary marine terraces and their deformation along the Altos de Talinay, north-Central Chile. Rev Géologica de Chile 22:89–102

Patzwahl R, Mechie J, Schulze A, Giese P (1999) Two-dimensional velocity models of the Nazca plate subduction zone between 19.5°S and 25°S from wide-angle seismic measurements during the CINCA95 project. J Geophys Res 104:7293–7317

Pelz K (2000) Tektonische Erosion am zentralandinen Forearc (20°– 24°S). GDZ Scientif Technical Report STR 00/20

Rabassa J, Clapperton CM (1990) Quaternary glaciations of the southern Andes. Quat Sci Rev 9:153–174

Ranero CR, von Huene R (2000) Subduction erosion along the Middle America convergent margin. Nature 404:748–752

Rauch K (2005) Cyclicity of Peru–Chile trench sediments between 36°S and 38°S: a footprint of paleoclimatic variations? Geophys Res Lett 32: doi 10.1029/2004GL022196

Resig J (1990) Benthic foraminiferal stratigraphy and paleoenvironments off Peru, Leg 112. Proceedings of the Ocean Drilling Program, Scientific Results 112:263–296

Reymer A, Schubert G (1984) Phanerozoic addition rates to the continental crust and crustal growth. Tectonics 3:63–77

Le Roux JP, Tavares Correa C, Alayza F (2000) Sedimentology of the Rímac-Chillón alluvial fan at Lima, Peru, as related to Plio-Pleistocene sea-level changes, glacial cycles and tectonics. J S Am Earth Sci 13:499–510

Le Roux JP, Gómez C, Venegas C, Fenner J, Middleton H, Marchant M, Buchbinder A, Frassinetti D, Marquardt C, Gregory-Wodzicki KM, Lavenu A (2005) Neogene-Quaternary coastal and offshore sedimentation in north central Chile: record of sea-level changes and implications for Andean tectonism. J S Am Earth Sci 19:83–98

Rötzler K, Naumann R, Wilke H-G (1998) Tectonic erosion of terrigenous rocks in northern Chile (19°S and 24°S) In: Hinz K (ed) Crustal investigations off- and onshore Nazca/Central Andes (CINCA). Rep BGR 117.613

Rutland RWR (1971) Andean orogeny and ocean floor spreading. Nature 233:252–255

Sallares V, Ranero CR (2005) Structure and tectonics of the erosional convergent margin off Antofagasta, north Chile (23°30'S). J Geophys Res 110: doi 10.1029/2004JB003418

Scheuber E, Bogdanic T, Jensen A, Reutter K-J (1994) Tectonic development of the north Chilean Andes in relation to plate convergence and magmatism since the Jurassic. In: Reutter K-J, Scheuber E, Wigger PJ (eds) Tectonics of the Southern Central Andes. structure and evolution of an active continental margin. Springer-Verlag, Berlin Heidelberg New York, pp 121–139

Scholl DW, Christensen MN, von Huene R, Marlow MS (1970) Peru-Chile trench sediments and sea-floor spreading. Geol Soc Am Bull 81:1339–1360

Seifert W, Rosenau M, Echtler H (2005) Crystallization depths of granitoids of South Central Chile estimated by AL-in-hornblende geobarometry: implications for mass transfer processes along the active continental margin. N Jb Geol Paläont 236:115–127

Sibson RH (1992) Implications of fault valve behaviour for rupture nucleation and recurrence. Tectonophysics 211:283–293

Sick C, Yoon M-K, Rauch K, Buske S, Lüth S, Araneda M, Bataille K, Chong G, Giese P, Krawczyk C, Mechie J, Meyer H, Oncken O, Reichert C, Schmitz M, Shapiro S, Stiller M, Wigger P (2006) Seismic images of accretive and erosive subduction zones from the Chilean margin. In: Oncken O, Chong G, Franz G, Giese P, Götze H-J, Ramos VA, Strecker MR, Wigger P (eds) The Andes – active subduction orogeny. Frontiers in Earth Science Series, Vol 1. Springer-Verlag, Berlin Heidelberg New York, pp 147–170, this volume

Somoza R (1998) Updated Nazca (Farallon)-South America relative motions during the last 40 Myr: implications for mountain building in the central Andean region. J S Am Earth Sci 11:211–215

Sosson M, Bourgois J, Mercier de Lépinay B (1994) SeaBeam and deep-sea submersible Nautile surveys in the Chiclayo canyon off Peru (7°S): subsidence and subduction-erosion of an Andean-type convergent margin since Pliocene times. Mar Geol 118:237–256

Stern CR (1989) Pliocene to present migration of the volcanic front, Andean southern volcanic zone. Rev Geol Chile 16:145–162

Stern CR (1991) Role of subduction erosion in the generation of Andean magmas. Geology 19:78–81

Suess E, von Huene R (1988) Proceedings of the Ocean Drilling Program, Initial Reports 112. College Station TX

Thomson SN (2002) Late Cenozoic geomorphic and tectonic evolution of the Patagonian Andes between latitudes 42°S and 46°S: An appraisal based on fission-track results from the transpressional intra-arc Liquiñe-Ofqui fault zone. Geol Soc Am Bull 114:1159–1173

Trumbull RB, Riller U, Oncken O, Scheuber E, Munier K, Hongn F (2006) The time-space distribution of Cenozoic volcanism in the South-Central Andes: a new data compilation and some tectonic implications. In: Oncken O, Chong G, Franz G, Giese P, Götze H-J, Ramos VA, Strecker MR, Wigger P (eds) The Andes – active subduction orogeny. Frontiers in Earth Science Series, Vol 1. Springer-Verlag, Berlin Heidelberg New York, pp 29–44, this volume

Vannucchi P, Scholl DW, Meschede M, McDougall-Reid K (2001) Tectonic erosion and consequent collapse of the Pacific margin of Costa Rica: combined implications from ODP Leg 170, seismic offshore data, and regional geology of the Nicoya Peninsula. Tectonics 20(5):649–668

Vannucchi P, Galeotti S, Clift PD, Ranero CR, von Huene R (2004) Long-term subduction-erosion along the Guatemalan margin of the Middle America Trench. Geology 32:617–620

Victor P, Oncken O, Glodny J (2004) Uplift of the western Altiplano plateau: evidence from the Precordillera between 20° and 21°s (northern Chile). Tectonics 23: doi 10.1029/2003TC001519

Vietor T, Echtler H (2006) Episodic Neogene southward growth of the Andean subduction orogen between 30° S and 40° S – plate motions, mantle flow, climate, and upper-plate structure. In: Oncken O, Chong G, Franz G, Giese P, Götze H-J, Ramos VA, Strecker MR, Wigger P (eds) The Andes – active subduction orogeny. Frontiers in Earth Science Series, Vol 1. Springer-Verlag, Berlin Heidelberg New York, pp 375–400, this volume

Völker D, Wiedicke M, Ladage S, Gaedicke C, Reichert C, Rauch K, Kramer W, Heubeck C (2006) Latitudinal variation in sedimentary processes in the Peru-Chile trench off Central Chile. In: Oncken O, Chong G, Franz G, Giese P, Götze H-J, Ramos VA, Strecker MR, Wigger P (eds) The Andes – active subduction orogeny. Frontiers in Earth Science Series, Vol 1. Springer-Verlag, Berlin Heidelberg New York, pp 193–216, this volume

von Huene R, Culotta R (1989) Tectonic erosion at the front of the Japan trench convergent margin. Tectonophysics 160:70–90

von Huene R, Lallemand S (1990) Tectonic erosion along the Japan and Peru convergent margins. Geol Soc Am Bull 102: 704–720

von Huene R, Ranero CR (2003) Subduction erosion and basal friction along the sediment-starved convergent margin off Antofagasta. J Geophys Res 108(B2): doi 10.1029/2001JB001569

von Huene R, Scholl DW (1991) Observations at convergent margins concerning sediment subduction, subduction erosion, and the growth of continental crust. Rev Geophys 29:279–316

von Huene R, Weinrebe W, Heeren F (1999) Subduction erosion along the north Chile margin. J Geodynamics 27:345–358

von Huene R, Ranero CR, Vannucchi P (2004) Generic model of subduction erosion. Geology 32:913–916

Wessel P, Smith WHF (1998) New, improved version of the Generic Mapping Tools released. EOS 79(47):579

Willner A, Glodny J, Gerya TV, Gogoy E, Massone H-J (2004) A counter-clockwise pTt-path of high pressure-low temperature rocks from the coastal Cordillera accretionary complex of south Central Chile: constraints for the earliest stage of subduction mass flow. Lithos 75:283–310

Yañez G, Ranero C, von Huene R, Diaz J (2001) Manetic anomaly interpretation across the southern-central Andes (32°–34°S): the role of the Juan Fernandez ridge in the late Tertiary evolution of the margin. J Geophys Res 106:6325–6345

Zachos J, Pagani H, Sloan L, Thomas E, Billups K (2001) Trends, rhythms, and aberrations in global climate 65 Ma to present. Science 292:686–693

Zazo C, Ortlieb L, Goy JL, Macharé J (1994) Fault tectonics and crustal vertical motions on the coastal area of southern Peru. Bull INQUA Neotectonics Commission 17:31–33

# Subduction Channel Evolution in Brittle Fore-Arc Wedges – a Combined Study with Scaled Sandbox Experiments, Seismological and Reflection Seismic Data and Geological Field Evidence

Jo Lohrmann · Nina Kukowski · Charlotte M. Krawczyk · Onno Oncken · Christof Sick · Monika Sobiesiak · Andreas Rietbrock

**Abstract.** With a series of scaled sandbox experiments, we investigated the mass-flux patterns at the interface of convergent plates, with emphasis on the upper (brittle) part of subduction channels. Analysis of the particle displacement field integrated over short time periods shows that both types of simulated subduction channels (accretive and tectonically erosive) are characterized by episodically active thrusts (roof thrusts) at the top and a continuously active basal detachment. The short-term material flux reveals a complex temporal and spatial variability in the active mass-transfer processes within the subduction channel, and is particularly influenced by the activity of fore-arc structures (e.g. reactivation of backthrusts or duplexes). In the subduction channel, the localization of deformation also shows temporal and spatial fluctuations, which range from periodic kinematic cycles to unpredictable, apparently chaotic behaviour involving the activation and reactivation of shear zones. However, the location of the roof thrusts and their reactivation pattern during the periodic cycles is indicative of either tectonically erosive or accretive mass-transfer modes. In contrast to the short-term observations, the long-term material flux integrated over one kinematic cycle exhibited diagnostic patterns for the location of sediment accretion and subduction erosion. The series of accretive experiments shows that the combination of several parameters (initial wedge thickness, absence/presence of upper-crustal structures, and depth-dependent softening of the top of the subduction channel) can cause the same bulk effect in the upper plate (i.e., the migration of the center of uplift as an indicator of the position of rearward accretion). This experimental result precludes the determination of controlling parameters in nature.

We demonstrate that comparison of the subduction channel-related structures detected in the sandbox experiments with the structures of the accretive south-central Chilean fore-arc (37–38° S) and the tectonically erosive north Chilean fore-arc (21–24° S) is possible, when the restrictions of the analogue experiments (strongly idealized set-up and simplified material behaviour) and observational methods for nature (observation time window, spatial resolution) are taken into account. We show that the differences between analogue models simulating accreting and tectonically eroding subduction channels can be applied to distinguish these end-member types of subduction channels in nature. In contrast to the data from the analogue simulations, the reflection seismic profiles and the seismological data only reveal the geometry of the currently active subduction channel, which might have fluctuated over time. The results imply that subduction channels in nature could have a particle-velocity pattern at least as complex as those seen in the analogue experiments.

## 11.1 Introduction

Tectonic mass transfer in subduction zones is nearly entirely effected through the so-called subduction channel (Cloos and Shreve 1988a; Beaumont et al. 1999; Ellis et al. 1999). The subduction channel is a narrow zone between the upper and lower plate defined by material that exhibits a velocity gradient with respect to both plates on a long-term basis. Material flowing within the subduction channel is derived from trench deposits, offscrapings from the base of the upper plate (by tectonic erosion), or from the top of the downgoing plate at depth. Occasional return-flow back to the surface may also occur. The concept was originally developed by Hsü (1974), Shreve and Cloos (1986), and Cloos and Shreve (1988a, 1988b) in order to explain the complex juxtaposition of rocks of different paleo-geographic provenance and metamorphic grade within the Franciscan subduction melange. It has been elaborated more recently from two perspectives in particular. One concerns field observations of mainly metamorphic rocks and the attempt to numerically model their pathways. The other is mainly based on marine geophysical observations of sediment input in the subduction zone and its transport below, or accretion at, the lower fore-arc slope.

The first, combined, field-based and numerical modeling approach (e.g. Gerya et al. 2002 and references therein) focused on petrological processes. Pressure-temperature-time ($PTt$) paths of exhumed melange rocks typically indicate material transport from the trench to depths often exceeding the crustal thickness of the continental upper plate, with subsequent uplift to the surface. In addition, geochemical evidence ($^{10}$Be and trace elements) documents the involvement of subducted crustal material (tectonically eroded material and sediments from the top of the oceanic crust) in the generation of volcanic arc magmas originating as deep as 100–200 km (e.g. Woodhead and Fraser 1985; Tera et al. 1986; Morris et al. 1990; Plank and Langmuir 1993). It seems apparent that these observations mainly pertain to material fluxed in the subduction channel below the brittle-ductile boundary, where viscous and creep processes are the dominant mechanisms governing rock rheology, and to compositional and petrological processes that affect physical rock properties.

While numerical modeling of material flux at greater depths, and its dependence on various intrinsic parameters, has significantly advanced our understanding of the subduction-channel system (Gerya et al. 2002), we lack understanding of the flux mode in the upper, brittle part of the fore-arc system (accretion versus subduction erosion and their sites). Although the basic rheology appears to be simpler – the constitutive law is commonly considered to be Mohr Coulomb failure under conditions of high (to lithostatic) pore pressure – transient changes in the kinematic behaviour have not yet allowed a systematic assessment of the mechanisms controlling this part of the subduction system. These kinematic changes include elastic deformation and relaxation in the seismic cycle, the ratio between seismic slip and aseismic creep, and the role of material input into the system and its rate.

Recent studies indicate that the general mode of mass flux in subduction channels (i.e. sediment accretion versus tectonic erosion) is chiefly controlled by parameters such as the roughness of the plate interface, sediment input into the trench, and convergence velocity. The change from subduction erosion to accretion, more particularly, seems to be linked to exceeding a critical trench-fill thickness (~1 km), which may bear on the transport capacity of subduction channels (e.g. von Huene and Scholl 1991; Lallemand et al. 1994; Clift and Vannucchi 2004). In their analytical models, Cloos and Shreve (1988b) additionally suggest that the capacity of the subduction channel is influenced by the physical properties of its material and the overlying wedge material, as well as by the geometry of the fore-arc. Although the various resulting mass-transfer modes and flow patterns have a substantial impact on the internal architecture of fore-arc wedges and on their vertical surface movements, the relative role of the above parameters, and of the interactions between them, are poorly known.

In this study, we investigated the mass-transfer processes and kinematics within the upper (brittle) part of accretive and erosive subduction channels by employing scaled, two-dimensional sandbox experiments. Previous studies have shown that this method is able to provide detailed insights into mass-transfer processes (e.g. Malavieille 1984; Mulugeta and Koyi 1992; Storti and McClay 1995; Gutscher et al. 1998; Leturmy 2000; Marques and Cobbold 2002). We used analogue materials that reproduce the plastic frictional behaviour of brittle rocks with a strain hardening and softening cycle that allowed shear localization, as observed in natural rocks (e.g. Ranalli 1987; Marone 1998). Our focus was entirely on long-term processes, unrelated to the seismic cycle, and on dry ma-

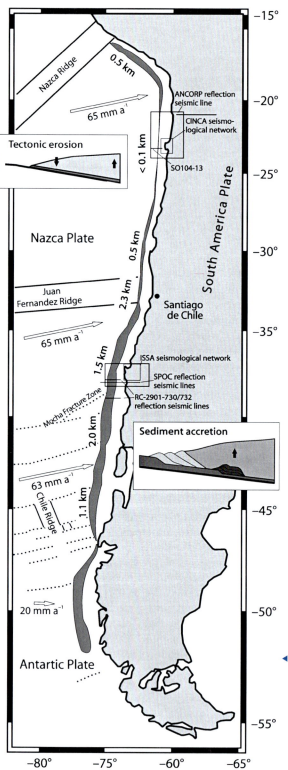

◄ **Fig. 11.1.** Location of the northern and southern study areas with the locations of available data acquired during geophysical field campaigns (Díaz-Naveas 1999; Hinz et al. 1998; ANCORP Working Group 1999, 2003; Bohm et al. 2002; Bohm 2004; Krawczyk et al. 2003, 2006, Chap. 8 of this volume). *Dark gray area* denotes the width of the trench fill with indicated thickness (Bangs and Cande 1997), which is proposed to be an important parameter controlling the mass-transfer mode (shown in the insets). *Dotted lines* – fracture zones in the oceanic crust. Current convergence rates after Angermann et al. (1999)

terials, neglecting the chemical and mechanical effects of fluids. As a consequence, our study only attempted to resolve fundamental kinematic aspects of deformation accumulation in the subduction channel and the fore-arc system, and to the related mass-flux evolution. We applied our experimental results to the Chilean fore-arc for which a multidisciplinary database is available (Kaizuka et al. 1973; ANCORP Working Group 1999, 2003; Boettcher 1999; Díaz-Naveas 1999; von Huene et al. 1999; Patzig et al. 2002; Buske et al. 2002; Krawczyk et al. 2003, 2006, Chap. 8 of this volume; Potent 2003; Rauch 2005; Rehak 2004; Sick et al. 2006, Chap. 7 of this volume). This plate margin is ideally suited for our attempt, as it is accretive in the south and tectonically erosive in the north, while plate kinematic boundary conditions remain the same (Fig. 11.1).

## 11.2 Set-up of the Physical Experiments

The analogue experiments were performed in a glass-sided box (3 200–3 800 mm length and 200 mm width, Lohrmann 2002; for details see Table 11.1 and Figs. 11.2 and 11.3; see also movies and diagram on DVD) at a scale factor of about $10^{-5}$ (Hubbert 1937). We used sand and glass beads, which deform similarly to brittle crustal rocks with a strain-hardening and strain-softening phase prior to stable sliding (Schellart 2000; Lohrmann et al. 2003). In the device, a conveyor belt, which simulates the top of the basement of the downgoing plate in a convergent margin scenario, was drawn towards a rigid back wall along a 5–10° dipping base. The conveyor belt was covered with sandpaper, which represents the rough top of the oceanic crust. In front of the back wall, a wedge-shaped body (i.e. deformable backstop; 1 100–1 350 mm initial length, 0–6° initial surface slope) acts as continental fore-arc wedge. Focussing on the Neogene timescale, an average convergence rate of 100 mm yr$^{-1}$ from the Miocene to Recent in Chile (DeMets and Dixon 1999) translates 10 mm of simulated convergence to ~10 000 years in nature.

Two series of accretive experiments and one erosive experiment were performed in this study. In the accretive experiments, a sand layer (20–35 mm thick) was sifted onto the conveyor belt to simulate trench input. The following parameters were varied (Table 11.1, Fig. 11.2):

- *Rearward material loss through the subduction channel.* This has been shown experimentally to strongly influence the mass-transfer mode in fore-arcs (e.g. Kukowski et al. 1994; Gutscher et al. 1998; Lohrmann 2002). In our experiments, it was simulated by lifting the rigid back wall (i.e. open set-up) which allowed the loss of 50–80% of the underthrust sediment. This would allow material to move towards greater depth as indicated by the isotopic compositions of arc magmas and high-pressure rocks exhumed in subduction melanges (e.g. Morris et al. 1990).
- *Vertical load.* This parameter was varied by changing the initial thickness of the backstop without changing its surface slope. Cloos and Shreve (1988b) showed that the vertical load, depending on fore-arc geometry and rock density, can significantly influence the amount and position of rearward material loss.
- *Friction contrast along the boundary between the subduction channel and the original wedge.* This was varied by choosing different material combinations as the use of the same sand in an experiment produces only a very minor friction contrast. Incorporating a thin layer of glass beads, which are significantly weaker than sand (Kukowski et al. 2002, Lohrmann 2002), in the incoming sand layer increased the friction contrast along the entire length of the subduction channel. A glass-bead layer present only in the rear of the deformable backstop simulates depth-dependent softening along the top of the subduction channel. This approach concurs with rheological studies indicating depth- and material-dependent changes in friction contrasts in nature (e.g. Hacker et al. 2003).

The first series of accretive experiments (KOL, NOL and KOI; see Table 11.1, Fig. 11.2 left column) focused on the fundamental processes within subduction channels. A strongly idealized, open set-up was chosen with a homogenous incoming sand layer and a deformable backstop. Both consisted of sifted sand (sifting height ~200 mm, sifting rate 26 mm min$^{-1}$). The experiments of this series underwent 3 000–3 900 mm of convergence. Here, we tested the influence of depth-dependent softening at the top of subduction channels and of the vertical load. The second experimental series (KOH, NOH and KIH; see Table 11.1 Fig. 11.2 right column) aimed at an application to the south-central Chilean fore-arc. In this case, we used a mechanically stratified, incoming sand/glass-bead layer and a deformable backstop consisting of mortar. The former enables decoupling between the upper and lower parts of the incoming sediment to be simulated, the latter represents the continental basement wedge. These experi-

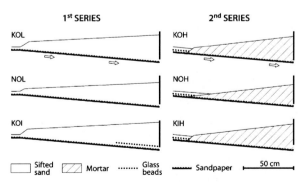

**Fig. 11.2.** Schematic set-up of the accretive experiment series. See DVD for a diagram of the sandbox apparatus

**Table 11.1.** Set-up conditions and nomenclature of the sandbox experiments performed in this study

| Experiment | | | Varied parameters | | | Stable-dynamic friction coefficient and cohesion (Pa) | | | | Thickness of sand layer (mm) | Backstop surface slope (deg) | Convergence (mm) |
|---|---|---|---|---|---|---|---|---|---|---|---|---|
| | | | Backstop thickness at backwall (mm) | Rear material loss (%) of underthrust sand | Friction contrast (%) between wedge and basal detachment | Sand paper/sand | Sand | Glass beads | Mortar | | | |
| Accretive | 1st series | KOL | 26 (thicK) | 50 (lOw) | 3.5 (Low) | 0.55 ±0.002; 44 ±10 | 0.57 ±0.002; 67 ±5 | ... | ... | 20 | 2 | 3 300 |
| | | NOL | 22 (thiN) | 50 (lOw) | 3.5 (Low) | 0.55 ±0.002; 44 ±10 | 0.57 ±0.002; 67 ±5 | ... | ... | 20 | 2 | 3 900 |
| | | KOI | 26 (thicK) | 50 (lOw) | 3.5–24.6 (Increase) | 0.55 ±0.002; 44 ±10 | 0.57 ±0.002; 67 ±5 | 0.43 ±0.014; 34 ±2 | ... | 20 | 2 | 3 000 |
| | 2nd series | KOH | 24 (thicK) | 50 (lOw) | 24.6 (High) | 0.55 ±0.002; 44 ±10 | 0.57 ±0.002; 67 ±5 | 0.43 ±0.014; 34 ±2 | 0.66 ±0.004; 24 ±13 | 14 + 20 | 6 | 1 800 |
| | | NOH | 20 (thiN) | 50 (lOw) | 24.6 (High) | 0.55 ±0.002; 44 ±10 | 0.57 ±0.002; 67 ±5 | 0.43 ±0.014; 34 ±2 | 0.66 ±0.004; 24 ±13 | 14 + 20 | 6 | 2 000 |
| | | KIH | 24 (thicK) | 80 (hIgh) | 24.6 (High) | 0.55 ±0.002; 44 ±10 | 0.57 ±0.002; 67 ±5 | 0.43 ±0.014; 34 ±2 | 0.66 ±0.004; 24 ±13 | 14 + 20 | 6 | 1 800 |
| Erosive | | | | 100 | 24.6 | 0.55 ±0.002; 44 ±10 | 0.57 ±0.002; 67 ±5 | ... | ... | 0 | 0 | 8 000 |

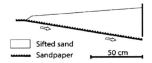

**Fig. 11.3.** Schematic set-up of the erosive experiment. See DVD for a diagram of the sandbox apparatus

ments underwent 1 800–2 000 mm of convergence. Both set-ups were similar to those used in earlier studies, but they also take into account specific features of the Chilean margin such as subduction dip or initial surface slope (see also comparable experimental features of Gutscher et al. 1996, Kukowski et al. 2002).

To simulate subduction erosion, there was no incoming sand layer on top of the sandpaper-covered conveyor belt (Table 11.1, Fig. 11.3; set-up according to Lohrmann 2002). This accords with observations that tectonically erosive margins are characterized by the absence, or only a very small thickness, of sediment entering the trench (Lallemand et al. 1994; von Huene and Scholl 1991, Clift and Vannucchi 2004). The geometrical conditions of the tectonically erosive experiment were adopted from the north Chilean fore-arc (von Huene et al. 1999; Buske et al. 2002). Since subduction erosion in physical experiments is much slower than accretion, we applied 8 000 mm of convergence. We used this experiment to investigate the general differences between tectonically erosive and accretive subduction channels. Therefore, it was designed to simulate a purely erosive subduction channel without basal accretion to ensure that the observed features were not influenced by other processes.

We monitored the experiments with a Particle Imaging Velocimetry (PIV) system (Adam et al. 2005). This image-correlation technique provides a measure of the incremental displacement field with high temporal and spatial resolution. We chose a temporal resolution of 1 second (equivalent to 0.9 mm of convergence in the experiment, ~0.090 km in nature or 900 years at 100 mm convergence yr$^{-1}$), which resulted in 1 000–3 000 time increments (PIV frames) per experiment depending on its duration. We calculated the displacement field with a spatial resolution of 1.4 mm to enable the identification of individual shear zones, which have a typical thickness of about 3 mm. From the displacement field, all components of the displacement gradient tensor were calculated. With these components we studied the influence of the aforementioned parameters on kinematics, growth mechanisms, and internal deformation patterns of the convergent sand wedges. For incremental analysis, we particularly used the vorticity (Eyx-Exy). The vorticity shows the degree of turbulence in the particle displacement field and illustrates the shear sense and magnitude of active shear zones in all directions. Furthermore, the summation of individual components of the displacement field accumulated in each time increment enables us to study long-term effects such as surface uplift or horizontal displacement.

## 11.3 Results of the Analogue Experiments

In the accretive experiments of the first series, all new thrusts were initiated in front of the previous deformation front (Fig. 11.4; e.g. KOL or NOL). The frontal ramp of the thrusts cut through the entire incoming sand layer (Fig. 11.4, left column). Depending on the specific boundary conditions of the individual experiments, the imbricates were accreted at the front (Fig. 11.4, KOL) or thrust sheets were transported beneath the deformable backstop (Fig. 11.4, NOL). During transport of these long thrust slices, duplexes were formed in their rear parts and left the subduction channel through accretion beneath the wedge (Fig. 11.4, NOL, KOI). In the accretive experiments of the second series, the new thrusts at the tip of the evolving accretionary wedge formed above the weak, glass-bead layer (Fig. 11.4, right column), accreting short imbricates in front of the backstop. The lower part of the incoming sand layer was underthrust and remained internally undeformed until duplex formation led to accretion more rearwards beneath the backstop (Fig. 11.4, right column, best seen in KOH).

The erosive wedge lost material by episodic slumping at the tip (see 'erosive movie' on DVD). The frontally eroded material was transported within the subduction channel (Fig. 11.5). Subsidence of the first 100–300 mm of the wedge tip (red area in Fig. 11.5) indicated that material was additionally offscraped from the base of the backstop when it was dragged downdip within the subduction channel.

## 11.4 Subduction Channel Kinematics and Surface Response

The kinematics of the subduction channel can be identified from the vorticity accumulated during short convergence increments (Fig. 11.6a–f). We tested different sampling rates and found that sampling at every 4 seconds (equivalent to 3.6 mm of convergence) was sufficient to unequivocally identify short-term fault patterns in the deforming sand wedges.

The accretive experiments showed highly variable subduction channel kinematics (cf. to Fig. 11.6a–f, see 'accretive movie' on DVD). Subduction channel kinematics was similar in all *accretive experiments of the first series* only when a new thrust was initiated in the incoming sand layer. During this stage, only one detachment on top of the sandpaper, with no material transport along the base of the wedge, was active. We name the interval from one thrust initiation to the next a "thrust cycle". During such thrust cycles, no systematic patterns of activity for the roof thrusts and internal thrusts of the subduction channel were observed (Fig. 11.7). For example, transport within the subduction

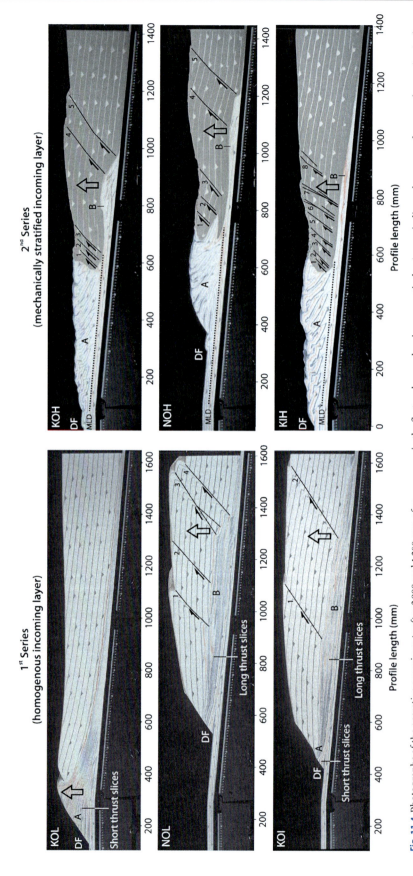

Fig. 11.4. Photographs of the accretive experiments after 2 900 and 1 200 mm of convergence in the first and second series, respectively. *Arrows* indicate the center of rearward accretion in these stages. *DF*: deformation front, *black lines* mark backthrusts, *numbers* indicate the successive initiation of backthrusts; *A*: frontal accretion of short slices; *B*: rearward accretion of formerly underthrust long slices; *MLD*: mid-level detachment. For nomenclature of experiments and parameters see Table 11.1

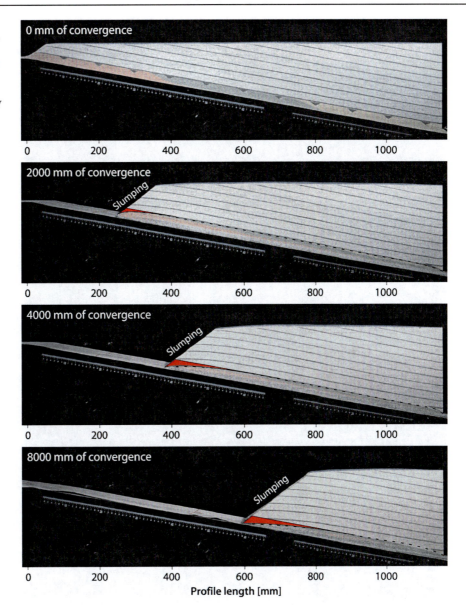

**Fig. 11.5.**
Successive stages of the tectonically erosive experiment. The *red area* marks subsidence that is mainly caused by basal erosion along the wedge base. The *dashed line* traces the boundary between material transported within the subduction channel and original backstop material

channel was subsequently initiated by reactivation of a roof thrust (Fig. 11.6b). Later during the same thrust cycle, the position of the slip planes at the top of the subduction channel fluctuated rapidly (Fig. 11.7b,d) and their vorticity magnitudes varied strongly (Fig. 11.6c–e). Further convergence initiated and reactivated branch thrusts (Figs. 11.6e and 11.7) with the formation of duplexes in the underthrust layer (Fig. 11.6c).

The location of initiation of deformation in the underthrust layer migrated towards the rear during one thrust cycle. The duplexes left the subduction channel by accretion to the base of the overlying wedge (Fig. 11.6c,e) causing its uplift and bending (i.e. of previously accreted material and original wedge material; arrows in Fig. 11.4). The uplift was accommodated by episodic activation of backthrusts (Figs. 11.6c,e). New backthrusts were initiated rearwards of former backthrusts creating a hinterland propagating series (Fig. 11.6e). Reactivation of earlier back-thrusts was often observed during duplex transport within the subduction channel. During some thrust cycles, the reactivation of backthrusts migrated towards the rear (Fig. 11.6c,d), whereas during other thrust cycles several backthrusts were activated simultaneously.

The *second accretive experimental series* with the mechanically stratified, incoming layer showed two types of accretion events. The first were thrust cycles involving frontally accreted material from the upper part of the incoming layer (comparable to the thrust cycles in the first experiment series). The second were duplex events in the underthrust, lower part of the incoming layer. These deformation events generated temporal and spatial uplift signals at the surface of the wedge (Fig. 11.8a).

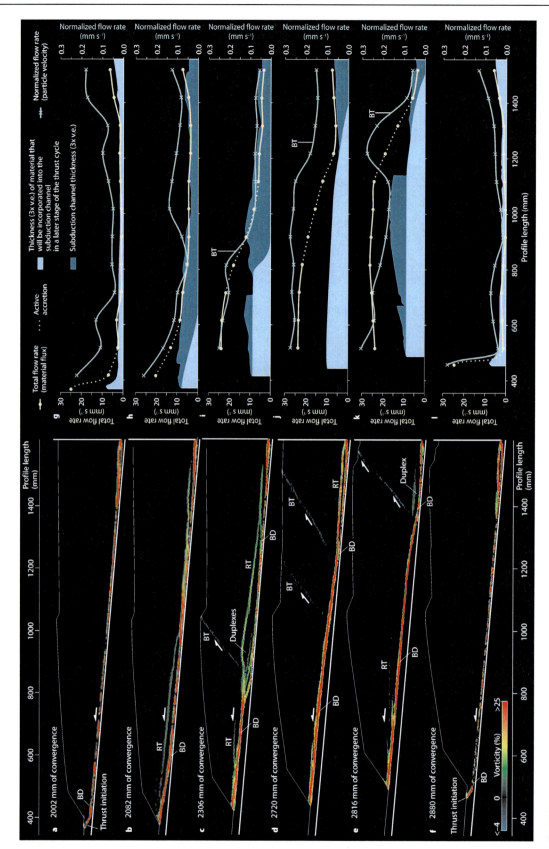

Fig. 11.6. Evolution of the third thrust cycle in the NOL experiment (see also Fig. 11.8c,d). a–f Incremental vorticity (Eyx-Exy) accumulated during 878 mm of convergence. The color scale indicates the vorticity magnitude and enables the visualization of shear-zone activity. Positive values mark sinistral shear sense, negative values denote dextral shear sense. BD – basal detachment, BT – backthrust, RT – roof thrust. g–l Diagrams of material flux along the wedge base during the respective time increments shown in (a–f). The material thickness is calculated from the highest active shear zone down to the conveyor belt

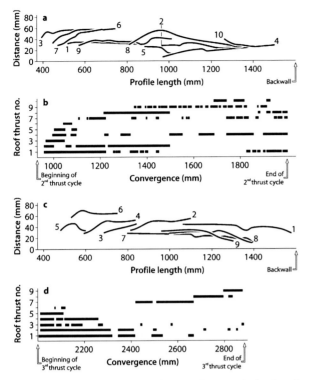

**Fig. 11.7.** Kinematics of all the roof thrusts that were (re-)activated in the second (**a, b**) and third (**c, d**) thrust cycle of the NOL experiment (see also Fig. 11.6). **a, c** Roof-thrust location along the experiment (abscissa) displayed as a function of vertical distance (ordinate, vertical exaggeration ×3) from the active shear zone to the conveyor belt. Numbers indicate the roof thrust numbers used in (**b**) and (**d**). **b, d** Time of roof-thrust activity. Time is displayed as millimeters of convergence. The roof-thrust numbers denote the individual shear zones as marked in (**a**) and (**c**)

The strongest uplift signal near the deformation front developed during the frontally accretive thrust cycles. At the beginning of these cycles, the uplift signal jumped towards the front and later expanded and migrated approximately 50 mm towards the rear with a decrease in uplift rate (Fig. 11.8a).

These well-distinguishable uplift cycles did not occur with the uplift events in the rear of the wedge. In the KOH experiment, shown in Fig. 11.8a, the rearward uplift signal showed three phases. The first phase (1 in Fig. 11.8a) was characterized by a continuously rearward-migrating uplift signal that affected a 150–200 mm-wide area. This phase lasted until duplex formation in the underthrust part of the incoming layer started and the first backthrust was formed. Thus, the migration of the uplift signal reflected the migration of the tip of the underthrust layer. The second phase (2 in Fig. 11.8a) was marked by episodic uplift pulses arching the uplifted region over an area of 200–250 mm, the front of which fluctuated by as much as 100 mm owing to the migration of the position of deformation onset in the underthrust layer. The duration of each uplift pulse was equivalent to 150–200 mm convergence. During this second phase, uplift rates were a composite of several horizontal shortening processes within the subduction channel that affected the wedge thickness: duplex initiation and accretion, as well as reactivation of earlier accreted duplexes. The third phase (3 in Fig. 11.8a) also had uplift pulses with width and duration similar to those observed in the second phase. However, in the third phase, the entire uplifted area showed laterally uniform uplift rates. During this phase, only individual duplexes were reactivated with little, or no more, shortening within the basally accreted material compared to the second phase.

The KIH experiment underscores the observation that the uplift signal is chiefly dependent on the deformation events within the subduction channel. In this experiment, only a small amount of the underthrust layer was basally accreted (Fig. 11.4, KIH) and no reactivation of large duplex complexes was observed. Accordingly, only a first- and third-type phase occurred in this experiment. When compared to the KOH experiments, this experiment indicates that lower uplift rates are caused by lower rates of basal accretion.

Detailed analysis shows that the uplift signal at a specific point on the wedge surface changed during the experimental run. These changes can be caused by two processes: a change in the shortening rate of the duplexes or duplex complexes; and by the migration of the position where deformation begins in the underthrust layer. These processes generate specific patterns of uplift rates. Figure 11.8b shows that the uplift rates varied strongly in the frontal area of the backstop, which was only uplifted when the underthrust layer was deformed close to the front (P1 in Fig. 8b), whereas the pulses of uplift owing to the variation in shortening rates in the underthrust layer caused only minor changes in the rearward uplift rates (P2, P3 in Fig. 11.8b). This analysis reveals that typical uplift rates caused by basal accretion reach a value of 2% of the flow rate of the internally undeformed and underthrust material. However, this value is a composite including the effects of incoming-layer thickness, the thickness of basally accreted material and the overlying wedge, as well as local subduction channel processes.

During the *erosive experiment*, the floor thrust of the subduction channel continuously acted as a basal detachment along the entire conveyor belt (Fig. 11.9). The top of the subduction channel was marked by (nearly) base-parallel, localized thrusts that were initiated and reactivated repeatedly. The roof thrusts developed from the rear to the front of the wedge (crosses in Fig. 11.9). The duration of these kinematic cycles varied greatly from 10 to 200 mm of convergence. Within individual kinematic cycles, the pattern of roof thrust activity was not systematic. For example, the absolute positions and lengths of

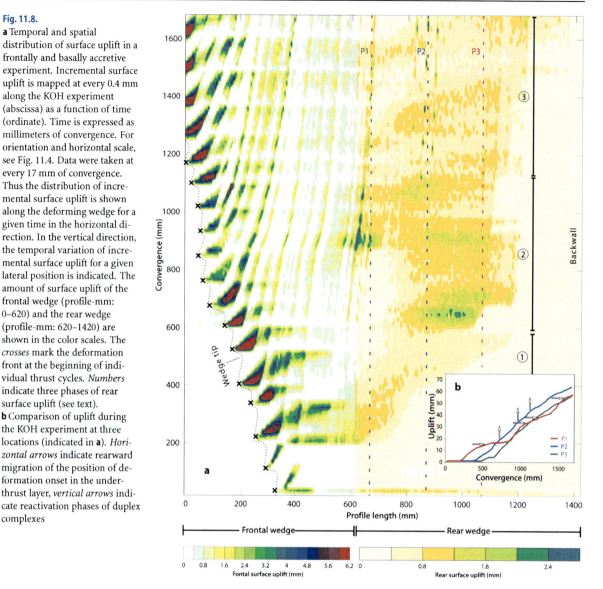

**Fig. 11.8.**
**a** Temporal and spatial distribution of surface uplift in a frontally and basally accretive experiment. Incremental surface uplift is mapped at every 0.4 mm along the KOH experiment (abscissa) as a function of time (ordinate). Time is expressed as millimeters of convergence. For orientation and horizontal scale, see Fig. 11.4. Data were taken at every 17 mm of convergence. Thus the distribution of incremental surface uplift is shown along the deforming wedge for a given time in the horizontal direction. In the vertical direction, the temporal variation of incremental surface uplift for a given lateral position is indicated. The amount of surface uplift of the frontal wedge (profile-mm: 0–620) and the rear wedge (profile-mm: 620–1420) are shown in the color scales. The *crosses* mark the deformation front at the beginning of individual thrust cycles. *Numbers* indicate three phases of rear surface uplift (see text).
**b** Comparison of uplift during the KOH experiment at three locations (indicated in **a**). *Horizontal arrows* indicate rearward migration of the position of deformation onset in the underthrust layer, *vertical arrows* indicate reactivation phases of duplex complexes

the thrusts were not the same from one cycle to the next. Also differing between cycles was the frequency of thrust reactivation and the number of coevally active faults (see 'erosive movie' on DVD).

The roof thrusts of the erosive subduction channel were located in material originating from the deformable backstop. Thus, the wedge base underwent basal erosion each time a roof thrust was reactivated. However, the basally eroded material in the rear of the wedge was immediately replaced by material that was previously eroded closer to the front. The vertical movements of the rear of the backstop depended on the geometry of the thrusts at the top of the subduction channel. Thrusts that were inclined towards the base or had a curved shape produced uplift. In the first phase of the experiment (up to 700 mm of convergence, 1 in Fig. 11.10), in which the backstop adjusted to the boundary conditions, the geometry of the thrusts at the top of the subduction channel strongly changed and parts of the backstop were permanently uplifted. During further convergence, thrust geometry changes at the top of the subduction channel became less frequent and related uplift in the overlying backstop was recorded as separate events occurring at every 2 to 35 mm of convergence (2 in Fig. 11.10). After 2000 mm of convergence, all thrusts at the top of the subduction channel were base-parallel and backstop uplift ceased (3 in Fig. 11.10).

At the front of the wedge, basal material loss was not replaced and, thus, only this part of the wedge underwent subsidence caused by basal erosion (red area in Figs. 11.5 and 11.10). The subsiding area continuously widened, up to 300 mm, during the first and second phases (1 and 2 in

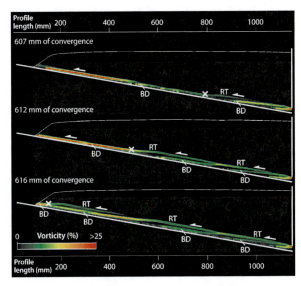

**Fig. 11.9.** Evolution of one kinematic cycle in the erosive experiment. The color scale indicates the magnitude of vorticity (Eyx-Exy) that was accumulated during 2.3 mm of convergence. The vorticity illustrates the activity of the shear zones. In this experiment, only positive values (sinistral shear) occurred. *BD* – basal detachment, *RT* – roof thrust. The *cross* indicates the tip of the most frontal roof thrust in each time increment

Fig. 11.10) of the experiment, and then maintained this width until the end of the experiment. The maximum width of the subsiding area was achieved after uplift ceased in the more rearward parts of the wedge, suggesting that these width changes were caused by an overlap of subsidence and uplift processes.

In conclusion, the accretive experiments and the erosive one showed that the top of a subduction channel is defined by episodically active thrusts, independent of the experimental set-up. However, the relative position of these thrusts differed between the accretive and erosive examples: the roof thrusts of the subduction channel in the accretive case occurred at the top of the material being continuously transferred into the subduction channel or they occurred in material that was previously accreted. Thus, subduction channel-related deformation in the accretive experiments did not penetrate through original backstop material, as was the case in the erosive experiment (Figs. 11.4 and 11.5). The accretive examples also differed from the erosive in that the interface between the conveyor belt and the accretive wedge was active along the entire base of the wedge only at the beginning of a thrust cycle (Fig. 11.6a,f). Later, during a thrust cycle, the basal detachment was progressively uplifted rearwards by the newly underthrust material attached to the conveyor belt.

Another key difference is the ability of the accretive and the erosive cases to form shear-zones that transect the overlying accretionary wedge or backstop. Steeply inclined shear zones were only formed in the overlying accretionary wedge and backstop during the accretive experiments. In the tectonically erosive experiment, no steeply inclined shear zones were formed in the backstop. No potential backthrusts were required to accommodate basal accretion as long as tectonically eroded material was subjected to rearward material loss.

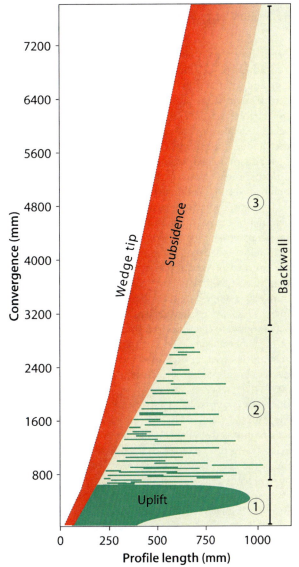

**Fig. 11.10.** Schematic diagram illustrating incremental subsidence and regions of uplift directly above the subduction channel along the tectonically erosive sand wedge (abscissa) as a function of time (ordinate). Time is expressed in millimeters of convergence (see also Fig. 11.5). The distribution of incremental, vertical particle movement is shown along the sand wedge for a given time in the horizontal direction. Temporal variations in incremental, vertical particle movements for a given lateral position are indicated in the vertical direction. *Numbers* indicate phases of vertical particle movements with different characteristics (see text)

## 11.5 Short-Term Material Flux

In our experiments, subduction channel material occurred beneath the base of the accreted material and/or the backstop and was transported faster towards the rear than the directly overlying material. Thus, the top of the subduction channel was marked by the strongest gradient in the vertical-displacement rate. The active roof thrusts defined the subduction-channel geometry and the rapid changes in subduction-channel kinematics that were observed in all experiments led to different subduction-channel geometries during one thrust cycle (Figs. 11.6g–l).

In this section, we focus on the accretive subduction channel of the NOL experiment to investigate the relationship between subduction-channel thickness and particle velocity defining the material flux. Material flux is the product of subduction-channel thickness and particle velocity, which is best seen in Figure 11.6k (profile mm: 650–1100) where the subduction channel thickness increases and particle velocity decreases. Areas of active accretion were characterized by changes in one or both of these parameters. For example, in Fig. 11.6j, particle velocity is constant and subduction channel thickness decreases in the area of active accretion (profile mm: 800–1200), whereas in Fig. 11.6h, subduction channel thickness is constant and particle velocity decreases (profile mm: 400–700). If a backthrust in the overlying wedge was highly active (Fig. 11.6i,k), an increase in particle velocity was related to a thinning of the subduction channel in the area of active accretion.

When the thickness of the incoming layer was constant, so too was the material flux at the deformation front. Rearwards, material flux varied with time and decreased where material was accreted beneath the base of the wedge and had left the subduction channel. However, at this stage, material accretion was not final and, thus, was not long-term accretion. Material could be reincorporated into the subduction channel in the later stages of a thrust cycle with reactivation of other roof thrusts (Fig. 11.6d,e). This active basal accretion is best seen by the decrease in material flux, which successively migrated rearwards over one thrust cycle (Fig. 11.6j,k). Because the area of active accretion changed during a thrust cycle, the short-term material flux pattern does not represent the long-term material flux. However, it is suitable to consider material flux as a continuous process over longer time spans because it was similar during the individual thrust cycles of this experiment type (Fig. 11.11).

## 11.6 Long-Term Material Flux

To determine the long-term material flux within the subduction channel, we integrated the short-term material flux over one thrust cycle. In this way, we could consider the whole range of subduction channel thicknesses without accounting for the displacement signal of material that had left the subduction channel by accretion. The material fluxes detected with this method show distinct patterns indicative for each of the accretion modes (Fig. 11.11).

In the first series, the experiments showed only one mode of accretion in one thrust cycle. In the case of the initially thick wedge (KOL), accretion of the main part of the incoming material occurred near the front during the entire experiment (Fig. 11.11a). We did not observe the position of accretion migrating. In contrast, in the case of an initially thin wedge (NOL), the main part of the incoming material was accreted in a rearward position (Fig. 11.11b). Here, the position of the onset of accretion migrated towards the rear during the experiment. The experiment with an initially thick wedge and weakening towards the rear (KOI) exhibited a more complex accretion behaviour (Fig. 11.11c). Long and short underthrusting phases alternated erratically. Phases of frontal accretion, which were similar to the KOL experiment, and phases of rearward accretion, similar to the NOL experiment, were observed. Again, the position of frontal accretion remained the same during the experiment. The position of the onset of rearward accretion was unique in each thrust cycle and did not show a systematic migration as observed in the NOL experiment.

In the accretive experiments of the second series (Fig. 11.4, right column), the presence of the weak, glass-bead layer within the incoming sand layer enabled decoupling of the upper and lower portions of the sand layer. Thus, long-term material accretion was not only dependent on the onset of deformation at the front of the accretionary wedge, but also on the onset of deformation beneath the backstop in the lower, underthrust part of the incoming sand layer (Fig. 11.11d). The deformation events did not show systematic alternation during the experiment. Material that had left the subduction channel, owing to one of these deformation events, had to be subtracted from the thickness of the subduction channel to detect the long-term material flux within the channel. To test whether the two modes of material accretion occur simultaneously or alternate during a thrust cycle, we investigated material flux for the time increment between different deformation events. For each time increment, the mode of material flux was basically the same (Fig. 11.11d), with decreasing material flux near the deformation front and in the rear. Therefore, accretion was simultaneous at the front and rear, and independent of the deformation events taking place within the subduction channel. This mode was similar in each experiment of the second series.

The sites of accretion between the individual experiments differed depending on the underthrust lengths. The underthrust units observed in the NOL experiment were longer than those of the KOL experiment (Fig. 11.4, left

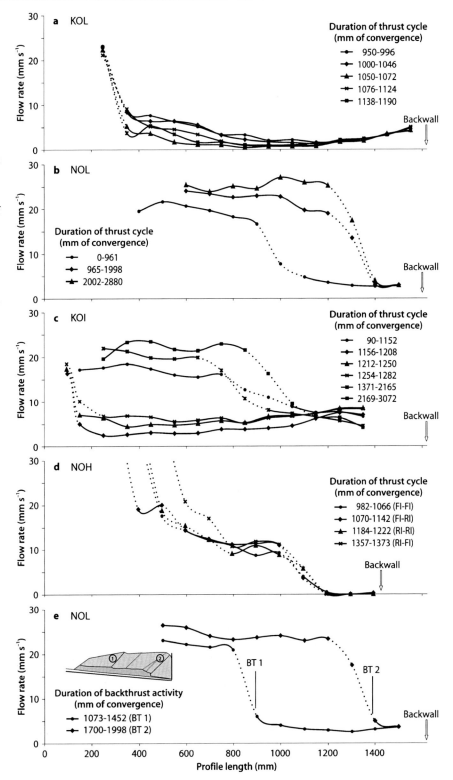

Fig. 11.11.
Comparison of material flux within subduction channels integrated over different thrust cycles in the accretive experiments (see also Fig. 11.4). *Dotted lines* – material loss within the subduction channel owing to long-term accretion, *solid lines* – no material loss within the subduction channel. **a–c** Material flux integrated from one frontal thrust initiation to the next. **d** Material flux integrated from one thrust initiation to the next. Position of thrust initiation as indicated (*FI* – frontal initiation, *RI* – rear initiation). **e** Comparison of long-term material flux within the subduction channel during two phases of backthrust activity in one thrust cycle of the NOL experiment. Position of backthrusts (*BT*) are marked in the inserted sketch of the experiment

column) and this feature was also seen in the second series (compare NOH and KOH experiments in Fig. 11.4). The only constraint, which was changed between the two experiments, was the thickness of the initial wedge and, thus, the vertical load acting on the subduction channel. In the case of an initially thin wedge, the low, normal load required lower shear traction for underthrusting, allowing displacement for longer distances. This conclusion concurs with the minimum work concept (e.g. Mitra and Boyer 1986; Masek and Duncan 1998), which was shown to be applicable to convergent sandbox experiments in terms of force balances (Gutscher et al. 1998).

Consequently, the increased thickness of the NOL experiment, owing to accretion, should have caused the onset of accretion to migrate towards the tip of the wedge. However, this was not observed in the experiments. The observed rearward migration of accretion onset indicates that an additional parameter facilitated underthrusting during the later stages of this experiment. We suggest that active backthrusts in the overlying wedge are able to decrease the force that is required for uplift of the overlying wedge. Once a backthrust is initiated, it forms a weak zone within the wedge (due to strain-softening of the analogue material) and it is easier to uplift the hanging wall. This suggestion is supported by the observation that back-thrusts significantly increased the particle velocity in the subduction channel each time they were reactivated (Fig. 11.6i,k). Also, the area of accretion was located in front of the active backthrust indicating that underthrusting occurred only beneath the hanging wall until the next backthrust was reactivated. Thus, the initiation of backthrusts further rearwards from one thrust cycle to the next, and the reactivation of previously formed backthrusts, control the rearward migration of the onset of basal accretion.

Comparison of the second-series experiments, KOH and KIH (Fig. 11.4, right column), in which the amount of rearward material loss was varied, shows that in both experiments the position of rearward accretion was the same, but that the amount of the accreted material differed. This suggests that rearward material loss has no influence on the position of basal accretion. Rearward material loss is the only parameter varied in our experiments that did not lead to a migration of the position of basal accretion.

The discussed differences in the experiments suggest that several parameters potentially cause the position of basal accretion to migrate. For example, the differences in accretion location between the NOL and KOL experiments were caused by different vertical loads and the presence or absence of backthrusts in the overlying wedge. This unequivocal correlation between accretion mode and controlling parameter is more difficult for the KOI experiment (Fig. 11.4). Here, two parameters (weakening of the top rear of the subduction channel and occurrence of backthrusts) support the underthrusting of long thrust sheets, whereas one parameter (large initial wedge thickness) acts against it. However, it might be possible to predict and interpret the changes between long and short underthrusting phases by monitoring the force balance throughout the experiment.

In summary, the analogue experiments have shown that the variation of several parameters can cause the same bulk effect in the upper plate: migration of the center of uplift as an indicator of the position of rearward accretion. In all cases, the fundamental property of subduction channel deformation that emerges is the substantial fluctuation in the localization of deformation over time, as well as within the channel. These variations resulted in a range of periodic, yet differing, cycles of unpredictable, apparently chaotic behaviour regarding the activation and reactivation of shear zones within the channel and at its top. Both end-member modes showed substantial variability in the length of their active periods, indicative of a continuum of short- to long-term transient behaviour.

## 11.7 Application to Nature

The analogue experiments provide detailed and continuous data for subduction channel processes equivalent to a low sampling rate when scaling for natural time – i.e. individual PIV frames chosen in the experimental set-up integrate up to ~$10^3$–$10^4$ years of deformation in nature (for convergence rates between 100 and 10 mm yr$^{-1}$). In this section, we compare these results with geophysical images and kinematic data from the Chilean margin, which reflect "snapshots" of time, depending on the data used (GPS and seismological: < 30 years; paleo-seismological and neotectonic: $10^2$–$10^7$ years), creating a conceptual geodynamic model. Because of the observational limitation in the experimental data, it is obvious that short-period events, and the data recording these, are not comparable with our results. Figure 11.12 shows the spectrum of activity periods observed for the various processes in the experiment with a tentative scaling up to natural time periods. It becomes clear that the size as well as the position of the observational time window in both experiment and nature are critical for assessing subduction-channel behaviour.

The short-period events that were inaccessible in our experiments include the seismic cycle and various transient events related to it, as well as other, short-term kinematics. Also, our results do not include the effects resulting from fluid system processes and the breakdown of mineral phases in the subduction zone. In particular, the mechanical properties of the subduction channel and overlying fore-arc are transiently affected by: (1) continuous or discontinuous fluid release from mineral reactions and pore space collapse; (2) subsequent hydraulic fracture with expulsion of the fluids; and (3) gradual fault sealing. In addition to changes caused by mineral breakdown, the nucleation of new phases and the continuous evolution of phyllosilicates – all affecting a rock's mechanical properties – will create additional complexity. This will affect the seismic cycle and provide depth-dependent gradients along the subduction channel with more down-dip changes, as analysed here. Apart from the

**Fig. 11.12.**
Comparison of the processes and structures that are indicative for accretive and erosive subduction channels. **a** Duration ranges of process events related to accretive (acc) and erosive (er) subduction channels derived from the analogue experiments (timescale below) and assumed for nature (timescale above). For details of the itemized processes and experiment phases see text and Figs. 11.9 and 11.10. **b** Fore-arc structures related to an accretive subduction channel. **c** Fore-arc structures related to an erosive subduction channel. Features in (**b**) and (**c**) were detected in the analogue experiments and observed in nature

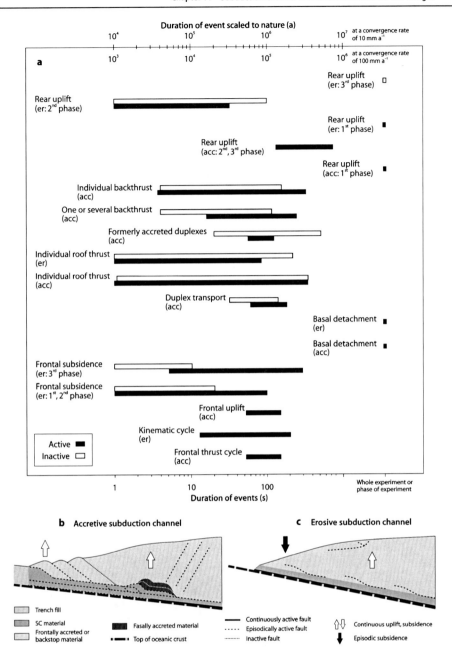

time-dependent aspects, the hypothetical mechanical influence of fluids in convergent analogue experiments would only generate lower tapers and, possibly, alter fault angles. Both would modify geometry but not kinematics. Our experiments, therefore, provide a simplified image of kinematic complexity at larger timescales, but do not fundamentally differ from similar studies performed under fluid-assisted conditions.

We compared our experimental results with two different parts of the Chilean subduction zone, for which an abundance of geophysical and geological data is available. The accretive fore-arc in south-central Chile (37–38° S) has a trench filled with thick, terrigeneous sediment (Bangs and Cande 1997; Díaz-Naveas 1999; Rauch 2005) and a subduction dip of 8–19° beneath the brittle part of the fore-arc wedge (Krawczyk et al. 2003, 2006, Chap. 8 of this volume). The upper part of the incoming sediments is frontally accreted and the lower part is underthrust (Díaz-Naveas 1999; Sick et al. 2006, Chap. 7 of this volume). In contrast, the tectonically erosive fore-arc in northern Chile (21–24° S) is characterized by the absence of sediments on the subducting, oceanic Nazca Plate, which dips at 10–20° under the frontal fore-arc (von Huene et al. 1999; ANCORP Working Group 2003).

## 11.8 South-Central Chile

Industry marine seismic reflection lines (Bangs and Cande 1997; Díaz-Naveas 1999) have shown that a thick trench fill (1.5–2.5 km) is only partly accreted to the tip of the upper plate, with most of the sediment being subducted (up to 70%). Another, more recent, marine survey provided better images of sediment entering the subduction system (Fig. 11.13b, Rauch 2005) and was complemented by an onshore reflection seismic profile at 38° S that imaged the plate interface from 20–40 km depth (project SPOC: Subduction Processes Off Chile; Krawczyk et al. 2003, 2006; Chap. 8 of this volume). The onshore line shows several groups of reflections that might be related to subduction-channel processes, as observed in the analogue experiments (Fig. 11.13c, Table 11.2). Wide-angle reflections show one group occurring at 20 to 30 km depth and some 3–7 km above the top of the subducting plate (1 in Fig. 11.13c). A second group consists of several upwardly convex, but flat-lying, prominent reflections that are imaged between 5 and 25 km depth within the fore-arc crust (2 in Fig. 11.13c). A third group (3 in Fig. 11.13c) occurs between profile-km 3 and 18 and these reflections seem to be dissected by steeply west-dipping structures. The structures of the third group can only be identified in the upper part of the fore-arc crust to 13 km depth.

Seismological data from the ISSA-2000 campaign (Integrated Seismological experiment in the Southern Andes, Bohm et al. 2002) provide kinematic constraints at depth. Figure 11.14 shows that recorded earthquakes mainly occur in two areas: NE of Isla Santa Maria and in the southern part of the Arauco Peninsula. In the southern area, seismic activity along the plate interface occurs from 20 to 30 km depth in a narrow zone less than 5 km wide. This zone of seismic activity matches the domain between the top of the downgoing plate and the strong reflections of group 1 (Fig. 11.13c). The projection of the northern earth-

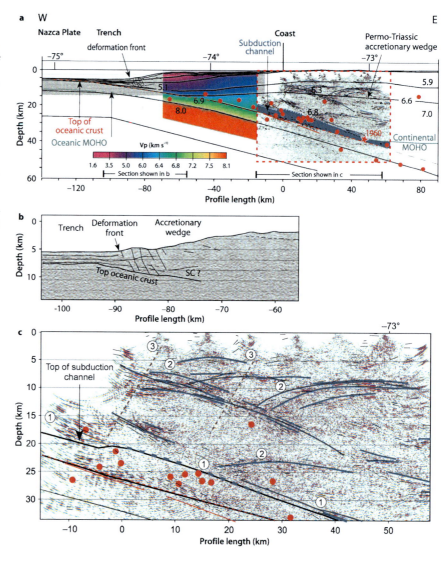

**Fig. 11.13.**
Architecture of the accretive south-central Chilean fore-arc at 38.2° S. **a** Compilation of offshore and onshore geophysical data imaging the subduction channel-related structures from the trench to 40 km depth (after Krawczyk et al. 2006, Chap. 8 of this volume). Reflection seismic data were acquired during the SPOC campaign (processing: Rauch 2005; Krawczyk et al. 2003). ISSA seismicity data were analysed by Bohm (2004). Details of the off-shore and onshore subduction channel are shown in (**b**) and (**c**). Numbers in (**c**) indicate reflectors that belong to the structural groups 1–3 (see text)

quake cluster onto a section (Fig. 11.14b) does not show seismic activity along the plate interface. The lack of seismic activity at this site is also apparent in the data recorded (since ~1970) by the National Earthquake Information Center. The crustal seismicity of both areas traces steeply west-dipping features from 2 km depth to the plate interface (Fig. 11.14b,c). The focal mechanisms of two events in the northern cluster reveal reverse shear sense in this seismic active zone, uplifting the seaward hanging wall.

The position and geometry of the first, plate interface-parallel, reflector group suggest that they might be either the boundary between actively underthrusting and basally accreted material or the interface between individual, accreted duplexes. At the tip of the accretionary wedge, between 37–38° S, the roof detachment, as identified in several reflection seismic profiles (RC-2901/730,732, Díaz-Naveas 1999), occurs an average distance of 2–3 km from the top of the lower plate. This value, defining the inlet capacity of the uppermost part of the subduction channel, is of the same order of magnitude as the distance between the top of oceanic basement and the strong, parallel reflections in the SPOC-profile between profile-km –18 and –6 (15–20 km depth, Fig. 11.13c). Therefore, we interpret these reflectors as the interface between the upper plate and the top of the underthrust material.

In the analogue experiments, the interface between the underthrust and basally accreted material formed the active detachment at the top of the subduction channel, except for the relatively short phases when a new duplex was formed or the previously accreted duplexes were reactivated. Eastwards of profile-km –6, the distance between oceanic plate and reflectors group 1 increases to 7 km (Fig. 11.13c). Also, the wide band of seismicity occurring in the southern working area along this segment of the plate interface indicates a widened, actively deforming zone. In accordance with the analogue experiments, these reflectors might image the upper limit of basally accreted duplexes that were underplated during the late Neogene subduction cycle and are currently being reactivated and deformed internally. In the area of basal accretion, the analogue experiments have shown strong fluctuations in subduction channel thickness and geometry over time, in line with the observations of seismically quiescent and active, plate-interface domains in these two neighbouring sections of south-central Chile (Fig. 11.14a).

Compared to the analogue experiments, the reflective bands of the second reflector group (2 in Fig. 11.13c) may mimic some of the basally accreted duplexes, which are uplifted by ongoing basal accretion and are not undergoing deformation. Isotopic dating of these structures, where they reach the surface, yielded Triassic peak metamorphic and uplift ages (Glodny et al. 2005) with a very minor Mesozoic/early Cenozoic component of exhumation. The steeply dipping structures of the third group (3 in Fig. 11.13c) are comparable to the backthrusts in the analogue experiments, which accommodated the uplift caused by basal accretion. One important feature of the backthrusts in the analogue experiments is that these

**Table 11.2.**
Comparison of the features of the accretive south-central Chilean and erosive north Chilean subduction channels (SC) based on reflection seismic and seismological data, as well as analogue experiments

| | Reflection seismic data | | |
|---|---|---|---|
| | SC thickness (km) | | Upper crustal structures |
| South-central Chile | 3–7[a] | | Suggested backthrusts[a] |
| North Chile | 5–7[b,c] | | Normal faulting[b,d] |
| | Seismological data | | |
| | SC thickness (km) | SC kinematics | Upper crustal structures |
| South-central Chile | 0–7[e] | Thrusting[e,f] | Backthrusts[e,f] |
| North Chile | ~3[g] | Thrusting[h] | ... |
| | Analogue experiments (scaled to nature) | | |
| | SC thickness (km) | SC kinematics | Structures within the backstop |
| Accretive SC | 2–5 | Thrusting | Backthrusts |
| Erosive SC | 2–4 | Thrusting | ... |

[a] Fig. 11.13 (Krawczyk et al. 2006, Chap. 8 of this volume).
[b] Fig. 11.15c (Sick et al. 2006, Chap. 7 of this volume).
[c] Fig. 11.15a (ANCORP Working Group 2003).
[d] von Huene and Ranero (2003).
[e] Fig. 11.14 (Rietbrock, pers comm.).
[f] Bohm (2004).
[g] Fig. 11.15d (Patzig et al. 2002).
[h] Engdahl et al. (1998).

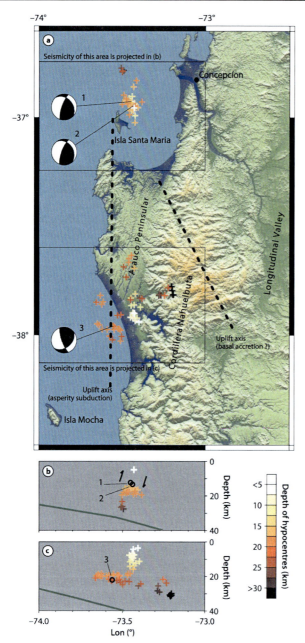

**Fig. 11.14. a** Relocalized hypocenters ($M < 5$) recorded during the ISSA-2000 campaign (Rietbrock, pers. comm). Neogene uplift axes of the south-central Chilean fore-arc as inferred from paleo-shorelines, paleoplains, marine and fluvial terraces (Kaizuka et al. 1973). **b** Depth distribution of the hypocenters in the Isla Santa Maria area. **c** Depth distribution of the hypocenters in the southern Arauco area. *Green solid lines* in (**b**) and (**c**) mark slab geometry derived from wide-angle seismic velocity models of the data acquired during the SPOC campaign in the respective regions (Krawczyk et al. 2006, Chap. 8 of this volume)

faults transected the entire backstop from the top of the subduction channel up to the wedge surface. Steep alignment of seismicity with reverse fault kinematics (see above) is similarly in accordance with the backthrusts in the analogue experiments. The presence of active backthrusts supports the aforementioned suggestion that the steeply west-dipping structures in the SPOC reflection seismic profile probably also represent backthrusts.

The currently active backthrust system might have migrated trenchwards during the Neogene. Recent uplift characterises this region, with two areas of surface uplift, each 50 km wide (Fig. 11.14a; Kaizuka et al. 1973): the area of the Arauco Peninsula (including Isla Mocha and Isla St. Maria) and the Cordillera Nahuelbuta, as part of the Coastal Cordillera. Paleo-shorelines and marine terraces document surface uplift of 0.8–5.5 mm yr$^{-1}$ in the Arauco Peninsula and its environs along an N-S-trending axis since the late Pleistocene (Kaizuka et al. 1973). The NW-SE-trending axis of surface uplift of the Cordillera Nahuelbuta is recorded by marine terraces, paleoplains and fluvial terraces (Kaizuka et al. 1973). Although geochronological data are not available for this region, it is assumed that uplift started earlier (Pliocene) and has been slower compared to the Arauco area (Kaizuka et al. 1973, Rehak 2004). Morphometric analysis of the Cordillera Nahuelbuta indicates a period of slow uplift followed by a phase of faster uplift (Rehak 2004). It should be noted that the widening, seismically active subduction channel observed in the reflection data underlies the Cordillera Nahuelbuta.

Based on geochronological data, it has been argued that the strong uplift of the Arauco area results from the subduction of the Mocha fracture zone accompanied by isostatic adjustments and subduction earthquakes (Kaizuka et al. 1973; Boettcher 1999, Potent 2003). However, the amount of uplift experienced by the Arauco Peninsula requires additional tectonic underplating. The underlying processes of the discontinuous, long-term, and slow uplift of the Cordillera Nahuelbuta is here explained by basal accretion of underthrust material, as observed in the analogue experiments. A landward backthrust confining the Cordillera Nahuelbuta to the Longitudinal Valley is not presently obvious. Hence, we may be facing two alternately active backthrust systems responsible for Neogene uplift in this part of the Chilean margin.

## 11.9 North Chile

The architecture of the erosive north Chilean margin at 23° S (Fig. 11.15) was investigated with the offshore reflection seismic profile SO104-13, acquired during the CINCA project (Crustal investigations off- and onshore Nazca Central Andes, Hinz et al. 1998) and its onshore continuation, the ANCORP'96 line (ANCORP Working Group 1999; 2003). Recent reprocessing of the marine data by Sick et al. (2006, Chap. 7 of this volume; Fig. 11.15c) resolved more details at the lower slope, including several reflections parallel to the plate interface that occur up to 5 km above the top of the downgoing plate and can be seen

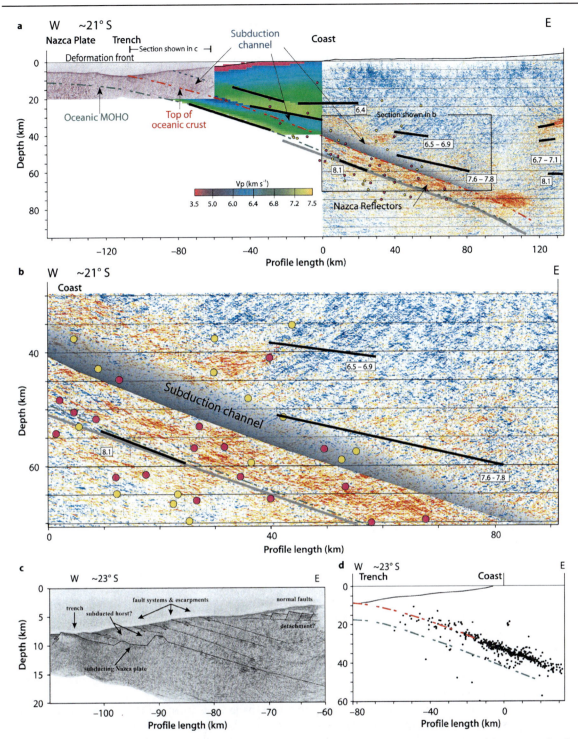

Fig. 11.15. Architecture of the north Chilean fore-arc. a Compilation of offshore and onshore geophysical data imaging the structures related to an erosive subduction channel from the trench to ~60 km depth. Offshore $v_p$ velocity model from Patzwahl et al. (1999). Onshore reflection seismic data, $v_p$ velocity model, and background seismicity were processed by the ANCORP Working Group (2003). *Thick black lines* with $v_p$ velocities denote reflector elements of the wide-angle reflection line. The *gray line* is the oceanic MOHO position as inferred from receiver function data (Heit 2005). *Yellow dots* – seismic events over 3 months of 1996 (ANCORP Working Group 2003), *Red dots* – relocalized hypocenters of events M > 4.5 recorded by the global teleseismic network over 30 years (Engdahl et al. 1998). b Details of the onshore ANCORP reflection seismic line. c Details of the reprocessed, offshore CINCA reflection seismic line SO104-13 (Sick et al. 2006, Chap. 7 of this volume). d Hypocenters from the aftershock sequence of the 1995 Antofagasta earthquake (Patzig et al. 2002) trace a ~3 km-thick damage zone along the plate interface

to a depth of ~10 km. These reflectors correlate with small escarpments at the sea floor that indicate eastward underthrusting of material that is now forming the frontal prism. The frontal prism consists of material slumped trenchwards from the middle slope (see also von Huene and Ranero 2003) as well as material from the upper plate. In the middle slope, several normal faults bend into a low-angle detachment at some 7 km depth. The related fault blocks are tilted and indicate the near-surface, gravitational mass flux that provides material to the frontal prism (von Huene and Ranero 2003).

In the erosive sandbox experiment, a comparable frontal prism developed only when the material volume of the erosional debris was larger than the capacity of the subduction channel. The north Chilean fore-arc may, therefore, currently receive more material from mass wasting than is gathered in its subduction channel. In the reflection seismic profile, the low-angle detachment beneath the normal faults separates the near-surface, extensional kinematics from the compressional domain near the base of the wedge. Similar gravitational mass flux in the analogue experiments affected only material at the surface and localized normal faults did not occur. From earlier experiments with mechanically stratified materials (Lohrmann 2002), we infer that the differences between nature and the sandbox experiments presented in this study are due to the absence of mechanical stratigraphy within the sand wedge and/or to the inability of cohesionless material to deform by localized failure when near its angle of repose. Moreover, it is clear that the kinematics observed near the surface in both nature and the experiment do not reflect the kinematic regime of the fore-arc subduction system, but are nearly entirely controlled by the gravitational potential and, hence, not diagnostic for deep mass flux.

Sick et al. (2006, Chap. 7 of this volume) suggest that the interface-parallel reflectors linking to small surface escarpments are active thrust faults within a 5 km-thick zone above the top of the oceanic basement (Fig. 11.15c). In contrast, von Huene and Ranero (2003) inferred that these structures might image the lithological boundary between the continental, Palaozoic basement and the Jurassic, volcanic La Negra Formation, rotated clockwise from its originally horizontal orientation. However, the sandbox experiments document that structures above the area of basal erosion undergo anti-clockwise rotation. The only structures in the sandbox experiments with a base-parallel orientation are the episodically active thrusts at the top of the subduction channel. The distance between these thrusts – defining the top of the subduction channel – and the conveyor belt varies between 20 and 40 mm (scaled to nature: 2–4 km), supporting the interpretation by Sick et al. (2006, Chap. 7 of this volume).

The offshore $v_p$ velocity model (Patzwahl et al. 1999) shows a low-velocity zone in continuation of the subduction channel identified in the off-shore reflection seismic line (Fig. 11.15a). This feature has been interpreted as serpentinized mantle material (ANCORP Working Group 2003). However, the low-velocity zone is directly located above the top of the oceanic crust and thus might be also interpreted as subduction channel material.

From correlations of the onshore reflection seismic profile with wide-angle observations, the Nazca reflector and its up-dip continuation has been identified as oceanic crust (ANCORP Working Group 2003). Above these reflections the reflectivity of the onshore seismic profile shows no evidence of subduction-channel related structures (Fig. 11.15b). However, background seismicity at this latitude – which last ruptured in 1877 – is scattered widely about these reflections. Further south, the 1995 Antofagasta earthquake ($M = 8$) generated a series of aftershocks that were recorded by a local network (Fig. 11.15d). Nearly all of the aftershock hypocenters were located within a 3 km-wide zone (Patzig et al. 2002) above the top of the subducting plate, ranging from 0 to 40 km depth (Figs. 11.15 and 11.16, Patzig et al. 2002; Sobiesiak 2004). This indicates current material movement between the upper and lower plate.

We interpret this combined feature as outlining the current subduction channel between the upper and lower plate. Whereas the analogue experiments revealed thrust kinematics in this domain, the aftershock focal mechanisms of the seismogenic zone at the north Chilean margin indicate thrusting, dip-slip and normal faulting (Fig. 11.16a). Sobiesiak (2004) showed that the area directly above the top of the oceanic plate suffers thrusting, whereas the dip-slip and normal faulting occurs in the higher parts of the seismogenic zone. Obviously, aftershock faulting in the subduction channel of the north Chilean fore-arc at this very short timescale is not mimicked in the analogue experiments, which have a temporal resolution corresponding to much longer time spans.

In contrast to south-central Chile, the permanent Antofagasta, and other temporary, seismological networks show that the erosive northern Chile domain has nearly no crustal fore-arc seismicity (Delouis et al. 1996; Sobiesiak 2000; Patzig et al. 2002). The lack of upper-crustal seismicity in the north Chilean fore-arc is in accordance with observations from the erosive sandbox experiments, where the compressive deformation within the subduction channel was accommodated at the base of the backstop and had no impact on the kinematics of the overlying material (Table 11.2). Yet, the eastern part of the north Chilean margin has been recording very slow surface uplift (< 0.1 mm yr$^{-1}$; Delouis et al. 1998, Pelz 2000), generating the Coastal Cordillera, as opposed to the subsiding marine part of the margin. Pelz (2000) demonstrated that this effect is probably largely, or entirely, owing to isostatic compensation caused by unloading the oceanic plate through basal erosion and thinning of the upper plate.

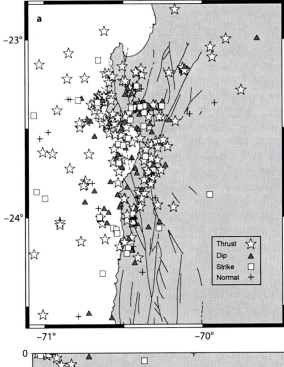

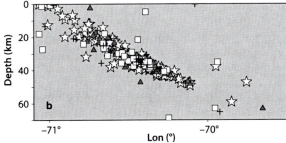

**Fig. 11.16. a** Absolute, relocalized hypocenters and (**b**) their depth distribution from the aftershock sequences of the 1995 Antofagasta earthquake (Sobiesiak 2004). The symbols indicate the structural kinematics of the events. The offshore events scatter at depth more widely than the single, relocalized hypocenters (see Fig. 11.15d) owing to the single-directed monitoring that was limited to the onshore network

Again, we note that the lateral variability of observed kinematics in these neighbouring domains in northern Chile also reflects the variability of subduction-channel behaviour observed in erosive experiments.

## 11.10 Discussion of the Results with Respect to Chile

The second series of the accretive analogue experiments shows a good first-order agreement with the features of the south-central Chilean margin. The experiments resulted in frontal accretion of the upper part and underthrusting of the lower part of the incoming sediments (Fig. 11.13b, Díaz-Naveas 1999; Rauch et al. 2005). The experiments also showed that the partial underthrusting of incoming material led to basal accretion beneath the fore-arc wedge and uplift, backthrusting with bending, and extension at the top of the respective backthrust hanging walls. This is corroborated by reflection seismic and seismological observations of structure and deep kinematics. In addition, an area of minor, Neogene uplift and near-surface extension in the south-central Chilean margin is observed between the shoreline and the Coastal Cordillera/Longitudinal Valley boundary (Lavenu and Cembrano 1999; Melnick et al. 2006). These observations suggest that basal accretion took, and is currently taking place beneath this region during the accretionary phases of this fore-arc. Moreover, from estimated surface uplift (< 2 km), only a fraction of the subducted sediment is accreted at this point and a large part must be subducted towards greater depth.

The tectonically erosive experiment showed the typical features of tectonically erosive fore-arcs in nature: strong subsidence of the marine fore-arc, trench retreat, large wedge taper, the absence of a frontal accretionary wedge, and lack of active backthrusts. The underlying, mass-transfer processes observed in the analogue experiments were frontal erosion by mass wasting of debris and basal erosion by the removal of material from the base of the wedge. The bathymetry of the north Chilean fore-arc shows comparable slumps involving huge volumes of material at the front of the offshore fore-arc (Li and Clark 1991; von Huene et al. 1999; von Huene and Ranero 2003), whereas the basally erosive process is indirectly inferred by the subsidence of a Miocene, subaerial unconformity to as much as 4 km depth (Kudrass et al. 1998; von Huene et al. 1999; Pelz 2000). In contrast to the analogue experiments, the north Chilean, offshore fore-arc shows a morphological zonation with a steep frontal slope (9°), a moderately inclined middle slope (6°) and a gentle upper slope (2.5–4°) (von Huene et al. 1999; Pelz 2000). These kinematic domains were often interpreted as indicating the underlying mass-transfer processes (von Huene et al. 1999; Adam and Reuther 2000; Pelz 2000). The analogue experiments lack a comparable morphological zonation, as, in plan view, the area of frontal erosion overlaps the area of basal erosion. Thus, the subsidence signal caused by basal erosion in the experiment was located within the area where the wedge is at the angle of repose and, therefore, controlled by gravitational potential which obliterates deeper-sourced signals.

This aforementioned comparison shows that a combination of geophysical and geological data is able to detect the presence of subduction channels in nature. Although both the geometry and position of a subduction channel determined from reflection seismic data at the Chilean margin are comparable to those observed in the analogue experiments, subduction channels in both nature and analogue experiment are fundamentally different. In the analogue experiments, we considered the whole range of geometric variability over time to define the long-term geometry of the subduction channel. The reflection seismic profiles and the seismological data chiefly reveal the geometry of the currently active subduction channel, which might fluctuate vertically over time, becoming as shallow as formerly accreted material. This variability is reflected by local differences in both study areas.

The analogue experiments indicate that the transport velocity within the subduction channel is equal to the convergence velocity only in that part of an accretive subduction channel that contains an undeformed, incoming sand layer. In the experiments, the transport velocity in deeper parts of the accretive, as well as erosive, subduction channels was much slower than the velocity of the conveyor belt. In nature, low transport velocities within the subduction channel are indirectly corroborated by the erosion rates of the north Chilean fore-arc between 22 and 23.5° S. Values range between ~40 km$^3$ Myr$^{-1}$ per kilometer of subduction (Pelz 2000; Kukowski and Oncken 2006, Chap. 10 of this volume) and 45–50 km$^3$ Myr$^{-1}$ km$^{-1}$ (von Huene and Ranero 2003), based on the analysis of the trench/lower-slope system. Also, a retreat rate of ~2 km Myr$^{-1}$ has been determined for the Neogene volcanic arc (Oncken et al. 2006, Chap. 1 of this volume). Using a convergence rate of 100 mm yr$^{-1}$ from the Miocene to Recent (DeMets and Dixon 1999) results in a subduction channel thickness of 0.5–2 km. This is less than the thickness obtained by various seismological data and might indicate that transport velocities in the subduction channel are lower than the convergence velocity.

While the analogue simulations suggest that analysis of the surface movements of fore-arc wedges could, indeed, provide semi-quantitative clues for assessing subduction-channel flow fields (such as basal erosion versus accretion, the migration of the position of basal accretion, or the migration of the position of the onset of deformation in the underthrust layer), it is not possible to draw conclusions about the controlling parameters. The reason is that the variation or combination of several of the here-analysed parameters can result in the same bulk effect with respect to surface kinematics.

## 11.11 Discussion with a Global Point of View

Geophysical data from other fore-arc systems suggest that a subduction channel is formed in nearly all subduction zones in nature. Investigations of the crustal fore-arc structures with wide-angle seismic data at several subduction zones reveal thin, low-velocity zones at the base of fore-arc wedges, hinting at the existence of subduction channels in north Chile (Patzig et al. 2002), Peru (Krabbenhöft et al. 2004), Costa Rica (Christeson et al. 1999), south-central Chile (Krawczyk et al. 2003, 2006; Chap. 8 of this volume), Nankai (Takahashi et al. 2003) and Makran (Kopp et al. 2000). While these data do not have the resolution to identify detailed subduction-channel geometries, reflection seismic profiles from some of these fore-arcs image plate interface-parallel reflectors that are comparable to those observed at the Chilean margin. Examples include Peru (von Huene and Suess 1988), Costa Rica (Ranero and von Huene 2000, von Huene et al. 2004), South Sandwich (Vaneste and Larter 2002), Nankai (Takahashi et al. 2003, Moore et al. 2001) and Makran (Kopp et al. 2000), but all of these examples only trace the subduction channel below the marine fore-arc to shallow depths (< 15 km), as opposed to our significantly deeper data.

To ensure that these reflectors are subduction-channel structures, they have to be tested against seismological data. Seismological data such as that from Nankai (Matsumura 1997) and Costa Rica (DeShon et al. 2003) show the occurrence of thin seismogenic zones along the base of the wedge, with a characteristic increase in thickness where basal accretion is suggested. Comparing the thicknesses of the potential subduction channels, based on plate interface-parallel reflectors, we see the entire range encountered in the analogue experiments. Accretive subduction channels may be < 100 m thin (Sunda, Kopp and Kukowski 2003) – corresponding to a stage of underthrusting of trench sediment – to as much as 4 km thick (Makran, Kopp et al. 2000), underscoring the experimental observation that the fluctuation of subduction channel thickness strongly depends on the position of the observational window with respect to the evolutionary cycle of the subduction channel. The potential subduction channels of erosive fore-arcs have a narrower range of thicknesses, from 1 km (NE Japan, von Huene et al. 1994) to 3 km (Costa Rica, von Huene et al 2004).

Where observed, vertical surface movements that help quantify the mass flux within a subduction channel corroborate the interpretation of geophysical data from the respective plate-interface zones. This is achieved by either reconstructing the subsidence signal at the front of an erosive fore-arc, as in Guatemala (Vannucchi et al. 2004), Peru (Clift et al. 2003) and Tonga (Clift and McLeod 1999), or by quantifying the uplift signal generated by basal accretion – compare the unresolvable, marine, surface uplift at Nankai (Moore et al. 2001) and Cascadia (Fisher, et al. 1999) with the resolvable uplift at Makran (Platt et al. 1985).

## 11.12 Conclusions

The analogue experiments show that both erosive and accretive subduction channels are marked by episodically active thrusts (roof thrusts) at the top, and a continuously active, basal detachment (floor thrust). Kinematic cycles are observed in the activity of the roof thrusts. A kinematic cycle of an accretive subduction channel is defined by the initiation of a new, frontally or basally accreted duplex, whereas the kinematic cycle of an erosive subduction channel is marked by the reactivation of the most-rearward roof thrust. Yet, the cycles are only mildly repetitive and may grade into entirely irregular and unpredictable fault activation patterns. The main differences between accretive and erosive subduction channels were (Fig. 11.12):

- *The location of the roof thrusts.* The roof thrusts in accretive experiments occurred within the subduction-channel material, whereas those in the erosive experiments also occurred in the original backstop material.
- *The length of the actively deforming floor thrust along the sandpaper interface* (which is equivalent to a detachment along the top of the oceanic crust). Accretive experiments resulted in a variable length of the actively deforming floor thrust that was dependent on the stage of the kinematic cycle. The erosive experiments had an actively deforming floor thrust along the whole wedge base.
- *The occurrence of faults in the backstop.* In the accretive experiments, backthrusts transected the whole backstop, possibly with extensional structures in the upper part, whereas no faults were observed in the erosive experiment. In the latter, extensional structures at the surface were entirely caused by the gravitational potential of the steep wedge slope.

The episodically active thrusts at the top of the subduction channel caused its geometry and particle velocity field to vary both temporally and spatially. As a consequence, the short-term material flux also shows strong fluctuations at the scale of 0.2 mm s$^{-1}$, which is equivalent to 20 mm yr$^{-1}$ in nature at 100 mm yr$^{-1}$ convergence. Investigation of the short-term material flux reveals the influence of the activity of fore-arc structures within the subduction channel and helps to identify areas of transient, active accretion. The short-term material flux does not provide information about the long-term, mass-transfer mode, owing to its nature. In contrast, the long-term material flux, which is integrated over one kinematic cycle, shows diagnostic patterns: frontal accretion, basal accretion or simultaneous frontal and basal accretion.

The analysis of accretive analogue experiments reveals that the variation of different parameters (initial wedge thickness, absence/presence of upper-crustal structures, depth-dependent softening of the top of the subduction channel) has the same bulk effect, i.e. a migration of the position of basal accretion. The simultaneous variation of two or more parameters results in forward and/or rearward migration of the position of basal accretion, hinting at the complexity of these processes in nature. The potential interaction of several parameters in nature, and the fact that the change of an individual parameter results in the same effect, preclude the determination of controlling parameters.

Moreover, the temporal complexity in subduction-channel, fault activation, with quite different patterns, underscores the relevance of long observation time windows in both experiment and nature for identifying underlying, mass-transfer patterns. On this basis, we show that it is possible to identify structures related to subduction channels in accretive and erosive fore-arc settings that are comparable to the structures observed in the sandbox experiments. We note, however, that geophysical and geodetic data may image a transient stage in the evolution of the subduction channel as typical observation windows (< 1 year) are several orders of magnitude shorter than the scaled, temporal observation window of our physical experiments. The latter show a wide range in the duration of structural activity in individual subduction channels (equivalent to $10^3$–$10^6$ years in nature) during the kinematic cycles of smaller range ($10^4$–$10^6$ years in nature).

Relating these experimental results to observational data from the south-central and north Chilean fore-arcs suggests the presence of typical structures found in accretive and erosive subduction channels, respectively. They also appear to exhibit the lateral variability with respect to the transient stages in subduction-channel evolution predicted from the experiments. These results imply that subduction channels in nature might have particle-velocity patterns that are, at least, as complex as those in the analogue experiments. In particular, we expect to see additional complexity in natural systems caused by short-term, transient changes in the strength properties of the subduction-channel material as well as of the base of the upper plate during the seismic cycle. Mechanisms for transient change include fluid production, hydraulic fracturing and fluid percolation at elevated pore pressure, fault and fracture sealing and precipitation, and mineral reactions involving changes in grain size and mechanical properties, with most of these processes involving time-dependence in their constitutive laws.

## Acknowledgments

This study was carried out in the SP F1 of the Collaborative Research Center 'Deformation Processes in the Andes' (SFB 267), financially supported by the German Science Foundation (DFG). Technical laboratory assistance came from Günter Tauscher. We gratefully acknowledge Heidrun Kopp for a helpful review. We thank A. Britt for improving the language of the manuscript.

# References

Adam J, Reuther C-D (2000) Crustal dynamics and active fault mechanics during subduction erosion. Application of frictional wedge analysis on to the North Chilean Forearc. Tectonophysics 321:297–325

Adam J, Urai JL, Wieneke B, Oncken O, Pfeiffer K, Kukowski N, Lohrmann J, Hoth S, Van der Zee W, Schmatz J (2005) Shear localisation and strain distribution during tectonic faulting – new insights from granular-flow experiments and high-resolution optical image correlation techniques. J Struc Geol 27(2):283–301

ANCORP Working Group (1999) Seismic reflection image revealing offset of Andean subduction-zone earthquake locations into oceanic mantle. Nature 397(6717):341–344

ANCORP Working Group (2003) Seismic imaging of a convergent continental margin and plateau in the central Andes (Andean Continental Research Project ANCORP 1996). J Geophys Res 108(B7): doi 10.1029/2002JB001771

Angermann D, Klotz J, Reigber C (1999) Space-geodetic estimation of the Nazca-South America Euler vector. Earth Planet Sci Lett 171:329–334

Bangs NL, Cande SC (1997) Episodic development of a convergent margin inferred from structures and processes along the southern Chile margin. Tectonics 16:489–503

Beaumont C, Ellis S, Pfiffner A (1999) Dynamics of sediment subduction-accretion at convergent margins: Short-term modes, long-term deformation, and tectonic implications. J Geophys Res 104(B8):17573–17602

Boettcher M (1999): Tektonik der Halbinsel Arauco und angrenzende Forearc-Bereiche (südliches Zentral-Chile). Mitt Geol-Paläont Inst Univ Hamburg 83:1–53

Bohm M (2004) 3-D Lokalbebentomographie der südlichen Anden zwischen 36° und 40°S. GFZ Scientific Technical Report 04/15, http://www.gfz-potsdam.de/bib/pub/str0415/0415.pdf

Bohm M, Lüth S, Echtler H, Asch G, Bataille K, Bruhn C, Rietbrock A, Wigger P (2002) The Southern Andes between 36° and 40°S latitude: seismicity and average seismic velocities. Tectonophysics 356:275–289

Buske S, Lüth S, Meyer H, Patzig R, Reichert C, Shapiro S, Wigger P, Yoon M (2002) Broad depth range seismic imaging of the subducted Nazca Slab, North Chile. Tectonophysics 350:273–282

Christeson GL, McIntosh KD, Shipley TH, Flueh ER, Goedde H (1999) Structure of the Costa Rica convergent margin, offshore Nicoya Peninsula. J Geophys Res 104:25443–25468

Clift PD, MacLeod CJ (1999) Slow rates of tectonic erosion estimated from the subsidence and tilting of the Tonga forearc basin. Geology 27:411–414

Clift PD, Vannucchi P (2004) Controls on tectonic accretion versus erosion in subduction zones: implications for the origin and recycling of the continental crust. Rev Geophys 42: doi 10.1029/2003RG000127

Clift PD, Pecher I, Kukowski N, Hampel A (2003) Tectonic erosion of the Peruvian forearc, Lima Basin, by subduction and Nazca Ridge collision. Tectonics 22:7/1–7/16

Cloos M, Shreve RL (1988a) Subduction-channel model of prism accretion, melange formation, sediment subduction, and subduction erosion at convergent plate margins: Part I, background and description. Pure Appl Geoph 128(3–4):455–500

Cloos M, Shreve RL (1988b) Subduction-channel model of prism accretion, melange formation, sediment subduction, and subduction erosion at convergent plate margins: Part II, implications and discussion. Pure Appl Geoph 128(3–4):501–545

Delouis B, Cisternas A, Dorbath L, Rivera L, Kausel E (1996) The Andean subduction zone between 22 and 25°S (northern Chile): precise geometry and state of stress. Tectonophysics 259:81–100

Delouis B, Phillip H, Dorbath L, Cisternas A (1998) Recent crustal deformation in the Antofagasta region (northern Chile) and the subduction process. Geophys J Int 132:302–338

DeMets C, Dixon TH (1999) New kinematic models for Pacific-North America motion from 3 Ma to present: I. Evidence for steady motion and biases in the NUVEL-1A model. Geophys Res Let 26:1921–1924

DeShon HR, Schwartz SY, Bilek SL, Dorman LM, Gonzalez V, Protti JM, Flueh ER, Dixon TH (2003) Seismogenic zone structure of the southern Middle America Trench, Costa Rica. J Geophys Res 108: doi 10.1029/2002JB002294

Díaz-Naveas JL (1999) Sediment subduction and accretion at the Chilean convergent margin. PhD thesis, Christian-Albrechts-Universität Kiel

Ellis S, Beaumont C, Pfiffner OA (1999) Geodynamic models of crustal-scale episodic tectonic accretion and underplating in subduction zones. J Geophys Res 104(B7):15169–15190

Engdahl ER, Van der Hilst RD, Buland RP (1998) Global teleseismic earthquake relocation with improved travel times and procedures for depth determination. Bull Seismol Soc Am 88:722–743

Fisher MA, Flueh ER, Scholl DW, Parsons T, Wells RE, Trehu A, TenBrink U, Weaver CS (1999) Geologic processes of accretion in the Cascadia subduction zone west of Washington State. J Geodyn 27:277–288

Gerya TV, Stöckhert B, Perchuk AL (2002) Exhumation of high-pressure metamorphic rocks in a subduction channel: a numerical simulation. Tectonics 21(6): doi 10.1029/2002TC001406

Glodny J, Lohrmann J, Echtler H, Gräfe K, Seifert W, Collao S, Figueroa O (2005) Internal dynamics of an accretionary wedge: insights from combined isotope tectonochronology and sandbox modelling of the South-Central Chilean forearc. Earth Planet Sci Lett 231:23–39

Gutscher MA, Kukowski N, Malavieille J, Lallemand S (1996) Cyclical behaviour of thrust wedges: insights from high basal friction sandbox experiments. Geology 24:135–138

Gutscher MA, Kukowski N, Malavieille J, Lallemand S (1998) Episodic imbricate thrusting and underthrusting: analogue experiments and mechanical analysis applied to the Alaskan accretionary wedge. J Geophys Res 103(B5):10161–10176

Hacker BR, Abers GA, Peacock SM (2003) Subduction factory 1. Theoretical mineralogy, densities, seismic wave speeds, and $H_2O$ contents. J Geophys Res 108(B1): doi 10.1029/2001JB001127

Heit B (2005) Teleseismic tomographic images of the southern Central Andes at 21°S and 25.5°S: an inside look at the Altiplano and Puna plateaus. PhD thesis, Free University of Berlin

Hinz K, Reichert C J, Flueh ER, Kudrass HR (1998) Final report: crustal investigations off- and onshore Nazca Central Andes (CINCA). Bundesanstalt für Geowissenschaften und Rohstoffe, Hannover

Hsü KJ (1974) Melanges and their distinction from olistostroms. Soc Econ Paleon Mineral Spec Pub 19:321–333

Hubbert MK (1937) Theory of scale models as applied to the study of geological structures. Geol Soc Am Bull 48:1459–1520

Kaizuka S, Matsuda T, Nogami M, Yonekura N (1973) Quarternary tectonics and recent seismic crustal movements in the Arauco Peninsular and its environs, Central Chile. Tokyo Metropolitan University Geographical Report 8:1–49

Kopp H, Kukowski N (2003) Backstop geometry and accretionary mechanics of the Sunda margin. Tectonics 22(6): doi 10.1029/2002TC001420

Kopp C, Fruehn J, Flueh ER, Reichert C, Kukowski N, Bialas J, Klaeschen D (2000) Structure of the Makran subduction zone from wide-angle and reflection seismic data. Tectonophysics 329:171–191

Krabbenhöft A, Bialas J, Kopp H, Kukowski N, Hübscher C (2004) Crustal structure of the Peruvian Continental Margin from wide-angle seismic studies. Geophys J Int 159:749–764

Krawczyk CM, SPOC Team (2003) Amphibious seismic survey images plate interface at 1960 Chile earthquake. EOS 84(32):301, 304–305

Krawczyk CM, Mechie J, Lüth S, Tašárová Z, Wigger P, Stiller M, Brasse H, Echtler HP, Araneda M, Bataille K (2006) Geophysical signatures and active tectonics at the south-central Chilean margin. In: Oncken O, Chong G, Franz G, Giese P, Götze H-J, Ramos VA, Strecker MR, Wigger P (eds) The Andes – active subduction orogeny. Frontiers in Earth Science Series, Vol 1. Springer-Verlag, Berlin Heidelberg New York, pp 171–192, this volume

Kudrass HR, Von Rad U, Seyfried H, Andruleit H, Heinz K, Reichert C (1998) Geological sampling. In: Hinz K, Reichert CJ, Flueh ER, Kudrass HR (eds) Final report: Crustal Investigations off- and onshore Nazca Central Andes (CINCA). Bundesanstalt für Geowissenschaften und Rohstoffe, Hannover, pp 44–52

Kukowski N, Oncken O (2006) Subduction erosion – the "normal" mode of fore-arc material transfer along the Chilean margin? In: Oncken O, Chong G, Franz G, Giese P, Götze H-J, Ramos VA, Strecker MR, Wigger P (eds) The Andes – active subduction orogeny. Frontiers in Earth Science Series, Vol 1. Springer-Verlag, Berlin Heidelberg New York, pp 217–236, this volume

Kukowski N, von Huene R, Malavieille J, Lallemand SE (1994) Sediment accretion against a buttress beneath the Peruvian continental margin at 12°S as simulated with sandbox modelling. Geol Rundsch 83:822–831

Kukowski N, Lallemand SE, Malavieille J, Gutscher MA, Reston TJ (2002) Mechanical decoupling and basal duplex formation observed in sandbox experiments with application to the Western Mediterranean Ridge accretionary complex. Marine Geology 186(1–2):29–42

Lallemand SE, Schnuerle P, Mallavieille J (1994) Coulomb theory applied to accretionary and nonaccretionary wedges: possible causes for tectonic erosion and/or frontal accretion. J Geophys Res 99(B6):12033–12056

Lavenu A, Cembrano J (1999) Compressional- and transpressional-stress pattern for Pliocene and Quaternary brittle deformation in fore-arc and intra-arc zones (Andes of Central and Southern Chile). J Struc Geol 21:1669–1691

Leturmy P, Mugnier JL, Vinour P, Baby P, Colletta B, Charbon E (2000) Piggyback basin development above a thin-skinned thrust belt with two detachment levels as a function of interactions between tectonic and superficial mass transfer: the case of the Subandean Zone (Bolivia). Tectonphysics 320(1):45–67

Li C, Clark AL (1991) SeaMARC II study of a giant submarine slump on the northern Chile continental slope. Marine Geotechnology 10:257–268

Lohrmann J (2002) Identification of parameters controlling the accretive and tectonically erosive mass-transfer mode at the South-Central and North Chilean Forearc using scaled 2D sandbox experiments. GeoForschungsZentrum Potsdam Scientific Technical Report STR02/10

Lohrmann J, Kukowski N, Adam J, Oncken O (2003) The impact of analogue material properties on the geometry, kinematics, and dynamics of convergent sand wedges. J Struc Geol 25(10):1691–1711

Malavieille J (1984) Modélisation expérimentale des chevauchementes imbriqués: application aux chaines de montagnes. Societé Géologique de France Bulletin 7:129–138

Marone C (1998) Labory-derived friction laws and their application to seismic faulting. Ann Rev Earth Planet Sci 26:643–696

Marques FO, Cobbold PR (2002) Topography as a major factor in the development of arcuate thrust belts: insights from sandbox experiments. Tectonophysics 348(4):247–268

Masek JG, Duncan CC (1998) Minimum-work mountain building. J Geophys Res 103(B1):907–918

Matsumura S (1997) Focal zone of a future Tokai earthquake inferred from the seismicity pattern around the plate interface. Tectonophysics 273:271–291

Melnick D, Bookhagen B, Echtler H, Strecker M (2006) Coastal Deformation and great subduction earthquakes: Isla Santa Maria, Chile (37°S). Geol Soc Am Bulletin, in press

Mitra G, Boyer SE (1986) Energy balance and deformation mechanisms of duplexes. J Struc Geol 8(3–4):291–304

Moore GF, Taira A, Klaus A, Shipboard Scientific Party (2001) Deformation and fluid flow processes in the Nankai trough accretionary prism. Proceedings of the Ocean Drilling Program, Initial Report 190:1–87

Morris JD, Leeman WP, Tera F (1990) The subducted component in island arc lavas: constraints from B-Be isotopes and Be systematics. Nature 344(6261):31–36

Mulugeta G, Koyi H (1992): Episodic accretion and strain partitioning in a model sand wedge. Tectonophysics 202(2–4):319–333

Oncken O, Hindle D, Kley J, Elger K, Victor P, Schemmann K (2006) Deformation of the central Andean upper plate system – facts, fiction, and constraints for plateau models. In: Oncken O, Chong G, Franz G, Giese P, Götze H-J, Ramos VA, Strecker MR, Wigger P (eds) The Andes – active subduction orogeny. Frontiers in Earth Science Series, Vol 1. Springer-Verlag, Berlin Heidelberg New York, pp 3–28, this volume

Patzig R, Shapiro S, Asch G, Giese P, Wigger P (2002) Seismogenic plane of the northern Andean Subduction Zone from aftershocks of the Antofagasta (Chile) 1995 earthquake. Geoph Res Let 29: doi 10.1029/2001GL013244

Patzwahl R, Mechie J, Schulze A, Giese P (1999) Two-dimensional velocity models of the Nazca plate subduction zone between 19.5°S and 25°S from wide angle seismic measurements during the CINCA95 project. J Geophys Res 104(B4):7293–7318

Pelz K (2000) Tektonische Erosion am zentralandinen Forearc (20°–24°S). Scientific Technical Report STR00/20

Plank T, Langmuir CH (1993) Tracing trace elements from sediment input to volcanic output at subduction zones. Nature 362:739–742

Platt JP, Leggett JK, Young J, Raza H, Alam S (1985) Large-scale sediment underplating in the Makran accretionary prism, southwest Pakistan. Geology 13:507–511

Potent S (2003) Kinematik und Dynamik neogener Deformationsprozesse des südzentralchilenischen Subduktionssystems, nördlichste Patagonische Anden (37–40°S). PhD thesis, Universität Hamburg

Ranalli R (1987) Rheology of the earth. Allen & Unwin, Wellington

Ranero CR, von Huene R (2000): Subduction erosion along the Middle America convergent margin. Nature 404:748–752

Rauch K (2005) Cyclicity of Peru-Chile trench sediments between 36° and 38° S: a footprint of paleoclimatic variations? Geophys Res Lett 32: doi 10.1029/2004GL022196

Rehak K (2004) Morphometrische Analyse eines aktiven Kontinentalrandes – Cordillera Nahuelbuta (Chile). Diploma thesis, Humboldt-Universität Berlin

Schellart WP (2000) Shear test results for cohesion and friction coefficients for different granular materials, scaling implications for their usage in analogue modelling. Tectonophysics 324(1–2): 1–16

Shreve LR, Cloos M (1986) Dynamics of sediment subduction, melange formation, and prism accretion. J Geophys Res 91(B10): 10229–10245

Sick C, Yoon M-K, Rauch K, Buske S, Lüth S, Araneda M, Bataille K, Chong G, Giese P, Krawczyk C, Mechie J, Meyer H, Oncken O, Reichert C, Schmitz M, Shapiro S, Stiller M, Wigger P (2006) Seismic images of accretive and erosive subduction zones from the Chilean margin. In: Oncken O, Chong G, Franz G, Giese P, Götze H-J, Ramos VA, Strecker MR, Wigger P (eds) The Andes – active subduction orogeny. Frontiers in Earth Science Series, Vol 1. Springer-Verlag, Berlin Heidelberg New York, pp 147–170, this volume

Sobiesiak M (2000) Fault plane structure of the Antofagasta, Chile earthquake of (1995) Geophys Res Lett 27:577–580

Sobiesiak M (2004) Fault plane structure of the 1995 Antofagasta earthquake (Chile) derived from local seismological parameters. PhD thesis, University of Potsdam

Storti F, McClay K (1995) Influence of syntectonic sedimentation on thrust wedges in analogue models. Geology 23(11): 999–1002

Takahashi N, Kodaira S, Park JO, Diebold J (2003) Heterogeneous structure of western Nankai seismogenic zone deduced by multichannel reflection data and wide-angle seismic data. Tectonophysics 364:167–190

Tera F, Brown L, Morris J, Sacks SI, Klein J, Middleton R (1986) Sediment incorporation in island-arc magmas: infernces from 10Be. Geochim Cosmochim Acta 50(4):535–550

Vaneste LE, Larter RD (2002) Sediment subduction, subduction erosion, and strain regime in the northern South Sandwich forearc. J Geophys Res 107: doi 10.1029/2001JB000396

Vannucchi P, Galeotti S, Clift PD, Ranero CR, von Huene R (2004) Long-term subduction-erosion along the Guatemalan margin of the Middle America Trench. Geology 32:617–620

von Huene R, Ranero C (2003) Subduction erosion and basal friction along the sediment-starved convergent margin off Antofagasta, Chile. J Geophys Res 108: doi 10.1029/2001JB001596

von Huene R, Scholl DW (1991) Observations at convergent margins concerning sediment subduction, subduction erosion, and the growth of continental crust. Rev Geophys 29(3):279–316

von Huene R, Suess E (1988) Ocean Drilling Program Leg 112, Peru continental margin: Part 1, tectonic history. Geology 16: 934–938

von Huene R, Klaeschen D, Cropp B, Miller J (1994) Tectonic structure across the accretionary and erosional parts of the Japan Trench margin. J Geophys Res 99(B11):22349–22362

von Huene R, Weinrebe W, Heeren F (1999) Subduction erosion along the Chile margin. In: Flueh ER, Carbonell R (eds) Lithospheric structure and seismicity at convergent margins. Journal of Geodynamics 27:345–358

von Huene R, Ranero CR, Vannucchi P (2004) Generic model of subduction erosion. Geology 32:913–916

Woodhead JD, Fraser DG (1985) Pb, Sr and 10Be isotopic studies of volcanic rocks from the northern Mariana Islands: Implications for magma genesis and crustal recycling in the western Pacific. Geochim Cosmochim Acta 49(9):1925–1930

# Part III
# Tectonics and Surface Processes – Responses to Change

**Chapter 12**
Tectonics, Climate, and Landscape Evolution of the Southern Central Andes: the Argentine Puna Plateau and Adjacent Regions between 22 and 30° S

**Chapter 13**
Exhumation and Basin Development Related to Formation of the Central Andean Plateau, 21° S

**Chapter 14**
The Salar de Atacama Basin: a Subsiding Block within the Western Edge of the Altiplano-Puna Plateau

**Chapter 15**
Upper-Crustal Structure of the Central Andes Inferred from Dip Curvature Analysis of Isostatic Residual Gravity

**Chapter 16**
Central and Southern Andean Tectonic Evolution Inferred from Arc Magmatism

**Chapter 17**
The Segmented Overriding Plate and Coupling at the South-Central Chilean Margin (36–42° S)

**Chapter 18**
Episodic Neogene Southward Growth of the Andean Subduction Orogen between 30° S and 40° S – Plate Motions, Mantle Flow, Climate, and Upper-Plate Structure

**Chapter 19**
Long-Term Geological Evolution and Mass-Flow Balance of the South-Central Andes

**Chapter 20**
Links between Mountain Uplift, Climate, and Surface Processes in the Southern Patagonian Andes

Regional studies presented in this part from various parts of the Andean system provide more detailed analysis of the mechanisms underlying deformation, and give particular emphasis on the relationships between tectonics and the erosional mass flux. The opening chapter by Alonso et al. (Chap. 12) explores the evolution of the Puna Plateau, its effect on atmospheric circulation, and the complex feedback mechanisms among deformation, topography build-up, and climate. They point out an intricate relationship between the formation, stabilization and destruction of basins, and the role of aridity in building the plateau.

In a parallel approach, E. Scheuber et al. (Chap. 13) study the southern Altiplano Plateau in terms of its sedimentary basins and exhumation history. The data reveal a close connection between early, and previously poorly-known, stages of deformation and the evolution of basins and their infill facies, both of which possibly reflect the influence of deeper processes.

K. J. Reutter et al. (Chap. 14) compiled a wide variety of data to address the enigmatic nature of the Salar de Atacama block, which stands out as a first-order anomaly in the Central Andes. They note a number of unusual geophysical properties of the Atacama block extending down as far as the plate interface, and demonstrate spatial coincidence with the long record of sedimentation.

In a complementary study, U. Riller et al. (Chap. 15) look at the nature of the Atacama and other large salars in the Central Andes using a novel dip-curvature analysis of the isostatic residual gravity signal. They suggest that the residual gravity field in the Central Andes is largely controlled by the distribution of late Cenozoic volcanism and by first-order tectonic features in the upper crust.

M. Haschke et al. (Chap. 16) study the variable geochemical signatures of magmatism from the Northern and Southern Chile arc to understand their underlying relationship with deformation. Long-term cyclicity in patterns of magmatic, geochemical, and tectonic features in Northern Chile are used to propose that rheologic weakening of the lithosphere and changing slab geometry (flat-slab episodes) may be typical features of Andean-type margins generally.

R. Hackney et al. (Chap. 17) show that the gravity signal of the Southern Chile fore-arc can be used to infer variations in the grade of seismic coupling at the plate interface. The patterns they present using this method may help to explain the present-day distribution of seismicity and vertical motion of the crust, and to identify the underlying physical causes.

T. Vietor et al. (Chap. 18) present time-series analyses of the Neogene deformation of the Southern Andes and relate them to climatic, plate kinematic and geological data. According to their results, the latitudinal and temporal variations of deformation can be explained by varying proportions of slab rollback, which are controlled by mantle flow and the respective strengths of the plate interface and of the upper plate.

Using isotopic data, J. Glodny et al. (Chap. 19) reveal that for most of its lifetime, the Southern Chile margin maintained a delicate balance between constructive and destructive processes. Over the long term, the continental margin has not been a site of net growth, but rather, of continental mass wasting, crustal recycling and rejuvenation.

Finally, T. Blisniuk et al. (Chap. 20) present a study linking climate and deformation in the Southern Andes. They show that the uplift and exhumation of the Southern Andes generated a pronounced rain shadow since the Middle Miocene. The subsequent acceleration of erosion, in turn, was a key factor in reducing the mechanical coupling at the plate interface, thus leading to decreased shortening rates.

# Tectonics, Climate, and Landscape Evolution of the Southern Central Andes: the Argentine Puna Plateau and Adjacent Regions between 22 and 30° S

Ricardo N. Alonso · Bodo Bookhagen · Barbara Carrapa · Isabelle Coutand · Michael Haschke · George E. Hilley
Lindsay Schoenbohm · Edward R. Sobel · Manfred R. Strecker* · Martin H. Trauth · Arturo Villanueva

**Abstract.** The history of the Puna Plateau and its marginal basins and ranges in the Eastern Cordillera and the northern Sierras Pampeanas structural provinces in northwestern Argentina impressively documents the effects of tectonics and topography on atmospheric circulation patterns, the successive evolution of orographic barriers, as well as their influence on erosion and sedimentation processes. In addition, this region exemplifies that there are several pathways by which tectonic activity may be coupled to the effects of climate and erosion. Apatite fission track and sedimentologic data indicate that distributed, diachronous uplift of ranges within the present Puna Plateau of NW Argentina began as early as Oligocene time, compartmentalizing a foreland region similar to tectonically active sectors along the current eastern plateau margins. However, fission track data from detrital apatite in sedimentary basins and vertical profiles along the eastern plateau margin document that wholesale plateau uplift probably affected this region in mid to late Miocene time, which may have been associated with mantle delamination. This coincided with the establishment of humid conditions along the eastern Puna margin and a sustained arid to hyper-arid climate within the plateau region. A common feature of the Puna Plateau is that its location corresponds to hyper-arid areas of the landscape in which channels fail to incise deeply into basin sediments or surrounding basement ranges. Importantly, the local base-level is hydrologically isolated from the foreland. This isolation occurs where the incising power of regional drainage systems has been greatly reduced due to a combination of diminished precipitation related to regional climate and local orography, and exposure of resistant bedrock. Hydrologic isolation of the plateau from the foreland permits deposition within basins as material is eroded from the surrounding ranges, reducing the relief between basins and adjacent peaks. While a variety of deformation styles and possibly combinations of different processes may have generated the high elevations observed in the Puna Plateau, the observed low-relief morphology requires evacuation of material via regional fluvial systems to be restricted. Therefore, the low-relief character of the orogenic plateau may be a geomorphic, rather than a tectonic phenomenon. At the eastern plateau margins similar basin histories can be observed in fault-bounded intramontane depressions that straddle the eastern Puna border. However, these basins remain only transiently isolated and internally drained due to their proximity to the high precipitation gradients which were established by orographic barriers in the course of Pliocene uplift. These outlying barriers focus precipitation, erosion, promote headward erosion, stream capture, and ultimately basin exhumation. This conspiring set of processes thus prevents these areas to become incorporated into the plateau realm, while the interior of the orogen conserves mass and may influence deformation patterns in the foreland due to high lithostatic stresses. Sustained aridity in the core of the orogen may thus be responsible for the creation, maintenance and potential for future lateral growth of the plateau, thus emphasizing the coupling between tectonics, climate and erosion.

---

\* Corresponding author.

## 12.1 Introduction

With a length of about 7000 km, peak elevations in excess of 6 km, pronounced tectonic activity, and different climatic regimes, the Andes are the ideal non-collisional orogen for studying the coupled processes of deformation, climate, and erosion. Here, meridionally oriented ranges built by Cenozoic tectonic activity are generally oriented perpendicular to moisture-bearing winds, and so as moist air masses impinge on their margins, large precipitation gradients result. In southern Bolivia and NW Argentina, the Subandean, Interandean, and Eastern Cordillera Ranges block moisture-bearing winds that originate in the Amazon Basin and the Atlantic, leading to pronounced contrasts between humid sectors along the eastern flanks of the orogen, and arid conditions within the Puna-Altiplano Plateau and the Western Cordillera (Fig. 12.1). Similarly, farther south where the southern hemisphere westerlies impinge on the Patagonian Andes, high precipitation on the windward western side of the orogen contrasts drastically with the arid climate to the east (WMO 1975; Lenters and Cook 1997). These differences create strong effects on erosion and sediment transport in both regions (e.g., Bangs and Cande 1997; Haselton et al. 2002; Hartley 2003; Stern and Blisniuk 2003; Blisniuk et al. 2005).

Given the large gradients in climate, and coupling between deformation and erosional efficiency that have been postulated for other orogens (e.g., Koons 1989; Isacks 1992; Willett 1999; Zeitler et al. 2001; Burbank 2002; Reiners et al. 2003; Hilley et al. 2004; Whipple and Meade 2004; Thiede et al. 2004), it is likely that the feedbacks between erosion and deformation could be important in the development of the Andes as well (Masek et al. 1994; Horton 1999; Montgomery et al. 2001; Sobel et al. 2003; Sobel and Strecker 2003; Lamb and Davis 2003). First, the topographic configuration and climatic conditions imply that ongoing tectonic uplift and migration of deformation in the Andes may have had a significant impact on the distribution of precipitation, and hence erosion, in space and time. For example, the eastward migration of tectonic activity in the central Andes (Strecker et al. 1989; Ramos et al. 2002) combined with an easterly moisture source focuses precipitation on the eastern windward slopes and successively starves the leeward western portions of the orogen of moisture (e.g., Kleinert and Strecker 2001; Sobel and Strecker 2003). Consequently, discharge within channels, and hence incision and landscape lowering rates will ultimately be reduced over geologic time in the arid interior of the orogen (e.g., Sobel et al. 2003). Second, as a consequence of the topographic configuration erosion will likely be strongest in those parts of the Andes where precipitation impinges on the eastern slopes or where topographic lows or gaps associated with structural discontinuities allow moisture to migrate farther into the orogen (Horton 1999; Sobel et al. 2003; Sobel and Strecker 2003; Coutand et al. 2006). In light of the long-term aridity of the Central Andes Lamb and Davis (2003) suggested an intimate link between aridity and tectonic uplift. In their view the rise of the orogen is closely related to the lack of sediment input into the Peru-Chile trench, which results in a high degree of plate coupling and increase in shear stresses that are ultimately responsible for supporting the high elevation of the Andes.

Here, we summarize our previous work in the Argentine Andes that strives to elucidate the relationships between tectonics and climate. We focus on the Puna Plateau and adjacent areas between about 22 and 27° S as well as the Argentine Precordillera in the area of flat-slab subduction at about 30° S. We have constrained variations in the loci of deformation, changes in erosional style, and Tertiary and Quaternary climate fluctuations using geologic mapping, geochronology, provenance, and geochemical analysis. We first review the geologic and long-term climatic development of the central Andes and show that tectonic uplift has had a significant impact on establishing orographic barriers and moderating the spatial distribution of precipitation. As a result, the associated concentration of precipitation has strongly moderated erosional exhumation in various parts of the orogen. In a second step, we evaluate the interplay between climate-driven surface processes and tectonism and their influence on developing intra-orogenic plateau topography, as well as their role in causing oscillations between transiently closed and externally drained intramontane basins. Finally, we review long-term climate conditions and deformation patterns in the northwest and central Argentine Andes to speculate that climate may influence tectonism in these sectors of the Andean orogen.

## 12.2 Geologic Setting

The Andes of western and northwestern Argentina are an integral part of the southern end of the Central Andes structural domain (Fig. 12.1). The key morphotectonic divisions in the area consist of, generally from west to east, the western slope of the Andes, the active magmatic arc,

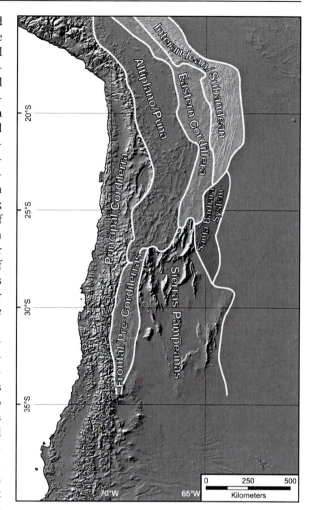

**Fig. 12.1.** Major geologic provinces of the southern central Andes superposed on shaded relief map derived from Shuttle Radar Topography Mission (SRTM). Boundaries of geologic provinces modified after Jordan et al. (1983)

the arid Puna Plateau, the Cordillera Oriental, and the Sierras Pampeanas reverse fault bounded ranges to the east. The Puna Plateau region has an average elevation of 3700 m with broad depocenters separated by meridionally trending mountain ranges, often in excess of 6000 m elevation. These sedimentary basins are presently internally drained and contain thick sequences of continental evaporites and clastic deposits (Jordan and Alonso 1987; Alonso et al. 1991). Contraction in the region of the present-day plateau and adjacent areas in the Eastern Cordillera and the northern Sierras Pampeanas has apparently contributed to the formation of closed depocenters on the present plateau and transiently closed intramontane basins within the other tectonic provinces of northwestern Argentina (Strecker et al. 1989; Alonso et al. 1991; Grier et al. 1991; Kraemer et al. 1999; Coutand et al. 2001; Marrett and Strecker 2000; Bossi et al. 2001; Hilley and Strecker 2005; Carrapa et al. 2005).

The Eastern Cordillera comprises a region of late Proterozoic to early Paleozoic metasedimentary and intrusive rocks (e.g., Jezek et al. 1985; Omarini 1983) that were uplifted along bivergent, north-northeast striking thrust faults, which were largely responsible for the deformation of Neogene intramontane and foreland sedimentary sequences (Mon and Salfity 1995). In the northern and central portion of this area east of the Cordillera Oriental, the Santa Barbara thrust belt involves Precambrian metagreywackes and Cretaceous and Tertiary sediments. These units are deformed by reverse faulting along pre-existing Cretaceous normal faults that define the southern termination of the Cretaceous Salta continental rift basin (e.g., Grier et al. 1991; Kley and Monaldi 2002; Kley et al. 2005; Marquillas et al. 2005; Gonzáles and Mon 1996; Mon et al. 2005).

The Sierras Pampeanas between 27 and 33° S are structurally transitional to the mountain ranges of the Puna and the Cordillera Oriental and comprise Laramide-style crystalline basement uplifts in a region, which coincides with the flat subduction of the oceanic Nazca Plate (e.g., Mon 1979; Jordan et al. 1983; Gonzáles and Mon 1996) that became active in late Miocene time after the last Tertiary transgression into the Andean foreland (e.g., Ramos and Alonso 1995; Ramos et al. 2002), accelerated after about 4 Ma, and culminated after 3 Ma, when intramontane basin deposits were folded and partly overthrust (Strecker et al. 1989; Bossi et al. 2001; Kleinert and Strecker 2001; Sobel and Strecker 2003).

To the west, the Sierras Pampeanas province is bordered by the thin-skinned Cordillera Principal, the Cordillera Frontal, and the Precordillera, a foreland fold and thrust belt that originated at approximately 20 Ma and which has progressed eastward, forming an orogenic wedge geometry with current activity concentrated in the Precordillera (Jordan et al. 1993; Ramos et al. 2002; Brooks et al. 2003).

## 12.3 Quaternary Climate Characteristics and Geomorphic Processes

Although this study emphasizes aspects of climate and tectonics on 100 kyr to 10 Ma timescales, it is important to characterize the present-day climate and its variability on timescales of kyrs and yrs to better evaluate long-term climate trends that have evolved in this part of the Andes. The arid highlands of the Puna and intramontane basins to the east are now one of the driest regions in the Andes. Although moisture-bearing winds impinge on the east-facing margin of the plateau, the high topography of the Eastern Cordillera and the northwestern extension of the Sierras Pampeanas borders the full length of the plateau, effectively shielding this region from significant amounts of eastwardly-derived precipitation (Fig. 12.2) (WMO 1975; Masek et al. 1994; Haselton et al. 2002). This situation

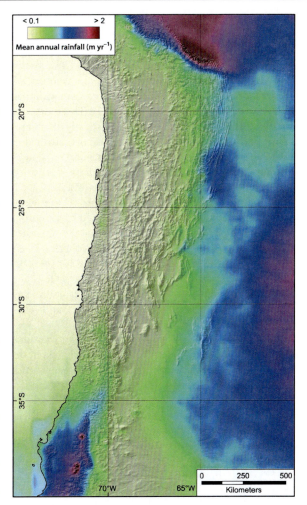

**Fig. 12.2.** Mean annual rainfall (m yr$^{-1}$) draped over shaded relief for the southern central Andes. Remotely-sensed rainfall data are derived from the TRMM (Tropical Rainfall Measurement Mission) satellite and have been calibrated using ground-control stations. Processing of the data is similar to the procedure described in Bookhagen and Burbank (2006). Note the pronounced high rainfall areas on the windward slopes in the Bolivian and Argentine Subandean belts, the Santa Barbara System and the northwestern Sierras Pampeanas vs. the interior and western flanks of the central Andes

inhibits the development of an effective bedrock fluvial system, isolating the high plateau from the foreland base-level, a setting reminiscent of conditions along the northeastern border of the Tibetan Plateau (e.g., Sobel et al. 2003).

The Puna/Altiplano region causes summertime climate and atmospheric circulation over northwestern Argentina to be governed by the South American monsoon system (e.g., Zhou and Lau 1998), characterized by an upper-air anticyclone (Bolivian High), a low-level trough (Chaco Low) and high precipitation (Garreaud et al. 2003). Approximately 80% of the annual precipitation falls within the summer months November–February (Bianchi and Yañez 1992), associated with southward moisture transport to the east of the Andes through the Andean low-level jet (e.g., Nogués-

Paegele and Mo 1997). Superposed on this circulation pattern are orographic effects determined by the uplift of ranges to the east of the Puna that shield intramontane basins to the west from easterly moisture. These basins receive less than 200 mm yr$^{-1}$ precipitation, whereas the regions along the orographic barrier receive more than 1 500 mm yr$^{-1}$ (Bianchi and Yañez 1992). These extreme spatial differences in precipitation drastically affect erosional processes. For example, the orographic rainshadow effect also reduces the long-term background erosion by a factor of four on an east-west transect in the transition between the northern Sierras Pampeanas and the Cordillera Oriental (Trauth et al. 2003a). Slopes of the windward flanks of the ranges are characterized by numerous small rotational slumps and earthflows that are generated annually, whereas rock avalanches with volumes in excess of $10^9$ m$^3$ and recurrence rates of $10^3$ to $10^4$ years occur in the semi-arid and arid intramontane valleys and basins (e.g., Fauqué and Strecker 1988; Hermanns and Strecker 1999; Hermanns et al. 2000, 2001; Trauth et al. 2000, 2003b; Trauth and Strecker 1999).

NW Argentina experiences significant intraannual fluctuations in precipitation. The seasonal change in the tropospheric temperature gradient between low and midlatitudes causes the subtropical westerly jet to extend farther north during the winter months, reaching its northernmost position around 27° S (Prohaska 1976; Hastenrath 1991). The resulting wintertime mean westerly flow, which prevails over the study region in the mid and upper troposphere, prevents regional moisture transport over the eastern slopes of the Andes and leads to a dry winter climate (Prohaska 1976; Hastenrath 1991; Bianchi and Yañez 1992). On interannual timescales, summer precipitation in the Central Andes is primarily related to changes in meridional baroclinicity between tropical and subtropical latitudes, which in turn is a response to sea surface temperature anomalies in the tropical Pacific (Ropelewski and Halpert 1987; Kiladis and Diaz 1989; Bianchi and Yañez 1992; Vuille et al. 2000; Garreaud and Aceituno 2001; Garreaud et al. 2003). The study region therefore shows a significant relationship with the El Niño Southern Oscillation (ENSO), featuring a weakened westerly flow with a significantly enhanced easterly moisture transport during La Niña summers and strengthened westerly flow with a significantly subdued easterly moisture transport during El Niño summers. As a result, the rainy season is much more active during La Niña episodes and less active during El Niño episodes, although this trend is spatially and temporally highly variable (Bianchi and Yañez 1992; Vuille 1999; Trauth et al. 2000, 2003b; Marwan et al. 2003).

Subtropical South America has also been subjected to important fluctuations in climate on timescales of $10^3$ to $10^4$ years. Quaternary lake-level changes, pollen data, and snowline changes document significant shifts in temperature and precipitation (e.g., Ammann et al. 2001; Haselton et al. 2002; Seltzer et al. 2003; Placzek et al. 2006). The general agreement between the timing of more humid conditions and periods of maximum summer insolation on the Puna-Altiplano Plateau supports a first-order orbital control of the intensity of the South American summer monsoon (Baker et al. 2001). In greater detail, however, the timing and spatial significance of the wetter/drier or cooler/warmer periods varies regionally, although the differences between records from various regions may reflect artefacts of dating uncertainties or ambiguous environmental proxies (e.g., Markgraf and Seltzer 2001). Periods of wetter climates and a stronger ENSO and/or regional ENSO-influence during the last 50 kyrs in the northwest Argentine Andes may also be responsible for increased landsliding activity, landslide-dammed lakes, and fluvial terraces (Trauth and Strecker 1999; Trauth et al. 2000, 2003b; Bookhagen et al. 2001; Robinson et al. 2005). Although large landslide deposits (*sturzstroms*) in NW Argentina occur along mountain fronts with known Quaternary tectonic activity (e.g., Fauqué and Strecker 1988; Hermanns and Strecker 1999; Hermanns et al. 2000, 2001; Fauqué and Tschilinguirian 2002), two significant landslide clusters at ~35 kyr BP ($^{14}$C age) and at 4.5 kyr BP (calibrated $^{14}$C ages) correlate with wet periods in the Altiplano, while apparently no landslides correlate with the Tauca (13.4–11.5 kyr BP) and Coipasa (17.8–13 kyr BP) humid phases (Trauth et al. 2003b). This observation suggests that in addition to orographic precipitation gradients, temporal variability of subtropical South American climate also strongly impacts erosional processes in the central Andes. In this context, the impact of ENSO-related precipitation fluctuations on erosion may be more important than the influence of orbitally-induced shifts in humidity (Trauth et al. 2003b).

## 12.4 Tertiary Climate Characteristics

Subtropical atmospheric subsidence and aridity in this part of the Andes have apparently been long-lived features as South America has remained at its present latitude during the last 18 Ma (Scotese et al. 1988; Hartley et al. 1992) and the present circulation system may have already been established during the Paleogene (e.g., Parrish et al. 1982). However, paleoclimate proxy indicators from this region are ambiguous, sometimes contradictory, and limited. According to Hartley (2003), the effects of the Hadley circulation, the presence of the cold Humboldt current, and the establishment of the Panama land bridge resulted in the transition from an arid to a hyper-arid climate after 4 Ma. Prior to this, the existence of the proto-Humboldt current (e.g., Zachos et al. 2001) may have aided aridity in this part of South America. This notion is supported by gypsiferous Oligocene units in the Salar de Antofalla region (Fig. 12.3) at about 26°15' S (Adelmann 2001; Carrapa et al. 2005). However, Gaupp et al. (1999) and Sáez et al. (1999) report sedimentary sequences in Chile from separate basins in the core region of the arid sector of the Andes that also record freshwater lake environments under semi-humid conditions in late Miocene time and a drastic switch to arid conditions between 6.4 and 3.7 Ma. This was followed by a return to more variable, semi-humid conditions that lasted until 2.6 Ma, after which the present hyper-arid conditions were established. The changeover to hyper-arid conditions at this time is in agreement with interpretations by Hartley and Chong (2002) about this region.

In NW Argentina limited sedimentologic, paleontologic, and stable isotope data sets can be combined to reconstruct Tertiary climatic conditions. In the Puna, aridification is commonly linked with the onset of internal drainage and the appearance of evaporites between 24 and 15 Ma (e.g., Alonso et al. 1991; Vandervoort et al. 1995), although arid environments may have existed already in the Eocene (Adelmann 2001). In line with these observations is the termination of supergene alteration and copper-sulfide enrichment at about 24° S in the Atacama desert of northern Chile between 14 and 8.7 Ma (Alpers and Brimhall 1988), at 27°10' S at 13.3 Ma, and at 27°47' S at 13.5 Ma (Sillitoe et al. 1991), in the volcanic arc immediately west of the Puna. Furthermore, based on the analysis of drainage patterns between 18 and 22° S, Hoke et al. (2004) suggest that a changeover to hyper-arid climate conditions may have occurred along the western slopes of the Andes between about 10 and 5.8 Ma.

Sedimentologic indicators documenting climatic change are also available from the intramontane basins parallel to the eastern Puna margin. For example, in the Santa Maria Valley, an intramontane basin between the Aconquija, Calchaquí and Quilmes Ranges in the northernmost Sierras Pampeanas (Fig. 12.3), vertic paleosols in sediments of the Santa María Group deposited in this basin between 12 and 10.7 Ma suggest a wet-dry seasonal climate (Kleinert and Strecker 2001). These latter soils were apparently waterlogged frequently and could have hosted salt-tolerant $C_3$-vegetation based on $\delta^{13}C$ analysis of paleosol carbonates (Kleinert and Strecker 2001). Authigenic clays in the Andalhuala Formation (7.1–3.4 Ma) of the Santa María Group suggest weathering under more humid conditions than now, but stable carbon isotope data from this formation document that $C_4$ plants appeared in the valley by approximately 7 Ma, similar to observations at Corral Quemado (CQ in Fig. 12.3) farther south (Latorre et al. 1997;

**Fig. 12.3.** Digital elevation map of the southern central Andes based on SRTM data, including important structures. Structures are modified from Sobel et al. (2003) and references therein. *Colored dots* show apatite fission-track (AFT) cooling ages and $^{40}Ar/^{39}Ar$ and K/Ar ages of supergene and hypogene mineralization. The plotted data are listed in Tables 12.1 and 12.2. AFT age distribution reflects the role of aridity in minimizing magnitude of erosion since tectonically driven exhumation occurred; argon ages reflect changes in ground water availability and hence timing of aridification. Locality abbreviations are: *A*: Angastaco Basin; *AB*: Salar de Arizaro; *AC*: Sierra Aconquija; *AD*: Atacama Basin; *C*: Sierra Calalaste; *CA*: Valle del Cajón; *CC*: Calchaquí Range; *CD*: Cordillera Domeyko; *CQ*: Puerta de Corral Quemado; *CR*: Sierra Chango Real; *F*: Bolsón de Fíambalá; *H*: Quebrada de Humahuaca; *HM*: Salar de Hombre Muerto; *PG*: Salar de Pastos Grandes; *QT*: Quebrada del Toro; *SA*: Salar de Antofalla/Antofalla Basin; *SG*: Salinas Grandes Basin; *SC*: Siete Curvas; *SP*: Sierra Pasha; *SM*: Santa María Valley; *SQ*: Sierra de Quilmes; *TC*: Tres Cruces Basin; *VC*: Calchaquí Valley. The sources for all reported ages are documented in Tables 12.1 and 12.2

**Table 12.1.** Compilation of apatite fission-track data used in Fig. 12.3. $Chi^2$: Probability of passing chi squared test; $P$ is pass; $f$ is fail; $pop$ indicates sample analyzed using population method. References: AR 1994: Andriessen and Reutter 1994; C 2005: Carrapa et al. 2005; C 1998: Coughlin et al. 1998; C 2001: Coutand et al. 2001; C 1999: Coutand 1999; D 2004: Deeken et al. 2004; D rev: Deeken et al. in review; E 2004: Ege 2004; J 1989: Jordan et al. 1989; MZ 1999: Maksaev and Zentilli 1999; M rev: Mortimer et al. in review; SS 2003: Sobel and Strecker 2003; T 1996: Tawackoli et al. 1996

| Age (Ma) | ±1σ | chi² | Long.W | Lat.S | Reference | Age (Ma) | ±1σ | chi² | Long.W | Lat.S | Reference | Age (Ma) | ±1σ | chi² | Long.W | Lat.S | Reference |
|---|---|---|---|---|---|---|---|---|---|---|---|---|---|---|---|---|---|
| 30.0 | 3.0 | pop | 65.92222 | 24.48056 | AR 1994 | 19.1 | 0.8 | p | 66.70455 | 25.00933 | D rev | 36.7 | 8.0 | p | 68.73889 | 22.09167 | MZ 1999 |
| 38.6 | 5.6 | pop | 68.27111 | 23.79361 | AR 1994 | 18.2 | 1.6 | p | 66.51028 | 25.33243 | D rev | 38.6 | 4.4 | p | 68.82778 | 22.13194 | MZ 1999 |
| 64.5 | 7.9 | pop | 68.65250 | 23.10750 | AR 1994 | 13.2 | 2.1 | f | 66.51370 | 25.32527 | D rev | 32.7 | 4.4 | p | 68.87778 | 22.26944 | MZ 1999 |
| 119.0 | 13.0 | pop | 70.42778 | 23.93472 | AR 1994 | 16.3 | 0.7 | p | 66.50465 | 25.32013 | D rev | 35.4 | 5.2 | p | 68.85417 | 22.33111 | MZ 1999 |
| 85.0 | 13.0 | pop | 70.31250 | 23.89861 | AR 1994 | 16.0 | 0.7 | p | 66.49410 | 25.31532 | D rev | 45.9 | 8.8 | p | 69.28472 | 26.35833 | MZ 1999 |
| 78.7 | 11.6 | pop | 70.22222 | 23.84167 | AR 1994 | 16.0 | 0.7 | p | 66.49202 | 25.31175 | D rev | 44.4 | 4.4 | p | 69.27222 | 26.36667 | MZ 1999 |
| 69.2 | 7.0 | pop | 70.05222 | 23.72972 | AR 1994 | 15.6 | 0.7 | p | 66.42565 | 25.24920 | D rev | 43.7 | 4.6 | p | 69.25417 | 26.34167 | MZ 1999 |
| 59.2 | 7.5 | pop | 69.51389 | 23.37778 | AR 1994 | 51.9 | 2.6 | p | 66.32022 | 25.66888 | D rev | 62.4 | 7.4 | p | 69.09722 | 22.33333 | MZ 1999 |
| 50.8 | 7.6 | pop | 69.03250 | 23.53222 | AR 1994 | 86.0 | 3.4 | f | 66.35378 | 25.65930 | D rev | 58.9 | 6.0 | p | 69.08056 | 22.32500 | MZ 1999 |
| 25.8 | 1.6 | p | 67.42150 | 26.14360 | C 2005 | 95.2 | 3.7 | f | 66.35378 | 25.66110 | D rev | 58.2 | 5.6 | p | 69.14861 | 22.34861 | MZ 1999 |
| 27.7 | 3.9 | p | 67.48190 | 26.16180 | C 2005 | 69.0 | 2.4 | f | 66.34952 | 25.65487 | D rev | 51.7 | 5.0 | p | 69.14444 | 22.33611 | MZ 1999 |
| 23.9 | 3.1 | p | 67.45160 | 26.13560 | C 2005 | 162.4 | 5.5 | f | 66.41838 | 25.56748 | D rev | 55.6 | 6.0 | p | 69.19167 | 22.31528 | MZ 1999 |
| 28.5 | 1.8 | p | 67.45270 | 26.13560 | C 2005 | 16.9 | 0.9 | p | 66.42355 | 25.57077 | D rev | 47.4 | 5.4 | p | 69.19167 | 22.28750 | MZ 1999 |
| 29.0 | 2.0 | p | 67.46060 | 26.15260 | C 2005 | 15.0 | 1.7 | p | 66.34063 | 25.02088 | D rev | 37.1 | 3.6 | p | 68.96389 | 22.28750 | MZ 1999 |
| 26.5 | 2.2 | p | 67.46090 | 26.15010 | C 2005 | 189.4 | 11.2 | p | 67.32317 | 24.59590 | D rev | 30.1 | 3.2 | p | 68.92500 | 22.23611 | MZ 1999 |
| 3.0 | 1.5 | p | 66.54000 | 27.35650 | C 1998 | 198.4 | 12.0 | p | 67.33080 | 24.59717 | D rev | 34.4 | 5.0 | p | 68.98056 | 22.23056 | MZ 1999 |
| 3.0 | 3.0 | p | 66.53983 | 27.35650 | C 1998 | 27.6 | 1.1 | p | 67.41948 | 21.01484 | E 2004 | 30.8 | 3.6 | p | 68.97500 | 22.30000 | MZ 1999 |
| 7.1 | 1.1 | p | 66.52567 | 27.45650 | C 1998 | 31.4 | 2.4 | p | 67.04153 | 20.79568 | E 2004 | 32.0 | 3.8 | p | 68.97639 | 22.31389 | MZ 1999 |
| 7.6 | 2.2 | p | 66.52083 | 27.46700 | C 1998 | 33.8 | 2.1 | p | 67.03553 | 20.79567 | E 2004 | 30.3 | 3.6 | p | 68.96667 | 22.34639 | MZ 1999 |
| 118.0 | 11.0 | f | 66.52133 | 27.47450 | C 1998 | 32.7 | 2.4 | p | 66.92051 | 20.76923 | E 2004 | 30.2 | 4.4 | p | 68.89583 | 22.27917 | MZ 1999 |
| 4.2 | 1.1 | p | 66.52133 | 27.48050 | C 1998 | 27.7 | 2.5 | p | 66.28170 | 21.35024 | E 2004 | 36.3 | 3.6 | f | 68.83333 | 21.91833 | MZ 1999 |
| 89.0 | 13.0 | f | 66.50700 | 27.53100 | C 1998 | 24.3 | 2.8 | p | 66.02907 | 21.11034 | E 2004 | 33.6 | 3.8 | p | 68.85139 | 21.92917 | MZ 1999 |
| 182.0 | 13.0 | f | 66.50583 | 27.51383 | C 1998 | 21.5 | 2.1 | p | 65.93013 | 21.17228 | E 2004 | 35.9 | 4.4 | p | 68.81389 | 21.91944 | MZ 1999 |
| 67.0 | 3.0 | p | 66.51150 | 27.56217 | C 1998 | 26.7 | 2.1 | p | 65.90618 | 21.43629 | E 2004 | 47.0 | 2.2 | f | 66.26830 | 26.48927 | M rev |
| 38 | 3 | f | 66.75667 | 26.85167 | C 2001 | 31.5 | 2.6 | p | 65.74974 | 21.41507 | E 2004 | 81.9 | 2.8 | p | 66.30317 | 26.62567 | M rev |
| 29 | 3 | f | 66.73667 | 26.86333 | C 2001 | 36.4 | 5.2 | p | 65.40043 | 21.42513 | E 2004 | 59.7 | 4.0 | p | 65.72108 | 26.53455 | SS 2003 |
| 31 | 3 | f | 66.73167 | 26.88000 | C 2001 | 20.1 | 2.4 | p | 65.16355 | 21.27165 | E 2004 | 65.2 | 4.5 | p | 65.71887 | 26.53118 | SS 2003 |
| 30 | 4 | p | 66.74000 | 26.88667 | C 2001 | 22.4 | 2.1 | p | 65.12304 | 21.29792 | E 2004 | 82.2 | 5.4 | p | 65.72418 | 26.51752 | SS 2003 |
| 87 | 4 | p | 65.89167 | 24.48500 | C 1999 | 28.2 | 2.4 | p | 65.05028 | 21.36039 | E 2004 | 73.9 | 5.0 | p | 65.73273 | 26.51293 | SS 2003 |
| 67 | 3 | p | 65.90667 | 24.48000 | C 1999 | 29.3 | 3.7 | p | 65.03707 | 21.36655 | E 2004 | 74.8 | 4.9 | p | 65.73938 | 26.50892 | SS 2003 |
| 52 | 3 | p | 65.91683 | 24.47500 | C 1999 | 24.5 | 2.0 | p | 65.00494 | 21.36091 | E 2004 | 74.0 | 6.4 | p | 65.75323 | 26.47710 | SS 2003 |
| 42 | 3 | f | 65.95500 | 24.44833 | C 1999 | 23.8 | 1.7 | p | 64.92073 | 21.48545 | E 2004 | 78.2 | 5.5 | p | 65.76530 | 26.46568 | SS 2003 |
| 93 | 5 | p | 65.99033 | 24.41533 | C 1999 | 23.9 | 1.6 | p | 64.90326 | 21.49423 | E 2004 | 79.7 | 5.8 | p | 65.77495 | 26.46423 | SS 2003 |
| 68 | 7 | f | 66.01000 | 24.33333 | C 1999 | 26.1 | 2.2 | p | 64.88323 | 21.46372 | E 2004 | 83.2 | 6.4 | p | 65.79672 | 26.47738 | SS 2003 |
| 24.5 | 1.3 | p | 66.29893 | 23.62935 | D 2004 | 28.8 | 1.9 | p | 64.84911 | 21.46167 | E 2004 | 81.6 | 5.5 | p | 65.81187 | 26.47738 | SS 2003 |
| 22.7 | 1.3 | p | 66.29198 | 23.63730 | D 2004 | 29.3 | 3.2 | p | 64.78190 | 21.43632 | E 2004 | 3.6 | 0.6 | p | 66.11088 | 27.19595 | SS 2003 |
| 21.3 | 2.4 | p | 66.29237 | 23.64092 | D 2004 | 16.4 | 1.4 | p | 64.48771 | 21.44957 | E 2004 | 5.1 | 0.7 | p | 66.11335 | 27.19392 | SS 2003 |
| 20.8 | 1.1 | p | 66.29340 | 23.64372 | D 2004 | 9.2 | 1.2 | p | 64.28410 | 21.41101 | E 2004 | 5.7 | 0.8 | p | 66.11325 | 27.18987 | SS 2003 |
| 23.0 | 3.0 | p | 65.68882 | 23.19240 | D 2004 | 237 | 16 |  | 65.54762 | 28.47368 | J 1989 | 5.9 | 0.6 | p | 66.11812 | 27.18178 | SS 2003 |
| 16.8 | 1.2 | p | 65.69335 | 23.21432 | D 2004 | 208 | 13 |  | 65.58730 | 28.47368 | J 1989 | 5.8 | 0.7 | p | 66.11527 | 27.17725 | SS 2003 |
| 64.4 | 2.7 | p | 65.45967 | 23.38313 | D 2004 | 45.3 | 6.2 | p | 68.70278 | 21.06944 | MZ 1999 | 6.0 | 0.8 | p | 66.10930 | 27.17292 | SS 2003 |
| 17.8 | 3.7 | f | 65.45495 | 23.38288 | D 2004 | 43.4 | 4.8 | p | 68.75833 | 21.16250 | MZ 1999 | 4.2 | 0.7 | p | 66.12052 | 27.16545 | SS 2003 |
| 15.0 | 1.2 | p | 65.45322 | 23.38078 | D 2004 | 43.5 | 6.6 | p | 68.80833 | 21.54167 | MZ 1999 | 4.9 | 0.7 | p | 66.12933 | 27.15758 | SS 2003 |
| 14.0 | 1.4 | p | 65.44888 | 23.37897 | D 2004 | 36.4 | 4.4 | p | 68.81389 | 21.50833 | MZ 1999 | 5.0 | 0.7 | p | 66.13080 | 27.15040 | SS 2003 |
| 13.1 | 0.9 | p | 65.44780 | 23.37892 | D 2004 | 41.6 | 5.8 | p | 68.79861 | 22.06667 | MZ 1999 | 5.1 | 0.6 | p | 66.13283 | 27.14450 | SS 2003 |
| 12.4 | 1.0 | p | 65.44162 | 23.38090 | D 2004 | 39.8 | 7.0 | p | 68.73750 | 22.05167 | MZ 1999 | 4.8 | 0.6 | p | 66.17917 | 27.15000 | SS 2003 |
| 56.4 | 2.0 | p | 66.61252 | 24.97978 | D rev | 39.9 | 5.4 | p | 68.77083 | 22.08056 | MZ 1999 | 32.1 | 4.5 | p | 65.70179 | 21.50476 | T 1996 |
| 20.3 | 0.9 | p | 66.61835 | 24.98103 | D rev | 39.8 | 7.0 | p | 68.83333 | 22.11667 | MZ 1999 | 29.7 | 2.7 | p | 65.59637 | 21.05903 | T 1996 |
| 18.7 | 0.9 | p | 66.62813 | 24.98423 | D rev | 38.9 | 8.6 | p | 68.82778 | 22.11944 | MZ 1999 | | | | | | |
| 25.8 | 1.6 | p | 66.67258 | 24.98708 | D rev | 39.7 | 6.2 | p | 68.82778 | 22.11944 | MZ 1999 | | | | | | |

Table 12.2. Compilation of Ar/Ar and K/Ar ages used in Fig. 12.3. *BC 2002:* Bouzari and Clark 2002; *SC 1998:* Sasso and Clark 1998; *AB 1988:* Alpers and Brimhall 1988; *SM 1996:* Sillitoe and McKee 1996; *S 1991:* Sillitoe et al. 1991

| Age (Ma) | ±2σ | Type | Long. W | Lat. S | Reference | Age (Ma) | ±2σ | Type | Long. W | Lat. S | Reference |
|---|---|---|---|---|---|---|---|---|---|---|---|
| 35.3 | 0.7 | Ar/Ar | 69.25972 | 20.04472 | BC 2002 | 20.4 | 0.6 | K/Ar | 68.91222 | 22.37556 | SM 1996 |
| 22.4 | 1.1 | Ar/Ar | 69.25972 | 20.04472 | BC 2002 | 20.8 | 0.6 | K/Ar | 68.91222 | 22.37556 | SM 1996 |
| 21.5 | 0.5 | Ar/Ar | 69.25972 | 20.04472 | BC 2002 | 21.0 | 1.3 | K/Ar | 69.24222 | 22.56417 | SM 1996 |
| 19.6 | 1.6 | Ar/Ar | 69.25972 | 20.04472 | BC 2002 | 14.1 | 0.6 | K/Ar | 69.34083 | 22.86500 | SM 1996 |
| 14.6 | 2.5 | Ar/Ar | 69.25972 | 20.04472 | BC 2002 | 20.0 | 1.0 | K/Ar | 69.07639 | 22.98778 | SM 1996 |
| 5.4 | 0.1 | Ar/Ar | 66.26416 | 27.43584 | SC 1998 | 20.8 | 0.8 | K/Ar | 69.51667 | 23.43333 | SM 1996 |
| 33.7 | 1.3 | K/Ar | 69.07864 | 24.24991 | AB 1988 | 31.0 | 1.0 | K/Ar | 69.60750 | 24.39833 | SM 1996 |
| 33.2 | 1.4 | K/Ar | 69.06990 | 24.27205 | AB 1988 | 23.1 | 0.7 | K/Ar | 69.56389 | 26.25944 | SM 1996 |
| 32.8 | 1.3 | K/Ar | 69.07088 | 24.27025 | AB 1988 | 14.5 | 0.4 | K/Ar | 69.26944 | 26.79056 | SM 1996 |
| 33.7 | 1.4 | K/Ar | 69.05336 | 24.28281 | AB 1988 | 14.2 | 0.6 | K/Ar | 60.87139 | 26.76611 | SM 1996 |
| 31.5 | 1.4 | K/Ar | 69.06399 | 24.27025 | AB 1988 | 23.2 | 0.7 | K/Ar | 69.17556 | 26.65056 | S 1991 |
| 31.0 | 1.4 | K/Ar | 69.06497 | 24.27025 | AB 1988 | 21.7 | 0.7 | K/Ar | 69.16556 | 26.65083 | S 1991 |
| 31.6 | 1.6 | K/Ar | 69.05720 | 24.27477 | AB 1988 | 20.7 | 0.7 | K/Ar | 69.17556 | 26.64222 | S 1991 |
| 17.7 | 0.7 | K/Ar | 69.06497 | 24.27025 | AB 1988 | 20.2 | 1.0 | K/Ar | 69.17306 | 26.64167 | S 1991 |
| 14.7 | 0.6 | K/Ar | 69.06990 | 24.26844 | AB 1988 | 23.4 | 0.7 | K/Ar | 69.25444 | 26.82917 | S 1991 |
| 16.4 | 0.7 | K/Ar | 69.06094 | 24.27479 | AB 1988 | 20.2 | 0.6 | K/Ar | 69.25833 | 26.81250 | S 1991 |
| 18.0 | 0.7 | K/Ar | 69.06445 | 24.27068 | AB 1988 | 17.3 | 0.5 | K/Ar | 69.26667 | 26.78333 | S 1991 |
| 34.3 | 1.1 | K/Ar | 69.28806 | 20.04333 | SM 1996 | 15.8 | 0.6 | K/Ar | 69.25389 | 26.80611 | S 1991 |
| 30.3 | 1.1 | K/Ar | 69.28139 | 20.04556 | SM 1996 | 23.0 | 0.7 | K/Ar | 69.23889 | 27.27361 | S 1991 |
| 15.2 | 0.5 | K/Ar | 68.63833 | 20.99611 | SM 1996 | 22.3 | 0.7 | K/Ar | 69.23472 | 27.27500 | S 1991 |
| 21.1 | 0.6 | K/Ar | 69.71389 | 21.92778 | SM 1996 | 22.8 | 0.6 | K/Ar | 69.31000 | 27.55778 | S 1991 |
| 16.3 | 0.6 | K/Ar | 68.90000 | 22.29444 | SM 1996 | 24.3 | 0.7 | K/Ar | 69.35333 | 27.68250 | S 1991 |
| 19.0 | 0.7 | K/Ar | 68.89306 | 22.26750 | SM 1996 | 24.1 | 0.8 | K/Ar | 69.34556 | 27.68611 | S 1991 |
| 16.5 | 0.5 | K/Ar | 68.89639 | 22.26972 | SM 1996 | 22.0 | 0.6 | K/Ar | 69.06028 | 27.43333 | S 1991 |
| 18.1 | 0.7 | K/Ar | 68.89639 | 22.26972 | SM 1996 | 13.6 | 0.4 | K/Ar | 69.00778 | 27.17306 | S 1991 |
| 17.6 | 0.6 | K/Ar | 68.89639 | 22.26972 | SM 1996 | 13.3 | 0.4 | K/Ar | 69.02667 | 27.17222 | S 1991 |
| 15.2 | 0.5 | K/Ar | 68.89639 | 22.26972 | SM 1996 | 12.9 | 0.4 | K/Ar | 69.02972 | 27.23361 | S 1991 |
| 16.8 | 1.2 | K/Ar | 68.89972 | 22.27361 | SM 1996 | 13.5 | 0.5 | K/Ar | 69.31667 | 27.78333 | S 1991 |

Kleinert and Strecker 2001). Yet the ecosystem also contained $C_3$ plants, which increased in proportion relative to $C_4$ plants between ~4.5 and 3 Ma (Strecker and Kleinert 2001). Fossils of *Paracacioxylon o'donelli* and *Mimosoxylon piptadensis* (Leguminosae; $C_3$ plants) were identified in the Andalhuala Formation, indicating semi-humid conditions (Menéndez 1962; Lutz 1987). Fossil vertebrates include racoons, giant running birds, ground sloths, and a variety of ungulates (Marshall and Patterson 1981). These faunal assemblages characterize a subtropical tree savanna, more humid than today and with a pronounced precipitation seasonality, similar to the present Chaco region in the undeformed foreland farther east (Pascual et al. 1985; Pascual and Ortiz Jaureguizar 1990; Pascual et al. 1996; Nasif et al. 1997). However, after 2.5 ± 0.6 Ma conditions became drastically different after the Santa María Group was folded and overthrust by basement rocks of the Aconquija and Calchaquí Ranges (Strecker et al. 1989). From this time onward aridity has been sustained in the valley and is reflected by thick $CaCO_3$-bearing paleosol horizons that cement conglomeratic gravels of fluvial terraces and pediments (Villanueva García and Strecker 1986; Strecker et al. 1989; Kleinert and Strecker 2001).

North of 26° S in the Calchaquí Valley (Fig. 12.3), Anzótegui (1998) and Starck and Anzótegui (2001) describe an arid paleoenvironment at approximately 13.4 ± 0.4 Ma (Grier et al. 1991), which heralds the deposition of the coarse sandstones of the Angastaco Formation in this intramontane basin near the village of Angastaco. These conditions were followed by the deposition of sandstones and siltstones rich in organic material during more humid conditions that were dominated by fluvial and lacustrine sedimentary environments recorded in the Palo Pintado formation by about 5.4 Ma (Starck and Anzótegui 2001; Coutand et al. 2006). The Palo Pintado and the lower section of the superceding San Felipe Formation contain a rich mammal fauna, pollen, leaves and tree-trunk remnants that all indicate a hot, semi-humid climate comparable to that found in the present Chaco lowlands (Anzótegui 1998; Starck and Anzótegui 2001). With the formation of thick calcretes in the upper part of the San Felipe Formation, however, humid conditions reverted back to aridity between about 3.4 and 2.4 Ma, identical to the situation reflected in the Santa María record.

Pronounced changes in available moisture are documented by fossil leaf assemblages from the semi-arid Quebrada del Toro, an intramontane arid valley immediately to the east of the Puna at about 24°30' S (Fig. 12.3). Plant fossils including *Typha sp.* (Thyphaceae), *Equisetum sp.* (Equisetaceae), and *Thelypteris sp.* (Thelypteridaceae) are embedded in lacustrine silt and claystones and crop out in the southernmost part of this basin. Based on lateral correlations within the basin, these deposits are < 8 Ma and > 4.6 Ma (Marrett and Strecker 2000; Hilley and Strecker 2005) and document an evaporite-free lacustrine environment. Living related plants presently occur east of the basin-bounding ranges and are part of the Subandean phytogeographic province, which receives between 700 and 1000 mm yr$^{-1}$ precipitation (Bianchi and Yañez 1992), hence a region with abundant water availability and acid pH conditions (Cabrera 1994). In particular, *Thelypteris* is a plant that cannot support any salinity and documents a humid environment with abundant freshwater. The interpretation of a freshwater environment is also supported by very low boron concentrations of 20ppm in these sediments. Further paleo-environmental information comes from Toxodontidae (*Notoungulata*) remains in the vicinity of these outcrops (Sirolli 1968) and support the interpretation of a paludal environment, as these mammals lived in swamp environments with access to abundant grass (e.g., Bond et al. 1995).

Farther east, the now semi-arid intramontane Quebrada de Humahuaca at approximately 23°30' S and 65°20' W long (Fig. 12.3) yields a rich record of herbivore fossils, including glyptodonts, sloths, armadillos, rodents, horses, giant lamas, peccaries, and deer (Reguero et al. 2003). This suggests humid conditions, probably with some wet-dry seasonality permitting an open forest environment. The presence of *Hydrochoeropsis dasseni*, a capybara-like rodent indicates the presence of permanent water bodies (Reguero et al. 2003). All fossils occur in the Uquía Formation, > 2.78-Myr-old (Marshall et al. 1982), which conformably overlies gypsum-bearing mud and sandstones of the Maimará Formation, which is older than 3.5 Ma (Walther et al. 1998). The latter unit is a foreland basin deposit and contains clasts of the eastern Puna margin. Both units are folded and faulted and are unconformably overlain by an early Pleistocene massive, calcrete-bearing conglomerate fill unit that once covered the paleo-topography of this basin.

Finally, Uba et al. (2005) report on paleoclimate indicators in sediments in the Subandean foreland fold and thrust belt of southern Bolivia. According to their investigations, aridity characterized this environment until about 10 Ma, which was followed by semi-arid conditions between 10 and 7 Ma, and finally humid conditions accompanied by the formation of coal after 4 Ma.

Although limited in spatial and temporal resolution, the different data sets from the western flanks and the interior of the southern central Andes show evidence for protracted Tertiary arid to semi-humid conditions followed by the Miocene initiation of hyper-arid conditions that are observed at present. In contrast, data from the late Miocene to early Pliocene intramontane valleys that straddle the Puna margin document early arid/semi-arid conditions that subsequently changed into more humid environments, and ultimately back to aridity after 4 Ma. This is in contrast with the global trend toward increased aridity at that time (e.g., Cerling et al. 1993). The change-over from arid to humid conditions along the eastern margin of the Puna, as well as sustained aridity within and in regions west of the plateau therefore points a tectonic origin. This change appears to have been associated with the establishment of effective orographic barriers that blocked moisture-bearing winds, which were forced to precipitate on the windward side of the developing central Andean Plateau and its flanking ranges.

Taking studies on the effects of orographic barriers on amount and distribution of precipitation from other regions into account (e.g., Stern and Blisniuk 2002; Blisniuk et al. 2005; Bookhagen et al. 2005; Blisniuk and Stern 2006), then the elevation of the present plateau region and its eastern border may have been between 2000 and 2500 m in late Miocene to early Pliocene time. This is in agreement with studies from the southern rim of the Puna Plateau at about 27° S, immediately north of the Fiambala Basin (Fig. 12.3) suggesting a minimum paleorelief of ~1800 m between the present plateau border and that basin by late Miocene time (Carrapa et al. 2006). This assessment is also compatible with paleo-elevation estimates farther north based on leaf morphology by Gregory-Wodzicki (2000), who suggested that the Bolivian Altiplano had possibly reached about half of its present elevation by 10 Ma. New paleo-altimetry studies by Garzione et al. (2006) using oxygen isotopes in paleosol carbonates in the Altiplano also indicate that the present elevation of the northern plateau region may had come into existence between 9 and 6 Ma.

## 12.5 Evidence for Tectonically-Induced Climatic Gradients: the Role of Orographic Barriers

To determine the timing of orographic barrier construction, it is important to document both the onset and the cause of aridity. The former is best determined from sedimentology, paleontology, and stable isotope studies. The latter may be due to regional climatic shifts or the construction of significant topography. Paleoaltimetry data is notoriously imprecise and scarce in the Andes (e.g., Blisniuk et al. 2005; Blisniuk and Stern 2006). However, under favorable circumstances, thermochronology data from range uplifts and studies of the adjacent basins can be combined to provide an estimate of the paleotopographic development of orographic barriers. In the following sections, the relations between the generation of orographic barriers and climate will be discussed in different regions of NW Argentina, including the Puna Plateau and its eastern margin, the Eastern Cordillera, including basins in the transition between the Eastern Cordillera and the Puna, and the northernmost Sierras Pampeanas basement uplifts.

Apart from the volcanic arc the western flank of the Puna also includes the Chilean Cordillera Domeyko (Chilean Precordillera), which extends 20 to 26° S with peak elevations between 3500 and 5000 m (Fig. 12.3). Apatite fission-track data demonstrate that this range experienced a strong pulse of exhumation during the Eocene when several kilometers of rock were eroded at rates of 0.1 to 0.2 mm yr$^{-1}$ and since ca. 30 Ma, the exhumation rate has decreased to ca. 0.05 mm yr$^{-1}$ (Maksaev and Zentilli 1999). Based on structural and thermochronologic data, Maksaev and Zentilli (1999) concluded that much of the present relief formed during the Eocene and has been preserved due to the extremely arid climate. New (U-Th)-He ages (Juéz-Larré et al. 2004) and cosmogenic nuclide dating (Dunai et al. 2005) from the western flank of the northern Andes of Chile also indicate minimal erosional modification of the landscape and sustained aridity since Oligo-Miocene time. Similarly, Coutand et al. (2001, 2006) report the onset of exhumation along the southeastern margin of the Puna in the Oligocene. Carrapa et al. (2005) document apatite fission track cooling ages for the Sierra de Calalaste Range (southern Puna Plateau) inferring range exhumation in the Oligocene followed by slow erosion within the present plateau region. In the Eastern Cordillera between 23 and 25.5° S, apatite fission-track data from a series of vertical profiles in basement rocks document moderately rapid exhumation between ca. 25 and 15 Ma (Deeken et al. 2004). The internally drained Salar de Pastos Grandes and Salar de Hombre Muerto Basins are on the lee side of these ranges and became aridified during this time (Alonso et al. 1991; Vandervoort et al. 1995). Analysis of additional fission track samples from the interior of the Puna Plateau, east of the Salar de Arizaro at about 25° S, yields Jurassic ages, also documenting minimal Cenozoic exhumation in this region (Deeken et al. 2004; Deeken et al. in press). The coincidence between low exhumation rates on the plateau, the onset of topographic construction, and the creation of internal drainage and deposition of evaporites suggests an efficient orographic barrier for moisture incursions within a region already dominated by aridity.

The northwestern Sierras Pampeanas show strong relations between evolving orographic barriers and exhumation patterns that constrain the timing of range uplift. Sierra Aconquija forms the windward (eastern) flank of the semiarid Santa María Basin (Fig. 12.3), which became aridified between 4 and 3 Ma (Kleinert and Strecker 2001).

This range towers more than 5000 m above the virtually undeformed foreland without other upwind topographic barriers. Apatite fission-track data demonstrates that rapid exhumation of the western flank of Sierra Aconquija commenced at ca. 6 Ma with ca. 1 mm yr$^{-1}$; the rate slowed significantly at about 3 Ma (Sobel and Strecker 2003). In the last 3 Ma, surface uplift rates have increased from 0.1–0.5 mm yr$^{-1}$ to ca. 1.1 mm yr$^{-1}$ while exhumation on the now arid western flank of the range has decreased. Today, there is a marked decrease in precipitation at elevations in excess of 2000–2500 m on the windward flank of the range. By extension, at about 3 Ma, the crest of the range must have exceeded this elevation. In this example, rapid exhumation preceded rapid surface uplift by one to three million years, because thick Miocene sediments that formerly covered the present range could be easily eroded. At present, the range exposes metamorphic rocks; the combination of leeward aridity and such resistant lithologies are thus jointly responsible for the low exhumation rate.

In contrast, the Cumbres Calchaquíes Range immediately north of Sierra Aconquija (Fig. 12.3) reveals a different history, which also appears to be profoundly influenced by climate. The sedimentary fill of the adjoining Santa María Basin and its deformation history suggest that these two ranges were exhumed at the same time and were previously covered by comparable thicknesses of Miocene sediments (Sobel and Strecker 2003). However, the Cumbres Calchaquíes Range preserves remnants of a regional Cretaceous erosion surface along its crest. This is in marked contrast to Sierra Aconquija, which has had up to 4300 m of basement rocks removed from above the highest peaks. The mean exhumation rate over the last 6 Ma for the Cumbres Calchaquíes was between 0.4 and 0.5 mm yr$^{-1}$. The maximum amount of precipitation along the windward side of the ranges varies dramatically. In contrast to Sierra Aconquija, the precipitation maximum at Cumbres Calchaquíes is reduced by 50% (Bianchi and Yañez 1992) and located on the crest of lower ranges to the east of the main range. The implication is that the differences in exhumation at these localities are due primarily to the varying amounts of precipitation which are available to drive erosion. These examples suggest that a positive feedback may be initiated in which uplifting areas with little incoming precipitation may be only slowly exhumed, allowing topography to be rapidly constructed and high elevations, which preserve old geomorphic surfaces, to be maintained. This topography may either become high or create a wide mountain range to accommodate the deformation. In either of these cases, the orographic precipitation gradients become steep, further starving the dry leeward areas of moisture. In contrast, the relations investigated in the northern Sierras Pampeanas underscore that when ample moisture is available, precipitation will be focused, exhumation will be rapid, and deformation may not migrate farther into the foreland.

## 12.6 Processes of Relief Reduction, the Role of Internal Drainage, and Generation of Plateau Morphology in the Puna Plateau

In contrast to the aridity of the plateau interior the plateau margins block moisture-bearing winds that create an effective erosional regime at the flanks. The low relief typical of the plateau regions may reflect isolation of the local base-level of plateau basins from the low-elevation foreland. During this process, basin aggradation replaces incision and transport, while erosion of the surrounding peaks reduces the internal relief of the region (e.g., Sobel et al. 2003).

Uplift of basement ranges in the low-elevation Andean foreland may cause steepening of channels that must incise through them to allow interior basins to continue to drain to the foreland. As steepening proceeds and topography is built, however, orographic precipitation gradients reduce moisture on the leeward interior basins. Consequently, as discharge is also reduced, this must result in further steepening of the channels and aggradation within the interior basins. In the course of this process, aggradation within the interior basins may be eventually outpaced by channel steepening, isolating them from the foreland. Simultaneously, the interior basins aggrade as long as there is sufficient sediment production to fill them, reducing the relief between the basin floor and surrounding peaks (e.g., Sobel and Strecker 2003; Sobel et al. 2003; Hilley and Strecker 2005).

Paleocurrent indicators, sediment provenance, and apatite fission track thermochronology indicate that flow within a once contiguous foreland basin (e.g., Jordan and Alonso 1987) in the vicinity of the present Antofalla Basin of the southern Puna was disrupted by uplift of this range to the east of this basin between 29 and 24 Ma (Carrapa et al. 2005). Late Oligocene sediments in this area indicate that an arid environment existed prior to and during the uplift of the Sierra de Calalaste (Adelmann 2001), and internal drainage was established in the Antofalla Basin possibly as early as late Oligocene to early- to mid-Miocene time (Kraemer et al. 1999; Adelmann 2001).

Fission-track ages and abundant volcanic lithologies from areas within and along the margin of the Puna provide a provenance signal that can be used to discern sediment source areas over time. The Angastaco Basin lies southeast of the Puna margin within the transition between the northern Sierras Pampeanas, the Santa Barbara System and the Cordillera Oriental at 25° 30' S (Fig. 12.3), and preserves a sedimentary sequence in excess of 6 km thickness and spanning the interval from 13.4 to 2.4 Ma (Grier et al. 1991; Strecker, unpub. data 2005). The majority of the section records deposition from westerly sources (Díaz et al. 1987; Starck and Anzótegui 2001; Coutand et al. 2006). Detrital apatite fission-track analysis and sedimentary petrography from throughout this section demonstrate that material from the interior of the Puna was apparently not deposited in this locality (Coutand et al. 2006). Much of the sediment deposited in the basin was sourced from the intervening ranges in the Eastern Cordillera and the northwestern Sierra de Quilmes (Fig. 12.3). Exhumation of these source areas commenced ca. 6 Ma prior to deposition of the base of the stratigraphic sequence (Deeken et al. 2004). Therefore, by at least 14 Ma, the Eastern Cordillera and its continuation at the Puna margin farther south (Strecker 1987; Coutand et al. 2001) had likely disrupted the fluvial system that may have once drained the Puna.

The transition from external to internal drainage within other basins of the Puna (e.g., Alonso 1986; Strecker 1987; Coira et al. 1993; Marrett 1990; Vandervoort 1993) appears to be a unifying feature of this region. For example, at about 22° S, the Salinas Grandes and Tres Cruces Basins contain evidence of deformation beginning in the late Eocene to early Oligocene (Coutand et al. 2001). At 24° S, internal drainage at Siete Curvas (SC in Fig. 12.3) may have formed as early as the late Oligocene and certainly by late Miocene time (Vandervoort et al. 1995), while to the west internal drainage within the Arizaro and Tolar Grande Basins commenced no later than early Miocene (Donato 1987; Coutand et al. 2001), and early- to mid-Miocene time (Vandervoort et al. 1995), respectively. Within the Salar de Pastos Grandes to the west of Siete Curvas, thick evaporites show that internal drainage had formed sometime between 11.2 Ma and perhaps as early as the late Eocene to early Oligocene (Alonso 1992). Within the Salar de Hombre Muerto area at 25° S, evaporite deposition and internal drainage was established by $15 \pm 1.2$ Ma, and perhaps as early as the Oligocene (Alonso et al. 1991; Vandervoort et al. 1995). In the same region the termination of supergene copper mineralization at 14.7 Ma provides a minimum age for the time when hyper-aridity was established (Alpers and Brimhall 1988). Finally, in the Sierra Chango Real along the present Puna margin at about 27° S (Fig. 12.3), apatite fission track data indicates that deformation and exhumation had begun between 29–38 Ma (Coutand et al. 2001). It is thus conceivable that the southern Puna Basins became already isolated at that time.

Deformation and the establishment of internal drainage within the Puna were linked, but diachronous and widespread in the present plateau region (e.g., Vandervoort et al. 1995; Coutand et al. 2001; Carrapa et al. 2005; Carrapa et al. 2006). The timing of the onset of internal drainage may be dependent on the details of the uplift of discrete mountain ranges as well as the construction of lateral barriers between north-south oriented basin margins by volcanic activity (e.g., Alonso et al. 1984) or faulting (e.g., Segerstrom and Turner 1972; Alonso 1992).

Taken together, the evolution of the present plateau area thus appears to have taken place in consecutive steps. First, widely distributed deformation and range uplift caused the breaking of former foreland regions east of the volcanic arc during the Oligocene, which was coupled with an intensification of arid conditions due to the evolving orographic barriers. Tectonism and arid climate conditions thus conspired in the defeat of the fluvial system and caused large amounts of material to be deposited in the adjacent closed depressions. These basin fills reach several kilometers of thickness (e.g., Jordan and Alonso 1987; Coutand et al. 2001). In some cases (e.g., the Salinas Grandes Basin and basins to the west), basins apparently filled past their spill-points into adjacent basins, causing the formerly isolated closed depressions to coalesce into a broad aggradational plain.

The combined observations concerning aridity from within the present plateau and the onset of the highly asymmetric climatic conditions along the plateau margins, indicate uplift of the plateau region in late Miocene to early Pliocene time. This second step in plateau evolution may have been coupled with mantle delamination, which caused wholesale surface uplift of the region that already had previously experienced distributed shortening and foreland compartmentalization into compressional, internally drained basins and ranges (e.g., Kay and Kay 1993; Kay et al. 1994; Allmendinger et al. 1997; Garzione et al. 2006).

## 12.7 Transient Relief Reduction in the Eastern Cordillera: Oscillatory Filling and Excavation of Marginal Basins, and the Voracity of Erosion

Structurally, the basins along the eastern margin of the Puna Plateau are virtually indistinguishable from those within the plateau. During their evolution these basins have alternated between internal drainage conditions similar to those of the plateau region and open drainage with a connection to the foreland plateau. Thus, their spatial location and the processes that control their aggradation and incision are transitional between these two environments. The intramontane basins east of the Puna currently drain into the foreland; however, many of these basins contain multiple thick conglomeratic fill units that likely formed in restricted drainage or closed basin setting (Hilley and Strecker 2005). Because thick sedimentary units that contain intercalated ash deposits are partially preserved in these basins, the timing and processes of basin infilling that are likely responsible for the characteristic topography of the Puna Plateau can be assessed here. The frequent occurrence and similarity of such basins along the plateau margin suggests that a common set of processes is responsible for their existence. In general, due to the easterly moisture sources, the establishment of orographic barriers by surface uplift concentrates erosion along the windward eastern range fronts, whereas the leeward basins become increasingly arid. Basin outlets in these environments always coincide with structurally complex parts of the orogen where along-strike changes in displacement allow the fluvial system to either remain or be easily reconnected to the foreland. Another characteristic of these sectors of the orogen is that they coincide with high precipitation gradients that may aid in ultimately regaining and maintaining external drainage conditions. However, changes in climate, tectonic rates or the unroofing of resistant units may have led repeatedly to reduced evacuation of sediment, increased sediment storage, and potentially to basin isolation.

A type example for such a setting is the small Toro Basin at about 24° S (Fig. 12.3). The regional tectonic and stratigraphic relationships in the Toro Basin area record the transition from a foreland basin to an intramontane basin setting. Today, the basin lies in an arid environment and is bounded on the west and east by high basement ranges, between 3 and 5 km high. The Toro Basin is drained by the Río Toro, which traverses the Sierra Pasha to the east through a constricted bedrock gorge. However, before about 6 Ma this region was an integral part of a foreland basin, with sediments sourced at the Puna margin transported eastward. Between 6 and 0.98 Ma exposed conglomerates in the Toro Basin record a changeover from a foreland to an intramontane, probably closed-basin setting (Marrett and Strecker 2000; Hilley and Strecker 2005). This basin-filling episode was in turn superseded by deformation, basin excavation, and removal of conglomerates sometime after 0.98 Ma, which produced a pronounced unconformity. The unconformity is overlain by an undeformed conglomeratic basin-fill unit, at least 600 m thick. Conglomeratic gravel from this filling episode 600 m above the present trunk stream records the top of this aggradational surface, and its geometry suggests that the basin became internally-drained again as it filled with sediments. This last fill episode may have occurred as the result of an increase in uplift rate within the southern Sierra Pasha that led to steepening and tectonic defeat of the Río Toro, a trend toward aridity in the basin region, the exposure of Precambrian rocks resistant to fluvial incision in the Río Toro gorge, an increase in the production of coarse sediment associated with a more erosive climate, or some combination of these factors. In any case, the landscape that was likely present during the time that the Toro Basin was internally-drained consisted of a broad, aggradational, low-relief surface surrounded by mountain peaks, identical to the landscape of the Puna Plateau today. However, rather than remaining internally-drained, the basin was subsequently recaptured, leading to the partial excavation of the conglomeratic basin fill units and a re-establishment of foreland-basin connectivity.

Similar observations of basin filling and re-excavation can be made in the Santa María Valley farther south. For example, following uplift of the Aconquija and Calchaquí Ranges and the Puna margin the Andean foreland was further segmented by the uplift of the southern part of Sierra de Quilmes after 5.4 Ma (Strecker et al. 1989), which created the intramontane Santa Maria and Cajón Basins and caused widespread deformation and erosion of the late Tertiary sedimentary sequences. This was followed by a phase of erosion and subsequent conglomerate deposition after 2.9 Ma that covered the erosional paleo-topography and probably created transient internal drainage conditions. The deposition of conglomerates, several hundred meters thick, was followed by episodic incision and formation of terraced pediments and fluvial terraces that record the stepwise drainage reintegration (Strecker et al. 1989).

Other impressive examples for this type of basin evolution along the eastern Puna can also be found in the Quebrada de Humahuaca, Calchaquí, Cajón, Hualfín, and Fíambalá Basins. All of them record multiple filling and isolation events that were followed by sediment evacuation. After the intramontane basin stage had been attained, such filling events occurred after 2.7 in the Humahuaca Basin, after 3.6 Ma in the Fiambalá, and after 2.4 Ma in the Calchaquí Basins (Strecker, unpubl. data). The evolution of these basins is thus highly diachronous, but tectonism and similar processes of erosion and sedimentation have led to indistinguishable tectonic deformation features and facies associations, respectively. These observations emphasize the complex interplay between tectonism, sedimentation, and subsequent efficient headward erosion processes at the periphery of a tectonically active orogen, located in proximity to pronounced climatic gradients.

## 12.8 Possible Feedbacks between Erosion and Tectonics in the Central Andes

When viewed at a regional scale, the southern central Andes comprise three principal morphotectonic settings, in which coupling between erosion and deformation may be important, but profoundly different between each individual province. The major morphotectonic divisions in this respect are based on mechanical and erosional differences throughout this region. First, we consider the internally-drained Puna-Altiplano and its adjacent externally drained basins to be governed by a distinctive set of erosional and mechanical conditions. Second, the externally-draining Precordillera fold-and-thrust belt of central Argentina (Fig. 12.1) is located where thick basin sediments with mechanically weak layers may allow low-angle decollements to develop (e.g., Ramos et al. 2002). Thirdly, the basement-cored uplift province of the Sierras Pampeanas has internally and externally drained sectors, but horizontal, mechanically weak layers are absent. In this setting deformation is primarily accommodated by high-angle structures generated during previous episodes of faulting, involving suturing of terranes to the South American craton (e.g., Allmendinger et al. 1983; Jordan and Allmendinger 1986; Ramos et al. 2002).

First, Sobel et al. (2003) and Hilley and Strecker (2005) investigated the conditions that promote basin isolation from the foreland base-level and the resulting mechanical implications. Based on bedrock incision, erosion and fluvial aggradation laws, they determined the controlling factors and conditions that lead to the disconnection of the downstream fluvial system from the upstream intramontane basin. In their formulation, the establishment of internal drainage is a threshold process in which internal drainage ensues if uplift rates are high, rocks are resistant to fluvial incision, and/or precipitation is low. In case of the Puna-Altiplano Plateau, changes in precipitation that result from the construction of topography may starve moisture from the headwaters of the trunk stream, favoring internal drainage and storage of material. Consequently, the trapped mass will increase the lithostatic load, which may cause regional faults to become unfavorable to accommodate further deformation (e.g., Royden 1996; Willett 1999). Thus, contractional deformation may be forced to migrate toward lower elevations in the foreland. Therefore, if the threshold of internal drainage is crossed, as basins fill and coalesce in the internal portions of the orogen, deformation may eventually cease there. Likewise, if recapture of these areas should occur and material were rapidly evacuated from them, deformation is expected to migrate back into the interior of the orogen as sediment load decreases.

Evidence for this process exists within the Puna Plateau and along its margins. In Oligocene through Miocene time, shortening was taken up within the present plateau and along its future margins. As average elevations increased due to shortening (e.g., Allmendinger et al. 1997), the establishment of internal drainage and coeval basin filling (Sobel et al. 2003), and perhaps thermal buoyancy effects associated with the delamination of the mantle lithosphere in the area (Kay et al. 1994), the lithostatic stresses exerted by these highlands may have favored a migration of deformation along its margins, at the expense of shortening in the plateau area. During the Pliocene and Quaternary shortening was negligible in this environment. In fact, during the ultimate 2 Ma strike-slip and normal faulting have prevailed in the Puna region, whereas only limited evidence for shortening can be found during that time (Strecker 1987; Allmendinger et al. 1986; Marrett et al. 1994; Cladhous et al. 1994; Schoenbohm and Strecker 2005). In contrast, since early Pliocene time all active shortening in the orogen has been focused along the plateau margin and the foreland. Therefore, it is plausible that the combination of tectonic, thermal, and geomorphic processes created and maintained the high topography of the Puna, but at the same time the creation of high topography may also moderate deformation in the area.

Further support for this assessment may be observed in the intramontane basins along the eastern Puna margin. Here, the timing of motion along faults within the basin appears to correspond to periods when basin fill had been largely removed. For example, the reactivation of basin-bounding faults temporally coincides with the excavation of basin fill in the Toro, Humahuaca, Santa María and Cajón Basins (Strecker et al. 1989; Hilley and Strecker 2005). Therefore, there may exist a causative relationship between phases of basin filling, variations in lithostatic loading, and activity along the basin-bounding structures, which is ultimately controlled by the overall climatic conditions.

The second type of coupling between erosion and tectonism can be observed in the Cordillera fold-and-thrust belt at about 30° S. Here, basin sediments with many detachment levels in the mechanical stratigraphy allow shortening to be accommodated along low-angle structures (Ramos et al. 2002). As foreland material is incorporated into the fold-and-thrust belt, the geometry of and deformation within the mountain belt must change to accommodate the volume increase caused by the added material. As the volume of the orogen increases, it may widen as the basal detachment deepens and surface slopes increase. The latter effect may change the lithostatic stresses within the thrust belt, promoting a change in the basal detachment geometry (e.g., Davis et al. 1983; Dahlen 1984). While tectonic deformation causes addition, erosion removes a portion or all of this material from the orogen (e.g., Dahlen and Suppe 1988; Dahlen and Barr 1989). Thus, erosion relative to the rate of tectonic accretion may play an important role in determining the stresses and deformation in the orogen (e.g., Dahlen and Suppe 1988; Hilley et al. 2004; Hilley and Strecker 2004; Whipple and Meade 2004).

As lithostatic stresses within the orogen change in the course of material removal, a response in the geometry of and deformation within the fold-and-thrust belt may be necessary. Therefore, unlike the case of the Puna Plateau in which internal drainage creates a threshold beyond which erosional mass export is prohibited, when externally-draining fold-and-thrust belts are being eroded, their deformation and geometry may change continuously as topography is being built. This creates a direct feedback between tectonic deformation and erosional processes (e.g., Beaumont et al. 1992; Willett et al. 1993; Willett 1999).

Using a combination of theoretical models and field data (e.g., Davis et al. 1983; Suppe 1981; Ramos et al. 2002), Hilley et al. (2004) and Hilley and Strecker (2004) presented the first-order relationships expected between tectonic accretion and the erosion of a fold-and-thrust belt, and tested these relationships in the Precordillera of Argentina, and the Himalayan and Taiwanese orogens. They found that as erosional removal of rock increased relative to tectonic accretion, the fold-and-thrust belts would be narrower and experience more out-of-sequence deformation (e.g., Dahlen 1984) than would be expected when erosion were inefficient. These general results quantitatively agree with the erosional rates of removal from and tectonic rates of addition to the Precordillera thrust belt in the flat-slab region of the Andes, and the deformation and geometry observed in that area. In addition, the geometry of the Sierras Subandinas in the Bolivian Andes qualitatively supports these predictions. For example, the northern portion of this fold-and-thrust belt is narrow, out-of-sequence deformation is common, and precipitation and erosional processes are efficient (e.g., Horton 1999). Conversely, the southern sector of this tectonic province corresponds with reduced precipitation a wider fold-and-thrust belt, regional erosion surfaces are preserved, and deformation has successively migrated eastward (e.g., Horton 1999). Therefore, as erosion becomes less efficient in the Subandean belt, systematic changes in geometry and deformation are observed.

In the third scenario of coupled erosion and deformation processes Hilley et al. (2005) analyzed the Sierras Pampeanas structural province where the absence of thick foreland sediments does not allow deformation to be taken up along low-angle detachment structures. Instead, deformation appears to be focused on reactivated crustal structures that predate the Cenozoic Andean orogeny. In the case that these high-angle structures are weak enough to be mechanically favored for accommodating plate convergence, the failure stresses along structures may greatly exceed failure stresses within the mountain belt (Dahlen 1984), allowing slip along the fault without internal deformation within the range. This contrasts with fold-and-thrust belts, where failure stresses throughout the orogen may be approximately equivalent (e.g., Davis et al. 1983). Because these conditions result in a mechanically stable condition in which basal sliding accommodates horizontal shortening along the weakest structure in the crust, in the absence of any additional pre-existing structures, deformation is expected to be exclusively taken up along this structure. However, if other pre-existing structures exist, the construction of topography above the growing mountain range may change the loading conditions along its base, potentially favoring failure of a different pre-existing structure with little or no topography above it. In a hypothetical landscape with no initial topography in which deformation commences along the frictionally weakest structure, movement along this structure changes the lithostatic loading along the fault plane. If topographic loading is large, deformation may be either forced to migrate to a frictionally stronger structure with no topographic load above it. Therefore, in tectonic environments such as the Sierras Pampeanas or structurally similar settings such as the Tien Shan in Central Asia or the Laramide uplifts in North America, over short timescales deformation may be localized on the most mechanically favored pre-existing structure. However, as topography is built, deformation may be distributed between many similarly stressed pre-existing structures. The unsystematic deformation and uplift history of the Sierras Pampeanas structural province underscores this assessment (Strecker et al. 1989).

When deformation within orogens is strongly influenced by the presence of pre-existing structures, surface slopes may steepen without triggering a change in the orogenic width or basal decollement angle as is the case for fold-and-thrust belts (Hilley et al. 2004; Hilley and Strecker 2004; Whipple and Meade 2004). As slopes steepen, erosional efficiency may increase (e.g., Ahnert 1970; Howard and Kerby 1983) and hence the proportion of rock removed from the orogen by erosion increases relative to that introduced by shortening. In the case that the rate of material removal from an orogen is balanced by the introduction of rock by deformation (Willett and Brandon 2001), topography may attain an equilibrium in which slopes and elevations remain constant with time (e.g., Hilley and Strecker 2005). If this equilibrium topography is insufficient to force deformation to move to stronger structures with no initial topography, erosional removal of rock from an orogen may allow deformation to remain concentrated along a single set of weak structures, despite the presence of other similarly weak structures in the crust. Therefore, while erosion in these tectonic environments may not be directly coupled to deformation as in fold-and-thrust belts, it ultimately moderates the slopes that may force deformation to migrate to other structures. The efficiency of erosion in relation to the tectonic uplift rate thus controls the threshold that determines if deformation may remain focused on a set of active mountain fronts or migrates to equally weak structures, similar to what Sobel and Strecker (2003) predicted for the different evolution of Sierra Aconquija and Cumbres Calchaquíes in the northernmost Sierras Pampeanas.

## 12.9 Concluding Remarks

The history of the Puna Plateau and its marginal basins and ranges in the Eastern Cordillera and the northern Sierras Pampeanas impressively documents the effects of tectonics and topography on atmospheric circulation pattern, the development of orographic barriers, and their influence on erosion and landscape evolution at various timescales. Available sedimentologic, paleontologic and stable isotope data sets show a major shift from aridity toward increased humidity along the eastern border of the Puna Plateau in late Miocene to early Pliocene time due to the establishment of orographic barriers, high enough to intercept moisture-bearing winds.

The evolution of this region also emphasizes that there are several pathways by which tectonic activity in an orogen may be coupled to the effects of climate and erosion. Our field studies and theoretical analyses indicate, however that the nature of this coupling may also vary strongly depending on the pre-orogenic geologic history. This is emphasized by the width of the present Puna Plateau and the adjacent tectonic provinces and their relation to the Cretaceous Salta Rift (e.g., Allmendinger et al. 1983; Grier et al. 1991; Kley and Monaldi 2002; Kley et al. 2005; Mon et al. 2005; Hilley et al. 2004) and those structures within the present plateau that were generated during a different subduction regime in Oligocene time (e.g., Kay et al. 1999; Kraemer et al. 1999; Carrapa et al. 2005). The width of these tectonic provinces in the Central Andes is thus a first-order result of structural inheritance and tectonic conditions and may not be primarily a function of the arid climate, as envisioned by Montgomery et al. (2001).

Nevertheless, the fact that the Puna region is located in an area of pronounced inherent aridity has helped to create and maintain the second largest orogenic plateau on Earth (Lamb and Davis 2003; Sobel et al. 2003), but the climatic conditions along the eastern Puna also prevent further eastward expansion due to cyclic intramontane basin filling, headward erosion, re-exhumation of basin fills, and renewed tectonic activity of basin bounding reverse faults (Hilley and Strecker 2005).

The broad Oligo-Miocene deformation of the present plateau region set the stage for the landscape that characterizes this region now. Aridity and tectonism conspired in isolating the local base levels of basins in the plateau region from the undeformed foreland, resulting in a reduction of erosional capacity, a redistribution of mass and filling of basins in the orogen interior. Wholesale plateau uplift due to mantle delamination as envisioned by Kay et al. (1994) and Garzione et al. (2006) in combination with shortening and filling of internally drained basins and conservation of material within the orogen as suggested herein are thus expected to have had a profound impact on the distribution of stresses in the orogen and tectonic deformation in adjacent regions. Due to an increased gravitational potential in the interior of the orogen it is therefore likely that deformation migrated toward foreland areas, affecting pre-existing weak structures that gave rise to new range bounding faults. This can be observed in the northern Sierras Pampeanas or in the Santa Barbara structural province in the transition between the Sierras Pampeanas and the Eastern Cordillera (Allmendinger et al. 1983; Grier et al. 1991; Sobel and Strecker 2003; Sobel et al. 2003; Hilley et al. 2005), where the late Miocene to Pliocene onset of range uplifts are in stark contrast to structurally similar ranges within the plateau.

If the majority of future faults has similar strikes to those observed in the compartmentalized foreland now and if the associated ranges are capable to intercept eastwardly derived moisture, the aridification of the NW Argentine Andes will successively migrate eastward and the plateau region with its internally drained basins may prevail as a morpho-structural entity, and even expand eastward over time. The inability of effectively removing mass from the interior of the orogen over long timescales leaves the southern Central Andes in a situation, in which it will be difficult to reach erosional steady-state conditions.

In conclusion, climate conditions have been profoundly influenced by the tectonic evolution of the NW Argentine Andes, which is expressed by the pronounced climate gradients and the different surface processes and rates on the windward flanks versus the arid interior of the orogen. However, there is also a complex influence of climate and related surface processes on tectonic style and tectonic activity in this orogen. Prevailing aridity in the core zone of the southern central Andes (e.g. Hartley and Chong 2002; Hartley 2003) has helped to create and maintain the Puna Plateau morphotectonic province, which resulted in low exhumation of basement uplifts, and caused the low erosional capacity of fluvial systems that eventually became isolated from the low-elevation foreland, while deformation stepped eastward.

## Acknowledgments

We thank the German Research Council (Deutsche Forschungsgemeinschaft) for generous financial support. Many colleagues and friends have contributed and helped during our research in the Andes, which made our work a successful and pleasant experience. In particular, we would like to thank R. Allmendinger, A. Ahumada, A. Bonvecchi, G. Bossi, L. Chiavetti, L. Fauque, K. Haselton, R. Herbst, R. Hermanns, T. Jordan, S. Kay, K. Kleinert, R. Marrett, R. Mon, E. Mortimer, V. Ramos, J. Sayago, J. Sosa Gomez, I. Vila, J. Viramonte, and all members of the SFB 267. We thank G. Chong, A. Hartley, and M. Suárez for helpful reviews and thank B. Fabian for artwork. M. Strecker also acknowledges the A. Cox Fund of Stanford University for financial support.

## References

Adelmann D (2001) Känozoische Beckenentwicklung in der südlichen Puna am Beispiel des Salar de Antofolla (NW-Argentinien). PhD thesis, Freie Universität Berlin

Ahnert F (1970) Functional relationship between denudation, relief, and uplift in large mid-latitude drainage basins. Am J Sci 268: 243–263

Allmendinger RW, Ramos VA, Jordan TE, Palma M, Isacks BL (1983) Paleogeography and Andean structural geometry, northwest Argentina. Tectonics 2:1–16

Allmendinger RW, Strecker MR, Eremchuk J, Francis P (1989) Neotectonic deformation of the southern Puna Plateau, northwestern Argentina, J S Am Earth Sci 2: 111–130

Allmendinger RW, Jordan TE, Kay SM, Isacks BL (1997) Evolution of the Puna-Altiplano Plateau of the central Andes. Ann Rev Earth Planet Sci 25:139–174

Andriessen, PAM, Reutter KJ (1994) K-Ar and fission track mineral age determinations of igneous rocks related to multiple magmatic arc systems along the 23 degrees S latitude of Chile and NW Argentina. In: Reutter KJ Scheuber E, Wigger PJ (eds) Tectonics of the southern Central Andes: structure and evolution of an active continental margin. Springer-Verlag, Berlin Heidelberg New York, pp 141–153

Alonso RN (1986) Occurencia, posición estratigráfica y génesis de boratos de la Puna Argentina. PhD thesis, Universidad Nacional de Salta

Alonso RN (1992) Estratigrafia del Cenozoico de la cuenca de Pastos Grandes (Puna Salteña) con énfasis en la Formación Sijes y sus boratos. Rev Asoc Geol Arg, 47: 189–199

Alonso R, Viramonte J, Gutiérrez R (1984) Puna Austral – Bases para el subprovincialismo Geológico de la Puna Argentina. Noveno Congreso Geológico Argentino, S.C. Bariloche, Actas, pp 43–63

Alonso RN, Jordan TE, Tabbutt KT, Vandervoort DS (1991) Giant evaporite belts of the Neogene central Andes. Geology 19:401–404

Alpers CN, Brimhall GH (1988) Middle Miocene climatic change in the Atacama Desert, northern Chile: evidence from supergene mineralization at La Escondida. Geol Soc Am Bull 100: 1640–1656

Ammann C, Jenny B, Kammer K, Messerli B (2001) Late Quaternary Glacier response to humidity changes in the arid Andes of Chile (18–29°S). Palaeogeo Palaeoclim Palaeoecol 172:313–326

Anzótegui LM (1998) Hojas de Angiospermas de la formación Palo Pintado, Mioceno Superior, Salta, Argentina. Parte I : Anacardiaceae, Lauraceae y Moraceae. Ameghiniana, (Rev. Asoc. Paleontol. Argent.) 35: 25–32

Baker PA, Rigsby CA, Seltzer GO, Fritz SC, Lowenstein TK, Bacher NP, Veliz C (2001) Tropical climate changes at millenial and orbital timescales on the Bolivian Altiplano. Nature 409:698–701

Bangs NL, Cande SC (1997) Episodic development of a convergent margin inferred from structures and processes along the southern Chile margin. Tectonics 16:489–503

Bianchi AR, Yañez CE (1992) Las precipitaciones en el noroeste Argentino. Instituto Nacional de Tecnologia Agropecuaria, Estacíon Experimental Agropecuaria Salta, Argentina: 393 pp

Blisniuk PM, Stern LA (2006) Stable isotope paleo-altimetry: a critical review. Am J Sci, in press

Blisniuk PM, Stern LA, Chamberlain CP, Idleman B, Zeitler PK (2005) Climatic and ecologic changes during Miocene uplift in the Southern Patagonian Andes. Earth Planet Sci Lett 230:125–142

Bond M, Cerdeño E, López G (1995) Los ungulados nativos de América del sur. Evolución biológica y climática de la región Pampeana durante los últimos cinco millones de años. In: Alberdi MT, Leone G, Tonni EP (eds) Museo Nacional de Ciencias Naturales de Madrid Monografias, chapter 12. pp 259–275

Bookhagen B, Burbank DW (2006) Topography, relief, and TRMM-derived rainfall variations along the Himalaya. Geophys Res Lett 33: doi 10.1029/2006GL026037

Bookhagen B, Haselton K, Trauth MH (2001) Hydrologic modelling of a Pleistocene landslide-dammed lake in the Santa María Basin, NW Argentina. Paleogeog Paleoclim Paleoecol 169: 113–127

Bossi GE, Georgieff SM, Gavriloff IJC, Ibañez LM, Muruaga CM (2001) Cenozoic evolution of the intramontane Santa Maria basin, Pampean Ranges, northwestern Argentina. J S Am Earth Sci 14: 725–734

Bouzari F, Clark AH (2002) Anatomy, evolution, and metallogenic significance of the supergene orebody of the Cerro Colorado porphyry copper deposit, I region, northern Chile. Econ Geol 97:1701–1740

Brooks BA, Bevis M, Smalley R Jr, Kendrick E, Manceda R, Lauria E, Maturana R, Araujo M (2003) Crustal motion in the Southern Andes (26°–36°S): Do the Andes behave like a microplate? Geochem Geophys Geosys 4:1–14

Burbank DW (2002) Rates of erosion and their implications for exhumation. Min Mag 66(1): 25–52

Cabrera A (1994) Regiones fitogegráficas de Argentina. ACME, Buenos Aires

Carrapa B, Adelmann D, Hilley GE, Mortimer E, Sobel ER, Strecker MR (2005) Oligocene range uplift and development of plateau morphology in the southern Central Andes. Tectonics 24: doi 10.1029/2004TC001762

Carrapa B, Strecker MR, Sobel E (2006) Sedimentary, tectonic and thermochronologic evolution of the southernmost end of the Puna Plateau (NW Argentina). Earth Planet Sci Lett in press

Cerling TE, , Wang Y, Quade J (1993) Expansion of C4 ecosystems as an indicator of global ecological changes in the late Miocene. Nature 361:344–345

Cladhous TT, Allmendinger RW, Coira B, Farrar E (1994) Late Cenozoic deformation in the Central Andes: fault kinematics from the northern Puna, northwestern Argentina and southwestern Bolivia. J S Am Earth Sci 7:209–228

Coira B, Kay SM, Viramonte J (1993) Upper Cenozoic magmatic evolution of the Argentina Puna – a model for changing subduction geometry. Int Geol Rev 35:677–720

Coughlin TJ, O'Sullivan PB, Kohn B, and Holcombe RJ (1998) Apatite fission-track thermochronology of the Sierras Pampeanas, central western Argentina: implications for the mechanism of plateau uplift in the Andes. Geology 26:999–1002

Coutand I, Cobbold PR, De Urreiztieta M, Gautier P, Chauvin A, Gapais D, Rossello EA, Lòpez-Gamundí O (2001) Style and history of Andean deformation, Puna plateau, Northwestern Argentina. Tectonics 20:210–234

Coutand I, Carrapa B, Deeken A, Schmitt AK, Sobel ER, Strecker MR (2006) Orogenic plateau formation and lateral growth of compressional basins and ranges: insights from sandstone petrography and detrital apatite fission-track thermochronology in the Angastaco Basin, NW Argentina. Basin Research 18:1–26

Dahlen FA (1984) Noncohesive critical Coulomb wedges: an exact solution. J Geophys Res 89:10125–10133

Dahlen FA, Barr TD (1989) Brittle frictional mountain building: 1. Deformation and mechanical energy budget. J Geophys Res 94:3906–3922

Dahlen FA, Suppe J (1988) Mechanics, growth, and erosion of mountain belts. Spec Pap Geol Soc Am 218:161–178

Davis D, Suppe J, Dahlen FA (1983) Mechanics of fold-and-thrust belts and accretionary wedges. J Geophys Res 88:1153–1172

Deeken A, Sobel ER, Haschke M, Strecker MR, Riller U (2004) Age of initiation and growth pattern of the Puna Plateau, NW-Argentina, constrained by AFT Thermochronology. International Fission Track Conference Amsterdam, Abstract Volume 82

Díaz JI, Malizia DC, and Bossi GE (1987) Análisis estratigráfico y sedimentológico del Grupo Payogastilla. X Congreso Geológico Argentino, Tucumán, Actas 113–117

Donato E (1987) Características estructurales del sector occidental de la Puna Salteña. Bol Inv Petrol 12:89–99

Dunai TJ, González López GA, and Juez-Larré J (2005) Oligocene/Miocene age of aridity in the Atacama Desert revealed by exposure dating of erosion sensitive landforms. Geology 33:321–324

Ege H (2004) Tectono-sedimentary evolution of the southern Altiplano: basin evolution, thermochronology and structural geology. PhD thesis, Freie Universität Berlin

Fauqué L, and Strecker MR (1988) Large rock avalanche deposits (Sturzströme, sturzstroms) at Sierra Aconquija, northern Sierras Pampeanas, Argentina. Ecl Geol Helv 81:579–592

Fauqué L, Tschilinguirian P (2002) Villavil rockslides, Catamarca Province, Argentina. In: Evans SG, DeGraff JV (eds) Catastrophic landslides: effects, occurrence, and mechanism. Geol Soc Am Rev Eng Geol, Boulder CL, pp 303–324

Garreaud R, Aceituno, P. (2001) Interannual rainfall variability over the South American Altiplano. J Clim 14:2779–2789

Garreaud R, Vuille M, Clement AC (2003) The climate of the Altiplano: observed current conditions and mechanisms of past changes. Palaeogeo Palaeoclim Paleoecol 194:5–22

Garzione CN, Molnar P, Libarkin JC, MacFadden BJ (2006) Rapid late Miocene rise of the Bolivian Altiplano: evidence for removal of mantle lithosphere. Earth Planet Sci Lett 241:543–556

Gaupp R, Kött A, Wörner G (1999) Palaeoclimatic implications of Mio-Pliocene sedimentation in the high-altitude intra-arc Lauca Basin of northern Chile. Palaeogeo Palaeoclim Paleoecol 151:79–100

Gregory-Wodzicki KM (2000) Uplift history of the central and northern Andes: a review. Geol Soc Am Bull 112:1091–1105

Grier ME, Salfity JA, Allmendinger RW (1991) Andean reactivation of the Cretaceous Salta rift, northwestern Argentina. J S Am Earth Sci 4:351–372

González O, Mon R (1996) Evolución tectónica del extremo norte de las Sierras Pampeanas y su transición a la Cordillera Oriental y a las Sierras Subandinas. XIII Congreso Geológico Argentino y III Congreso de Exploración de Hidrocarburos II, Buenos Aires, pp 149–160

Gutmann GJ, Schwerdtfeger W (1965) The role of latent and sensible heat for the development of a high pressure system over the tropical Andes in the summer. Meteor Rundsch 18:69–75

Hartley AJ (2003) Andean uplift and climate change. J Geol Soc Lond 160:7–10

Hartley AJ, Chong G (2002) Late Pliocene age for the Atacama Desert: implications for the desertification of western South America. Geology 30: 43–46

Hartley AJ, Flint S, Turner P, and Jolley EJ (1992) Tectonic controls on the development of a semi-arid, alluvial basis as reflected in the stratigraphy of the Purilactis Group (Upper Cretaceous-Eocene), northern Chile. J S Am Earth Sci 5:275–296

Haselton K, Hilley G, Strecker MR (2002) Average Pleistocene climatic patterns in the southern Central Andes: controls on mountain glaciation and palaeoclimate implications. J Geol 110:211–226

Hastenrath S (1991) Climate dynamics of the Tropics. Kluwer, Dordrecht

Hermanns RL, Strecker MR (1999) Structural and lithological controls on large Quaternary rock avalanches (sturzstroms) in arid northwestern Argentina. Geol Soc Am Bull 111:934–948

Hermanns RL, Trauth MH, Niedermann S, McWilliams M, Strecker MR (2000) Tephrochronologic constraints on temporal distribution of large landslides in NW-Argentina. J Geol 108:35–52

Hermanns RL, Niedermann S, Villanueva García A, Gomez JS, Strecker MR (2001) Neotectonics and catastrophic failure of mountain fronts in the southern intra-Andean Puna Plateau, Argentina. Geology 29:619–623

Hilley GE, Strecker MR (2004) Steady state erosion of critical Coulomb wedges with applications to Taiwan and the Himalaya. J Geophys Res 109(B1): doi 10.1029/2002 JB002284

Hilley GE, Strecker MR (2005) Processes of oscillatory basin infilling and excavation in a tectonically active orogen: Quebrada del Toro Basin, NW Argentina. Geol Soc Am Bull 117:887–901

Hilley GE, Strecker MR, Ramos VA (2004) Growth and erosion of fold-and-thrust belts with an application to the Aconcagua fold-and-thrust belt, Argentina. J Geophys Res 109(B1) doi 10.1029/2002 JB002282

Hilley GE, Blisniuk PM, Strecker MR (2005) Mechanics and erosion of basement-cored uplift provinces. J Geophys Res 110: doi 101029/2005JB003704

Hoke GD, Isacks BL, Jordan TE and Yu JS (2004) Groundwater-sapping origin for the giant quebradas of northern Chile. Geology 32:605–608

Horton BK (1999) Erosional control on the geometry and kinematics of thrust belt development in the central Andes. Tectonics 18:1292–1304

Howard AD, Kerby G (1983) Channel changes in badlands. Geol Soc Am Bull 94:739–752

Jezek P, Willner AP, Aceñolaza FG, Miller H (1985) The Puncoviscana trough. A large basin of Late Precambrian to Early Cambrian age on the Pacific edge of the Brazilian shield. Geol Rundsch 74:573–584

Jordan TE, Allmendinger RW (1986) The Sierras Pampeanas of Argentina: a modern analogue of Laramide deformation. Am J Sci 286:737–764

Jordan TE, Alonso RN (1987) Cenozoic stratigraphy and basin tectonics of the Andes mountains, 20°–28° South Latitude. Am Assoc Petr Geol Bull 71:49–64

Jordan TE, Isacks BL, Allmendinger RW, Brewer JA, Ramos VA, Ando CJ (1983) Andean tectonics related to the geometry of the subducted Nazca Plate. Geol Soc Am Bull 94:341–361

Jordan TE, Zeitler P, Ramos V, Gleadow AJW (1989) Thermochronometric data on the development of the basement peneplain Sierras Pampeanas, Argentina. J S Am Earth Sci 2:207–222

Jordan T E, Allmendinger R W, Damanti J F, Drake, R (1993) Chronology of motion in a complete thrust belt: the Precordillera, 30–31°S, Andes Mountains. J Geol101:135–156

Juez-Larré J, Dunai T, González López GA (2005) Unraveling the link between climate and tectonic forces along the Andean margin of Chile, by means of thermochronological and exposure age dating. Geol Soc Am Ann Meeting, Salt Lake City, Abstract Volume, pp 83–11

Kay RW, Kay SM (1993) Delamination and delamination magmatism. Tectonophysics 219:177–189

Kay SM, Coira B, Viramonte J (1994) Young mafic back-arc volcanic rocks as indicators of continental lithospheric delamination beneath the Argentine Puna plateau, Central Andes. J Geophys Res 99:24323–24339

Kay SM, Mpodozis C, Coira B (1999) Magmatism, tectonism, and mineral deposits of the Central Andes (22°–33°S latitude). In: Skinner BJ (ed) Geology and Ore Deposits of the Central Andes. Soc Econ Geol Spec Pub 7:27–59

Kiladis GN, Diaz H (1989) Global climatic anomalies associated with extremes in the Southern Oscillation. J Clim 2:1069–1090

Kleinert K, Strecker MR (2001) Changes in moisture regime and ecology in response to late Cenozoic orographic barriers: the Santa Maria Valley, Argentina. Geol Soc Am Bull 113:728–742

Kley J, Monaldi CR (2002) Tectonic inversion in the Santa Barbara System of the central Andean foreland thrust belt, northwestern Argentina. Tectonics 21:1–18

Kley J, Rossello EA, Monadi CR, Habighorst B (2005) Seismic and field evidence for selective inversion of Cretaceous normal faults, Salta rift, northwest Argentina. Tectonophysics 399:155–172

Kraemer B, Adelmann D, Alten M, Schnurr W, Erpenstein K, Kiefer E, Van den Bogaard P, Görler K (1999) Incorporation of the Paleogene foreland into Neogene Puna plateau: the Salar de Antofolla, NW Argentina. J S Am Earth Sci 12:157–182

Koons PO (1989) The topographic evolution of collisional mountain belts: a numerical look at the Southern Alps, New Zealand. Am J Sci 289:1041–1069

Lamb S, Davis P (2003) Cenozoic climate change as a possible cause for the rise of the Andes. Nature 425:792–797

Latorre C, Quade J, McIntosh WC (1997) The expansion of C-4 grasses and global change in the late Miocene: stable isotope evidence from the Americas. Earth Planet Sci Lett 146:83–96

Lenters JD, Cook KH (1995) Simulation and diagnosis of the regional summertime precipitation climatology of South America. J Clim 8:2988–3005

Lenters JD, Cook KH (1997) On the origin of the Bolivian High and related circulation features of the South American climate. J Atm Sci 54:656–677

Lutz AI (1987) Estudio anatómico de maderas terciarias del valle de Santa María (Catamarca-Tucumán). Facena 7:125–143

Maksaev V, Zentilli M (1999) Fission track thermochronology of the Domeyko Cordillera, northern Chile: implications for Andean tectonics and porphyry copper metallogenesis. Explor Min Geol 8:65–89

Markgraf V, Seltzer GO (2001) Pole-equator-pole paleoclimates of the Americas integration: toward the big picture. In: Markgraf V (ed) Interhemispheric climate linkages. Academic Press, San Diego, pp 433–442

Marquillas RA, Del Papa C, Sabino IF (2005) Sedimentary aspects and paleoenvironmental evolution of a rift basin: Salta Group (Cretaceous-Paleogene), northwestern Argentina. Int J Earth Sci 94:94–113

Marrett RA (1990) The late Cenozoic tectonic evolution of the Puna plateau and adjacent foreland, northwestern Argentine Andes. Cornell University, Ithaca NY

Marrett R, Strecker MR (2000) Response of intracontinental deformation in the central Andes to late Cenozoic reorganization of South American Plate motions. Tectonics 19:452–467

Marrett RA, Allmendinger RW, Alonso RN, Drake R (1994) Late Cenozoic tectonic evolution of the Puna Plateau and adjacent foreland, northwestern Argentine Andes. J S Am Earth Sci 7:179–207

Marshall LG, Patterson B (1981) Geology and geochronology of the mammal-bearing Tertiary of the Valle de Santa Maria and Rio Corral Quemado, Catamarca Province, Argentina. Fieldiana. Geology 9:1–80

Marshall LG, Butler RF, Drake RE, Curtis GH (1982) Geochronology of type Uquian land mammal age, Argentina. Science 216:986–989

Marwan N, Trauth MH, Vuille M, Kurths J (2003) Nonlinear time-series analysis on present-day and Pleistocene precipitation data from the NW Argentine Andes. Clim Dyn 21:317–326

Masek JG, Isacks BL, Gubbels TL, Fielding EJ (1994) Erosion and tectonics at the margins of continental plateaus. J Geophys Res 99:13941–13956

Menéndez C (1962) Leño petrificado de una Leguminosa del Terciario de Tiopunco, Prov. de Tucumán. Ameghiniana (Rev. Asociación Paleontológica Argentina) II:121–126

Mon R (1979) Esquéma estructural del Noroeste Argentino. Rev Asoc Geol Arg 35:53–60

Mon R, Salfity JA (1995) Tectonic evolution of the Andes of northern Argentina. In: Tankard AJ, Suárez Soruco R, Welsink HJ (eds) Petroleum basins of South America. Am Assoc Petr Geol Mem 62:269–283

Mon R, Monaldi CR, Salfity JA (2005) Curved structures and interference fold patterns associated with lateral ramps in the Eastern Cordillera, Central Andes of Argentina. Tectonophysics 399:173–179

Montgomery DR, Balco G, Willett SD (2001) Climate, tectonics, and the morphology of the Andes. Geology 29:579–582

Mortimer E, Schoenbohm L, Carrapa B, Sobel ER, Sosa Gomez J, Strecker MR (in rev) Compartmentalization of a foreland basin in response to plateau growth and diachronous thrusting: El Cajón-Campo Arenal basin, NW Argentina. Geol Soc Am Bull

Nasif N, Musalem S, Esteban G, Herbst R (1997) Primer registro de vertebrados para la Formación Las Arcas (Mioceno tardio), Valle de Santa María, Provincia de Catamarca, Argentina. Ameghiniana 34:538

Nogués-Paegele J, Mo KC (1997) Alternating wet and dry conditions over South America during summer. Monthly Weather Rev 125: 279-291

Omarini RH (1983) Caracterización litológica, diferenciación y génesis de la Formación Puncoviscana entre el Valle de Lerma y la Faja Eruptiva de la Puna. PhD thesis, Salta

Parrish JT, Ziegler AM, Scotese CR (1982) Rainfall patterns and the distribution of coals and evaporites in the Mesozoic and Cenozoic. Palaeogeogr Palaeoclimatol Palaeoecol 40:67-101

Pascual R, Ortiz Jaureguizar E (1990) Evolving climates and mammal faunas in Cenozoic South America. J Human Evol 19:23-60

Pascual R, Vucetich MG, Scillato-Yané GJ, Bond M (1985) Main pathways of mammalian diversification in South America. In: Stehli F, Webb SD (eds) The Great American Biotic Interchange. New York, pp 219-247

Pascual R, Ortiz Jaureguizar E, Prado JL (1996) Land Mammals: paradigm for Cenozoic South American geobiotic evolution. Münchner Geowiss Abh 30:265-319

Placzek C, Quade J, Patchett J (2006) Geochronology and stratigraphy of late Pleistocene lake cycles on the southern Bolivian Altiplano: implications for causes of tropical climate change. Geol Soc Am Bull 118:515-532

Prohaska FJ (1976) The climate of Argentina, Paraguay and Uruguay. In: Schwerdtfeger W (ed) Climates in Central and South America. World Survey of Climatology 12:13-73

Ramos VA, Alonso RN (1995) El Mar Paranense en la provincia de Jujuy. Rev Geol Jujuy 10:73-80

Ramos VA, Cristallini EO, Pérez DJ (2002) The Pampean flat-slab of the Central Andes. J S Am Earth Sci 15:59-78

Rao GV, Erdogan S (1989) The atmospheric heat source over the Bolivian plateau for a mean January. Bound Layer Met 17:45-55

Reguero MA, Candela AM, Alonso RN (2003) Biochronology and biostratigraphy of the Uquia Formation (Pliocene-early Pleistocene, NW of Argentina) and its significance in the Great American Biotic Interchange. Ameghiniana, 40:69R

Reiners PW, Ehlers TA, Mitchell SG, Montgomery DR (2003) Coupled spatial variations in precipitation and long-term erosion rates across the Washington Cascades. Nature 426:645-647

Reynolds JH, Galli CI, Hernández RM, Idleman BD, Kotila JH, Hilliard RV, Naeser CW (2000) Middle Miocene tectonic development of the Transition Zone, Salta Province, northwest Argentina: magnetic stratigraphy from the Metán Subgroup, Sierra de González. Geol Soc Am Bull 112:1736-1751

Robinson RAJ, Spencer J, Strecker MR, Richter A, Alonso RN (2005) Luminescence dating of alluvial fans in intramontane basins of NW Argentina. Geol Soc Lond Spec Pub 251:153-168

Ropelewski CF, Halpert MS (1987) Global and regional scale precipitation patterns associated with the El Niño/Southern Oscillation. Monthly Weather Rev 115:1606-1626

Royden L (1996) Coupling and decoupling of crust and mantle in convergent orogens: implications for strain partitioning in the crust. J Geophys Res 101:17679-17705

Sáez A, Cabrera L, Jensen A, Chong G (1999) Late Neogene lacustrine record and Palaeogeography in the Quillagua-Llamara basin, Central Andean fore-arc (northern Chile). Palaeogeo Palaeoclim Palaeoecol 151:5-37

Sasso AM, Clark AH (1998) The Farallón Negro Group, northwest Argentina: magmatic, hydrothermal and tectonic evolution and implications for Cu-Au metallogeny in the Andean back-arc. Soc Econ Geol Newsletter 34:6-18

Schoenbohm L, Strecker MR (2005) Extension in the Puna-Altiplano plateau since 1-2 Ma: lithospheric delamination, lower crustal flow or gravitational collapse? Geol Soc Am Ann Meeting, Salt Lake City, Abstract Volume 121:5

Scotese CR, Gahagan LM, Larson RL (1988) Plate tectonic reconstructions of the Cretaceous and Cenozoic ocean basins. Tectonophysics 155:27-48

Segerstrom K, Turner JCM (1972) A conspicuous flexure in regional structural trend in the Puna of northwest Argentina. US Geol Surv Prof Pap 800B:205-209

Seltzer GO, Rodbell DT, Wright (2003) Late-quaternary paleoclimates of the southern tropical Andes and adjacent regions. Palaeogeo Palaeoclim Paleoecol 194:1-3

Sillitoe RH, McKee EH, Vila T (1991) Reconnaissance K-Ar Geochronology of the Maricunga Gold-Silver Belt, Northern Chile, Econ Geol 86:1261-1270

Sirolli AR (1968) El Toxodon de la finca El Gólgota. El Tribuno, Salta, Argentina, Revista 28:6

Sobel ER, Strecker MR (2003) Uplift, exhumation and precipitation: tectonic and climatic control of late Cenozoic landscape evolution in the northern Sierras Pampeanas, Argentina. Basin Research 15:431-451

Sobel ER, Hilley GE, Strecker MR (2003) Formation of internally-drained contractional basins by aridity-limited bedrock incision. J Geophys Res 108(B7):6/1-6/23

Starck D, Anzótegui LM (2001) The late Miocene climatic change – persistence of a climate signal through the orogenic stratigraphic record in northwestern Argentina. J S Am Earth Sci 14: 763-774

Stern LA, Blisniuk PM (2002) Stable isotope composition of precipitation across the Southern Patagonian Andes. J Geophys Res 107(D23): doi 1029/2002jd002509

Strecker MR (1987) Late Cenozoic landscape development, the Santa Maria Valley, Northwest Argentina. Cornell University, Ithaca NY

Strecker MR, Cerveny P, Bloom AL, Malizia D (1989) Late Cenozoic tectonism and landscape development in the foreland of the Andes: Northern Sierras Pampeanas, Argentina. Tectonics 8:517-534

Suppe J (1981) Mechanics of mountain building and metamorphism in Taiwan. Mem Geol Soc China 4:67-89

Tawackoli S, Jacobshagen V, Wemmer K, Andriessen PAM (1996) The Eastern Cordillera of southern Bolivia: a key region to the Andean backarc uplift and deformation history. Third International Symposium on Andean Geodynamics (ISAG), St. Malo, France, pp 505-508

Thiede R, Bookhagen B, Arrowsmith JR, Sobel E, Strecker M (2004) Climatic control on rapid exhumation along the southern Himalayan front. Earth Planet Sci Lett 222:791-806

Trauth MH, Strecker MR (1999) Formation of landslide-dammed lakes during a wet period between 40,000 – 25,000 yr bp in northwestern Argentina. Palaeogeo Palaeoecol 153:277-287

Trauth MH, Alonso RA, Haselton KR, Hermanns RL, Strecker MR (2000) Climate change and mass movements in the northwest Argentine Andes. Earth Planet Sci Lett 179:243-256

Trauth MH, Bookhagen B, Mueller A, Strecker MR (2003a) Erosion and climate change in the Santa Maria Basin, NW Argentina during the last 40,000 yrs. J Sed Res 73:82-90

Trauth MH, Bookhagen B, Marwan N, Strecker MR (2003b) Multiple landslide clusters record Quaternary climate changes in the NW Argentine Andes. Palaeogeo Palaeoclim Paleoecol 194:109-121

Uba CE, Heubeck C, Hulka C (2005) Facies analysis and basin architecture of the Neogene Subandean synorogenic wedge, southern Bolivia. Sed Geol 180:91-123

Vandervoort DS (1993) Non-marine evaporite basin studies, southern Puna Plateau, central Andes. PhD thesis, Cornell University, Ithaca NY

Vandervoort DS, Jordan TE, Zeitler PK, Alonso RN (1995) Chronology of internal drainage development and uplift, southern Puna plateau, Argentine Central Andes. Geology 23:145–148

Vera C, Higgins W, Amador J, Ambrizzi T, Garreaud R, Gochis D, Gutzler D, Lettenmaier D, Marengo J, Mechoso CR, Nogues-Paegle J, Silvas Dias PL, Zhang C (in press) J Climate Spec Issue on 1st International CLIVAR Science Conference

Villanueva García A, Strecker MR (1986) Origen de las toscas (calcrete) en el Valle de Santa María, Noroeste Argentino. Actas, Primera reunión Argentina de sedimentología, La Plata, Argentina, pp 81–84

Viramonte JG, Reynolds JH, Del Papa C, Disalvo A (1994) The Corte Blanco garnetiferous tuff: a distinctive late Miocene marker bed in northwestern Argentina applied to magnetic polarity stratigraphy in the Rio Yacones Salta Province. Earth Planet Sci Lett 121:519–531

Vuille M (1999) Atmospheric circulation over the Bolivian Altiplano during dry and wet periods and extreme phases of the Southern Oscillation. International J Climat 19:1579–1600

Vuille M, Bradley RS, Keimig F (2000) Interannual climate variability in the Central Andes and its relation to tropical Pacific and Atlantic forcing. J Geophys Res 105:12447–12460

Walther AM, Orgeira MJ, Reguero MA, Verzi DH, Vilas JF, Alonso RN, Gallardo E, Kelley S, Jordan T (1998) Estudio paleomagnético, paleontológico y radimétrico de la Formación Uquía (Plio-Pleistoceno) en Esquina Blanca (Jujuy). X Congreso Latinoamericano de Geología y VI Congreso Nacional de Geología Económica, Actas 1:77

Whipple KX, Meade BJ (2004) Controls on the strengh of coupling among climate, erosion, and deformationin two-sided, frictional orogenic wedges at steady-state. J Geophys Res 109: doi 10.1029/2003JF000019

Willett SD (1999) Orogeny and orography: the effects of erosion on the structure of mountain belts. J Geophys Res 104:28957–28981

Willett SD, Brandon MT (2002) On steady states in mountain belts. Geology 30:175–178

Willett SD, Beaumont C, Fullsack P (1993) Mechanical model for the tectonics of doubly vergent compressional orogens. Geology 21:371–274

WMO (1975) Climatic atlas of South America. World Meteorological Organization, Geneva

Zachos J, Pagani M, Sloan L, Thomas E, Billups K (2001) Trends, rhythms, and aberrations in global climate 65 Ma to present. Science 292:686–693

Zeitler PK, Meltzer AS, Koons PO, Craw D, Hallet B, Chamberlain CP, Kidd WSF, Park SK, Seeber L, Bishop M, Shroder J (2001) Erosion, Himalayan geodynamics, and the geomorphology of metamorphism. GSA Today 11:4–9

# Chapter 13

# Exhumation and Basin Development Related to Formation of the Central Andean Plateau, 21° S

Ekkehard Scheuber · Dorothee Mertmann · Harald Ege · Patricio Silva-González · Christoph Heubeck
Klaus-Joachim Reutter · Volker Jacobshagen

**Abstract.** Thermochronological (apatite fission track, AFT) and sedimentological data from the central Andean high plateau and its eastern foreland reflect the Tertiary tectonic evolution of this plateau. AFT data define several stages of exhumational cooling in the plateau and its foreland: A first stage (AFT dates of 40 and 36 Ma), restricted to the central Eastern Cordillera, can be attributed to initial thrusting increments following the Incaic Phase. Around 33–30 Ma, cooling occurred along basement highs over the entire plateau; kinematics were thrusting (eastern Cordillera) as well as normal faulting (Altiplano). From 17 Ma onwards, exhumational cooling took place in the eastern foreland (Interandean-Subandean), reflecting the eastward migration of the thrust front into the Chaco foreland basin.

Sedimentary data agree with the AFT data. Before 32 Ma, fine-grained playa-mudflat deposits indicate tectonic quiescence in most parts of the endorheic basin. Around 32 Ma, a strong change to coarse clastics took place. Between 32 and 18 Ma, alluvial fans and fluvial deposits formed adjacent to fault-bound basement highs. Later growth strata indicate thrusting activity. Since about 10 Ma, flat-lying continental deposits indicate tectonic quiescence from the Altiplano to the Eastern Cordillera, whereas in the Chaco the present foreland was created.

The data are discussed in terms of the development of the active continental margin and a possible scenario is suggested: The Incaic Phase (45–38 Ma) might have been linked to a slab breakoff and subsequent flat slab conditions. During the associated gap in volcanism (38–28 Ma), flat slab subduction may have caused hydration of the upper plate mantle and the initial formation of the plateau. Later plateau uplift could have been caused by crustal thickening owing to tectonic shortening, first within the plateau and its foreland and, since about 10 Ma, in the foreland only.

## 13.1 Introduction

It is generally accepted that the formation of the Altiplano-Puna Plateau resulted from the convergence of the Farallon and Nazca Plates with the South American Plate. Gephart (1994), for example, has shown that both Andean topography and slab geometry are highly symmetrical around a vertical NNE-trending plane that corresponds to the Euler equator and to the Farallon Euler pole of approximately 35 million years ago. Small-scale asymmetries in topography and slab geometry have been attributed to the shifting of the Euler pole after 20 Ma. Thus, it is concluded that the initial stages of plateau formation occurred during the late Eocene to early Oligocene.

The causes of plateau formation and the single stages of its evolution have, however, remained a matter of debate for a long time. In his widely accepted model, Isacks (1988) assumed that plateau formation started approximately 28 million years ago, caused by processes in the asthenospheric wedge that were effective in thinning the lithospheric mantle of the upper plate. By contrast, James and Sacks (1999) attributed these early stages to a time of flat slab (nearly horizontal) subduction which caused a magmatic pause during much of the Oligocene. Flat slab subduction led to a low geothermal gradient and hydration (serpentinization) of the upper plate mantle, resulting in an early uplift, and extensive shortening deformation in the Eastern Cordillera.

The resumption of volcanism some 28 million years ago indicates the installation of a new asthenospheric wedge beneath the upper plate and, following James and Sacks (1999), a steepening of subduction. In this stage, plateau uplift is assumed to have occurred owing to lithospheric thinning and distributed shortening (pure shear mode) of the upper plate over a "heated and weakened zone" (Isacks 1988). Since 18–12 Ma, the Brazilian shield has been thrust underneath the Andean orogen by transferring the principal location of shortening, first into the Interandean and later into the Subandean Belt (simple shear mode; Baby et al. 1992, 1997; Dunn et al. 1995; Kley 1996; Moretti et al. 1996; DeCelles and Horton 2003; Gubbels et al. 1993). The model of Isacks has been modified by Lamb et al. (1997) and Lamb (2000) for the present stage of the plateau. These authors suggested that the thin-skinned thrusting in the upper crustal foreland is accommodated by ductile distributed shortening beneath the plateau.

The objective of our work is to illuminate the plateau-forming processes in the Central Andes by investigating the geological record within the plateau and its eastern foreland. We have carried out thermochronological, sedimentary and structural work in the plateau and foreland, and have particularly focused on the Tertiary strata of the Altiplano, which we compare with those of the Subandean Zone (Figs. 13.1 and 13.2). The Tertiary Altiplano sediments reflect the tectonic activity of the early

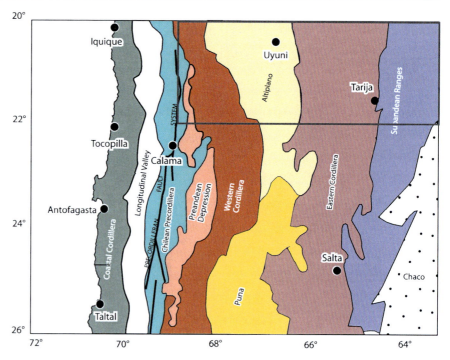

**Fig. 13.1.**
Morphostructural units of the Central Andes. (The *box* locates the study area of Fig. 13.2)

stages of plateau formation, whereas the basin sediments of the Chaco foreland reflect lithospheric loading of the foreland lithosphere resulting from the thrusting of the Brazilian shield beneath the Andean orogen. Our sedimentological work was accompanied by apatite fission track (AFT, e.g. Wagner and van der Haute 1992) analyses along an E-W traverse at about 21° S, from the Western Cordillera to the Subandean, in order to date the exhumation and uplift history.

## 13.2 Geological Setting

The Central Andes formed in a long-lived convergent setting in which, since the Jurassic, several oceanic plates were subducted beneath the leading edge of the South American Plate. Subduction resulted in the formation of a magmatic arc that, due to tectonic erosion, has migrated about 200 km eastwards since 120 Ma, moving from the Coastal Cordillera to its present position in the Western Cordillera (Fig. 13.1). This volcanic arc is located along the western margin of the Altiplano-Puna Plateau, which (with a mean 3.8 km elevation) mainly belongs to the back-arc region of the Andean arc system, but also comprises the Chilean Precordillera at about 21° S where it forms the western monocline of the plateau (Isacks 1988).

Eastwards, the Altiplano-Puna Plateau is limited by the Eastern Cordillera with elevations exceeding 5 000 m. On the eastern slope of the plateau, the (west to east) Interandean and Subandean Belts represent detached and deformed foreland sediments, whereas the Chaco represents the present foreland basin. The Interandean to Chaco units reflect the lateral growth of the Andean orogen and its thrusting onto the stable Precambrian Brazilian Shield.

In the north Chilean Precordillera, along the western margin of the plateau, rocks and structures of the Paleogene magmatic arc crop out in orogen-parallel, bivergent ridges and are underlain by Mesozoic formations and a Paleozoic basement (Günther 2001). Its major structure is the Precordilleran Fault System (PFS), a bivergent, transpressional zone that originated as a trench-linked, strike-slip system within the Eocene magmatic arc (Döbel et al. 1992). Several world-class ore deposits (e.g. Chuquicamata) exist along this fault system.

The Altiplano forms a wide, low-relief and mainly internally-drained basin that is covered by undeformed Upper Miocene to Quaternary volcanics and sediments (Fig. 13.2). This basin is morphologically structured by isolated, N-S- to NNE-SSW-trending ridges with elevations ranging from 4 000 m to 5 350 m. These are built from Paleozoic to Neogene strata that, in mid to late Miocene times, underwent dextral transpressional deformation with thrusting toward the east, and have dextral, strike-slip displacements along the Uyuni-Khenayani Fault Zone (UKFZ, Fig. 13.3). The eastern margin of the Altiplano is marked by the San Vicente Fault System (SVFS) where Ordovician sedimentary rocks of the Eastern Cordillera were thrust westwards onto Paleogene deposits of the Altiplano in the early Miocene (Müller 2000; Müller et al. 2002).

The Eastern Cordillera forms a N-S-trending, bivergent thrust system centered near 65.5° W (Baby et al.

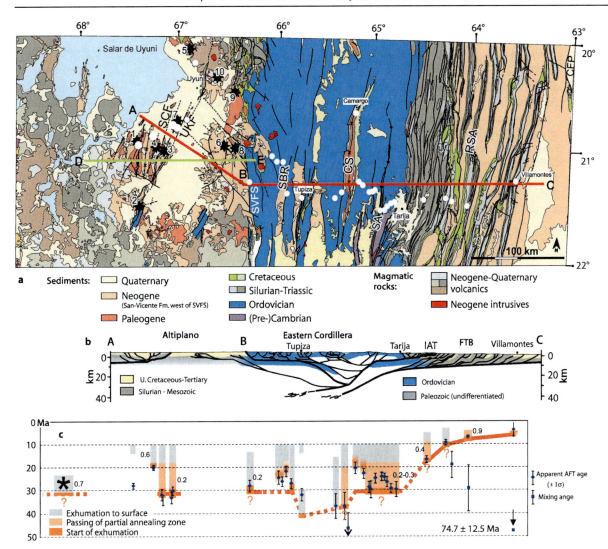

**Fig. 13.2. a** Geological map of the study area: *ABC*: location of the Apatite fission track thermochronology (AFT) traverse; *white dots*: AFT sample locations; *DE*: section of Fig. 13.3; *SCF*: San Cristobal Fault; *UKF*: Uyuni-Khenayani Fault; *SVFS*: San Vicente Fault System; *SBR*: Santa Barbara Ranges; *CS*: Camargo Syncline; *SA*: Sama Anticline; *RSA*: Rio Salado Anticline; *CFP*: Cabalgimiento Frontal Principal. Numbered black stars: *1* – Santa Ines, *2* – Cerro Gordo, *3* – San Cristobal 2, *4* – San Cristobal 1, *5* – Chita, *6* – Animas 2, *7* – Corregidores, *8* – Animas 1, *9* – Ubina, *10* – Pulacayo. **b** Cross section along the AFT traverse ABC. *IAT*: Interandean Thrust; *FTB*: Foreland fold-and-thrust belt. **c** AFT data and modeled cooling and exhumation paths along the cross section: (*: westernmost data point, assumed source area of crystalline basement pebbles in San Vicente Formation. *Numbers* refer to averaged exhumation rates in mm yr$^{-1}$.

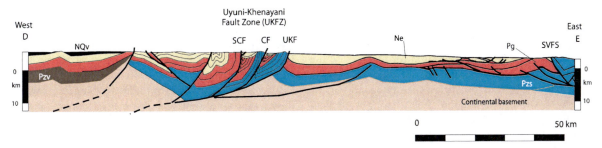

**Fig. 13.3.** Schematic section across the central and eastern Altiplano at ~21° S. For location (D-E) see Fig. 13.2. *NQv*: Neogene-Quaternary volcanics, *Ne*: Neogene sediments of the San Vicente Formation, *Pg*: Paleogene sediments with Upper Cretaceous at the base, *PzV*: Paleozoic volcanics, *Pzs*: Paleozoic sediments, *SCF*: San-Cristobal Fault, *CF*: Corregidores Fault, *UKF*: Uyuni-Khenayani Fault

1992, 1996). It is predominantly composed of clastic, metasedimentary rocks of Ordovician age, up to 10 km thick. These rocks are disconformably overlain by either Cretaceous to Paleocene or mid Tertiary to Neogene sequences. In the late Cretaceous, after mid Cretaceous rifting, a wide, post-rift basin developed which covered large parts of the Eastern Cordillera and the adjacent Altiplano (Sempere 1994; Marquillas and Salfity 1994; Fiedler et al. 2003). Marine deposits of the late Cretaceous and early Tertiary (El Molino Fm.) provide the last regional paleoelevation record prior to Andean deformation (Kley et al. 1997).

During the Eocene and early Oligocene, the central Eastern Cordillera was the source area (Proto-Eastern Cordillera, Lamb et al. 1997) of continental, upwardly coarsening deposits (Potoco and Camargo Fms., Horton and DeCelles 2001; Silva-González 2004) in the eastern Altiplano and in the area of the Camargo syncline (CS in Fig. 13.2). Late Oligocene to Miocene coarse-grained sediments disconformably cover older strata in several N-S-trending intramontane basins (Horton 1998; Kley et al. 1997) that were formed during intense shortening (Horton 1998; Tawackoli 1999, Müller et al. 2002). The formation of a regionally extensive, erosional surface, dating from 13 to 9 Ma, which is covered by undeformed strata, documents the end of deformation in the entire Eastern Cordillera (Gubbels et al. 1993).

Descending to the east, the mean elevation diminishes by about 1500 m to the Interandean and Subandean zones, which form an east-vergent, fold-and-thrust belt and expose Paleozoic to Mesozoic units. The widely-spaced, thrust-bounded anticlines of the Subandean zone progressively include the Neogene foreland basin sequences of the Chaco and continue eastwards into the undeformed Recent Chaco Basin (Gubbels et al. 1993; Isacks 1988; Kley et al. 1997). At present, approximately half of the original width of the Chaco foreland basin, between the Cabalgimiento Frontal Principal (CFP) in the west and the gentle onlap of flat-lying sediments over the crystalline rocks of the Izozog High (part of the Brazilian Shield) in the east, has been incorporated into the Subandean fold-and-thrust belt; there, shortening reaches approximately 38% (Baby et al. 1992, 1996).

Tectonic shortening thickened the crust to 70 km (Wigger et al. 1994) beneath the Central Andean Plateau. Quantitative analyses of structural profiles across the whole back-arc have revealed that tectonic shortening predominantly occurred during the Cenozoic. For the southern Altiplano, 60–80 km of shortening was calculated (Elger et al. 2005), for the Eastern Cordillera, 90–95 km (Müller et al. 2002), and for the Interandean and Subandean zones, 130–140 km (citations in Müller et al. 2002), although some lower crustal flow out of the section has to be considered (Hindle et al. 2002, 2005).

## 13.3 Exhumation History (Apatite Fission Track Thermochronology)

The exhumation history of the central Andean Plateau has been addressed using AFT thermochronology. A detailed description and interpretation of the AFT data is the focus of another paper (Ege et al. submitted). To date, only a few fission track studies exist from the Altiplano-Puna segment (Benjamin et al. 1987; Kontak et al. 1990; Andriessen and Reutter 1994; Moretti et al. 1996).

Fission tracks in apatite become progressively shorter in the broad temperature range (i.e. 125–60 °C for Durango apatite, Green et al. 1989) of the partial annealing zone (PAZ); a representative number of track lengths measured in a sample can be compared with forward annealing models to constrain the thermal history (e.g. Gallagher 1995). Mean track lengths (MTL) of 14–15 μm indicate rapid cooling, whereas reduced lengths are indicative of longer residence in the PAZ. To address the annealing kinetics of the detrital apatites, the etching parameter "Dpar" was used (Carlson et al. 1999; Ketcham et al. 1999) and thermal models were made with the program "AFTSolve" (Ketcham et al. 2000). For analytical details and modeling parameters, the reader is referred to Ege et al. (2003) and Ege (2004).

Samples from mostly pre-Tertiary sedimentary rocks were collected along a 400 km long transect at ~21° S, extending from the central Altiplano to the Subandean belt (Fig. 13.2), with the aim of characterizing the regional pattern of exhumation. Structurally deep locations, e.g. hanging walls of major thrusts, were preferentially sampled, but, for comparison purposes, some samples were taken from shallower structural positions. A common approach in fission track studies is to take a suite of samples over a range of elevations with the goal of encountering a kink in the age-elevation gradient, marking the time when exhumational cooling began (Benjamin et al. 1987; Fitzgerald et al. 1995). However, owing to low relief and intense deformation within the plateau, such profiles are absent.

Isolated fission track data from buried and subsequently exhumed rocks can provide two different types of information. Where rocks were buried deep enough to exceed the total annealing temperature ($T_A$), the fission track clock was reset to zero and the data record information about the following cooling path. Where $T_A$ was not achieved, ages and track lengths reflecting the anterior thermal history will be preserved, modified by the final cooling from a maximum temperature lower than $T_A$. Small portions of strongly shortened tracks encountered in most samples suggest that they are of the latter type. To distinguish between the two possibilities, the maximum temperature is decisive and is estimated from stratigraphic and structural information, and the assumed paleo-geothermal gradient (see below).

## 13.4 Thermal State of the Crust

Thermal processes might have influenced the observed cooling record. The recent thermal state of the crust is characterized by an elevated heat flow of 100–80 mW m$^{-2}$ in the plateau, decreasing to 40 mW m$^{-2}$ in the Subandean zone (Springer and Förster 1998). This thermal anomaly has been created by magmatism and crustal thickening since the late Oligocene (Babeyko et al. 2002) and, hence, is coeval with the observed cooling. However, as volcanism started around 28–25 Ma (see below), i.e. later than the observed cooling ages (Fig. 13.2c), a magmatic event that reset the fission track ages can be excluded. Thus, the cooling solely reflects exhumation.

In order to interpret thermochronological data in the spatial frame of reference, the paleo-geothermal gradient has to be assumed. Today, temperature gradients are $39 \pm 9$ °C km$^{-1}$ ($\pm 1\sigma$) in the Altiplano, $25 \pm 8$ °C km$^{-1}$ in the Eastern Cordillera and $22 \pm 3$ °C km$^{-1}$ in the Subandean-Chaco (Springer and Förster 1998). A constant geothermal gradient of $25 \pm 5$ °C km$^{-1}$ was assumed for the calculation of exhumation rates (Ege et al. submitted). Given the low amounts and rates of exhumation within the plateau, heat advection by erosion (Brown and Summerfield 1997) is neglected. To relate sediment thickness to temperature, thermal conductivity data from Bolivia were used (Henry and Pollack 1988; Springer 1996).

## 13.5 General Results

In the plateau domain (Altiplano and Eastern Cordillera), apparent cooling ages range between 37 and 20 Ma, with the majority around 30 Ma (Fig. 13.2). The apatite fission tracks exhibit variable amounts of moderate length reduction (< 20%). By contrast, exhumational cooling in the eastern foreland fold-and-thrust belt started much later with cooling ages of 16–9 Ma in the Interandean zone. Mixed ages are obtained in the Subandean zone. Our interpretation of the exhumation history (for details see Ege et al., in prep) is presented below in temporal order.

### 13.5.1 Late Eocene – Early Oligocene

The earliest cooling along the transect is observed in the central part of the Eastern Cordillera (Fig. 13.2). Near Tupiza, an Ordovician sample (AFT age of 31.5 Ma and MTL of 11.9 μm) was taken below an unconformity with sediments of possible Oligocene age. These sediments are more than 1.2 km thick and were folded in the mid Miocene together with the Ordovician substrate (Kley et al. 1997; Müller et al. 2002). Modeling with corresponding constraints yielded good fits for a two-phase thermal history with initial cooling since ca. 40 Ma, reheating by burial to about 80 °C, and final cooling to surface temperature after ca. 15 Ma.

The sample from the hanging wall of the Camargo-Yavi Thrust (CYT) which is the western border of the Camargo Syncline (CS, Fig. 13.2) slowly cooled from at least 36 Ma, most likely owing the to initiation of the CYT. Together with an apparent age of $36.4 \pm 5.2$ Ma from a sample further west, the AFT data from the central Eastern Cordillera delineate a zone of initial and, suggestively, eastwardly propagating uplift in the late Eocene to early Oligocene. Our results are similar to those from the Eastern Cordillera of northern Bolivia where zircon fission track ages indicate the onset of cooling at 45–40 Ma (Benjamin et al. 1987; Lamb and Hoke 1997). We interpret late Eocene exhumation to be contractionally-driven uplift of a narrow proto-Eastern Cordillera and the related inversion of Cretaceous rift structures (Kley et al. 1997; Lamb and Hoke 1997; Hansen and Nielsen 2003).

### 13.5.2 Oligocene

Widespread cooling occurred within the Oligocene (33–27 Ma) from the Western Cordillera to the eastern margin of the Eastern Cordillera. Apatite crystals from basement pebbles, collected from an early Miocene conglomerate of the San Vicente Fm. in the western *Altiplano* (Fig. 13.2), yield a uniform cooling age of $28.0 \pm 1.2$ Ma. The pebbles comprise west-derived, Proterozoic-Paleozoic, basement rocks (Bahlburg et al. 1986; Wörner et al. 2000). Post-depositional burial of ca. 1 km (Elger et al. 2005) did not heat the sample above 60 °C; the AFT data thus reflect cooling in the source areas of the western Altiplano. In particular, thermal models indicate rapid cooling to less than 145–130 °C since 32–30 Ma.

Four Paleozoic samples from the hanging walls of two east-vergent thrusts in the Uyuni-Khenayani fault zone (UKFZ, Fig. 13.3) yielded similar AFT cooling ages of between 34 and 30 Ma, but had variable amounts of track shortening. Thermal modeling and the reconstruction of pre-exhumational burial depth indicate maximum temperatures slightly below $T_A$ with the onset of exhumation occurring between 33 to 29 Ma. Modeled maximum temperatures still require burial heating beneath a thickness of $1.25 \pm 0.75$ km of early Tertiary sediments (Potoco Fm.).

In principle, the observed exhumational cooling along the UKFZ is compatible with uplift in the footwall of a west-dipping, normal fault (the future San Cristobal thrust, SCF, Figs. 13.2 and 13.3), and also with uplift in the hanging walls of east-vergent, reverse faults (Elger et al. 2005). However, the coarseness of the San Vicente Fm., the abundance of Paleozoic clasts, even in the basal parts of the San Vicente Fm., and the high sediment thickness west of the fault rather argue for normal faulting along the SCF during early exhumation (see below).

Until ca. 18 Ma, the Paleozoic rocks in the UKFZ cooled to a temperature below ~60 °C as exhumational unroofing of 2–3 km occurred, while the Tertiary sample from the basin to the west was continuously heated by sedimentary burial (Fig. 13.2, for details see Ege et al., submitted). Exhumation in this basin started by 20–18 Ma and most likely reflects that the SCF became active as a reverse fault. All samples reached the surface 11–7 million years ago.

In the west-vergent part of the *Eastern Cordillera*, apparent AFT ages range between 28 and 22 Ma. The precise timing of cooling is restricted by incomplete length data and poor constraints on pre-exhumational burial. The stratigraphic throw along the thrust of the San Vicente Fault System (SVFS) of at least 3.8 km (Kley et al. 1997) and the ~2 km thick Paleogene Potoco Fm. in the footwall (Elger et al. 2005) indicate that the hanging wall sample cooled by about 50 °C prior to its apparent AFT age of 27.7 Ma.

In the western part of the Eastern Cordillera, the fault-bounded Santa Barbara Range (SBR, Fig. 13.2) was exhumed around the same time (AFT age of 26.7 Ma), possibly due to the formation of a fault bend fold at depth (Müller et al. 2002). Samples west of the SBR had shallow positions in the Ordovician stratigraphy (< 1 km), except the sample in the core of the Tres Palcas anticlinal structure (3.9 km). However, the apparent AFT age of the core sample is similar to the sample from the flank (AFT ages of 24.3 Ma and 25.8 Ma, respectively), suggesting that the anticline had mostly formed during the late Paleozoic. Therefore, samples west of the SBR most likely cooled from a temperature slightly below $T_A$ some time after their apparent AFT ages, i.e. during the early Miocene.

The exhumation pattern, consistent with relative faulting ages (Müller et al. 2002), indicates that the west-vergent thrust system did not form by in-sequence, propagating deformation, despite the thin-skinned geometry (for discussion see Müller et al. (2002). Earliest exhumation along the SVFS may correlate with the deposition of the alluvial conglomerates of the basal San Vicente Fm., or with the upward coarsening of sediment in the underlying Potoco Fm.

In the eastern part of the Eastern Cordillera, a series of ten lower Ordovician samples have been taken across 1 500 m of local relief in the Sama anticlinal structure (SA in Fig. 13.2). These samples yielded cooling ages between 29 and 20 Ma. These data are interpreted to represent the base of a fossil partial annealing zone (for details see Ege et al. submitted). This zone was exhumed by at least 28 Ma, and most likely by 30 Ma. Until ca. 18 Ma, block-like exhumation of 2.8 km was probably achieved by uplift over the ramp of the active main Interandean thrust (IAT).

### 13.5.3 Miocene – Pliocene

Tectonic shortening of the central Altiplano and the Eastern Cordillera occurred until 11–7 Ma (Lamb and Hoke 1997; Müller et al. 2002; Victor et al. 2004; Elger et al. 2005), with bulk shortening rates increasing during the Miocene (Elger et al. 2005). However, the associated rock uplift did not exhume rocks from below the PAZ after the early Miocene.

The onset of exhumational cooling migrated from the eastern margin of the Eastern Cordillera (Sama anticline) to the east from about 30 Ma, and is clearly linked to the propagation of the deformation front into the eastern foreland. Cooling ages of 16.4 and 9.2 Ma in the central and easternmost part of the *Interandean* zone, respectively, indicate that the exhumational front propagated through this region until about 10 Ma. A sample from the westernmost anticline (mixed age of 28.6 Ma, Fig. 13.2) in the western *Subandean* zone consists of two kinetically different apatite populations; the younger population has a completely reset AFT cooling age of 6.8 ± 0.6 Ma and a MTL of 13.9 ± 2.3 μm that dates the onset of exhumational cooling by 8 Ma, owing to the formation of the anticline. AFT data from the easternmost Subandean anticline near Villamontes (mixed age of 74.7 Ma, Fig. 13.2), cannot constrain the onset of cooling, but suggest that cooling started considerably after 8 Ma. The onset of cooling in the Subandean is clearly reflected in a change in sedimentation (see below).

### 13.5.4 Exhumation Rates and Surface Uplift

Apparent mean exhumation rates were calculated either from the portion of modeled cooling paths hotter than 60 °C or with respect to the surface temperature (15 °C), but only where the timing is constrained and assuming a constant geothermal gradient of 25 ± 5 °C km$^{-1}$ (Fig. 13.2). Within the plateau, mean apparent exhumation rates were predominantly low (0.1–0.3 mm yr$^{-1}$) between 33 and 18 Ma (Fig. 13.2). Exceptionally high apparent exhumation rates of 0.6–0.9 mm yr$^{-1}$ are observed in samples from the basement of the western Altiplano/Western Cordillera. There, notable late Eocene-Oligocene exhumation (Maksaev and Zentilli 1999) could have created an elevated thermal gradient; therefore, a value closer to 0.6 mm yr$^{-1}$ is considered to be appropriate. Cooling from 60 °C to surface temperature occurred in most of these samples within the time span of 18–10 Ma; this corresponds to bulk exhumation rates of ca. 0.2 mm yr$^{-1}$. Significant changes in exhumation rates are not observed or were not within the resolution of our data.

Widespread Oligocene exhumation from the western Altiplano to the Eastern Cordillera most likely produced moderately elevated mountain ranges between which the

Altiplano Basin was filled. However, this surface uplift cannot be quantified.

Removal of Tertiary cover sediments and exposure of Paleozoic rocks has increased erosional resistance since the Oligocene. Several indicators in the Eastern Cordillera south of ~18° S suggest that erosion rates decreased around ~15 Ma (see Horton (1999) for a summary) and, at this time, arid climatic conditions were established in the area west of the plateau (Alpers and Brimhall 1988; Sillitoe and McKee 1996).

Uplift was largely a function of tectonic shortening (cf. Isacks 1988), which continued after the end of internal deformation around 10 Ma due to underthrusting of the foreland (Allmendinger and Gubbels 1996; Kley et al. 1999). Probably, elevations along the eastern margin of the plateau became sufficiently high by 15–10 Ma to form a rain shadow, thus strongly decreasing erosional exhumation (Alpers and Brimhall 1988). However, paleo-geomorphological studies predict lower mean elevations of 1 000–1 500 m around 10 Ma (Kennan et al. 1997; Gregory-Wodzicki 2000; Kennan 2000).

## 13.6 Sedimentary History of the Altiplano Basin and Adjacent Areas

In the southern Altiplano, an endorheic (internally-drained) basin was formed, presumably during the late Paleocene, and persists today. Sediments and magmatic rocks, up to 8 km thick, were deposited and emplaced mainly during the Eocene to Miocene (Mertmann et al. 2003). The Tertiary sediments of the southern Altiplano were investigated in 10 localities south of the Salar de Uyuni (Fig. 13.2) and the stratigraphic columns, sedimentological features, and the results of isotope dating are depicted in Figs. 13.4 and 13.5, and Table 13.1.

A major change in basin configuration occurred with the onset of widespread, coarse-clastic sedimentation in the area of the present central Andean Plateau during the early Oligocene. This change is coeval with the onset of exhumational cooling at ca. 32 Ma (see above) within the plateau and was followed by the onset of volcanism at ~28 Ma. In contrast to the widespread playa-mudflat environments recorded in the Potoco Fm., the sedimentary characteristics of the overlying San Vicente Fm. indicate high relief, fault-bounded basins, and a coincidence between sedimentation, tectonism, and (since ~28 Ma) magmatism. The sedimentation forming the San Vicente Fm. probably began during the initial uplift of the plateau. A brief outline of the basin development implies four time intervals: Pre-32 Ma, 32–17 Ma, 17–8 Ma, and 8 Ma–Recent.

### 13.6.1 Pre-32 Ma

The rocks underlying the Mesozoic to Cenozoic strata of the southern Altiplano Basin belong to different basement blocks. In the west, evidence for a shallow, but hidden, basement comes from pebbles within the Oligocene-Miocene San Vicente Fm. The clasts are mainly granites and gneisses of probable Precambrian age and Permian igneous rocks (Figs. 13.4 and 13.5, Table 13.1). Comparable

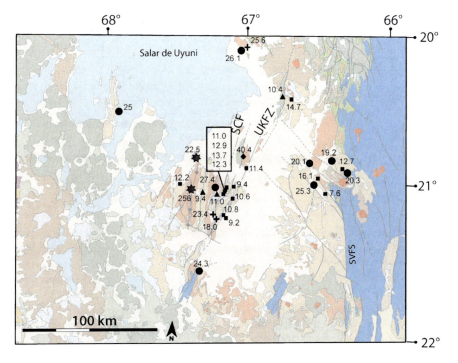

**Fig. 13.4.**
Distribution of sample localities and age values for isotope age determinations (cf. Table 13.1). The map shows the western part of Fig. 13.2

**Table 13.1.**
K-Ar age data from the southern Altiplano

| Sample | Age (Ma) | Mineral dated | Rock type | Position of sample | |
|---|---|---|---|---|---|
| | | | | deg W | deg S |
| **Flat-lying strata, unconformable over tilted San Vicente strata** | | | | | |
| ED00-41 | 9.4 ±0.3 | Biotite | Dacitic tuff | -67.3305 | -21.0318 |
| SR99-15 | 10.4 ±0.3 | Biotite | Dacitic tuff | -66.7339 | -20.3934 |
| SR99-10 | 11.0 ±0.5 | Hornblende | Andesitic lava | -67.2223 | -21.0076 |
| **Upper San Vicente Formation, from growth strata** | | | | | |
| ED 00-17 | 7.6 ±0.2 | Biotite | Ignimbrite | -66.4193 | -21.0481 |
| ED 00-40 | 9.2 ±0.3 | Biotite | Dacitic tuff | -67.1636 | -21.1979 |
| ED 00-07 | 9.4 ±0.3 | Biotite | Dacitic tuff | -67.0998 | -20.9969 |
| ED 00-02 | 10.6 ±0.3 | Biotite | Dacitic tuff | -67.1080 | -21.0776 |
| ED 00-33 | 10.8 ±0.3 | Biotite | Dacitic tuff | -67.1657 | -21.1940 |
| PS 99/26 | 11.0 ±0.3 | Biotite | Dacitic tuff | -67.1678 | -21.0324 |
| SR 99-06 | 11.4 ±0.3 | Biotite | Dacitic tuff | -67.0085 | -20.8749 |
| ED 00-42 | 12.2 ±0.3 | Biotite | Dacitic tuff | -67.4997 | -20.9759 |
| PS 99/47 | 12.3 ±0.4 | Biotite | Dacitic tuff | -67.1437 | -21.0288 |
| ED 00-14 | 12.7 ±0.4 | Biotite | Dacitic tuff | -66.2952 | -20.8827 |
| SR 99-13 | 12.9 ±0.3 | Biotite | Dacitic tuff | -67.1755 | -21.0376 |
| PS 99/17 | 13.7 ±0.4 | Biotite | Dacitic tuff | -67.1760 | -21.0373 |
| SR 99-01 | 14.7 ±0.4 | Biotite | Dacitic tuff | -66.6676 | -20.4152 |
| DLP 02/2000 | 16.1 ±0.4 | Biotite | Dacitic tuff | -66.4707 | -20.9451 |
| **Lower San Vicente Formation** | | | | | |
| ED 00-12 | 19.2 ±0.5 | Biotite | Dacitic tuff | -66.3669 | -20.8284 |
| ED 00-10 | 20.1 ±0.5 | Biotite | Dacitic tuff | -66.5333 | -20.8427 |
| PS 99/01 | 20.3 ±0.5 | Biotite | Dacitic tuff | -66.2702 | -20.8963 |
| SR 99-14 | 24.3 ±1.2 | Biotite | Andesitic tuff | -67.3566 | -21.5611 |
| SR 99-16 | 25.0 ±0.6 | Biotite | Ignimbrite | -67.9466 | -20.4903 |
| AN 14/2000 | 25.3 ±1.0 | Hornblende | Dacitic tuff | -66.4989 | -20.9928 |
| Chi 31/2000 | 26.1 ±0.7 | Biotite | Dacitic tuff | -67.0334 | -20.0702 |
| SR 99-12 | 27.4 ±0.7 | Biotite | Dacitic tuff | -67.2316 | -21.0024 |
| **Intrusive rocks** | | | | | |
| ED 00-31 | 18.0 ±0.5 | Biotite | Granodiorite | -67.2372 | -21.1956 |
| ED 00-25 | 23.4 ±1.1 | Hornblende | Diorite | -67.2407 | -21.1933 |
| ED 00-46 | 25.6 ±0.9 | Hornblende | Andesite | -67.0220 | -20.0640 |
| **Intrusive-derived clast from the lower San Vicente Formation** | | | | | |
| ED 00-44 | 22.5 ±0.8 | Hornblende | Microdiorite | -67.3783 | -20.8015 |
| ED 00-19 | 256 ±7 | Hornblende | Granodiorite | -67.4194 | -21.0144 |
| **Tuff from the Potoco Formation** | | | | | |
| SR 99-03 | 40.4 ±1.1 | Biotite | Dacitic tuff | -67.0295 | -20.7959 |

rocks are exposed north of the Salar de Uyuni at Cerro Uyarani (Wörner et al. 2000). In the eastern part of the Altiplano, the basement is composed mainly of Ordovician to Devonian sedimentary rocks exposed along the UKFZ and in the Eastern Cordillera.

In the Altiplano, a significant angular unconformity is present between intensely foliated and folded Ordovician rocks and non-foliated late Ordovician (Cancaniri Fm.) and Silurian rocks. Cretaceous to Paleocene rock sequences rest conformably on the Silurian strata. By contrast, in the Eastern Cordillera a marked angular unconformity of Carboniferous age (Müller et al. 2002) separates early Paleozoic from Meozoic and younger strata.

The Cretaceous to Paleocene strata consist of terrigenous sediments, subordinate mafic volcanics and carbonates. In the early Cretaceous, deposition started in a small rift basin near Tupiza in the Eastern Cordillera. During the Maastrichtian and the Paleocene, former ar-

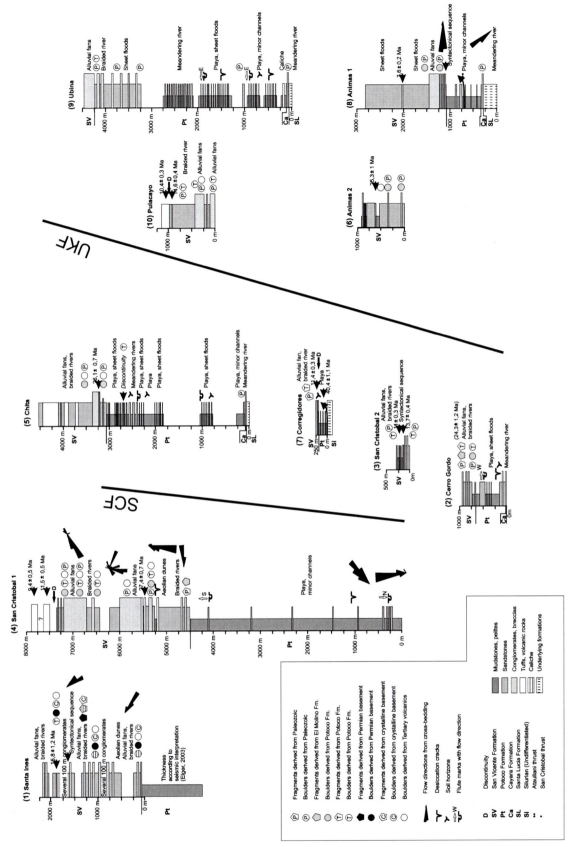

**Fig. 13.5.** Stratigraphic sections from the Altiplano with their respective position to major faults. For the location numbers and abbreviations refer to Fig. 13.2

eas of non-deposition were integrated into a post-rift basin extending from the Eastern Cordillera over the Altiplano to the Salar de Atacama region in Chile (Sempere 1995; Sempere et al. 1997, Fiedler 2001, Reutter et al. 2006, Chap. 14 of this volume). The limestones of the El Molino Formation and its equivalents are a significant, shallow marine, marker horizon indicating that the future plateau lay near sea level at the beginning of the Paleocene. There are first indications of sediment transport from the central Eastern Cordillera towards the west, as well as to the east, (Horton and DeCelles 2001) in the Paleocene-early Eocene Cayara Fm. (Silva-González 2004).

The Eocene Potoco Fm. shows significant differences between the eastern/central and the western Altiplano. To the east, an upwardly thickening and coarsening sequence (Lipez Basin, thickness > 3000 m) documents increasing relief along the eastern basin margin and shows a succession from the individual channels of a floodplain to a braided river environment. Flute casts and cross beds indicate an east to west transport direction (Fig. 13.5: Animas 1). To the west, the Potoco Fm. is mainly made up of red, thin-bedded to laminated pelites with frequent gypsum layers. The facies association points to deposition at the bottom of a quiet water column in a playa lake to playa-mudflat environment.

Provenance analyses of sandstones (Fig. 13.6) have revealed a predominantly recycled orogenic source, probably derived from the Paleozoic basement of the Eastern Cordillera, but also, locally, a magmatic arc source which was situated in the Western Cordillera and in the Chilean Precordillera. In the western Altiplano, east-directed sediment transport has been reported by Almendras-Alarcón et al. (1997). Thus, we found evidence for the eastern as well as the western basin margins. This contrast with the findings of Horton et al. (2001) who described a persistent sediment source for the Potoco Formation only to the west associated with a paleocurrent reversal across the Santa Lucia–Potoco contact.

The Potoco Fm. is distributed throughout the Altiplano and, as revealed by seismic interpretations across the UKFZ, thicknesses vary significantly, probably owing to extensional tectonics (Elger 2003). The age of the Potoco Fm. is poorly constrained, however, a wind-blown tuff found near Corregidores provided an age of 40.4 ± 1.1 Ma (K-Ar in biotite, Table 13.1, Fig. 13.4). This is in agreement with the palynomorph data published by Horton et al. (2001) from sections north of the Salar de Uyuni, which gave a late Eocene to Oligocene age for the majority of the Potoco strata. Horton et al. (2001) also inferred from age dating of the underlying Santa Lucia Fm. (~58 Ma, Sempere et al. 1997) that the lowermost Potoco Fm. was deposited during the late Paleocene to middle Eocene. In general, the sedimentary characteristics of the Potoco Fm. indicate a relatively low relief (except for the eastern basin margin) and little tectonic activity.

## 13.6.2 The Early Oligocene Turnover and the 32–17 Ma Interval

In the center of the Altiplano Basin, the Potoco Fm. is overlain by the San Vicente Fm. which spans an age interval of about 32–10 Ma. The boundary between both formations is marked by the first appearance of breccias and conglomerates containing Paleozoic clasts. The contact between the two formations is conformable west of the San Cristobal fault, which is the westernmost branch of the UKFZ (SCF in Figs. 13.2 and 13.3), but angular unconformities have been observed towards the east. Moreover, in several localities the San Vicente Fm. directly rests on Silurian or Ordovician strata along the UKFZ, indicating a considerable erosional phase prior to its deposition and a strong paleo-relief. For example, in the Cerro Gordo area, the San Vicente Fm. lies on foliated Ordovician strata. Basal tuffs are 24.3 ± 1.2 Ma old, indicating

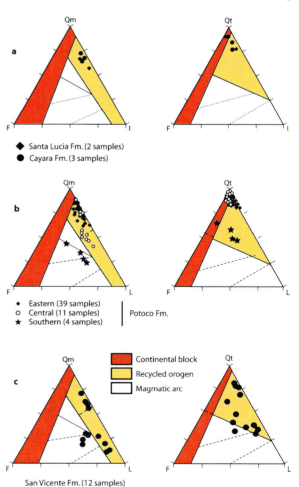

**Fig. 13.6.** Provenance of sandstones from the Altiplano according to Dickinson and Suczek (1979). **a** Paleocene (Cayara and Santa Lucia Fms.); **b** Eocene (Potoco Fm.); **c** Oligocene-Miocene (San Vicente Fm.). *Qm*: monocrystalline quartz; *Qt*: total quartz; *F*: feldspar; *L*: lithics

that this phase of erosion had taken place in the late Oligocene or earlier.

The age of basal San Vicente sediments varies considerably. West of the UKFZ, deposition started in the early Oligocene, but east of it, sedimentation began in the late Oligocene to early Miocene (between 20 and 25 Ma). In the San Cristobal section, a complete profile of the Potoco and San Vicente deposits has been investigated. Here, the first volcanic intercalations (K-Ar in biotite: 27.4 ± 0.7 Ma) have been found some 900 m above the base of this ~4000 m thick formation, but, owing to the presence of volcanic boulders in the conglomerates 750 m above the base, the onset of volcanism should have occurred somewhat earlier.

To estimate when deposition of the San Vicente Fm. began, we tentatively assumed a constant sedimentation rate for the entire formation, thus implying an age of ~32 Ma for the base of the San Vicente strata. This is significantly earlier than the onset of volcanism. The first occurrence of Paleozoic clasts in the basal San Vicente would then be coeval with the onset of exhumational cooling (33–29 Ma) along the UKFZ to the east (see above). In the Chita section (Fig. 13.5), where the San Vicente Fm. also lies conformably upon Potoco strata, the basal San Vicente conglomerates contain volcanic clasts, besides Paleozoic and El Molino (late Cretaceous – Paleocene) clasts. An age of 26.1 ± 0.7 Ma was derived from the lowermost tuff, approximately 200 m above the base.

The San Vicente Fm. differs considerably from the Potoco Fm. It is composed of breccias, conglomerates, sandstones (Fig. 13.5), pelites, tuffaceous sandstones, tuffs and lavas in a complicated lateral and vertical facies amalgamation. Thickness varies considerably within the basin, reaching a maximum of ca. 4000 m west of the UKFZ in the San Cristobal area. Towards the west and the south, thicknesses diminish gradationally, whereas to the east, across the UKFZ, there is a remarkable change in the San Vicente thickness from these maximum values to only several tens of meters.

Facies associations of the San Vicente Fm. comprise alluvial fans, aeolian deposits, fluvial systems, and subordinate lacustrine and playa associations (Mertmann et al. (2003). Conglomerates and breccias of the western Altiplano region are derived from a granitic and metamorphic basement of probable Precambrian age, as well as from Permian volcanics that were exhumed around 28 Ma (see above).

Detrital AFT cooling ages of these components record rapid exhumation from ~32 Ma in the western source areas. In the central and eastern parts, all sections show a predominance of clasts derived from the early Paleozoic sedimentary basement mixed with minor amounts of clasts derived from the Potoco Fm. and the late Cretaceous – Paleocene El Molino Fm. It is important to note that in the vicinity of the SCF, the basal San Vicente Fm. consists of breccias containing Paleozoic and El Molino clasts indicating the existence of nearby basement highs when deposition of the San Vicente Fm. began. The same clast compositions exist in Cerro Gordo as well as breccia horizons which, however, also contain Potoco clasts.

Volcanic rocks, tuffs and lahar suites indicate intense volcanism during much of the late Oligocene to Miocene. This corresponds to the magmatic arc source indicated by the provenance analyses diagrams of the San Vicente Fm. (Fig. 13.6).

### 13.6.3 17–8 Ma (Miocene)

In the upper part of the San Vicente Fm. (Pilkhaua subsequence, Elger 2003), growth strata are developed along thrusts indicating deposition during shortening deformation. The age data from the upper part of the San Vicente Fm. Range between 16.1 and 7.6 Ma (Table 13.1). Synthrusting sequences with progressive unconformities have been observed in the Pulacayo, Santa Ines, Animas 1, and San Cristobal sections. The alluvial fans and fluvial systems are composed once more of conglomerates and sandstones, tuffaceous sandstones, tuffs and lavas. In younger parts of the San Vicente Fm., Potoco-derived clasts clearly become more frequent than in the lower part, indicating a significant change in the composition of the source area. The overall tendency of the associated sediment is an upwardly coarsening cycle, which may point to a thrust source (see below). Exhumation in the hanging wall of the San Cristobal fault started around 20 Ma. Hence, the thick strata of the former half-graben were thrust to the east, tilted, and eroded, plausibly explaining these significant sedimentological changes in the upper San Vicente. The San Vicente strata are overlain by late Miocene volcanics and volcaniclastic sediments with an angular unconformity. The age of the unconformity varies between approximately 11 and 8 Ma (Table 13.1).

### 13.6.4 8 Ma–Recent

After the late Miocene deformation events, the southern Altiplano remained an endorheic basin developing during a tectonic quiescence that has characterized both the Altiplano and the Eastern Cordillera over the last 8 million years. Strata younger than 8 Ma are flat-lying, and no significant faults have been observed. Undeformed sediments covering the San Juan del Oro peneplain are distributed over SW Bolivia and the Puna of NW Argentina (Müller et al. 2002). Our AFT results suggest that the inner part of the foreland-thrust belt (Interandean zone) had largely formed by this time, and that the eastern deformation front stepped into the Subandean zone. The zone of active shortening and uplift is located approximately 10 to 20 km east of the easternmost frontal range of the Subandean belt, below the Chaco plain.

The sedimentary sequences of the Interandean, Subandean and Chaco area document the eastward migration of the Andean orogenic front. This can be shown by the development of the foreland basin infill of the Chaco Basin. This infill can be subdivided into five formations: the Petaca, Yecua, Tariquia, Guandacay and Emborozu (Hulka et al. in press, Uba et al. in press).

The *Petaca Fm.* (approximately 60 to and 15 Ma, Marshall and Sempere 1991; Marshall et al. 1993) is up to 50 m thick and consists of duricrusts interbedded with channel fills of poorly-defined, sandy, ephemeral streams deposited in an arid, braidplain environment. During the mid to late Miocene, several short-lived, weak, marine transgressions occurred and the marine lagoonal to tidal facies of the *Yecua Fm.* may represent the underfilled stage and distal position of the eastward-advancing, Andean foredeep. The Yecua Fm. is overlain by fluvial redbeds, up to 2 km thick, of the *Tariquia Fm.* The provenance of this formation is from the west; the medium-grained, multiply-recycled quartz grains clearly record the uplift and unroofing of major continental sedimentary sequences to the west.

The base of the overlying *Guandacay Fm.* (Miocene – Pliocene, more than 100 m thick) is conventionally drawn at the first sustained occurrence of pebble conglomerates. The contact is gradational and marks the passing from lowland, anastamosing streams and semiarid floodplains to gravelly, braided fluvial systems, and suggests an approaching source area of high relief. The base of the youngest unit of the Chaco Basin fill, the *Emborozu Fm.* (Pliocene – Pleistocene), is marked by the first sustained occurrence of cobble and boulder conglomerates. These mark alluvial fan facies and the proximity of high morphological relief. The Guandacay and Emborozu Fms. are almost exclusively exposed near the present mountain front and grade eastwards, presumably rapidly, into finer-grained deposits, overlain by the Quaternary floodplain and aeolian deposits of the Chaco plain. As stated above, the Miocene to Pleistocene sedimentary sequences of the Chaco indicate the approach of the Andean thrust front, which is also reflected in the AFT data.

## 13.7 Discussion

Our sedimentary and exhumational data have tectonic implications for plateau evolution and, furthermore, are relevant to the evolution of the Andean active margin. Our results show four stages in the tectonic evolution of the southern Altiplano plateau and adjacent areas:

1. *Pre 32 Ma*: Relative tectonic quiescence in the Altiplano Basin, Incaic shortening in the Precordillera (45–38 Ma) and initial thrusting and uplift in the Proto Eastern Cordillera (40–36 Ma).

2. *32–17 Ma*: Exhumation, creation of strong relief, construction of a magmatic arc (from ~28 Ma). Deformation restricted to the present plateau, extension in the Altiplano, shortening in the Eastern (and Western) Cordillera.

3. *17–8 Ma*: Shortening in the entire plateau, beginning of thrusting onto the foreland.

4. *8–0 Ma*: Shortening in the foreland, tectonic quiescence in the plateau.

### 13.7.1 Tectonic Implications of the Sedimentary and Exhumational Result

The Andean high plateau and adjacent areas are characterized by surface uplift and exhumation, as well as by strong subsidence of sedimentary basins since the Paleogene. In areas, exhumed along major faults, the Paleozoic and Precambrian basement is eroded, whereas basinal areas receive continental detritus, in some places up to 8 km thick (Oligo-Miocene San Vicente Fm.: 4 km). Today, these basins are thrust-bound. However, the observations of Elger (2003) indicate that normal faulting has also played an important role during the basin evolution in the Altiplano. Our AFT data show that inversion of normal faults occurred around 17 Ma, during deposition of the San Vicente Fm. (see below). This inversion contrasts with the evolution of the Eastern Cordillera where, according to Müller et al. (2002), shortening has prevailed since the late Eocene.

At the beginning of the Tertiary and before ~32 Ma, the Altiplano was near sea level, evidenced by the marine strata of the El Molino Fm. (Sempere et al. 1997). This does not seem to have changed fundamentally until the early Oligocene. In the central Eastern Cordillera, however, exhumation started in the late Eocene (see above and cf. Lamb and Hoke 1997). Sandstone composition and the upward coarsening of sediment in the Potoco Fm., shed from the Proto-Cordillera, indicate growing relief at the basin margin. The thicknesses of Potoco sediments vary strongly from some hundred meters to (locally) 4000 m.

In general, sedimentary thicknesses can be attributed to some crustal thickening and, thus, some uplift. On the other hand, Potoco extension of about 5% (Elger 2003) caused crustal thinning. A quantitative estimate of these concurring processes is difficult since the original thickness of the Potoco Fm. is known only from a few conformable successions into the San Vicente Fm. It seems, however, that there was only little uplift in the central Altiplano before the early Oligocene. This concurs with the uplift history data of Gregory-Wodzicki (2000) which state no uplift before the early Miocene.

With the beginning of San Vicente deposition, a major change in basin configuration took place. Differential uplift produced local basement highs where Cretaceous and Paleozoic units became exposed. The distribution of sediments, for example, across the west-dipping SCF is such that to the east, i.e. in the footwall of this fault, thicknesses of the Potoco and San Vicente Fms. are only 400 m and 200 m, respectively (Elger 2003; Silva-González 2004). West of the fault, each formation reaches a maximum thickness of 4000 m. This indicates that the basement high was already bounded by a precursor to the San Cristobal fault.

Questions arise about the kinematic regime of this precursor fault during early San Vicente time. We suggest that the variations in sedimentary thickness across the SCF is more indicative of normal faulting than of thrusting, although, today, this steeply westward-dipping fault shows thrusting kinematics which, according to the AFT data, took place in the mid to late Miocene (20–11 Ma). If these expansive thicknesses were accommodated in the back of a thrust fault, this would imply an exceptionally thick, piggyback basin fill on top of the thrusted strata, in contrast to a very thin foreland infill.

Furthermore, if this maximum thickness of 4000 m was accommodated behind the thrust, the components should represent an inverse lithostratigraphical succession. This, however, contrasts with the spectrum of components in the San Vicente Fm., which predominantly contains Paleozoic clasts from the base upward. The Paleozoic components of this formation are only poorly rounded indicating the existence of a nearby source area at approximately 32 Ma. East-west extension is also in accordance with the orientation of N-S-trending mafic dikes (25.6 ± 0.9 Ma) that have been found in the Chita area.

To the west, the San Vicente Basin was bounded by a basement high composed of Proterozoic to Permian rocks, representing the western source area of the Tertiary strata. However, due to poor outcrop data we cannot decide whether the San Vicente depocenter was a half-graben or graben setting. The seismic interpretations of Elger (2003) rather indicate a half-graben setting.

Since ca. 28 Ma, widespread volcanic activity created further relief and an additional source of sediment supply. West of the UKFZ, the depocenter persisted and formed a local intra-arc basin, probably still bounded by a normal fault. This extensional basin was inverted, but the onset of contraction is poorly constrained. Exhumation of the depocenter west of the San Cristobal fault yields a minimum age of 20 Ma. Clearly, the major phase of shortening in the central Altiplano began by 20–19 Ma (Elger 2003).

### 13.7.2 Interpretation in the Context of the Andean Active Continental Margin Evolution

Before 38 Ma, the Altiplano occupied a back-arc position in relation to the magmatic arc in the Chilean Precordillera (Fig. 13.7a; Charrier and Reutter 1994). At the beginning of the Paleogene, this back-arc area formed one single basin and the generally fine-grained sediments in the basin center seem to indicate weak relief. However, the original basin progressively divided into smaller basins owing to the uplift of the central Eastern Cordillera (Protocordillera, Lamb and Hoke 1997), which became the source area for fringing river systems.

During this time, internal normal faults might have been active (Elger 2003), thus already indicating areas of, later accentuated, tectonic basin segmentation. AFT data indicate that uplift occurred in the central Eastern Cordillera caused by initial thrusting increments between 40 and 36 Ma (Fig. 13.7b). These thrusting movements partly overlapped with the Incaic phase transpressional movements that affected the magmatic arc in the north Chilean Precordillera between 45 and 38 Ma (Haschke and Günther 2003). Thus, there were deformations in two zones separated by an approximately 430 km wide zone of tectonic quiescence and some normal faulting.

Around 38 Ma, volcanism ceased in the Chilean Precordillera for some 10 million years (Döbel et al. 1992). This period was crucial for the early evolution of the central Andean Plateau: Exhumational cooling started around 32 Ma along discrete faults (Fig. 13.7c) distributed between the eastern margin of the Eastern Cordillera and the western Altiplano, and possibly continued into the north Chilean Precordillera (Maksaev and Zentilli 1999), i.e. over the future plateau area which was > 400 km wide before the onset of Andean shortening. Exhumation occurred together with a fundamental change in the sedimentary record in adjacent basins prior to the resumption of volcanism at ~28 Ma, as the first (and oldest) volcanic intercalations in the San Vicente Fm. have been found 900 m above its base.

Our ages for the described tectonic events of between 38 and 32 Ma is in good agreement with Gephart's (1994) age for the initial uplift of the central Andean Plateau. Gephart considered topographic and slab symmetry and suggested the beginning of plateau formation at about 35 Ma. Given the uncertainties in age dating, this seems a rather good accordance.

A clue for understanding this early plateau comes from inferences about the subduction geometry in the early Oligocene when the downgoing slab was probably nearly horizontal at 100 km depth (James and Sacks 1999; Haschke et al. 2002, Reutter 2001). This so-called flat subduction probabaly was caused by slab detachment which has been suggested by Fukao et al. (2001) to have ocurred in the circum-Pacific subduction zones in the Eocene epoch. Flat subduction has severe consequences for the tectonic regime in the upper plate. According to Gutscher et al. (2000), flat subduction leads to a pause in volcanism and to strong intraplate coupling, which leads to strong deformation in the cold seismogenic crust of the

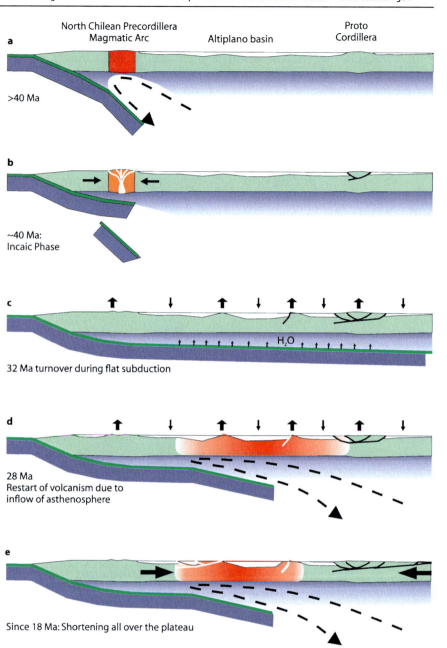

**Fig. 13.7.**
Cartoons showing the tectonic evolution of the central Andean Plateau around 21° S

upper plate. Deformation is concentrated on the zone above the point where the slab resteepens, and this could explain how shortening deformation jumped from the north Chilean Precordillera to the Eastern Cordillera.

A further consequence of flat subduction should be the hydration of the upper plate mantle. The water carried by the subducting slab might have migrated into the upper plate mantle, partially altering the mantle peridotite into serpentinite. Serpentinization, in turn, increased the volume of the mantle by decreasing its density, consequently making a considerable contribution to the initial uplift of the future plateau. The inferred extensional movements might be caused by variations in the degree of coupling between the plates from the central Altiplano to the Eastern Cordillera. They might also be due to local and near-surface effects of differential uplift, perhaps owing to variations in the degree of hydration in the lithospheric mantle.

Around 28 Ma, subsequent to the beginning of exhumation and faulting, magmatism started again in the Central Andes over a very wide area, reaching from the Western Cordillera over the Altiplano to the Eastern Cordillera (Fig. 13.7d). Intense and widespread magmatic activity, which is strongly reflected in the composition of sandstones (Fig. 13.6), began with basaltic-andesitic to silicic lavas, tuffs, and shallow intrusions. The southern Altiplano also became part of the Central Andean magmatic arc. Such a broad arc indicates a rather low dip angle for the subducting slab, which, however, must have been steep enough to allow the formation of an asthenospheric wedge (Tatsumi and Eggins 1995). Thus, since the restart of volcanism, uplift should have been accompanied by lithospheric thinning together with crustal thickening because of distributed shortening within the plateau, as has been suggested by Isacks (1988).

Around 18 Ma, shortening started in the Altiplano (Fig. 13.7e), and the Andean orogen was progressively thrust along the Interandean thrust onto the Brazilian shield (Jordan et al. 1997). Thus, crustal thickening has been the major plateau-forming process since that time. A further change occurred in the Andean system around 10 Ma. Before this time, shortening occurred in the plateau and in the western foreland, but after 10 Ma, deformation ceased in the plateau (Gubbles et al. 1993), becoming stronger in the foreland. The termination of deformation within the plateau was accompanied by the underthrusting of cold Brazilian shield lithosphere, causing a narrowing of the magmatic arc and restricting it to the Western Cordillera. The Altiplano and the Eastern Cordillera are now underlain by a cold thick lithosphere mantle (Whitman et al. 1992).

To conclude, the style of deformation changed through the stages outlined above: shortening and extension operated simultaneously during initial plateau formation (32–19 Ma); between 19 and 8 Ma, shortening occurred within the plateau and also in the eastern foreland; and since 8 Ma, shortening has been localized to the thin-skinned fold-and-thrust belt in the eastern foreland. Furthermore, uplift during the first two stages occurred together with subsidence over wide parts of the plateau. Neither a migration of an exhumation front nor an eastward propagation of a thrust front operated before 18 Ma. A further important result is that initial plateau formation occurred significantly earlier than the onset of volcanism.

## Acknowledgments

Research was supported by the Deutsche Forschungsgemeinschaft and the Freie Universität Berlin (Sonderforschungsbereich 267 "Deformation Processes in the Andes"). We thank O. Oncken for helpful discussions. Fieldwork was facilitated by Umberto Castro, Reinhard Rössling and Heriberto Salamanca. The manuscript was improved by the constructive rewies of Teresa Jordan, (Cornell University, Ithaca New York) and Fritz Schlunegger (University of Bern).

## References

Allmendinger RW, Gubbels T (1996) Pure and simple shear plateau uplift, Altiplano-Puna, Argentina and Bolivia. Tectonophysics 259: 1–14

Almendras-Alarcón OD, Baldelón EG, López R (1997) Hoja geológica Volcán Ollague/San Agustín. Programa Carta Geológica de Boliva 1: 100 000, Publ. SGM Serie I-CGB-46 SG

Alpers CN, Brimhall GH (1988) Middle Miocene climatic change in the Atacama Desert, northern Chile: evidence from supergene mineralization at La Escondida. Geol Soc Am Bull 100:1640–1656

Andriessen PAM, Reutter K-J (1994) K-Ar and fission track mineral age determinations of igneous rocks related to multiple magmatic arc systems along the 23°S latitude of Chile and NW Argentina. In: Reutter K-J, Scheuber E, Wigger P (eds) Tectonics of the Southern Central Andes. Springer-Verlag, Berlin Heidelberg New York, pp 141–153

Babeyko AY, Sobolev SV, Trumbull RB, Oncken O, Lavier LL (2002) Numerical models of crustal scale convection and partial melting beneath the Altiplano-Puna plateau. Earth Planet Sci Lett 199: 373–388

Baby P, Herail H, Salinas R, Sempere T (1992) Geometry and kinematic evolution of passive roof duplexes deduced from cross-section balancing: Example from the foreland thrust system of the southern Bolivian Subandean Zone. Tectonics 11:523–536

Baby P, Rochat P, Herail G, Mascle P, Paul A (1996) Neogene thrust geometry and crustal balancing in the northern and southern branches of the Bolivian oroline (Central Andes). International Symposium on Andean Geodynamics 3, Saint Malo

Baby P, Rochat P, Herail H, Mascle G (1997) Neogene shortening contribution to crustal thickening in the back arc of the Central Andes. Geology 25:883–886

Bahlburg H, Breitkreuz C, Zeil W (1986) Paläozoische Sedimente Nordchiles. Berliner Geowiss Abh A66, pp 147–168

Benjamin MT, Johnson NM, Naeser CW (1987) Recent rapid uplift in the Bolivian Andes: evidence from fission-track dating. Geology 15:680–683

Brown RW, Summerfield MA (1997) Some uncertainties in the derivation of rates of denudation from thermochronologic data. Earth Surf Process Landforms 22:239–248

Carlson WD, Donelick RA, Ketcham RA (1999) Variability of apatite fission-track annealing kinetics: I. Experimental results. Am Mineralogist 84:1213–1223

Charrier R, Reutter K-J (1994) The Purilactis Group of Northern Chile: boundary between arc and backarc from late Cretaceous to Eocene. In: Reutter K-J, Scheuber E and Wigger P (eds) Tectonics of the Southern Central Andes. Springer-Verlag, Berlin Heidelberg New York, pp 189–202

DeCelles PG, Horton BK (2003) Early to middle Tertiary foreland basin development and the history of Andean crustal shortening in Bolivia. Geol Soc Am Bull 115:58–77

Dickinson WR, Suczek CA (1979) Plate tectonics and sandstone compositions. Am Ass Petrol Geol Bull 63:2164–2182

Döbel R, Friedrichsen H, Hammerschmidt K (1992) Implication of $^{40}Ar/^{39}Ar$ dating of early Tertiary volcanic rocks from the North Chilean Precordillera. Tectonophysics 202:55–81

Dunn JF, Hartshorn KG, Hartshorn PW (1995) Structural styles and hydrocarbon potential of the Subandean Belt of southern Bolivia. In: Tankard AJ, Suarez R, Welsink HJ (eds) South American Petroleum Basins. AAPG Memoir 62, pp 523–543

Ege H (2004) Exhumations- und Hebungsgeschichte der zentralen Anden in Südbolivien (21°S) durch Spaltspur-Thermochronologie an Apatit. PhD thesis, Freie Universität Berlin

Ege H, Sobel E, Jacobshagen V, Scheuber E, Mertmann D(2003) Exhumation history of the Central Andes of southern Bolivia by apatite fission track dating. Revista Tecnica YPFB 21:165–172

Ege H, Sobel ER, Scheuber E, Jacobshagen V (submitted) Exhumation history of the Central Andean plateau (southern Bolivia) constrained by apatite fission track thermochronology. Submitted to Tectonics

Elger K (2003) Analysis of deformation and tectonic history of the Southern Altiplano Plateau (Bolivia) and their importance for plateau formation. GeoForschungsZentrum Potsdam, Scientific technical report STR 03/05, pp 1–152

Elger K, Oncken O, Glodny J (2005) Plateau-style accumulation of deformation – the Southern Altiplano. Tectonics (in press)

Fiedler K (2001) Die kretazisch-alttertiäre Entwicklung des südlichen Potosí-Beckens (Süd-Bolivien). Berliner Geowiss Abh A215:1–185

Fiedler K, Mertmann D, Jacobshagen V (2003) Cretaceous marine ingressions in the southern Potosi basinof southern Bolivia: tectonic and eustatic control. Revista Tecnica YPFB 21:157–164

Fitzgerald PG, Sorkhabi RB, Redfield TF, Stump E (1995) Uplift and denudation of the central Alaska Range: a case study in the use of apatite fission track thermochronology to determine absolute uplift parameters. J Geophys Res 100:20175–20191

Fukao Y, Widiyantoro S, Obayashi M (2001) Stagnant slabs in the upper and lower mantle transition region. Rev Geophys 39(3): 291–323

Gallagher K (1995) Evolving temperature histories from apatite fission-track data. Earth Planet Sci Lett 136:421–435

Gephart JW (1994) Topography and subduction geometry in the central Andes: clues to mechanics of a noncollisional orogen. J Geophys Res 99(B6):12279–12288

Green PF, Duddy IR, Laslett GM, Hegarty KA, Gleadow AJW, Lovering JF (1989) Thermal annealing of fission tracks in apatite. 4. Quantitative modelling techniques and extension to geological time-scales. Chem Geol (Isotope Geosci Sec ) 79:155–182

Gregory-Wodzicki KM (2000) Uplift history of the Central and Northern Andes: a review. Geol Soc Am Bull 112(7):1091–1105

Gubbels TL, Isacks BL, Farrar E (1993) High-level surfaces, plateau uplift, and foreland development, Bolivian central Andes. Geology 21:695–698

Günther A (2001) Strukturgeometrie, Kinematik und Deformationsgeschichte des oberkretazisch-alttertiären magmatischen Bogens (nord-chilenische Präkordillere, 21.7°–23°S). Berliner Geowiss Abh A 213

Gutscher ME, Spakman W, Bijwaard H, Engdahl ER (2000) Geodynamics of flat subduction: seismicity and tomographic constraints from the Andean margin. Tectonics 19(5):814–833

Hansen DL, Nielsen SB (2003) Why rifts invert in compression. Tectonophysics 373:5–24

Haschke M, Günther A (2003) Balancing crustal thickening in arcs by tectonic vs. magmatic means. Geology 31(11):933–936

Haschke M, Scheuber E, Günther A, Reutter K-J (2002) Evolutionary cycles during the Andean orogeny. repeated slab breakoff and flat subduction? Terra Nova 14(1):49–55

Henry SG, Pollack HN (1988) Terrestrial heat flow above the Andean subduction zone in Bolivia and Peru. J Geophys Res 93: 15153–15162

Hindle D, Kley J, Klosko E, Stein S, Dixon T, Norabuena E (2002) Consistency of geologic and geodetic displacements in Andean orogenesis. Geophys Res Lett 29: doi 10.1029/2001GL013757

Hindle D, Kley J, Oncken O, Sobolev S (2005) Crustal flux and crustal balance from shortening estimates in the Central Andes. Earth Planet Sci Lett 230:113–124

Horton BK (1998) Sediment accumulation on top of the Andean orogenic wedge: Oligocene to Late Miocene basins of the Eastern Cordillera, Southern Bolivia. Geol Soc Am Bull, 110(9):1174–1192

Horton BK (1999) Erosional control on the geometry and kinematics of thrust belt development in the central Andes. Tectonics 18:1292–1304

Horton BK, DeCelles PG (2001) Modern and ancient fluvial megafans in the foreland basin systems of the central Andes, southern Bolivia: implications for drainage network evolution fold-thrust belts. Basin Research 13:43–63

Horton BK, Hampton BA, Waanders GL (2001) Paleogene synorogenic sedimentation in the Altiplano plateau and implications for initial mountain building in the central Andes. Geol Soc Am Bull 113(11):1387–1400

Hulka C, Gräfe K-U, Sames B, Uba CE, Heubeck C (in press) Depositional setting of the middle to late Miocene Yecua Formation of the Chaco foreland basin, southern Bolivia. J S Am Earth Sci

Isacks BL (1988) Uplift of the Central Andean Plateau and bending of the Bolivian Orocline. J Geophys Res 93:3211–3231

James DE, Sacks IS (1999) Cenozoic formation of the Central Andes: a geophysical perspective. In: Skinner BJ (ed) Geology and ore deposits of the Central Andes. Spec Pub Soc Econ Geol 7, pp 1–25

Jordan TE, Reynold JH III, Erikson JP (1997) Variability in age of initial shortening and uplift in the Central Andes, 16–33°30'S. In: Ruddiman WF (ed) Tectonic uplift and climate change. Plenum Press, pp 41–61

Kennan L (2000) Large-scale geomorphology of the Andes: interrelationships of tectonics, magmatism and climate. In: Summerfield MA (ed) Geomorphology and global tectonics. John Wiley & Sons, Chicester, pp 167–200

Kennan L, Lamb SH, Hoke L (1997) High altitude palaeosurfaces in the Bolivian Andes: evidence for late Cenozoic surface uplift. In: Widdowson M (ed) Palaeosurfaces: recognition, reconstruction and interpretation. Geol Soc London Spec Pub 120, pp 307–324

Ketcham RA, Donelick RA, Carlson WD (1999) Variability of apatite fission-track annealing kinetics: III. Extrapolation to geological time scales. Am Mineralogist 84(9):1235–1255

Ketcham RA, Donelick RA, Donelick MB (2000) AFT-Solve: a program for multi-kinetic modeling of apatite fission-track data. Geol Materials Res 2(1):1–32

Kley J (1996) Transition from basement-involved to thin-skinned thrusting in the Cordillera Oriental of southern Bolivia. Tectonics 15:763–775

Kley J, Müller J, Tawackoli S, Jacobshagen V, Manutsoglu E (1997) Pre-Andean and Andean-age deformation in the Eastern Cordillera of Southern Bolivia. J S Am Earth Sci 10(1):1–19

Kley J, Monaldi CR, Salfity JA (1999) Along-strike segmentation of the Andean foreland: causes and consequences. Tectonophysics 301:75–94

Kontak DJ, Farrar E, Clark AH, Archibald DA (1990) Eocene tectonothermal rejuvenation of an upper Palaeozoic-lower Mesozoic terrane in the Cordillera de Carabaya, Puno, southeastern Peru, revealed by K-Ar and $^{40}Ar/^{39}Ar$ dating. J S Am Earth Sci 3(4):231–246

Lamb S (2000) Active deformation in the Bolivian Andes, South America. J Geophys Res 105(11):25627–25653

Lamb S, Hoke L (1997) Origin of the high plateau in the Central Andes, Bolivia, South America. Tectonics 16:623–649

Lamb S, Hoke L, Kennan L, Dewey J (1997) Cenozoic evolution of the Central Andes in Bolivia and northern Chile. In: Burg JP, Ford M (eds) Orogeny through time. Geol Soc Spec Pub 121, pp 237–264

Maksaev V, Zentilli M (1999) Fission track thermochronology of the Domeyko Cordillera, northern Chile: implications for Andean tectonics and porphyry copper metallogenesis. Expl Mining Geol 8:65–89

Marquillas RA, Salfity JA (1994) Tectonic and sedimentary evolution of the Cretaceous-Eocene Salta Group, Argentina. In: Salfity JA (ed) Cretaceous Tectonics of the Andes. Earth Evolution Science 6, pp 266–315

Marshall LG, Sempere T (1991) The Eocene to Pleistocene vertebrates of Bolivia and their stratigraphic context: a review. In: Suarez-Soruco R (ed) Vertebrados. Revista técnica de YPFB 12, pp 631–652

Marshall LG, Sempere T, Gayet M (1993) The Petaca (Late Oligocene–Middle Miocene) and Yecua (Late Miocene) formations of the Subandean-Chaco basin Bolivia and their tectonic significances. Documents du laboratoire de Lyon, pp 291–301

Mertmann D, Scheuber E, Silva-González P, Reutter K-J (2003) Tectono-sedimentary evolution of the southern Altiplano: basin evolution, thermochronology and structural geology. Revista técnica de YPFB, 21: 17–22

Moretti I, Baby P, Mendez E, Zubieta D (1996) Hydrocarbon generation in relation to thrusting in the Subandean zone from 18° to 22°S, South Bolivia. Petrol Geosci 2:17–28

Müller J (2000) Tektonische Entwicklung und Krustenverkürzung der Ostkordillere Südboliviens (20.7°S–21. 5°S). PhD thesis, Freie Universität Berlin

Müller JP, Kley J, Jacobshagen V (2002) Structure and Cenozoic kinematics of the Eastern Cordillera, southern Bolivia (21°S). Tectonics 21: doi 10.1029/2001TC001340

Reutter K-J (2001) Le Ande centrali: elementi di un' orogenesi di margine continentale attivo. Acta Naturalia de "L' Ateneo Parmense" 37(1/2):5–37

Reutter K-J, Charrier R, Götze H-J, Schurr B, Wigger P, Scheuber E, Giese P, Reuther C-D, Schmidt S, Rietbrock A, Chong G, Belmonte-Pool A (2006) The Salar de Atacama Basin: a subsiding block within the western edge of the Altiplano-Puna Plateau. In: Oncken O, Chong G, Franz G, Giese P, Götze H-J, Ramos VA, Strecker MR, Wigger P (eds) The Andes – active subduction orogeny. Frontiers in Earth Science Series, Vol 1. Springer-Verlag, Berlin Heidelberg New York, pp 303–326, this volume

Sempere T (1994) Kimmeridgian to Paleocene tectonic evolution of Bolivia. In: Salfity JA (ed) Cretaceous Tectonics in the Andes. Earth Evolution Science 6, pp 169–212

Sempere T (1995) Phanerozoic evolution of Bolivia and adjacent regions. In: Tankard AJ, Suárez SR, Welsink, HJ (eds) Petroleum Basins of South America. AAPG Memoir 62, pp 207–230

Sempere T, Butler RF, Richards DR, Marshall LG, Sharp W, Swisher CC III (1997) Stratigraphy and chronology of Upper Cretaceous-lower Paleogene strata in Bolivia and northwest Argentina. Geol Soc Am Bull 109(6):709–727

Sillitoe RH, McKee EH (1996) Age of supergene oxidation and enrichment in the Chilean Porphyry Copper Province. Econ Geol, 164–179

Silva González P (2004) Das Süd-Altiplano-Becken (Bolivien) im Tertiär: sedimentäre Entwicklung und tektonische Implikationen. PhD thesis, Freie Universität Berlin

Springer M (1996) Die regionale Oberflächenwärmeflußdichte-Verteilung in den zentralen Anden und daraus abgeleitete Temperaturmodelle der Lithosphäre. PhD thesis, Freie Universität Berlin

Springer M, Förster A (1998) Heat-flow density across the Central Andean subduction zone. Tectonophysics 291:123–139

Tatsumi Y, Eggins S (1995) Subduction zone magmatism. Blackwell Scientific

Tawackoli S (1999) Andine Entwicklung der Ostkordillere in der Region Tupiza (Südbolivien). Berliner Geowiss Abh A203, p 116

Uba CE, Heubeck C, Hulka C (in press) facies analysis and basin architecture of the Neogene Subandean synorogenic wedge, southern Bolivia. Sed Petrol

Victor P, Oncken O, Glodny J (2004) Uplift of the western Altiplano plateau: evidence from the Precordillera between 20° and 21°S (northern Chile). Tectonics 23: doi 10.1029/2003TC001519

Wagner G, Van der Haute P (1992) Fission-track dating. Kluwer Academic Publishers

Whitman D, Isacks BL, Chatelain JL, Chiu JM, Perez A (1992) Attenuation of high-frequency seismic waves beneath the Central Andean plateau. J Geophys Res 97(B13):19929–19947

Wigger PJ, Schmitz M, Araneda M, Asch G, Baldzuhn S, Giese P, Heinsohn, WD, Eloy M, Ricaldi E (1994) Variation in the crustal structure of the Southern Central Andes deduced from seismic refraction investigations. In: Reutter K-J, Scheuber E, Wigger P (eds) Tectonics of the Southern Central Andes. Springer-Verlag, Berlin Heidelberg New York, pp 23–48

Wörner G, Lezaun J, Beck A, Heber V, Lucassen F, Zinngrebe E, Rössling R, Wilke HG (2000) Precambrian and early Paleozoic evolution of the Andean basement at Belén (northern Chile) and Cerro Uyarani (western Bolivian Altiplano). J S Am Earth Sci 13:717–737

# The Salar de Atacama Basin: a Subsiding Block within the Western Edge of the Altiplano-Puna Plateau

Klaus-J. Reutter · Reynaldo Charrier · Hans-J. Götze · Bernd Schurr · Peter Wigger · Ekkehard Scheuber · Peter Giese †
Claus-Dieter Reuther · Sabine Schmidt · Andreas Rietbrock · Guillermo Chong · Arturo Belmonte-Pool

**Abstract.** The internally drained Salar de Atacama (SdA) Basin, located in the proximal fore-arc between the present magmatic arc (Western Cordillera) to the east and the North Chilean Precordillera (Cordillera de Domeyko) to the west, represents a prominent morphological anomaly in the Central Andean Plateau. The basin is a post-Incaic feature that developed contemporaneously with the initial plateau uplift. Before 38 Ma, the magmatic arc was positioned in the present-day Precordillera; as a result, the Cretaceous to Eocene sequences that underlie younger SdA sediments were deposited in a proximal back-arc location, where a westward extending arm of the Salta Rift interfered with the magmatic arc.

The lithospheric block underlying the SdA area has different physical properties to its surroundings. Strong, positive, isostatic residual gravity anomalies in the SdA area are caused by dense bodies that were magmatically or tectonically emplaced at depths of 10 to 15 km and 4 to 6 km, probably during Cretaceous rift processes. They contrast with gravity minima in the Precordillera and Western Cordillera. Seismic refraction profiles in the SdA lithospheric block show velocities of > 6.5 km s$^{-1}$ and 7.5 km s$^{-1}$ at depths of < 10 km and ~30 km, respectively, but no clear Moho. A crustal thickness of 67 km, obtained from receiver functions, is reported, however Mesozoic and Tertiary structures do not indicate crustal thickening. Seismic tomography shows $v_p/v_s$ ratios that do not reveal mantle serpentinization, and high $v_p$ and high $Q_p$ values (> 2000) point to a dense and strong lithosphere down to the plate interface.

It is proposed that the SdA lithospheric block which, as part of the Salta Rift system in the pre-Incaic back-arc, had been hotter than the adjacent lithosphere was not cooled sufficiently to be serpentinized after the Incaic Event. Instead, mantle rocks hydrated at 600–800 °C (chlorite lherzolite) fit the measured physical properties. An episode of flat or low-angle subduction from 38 to 28 Ma, that preceded the emplacement of the modern magmatic arc in the Western Cordillera, was probably an important starting condition that allowed these processes to occur. Slab resteepening caused subsidence of the heavy SdA lithospheric block and its contact with the lower plate resisted the westward advancement of the asthenospheric wedge, thus causing the eastwardly convex curvature of the volcanic front around the east side of the SdA Basin.

## 14.1 Introduction

In satellite images and maps, the area of the Salar de Atacama (SdA) appears to be a first-order morphological anomaly in the long-wavelength topographic form of the Central Andean edifice (Fig. 14.1). The salt lake (salar), measuring 90 km from north to south and 55 km from east to west, forms the center of a deep fore-arc basin that is located in a wide, eastward deflection of the volcanic arc in the Western Cordillera. The SdA is, by far, the largest and widest in a series of basins and topographic lows in northern Chile that are somewhat irregularly aligned along strike from the upper Loa Valley to the Salar de Maricunga (21–27° S; Fig. 14.2). This Preandean Depression morphologically separates the Western Cordillera to the east from the north Chilean Precordillera to the west, which represents the western glacis of the Altiplano-Puna Plateau. Near Calama, a transverse depression (Calama Basin) forms a gap in the Precordillera determining a southern segment (Cordillera de Domeyko) and a northern segment (Sierra de Moreno). As the SdA basin extends quite far to the east, it exhibits stratigraphic sequences and structures with prolongations extending to the north and south that are hidden beneath the volcanic rocks of the Western Cordillera.

As a consequence of tectonics and an arid climate, the Salar de Atacama is an endorheic (internally-drained) basin, like all other salars and many lakes of the Central Andes. The present surface of the salt lake lies at 2 350 m, more than 700 m below the lowest point of its western rim in the Precordillera. Along the western margin (El Bordo Escarpment), a narrow fold-and-thrust belt exists within a thick sequence of Cretaceous to Eocene sediments. The Cordillera de la Sal, a narrow and low fold belt developed in thick clastic and evaporitic sediments of Miocene and Pliocene age, traverses the SdA near this western border. The eastern border, in contrast, is a monoclinal surface blanketed by Neogene ignimbrites that gently dip from the volcanoes bordering the Altiplano and Puna at 4 200 m down to the salt lake surface.

The SdA, though representing an actively subsiding sedimentary basin, is emplaced – together with the Calama Basin – in an area with strong, positive gravity anomalies in the isostatic residual field (Fore-arc Gravity High). A genetic relationship between other relatively depressed areas and positive gravity anomalies is also suggested for the Altiplano-Puna back-arc region, e.g. the salars of Uyuni, Arizaro, and Olaroz-Cauchari. Also, these internally drained basins represent areas whose pre-Oligocene floor beneath the accumulating younger sediment is relatively,

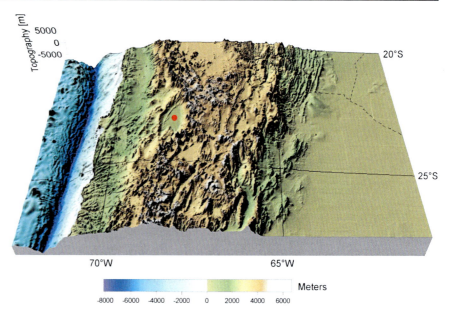

**Fig. 14.1.**
Elevation model of the Central Andean segment between 21° S and 26° S showing the prominent morphological anomaly of the Salar de Atacama (*red point*)

if not absolutely, subsiding with respect to the surrounding uplifting parts of the Altiplano-Puna Plateau.

The SdA Basin, however, is unique in the Central Andes for its strong, negative topographic expression in the westernmost Atiplano-Puna Plateau, and for its relationship to the eastwardly deflected volcanic front. Gravimetric and seismic investigations showed that the physical properties of the lithosphere under the SdA area are different from those of the surrounding lithosphere. For this reason, a peculiar, somewhat exotic nature, supposedly inherited from the pre-Mesozoic geological past, was repeatedly postulated for this lithospheric block (e.g. Schurr and Rietbrock 2004, Götze and Krause 2002; Schurr 2000, Giese et al. 1999).

Such assumptions, however, are in contrast with the late Paleozoic and early Mesozoic geological record, which does not reveal significant variations over the area. Only since the Cretaceous, and especially since the late Eocene, has the geology of SdA area developed in a characteristic way, thus suggesting that the peculiarities of the SdA lithospheric block were also acquired during this time. This leads us to re-analyse the gelogical record with special regard for its evolution since the Cretaceous and to reconsider the geophysical anomalies in order to elaborate a consistent model for the evolution of the SdA lithosperic block that satisfies both the geological and the geophysical data.

## 14.2 Geology of the SdA Basin

The upper crustal lithology and structure of the SdA Basin are well known from several publications based on industry seismic data and a deep exploration well drilled

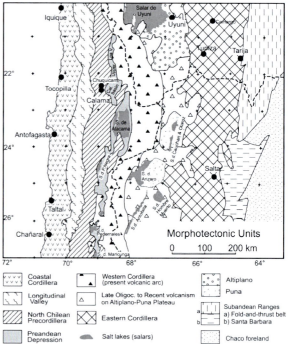

**Fig. 14.2.** Morphotectonic units of the southern Central Andes

close to the basin center (e.g. Townsend 1988, Macellari et al. 1991, Muñoz and Townsend 1997, Muñoz et al. 2002, Pananont et al. 2004, Arriagada et al. 2006). The interpretation of these data is supported by numerous geological field studies in the surroundings of the SdA (Fig. 14.3). The main geological units involved in the development of the SdA area are: (*1*) Pre-Cretaceous basement rocks that follow, more or less regularly, the structural grain of the Andes over long distances; (*2*) Cretaceous to Eocene

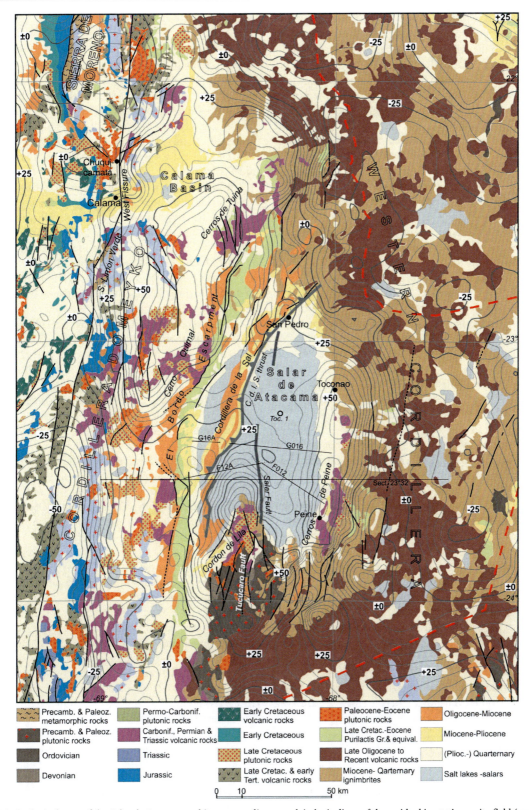

Fig. 14.3. Geological map of the Salar de Atacama and its surroundings overlain by isolines of the residual isostatic gravity field (map base Reutter et al. 1994). Note the good correlation between major geological structures and gravity anomalies

deposits, related to the Salta Rift, which are irregularly distributed and extend from the distal back-arc over the SdA Basin to the volcanic arc, located at that time in the North Chilean Precordillera; and (3) late Eocene to Recent sediments, restricted to the SdA depocenter, but with disconnected time-equivalents in other depocenters of the developing Altiplano-Puna Plateau and its western slope.

## 14.2.1 Pre-Cretaceous Basement Rocks

The basement of the thick Cretaceous and Cenozoic sediments of the SdA Basin crops out in vast areas at its margins. It consists of early Paleozoic sedimentary and igneous rocks in the southern Cordón de Lila peninsula and widely distributed, stratified volcanic and sedimentary rocks of late Paleozoic to Triassic age. Here, the basement rocks are only summarized: for more detail see the discussion on the origin of the Fore-arc Gravity High and Table 14.1.

The early Paleozoic in the Cordón de Lila peninsula and further south is represented by a sequence of clastic sediments with intercalations of pillow-lavas of probable Ordovician age (Niemeyer 1989: Complejo Igneo y Sedimentario de Lila; Moraga et al. 1974: Lila Fm.) and Ordivician to Carboniferous plutonic rocks (Mpodozis 1983). Devonian and early Carboniferous clastic sediments unconformably overlie the older units (Niemeyer et al. 1997).

During the Permo-Triassic, the Cordon de Lila area probably formed a horst structure because coeval volcanic and sedimentary rocks form only thin patches upon the older basement. To the east and west of the Cordon de Lila, however, Permo-Triassic volcanic sequences, intercalated with sediment, comprise formations more than 2000 m thick (Coira et al. 1982; Breitkreuz and van Schmus 1995). Mpodozis and Mahlburg Kay (1990) attributed these late Permian to Triassic sequences to the not subduction-controlled Choiyoi Group (Table 14.1). Normal faults of the extensional tectonic regime of that time were, at least in part, reactivated during the Mesozoic and Cenozoic (Günther 2001).

In most parts of the SdA area, the pre-Cretaceous basement consists of Carboniferous and/or Permian to early Triassic volcanic and sedimentary successions, such as the La Tabla (Marinovic et al. 1992), Tuina, Cas, Peine, Estratos El Bordo, Estratos de Cerros Negros, and, in part, Aguadulce Formations (Marinovic and Lahsen 1984; Gardeweg and Ramirez 1985). At the western, southern, and southeastern SdA borders, these basement rocks are partly penetrated by late Cretaceous plutons (e.g. Cerro Quimal). In outcrop, a very low-angle unconformity separates the Permian and Triassic sequences from the overlying Cretaceous sediments. As a result, the basement-cover boundary is not clearly detectable in the industry seismic records of the SdA interior (e.g. seismic line Z1F012 in Flint et al. 1993; Townsend 1988).

## 14.2.2 Cretaceous to Eocene Development

The Cretaceous to Eocene Purilactis Group (Brüggen 1942; Dingman 1963; Ramirez and Gardeweg 1982; Marinovic and Lahsen 1984; Hartley 1988, Hartley et al. 1992, Flint et al. 1993, Charrier and Reutter 1994) underlies the late Eocene to Recent sedimentary fill of the SdA depocenter. These red, clastic sediments, varying in thickness between 2000 and 4000 m, are well exposed along the narrow fold-and-thrust belt of the El Bordo Escarpment (Fig. 14.3).

The lowermost formation of the Purilactis Group (Tonel Fm.) shows, in its basal part, local limestones, gypsum, and halite (Charrier and Reutter 1994) and was therefore considered an equivalent of the Maastrichtian marine sediments in the Puna (Yacoraite Fm.) and Altiplano regions (El Molino Fm.). However, Mpodozis et al. (1999, 2005), based on paleomagnetic results and regional relationships, put the onset of the Purilactis Group in the Cenomanian.

Conglomerates prevail in the upper part of the Purilactis Group, but their age of onset seems to vary along the El Bordo escarpment. Mpodozis et al. (1999, 2005) and Arriagada et al. (2006) mention two lithologically similar formations of presumed Paleogene age that unconformably (not clearly constrained by isotope age data!) overlie the early Paleocene top of the Purilactis Group. Basaltic to andesitic lavas, dykes and sills, as well as subvolcanic intrusives, have been found throughout the Cretaceous to Eocene succession. The Purilactis Group shows an angular unconformity with the overlying, Oligocene, Tambores Formation, which to the east gradually passes into the San Pedro Formation of the SdA Basin.

In the exploration well Toconao X1, which penetrated an anticlinal structure near the center of the SdA Basin (Fig. 14.3), the Purilactis Group was encountered between 2800 and 3900 m depth (Muñoz et al. 2002; Pananont et al. 2004). There, above volcanic and volcaniclastic rocks of the Permo-Triassic basement, occurs a sequence of sandstones, limestones, and evaporitic beds attributed to the late Cretaceous part of the Purilactis Group. Sedimentation continued with a thick succession of siltstones and sandstones, probably extending into the early Tertiary.

In a reinterpretation of seismic data for the area, Muñoz et al. (2002) detected two anticlinal structures in Purilactis Group sediments that appear to be filled half-grabens, active since the Cretaceous, whose bordering, west-dipping normal faults inverted during subsequent shortening. These structures are unconformably covered by a younger formation of volcaniclastic sandstones and conglomerates which Muñoz et al. (2002: "Sequence B"), Pananont et al. (2004: "Sequence H") and Arriagada et al. (2006) consider to be a Paleocene part of the Purilactis Group.

Our interpretations of the industry seismic sections G016 – G16A (Fig. 14.4) and F012 – F12A (Fig. 14.5) also

**Table 14.1.** Crustal evolution of the Salar de Atacama and adjacent areas. *Arrows* symbolize shortening, extension and subsidence, as well as their intensities. For references see Sect. 14.2 and 14.4.2

| Time (Ma) | Magmatic arc system and tectonic events | Position of SdA block | Tectonics, magmatism and sedimentation in SdA basin and surroundings | SdA crustal development |
|---|---|---|---|---|
| 10 – 0 | Neogene-Recent arc – Western Cordillera | Proximal fore-arc | Very slight shortening, subsidence | ⇓ ⇒⇐ |
| 17 – 10 | Neogene-Recent arc; *Tectonic event:* Quechua Phase, mainly Altiplano-Puna Plateau; closure of magmatic gap E of SdA at 17 Ma | Proximal fore-arc since 17 Ma | Shortening at western border of SdA, Cordillera de la Sal thrust, crustal thickening west of SdA, basin gets modern outlines. Since 17 Ma: volcanic front develops eastwardly concave curvature around E-side of SdA | ⇓ ⇒⇐ |
| 28 – 17 | Neogene-Recent arc gets active at 28 Ma in Altiplano-Puna Plateau and Western Cordillera, volcanic gap between 22° and 25° S | SdA within volcanic gap | Subsidence, San Pedro and Tambores Fms. since 28 Ma. First deposits in SdA centre possibly since 33 Ma? | ⇓ |
| 38 – 28 | *Amagmatic interval*, but last plutonism and Cu mineralization in Precordillera; uplift of Altiplano-Puna Plateau begins; shortening in Precordillera, Eastern Cordillera and Altiplano-Puna Plateau (locally) | No arc | N-S sinistral strike slip in El Bordo Escarpment, local shortening after Incaic Event. Subsidence in SdA basin | ⇓ |
| 45? – 38 | *Tectonic event:* (Incaic Event), terminates arc magmatism in Precordillera | Proximal back-arc | Reverse faulting, folding, and N-S dextral strike slip in Precordillera and El Bordo Escarpment. Inversion tectonics in SdA basin? Crustal thickening beneath Precordillera, minor thickening beneath SdA? | ⇒⇐ |
| 75 – 38 | Late Cretaceous-Eocene magmatic arc in Precordillera; Salta Rift: post-rift stage, but magmatism until 65 Ma; (Early Palaeocene shortening event?) | Proximal back-arc | Normal faulting, strong subsidence; onset (if not earlier) of Purilactis group sedimentation, basaltic-andesitic volcanism. Around 65 Ma, strong, alkaline magmatism, probably causing low wave-length gravity anomalies, and uplift of Cordillera de Domeyko. Crustal thinning from SdA to Precordillera N of Calama | ⇓ ⇐⇒ |
| 85 – 75 | *Amagmatic interval* | No arc | Early rift tectonics or slow uplift and erosion? | ? |
| 85 | *Tectonic event:* (Peruvian Event), shortening in magmatic arc, end of arc volcanism | | No shortening | ? |
| 125 – 85 | Mid-Cretaceous volcanic arc, W of Precordillera and Longitudinal Valley; Salta Rift: syn-rift stage | Back-arc | Early Salta Rift tectonics in SdA region and Precordillera N of Calama and possible onset of Purilactis Group sedimentation | ⇐⇒? |
| 155 – 125 | Plutonism in Coastal Cordillera; *Tectonic event:* Araucanian Phase ~155 Ma in Coastal Cordillera | Back-arc | Uplift and erosion? Gentle deformation (possibly younger) | ⇑? |
| 200 – 155 | Magmatic arc (basaltic andesites) in the Coastal Cordillera under strongly extensional conditions | Back-arc | Thick marine sediments in Precordillera, no sediments preserved in SdA surroundings, slight crustal thinning? | ⇐⇒? |
| 220 – 205 | Acid and intermediate volcanism from Coastal Cordillera to Precordillera | Proximal back-arc | Coaly sandstones and acid volcanics from Precordillera to coast | ⇐⇒? |
| ~270 – 240 | Choiyoi magmatism (crustal melting or extensional arc), strong extension from Precordillera to Western Cordillera | No arc or magmatic arc-graben? | Normal faulting from Precordillera to Western Cordillera, up to 3 km of volcanics and sediments in graben-type depocentre, strong crustal thinning, intrusions, basaltic underplating | ⇓ ⇐⇒ |
| ~300 – 280 | Arc magmatism in Precordillera, shortening in magmatic arc? | Proximal back-arc | Probably no back-arc shortening. | ?? |
| ~450 | Ocloic orogeny | No arc | Suture of Arequipa-Gondwana continental collision. Strong crustal thickening | ⇒⇐ |
| ~495 – 470 | Arc magmatism in Faja eruptiva occidental | Magmatic arc | Faja Eruptiva Occidental, arc magmatism, crustal thinning? | ⇐⇒ |

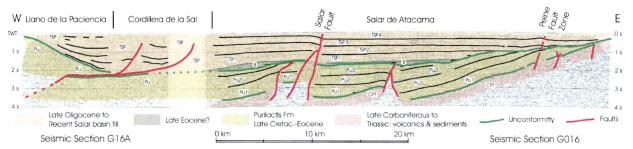

**Fig. 14.4.** Interpretation of the industry seismic sections *G016* and *G16A* through the Salar de Atacama, kindly made available by Empresa Nacional del Petroleo (ENAP); location in Fig. 14.2. The Tambores-San Pedro Fm. (*TSP*) unconformably overlies Purilactis Group (*PU*) sediments with increasing thickness from east to west. The western border zone of the *TSP* shows a long-lasting growth structure that was thrust eastwards during the mid Miocene, leading to the formation of the Cordillera de la Sal. Probable coeval uplift at the eastern border of SdA led to the truncation of Purilactis sediments and onlapping sedimentation of the *TSP* and younger rocks. The Purilactis Group unconformably overlies (mostly with low angle) volcanic and sedimentary rocks of late Carboniferous to early Triassic age (*CPT*). Folding and reverse faulting of the Purilactis Group is probably of mid or late Eocene age. *B* is a basal unit of the *TSP* that filled the relief formed on top of the *PU* during folding. The Salar and the Peine fault zones are young or active reverse thrusts

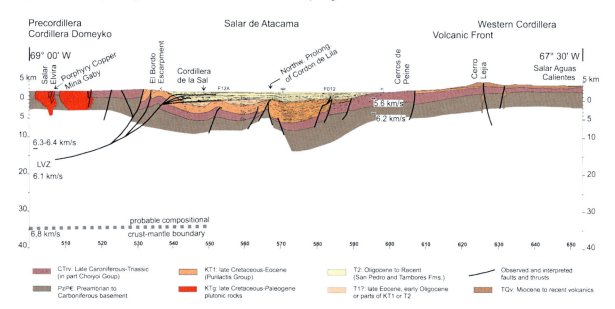

**Fig. 14.5.** Crustal section from the Precordillera (*W*) through the southern part of Salar de Atacama (23° 32') into the Western Cordillera (*E*) based on geological data, gravity studies, and seismic experiments in the interior of the SdA Basin; industry data kindly made available by ENAP. The western border shows structures resulting from strong shortening during the Eocene and the mid Miocene. Most faults in the Salar area are inverted normal faults. Crustal shortening of the Salar area cannot account for the present crustal thickness in this region of > 60 km

show, beneath the very obvious unconformity, gently folded Purilactis rocks with the anticlines controlled by reverse faults. Here, again, their steep dip suggests they originated as normal faults that were inverted when the late Cretaceous tensional stress regime was superseded by a compressional one during early Tertiary times. Different apparent thicknesses on both sides of the normal faults point to synsedimentary tensional tectonics.

Muñoz et al. (2002), Pananont et al. (2004), and Arriagada et al. (2006) postulated that tectonic inversion of the Cretaceous normal faults had already occurred during the Paleocene, which is similar to the findings of Mpodozis et al. (2005) and Arriagada et al. (2006) in the El Bordo Escarpment. This age is corroborated by K/Ar and Fission Track dating of biotite, sphene, zircon and apatite from granitoids rocks of the Cerro Quimal pluton by Andriessen and Reutter (1994). Grouping of these data around 63 Ma indicates that the cooling pluton and adjacent parts of the Cordillera de Domeyko to the west of the SdA were rapidly uplifted and exhumed during the early Paleocene by at least 2 km.

However, there is also evidence that shortening of the Purilactis Group in the SdA Basin and in the El Bordo Escarpment occurred later, i.e. during the Incaic Event (around 45–38 Ma). This evidence consists of: (*1*) an Ordovician pluton near the southern border of the SdA Basin that has yielded an Apatite Fission Track (AFT) exhumation age of 38.6 ± 5.6 Ma (Andriessen and Reutter

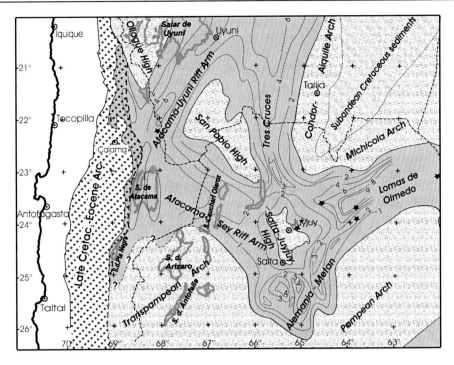

Fig. 14.6.
The Maestrichtian sedimentary basins in the arc and back-arc (Tolar Fm. and Purilactis Group, respectively) and in the Salta Rift (post-rift stage, Balbuena Subgroup; isopachs in 100 m, asterisks show locations of magmatism) after Marquillas and Salfity (1988) with modifications in the SdA and Calama regions

1994); and (2) in the El Bordo Escarpment, tuffaceous layers of the folded Purilactis Group overlain by late Paleogene conglomerates have yielded radiometric ages of 44.0 ± 0.5 Ma (Ar/Ar by Hammerschmidt et al. 1992) and 39.9 ± 3.0 Ma (K/Ar by Ramirez and Gardeweg 1982). In the Toconao X1 well, AFT ages presented by Muñoz et al. (2002) show values varying between 73 and 46 Ma, independent of their location below or above the unconformity. These age variations presumably reflect different exhumation ages in the source area of the clastic components with these sediments suffering no (or only very incomplete) resetting by burial in the SdA Basin. Thus, in the light of these data, Purilactis rocks beneath the unconformity would stratigraphically also contain the Eocene.

In the sections represented in Figs. 14.4 and 14.5, thicknesses increase from the eastern margin of the basin towards its center near the Salar Fault. According to Fig. 14.4, a maximum thickness of > 3.5 km (1.8–3.6 seconds TWT) is likely. Sedimentary rocks, twice as thick, appear in section F012 (Fig. 14.5) in the basin to the east of the Salar Fault Zone. In this case, it is possible that lithologically similar formations of early and/or mid Cretaceous age exist beneath the late Cretaceous Purilactis sediments. These later sediments would have been deposited in a fault-controlled basin corresponding to the synrift depocenters of the Salta Rift (Salfity and Marquillas 1994; Marquillas et al. 2005).

The distribution of (late) Cretaceous to mid Eocene sediments far exceeds the extent of the Recent SdA Basin. Purilactis Group sediments can be traced southwards along the Precordillera to the Salar Punta Negra and towards the NNW, over Cerros de Tuina and the Calama Basin, into the Precordillera north of Calama (Tolar Fm. or Eastern Sequence; Bogdanic 1990, Tomlinson et al. 2001). Also, towards the east, the Purilactis depocenter reached far beyond the present border of the SdA Basin. At its eastern side, a cross section through the SdA Basin (Fig. 14.4) reveals truncated Purilactis sediments overlain by onlapping sequences. Furthermore, equivalents of the Purilactis Group are exposed on the eastern monoclinal slope of the SdA Basin (Estratos de Quepe, Ramirez and Gardeweg 1982), near the international boundary (Gardeweg and Ramirez 1985), and in the adjacent Argentine Puna, where they are widely distributed (Marquillas and Salfity 1988).

This distribution suggests that the depocenter of the Cretaceous to Eocene sediments of the SdA and its surroundings was located in a continuation of a westwardly extending arm of the Salta Rift, as originally proposed by Marquillas and Salfity (1988) and represented in Fig. 14.6 as "Atacama-Sey Arm" (Salfity and Marquillas 1994; Marquillas et al. 2005). Mpodozis et al. (2005) and Arriagada et al. (2006) also recognized the connection between the SdA area and the Salta Rift since the Campanian or Maastrichtian, but proposed that the Purilactis depocenter, independent of back-arc rifting, developed during the mid-Cretaceous as a foreland basin to the east of a Cretaceous Cordillera.

Strong magmatism around the Cretaceous-Tertiary boundary is evident in the surroundings of the Purilactis Basin. Figure 14.3 shows a great number of plutonic rocks west, south and southeast of the SdA Basin. They are probably of late Cretaceous age, similar to those of Cerro Quimal, immediately west of El Bordo, whose rapid uplift and exhumation around 63 Ma cannot have been much younger than its intrusion (Andriessen and Reutter 1994). Mpodozis et al. (2005) also report lavas and dykes of this age in the El Bordo area and emphasise the alkaline character of this magmatism. This last feature points to low-degree partial melting of the mantle, consistent with an extensional regime at the end of the Cretaceous. Eocene magmatism, however, appears to be confined to the western side of the SdA, i.e. to the transition zone between arc and back-arc.

## 14.2.3 Incaic Event and Late Eocene to Recent Development of the SdA Basin

As mentioned above, Paleocene or mid to late Eocene shortening caused reverse faulting and some fold structures in the SdA Basin, but subsidence continued and further sedimentation took place over an angular unconformity. Based on the importance of the Incaic Event for the region (see below) and on the above mentioned arguments, we tend to attribute a post-Incaic age (i.e. not older than the late Eocene) to the sequences overlying the unconformity at the top of the Purilactis Group in the SdA Basin. We are aware, however, that this timing still represents an unsolved problem.

The structures in the SdA Basin do not reveal the supraregional importance of the tectonic processes occurring around the mid to late Eocene boundary, which in the Andes are commonly associated with the Incaic Phase or Event of Andean mountain building (Steinmann 1929; Mégard 1984). In the vicinity of the SdA, tectonic movements of this age seem to have been concentrated in the thermally weakened crust of the waning Eocene magmatic arc, which had been located in the present Precordillera since the late Cretaceous, and also affected the El Bordo Escarpment at the boundary to the back-arc. The Precordillera north of Calama was strongly shortened at 38.5 Ma, as revealed by a prominent and well-constrained unconformity (Ar/Ar dating by Hammerschmidt et al. 1992). However, as there were also precursory and later movements, the duration of the Incaic Event probably extended over some millions of years, possibly from 45 to 38 Ma (Günther 2001; Scheuber et al. 2006, Chap. 13 of this volume).

After E-W-directed shortening, the Precordillera, as well as the El Bordo Escarpment (Charrier and Reutter 1994), was subject to longitudinal strike-slip displacement (Mpodozis et al. 1993, 2005; Reutter et al. 1991, 1996; Tomlinson and Blanco 1997). Furthermore, the Precordillera and the El Bordo area south of Cerro Quimal, between 23°10' and 23°20' S, underwent up to 60° of clockwise block rotation about a vertical axis. That is about 30° more than most other paleomagnetic measurements for pre-Miocene rocks from elsewhere in northern Chile (Mpodozis et al. 1999; Arriagada et al. 2000, 2003). These authors, as well as Somoza and Tomlinson (2002) and Somoza et al. (1999), ascribed block rotation in the immediate vicinity of the SdA to the Incaic Event. Incaic tectonics led to crustal thickening in the Precordillera (Günther 2001; Haschke and Günther 2003) and the formation of a high mountain range (Maksaev and Zentilli 1999), which became a source area for the clastic sediments of post-Incaic basins.

Two further effects of the Incaic Event were important in the area under consideration. Firstly, except for minor hypabyssal intrusions related to the famous porphyry copper deposits of the region, magmatic arc activity in the Precordillera ceased at about 38 Ma. Consequently, the SdA area was no longer in a back-arc position. It became part of the Andean fore-arc when arc magmatism resumed its activity in the areas of the present-day Altiplano-Puna Plateau and Western Cordillera after a 10 million year amagmatic interval (38–28 Ma) and the closure of a volcanic gap east of the SdA area at about 17 Ma (Fig. 14.7).

The second effect was a reorganization of the whole pre-Incaic back-arc area, which was the first step towards the formation of the Altiplano-Puna Plateau. Incaic tectonic shortening, uplift and exhumation affected not only the north Chilean Precordillera, but also parts of the modern Eastern Cordillera, i.e. an area located at a present-day (i.e. post-shortening) distance of some 400 km farther to the east, where a "Protocordillera" developed (Lamb and Hoke 1997; Coutand 2001; Scheuber et al. 2006, Chap. 13 of this volume), thus delineating the outlines of the future plateau. The principal plateau uplift, however, is younger; it started in the late Oligocene around 32 Ma, somewhat before the end of the amagmatic interval (28 Ma; Scheuber et al. 2006, Chap. 13 of this volume).

The reorganization of the former back-arc area, in turn, affected the post-Incaic sedimentary basins in the area. New depocenters developed in the late Eocene, in part with sedimentary continuity, over the Salta post-rift basins (e.g. in the southern Altiplano), while others formed outside the now inactive rift arms over a much older basement (Kraemer et al. 1999; Carrapa et al. 2005). DeCelles and Horton (2003) interpreted early Tertiary sediments in the Southern Altiplano and Eastern Cordillera as deposits of a vast, eastwardly migrating foreland basin in front of the folded and upwarped Precordillera (Carrapa et al. 2005 applied this model also to the Puna). However, based on structural, sedimentological and thermochronological data, Elger (2003), Elger et al. (2005) and Scheuber et al. (2006, Chap. 13 of this volume) have shown that deformation was distributed irregularly over the Altiplano and that no distinct migration direction of the deformation front could be inferred. Moreover, the Altiplano was not only subject to shortening but also to Oligocene extension.

Incaic tectonics did not stop the subsidence in the SdA area which, close to the uprising Precordilleran moun-

**Fig. 14.7.**
Distribution of magmatic ages in the Central Andes since the Oligocene. Note the temporal volcanic gap to the east of the SdA region from 28 to 17 Ma

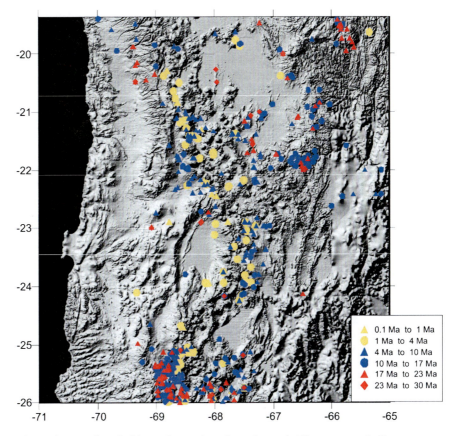

tain range, developed with an independent, and probably isolated, depocenter. The age of the first post-Incaic deposits in the center of the new basin is uncertain. At the El Bordo Escarpment, the onset of coarse-grained clastic (Tambores Fm.) and equivalent fine-grained clastic and evaporitic sedimentation (San Pedro Fm.) over truncated Purilactis rocks is generally attributed to the Oligocene (Naranjo et al. 1994; Flint et al. 1989, 1993; Marinovic and Lahsen 1984; Ramirez and Gardeweg 1982). The dimensions of the Oligocene depocenter were larger than those of the present SdA Basin. Outcrops of the Tambores Fm. along the El Bordo Escarpment (Ramirez and Gardeweg 1982; Charrier and Reutter 1994) reveal that the Oligocene SdA Basin extended farther to the west and northwest than the present basin, and outcrops of the San Pedro formation along the continuation of the Cordillera de la Sal north of the present SdA Basin also document the larger size of the fossil basin.

As revealed by the industry seismic sections, large areas of the SdA Basin show an almost undisturbed stratification of Oligocene to Recent clastic and evaporitic filling (Figs. 14.4 and 14.5; Flint et al. 1993; Muñoz et al. 2002). The most important feature is an increase in thickness from almost zero in the east to more than 4 km at the western margin, where these sequences thin out in a spectacular progressive unconformity, similar to that in the outcrops described by Wilkes and Görler (1994; see also Arriagada et al. 2006). The structure indicates a synsedimentary steepening of the western basin border leading to a basinward divergence of the growth strata by up to 25° (Fig. 14.4). This tectonic activity culminated in a thrust that displaced the western basin margin eastwards by approximately 5 km and caused the folds and faults of the Cordillera de la Sal (Fig. 14.3), thus establishing a basin configuration similar to the present one. These movements also caused the reactivation of the Incaic structures in the El Bordo Escarpment which increased shortening (though, the amount is not calculated here). Volcanics associated with the unconformably overlying San Bartolo Group constrain this shortening event to between 17 and 10 Ma (Flint et al. 1993).

The surface structures near the western border of the SdA Basin apparently indicate an essentially E-W or ESE-WNW compressive stress regime during the early and middle Miocene. However, the findings of Pananont et al. (2004) point to more complex conditions. These authors interpreted a very important ESE-dipping, synsedimentary normal fault in the northwestern sector of the SdA and proposed its development by N-S-oriented strike-slip.

At the eastern border of the SdA Basin, the E-W seismic sections (Figs. 14.4 and 14.5) show a progressive eastward onlapping upon the erosionally truncated, unconformable Purilactis Group and the Permo-Triassic basement. This structure suggests that these Purilactis rocks were already uplifted during the Incaic event, followed by further uplift in the area of the present Western Cordillera, as well as relative subsidence of the Oligo-Miocene basin near the eastwardly shifting eastern border. The asymmetric thickness and the synsedimentary tectonic features at the western border, and also the apparently continuous eastward onlapping at the eastern border, are features similar in character to those of a foreland basin. However, it appears that this basin had no connections with the late Eocene to middle Miocene basins of the present Puna and Altiplano.

As mentioned, arc magmatism resumed at 28 Ma in the region of the present-day Altiplano-Puna Plateau; however, between 22 and 25° S, volcanism only started at 17 Ma, delineating the westwardly concave shape of the arc around the SdA area (Fig. 14.7). In this part of the modern Western Cordillera, crustal uplift caused by magmatism and crustal shortening was at a maximum probably between 17 Ma and the onset of intensive ignimbrite magmatism at 10 Ma (Ramirez and Gardeweg 1982; Marinovic and Lahsen 1984).

The emplacement and activity of the volcanic arc in the Western Cordillera was hitherto unclear in its relationship to the development of the Preandean Depression in the areas adjacent to the SdA (Salar de Punta Negra, Calama Basin and upper Loa Valley; Fig. 14.2). The Calama Basin, separated from the SdA Basin by the Cordon de Barros Arana, formed as part of the Preandean Depression within the Precordillera (Fig. 14.3) during the late Miocene (May et al. 1999; Somoza et al. 1999; Hartley et al. 2000, Panannont et al. 2004), i.e. after the end of Eocene arc magmatism and Incaic deformation, and probably also after mid Miocene shortening (Cordillera de la Sal thrust). These relatively young processes determined the definitive shaping of the whole region.

## 14.3 Neotectonics of the SdA Area and Crustal Seismicity

Abundant structures within Plio-Pleistocene ignimbrites and clastic sediments at the margins of the SdA Basin, as well as in the interior of the basin and in its two southern sub-basins (divided by the Cordón de Lila; Fig. 14.3), indicate continued, but not intense, shortening during the last 10 million years. This shortening is expressed by E- and W-verging thrusts, reverse faults and folds. Most of the thrusts and fault-bend folds are of minor (dm to m) throw. The northern border of the SdA is affected by NW-verging thrusts and fault-bend folds that deform Plio-Pleistocene lacustrine limestones (Kuhn 1996).

Seismic sections through the SdA reveal neotectonic E-W shortening. There is evidence along the "Peine Fault Zone" in the southwestern SdA that inversion of Cretaceous normal faults occurred not only during the Incaic Event, but also during the Neogene (Figs. 14.4 and 14.5; Muñoz et al. 2002). The "Salar Fault Zone", extending from the SdA Basin center southwards to the Tucucaro fault (Niemeyer 1984) of the Cordon de Lila, is developed in the upper evaporitic sequences as a narrow flexure over a steeply dipping reverse fault (Figs. 14.4 and 14.5; Jordan et al. 2002) that uplifts the western block. A similar fault in the northern part of the basin interferes with the Cordillera de la Sal near San Pedro and strikes northward into the Río Salado fault, where neotectonic deformation is expressed by sinistral movements, block-rotations, and displaced fold axes (Kuhn and Reuther 1999).

Generally, the axial trends of folds and thrust planes suggest maximum horizontal stress varying between E-W and NW-SE. This stress regime is also reflected in N-S-trending, sinistral, strike-slip faulting that has reactivated pre-existing faults at the western and eastern border of the SdA Basin (Kuhn and Reuther 1996). W- and E-verging thrust planes and pop-up structures with N-S-oriented axes in Plio-Pleistocene ignimbrites are common along the eastern slope of the SdA Basin and up to the Western Cordillera. The eastward vergence of these structures demonstrates that compression was not induced by gravity (Adam and Reuther 2000).

Shallow crustal seismicity beneath the SdA area is weak. Fault-plane solutions appear to indicate both horizontal tension and compression (Belmonte 2002). However, the solutions are unreliable because of the scarcity of data, the weakness of first arrivals and unfavorable ray geometry. Similarly, few well-located events have occurred in the crust at depths between 30 and 40 km, indicating either an unusually brittle rheology, owing to extremely low temperatures, or active metamorphic processes such as dehydration embrittlement. In this context, Muñoz et al. (2002) found that samples taken from depths between 1 800 m and 2 900 m in the Toconao X1 well showed no detectable AFT age resetting, which indicates temperatures below the partial annealing zone (60 °C). This corresponds to the modest heat flow density of $< 60$ mW m$^{-2}$ determined by Springer and Förster (1998) for the north Chilean Precordillera.

## 14.4 The Fore-Arc Gravity High

### 14.4.1 Gravity Field and Resulting Structural Constraints

Owing to the thick Central Andean crustal root, the Bouguer anomalies of the Central Andes are extremely negative. Along 23.5° S, they drop from $-250 \times 10^{-5}$ m s$^{-2}$ at 69° W to almost $-450 \times 10^{-5}$ m s$^{-2}$ at 67° W (Götze et al. 1994, Götze and Kirchner 1997, Götze and Krause 2002). Gravity sources predominantly correspond to regional, deep-seated density inhomogeneities whose effects outweigh those of near-surface geological bodies. In order to visualize short-wavelength effects, a regional isostatic gravity field was calculated and subtracted from the Bouguer anomaly map. This Vening-Meinesz model was calculated assuming a normal crustal thickness of 40 km, a rigidity of the underlying plate of $10^{23}$ N m and a crust-mantle density contrast of 0.4 Mg m$^{-3}$. The resulting short-wavelength anomalies are thought to be generated by upper-crustal density inhomogeneities.

The isostatic residual gravity map (Fig. 14.8) shows an area of positive anomalies in the fore-arc region some 100 km wide and 260 km long. This, here-named, Fore-arc Gravity High, which is the fore-arc part of the "Central Andean Gravity High" of Götze and Krause (2002), comprises the Calama Basin and the the SdA area. To the southeast, it is weakly separated by the reduced values of the Western Cordillera from the back-arc areas of high gravity in the Salar de Arizaro region of the Puna and further south. In contrast, strong negative anomalies in the Western Cordillera (isostatic residual minimum $> -50 \times 10^{-5}$ m s$^{-2}$) separate the Fore-arc Gravity High (isostatic residual maximum $> +70 \times 10^{-5}$ m s$^{-2}$) from the Uyuni back-arc area of positive anomalies (isostatic residual maximum $> +50 \times 10^{-5}$ m s$^{-2}$) in the southern Altiplano (northeastern sector of Fig. 14.8). The Fore-arc Gravity High is more intense than the positive anomalies of the back-arc region, excluding the gravity high of the Eastern Cordillera. Minima from 0 to $-50 \times 10^{-5}$ m s$^{-2}$ located in the Precordillera limit the Fore-arc Gravity High to the west.

Within the Fore-arc Gravity High, three local maxima exceeding $+50 \times 10^{-5}$ m s$^{-2}$ can be recognized (Fig. 14.3):

1. the 'Peine High', an N-S-elongated maximum with an axis running along the eastern border of the Salar de Atacama;
2. the 'Lila High', also with a N-S-oriented axis, lying in the Cordon de Lila peninsula and extending northwards in the SdA Basin, west of the Salar-Tucucaro fault zone; and
3. the 'Quimal-Tuina High', the largest and most intense maximum, occurring in the Cordillera de Domeyko, immediately west of the El Bordo Escarpment.

The Quimal-Tuina High extends northwards from a shallow, longitudinal, morphological depression at 23.5° S and enters the Calama Basin west of Cerros de Tuina, where it widens to more than 50 km while decreasing in intensity. Spectral analyses show that the gravity sources of the short-wavelength, positive anomalies should be situated at depths of 4 to 6 km, while those of the broad, regional anomaly are inferred to be at 10 to 15 km depth. As the Calama Basin is filled by not more than 700 m of Oligocene to Quaternary sediments (May et al. 1999; Somoza et al. 1999; Hartley et al. 2000, Panannont et al. 2004), the Quimal Tuina High is more intense than the positive anomalies beneath the SdA Basin (Fig. 14.8).

The three short-wavelength anomalies coincide well with geological structure (Figs. 14.3, 14.8 and 14.9). The Peine High correlates with a major area of pre-Mesozoic basement outcrop of the late Permian Peine Group and its northward and southward prolongations, whereas the Lila High corresponds to a basement uplift of early and late Paleozoic sedimentary and plutonic rocks, partly overlain by Permo-Triassic volcanic deposits and intruded by younger plutonic stocks. The area of the Quimal-Tuina High, to the south of the Calama Basin, shows small and

**Fig. 14.8.**
Isostatic residual gravity field showing the strong positive anomalies of the Fore-arc Gravity High and the less intense back-arc gravity highs, which mostly correlate with subsiding basins of the Altiplano-Puna. The strong negative anomalies of the magmatic arc (Western Cordillera) abruptly separate the two areas of gravity highs

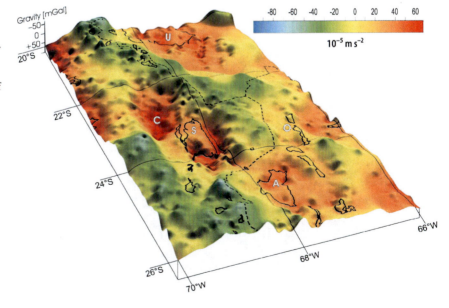

large plutonic intrusions of late Cretaceous to early Tertiary age. The Calama Basin, which occupies the northern part of this gravity high, developed at the northern termination of the Cordillera de Domeyko. This is where it is thought that the Purlactis depocenter and that of the Tolar Fm. in the Precordillera north of Calama were connected in the late Cretaceous and early Tertiary (Tomlinson et al. 2001; Bogdanic 1990: Eastern Sequence). Belts of reduced positive anomalies mark the Cordillera de la Sal and the inferred axis of very thick sediments (Purilactis Group and other formations) between the Peine High and the Lila High. Figure 14.9 shows a gravity and density model across the southern part of the SdA along 23°32' S (see Fig. 14.5 for comparison).

## 14.4.2 Origin of the Fore-Arc Gravity High

The dense bodies that cause the relatively shallow, local, positive gravity anomalies must have been emplaced either by tectonic or magmatic processes. When checking the geological history of the SdA for periods of high tectonic activity or strong basic magmatism, three paleotectonic situations should be considered: (*1*) Ordovician orogeny, (*2*) Permo-Triassic processes associated with the Choiyoi Group, and (*3*) Salta Rift processes.

### 14.4.2.1 Ordovician Orogeny

During the Ordovician, abundant magmatism and tectonics, culminating in the Ocloyic orogeny, affected the crust of the present Puna from Argentina to the SdA. Coira et al. (1999) distinguish a western magmatic belt that became active as a magmatic arc at the Cambrian-Ordovician boundary ("Faja Eruptiva Occidental"; Palma et al. 1986) from a younger eastern belt ("Faja Eruptiva Oriental"). According to Bahlburg and Hervé (1997), the western magmatic arc developed within the continental Arequipa-Antofalla Terrane over an eastwardly subducting, oceanic lithosphere. This terrane is thought to have been welded to the South American continent during the late Ordovician Ocloyic orogeny after overthrusting (northern Puna) or eastward subduction (southern Puna) of intermediate, partially oceanic zones.

In the SdA area, basalts (Niemeyer 1989) and granitoid plutons (Mpodozis et al. 1983) representing the Faja Eruptiva Occidental can be traced to the Salar de Arizaro region (Fig. 14.2) in isolated patches that outcrop from beneath the young volcanics of the Western Cordillera. This apparently SSE-NNW-trending connection has been repeatedly proposed as a probable source of the gravity anomalies (i.e. the Central Andean Gravity High of Götze and Krause (2002); Götze and Kirchner 1997; Schurr and Rietbrock 2004). However, as the basic rocks of this zone are not abundant, and ultrabasic rocks are actually lacking, surface geology does not support this hypothesis. Neither do post-Ordovician sedimentary formations, as there is no evidence for extraordinary subsidence that might have been caused by a high density portion of the crust.

### 14.4.2.2 Permo-Triassic Extension and Magmatism

South of 28° S, Mpodozis and Kay (1990) showed that magmatic rocks of Carboniferous to Triassic age have to

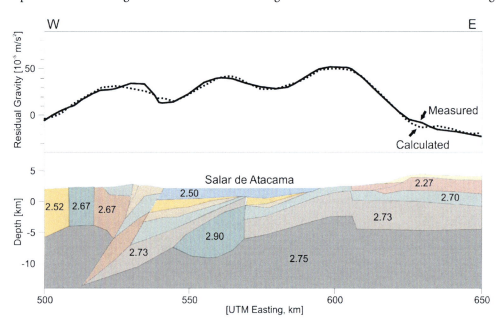

**Fig. 14.9.** Gravity profile and density model along a section through the SdA at 23°32' S (Choi 1996)

be subdivided into two superunits of very different origin. Whereas the late Carboniferous to Permian "Elqui Superunit" was subduction-controlled and is associated with tectonic shortening, the Permian to Triassic "Ingaguas Superunit" originated from crustal melting in an extensional environment following the cessation of subduction at the Gondwana margin. According to these authors, this last magmatic superunit represents the magmatic Choiyoi province that extends along the Andes from 43° to 20° S. In the area under consideration, the Elqui Superunit is possibly restricted to the plutons in the Precordillera, whereas most of the volcanic formations of Permo-Triassic age (e.g. Cas Fm., Peine Fm., Aguadulce Fm. and others; Ramirez and Gardeweg 1982; Marinovic and Lahsen 1984; Table 14.1) probably pertain to the Inguaguas Superunit, i.e. the Choiyoi province.

Areal distribution and variations in thickness of the Permo-Triassic volcanic formations also point to crustal extension (Coira et al. 1982). In many places (e.g. Cerros de Peine and Cerros de Tuina) sequences more than 2 km thick have been described (Breitkreuz and van Schmus 1995), but in other places, such as the Cordon de Lila peninsula of the SdA, thicknesses are strongly reduced, locally even to zero. Breitkreuz (1995) recognized the extensional depositional environment of these volcanic units and their sedimentary intercalations. In contrast to the interpretation given by Mpodozis and Kay (1990) for the Choiyoi province, Breitkreuz and van Schmus 1995, as well as Breitkreuz (1995), inferred a subduction-controlled arc-graben setting for the late Permian volcanic suites at the southeastern SdA border. Evidence for extension is seen in the Precordillera north of Calama where, along a significant fault zone (Günther et al. 1997, Günther 2001: Sierra Moreno Border Fault), Permo-Triassic volcanics abruptly border on the Paleozoic mica schists of Sierra de Moreno, both forming the basement to Mesozoic sedimentary rocks. In the Precordillera southwest of Calama (Sierra Limon Verde, Fig. 14.3), Lucassen et al. (1999) found lower-crustal rocks that indicate Permian high-pressure metamorphism and subsequent rapid uplift at the Permian-Triassic boundary, thus proving strong crustal thinning.

Besides lithospheric thinning, intracontinental rifting, associated with magmatism, involves the addition of basic material to the base of the crust ("basaltic underplating"; Henk et al. 1997) and the creation of a new crust-mantle boundary. By such processes, older crustal structures become strongly overprinted, or even obliterated. However, Permo-Triassic magmatism and tectonics cannot be the unique cause of the Fore-arc Gravity High for two reasons: (*1*) the outcrops of the Choiyoi-type rocks extend far beyond the limits of this anomaly without causing similar effects along strike, and (*2*) the SdA region has not acted as an important depocenter since the late Triassic; on the contrary, it was slightly folded and uplifted before the mid or late Cretaceous.

### 14.4.2.3 Salta Rift Tectonics

During the early Cretaceous, the South American continent was affected by thermal mantle perturbations related to the initial opening of the South Atlantic (e.g. Paraná flood basalts). The Salta Rift documents the activity of such a plume-like process close to the western, active continental margin. Owing to this origin, the rift arms and, hence, the depocenters and the associated magmatism do not, or only in part, follow the structural grain of the Andes (Fig. 14.6). The predominantly alkaline magmatism was active in three phases between 130 and 60 Ma (Galliski and Viramonte 1988; Viramonte et al. 1999). Magmatic activity overlapped with post-rift sedimentation beginning at ~70 Ma, according to Marquillas et al (2005), and lasting until the late Eocene or Oligocene (Marquillas and Salfity 1988).

Possibly the SdA area had already been affected by extensional tectonics since the early or mid Cretaceous, i.e. the Salta pre- and syn-rift phase (Marquillas and Salfity 1988), though alkaline magmatic rocks of this age are unknown there. A coherent depocenter from the SdA to Salta might have developed only during the post-rift phase (Salfity and Marquillas 1994; Marquillas et al. 2005, Mpodozis et al. 2005). Nevertheless, normal faulting affected the Purilactis Basin during the late Cretaceous (Muñoz et al. 2002; Figs. 14.4 and 14.5) and possibly, according to the stratigraphic interpretation, during the Tertiary before Incaic shortening. Intense alkaline magmatism in the SdA area at the Cretaceous-Tertiary boundary (early stage, back-arc magmatism of Haschke 1999; Haschke et al. 2002; Mpodozis et al. 2005) is coeval with the last phase of Salta Rift magmatism, whereas younger magmatism is restricted to the Paleocene-Eocene arc.

Cretaceous plutonic rocks are associated with the Peine and Lila Highs and, in the latter case, occur along important faults that were active during the Cretaceous (Salar-Tucucaro Fault; Fig. 14.3). Additionally, the axis of the Quimal-Tuina High follows an alignment of late Cretaceous plutons along a zone of Cretaceous (and younger) faults, thus suggesting that these gravity anomalies could be caused by basic plutons at shallow depth. The high gravity anomalies of the Calama Basin, as a whole, may be another effect of the late Cretaceous tectonics and magmatism.

The interference of the Atacama-Sey arm of the Salta Rift (Fig. 14.6) with the Cretaceous magmatic arc can be considered to have determined not only the development of the Purilactis depocenter, but also the structure of the Precordillera south (Cordillera de Domeyko) and north (Sierra de Moreno) of Calama. In the north, extensional tectonics and the formation of Cretaceous sedimentary basins advanced to the NW over the present-day Calama Basin into the Sierra de Moreno, which was subsequently affected by mid Eocene tectonic inversion. In contrast, along the Cordillera de Domeyko south of 23° S, the original western border of the Purilactis depocenter was probably not much farther west than the El Bordo Escarpment (Fig. 14.3).

For all of these reasons, the crustal structure acquired by tectonic and magmatic processes during the mid and late Cretaceous is the most likely source of the low-wavelength, positive gravity anomalies. It has to be assumed that these back-arc processes were once related to lithospheric thinning and modifications at the crust-mantle boundary (Folguera et al. 2005).

## 14.5 Crustal Structure and Thickness from Refraction Seismic Data

Further insight into the deep crustal structure of the north Chilean Andes comes from active source seismic experiments (Wigger et al. 1994; Giese et al. 1999; Schmitz et al. 1999). Four seismic refraction sections, though not designed to study the SdA Basin in detail, partly cover the SdA area and provide important data for our assessment of the region. These sections are: the short San Pedro – Peine section along the eastern border of the SdA; the N-S-oriented

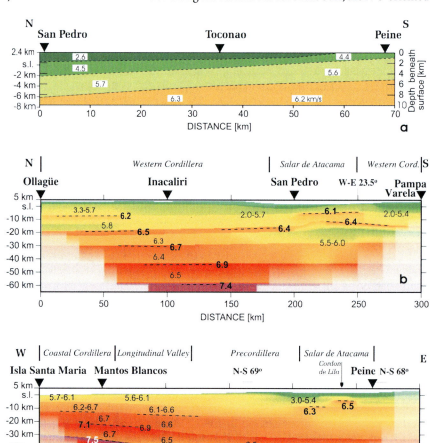

**Fig. 14.10.** Refraction seismic profiles through the SdA area, modeled by Lessel (1997). **a** Short section along the eastern SdA border showing velocities that can be interpreted as Neogene to Recent Salar evaporitic and clastic sediments (2.6 km s$^{-1}$), Cretaceous – early Tertiary Purilactis Group (4.5–4.4 km s$^{-1}$), volcanic-sedimentary Peine Group (5.7–5.6 km s$^{-1}$), and the underlying basement (Ordovician granites?) (6.3–6.2 km s$^{-1}$); **b** N-S section from Ollagüe in the Western Cordillera, along the east side of the SdA and towards the SSE to the border of the Western Cordillera (Pampa Varela); **c** E-W section from the coast to the eastern border of the SdA. In (**b**) and (**c**), increasing velocities beneath the low-velocity layer are interpreted as lower crust near the compositional crust-mantle boundary. Note elevated position of high density material in the SdA area

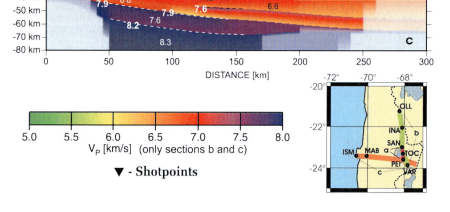

Ollagüe – Pampa Varela section; the E-W-oriented Isla Santa Maria – Peine section (Fig. 14.10a–c); and the NW-SE-oriented Chuquicamata – Huaitiquina section (not shown).

The SdA area shows considerable lateral variation in the $v_p$-depth distribution in the uppermost crust. Relatively high $v_p$ of > 6.3 km s$^{-1}$ is found at two levels separated by a low velocity zone: The upper level is observed at a depth of 5 km in the Cordón de Lila region (Fig. 14.10c). In the San Pedro–Peine section (Fig. 14.10a), the top of this high-velocity zone rises from 9 km depth near San Pedro to 5 km beneath the surface near Peine. It is most probable that this upper level corresponds to the high-density body causing the Lila and Peine gravity highs.

The question is whether this material corresponds to voluminous, late Paleozoic or Cretaceous, basic intrusions or to dense, lower crust uplifted to shallow levels by extensional detachment faulting. A low-velocity layer ($v_p$ 5.9–6.1 km s$^{-1}$) occurring at < 20 km depth, beneath the zone of high $v_p$ values (Fig. 14.10b,c), corroborates the former assumption. Similar results were obtained from the Calama Basin and from the Quimal-Tuina part of the Fore-arc Gravity High; Wigger (1988) calculated high p-wave velocities exceeding 6.5 km s$^{-1}$ that also occurred at a depth of < 10 km.

The lower crustal structure in the SdA area and its velocities, shown in the seismic refraction sections of Fig. 10b,c, are not well constrained and no prominent waves indicating the crust-mantle boundary could be observed to a depth of 60 km. The Moho values beneath the Precordillera, at 70 km depth (Fig. 14.10c), are signals from the crust-mantle boundary of the subducting plate. By means of receiver functions, Yuan et al. (2002) obtained a crustal thickness of 67 km in the SdA area, ~60 km beneath the Western Cordillera, and 50–55 km beneath the Precordillera. Thus, in a cross section through the Andes based on this technique, Schurr (2000) and Schurr and Rietbrock (2004) assigned maximum values for crustal thickness beneath the SdA depression (Fig. 14.13).

Interestingly, these results are inconsistent with estimates of crustal thickness based on the geological record (Table 14.1). The geological evolution of the SdA area since the early Paleozoic shows alternating periods of crustal contraction and extension. After probable crustal thickening during the Ordovician by shortening and/or subcrustal magmatic accretion, extension prevailed during the late Paleozoic and Mesozoic. The inversion of Cretaceous normal faults in the interior of the SdA during the Paleogene (Incaic Event or earlier) produced only minor crustal thickening because of the steep dip of the faults. Late Eocene shortening in the El Bordo Escarpment and mid Miocene eastward thrusting (Cordillera de la Sal) caused thickening only beneath the El Bordo area and the Precordillera, while shortening since 10 Ma has been concentrated at the marginal zones of the SdA block (Figs. 14.4 and 14.5).

Therefore, there are no geological grounds to assume that the felsic crust beneath the SdA Basin is thicker than a normal felsic crust. However, the apparent contradiction between the geophysically- and geologically-derived crustal thickness estimates can be reconciled if partial hydration of the mantle is assumed (Schmitz et al. 1999). In this case, the high velocities of about 7 kms$^{-1}$ may be attributed to the lowermost crust near the compositional crust-mantle boundary, at about 40 km beneath the SdA.

## 14.6 Seismic Tomography

Schurr and Rietbrock (2004) studied the deep structure of the SdA Basin using local earthquake tomography. Travel times and $t^*$ parameters from nearly 1 600 earthquakes of predominantly intermediate depth (80–250 km) were used to determine the spatial distribution of $v_p$, $v_p/v_s$ and $Q_p$ values (quality factor = attenuation$^{-1}$). The effect of intrinsic seismic attenuation is mostly influenced by temperature, allowing important inferences about rheology. Resolution of the models is approximately 20–30 km in all directions, which enables three-dimensional imaging of the major tectonic units from the surface to the subducting slab.

The tomographic models reveal a quite anomalous fore-arc lithosphere between the SdA Basin and the plate interface. Figure 14.11 shows $v_p$, $v_p/v_s$, and $Q_p$ horizontal slices (maps) at 40 km, 60 km, and 85 km depth. If the deductions from the geological evolution are applied, all these planes lie within the mantle, only the lowermost plane at its western side intersects the subducting plate. However, according to the lithospheric model from receiver functions (Fig. 14.13; Schurr 2000, Yuan et al. 2002, Schurr and Rietbrock 2004), the planes at 40 and 60 km intersect the middle and lower crust of the SdA block, respectively.

In the $v_p$ map at 40 km depth, a band of high velocity follows the outlines of the Preandean Depression, from the Calama Basin over the SdA to the Salar de Punta Negra (Fig. 14.2). This high velocity zone is bordered to the east by low velocities underneath the Western Cordillera, and to the west and north by a narrow, low-velocity zone (~6 kms$^{-1}$) beneath the Cordillera de Domeyko, its prolongation into the Cordón de Barros Arana, and the Sierra de Moreno, somewhat displaced to the NW. The high-velocity zone is, in parts, congruent with the Fore-arc Gravity High (Fig. 14.8). In the $v_p$ map at 85 km depth, mantle velocities beneath the arc and back-arc are very low, but are regular beneath the SdA area. In the west, the map intersects subducted oceanic lithosphere which is characterized by a pronounced high-velocity anomaly.

**Fig. 14.11.**
Seismic tomography of the SdA lithospheric block: Horizontal slices at depths of 40, 60 and 85 km

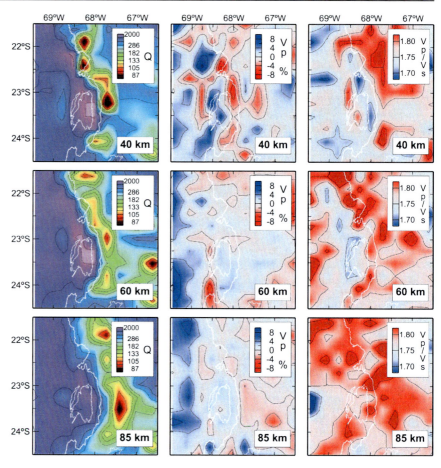

The strongest signal in the $v_p/v_s$ map at 40 km is the occurrence of very high ratios (> 1.85) beneath the magmatic arc. This region is clearly separated from the fore-arc by a band of low $v_p/v_s$ ratios that follow the eastern edge of the SdA Basin. The situation is simlar at 60 km depth, but appears to be displaced to the west. At 85 km depth, $v_p/v_s$ ratios are slightly reduced beneath the SdA Basin, compared to the high ratios measured beneath the fore-arc north of 23° S. These high ratios in the fore-arc, at least west of 69° W, correspond to the mafic crust of the slab. For seismic attenuation, high $Q_p$ values in the fore-arc contrast with very low $Q_p$ values beneath the magmatic arc and back-arc (Fig. 14.11: $Q_p$ maps at all three levels; Fig. 14.12). The lithosphere underlying the SdA and the Calama Basins shows the highest $Q_p$ values (> 2000). The SdA is sharply bounded by extremely low values (< 150) outlining the Western Cordillera and by moderately reduced values (> 350) beneath the Precordillera.

The high $Q_p$ zone extending to the plate interface characterizes the particular nature of the SdA lithospheric block. It is encased by regions of low velocities and low $Q_p$. To the east, beneath the Western Cordillera, ongoing magmatic processes find a pronounced expression as regions of low $v_p$, high $v_p/v_s$, and low $Q_p$, suggesting significant amounts of partial melting in the crust and mantle (Haberland and Rietbrock 2001; Schurr et al. 2003). The zone of low $v_p$ and low $Q_p$ that follows the trend of the Cordillera de Domeyko along the western and northern limits of the SdA Basin (Fig. 14.11, map at 40 km depth) was interpreted by Schurr and Rietbrock (2004) as a deep-reaching zone of deformation. Its main activity from the late Eocene to the mid Miocene might have been induced by heat accumulated during the Cretaceous to Eocene activity of the Precordilleran magmatic arc system and the resulting crustal weakness.

It has been mentioned that the apparently overthickened crust beneath the SdA, as obtained from receiver functions (Fig. 14.13; Yuan et al. 2002), can be readily explained assuming hydrated mantle rocks beneath a felsic upper and a, possibly, basic lower crust. Hydration reduces P-wave velocities in the mantle and obscures the compositional crust-mantle boundary. According to Hyndman and Peacock (2003), extensive serpentinization of the fore-arc mantle is revealed by high values of the $v_p/v_s$ ratio (> 1.80). From the $v_p/v_s$ map at 60 km (Fig. 14.11) and the $v_p/v_s$ section (Fig. 14.12), mantle serpentinization can be interpreted beneath the Precordillera and further to the west. However, the low $v_p/v_s$ values beneath the SdA exclude detectable amounts of serpentinization, underlining once again the apparent "exotic" nature of the SdA lithospheric block.

## 14.7 Discussion

### 14.7.1 Was the Strong SdA Lithospheric Block Cold?

The above mentioned features (morphological anomaly, basin development in the uprising plateau, contractional neotectonics, interrelation with the magmatic arc, association with high gravity, preceding rift and back-arc history, special crustal structure, and high strength lithosphere) distinguish the SdA block as an exceptional element within the Central Andean edifice. The crustal thickness of 67 km obtained from receiver functions (Fig. 14.13; Yuan et al. 2002) is perhaps the strangest of these peculiarities.

Schurr and Rietbrock (2004), referring to the high P-wave velocities and the low attenuation (Figs. 14.11 and 14.12), described the lithosphere of the SdA lithospheric block as "strong and cold". Yuan et al. (2002) calculated that the height of the SdA area is more than 1 km below the isostatic equilibrium expected from the thick crustal root and concluded that the SdA block has an extremely cold lithospheric mantle. However, these statements cannot be reconciled with the geological history.

On the contrary, temperatures in the crust and mantle remained high as long as the future SdA area was part of the back-arc and Salta rift system, the latter having formed above a positive thermal mantle anomaly that supplied the crust with basic melts. Cooling of the hot lithosphere only began at the end of the Incaic Event (~38 Ma), especially during a probable flat-slab episode from 38 to 28 Ma (Haschke et al. 2002; James and Sacks 1999; Reutter 2001; Scheuber et al. 2006, Chap. 13 and Fig. 13.7 of this volume). Furthermore, since 17 Ma, at least the eastern side of SdA lithospheric block must have been additionally heated by lateral conduction from the hot lithosphere and the asthenopheric wedge beneath the magmatic arc. According to simplistic calculations, the temperature would have risen about 125 °C at a horizontal distance of 30 km from the magma chambers and an original temperature difference of 400 °C. It has been mentioned, however, that independent of possible lateral heating, the temperature gradient in the Toconao X1 well is rather low (Apatite FT values by Muñoz et al. 2002).

### 14.7.2 A Tentative Solution

In the following discussion, we consider the development of the SdA over the last 38 million years under conditions of a cooling lithosphere. It is generally accepted that heat flow in the upper plate is much higher in the back-arc than in the cool fore-arc. In the fore-arc, where serpentine is stable, the mantle is hydrated and serpentinized to different degrees by percolating subduction fluids, whereas in the back-arc, much higher tempera-

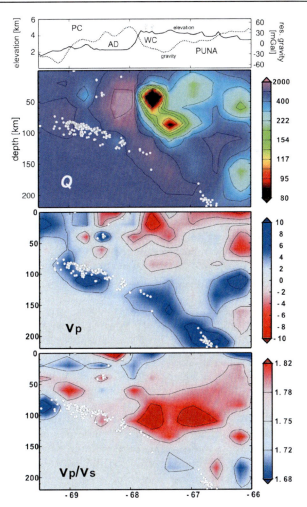

**Fig. 14.12.** Seismic tomography of the of the SdA lithospheric block: E-W sections along latitude 23.2° S

tures and a scarcity of fluids favor a dry, peridotitic mantle (Hyndman and Peacock 2003). Thus, during the back-arc history of the future SdA area, i.e. from the beginning of the Jurassic (205 Ma) to the end of the mid Eocene (38 Ma), we expect high heat flow. Particularly high temperatures, as indicated by intrusive and extrusive magmatism, prevailed during the Cretaceous when the SdA lithospheric block was an integral part of the Salta Rift system. The mantle lithosphere beneath the rift must have been thin and was probably penetrated by swarms of dykes filled with basic melts rising from the asthenosphere.

Cooling might have started in the early Tertiary but became influential only after the Incaic Event, i.e. during the amagmatic interval (38–28 Ma) and the following incorporation of the SdA area into the fore-arc realm of the new magmatic arc that developed thereafter. Since the beginning of the amagmatic interval, considered by many as a time of low-angle subduction (Isacks 1988) or even flat-slab subduction (Haschke et al. 2002; James and Sacks

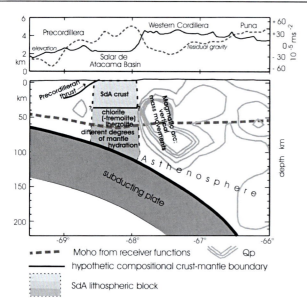

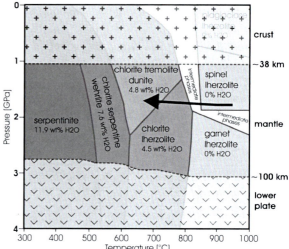

**Fig. 14.13.** Sketch adapted from Schurr and Rietbrock (2004) showing outlines of the Salar de Atacama lithospheric block, Moho discontinuity (obtained from receiver functions), presumed compositional crust-mantle boundary (based on slight or absent contraction in SdA crust), and interpretation of SdA compositional mantle. Note crustal thickening in the Precordillera owing to Eocene and Miocene east-vergent thrusting

**Fig. 14.14.** Detail of the metamorphic phases of lherzolite (after Hacker and Peacock 2004) for a mantle segment of the upper plate corresponding to the present Atacama lithospheric block. Incorporation of this block from the back-arc into the fore-arc caused retrograde metamorphism owing to cooling in the presence of water (from the slab?). *Arrow* visualizes a possible metamorphic path of the Atacama block at a depth of ~ 50 km

1999; Reutter 2001; Scheuber et al. 2006, Chap. 13 of this volume), the mantle of the still hot SdA lithospheric block became exposed to rising subduction fluids. According to the phase diagrams of Hacker et al. (2003), cooling lherzolitic mantle rocks start to metamorphose in the presence of water at temperatures < 800 °C to chlorite lherzolite or chlorite-tremolite dunite, which contain between 2% and 4.5% water, whereas the serpentinization reaction starts at < 625 °C and becomes intense below ~500 °C (Fig. 14.14). According to Hyndman and Peacock (2003), fluid flow in the fore-arc mantle and, hence, hydration are controlled by fractures. Faulting, however, became possible only once the temperature in the dry mantle rock had dropped beneath the brittle-ductile boundary of 600–800 °C. Mantle cooling under conditions of hydration is slow because this type of metamorphism is an exothermic process.

Hacker et al. (2003) published experimental $v_p$ and $v_p/v_s$ values for chlorite lherzolite and chlorite-tremolite dunite, both are metamorphic products from a lherzolite composition that match well the tomographic results for the SdA lithospheric block (Figs. 14.11 and 14.12; Schurr 2000; Schurr and Rietbrock 2004). The physical properties are almost the same for weakly hydrated, lherzolitic mantle rocks, unmetamorphosed basic rocks of the lower crust (Hyndman and Peacock 2003) and also, presumably, mantle rocks with small degrees of serpentinization in patches or along faults. Under these circumstances, a compositional crust-mantle boundary is almost undetectable. A weakly hydrated mantle with properties similar to those of the overlying lower crust would lead to measurements of extreme crustal thickness in a lithospheric block not affected by crustal thickening (Fig. 14.13). Besides the strength of the SdA block, as derived from the high $v_p$, high $Q_p$ and low $v_p/v_s$ values, its mechanical behavior as a compact block is also best explained by assuming slow lithospheric cooling to the brittle-ductile boundary because deep faulting could occur only relatively late in the cooling history.

### 14.7.3 Plateau and SdA Development and a Possible Flat-Slab Episode

The evolution of the SdA Basin is intimately related to the formation of the Altiplano-Puna Plateau. Many authors have attributed the origin of this plateau to a period of flat-slab subduction (James and Sacks 1999; Haschke et al. 2002; Reutter 2001; Scheuber et al. 2006, Chap. 13 of this volume). Flat-slab subduction is concluded from the lull of arc volcanism between 38 Ma and 28 Ma and the emplacement and changing distribution of the Oligocene to Recent magmatic arc system that followed from the resteepening of the slab dip and the creation of a new asthenospheric wedge. Such a scenario of strongly varying slab dip appears to be useful for understanding the evolution of the SdA Basin and its relationships to the adjacent morphostructural units.

It is thought that, starting with the Incaic Event, possibly after a slab break-off (Fukoa et al. 2001; Haschke 1999; Haschke et al. 2002; Reutter 2001; Scheuber et al. 2006, Chap. 13 of this volume), the slab began to travel a subhorizontal path several hundred kilometers long at a depth between 50 and 100 km (Barazangi and Isacks 1976; Gutscher 2002). As temperatures beneath the late Cretaceous-Eocene arc and back-arc became too low for melting and arc magmatism, fluids released from the slab caused metamorphic reactions in the overlying lithosphere. Mantle hydration led to an increase in volume and, hence, is considered to be responsible for initial vertical movements in the area of the present Altiplano-Puna Plateau (Ege 2004). According to Hacker et al. 2003, this type of mantle rock metamorphism, as well as the degree of hydration and volume increase, is temperature-dependent.

In the former back-arc lithosphere overlying the flat slab, temperatures must have been irregularly distributed as a consequence of the preceding rift tectonics. Those parts of the lithospheric mantle exceeding temperatures of 625 °C were subject to chloritization, while cooler parts underwent serpentinization, the intensity of which depended on the availability of water (Fig. 14.14). This may have already caused irregular topography during the first stage of plateau formation. When the slab dip steepened again, the resulting density differences in the upper plate presumably influenced the advance of hot mantle material, i.e. the restoration of an asthenospheric wedge and, hence, the next stage of plateau uplift, as well as the emplacement of a new magmatic arc.

## 14.7.4 The SdA Lithospheric Block and the West Cordilleran Magmatic Arc

Arc magmatism resumed at 28 Ma. Two features produce evidence of a causal relationship between the formation of both the basin and the magmatic arc: (1) the existence of a volcanic gap from 28 to 17 Ma between 22 and 25° S, i.e. east of the SdA area, and the reduced width of the volcanic arc that developed thereafter with respect to the arc segments to the north and south (Fig. 14.7); and (2) the wide eastward embayment of the volcanic arc along the SdA.

According to Schurr and Rietbrock (2004), the strong SdA lithospheric block may have been ultimately responsible for obstructing the asthenospheric flow and, hence, the emplacement of a rectilinear volcanic front. When the lower plate resteepened, the dense SdA lithospheric block sank relative to the surrounding, less dense regions. This assessment is supported by the unusually deep plate interface at 125 km beneath the volcanic front to the east of the SdA (instead of ~100 km; ANCORP 2003; Yuan et al. 2002).

If the resteepening of the slab at the end of the inferred flat-slab episode took place in a "rollback" style, it must have occurred earlier under the present Altiplano and Puna than in the SdA area. Arc volcanism then spread over the wide deformation zone which now constitutes the plateau, even though it contained subsiding "heavy" basins located on the former Salta Rift arms. In the western Cordillera to the east of the SdA, no residual positive gravity anomaly marks the Cretaceous Atacama-Sey arm of the Salta Rift. However, a gravity high in the Puna, to the east of the present intra-arc Olaroz-Cauchari Basin, points to relics of a lithosphere similar to that of the SdA area (Figs. 14.6 and 14.8). We therefore speculate that, in the course of slab resteepening, the Atacama-Sey parts of dense lithosphere obstructed and retarded the westward propagation of the asthenospheric wedge, thus causing the magmatic gap from 28 to 17 Ma (Fig. 14.7).

The processes related to arc magmatism caused fundamental transformations in the lithosphere of the former, late Cretaceous-Eocene back-arc, especially in the western Cordillera. We envision a scenario where pre-existing structures and primary compositions were overprinted or even obliterated (see also Riller et al. 2006, Chap. 15 of this volume). Vertical mass movements in the form of rising magmas and sinking dense rocks, as well as metamorphism caused by heat and contaminating melts and fluids, reduced the bulk density to such a degree that a zone of extremely negative, residual isostatic gravity anomalies now characterizes the present-day volcanic arc in the Western Cordillera (Fig. 14.8). In the back-arc, these younger processes may have partly blurred the lateral differences between the lithoshere within and out of the Cretaceous rift arms.

## 14.7.5 Further Implications

### 14.7.5.1 Inclination of the SdA Basement

Yuan et al. (2002) and Schurr and Rietbrock (2004) draw attention to a probable coupling of the heavy SdA lithospheric block to the lower plate. In this respect, the westward thickening of the SdA deposits and the foreland-like, west-dipping inclination of their basement can be attributed to the tilting of the strong SdA lithospheric block induced by the lower plate, or to corresponding shear movement in the upper plate (Figs. 14.4 and 14.5). In this scenario, the still active Salar Fault would separate two blocks undergoing book-shelf tilting. It remains an open question whether the SdA area can be considered a thrust-determined back-arc type foreland to the growing Precordillera.

#### 14.7.5.2 The Uyuni-Khenayani Thrust

The relative downward motion of the dense SdA lithospheric block between the encasing lithospheres of the Western Cordillera and the Precordillera, in conjunction with high convergence rates in the Miocene, presumably triggered the mid Miocene (Quechua Event), east-vergent thrusting towards the SdA that remodeled the El Bordo Escarpment and created the Cordillera de la Sal. These thrust structures, however, trend to the NE, towards the Khenayani-Uyuni Fault Zone of the southern Altiplano (Elger 2003; Scheuber et al. 2006, Chap. 13 of this volume). As this fault zone runs into the Salar de Uyuni, which is a subsiding back-arc basin, the question arises as to whether the distribution of dense material, as indicated by the positive gravity anomalies of the Salar de Uyuni and the SdA (Fig. 14.8), in combination with Cretaceous normal faults in the Atacama-Uyuni Rift arm (Fig. 14.6) controlled the emplacement and activity of the Uyuni-Khenayani Fault System.

#### 14.7.5.3 Subsidence and Lateral Contraction

The transition from a "flat-slab" to a "normal-slab" profile by a rollback-like mechanism, and the consequent filling of the opening space by asthenospheric material, must have caused vertical movements in the overlying weakened lithosphere. This means that old lateral contrasts in strength and density, as well as new ones created by uprising magmas, were controlled by buoyancy rather than by a mechanically supporting slab. The resulting vertical movements might have been combined with horizontal shortening in stress fields of changing intensity. In general, subsiding sedimentary basins bordered by steeply dipping normal faults that reactivated as reverse faults are a characteristic feature of Central Andean tectonics (e.g. Charrier and Reutter 1994; Günther et al. 1997; Scheuber et al. 2006, Chap. 13 of this volume).

### 14.8 Conclusions

The following factors contributed to the formation of the SdA depression:

1. The area was part of the Cretaceous Salta Rift, where the crust was thinned, basic material added at all crustal levels, and the deep lithosphere renewed by hot, uprising, mantle material.
2. Its location at the westernmost extension of this rift system, immediately behind the Cretaceous to Eocene magmatic arc in the north Chilean Precordillera, led to its incorporation into the fore-arc when the new late Oligocene to Recent magmatic arc was generated further east in the Western Cordillera and the growing Altiplano-Puna Plateau.
3. Relatively slow cooling ensured that the mantle of the SdA lithospheric block suffered only a low degree of hydration (probably chloritization) and thus remained strong and dense.
4. Its high bulk density relative to the Western Cordillera in the east, and to the Precordillera in the west, prevented the SdA lithoshperic block from participating in the Central Andean Plateau uplift.
5. Slab steepening under the SdA lithospheric block favored its downward motion.
6. The density and compactness of the block has obstructed the westward advancement of the asthenospheric mantle wedge and the straightening of the embayed volcanic front.
7. Downward motion and westward tilting of the block by subduction drag is thought to have induced eastward thrusting of the Precordillera (Cordillera de Domeyko) onto the SdA area during the Miocene.

### Acknowledgments

The research work in the SFB 267 project "Deformation Processes of the Central Andes" was mainly supported by the Deutsche Forschungsgemeinschaft in cooperation with the Freie Universität Berlin, the Technische Universität Berlin, the University Potsdam, and the Geoforschungszentrum. The authors E. S., H. J. G., K. J. R., P. G., P. W., and S. S. obtained funding by the Deutsche Forschungsgemeinschaft and Freie Universität Berlin not only for their SFB activities, but also for preceding smaller research projects in the southern Central Andes. The cooperation with the Chilean geoscientists was supported by the Deutscher Akademischer Austauschdienst (A.B.), the Humboldt Foundation (G.C. and R.C.) and by their national institutions. We thank Christoph Heubeck for first critical reading of the manuscript and the referees Gregory Hoke and Manfred Strecker for their thorough revision and helpful criticism.

### References

Adam J, Reuther CD (2000) Crustal dynamics and active fault mechanics during subduction erosion. Application of frictional wedge analysis onto the North Chilean Forearc. Tectonophysics 321:297–325

ANCORP Working Group (2003) Seismic imaging of a convergent continental margin and plateau in the Central Andes (Andean Continental Research Project (ANCORP) 1996). J Geophys Res 108(B7): doi 10.1029/2002JB001771

Andriessen PAM, Reutter K-J (1994) K-Ar and fission track mineral age determination of igneous rocks related to multiple magmatic arc systems along the 23°S latitude of Chile and NW Argentina. In: Reutter K-J, Scheuber E, Wigger P (eds) Tectonics of the Southern Central Andes. Springer-Verlag, Berlin Heidelberg New York, pp 141–154

Arriagada C, Roperch P, Mpodozis C (2000) Clockwise block rotations along the eastern border of the Cordillera Domeyko, Northern Chile (22°45'–23°30' S). Tectonophysics 326:153–171

Arriagada C, Roperch P, Mpodozis C, Dupont-Nivet G, Cobbold PR, Chauvin A, Cortés J (2003) Paleogene clockwise tectonic rotations in the forearc of central Andes, Antofagasta region, northern Chile. J Geophys Res 108(B1): doi 10.1029/2001JB001598

Arriagada C, Cobbold PR, Roperch P (2006) Salar de Atacama basin: a record of compressional tectonics in the central Andes since the mid-Cretaceous. Tectonics 25: doi 1029/2004TC001770

Bahlburg H, Hervé F (1997) Geodynamic evolution and tectonostratigraphic terranes of northwestern Argentina and northern Chile. Geol Soc Am Bull 109:869–884

Barazangi M, Isacks B (1976): Spatial distribution of earthquakes and subduction of the Nazca plate beneath South America. Geology 4:686–692

Belmonte-Pool A (2002) Krustale Seismizität, Struktur und Rheologie der Oberplatte zwischen Präkordillere und dem magmatischen Bogen in Nordchile. PhD thesis, Freie Universität Berlin

Bogdanic T (1990) Kontinentale Sedimentation der Kreide und des Alttertiärs im Umfeld des subduktionsbedingten Magmatismus in der chilenischen Präkordillere. Berliner Geowiss Abh A123

Breitkreuz C (1995) The late Permian Peine and Cas Formations at the eastern margin of the Salar de Atacama, northern Chile: stratigraphy, volcanic facies, and tectonics. Rev Geol Chile 22(1):3–23

Breitkreuz C, Van Schmus WR (1995) U/Pb geochronology and significance of late Permian ignimbrites in northern Chile. J S Am Earth Sci 9(5/6):281–293

Brüggen J (1942) Geología de la Puna de San Pedro de Atacama y sus formaciones de areniscas y arcillas rojas. Primer Congreso Panamericano Ing. Minas y Geología, Santiago, Anales 2:342–367

Carrapa B, Adelmann D, Hilley GE, Mortimer E, Sobel ER, Strecker MR (2005) Oligocene range uplift and development of plateau morphology in the southern central Andes. Tectonics 24: doi 10.1029/2004TC001762 (2005)

Charrier R, Reutter K-J (1994) The Purilactis Group of Northern Chile: boundary between arc and backarc from Late Cretaceous to Eocene. In: Reutter K-J, Scheuber E, Wigger P (eds) Tectonics of the Southern Central Andes, Springer-Verlag, Berlin Heidelberg New York, p 189–202

Choi S (1996) Ein 3D-Dichtemodell der Oberkruste in den Zentralanden zwischen 21° und 24° S. Diploma thesis, Freie Universität Berlin

Coira BL, Davidson J, Mpodozis C, Ramos V (1982) Tectonic and magmatic evolution of the Andes of northern Argentina and Chile. Earth Sci Rev 18:303–332

Coira BL, Mahlburg Kay S, Pérez B, Woll B, Hanning M, Flores P (1999) Magmatic sources and tectonic setting of Gondwana margin Ordovician magmas, northern Puna of Argentina and Chile. Geol Soc Am Spec P 336:145–170

DeCelles PG, Horton BK (2003) Early to middle Teriary foreland basin development and the history of the Andean crustal shortening in Bolivia. Geol Soc Am Bull 115:58–77

Dingman RJ (1963) Cuadrángulo Tulor. Carta Geológica de Chile 1:50 000, 11. Instituto de Investigaciones Geológicas, Santiago de Chile, p 35

Ege H (2004) Exhumations und Hebungsgeschichte der zentralen Anden in Südbolivien (21°S) durch Spaltspur-Thermochronologie an Apatit. PhD thesis, Freie Universität Berlin

Ege H, Sobel E, Jacobshagen V, Scheuber E, Mertmann D (2003) Exhumation history of the Central Andes of Southern Bolivia by apatite fisson track dating. Revista Tecnica YPFB 21:165–172

Elger K (2003) Analysis of deformation and tectonic history of the southern Altiplano Plateau (Bolivia) and their importance for plateau formation. GeoForschungsZentrum Potsdam Scientific Technical Report STR03/05

Elger K, Oncken O, Glodny J (2005) Plateau-style accumulation of deformation: Southern Altiplano. Tectonics 24: doi 10.1029/2004TC001675

Flint S, Hartley AJ, Rex DC, Guise P, Turner P (1989) Geochronology of the Purilactis Formation, Northern Chile: an insight into Late Cretaceous/Early Tertiary basin dynamics of the Central Andes. Rev Geol Chile 16:241–246

Flint S, Turner P, Jolley EJ, Hartley AJ (1993) Extensional tectonics in convergent margin basins: an example from the Salar de Atacama, Chilean Andes. Geol Soc America Bull 105:603–617

Folguera A, Ramos VA, Zapata T, Spagnuolo M, Miranda F (2005) Pliocene to Quaternary retro-arc extension in the Andes at 35°–37°30' S. 6th International Symposium on Andean Geodynamics, Barcelona, Extended Abstracts, pp 277–280

Galliski MA, Viramonte JG (1988) Cretaceous paleorift in northwestern Argentina – petrological approach. J S Am Earth Sci 1:329–342

Gardeweg M, Ramirez CF (1985) Hoja Río Zapaleri, II Región de Antofagasta. Carta Geológica de Chile 1:250000, 66. Servicio Nacional de Geología Minería, Santiago de Chile, p 89

Giese P, Scheuber E, Schilling F, Schmitz M, Wigger P (1999) Crustal thickening processes in the Central Andes and the different natures of the Moho-discontinuity. J S Am Earth Sci 12(2):201–220

Götze H-J, Kirchner A (1997) Interpretation of gravity and geoid in the Central Andes between 20° and 29° S. J S Am Earth Sci 10(2):179–188

Götze H-J, Krause S (2002) The Central Andean gravity high, a relic of an old subduction complex? J S Am Earth Sci 14(8):799–811

Götze H-J, Lahmeyer B, Schmidt S, Strunk S (1994) The lithospheric structure of the Central Andes (20°–26°S) as inferred from interpretation of regional gravity. In: Reutter K-J, Scheuber E, Wigger P (eds) Tectonics of the Southern Central Andes. Springer-Verlag, Berlin Heidelberg New York, p. 7–22

Günther A (2001) Strukturgeometrie, Kinematik und Deformationsgeschichte des oberkretazisch-alttertiären magmatischen Bogens (nordchilenische Präkordillere, 21.7–23°) Berliner Geowiss Abh A213, pp 1–170

Günther A, Haschke M, Reutter, KJ, Scheuber E (1997) Repeated reactivation of an ancient fault zone under changing kinematic conditions: the Sierra de Moreno fault system (SMFS) (North–Chilean Precordillera). 8th Congreso Geologico Chileno, Actas 1:85–89

Haberland C, Rietbrock A (2001) Attenuation tomography in the western central Andes: a detailed insight into the structure of a magmatic arc. J Geophys Res 106:11151–11167

Hacker BR, Abers GA, Peacock SM (2003) Subduction factory 1. Theoretical mineralogy, densities, seismic wave speeds, and $H_2O$ contents. J Geophys Res 108(B1) doi 10.1029/2001JB001127

Hammerschmidt K, Döbel R, Friedrichsen H (1992) Implication of $^{40}Ar/^{39}Ar$ dating of Early Tertiary volcanic rocks from the north-Chilean Precordillera. Tectonophysics 202:55–81

Hartley AJ, Flint S, Turner P (1988) A proposed lithostratigraphy for the Cretaceous Purilactis Formation, Antofagasta Province, Northern Chile. 5th Congreso Geológico Chileno, Actas 3:83–99

Hartley AJ, Flint S, Turner P, Jolley EJ (1992) Tectonic controls on the development of a semi-arid, alluvial basin as reflected in the stratigraphy of the Purilactis Group (Upper Cretaceous–Eocene), northern Chile. J S Am Earth Sci 5(3–4):257–296

Hartley AJ, May G, Chong G, Turner P, Kape SJ, Jolley EJ (2000) Development of a continental forearc: a Cenozoic example from the Central Andes, northern Chile. Geology 28:331–335

Haschke MR (1999) Variationen der Magmengenese oberkretazischer bis obereozäner Magmatite in Nordchile (21,5–26°S). Berliner Geowiss Abh A202, pp 1–120

Haschke MR, Günther A (2003) Balancing crustal thickening in arcs by tectonic vs. magmatic means. Geology 31(11):933–936

Haschke MR, Scheuber E, Günther A, Reutter KJ (2002) Evolutionary cycles during the Andean Orogeny: repeated slab breakoff and flat subduction? Terra Nova 14:49–55

Henk A, Franz L, Teufel S, Oncken O (1997) Magmatic underplating, extension, and crustal reeqilibration: insights from a cross-section through the Ivrea Zone and Strona-Ceneri Zone, Northern Italy. J Geol 105(3):367–377

Hyndman RD, Peacock SM (2003) Serpentinization of the forearc mantle. Earth Planet Sci Lett 212:417–432

Isacks BL (1988) Uplift of the Central Andean plateau and bending of the Bolivian orocline. J Geophys Res 93:3211–3231

James DE, Sacks S (1999) Cenozoic formation of the Central Andes: a geophysical review. In Skinner BJ (ed) Geology and ore deposits of the Central Andes. Soc Econ Geol Spec Pub 7:1–25

Jordan TE, Muñoz N, Hein M, Lowestein T, Godfrey L, Yu J (2002) Active faulting and folding without topographic expression in an evaporite basin, Chile. Geol Soc Am Bull 114:1406–1421

Kraemer B, Adelmann D, Alten M, Schnurr W, Erpenstein K, Kiefer E, Van Den Bogaard P, Görler K (1999) Incorporation of the Paleogene foreland into the Neogene Puna plateau: the Salar de Antofalla area, NW Argentina. J S Am Earth Sci 12:157–182

Kuhn D (1996) Deformationsanalyse der Salar de Atacama Region (Nordchile) und Interpretation des neotektonischen Spannungsfeldes. PhD thesis, Technische Universität Berlin

Kuhn D, Reuther CD (1996) Sinistrale Transpression im Bereich des Salar de Atacama, Nordchile: Ausdruck einer neogenen Forearc Rotation? Terra Nostra 8:79–80

Kuhn D, Reuther CD (1999) Strike-slip faulting and nested block rotations: structural evidence from the Cordillera Domeyko, northern Chile. Tectonophysics 331:383–398

Lamb S, Hoke L (1997) Origin of the high plateau in the Central Andes, Bolivia, South America. Tectonics 16(4):623–649

Lessel K (1997) Die Krustenstruktur der Zentralen Anden in Nordchile (21–24°S), abgeleitet aus 3D-Modellierungen refraktionsseismischer Daten. PhD thesis, Freie Universität Berlin

Lucassen F, Franz G, Laber A (1999) Permian high pressure rocks – the basement of the Sierra de Limòn Verde in northern Chile. J S Am Earth Sci 12:183–199

Macellari CE, Su MJ, Townsend F (1991) Structure and seismic stratigraphy of the Atacama basin, northern Chile. 6th Congreso Geológico Chileno Actas 1:133–137

Marinovic N, Lahsen A (1984) Hoja Calama. Carta Geológica de Chile 1:250.000 N°58, Servicio Nacional Geología Minería, p 140

Marquillas RA, Salfity JA (1988) Tectonic framework and correlations of the Cretaceous-Eocene Salta Group: Argentina. In: Bahlburg H, Breitkreuz C, Giese P (eds) The Southern Central Andes. Lecture Notes in Earth Sciences 17, Springer-Verlag, Berlin Heidelberg New York, pp 119–136

Marquillas RA, Del Papa C, Sabino IF (2005) Sedimentary aspects and paleoenvironmental evolution of a rift basin: Salta Group (Cretaceous-Paleogene), northwestern Argentina. Int J Earth Sci 94:94–113

May G, Hartley AJ, Stuart FM, Chong G (1999) Tectonic signature in arid continental basins: an example from the Upper Miocene–Pleistocene Calama Basin, Andean forearc, northern Chile. Palaeogeog Palaeoclim Palaeoecol 151:55–77

Mégard F (1984) The Andean orogenic period and its major structures in central and northern Peru. J Geol Soc London 141: 893–900

Moraga A, Chong G, Fortt MA (1974) Estudio geológico del Salar de Atacama, provincia de Antofagasta, Chile. Instituto Investigaciones Geológicas Boletino 29

Mpodozis C, Mahlburg Kay S (1990) Provincias magmáticas ácidas y evolución tectónica de Gondwana: Andes chilenos (28–31°S). Rev Geol Chile 17(2):153–180

Mpodozis C, Hervé F, Davidson J, Rivano S (1983) Los granitoides de Cerro Lila, manifestaciones de un episodio intrusivo y termal del Paleozoico inferior en los Andes del norte de Chile. Rev Geol Chile 18:3–14

Mpodozis C, Marinovic N, Smoje I, Cuitiño G (1993) Estudio geológico estructural de la Cordillera de Domeyko entre Sierra Limón Verde y Sierra Mariposas, Región de Antofagasta. Servicio Nacional de Geología y Minería – Corporación del Cobre de Chile, unpublished report

Mpodozis C, Arriagada C, Roperch P (1999) Cretaceous to Paleogene geology of the Salar de Atacama, northern Chile: a reappraisal of the Purilactis Group stratigraphy. Proc 4th International Symposium on Andean Geodynamics Göttingen, pp 523–526

Mpodozis C, Arriagada C, Basso M, Roperch P, Cobbold P, Reich M (2005) Late Mesozoic to Paleogene stratigraphy of the Salar Atacama Basin, Northern Chile: tectonic implications for the tectonic evolution of the Andes. Tectonophysics 339:125–154

Muñoz GN, Townsend GF (1997) Estratigrafía de la Cuenca Salar de Atacama. Resultados del pozo exploratorio Toconao. 1st Implicaciones Regionales. 8th Congreso Geologico Chileno, Actas 1:555–559

Muñoz N, Charrier R, Jordan T (2002) Interactions between basement and cover during the evolution of the Salar de Atacama basin, northern Chile. Rev Geol Chile 29(1):55–80

Naranjo JA, Ramírez CF, Paskoff R (1994) Morphostratigraphic evolution of the northwestern margin of the Salar de Atacama basin. Rev Geol Chile 21(1):91–104

Niemeyer H (1984) La Megafalla Tucucaro en el extremo sur del Salar de Atacama: una antigua zona de cizalle reactivada en el Cenozoico. Comunicaciones 34:37–45

Niemeyer H (1989) El Complejo Igneo-Sedimentario del Cordón de Lila, Región de Antofagasta: significado tectónico. Rev Geol Chile 16(2):163–181

Niemeyer H, Urzúa F, Rubinstein C (1997) Nuevos antecedentes estratigráficos y sedimentológicos de la Formación Zorritas, Devónico–Carbonífero de Sierra Almeida, Región de Antofagasta, Chile. Rev Geol Chile 24(1):25–43

Palma MA, Parica PD, Ramos VA (1986) El Granito Archibarca: Su edad y significado tectonico, provincia de Catamarca. Asociacion Geologica Argentina XLI(3–4):414–419

Pananont P, Mpodozis C, Blanco N, Jordan TE, Brown LD (2004) Cenozoic evolution of the northwestern Salar de Atacama Basin, northern Chile. Tectonics 23: doi 10.1029/2003TC001595

Ramirez CF, Gardeweg M (1982) Hoja Toconao. Carta Geológica de Chile 1:250.000 N°54, Servicio Nacional de Geología y Minería, pp 122

Reutter KJ (2001) Le Ande centrali: elementi di una orogenesi di margine continentale attivo. Acta Naturalia de l'Ateneo Parmense 37(1/2):5–37

Reutter K-J, Scheuber E, Helmcke D (1991) Structural evidence of orogen-parallel strike-slip displacements in the Precordillera of Northern Chile. Geol Rdsch 80:135–153

Reutter KJ, Doebel R, Bogdanic T, Kley J (1994) Geological Map of the Central Andes between 20°S and 26°S, 1:1000000. In: Reutter K-J, Scheuber E, Wigger P (eds) Tectonics of the Southern Central Andes. Springer-Verlag, Berlin Heidelberg New York

Reutter K-J, Scheuber E, Chong G (1996) The Precordilleran Fault System of Chuquicamata, Northern Chile: evidence for reversals along arc-parallel strike-slip faults. Tectonophysics 259:213–228

Riller U, Götze H-J, Schmidt S, Trumbull RB, Hongn F, Petrinovic IA (2006) Upper-crustal structure of the Central Andes inferred from dip curvature analysis of isostatic residual gravity. In: Oncken O, Chong G, Franz G, Giese P, Götze H-J, Ramos VA, Strecker MR, Wigger P (eds) The Andes – active subduction orogeny. Frontiers in Earth Science Series, Vol 1. Springer-Verlag, Berlin Heidelberg New York, pp 327–336, this volume

Salfity J, Marquillas R (1994) Tectonic and sedimentary evolution of the Cretaceous–Eocene Salta Group Basin, Argentina. In: Salfity J (ed) Cretaceous tectonics of the Andes. Vieweg, Braunschweig, pp 266–315

Scheuber E, Mertmann D, Ege H, Silva-González P, Heubeck C, Reutter K-J, Jacobshagen V (2006) Exhumation and basin development related to formation of the Central Andean Plateau, 21° S. In: Oncken O, Chong G, Franz G, Giese P, Götze H-J, Ramos VA, Strecker MR, Wigger P (eds) The Andes – active subduction orogeny. Frontiers in Earth Science Series, Vol 1. Springer-Verlag, Berlin Heidelberg New York, pp 285–302, this volume

Schmitz M, Lessel K, Giese P, Wigger P, Araneda M, Bribach J, Graeber F, Grunewald S, Haberland C, Lüth S, Röwer P, Ryberg T, Schulze A (1999) The crustal structure beneath the Central Andean forearc and magmatic arc as derived from seismic studies – the PISCO 94 experiment in northern Chile (21°–23°S). J S Am Earth Sci 12:237–260

Schurr B (2000) Seismic structure of the Central Andean subduction zone from local earthquake data. GeoForschungsZentrum Potsdam Scientifc Technical Report STR01/01

Schurr B, Rietbrock A (2004) Deep seismic structure of the Atacama basin, Northern Chile. Geophys Res Lett 31: doi 10.1029/2004GL019796

Schurr B, Asch G, Rietbrock A, Trumbull R, Haberland C (2003) Complex pattern of fluid and melt transport in the central Andean subduction zone revealed by attenuation tomography. Earth Planet Sci Lett 215:105–119

Somoza R, Tomlinson A (2002) Paleomagnetism in the Precordillera of northern Chile (22°30' S): implications for the history of tectonic rotations in the Central Andes. Earth Planet Sci Lett 194:369–381

Somoza R, Singer S, Tomlinson A (1999) Paleomagnetic study of upper Miocene rocks from northern Chile: implications for the origin of late Miocene–Recent tectonic rotations in the southern Central Andes. J Geophys Res 104(B10):22923–22936

Springer M, Förster A (1998) Heat-flow density across the Central Andean subduction zone. Tectonophysics 291:123–139

Steinmann G (1929) Geologie von Peru. Karl Winter, Heidelberg

Tomlinson AJ, Blanco N (1997) Structural evolution and displacement history of the West Fault system, Precordillera, Chile: Part 1, syn-mineral history. 8$^{th}$ Congreso Geológico Chileno, Actas 3:1873–1877

Tomlinson AJ, Blanco N, Maksaev V, Dilles JH, Grunder AL, Ladino M (2001) Geología de la Precordillera Andina de Quebrada Blanca-Chuquicamata, Región I y II (20°30'–22°30') 1:50000, mapas 1–20. Informe Registrado, Sernageomin

Townsend F (1988) Exploracion petrolera en la cuenca del Salar de Atacama, Region de Antofagasta, Chile, Antofagasta. Vertiente 4:45–55

Viramonte JG, Mahlburg Kay S, Becchio R, Escayola M, Novitsky I (1999) Cretaceous rift_related magmatism in central-western South America. J S Am Earth Sci 12:109–121

Wigger P (1988) Seismicity and crustal structure of the Central Andes. In: Bahlburg H, Breitkreuz C, Giese P (eds) The Southern Central Andes – contributions to structure and evolution of an active continental margin. Lecture Notes in Earth Sciences 17, Springer-Verlag, Berlin Heidelberg New York, pp 209–229

Wigger PJ, Schmitz M, Araneda M, Asch G, Baldzuhn S, Giese P, Heinsohn WD, Martinez E, Ricaldi E, Röwer P, Viramonte J (1994) Variation in the crustal structure of the southern Central Andes. In: Reutter K-J, Scheuber E, Wigger P (eds) Tectonics of the Southern Central Andes. Springer-Verlag, Berlin Heidelberg New York, pp 23–48

Wilkes E, Görler K (1994) Sedimentary and structural evolution of the Salar de Atacama depression. In: Reutter K-J, Scheuber E, Wigger P (eds) Tectonics of the Southern Central Andes. Springer-Verlag, Berlin Heidelberg New York, pp 171–188

Yuan X, Sobolev SV, Kind R (2002) Moho topography in the Central Andes, and its geodynamic implications. Earth Planet Sci Lett 199:389–402

# Upper-Crustal Structure of the Central Andes Inferred from Dip Curvature Analysis of Isostatic Residual Gravity

Ulrich Riller · Hans-Jürgen Götze · Sabine Schmidt · Robert B. Trumbull · Fernando Hongn · Ivan Alejandro Petrinovic

**Abstract.** The relationship between Bouguer gravity, isostatic residual gravity and its dip curvature, first-order structural elements and distribution of Neogene volcanic rocks was examined in the southern Altiplano and Puna Plateau. In the southern Altiplano, strong positive Bouguer gravity corresponds to areas affected by late Cenozoic faulting and large-scale folding of upper crustal rocks. Dip curvature analysis of isostatic residual gravity shows that elongate zones of maximum curvature correspond remarkably well with the structural grain defined by first-order folds and faults. Similarly, isostatic residual gravity in the Puna is largely controlled by prominent, upper-crustal structures and also by the distribution of Miocene and younger volcanic rocks. In particular, the Central Andean Gravity High, one of the most prominent features of the residual gravity field, corresponds with domains of low topography, i.e., internally-drained basins, which are surrounded by zones of Neogene faults and abundant felsic volcanic rocks. Dip curvature analysis of the isostatic residual gravity field shows that elongate zones of maximal curvature correlate with the strike of prominent Neogene faults. Our study suggests that such analysis constitutes an important tool for imaging upper-crustal structures, even those that are not readily apparent at surface. For example, upper-crustal faults in the Salar de Atacama area, the presence of which is suggested by the dip curvature of residual gravity, offers a plausible explanation for the pronounced angular departure of the volcanic belt from its overall meridional trend and its narrowing south of the salar. In contrast to previous interpretations, our study suggests that gravity anomalies of the Central Andes are largely controlled by the distribution of late Cenozoic volcanism and tectonism. Dip curvature analysis of gravity fields bear great potential for elucidating first-order structural elements of deformed, upper-crustal terrains such as the modern Andes.

## 15.1 Introduction

The central and southern Andes have been the focus of the collaborative research program (SFB 267) in its goal to understand orogenic processes at convergent margins – processes such as continental plateau formation, the relationships between tectonism and magmatism as well as those between surface uplift, rock exhumation and climate. Central to understanding these relationships is knowledge of the crustal structure as revealed, for example, by seismic studies (e.g., Wigger et al. 1994; Whitman et al. 1996; Zandt et al. 1994; Yuan et al. 2000, 2002), gravity studies (e.g., Lyon-Caen et al. 1985; Götze et al. 1994; Götze and Kirchner 1997; Götze and Krause 2002) and field analyses of upper-crustal deformation (e.g., Kley 1999; Kley and Monaldi 1998). However, only a few studies have specifically addressed the correspondence between structures imaged by geophysical methods and those observed at surface (e.g., Gangui 1998; Schmitz et al. 1999). This paper provides an example of such a collaborative study combining the analysis of the gravity field, prominent upper-crustal structures and distribution of volcanic centers in the southern Altiplano and Puna Plateau regions of the Central Andes (Fig. 15.1a). The emphasis is on the importance of the gravity field and its derivatives – isostatic residual gravity and its dip curvature – for imaging upper-crustal structures, and using these to better understand the magmatic and tectonic characteristics of this area.

The Central Andes are dominated by the ~4000 m-high Altiplano-Puna Plateau, which is bounded by the Miocene to Recent magmatic arc to the west, and the Eastern Cordillera and Subandean foreland fold-and-thrust belt to the east (Fig. 15.1a). The plateau formed chiefly by E-W crustal shortening (e.g., Isacks 1988; Allmendinger et al. 1997; Kley and Monaldi 1998; Elger et al. 2005), which led to the development of internally-drained, contractional sedimentary basins (Kraemer et al. 1999; Riller and Oncken 2003; Sobel et al. 2003; Sobel and Strecker 2003). The basins are structurally limited by orogen-parallel reverse faults and by NE-SW-striking, strike-slip faults.

Shortening and associated basin formation in the plateau area commenced during the Eocene-Oligocene (Jordan and Alonso 1987; Allmendinger et al. 1997; Kraemer et al. 1999; Scheuber et al. 2006, Chap. 13 of this volume; Oncken et al. 2006, Chap. 1 of this volume) and, overall, may have propagated southwards in the south-central Andes (Riller and Oncken 2003). Coarse, clastic foreland deposits in the Eastern Cordillera that date from 13–10 Ma (Viramonte et al. 1994; Grier et al. 1991) indicate that the plateau area must have been already elevated well above its foreland by this time. This onset of vertical plateau growth agrees well with (*1*) the age of uplift of the Eastern Cordillera, which formed an orographic barrier to the evolving plateau (Sobel et al. 2003; Sobel and Strecker 2003); (*2*) aridization of the plateau area in the late Miocene (Vandervoort et al. 1995; Gaupp et al. 1999); (*3*) an abrupt east-

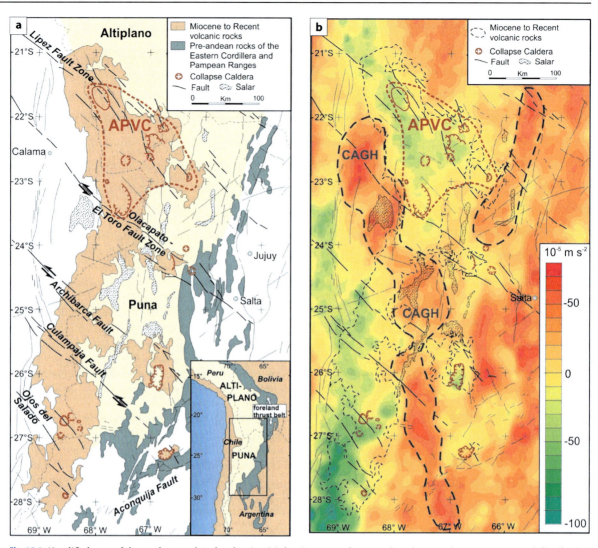

**Fig. 15.1.** Simplified map of the south-central Andes showing (**a**) dominant morpho-tectonic units, prominent structures and distribution of late Cenozoic collapse calderas as well as (**b**) isostatic residual gravity. Gravity highs in the Puna, outlined by *stippled gray lines* in (**b**), correspond with internally-drained basins, i.e., salars. In contrast, gravity lows are spatially associated with Neogene volcanic complexes, specifically the active magmatic arc, transverse volcanic belts, the Altiplano Puna Volcanic Complex (APVC) and individual, large collapse calderas. *CAGH*: Central Andean Gravity High (Götze and Krause 2002)

ward shift in upper-crustal deformation at about 10 Ma into the Subandean fold-and-thrust belt (Gubbels et al. 1993); and (*4*) initiation of extensive ignimbrite volcanism in the south-central Andes (de Silva 1989; Riller et al. 2001).

Between about 10 Ma and 1 Ma, large volumes of dacitic to rhyolitic ignimbrite deposits were emplaced in the Puna and the southern Altiplano (Fig. 15.1a). These ignimbrites erupted mainly from collapse calderas that comprise the so-called Altiplano Puna Volcanic Complex (de Silva 1989). Many of the calderas are located in, or near, prominent NW-SE-trending transverse volcanic belts (Viramonte and Petrinovic 1990). The transverse volcanic belts, in turn, are spatially associated with major NW-SE-striking fault zones, such as the Olacapato-El Toro fault zone (Fig. 15.1a), which has been intermittently active since the Paleozoic and has caused left-lateral displacement of the order of 20 km in the Neogene (Allmendinger et al. 1983). Orogen-parallel dilation on NW-SE-striking fault zones may well have facilitated magma ascent and caldera formation in the Puna (Riller et al. 2001; Caffe et al. 2002; Chernicoff et al. 2002; Petrinovic et al. 2005; Ramelow et al. 2006).

The gravity field of the Central Andes (Fig. 15.1b) has been examined with regard to the isostatic state, crustal density structure and the rigidity of the lithosphere (Whitman 1994; Götze and Kirchner 1997; Romanyuk et al. 1999). Field measurements and satellite altimetry have contributed to the gravity database. Overall, low gravity is evident in the Miocene to Recent magmatic arc, whereas positive gravity is observed in the Eastern Cordillera and Pampean Ranges. This can be explained by the difference in the thermo-mechanical states of the two morphotectonic provinces. More specifically, the volcanic belt is hotter and, thus, less rigid than the Eastern Cordillera and the Pampean Ranges (Tassara 2005). However, the plateau area between the volcanic belt and the Eastern Cordillera displays both positive and negative gravity anomalies.

In the Puna Plateau, a major characteristic of the gravity field is the presence of two conspicuous zones of positive isostatic residual gravity: a western zone trending NW-SE between 28.5° S and 22° S and a less prominent one to the east, trending NNE-SSW between 26° S and 22° S (Fig. 15.1b). The western gravity anomaly is known as the Central Andean Gravity High (CAGH) and has been interpreted as being caused by a dense rock mass measuring about 400 km × 120 km occurring between 10 km and 38 km depth (Götze and Krause 2002). The CAGH coincides spatially with Paleozoic granitoid rocks of the Faja Eruptiva Occidental, whereas the eastern gravity anomaly corresponds to the Faja Eruptiva Oriental (Omarini et al. 1999), which contains mafic rocks of Cretaceous and Ordovician age. Accordingly, the two gravity anomalies have been thought to largely reflect early Paleozoic features (Omarini et al. 1999).

Here, we present an alternative explanation for the existence of the CAGH that is linked with upper-crustal deformation and late Cenozoic, fault-controlled magmatism in this area. Moreover, we demonstrate that dip curvature analysis of the gravity field can effectively contribute to the imaging of prominent upper-crustal discontinuities, even if they are obscured by sedimentary or volcanic cover. The results also have important implications as to the cause of the deviation of the present-day volcanic belt from its overall meridional trend in the Salar de Atacama area.

## 15.2 Dip Curvature as a Tool for Gravity Image Processing

The concept of calculating surface curvature is known from the work by Gauss in the 1820s, but practical applications have only recently been made possible with the advent of powerful workstations (e.g., Sigismondi and Soldo 2003; Roberts 2001). Those working with three-dimensional (3D) seismic data have been deriving various surface attributes (e.g., dip and azimuth) for several years. Typically, these attributes are obtained in order to locate subtle discontinuities that are not readily apparent on seismic profiles (e.g., Hesthammer and Fossen 1997; Townsend et al. 1998) and may be important for localizing natural reservoirs, in particular hydrocarbons. Roberts (2001) demonstrated the use of curvature attributes, many of them developed in the field of terrain analysis,

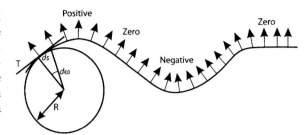

**Fig.15.2.** Cartoon illustrating quantities of dip curvature. $T$ denotes the tangent and $R$ the radius of a curve at a given point. The curvature of a line segment is related to the length of an arc segment, $s$, and its angle $\omega$. See text for an explanation of $d\omega$, $ds$ and the dip curvature concept

for detection of subtle structural elements. Among many others, important curvature attributes include strike curvature, dip curvature, maximum curvature and minimum curvature.

In terms of methodology, curvature is related to the second derivative of a surface. In addition to applications concerning 3D seismic arrays (Roberts 2001), curvature analysis has recently been tested and applied to gravity data (e.g., Schmidt and Götze 2003; Tašárová 2004). Curvature analysis is particularly useful for enhancing linear elements, more specifically, gravity lineaments of the residual field, and it helps to identify the orientations of such elements. Mathematically, curvature is an attribute of a curve or a surface in 3D space, and quantifies its angular departure from a straight line, or planar surface, respectively, as well as the magnitude of bending at a specific point (Fig. 15.2). The power of curvature analysis lies in the fact that it is a surface-dependent attribute and, thus, provides insight into surface properties that are intrinsic to that surface and difficult to express otherwise.

To aid understanding of curvature calculations with respect to the interpretation of gravity fields, some basic concepts are given here. A two-dimensional curve in the $xy$-plane can be thought of as a consecutive string of arcs of a circle characterized by variable centers and radii (Fig. 15.2). Therefore, the curvature, $k$, at any given point on this curve can be defined as the reciprocal of the radius, $R$, of the particular arc segment at that point. It can also be defined as the derivative of the curve's tangent, $T$, with respect to its position on the curve at that point. In other words, if $\omega(s)$ is the angle of the curve with respect to a reference axis as a function of the path length, $s$, of a specified curve segment, then $k = d\omega/ds$. In terms of Cartesian coordinates $X$ and $Y$, $\tan\omega = dY/dX$. A plane curve can be represented as a function $Y = f(X)$ where $X$ and $Y$ are Cartesian coordinates. The Pythagoras theorem results in $(ds)^2 = (dX)^2 + (dY)^2$, which can be rearranged as $ds/dX = [1 + (dY/dX)^2]^{1/2}$. The curvature $k$ can then be determined directly by evaluating the derivative $d\omega/ds$ as follows:

$$k = \frac{\dfrac{d^2Y}{dX^2}}{\sqrt{\left(1+\left(\dfrac{dY}{dX}\right)^2\right)^3}}$$

If the $X$ axis is the tangent to the curve at a specific point, then $\tan\omega$ approaches $d\omega$ and $dX$ approaches ds (i.e., the zero dip situation). From the equation above, the curvature can be simply defined as the second derivative, $\kappa = d^2Y/dX^2$. A sense for both the sign and magnitude of curvature of any curve can be obtained by replacing the radii by vectors normal to the curve (Fig. 15.2). Following the convention proposed by Roberts (2001), the configuration of normal vectors illustrates the curvature, i.e., where a given set of vectors diverge, converge or are parallel to each other. It follows that planar surfaces have zero curvature, antiforms have positive curvature and synforms display negative curvature. Another useful property of the curvature attribute is its independence on surface orientation, which does not apply for the first-derivative attributes such as dip or edge. The equation above shows that curvature depends on the second derivative of a function. Therefore, it is recommended to first filter data sets to reduce or remove any noise or small-wavelength anomalies, i.e., to enhance the gravity surface (gravity field) before analysing the data in terms of curvature.

Curvature analysis can be easily extended to three dimensions as it is possible to calculate the curvature of a line that results from the intersection of a surface with any plane. For example, if a surface is cut by a plane orthogonal to the surface, the curvature of their intersection is termed normal (Roberts 2001). From any two sets of orthogonal, normal curvature values, it is possible to calculate other curvature attributes such as mean curvature, Euler's curvature and Gaussian curvature. The curvature of isostatic residual gravity fields can be used to quantitatively display field gradients that are due to density contrasts caused by lithological variation or distinct structural dislocations. Consequently, the comparison of curvature with crustal structure can be expected to furnish information as to the origin of gravity anomalies. Such a comparison was conducted in this study for two areas of the Central Andes, the southern Altiplano (Fig. 15.3) and the Puna (Fig. 15.4), for which sufficient gravity measurements and upper-crustal structural information are available.

## 15.3 Correlation of Upper-Crustal Structure with Dip Curvature in the Southern Altiplano

The relationship between prominent geological structures and the gravity field in the southern Altiplano is examined by superimposing mapped structures on the Bouguer gravity field (Fig. 15.3b) and on the dip curvature map of the isostatic residual gravity (Fig. 15.3c). The isostatic residual field was obtained by subtracting the isostatic (Vening Meinesz) regional field from the Bouguer gravity field (Fig. 15.3b). The curvature of the isostatic residual gravity field (Fig. 15.3c) was calculated using an in-house JAVA program written by S. Schmidt and based on algorithms published by Roberts (2001).

The structural grain of the southern Altiplano is dominated by NNE-SSW-striking reverse faults and prominent folds associated mostly with the Uyuni-Khenayani fault

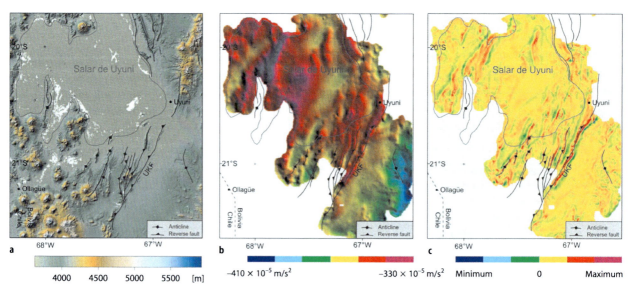

**Fig. 15.3.** Prominent upper-crustal structures of the southern Altiplano superimposed on (**a**) a shuttle radar topography model, (**b**) the Bouguer gravity field and (**c**) the dip curvature of the isostatic residual gravity. Note the remarkable correspondence between structures and positive gravity anomalies in (**b**) and elongate zones of maximum dip curvature in (**c**). *UKF*: Uyuni-Khenayani fault

(Fig. 15.3a). This fault and its secondary strands displaced Paleozoic and Mesozoic sedimentary rocks as well as syn-orogenic, clastic, continental deposits during crustal shortening in the Neogene (Elger et al. 2005). Where exposed, the reverse faults and associated hanging-wall anticlines generated elongate morphological ridges (Fig. 15.3a). Large parts of the area are, however, covered by young, undeformed salt deposits, notably those of the Salar de Uyuni, and Cenozoic structures are obscured.

Elger (2003) pointed out the strong correspondence between the location of first-order geological structures and elongate, positive anomalies of the Bouguer gravity field (Fig. 15.3b), and used this correspondence to successfully locate buried structures in the Salar de Uyuni area. Specifically, anticlines correlate well with positive Bouguer anomalies, which can be explained by an elevated position of excess mass in uplifted basement rocks or the hanging walls of major reverse faults, notably the Uyuni-Khenayani fault and its western fault splays. North-south-trending, positive gravity anomalies are also prominent in the Salar de Uyuni area and they can be traced to exposed faults north of the salar as well as to reverse faults and anticlines associated with the Uyuni-Khenayani fault south of the salar (Fig. 15.3b). Conversely, gravity lows correspond to areas devoid of prominent, upper-crustal structures.

The influence of upper-crustal structures on the gravity field is even more apparent in a map displaying variations in dip curvature of the isostatic residual gravity field (Fig. 15.3c). There is a remarkable coincidence between the gravity lineaments indicated by elongate zones of maximum dip curvature and the surface trace of known geological structures. Furthermore, as with the positive Bouguer anomalies, zones of maximum dip curvature in the Salar de Uyuni, where structures are fully buried, can be clearly linked with exposed structures to the north and south of the salar. Finally, the areas with low or zero curvature are devoid of obvious structures. This example demonstrates the power of curvature analysis for identifying and mapping upper-crustal structures that are buried by post-tectonic, sedimentary or volcanic deposits.

## 15.4 Correlation of Upper-Crustal Structure with Dip Curvature in the Puna

In order to determine the relationship between upper-crustal structures and the pattern of dip curvature of a larger region, these properties were examined in the Puna Plateau and adjacent regions (Fig. 15.4). The fault pattern of the Puna Plateau is dominated by orogen-parallel and NE-SW-striking fault systems, mostly made up of Neogene reverse faults that follow the mountain fronts of uplifted ranges (Fig. 15.4a). The ranges constitute large-amplitude, often basement-cored, ramp anticlines (Jordan and Alonso 1987; Strecker et al. 1989; Riller et al. 1999) that formed by displacement on reverse faults affecting pre-Cenozoic basement rocks (Allmendinger and Zapata 1996; Gangui 1998). The curvature of fault termini of the NE-SW system towards parallelism with the orogen-parallel system indicates that the two systems are geometrically and kinematically linked. The consequence of such localization of upper-crustal deformation is a segmentation of the upper crust into domains which are rhombic in plan view and coincide with topographic depressions such as the Atacama and Arizaro Basins (see also Riller and Oncken 2003). Owing to upper-crustal segmentation, the salar areas escaped pervasive deformation and, therefore, remained largely free of fault-controlled volcanic centers (see next section). This, in turn, may account for the enhanced rigidity of the crust underlying the Salar de Atacama as observed by seismic tomography (Schurr and Rietbrock 2004; Reutter et al. 2006, Chap. 14 of this volume).

The isostatic residual gravity field and its dip curvature in the Puna region are shown in Fig. 15.4b,c, respectively. The pattern of dip curvature is dominated by well-defined zones of maximum curvature trending NE-SW (bright colors in Fig. 15.4c), which are particularly prominent north and south of the Salar de Atacama, south of the Salar de Arizaro, at the Salar de Antofalla and south of the Cerro Galan caldera. In contrast, curvature maxima that trend parallel to the orogen are evident in the eastern portion of the study area, the Eastern Cordillera. Although less prominent, NW-SE-trending zones of curvature maxima are also apparent. A particularly prominent example of these is the maximum curvature associated with the Olacapato-El Toro fault system next to the Aguas Calientes caldera and north of the Salar de Atacama (Fig. 15.4c,d). Overall, the pattern of dip curvature corresponds well to that of prominent, upper-crustal structures in the Puna (Fig. 15.4d).

It must be emphasized that the presence of curvature maxima depends to some degree on the distribution and station density of primary gravity measurements (Fig. 15.4c). More specifically, curvature maxima are more pronounced in areas with a high spatial density of measurements. Conversely, the presence of large crescent-shaped curvature maxima in regions devoid of gravity measurements, e.g., the areas north of the La Pacana caldera and east of the Cerro Galan caldera, should be regarded with caution as these may be artefacts. Nonetheless, in areas where both well-developed maxima of dip curvature and prominent faults are obvious, there is a close correspondence between the two.

Although the correspondence between zones of maximum dip curvature of the residual gravity field and prominent upper-crustal faults is more striking in the southern Altiplano, i.e., on the local scale, it is also apparent regionally in the Puna. As demonstrated for the Salar de Uyuni area, this may permit the detection of shallow

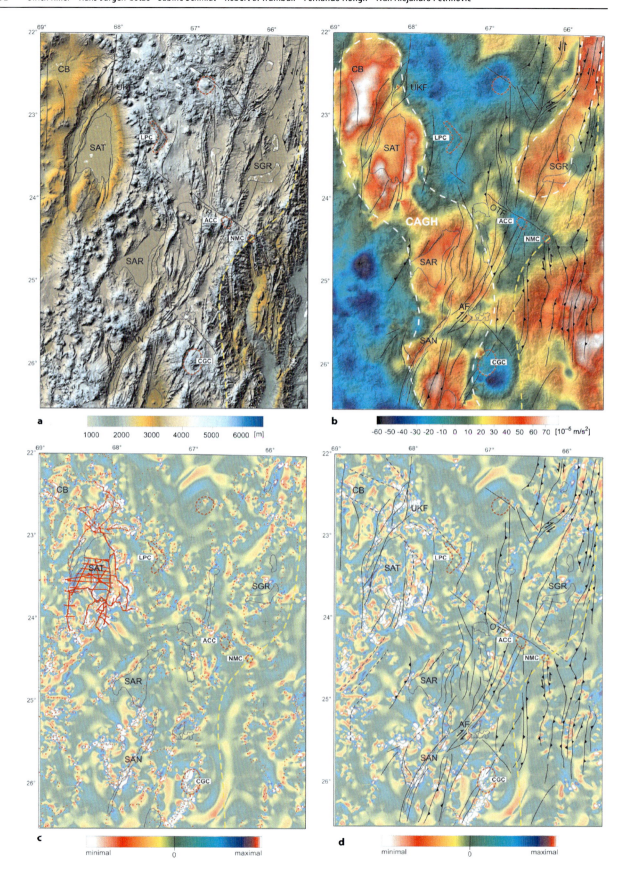

**Fig. 15.4.** Maps showing the spatial relationships between topography, first-order upper-crustal structures, isostatic residual gravity and its dip curvature of the Puna Plateau and adjacent morphotectonic units. *Stippled yellow line* denotes the boundary between the Puna Plateau and the Eastern Cordillera. *CB*: Calama Basin, *SAT*: Salar de Atacama, *SAR*: Salar de Arizaro, *SAN*: Salar de Antofalla, *SGR*: Salinas Grandes, *AF*: Acazoque fault, *UKF*: Uyuni-Khenayani fault, *OTF*: Olacapato – El Toro fault zone, *LPC*: LaPacana Caldera, *ACC*: Aguas Calientes Caldera, *NMC*: Negra Muerta Caldera, *CGC*: Cerro Galan Caldera. **a** Upper-crustal structures superimposed on a shuttle radar topography mission model. **b** Isostatic residual gravity field delineating the Central Andean Gravity High (CAGH: Götze and Krause 2002). The NW-trending gravity high is morphologically characterized by topographic depressions. Note the prominent NE-SW-trending and orogen-parallel structural discontinuities separating the CAGH into individual domains, elliptical in plan view. **c** Location of gravity measurement stations (*red dots*) superimposed on the dip curvature of the isostatic residual gravity. **d** Upper-crustal structures superimposed on the dip curvature of the isostatic residual gravity. NE- and NW-trending zones of maximum dip curvature delineated by stippled lines are well apparent south and north of the Salar de Atacama, respectively

crustal structures obscured beneath sedimentary or volcanic cover. For example, strong NE-SW-trending curvature maxima are evident to the south of the Salar de Atacama (Fig. 15.4c,d). Corresponding faults or lineaments that might explain the linear curvature maxima are, however, not apparent at surface. Since this area is covered by young volcanic rocks of the magmatic arc, it is conceivable that buried faults (stippled lines in Fig. 15.4d) are present and, furthermore, that the distribution of volcanic centers in this area may be controlled by NE-SW-striking, upper-crustal discontinuities.

Similarly, there are distinct NW-SE-trending zones of maximum dip curvature north of the Salar de Atacama. These are subparallel to the northwestward prolongation of maximum curvature associated with the Olacapato-El Toro fault zone but do not seem to be marked by faults at surface. However, the zones may indicate the presence of upper-crustal faults obscured by volcanic cover. We speculate that NE-SW-striking, upper-crustal discontinuities to the south of the Salar de Atacama, and NW-SE-striking discontinuities to the north have localized volcanic activity and may be the cause for the conspicuous deflection of the magmatic arc around the eastern portion of the salar (see also Reutter et al. 2006, Chap. 14 of this volume).

## 15.5 Morpho-Tectonic and Lithological Characteristics of the Cagh

As mentioned above, the CAGH is the most conspicuous anomaly of the isostatic residual gravity field in the Puna (Götze and Krause 2002). It is segmented into at least four individual domains, each of which has elliptical geometry in plan view and corresponds to the rhombic topographic depressions described above (Figs. 15.1b and 15.4a,b). From SE to NW, these depressions are: a topographic low southwest of the Cerro Galan caldera, the Salar de Arizaro, the Salar de Atacama and the Calama Basin within the Cordillera de Domeyko. The isostatic residual gravity in individual domains of the CAGH reaches values of the order of $60 \times 10^{-5}$ m s$^{-2}$.

The rhombic domains are separated by narrow, NE-SW-trending zones of lower gravity. In some cases, there is surface evidence that links these intermittent gravity lows to NE-SW-striking faults, notably the Acazoque fault and the southern strand of the Uyuni-Khenayani fault between the Salar de Atacama and Domeyko Cordillera domains (Fig. 15.4a,b). Between the Salar de Atacama and Salar de Arizaro domains, prominent NE-SW-striking faults are not evident at surface, but are suggested by the presence of the two NE-trending curvature maxima in this area (Fig. 15.4c,d). Here, the active volcanic arc constricts to less than 50 km width and departs from its overall N-S trend to a NE-SW trend, similar to that of the Acazoque and Uyuni-Khenayani faults. Similarly, north-trending curvature maxima border the topographic depression southwest of the Cerro Galan caldera, whereby the western part matches a prominent reverse fault segment of the Acazoque fault (Fig. 15.4c,d).

The CAGH is confined to the east and west by orogen-parallel faults which are also evident by the pattern of curvature maxima, e.g., on either side of the Calama Basin, the Salar de Atacama and the Salar de Arizaro, respectively (Fig. 15.4a–c). Regionally, these faults merge with NE-SW-striking faults to form an anastomosing pattern that envelops largely undeformed and topographically lower domains (Riller and Oncken 2003). Although less well expressed than the CAGH, the gravity high to the east of the CAGH is also spatially associated with topographic depressions such as the Salinas Grandes Basin (Figs. 15.1b and 15.4b). This NNE-SSW-trending gravity anomaly has been correlated with the Faja Eruptiva Oriental (Omarini et al. 1999) and is related to the same morpho-tectonic features as the individual domains of the CAGH (Fig. 15.4b,d). Thus, gravity highs in the Puna Plateau are associated mainly with fault-bound topographic depressions.

In contrast, the broad areas of low gravity in the south-central Andes (–30 to –50 × 10$^{-5}$ m s$^{-2}$) correspond well with the location of the Miocene to Recent magmatic arc and the post-10 Ma, transverse volcanic belts (Figs. 15.1b and 15.4b). There is also a good correspondence between areas of low gravity and the distribution of collapse calderas, such as those of the Altiplano Puna Volcanic Complex (APVC, Fig. 15.1b), Cerro Galan and Aguas Calientes (Figs. 15.1b and 15.4b). These collapse calderas are the eruptive sources for extremely large volumes of felsic ignimbrites, which have been shown to be derived dominantly from crustal melts (e.g., de Silva 1989). Lindsay et al. (2001) estimated that the magma chambers beneath the largest Central Andean caldera complex, La Pacana, resided at about 5–8 km depth. Seismic and magnetotelluric evidence points to still-active zones of partial melting beneath the APVC (Schilling et al. 2006, Chap. 22 of this volume, and references therein). Thus, it is likely that much of the crust underlying the late Cenozoic volcanic centers and their erupted material is felsic and weak, consequently, of low density and low rigidity, which accounts well for the observed negative gravity anomalies.

Based on the composition of Paleozoic basement rocks and of lower crustal xenoliths contained in Cretaceous and Cenozoic volcanic rocks, Lucassen et al. (2001) argued that much of the crust underlying the south-central Andes is felsic and supported this with gravity models. There is evidence for local presence of higher-density basement rocks below the CAGH, notably at its northwestern terminus (Lucassen et al. 1999; Schmitz et al. 1999; Götze and Krause 2002), but the combined evidence from basement composition, ignimbrite source constraints, and inferences from seismic studies (Zandt et al. 1994; Yuan et al. 2002), all but rule out large volumes of mafic rocks underlying the CAGH. An alternative explanation, based on the distribution of ignimbrite centers and inferred areas of partial melting in the crust, is that the CAGH domains reflect areas relatively lacking in felsic volcanism compared to their surroundings, and not to the presence of particularly mafic rocks.

The general lack of felsic volcanism in the CAGH domains underlying the topographic depressions could be explained in terms of localization of upper-crustal deformation. In the south-central Andes volcanism is partly fault-controlled, particularly along the transverse volcanic belts which follow prominent NW-SE striking lineaments. In contrast, the topographic depressions are less affected by upper-crustal deformation than adjacent regions (Fig. 15.4a) and are also largely devoid of felsic volcanic rocks (Fig. 15.1a). Thus, the field of isostatic residual gravity in the south-central Andes may portray several phenonema that are interrelated – the distribution of late Cenozoic volcanic rocks, the felsic character of the crust, and the variable degree of upper crustal fracturing.

## 15.6 Conclusions

Analysis and interpretation of the gravity field in the southern Altiplano and the Puna Plateau regions can be greatly enhanced by mapping the dip curvature of the isostatic residual gravity field. Comparison of the dip curvature maps with the regional geology shows that the variation in the gravity field is largely affected by the distribution of the late Cenozoic volcanic arc and back-arc ignimbrite fields and by first-order deformational structures in the upper crust.

Our example from the Salar de Uyuni area, southern Altiplano, demonstrates that dip curvature analysis is effective in revealing upper-crustal structures, even where buried by post-tectonic sedimentary or volcanic cover. Maps of the dip curvature in the Puna region show that elongate zones of maximum dip curvature correspond to prominent Neogene faults, forming a pattern characterized by NE-SW-, N-S- and NW-SE-striking fault zones.

The localization of upper-crustal deformation led to the formation of domains with rhombic geometry in map view, which generally correspond to internally-drained basins, notably the salars of the NNW-SSE-trending Central Andean Gravity High (CAGH). Previous interpretations attributed the CAGH to domains of relatively mafic crust. We suggest an alternative explanation that the CAGH constitutes a late Cenozoic feature and that the domains of relatively high gravity reflect an absence of felsic volcanism and crustal melting relative to their surroundings.

Dip curvature analysis suggests the presence of buried, upper-crustal discontinuities bounding the eastern margin of the Salar de Atacama to the south and to the north. These discontinuities may well represent prominent fault zones that localized Cenozoic volcanic activity and, thus, provide an explanation for the conspicuous angular departure of the volcanic arc from its overall meridional trend in the south-central Andes.

Our study suggests that dip curvature analysis of isostatic residual gravity bears great potential for elucidating first-order structural elements of deformed upper-crustal terranes such as the modern Andes.

## Acknowledgments

This work was funded by the German Science Foundation as part of the Collaborative Research Center 267 "Deformation Processes in the Andes". Insightful reviews of the manuscript by John Dehls and Jonas Kley are greatly appreciated. We are indebted to Allison Britt for her editorial and linguistic efforts in improving the quality of the text.

## References

Allmendinger RW, Zapata TR (1996) Imaging the Andean structure of the Eastern Cordillera on reprocessed YPF seismic reflection data. XIII Congreso Geológico Argentino, Actas 2:125–134

Allmendinger RW, Ramos VA, Jordan TE, Palma M, Isacks BL (1983) Paleogeography and Andean structural geometry, northwestern Argentina. Tectonics 2:1–16

Allmendinger RW, Jordan TE, Kay SM, Isacks BL (1997) The evolution of the Altiplano-Puna Plateau of the Central Andes. Ann Rev Earth Planet Sci 25:139–174

Caffe PJ, Trumbull RB, Coira BL, Romer RL (2002) Petrogenesis of early volcanic phases in Northern Puna Cenozoic magmatism. Implications for magma genesis and crustal processes in the Central Andean Plateau. J Petrol 43:907–942

Chernicoff CJ, Richards JP, Zappettini EO (2002) Crustal lineament control on magmatism and mineralization in northwestern Argentina: geological, geophysical and remote sensing evidence. Ore Geol Rev 21:127–155

De Silva SL (1989) Altiplano-Puna volcanic complex of the central Andes. Geology 17:1102–1106

Elger K (2003) Analysis of deformation and tectonic history of the Southern Altiplano Plateau (Bolivia) and their importance for plateau formation. GeoForschungsZentrum Potsdam Scientific Technical Report STR03/07

Elger K, Oncken O, Glodny J (2005) Plateau-style accumulation of deformation: Southern Altiplano. Tectonics 24: doi 10.1029/2004TC001675

Gangui AH (1998) A combined structural interpretation based on seismic data and 3-D gravity modelling in the northern Puna / Eastern Cordillera, Argentina. PhD thesis, Freie Universität Berlin

Gaupp R, Kött A, Wörner G (1999) Paleoclimatic implications of Mio-Pliocene sedimentation in the high-altitude intra-arc Lauca Basin of northern Chile. Paleogeog Paleoclimat Paleoecol 151:79–100

Götze H-J, Kirchner A (1997) Interpretation of gravity and geoid in the Central Andes between 20° and 29°S. J S Am Earth Sci 10:179–188

Götze H-J, Krause S (2002) The Central Andean Gravity High, a relic of an old subduction complex? J S Am Earth Sci 14:799–811

Götze H-J, Lahmeyer B, Schmidt S, Strunk S (1994) The lithospheric structure of the Central Andes (20–26°S) as inferred from interpretation of regional gravity. In: Reutter K-J, Scheuber E, Wigger P (eds) Tectonics of the Southern Central Andes. Springer-Verlag, Berlin Heidelberg New York, pp 7–23

Grier ME, Salfity JA, Allmendinger RW (1991) Andean reactivation of the Cretaceous Salta rift, northwestern Argentina. J S Am Earth Sci 4:351–372

Gubbels T, Isacks B, Farrar E (1993) High-level surfaces, plateau uplift, and foreland development, Bolivian central Andes. Geology 21:695–698

Hesthammer J, Fossen H (1997) Seismic attribute analysis in structural interpretation of the Gullfaks Field, northern North Sea. Petroleum Geoscience 3:13–26

Isacks B (1988) Uplift of the central Andean plateau and bending of the Bolivian orocline. J Geophys Res 93:3211–3231

Jordan TE, Alonso RN (1987) Cenozoic stratigraphy and basin tectonics of the Andes Mountains, 20–28°S. Am Assoc Petr Geol Bull 71:49–64

Kley J (1999) Geologic and geometric constraints on a kinematic model of the Bolivian orocline. J S Am Earth Sci 12:221–235

Kley J, Monaldi CR (1998) Tectonic shortening and crustal thickness in the central Andes: how good is the correlation? Geology 26:723–726

Kraemer B, Adelmann D, Alten M, Schnurr W, Erpenstein K, Kiefer E, Van den Bogaard P, Görler K (1999) Incorporation of the Paleogene foreland into the Neogene Puna plateau: the Salar de Antofalla area, NW Argentina. J S Am Earth Sci 12:157–182

Lindsay JM, Schmitt AK, Trumbull RB, De Silva SL, Siebel W, Emmermann R (2001) Magmatic evolution of the La Pacana Caldera system, central Andes, Chile: compositional variation of two cogenetic, large-volume felsic ignimbrites and implications for contrasting eruption mechanisms. J Petrol 42:459–486

Lucassen F, Franz G, Laber A (1999) Permian high-pressure rocks – the basement of the Sierra de Limón Verde in Northern Chile. J S Am Earth Sci 12:183–199

Lucassen F, Becchio R, Harmon R, Kasemann S, Franz G, Trumbull R, Romer RL, Dulski P (2001) Composition and density model of the continental crust in an active continental margin – the Central Andes between 18° and 27°S. Tectonophysics 341:195–223

Lyon-Caen H, Molnar P, Suárez G (1985) Gravity anomalies and flexure of the Brazilian Shield beneath the Bolivian Andes. Earth Planet Sci Lett 75:81–92

Omarini RH, Sureda RJ, Götze H-J, Seilacher A, Pflüger F (1999) Puncoviscana fold belt in Northwestern Argentina: testimony of Late Proterozoic Rodinia fragmentation and pre-Gondwana collisional episodes. Int J Earth Sci 88:76–97

Oncken O, Hindle D, Kley J, Elger K, Victor P, Schemmann K (2006) Deformation of the central Andean upper plate system – facts, fiction, and constraints for plateau models. In: Oncken O, Chong G, Franz G, Giese P, Götze H-J, Ramos VA, Strecker MR, Wigger P (eds) The Andes – active subduction orogeny. Frontiers in Earth Science Series, Vol 1. Springer-Verlag, Berlin Heidelberg New York, pp 3–28, this volume

Petrinovic IA, Riller U, Brod JA (2005) The Negra Muerta Volcanic Complex, southern Central Andes: geochemical characteristics and magmatic evolution of an episodically active volcano-tectonic complex. J Volcanol Geotherm Res 140:295–320

Ramelow J, Riller U, Romer RL, Oncken O (2006) Kinematic link between episodic caldera collapse of the Negra Muerta Collapse Caldera and motion on the Olacapato – El Toro Fault Zone, NW-Argentina. Inter J Earth Sci 95:529-541. doi 10.1007/s00531-005-0042-x

Reutter K-J, Charrier R, Götze H-J, Schurr B, Wigger P, Scheuber E, Giese P, Reuther C-D, Schmidt S, Rietbrock A, Chong G, Belmonte-Pool A (2006) The Salar de Atacama Basin: a subsiding block within the western edge of the Altiplano-Puna Plateau. In: Oncken O, Chong G, Franz G, Giese P, Götze H-J, Ramos VA, Strecker MR, Wigger P (eds) The Andes – active subduction orogeny. Frontiers in Earth Science Series, Vol 1. Springer-Verlag, Berlin Heidelberg New York, pp 303–326, this volume

Riller U, Oncken O (2003) Growth of the central Andean Plateau by tectonic segmentation is controlled by the gradient in crustal shortening. J Geol 111:367–384

Riller U, Greskowiak J, Ramelow J, Strecker M (1999) Dominant modes of Andean deformation in the Calchaquí River Valley, NW-Argentina. XIV Argentine Geological Congress, Actas I, Salta, pp 201–204

Riller U, Petrinovic I, Ramelow J, Strecker M, Oncken O (2001) Late Cenozoic tectonism, caldera and plateau formation in the central Andes. Earth Planet Sci Lett 188:299–311

Roberts A (2001) Curvature attributes and their application to 3D interpreted horizons. First Break 19:85–99

Romanyuk TV, Götze H-J, Halvorson PF (1999) A density model of the Andean subduction zone. The Leading Edge 18:264–268

Scheuber E, Mertmann D, Ege H, Silva-González P, Heubeck C, Reutter K-J, Jacobshagen V (2006) Exhumation and basin development related to formation of the Central Andean Plateau, 21° S. In: Oncken O, Chong G, Franz G, Giese P, Götze H-J, Ramos VA, Strecker MR, Wigger P (eds) The Andes – active subduction orogeny. Frontiers in Earth Science Series, Vol 1. Springer-Verlag, Berlin Heidelberg New York, pp 285–302, this volume

Schilling FR, Trumbull RB, Brasse H, Haberland C, Asch G, Bruhn D, Mai K, Haak V, Giese P, Muñoz M, Ramelow J, Rietbrock A, Ricaldi E, Vietor T (2006) Partial melting in the Central Andean crust: a review of geophysical, petrophysical, and petrologic evidence. In: Oncken O, Chong G, Franz G, Giese P, Götze H-J, Ramos VA, Strecker MR, Wigger P (eds) The Andes – active subduction orogeny. Frontiers in Earth Science Series, Vol 1. Springer-Verlag, Berlin Heidelberg New York, pp 459–474, this volume

Schmidt S, Götze H-J (2003) Pre-interpretation of potential fields by the aid of curvature attributes. Geophys Res Abs 5:07689

Schmitz M, Lessel K, Giese P, Wigger P, Araneda M, Bribach J, Graeber F, Grunewald S, Haberland C, Lüth S, Röwer P, Ryberg T, Schulze A (1999) The crustal structure beneath the Central Andean forearc and magmatic arc as derived from seismic studies – the PISCO 94 experiment in northern Chile (21°–23°S). J S Am Earth Sci 12:237–260

Schurr B, Rietbrock A (2004) Deep seismic structure of the Atacama basin, northern Chile. Geophys Res Lett 31: doi 10.1029/2004GL019796

Sigismondi ME, Soldo JC (2003) Curvature attributes and seismic interpretation: case studies from Argentina basins. The Leading Edge 22:1122–1126

Sobel ER, Strecker MR (2003) Uplift, exhumation, and precipitation: tectonic and climatic control of Cenozoic landscape evolution in the northern Sierras Pampeanas, Argentina. Basin Res 15:431–451

Sobel ER, Hilley GE, Strecker MR (2003) Formation of internally-drained contractional basins by aridity-limited bedrock incision. J Geophys Res 108: doi 10.1029/2002JB001883,2003

Strecker MR, Cerveny P, Bloom AL, Malizia D (1989) Late Cenozoic tectonism and landscape development in the foreland of the Andes: Northern Sierras Pampeanas (26°–28°S). Tectonics 8:517–534

Tašárová Z (2004) Gravity data analysis and interdisciplinary 3D modelling of a convergent plate margin (Chile, 36–42°S). PhD thesis, Freie Universität Berlin, http://deposit.ddb.de/cgi_bin/dokserv?idn=973534729

Tassara A (2005) Interaction between the Nazca and South American plates and formation of the Altiplano-Puna plateau: review of a flexural analysis along the Andean margin (15°–34°S). Tectonophysics 399:39–57

Townsend C, Firth IR, Westerman R, Kirkevollen L, Hårde M, Andersen T (1998) Small seismic-scale fault identification and mapping. In: Jones G, Fisher QJ, Knipe RJ (eds) Faulting, fault sealing and fluid flow in hydrocarbon reservoirs. Geol Soc Spec Publication 147:1–25

Vandervoort DS, Jordan TE, Zeitler PK, Alonso RN (1995) Chronology of internal drainage and uplift, southern Puna plateau, Argentine central Andes. Geology 23:145–148

Viramonte JG, Petrinovic IA (1990) Cryptic and partially buried calderas along a strike-slip fault system in the central Andes. ISAG Grenoble, pp 317–320

Viramonte JG, Reynolds JH, Del Papa C, Disalvo A (1994) The Corte Blanco garnitiferous tuff: a distinctive late Miocene marker bed in Northwestern Argentina applied to magnetic polarity stratigraphy in the Rio Yacones, Salta Province, Earth Planet Sci Lett 121:519–531

Whitman D (1994) Moho geometry beneath the eastern margin of the Andes, northwestern Argentina, and its implications to the effective elastic thickness of the Andean foreland. J Geophys Res 99:15277–15289

Whitman D, Isacks B, Kay S (1996) Lithospheric structure and along-strike segmentation of the Central Andean Plateau: seismic Q, magmatism, flexure, topography and tectonics. Tectonophysics 259:29–40

Wigger P, Schmitz M, Araneda M, Asch G, Baldzuhn S, Giese P, Heinsohn WD, Martinez E, Ricaldi E, Röwer P, Viramonte J (1994) Variations of the crustal structure of the Southern Central Central Andes deduced from seismic refraction investigations. In: Reutter K-J, Scheuber E, Wigger P (eds) Tectonics of the Southern Central Andes. Springer-Verlag, Berlin Heidelberg New York, pp 23–48

Yuan X, Sobolev SV, Kind R, Oncken O, Bock G, Asch G, Schurr B, Graeber F, Rudloff A, Hanka W, Wylegalla K, Tibi R, Haberland C, Rietbrock A, Giese P, Wigger P, Röwer P, Zandt G, Beck S, Wallace T, Pardo M, Comte D (2000) Subduction and collision processes in the Central Andes constrained by converted seismic phases. Nature 408:958–961

Yuan X, Sobolev SV, Kind R (2002) Moho topography in the central Andes and its geodynamic implications. Earth Planet Sci Lett 199:389–402

Zandt G, Velasco AA, Beck SL (1994) Composition and thickness of the southern Altiplano crust, Bolivia. Geology 22:1003–1006

# Central and Southern Andean Tectonic Evolution Inferred from Arc Magmatism

Michael Haschke · Andreas Günther · Daniel Melnick · Helmut Echtler · Klaus-Joachim Reutter · Ekkehard Scheuber · Onno Oncken

**Abstract.** Patterns of spatial distribution, and geochemical and isotopic evolution from subduction-related igneous rocks provide tools for scaling, balancing and predicting orogenic processes and mechanisms. We discuss patterns from two Andean key arc segments, which developed into fundamentally different types of orogens: (1) A plateau-type orogen with thick crust in the central Andes, and (2) a non-plateau orogen with normal crust in the southern Andes.

*Northern Chile* (21–26° S) shows a collage of stepwise, eastward-migrating arc axes from 200 Ma to the Present. Each arc is characterized by a repeating sequence of magmatic-tectonic events: Magmatism for 30–40 million years with increasing REE fractionation (increasing La/Yb, La/Sm and Sm/Yb ratios); increasing crust-like initial Sr and Nd isotopes; early-stage, back-arc, alkaline magmatism; and late-stage tectonic activity and (mainly) crustal shortening followed by intra-arc strike-slip fault motion, followed by mineralization and magmatic quiescence for 5–12 million years, before the next main-arc evolved up to 100 km further east. Increasing REE fractionation and crust-like Sr and Nd isotopes correlate with crustal thickening by tectonic shortening and magmatic underplating, from 30–35 km (Jurassic) to 45 km (Eocene) to 70 km thick in the modern central Andes. Episodes of magmatic quiescence for 5–12 million years reflect episodes of flat subduction; the repeated nature of these episodes reflects dynamic subduction cycles including flat subduction, slab steepening, and slab breakoff.

*Southern Chile* (41–46° S) shows stationary arc magmatism from 200 to 50 Ma; followed by trench retreat and arc widening from 50–28 Ma; arc narrowing from 28–8 Ma; and magmatic quiescence from 8–3 Ma. Volcanism from the Pliocene to the Present was concentrated in a narrow volcanic arc. Moderate crustal shortening occurred from 70–55 Ma (mainly back-arc) and ~9–8 Ma (intra-arc). REE fractionation patterns (low and constant La/Yb ratios) are similar to those of Jurassic rocks from northern Chile, consistent with crustal thicknesses of 30–35 km. Initial Sr and Nd isotopes between 200 and 20 Ma evolved from crust-like to mantle-like ratios, with a reversal at 20 Ma to more diffuse and crust-like ratios. This pattern can be related to successive isotopic shielding and/or asthenospheric depletion (200–20 Ma), and increasing crustal assimilation (20 Ma to Recent) due to moderate crustal thickening. Similarly to northern Chile, magmatic quiescence from 8–3 Ma may reflect an episode of flat subduction.

The cyclicity of magmatic, isotopic, and tectonic features in northern Chile suggests that rheologic weakening of the lithosphere plays an important role. Shared magmatic-tectonic features of paleo-arcs in northern Chile and regions of subhorizontal in southern Chile suggests that flat slab episodes may be a typical feature of Andean-type margins.

## 16.1 Introduction

Among the bulk global occurrence of subduction zones, the Andean convergent margin is the most commonly referenced classic type of orogen induced by non-collisional subduction of oceanic beneath continental lithosphere. Yet, plate convergence and subduction-induced magmatism since at least 200 Ma generated two fundamentally different types of orogens along western South America: (1) a plateau-orogen with anomalous thick orogenic crust in the Central Andes, and (2) a non-plateau orogen in the Southern Andes with thinner crust. Both arc systems show evidence of contemporaneous igneous activity for at least 200 Myr, yet it is unclear why they developed into fundamentally different arc orogens. Although much previous work focused on the Central Andean orogenic arc, few hints exist on which segment the evolution of the Andean orogen *is* typical, and which one is not.

We address this issue by exploring long-term evolutionary tectonic, magmatic, geochemical and isotopic patterns of two contrasting Andean arc segments (Fig. 16.1): (1) the foreland and modern volcanic arc of the Central Andes in north Chile (21–26° S), and (2) the Southern Volcanic Zone (SVZ) in south Chile (41–46° S).

The north Chilean arc segment is a wide mature volcanic arc system with shortened, thick orogenic crust (up to 70 km, Wigger et al. 1994), whereas the south Chilean arc segment is a narrow segment with relatively normal crustal thicknesses (30–40 km, Hildreth and Moorbath 1988, Yuan et al., in review). Spatial and temporal changes in evolutionary magmatic, geochemical and isotopic patterns during this time presumably recorded changes in both the geometry of the subducting slab and the tectonic conditions in the overriding plate. Hence they provide ideal opportunities to explore the cause and effect relationships of tectonics and magmatism along this classic non-collisional convergent margin. Moreover, these patterns offer efficient yet underestimated tools for scaling and balancing orogenic and magmatic processes, including subduction erosion (Scholl et al. 1980; von Huene and Scholl 1991), crustal thickening (e.g. Hildreth and Moorbath 1988; McMillan et al. 1989; Kay et al. 1991; Kay and Mpodozis 2001; Haschke et al. 2002a, 2002b), changing slab geometry (Kay et al. 1987, 1994; Haschke et al. 2002b) and crustal growth (Parada et al. 1999; Pankhurst et al. 1999; Haschke and Günther 2003). We argue that such patterns hold important constraints on the discussion of

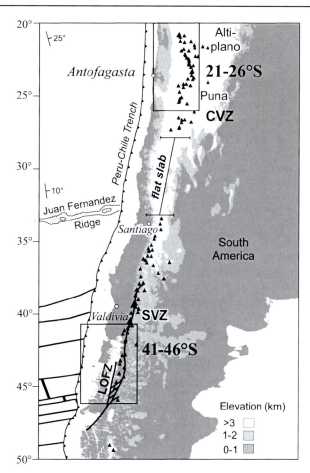

**Fig. 16.1.** Distribution of present-day volcanism (*triangles*) in the Andes between 20° and 50° S. Topography is shaded at 1 000 and 3 000 m elevation. Obliquity is relative to the Peru-Chile trench axis after Somoza (1998). *Rectangles* indicate studied arc segments in north and south Chile

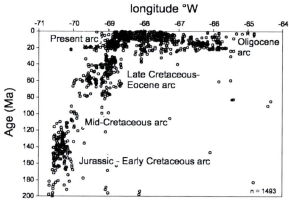

**Fig. 16.2.** Longitude °W versus radiometric ages of rocks from the north Chilean arc segment showing stepwise eastward migration of magmatic units

the causes and mechanisms that are responsible for generating either a plateau or non-plateau orogen. For comparison, we selected two representative arc segments with high geochemical rock data density and sufficient tectonic constraints to derive a base for interpretation of driving forces.

For the north Chilean arc segment (21–26° S), we compiled major and trace element, and Sr-Nd isotopic data (649 samples) from basaltic andesites to dacites and igneous equivalents with radiometric ages from 200 Ma to Recent, including new data from the Jurassic arc by Bartsch (2004). Primitive basalts and rhyolites and/or its intrusive equivalents are rare, which is diagnostic of continental arcs. The complete database for north Chile is available upon request.

For south Chile (41–46° S), we compiled data from 125 samples with ages of the same time frame comprising late Jurassic to late Cenozoic rocks (70 samples) from Pankhurst et al. (1999), Eocene-Miocene rocks (12 samples) from Lopez-Escobar and Vergara (1997), Upper Oligocene-Miocene rocks (28 samples) from Munoz et al. (2000), and Pliocene-Pleistocene rocks (15 samples) from Lara et al. (2001). Our compilation includes geochemical and geochronological constraints from Pankhurst et al. (1992, 42–42°30' S) and Pankhurst and Hervé (1994, 44–47° S).

## 16.2 Tectonic Evolution: North vs. South Chile

North Chile shows the most complete picture of Andean igneous history to date. Magmatic activity between 21° and 26° S can be separated into a geochemically and isotopically distinct, eastward-younging collage of four largely parallel north-south-trending magmatic arcs (Scheuber and Reutter 1992; Haschke et al. 2002a; Fig. 16.2):

1. the Jurassic arc along the present Coastal Range (195–130 Ma),
2. the mid-Cretaceous arc (125–90 Ma), aligning with the Longitudinal Valley,
3. the late Cretaceous-Eocene arc (78–37 Ma), largely situated in the Chilean Precordillera, and
4. the Neogene and modern Western Cordillera volcanic arc (26 Ma–Recent).

Each magmatic episode is spatially separated by up to 100 km wide gaps with few or no igneous arc rocks, and temporally separated by 5–12 million years of relative magmatic quiescence (130–125 Ma, 90–78 Ma, and 37–26 Ma, Haschke et al. 2002b). The present-day lack of a Jurassic fore-arc in the Coastal Cordillera attest to a continuous landward retreat of the trench associated with subduction erosion in the order of 150–200 km (Scholl et al. 1980; von Huene and Scholl 1991).

Most magmatic activity between 200 Ma and the Present occurred during tectonic extension to transtension (Scheuber et al. 1994; Charrier and Reutter 1994; Scheuber and Gonzales 1999) as indicated by large-scale

**Fig. 16.3.**
Position of the main arc (*thick black line*) in the north Chilean arc segment at: (**a**) 195–130 Ma, (**b**) 125–90 Ma, (**c**) 78–36 Ma, and (**d**) 26–Present, with position of major intra-arc strike-slip fault zones (see text for references)

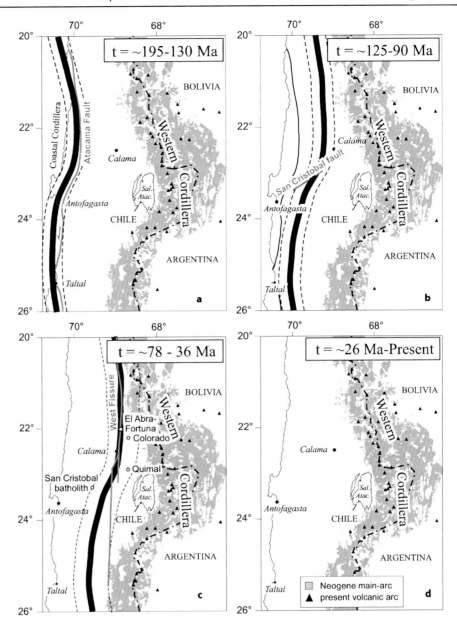

dyke injections in the Jurassic arc, intra-arc basins with deposits > 3 000 m thickness in the mid-Cretaceous and Cretaceous-Eocene arcs, and local extension-related alkaline magmatism.

Evidence for transpressive crustal shortening exists in the mid-Cretaceous arc (~100 Ma Peruvian; Megard 1987; Munoz et al. 1989; Charrier and Reutter 1994; Arancibia 2004), the Cretaceous-Eocene arc (45–38 Ma Incaic, Döbel et al. 1992; Haschke and Günther 2003) and the modern Western Cordillera (26–6 Ma, Allmendinger 1986; Isacks 1988). Episodes of contractional deformation tend to be concentrated during late-stage main-arc magmatism. In the Neogene arc and back-arc, large-scale, mid to late Miocene shortening (300 km, Allmendinger et al. 1997) occurred in an already thickened crust (45 km, Haschke and Günther 2003) and produced the over-thickened orogenic crust with the Altiplano-Puna Plateau.

All Andean transtensional and transpressional tectonic episodes were followed by major intra-arc strike-slip faulting (125 Ma Atacama Fault Zone: 450 km dextral, St. Amand and Allen 1960; min. 100 km sinistral, Scheuber and Reutter 1992; Scheuber et al. 1994; Scheuber and Gonzales 1999; 85 Ma San Cristobal Fault Zone: unknown offset; 32–20 Ma West Fissure: 2–10 km dextral and 35–37 km sinistral, Reutter et al. 1991; Charrier and Reutter 1994; Reutter et al. 1996; Tomlinson and Blanco 1997; Fig. 16.3a–d), which accommodated margin-parallel components of oblique convergence by lateral displacement at, or near the rheologically-weakened arc axis. Tectonic activity along these faults occurred contemporaneously with the cessation of arc activity. The resulting episodes of magmatic quiescence can be related to: (*1*) a decrease in the angle of subduction (Barazangi and Isacks 1976; Cahill and Isacks 1992; Kay et al. 1999), (*2*) discontinuous eastward migration of the Chile trench due to tectonic erosion of the continental margin, and (*3*) cooling of the mantle wedge overlying the subducted plate, resulting in a deepening zone of magma generation below the volcanic front. As discussed below, it is likely that these three processes occurred and were linked within dynamic subduction cycles.

## 16.2.1 South Chile

Is the magmatic and tectonic evolution of the arc segment in southern Chile compatible with that seen in northern Chile? The tectono-magmatic history of southern Chile (41–46° S) is less well constrained, and debate centers on some different yet interrelated issues:

1. Was main-arc magmatism between 200 Ma and Present stationary in southern Chile, and what are the implications for the geometry of the subducting slab during this time?
2. Did the arc continental crust in southern Chile thicken, and, if so, how thick was the crust, and how much can be attributed to basaltic underplating?

The southern Andean margin lacks evidence of crustal shortening between 200 Ma and 90 Ma, and little spatial variation in the distribution of main-arc rocks during this time suggest a largely stationary arc. Between 90 and 70 Ma, the volcanic arc started to widen by trench-retreat into a broad volcanic zone (from 150 km to 350 km width), followed mainly by crustal shortening in the back-arc (70–55 Ma, Neuquen Basin; Ramos 1989; Melnick et al. 2004, Ramos 2005). From 55–11 Ma the main-arc narrowed (Fig. 16.4), and from 11–6 Ma, the previously wide magmatic zone merged into a narrow volcanic chain (< 75 km width, Lavenu and Cembrano 1999; Fig. 16.5).

Igneous activity in this narrowed arc developed into ignimbrite volcanism by 9–8 Ma, coeval with some shortening and thickening of the arc crust, tectonic inversion and uplift of the former intra-arc basins in the Neuquen region (Melnick et al. 2004), which generated the Main Cordillera. Intra-arc deformation between the Miocene and the Present (42–46° S) can be divided into two tectonic events: (*1*) transpressive (dextral-oblique) deformation at 5.4–1.6 Ma (as early as 13.3 Ma at 44°29' S; Munizaga et al. 1988; Pankhurst and Herve 1994; Lavenu and Cembrano 1999), and (*2*) NE-SW shortening at 1.6 Ma (Lavenu and Cembrano 1999), coeval with magmatism along the Liquine-Ofqui Fault Zone since 5 Ma (Fig. 16.5).

### 16.2.1.1 Plate Convergence Parameters

Most previous workers attributed the changing tectonic settings in north and south Chile with changing conver-

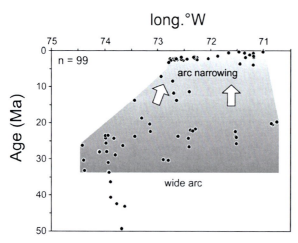

**Fig. 16.4.** Western longitude (corrected for 10° N strike of the arc) versus radiometric ages of rocks from the south Chilean arc segment (41–46° S) showing ~150 km arc narrowing between 50 and 5 Ma

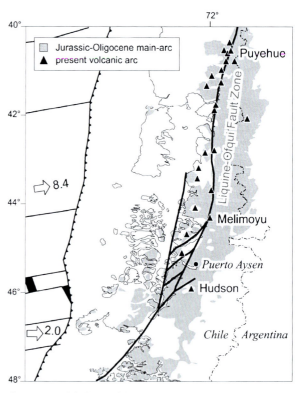

**Fig. 16.5.** Simplified map of the south Chilean arc segment (40–48° S) showing the wide Neogene arc (*gray shaded*), the present-day distribution of active volcanoes in the modern Southern Volcanic Zone (SVZ, *black triangles*) and the major Liquine-Ofqui intra-arc fault zone (*black line*) in relation to the position of the trench (*line with tick marks*). *Arrows with numbers* indicate change in convergence rate after subduction of the Chile Rise (after Cande et al. 1986)

gence parameters through time. Cretaceous-Tertiary plate convergence parameters for north and south Chile were established by Pilger (1984), Pardo-Casas and Molnar (1987), Tebbens and Cande (1997), and Somoza (1998). Pre-Cretaceous convergence parameters are avoided because the evidence for plate motion is more obscure, and reconstructions increasingly conjectural the farther back we look in time. Reconstructions of Cretaceous to modern convergence parameters indicate low convergence rates ($< 5$ cm yr$^{-1}$) and highly oblique subduction from 78–49 Ma (Pardo-Casas and Molnar 1987, Somoza 1998), correlating with extensional tectonics in both north and south Chile (Fig. 16.6).

In north Chile, convergence is nearly normal to the arc and the convergence rate between 49 and 36 Ma more than doubled, accompanied by transpression, and crustal shortening and thickening (Fig. 16.6). South Chile lacks any indication of crustal shortening during this time. After an episode with low convergence rates (6 cm yr$^{-1}$) and high obliquity (55°) between 36 and 28 Ma, convergence rates nearly tripled between 28 and 26 Ma in north Chile and approximately doubled in south Chile, and the angle of convergence became nearly orthogonal to the continental margin (Fig. 16.6). The increase in convergence rates from 28–26 Ma is a general feature of the western South American plate margin south of the bend at 18° S (Pilger 1984; Pardo-Casas and Molnar 1987).

One possible explanation for the rough correlation of Neogene contractional deformation between north and south Chile may be the coeval increase in westward motion of continental South America since the Eocene (Silver et al. 1998). Although the absolute ages of any changes in convergence parameters are only as accurate as the uncertainties in seafloor spreading history, the overall pattern is clear. It suggests that convergence was much slower prior to, and much faster after, 28 Ma (Somoza 1998; Silver et al. 1998), and possibly due to enhanced mechanical coupling between the plates.

The changes in convergence rates cannot be attributed to changes in age, composition and/or buoyancy of the subducting oceanic lithosphere at 28–26 Ma (Jordan et al. 2001). The buoyancy of the slab might have changed as the slab descended, resulting in rock flux through gabbro-eclogite and olivine-spinel transitions (Cloos 1993), yet it would take millions of years to reach a state of equilibrium within the slab and resisting asthenosphere (see discussion in Jordan et al. 2001). Meanwhile, the asthenosphere's resistance to a more rapidly subducting, and hence more buoyant slab, together with an accelerated horizontal component of subduction, would induce slab bending and kinking, and hence subhorizontal subduction (Haschke et al. 2002b).

Another possible mechanism for changing slab buoyancy is the subduction of an oceanic plateau and/or ridge

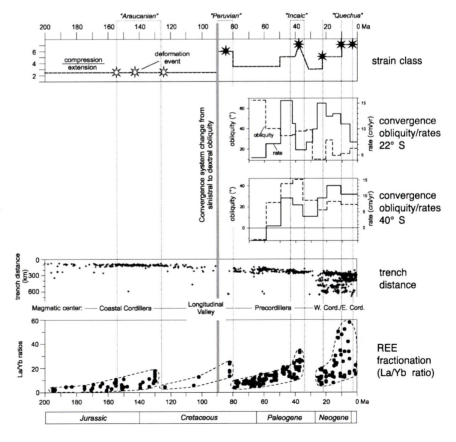

**Fig. 16.6.**
Compilation of magmatic-tectonic convergence parameters for the Andean margin in north (22° S) and south Chile (40° S, Günther 2001) showing: changing strain classes (Jaillard 1986; Scheuber et al. 1994); relative convergence rates and obliquity; the ratio of arc-normal/parallel vectors (Pardo-Casas and Molnar 1987, Somoza 1998, Silver et al. 1998); spatial-temporal distribution patterns of Andean subduction-related igneous rocks; and evolutionary REE patterns of arc rocks from north Chile

(Gutscher et al. 2000), as the basalt/gabbro to eclogite transformation is delayed when subducting thickened, young to middle-aged oceanic crust. If flat subduction of an oceanic plateau caused the central Andean Eocene-Oligocene amagmatic episode (38–26 Ma), it must date prior to the Juan Fernandez Ridge, subducting since 22 Ma (Yanez et al. 2002), and prior to the Nazca Ridge and the postulated "Inca Plateau", both subducting since 12–10 Ma (Gutscher et al. 2000). Given the observed systematic geochemical Andean patterns (see below), this would require repeated generation and subduction of oceanic plateaus large enough to cause sufficient buoyancy for flat subduction every 30–40 million years. This is difficult to explain and remains, at best, speculative, as the record of subducted oceanic features along the Pacific margin of South America is not well understood.

However, as the resolution of temporal and structural studies improves, evidence is increasing that there may be no direct link between increasing convergence rates and the onset of shortening in the upper plate. Recent results from south Chile (33–45° S) challenge this generally inferred correlation by showing moderate extensional tectonics and basin formation during increasing convergence rates (Jordan et al. 2001). McCaffrey (1997) showed that the correlation of convergence rate and plate coupling intensity (i.e. stress state, approximated by earthquake size) does not withstand statistical tests. Other authors have reported delays in shortening of 8–12 million years (Kley et al. 1997; Jordan et al. 1997; Horton 1998). Therefore, increasing convergence rates might not have directly triggered shortening in the upper plate, but they might have changed the slab geometry in the asthenospheric wedge causing flat subduction, which helped to drive contractional deformation.

### 16.2.1.2 Chemical Clues to Temporal Changes in Magmatic and Tectonic Processes

Rare Earth Elements (REE) provide diagnostic clues for scaling, balancing and predicting orogenic processes and mechanisms. REE fractionation (La/Yb, La/Sm and Sm/Yb ratios) provides information about the evolution of pressure-sensitive residual minerals and serves as a first-order guide for determining relative crustal thicknesses. These REE correlations assume that a melt is equilibrated with, and reflects, pressure-dependent changes ranging from clinopyroxene to amphibole to garnet transitions in the residual or source mineralogy (Kay and Kay 1991). In the boundary region between mantle and a hydrous mafic crust these changes include processes of melting, assimilation, storage and homogenization (Hildreth and Moorbath 1988).

As a rough guide for mafic source mineralogies, gabbroic (clinopyroxene-bearing) source mineralogies dominate at depths of 30–35 km, amphibolitic (amphibole-bearing) at ~40 km (Rapp and Watson 1995), and eclogitic (garnet-bearing) at $\geq 45$–50 km (or $\geq 12$–15 kbar; Rushmer 1993; Rapp and Watson 1995). These inferred depths are approximations as mineral breakdown pressures are influenced by bulk composition and temperature. Lower ratios of La/Yb ($< 20$), La/Sm ($<4$) and Sm/Yb ($< 3$) generally indicate clinopyroxene-dominated mineral residues, higher ratios (La/Yb $> 20$–30, La/Sm $> 4$, Sm/Yb 3–5) indicate middle REE retention (such as Sm) by amphibole, and high ratios (La/Yb $> 30$, Sm/Yb $> 5$) reflect heavy REE retention (such as Yb) by garnet in the residue (e.g. Van Westrenen et al. 1999; Barth et al. 2002). Garnet-bearing basaltic sources transform to eclogite at 15 kbar (50 km; Rapp and Watson 1995) and these melts retain residues with high La/Yb ($> 30$) and Sm/Yb ($> 5$) ratios.

Such a guide can be considered a rough approximation, as the choice of distribution coefficients is difficult and somewhat arbitrary. Recent studies on trace element partitioning in clinopyroxene, amphibole and garnet revealed a range of partly overlapping partition coefficients (KD), respectively, depending on melt temperature, pressure, and mineral composition. Furthermore, distinguishing between clinopyroxene and amphibole involvement may be difficult given the small differences in KD of REE such as Sm ($KD_{Sm}$ for clinopyroxene: 0.59–1.17, Barth et al. 2002; $KD_{Sm}$ for amphibole: 1.37–2.01, Klein et al. 1997), so that bowl-shaped REE patterns (as seen in Eocene arc rocks from north Chile, see Haschke et al. 2002a) may be more diagnostic of amphibole involvement. More important for our discussion, however, is the trace element partitioning of the heavy REE (such as Yb) in garnet as they are ~5 orders of magnitude higher than those of clinopyroxene and/or amphibole ($KD_{Yb}$ for garnet: 8.86–10, Barth et al. 2002), and hence much more diagnostic.

Melts migrating through thickened Andean continental crust also tend to show higher ('crust-like') initial $^{87}Sr/^{86}Sr$ and lower $^{143}Nd/^{144}Nd$ ratios relative to island-arc melts. In addition, isotope enrichment of Miocene to Recent rocks tends to correlate with increasing $SiO_2$ and decreasing Sr content, suggesting upper crustal contamination during low pressure differentiation. In contrast, the lack of such correlations in $SiO_2$-rich rocks may point to higher pressure (deep crustal) enrichment (e.g. Hildreth and Moorbath 1988; Rogers and Hawkesworth 1989; Davidson et al. 1991).

## 16.2.2 North Chile

A prominent feature of the north Chilean patterns of geochemical and isotopic evolution (200 Ma-Recent) are increasing La/Yb and initial $^{87}Sr/^{86}Sr$, and decreasing $^{143}Nd/^{144}Nd$ ratios (Fig. 16.7a–c). Detailed discussions of

**Fig. 16.7.**
Ages of igneous rocks from the north Chilean arc (21–26° S) versus (**a**) REE fractionation (here as La/Yb ratios), (**b**) initial $^{87}Sr/^{86}Sr$ ratios, and (**c**) initial $^{143}Nd/^{144}Nd$-isotopic ratios (data sources in Haschke et al. 2002a, and references therein). Steepening REE patterns and increasing isotopic enrichment correlate with younging and increasing crustal thickening of the arc through time, and within each individual arc system

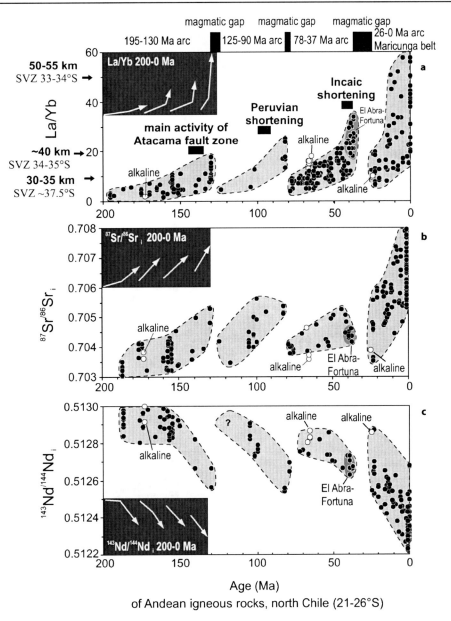

petrological and isotopic characteristics, petrogenetic mechanisms and geodynamic implications from these patterns are published in Haschke et al. (2002a, 2002b), Haschke (2002), and Haschke and Günther (2003). The anchorpoints of our interpretation are (*1*) the high geochemical data density of the Eocene arc (Fig. 16.7a), (*2*) die timing of shortening and/or tectonic activity at the end of each magmatic episode, (*3*) the pattern of stepwise arc migration every 30–40 Myr, and (*4*) the pronounced magmatic gap of 5–12 Myr between the Eocene and Neogene arc. The geochemical characteristics of the paleo-Andean arcs are also consistent with structural and sedimentologic constraints, which reflect a thickening crust, from the Jurassic (30–35 km), to the mid-Cretaceous (35 km), to the Eocene (37–45 km, Haschke and Günther 2003), to the Recent arc with 70 km of thickness (Wigger et al. 1994; Zandt et al. 1994; Yuan et al. 2000). On a smaller scale, these patterns are repeated during increasing arc maturity, reflecting gradual intra-arc crustal thickening (Fig. 16.7a–c).

In addition, the highest La/Yb ratios in each arc always correlate with intra-arc tectonic activity, either transtensional (Jurassic) and/or transpressional (mid-Cretaceous, Eocene, Miocene) with shortening of the arc crust. Each tectonic episode was followed by 5–12 million years with little or no magmatism in the arc, before the next main-arc was established up to 100 km further east. These cyclic geochemical and isotopic patterns are best explained as reflecting melting (recycling) of an underplated basaltic base of a thickening orogenic crust (e.g. Kay et al. 1991; Haschke and Günther 2003). The overall trend to more crust-like isotope compositions through time resulted from the increased assimilation of the thickening overlying crust (e.g. Hildreth and Moorbath 1988; Rogers and Hawkesworth 1989; Davidson et al. 1991 and discussions therein).

### 16.2.3 South Chile

Igneous arc rocks between 41 and 46° S show different REE fractionation and isotopic patterns to those from north Chile (Fig. 16.8a–c). Little is known about the igneous evolution of the pre-150 Ma arc, but the post-150 Ma evolutionary trend is better constrained.

Rocks from ~110 Ma to the Present generally show low La/Yb ratios (< 10), similar to those from Jurassic rocks in north Chile, suggesting a crustal thickness in the order of 30–35 km. The lack of increasing La/Yb ratios suggests that this crustal thickness remained constant throughout most of the time that the arc existed.

Between ~150 and ~20 Ma, initial Sr isotope ratios of andesitic to dacitic rocks (and intrusive equivalents) tend to decrease from higher (crust-like) to lower (mantle-like < 0.704) $^{87}Sr/^{86}Sr$ ratios (Pankhurst et al. 1999), where-

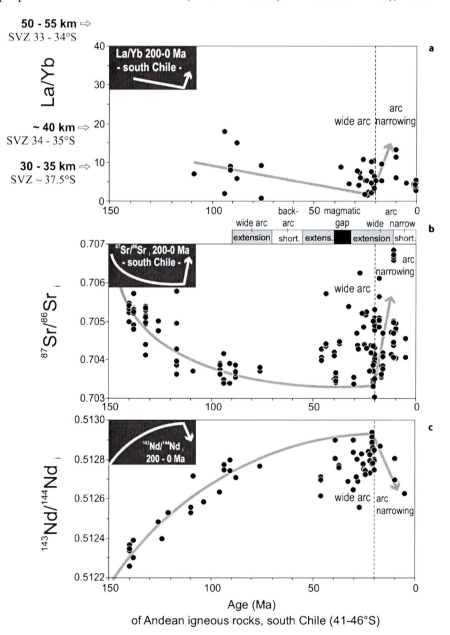

**Fig. 16.8.**
Ages of igneous rocks from south Chile (41–46° S) versus (**a**) La/Yb ratios, (**b**) initial $^{87}Sr/^{86}Sr$ ratios, and (**c**) initial $^{143}Nd/^{144}Nd$-isotopic ratios (data sources in Lopez-Escobar and Vergara 1997, Pankhurst et al. 1999, Munoz et al. 2000, Lara et al. 2001). Generally low La/Yb ratios are consistent with relatively thin arc crust (30–40 km). Both initial Sr and Nd isotope patterns show a deflection in the evolutionary patterns at ~20 Ma (see text for discussion), followed by arc narrowing. *Gray arrows* indicate interpreted evolutionary La/Yb and Sr-Nd isotope trends (shown simplified in insets)

as rocks between ~20 Ma and Present lack mantle-like $^{87}Sr/^{86}Sr$ ratios (Fig. 16.8b). Initial $^{143}Nd/^{144}Nd$ isotope ratios between 150 Ma and Present show a reverse evolution yet a similar overall pattern: increasing ratios between 150 Ma and 20 Ma, followed by lower $^{143}Nd/^{144}Nd$ ratios from 20 Ma to the Present (Fig. 16.8c).

Possible explanations for the gradual Sr- and Nd-isotopic evolution between 150 and 20 Ma include: (*a*) increasing isotopic shielding of younger plutons, with the result that formerly emplaced plutonic rocks were assimilated in a stationary magmatic arc (Pankhurst et al. 1999), (*b*) melting of a reservoir of successively depleting lithospheric mantle (Bruce et al. 1991), (*c*) gradually decreasing assimilation of more radiogenic Sr- and Nd-isotopes due to crustal thinning (e.g. Hildreth and Moorbath 1988; Haschke et al. 2002a), and/or (*d*) mixing of mantle basalt with increasing proportions of crustal melt from (previously underplated) mantle basalt (Pankhurst et al. 1999). However, these mechanisms do not provide a satisfactory explanation for the late Miocene change in isotopic signatures. Isotopic shielding should increase (not decrease) during arc narrowing, thus providing more primitive isotopic signatures (rather than crust-like ratios, see Fig. 16.8b,c). Similarly, melting of a successively depleting mantle should not be affected by arc narrowing and should further generate more mantle-like isotopes between 20 Ma to the Present (as opposed to crust-like ratios). Crustal thinning is consistent with the overall low La/Yb ratios and the mainly extensional tectonic setting in southern Chile between 150 and 10 Ma (Fig. 16.8b). Consequently, the changing trend of isotopic signatures would indicate crustal thickening at 20 Ma, yet evidence of Neogene tectonic shortening in southern Chile is restricted to 9–8 Ma, coeval with arc narrowing, therefore postdating crustal thickening. Pankhurst et al. (1999) reported slightly higher initial $^{87}Sr/^{86}Sr$ ratios for late Miocene and Pliocene plutons near the Liquine-Ofqui fault zone, which is roughly consistent with an increase in crustal assimilation upon shortening at 9–8 Ma. However, the overall low La/Yb ratios of rocks before and after 20 Ma (Fig. 16.8a) suggest that any crustal shortening – regardless if at 20 Ma or 9–8 Ma – did not cause any major changes in the melt source and/or residual mineralogy, and therefore in crustal thickness.

Contamination of the mantle source by subducted sediments (Kilian and Behrmann 2003) could also explain the changes in isotopic signatures seen in southern Chile. However, Pankhurst et al. (1999) showed that this explanation is unlikely. $^{87}Sr/^{86}Sr$ and $^{143}Nd/^{144}Nd$ ratios show a clear mixing relation between the mantle reservoir and isotopes from the North Patagonian batholith (yet not with the Chilean sediments), whose lower crust probably served as contaminating component for the Neogene arc rocks.

Analogous to north Chile, the isotopic evolution in south Chile rather reflects a combination of petrologic processes. Isotopic shielding and gradual lithospheric depletion might have controlled the isotopic evolution in the stationary arc prior to ~20 Ma, but slab-induced mechanisms like arc widening and narrowing, subsequent shortening of arc crust, the lack of arc migration and isotopic evidence of hybrid melts (mantle basalts with increasing lower crustal assimilation) suggest that the 20 Ma to present-day geochemical and isotopic evolution in south Chile may have been dominated by tectonic processes. This line of arguments would be similar to north Chile, where interacting tectono-magmatic processes are better constrained.

## 16.3 Balancing Tectonic Versus Magmatic Crustal Thickening

The geochemical and isotopic constraints suggest that recycling melting of basaltic underplated crust at or near the arc crustal base played an important role during the evolution of arc magmatism in both north and south Chile.

Throughout the Andean orogeny (200 Ma to Recent), mantle-derived hydrous melts sampled and equilibrated with different lower crustal mineralogies at different crustal depths. As basaltic melts from the hydrous mantle wedge underplated and intruded the thickening lower crust, a mafic arc crustal keel developed and transformed from anhydrous clinopyroxene-dominated gabbroic (~35 km) to dominantly hydrous amphibolitic (~40 km) into anhydrous eclogitic mineralogies (= 50 km). The formation of major ore deposits in the Andes appears to be closely related to arc crustal thickening. In north Chile, the occurrence of ore mineralization is concentrated on the late Oligocene and Miocene episode, specifically at the beginning and end of volcanic episodes (e.g. El Abra-Fortuna pluton at 35–31 Ma, Boric et al. 1990; Fig. 16.9). Igneous rocks from this episode were in equilibrium with a mafic, garnet-amphibolite residue (Haschke et al. 2002a) indicating that these melts erupted through a thickened orogenic crust (45 km in the late Eocene/Oligocene, Haschke and Günther 2003). Arc magmas erupting through such thickening crust reflect the transition in mineralogy by increasing La/Yb, La/Sm and Sm/Yb ratios through time (Fig. 16.9).

### 16.3.1 North Chile

The low La/Sm and Sm/Yb ratios of rocks from the Jurassic arc (195–130 Ma field in Fig. 16.9) imply clinopyroxene-dominated, gabbroic mineral assemblages (30–35 km crustal thickness), whereas rocks from the mid-Cretaceous arc (125–90 Ma field) show higher La/Sm ratios, indicat-

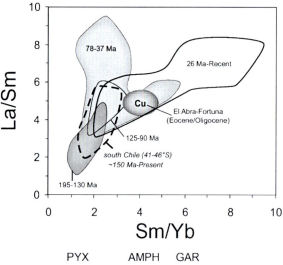

**Fig. 16.9.** Sm/Yb versus La/Sm ratios for more than 500 igneous rocks from north Chile (21–26° S) showing increasing La/Sm and Sm/Yb through time. Jurassic rocks (195–130 Ma) correspond to clinopyroxene-dominated (PYX) residual mineralogies, mid-Cretaceous rocks (125–90 Ma) are consistent with amphibole-dominated (AMPH) mineral assemblages, and Eocene rocks (78–37 Ma) show the first substantial increase in Sm/Yb ratios (> 4) indicating garnet stability (GAR) in the residual mineralogy. The Cu-field encloses magmas that produced the major copper (Cu) mineralization of the El Abra-Fortuna plutonic complex (Chuquicamata mine) during the Eocene, and also La/Sm and Sm/Yb ratios that are in equilibrium with amphibole and garnet-bearing residual mineral assemblages (data source in Haschke et al. 2002a and references therein). The field of igneous rocks from the south Chilean arc segment (150 Ma-Present, *dashed line*) overlaps with the Jurassic and mid-Cretaceous rocks from north Chile, implying transitional clinopyroxene-amphibole-dominated residual mineral assemblages, consistent with a 30–35 km crustal thickness for south Chile

ing the presence of amphibole. The Eocene arc (78–37 Ma field) shows high La/Sm *and* high Sm/Yb ratios after late Eocene (Incaic) compressional deformation and thickening of the crust, attesting to the presence of garnet as a residual mineral at, or near, the crust-mantle boundary upon crustal shortening and that these melts erupted through a thickened orogenic crust (45 km in the late Eocene/Oligocene, Haschke et al. 2002a). The Neogene arc and the recent Western Cordillera (26 Ma to the Present field) show high La/Sm and the highest Sm/Yb ratios, indicating garnet-dominated (eclogitic) mineral assemblages (Fig. 16.10), which is consistent with present arc crustal thickness of 70 km.

This link between the evolutionary REE patterns and amount of crustal shortening can be used to balance the relative amount of tectonic and magmatic crustal thickening. In north Chile, Haschke and Günther (2003) used combined structurally balanced arc crustal sections and independent petrologic trace element data from the Eocene arc (Fig. 16.7a) to show that the amount of crustal shortening alone was insufficient to explain the 45 km thick crust indicated by the trace element signatures seen in post-deformational melts (onset of garnet in the source and/or residual mineralogy). Basaltic underplating in north Chile produced up to 3 km of lower continental crust during the 40 million years that this arc was active, with the relative amount of tectonic versus magmatic thickening in the order of 2 : 1 (Haschke and Günther 2003). This amount of magmatically underplated crust is significantly less than the 25 km proposed by Furlong and Fountain (1986), and the results support earlier models suggesting that shortening of the crust is, and *was,* the most efficient mechanism for crustal thickening in the Neogene and pre-Neogene Andean arcs from 200 Ma to the Present (Allmendinger et al. 1997).

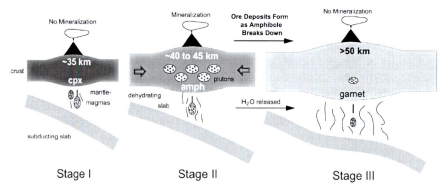

**Fig. 16.10.** Scheme for the generation of major Andean ore deposits (after Kay et al. 1999; Kay and Mpodozis 2001). *Stage I*: magmas equilibrate with pyroxene-bearing (*cpx*) mineral residues in continental crust of normal thickness over a relatively steep subduction zone. *Stage II*: magmas equilibrate with an amphibole-dominated (*amph*) lower crust. *Stage III*: magmas equilibrate with garnet-bearing (*garnet*) mineral residues in of the lower thickening crust over a shallowing subduction zone. Mineralization occurs between stages II and III. Critical ingredients are: a hydrated mantle above a shallowing subduction zone; storage of fluid in amphibole and hydrous magmas in the deep crust; and release of that fluid in conjunction with a breakdown of amphibole-bearing crustal assemblages in zones of crustal melting during shortening and thickening of ductile, magma-injected crust

## 16.3.2 South Chile

Rocks from the south Chilean arc (dashed field in Fig. 16.9) show La/Sm and Sm/Yb ratios overlapping with the fields of Jurassic and mid-Cretaceous rocks from north Chile. However, they do not overlap with that of the El Abra-Fortuna plutonic rocks (Cu-field in Fig. 16.9), consistent with clinopyroxene/amphibole-bearing residual mineralogies and an arc crustal thickness of 30–40 km (Hildreth and Moorbath 1988; Yuan et al. 2006), therefore failing to provide the minimum crustal thickness required for stabilizing garnet in the crustal base. In fact, the trace element patterns from southern Chile lack any evidence for major changes in crustal thickness, in spite of 200 Myr of mainly stationary arc magmatism. Given the tectonomagmatic constraints from the north Chilean Eocene arc (3 km underplated basaltic crust in 40 million years, Haschke and Günther 2003), one would expect a minimum of 15 km of underplated mafic lower crust and a total crustal thickness of ~50 km in southern Chile, yet these values are inconsistent with regional geochemical, geophysical and structural data. Consequently, either the concept of underplating is wrong, or there is another, previously underestimated, mechanism capable of removing newly underplated basalts from the arc and preventing magmatic thickening of the crust.

One possible explanation may be delamination of the arc crustal keel (Kay and Kay 1991), but this requires thick orogenic crust (= 50 km, Rapp and Watson 1995), which can be ruled out for southern Chile (Yuan et al. 2006). Another possibility is that basaltic underplating may have been removed by lower crustal flow (Haschke et al. 2003). Flow of hot and viscous lower crust can be driven by lateral contrasts in crustal thickness between the thickened crustal keel and the non-thickened fore-arc and back-arc crust. The resulting Moho topography generates buoyancy forces that act to equalize crustal thicknesses. As the lower crust flows, the flow is deformed to smooth the Moho topography along crustal channels on a scale of 100–200 km (Royden 1996; McKenzie et al. 2000). Creeping flow at geologically important rates may occur when the absolute temperature of a solid exceeds ~70% of the solidus temperature (700–900 °C, Stocker and Ashby 1973). Thus, at temperature gradients of 20 °C km$^{-1}$, the lower 5–10 km of the crust may flow (McKenzie et al. 2000). Active volcanic arcs have substantially higher temperature gradients (30–60 °C km$^{-1}$) and the high temperature of mantle-derived basalt ponding at, or near, the crust-mantle boundary is likely to be one of the *main* causes leading to flow in the lower crust, possibly maintaining the largely constant crustal thicknesses in southern Chile.

As much of the Andean evolution appears to depend on the presence of shortened and, hence, thickened crust, caused by episodes of flat subduction, the next logical consideration is how to induce flat subduction.

## 16.4 Flat Subduction and Slab Geometry in the Andes

The present-day Andean subduction zone shows at least two spatial arc volcanic gaps: in central Chile (28–33° S, Fi. 16.1, Barazangi and Isacks 1976; Sacks 1983; Cahill and Isacks 1992; Kay et al. 1999) and Peru (3–15° S, Hasegawa and Sacks 1981; Boyd et al. 1984). These gaps in the volcanic chain correlate with seismic data showing subhorizontally subducting oceanic lithosphere (flat slab). Flat subduction disturbs the thermal structure in the subduction zone by preventing the development of an asthenospheric mantle wedge, and reducing the mantle temperatures below the solidus of hydrous peridotite. In spite of continued fluid release, the mantle peridotite is too cold to melt. In both cases of volcanic gaps in the Andes, flat subduction appears to be preceded by ceasing volcanism, crustal shortening and thickening, increasing La/Yb ratios, and enrichment of Sr and Nd isotope ratios (e.g. Kay et al. 1994, 1999). Interestingly, this sequence is seen repeatedly in the central Andean evolutionary geochemical and isotopic patterns, hence suggesting similar geodynamic processes. Given the present-day tectonic, magmatic, geochemical and isotopic patterns associated with modern flat subduction, we hypothesize about earlier episodes of flat subduction during the Andean orogeny.

### 16.4.1 North Chile

Evolutionary geochemical patterns of north Chilean arc rocks show repeated temporal gaps in arc magmatism. Every 30–40 million years, arc volcanism ceased for 5–12 Myr, coeval with arc crustal shortening and thickening, arc migration and resetting the geochemical and isotopic characteristics from mature to more immature signatures. This cyclicity in arc migration and tectonic activity implies a cause-and-effect relationship between plate convergence, mechanical interaction of both plates and the existence of a critical rheologic threshold of the overriding, magmatically weakened continental plate.

These patterns correspond to those seen in modern flat-slab scenarios, and imply that the 5–12 Myr temporal magmatic gaps may have been episodes of subhorizontal subduction (Haschke et al. 2002b). If accepted, then the next logical kinematic issues are: (*1*) How does the slab steepen, and perhaps more importantly (*2*) how does the slab re-shallow?

Steepening of the slab is inferred to be density-driven (e.g Von Blanckenburg and Davies 1995) due to eclogitization of the slab tip, increasing slab pull forces, and slab rollback. Both slab steepening and rollback cause a westward-prograding mantle wedge, typically at velocities of ~10% of the plate convergence rate (Garfunkel et al. 1986). Possible slab steepening in the Cretaceous-Eocene arc in north Chile (78–41 Ma) may be reflected by narrowing of the Eocene main-arc (Fig. 16.3b), and incipient back-arc rifting and related alkaline magmatism (65–67 Ma: Cerro Colorado and Quimal, Fig. 16.3c) during the early stage of the arc (Haschke et al. 2002b). A similar sequence of extension-related, alkaline magmatism is seen in the late Oligocene back-arc of the central Andes (Kay et al. 1994).

Relative plate convergence rates during this time (78–49 Ma, Pardo-Casas and Molnar 1987; Somoza 1998) increased, coeval with an increase in the westward motion of continental South America since Eocene times (Silver et al. 1998) and decreasing convergence obliquity (Fig. 16.6). One possible mechanical consequence of accelerated subduction without contemporaneously increasing the slab pull, is the development of an eastward-bending forebulge of the shallow slab which would result in overriding of the dense and already steepened eclogitic slab tip (Fig. 16.11a), causing slab bending and kinking (Haschke et al. 2002b).

The next step is to understand, how the slab shallows again, in order to explain the ceasing magmatic activity and following 5–12 Myr of magmatic quiescence. It is unlikely that the deep and dense eclogitic portion of the slab will return to shallow depths after the onset of rapid convergence, as this causes mantle mass transfer problems. Slab shallowing would require the 'squeezing out' of asthenospheric material between the two interacting plates, versus a corner flow from the opposite direction (Kincaid and Griffith 2004). One possible explanation is that the continuous overriding of the rapidly and shallowly subducting portion of the upper slab over the steepened, bent and kinked tip of the slab eventually resulted in slab breakoff and subsequent subhorizontal subduction (Reutter 2001; Haschke et al. 2002b; Fig. 16.11b,c).

The loss of the deep and dense eclogitic slab tip substantially decreased the remaining pulling forces on the slab, leading to an episode of buoyancy-driven, subhorizontal subduction (Fig. 16.11c). Although a torn and shallowly subducting slab edge adjacent to warm oceanic lithosphere is a potential site for slab melting, the relatively cool mantle wedge associated with shallow subduction might have prevented melting at the slab edge. In addition, this lack of melting might have been due to rapid convergence rates at this time, as rapid subduction produces lower slab surface temperatures than slow subduction (Kincaid and Sacks 1997; Kincaid and Griffith 2004).

The following flat-slab episode (37–26 Ma, Fig. 16.11c–e) correlates with the onset of shortening and thickening of the crust. If correct, then such compressional tectonic episodes (Peruvian: ~100 Ma, Incaic: 45–38 Ma, Quechua: 26–6 Ma) reflect intervals of slab-breakoff followed by flat

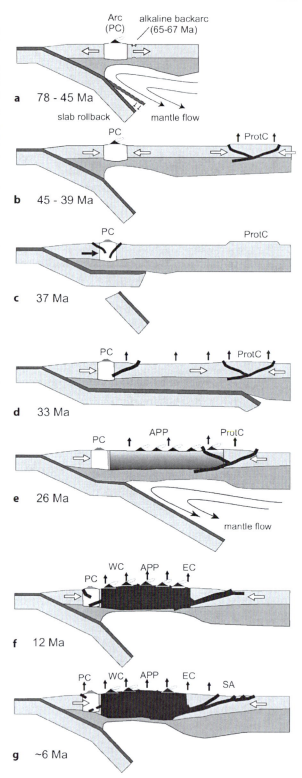

**Fig. 16.11. a–g** Model of repeated slab shallowing and breakoff during the Andean orogeny (Reutter 2001; Haschke et al. 2002b). *PC* = Precordillera, *APP* = Altiplano-Puna Plateau, *ProtC* = Proto-Eastern Cordillera, *WC* = Western Cordillera, *EC* = Eastern Cordillera, *SA* = South America

subduction. One first-order consequence of flat subduction and shortening may be enhanced uplift as the buoyant and shallow tip of the subducting slab advances eastward, such as shown after Incaic deformation at ~33 Ma in the Proto-Eastern Cordillera (Fig. 16.11d; Andriessen and Reutter 1994). As the slab begins to steepen, asthenospheric flow is allowed to re-establish arc magmatism (at 28–26 Ma, Fig. 16.11e) at threshold depths for genesis of arc magma (> 85 km, England et al. 2004). The propagation rate of asthenospheric wedge material is generally unknown but may be in the order of ~5 cm yr$^{-1}$, as suggested for delamination (Bird 1979). Magmatic quiescence of 5–12 Myr would allow the corner flow to propagate about 250–500 km, which is sufficient to recharge the asthenospheric wedge. As the tip of the slab passed through the threshold depth of magma genesis, the recharged mantle wedge partially melted (28–26 Ma) and generated a broad and diffuse magmatic zone across an area which is now the Altiplano-Puna Plateau (Fig. 16.11e,f). As the slab steepened, the former broad magmatic zone narrowed westward, accompanied by local, alkaline mafic back-arc flows during the Oligocene (e.g. at 26–28° S, Kay et al. 1994). Early-stage, mafic, alkaline back-arc magmatism is also characteristic of the Jurassic and late Cretaceous back-arc (e.g. 65–67 Ma, Cerros Colorado and Quimal, Fig. 16.3c) and may generally be diagnostic of slab steepening after an episode of flat subduction.

The broad magmatic zone eventually merged westward into the modern narrow volcanic arc of the Western Cordillera (Fig. 16.3d) and was accompanied by major shortening of the crust during the Miocene, generating the modern Andean Altiplano-Puna Plateau orogen (Fig. 16.11g).

## 16.4.2 South Chile

Unlike north Chile, the south Chilean arc segment (~41–46° S) lacks patterns of repeated or systematic arc migration. Existing patterns suggest a generally stationary arc axis between ~200 and 50 Ma, followed by a more complex history (Fig. 16.4).

Between 50 and 28 Ma, main-arc magmatism widened mainly towards the trench by 150 km into a broad (300 km) volcanic zone (Fig. 16.12a,b). Between 28 and 8 Ma, this broad volcanic zone narrowed eastward into a typical 100–150 km wide volcanic arc (Fig. 16.12c). Another pulse of arc-narrowing in the late Miocene (11–8 Ma) occurred coeval with modest crustal shortening and deformation of the arc at ~46° S (Ramos 2005, Fig. 16.12d). Suarez and de la Cruz (2001) found evidence of Paleogene and even older deformation in several areas of the Patagonian Cordillera.

In the Neuquen region (37–39° S), arc crustal shortening (11–6 Ma, Melnick et al. 2004) was followed by an episode of magmatic quiescence in the Main Cordillera

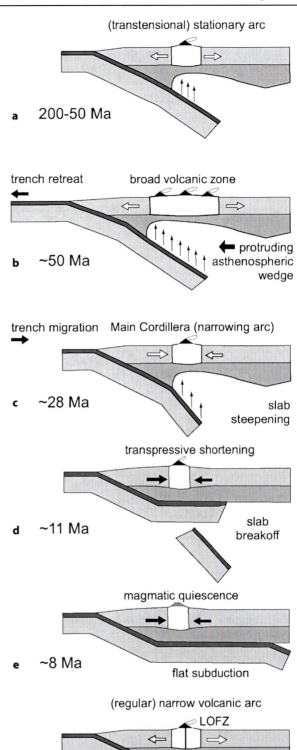

Fig. 16.12. a–f Model of changing subduction geometry between 200 Ma and the Present in south Chile (41–46° S)

(8–3 Ma), Fig. 16.12e). Arc volcanic activity resumed from Pliocene to early Pleistocene in a transtensional tectonic setting, and from the late Pleistocene to the Present, the reactivated volcanic arc narrowed along the Liquine-Ofqui strike-slip fault zone (Fig. 16.12f). Previous workers interpreted the changing igneous distribution pattern and magmatic gap as reflecting an episode of slab steepening followed by flat subduction (Kay and Mancilla 2001), similar to the flat-slab subduction cycles inferred for north Chile. In particular the late Miocene onset of shortening and the lack of volcanism in the Main Cordillera in the Neuquen Andes between 37° and 39° S (Kay 2002) are processes which can be linked to the flattening of the subduction angle.

However, the evolution of a flat subduction cycle in south Chile may be obscured by the westward trench retreat prior to ~28 Ma. Because there is no evidence of continuous arc migration (as in north Chile), pre-Oligocene magmatic zones are largely overlapping and difficult to distinguish. The episode of westward trench retreat and widening of the arc by ~150 km is complex and cannot be explained by simple variations in slab dip, but requires a rapidly protruding heat source in order to widen the melting zone into the former fore-arc (Fig. 16.9). Such a heat source could be (*a*) a hot asthenospheric wedge protruding westwards into the relatively cool and previously metasomatized, hydrous mantle wedge, or (*b*) decompression melting by return flow beneath the tip of the slab between 28 and 20 Ma (Kincaid and Griffiths 2004).

Kincaid and Griffiths (2004) simulated such a geodynamic scenario by analogue modeling of rollback subduction, i.e. sinking of the slab together with trench retreat. In their model, rollback subduction generates a major return flow from the ocean side of the slab into the slab wedge, which enhances the efficiency of decompression melting in the previous fore-arc mantle. Return flow can be an important parameter for tapping geochemically distinct mantle from the ocean side, which may flow around the plate edges into the mantle wedge transporting distinct geochemical and isotopic signatures (Wendt et al. 1997; Turner and Hawkesworth 1998; Pearce et al. 2001).

Yet if the south Chilean arc segment *was* influenced by flat subduction, then the problem of mass transfer into the mantle remains: How does the angle of subduction become re-shallow? Slab breakoff and subsequent subduction is one possible explanation, and is therefore considered as a possible scenario for south Chile (Fig. 16.12a–f). The actual differences in the onset and duration of magmatic quiescence and contractional deformation between the north and south Chilean arc segments may reflect the different variables and parameters involved in flat subduction, such as the timing and geometry of slab breakoff, the nature of the coupling between the interacting plates, and the time of onset and intensity of eclogitization.

## 16.5 Summary

The shared tectonic and magmatic features of Andean arc segments in north and south Chile, and the modern Western Cordillera volcanic arc support existing models on Andean deformation and changing slab dip (e.g. Megard 1987; Isacks 1988; Kay et al. 1991; Sandeman et al. 1995; Kay and Abruzzi 1996; James and Sacks 1999). They also suggest that the change from normal to flat subduction of the downgoing slab played a far more significant role in the formation of the modern Andes than is generally acknowledged. In south Chile, some earlier magmatic signals of flat subduction might have been obscured by the overlying magmatic episodes that followed.

In summary, the shared tectonic and magmatic features between the north and south Chilean arc segments can be divided into early and late stage features. Early Andean arc features are (*1*) back-arc extensional tectonics, (*2*) alkaline back-arc magmatism, (*3*) basaltic underplating and crustal thickening, and (*4*) lower crustal flow. Late stage arc features are: (*1*) crustal shortening and thickening, (*2*) arc-parallel strike-slip faulting, (*3*) episodes of flat subduction, and (*4*) mineralization. Each arc segment does not necessarily exhibit all features, and the order of features may vary, but the overall pattern is robust and clear.

Another important implication may be the analogy between modern and paleo-flat-slab episodes. Episodes of subhorizontal subduction may be a characteristic feature in the evolution of Andean-type margins (Kay 2002). A compilation of worldwide volcanic arc systems by McGeary et al. (1985) revealed at least 24 sites with major spatial gaps in the volcanic chains (> 200 km), indicating that flat subduction is a widespread worldwide phenomenon occurring in ~10% of modern convergent margins (e.g. Costa Rica, Mexico, S Alaska, SW Japan, W New Guinea). The paleo-flat slab episode between ~36 and 26 Ma in north Chile is particularly well constrained. Volcanic evolution during this time kept pace with the changes in slab geometry: between 18° and 45° S, there was a 10 Myr period with little or no volcanic activity which ended at ~27 Ma (Lahsen 1982; Mpodozis et al. 1995; Maksaev 1988; Munoz et al. 2000). This 10 Myr period might have been a transient state of slab reconfiguration, including subhorizontal subduction and subsequent slab steepening.

Episodes of flat subduction may be difficult to recognize in stationary arc segments like that of south Chile. We do not anticipate an identical onset of subhorizontal subduction in north and south Chile due to differences in plate coupling, asthenospheric circulation, the age of the subducting slab, the seafloor structures of the subducting slab, and local mantle anomalies (such as the Somun

Cura hot spot, Kay et al. 1993). But we do suggest that the subducting slab geometry may be a previously underestimated orogenic parameter, which preconditioned the Andean lithosphere to enhanced crustal thickening in the Neogene (James and Sacks 1999).

The cyclic magmatic, geochemical, isotopic and tectonic patterns, clear in north Chile yet obscured in south Chile, provides an estimate of the "Andean orogenic clock", at least for the north Chilean arc segment. Every 30-40 million years, arc magmatism migrated, coeval with major tectonic activity and a reset of geochemical and isotopic signatures in arc magmatism. What are the driving forces for the Andean cyclicity? Possible explanations include (a) cyclic changes in the geometry of the subducting slab (see discussion above), and, perhaps more likely (b) cyclic changes in the rheologic behavior of the upper continental lithosphere. Accordingly, the cyclicity would be the cumulative response to the same processes exceeding a critical threshold value of the upper plate's strength (weakened by arc magmatism) before yielding to shortening and thickening of the lithosphere.

## Acknowledgments

We gratefully acklowledge constructive reviews by H.A. Seck and S. Foley. This work was supported by the DFG Sonderforschungsbereich 267 "Deformationsprozesse in den Anden".

## References

Allmendinger RW (1986) Tectonic development, southeastern border of the Puna Plateau, northwestern Argentine Andes. Geol Soc Am Bull 97:1070–1082

Allmendinger RW, Jordan TE, Kay SM, Isacks BL (1997) The evolution of the Altiplano-Puna plateau of the Central Andes. Ann Rev Earth Planet Sci 25:139–174

Andriessen PA, Reutter K-J (1994) K-Ar and fission-track mineral age determination of igneous rocks related to multiple magmatic arc systems along the 23°S lat. of Chile and NW Argentina. In: Reutter K-J, Scheuber E, Wigger P (eds) Tectonics of the Southern Central Andes. Springer-Verlag, Berlin Heidelberg New York, pp 141–153

Arancibia G (2004) Mid-Cretaceous crustal shortening: evidence from a regional-scale ductile shear zone in the Coastal range of central Chile (32°S). J South Am Earth Sci 17:209–226

Barazangi M, Isacks B (1976) Spatial distribution of earthquakes and subduction of the Nazca plate beneath South America. Geology 4:686–692

Barnes HL (1997) Geochemistry of hydrothermal ore deposits. John Wiley & Sons, New York, p 972

Barth MG, Foley SF, Horn I (2002) Partial melting in Archean subduction zones: constraints from experimentally determined trace elements partition coefficients between eclogitic materials and tonalitic melts under upper mantle conditions. Precambrian Res 113:323–340

Bartsch V (2004) Magmengenese der obertriassischen bis unterkretazischen Vulkanite der mesozoischen Vulkanite in der Küstenkordillere von N-Chile zwischen 24° und 27°S: Petrographie, Mineralchemie, Geochemie und Isotopie. PhD thesis, Technische Universität Berlin

Bird P (1979) Continental delamination and the Colorado Plateau. J Geophys Res 84(B13):7561–7571

Boric R, Diaz F, Maksaev V (1990) Geologia y yacimientos metaliferos de la Region de Antofagasta, Serv Nac Geol Miner Boletin 40

Boyd TM, Snoke JA, Sacks IS, Rodriguez AB (1984) High resolution determination of the Benioff zone geometry beneath southern Peru, Bull Seismol Soc Am 74:559–568

Bruce RM, Nelson EP, Weaver SG, Lux DR (1991) Temporal spatial variation in the southern Patagonian batholith: constraints on magmatic arc development. In: Harmon RS, Rapela CW (eds) Andean magmatism and its tectonic setting. Geol Soc Am Spec P 265:1–12

Cahill TA, Isacks BL (1992) Seismicity and shape of the subducted Nazca plate. J Geophys Res 97:17503–17529

Cande SC, Leslie RB, Parra JC, Hobart M (1986) Interaction between the Chile ridge and Chile trench; Geophysical and geothermal evidence. J Geophys Res B92:495–520

Charrier R, Reutter K-J (1994) The Purilactis Group of northern Chile: boundary between arc and backarc from late Cretaceous to Eocene. In: Reutter K-J, Scheuber E, Wigger P (eds) Tectonics of the Southern Central Andes. Springer-Verlag, Berlin Heidelberg New York, pp 189–202

Cloos M (1993) Lithospheric buoyancy and collisional orogenesis: Subduction of oceanic plateaus, continental margins, island arcs, spreading ridges, and seamounts. Geol Soc Am Bull 105:715–737

Davidson JP, Harmon RS, Wörner G (1991) The source of central Andean magmas: some considerations. In: Harmon RS, Rapela CW (eds) Andean magmatism and its tectonic setting. Geol Soc Am Spec P 265:233–243

Döbel R, Hammerschmidt K, Friedrichsen H (1992) Implication of $^{40}Ar/^{39}Ar$ dating of Early Tertiary volcanic rocks from the north-Chilean Precordillera. Tectonophysics 202:55–81

England P, Engdahl R, Thatcher W (2004) Systematic variation in the depths of slabs beneath arc volcanoes. Geophys J Int 156:377–408

Furlong K, Fountain Dm (1986) Continental crustal underplating: thermal considerations and seismic–petrological consideration. J Geophys Res 91:8285–8294

Garfunkel Z, Anderson CA, Schubert G (1986) Mantle circulation and the lateral migration of subducted slabs. J Geophys Res 91(B7): 7205–7223

Grocott J, Brown M, Dallmeyer RD, Taylor GK, Treloar TJ (1994) Mechanisms of continental growth in extensional arcs: an example from the Andean plate-boundary zone, Geology 22:391–394

Günther A (2001) Strukturgeometrie, Kinematikanalyse und Deformationsgeschichte des oberkretazisch–alttertiären magmatischen Bogens (21.5–23°S), Berliner Geowiss Abh 213

Gutscher MA, Spakman W, Bkjwaard H, Engdahl ER (2000) Geodynamics of flat subduction: seismicity and tomographic constraints from the Andean margin. Tectonics 19:814–833

Haschke (2002) Evolutionary geochemical patterns of Late Cretaceous to Eocene arc magmatic rocks in North Chile: implications for Archean crustal growth. EGU Stephan Mueller Spec Pub Ser 2, pp 207–218

Haschke M, Günther A (2003) Balancing crustal thickening in arc by tectonic vs. magmatic means. Geology 31(11):933–936

Haschke M, Siebel W, Günther A, Scheuber E (2002a) Repeated crustal thickening and recycling during the Andean orogeny in North Chile (21°–26°S). J Geophys Res 107(B1): 10.1029/2001JB000328

Haschke M, Scheuber E, Günther A, Reutter K-J (2002b) Evolutionary cycles during the Andean orogeny: repeated slab breakoff and flat subduction? Terra Nova 14(1):49–56

Haschke M, Echtler H, Oncken O (2003) Fate of Basaltic Underplate in North and South Chile. Geol Soc Am Abs Prog 64815, 35(6)

Hasegawa A, Sacks IS (1981) Subduction of the Nazca Plate beneath Peru as determined from seismic observations. J Geophys Res 86:4971–4980

Hildreth W, Moorbath S (1988) Crustal Contributions to arc magmatism in the Andes of Central Chile, Contrib Mineral Petrol 98: 455–489

Horton BK (1998) Sediment accumulation on top of the Andean orogenic wedge: Oligocene to late Miocene basins of the Eastern Cordillera, southern Bolivia. Geol Soc Am Bull 110:1174–1192

Isacks BL (1988) Uplift of the Central Andean Plateau and bending of the Bolivian orocline. J Geophys Res 93:3211–3231

Jaillard RD (1986) Relations among subduction parameters, Rev Geophys 24:217–284

James DE, Sacks IS (1999) Cenozoic formation of the Central Andes: a geophysical perspective. In: Skinner BJ (ed) Geology and ore deposits of the Central Andes. Soc Econ Geol Spec Pub 7, pp 1–25

Jordan TE, Reynolds JH, Erikson JP (1997) Variability in age of initial shortening and uplift in the central Andes, 16–33°30'S. In: Ruddiman W (ed) Tectonic uplift and climate change. Plenum, New York, pp 41–61

Jordan TE, Burns WM, Veiga R, Pangaro F, Copeland P, Kelley S, Mpodozis C (2001) Extension and basin formation in the southern Andes caused by increased convergence rate: a mid-Cenozoic trigger for the Andes. Tectonics 20(3):308–324

Kay SM (2002) Tertiary to Recent transient shallow subduction zones in the Central and Southern Andes. XV Congreso Geologico Argentino, Calafata, Argentina, Abstract 237

Kay SM, Abbruzzi JM (1996) Magmatic evidence for Neogene lithospheric evolution of the central Andean "flat–slab" between 30°S and 32°S. Tectonophysics 259:15–28

Kay RW, Kay SM (1991) Creation and destruction of lower continental crust. Geol Rundsch 80(2):259–278

Kay SM, Mancilla O (2001) Neogene shallow subduction segments in the Chilean/Argentine Andes and Andean-type margins. Geol Soc Am Abs Prog 33, A156

Kay SM, Mpodozis C (2001) Central Andean ore deposits linked to evolving shallow subduction systems and thickening crust, GSA Today 11(3):4–9

Kay SM, Maksaev V, Mpodozis C, Moscoso R, Nasi C (1987) Probing the evolving Andean lithosphere: Mid–late Tertiary magmatism in Chile (29°–30.5°S) over the zone of subhorizontal subduction. J Geophys Res 92:6173–6189

Kay SM, Mpodozis C, Ramos VA, Munizaga F (1991) Magma source variations for Mid-Tertiary magmatic rocks associated with a shallowing subduction zone and a thickening crust in the Central Andes (28°–33°S). In: Harmon RS, Rapela CW (eds) Andean magmatism and its tectonic setting. Geol Soc Amer Spec P 265: 113–137

Kay SM, Ardolino AA, Franchi M, Ramos VA (1993) El origen de la meseta de Somun Cura: distribucion y geoquimica de sus rocas volcanicas maficas, XII Congr. Geol. Argentino y II. Congr. Explor. Hidrocarburos, 4, pp 236–248

Kay SM, Mpodozis C, Tittler A, Cornejo P (1994) Tertiary magmatic evolution of the Maricunga Mineral Belt in Chile. Int Geol Rev 36:1079–1112

Kay SM, Mpodozis C, Coira B (1999) Neogene magmatism, tectonism, and mineral deposits of the central Andes (22° to 33°S latitude). In: Skinner BJ (ed) Geology and ore deposits of the Central Andes. Soc Econom Geol Spec Pub 7, pp 27–59

Kincaid C, Griffiths RW (2004) Variability in flow and temperatures within mantle subduction zones. Geochem Geophys Geosyst 5(6): doi 10.1029/2003GC000666

Kincaid C, Sacks S (1997) The thermal and dynamical evolution of the upper mantle in subduction zones. J Geophys Res 102:12295–12315

Kilian R, Behrmann JH (2003) Geochemical constraints on the sources of Southern Chile Trench sediments and their recycling in arc magmas of the Southern Andes. J Geol Soc London 160: doi 10.1144/0016-764901–143

Klein M, Stosch HG, Seck HA (1997) Partitioning of high field-strength and rare-earth elements between amphibole and quartz-dioritic to tonalitic melts: an experimental study. Chem Geol 138:257–271

Kley J, Müller J, Tawackoli S, Jacobshagen V, Manutsoglu E (1997) Pre-Andean and Andean-age deformation in the Eastern Cordillera of southern Bolivia. J S Am Earth Sci 10:1–19

Lahsen A (1982) Upper Cenozoic volcanism and tectonism in the Andes of northern Chile. Earth Sci Rev 18:258–302

Lara L, Rodriguez C, Moreno H, Arce C (2001) Geocronologia K-Ar y geoquimica del volcanismo plioceno superior-pleistoceno de los Andes del sur (39–42°S). Rev Geol Chile 28(1):67–90

Lavenu A, Cembrano J (1999) Compressional- and transpressional-stress pattern for Pliocene and Quaternary brittle deformation in fore arc and intra-arc zones (Andes of Central and Southern Chile). J Struct Geol 21:1669–1691

Lopez-Escobar L, Vergara M (1997) Eocene–Miocene longitudinal depression and Quaternary volcanism in the southern Andes, Chile (33–42.5°S): a geochemical comparison. Rev Geol Chile 24(2):227–244

Maksaev V (1988) Metallogenic implications of K-Ar, $^{40}$Ar-$^{39}$Ar, and fission-track dates of mineralized areas in the Andes of northern Chile. V Congreso Geologíco Chileno, Actas 1:B65–B86

McCaffrey R (1997) Influences of recurrence times and fault temperatures on the age-rate dependence of subduction zone seismicity. J Geophys Res 102:22839–2285

McGeary S, Nur A, Ben-Avraham Z (1985) Spatial gaps in arc volcanism: the effect of collision or subduction of oceanic plateaus. Tectonophysics 119:195–221

McKenzie D, Nimmo F, Jackson JA (2000) Characteristics and consequences of flow in the lower crust. J Geophys Res 105(B5):11029–11046

McMillan N, Davidson J, Wörner G, Harmon RS, Lopez-Escobar L, Moorbath S (1993) Mechanism of trace element enrichment related to crustal thickening: the Nevados de Payachata region, Northern Chile. Geology 21:467–470

Megard F (1987) Cordilleran Andes and marginal Andes: a review of Andean geology north of the Arica elbow (18°S). In: Monger JW, Francheteau J (eds) Circum-Pacific orogenic belts and evolution of the Pacific Ocean Basin. AGU Geodyn Ser 18, pp 71–95

Melnick D, Rosenau M, Folguera A, Echtler H (2006) Neogene tectonic evolution of the Neuquén Andes western flank. In: Kay SM, Ramos VA (eds) Evolution of an Andean margin: a tectonic and magmatic view from the Andes to the Neuquén Basin (35°–39° S lat). Geol Soc Am Spec P 407:73–95

Mpodozis C, Cornejo P, Kay SM (1995) La franja de Maricunga: sintesis de la evolucion del frente volcanico Oligocene-Miocene de la zona sur de los Andes centrales. Rev Geol Chile 22:273–314

Munizaga F, Herve F, Drake R, Pankhurst RJ, Brook M, Snelling N (1988) Geochronology of the Lake Region of south-central Chile (39°–42°S): preliminary results. J S Am Earth Sci 1:309–316

Munoz J, Duhart P, Crignola P, Farmer GL, Stern CR (1989) Alkaline magmatism within the segment 38–39°S of the Plio-Quaternary volcanic belt of the southern South American continental margin. J Geophys Res 94:4545–4560

Munoz J, Duhart P, Farmer L (2000) The relation of the mid-Tertiary coastal magmatic belt in South-Central Chile to the late Oligocene increase in plate convergence rate. Rev Geol Chile 27(2):177–203

Pankhurst RJ, Rojas L, Cembrano J (1992) Magmatism and tectonics in continental Chile, Chile (41°–42°30'S). Tectonophysics 205: 283–294

Pankhurst RJ, Herve F (1994) Granitoid age distribution and emplacement control in the North Patagonian Batholith, Aysen, Southern Chile. 7. Congreso Geológico Chileno, Actas 2:1409–1413

Pankhurst RJ, Weaver SD, Herve F, Larrondo P (1999) Mesozoic-Cenozoic evolution of the North Patagonian Batholith in Aysen, southern Chile. J Geol Soc London 156:673–694

Parada MA, Nyström JO, Levi B (1999) Multiple sources for the Coastal Batholith of central Chile (31–34°S): geochemical and Sr-Nd isotopic evidence and tectonic implications. Lithos 46:505–521

Pardo Casas F, Molnar P (1987) Relative motion of the Nazca (Farallon) and South American plates since Late Cretaceous time. Tectonics 6(3):233–248

Pearce JA, Leat PT, Barker PF, Millar IL (2001) Geochemical tracing of Pacific-to-Atlantic upper mantle flow through the Drake Passage. Nature 410:457–461

Pilger RH (1984) Cenozoic plate kinematics, subduction and magmatism: South American Andes. J Geol Soc London 141(5):793–802

Ramos VA (1989) Andean foothills structures in northern Magellanes Basin, Argentina. AAPG 73:887–903

Ramos VA (2005) Seismic ridge subduction and topography: foreland deformation in the Patagonian Andes. Tectonophysics 399:73–86

Rapp RP, Watson EB (1995) Dehydration melting of metabasalt at 8–32 kbar: implications for continental growth and crust-mantle recycling. J Petrol 36(4):891–931

Reutter K-J (2001) Le Ande centrali: elementi di un'orogenesi di margine continental attivo, Acta Naturalia L'Ateneo Parmense 37(1–2): 5–37

Reutter K-J, Scheuber E, Helmcke D (1991) Structural evidence of orogen-parallel strike-slip displacements in the Precordillera of northern Chile. Geol Rundsch 80:135–153

Reutter KJ, Scheuber E, Chong G (1996) The Precordilleran fault system of Chuquicamata, Northern Chile: evidence for reversals along arc-parallel strike-slip faults. Tectonophysics 259:213–228

Rogers G, Hawkesworth CJ (1989) A geochemical traverse across the North Chilean Andes: evidence for crust generation from the mantle wedge. Earth Planet Sci Lett 91:271–285

Royden L (1996) Coupling and decoupling of crust and mantle in convergent orogens: implications for strain partitioning in the crust. J Geophys Res 101(B8):17679–17705

Rushmer T (1993) Experimental high pressure granulites: some applications to natural mafic xenolith suites and Archean granulite terranes. Geology 21:411–414

Sacks IS (1983) The subduction of young lithosphere. J Geophys Res 88:3355–3366

Sandeman HA, Clark AH, Farrar E (1995) An integrated tectonomagmatic model for the evolution of the southern Peruvian Andes (13–20°S) since 55 Ma. Int Geol Rev 37:1039–1073

Scheuber E, Gonzales G (1999) Tectonics of the Jurassic–Early Cretaceous magmatic arc of the north Chilean Coastal Cordillera (22°–26°S): a story of crustal deformation along a convergent plate margin. Tectonics 18(5): 895–910

Scheuber E, Reutter K-J (1992) Magmatic arc tectonics in the central Andes between 21° and 25°S. Tectonophysics 205:127–140

Scheuber E, Bogdanic T, Jensen A, Reutter K-J (1994) Tectonic development of the North Chilean Andes in relation to plate convergence and magmatism since the Jurassic. In: Reutter K-J, Scheuber E, Wigger P (eds) Tectonics of the Southern Central Andes. Springer-Verlag, Berlin Heidelberg New York, pp 121–139

Scholl WD, von Huene RE, Vallier T, Howell DG (1980) Sedimentation masses and concepts about tectonic processes at underthrust ocean margins. Geology 8:564–568

Silver PG, Russo RM, Lithgow-Bertelloni C (1998) Coupling of South American and African plate motion and plate deformation. Science 279:60–63

Somoza R (1998) Updated Nazca (Farallon)-South America relative motions during the last 40 Ma: implications for mountain building in the central Andean region. J South Amer Earth Sci 11(3):211–215

St Amand P, Allen CR (1960) Strike-slip faulting in northern Chile. Geol Soc Am Bull Abs 71:8965

Stocker RL, Ashby MF (1973) On the rheology of the upper mantle. Rev Geophys 11:391–426

Suarez M, De La Cruz R (2001) Jurassic to Miocene K-Ar dates from eastern central Patagonian Cordillera plutons, Chile (45–48°S). Geol Mag 138(1):53–66

Tebbens SF, Cande SC (1997) Southeast Pacific tectonic evolution from early Oligocene to Present. J Geophys Res 102(B6):12061–12084

Tomlinson AJ, Blanco N (1997) Structural evolution and displacement history of the West Fault system, Precordillera, Chile. VIII Congreso Geológico Chileno, Actas 4, pp 1873–1882

Turner S, Hawkesworth C (1998) Using geochemistry to map mantle flow beneath the Lau Basin. Geology 26:1019–1022

Van Westrenen W, Blundy JD, Wood BJ (2001) High field strength element/rare earth element fractionation during partial melting in the presence of garnet: implications for identification of mantle heterogeneities. Geochem Geophys Geosyst 2: doi 2000GC000133

Von Blanckenburg F, Davies JH (1995) Slab breakoff: a model for syncollisional magmatism and tectonics in the Alps. Tectonics 14:120–131

von Huene RE, Scholl DW (1991) Observations at convergent margins concerning sediment subduction, subduction erosion, and the growth of continental crust. Rev Geophys 29:279–316

Wendt JL, Regelous M, Collerson KD, Ewart A (1997) Evidence for a contribution from two mantle plumes to the island-arc lavas from northern Tonga. Geology 25:611–614

Wigger PJ, Schmitz M, Araneda M, Asch G, Baldzuhn S, Giese P, Heinsohn WD, Martinez E, Ricaldi E, Rüwer P, Viramonte J (1994) Variation in the crustal structure of the southern central Andes deduced from seismic refraction investigations. In: Reutter K-J, Scheuber E, Wigger P (eds) Tectonics of the Southern Central Andes. Springer-Verlag, Berlin Heidelberg New York, pp 23–48

Yanez G, Cembrano J, Pardo M, Ranero C, Selles D (2002) The Challenger-Juan Fernandez-Maipo major transition of the Nazca-Andean subduction system at 33–34°S: geodynamic evidence and implications. J S Am Earth Sci 15:23–38

Yuan X, Sobolev SV, Kind R, Oncken O, Bock G, Asch G, Schurr B, Graeber F, Rudloff A, Hanka W, Wylegalla K, Tibi R, Haberland C, Rietbrock A, Giese P, Wigger P, Rower P, Zandt G, Beck S, Wallace T, Pardo M, Comte D (2000) New constraints on subduction and collision processes in the Central Andes from P-to-S converted seismic phases. Nature 408(6815):958–961

Yuan X, Asch G, Bataille K, Bock G, Bohm M, Echtler H, Kind R, Oncken O, Wölbern I (2006) Deep seismic images of the southern Andes. In: Kay SM, Ramos V (eds) Evolution of an Andean margin: a tectonic and magmatic view from the Andes to the Neuquén Basin (35°–39° S lat). Geol Soc Am Spec P 407

Zandt G, Velasco AA, Beck SL (1994) Composition and thickness of the southern altiplano crust, Bolivia. Geology 22:1003–1006

# Chapter 17

# The Segmented Overriding Plate and Coupling at the South-Central Chilean Margin (36–42° S)

Ron I. Hackney · Helmut P. Echtler · Gerhard Franz · Hans-Jürgen Götze · Friedrich Lucassen · Dmitriy Marchenko
Daniel Melnick · Uwe Meyer · Sabine Schmidt · Zuzana Tašárová · Andrés Tassara · Susann Wienecke

**Abstract.** Identifying the parts of subduction zones that are susceptible to great earthquakes is a challenge that warrants considerable attention. In south-central Chile, where the 1960 $M_w$ 9.5 Valdivia earthquake occurred, we have combined surface geology and gravity data into a three-dimensional density model that helps to identify trench-parallel changes in fore-arc properties between 36 and 42° S. In light of suggestions that gravity data predict the seismogenic behavior of subduction zones, we use the gravity data and geological observations to separate the fore-arc in this region into three segments.

The northern Arauco-Lonquimay segment, where gravity anomalies are strongly positive, should be characterized by low coupling. In contrast, the plate interface under the Valdivia-Liquiñe and Bahia Mansa-Osorno segments to the south, where anomalies are negative or near-zero, should be highly coupled. The inferred differences in coupling are consistent with the extent of rupture during the Valdivia earthquake, which initiated under the southern part of the Arauco-Lonquimay segment but propagated southwards through the zone of inferred high coupling under the Valdivia-Liquiñe and Bahia Mansa-Osorno segments.

A three-dimensional gravity model of this region, constrained by surface geology and, in part, by independent seismic information, shows that one major control on the changing gravity-anomaly characteristics is the depth to the slab below the fore-arc. The model suggests that a north-to-south increase in the depth to the slab of about 5 km is possible. This increasing depth to the slab (i.e. increasing fore-arc thickness) can account for the inferred increase in coupling under the southern segments. A deeper slab would lead to greater shear stress and an increased coupling force at the plate interface. This increase is a result of: (1) greater normal stress acting on the plate interface induced by a thicker fore-arc, (2) slab buoyancy effects related to plate age (which also decreases from north to south), and (3) a shallower onset of sediment consolidation that increases the rigidity of material at the plate interface, thereby increasing the width of the frictionally coupled (unstably sliding) part of the subduction interface. Differences in fore-arc rheology, reflected in along-strike compositional differences and seismicity patterns, also have an important influence on coupling.

Other parameters that are often invoked to explain coupling differences (e.g. changes in trench sediment, convergence rate and seafloor texture) cannot explain differences in coupling here because these parameters do not change over the length of the trench examined, or they cause an effect that is contrary to the inferences based on gravity anomalies. The inferred coupling differences are also consistent with observed seismicity. In the Arauco-Lonquimay segment, where we infer coupling to be low, prominent seismicity is evident, indicating that the fore-arc is releasing the strain built up during convergence. In the Valdivia-Liquiñe and Bahia Mansa-Osorno segments, where we infer coupling to be high, fore-arc seismicity is limited. This suggests that strain is accumulating as the result of a locked plate interface, although aseismic slip cannot be ruled out.

## 17.1 Introduction

The convergent Andean margin, susceptible to some of the largest earthquakes on Earth, is an important region for understanding the mechanisms associated with, and the consequences of, plate coupling in subduction zones. Determining the degree of plate coupling gives a measure of how well the subducting and overriding plates are "stuck" and assists the identification of regions susceptible to great subduction-zone earthquakes. It has been suggested that plate coupling, or "the level of long-term shear stress along the plate interface" (Wang and Suyehiro 1999), depends on factors such as: the age of the subducting plate (thermal regime); subduction velocity and slab dip (e.g. Jarrad 1986; Kanamori 1986; Scholz 1990; Scholz and Campos 1995; Yáñez and Cembrano 2004); the amount of trench sediment (Ruff 1989; Lamb and Davis 2003); fore-arc rheology (McCaffrey 1993); properties of material at the subduction interface (Kanamori 1986; Pacheco et al. 1993; Scholz 1990); and the degree to which the incoming oceanic plate is attached to the slab descending in the upper mantle (Conrad et al. 2004).

Compared to other subduction zones (e.g. Marianas, Izu Bonin), the Andean subduction interface is considered to be highly coupled (e.g. Scholz and Campos 1995; Conrad et al. 2004). This high coupling is associated with the occurrence of large mega-thrust earthquakes with high seismic moment release (e.g. the 1960 $M_w$ 9.5 Valdivia and 1995 $M_w$ 8.0 Antofagasta earthquakes: Plafker and Savage 1970; Cifuentes 1989; Ruegg et al. 1996). Despite the generally high coupling inferred along the Andean margin, along-strike coupling variations are evident (Yáñez and Cembrano 2004; Hoffmann-Rothe et al. 2006, Chap. 6 of this volume). These along-strike variations reflect, among other factors, the age and convergence velocity of the subducting plate (Yáñez and Cembrano 2004), the down-dip width of the inter-seismic locked zone required to match onshore GPS-determined velocities (Khazaradze and Klotz 2003; Klotz et al. 2006, Chap. 4 of this volume), and the presence or absence of trench sediments (Lamb and Davis 2003).

Smaller-scale variations in plate coupling (over less than a few hundred kilometers) are also likely. Such variations have been inferred for subduction zones from estimates of seismic moment release and coseismic slip during earthquakes (e.g. Pacheco et al. 1993; Yamanaka and Kikuchi 2004), stress inhomogeneity inferred from fault-plane solutions (e.g. Lu and Wyss 1996), GPS monitoring (e.g. Nishimura et al. 2004; Bürgmann et al. 2005), and fore-arc seismic tomography images (e.g. Mishra et al. 2003).

Identifying and understanding differences in plate coupling are essential if processes related to great subduction-zone earthquakes are to be better understood. Seismic methods are typically used to identify regions of high coupling (asperities). However, an apparent global correlation between trench-parallel, negative gravity anomalies, fore-arc sedimentary basins and regions of high slip during earthquakes (Song and Simons 2003; Wells et al. 2003) suggests that gravity data could also be a useful indication of where slip is likely to occur during great subduction-zone earthquakes.

In this paper, we focus on the Chilean convergent margin between 36 and 42° S (Fig. 17.1) where the globally observed correlation between gravity anomalies and rupture of the plate interface is particularly evident (Song and Simons 2003; Wells et al. 2003). We seek to determine the factors controlling the differences in coupling that can be inferred from gravity anomalies. These factors may result from subducting plate characteristics or, because gravity anomalies and the geoid also provide a clear indication for regional-scale segmentation of the overriding (continental) plate, from the influence of the overriding plate.

To aid the interpretation of stress-state on the subduction interface in south-central Chile, we describe the substantial geological and geophysical data that are now available for constraining slab and fore-arc geometry in this region, some of which has not been previously published. Gravity data and the geoid clearly define a segmented south-central Chilean fore-arc, and present geological knowledge allows the geological characteristics of these segments to be determined. A three-dimensional density model constrained by potential field data, geological information and other geophysical data suggests that segmentation of the overriding plate coincides with variations in the depth to the subducting Nazca Plate under

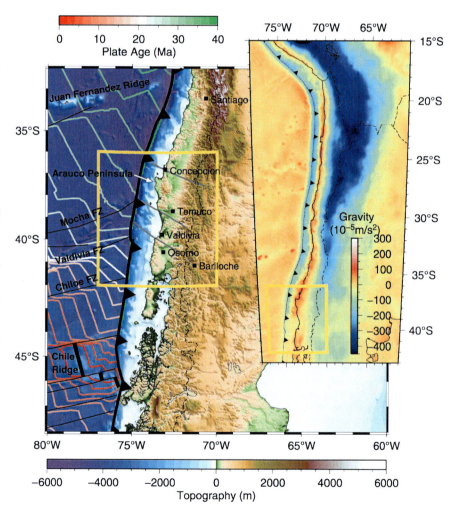

**Fig. 17.1.**
Topographic and bathymetric map of the south-central Andes from the GEBCO database (http://www.ngdc.noaa.gov/mgg/gebco/gebco.html). *Colored contours* offshore show the age of the subducting Nazca Plate (Müller et al. 1997). The *yellow box* marks the region examined in this paper and the WNW-trending *gray lines* within this box mark the boundaries of the arc/fore-arc segments discussed in the text. The *inset map* shows gravity anomalies for part of the South American convergent margin. Free-air anomalies are shown offshore (from the 2001 version of the satellite altimetry-derived database described by Andersen and Knudsen 1998) and Bouguer anomalies onshore (data from Araneda et al. 1999; Blitzkow, pers. comm. 2003; Götze et al. 1994, 1996; Tašárová 2004)

the fore-arc. The associated variations in the thickness of the fore-arc are likely to lead to fore-arc-parallel variations in normal stress and differences in the material properties of rocks at, and near, the subduction interface. The ability of the fore-arc to accumulate elastic strain should also be an important control on where slip during earthquakes occurs. All of these differences must then exert some control on the level of shear stress at the plate interface, on the degree of coupling between the subducting and overriding plates, and on the likely location of rupture during great subduction-zone earthquakes.

## 17.2 Arc/Fore-Arc Geology in South-Central Chile

A morphotectonic map of the south-central Andean arc and fore-arc between 36 and 42° S is shown in Fig. 17.2. This map is based on more detailed compilations (Melnick et al. 2006a,b; Melnick and Echtler 20a06, Chap. 31 of this volume; Götze et al. 2006, Chap. 26 of this volume) of geological data sourced from 1 : 250 000- and 1 : 100 000-scale maps of Chile published by the Servicio Nacional de Geología y Minería. Regional information was also partly taken from 1 : 1 000 000 (Chile) and 1 : 2 500 000 (Argentina) maps produced by SERNAGEOMIN (2003) and SEGEMAR (1998), respectively. This earlier information has been supplemented with new field mapping (e.g. Melnick et al. 2006a,b).

In general, the Andean margin between 36 and 42° S is characterized by a coastal range, a longitudinal depression and a main cordillera where the active volcanic arc is located. In detail, the margin can be subdivided into nine main morphotectonic units (Fig. 17.2) on the basis of geological, geophysical and geomorphological data.

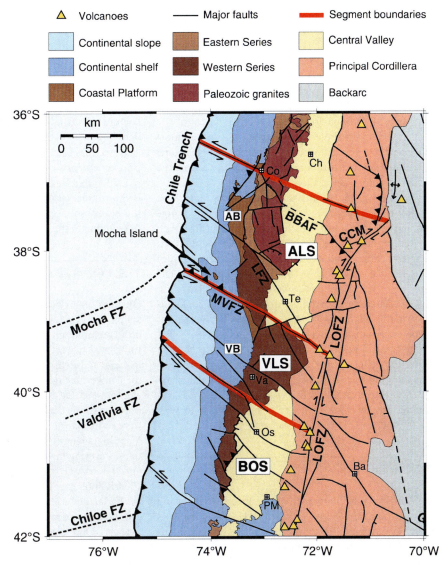

**Fig. 17.2.**
Map of the arc and fore-arc in south-central Chile showing major morphotectonic units and structural features. The *thick red lines* separate arc/fore-arc segments defined on the basis of gravity and geoid data (Fig. 17.3). ALS is Arauco-Lonquimay, VLS is Valdivia-Liquiñe, and BOS is Bahia Mansa-Osorno segment. The location of the Chile trench is interpreted from bathymetric data obtained during the SPOC project (see Völker et al. 2006, Chap. 9 of this volume). *Yellow triangles* show active volcanoes and *dashed lines* offshore mark fracture zones (*FZ*). Offshore fore-arc basins discussed in the text are also indicated: *AB*: Arauco Basin; *VB*: Valdivia Basin. Major fault structures are: *BBAF*: Bío-Bío-Aluminé Fault; *CCM*: Callaqui-Copahue-Mandolegüe transfer zone; *LOFZ*: Liquiñe-Ofqui Fault Zone; *LFZ*: Lanalhue Fault Zone; *MVFZ*: Mocha-Villarica Fault Zone. Locations marked include: *Co*: Concepción; *Ch*: Chillan; *Te*: Temuco; *Va*: Valdivia; *Os*: Osorno; *PM*: Puerto Montt; and *Ba*: Bariloche

## 17.2.1 Fore-Arc

The fore-arc comprises a coastal platform, Coastal Cordillera and Central Valley (Fig. 17.2). The coastal platform comprises uplifted shelf deposits consisting of late Cretaceous to Quaternary marine and minor continental deposits (e.g. Pineda 1986) that outcrop in coastal flatlands and low altitude ranges.

The Coastal Cordillera is a narrow mountain belt located between the coast and the Central Valley. It is ~10–100 km wide and ranges in elevation from 1 500 m at Nahuelbuta (38° S) to only a few hundred meters in the Valdivia region (40–41° S). Rocks of the Coastal Cordillera include those from a Permo-Triassic accretionary complex and a Permo-Carboniferous magmatic arc (Hervé 1977; 1988; Hervé et al. 1988; Martin et al. 1999; Duhart et al. 2001; Willner et al. 2004).

The accretionary complex is generally recognised as comprising two major units whose composition is summarized by Willner et al. (2004). The low-grade Western Series is dominated by metagraywacke and metapelite, but contains 15 to 20% metabasite and other rocks of oceanic origin. These rocks are penetratively deformed and are characterized by an intense planar transposition and schistocity. In comparison, the metagraywackes of the Eastern Series were subjected to very low-grade metamorphism and lack any strong foliation. These deformational and compositional differences suggest variance in the fore-arc rheology along the margin.

The late Carboniferous to Permian magmatic arc, which forms a prominent batholith in the Coastal Range (Nahuelbuta Mountains) north of about 38° S (Fig. 17.2), resulted from a major phase of granitoid magmatism along the entire western margin of Gondwana (Rapela and Kay 1988; Hervé et al. 1988; Lucassen et al. 1999). The geochemistry and isotopic composition of this batholith and associated metamorphic and sedimentary rocks indicate major recycling of continental crust by the late Paleozoic magmatism (Lucassen et al. 2004).

The Central Valley contains about two kilometers of Pliocene and Quaternary sediments unconformably overlying late Oligocene to Miocene volcano-sedimentary basins (Muñoz et al. 2000; Jordan et al. 2001). It extends discontinuously throughout the study area, from 36 to 39° S and from 40 to 42° S (where it is partly submarine).

## 17.2.2 Intra-Arc

The Principal Cordillera, with mean altitudes decreasing from ~2 700 m at 36° S to ~1 000 m at 42° S (Fig. 17.1), is the main morphologic component of this region. It includes the southern active volcanic arc (the Southern Volcanic Zone of Hildreth and Moorbath 1988) that coincides with the axis of a long-lived magmatic arc that was constructed in the Jurassic and has occupied a relatively stationary position since (Mpodozis and Ramos 1989). In general, only Pliocene to Recent subaerial volcanic complexes are preserved. South of 39° S, higher exhumation rates related to glacial erosion and transpressional tectonics have lead to exposure of the ~5–15 km-deep roots of the magmatic arc (Seifert et al. 2005). North of 39° S, the Mesozoic–Cenozoic volcano-sedimentary cover is still preserved.

Voluminous, Jurassic to Recent, subduction-related magmatism is restricted to the Principal Cordillera (López-Escobar 1984; Hildreth and Moorbath 1988; Lucassen et al. 2004; North Patagonian Batholith south of 40° S, see Pankhurst et al. 1999). Most intrusions between 36 and 42° S are late Cretaceous to late Tertiary in age (Suárez and Emparan 1997; Lucassen et al. 2004, and references therein). West of the Cordillera, isolated Mesozoic magmatism at the coast (Martin et al. 1999) and a mid-Tertiary belt (Muñoz et al. 2000) of scattered volcanic and intrusive rocks are also present.

## 17.3 Potential Field Data and Fore-Arc Segmentation

Segmentation of the fore-arc in the southern Andes is evident in gravity data and the geoid. While segmentation is also evident in surface geology, the gravity and geoid data provide a depth-integrated image of fore-arc characteristics. Examining the segmentation of the overriding plate is important because if it is more than surficial, then it might play some role in causing differences in the degree of coupling at the plate interface.

## 17.3.1 Gravity Data and Geoid Model

Onshore gravity data are derived from several different sources and comprise data acquired over the last 30 years or so (Schmidt and Götze 2006, Chap. 28 of this volume). Industry data were provided under agreements with ENAP in Chile and Repsol-YPF in Argentina. Data measured by the Universidad de Chile, covering a large part of the Chilean fore-arc that is the main focus of this paper, are also included in the database (Araneda et al. 1999). Other data were sourced from the United States National Imagery and Mapping Agency. A large amount of new data (~2000 stations) covering the region in Argentina between the Andes and the Atlantic coast were measured in 2000 (Wienecke 2002; Ramos et al. 2002). Figure 17.3 shows gravity anomalies based on the above data sources and, for comparison, a geoid model computed as part of a project to determine a new geoid model for South America.

The variety of methods and instrumentation used to acquire the different onshore gravity data meant that data

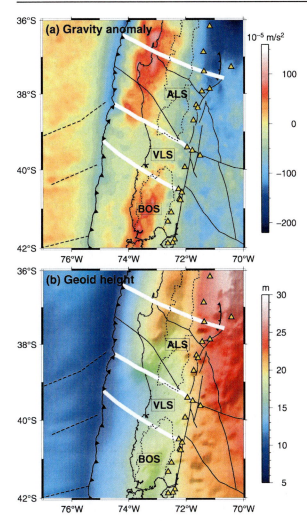

**Fig. 17.3. a** Gravity anomaly map for the south-central Chilean convergent margin: Bouguer anomalies onshore and free-air anomalies offshore. **b** Geoid model for the same region. The images are overlain with interpreted arc/fore-arc segment boundaries (*thick white lines*), and major structural features (*thin black lines*; cf. Fig. 17.2). The dashes outline the Central Valley and volcanoes are shown with *yellow triangles*. *Dashed lines* offshore mark the location of fracture zones

covering Chile were reprocessed and checked against new control measurements made during 2002 (Tašárová 2004). Offshore data are sourced from shipbourne measurements of the SPOC (Subduction Processes Off Chile) project (Flüh et al. 2002; Reichert et al. 2002), and these data are integrated with data from the 2001 version of the KMS global free-air anomaly database (Andersen and Knudsen 1998).

All measurements are tied to the IGSN71 gravity datum. Bouguer anomalies were computed using the normal gravity formula of 1967 and a spherical Bouguer cap correction (cap radius 167 km, density 2670 kg m$^{-3}$). Terrain corrections on land were computed using a method that incorporates triangular facets to approximate topography to a distance of 167 km from the station (Götze and Lahmeyer 1988; Müller 1999). These terrain corrections were applied using the 1×1 km GLOBE (onshore) and ETOPO5 (offshore) digital elevation models (Hastings et al. 1999; NOAA 1988). Processing was conducted using an in-house Java-based program, DBGrav.

The resulting gravity anomaly map is shown in Fig. 17.3a. The most obvious features in the combined free-air and Bouguer gravity field are the anomaly low associated with the Chile trench, the negative anomalies associated with the Andean mountain range, and the strong positive anomaly that parallels the coastline but is disrupted around 39° S. The general trend of gravity anomalies is north–south, reflecting the dominant orientation of the major morphotectonic elements of the south-central Andes (cf. Fig. 17.2). The prominent coastal gravity high evident in Fig. 17.3a is as much a feature of the southern Chile margin as it is of the entire Pacific coast of the Andes (Fig. 17.1). The nature of this coastal high is also typical of most convergent continental margins. The positive coastal anomaly of about $100 \times 10^{-5}$ m s$^{-2}$ ($1 \times 10^{-5}$ m s$^{-2}$ = 1 mGal) is disrupted at about 39° S where the Central Valley is absent and the Valdivia Basin (VB in Fig. 17.2) dominates the offshore fore-arc.

In contrast to the central Andes, the Andean mountain range in this region is generally not associated with an extreme Bouguer anomaly low (cf. Götze et al. 1994; Tassara and Yáñez 2003) (Fig. 17.1). North of 38° S, the minimum Bouguer anomalies reach about $-180 \times 10^{-5}$ m s$^{-2}$ (cf. $-450 \times 10^{-5}$ m s$^{-2}$ in the central Andes). This less-negative anomaly reflects the lower elevation of the Andes in this region and the associated reduction in crustal thickness (up to 80 km in the central Andes versus about 40–45 km in this region: e.g. Bohm et al. 2002; Yuan et al. 2002, 2006; Lüth et al. 2003; Tassara and Yáñez 2003).

Offshore, the free-air anomaly low associated with the Chile trench reflects the mass deficit associated with its substantial depth (4–5 km; Fig. 17.1). The gravity expression of fracture zones and ridges in the oceanic plate is also visible.

Variations in the high-resolution geoid model covering south-central Chile (Fig. 17.3b) also reflect the features evident in gravity anomalies. Because the geoid still contains the effects of topography, relatively positive geoid heights correlate with the Coastal Cordillera and the higher Andes in the north of the area (Figs. 17.2 and 17.3b). Offshore, where free-air anomalies closely mimic bathymetry, the geoid better reflects mass distribution within the oceanic lithosphere. The Chile trench is an obvious feature, but positive geoid height anomalies (~5 m) correlate with the flexural bulge seawards of the trench that results from the bending of the Nazca Plate as it subducts. The smooth pattern related to this bulge is interrupted by the effects of the ENE-trending oceanic ridges and fracture zones discussed above.

### 17.3.2 Segmentation of the Overriding Plate

The general trend of gravity anomalies is north–south, reflecting the dominant structural trend of the margin. However, the south-central Andes are characterized by a prominent along-strike segmentation that is evident in both geological and geophysical data (Figs. 17.2 and 17.3). Much of the along-strike segmentation is controlled by pre-existing structures, particularly the NW–SE-trending fault zones shown in Fig. 17.2. The Bío-Bío-Aluminé fault (BBAF in Fig. 17.2) separates two major tectonic domains. The block north of the Bío-Bío-Aluminé fault is characterized by Quaternary shortening and compression in the back-arc and significant late Miocene inversion (Folguera et al. 2003). The region south of the Bío-Bío-Aluminé fault is characterized by along–strike transtension in the intra-arc since the Pliocene (Folguera et al. 2002; Melnick et al. 2006a), westward migration and narrowing of the volcanic arc since the early Pleistocene (Stern 1989; Lara et al. 2001), and uplift and contractional deformation in the fore-arc (Melnick and Echtler 2006b).

In detail, based mainly on observations of the fore-arc and Principal Cordillera, we discuss three domains along the south-central Andes, namely the Arauco-Lonquimay, Valdivia-Liquiñe and Bahía Mansa-Osorno segments (Figs. 17.2 and 17.3). These segments are characterized by distinct gravity and geoid signatures (Fig. 17.3) and their different characteristics may also reflect differences in coupling at the plate interface below the fore-arc.

#### 17.3.2.1 Arauco-Lonquimay Segment

The Arauco-Lonquimay segment and the region north of the Bío-Bío-Aluminé fault are characterized by a prominent north–south-trending positive gravity anomaly (up to $100 \times 10^{-5}$ m s$^{-2}$) centered on the Coastal Cordillera that extends roughly from the coastline to the Central Valley (Fig. 17.3a). Over the trench, free-air anomalies are more negative than further to the south ($-100 \times 10^{-5}$ m s$^{-2}$ versus $-25 \times 10^{-5}$ m s$^{-2}$). In the arc, negative Bouguer anomalies reaching $-200 \times 10^{-5}$ m s$^{-2}$ are associated with the Andean Range. Relatively positive geoid heights (~5 m) also coincide with the Coastal Cordillera. Geoid height anomalies corresponding to the Central Valley are small (Fig. 17.3b).

Based on a major change in tectonic style, this segment is bound to the north by the Bío-Bío-Aluminé fault in the fore-arc and in the intra-arc zone by the Callaqui-Copahue-Mandolegüe transfer zone (CCM in Fig. 17.2; Folguera and Ramos 2000; Melnick et al. 2006a). This transfer zone decouples Plio-Quaternary back-arc shortening to the north from the dominant intra-arc transtension and strike-slip motion associated with the Liquiñe-Ofqui fault zone (LOFZ in Fig. 17.2) to the south. Despite this apparent tectonic segmentation, fore-arc gravity anomalies across this boundary are very similar (Fig. 17.3a). Therefore, on the basis of gravity anomalies alone, the Arauco-Lonquimay segment could be considered to extend further north.

Within the Arauco-Lonquimay segment, a clear lithological and tectonic change across the NW–SE-trending Lanalhue fault zone (LFZ in Fig. 17.2) is also present in the Coastal Range at about 38° S. This is where the subduction-related accretionary complex, with its so-called Western Series high-pressure/low-temperature metamorphism (Hervé 1988; Martin et al. 1999; Willner et al. 2000; 2004), is juxtaposed against the late Paleozoic coastal batholith (Fig. 17.2). The lack of a corresponding gravity gradient across this boundary in the Coastal Cordillera suggests that fore-arc compositional differences are not the major cause of fore-arc gravity differences.

To the south, the Arauco-Lonquimay segment is bound by the Mocha-Villarica fault zone (MVFZ in Fig. 17.2), which crosses the entire active margin. The Mocha-Villarica fault zone is defined by Melnick et al. (2003) on the basis of structural data from Mocha Island, bathymetry from the SPOC project (Völker et al. 2006, Chap. 9 of this volume) and regional morphology and structure within the Central Valley and Principal Cordillera. This boundary is clearly marked by the change in erosion level across the Mocha-Villarica fault zone. South of the Mocha-Villarica fault zone, the deeper parts of the magmatic arc are exposed, whereas north of the fault zone, Mesozoic–Cenozoic volcano-sedimentary cover rocks are still preserved. In the fore-arc, the Mocha-Villarica fault zone controls the uplift that produced Mocha Island and it is evident as a 1 500 m-high, NW-trending fault-scarp in the continental slope that sinestrally offsets the trench (Fig. 17.2). In the arc, the alignment over some 50 km of the Villarica, Quetrupillan and Lanin volcanoes is the Quaternary expression of the Mocha-Villarica fault zone.

The main morphological feature of this segment is the Arauco Peninsula (Fig. 17.1), where the South American coastline extends furthest to the west, resulting in the minimum distance to the trench (~70 km) and hence the broadest part of the onshore fore-arc. The Arauco Peninsula is an uplifted block of continental shelf that records temporally and spatially discontinuous marine and continental fore-arc basin formation since the late Cretaceous. Episodes of uplift and erosion alternate with subsidence, sedimentation and sea-level change (Pineda 1986; LeRoux and Elgueta 1997; Melnick and Echtler 2006b). Since the early Quaternary, active fore-arc deformation was dominated by regional coastal uplift (Plafker and Savage 1970; Kaizuka et al. 1973; Nelson and Manley 1992). This suggests that significant basal accretion of sediments is taking place under the margin (Glodny et al. 2006, Chap. 19 of this volume). Evidence for basal accretion is interpreted in seismic data at 38°15' S (Krawczyk et al. 2006, Chap. 8 of this volume) and also includes a zone of enhanced seismicity (Bohm et al. 2002) and relatively low p-wave seismic velocities (Bohm 2004) in the Arauco area.

### 17.3.2.2 Valdivia-Liquiñe Segment

The Valdivia-Liquiñe segment is anomalous with respect to adjacent segments. In stark contrast to the gravity signature of the Arauco-Lonquimay segment, it is characterized by negative gravity anomalies, ranging from about 0 to $-50 \times 10^{-5}$ m s$^{-2}$ (Fig. 17.3a), that are unique along almost the entire Andean margin (Fig. 17.1). Geoid height differences are also subdued (Fig. 17.3b).

This segment is bound to the south by increasingly positive gravity anomalies associated with the onshore fore-arc. Morphologically, this segment is characterized by a prominent retreat of the coastline (maximum distance to the trench ~150 km). In this segment, marine and continental fore-arc basin formation started in the Eocene and the Valdivia Basin dominates the offshore fore-arc (Mordojovich 1981; Fig. 17.2). Late Quaternary minor uplift is limited to the coastal area.

The topography of the Coastal Cordillera in this segment does not exceed a few hundred meters. The most notable morphological feature is the absence of a Central Valley, meaning that the low-relief Coastal Cordillera has a direct morphologic transition to the Principal Cordillera (Fig. 17.2). In the Liquiñe area, Neogene to Recent deformation along the intra-arc reflects strike-slip kinematics related to the Liquiñe-Ofqui Fault Zone (Rosenau et al. 2006) and Pliocene-Quaternary fission-track-cooling ages indicate high exhumation rates along the axis of the Principal Cordillera (Gräfe et al. 2002).

### 17.3.2.3 Bahía Mansa-Osorno Segment

The Bahía Mansa-Osorno segment is characterized by positive gravity anomalies similar to those observed in the Arauco-Lonquimay segment (Fig. 17.3a). However, these positive anomalies are centered on the Central Valley rather than over the Coastal Range. Anomalies over the offshore fore-arc are dominantly negative. A relative geoid high of a few meters is also centered on the Central Valley (Fig. 17.3b). It lies adjacent to a relative geoid low that coincides with the glacial lakes at the western edge of the Principal Cordillera.

The geometry of this segment is intermediate to the northern segments. The trench–coastline separation is about 100 km and the topography of the narrower Coastal Cordillera, as in the Valdivia-Liquiñe segment, is moderate. The most outstanding feature is the well-developed sedimentary basin of the Central Valley. Late Miocene, partly inverted, margin-parallel normal faults along the western contact with the Coastal Cordillera have controlled the formation of the basin since the late Oligocene (Muñoz et al. 2000; Jordan et al. 2001; SERNAGEOMIN 2002). The intra-arc zone, comprised mainly of the ~10 km-deep roots of the Cretaceous to Miocene volcanic arc (Cingolani et al. 1991; Seifert et al. 2005), also has low average topography (~1 500 m) and is dominated by dextral transpression along the Liquiñe-Ofqui Fault Zone (Lavenu and Cembrano 1999).

## 17.4 3D Density Model

To further investigate the factors related to and controlling the segmentation of the arc and fore-arc in south-central Chile, a three-dimensional (3D) density model was constructed (Fig. 17.4; Tašárová 2004). This model is constrained by the geological observations and gravity data already discussed, but also by independent information derived from seismic studies. The gravity model provides a means to examine the cause of the prominent differences in gravity anomalies for each segment. Possible causes are variations in fore-arc density, changes in the depth to the slab below the fore-arc and, to a lesser extent, the geometry of the fore-arc at depth and crustal thinning under the Central Valley.

### 17.4.1 Model Construction

The 3D density model (Fig. 17.4), discussed in detail by Tašárová (submitted), is focussed on the fore-arc region and includes the accretionary complex, the crust of the overriding plate, the mantle wedge and the subducted plate. It was constructed using the IGMAS forward-modeling software, which represents geological bodies with triangular-faceted polyhedra (e.g. Schmidt and Götze 1998). The gravity effect of these polyhedra is computed using the method described by Götze and Lahmeyer (1988). The 3D geometry is achieved by triangulating between the common interfaces of 29 separate cross sections that have an average separation of 20 km in areas where constraining data are abundant (36–39° S), and 30 km elsewhere.

Where possible, the density model is constrained by geological and geophysical observations. Substantial new geophysical data have been acquired between 36 and 39° S as part of the projects ISSA 2000 (Bohm et al. 2002; Bohm 2004; Lüth et al. 2003; Yuan et al. 2006) and SPOC (Krawczyk et al. 2003, 2006, Chap. 8 of this volume; Völker et al. 2006, Chap. 9 of this volume). These onshore and offshore seismic stud-

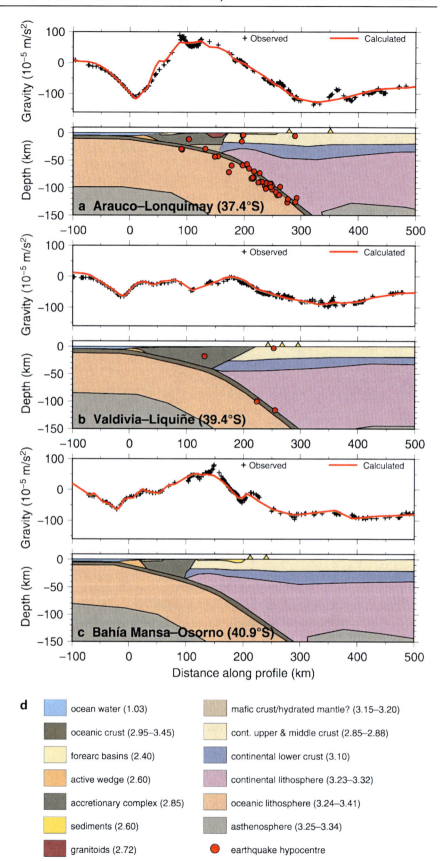

**Fig. 17.4.**
Cross sections through the 3D density model that are representative of each arc/fore-arc segment (see Fig. 17.5 for locations). **a** Arauco-Lonquimay segment; **b** Valdivia-Liquiñe segment; **c** Bahía Mansa-Osorno segment. A legend for the cross sections, including the range of densities for each unit (in Mg m$^{-3}$), is shown in (**d**). *Red circles* represent seismicity within 20 km of each profile recorded during the ISSA 2000 campaign (Bohm et al. 2002) and *yellow triangles* mark volcanoes within about 50 km of each profile. Observed gravity data (+ symbols) are from measurements within 2 km of each profile

ies provide important constraints on the geometry, composition and tectonics of the fore-arc region. The geophysical data are complemented by the geological and structural information gathered both in the field and from published work. Densities in the model were assigned on the basis of the ISSA 2000 and SPOC seismic velocity and local earthquake tomography models (e.g. Lüth et al. 2003; Bohm 2004; Krawczyk et al. 2006, Chap. 8 of this volume) and published relationships between seismic velocity and density for appropriate rock compositions (Sobolev and Babeyko 1994). A procedure using data and methods described by Hacker and Abers (2004) was also used in part.

## 17.4.2 Main Features of the Model

### 17.4.2.1 Continental Crust

The continental crust beneath the Central Valley, volcanic arc and back-arc was modeled with three layers: an upper, middle and lower crust. The upper crust is characterized by densities that change according to geological observations, whereas the middle and lower crust have constant densities of 2.85 Mg m$^{-3}$ and 3.10 Mg m$^{-3}$, respectively, throughout the entire model. These densities are constrained by seismic velocities in the Arauco-Lonquimay segment. With upper-crustal density constrained by surface geology, the Moho depth along the ISSA 2000 receiver function profile at 39° S (Yuan et al. 2006) also provides a constraint on densities in the deeper crust.

The volcanic arc is characterized by a crustal thickness of 45 to 50 km in the Arauco-Lonquimay segment, which is consistent with available interpretations (Tassara and Yáñez 2003; Bohm 2004; Yuan et al. 2006). The modeled crustal thickness gradually decreases to less than 40 km in the Bahía Mansa-Osorno segment, but crustal thickness in the southern segments and in the back-arc is poorly constrained. In the back-arc, only receiver function data on the ISSA 2000 profile give an indication of Moho geometry (Yuan et al. 2006).

The ISSA 2000 local earthquake tomography model suggests significant crustal thinning below the Central Valley in the Arauco-Lonquimay segment (Bohm 2004). Based on the tomography results, the crustal thickness beneath the Central Valley is about 35 km and the excess mass associated with shallower mantle contributes to the observed positive gravity anomalies. In the Bahía Mansa-Osorno segment, which has a similar morphology to the Arauco-Lonquimay segment, thinned crust of about 28 km thickness is also inferred. In this case, the thinned crust reproduces the positive anomalies centered on the Central Valley (Figs. 17.3 and 17.4). In the Valdivia-Liquiñe segment, where the Central Valley is absent and gravity anomalies are more negative, crustal thickness is modeled as 38–43 km, similar to that modeled under the volcanic arc.

### 17.4.2.2 Fore-Arc Accretionary Complex

In all three segments, the fore-arc accretionary complex is represented by a body that is triangular in cross section and that has a constant density of 2.85 Mg m$^{-3}$. The choice of density for the accretionary complex strongly influences modeled gravity anomalies and, where no constraints are available, also the position of the slab beneath the fore-arc. A density of 2.85 Mg m$^{-3}$ provides the best match between observed and calculated anomalies in the Arauco-Lonquimay segment where slab geometry is relatively well constrained by seismic results.

The modeled structure of the accretionary complex in the Arauco-Lonquimay segment is more complex than in the other segments. It contains a body representing the granitoids of the Nahuelbuta Mountains (red body, Fig. 17.4a), which are not present south of about 38° S (Fig. 17.2). Moreover, a high-density body (3.15 Mg m$^{-3}$ in the upper part and 3.20 Mg m$^{-3}$ in the lower part) at the base of the accretionary complex is required as a shorter-wavelength contribution to the pronounced gravity high in the Arauco-Lonquimay segment. This body could represent mafic lower crust or serpentinized/hydrated mantle, but its exact nature is unclear (Krawczyk et al. 2006, Chap. 8 of this volume).

### 17.4.2.3 The Subducting Plate and Its Geometry

The densities in the subducting plate are based on the petrological models of Hacker et al. (2003) and its temperature is related to plate age. Its thickness, which decreases from north to south, is calculated from a simple age-thickness relationship (Turcotte and Schubert 2002, p 159). North of about 39° S, the depth to the subducting plate is constrained by earthquake hypocenters and seismic results (Bohm et al. 2002; Lüth et al. 2003; Krawczyk et al. 2006, Chap. 8 of this volume). At greater depths, the top of the slab is simply assumed to be about 100 km below the volcanic arc. However, given the globally wide range of slab depths below arcs (65–120 km; England et al. 2004) and the complicated relationship between depth of melting, melt paths and arc location (e.g. Schurr et al. 2003), this is a grossly simplified assumption.

Unlike the Arauco-Lonquimay segment, the slab dip below the Valdivia–Liquiñe segment is not constrained. Therefore, the anomalously negative gravity anomalies in this region could be explained by either a deeper slab or a reduced density within the fore-arc accretionary complex (i.e. density less than 2.85 Mg m$^{-3}$). A reduced density is not consistent with more mafic fore-arc composition in the south, nor is it consistent with densities measured from surface samples (Tašárová 2004). As a consequence, we assume that the density in the accretionary complex is constant and that the negative gravity anomalies in this region are explained by a deeper slab below the fore-arc. If the density in the accretionary complex was increased to better reflect geology then, to explain the negative gravity anomalies, the slab would be even deeper.

In the Arauco-Lonquimay segment, the model slab is at shallower depth than in the two segments to the south (Figs. 17.4 and 17.5). At a distance of 100 km from the trench axis, the seismically-constrained top of the slab lies at about 25 km depth. In contrast, at the same distance from the trench in the Bahía Mansa–Osorno segment, the top of the slab lies at about 30 km depth. In addition to the effects from crustal thinning under the Central Valley and denser material near the base of the accretionary complex, the shallower slab under the Arauco–Lonquimay segment makes a significant contribution to the positive anomalies associated with the Coastal Cordillera (especially at longer-wavelengths). The lack of positive anomaly in the Valdivia–Liquiñe segment largely reflects the increased depth to the slab, but also the absence of the Central Valley and its corresponding thinned crust. It is important to reiterate that the depth to the subducting plate is not constrained south of 39° S. However, significant variations in the depth to the Nazca Plate are evident in SPOC seismic data (Fig. 17.6), which means that the depth-to-slab variations inferred south of 39° S are plausible.

## 17.5 Discussion

The geological and geophysical data discussed above suggest that the south-central Chilean margin can be separated into distinct segments characterized by different fore-arc thickness and depth to the subducting plate. In the following discussion, we will argue that these differences, together with differences in fore-arc rheology, are likely to influence variations in shear stress at the plate interface and have implications for the degree of plate coupling, deformation of the overriding plate, and the occurrence of great earthquakes.

### 17.5.1 What Is Plate Coupling?

Plate (or fault) coupling can be an ambiguous term that gives only a vague indication of the mechanical interaction between rocks on either side of a fault (e.g. Wang and Dixon 2004; Lay and Schwartz 2004). Therefore, a

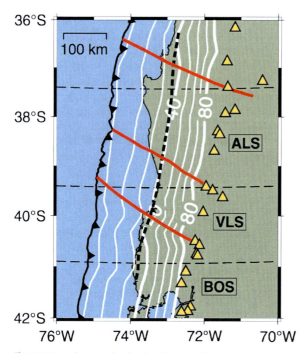

**Fig. 17.5.** Map showing the depth to the top of the subducting Nazca Plate from the 3D density model. Contours are only shown to a depth of 80 km (contour interval 10 km). *Triangles* mark volcanoes and *fine dashed lines* show the location of the cross sections plotted in Fig. 17.4. The *thick black dashed line* shows the intersection between the slab and continental Moho. The *contours* suggest that to depths of about 30 km, the slab is shallower beneath the Arauco-Lonquimay segment than beneath the Valdivia-Liquiñe and Bahía Mansa-Osorno segments

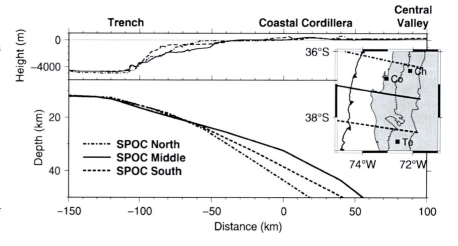

**Fig. 17.6.** Profiles showing bathymetry/topography (*above*) and depth to the Moho of the subducting Nazca Plate (*below*) from SPOC wide-angle seismic profiles (Krawczyk et al. 2006, Chap. 8 of this volume). Moho variations are assumed to parallel the depth to the top of the slab. Depth variations of up to 10 km exist beneath the Coastal Cordillera and the slab is shallowest under the Arauco Peninsula (SPOC Middle). The *inset map* shows the approximate location of each profile, main cities (as in Fig. 17.2) and the outline of the Central Valley (*dotted line*)

short clarification of the meaning of plate coupling in the context of this paper is warranted before discussing the implications of the inferred segmentation.

In the simplest terms, plate coupling can be thought of as the degree to which the overriding plate is "stuck" to the subducting plate – high coupling means that the plates are "well stuck" at all times other than during earthquakes. From a seismological perspective, high coupling is related to a high seismic-coupling coefficient, the ratio of the amount of slip during earthquakes to plate convergence velocity (e.g. Ruff and Kanamori 1983; Pacheco et al. 1993; Scholz and Campos 1995). A high seismic-coupling coefficient means that almost all accumulated strain is released during earthquakes and that the subduction interface is likely to be 100% locked during inter-seismic periods, as might be inferred from the modeling of GPS displacements.

These definitions of coupling are commonly explained by the distribution of asperities on the plate interface (e.g. Kanamori 1986; Pacheco et al. 1993; Scholz 1998; Lay and Schwartz 2004). Between earthquakes, asperities are locked zones where inter-seismic strain accumulates, eventually leading to rupture and high seismic moment release. The properties of the parts of the plate interface associated with asperities are conducive to stick-slip movement (unstable frictional sliding), which means that high coupling reflects a situation in which a large part of the plate interface is susceptible to unstable frictional sliding. Extensive asperities may be caused by the subduction of large amounts of sediment (e.g. Ruff and Kanamori 1983; Ruff 1989), whereas more isolated asperities may be caused by subducted seamounts (Scholz and Small 1997) or horst-and-graben structures (Ruff and Kanamori 1983).

The rheology, geometry and structure of the overriding plate may also be an important control on plate coupling (e.g. McCaffrey 1993; Reyners 1998). Fore-arc rheology is especially important because it will determine whether or not the shear stress acting across the subduction interface can be sustained (McCaffrey 1993), thereby governing whether or not extensive asperities can be formed. In this case, high coupling requires that the fore-arc is capable of accumulating significant strain.

In this discussion, we refer to plate coupling in terms of a "coupling force" (Wang and He 1999) that can be related to the different definitions and explanations outlined above and to parameters that can be inferred from geophysical observations. When we refer to high coupling, we mean a large coupling force. With this definition, high coupling corresponds to: a situation in which the subducting and overriding plates are well stuck; a seismic coupling coefficient close to one; a plate interface dominantly in the unstable frictional sliding regime; an interface characterized by many asperities that are possibly interlinked; or a situation in which the fore-arc can accumulate large amounts of convergence-induced strain.

### 17.5.2 Inferred State of Plate Coupling in South-Central Chile

Gravity anomalies can be used to infer the degree of plate coupling in subduction zones. Song and Simons (2003) showed that there is a strong global correlation between high seismic moment release and negative, trench-parallel, gravity anomalies associated with the fore-arc (see also Wells et al. 2003). Their interpretation of this correlation is that the parts of the fore-arc characterised by negative gravity are regions of high shear stress at the plate interface. They argued that this high shear stress induces a deep trench and subdued fore-arc topography because the overriding plate is dragged down with the subducting plate. The resulting deep trench and subdued fore-arc topography are reflected in negative gravity anomalies. In contrast, regions with positive fore-arc gravity are interpreted to reflect reduced shear stress at the plate interface, which results in a shallower trench and higher fore-arc. This geometry results because the subducting slab slides more easily under the fore-arc wedge.

The observed differences in fore-arc gravity anomalies in south-central Chile (Fig. 17.3) can be used together with the Song and Simons (2003) interpretation to infer the state of shear stress on the plate interface. Shear-stress differences can be related to differences in plate coupling (coupling force), so gravity anomalies can be used to infer the degree of coupling at the plate interface. This means that under the Arauco-Lonquimay segment, where the onshore fore-arc is widest and highest and gravity anomalies are positive, the plate interface is characterized by lower shear stress and reduced coupling. In contrast, the Valdivia-Liquiñe segment, which has a narrower and deeper onshore fore-arc and negative gravity anomalies, is characterized by higher shear stress and greater coupling. If these gravity-based inferences are correct, then the causes of coupling differences in south-central Chile can be examined.

### 17.5.3 Explanations for Inferred Coupling Differences

Margin-parallel differences in the degree of plate coupling could arise from changes in one or more different parameters. These parameters are summarized in Tables 17.1 and 17.2, and their potential influence on coupling is discussed below in the context of the shear stress and plate coupling force acting at the plate interface. The discussion concentrates on the Arauco-Lonquimay and Valdivia-Liquiñe segments because these segments are the best constrained by geophysical measurements and both display distinct geological and geophysical contrasts.

### 17.5.3.1 Differences in Shear Stress

In simple, illustrative terms, shear stress, $\tau$, on an inclined fault plane depends on the effective coefficient of friction, $\mu'$ (the friction coefficient minus the effects of pore pressure), and the normal stress acting on the plate interface, $\sigma_N$ (e.g. Wang and Suyehiro 1999; Turcotte and Schubert 2002, p 342):

$$\tau = \mu' \sigma_N \qquad (17.1)$$

Equation 17.1 demonstrates that increases in shear stress at the plate interface could result from increases in either: (*a*) normal stress, through factors such as slab buoyancy, seamount subduction and fore-arc loading, or (*b*) the effective coefficient of friction through changes in plate texture, the presence or absence of sediments, or changes in the material properties of subducting material. The composition of the fore-arc, and hence its rheology, will also influence the shear stress at the plate interface because the shear stress is related to the capacity of the fore-arc to sustain the stresses associated with plate convergence (e.g. McCaffrey 1993). The effects of these parameters are summarised in Table 17.1 and discussed below. Of the parameters listed in Table 17.1, only differences in plate age, slab depth and fore-arc rheology can explain the shear stress under the Arauco-Lonquimay (low $\tau$) and Valdivia-Liquiñe (high $\tau$) segments inferred from gravity anomalies.

The convergence rate does not vary between the segments (e.g. Kendrik et al. 2003) and trench-fill sediments are thick enough to largely cover any seafloor features (e.g. seamounts) that are present (Völker et al. 2006, Chap. 9 of this volume). Therefore, these parameters cannot explain the inferred coupling differences ('a' and 'b' in Table 17.1). The seafloor in this region is cut by several fracture zones (Fig. 17.1). These fracture zones could influence shear stress on the plate interface but, because fracture zones tend only to separate the plate interface into distinct asperities (e.g. Lu and Wyss 1996; Wells et al. 2003), it is unlikely that the subduction of these fracture zones can explain the large differences in shear stress inferred to exist between the Arauco-Lonquimay and Valdivia-Liquiñe segments ('c' in Table 17.1).

Age-related buoyancy effects induced by the slab are important, at least at large length scales (e.g. Jarrad 1986; Kanamori 1986; Scholz and Campos 1995; Yáñez and Cembrano 2004). A younger plate, being more buoyant, would tend to increase the normal stress acting at the plate interface and, for constant frictional properties, lead to an increase in shear stress. In south-central Chile, there is a north-to-south decrease in the age of the subducting plate, with a step of about 10 million years at the boundary between the Arauco-Lonquimay and Valdivia-Liquiñe segments (Fig. 17.1). The increased buoyancy of the plate under the Valdivia-Liquiñe segment should lead to higher shear stress on the plate interface there than on the interface further to the north. Thus, increased buoyancy related to decreasing plate age can explain the north-to-south increase in shear stress inferred on the basis of gravity anomalies ('d' in Table 17.1). However, fore-arc topography is not consistent with increasing buoyancy.

In this region, fore-arc topography is most subdued and the 3D gravity model suggests that the plate is deepest (Fig. 17.4) where the slab is youngest. This means that while plate age can explain the increased shear stress associated with the Valdivia-Liquiñe segment, it cannot account for the observed fore-arc topography and associated gravity anomalies. This is an indication that the rheology of the fore-arc may be helping to prevent the formation of elevated fore-arc topography.

The geometry and properties of the overriding plate could also influence shear stress at the plate interface and the degree of plate coupling. The 3D gravity model presented here suggests differences in slab depth of several kilometers along the south-central Chilean margin (Figs. 17.4, 17.5 and 17.6), which also means that the thickness of the fore-arc varies. The Arauco-Lonquimay segment, with positive anomalies at the coastline, has the shallowest slab under the fore-arc and the thinnest fore-arc wedge (despite the higher topography). In contrast, the slab under the fore-arc of the Valdivia-Liquiñe segment, where negative anomalies are dominant, is about 5 km deeper and the fore-arc is thicker. The increased loading associated with a thicker fore-arc would lead to increased normal stress and, in turn, increased shear stress acting on the plate interface ('e' in Table 17.1). Therefore, normal stress changes over length scales of 100s rather than 1000s of kilometers may have an important influence on shear stress and plate coupling, despite the apparent insignificance of such changes on the scale of the entire margin (e.g. Lamb and Davis 2003).

Changes in fore-arc thickness would also contribute to shear-stress differences through variations in the state of friction at the interface. These variations result from along-strike differences in the degree of consolidation and metamorphism of subducted sediments (just as these properties change down-dip in response to increasing depth and temperature: e.g. Bilek and Lay 1999; Scholz 1990, p320). Where the fore-arc is thicker (i.e. the slab is deeper), consolidation of subducted sediments and associated increases in sediment rigidity (Bilek and Lay 1999) would be initiated at shallower depth, thereby leading to an increase in the effective coefficient of friction and, subsequently, to higher shear stress ('f' in Table 17.1).

Because the shear stress at the plate interface is directly related to the capacity of the fore-arc to sustain the stresses associated with plate convergence, the effect of differences in fore-arc rheology should also be considered. McCaf-

**Table 17.1.** Summary of the effects of various parameters on shear stress ($\tau$) at the plate interface. $\sigma_N$ is normal stress and $\mu'$ is the effective coefficient of friction at the plate interface. The relationship between these parameters is expressed by Eq. 17.1

| | Parameter | Effect on | Example cause | Resulting effect | Effect on $\sigma_N$ | Effect on $\mu'$ | Effect on $\tau$ | Applicable to implied shear-stress differences in south-central Chile? |
|---|---|---|---|---|---|---|---|---|
| a | Convergence rate | Stress | Increased velocity | Increased stress, cooler fore-arc | Increase | – | Increase | NO, constant for both segments |
| b | Seafloor fabric | Roughness | Seamounts | Increased resistance | Increase | – | Increase | NO, negated by trench sediment |
| c | Fracture zones | Pore pressure | | Increased pore pressure | – | Decrease | Decrease | YES, but only locally |
| d | Plate age | Buoyancy | Decreasing age | Increasing buoyancy | Increase | – | Increase | YES, but slab is deepest where buoyancy is greatest |
| e | Slab dip | Fore-arc loading | Deeper slab | Thicker fore-arc | Increase | – | Increase | YES |
| | | | Shallower slab | Thinner fore-arc | Decrease | – | Decrease | |
| f | Slab dip | Sediment properties | Thicker fore-arc | Shallower onset of sediment consolidation | – | Increase | Increase | YES |
| | | | Thinner fore-arc | Deeper onset of sediment consolidation | – | Decrease | Decrease | |
| g | Fore-arc rheology | Strain accumulation | Strong fore-arc | Fore-arc sustains high stresses | – | – | Increase | YES |
| | | | Weak fore-arc | Fore-arc cannot sustain high stresses | – | – | Decrease | |

**Table 17.2.** Effects of various parameters on the coupling force acting across the plate interface. The coupling force, $F_0$, is defined in Eq. 17.2. $\mu'$ and $W$ are the effective coefficient of friction and the width of the frictionally coupled zone (cf. Wang and He 1999)

| | Parameter | Effect on | Caused by | Resulting effect | Effect on $\mu'$ | Effect on $W$ | Effect on $F_0$ | Inferred coupling state | |
|---|---|---|---|---|---|---|---|---|---|
| | | | | | | | | Arauco-Lonquimay | Valdivia–Liquiñe |
| h | Convergence rate | Temperature | Increased velocity | Cooler fore-arc | – | Increase | Increase | Not applicable, convergence constant | |
| i | Fracture zones | Pore pressure | | Increased pore pressure | Decrease | – | Decrease | Locally lower | |
| j | Trench sediment | Stress coherency | Presence | Coherent asperities | – | Increase | Increase | Not applicable, trench sediment everywhere | |
| | | | Absence | Dispersed asperities | – | Decrease | Decrease | | |
| k | Plate texture | Stress coherency | Rough plate | Isolated asperities | – | Decrease | Decrease | Not applicable, trench sediment everywhere | |
| | | | Smooth plate | Coherent asperities | – | Increase | Increase | | |
| l | Slab dip | Fore-arc loading | Deeper slab | Thicker fore-arc (increase $d_1, d_2$) | Increase | – | Increase | Lower | Higher |
| | | | Shallower slab | Thinner fore-arc (decrease $d_1, d_2$) | Decrease | – | Decrease | | |
| m | Slab dip | Sediment properties | Thicker fore-arc | Wider coupled zone (shallower onset of unstable sliding) | Increase | Increase | Increase | Lower | Higher |
| | | | Thinner fore-arc | Narrower coupled zone (deeper onset of unstable sliding) | Decrease | Decrease | Decrease | | |
| n | Plate age | Temperature | Decreasing age | Increase in width of the unstable sliding zone (up-dip end) | – | Increase | Increase | Lower | Higher |
| o | Fore-arc rheology | Strain accumulation | Strong fore-arc | Fore-arc sustains high stresses | – | Increase | Increase | Lower | Higher |
| | | | Weak fore-arc | Fore-arc cannot sustain high stresses | – | Decrease | Decrease | | |

frey (1993) suggests that the rheology of material either side of a subduction fault is an important control on where coupling is high. This provides an explanation for high coupling on a global scale, but given the differences in fore-arc composition and seismicity along the south-central Chilean margin, his explanation of large-scale differences in coupling could also account for smaller-scale differences. Differences in rheology between the Arauco-Lonquimay and Valdivia-Liquiñe segments are suggested by differences in fore-arc composition (Fig. 17.2) and observed seismicity (Fig. 17.7).

The prevalent seismicity of the Arauco-Lonquimay segment (Fig. 17.7) suggests that it is weaker and releasing elastic strain, rather than storing it, meaning that shear stress at the plate interface should be lower ('g' in Table 17.1). In contrast, the Valdivia-Liquiñe segment lacks seismicity and, assuming that this lack of seismicity doesn't reflect aseismic slip, appears to be stronger and accumulating strain elastically.

This inferred difference in rheology may be related to the differences in fore-arc composition discussed previously (i.e. a more mafic Valdivia-Liquiñe segment), but may also reflect the thermal state of the different parts of the fore-arc. Along-strike differences in subducting plate age (Fig. 17.1) mean that between 36 and 42° S, heat flow at the trench increases from about 65 to 75 mW m$^{-2}$ (see Fig. 5b of Hoffmann-Rothe et al. 2006, Chap. 6 of this volume). The higher heat flow beneath the Valdivia-Liquiñe segment could lead to a warmer and weaker fore-arc, in contrast to our interpretation inferred from seismicity. However, the true thermal state can only be deduced by quantifying the insulating effect of subducted sediments and the relative contributions of heat flow from the subducting plate, shear heating on the subduction interface, and radiogenic heat production within the crust (cf. McCaffrey 1993; Molnar and England 1990).

Differences in fore-arc rheology can also be inferred on the basis of fore-arc deformation. For example, Reyners (1998) has shown that on the east coast of New Zealand's South Island, the rapidly-deforming and uplifting Kaikoura Ranges are associated with low seismic velocities in the overriding plate caused by fluids driven from the subducting slab. The deformation of the overriding plate is thought to be taking place because of its fluid-induced weakness. A similar situation is possible in the Arauco-Lonquimay segment, though a comparison between fore-arc seismic velocities for the Arauco-Lonquimay and Valdivia-Liquiñe segments is not possible. The Arauco-Lonquimay segment is characterized by uplift and active faulting (see earlier description) and, regardless of the degree of coupling at the plate interface, this deformation is an indication of the weakness of the overriding plate in this segment.

Deformation of the fore-arc can also affect coupling in other ways. For example, the more frequent seismicity within the fore-arc of the Arauco-Lonquimay segment suggests that it is experiencing more deformation through faulting than is the Valdivia-Liquiñe segment. These faults probably extend to the slab interface (Melnick et al. 2003; 2006b) where they can also increase the stress heterogeneity on the interface and inhibit rupture propagation to adjacent parts of the interface (McCaffrey 1993). The Arauco-Lonquimay segment also lies adjacent to the northerly termination of the Liquiñe-Ofqui Fault Zone. It is possible, therefore, that the apparent weakness of this segment may be related to the buttressing effect induced by the northward movement of the fore-arc block, which is reflected in the dextral displacement on the Liquiñe-Ofqui Fault Zone (cf. Reuther et al. 2003, Rosenau et al. 2006).

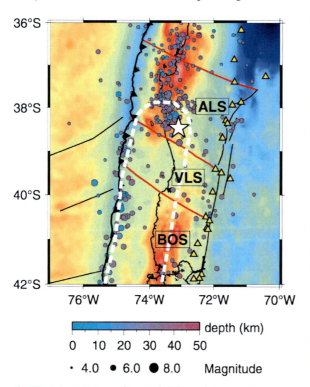

**Fig. 17.7.** Seismicity in south-central Chile overlaying gravity anomalies (Fig. 17.3a). Epicenters are plotted for earthquakes with magnitudes greater than 4 recorded between 1973 and 2004 (data from the National Earthquake Information Center) and only for events above 50 km depth (i.e. only crustal and interplate earthquakes). The epicenter of the 1960 $M_w$ 9.5 Valdivia earthquake, relocated by Krawczyk et al. (2003), is marked by the *white star* and the rupture extent is roughly indicated by the *white dashed line*

### 17.5.3.2 Differences in Coupling Force

Further insight into the factors influencing inferred coupling differences in south-central Chile can be gained from the coupling force that acts at the plate interface, the quantity that we use to explain the degree of coupling. This force results from the shear stress acting over the frictionally coupled part of the subduction interface. The coupling force, $F_0$, is given by Wang and He (1999) as:

$$F_0 = \frac{1}{2}\mu'W(d_1 + d_2)\rho g \qquad (17.2)$$

where $W$ is the width of the frictionally coupled part of the plate interface, $d_1$ and $d_2$ are the depth of the up-dip and down-dip ends of this coupled zone, $\rho$ is the density of the overriding plate, and $g$ is acceleration due to gravity. Equation 17.2 shows that the degree of coupling is affected by the width of the frictionally unstable zone and the composition (through density differences) and thickness of the fore-arc. These parameters that can be inferred from geophysical observations: $W$ can be determined on the basis of GPS observations or thermal modeling; $d_1$ and $d_2$ can be determined from seismic observations (if $W$ is known), and $\rho$ can be inferred from gravity modeling.

Table 17.2 summarizes the influence of various parameters on the friction coefficient and the width of the frictionally coupled zone. Only differences in slab depth and fore-arc rheology impart a coupling regime consistent with that inferred from the gravity anomaly differences. The arguments concerning convergence rate and seafloor fabric discussed in relation to shear stress are equally applicable here – neither can explain north-to-south differences in the coupling force ('h'–'k' in Table 17.2).

Equation 17.2 shows that fore-arc thickness also affects the coupling force. The deeper slab under the fore-arc of the Valdivia-Liquiñe segment leads to an increased coupling force through increases in $d_1$, $d_2$, $\mu'$ and $W$. The increase in $d_1$ and $d_2$ is a consequence of the deeper slab ('l' in Table 17.2), but the increase in $\mu'$ and $W$ is a consequence of the shallower onset of sediment consolidation after it is subducted ('m' in Table 17.2).

The shear stress differences expected from the age-related buoyancy induced by the slab are consistent with those inferred from gravity anomalies, as is the effect of subducting plate age on the coupling force ('n' in Table 17.2). Temperature affects plate coupling through its control on the width of the frictionally coupled zone (e.g. Oleskevich et al. 1999). The up-dip transition from stable to stick-slip sliding occurs at about 150 °C, while the down-dip transition from stick-slip to stable sliding on the plate interface takes place between 350 °C and 450 °C. If the critical down-dip temperatures are not reached until below the depth where the continental Moho meets the slab, then the stable-sliding regime below the seismogenic zone can be reached at lower temperatures. Based on the model of Oleskevich et al. (1999) at 44° S, the critical down-dip isotherms in south-central Chile probably occur landwards of the contact between the slab and the continental Moho (Fig. 17.5). Therefore, the north-to-south decrease in plate age, with a jump at the Valdivia fracture zone (Fig. 17.1), should lead to southward shallowing of the 150°C isotherm and an increase in the width of the frictionally coupled zone. This means that a shallower up-dip stable to stick-slip sliding transition (increased $W$ in Eq. 17.2) and increased coupling force would be expected under the Valdivia-Liquiñe segment ('n' in Table 17.2).

In addition to thickness-related differences, the composition of the fore-arc would also affect the normal stress, or load, acting on the plate interface (through $\rho$ in Eq. 17.2). In the Arauco-Lonquimay segment, less-dense granitoid rocks and low-grade metasediments dominate the fore-arc, whereas in the Valdivia-Liquiñe segment, granites only outcrop locally and metamorphosed mafic rocks are more common. However, the gravity model suggests that to explain along-strike gravity differences, varying fore-arc density is not essential. This means that the fore-arc compositional variations evident at the surface don't necessarily make a contribution to along-strike loading differences. Despite this, the differences in fore-arc composition would also influence the coupling force through differences in fore-arc rheology ('o' in Table 17.2). This is because the coupling force, like shear stress, is directly related to the ability of the fore-arc to sustain convergence-related stresses and accumulate strain.

### 17.5.3.3 A Dominant Control on Shear Stress and Plate Coupling?

The qualitatively inferred state of shear stress and the degree of plate coupling (coupling force $F_0$) for the south-central Chilean margin are summarized in Table 17.3. The interpretation suggests that both shear stress and coupling force are affected by fore-arc thickness (depth to the slab), slab buoyancy and fore-arc rheology. The thicker fore-arc of the Valdivia-Liquiñe segment would lead to increased normal stress, to related changes in frictional properties of the material at the subduction interface, and to a wider frictionally coupled zone. In combination with a more mafic composition and correspondingly stronger rheology, this equates to increased coupling in the Valdivia-Liquiñe segment. Higher buoyancy forces associated with the younger slab in the Valdivia-Liquiñe segment should also lead to increased shear stress, but the rheology of the fore-arc may prevent these buoyancy effects from producing the expected, elevated, fore-arc topography. Thermal effects related to the age of the slab should also act to increase the coupling force under the Valdivia-Liquiñe segment by shallowing the up-dip limit of the frictionally coupled zone. Thermal effects at the down-dip end are not important because the critical isotherms are probably deeper than the intersection between slab and continental Moho.

**Table 17.3.** Inferred state of shear-stress and plate coupling in the Arauco-Lonquimay and Valdivia-Liquiñe segments. In combination with observations of seismicity, this interpretation suggests that the effects of fore-arc thickness complement thermally-induced effects and that in south-central Chile there is a direct correlation between shear stress and coupling force over length scales of a few hundred kilometres

|  | **Arauco-Lonquimay** | **Valdivia-Liquiñe** | **Cause of difference** |
|---|---|---|---|
| Inferred state of shear stress | Lower | Higher | Fore-arc thickness |
|  | Lower | Higher | Age-related slab buoyancy |
|  | Higher | Lower | Fore-arc rheology |
| Inferred degree of coupling | Lower | Higher | Thermal effects on W (up-dip) related to slab age |
|  | Lower | Higher | Fore-arc thickness |
|  | Lower | Higher | Fore-arc rheology |

**Table 17.4.** Summary of characteristic features and expected coupling characteristics of the Arauco-Lonquimay and Valdivia-Liquiñe segments as inferred from the Song and Simons (2003) interpretation of gravity anomalies, fore-arc topography and seismic moment release

|  |  | **Arauco-Lonquimay segment** | **Valdivia-Liquiñe segment** |
|---|---|---|---|
| Observations | Gravity anomalies | Positive | Negative |
|  | Fore-arc topography | Elevated | Subdued |
|  | Fore-arc seismicity | More frequent | Less frequent |
|  | Seismic moment release | Low | High |
| Expected state | Shear stress at interface | Lower | Higher |
|  | Inter-seismic locking | <100% | ~100% |
|  | Seismic-coupling coefficient | <1 | ~1 |
|  | Dominant friction state | Conditionally stable | Unstable sliding |
|  | Asperities | Smaller and isolated | Extensive, interlinked |
|  | Seismicity | More frequent moderate earthquakes | Infrequent large earthquakes |
|  | Description | Slab slides more easily past overriding plate | Slab and overriding plate are largely stuck |
|  | Degree of coupling | Lower | Higher |

The lack of seismicity in the Valdivia-Liquiñe segment (Fig. 17.7) could reflect either aseismic subduction (stable sliding) or a long recurrence time between earthquakes (stick slip). Given that rupture during the largest ever recorded earthquake, the $M_w$ 9.5 1960 Valdivia earthquake, dominantly occurred under and to the south of the Valdivia-Liquiñe segment (Fig. 17.7), it seems unlikely that the lack of seismicity associated with the Valdivia-Liquiñe segment reflects aseismic slip. With this assumption, the pattern of seismicity suggests that the Arauco-Lonquimay segment is more frequently releasing the elastic strain built up during convergence, whereas the distinct lack of earthquakes in the fore-arc of the Valdivia-Liquiñe segment suggests that elastic strain is accumulating there over longer time periods. Stick-slip motion in the Valdivia-Liquiñe segment is consistent with higher coupling, which suggests that fore-arc thickness and rheology, in combination with thermal effects related to plate age, are important controls on plate coupling.

In summary, the gravity anomaly pattern in south-central Chile is consistent with higher shear stress at the plate interface in the Valdivia-Liquiñe segment than in the Arauco-Lonquimay segment. This shear stress difference is reflected in the degree of coupling across the plate interface (coupling force), which is higher in the Valdivia-Liquiñe segment. This contrast in coupling should also be evident in differences in inter-seismic locking, seismic coupling, friction state, asperity distribution and seismicity (Table 17.4).

### 17.5.4 Coupling Differences and Rupture Characteristics of the 1960 Valdivia Earthquake

The 1960 $M_w$ 9.5 Valdivia earthquake initiated under the southern part of the Arauco Pensinsula within the Arauco-Lonquimay segment (Fig. 17.7). Rupture from this earthquake propagated southwards into the Valdivia-Liquiñe segment, rather than northwards and further into the Arauco-Lonquimay segment. This pattern of rupture propagation can be explained by the contrasting coupling inferred for each segment.

The arguments presented previously suggest that the plate interface under the Arauco-Lonquimay segment is characterized by low coupling that reflects the weaker and thinner fore-arc (shallower slab) and incoherent stress distribution on the plate interface caused by faulting in the fore-arc. In contrast, the stronger and thicker fore-arc of the Valdivia-Liquiñe segment and a more uniform stress distribution at the plate interface mean that the plate interface is more highly coupled. High coupling in the Valdivia-Liquiñe segment means that the slip resulting from large earthquakes should dominantly occur on the plate interface under this segment, but this does not necessarily require that rupture initiates in this region of high coupling. If the parts of the subduction interface characterized by low coupling are considered to comprise small portions of highly coupled interface (asperities), then slip can be initiated in a region of low overall coupling and subsequently trigger rupture in surrounding regions that are within the conditionally stable frictional regime. Once this rupture reaches highly coupled parts of the plate interface (i.e. regions with a high portion of material within the unstable sliding field), then it rapidly leads to slip over a large area.

Triggering of slip from outside or near the edge of major asperities seems to be a common feature of subduction interfaces (e.g. Kanamori 1986). For example, Yamanaka and Kikuchi (2004) demonstrate that slip during large earthquakes off north-eastern Japan nearly always propagates away from the focus of main shocks and that these main shocks occur at the edge of, not within, subduction zone asperities. Aftershocks from these large events also dominantly occur outside the asperities.

## 17.6 Conclusions

In south-central Chile, gravity anomalies and surface geology provide a clear indication of fore-arc segmentation. Three-dimensional gravity modeling suggests that this segmentation coincides with differences in slab dip below the fore-arc. Regions where the slab is deeper (Valdivia-Liquiñe segment) correspond with negative gravity anomalies and, based on a global correlation between such negative anomalies, subdued fore-arc topography and high seismic moment release, the plate interface under this part of the fore-arc is inferred to be characterized by high coupling. In contrast, where the fore-arc wedge is thin (Arauco-Lonquimay segment), broad and positive gravity anomalies reflect the shallower depth to the dense subducting plate. These regions of positive gravity correlate with lower coupling at the plate interface.

A qualitative investigation of the characteristics of the south-central Chilean margin between 36 and 42° S suggests that the variations in coupling are dominantly controlled by slab buoyancy, fore-arc thickness and fore-arc rheology. Fore-arc thickness contributes to high plate coupling (i.e. high coupling force) under the Valdivia-Liquiñe segment where seismicity is reduced and the plate interface ruptured during the $M_w$ 9.5 Valdivia earthquake. Thermal effects related to plate age would further increase coupling under the Valdivia-Liquiñe segment. Therefore, thermally-induced effects on coupling are probably complemented by the geometrically- and compositionally-induced effects that change the amount of fore-arc loading, the width of the frictionally coupled zone and the ability of the fore-arc to accumulate strain.

In the Valdivia–Liquiñe segment, the inferred high coupling force (i.e. substantial stress transfer across the interface), stronger fore-arc (reduced seismicity, more mafic composition), and sediment-smoothed interface mean that conditions are suitable for the extensive plate-interface rupture that leads to great earthquakes such as the $M_w$ 9.5 Valdivia earthquake of 1960.

The features observed along the south-central Chilean margin cause stark contrasts in gravity anomalies and fore-arc thickness, and major differences in the characteristics of the plate interface are expected. Therefore, this area is particularly suitable for detailed studies of the fore-arc and plate-interface characteristics that lead to great subduction zone earthquakes. Making high-density measurements of heat flow and establishing dense seismic and GPS networks, like those already established in other subduction zones (e.g. Nishimura et al. 2004; Mishra et al. 2003), would provide significant new insight into the processes that control plate coupling and the occurrence of great subduction zone earthquakes.

## Acknowledgments

This work is a part of SFB267 "Deformation Processes in the Andes" funded by the Deutsche Forschungsgemeinschaft. We thank Muriel Gerbault for her thorough review, Charlotte Krawczyk, Jo Lohrmann, Arne Hoffmann-Rothe, Mark Handy and Nina Kukowski for discussions, Stefan Lüth for providing the data plotted in Fig. 17.6, Kirsten Borchardt and Denise Wilhelmsen for assistance in preparing the manuscript, and our many colleagues in the SFB267 and in South America for their contributions. For data, we are grateful to ENAP in Chile, Repsol-YPF in Argentina, and to Denizar Blitzkow for providing his South American gravity database. Figures were prepared using the freely available GMT software (Wessel and Smith 1998).

# References

Andersen O, Knudsen P (1998) Global marine gravity field from the ERS-1 and Geosat geodetic mission altimetry. J Geophys Res 103(C4):8129–8138

Araneda M, Avendaño MS, Götze H-J, Schmidt S, Munoz J, Schmitz M (1999) South central Andes gravity, new data base. 6th International Congress of the Brazilian Geophysical Society

Bilek SL, Lay T (1999) Rigidity variations with depth along interplate megathrust faults in subduction zones. Nature 400:443–446

Bohm M (2004) 3-D Lokalbebentomographie der südlichen Anden zwischen 36° und 40°S. PhD thesis, Freie Universität Berlin, http://www.diss.fu-berlin.de/2005/7/

Bohm M, Lüth S, Echtler H, Asch G, Bataille K, Bruhn C, Rietbrock A, Wigger P (2002) The Southern Andes between 36° and 40°S latitude: seismicity and average seismic velocities. Tectonophysics 356:275–289

Bürgmann R, Kogan MG, Steblov GM, Hilley G, Levin VE, Apel E (2005) Interseismic coupling and asperity distribution along the Kamchatka subduction zone. J Geophys Res 110: doi 10.1029/2005JB003648

Cifuentes IL (1989) The 1960 Chilean Earthquakes. J Geophys Res 94(B1):665–680

Cingolani C, Dalla Salda L, Hervé F, Munizaga F, Pankhurst RJ, Parada MA, Rapela CW (1991) The magmatic evolution of northern Patagonia: new impressions of pre-Andean and Andean tectonics. In: Harmon RS, Rapela CW (eds Andean magmatism and its tectonic setting. Geol Soc Am Spec P 265:29–44

Conrad CP, Bilek S, Lithgow–Bertelloni C (2004) Great earthquakes and slab pull: interaction between seismic coupling and plate-slab coupling. Earth Planet Sci Lett 218:109–122

Duhart P, McDonough M, Muñoz J, Martin M, Villeneuve M (2001) El complejo metamórfico Bahía Mansa en la Cordillera Costa del centro-sur de Chile (39°30'–42°00'S): geocronologia K-Ar, $^{40}Ar/^{39}Ar$ y U-Pb e implicancias en la evolución del margen sur-occidental de Gondwana. Rev Geol Chile 28:179–208

England P, Engdahl R, Thatcher W (2004) Systematic variation in the depths of slabs beneath arc volcanoes. Geophys J Int 156:377–408

Flüh ER, Kopp H, Schreckenberger B (2002) FS Sonne Cruise Report SO161-1&4, Subduction Processes Off Chile. GEOMAR Report 102

Folguera A, Ramos V (2000) Structural control of the Copahue volcano. Tectonics implications for the Quaternary volcanic arc (36°–39°S). Rev Asoc Geol Argentina 55(3):229–244

Folguera A, Ramos V, Melnick D (2002) Partición de la deformación en la zona del arco volcánico de los Andes neuquinos (36–39°S) en los últimos 30 millones de años. Rev Geol Chile 29(2):151–165

Folguera A, Ramos VA, Melnick D (2003) Recurrencia en el desarrollo de cuencas de inraarco. Cordillera Neuquina (37°30' – 38°S). Rev Asoc Geol Argentina 58(1):3–19

Glodny J, Echtler H, Figueroa O, Franz G, Gräfe K, Kemnitz H, Kramer W, Krawczyk C, Lohrmann J, Lucassen F, Melnick D, Rosenau M, Seifert W (2006) Long-term geological evolution and mass-flow balance of the South-Central Andes. In: Oncken O, Chong G, Franz G, Giese P, Götze H-J, Ramos VA, Strecker MR, Wigger P (eds) The Andes – active subduction orogeny. Frontiers in Earth Science Series, Vol 1. Springer-Verlag, Berlin Heidelberg New York, pp 401–428, this volume

Götze H-J, Lahmeyer B (1988) Application of three-dimensional interactive modeling in gravity and magnetics. Geophysics 53(8):1096–1108

Götze H-J, Lahmeyer B, Schmidt S, Strunk S (1994) The lithospheric structure of the Central Andes (20°–26°S) as inferred from quantitative interpretation of regional gravity. In: Reutter K-J, Scheuber E, Wigger P (eds) Tectonics of the Southern Central Andes. Springer-Verlag, Berlin Heidelberg New York, pp 7–21

Götze H-J, Alvers M, Goltz G, Kirchner A, Müller A, Schäfer U, Schmidt S, Araneda M, Ugalde H, Chong D, Barrio GL, Lopez N, Omarini R (1996) Group updates gravity database for Central Andes. EOS 77(19):181

Götze H-J, Alten M, Burger H, Goni P, Melnick D, Mohr S, Munier K, Ott N, Reutter K, Schmidt S (2006) Data management of the SFB 267 for the Andes – from ink and paper to digital databases. In: Oncken O, Chong G, Franz G, Giese P, Götze H-J, Ramos VA, Strecker MR, Wigger P (eds) The Andes – active subduction orogeny. Frontiers in Earth Science Series, Vol 1. Springer-Verlag, Berlin Heidelberg New York, pp 539–556, this volume

Gräfe K, Glodny J, Seifert W, Rosenau M, Echtler H (2002) Apatite fission track thermochronology of granitoids at the south Chilean active continental margin (37°S–42°S): implications for denudation, tectonics and mass transfer since the Cretaceous. 5th Int Symposium Andean Geodynamics, Toulouse, Extended Abs, pp 275–278

Hacker BR, Abers GA (2004) Subduction factory 3: an Excel worksheet and macro for calculating the densities, seismic wave speeds, and water contents of minerals and rocks at pressure and temperature. Geochem Geophys Geosyst 5(1): doi 10.1029/2003GC000614

Hacker BR, Abers GA, Peacock SM (2003) Subduction factory 1: theoretical mineralogy, density, seismic wave speeds, and $H_2O$ content. J Geophys Res 108(B1): doi 10.1029/2001JB001127

Hastings DA, Dunbar PK, Elphingstone GM, Bootz M, Murakami H, Maruyama H, Masaharu H, Holland P, Payne J, Bryant NA, Logan TL, Muller JP, Schreier G, MacDonald JS (1999) The Global Land One-kilometer Base Elevation (GLOBE) Digital Elevation Model, Version 1.0. National Oceanic and Atmospheric Administration, National Geophysical Data Center, Boulder, Colorado, http://www.ngdc.noaa.gov/mgg/topo/globe.html

Hervé F (1977) Petrology of the crystalline basement of the Nahuelbuta Mountains, South-Central Chile. In: Ishikawa T, Aguirre L (eds) Comparative studies on the geology of the Circum–Pacific orogenic belt in Japan and Chile, 1st Report. Japan Soc Prom Sci, Tokyo, pp 1–51

Hervé F (1988) Late Paleozoic subduction and accretion in southern Chile. Episodes 11:183–188

Hervé F, Munizaga F, Parada MA, Brook M, Pankhurst RJ, Snelling NJ, Drake R (1988) Granitoids of the Coast Range of central Chile: geochronology and geologic setting. J S Am Earth Sci 1:185–194

Hildreth W, Moorbath S (1988) Crustal contributions to arc magmatism in the Andes of Central Chile. Contrib Miner Petrol 98:455–489

Hoffmann-Rothe A, Kukowski N, Dresen G, Echtler H, Oncken O, Klotz J, Scheuber E, Kellner A (2006) Oblique convergence along the Chilean margin: partitioning, margin-parallel faulting and force interaction at the plate interface. In: Oncken O, Chong G, Franz G, Giese P, Götze H-J, Ramos VA, Strecker MR, Wigger P (eds) The Andes – active subduction orogeny. Frontiers in Earth Science Series, Vol 1. Springer-Verlag, Berlin Heidelberg New York, pp 125–146, this volume

Jarrard RD (1986) Relations among subduction parameters. Rev Geophys 24(2):217–284

Jordan TE, Burns WM, Veiga R, Pángaro F, Copeland P, Kelley S, Mpodozis C (2001) Extension and basin formation in the southern Andes caused by increased convergence rate: a mid-Cenozoic trigger for the Andes. Tectonics 20(3):308–324

Kaizuka S, Matsuda T, Nogami M, Yonekura N (1973) Quaternary tectonics and recent seismic crustal movements in the Arauco Peninsula and its environs, Central Chile. Tokyo Metropolitan University Geographical Reports 8:1–49

Kanamori H (1986) Rupture process of subduction-zone earthquakes. Ann Rev Earth Planet Sci Lett 14:293–322

Kendrick E, Bevis M, Smalley R, Brooks B, Barriga R, Lauría E, Souto L (2003) The Nazca-South America Euler vector and its rate of change. J S Am Earth Sci 16:125–131

Khazaradze G, Klotz J (2003) Short- and long-term effects of GPS measured crustal deformation rates along the south-central Andes. J Geophys Res 108(B6): doi 10.1029/2002JB001879

Klotz J, Abolghasem A, Khazaradze G, Heinze B, Vietor T, Hackney R, Bataille K, Maturana R, Viramonte J, Perdomo R (2006) Long-term signals in the present-day deformation field of the Central and Southern Andes and constraints on the viscosity of the Earth's upper mantle. In: Oncken O, Chong G, Franz G, Giese P, Götze H-J, Ramos VA, Strecker MR, Wigger P (eds) The Andes – active subduction orogeny. Frontiers in Earth Science Series, Vol 1. Springer-Verlag, Berlin Heidelberg New York, pp 65–90, this volume

Krawczyk C, SPOC Team (2003) Amphibious seismic survey images plate interface at 1960 Chile earthquake. EOS Transact 84(32):301,304–305

Krawczyk CM, Mechie J, Lüth S, Tašárová Z, Wigger P, Stiller M, Brasse H, Echtler HP, Araneda M, Bataille K (2006) Geophysical signatures and active tectonics at the south-central Chilean margin. In: Oncken O, Chong G, Franz G, Giese P, Götze H-J, Ramos VA, Strecker MR, Wigger P (eds) The Andes – active subduction orogeny. Frontiers in Earth Science Series, Vol 1. Springer-Verlag, Berlin Heidelberg New York, pp 171–192, this volume

Lamb S, Davis P (2003) Cenozoic climate change as a possible cause for the rise of the Andes. Nature 425:792–797

Lara L, Rodríguez C, Moreno H, Pérez de Arce C (2001) Geocronología K-Ar y geoquímica del volcanismo plioceno superior-pleistoceno en los Andes del sur (39°–42°S). Rev Geol Chile 28(1):67–91

Lavenu A, Cembrano J (1999) Compressional and transpressional stress pattern for Pliocene and Quaternary brittle deformation in fore-arc and intra-arc zones (Andes of Central and Southern Chile). J Struct Geol 21:1669–1691

Lay T, Schwartz SY (2004) Comment on "Coupling Semantics and Science in Earthquake Research" by Wang K and Dixon T. EOS 85(36):339–340

LeRoux JP, Elgueta S (1997) Paralic parasequences associated with Eocene sea-level oscillations in an active margin setting: Trihueco Formation of the Arauco Basin, Chile, Sediment Geol 110(3–4):257–276

López-Escobar L (1984) Petrology and chemistry of volcanic rocks of the southern Andes. In: Harmon RS, Barreiro BA (eds) Andean Magmatism, chemical and isotopic constraints. Shiva Publ Co, Cheshire UK, pp 47–71

Lu Z, Wyss M (1996) Segmentation of the Aleutian plate boundary derived from stress direction estimates based on fault plane solutions. J Geophys Res 101(B1):803–816

Lucassen F, Franz G, Thirlwall MF, Mezger K (1999) Crustal recycling of metamorphic basement: Late Paleozoic granites of the Chilean Coast Range and Precordillera at 22°S. J Petrol 40:1527–1551

Lucassen F, Trumbull R, Franz G, Creixell C, Vásquez P, Romer RL, Figueroa O (2004) Distinguishing crustal recycling and juvenile additions at active continental margins: the Paleozoic to Recent compositional evolution of the Chilean Pacific margin (36–41°S). J S Am Earth Sci 17:103–119

Lüth S, Wigger P, ISSA Research Group (2003) A crustal model along 39°S from a seismic refraction profile–ISSA (2000). Rev Geol Chile 30(1):83–101

Martin MW, Kato TT, Rodríguez C, Godoy E, Duhart P, McDonough M, Campos A (1999) Evolution of the late Paleozoic accretionary complex and overlying forearc-magmatic arc, south central Chile (38°–41°S): constraints for the tectonic setting along the southwestern margin of Gondwana. Tectonics 18:582–605

McCaffrey RM (1993) On the role of the upper plate in great subduction zone earthquakes. J Geophys Res 98(B7):11953–11966

Melnick D, Echtler HP (2006a) Morphotectonic and geologic digital map compilations of the South-Central Andes (36°–42° S). In: Oncken O, Chong G, Franz G, Giese P, Götze H-J, Ramos VA, Strecker MR, Wigger P (eds) The Andes – active subduction orogeny. Frontiers in Earth Science Series, Vol 1. Springer-Verlag, Berlin Heidelberg New York, pp 565–568, this volume

Melnick D, Echtler HP (2006b) Inversion of forearc basins in South-Central Chile caused by rapid glacial age trench fill. Geology 4(9):709–712

Melnick D, Sanchez M, Echtler HP, Pineda V (2003) Geológica structural de la Isla Mocha, centro-sur de Chile (38°30'S, 74°W): implicancias en la tectónica regional. X Congreso Geológico Chileno, Extended Abstracts

Melnick D, Rosenau M, Folguera A, Echtler HP (2006a) Neogene tectonic evolution of the Neuquén Andes western flank (37–39°S). In: Kay SM, Ramos VA (eds) Evolution of an Andean margin: a tectonic and magmatic view from the Andes to the Neuquén Basin (35°–39° S lat). Geol Soc Am Spec P 407:73–95, doi 10.1130/2006.2407(04)

Melnick D, Bookhagen B, Echtler HP, Strecker MR (2006b) Coastal deformation and great subduction earthquakes, Isla Santa María, Chile (37°S). Geol Soc Am Bull 118(9), in press

Mishra OP, Zhao D, Umino N, Hasegawa A (2003) Tomography of northeast Japan forearc and its implications for interplate seismic coupling. Geophys Res Lett 30(16): doi 10.1029/2003GL017736

Molnar P, England P (1990) Temperatures, heat flux, and frictional stress near major thrust faults. J Geophys Res 95(B4):4833–4856

Mordojovich C (1981) Sedimentary basins of Chilean Pacific Offshore. In: Halbouty MT (ed) Energy Resources of the Pacific Region. Am Assoc Petr Geologists, Stud Geol 12:63–82

Mpodozis M, Ramos V (1989) The Andes of Chile and Argentina. In: Geology of the Andes and its relation to hydrocarbon and mineral resources. In: Ericksen GE, Pinochet MT, Reinemund JE (eds) Circum-Pacific Council for Energy and Mineral Resources, Earth Sci Ser 11:59–90

Müller A (1999) Ein EDV-orientiertes Verfahren zur Berechnung der topographischen Reduktion im Hochgebirge mit digitalen Geländemodellen am Beispiel der Zentralen Anden. Berliner Geowiss Abh B34

Müller RD, Roest WR, Royer JY, Gahagan LM, Sclater JG (1997) Digital isochrons of the world's ocean floor. J Geophys Res 102(B2): 3211–3214

Muñoz J, Troncoso R, Duhart P, Cringnola P, Farmer L, Stern CR (2000) The relation of the mid-Tertiary coastal magmatic belt in South-Central Chile to the late Oligocene increase in plate convergence rate. Rev Geol Chile 27(2):177–203

Nelson R, Manley W (1992) Holocene coseismic and aseismic uplift of the Isla Mocha, South-Central Chile. Quaternary International 15/16:61–76

Nishimura T, Hirasawa T, Miyazaki S, Sagiya T, Tada T, Miura S, Tanaka K (2004) Temporal change of interplate coupling in northeastern Japan during 1995–2002 estimated from continuous GPS observations. Geophys J Int 157:901–916

NOAA (1988) Data Announcement 88-MGG-02, Digital relief of the surface of the Earth. National Oceanic and Atmosphere Administration, National Geophysical Data Center, Boulder, Colorado

Oleskevich DA, Hyndman RD, Wang K (1999) The updip and downdip limits to great subduction earthquakes: thermal and structural models of Cascadia, South Alaska, SW Japan and Chile. J Geophys Res 104(B7):14965–14991

Pacheco JF, Sykes LR, Scholz C (1993) Nature of seismic coupling along simple plate boundaries of the subduction type. J Geophys Res 98(B8):14133–14159

Pankhurst RJ, Weaver SD, Hervé F, Larrondo P (1999) Mesozoic–Cenozoic evolution of the North Patagonian Batholith in Aysén, southern Chile. J Geol Soc 156:673–694

Pineda V (1986) Evolución paleogeográfica de la cuenca sedimentaria Cretácico-Terciaria de Arauco. In: Frutos J, Oyarzún R, Pincheira M (eds) Geología y Recursos Minerales de Chile, Tomo 1. Universidad de Concepción, pp 375–390

Plafker G, Savage JC (1970) Mechanism of the Chilean earthquake of May 21 and 22, 1960. Geol Soc Am Bull 81:1001–1030

Ramos VA, Wienecke S, Götze H-J (2002) El basamento de la Cuenca Neua y regions adjacentes: datos gravimétricos preliminares. XV Congreso Geológico Argentino 2002, El Calafate, Santa Cruz, Argentina. Published on CDROM, ISBN 987-20190-1-0

Rapela CW, Kay SM (1988) Late Paleozoic to Recent magmatic evolution of northern Patagonia. Episodes 11:175–182

Reichert C, et al. (2002) Cruise Report SONNE Cruise 161 Leg 2&3, "Subduction Processes Off Chile". Bundesanstalt für Geowissenschaften und Rohstoffe, Hannover

Reuther CD, Potent S, Bonilla R (2003) Crustal stress history and geodynamic processes of a segmented active plate margin, south-central Chile: the Arauco Bío-Bío trench arc system. 10th Chilean Geological Congress, Concepción, Chile, Extended Abstracts

Reyners M (1998) Plate coupling and the hazard of large subduction thrust earthquakes at the Hikurangi subduction zone, New Zealand. N Zealand J Geol Geophys 41(4):343–354

Rosenau M, Melnick D, Echtler H (2006) Kinematic constraints on intra-arc shear and strain partitioning in the Southern Andes between 38°S and 42°S latitude. Tectonics 25(4), TC4013, doi 10.1029/2005TC001943

Ruegg JC, Campos J, Madariaga R, Kausel E, De Chabalier JB, Armijo R, Dimitrov D, Georgiev I, Barrientos S (1996) The Mw = 8.1 Antofagasta (North Chile) Earthquake of July 30 (1995) first results from teleseismic and geodetic data. Geophys Res Lett 23(9): 917–920

Ruff LJ (1989) Do trench sediments affect great earthquake occurrence in subduction zones? Pure Appl Geophys 129:263–282

Ruff LJ, Kanamori H (1983) Seismic coupling and uncoupling at subduction zones. Tectonophysics 99:99–117

Schmidt S, Götze H-J (1998) Interactive visualization and modification of 3D-models using GIS-functions. Phys Chem Earth 23(3): 289–295

Schmidt S, Götze H-J (2006) Bouguer and isostatic maps of the Central Andes. In: Oncken O, Chong G, Franz G, Giese P, Götze H-J, Ramos VA, Strecker MR, Wigger P (eds) The Andes – active subduction orogeny. Frontiers in Earth Science Series, Vol 1. Springer-Verlag, Berlin Heidelberg New York, pp 559–562, this volume

Scholz CH (1990) The Mechanics of Earthquakes and Faulting. Cambridge University Press

Scholz CH (1998) Earthquakes and friction laws. Nature 391:37–42

Scholz CH, Campos J (1995) On the mechanism of seismic decoupling and back arc spreading at subduction zones. J Geophys Res 100(B11):22103–22115

Scholz CH, Small C (1997) The effect of seamount subduction on seismic coupling. Geology 24(6):487–490

Schurr B, Asch G, Rietbrock A, Trumbull R, Haberland C (2003) Complex patterns of fluid and melt transport in the central Andean subduction zone revealed by attenuation tomography. Earth Planet Sci Lett 215:105–119

Seifert W, Rosenau M, Echtler H (2005) Crystallisation depths of granitoids of South Central Chile estimated by Al-in-hornblende geobarometry: implications for mass transfer processes along the active continental margin. In Miller H (ed) Contributions to Latin American Geology. Schweizerbart, pp 115–127

SEGEMAR (1998) Mapa geológico de la Republica Argentina. Servicio Geológico Minero Argentina, Buenos Aires, Argentina

SERNAGEOMIN (2003) Mapa Geológico de Chile: versión digital, N°4, CD-ROM version 1.0. Servicio Nacional de Geología y Minería, Publicación Geológica Digital, Santiago, Chile

Sobolev SV, Babeyko AY (1994) Modeling of mineralogical composition, density and elastic wave velocities in anhydrous magmatite rocks. Surveys Geophys 15:515–544

Song TA, Simons M (2003) Large trench-parallel gravity variations predict seismogenic behavior in subduction zones. Science 301: 630–633

Stern C (1989) Pliocene to present migration of the volcanic front, Andean Southern Volcanic Front. Rev Geol Chile 16(2): 145–162

Suárez M, Emparan C (1997) Hoja Curacautín, Regiones de la Araucanía y del BioBio. Carta Geol. de Chile 71, escala 1:250000. Servicio Nacional de Geología y Minería, Santiago, Chile

Tašárová Z (2004) Gravity data analysis and interdisciplinary 3D modelling of a convergent plate margin (Chile, 36–42°S). PhD thesis, Freie Universität Berlin, http://www.diss.fu-berlin.de/2005/19/indexe.html

Tašárová Z (submitted) An improved gravity database and a three-dimensional density model of a convergent plate margin (Chile, 36°–42 °S). Geophys J Int

Tassara A, Yáñez G (2003) Relación entre el espesor elástico de la litosfera y la segmentación tectónica del margen andino (15°–47°). Rev Geol Chile 30(2):159–186

Turcotte DL, Schubert G (2002) Geodynamics: application of continuum physics to geological problems. John Wiley, Hoboken NJ

Völker D, Wiedicke M, Ladage S, Gaedicke C, Reichert C, Rauch K, Kramer W, Heubeck C (2006) Latitudinal variation in sedimentary processes in the Peru-Chile trench off Central Chile. In: Oncken O, Chong G, Franz G, Giese P, Götze H-J, Ramos VA, Strecker MR, Wigger P (eds) The Andes – active subduction orogeny. Frontiers in Earth Science Series, Vol 1. Springer-Verlag, Berlin Heidelberg New York, pp 193–216, this volume

Wang K, Dixon T (2004) Coupling semantics and science in earthquake research. EOS 85(18):180

Wang K, He J (1999) Mechanics of low-stress forearcs: Nankai and Cascadia. J Geophys Res 104(B7):15191–15205

Wang K, Suyehiro K (1999) How does plate coupling affect crustal stresses in Northeast and Southwest Japan? Geophys Res Lett 26(15):2307–2310

Wells RE, Blakely RJ, Sugiyama Y, Scholl DW, Dinterman PA (2003) Basin-centred asperities in great subduction zone earthquakes: a link between slip, subsidence, and subduction erosion. J Geophys Res 108(B10): doi 10.1029/2002JB002072

Wessel P, Smith WHF (1998) New, improved version of Generic Mapping Tools released. EOS 79(47):579

Wienecke S (2002) Homogenisation and interpretation of the gravity field along the SALT-traverse between 36°–42°S. Diploma thesis, Freie Universität Berlin (in German)

Willner AP, Hervé F, Massonne HJ (2000) Mineral chemistry and pressure-temperature evolution of two contrasting high-pressure-low-temperature belts in the Chonos Archipelago, southern Chile. J Petrol 41:309–330

Willner AP, Glodny J, Gerya TV, Godoy E, Massonne HJ (2004) A counterclockwise $PTt$-path of high pressure-low temperature rocks from the Coastal cordillera accretionary complex of South Central Chile: constraints for the earliest stage of subduction mass flow. Lithos 75:283–310

Yáñez G, Cembrano J (2004) Role of viscous plate coupling in the late Tertiary Andean tectonics. J Geophys Res 109: doi 10.1029/2003JB002494

Yamanaka Y, Kikuchi M (2004) Asperity map along the subduction zone in northeastern Japan inferred from regional seismic data. J Geophys Res 109: doi 10.1029/2003JB002683

Yuan X, Sobolev SV, Kind R (2002) Moho topography in the central Andes and its geodynamic implications. Earth Planet Sci Lett 199:389–402

Yuan X, Asch G, Bataille K, Bock G, Bohm M, Echtler H, Kind R, Oncken O, Wöhlbern I (2006) Deep seismic images of the southern Andes. In: Kay SM, Ramos VA (eds) Evolution of an Andean margin: a tectonic and magmatic view from the Andes to the Neuquén Basin (35°–39° S lat). Geol Soc Am Spec P 407:61–72, doi 10.1130/2006.2407(03)

# Chapter 18

# Episodic Neogene Southward Growth of the Andean Subduction Orogen between 30°S and 40°S – Plate Motions, Mantle Flow, Climate, and Upper-Plate Structure

Tim Vietor · Helmut Echtler

**Abstract.** The tectonic shortening of the South American upper plate that formed the Andean subduction orogen shows significant along-strike variations in magnitude and timing. Shortening in the Central Andes started in the Eocene, whereas the contraction of the Southern Central Andes at 30° S started only in the Early Miocene and then migrated southwards. Upper-plate shortening reached 38° S in the Late Miocene and tapered off at around 40° S. At the beginning of the Pliocene, contraction in the Southern Central Andes significantly slowed and eventually changed to transtensional conditions south of about 33.5° S, where the intersection of the Juan Fernandez Ridge results in a radical change in subduction mechanics. A southward decrease of both the duration and the average rate of contraction explains the observed decrease of the absolute Neogene shortening from 160 km at 30° S to 15 km at 38° S.

The southward growth of the subduction orogen in the Mid-Late Miocene is not related to an increase of the relative Nazca-South America convergence but takes place during an acceleration of the westward motion of the South American Plate (overriding) with respect to the hot-spot reference frame. This suggests that overriding is the main plate kinematic driving mechanism for subduction orogeny.

Kinematical considerations show that overriding is converted either into plate boundary deformation or into subduction-zone rollback. From the correlation of the deformation time-series with climatic, plate kinematic and geological data we suggest that the latitudinal and temporal variations of three factors can explain the varying proportions of rollback and upper plate shortening along the orogen. These are (*1*) the distance to the southern edge of the subducting Nazca slab, (*2*) the strength of the plate interface and (*3*) the strength of the upper-plate.

From the fact that subduction orogeny expands southward during an acceleration of overriding we conclude that contraction of the upper plate in the more southerly regions required faster overriding velocities than in the north. This is most likely the result of easier rollback in the south which is enabled by the proximity of the Drake Passage or – after initiation of the Chile Rise triple junction at 14 Ma – the opening of slab windows under Patagonia that facilitate the transfer of mantle material from the sub-slab region to the slab wedge.

The almost simultaneous post-Miocene slowing of upper-plate shortening south of the Juan Fernandez Ridge intersection can be best explained by Late Cenozoic increased sediment flux to the trench that ultimately reduced the strength of the plate interface. The initial Mid-Miocene exhumation of the main source area for trench sediments in the Patagonian Andes was tectonically controlled by the northward migrating Chile rise collision and not related to global cooling. However, the rapid spreading of Patagonian ice sheets after 7 Ma on the previously uplifted topography most likely led to a sudden increase of the average erosion rates and sediment flux into the trench. This increase and the blocking of the sediment dispersal within the trench by the Juan Fernandez Ridge enabled the accumulation of a thick trench fill. Hence, the pronounced Pliocene slowing of upper-plate contraction by sediment induced lubrication and weakening of the plate interface in this region is most likely the combined result of global climate change, tectonic uplift of the sediment source region and limited sediment dispersal in the trench.

Once contraction started, deformation in the upper plate became focused in pre-Andean sedimentary basins and shortening rates in these basins were twice those in thick-skinned belts. Hence, we confirm that inherited soft regions in the upper-plate accelerate orogenic shortening. Subduction zone shallowing as a result of convergence-parallel ridge subduction weakens the upper plate thermally and has a similar accelerating effect on subduction orogeny.

## 18.1 Introduction

The Andes span more than 7500 km along the western margin of the South American Plate and they are the only active example of subduction orogens. Their central part forms the second largest plateau on Earth, resulting from Cenozoic tectonic shortening of the upper South American Plate over the subducting oceanic Nazca Plate (Isacks 1988, Kley and Vietor 2005). However, non-collisional orogeny is an exceptional phenomenon even along the Andean convergent margin. It has been concentrated to the Central Andes where the recent phase of shortening and subduction orogeny started in the Eocene. To the north and south of the plateau region, tectonic shortening has been only episodic and of smaller magnitude (Kley and Monaldi 1998; Mpodozis and Ramos 1989).

Various controlling factors for upper-plate shortening have been proposed in the past. In the earliest studies on tectonic deformation of the Andean margin it has recognized that contraction accelerated within distinct phases that were largely coeval along the Andean margin (Steinmann 1929). Some of these tectonic phases have been correlated with intervals of faster relative Nazca-South America convergence (Pardo-Casas and Molnar 1987). On the other hand, the absolute motions of the plates with respect to the mantle may play an important role for the timing of upper-plate shortening (Marrett and Strecker 2000; Russo and Silver 1996; Silver et al. 1998; Somoza 1998). In order to test this hypothesis detailed time series of shortening rates and plate kinematics are required.

In the Central Andes, back-arc shortening started almost 50 Ma ago (Oncken et al. 2006, Chap. 1 of this volume) and since then has involved a region up to 800 km wide (Fig. 18.1). Recent volcanics or post-tectonic sediments cover extensive areas and consequently there is only one transect at 21° S with complete coverage of detailed balanced sections and sufficient age data to construct a shortening rate evolution (Elger et al. 2005; Oncken et al. 2006, Chap. 1 of this volume). In contrast, in the Southern Central Andes, between 30° and 40° S, the orogen is in most regions less than 200 km wide, absolute contraction is much smaller, and has been restricted to the last 20 Ma. This facilitates the study of finite shortening and the timing of tectonic events over a large region. Here we review the available data for the upper-plate shortening of the Southern Central Andes between 30° and 40° S in order to decipher the temporal and spatial evolution of subduction orogeny south of the Central Andean Plateau. Finally, we compare this evolution with other time series (e.g. plate kinematics, climate evolution) and discuss possible controlling factors for the observed growth pattern of the Southern Central Andes.

## 18.2 Regional Setting

The Southern Central Andes between 30° and 40° S form the link between the Central Andean Plateau and the Northern Patagonian Andes (Figs. 18.1 and 18.2). Upper-plate contraction related to the Andean subduction orogeny shows large variations among the three segments. Contraction in the Central Andes, the Southern Central Andes at 30° S and the Northern Patagonian Andes at 42° S amounts to 300 km, 170 km and 15 km, respectively (Allmendinger et al. 1990; Diraison et al. 1998; Kley et al. 1996).

In contrast to the Central Andes, where plateau formation by tectonic shortening started in the Eocene, the region between 30° S and 40° S has undergone across-strike shortening only since the Early Miocene (Jordan et al. 1993). But even within the Southern Central Andes the onset of shortening has been diachronic and the duration of upper-plate contraction as well as the finite tectonic shortening decrease southwards (Gonzalez-Bono-

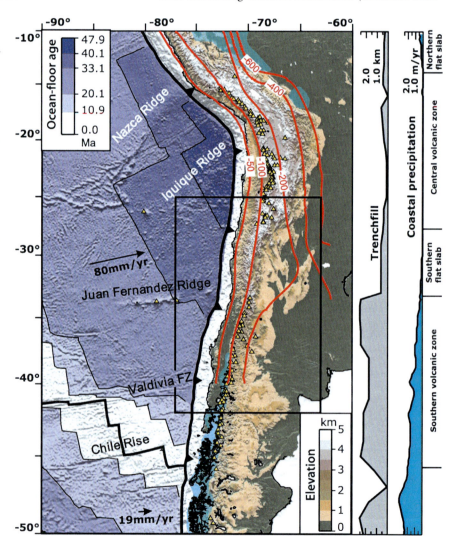

**Fig. 18.1.**
Shaded image of the topography of the Andean subduction orogen and the adjacent ocean floor. *Onshore contours* show the depth to the Wadati Benioff zone in km (Creager et al. 1995). *Triangles* mark Holocene volcanoes (Smithsonian Institution, Global Volcanism Program). Latitudinal variation of thickness of trench-fill (Hoffmann-Rothe et al. 2006, Chap. 6 of this volume) and of annual precipitation along the coast (New et al. 2002) are depicted at right. *Shading* on the South American Plate indicates regions with annual precipitation in excess of 2 000 mm yr$^{-1}$. Age contours for the oceanic lithosphere from Mueller (1997). *Box* shows outline of Figs. 18.2, 18.3 and 18.10a

rino et al. 2001; Ramos et al. 1996). At the northern end of the Southern Central Andes, at 30° S, Andean deformation started at 21 Ma (Jordan et al. 1993; Jordan et al. 2001b) and has accumulated a total upper-plate contraction of 150 to 200 km (Allmendinger et al. 1990; Kley 1999). While shortening had already started at 30° S an extensive basin system between 37° S and 39° S indicates persisting extensional conditions during the early Miocene further south (Jordan et al. 2001a; Radic et al. 2002). Subsequent Late Miocene trench-normal compression in the southern part of the Southern Central Andes accumulated less than 20 km of crustal shortening and in the Early Pliocene shortening even stopped (Folguera et al. 2002; Kley et al. 1999).

The differences in finite shortening have been correlated with along-strike-variations of both upper- and lower-plate properties. Within the plateau region the correlation between the amount of finite shortening and the thickness of the Pre-Andean sediments in the eastern foreland suggests that the basins enable the formation of thin-skinned thrust belts and thus facilitate shortening (Allmendinger and Gubbels 1996; Kley et al. 1999). On a larger scale, the differences between the modeled elastic thickness of the Central and the Southern Central Andes have lead to the conclusion that the southward decrease of contraction is due to a more mafic bulk composition and therefore a stronger lithosphere in the Southern Central Andes (Tassara and Yánez 2003).

Indeed, there are marked along-strike differences in the pre-Andean upper-plate geology. The pre-Andean Mesozoic basement between 30° S and 40° S owes its tectonic grain to deformational and accretionary processes during the Paleozoic Pampean – Famatinian and Gondwanide orogenic cycles and subsequent Mesozoic crustal extension phases. During this entire evolution, the region was strongly modified by tectonic and magmatic activity of Paleozoic to Cenozoic magmatic arcs (Herve 1988; Herve et al. 1988; Hildreth and Moorbath 1988; Parada 1990).

The western edge of the South American continent between 27° S and 42° S has been subdivided in several crustal segments (Mpodozis and Ramos 1989; Ramos et al. 1986). The segmentation comprises from the stable craton to the west: (*1*) the Famatina belt, (*2*) the Precordillera or Cuyana Terrane, (*3*) the Late Paleozoic to Early Mesozoic arc-fore-arc complex and Chilenia Terrane (Fig. 18.3). The Patagonian Massif is located south of these segments. In the Famatinian belt, Cambrian ('Pampean') and Ordovician ('Famatinian') metamorphic and magmatic phases are connected in a magmatic arc setting (Lucassen and Becchio 2003; Pankhurst et al. 1998; Rapela et al. 1998). The Precordillera Terrane is interpreted as an exotic fragment of Laurentian origin with a basement of Grenvillian age (Astini et al. 1995; Ramos and Keppie 1999; Sato et al. 2000; Vujovich and Kay 1998, and references therein). On top of this basement an up to 10 km thick cover of Paleozoic sediments developed in a platform setting and the terrane was accreted to the active margin (Famatinia Belt) in the Ordovician Ocloyic orogeny (Astini and Dávilla 2004). The Chilenia Terrane was accreted to the Precordillera Terrane along a north-striking suture in the latest Devonian to Early Carboniferous (Davies et al. 1999; Mpodozis and Ramos 1989). The North Patagonian Massif is delineated by south-east aligned Permian to Permo-Triassic magmatic rocks and the occurrence of Paleozoic metamorphic rocks (Caminos et al. 1988; Cingolani et al. 1991). The North Patagonian Massif amalgamated with South America during the Early Triassic Gondwanide orogeny (Ramos 1988; Ramos et al. 1986; von Gosen 2003).

The existence of exotic terranes and Paleozoic suture zone is well established in the Southern Central Andes. In contrast, north of about 27° S the isotopic composition of sediments and basement rocks practically rule out the possibility of large scale terrane accretion in the Paleozoic (Bock et al. 2000; Egenhoff and Lucassen 2003; Lucassen et al. 2002). The two regions are separated by a sinistral shear zone at 27° S that offsets the rocks of the Famatina belt from their coeval equivalents in the west-

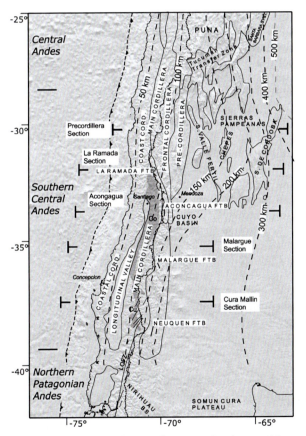

**Fig. 18.2.** Structural units of the Southern Central Andes 25–42° S redrawn and modified after Mpodosis and Ramos (1989)). Topography shown by shading. *Dashed lines* indicate depth to Wadati-Benioff zone from Creager et al. (1995). Named transects are discussed in the text. *Darker shading*: extent of Oligo-Miocene intra-arc basin system (*Co*: Coya Machali, *Cu*: Cura Mallin)

ern Puna to the north ("Faja eruptiva de la Puna Occidental") (Astini and Dávilla 2004).

In the Mesozoic repeated extension of the Gondwana margin led to the formation of a complex pattern of continental rifts and marine basins. The Neuquen basin of Argentina and its northern extensions, the Aconcagua and La Ramada depocenters in the Main Cordillera, as well as the Cuyo Basin in the eastern foreland, originated in segmented, predominantly NNW-striking rift systems (Fig. 18.3). Two phases of extension and rifting during the Triassic to Early Jurassic and during the Late Jurassic were both followed by phases of thermal subsidence that led to the deposition of extensive marine strata during the Middle Jurassic and Early Cretaceous (Giambiagi et al. 2003; Jacques 2004; Uliana et al. 1989). During the Cenozoic these basins were preferentially inverted and accommodated major fractions of the upper-plate contraction. Their normal faults controlled the geometry of Andean contractional structures on a local and regional scale (Manceda and Figueroa 1995; Ramos et al. 1996; Zapata et al. 1999).

Latitudinal changes in the amount and style of Andean contraction have not only been related to variations in upper-plate geology but also to lower-plate features. The dip of the Nazca Plate changes significantly along strike of the Andean margin (Barazangi and Isacks 1976; Cahill and Isacks 1992) and most authors agree that the regions of flat-slab subduction are supported by low-density anomalies within the downgoing plate (Gutscher et al. 2000b; Pilger 1981; Yanez et al. 2001). In the southern flat-slab section, which is associated with the Juan Fernandez Ridge of the lower plate, the anomalous width of the subduction orogen in the upper plate has been related to the displacement of the mantle wedge by subduction zone shallowing and subsequent changes in the thermal and mechanical conditions of upper-plate shortening (Jordan and Allmendinger 1986; Ramos et al. 2002).

Other transitions that coincide at the same latitude have been related to the varying trench fill which is thicker than 1.5 km and thinner than 0.5 km to the south and north of the Juan Fernandez Ridge, respectively (Bangs and Cande 1997) (Fig. 18.1). For example, Lamb and Davies (2003) have argued that the variations in coupling zone strength arising from the different thickness of the subducted sediment layer is responsible for the southward decrease in upper-plate shortening (Brooks et al. 2003; Dewey and Lamb 1992; Klotz et al. 2001) and the differences in interplate (Yanez et al. 2001) and upper-plate seismicity (Barrientos et al. 2004). The sediment thickness within the trench is primarily controlled by the flux of sediment resulting from upper-plate denudation. The latter strongly depends on precipitation and hence, climate may play a role for upper-plate contraction.

At present, the precipitation distribution along the Chilean margin is controlled by latitudinal changes of the wind pattern (Campetella and Vera 2002). In the Central Andes the easterlies loose their moisture on the eastern flank, so the western flank, the coastal regions of northern Chile, receive less than 50 mm yr$^{-1}$ of precipitation. In southern Chile the westerlies drive storm systems against the western flank of the Andes and here precipitation locally exceeds 4000 mm yr$^{-1}$ (New et al. 2002). Global climate simulations suggest that this contrast between the northern and southern Chilean margin prevailed throughout the Cenozoic (Bice et al. 2000) but in the Tertiary, global average temperatures were higher and the latitudinal temperature gradient was more subdued than in the Quaternary (Crowley and Zachos 2000).

Numerical simulations for the Tortonian climate that reproduce the warmer temperatures indicated by local climate proxies (Nielsen et al. 2004) show significantly reduced precipitation along the western flank of the Andes south of 35° S (Steppuhn 2002). Similar to the Holocene, this modified precipitation pattern is due to a southward shift of the westerly wind zone during global warming intervals (Lamy et al. 2004) and more southerly tracks of the westward moving storm systems. Hence, the tempera-

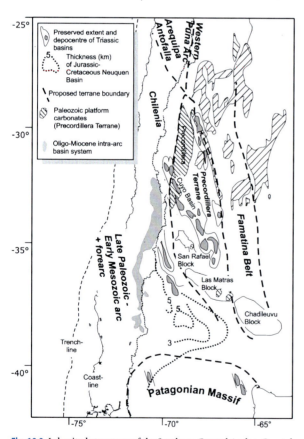

**Fig. 18.3.** Inherited structure of the Southern Central Andes. Crustal provinces, suspect terranes and pre-Andean sedimentary basins of the Southern Central Andes. Based on Astini and Davilla (2004), von Gosen et al. (2003), Uliana et al. (1989), Jacques (2004)

ture-dependent shifts of the westerly wind region control the precipitation pattern and thus influence the exhumation pattern and sediment flux into the trench.

## 18.3 Cenozoic Upper Plate Contraction between 30° S and 40° S

Within the Southern Central Andes, recent tectonic and geochronological studies allow a detailed reconstruction of the Neogene upper-plate deformation field. In this section we compile the available data along five transects (Fig. 18.2) and construct shortening-rate time series for four of them (Fig. 18.4) in order to identify possible controlling factors for upper-plate shortening.

### 18.3.1 Precordillera Transect (30.5° S)

Between 30° and 31°S, within the southern flat-slab region, the Andes are subdivided from west to east into the Coastal Cordillera, the Main Cordillera, the Frontal Cordillera, the Argentine Precordillera and the Sierras Pampeanas (Fig. 18.2). Neogene to recent Andean shortening has been mostly accommodated by thick-skinned thrusting or the local inversion of normal faults. Low-angle thrusting is restricted to the Argentinean Precordillera.

Finite shortening in the coastal area of this transect is probably less than 2 to 3 km (Allmendinger et al. 1990). For the western slope and the highest regions of the Main Cordillera shortening estimates are available from field studies (Maksaev et al. 1984; Reutter 1974) and fault length measurements in published maps. The latter use the correlation of fault length and fault displacement to estimate the finite slip from the surface trace of faults. For the western slope and the highest regions of the Main Cordillera these estimates are consistently higher than the field data (Allmendinger et al. 1990). The field data suggest a total Neogene contraction of the Main Cordillera in the range of 15 ± 5 km while the fault trace lengths indicate 35 ± 5 km. Similarly, the 25 ±5 km of shortening along the Vicuna fault system on the western slope calculated from a fault length of 230 km (Allmendinger et al. 1990) are contrasted by a field estimate of not much more than 2 km (Maksaev et al. 1984). Field studies in the Frontal Cordillera of Argentina attributed 10% of relative shortening to Andean deformation yielding an absolute contraction of 8 km (Heredia et al. 2002). Most likely, the shortening estimates from fault-length ratios suffer from the inevitable generalizations of fault traces in published maps which render the total displacement too high. Therefore, we assume that the field estimates, which sum up to a total shortening of 28 km for the Coastal, the Main, and the Frontal Cordilleras, are more reliable.

Shortening in the Main Cordillera probably started in the Early Miocene (Figs. 18.4 and 18.5). The earliest syntectonic units appear in the eastern foreland at ca. 21 to 20 Ma and provenance analysis indicates source regions in the Main and Frontal Cordilleras (Jordan et al. 1993; Reynolds et al. 1990). At about 16 Ma, most thrusts in the Chilean Main Cordillera can be considered inactive as they are covered by undeformed volcanics dated as 16.7 Ma (Maksaev et al. 1984). Hence, the available field estimates

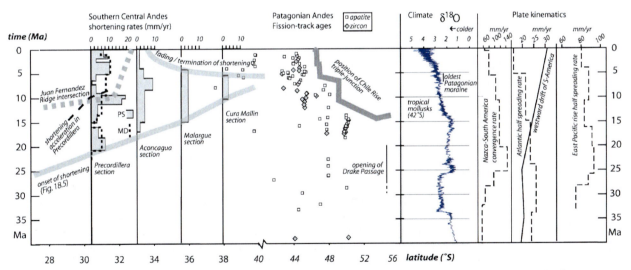

**Fig. 18.4.** Composite diagram showing the correlation of upper-plate shortening rates in the Southern Central Andes (*left*) with rock exhumation in Northern Patagonia (*center*), global climate evolution (*right*) and plate kinematic parameters (*far right*). Locations of transects used for shortening rate compilations are shown in Fig. 18.2; *PS, MD* rate curves based on "preferred" and "maximum duration" scenarios of Jordan et al. (1993). Fission-track ages from the Southern Central Andes and the Northern Patagonian Andes were compiled from Thomson et al. (2001, 2002) and Gräfe (2002). Position of Chile rise triple junction from Cande and Lisle (1986) and Cande (1987). Global $\delta^{18}O$ marine isotope record redrawn after Zachos et al. (2001). Plate kinematic data from Somoza (1998), Silver et al. (1998) and Tebbens and Cande (Tebbens and Cande 1997). Lines for onset and fading/termination of shortening from Fig. 18.5

of finite shortening and the radiometric data suggest that 28 ± 10 km of contraction were accommodated in the Frontal and Main Cordilleras from 21 Ma to 15 Ma, giving an interpolated rate of 4.7 ± 2 mm yr$^{-1}$ (Fig. 18.4).

At 30.5° S most of the Andean upper-plate shortening is taken up by the Argentinean Precordillera. Cambro-Ordovician limestones at the base of an up-to 10 km thick Paleozoic sequence constitute a regional detachment and enabled the accumulation of 117 km contraction during the Neogene by low-angle thrusting (Allmendinger et al. 1990; Kley and Monaldi 1998; Zapata and Allmendinger 1996). Similar magnitudes, corresponding to a relative shortening of more than 60%, have been determined at 29.5° S and at 31.5° S where estimates are 79 ± 9 km (Jordan et al. 2001b) and between 136 km and 88 km (Ramos and Cristallini 1995; von Gosen 1992), respectively.

The syntectonic sediments in the eastern foreland show that contraction in the Precordillera at 30.5° S started probably around 21 Ma but no later than 20 Ma (Jordan et al. 1993) (Fig. 18.5). Subsequently, the shortening rates in the Precordillera, based on the stratigraphy of the foreland basin and the kinematics of the thrusts, increased unsteadily (Jordan et al. 2001b) (Fig. 18.4). Between 20 and 11 Ma shortening rates varied between 0 and 4 mm yr$^{-1}$, followed by a phase of accelerated contraction at rates of 15 to 20 mm yr$^{-1}$. At 8 Ma, shortening rates dropped again below 5 mm yr$^{-1}$. Finally, at ca. 5 Ma contraction accelerated to the recent rate of 7 to 10 mm yr$^{-1}$. However, these strong variations of the shortening rate are smoothed when using the maximum duration of the different thrusts (Jordan et al. 1993). In that case, slow shortening of the Argentine Precordillera between 22 and 15 Ma (2 mm yr$^{-1}$) is followed by an acceleration to 5 mm yr$^{-1}$ until about 10 Ma.

In the 500 km wide Sierras Pampeanas the uplift of the individual ranges over high-angle reverse faults (Jordan and Allmendinger 1986) required only little crustal contraction. However, precise shortening data are not at hand because the inclination of the bounding faults of the ranges is only weakly constrained (Chinn and Isacks 1983; Schmidt et al. 1994). Jordan and Allmendinger (1986) estimated 2%, or 15 ± 5 km, of shortening across the entire Sierras Pampeanas Province.

The low temperature thermochronology of the Sierras Pampeanas indicates that uplift and contraction of the province started no later than 10 Ma (Coughlin et al. 1998) and, during continuous thrusting in the Precordillera, the onset of contraction in the Sierras Pampeanas resulted in another acceleration of the upper-plate shortening rate in the last 8–6 Ma to approximately 8–10 mm yr$^{-1}$ (Fig. 18.4). However, detailed data of individual ranges show that the shortening rate of the Sierras Pampeanas Province has not been constant since the initiation of uplift but that during episodic contraction individual ranges were uplifted at increased rates. For example, seismic data and the ages of the growth strata in the adjacent basins indicate a shortening rate for the Sierra Valle Fertil of 5 mm yr$^{-1}$ during the last 2 Ma (Zapata and Allmendinger 1996). Further to the east the deformation history of the Sierras de Chepes and the northern Sierras de Cordoba are poorly constrained. Only the onset of shortening of the Sierras de Cordoba at the easternmost end of the province has been dated to 6.5 Ma (Ramos et al. 2002).

In summary, the total Andean contraction along the transect at 30.5° S amounts to 160 ± 27 km which is based on the balanced section of the Precordillera and the less well constrained field estimates for the Frontal Cordillera, the High Andes and the Sierras Pampeanas (Table 18.1). Within the available constraints it remains undecided if the generally accelerating curve was punctuated by episodic fluctuations of more than 10 mm yr$^{-1}$ (preferred scenario of Jordan et al. 2001) or not. However, the two accelerations between 15 and 10 Ma and between 8 and 6 Ma are robust features, even if one assumes that the upper-plate contraction rate followed a rather smooth path (maximum duration scenario of Jordan et al. (1993) (Fig. 18.4).

## 18.3.2 La Ramada Transect (32° S)

This transect is still well within the southern flat-slab region of the downgoing Nazca Plate, and the Neogene Andean upper-plate contraction at this latitude is accommodated in the Main and Frontal Cordilleras, the Precordillera and the Sierras Pampeanas (Fig. 18.2).

The western flank of the Main Cordillera at 32° S is covered by continental clastics of the Coya Machali Basin (Abanico formation) and overlying arc volcanics (Farellones Formation). These deposits are part of the Eocene to Miocene intra-arc and fore-arc basin system of the Southern Central Andes. During the Neogene Andean orogeny the basins were inverted and uplifted. At 32° S contraction in the Coya Machali Basin, on the western slope of the Andes is estimated to be in the range of 7 ± 3 km (Cristallini and Ramos 2000).

Within the High Andes, thick Mesozoic sediments of the La Ramada Basin favored the concentration of shortening. The sediment fill allowed local thin-skinned thrusting in combination with inversion of Mesozoic extensional structures. Relative shortening in the Main Cordillera and the Frontal Cordillera thus reaches 30%, corresponding to 24 ± 2 km (Cristallini and Ramos 2000).

Similar to the transect at 30.5° S, contraction at 32° S is mostly concentrated in the Precordillera, where thin-skinned deformation accommodates between 88 km (von Gosen 1992) and 136 km (Ramos and Cristallini 1995) of contraction. In the adjacent Sierras Pampeanas, Ramos and Vujovich (2000) estimated 12 km of contraction in

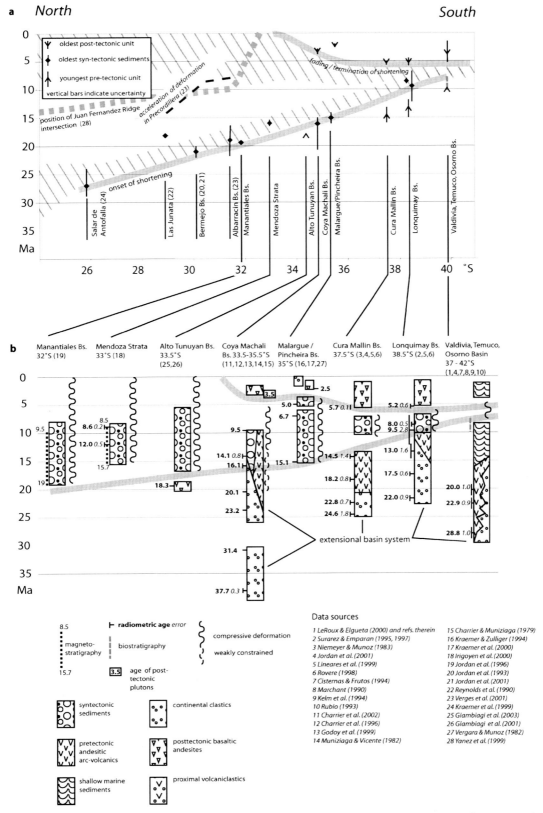

**Fig. 18.5.** Inferred onset of contraction along the strike of the Southern Central Andes (**a**) based on the constraints from syntectonic sediments, pre- and post-tectonic units (**b**)

**Table 18.1.**
Andean finite shortening estimates South of the Arica-Cochabamba bend. If not given in the original studies the errors of the shortening estimates are assumed to be 10% for geometrically balanced sections and 30% for all other methods

| Latitude (deg S) | Shortening from balanced sections (km) 10% error | Shortening from other methods (km) 30% error | Sum (km) Sum of errors |
|---|---|---|---|
| 21 | 297 ±30[a] | | 297 ±30 |
| 24 | 50 ±5[b] | | 102 ±13 |
|  | 52 ±8[c] | | |
| 30.25 | 117 ±12[d] | 20 ±7[e] | 160 ±27 |
|  |  | 8 ±3[f] | |
|  |  | 15 ±5[g] | |
| 31.5 – 32.5 | 24 ±2[h] | 12 ±4[k] | 180 ±20 or |
|  | 88 ±9[i] or | 8 ±2[h] | 132 ±17 |
|  | 136 ±12[j] | | |
| 33.5 | 70 ±7[l] | 5 ±2[m] | 75 ±9 |
| 35.5 | 38 ±4[n] | 20 ±6[o] | 58 ±10 |
| 38 | 5 ±1[p] | 4 ±1[q] | 14 ±3 |
|  |  | 5 ±1[r] | |
| 42.0 |  | 10 ±3[s] | 10 ±3 |

[a] Entire section at 21° S: Oncken et al. (2006, (Chap. 1, this volurme). [b] Puna and Western Cordillera: Cladouhos et al. (1994). [c] Kley and Monaldi (1998). [d] Precordillera: Kley et al. (1998). [e] High Andes, Western slope and Coastal Range: compiled from field data by Maksaev et al. (1984), Reutter (1974) and Allmendinger et al. (1990), see text. [f] Frontal Cordillera: Heredia et al. (2002). [g] Sierras Pampeanas: Jordan and Allmendinger (1986). [h] La Ramada Belt: Cristallini and Ramos (2000). [i] Precordillera: von Gosen (1992). [j] Argentine Precordillera: Cristallini and Ramos (1995). [k] Sierras Pampeanas (Pie de Palo): Ramos and Vujovich (2000). [l] Southern Aconcagua Fold Belt, Frontal Cordillera and Cuyo Basin: Giambiagi and Ramos (2002). [m] Coya Machali Basin: estimate from sections in Charrier et al. (2002). [n] Malargue Belt: Manceda and Figueroa (1995). [o] San Rafael Block and Chilean Andes: Ramos et al. (1996). [p] Cura Mallin: see Fig. 7. [q] covered Neuquen: see text. [r] Neuquen Belt: Zapata (1999). [s] Chilean Andes: Diraison et al. (1998).

the Sierra Pie de Palo from the combination of surface data and the available hypocenter distribution on a profile at 31.5° S. The total shortening along 32° latitude amounts to approximately 130 to 180 km. This estimate excludes the uplift of the easternmost Sierras Pampeanas that have accumulated probably less than 20 km of contraction.

Timing constraints for the shortening along the profile at 32° S are available from the syntectonic basins on the Frontal Cordillera, within the Argentine Precordillera and pre- and post-tectonic volcanics along the transect. Within the sediments on top of the Frontal Cordillera the onset of shortening in the Main Cordillera is recorded by conglomerates which have been assigned an age of 19 Ma based on paleomagnetic and radiometric dating (Jordan et al. 1996). Analysis of the clast contents of these sediments indicates initial thrusting in the contemporaneous arc on the western slope of the Main Cordillera and subsequent activation of the La Ramada belt and the Frontal Cordillera further to the east. Shortening in the Main and Frontal Cordillera proceeded at an average rate of 3 mm yr$^{-1}$ until about 10 or 9 Ma. At 9.2 ± 0.3 Ma the Cerro Piramides volcano sealed the contractional structures (Cristallini and Cangini 1993). Later deformation of the Main and Frontal Cordillera has been restricted to regional tilting of the western flank of the Andes (Ramos et al. 1996).

After about 10 Ma the deformation has been entirely focused on the Precordillera and the eastern foreland. Locally-derived conglomerates at the base of a thrust-top basin 30 km to the north suggest that deformation in the Precordillera started no later than 15.4 Ma (Bercowski et al. 1993). This is in agreement with the results of Cristallini and Ramos (2000) who propose that shortening in the Precordillera started at around 18 Ma. From these data Vergés et al. (2001) calculated a linear deformation rate for the Precordillera of 6.8 mm yr$^{-1}$ based on a shortening of 136 km. A strong increase of sediment accumulation rates observed in several sections along the strike of the Precordillera between 30 and 32° S suggest a southward migrating acceleration of shortening between 12 and 8 Ma (Vergés et al. 2001) (Fig. 18.5).

Deformation in the Sierras Pampeanas around 32° S most likely started approximately 6.5 Ma ago and shows a clear eastward progression of the deformation front following the eastward migration of magmatic activity with a delay of about 4 Ma (Ramos et al. 2002).

As the chronology of upper-plate shortening is far from complete along this transect we cannot attempt to construct a shortening rate evolution for the Neogene.

## 18.3.3 Aconcagua Transect (33.5° S)

Along the Aconcagua transect at 33.5° S (Fig. 18.2), at the southern limit of the southern flat-slab region, the onshore margin is subdivided into Coastal Cordillera, Longitudinal Valley, Main Cordillera (including the Aconcagua fold belt), the Frontal Cordillera and the Sierras Pampeanas. In the Main Cordillera, Neogene compression resulted in the inversion of the Coya-Machali intra-arc basin on the western slope and thick-skinned and locally thin-skinned contraction in the Aconcagua fold belt which forms the highest parts of the Main Cordillera. Thick-skinned contraction affected the Frontal Cordillera on the eastern slope of the Andes and the Cuyo Basin in the eastern foreland.

From the thickness distribution of the Coya Machali deposits and the tectonic style of the Andean structures Charrier et al. (2002) and Godoy et al. (1999) inferred syn-depositional normal faulting and concluded extensional to transtensional conditions lasting from the Oligocene until the Early Miocene (Fig. 18.5). On the grounds of intra-basin unconformities Godoy et al. (1999) suggested that the subsequent inversion started in the Middle Miocene and growth strata in the footwall of a thrust fault at 34.5° S have been dated at 16.1 Ma (Charrier et al. 2002). Thrusting in the Coya Machali Basin prevailed during the late Miocene, as 9 Ma old lavas are still cut by reverse faults (Godoy et al. 1999). The amount of Neogene shortening in the Coya Machali Basin is unknown, but intense deformation occurs only locally (Godoy et al. 1999) and we assume that total shortening is less than 10 km in analogy to around 32° S (Cristallini and Ramos 2000). Dioritic plutons with an age of 3.4 ± 0.4 Ma intruding the thrusts in the High Andes provide a minimum age for the termination of shortening (Ramos et al. 1997).

The Aconcagua thrust belt immediately east of the Coya Machali Basin formed within the sediments of the northern continuation of the Mesozoic retro-arc Neuquen Basin (Figs. 18.2 and 18.3). Here, locally thick marine Mesozoic sediments provided detachment horizons that allowed thin-skinned thrusting. Total shortening in the Aconcagua belt amounts to 47 km (Giambiagi and Ramos 2002). The chronology of shortening is recorded by the syn-tectonic sediments between the Main and Frontal Cordilleras which are underlain by pre-tectonic retro-arc volcanics dated as 18.3 Ma (Giambiagi et al. 2001). Based on correlations with more distal units of the foreland system, the first conglomeratic unit, which was mainly fed by the uplifting and contracting Aconcagua fold belt in the Main Cordillera to the West, is thought to span ages from 18 or 17 Ma to 10 Ma (Giambiagi et al. 2003). Paleocurrent indicators and the clast-content of the overlying volcaniclastics and epiclastics with an age range from 8.5 to 6 Ma indicate a main contribution from the thick-skinned Frontal Cordillera to the east. The final infill of the basin contains contributions from the east and west. The age of the overlying andesitic volcanics (5.9 Ma) suggests deposition between 7 and 6 Ma (Giambiagi and Ramos 2002). After the onset of shortening in the Main Cordillera at around 18–17 Ma the upper-plate contraction rate (Fig. 18.4) remained at around 4.3 mm yr$^{-1}$ until about 8.5 Ma. Between 8.5 and 6 Ma shortening accelerates to 8.3 mm yr$^{-1}$ due to the activation of thrusting in the Frontal Cordillera. Subsequently (6–4 Ma), contraction in the Main Cordillera faded as indicated by the mild deformation of the overlying 5.9 Ma volcanics and numerous Pliocene post-tectonic stocks (Ramos et al. 1997). At around 4 Ma deformation shifted east to the NNW-striking Cuyo Basin (Irigoyen et al. 1993) (Fig. 18.3). Subsequent shortening in these Triassic sediments and volcanics (Ramos and Kay 1991) amounts to approximately 6 km which yields a contraction rate of 1.5 mm yr$^{-1}$ (Giambiagi and Ramos 2002). Active fault scarps and seismically triggered rock-avalanches indicate ongoing contraction in the basin (Fauqué et al. 2000).

Summing up the finite shortening estimates within the Aconcagua fold belt, the Frontal Cordillera and the Cuyo Basin along the Aconcagua transect at 33.5° S leads to a cumulative across-strike shortening of approximately 75 km since the onset of shortening at around 18 or 17 Ma (Table 18.1).

## 18.3.4 Malargue Transect (35.5° S)

At 35.5° S the Andes show a cross-strike subdivision into Coastal Cordillera, Longitudinal Valley, Main Cordillera, Malargue fold belt and eastern foreland. Neogene Andean upper-plate contraction is concentrated on the western slope of the Main Cordillera in the Coya-Machali Basin and along the eastern margin of the orogen in the Malargue fold belt that developed within the Mesozoic Neuquen Basin.

Balanced cross-sections through the Malargue fold belt between 34.6° S and 36.1° S, which integrate surface geology, reflection seismics and well data, show an average shortening of 38 km (Manceda and Figueroa 1995) that decreases southward (Zapata et al. 1999). The thickness distribution of the Mesozoic units and the changing vergences within the fold belt indicate a strong control of the basin structure on the Andean deformation style by reactivation of the Mesozoic rift structures (Zapata et al. 1999). Outside of the Malargue fold belt shortening within the San Rafael block (Fig. 18.3) and on the Chilean side of the Andes is supposed to be less than 20 km (Ramos et al. 1996). Hence, total contraction at 35.5° S is in the range of 58 ± 10 km (Table 18.1).

The chronology of shortening in the Neuquen Basin is preserved within the thrust-top basins of the Malargue

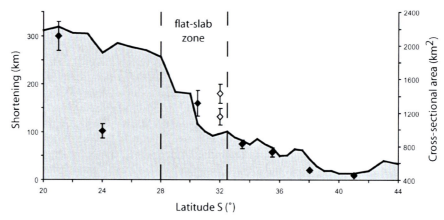

**Fig. 18.6.** Variation of finite shortening (*diamonds*) from Table 18.1 compared with cross sectional area (*solid line, shading*) along the western margin of South America. Cross sectional area of the topography above sea-level was extracted from the GEBCO dataset parallel to the average Neogene Nazca-South America convergence vector. Note the almost linear trends of the area curve between 20° S and 28° S and between 30° S and 39° S. The offset of 700–800 km² between the two segments marks the southern end of the plateau region at the northern limit of the southern flat slab region and does not coincide with the increase of finite shortening at around 32° S. See further explanation in the text. The Salar de Atacama depression at 24° S and regions with abundant back-arc volcanics at around 35° S and 42° S result in local anomalies

fold belt (Kraemer et al. 2000; Kraemer and Zulliger 1994). Volcanics close to the base of the growth strata indicate an onset of contraction just prior to 15.1 Ma which is in agreement with the Middle Miocene onset of contraction documented in the Coya Machali Basin at 34.5° S approximately 100 km to the north (Charrier et al. 2002). In the easternmost and youngest syntectonic thrust-top basin of the Malargue fold belt sedimentation started around 5 Ma (Kraemer et al. 2000). Overlying undeformed sediments of Quaternary age indicate a subsequent fading of contraction along this profile. This is corroborated by the undeformed Late Pliocene volcanics of the Cola de Zorro Formation which unconformably cover the deformed units of the Coya Machali Basin on the western slope of the Main Cordillera (Charrier et al. 2002). At 35.7° S the age span for the Cola de Zorro Formation is 2.47 to 0.96 Ma (Drake 1976; Vergara and Munoz 1982). Further to the south, at 37° S to 39° S, the base of the post-tectonic volcanics appears to be older (Fig. 18.5) and the ages for the lowermost flows range from 5.67 to 5.1 Ma (Linares et al. 1999).

The limited chronological constraints and the finite shortening estimate along the profile allow calculation of only a single interpolated upper-plate shortening rate. The total contraction of 58 ± 10 km and the duration of the deformation from 15 Ma to approximately 2.5 Ma yield a shortening rate of 4.1 mm yr$^{-1}$.

### 18.3.5 Cura Mallin Transect (38° S)

At around 38.5° S the Late Cenozoic Andean contraction resulted in inversion and shortening of the Mesozoic Neuquen Basin and the Oligo-Miocene Cura Mallin Basin (Fig. 18.7). The latter is comparable in its tectonic position to the Coya Machali Basin further to the north.

Likewise, it represents a pre-shortening extensional phase. However the upper-most units in the Cura Mallin Basin deposits are significantly younger than those in the Coya Machali Basin (Carpinelli 2000; Jordan et al. 2001a; Kemnitz et al. 2005; Niemeyer and Munoz 1983; Radic et al. 2002; Suarez and Emparan 1995). The available age constraints and sedimentological data show that the extension and subsidence in the northern Cura Mallin subbasin at around 37° S started during the Late Oligocene – Lower Miocene and lasted for about 5 Ma until the Middle Miocene (Jordan et al. 2001a; Radic et al. 2002). Within the southern subbasin in the Main Cordillera of Chile at 38.5° S ages of 17.5 ± 0.6 Ma and 13.0 ± 1.6 Ma from volcanic layers within sedimentary units and ages spanned from 19.9 ± 1.4 to 10.7 ± 1.1 Ma for the interfingering volcanic rocks (Suarez and Emparan 1995) suggest that semicontinuous sedimentation and coeval magmatism lasted until the Late Miocene. These pre-tectonic rocks are paraconformably overlain by fluviatile conglomerates of the Mitrauquen Formation. Interbedded ignimbrites yielded K/Ar ages younger than 9.5 ± 2.6 Ma (Suarez and Emparan 1995). Thickening and the decreasing dip of individual members towards the east indicate syntectonic deposition of the unit in the footwall of the east-dipping Pino Solo thrust at the Chile-Argentine border. The onset of shortening in the southern subbasin thus postdates 10.7 ± 1.1 Ma and predates 9.5 ± 2.6 Ma (Fig. 18.5).

Our balanced cross section at 38° S (Fig. 18.7) yields a minimum shortening of 5.5 km (15%), for the western part of the basin where structures dominantly verge to the west. East of the drainage divide, the Cura Mallin Basin is largely covered by the Plio-Pleistocene Cola de Zorro volcanics and younger deposits. From seismic section interpretation and geological field work Jordan et al. (2001a) concluded that the Late Miocene deformational

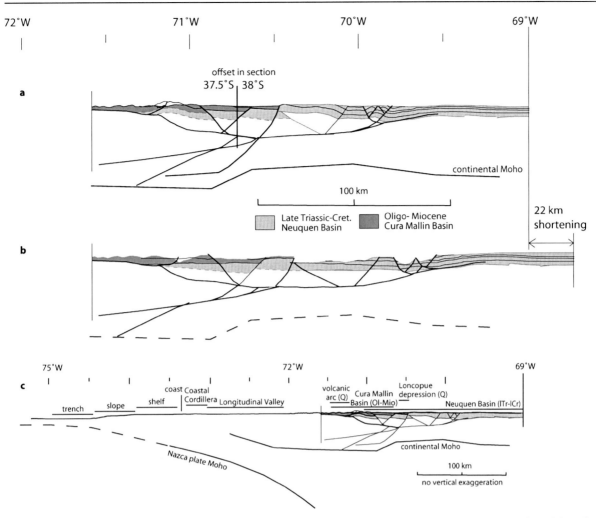

**Fig. 18.7. a** Restorable deformed composite section of the region between 37.5° S and 38° S constructed from seismic and geological data of (from E to W) Cobbold and Rosello (2003), Zapata (1999) and Jordan et al. (2001a), field mapping of Niemeyer and Munoz (1983) and supplemented by own data. **b** Restoration to pre-compressional stage (Middle Miocene in the Cura Mallin Basin and pre-Late Cretaceous in the Neuquen Basin). Note that geometry of Mesozoic strata in the foreland is a result of pre-compressional graben formation. Late Cretaceous and Neogene shortening inverts these grabens. **c** Deformed crustal section between the oceanic trench and the back-arc. Moho geometry (Yuan et al. 2005) shows crustal thinning in the region of thickest Mesozoic deposits of Neuquen Basin

structures in the eastern subbasin are mostly east-vergent and shortening is accommodated by inversion of normal faults. The inverted Cura Mallin Basin with the syntectonic units at the top is unconformably covered by the post-tectonic basalts and basaltic andesites of the Cola de Zorro Formation whereby in this region the oldest flows have an age of 5.7 Ma (Linares et al. 1999). Subsequent structures that cut the Early Pliocene Cola de Zorro volcanics of the Main Cordillera indicate across-strike extension and dextral transtensional faulting associated with the activity of the Liquine Ofqui intra-arc fault zone (Fig. 18.2) (Melnick et al. 2005).

Within the Neuquen Basin east of the Main Cordillera, contraction of the folded Mesozoic sediments is evident in the field and well documented in seismic sections. Similar to the Malargue transect, shortening is accommodated by thick-skinned structures and the thickness distribution of the Mesozoic units indicates that the shortening preferentially inverted pre-existing normal faults (Cobbold and Rosello 2003; Zapata et al. 1999). The balanced section at 38° S (Fig. 18.7) shows a finite shortening of 17 ± 2 km in the Neuquen Basin. However, increments in the Late Cretaceous and the Eocene have contributed to the contraction. Tertiary units in the Neuquen Basin are only locally developed and generally thinner than 200 m. Hence, the observed shortening represents an upper limit for the Andean increment in the back arc at 38° S. For simplicity we attribute half of the contraction in the Neuquen Basin to the Late Neogene shortening episode corresponding to 8.5 ± 2 km. Together with the 5.5 ± 1 km of the Cura Mallin Basin this yields a total Andean contraction of 14 ± 3 km at 38° S.

The timing of Andean deformation in the back-arc region of the Neuquen is only loosely constrained by the ages of syntectonic and posttectonic volcanic structures and syn-tectonic sediments are only locally developed. East-southeast striking dikes in the Huantrico syncline at 37.5° S, with ages of 15.2–18.9 Ma, have been interpreted to be the oldest indications for Neogene compressive conditions in the upper-plate (Cobbold and Rossello 2003). Seismic data in the same syncline indicate growth structures in Miocene lavas in the footwall of the Tromen thrust (Vines 1990).

Pliocene and Quaternary volcanics east of the Main Cordillera show only minor across-strike contraction. Faults in the Plio-Pleistocene Cola de Zorro volcanics on the eastern slope of the Andes have either strike-slip or normal offsets which do not exceed a few hundred meters. These structures are related to the northern termination of the Liquine-Ofqui intra-arc fault zone (Folguera and Iannizzotto 2004). The Loncopue depression (Fig. 18.7) along the western margin of the Neuquen Basin probably originated as a compressional basin in the footwall of a west-directed Late Cretaceous or Neogene back-thrust (Cobbold and Rossello 2003). However, the basin fill of Plio-Pleistocene volcanics of unknown thickness and distributed Quaternary basalt cones within the basin possibly signify ongoing extension. The depression is situated at the western edge of a Moho uplift (Yuan et al. 2005) that coincides with the thickest deposits of the Neuquen back-arc basin (Fig. 18.7) and thus may be a Mesozoic feature which has been reactivated during the ongoing extensional phase since the Pliocene.

Hence, the Andean contraction at around 38° S with a maximum total shortening of 5.5 km in the Cura Mallin Basin and 8.5 km in the Neuquen Basin can be constrained to the interval between 10 and 5 Ma, yielding a shortening rate of $3 \pm 0.5$ mm yr$^{-1}$.

## 18.3.6 Upper-Plate Contraction South of 39° S

Andean contraction of the upper-plate decreases southward and Cobbold and Rossello (2003) speculate that Neogene shortening in the back-arc region is small south of about 37.5° S. Some contraction is evident in a profile from the Nirihuau region at 42° S that crosses the eastern slope and foreland of the Andes and shows mainly thick-skinned thrusting in Paleozoic, Mesozoic and Oligo-Miocene sediments (Diraison et al. 1998). Similar to the structures in the Neuquen Basin, there is evidence for inversion of Mesozoic normal faults. Measuring the line length of the basement top yields a minimum contraction of 8.5 km in the sedimentary basins which extend over 80% of the cross section suggesting that contraction along the profile is in the range of 10 km.

The Tertiary sedimentary sequence in the Nirihuau region, at 42° S, comprises pre-tectonic Oligocene to Middle Miocene volcaniclastics with intermittent lake deposits (Nirihuau Formation) that are similar in age and facies to the Cura Mallin Basin. The Nirihuau Formation is unconformably overlain by Late Miocene syntectonic deposits of the Collón Curá Formation (Diraison et al. 1998) which are in turn covered by little-deformed basalts of Pliocene to Quaternary age. This suggests that some of the contraction observed in the back-arc is of Late Miocene to Pliocene age. This age range of upper-plate contraction derives support from the geology of the fore-arc basins between 39° and 42° S. In the onshore Valdivia Basin, LeRoux and Elgueta (2000) documented semi-continuous sedimentation from the Late Oligocene to the Late Miocene. The end of deposition in the Tortoninan (< 11.2 Ma) due to uplift of the Valdivia Basin is broadly coeval with the first occurrence of syntectonic sediments in the southern Cura Mallin subbasin at 38.5° S and thus a likely indication of tectonic shortening in the fore-arc.

Between 42° S and the Chile-Rise triple junction at 46.5° S the Andean upper-plate deformation involves a major strike-slip component which is accommodated by dextral motion along the intra-arc Liquine-Ofqui fault zone (LOFZ) (Cembrano 1992, Cembrano et al. 1996). Transpression at the southern end of the fault since the Middle Miocene is generally attributed to the collision of the Chile Rise (Forsythe and Nelson 1985; Nelson et al. 1994). Due to the deep exhumation of the Patagonian Andes and the absence of marker units the Neogene displacement of the LOFZ is poorly constrained. The most recent estimates decrease from about 80 km at the southern end (Rosenau 2004) to a few hundred meters along the splays of transtensional horsetail that forms its northern end at around 37° S (Folguera et al. 2004). In contrast to the Pliocene slowing of the margin normal shortening (see above) the transcurrent motion along the LOFZ has accelerated since the Miocene (Cembrano et al. 2002; Cembrano et al. 2000; Thomson 2002; Thomson et al. 2001).

South of the Chile Rise triple-junction (46.5° S) the magnitude of Neogene shortening increases again and the average topographic relief of the Main Cordillera is almost 2000 m higher than in the north (Ramos 2005). Most of the back-arc shortening however has to be attributed to pre-Neogene deformation phases, predominantly in the Cretaceous when up to 560 km of contraction and vertical axis rotations of up to 90°at the southern tip of South America formed the Patagonian orocline (Kraemer 2003).

After a period of tectonic quiescence in the Late Eocene incipient marine molasse sediments of the Centinela Formation signify renewed uplift and erosion in the Oligocene (Malumian and Ramos 1984; Ramos and Kay 1992) coeval with separation of Antarctica and South America and the opening of the Drake Passage (Barker 2001; Barker and Burrell 1977). Hence, the onset of Neogene deformation in the Magellan fold belt predates the collision and subsequent northward migration of the Chile rise triple junction (Lagabrielle et al. 2004; Ramos 1989) which took place at around 14 Ma (Cande and Leslie 1986; Cande et al. 1987; Gorring et al. 1997). However, the Neogene formation of upper-plate contractional structures in the Patagonian Andes reflects the impact of the northward migrating ridge-collision (Ramos 1989; Ramos 2005). Total Neogene shortening in the Magellan back-arc fold-belt ranges between 22 and 45 km at 47 to 49° S and increases to about 80 km at the southern tip of South America (Kraemer 2003; Ramos 1989). The present position of the Chile Rise triple junction was reached 6 Ma ago and since then, the subduction of convergence-parallel transform faults has kept it stationary.

South of the triple junction a conspicuous slowing of the back-arc contraction after the passage of the ridge intersection is indicated by the Strobel Basalts with an age of 8 to 9 Ma that cover the deformed Neogene structures and which are only slightly tilted (Lagabrielle et al. 2004; Ramos 1989). Here, the Andean margin has been subject to slow (2 cm yr$^{-1}$) and sinistrial oblique subduction of the Antarctic plate and reacts by mild left-lateral transtension sub-parallel to the convergent plate margin (Diraison et al. 1997).

In summary, the Neogene structural evolution of Southern Patagonia suggests that the passage of the triple-junction played an important role for the uplift and back-arc shortening. However, the onset of deformation as well as the strike-slip deformation subsequent to the ridge collision rather reflect the changes in the configuration of the subduction zone, e.g. the formation of the Drake Passage and the modifications of the subduction velocity and direction subsequent to the passage of the triple junction.

## 18.3.7 The Andean Subduction Orogeny between 30° S and 40° S

### 18.3.7.1 Distribution of Finite Shortening and Crustal Volume

Our compilation shows that Neogene finite shortening in the Southern Central Andes decreases monotonously from north to south (Table 18.1, Fig. 18.6). Between 33.5° S and 38° S the shortening diminishes from 75 to 14 km which corresponds to a gradient of approximately 11 km per ° latitude. Further south the gradient decreases to very low values because the finite shortening of the upper plate remains below 20 km up to about 46° S. Between 33.5° S and 32.25° S, the position of the Juan Fernandez Ridge intersection for the last 10 Ma, finite Andean shortening increases with a much higher gradient of ~70 km per °latitude to about 160 km and remains approximately constant up to the northern end of the Southern Central Andes at 30° S.

In contrast to the general northward increase observed between 40° S and 30° S upper-plate shortening most likely declines again north of 30° S. Cross section balancing across the Santa Barbara System and the Eastern Cordillera only revealed a maximum finite contraction of 70 km (Grier 1991; Kley 2002). North of 25° S finite contraction increases up to 21° S where it reaches the maximum along the Andean margin of approximately 300 km (Kley and Monaldi 1998). North of the Cochabamba-Arica bend (Fig. 18.1), which nearly coincides with the symmetry axis of the plateau in plan view and the average Tertiary Euler equator of the Nazca-South America convergence (Gephart 1994), finite shortening declines again on the opposite northern limb of the orocline (Kley and Monaldi 1998).

In Fig. 18.6 the shortening data for the southern limb of the Andean orocline are complemented with the variation of their cross-sectional area. The latter is generally accepted to be an approximation of tectonic shortening (Isacks 1988; Kley and Monaldi 1998; Montgomery et al. 2001). Areas above sea-level were calculated along sections parallel to the average Neogene Nazca-South America convergence vector because the geological data, the GPS vector field and the symmetry of the topography suggest that the deformation field of the Central Andes is parallel to that direction (Gephart 1994; Hindle et al. 2002; Kley 1999). Using assumptions for initial crustal thickness and the density distribution within the lithosphere it is possible to convert the topographic cross section into shortening but here we only discuss the general correlation of the along-strike variation of the area curve with the shortening observed on the geological sections.

Between 38° S and 31° S the cross-sectional area of the orogen increases linearly towards the north (Fig. 18.6) and is highly correlated with the geological shortening up to the position of the Juan Fernandez Ridge at 33.5° S. This suggests that shortening and topographic cross-section are strongly related south of 33.5° S. Local deviations, for example at 42° S, can be explained by the addition of magmatic material in the Somun Cura Plateau (Fig. 18.2). However, the area curve does not reproduce the sharp northward increase in geological shortening at around 33° S. Instead, the area increases further only with a linear trend up to about 30°S where it sharply increases into the plateau region. Obviously the intense shortening of the Argentinean Precordillera is not associated with a corresponding creation of topography. A weak correlation between geological shortening and cross sectional area north of 33.5° S is also supported by the data of the southern Puna Plateau. Here, the northward increase of the cross sectional area from 30.5° S to 28° S coincides with a drop of upper-plate contraction (Fig. 18.6).

Differential removal of crustal area by erosion is unable to account for these strong deviations as Neogene exhumation is generally low in the Central and Southern Central Andes (Montgomery et al. 2001) and weakly consolidated Tertiary sediments are equally widespread on both flanks of the orogen north and south of the region with the high shortening gradient at 33° S (Giambiagi et al. 2003; Jordan et al. 2001a).

These comparison of the cross-sectional area of the orogen and the geological shortening between 44° S and 20° S shows that the two data sets are closely correlated where crustal contraction is less than ~100 km. This relation breaks down in regions that have experienced stronger shortening. This may be either due to variations in the initial crustal or lithospheric thicknesses or a result of processes which are specific to plateau formation, e.g. upper-mantle thinning (Babeyko et al. 2006, Chap. 24 of this volume). Furthermore, steep along-strike shortening gradients may give rise to orogen-parallel flow in the lower crust and thus redistribute crustal volume of the Andean subduction orogen in north-south direction (Oncken et al. 2006, Chap. 1 of this volume).

### 18.3.7.2 Pre-Shortening Extension and the Middle/Late Cenozoic Onset of Subduction Orogeny in the Southern Central Andes

Similar to the southward declining finite shortening estimates, the ages of the oldest syntectonic sediments that document the initial tectonic uplift also decrease towards the south (Fig. 18.5). In the southeastern Puna, in the Salar de Antofalla, (25.45° S) the onset of deformation has been constrained to 28–25 Ma (Kraemer et al. 1999). At 30.5° S shortening started at around 21 to 20 Ma (Jordan et al. 1993). Finally, close to the southern end of the Southern Central Andes, at 38.5° S, deformation did not start before 10 Ma (Suarez and Emparan 1995). In between, the ages of the oldest syntectonic sedimentation satisfy a roughly linear trend with a progression of ca. 0.8° latitude per Myr or 90 km Myr$^{-1}$ (Fig. 18.5a).

Support for a southward younging onset of deformation can also be derived from the pre-shortening geology of the Andean margin. In the Southern Central Andes the Late Cenozoic intra-arc basins document that extensional conditions prevailed in the more southerly parts of the margin when shortening had already started to the north (Fig. 18.5b). In the Coya Machali Basin between 33° and 35.5° S extension has been documented to last from about 36 Ma into the Late Oligocene (Charrier et al. 2002; Godoy et al. 1999). At 37° S normal faulting is evident from seismic sections of the Cura Mallin Basin along the eastern margin of the Main Cordillera and their age range from 23 to 20 Ma is significantly younger than in the Coya Machali (Jordan et al. 2001a)). In the southern subbasins of the Cura Mallin at 38° S pre-contraction sedimentation lasted until about 11 Ma (Suarez and Emparan 1995). Still further south the Valdivia Basin (40° S) and the Osorno Basin (42° S) record renewed or accelerated subsidence at around 16 Ma (Kelm et al. 1994; LeRoux and Elgueta 2000) which are likely related to extension (Jordan et al. 2001a). This underlines the conclusion that shortening in the interior parts of the Central Andes that started in the Eocene and progressively migrated southwards was coeval with extension in the more southerly parts of the margin for most of the subduction orogeny (Fig. 18.5b).

### 18.3.7.3 Late Cenozoic Termination of Shortening South of the Juan Fernandez Ridge

Upper-plate shortening has been continuous since the southward migrating onset of deformation only north of 33° S, the present position of the Juan Fernandez Ridge (Giambiagi et al. 2003; Ramos et al. 2002). In contrast, the widespread sealing of deformation structures by Pliocene volcanics south of this latitude has lead to the conclusion that, in the last few Ma, upper plate margin-normal contraction has been slow or absent (Folguera and Iannizzotto 2004; Ramos 1989; Rapela and Kay 1988) (Fig. 18.5). At 38° S the oldest flows of the post-orogenic Cola de Zorro volcanics have an age of 5.67 ± 0.14 Ma (Linares et al. 1999). Correlative but slightly younger volcanics (2.5 Ma) also cover the deformation structures in the Coya-Machali Basin at 35.5° S (Charrier et al. 2002); and on the eastern flank of the Andes, thrust top basins in the Malargue fold belt at the same latitude are covered by undated but probably Quaternary post-tectonic sediments (Kraemer et al. 2000; Kraemer and Zulliger 1994). Even at the southern rim of the flat-slab zone along the Aconcagua transect shortening shows a conspicuous slowing at around 4 Ma (Fig. 18.4). The distribution of post-tectonic rocks suggests that possibly the fading of deformation migrated from south to north (Fig. 18.5). However, the progression remains poorly constrained.

The slowing of margin-normal contraction south of 33 or 34° S at around 5 Ma is corroborated by the reflection seismic surveys of the fore-arc shelf basins. Pre-Pliocene units are affected by the Late Miocene deformation that resulted in widespread inversion of pre-existing normal faults. In contrast, the para-conformably overlying Pliocene to Quaternary sediments are essentially undeformed (Gonzalez 1989; Jordan et al. 2001a; Mordojovich 1981).

Finally, the geodetic deformation field of the Southern Central Andes based on GPS data shows a sharp gradient of back-arc contraction rate at around 34° S. To the north, active deformation is most likely concentrated within the Precordillera fold belt and accommodates across-strike shortening at a rate of 4.5 mm yr$^{-1}$ (Brooks et al. 2003) or 6 mm yr$^{-1}$ Klotz et al. (2006, Chap. 4 of this volume). In contrast, south of 34° S the GPS permanent back-arc deformation rates are insignificant (Klotz et al. 2006, Chap. 4 of this volume).

The Plio-Quaternary slowing or termination of the subduction orogeny south of the Juan Fernandez Ridge at 33° S explains some of the along-strike variation of the shortening magnitude (Fig. 18.6). However, not only the duration but also the rates of contraction decrease southwards (Fig. 18.4), which indicates that the mechanical conditions for upper-plate deformation varied along the convergent margin.

## 18.4 Controlling Factors of Upper-Plate Shortening

The Neogene southward younging of the onset of upper-plate shortening, the Pliocene fading of contraction restricted to the region south of the Juan Fernandez Ridge and the southward diminishing of the magnitude of upper-plate shortening require a temporal and spatial variation of the controlling parameters for subduction orogeny along the margin. In this section we compare the evolution of possible controlling factors for upper-plate contraction with the observed deformation pattern in the Southern Central Andes.

### 18.4.1 Plate Motions and Mantle Flow

Tectonic shortening along the Andean margin is ultimately driven by the relative motions of the lithospheric plates. Consequently, the variations in plate velocities have been repeatedly used to explain the evolution of upper-plate shortening. On the one hand, the phases of accelerated deformation of the upper plate have been correlated with the varying convergence velocity between the subducting Nazca and the overriding South American Plate (Jordan et al. 1993; Jordan et al. 2001a; Pardo-Casas and Molnar 1987; Somoza 1998). On the other hand, the correlation of Andean shortening with the westward drift of South America in the hot-spot reference frame emphasizes the role of the plate motions with respect to the mantle (Russo and Silver 1996; Silver et al. 1998).

Figure 18.7 shows the shortening rates in the Southern Central Andes, the convergence velocity of the Nazca and the South American Plates, the spreading rates of the Mid-Atlantic Ridge and the East pacific rise, and the absolute motion rate of South America with respect to the Atlantic hot-spots during the last 40 Ma. It confirms that the onset of deformation in the northern part of the Southern Central Andes falls into the Late Oligocene-Early Miocene interval of accelerated Nazca – South America convergence after the break-up of the Farallon plate (Handschumacher 1976; Pardo-Casas and Molnar 1987). Subsequently however, while the southward-migrating onset of deformation led to contraction of increasing portions of the Southern Central Andes, the relative convergence rate between Nazca and South America decreased. Similarly, the spreading rates in the South Atlantic and at the East Pacific Rise decelerate throughout the Miocene and most of the Pliocene (DeMets et al. 1994; Tebbens and Cande 1997; Wilson 1993) (Fig. 18.4).

The only plate kinematic parameter that is positively correlated with the Neogene lateral expansion of the subduction orogen in the Southern Central Andes is the westward upper-plate motion (absolute motion of South America in the hotspot reference frame or "overriding"). After a phase of stagnation between 35 and 25 Ma, in which the extensional Coya Machali and Cura Mallin Basins were initiated, the overriding velocity increased throughout the Miocene from about 20 mm yr$^{-1}$ to approximately 30 mm yr$^{-1}$ (Fig. 18.4). During this acceleration of overriding the upper-plate contraction expanded southward. This may indicate that more southerly regions along the Central and Southern Central Andes require higher overriding velocities in order to shorten the upper plate.

Figure 18.8 shows the kinematic balance of the plate motions in the hot-spot reference frame. If the deformation of South America is restricted to the plate boundary zone along the Andean margin than the overriding rates and magnitudes are coupled and almost identical along the entire plate boundary. It must be compensated either by shortening in the plate boundary zone (subduction erosion or tectonic shortening) or by westward motion of the lower plate hinge in the hot-spot reference frame i.e. absolute rollback. The rollback of the subduction zone

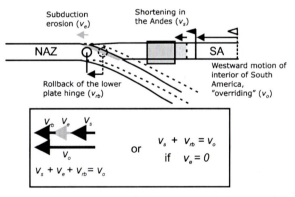

**Fig. 18.8.** Kinematical balance of plate motions in the hot-spot reference frame. Overriding or absolute westward motion of South America ($v_o$) is balanced by the sum of upper-plate shortening ($v_s$), subduction erosion ($v_e$) and absolute westward motion (rollback) of the lower-plate hinge ($v_{rb}$). The internal deformation of South America away from the plate boundary is low and consequently overriding is constant along the Andean margin. Hence, the along-strike variations of the plate boundary deformation must be mirrored by complementary changes of the lower-plate rollback

hinge with respect to the hot-spot reference frame requires lateral motion of the oceanic slab through the mantle and the transfer of mantle material from the sub-slab region to the mantle wedge (Fig. 18.9a). Russo and Silver (1996) and Silver et al. (1998) have suggested that the flow pattern of the mantle strongly influences the ability of the subducting plate to move towards the west. The bilateral symmetrical distribution of shortening along the Andean margin (Gephart 1994) suggests the intensity of the collision between the upper-plate and the subduction zone is a function of the distance from lateral slab-edges. This is confirmed by analog models which show that the mantle flow around the slab largely controls the motion of the subduction zone with respect to a mantle reference frame (Kincaid and Griffiths 2004). A circulation of mantle material around the slab is possibly supported by shear-wave splitting experiments that indicate horizontal flow in the Caribbean at the northern end of the Andean subduction zone (Russo and Silver 1994). At the opposite southern end of the convergent margin, the geochemical composition of the ocean floor suggests that a flow of mantle material around the southern end of the subduction zone started immediately after the opening of the Drake passage at around 30 Ma (Pearce et al. 2001). Russo and Silver (1996) and Silver et al. (Silver et al. 1998) argue that this outflow of subslab mantle enables faster rollback near the tips of the subduction zone. Large subduction windows that allow flow from the subslab into the mantle wedge above the slab may have a similar effect. In Patagonia Neogene flood basalts show a contribution from the subslab-region after the initiation of the Chile Rise triple junction and the opening of slab windows (Gorring et al. 2003; Ramos and Kay 1992). This suggests that mantle material from the Pacific flows around the lateral ends of the Nazca Plate.

A connection of upper plate deformation with the mantle circulation pattern can also be concluded from the geology of the Southern Patagonian Andes. Here, a change in the deformation regime coincides with the opening of the Drake passage in the Oligocene (Barker 2001; Barker and Burrell 1977; Eagles et al. 2005; Livermore et al. 2005). Prior to the opening several periods of back-arc contraction in the Cretaceous and Early Cenozoic that were coeval with the opening of the South Atlantic and which involved strong differential shortening lead to the formation of the Patagonian orocline with up to 560 km of contraction in southernmost South America (Kraemer 2003). After a period of tectonic quiescence in the Late Eocene, renewed deformation started in the Late Oligocene or Early Miocence (Ghiglione and Ramos 2005) subsequent to the opening of the Drake Passage. However, in contrast to the Cretaceous and Paleogene phases of deformation with strong back-arc contraction, this last phase of deformation has been dominated by sinistral wrenching along the southern tip of South America (Lodolo et al. 2003) and back-arc contraction has been subordinate (Ghiglione and Ramos 2005).

Geochemical, geophysical and geological data thus provide indications for a flow of mantle material around the slab tips and through slab windows of the Andean subduction zone. We suggest that this flow is the result of the westward motion of the subduction system pushed by the upper plate and that the rate of the flow limits the rollback velocity (Fig. 18.9a,b). In this context, displacement of the subslab mantle and a resulting rollback of the subduction zone are facilitated by pathways through or around the slab, especially close to slab windows or the lateral ends of the downgoing plate (Fig. 18.9b). In these regions, for example in the Southern Central Andes, shortening of the upper plate requires high overriding velocities. In the absence of such pathways, which is the case at large distance from the slab ends, the displacement of the subslab mantle by flow is inhibited. Here, rollback rates are slow and shortening rates of the upper-plate are high during overriding. This explains the situation in the Central Andes.

These considerations suggest that in the Southern Central Andes the southward increasing proximity to the southern slab edge enhances slab rollback and thus higher overriding velocities are required in more southerly regions to achieve upper-plate contraction (Fig. 18.9). The necessary higher overriding velocities were attained successively during the Early-Middle Miocene leading to a southward migrating onset of contraction in the Southern Central Andes between 21 and 10 Ma. Therefore overriding is a prime condition for the observed tectonic evolution of Southern Central Andes.

### 18.4.2 Pre-Andean Structures

The distribution of upper crustal volume in the Central Andes is almost perfectly symmetric with respect to the average Cenozoic convergence direction of South America and the Nazca Plate. Assuming that this distribution has been largely controlled by tectonic shortening, Gephart (1994) suggested that the finite contraction of South America is largely independent of the inherited geology of the upper plate. The southward decrease of the upper-plate shortening and the cross sectional area of the orogen described above (Fig. 18.6) suggest that this general assumption may also hold for the Southern Central Andes. In detail however, the steep along-strike gradients in both datasets indicate that they are can be significantly modified by local factors. This is underlined by the clear correlation of the magnitude of shortening and the thrust style in the back-arc of the Southern Central Andes which in turn is controlled by the pre-Andean structure of the upper-plate (Fig. 18.10).

**Fig. 18.9.**
**a** Conceptual model of the interaction of the west-moving South American Plate, the subducting oceanic Nazca slab and the surrounding mantle.
**b** The three elements that control if and to what extent overriding of the upper plate leads to upper-plate shortening:
(*1*) Flow conditions of the subslab mantle, (*2*) strength of the plate interface, (*3*) strength of the upper plate. Rollback is favored by a weak plate interface, a strong upper plate and the proximity of either a slab-tip or a large slab windows which enable a flow of subslab mantle into the supra-slab region. This scenario (*lower left*) impedes upper-plate shortening and overriding is completely converted into rollback of the slab. Subduction orogeny is favored by a weak upper plate, a strong plate interface and restricted flow of the subslab mantle in regions isolated from the lateral slab-edge (Central Andes). In this case (*lower right*) overriding is converted into upper-plate shortening

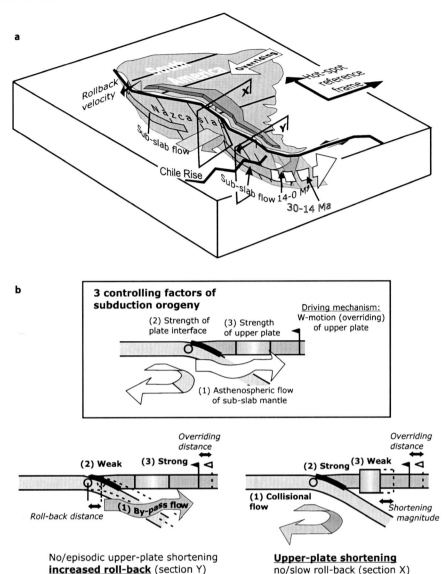

Shortening is generally concentrated in the pre-Andean basins which facilitate underthrusting of the foreland along low-angle detachments within or at the base of the sediments (Figs. 18.3 and 18.8). In the Precordillera belt of Argentina between 30° S and 32° S Paleozoic platform sediments are up to 10 km thick. A mechanically weak detachment at the base of the sequence enables the a relative shortening of 65–70% (Allmendinger et al. 1990). At both along-strike terminations of the Precordillera belt the total upper-plate contraction strongly decreases which suggests a control of the extent of the basin on the magnitude of contraction.

South of the Argentinean Precordillera, near 33.5° S, the strong shortening of the northern Aconcagua foldbelt indicates that some of the contraction of the Precordillera is transferred into the Main Cordillera (Figs. 18.2 and 18.8a) (Giambiagi and Ramos 2002). Here, the increased relative shortening (57%) was facilitated by the locally thick Mesozoic sediments (Fig. 18.10a). The thinning of sediments to the north and south, in the La Ramada fold belt and the southern Aconcagua fold belt resulted in thick-skinned or mixed mode thrusting with less intense shortening of approximately 30% (Cristallini and Ramos 2000; Giambiagi and Ramos 2002).

Shortening in the Coya Machali Basin west of the Aconcagua fold belt involved inversion of normal faults and thick-skinned structures. This resulted in a relative shortening of less than 10% (10 km) (Ramos et al. 1996). Similarly, the relative contraction of the Frontal Cordillera to the east amounts to only 20% (Giambiagi and Ramos 2002).

The same control on the deformation style and relative shortening is observed in the Neuquen Basin where

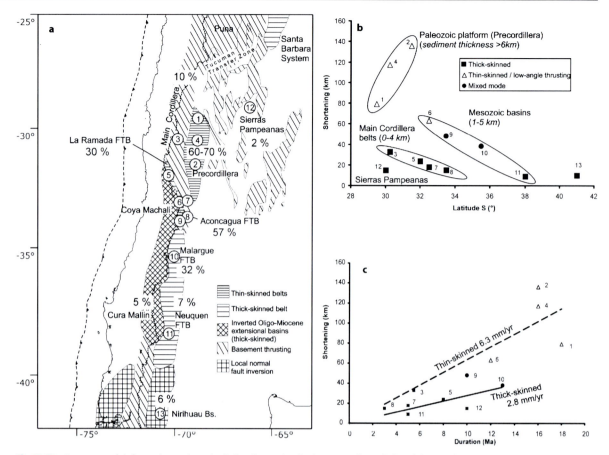

**Fig. 18.10. a** Summary of deformation style and relative shortening in the major thrust belts of the Southern Central Andes. **b** Shortening magnitude and deformation style. Paleozoic and Mesozoic marine basins with thick sedimentary cover absorb most of the upper-plate contraction. **c** Shortening and average duration of deformation. Linear fits show that shortening by low-angle thrusting is roughly twice as fast as in the thick-skinned belts. Data sources: *1*: Jordan et al. (1993), *2*: Ramos and Cristallini (1995), *3*: Heredia et al. (2002), *4*: Jordan et al. (1996), *5*: Cristallini and Ramos (2000), *6–9*: Giambiagi and Ramos (2002), *10*: Manceda and Figueroa (1996), *11*: this study, *12*: Jordan and Allmendinger (1986), *13*: Diraison et al. (1998)

the reduced thickness of the Mesozoic sediments and a wide strip of basement affected by Mesozoic extension prevented thin-skinned shortening and enforced the formation of the dominantly thick-skinned Malargue and Agrio fold-and-thrust belts (Kley et al. 1996).

Preferential shortening of the Mesozoic basins is also evident in the localization of shortening along the strike of the Main Cordillera. At 33° S relative contraction peaks in the Aconcagua belt (Fig. 18.10a) close to drainage divide of the Andes. Towards the south however, the relative shortening is strongest on the eastern flank of the chain and in the Neuquen Basin (Fig. 18.10b) because the strike of the Mesozoic rift is slightly oblique to the Main Cordillera (Figs. 18.3 and 18.10a).

A major control of the mechanical properties of the upper-plate on the distribution of finite deformation is also indicated by the deformation rates within the various belts of the Southern Central Andes. Average shortening rates in the thin-skinned belts are twice as high as in the thick-skinned regions and the contraction in the thin-skinned belts lasts longer by a factor of two than in their thick-skinned counterparts (Fig. 18.10c).

Hence, the available data on the Andean shortening of the Paleozoic and Mesozoic basins support the view of Allmendinger et al. (1983) that they represent soft regions within the upper plate that increased the magnitude and the rate of contraction during the Andean subduction orogeny. In consequence, the inherited distribution of pre-Andean basins represents one major controlling parameter of subduction orogeny.

### 18.4.3 Subduction geometry

The present flat-slab regions to the north and south of the Central Andean Plateau comprise the vicinities of the oceanic Nazca and the Juan Fernandez Ridges (Fig. 18.1). Most authors agree that these intra-slab low density features provide the buoyancy for slab shallowing and allow the displacement of the mantle wedge that enables the mechanical interaction of the upper and lower plates in the back-arc region (Gutscher 2002; Gutscher et al. 2000b; Jordan and Allmendinger 1986; Yanez et al. 2001). The shallowing of the slab in both regions is temporally associated with significant changes in magmatism and upper-plate tectonics and it has been suggested that phases of accelerated upper-plate shortening can be correlated with the migration of the ridge intersection along the margin (Ramos et al. 2002; Rousse et al. 2003).

Both the plate-motion reconstruction for the intersection point of the Juan Fernandez Ridge and the eastward migration of the arc magmatism document the arrival of the Juan Fernandez Ridge at its present position at approximately 12 to 10 Ma. The initial collision of the ridge at around 12 Ma coincides with a southward-migrating acceleration of thrusting and sediment accumulation along the Precordillera between 30 and 32° S (Jordan et al. 1993; Vergés et al. 2001) and the Frontal Cordillera between 32 and 34° S (Giambiagi et al. 2003). The subsequent and ongoing subduction of a convergence-parallel ridge segment has lead to progressive shallowing of the slab expressed in the eastward migration of arc magmatism and finally its termination at around 2 Ma (Ramos et al. 2002). Laramide-style thrusting in the Sierras Pampeanas followed the broadening of the magmatic-arc and affected the Andean foreland more than 800 km E of the trench (Fig. 18.1). South of the flat-slab region, upper-plate shortening is restricted to a much narrower belt predefined by the Mesozoic Neuquen and the Oligo-Miocene Cura Mallin Basins (Manceda and Figueroa 1996; Kley et al. 1999; Jordan et al. 2001) (Figs. 18.2 and 18.10).

The comparison of the migration of the Juan Fernandez Ridge intersection with the onset of upper-plate deformation (Fig. 18.4) shows that between 26 and 33° S the onset of contraction precedes the arrival of the ridge by more than 8 Ma at each point of the plate boundary. That means that the onset of upper plate shortening has a minimum lateral distance of 5° latitude or 550 km to the intersection of the Juan Fernandez Ridge. This excludes triggering of the upper-plate contraction by slab-shallowing. The rate evolution in the Precordillera transect (PS in Fig. 18.4) and in the Aconcagua transect show that the accelerations during the ridge passage are short-lived events which last no more than 4 Ma. This phase coincides with accelerated subduction erosion at the plate boundary probably associated with higher shear stresses along the plate interface (Laursen et al. 2002). Subsequently, the upper-plate contraction rate in the Aconcagua transect, at 33.5° S, declines below the Middle Miocene values. Apparently, the passing ridge leaves only an episodic signal in the shortening record. However, close to the center of the present flat-slab section, at 30.5° S, short-ening accelerates again in the Late Miocene due to the intermittent activity of the Sierras Pampeanas thrusts during ongoing shortening in the Precordillera.

It has been proposed that the modified upper-plate deformation style and the expansion of the deforming zone over the flat-slab section is due to the direct mechanical interaction of the upper- and lower-plates in the back arc region (Gutscher et al. 2000a; Jordan and Allmendinger 1986). However, the easternmost Sierras Pampeanas Range (Sierra de Cordoba) lies east of the flat slab (Fig. 18.2), precluding a direct mechanical influence of the lower plate. From the correlation of Sierras Pampeanas uplift with a preceding magmatic activity, Ramos (2002) proposed that the shallowing of the slab led to localized magmatic weakening which then facilitated shortening.

In summary, the effects of ridge subduction on upper-plate contraction appear to be two-fold: (*1*) It accelerates upper-plate shortening for short periods, which is due to the changes of the mechanical properties of the interplate surface. (*2*) Shallowing of the subduction angle leads to the eastward migration of the magmatic front which causes thermal weakening of a wide region within the upper plate that subsequently allows accelerated contraction.

### 18.4.4 Climate and the Strength of the Plate Interface

The Plio-Pleistocene decrease of the upper-plate shortening rate is limited to the region south of the Juan Fernandez Ridge where the sediment fill in the trench is thicker than 1 km (Bangs and Cande 1997) (Fig. 18.1). Thick layers of subducted sediment lubricate and thus weaken the plate-interface (Lamb and Davis 2003), and ultimately, such a mechanism links the strength of the plate interface to the long-term sediment flux into the trench and to exhumation along the Andean margin.

The distribution of exhumation along the Southern Central and the Patagonian Andes is highly correlated with recent precipitation pattern (Figs. 18.1 and 18.4). While the Neogene exhumation is low north of about 39° S, further south the absence of Neogene cover sequences in the Main Cordillera (Jordan et al. 2001a), lower- to mid-crustal intrusion depths for outcropping Neogene arc granitoids (Herve et al. 1996; Parada et al. 2000; Seifert et al. 2005) (Fig. 18.11) and finally abundant Neogene fission-track data (Thomson 2002; Thomson et al. 2001; Glodny et al. 2006, Chap. 19 of this volume) (Fig. 18.4) indicate substantial surface erosion.

Sediments from this source region between 40° S and 46° S which are delivered to the trench mostly along canyons that incise the offshore fore-arc are re-distributed northward along the plate boundary and fill the trench up to the position of the Juan Fernandez Ridge (Fig. 18.1). This has been shown by analysis of the trench fill struc-

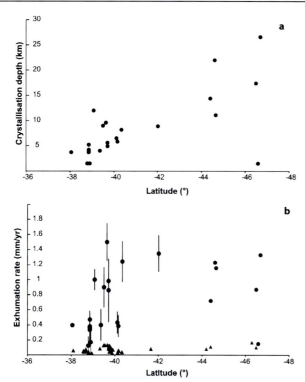

**Fig. 18.11. a** The crystallization depths (Al-in hornblende geobarometry) of Neogene granitoids in the Northern Patagonian Andes (39° S to 46° S) increase towards the south. **b** Linear exhumation rates of the Neogene granitoids calculated from intrusion depths and ages. *Vertical bars* indicate errors resulting from uncertainties of age determinations. Note the sharp increase of the exhumation rate to approximately 1 mm yr$^{-1}$ at 39° S. *Triangles* are pre-Neogene values. Data sources: Seifert et al. (2005), Herve et al. (1996), Parada et al. (2000), Pankhurst and Herve (1994)

ture and the northward dip of the axial channel within its surface (Bangs and Cande 1997; Völker et al. 2006, Chap. 9 of this volume).

Within the main sediment source area Neogene fission-track ages are generally restricted to the limits of the late Neogene and Quaternary ice-fields (Thomson 2002; Thomson et al. 2001), which suggests a connection of exhumation and the Northern Patagonian glaciation that started at around 7 Ma (Mercer and Sutter 1982). The large scatter of the fission-track ages within small distances shows that the exhumation pattern is strongly influenced by differential vertical motions along faults (Thomson 2002; Thomson et al. 2001). Outside the ice cover and in the back-arc region, fission track ages mostly fall in the Late Cretaceous or Early Tertiary (Glodny et al. 2006, Chap. 19 of this volume) and during the Miocene the fission track data show a conspicuous northward younging trend (Fig. 18.4).

A comparison of the low-temperature thermochronology data with the motion of the Chile rise triple junction shows that the exhumation is closely linked to and precedes the arrival of the triple junction at each point of the Patagonian margin (Fig. 18.4). The importance of tectonic uplift in connection with the triple junction as a prerequisite for denudation is underlined by the absence of young fission-track ages south of the Chile Rise intersection (Fig. 18.4). This shows that even thickly glaciated regions were unable to sustain high erosion rates once the tectonic uplift pulse associated with the passing ridge had ceased.

A close connection of the triple-junction migration with upper-plate contraction and the creation of topography is also evident from the Neogene shortening evolution in the Patagonian Andes. Wile the onset of contraction in the Late Oligocene or Early Miocene precedes the triple-junction initiation at 14 Ma (Ramos 2005) (Ghiglione and Ramos) the subsequent intense shortening pulses in the Main Cordillera and the eastern foreland of the Patagonian Andes clearly reflect the north-stepping collision of Chile rise segments (Lagabrielle et al. 2004; Ramos 1989). We conclude that the exhumation in the main sediment source region for the Chilean trench was mainly driven by the topographic evolution in response to the Chile Rise collision that started at 14 Ma.

The comparison of the upper-plate shortening chronology in the Southern Central Andes with the exhumation data from the Patagonian Andes (Fig. 18.4) shows that the southward expansion of upper-plate shortening between 30° S and 40° S during the Miocene is coeval with a northward migrating exhumation pulse controlled by the Chile rise collision in the Patagonian Andes. This suggests that the sediment flux resulting from the Middle Miocene exhumation in Patagonia was unable to sufficiently lubricate and weaken the plate interface along the Southern Central Andes as to stop upper-plate shortening.

In the Plio-Quaternary fission-track ages there is no apparent along-strike trend but the young ages extend far to the north, up to about 39° S. Hence, the region undergoing glacial erosion increased in the Late Miocene – Early Pliocene, which probably led to an increase in the sediment flux to the trench. This northward expansion of exhumation in the Late Miocene and Early Pliocene is closely followed by the slowing of upper-plate contraction between 40° S and 33° S (Fig. 18.4). The delay of about 2 Ma between the onset of glaciation at 7 Ma and the termination of shortening between 6 and 5 Ma at 38° S may be due to the transport time of the sediments to the stress supporting deeper parts of the plate interface. Today, the distance from the trench to the coastline, which generally coincides with the lower end of the coupling zone, is approximately 140 km (Hoffmann-Rothe et al. 2006, Chap. 6 of this volume). With a subduction rate of between 50 and 100 mm yr$^{-1}$ the complete lubrication of the plate interface requires between 1.4 and 2.8 Ma after delivery of the sediment to the trench.

In summary, these exhumation data suggest that the uplift of the sediment source region in the Patagonian Andes was controlled by tectonic parameters, i.e. the collision of the Chile Rise with the Andean margin and possibly the opening of the Drake Passage. In the middle Miocene, the resulting sediment flux was insufficient to stop upper-plate shortening by weakening the plate interface. However, the close temporal association of (*1*) north-ward expansion of Pliocene exhumation shown by fission-track data, (*2*) appearance of the first moraines in Patagonia at 7 Ma (Mercer and Sutter 1982) and (*3*) Pliocene slowing of upper plate shortening as far north as the Juan Fernandez Ridge indicate that global cooling in the Late Neogene increased the sediment flux sufficiently to suppress upper-plate contraction (Fig. 18.9b).

In this context, the contrasting nature and evolution of subduction orogeny north and south of the Juan Fernandez Ridge is a consequence of the long-lasting (> 10 Ma) stationary character of the ridge collision between 32° and 33° S. To the north, ridge subduction increased the upper plate contraction-rate by changing the mechanical properties of the plate boundary and by shallowing the subduction angle which led to upper-plate weakening. To the south the Juan Fernandez Ridge prevented the dispersal of sediment within the trench and during the Plio-Quaternary allowed the accumulation of a sediment layer thick enough to substantially weaken the plate-interface and inhibit upper-plate shortening.

## 18.5 Conclusions

In this study we compiled finite shortening data and available timing constraints for the Andean upper plate deformation from 30° S to 40° S. From that data we are able to better separate the potential geodynamic controls on subduction orogeny in this zone. Key observations derived from our compilation are:

1. The sediment record of Miocene basins in the Southern Central Andes shows that between 22 and 10 Ma the onset of shortening migrates southwards from 30° S to about 38.5° S.
2. South of the Juan Fernandez Ridge the upper-plate contraction drastically slows in the latest Miocene and even terminates in the Pliocene.
3. The deformation field shows a clear southward decrease in the magnitude of upper-plate shortening. This is a result of the southward decrease of shortening duration and average deformation rate.

Analysis of the spatial and temporal growth pattern of the subduction orogen in conjunction with ancillary data sets shows the following correlations:

1. Upper-plate shortening rates are increased in regions of thick sedimentary cover (Aconcagua belt, Argentine Precordillera) which enable low-angle thrusting. Acceleration of shortening is also observed in the flat-slab region that shows evidence for thermal weakening.
2. The southward expansion of the subduction orogen in the Middle and Late Miocene is not related to the Nazca-South America convergence velocity, which slowed throughout the Miocene. A positive correlation of the growth of the subduction orogen with the westward velocity of South America suggests that more southerly regions require higher overriding velocities to start and maintain upper-plate contraction. This shows that the conditions for upper-plate contraction vary with latitude. Apparently, the southward increasing proximity of the slab-edge in the Drake Passage or large subduction windows associated with the Chile rise intersection inhibit subduction orogeny and enhance rollback of the subduction zone.
3. The Plio-Pleistocene waning of shortening south of the Juan Fernandez Ridge coincides with the appearance of the first ice-sheets and the related increase of exhumation rates in the Patagonian Andes. However, the uplift and exhumation of the sediment source region in the Patagonian Andes following the opening of the Drake Passage was tectonically controlled by the northward migration of the Chile Rise triple-junction and appears to be a precondition of strong upper-plate denudation during Late Neogene glaciations.

From these relations we conclude that subduction orogeny is driven by the westward motion of South America and its collision with the subduction zone. The partitioning of the driving velocity into upper-plate shortening and trench rollback is controlled by (*1*) mantle flow conditions that change with the distance to the slab-edge (*2*) the strength of the plate interface which is controlled by the thickness of the trench fill (*3*) the strength of the upper plate which is determined by its inherited structure and heat flow.

## Acknowledgments

This work was financially supported by the DFG (Deutsche Forschungsgemeinschaft) in the frame of the Collaborative Research Center 267 – Deformation Processes in the Andes (Projects 2 and F4). The manuscript benefited from discussions with Onno Oncken, Stephan Sobolev, Daniel Melnick, Arne Hoffmann-Rothe and Friedrich Lucassen. We thank Jonas Kley and Javier Quinteros for constructive reviews. Figs. 18.1 and 18.2 were produced with the help of the GMT software package (Wessel and Smith 1998).

# References

Allmendinger RW, Gubbels T (1996) Pure and simple shear plateau uplift, Altiplano-Puna, Argentina and Bolivia. Tectonophysics 259(1–3):1–13

Allmendinger RW, Figueroa D, Snyder D, Beer J, Mpodozis C, Isacks BL (1990) Foreland shortening and crustal balancing in the Andes at 30°S latitude. Tectonics 9(4):789–809

Astini RA, Dávila FM (2004) Ordovician back arc foreland and Ocloyic thrust belt development on the western Gondwana margin as a response to Precordillera terrane accretion. Tectonics 23:1–19

Astini RA, Benedetto JL, Vaccari NE (1995) The early Paleozoic evolution of the Argentine Precordillera as a Laurentian rifted, drifted, and collided terrane: a geodynamic model. Geol Soc Am Bull 107(3):253–273

Götze H-J, Alten M, Burger H, Goni P, Melnick D, Mohr S, Munier K, Ott N, Reutter K, Schmidt S (2006) Data management of the SFB 267 for the Andes – from ink and paper to digital databases. In: Oncken O, Chong G, Franz G, Giese P, Götze H-J, Ramos VA, Strecker MR, Wigger P (eds) The Andes – active subduction orogeny. Frontiers in Earth Science Series, Vol 1. Springer-Verlag, Berlin Heidelberg New York, pp 539–556, this volume

Bangs NL, Cande SC (1997) Episodic development of a convergent margin inferred from structures and processes along the southern Chile margin. Tectonics 16(3):489–503

Barazangi M, Isacks BL (1976) Spatial distribution of earthquakes and subduction of the Nazca Plate beneath South America. Geology 4(11):686–692

Barker PF (2001) Scotia Sea regional tectonic evolution: implications for mantle flow and palaeocirculation. Earth Sci Rev 55(1–2):1–39

Barker PF, Burrell J (1977) The opening of Drake Passage. Marine Geol 25(1–3):15–34

Barrientos S, Vera E, Alvarado P, Monfret T (2004) Crustal seismicity in central Chile. J S Am Earth Sci 16(8):759–768

Bercowski F, Ruzycki L, Jordan TE, Zeitler P, Caballero MM, Perez I (1993) Litofacies y Edad Isotopica de la Secencia la Chilca y su Significado Paleografico para el Neogeno de Precordillera. XII Congreso Geologico Argentino, Actas 1:212–217

Bice KL, Scotese CR, Seidov D, Barron EJ (2000) Quantifying the role of geographic change in Cenozoic ocean heat transport using uncoupled atmosphere and ocean models. Palaeogeog Palaeoclimat Palaeoecol 161(3–4):295–310

Bock B, Bahlburg H, Woerner G, Zimmermann U (2000) Tracing crustal evolution in the southern Central Andes from late Precambrian to Permian with geochemical and Nd and Pb isotope data. J Geol 108(5), 515–535

Brooks BA, Bevis M, Smalley R Jr, Kendrick E, Manceda R, Lauria E, Manturana R, Araujo M (2003) Crustal motion in the Southern Andes (26°–36°S): do the Andes behave like a microplate? Geochem Geophys Geosyst 4(10): doi 10.1029/2003GC000505

Cahill TA, Isacks BL (1992) Seismicity and shape of the subducted Nazca Plate. J Geophys Res B Solid Earth Planet 97(12):17503–17529

Caminos R, Llambias EJ, Rapela CW, Parica CA (1988) Late Paleozoic-Early Triassic magmatic activity of Argentina and the significance of new Rb-Sr ages from northern Patagonia. J S Am Earth Sci 1(2):137–145

Campetella CM, Vera CS (2002) The influence of the Andes mountains on the South American low-level flow. Geophys Res Lett 29(17): doi 10.1029/2002GL015451

Cande SC, Leslie RB (1986) Late Cenozoic tectonics of the southern Chile Trench. J Geophys Res B 91(1):471–496

Cande SC, Leslie RB, Parra JC, Hobart M (1987) Interaction between the Chile Ridge and Chile Trench: geophysical and geothermal evidence. J Geophys Res B Solid Earth Planet 92(1):495–520

Carpinelli A (2000) Analisis estratigrafico, paloeambiental, estructural, y modelo tectono-estratigrafico de la Cuenca de Cura-Mallin VIII y IX Region, Chile, Provincia del Neuquen, Argentina. Master thesis, Universidad de Concepcion, Chile

Cembrano J (1992) The Liquine-Ofqui fault zone (LOFZ) in the Province of Palena: field and microstructural evidence of a ductile-brittle dextral shear zone. Departamento de Geologia, Facultad de Ciencias Fisicas y Matematicas, Universidad de Chile, Serie Comunicaciones 43:3–27

Cembrano J, Herve F, Lavenu A (1996) The Liquine Ofqui fault zone: a long-lived intra-arc fault system in southern Chile. In: Geodynamics of the Andes 259:1–3. Elsevier, Amsterdam, pp 55–66

Cembrano J, Schermer E, Lavenu A, Sanhueza A (2000) Contrasting nature of deformation along an intra-arc shear zone, the Liquine-Ofqui fault zone, southern Chilean Andes. Tectonophysics 319(2): 129–149

Cembrano J, Lavenu A, Reynolds P, Arancibia G, Lopez G, Sanhueza, A (2002) Late Cenozoic transpressional ductile deformation north of the Nazca-South America-Antarctica triple junction. Tectonophysics 354(3–4):289–314

Charrier R, Baeza O, Elgueta S, Flynn JJ, Gans P, Kay SM, Munoz N, Wyss AR, Zurita E (2002) Evidence for Cenozoic extensional basin development and tectonic inversion south of the flat-slab segment, southern Central Andes, Chile (33°–36°S). J S Am Earth Sci 15(1):117–139

Chinn DS, Isacks BL (1983) Accurate source depths and focal mechanisms of shallow earthquakes in western South America and in the New Hebrides island arc. Tectonics 2(6):529–563

Cingolani C, Dalla Salda LH, Herve F, Munizaga F, Pankhurst RJ, Parada MA, Rapela CW (1991) The magmatic evolution of northern Patagonia: new impressions of pre-Andean and Andean tectonics. In: Harmon RS, Rapela CW (eds) Andean magmatism and its tectonic setting. Geol Soc Am, Boulder CO, pp 29–44

Cladouhos TT, Allmendinger RW, Coira B, Farrar E (1994) Late Cenozoic deformation in the central Andes: fault kinematics from the northern Puna, northwestern Argentina and southwestern Bolivia. J S Am Earth Sci 7(2):209–228

Cobbold PR, Rossello EA (2003) Aptian to Recent compressional deformation, foothills of the Neuquen Basin, Argentina. Marine Petrol Geol 20(5):429–443

Coughlin TJ, O'Sullivan PB, Kohn BP, Holcombe RJ (1998) Apatite fission-track thermochronology of the Sierras Pampeanas, central western Argentina: implications for the mechanism of plateau uplift in the Andes. Geology 26(11):999–1002

Creager KC, Chiao LY, Winchester JP, Engdahl ER (1995) Membrane strain rates in the subducting plate beneath South America. Geophys Res Lett 22(16):2321–2324

Cristallini EO, Cangini A (1993) Estratigrafia y estructura de las nacientes del rio Volcan, Alta Cordillera de San Juan. XII Congreso Geologico Argentino, Actas 3, pp 85–92

Cristallini EO, Ramos VA (2000) Thick-skinned and thin-skinned thrusting in the La Ramada fold and thrust belt: crustal evolution of the High Andes of San Juan, Argentina (32°S). Tectonophysics 317(3–4):205–235

Crowley TJ, Zachos JC (2000) Comparison of zonal temperature profiles of past warm time periods. In: Huber B, MacLeod K, Wing S (eds) Warm climates in Earth history. Cambridge University Press,

Davies JS, Roeske SM, McClelland WC, Snee LW (1999) Closing the ocean between Precordillera terrane and Chilenia: Early Devonian ophiolites emplacement and deformation in the southwest Precordillera. In: Ramos VA, Keppie JD (eds) Laurentia-Gondwana connections before Pangea. Geol Soc Am Spec P 336:115–138

DeMets C, Gordon RG, Argus DF, Stein S (1994) Effect of recent revisions to the geomagnetic reversal time scale on estimates of current plate motions. Geophys Res Lett 21(20):2191–2194

Dewey JF, Lamb SH (1992) Active tectonics of the andes. Tectonophysics 205:79–95
Diraison M, Cobbold PR, Gabais D, Rosello EA (1997) Magellan Strait: part of a Neogene rift system. Geology 25:703–706
Diraison M, Cobbold PR, Rossello EA, Amos AJ (1998) Neogene dextral transpression due to oblique convergence across the Andes of northwestern Patagonia, Argentina. J S Am Earth Sci 11(6):519–532
Drake RE (1976) The chronology of Cenozoic igneous and tectonic events in the central Chilean Andes. In: Proceedings of the symposium on Andean and Antarctic volcanology problems, Santiago, Chile, September 1974. Int Assoc Volcanol and Chem Earth's Interior, Santiago, Chile, pp 670–697
Eagles G, Livermore RA, Fairhead JD, Morris P (2005) Tectonic evolution of the west Scotia Sea. J Geophys Res B110: doi 10.1029/2004JB0003154
Egenhoff SO, Lucassen F (2003) Chemical and isotopic composition of Lower to Upper Ordovician sedimentary rocks (Central Andes/South Bolivia): implications for their source. J Geol 111(4):487–497
Elger K, Oncken O, Glodny J (2005) Plateau-style accumulation of deformation: Southern Altiplano, Tectonics 24(4): doi TC4020
Fauqué L, Cortés JM, Folguera A, Etcheverria M (2000) Avalanchas de rocas asociadas a neotectonica en el valle del rio Mendoza, al sur de Uspallata. Rev Asoc Geol Argentina 55:419–423
Flynn JJ, Novacek MJ, Dodson HE, Frassinetti D, McKenna MC, Norell MA, Sears KE, Swisher CC, Wyss AR (2002) A new fossil mammal assemblage from the southern Chilean Andes: implications for geology, geochronology, and tectonics. J S Am Earth Sci 15(3):285–302
Folguera A, Iannizzotto NF (2004) The lagos La Plata and Fontana fold-and-thrust belt: Long-lived orogenesis at the edge of western Patagonia. J S Am Earth Sci 16(7):541–566
Folguera A, Melnick D, Ramos VA (2002) Particion de la deformacion en la zona del arco volcanico de los Andes neuquinos (36–39°S) en los ulitmos 30 millones des anos. Rev Geol Chile 29(2):227–240
Folguera A, Ramos VA, Hermanns R, Naranjo J (2004) Neotectonics in the foothills of the southernmost central Andes (37°–38°S): evidence of strike-slip displacement along the Antinir-Copahue fault zone. Tectonics 23: doi 10.10029/2003TC001533
Forsythe R, Nelson E (1985) Geological manifestations of ridge collision: evidence from the Golfo de Penas-Taitao basin, southern Chile. Tectonics 4(5):477–495
Gephart JW (1994) Topography and subduction geometry in the Central Andes: clues to the mechanics of a noncollisional orogen. J Geophys Res 99(B6):12279–12288
Ghiglione MC, Ramos VA (2005) Progression of deformation and sedimentation in the southernmost Andes. Tectonophysics 405(1-4):25–46
Giambiagi LB, Alvarez PP, Godoy E, Ramos VA (2003) The control of pre-existing extensional structures on the evolution of the southern sector of the Aconcagua fold and thrust belt, southern Andes. Tectonophysics 369(1-2):1–19
Giambiagi LB, Ramos VA (2002) Structural evolution of the Andes in a transitional zone between flat and normal subduction (33°30'-33°45'S), Argentina and Chile. J S Am Earth Sci 15(1):101–116
Giambiagi LB, Tunik MA, Ghiglione M (2001) Cenozoic tectonic evolution of the Alto Tunuyan foreland basin above the transition zone between the flat and normal subduction segment (33°30'-34°S), western Argentina. J S Am Earth Sci 14(7):707–724
Glodny J, Echtler H, Figueroa O, Franz G, Gräfe K, Kemnitz H, Kramer W, Krawczyk C, Lohrmann J, Lucassen F, Melnick D, Rosenau M, Seifert W (2006) Long-term geological evolution and mass-flow balance of the South-Central Andes. In: Oncken O, Chong G, Franz G, Giese P, Götz H-J, Ramos VA, Strecker MR, Wigger P (eds) The Andes – active subduction orogeny. Frontiers in Earth Science Series, Vol 1. Springer-Verlag, Berlin Heidelberg New York, pp 401–428, this volume

Godoy E, Yañez G, Vera E (1999) Inversion of an Oligocene volcano-tectonic basin and uplifting of its superimposed Miocene magmatic arc in the Chilean Central Andes: first seismic and gravity evidences. Tectonophysics 306(2):217–236
Gonzalez E (1989) Hydrocarbon Resources in the Coastal Zone of Chile. In: Ericksen GE, Cañas Pinochet MT, Reinemund JA (eds) Geology of the Andes and its relation to hydrocarbon and mineral resources. Circum–Pacific Council for Energy and Mineral Resources Earth Science Series, Houston, Texas, pp 383–404
Gonzalez-Bonorino G, Kraemer P, Re G (2001) Andean Cenozoic foreland basins: a review. J S Am Earth Sci 14(7):651–654
Gorring ML, Kay SM, Zeitler PK, Ramos VA, Rubiolo D, Fernandez MI, Panza JL (1997) Neogene Patagonian plateau lavas: continental magmas associated with ridge collision at the Chile Triple Junction. Tectonics 16(1):1–17
Gorring M, Singer B, Gowers J, Kay SM (2003) Plio-Pleistocene basalts from the Meseta del Lago Buenos Aires, Argentina: evidence for asthenosphere-lithosphere interactions during slab window magmatism. Chem Geol 193(3–4):215–235
Graefe K, Glodny J, Seifert W, Rosenau M, Echtler H (2002) Apatite fission track thermochronology of granitoids at the south Chilean active continental margin (37°–42°S): implications for denudation, tectonics and mass transfer since the Cretaceous. In: 5th International Symposium on Andean Geodynamics, Toulouse, France, pp 275–278
Grier ME, Salfity JA, Allmendinger RW (1991) Andean reactivation of the Cretaceous Salta Rift, northwestern Argentina. J S Am Earth Sci 4(4):351–372
Gutscher MA (2002) Andean subduction styles and their effect on thermal structure and interplate coupling. J S Am Earth Sci 15:3–10
Gutscher MA, Maury R, Eissen JP, Bourdon E (2000a) Can slab melting be caused by flat subduction? Geology 28(6):535–538
Gutscher MA, Spakman W, Bijwaard H, Engdahl ER (2000b) Geodynamics of flat subduction: seismicity and tomographic constraints from the Andean margin. Tectonics 19(5):814–833
Handschumacher D (1976) Post-Eocene plate tectonics of the eastern Pacific. Geophysical Monograph 19: the geophysics of the Pacific Ocean basin and its margin. 177–202
Heredia N, Rodriguez Fernandez LR, Gallastegui G, Busquets P, Colombo F (2002) Geological setting of the Argentine Frontal Cordillera in the flat-slab segment (30°00'–31°30' S latitude). J S Am Earth Sci 15(1):79–99
Herve F (1988) Late Paleozoic subduction and accretion in southern Chile. Episodes 11(3):183–188
Herve F, Munizaga F, Parada MA, Brook M, Pankhurst RJ, Snelling NJ, Drake R (1988) Granitoids of the Coast Range of central Chile: geochronology and geologic setting. J S Am Earth Sci 1(2):185–194
Herve F, Pankhurst RJ, Demant A, Ramirez E (1996) Age and Al-in-Hornblende geobarometry in the north patagonian batholith, Aysen, Chile. Third International Symposium on Andean Geodynamics, St Malo 1996, pp 579–582
Hildreth W, Moorbath S (1988) Crustal contributions to arc magmatism in the Andes of central Chile. Contrib Mineral Petrol 98(4):455–489
Hindle D, Kley J, Klosko E, Stein S, Dixon T, Norabuena E (2002) Consistency of geologic and geodetic displacements during Andean orogenesis. Geophys Res Lett 29(8): doi 10.1029/2001GL013757
Hoffmann-Rothe A, Kukowski N, Dresen G, Echtler H, Oncken O, Klotz J, Scheuber E, Kellner A (2006) Oblique convergence along the Chilean margin: partitioning, margin-parallel faulting and force interaction at the plate interface. In: Oncken O, Chong G, Franz G, Giese P, Götze H-J, Ramos VA, Strecker MR, Wigger P (eds) The Andes – active subduction orogeny. Frontiers in Earth Science Series, Vol 1. Springer-Verlag, Berlin Heidelberg New York, pp 125–146, this volume

Irigoyen MV, Buchan KL, Brown RL (1993) Magnetostratigraphy of Neogene Andean foreland-basin strata, lat 33°S, Mendoza Province, Argentina. Geol Soc Am Bull 112(6):803–816

Isacks B (1988) Uplift of the Central Andean Plateau and bending of the Bolivian Orocline. J Geophys Res 93(B4):3211–3231

Jacques JM (2004) The influence of intraplate structural accommodation zones on delineating petroleum provinces of the Sub-Andean foreland basins. Petrol Geosci 10(1):1–19

Jordan TE, Allmendinger RW (1986) The Sierras Pampeanas of Argentina: a modern analogue of Rocky Mountain foreland deformation. A J Sci 286(10):737–764

Jordan TE, Allmendinger RW, Damanti JF, Drake RE (1993) Chronology of motion in a complete thrust belt: the Precordillera, 30–31°S, Andes Mountains. J Geol 101(2):135–156

Jordan TE, Tamm V, Figueroa G, Flemings PB, Richards D, Tabbutt KD, Cheatham T (1996) Development of the Miocene Manantiales forealnd basin, Principal Cordillera, San Juan, Argentina. Rev Geol Chile 23(1):43–79

Jordan TE, Burns WM, Veiga R, Pangaro F, Copeland P, Kelley S, Mpodozis C (2001a) Extension and basin formation in the southern Andes caused by increased convergence rate: a mid-Cenozoic trigger for the Andes. Tectonics 20(3):308–324

Jordan TE, Schlunegger F, Cardozo N (2001b) Unsteady and spatially variable evolution of the Neogene Andean Bermejo foreland basin, Argentina. J S Am Earth Sci 14(7):775–798

Kelm U, Helle S, Cisternas ME, Méndez D (1994) Diagenetic character of the Tertiary basin between Los Angeles and Osorno, southern Chile. Rev Geol Chile 21(2):241–252

Kemnitz H, Kramer W, Rosenau M (2005) Jurassic to Tertiary tectonic, volcanic, and sedimentary evolution of the southern Andean intra-arc zone, Chile (38–39°S): a survey. N Jb Geol Paläont Abh 236(1/2):19–42

Kincaid C, Griffiths RW (2004) Variability in flow and temperatures within mantle subduction zones. Geochem Geophys Geosyst

Kley J (1999) Geologic and geometric constraints on a kinematic model of the Bolivian orocline. J S Am Earth Sci 12(2):221–235

Kley J, Monaldi CR (1998) Tectonic shortening and crustal thickness in the Central Andes: how good is the correlation? Geology 26(8):723–726

Kley J, Vietor T (2005) Subduction and mountain building in the Central Andes. In: Dixon T, Moore J (eds) Interplate subduction zone seismogenesis. MARGINS Theoretical and Experimental Earth Science Series, Vol 2

Kley J, Monaldi C, Salfity JA (1996) Along-strike segmentation of the Andean foreland. Third International Symposium on Andean Geodynamics, St Malo, pp 403–406

Kley J, Monaldi CR, Salfity JA (1999) Along-strike segmentation of the Andean foreland: causes and consequences. Tectonophysics 301(1–2):75–94

Klotz J, Khazaradze G, Angermann D, Reigber C, Perdomo R, Cifuentes O (2001) Earthquake cycle dominates contemporary crustal deformation in Central and Southern Andes. Earth Planet Sci Lett 193(3–4):437–446

Klotz J, Abolghasem A, Khazaradze G, Heinze B, Vietor T, Hackney R, Bataille K, Maturana R, Viramonte J, Perdomo R (2006) Long-term signals in the present-day deformation field of the Central and Southern Andes and constraints on the viscosity of the Earth's upper mantle. In: Oncken O, Chong G, Franz G, Giese P, Götze H-J, Ramos VA, Strecker MR, Wigger P (eds) The Andes – active subduction orogeny. Frontiers in Earth Science Series, Vol 1. Springer-Verlag, Berlin Heidelberg New York, pp 65–90, this volume

Kraemer PE (2003) Orogenic shortening and the origin of the Patagonian orocline (56° S.Lat). J S Am Earth Sci 15:731–748

Kraemer PE, Zulliger GA (1994) Sedimentacion Cenozoica sinorogenica en la faja plegada Andina a los 35 degrees S Malargue, Mendoza, Argentina. VII Congreso Geológico Chileno, Actas 7(1), pp 460–464

Kraemer B, Adelmann D, Alten M, Schnurr W, Erpenstein K, Kiefer E, Van den Bogaard P, Gorler K (1999) Incorporation of the Paleogene foreland into the Neogene Puna plateau: the Salar de Antofalla area, NW Argentina. J S Am Earth Sci 12(2):157–182

Kraemer PE, Silvestro J, Davila F (2000) Kinematic of the Andean fold-belt infered from the geometry and age of syntectonic sediments, Malarguee (35°30'S), Mendoza, Argentina. Annual Meeting of the Geological Society of America 2000, 32:7

Lagabrielle Y, Suarez M, Rossello EA, Herail G, Martinod J, Regnier M, De la Cruz R (2004) Neogene to Quaternary tectonic evolution of the Patagonian Andes at the latitude of the Chile Triple Junction. Tectonophysics 385(1–4):211–241

Lamb S, Davis P (2003) Cenozoic climate change as a possible cause for the rise of the Andes. Nature 425(6960):792–797

Lamy F, Kaiser J, Ninnemann U, Hebbeln D, Arz HW, Stoner J (2004) Antarctic timing of surface water changes off Chile and Patagonian ice sheet response. Science 304:1959–1962

Laursen J, Scholl DW, von Huene R (2002) Neotectonic deformation of the central Chile margin: deepwater forearc basin formation in response to hot spot ridge and seamont subduction. Tectonics 21(5):2/1–2/27

LeRoux JP, Elgueta S (2000) Sedimentologic development of a late Oligocene–Miocene forearc embayment, Valdivia Basin Complex, southern Chile. Sedimentary Geology 130(1–2):27–44

Linares E, Ostera HA, Mas LC (1999) Cronologia potasio-argon del complejo efusivo Copahue-Caviahue, Provincia del Neuquen. Rev Asoc Geol Argentina 54(3):240–247

Livermore R, Nankivell A, Eagles G, Morris P (2005) Paleogene opening of Drake Passage. Earth Planet Sci Lett 236(1–2):459–470

Lodolo E, Menichetti M, Bartole R, Ben AZ, Tassone A, Lippai H (2003) Magallanes-Fagnano continental transform fault (Tierra de Fuego, southernmost South America). Tectonics 22(6): doi 10.1029/2003TC001500

Lucassen F, Becchio R (2003) Timing of high-grade metamorphism: Early Palaeozoic U-Pb formation ages of titanite indicate long-standing high-T conditions at the western margin of Gondwana (Argentina, 26–29°S). J Metamorph Geol 21(7):649–662

Lucassen F, Harmon R, Franz G, Romer RL, Becchio R, Siebel W (2002) Lead evolution of the pre-Mesozoic crust in the Central Andes (18–27°): progressive homogenisation of Pb. Chem Geol 186(3–4):183–197

Maksaev V, Moscoso R, Mpodozis C, Nasi C (1984) Las unidades volcanicas y plutonicas del Cenozoico Superior en la alta Cordillera del Norte Chico (29°–31°S): Geologia, Alteracion Hidrothermal y Mineralizacion. Rev Geol Chile 21:11–51

Malumian N, Ramos VA (1984) Magmatic intervals, transgression-regression cycles and oceanic events in the Cretaceous and Tertiary of southern South America. Earth Planet Sci Lett 67(2):228–237

Manceda R, Figueroa D (1995) Inversion of the Mesozoic Neuquen Rift in the Malarguee fold and thrust belt, Mendoza, Argentina. In: Petroleum basins of South America 62. Am Assoc Petr Geologists, Tulsa OK, pp 369–382

Marrett R, Strecker MR (2000) Response of intracontinental deformation in the Central Andes to late Cenozoic reorganization of South American plate motions. Tectonics 19(3):452–467

Melnick D, Rosenau M, Folguera A, Echtler H (2006) Neogene tectonic evolution of the Neuquén Andes western flank. In: Kay SM, Ramos VA (eds) Evolution of an Andean margin: a tectonic and magmatic view from the Andes to the Neuquén Basin (35°–39° S lat). Geol Soc Am Spec P 407:73–95

Mercer JH, Sutter JF (1982) Late Miocene–earliest Pliocene glaciation in southern Argentina: implications for global ice-sheet history. Palaeogeog Palaeoclimat Palaeoecol 38(3–4):185–206

Montgomery DR, Balco G, Willett SD (2001) Climate, tectonics, and the morphology of the Andes. Geology 29(7):579–582

Mordojovich C (1981) Sedimentary basins of Chilean Pacific offshore. In: Energy resources of the Pacific region. Studies in Geology 12. Am Assoc Petr Geologists, Tulsa OK, pp 63–82

Mpodozis C, Ramos VA (1989) The Andes of Chile and Argentina. In: Ericksen G, Canas Pinochet M, Reinemund J (eds) Geology of the Andes and its relation to hydrocarbon and mineral resources. Earth Science Series 11, Circum-Pacific Council fo Energy and Mineral Resources, Houston TX, pp 59–89

Mueller RD, Roest WR, Royer JY, Gahagan LM, Sclater JG (1997) Digital isochrons of the world's ocean floor. J Geophys Res 102: 3211–3214

Nelson E, Forsythe R, Arit I (1994) Ridge collision tectonics in terrane development. J S Am Earth Sci 7(3–4):271–278

New M, Lister D, Hulme M, Makin I (2002) A high-resolution data set of surface climate over global land areas. Climate Research 21:1–25

Nielsen SN, Frassinetti D, Bandel K (2004) Miocene Vetigastropoda and Neritimorpha (Mollusca, Gastropoda) of central Chile. J S Am Earth Sci 17(1):73–88

Niemeyer H, Munoz JA (1983) Carta Geologica de Chile 1:250.000, Hoja Laguna de la Laja. Servicio Nacional de Geologia y Mineria, Santiago

Oncken O, Hindle D, Kley J, Elger K, Victor P, Schemmann K (2006) Deformation of the central Andean upper plate system – facts, fiction, and constraints for plateau models. In: Oncken O, Chong G, Franz G, Giese P, Götze H-J, Ramos VA, Strecker MR, Wigger P (eds) The Andes – active subduction orogeny. Frontiers in Earth Science Series, Vol 1. Springer-Verlag, Berlin Heidelberg New York, pp 3–28, this volume

Pankhurst RJ, Herve F (1994) Granitoid age distribution and emplacement control in the North Patagonian batholith in Aysen (44°–47°S). VII Congreso Geologico Chileno, Actas 2, pp 1409–1413

Pankhurst RJ, Rapela CW, Saavedra J, Baldo E, Dahlquist J, Pascua I, Fanning CM (1998) The Famatinian magmatic arc in the central Sierras Pampeanas: an Early to Mid-Ordovician continental arc on the Gondwana margin. In: Pankhurst RJ, Rapela CW (eds) The proto-Andean margin of Gondwana. Geol Soc London Spec Pub 142, pp 343–367

Parada MA (1990) Granitoid plutonism in central Chile and its geodynamic implications: a review. In: Plutonism from Antarctica to Alaska 241. Geol Soc Am, Boulder CO, pp 51–66

Parada MA, Lahsen A, Palacios C (2000) The Miocene plutonic event of the Patagonian Batholith at 44°30'S: thermochronological and geobarometric evidence for melting of a rapidly exhumed lower crust. Trans Royal Soc Edinburgh Earth Sci 91(1–2):169–179

Pardo Casas F, Molnar P (1987) Relative motion of the Nazca (Farallon) and South American plates since late Cretaceous time. Tectonics 6(3):233–248

Pearce JA, Leat PT, Barker PF, Millar IL (2001) Geochemical tracing of Pacific-to-Atlantic upper-mantle flow through the Drake Passage. Nature 410(6827):457–460

Pilger RH (1981) Plate reconstructions, aseismic ridges, and low-angle subduction beneath the Andes. Geol Soc Am Bull 92(7-1): 448–456

Radic JP, Rojas L, Carpinelli A, Zurita E (2002) Evolucion tectonica de la Cuenca Terciaria de Cura-Mallin, region cordillerana Chileno Argentina (36°30'-39°00' S). Actas Del XV Congreso Geologica Argentino, El Calafate

Ramos VA (1988) Late Proterozoic–Early Paleozoic of South America – a collisional history. Episodes 11(3):168–173

Ramos VA (1989) Andean Foothills structures in northern Magallanes Basin, Argentina. AAPG Bulletin 73(7):887–903

Ramos VA (2005) Seismic ridge subduction and topography: Foreland deformation in the Patagonian Andes. Tectonophysics 399(1–4):73–86

Ramos VA, Cristallini EO (1995) Perfil estructural de la Precordillera a lo largo del Rio San Juan. In: Andean Thrust Tectonics Symposium Field Guide, pp 1–42

Ramos VA, Kay SM (1991) Triassic rifting and associated basalts in the Cuyo Basin, central Argentina. In: Andean magmatism and its tectonic setting. Geol Soc Am Spec P 265, pp 79–91

Ramos VA, Kay SM (1992) Southern Patagonian plateau basalts and deformation: backarc testimony of ridge collisions. Tectonophysics 205(1–3):261–282

Ramos VA, Keppie JD (1999) Laurentia-Gondwana connections before Pangea. Geol Soc Am Spec P 336

Ramos VA, Vujovich G (2000) Hoja Geologica San Juan, escala 1:250.000. Servicio Geologico Minero Argentino

Ramos VA, Jordan TE, Allmendinger R, Mpodozis C, Kay SM, Cortes JM, Palma M (1986) Paleozoic Terranes of the Central Argentine-Chilean Andes. Tectonics 5(6):855–880

Ramos VA, Cegarra M, Cristallini EO (1996) Cenozoic tectonics of the high Andes of west-central Argentina (30–36°S latitude). Tectonophysics 259(1–3):185–200

Ramos VA, Alvarez PP, Aguirre Urreta MB, Godoy E (1997) La Cordillera Principal a la latitud del paso Nieves Negras (22°50'S), Chile-Argentina. VIII Congreso Geológico Chileno, Actas 3, pp 1704–1708

Ramos VA, Cristallini EO, Perez DJ (2002) The Pampean flat-slab of the Central Andes. J S Am Earth Sci 15(1):59–78

Rapela CW, Kay SM (1988) Late Paleozoic to Recent magmatic evolution of northern Patagonia. Episodes 11(3):175–182

Rapela CW, Pankhurst RJ, Casquet C, Baldo E, Saavedra J, Galindo C, Fanning CM (1998) The Pampean Orogeny of the southern proto-Andes: Cambrian continental collision in the Sierras de Cordoba. In: Pankhurst Rj, Rapela CW (eds) The proto-Andean margin of Gondwana. Geol Soc Spec Pub 142, pp 181–217

Reutter K-J (1974) Entwicklung und Bauplan der chilenischen Hochkordillere im Bereich 29° suedlicher Breite. N Jb Geol Paläont 146(2):153–178

Reynolds JH, Jordan TE, Johnson NM, Damanti JF, Tabbutt KD (1990) Neogene deformation of the flat-subduction segment of the Argentine-Chilean Andes: magnetostratigraphic constraints from Las Juntas, La Rioja province, Argentina. Geol Soc Am Bull 102(12):1607–1622

Rosenau M (2004) Tectonics of the Southern Andean Intra-arc Zone (38°–42°S). PhD thesis, Freie Universität Berlin, http://www.diss.fu-berlin.de/2004/280/

Rousse S, Gilder S, Farber D, McNulty B, Patriat P, Torres V, Sempere T (2003) Paleomagnetic tracking of mountain building in the Peruvian Andes since 10 Ma. Tectonophysics

Russo RM, Silver PG (1994) Trench-parallel flow beneath the Nazca Plate from seismic anisotropy. Science 263(5150):1105–1111

Russo RM, Silver PG (1996) Cordillera formation, mantle dynamics, and the Wilson cycle. Geology 24(6):511–514

Sato AM, Tickyj H, Llambias EJ, Sato K (2000) The Las Matras tonalitic-trondhjemitic pluton, central Argentina: Grenvillian-age constraints, geochemical characteristics, and regional implications. J S Am Earth Sci 13(7):587–610

Schmidt CJ, Astini RA, Costa CH, Gardini CE, Kraemer PE (1994) Cretaceous rifting, alluvial fan sedimentation, and Neogene inversion, southern Sierras Pampeanas, Argentina. In: Tankard AJ, Suarez Soruco R, Welsing HJ (eds) Petroleum basins of South America. Am Assoc Petr Geologists Mem 62, pp 341–35

Seifert W, Rosenau M, Echtler H (2005) The evolution of the South Central Chile magmatic arcs: crystallization depths of granitoids estimated by hornblende geothermobarometry – implications for mass transfer processes along the active continental margin. N Jb Geol Paläont Abh 236(1/2):115–127

Silver PG, Russo RM, Lithgow BC (1998) Coupling of South American and African Plate motion and plate deformation. Science 279(5347):60–63

Somoza R (1998) Updated Nazca (Farallon)-South America relative motions during the last 40 My: implications for mountain building in the central Andean region. J S Am Earth Sci 11(3):211–215

Steinmann G (1929) Geologie von Peru. Karl Winter, Heidelberg

Steppuhn A (2002) Climate and climate processes during the Upper Miocene – sensitivity studies with coupled general circulation models. PhD thesis, Universität Tübingen, http://w210.ub.uni-tuebingen.de/dbt/volltexte/2002/533/

Suarez M, Emparan C (1995) The stratigraphy, geochronology and paleophysiography of a Miocene fresh-water interarc basin, southern Chile. J S Am Earth Sci 8(1):17–31

Tassara A, Yáñez C (2003) Relación entre el espesor elástico de la litofera y la segmentación tectónica del margen andino (15–47°S). Rev Geol Chile 30:159–186

Tebbens SF, Cande SC (1997) Southeast Pacific tectonic evolution from early Oligocene to present. J Geophys Res B 102(6):12061–12084

Thomson SN (2002) Late Cenozoic geomorphic and tectonic evolution of the Patagonian Andes between latitudes 42 and 46°S: an appraisal based on fission-track results from the transpressional intra-arc Liquine-Ofqui fault zone. Geol Soc Am Bull 114(9):1159–1173

Thomson SN, Herve F, Stoeckhert B (2001) Mesozoic-Cenozoic denudation history of the Patagonian Andes (southern Chile) and its correlation to different subduction processes. Tectonics 20(5):693–711

Uliana MA, Biddle KT, Cerdan J (1989) Mesozoic extension and the formation of Argentine sedimentary basins. In: Tankard A, Balkwill H (eds) Extensional tectonics and stratigraphy of the North Atlantic margins. Am Assoc Petr Geol Mem 46, pp 599–614

Vergara MM, Munoz BJ (1982) La formacion Cola de Zorro en la Alta Cordillera sudina Chilena (36–39°S): sus caracteristicas petrograficas y petrologicas: una revision. Rev Geol Chile 17:31–46

Vergés J, Ramos E, Seward D, Busquets P, Colombo F (2001) Miocene sedimentary and tectonic evolution of the Andean Precordillera at 31°S, Argentina. J S Am Earth Sci 14:735–750

Vines RF (1990) Productive duplex imbrication at the Neuquina Basin thrust belt front, Argentina. In: Letouzey, J (ed) Petroleum and tectonics in mobile belts. Proceedings of the IFP exploration and production research conference, Paris, Ed Technip, pp 69–79

Völker D, Wiedicke M, Ladage S, Gaedicke C, Reichert C, Rauch K, Kramer W, Heubeck C (2006) Latitudinal variation in sedimentary processes in the Peru-Chile trench off Central Chile. In: Oncken O, Chong G, Franz G, Giese P, Götze H-J, Ramos VA, Strecker MR, Wigger P (eds) The Andes – active subduction orogeny. Frontiers in Earth Science Series, Vol 1. Springer-Verlag, Berlin Heidelberg New York, pp 193–216, this volume

Von Gosen W (1992) Structural evolution of the Argentine Precordillera: the Rio San Juan section. J Struct Geol 14(6):643–667

Von Gosen W (2003) Thrust tectonics in the North Patagonian Massif (Argentina): implications for a Patagonia plate. Tectonics 22(1):5/1–5/33

Vujovich GI, Kay SM (1998) A Laurentian Grenville-age oceanic arc/back-arc terrane in the Sierra de Pie de Palo, Western Sierras Pampeanas, Argentina. In: Pankhurst RJ, Rapela, CW (eds) The proto-Andean margin of Gondwana. Geol Soc Spec Pub 142:159–179

Wessel P, Smith WHF (1998) New, improved version of Generic Mapping Tools released. EOS 79(47):579

Wilson DS (1993) Confirmation of the astronomical calibration of the magnetic polarity timescale from sea-floor spreading rates. Nature 364(6440):788–790

Yañez GA, Ranero CR, von Huene R, Diaz J (2001) Magnetic anomaly interpretation across the southern Central Andes (32°–34°S): the role of the Juan Fernandez Ridge in the late Tertiary evolution of the margin. J Geophys Res B 106(4):6325–6345

Yuan X, Asch G, Bataille K, Bock G, Bohm M, Echtler H, Kind R, Oncken O, Wölbern I (2006) Deep seismic images of the Southern Andes. In: Kay SM, Ramos VA (eds) Evolution of an Andean margin: a tectonic and magmatic view from the Andes to the Neuquén Basin (35°–39° S lat). Geol Soc Am Spec P 407, in press

Zachos J, Pagani H, Sloan L, Thomas E, Billups K (2001) Trends, rhythms, and aberrations in global climate 65 Ma to present. Science 292(5517):686–693

Zapata TR, Allmendinger RW (1996) Growth stratal records of instantaneous and progressive limb rotation in the Precordillera thrust belt and Bermejo Basin, Argentina. Tectonics 15(5):1065–1083

Zapata TR, Brissón I, Dzelalija F (1999) The role of the basement in the Andean fold and thrust belt of the Neuquén Basin, Argentina. In: Thrust Tectonics Conference. Royal Holloway, University of London, pp 122–124

# Chapter 19

# Long-Term Geological Evolution and Mass-Flow Balance of the South-Central Andes

Johannes Glodny · Helmut Echtler · Oscar Figueroa · Gerhard Franz · Kirsten Gräfe · Helga Kemnitz · Wolfgang Kramer
Charlotte Krawczyk · Jo Lohrmann · Friedrich Lucassen · Daniel Melnick · Matthias Rosenau · Wolfgang Seifert

**Abstract.** In south-central Chile (36–42° S), the western edge of South America has evolved as an active margin since the Pennsylvanian (~305 Ma). Active margins are considered as sites of both potential continental growth and continental destruction. Continental growth in a margin setting can proceed by accretionary offscraping of juvenile material from the oceanic plate and by magmatic additions, whereas net mass loss can be achieved by subducting continental material, delamination, and chemical weathering. In south-central Chile, margin evolution was never interrupted by island-arc accretion or continental collision. Thus, the area provides an excellent field laboratory for studying mass flux through a long-term, persistent, convergent-margin system.

Using new isotopic age data, we summarize the current knowledge of the geological evolution of the south-central Chilean margin, from subduction initiation to the ongoing Andean morphotectonic processes, with emphasis on mechanisms of mass transfer. It is inferred that net crustal growth and mass losses alternated in time and space, and that dominance of one of the other process might have even occurred contemporaneously within a short distance along the same margin, controlled by factors such as sediment availability in the trench, lower-plate morphology, upper-plate tectonics, and climate.

In south-central Chile, the margin north of 38° S is characterized by a landward trench migration of ~100 km that occured mainly in the early Permian, whereas farther south, the modern and the late Paleozoic magmatic arcs are superimposed. For most of its lifetime, the margin evolved in a delicate balance between constructive and destructive processes. Over the long term, the south-central Chilean continental margin has not been a site of net growth, but, rather, a site of continental mass wasting, crustal recycling and crustal rejuvenation.

## 19.1 Introduction

How, and how fast, continental crust grows is of ongoing interest in the world of geodynamics. Usually, concepts are developed by identifying particular growth mechanisms, such as tectonic accretion of igneous oceanic crust, various mechanisms for adding basaltic plume material, and continent versus juvenile arc collision, and then combining them with geochemical, isotope-geochemical and geochronological inferences (cf. Albarède 1998). Most growth concepts imply an overall increase in the total volume of continental crust with time (Hurley and Rand 1969; De Paolo and Wasserburg 1979; Reymer and Schubert 1984; see compilation in Tarney and Jones 1994).

Whereas oceanic crust recycling into the mantle is inherent in the concept of plate tectonics, and has geochemically been proven (Hofmann and White 1980; Sobolev et al. 2005), destruction of the continental crust and net mass loss has been demonstrated at convergent margins using mass balance studies (von Huene and Scholl 1991). A very influential hypothesis for continental growth is the 'andesite model', introduced by Taylor (1967), which suggests significant horizontal and volumetric growth by the input of andesitic arc material at active continental margins. This concept has later been modified (Taylor and McLennan 1985), largely dismissed (Reymer and Schubert 1984), or restricted to ridge-subduction settings (Defant and Drummond 1990; Drummond and Defant 1990). However, the general idea of regarding active margins as sites of continental growth appears to be persistent (cf. Clift and Vannucchi 2004).

Recent studies have shown that active margins can episodically, within less than a few million years, shift between modes of accretion and tectonic erosion (Bangs and Cande 1997). However, the short-term net effects of convergence are difficult to assess as continent-derived sediment is lost into the mantle even during accretion. Geological indications for dominance of either continental growth or tectonic erosion are sparse. They comprise mainly of relative shifts in the position of the magmatic arc or of fixed points of the fore-arc, compared to the trench (e.g., Mpodozis and Ramos 1989; Meschede et al. 1999; Clift and Vannucchi 2004), or of basin formation by fore-arc subsidence caused by tectonic erosion (von Huene and Lallemand 1990; Laursen et al. 2002; Wells et al. 2003). Other convergence-related tectonic effects at active margins are difficult to interpret in terms of crustal growth. In particular, the emergence of morphological fore-arc highs and their extensional deformation may occur at both accretive and erosive margins (von Huene and Scholl 1991; Hartley et al. 2000).

Long-term, average rates of mass flux (over the 100 Myr timescale) are more relevant for estimating the overall behavior of a margin segment than short-term events and fluctuations. Therefore, when assessing the relevance of active, continental margin systems for crustal growth, the effect of long-term, undisturbed, oceanic-plate subduction beneath a continent needs to be isolated from all other possible mechanisms of crustal growth, in particular from continent-plume interaction and from arc-continent collisions. Long-term subduction at the persistent, active margin of south-central Chile (36–42° S) has never led to any significant collision processes (similarly to northern Chile; Franz et al. 2006, Chap. 3 of this volume), and indications for margin-parallel terrane transport are absent. The region is, thus, an ideal site for studying continental margin evolution.

In south-central Chile, oceanic-plate subduction began in the Pennsylvanian (~305 Ma; Willner et al. 2004). Traditionally, distinction is made between two subduction-related 'tectonic cycles' – the "Gondwanide" (Pennsylvanian to Triassic) and "Andean" (Jurassic to Holocene) cycles (e.g., Mpodozis and Ramos 1989). This distinction is based mainly on: (*1*) the abundance of rocks with specific age signatures; (*2*) a geochemical change from crust-dominated to more mantle-dominated magma sources; and (*3*) indications for major, Mesozoic, tectono-magmatic and paleogeographic changes. However, despite the Triassic-Jurassic margin reorganization, the overall convergent-margin setting persisted. Subduction has continued from the Pennsylvanian until the present day, generating a variety of geological features through time that reflect periods of magmatic addition, accretionary-wedge growth, tectonic erosion, and surface mass transfer (Mpodozis and Ramos 1989; Lucassen et al. 2004; Glodny et al. 2005).

In this contribution, we outline the geological evolution of the south-central Chilean fore-arc and arc since the Paleozoic, with emphasis on processes relevant for mass transfer. In our survey of the existing literature, we focus on the more recent contributions. We also present new data and interpretations on the late Paleozoic to Triassic accretion history, which is relevant in the context of this paper but has, so far, not been completely understood. We aim at both a review of the geological evolution of the margin between 36 and 42° S and toward qualitative estimates of volumes and rates of mass transfer. We show that, at the active margin, continental material is reworked, there is mass loss into the mantle, and the crust is 'rejuvenated' by the input of juvenile, mantle- or oceanic-plate-provided igneous additions. Over the long term, continental growth rates appear to be negligible or even negative.

## 19.2 Morphotectonic and Lithological Margin Architecture

Across a transect from west to east, the present-day, south-central Chilean active margin is built by five primary geomorphological units (Fig. 19.1). The submerged part of the fore-arc is made up of a steep slope in the west, associated with a 20 to 50 km wide, only shallowly submerged continental shelf area. The Coastal Cordillera fore-arc high, in most areas, emerges directly from the shoreline. Its crest is located about 130 km from the trench, and reaches altitudes of up to 1500 m in the Cordillera de Nahuelbuta at 38° S. Further east, the Central Depression (Longitudinal Valley) constitutes a sediment depository varying from only a few kilometers wide to nearly 100 km. The mountainous Andean Main Cordillera is ~200 km wide in the study area and is generally less than 1200 m high. Here, however, the magmatic arc, at a fairly constant distance of 280 km from the trench, currently generates individual volcanic edifices up to 3700 m high. These accumulated over Paleozoic to Miocene crystalline basement and Mesozoic-Cenozoic volcano-sedimentary sequences. The Andean back-arc, east of the Main Cordillera (Fig. 19.1), is characterized by mostly gentle, weakly dissected, plateau-like morphology.

The morphological architecture of the area is not entirely matched by the distribution of lithological units. The offshore fore-arc domain is, as inferred from drillcore and seismic reflection data, mainly made up of Pennsylvanian to Triassic accretionary-wedge material, with greenschist- to blueschist-facies and high P/T signature ('Western Series', Aguirre et al. 1972; Fig. 19.1), and contains a small Neogene-Quaternary wedge at the toe of the upper plate (Hervé 1988; Bangs and Cande 1997; Glodny et al. 2005). Late Cretaceous to Holocene sedimentary shelf basins also occur (Mordojovich 1981). The Coastal Cordillera north of 38° S contains late Paleozoic, arc granitoids that are associated with low P/T metasediments ('Eastern Series'), whereas further south, the Cordillera is mainly built by the Western Series paleo-wedge complex. The sedimentary basin in the Central Depression (Fig. 19.1) is an entirely young feature, filled with late Oligocene to Holocene deposits (Jordan et al. 2001, and references therein). In contrast, within the Andean Main Cordillera, and in the back-arc, a multitude of magmatic, metasedimentary, sedimentary and volcanic rocks occur, ranging in age from the Paleozoic to Holocene. The geological evolution toward the present-day situation is outlined below.

## 19.3 Pre-Pennsylvanian (Pre-Gondwanide) Margin Setting

The pre-Pennsylvanian history of the arc and fore-arc is widely obliterated by younger processes. No reliable magmatic or metamorphic ages for autochthonous, pre-Pennsylvanian processes are known from Chile between 36 and 42° S. However, older continental basement is exposed in the back-arc region in Argentina (e.g., Linares et al. 1988; Pankhurst et al. 2003). Along the eastern flank of the

**Fig. 19.1.**
a Digital elevation model of the south-central Chilean margin, compiled with TOPEX (Smith and Sandwell 1997) and SRTM data. b Morphotectonic and selected geological units of south-central Chile, after Melnick and Echtler (2006, Chap. 30 of this volume). *LOFZ* – Liquiñe-Ofqui Fault Zone

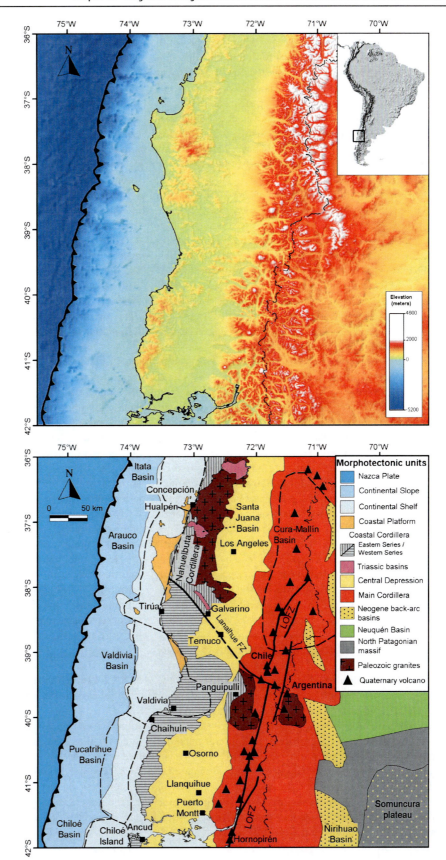

Andes, our study area includes the hypothetical transition between the North Patagonian Massif to the south and Gondwanan South America to the north. Although Mpodozis and Ramos (1989) suggested that Patagonia was accreted to South America in the late Paleozoic, there are various lithological, paleomagnetic, geochronological and isotopic indications that Patagonia was already occupying an autochthonous position during the Cambrian (Cingolani et al. 1991; Dalla Salda et al. 1994, and references therein; Rapalini 1998; Pankhurst et al. 2003; Lucassen et al. 2004).

Neodymium model ages for various pre-Pennsylvanian magmatic, metamorphic and sedimentary rocks are commonly between 2.0 and 1.4 Ga along the southern Andes (18–49° S; Sato et al. 2000, Lucassen et al. 2001, Augustsson and Bahlburg 2003, Lucassen et al. 2004). Thus, variably reworked, Proterozoic crust seems to have formed the pre-subduction basement. In our study area, the latest pre-Carboniferous, metamorphic episode recorded in rocks of the eastern Andean flank is Devonian (Lucassen et al. 2004, and references therein), which suggests a continuation of the Cambrian to Devonian Famatinian orogen of NW Argentina into NW Patagonia and, perhaps, even further southwards (Dalla Salda et al. 1994; Pankhurst et al. 2003).

All along the margin south of ~34° S, thick, turbiditic, Paleozoic (meta)sedimentary sequences are found. Among these are the Eastern Series rocks of the study area (Fig. 19.1), as well as successions of the Eastern Andean Metamorphic Complex (47–49° S). Despite their Permo-Carboniferous metamorphic overprints, several of these occurrences have independently been interpreted as Devonian-Carboniferous, passive-margin successions that were deposited on continental crust (Kato 1976; Hervé et al. 1987; Bahlburg and Hervé 1997; Augustsson and Bahlburg 2003), an interpretation that can be generalized. This suggests a coherent and continuous, passive-margin setting in south-central Chile during the early Carboniferous, prior to the onset of Pennsylvanian subduction.

## 19.4 Gondwanide Evolution (Carboniferous to Triassic)

Rocks from the Gondwanide orogeny predominate in the fore-arc region of south-central Chile. All elements of an evolved, accretive, active margin are present, including a magmatic arc and a complex accretionary wedge. The Gondwanide orogen in south-central Chile was part of a much larger active margin, oriented parallel to the present day margin, that extended along the southwestern edge of Gondwana. Evidence for this includes, e.g., a belt of Pennsylvanian-early Permian, magmatic-arc granitoids that occurs within the central and southern Andes of Chile and Argentina (Rapela and Kay 1988; Hervé et al. 1988; Parada 1990)

### 19.4.1 Incipient Subduction (Pennsylvanian)

#### 19.4.1.1 Onset of Subduction, Early Arc Magmatism

In the modern fore-arc, late Paleozoic to Triassic metasedimentary rocks are predominant together with Permo-Carboniferous, arc granitoids; no outcrops of older basement rocks have been described so far. This points to a voluminous sedimentary cover at the pre-subduction continental margin. We, therefore, hypothesize that the Pennsylvanian conversion of a passive margin into an active one might have been triggered by sediment overload (cf. Branlund et al. 2000).

The onset of subduction in south-central Chile has been constrained, by isotopic dating, to slightly older than 305 Ma (Willner et al. 2004). The rocks dated show a anticlockwise P/T path and form rafts within the Western Series accretionary complex. This age agrees with age data for magmatic-arc granitoids. In the Cordillera Nahuelbuta (Coastal Cordillera north of ~38° S; Figs. 19.1 and 19.2), these are calc-alkaline granodiorite, tonalite, diorite and granite plutons, with crystallization ages clustering tightly at around 295–305 Ma (Hervé et al. 1988; Lucassen et al. 2004). Equivalents of the Nahuelbuta granitoids are found in the western flank of the Andes at ~40° S, east of Valdivia (Fig. 19.2), and in the Argentinian Andes between 39 and 40° S, and are isotopically dated to between 316 and 285 Ma (Beck et al. 1991; Varela et al. 1994; Martin et al. 1999; Lucassen et al. 2004, and references therein). Indications from detrital zircon age populations for onset of arc magmatism in Pennsylvanian to Early Permian times have also been described from the Southern Andes (~50° S; Hervé et al. 2003).

This magmatic pulse, related to the early stages of subduction, compositionally resembles a mixture between mantle-derived melts and reworked crustal material. Strontium-Nd-Pb isotope data suggest that crustal melts predominate. Minimum estimates for the proportion of reworked continental crust in various granitoids from this early arc are between > 30 and 80% (Lucassen et al. 2004).

#### 19.4.1.2 Pennsylvanian to Early Permian Accretion: Eastern and Western Series

Along the margin, two voluminous, metamorphic units are interpreted as being parts of a complex, partly accretive, fore-arc fold and thrust belt. The *Eastern Series* (Fig. 19.2) consists entirely of continent-derived, mostly turbiditic successions (e.g., Hervé 1988). Metabasites are absent. Major occurrences of this series are seen in: the Coastal Cordillera north of 38° S, as a belt in the western flank; in the Andean Main Cordillera at ~40° S, also in the western flank (Fig. 19.2; here also known as the Trafún sequence, cf. Mar-

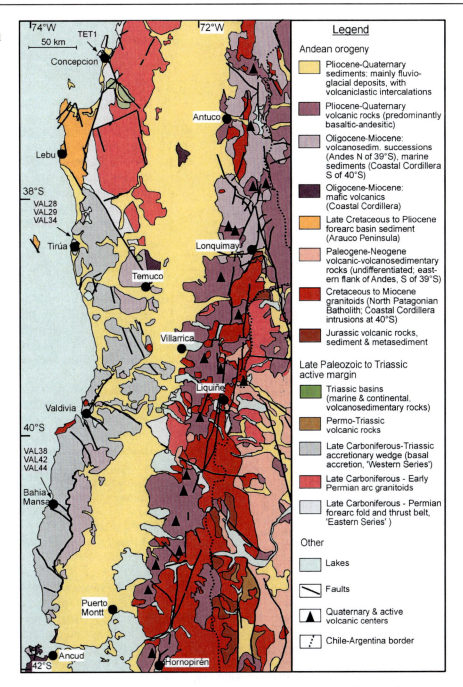

**Fig. 19.2.**
Geological map of south-central Chile and adjacent parts of Argentina (simplified, mainly after SERNAGEOMIN 2003), between 36° 30' S and 42° S

tin et al. 1999); and within the Andes, as relics on top of Mesozoic-Cenozoic granitoids (SERNAGEOMIN 2003; Rosenau 2004). Minor amounts of contemporaneous, possibly correlative, Carboniferous to early Permian, unmetamorphosed sediments occur in the eastern flank of the Andes, south of 41° S (the marine to deltaic Tepuel group; López-Gamundí and Rossello 1993, López-Gamundí et al. 1992).

The Eastern Series has been synkinematically intruded by Pennsylvanian to early Permian, magmatic-arc granitoids. Its metamorphic grade is highly variable but with low-P/T signature. It increases toward the intruding granitoids, starting with biotite grade leading into an andalusite zone, followed by a sillimanite zone in amphibolite- transitional to granulite-facies (González-Bonorino and Aguirre 1970; Hervé 1977). This metamorphism and internal deformation also occurred during the Pennsylvanian (Lucassen et al. 2004).

The structural inventory of the Eastern Series is that of a fold and thrust belt. This, together with the composition, the metamorphic signature and the geotectonic position of this series, implies that the Eastern Series was

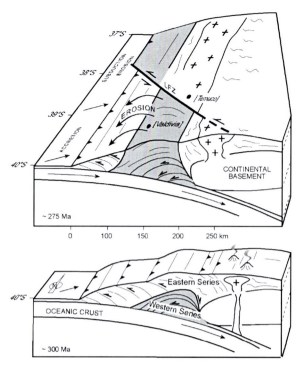

**Fig. 19.3.** Model of Pennsylvanian to early Permian margin evolution and formation of the Lanalhue fault, using vertically exaggerated cross sections of the continental margin at ~40° S. *Lower part*: Status at ~300 Ma, ~5–10 Ma after onset of subduction. Ongoing deformation of former passive-margin sediments (Eastern Series) is accompanied by initial growth of a basal-accretionary complex (Western Series), and arc magmatism. Note the divergent fault contact between Eastern and Western Series in the eastern flank of the Western Series. *Upper part*: Situation at ~275 Ma. Upwarping of the Western Series and surface erosion led to exposure of the Western Series in a fore-arc high, while arc magmatism ceased. Ongoing basal accretion south of 38° S contrasts with subduction erosion and fore-arc mass wasting further north. Differential margin behavior induces margin segmentation, with reworking of the Western-Eastern Series contact as a sinistral transform fault (the *LFZ*, Lanalhue Fault Zone)

formed by folding and thrusting of former passive-margin sediments (300 Ma situation in Fig. 19.3). A possible reason for folding and thrusting are either contemporaneous development of the Western Series accretionary complex farther west and down at the plate interface (see Fig. 19.3, and below). Alternatively, collision with a major oceanic structure may have caused deformation, as described e.g. for the collision of the Hikurangi Plateau, New Zealand (Collot et al. 1996). In the western foothills of the Main Cordillera at ~40° S, structural observations point to partly syndepositional, early deformation of the Eastern Series (Parada 1975), implying that sedimentation lasted until the establishment of the subduction regime. Chemical composition and paleo-Proterozoic Nd model ages (cf. Lucassen et al. 2004) characterize the Eastern Series as reworked, pre-Pennsylvanian, regional, continental basement.

The *Western Series* (Fig. 19.2) is a metamorphic unit consisting of meta-turbidite (metapelites, metapsammites) and metabasite in variable proportions, with local occurrences of ribbon chert, serpentinite and sulphide ore bodies. Geochemical signatures indicate the metabasite is similar to ocean floor basalt (Hervé et al. 1988). The series forms a continuous belt of variable width along the edge of the continent. South of ~38° S, it stretches from the landward trench slope to somewhere underneath the Central Depression, whereas further north, the Western Series is not exposed but known from drilling and is present only beneath the coastal platform and offshore fore-arc area (Fig. 19.2).

The metamorphic imprint has a high P/T signature – transitional greenschist – to blueschist-facies with conditions of ~400 °C and 8–9 kbar throughout the unit (Willner et al. 2001; Glodny et al. 2005, and references therein). Structurally, the Western Series complex is characterized by a mostly subhorizontal, near-penetrative, transposition foliation that is related to recumbent nappes. In places, this foliation is overprinted by folding and by zones of subhorizontal, mylonitic shear. Mineral-stretching lineations and fold axes, as well as compositional variations (e.g., lens-shaped metabasite bodies), generally trend NW-SE (Kato 1985; Godoy and Kato 1990; Martin et al. 1999; Glodny et al. 2005).

The complex has been shown, by reflection seismic imaging at 38°15' S (Krawczyk and the SPOC Team 2003), to continue down to the plate interface. It constitutes an extinct, fore-arc accretionary prism that was assembled by basal accretion (Willner et al. 2000; Glodny et al. 2005). Initial build-up of the complex must have started as early as ~305 Ma (Willner et al. 2004), soon after subduction was initiated (Fig. 19.3). It is worth noting that the short distance between the Western Series accretionary complex and the Paleozoic magmatic arc (in places less than 50 km, Fig. 19.2) requires that the Western Series originally was assembled at a large distance (of ~100 km and more) from the trench. This implies that the Western Series material must have travelled a considerable distance in a subduction channel at the plate interface before it was finally basally accreted, as described in analogy e.g. for the Orocopia-Pelona schists of southern California (Jacobson et al. 1996; Grove et al. 2003)

Basal growth increased the volume of the Western Series complex, most likely at the expense of the pre-existing margin sediment of the Eastern Series. In the early Permian, at around 275 Ma, erosion had already exposed the divergent, fault-zone contact between the Western and Eastern Series (Fig. 19.3). Locally, the Western Series continued to grow until the late Triassic, as explained below.

Neodymium model ages for Western Series rocks (cf. Hervé et al. 1990; Duhart et al. 2001; Lucassen et al. 2004; Glodny et al. 2005) show marked contrasts between metabasite and metasedimentary rocks. The metasediment signatures are indistinguishable from those of the Eastern Series and regional, pre-Pennsylvanian continental basement. Therefore, a terrigenous, turbiditic origin for these metasediments is indicated. The metabasite signatures, however, resemble Paleozoic-Mesozoic, ocean-floor rocks. This suggests that the metabasite is lower-plate material that has transferred into the upper-plate accretionary wedge. The proportion of metabasite and other lower-plate rocks (e.g., metachert, serpentinite) increases from north to south in the study area, from less than 10% at 38° S, to 50–60% at 42° S (Kato 1976).

### 19.4.2 Early Permian Margin Segmentation: Accretion versus Tectonic Erosion

The Coastal Cordillera of south-central Chile, at ~38° S, displays marked, lithological contrasts along strike. Here, Pennsylvanian to early Permian, magmatic-arc granitoids, and the associated Eastern Series in the north, are juxtaposed with the Western Series further south (Fig. 19.2). The contact between the Western and Eastern Series has long been recognized as a major crustal discontinuity (Ernst 1975; Kato 1976, 1985; Hervé 1977; Martin et al. 1999). The NW-SE-striking segment of this contact at ~38° S, termed the Lanalhue Fault Zone (Fig. 19.1), is unique; it is the only segment where this contact significantly deviates from a trench-parallel, NNE-SSW orientation. Given the iuxtaposition of magmatic arc granitoids with fore-arc HP/LT accretionary wedge rocks, the NW-SE strike of the contact here must be tectonic in origin, post-dates magmatism and accretion, and does not reflect an original curvature of the continental margin. The Lanalhue fault is interpreted as the former, divergent, fault-zone contact between the Western and Eastern Series, reactivated as a crustal-scale, sinistral, strike-slip fault (Fig. 19.3). Late shear increments at semibrittle to brittle conditions have been isotopically dated at 280–270 Ma (Burón 2003; Ardiles 2003; Glodny et al., unpublished data).

North of the Lanalhue fault trace, the late Paleozoic fore-arc zonation pattern, with markers like the western edge of the late Paleozoic magmatic arc, is displaced by nearly 100 km toward the present-day Andean trench (Fig. 19.3). This dissection of the Paleozoic margin architecture by the Lanalhue Fault Zone is interpreted as an effect of differential, early Permian, fore-arc evolution. North of 38° S, there was an early Permian period of subduction erosion, contrasting with the continuous accretion further south (see below). Assuming there were initially similar, fore-arc geometries north and south of the later Lanalhue Fault Zone, subduction erosion in the period between ~290 and 270 Ma (or an alternative mechanism of northward, trench-parallel, mass translation; cf. Martin et al. 1999) has truncated nearly 100 km of the Western Series rocks in the margin segment north of 38° S. This is only a minimum estimate of total mass removal for the area as it remains unknown whether another, frontally accreted wedge domain existed west of today's Western Series prior to 270 Ma.

### 19.4.3 Permo-Triassic Segmented Margin

After the margin segmentation of the early Permian Lanalhue Fault Zone, the evolution of the northern and southern margin segments differed markedly during the Permian and Triassic.

North of 38° S, internal ductile deformation of onshore, fore-arc lithologies waned prior to 290 Ma (Lucassen et al. 2004, Glodny et al., unpublished data; Willner et al. 2005). Around the Lanalhue Fault Zone (38° S), fluid-inclusion studies (Collao et al. 2003) show that today's surface had already been exhumed to upper-crustal levels of less than 3–4 km in the early Permian. North of 38° S, total cumulative exhumation of rocks in the Coastal Cordillera is estimated to be less than ~10 km (Martin et al. 1999). This is indicated by the virtual absence of muscovite in the arc granitoids, the metamorphic conditions of the Eastern Series rocks, and the local preservation of the anchimetamorphic Eastern Series.

In the Triassic, and possibly since the mid Permian, the area north and northeast of the Lanalhue Fault Zone was affected by extension and rifting (Dalziel et al. 1987; Kay et al. 1989), the early stages of which are related to waning Choiyoi volcanism (not exposed in the study area; cf. Ramos 2000 and references therein). An array of NW-SE-oriented, mid to late Triassic, sedimentary basins formed north of 38° S in wide areas of central Chile and adjacent Argentina (Charrier 1979; Suárez and Bell 1992; Franzese and Spalletti 2001). Basins such as the Galvarino Basin, NNW of Temuco (38° 30' S), and the Santa Juana Basin, SE of Concepción (Figs. 19.1 and 19.2), are filled with continental and marine successions. They contain some volcanic intercalations and, locally, coarse-grained clasts of the local basement (Hervé et al. 1976; Charrier 1979; Nielsen, in press). Deposition of the Triassic successions on top of Permo-Carboniferous granitoid basement is evidence for exhumation of the basement to a near-surface position already in the Triassic. Zircon fission-track data further show that the Coastal Cordilleran granitoid basement north of 38° S was cooled to generally less than ~200 °C when these Triassic basins were forming (Glodny et al. in press; Willner et al. 2005).

Triassic rifting was associated with widespread, bimodal magmatism (Kay et al. 1989; Ramos and Kay 1991; Suárez and Bell 1992) that, at least in our study area, is predominantly recycled, Precambrian, continental crust

**Table 19.1.** Rb/Sr analytical data

| Sample No. Analysis No. | Material | Rb (ppm) | Sr (ppm) | $^{87}Rb/^{86}Sr$ | $^{87}Sr/^{86}Sr$ | $^{87}Sr/^{86}Sr\ 2\sigma_m$ (%) |
|---|---|---|---|---|---|---|
| **Hualpén miarolitic granite** | | | | | | |
| TET1 (227 ±7 Ma, MSWD = 5.0, $Sr_i$ = 0.34 ±0.88) | | | | | | |
| PS782 | Muscovite (cm sized) | 5080 | 4.73 | 233000 | 755.6 | 0.1562 |
| PS874 | Muscovite + feldspar 1 | 680 | 2.94 | 830 | 3.189426 | 0.0026 |
| PS920 | K-feldspar | 798 | 6.59 | 386 | 1.748434 | 0.0012 |
| PS921 | Muscovite + feldspar 2 | 358 | 4.74 | 6510 | 20.97449 | 0.0034 |
| **Main deformation: Quidico** | | | | | | |
| VAL28 (289.3 ±2.2 Ma, MSWD = 0.98, $Sr_i$ = 0.713661 ±0.000046) | | | | | | |
| PS666 | wm 355–250 µm, m 0.45–0.8 A | 517 | 134 | 11.2 | 0.760114 | 0.0018 |
| PS673 | wm 250–160 µm, m 0.45–0.8 A | 512 | 133 | 11.2 | 0.759737 | 0.0018 |
| PS670 | wm 160–126 µm, m 0.45–0.8 A | 509 | 133 | 11.1 | 0.759612 | 0.0010 |
| PS669 | White mica nm 1.7 A | 485 | 171 | 8.23 | 0.747193 | 0.0014 |
| PS696 | Plagioclase | 7.20 | 52.9 | 0.394 | 0.715283 | 0.0014 |
| VAL29 (282.5 ±5.3 Ma, MSWD = 14, $Sr_i$ = 0.71234 ±0.00046) | | | | | | |
| PS643 | White mica 355–180 µm | 498 | 123 | 11.8 | 0.759935 | 0.0014 |
| PS642 | White mica 180–125 µm | 540 | 121 | 12.9 | 0.763838 | 0.0012 |
| PS641 | White mica 125–80 µm | 511 | 123 | 12.1 | 0.760975 | 0.0014 |
| PS681 | Apatite | 1.28 | 1890 | 0.00196 | 0.712216 | 0.0016 |
| PS702 | Albite | 9.85 | 29.0 | 0.982 | 0.716428 | 0.0014 |
| **Tension gash, Quidico/Tirúa** | | | | | | |
| VAL34 (249 ±10 Ma, MSWD = 473, $Sr_i$ = 0.7189 ±0.0010) | | | | | | |
| PS672 | White mica 1, ground in ethanol | 217 | 58.1 | 10.9 | 0.757000 | 0.0016 |
| PS662 | White mica 2, ground, etched | 332 | 87.4 | 11.0 | 0.758014 | 0.0014 |
| PS660 | White mica 3, big crystals | 526 | 140 | 10.9 | 0.757707 | 0.0016 |
| PS688 | Quartz 2 | 0.09 | 0.36 | 0.716 | 0.721175 | 0.0032 |
| PS691 | Quartz 1 | 0.09 | 0.44 | 0.620 | 0.720295 | 0.0020 |
| PS690 | Feldspar | 1.49 | 595 | 0.00727 | 0.719822 | 0.0018 |
| **Main deformation: Bahía Mansa** | | | | | | |
| VAL42 (236.4 ±1.7 Ma, MSWD = 1.4, $Sr_i$ = 0.708686 ±0.000027) | | | | | | |
| PS639 | White mica 90–125 µm | 330 | 106 | 9.00 | 0.738877 | 0.0014 |
| PS637 | White mica 125–160 µm | 388 | 26.0 | 43.7 | 0.854154 | 0.0018 |
| PS635 | White mica 160–250 µm | 372 | 52.7 | 20.5 | 0.778206 | 0.0014 |
| PS667 | White mica >250 µm | 321 | 47.9 | 19.5 | 0.774288 | 0.0014 |
| PS676 | Apatite | 3.78 | 823 | 0.0133 | 0.708711 | 0.0014 |
| PS693 | Titanite (impure separate) | 14.3 | 87.2 | 0.475 | 0.710311 | 0.0018 |
| VAL44 (228.6 ±5.2 Ma, MSWD = 338, $Sr_i$ = 0.7171 ±0.0022) | | | | | | |
| PS638 | Garnet | 2.72 | 20.0 | 0.395 | 0.719472 | 0.0036 |
| PS678 | Titanite-conc. | 7.98 | 131 | 0.176 | 0.717763 | 0.0018 |
| PS682 | Apatite | 1.38 | 259 | 0.0155 | 0.717624 | 0.0014 |
| PS831 | Whole rock | 75.1 | 65.6 | 3.32 | 0.726784 | 0.0014 |
| PS704 | White mica >250 µm | 369 | 14.6 | 75.1 | 0.964024 | 0.0020 |
| PS832 | White mica 200–160 µm | 312 | 25.2 | 36.3 | 0.836076 | 0.0016 |
| PS830 | White mica 160–100 µm | 315 | 18.0 | 51.5 | 0.881040 | 0.0014 |
| **Tension gashes, Bahía Mansa** | | | | | | |
| VAL38 (197 ±16 Ma, MSWD = 25, $Sr_i$ = 0.71518 ±0.00099) | | | | | | |
| PS671 | White mica 2 | 554 | 266 | 6.05 | 0.732332 | 0.0018 |
| PS674 | White mica 1 (+ quartz) | 200 | 98.2 | 5.90 | 0.732021 | 0.0012 |
| PS683 | Whole rock: host rock | 3.25 | 55.4 | 0.170 | 0.715853 | 0.0016 |
| PS687 | Quartz fragments | 0.30 | 0.70 | 1.23 | 0.718517 | 0.0036 |
| PS699 | Chlorite | 80.8 | 48.8 | 4.80 | 0.728077 | 0.0024 |

Errors are reported at the $2\sigma$ level. An uncertainty of ±1.5% is assigned to Rb/Sr ratios. 3% errors in Rb/Sr apply to analyses PS688 and PS691, owing to unfavourable spike/sample and high Rb blank/sample ratios. *wm:* White mica; *m/nm:* magnetic/nonmagnetic on Frantz magnetic separator; 13° tilt, at electric current as indicated. See Appendix for sample characterization and exact sample localities.

(Lucassen et al. 2004). A Triassic, leucogranitic, granophyric, shallow-level intrusion at the Hualpén Peninsula near Concepción, close to the contact between the Western and Eastern Series, has yield crystallization ages of 222 ± 2 Ma, 220 ± 5 Ma, and 227 ± 7 Ma (Rb/Sr mineral isochrons; Lucassen et al. 2004, and this work, sample TET1, Table 19.1, Fig. 19.2 and Appendix of this chapter). This age coincides with the age of a similar granitoid near Constitución, 35°20' S (Willner et al. 2005). Together, these ages provide a minimum estimate for the onset of extension-related magmatism. This extensional environment and associated magmatism is thought to be caused by subducting-plate hinge rollback toward the ocean, and from strike-slip movement owing to probable dextral-oblique convergence (cf. Martin et al. 1999; Franzese and Spalletti 2001) on a continuously active margin (Suárez and Bell 1992; Willner et al. 2005).

South of 38° S, the Coastal Cordilleran fore-arc behaved entirely differently during the Permian and Triassic. Numerous K-Ar-based mineral ages, mainly of muscovite, clustered between 260 and 220 Ma (Duhart et al. 2001) indicate Permo-Triassic metamorphism and exhumation of the Western Series. Glodny et al. (2005) have shown that in the Valdivia area (39°45' S), the Western Series was a cyclic system: Frontally underthrusted material was subducted, basally accreted, incorporated into an evolving, antiformal, duplex complex, and then passively uplifted by further progressive underplating, and, finally, eroded (Fig. 19.3).

Isotopic dating of various structural features that developed during different increments of the prograde and retrograde metamorphic history enabled us to use the present-day surface rocks as 'flight monitors' through the evolving wedge. Near Valdivia, prograde metamorphism at ~250 Ma was immediately followed by the main, penetrative deformation at 250–245 Ma (and 8–9 kbar) when subduction-channel material was attached to the upper plate. Nappe stacking ended at ~240 Ma (and 7–8 kbar). Within higher levels of the complex, extensional shearzones developed in a semiductile regime at ~235 Ma (5–6 kbar) and brittle extension in the upper crust occurred at ~210 Ma (2 kbar), resulting in tension gashes. From this record of pressure, temperature, time and depth, a long-term, average exhumation rate of 0.6 ± 0.2 mm yr$^{-1}$ can be calculated. Basal accretion ended in the Valdivia area at ~210 Ma, resulting in a drastic slowdown of long-term exhumation rates by more than an order of magnitude (Glodny et al. 2005).

To check for possible latitudinal variation in the evolutionary pattern of the Western Series south of 38° S, as described above, we made additional geochronological investigations at Quidico-Tirúa (38°15' S) and near Bahía Mansa (40°35' S) (Fig. 19.2). Similarly to the investigations in the Valdivia area, we used Rb/Sr multimineral isochrons to date different structural features from these sites.

For simplicity, we only determined ages for the main, penetrative deformation (Fig. 19.4a), and for the formation of tension gashes (Fig. 19.4b). The analytical procedures were those of Glodny et al. (2002), sample sites are shown in Fig. 19.2, and the data are presented in Table 19.1. The results are summarized and combined with data for the Quidico area (Glodny et al., unpublished results) and the Valdivia region (Glodny et al. 2005) in Table 19.2.

The new data show that near Quidico-Tirúa (38°15' S), rocks that were mainly deformed in the earliest Permian (~285 Ma) are now at the surface, whereas near Bahía Mansa (40°35' S), the main deformation was considerably younger (229–236 Ma). The passage of material from the base of the upper plate to the near-surface, as defined by the data for main deformation and tension-gash formation (Table 19.2), took around 40 million years at all sites. This suggests that growth mechanisms, growth velocities and the structural evolution were similar all along the Western Series complex. However, the absolute ages differ and they document a significant, southward younging. If we consider that the formation of tension gashes mark the termination of basal-accretionary mass flow, then

**Fig. 19.4.** Field aspects of ductile and brittle deformation structures within the Western Series, at coastal cliffs near Tirúa. **a** Quartz-rich mica schist with subhorizontal foliation, related to basal accretion, enclosing a lens-shaped metabasite body (in the *upper part* of photo) Hammerhead (14 cm long) for scale. **b** Typical aspect of tension gashes in chlorite schists. Hammerhead (14 cm long) for scale. *Arrows* indicate open cavities

**Table 19.2.**
Rb/Sr mineral isochron age data, Western Series

| Site | Main penetrative deformation (Ma) | Semiductile shear zones (Ma) | Tension gashes (Ma) |
|---|---|---|---|
| Quidico/Tirua (38°15' S) | 289.3 ±2.2<br>282.5 ±5.3 | 255.8 ±2.7[a] | 249 ±10 |
| Valdivia[b] (39°45' S) | 249 ±5 ($n = 5$) | 235 ±2 ($n = 2$) | ~210 ($n = 3$) |
| Bahía Mansa (40°35' S) | 236.4 ±1.7<br>228.6 ±5.2 | n.d. | 197 ±16 |

[a] Unpublished data of Glodny et al.
[b] Valdivia data from Glodny et al. 2005.
$n$: Number of independent age determinations.

shutdown of the accretion system occurred at ~250 Ma near Quidico, at ~210 Ma near Valdivia, and at roughly 200 Ma at Bahía Mansa. Zircon fission-track data further indicate that this differential evolution had finished by the earliest Jurassic. At ~200 Ma, present-day surface rocks were exhumed to paleodepths of < 7 km over the entire Western Series complex south of 38° S (Glodny et al., in press).

In the southern part of our study area, Permo-Triassic subduction and accretionary wedge growth was accompanied by magmatic activity. Scattered, calc-alkaline, granitoid intrusions aged between 280 and 200 Ma occur south of 39° S in the modern back-arc in Argentina, in the Somuncura area (Caminos et al. 1988; Pankhurst et al. 1992a), and in other locations within Northern Patagonia (Linares et al. 1988; Rapela et al. 1992; Rapela and Pankhurst 1992; Dalla Salda et al. 1994; Rapela and Pankhurst 1996; López de Luchi et al. 2000; von Gosen and Loske 2004). Most of these intrusions are probably related to subduction beneath the Western Patagonian margin (e.g., Rapela and Pankhurst 1996; von Gosen and Loske 2004). Scarce isotopic data (Sr, Nd; Pankhurst 1990, and references above) indicate that there is a major component of reworked continental crust in most of these intrusions, similar to those of the Pennsylvanian-early Permian arc.

The reason for the differential, early Permian margin behavior (i.e. mass loss north of 38° S and accretion further south that later became progressively extinct southwards) is a matter of speculation. It is well known that sediment-starved margins tend to be tectonically erosive, whereas a large amount of trench sediment is required for long-term accretion (von Huene and Scholl 1991; Lohrmann 2002) and, as observed in the Western Series, continental-clastic, turbiditic material dominates in many major accretionary wedges (Karig and Sharman 1975). Terrigenous sediment supply in our study area might well have been increasingly important toward the south because the area was already located and oriented similarly to today's situation, facing west at mid latitudes in the early Permian (Torsvik and Cocks 2004).

Because turbiditic currents can travel for more than 1 000 km downslope on the sea floor and along trench axes (Lewis et al. 1998; Wynn et al. 2002), local, terrigenous sediment supply is probably not the only factor governing the differential Permo-Triassic margin evolution in south-central Chile. Instead, a possible explanation can be found in a combined control of margin behavior by terrestrial sediment supply and lower-plate morphology. Seamount chains and ridges can effectively block trench-parallel turbidity currents. Recent examples include the dammed Hikurangi Trough, New Zealand (Lewis et al. 1998) and the Juan Fernández Ridge which segments the Chile Trench (Bangs and Cande 1997; von Huene et al. 1997; Yáñez et al. 2001). These lower-plate, aseismic ridges separate sediment-filled from sediment-starved trench domains and, thus, potentially seperate accretive from potentially non-accretive or erosive margin segments.

We, therefore, hypothesize that ridge-trench interaction controlled Permo-Triassic margin dynamics. Studies of the Juan Fernández Ridge have shown that ridge-trench interaction can persist for several tens of millions of years, with a slowly migrating, or even immobile, collision point (Yáñez et al. 2001). A scenario of Permo-Triassic sediment supply predominantly south of 38° S, combined with a stable, ridge-trench collision site at ~38° S (~280–250 Ma), followed by the slow migration (5–10 km Myr$^{-1}$) of the collision site would elegantly explain the inferred, differential, margin evolution.

## 19.5 Andean Evolution (Jurassic to Holocene)

The initiation of the Andean evolution in southern South America is conventionally defined by processes along the Pacific margin in the early Jurassic, in particular by the onset of 'Andean', widespread magmatism in a N-S arc system along the entire margin (cf. Mpodozis and Ramos 1989), which persists today. The Andean period also encompasses the break-up of Gondwana and the opening of the southern Atlantic Ocean, accompanied by the establishment of a volcanic, passive margin along the Atlantic coast. The Jurassic, voluminous, Chon Aike silicic magmatism, found in Argentina south of 37° S, has been controversially assigned either to back-arc extension (Féraud et al. 1999) or to the opening of the South Atlantic (Riley et al. 2001). Between the southern Andean arc and the Atlantic margin, major sedimentary basins were formed, in particular the late Triassic to Cenozoic Neuquén Basin (Vergani et al. 1995).

In northern Chile, Triassic, dextral-oblique convergence (see above) continued into the Jurassic. Magmatism there resumed in the early Jurassic, with clear, subduction-related signatures (e.g., Kramer et al. 2005). In south-central Chile and adjacent Argentina, the Andean convergence is lithologically manifested in Jurassic to Holocene, subduction-related, igneous rocks that are located mostly within the Main Andean Cordillera and also in minor Late Cretaceous and Miocene, fore-arc, magmatic centers.

Andean magmatism has generated the North Patagonian Batholith, a huge, composite, granitoid body, within the Main Cordillera. The exposed, E-W dimension of the batholith widens toward the south, from discontinuous outcrops at 38°30' S to about 100 km at 42° S (Fig. 19.2). Available age data of igneous rocks from the North Patagonian Batholith are clustered at around, 130–80 Ma (Cretaceous) and 20–5 Ma (Miocene), with subordinate magmatic episodes in other periods since the Early Jurassic, and abundant volcanism in the Plio-Pleistocene (cf. Rapela and Kay 1988; Pankhurst et al. 1999; SERNAGEOMIN 2003; Rapela et al. 2005). Andean evolution further comprised the formation of sedimentary basins in various fore-arc, intra-arc and back-arc settings, as presented below.

### 19.5.1 Jurassic: Incipient Andean Orogeny

Jurassic, magmatic activity in south-central Chile was confined to the Main Andean Cordillera and the area of today's back-arc. However, an Upper Triassic to Lower Jurassic age attributed to scarce plutonic bodies in the eastern flank of the Coastal Cordillera at 36–37°15' (e.g., Hervé and Munizaga 1978) awaits confirmation. Outcropping remnants of Jurassic magmatism consist of volcanic and volcaniclastic intercalations in the preserved, western flank of the Neuquén Basin. They occur within the Lower Jurassic Formación Nacientes del Bíobío at ~38–39° S, near the town of Lonquimay (Fig. 19.2) (Suárez and Emparán 1997; SERNAGEOMIN 2003). The mainly calcalkaline basalts and basaltic andesites have clear, arc-related, geochemical characteristics (Kemnitz et al. 2005).

Martin et al. (1999) confirmed the presence of Jurassic granites (~177 Ma zircon crystallization age) in the Panguipulli Area of Chile at 39°40' S. This observation, together with the occurrence of Jurassic K-Ar ages in Palaozoic rocks of that region (Munizaga et al. 1988; Martin et al. 1999), suggests the Panguipulli area occupied an intra-arc position in the Jurassic. Jurassic, basaltic-andesitic rocks also occur south of 41° S in scattered outcrops along the present-day magmatic arc and continue to ~44° S, suggesting that the western front of the Jurassic magmatic arc was close to the current position of the arc (see also Legarreta and Uliana 1996).

Jurassic, subduction-related rocks in Patagonia also exist further east and south. Among other instances, they occur as a belt in the eastern flank of the Andes, south of 41° S (Fig. 19.2), in particular as part of the late Triassic to mid Jurassic Sub-Cordilleran Patagonian Batholith, 41°–42°30' S (Rapela and Kay 1988; Gordon and Ort 1993; Rapela and Pankhurst 1996; Haller et al. 1999; Rapela et al. 2005). Geochemical and isotopic signatures of early Andean magmatism indicate that during this time magma sources were increasingly dominanted by mantle melts (Pankhurst 1990; Lucassen et al. 2004) and, thus, juvenile magmatic additions to the crust were increasingly important.

Jurassic sedimentary rocks or metasediments are not known between 36 and 42° S in the Chilean fore-arc, suggesting that large topographic contrasts from previous rifting or mountain-building processes did not persist there. North of 35° S, the arc – fore-arc region was extending during the Jurassic (Charrier and Muñoz 1994; Willner et al. 2005). The Neuquén Basin, in the eastern flank of the Andes in Argentina and easternmost Chile (33–40° S and 67–71° W), is a major depocenter which nucleated from late Triassic rifts that were filled with continental, pyroclastic deposits. From the early Jurassic, it further evolved as a marine basin in a back-arc setting, driven by trench roll-back (Vergani et al. 1995; Legarreta and Uliana 1991, 1996).

From its sedimentary history, it appears that during the Jurassic, the Andean magmatic arc and the fore-arc basement were partly and episodically submerged. This created open connections between the Neuquén back-arc basin and the Pacific Ocean at these times. This island-arc system, related to eastward, Andean subduction, was fully developed by the mid-Jurassic and supplied immature, volcaniclastic material to the basin (Legarreta and Uliana 1996; Eppinger and Rosenfeld 1996; Burgess et al. 2000). However, the productivity of the Jurassic arc was probably not particularly high because the arc was only a secondary source area for Jurassic Neuquén sediments (Legarreta and Uliana 1996).

The low average topography and episodic submergence during the Jurassic of today's arc and fore-arc indicate that no major erosion occurred in our study area at that time. This supposition agrees with inferences from fore-arc geology, tectonochronology and zircon fission-track data (Martin et al. 1999; Glodny et al. 2005; Glodny et al., in press).

### 19.5.2 Cretaceous

The Cretaceous was a period of intense, subduction-related plutonism all along the central and southern Andes.

In our study area, Cretaceous granitoids outcrop within the North Patagonian Batholith south of 38°30' S, and are also present in subordinate volumes in the Coastal Cordillera south of 39° S.

In the Main Cordillera between 38°30' S and 41° S, outcrops of Cretaceous, generally calc-alkaline granitoids are patchy and unsystematically distributed, whereas south of 41° S, they make up a continuous belt (up to 80 km broad) within the center and eastern flank of the North Patagonian Batholith (SERNAGEOMIN 2003; Pankhurst et al. 1999). Batholith formation was apparently episodic, with periods of relatively low activity interrupted by intense pulses. This pattern was possibly governed by subduction rates and variable convergence obliquity (Pankhurst et al. 1999). The mid Cretaceous magmatic pulse appears to have been particularly important and productive, in both our study area and further to the south, where numerous mid Cretaceous ages have been reported (K-Ar mineral and Rb/Sr whole rock ages between ~110 and 80 Ma; e.g., Emparán et al. 1992; Pankhurst et al. 1992; Lara and Moreno 1998; Pankhurst et al. 1999; Latorre et al. 2001; Rolando et al. 2002; and age data compiled by Adriasola 2003).

Deformation of late Paleozoic metamorphic rocks near Liquiñe (Fig. 19.2; 39°45' S), associated with an important thermal overprint, occurred between ~100 and 80 Ma (Rosenau 2004), signifying intense heating of the upper crust by magmatic input at that time. Nearly all magmatism within the North Patagonian Batholith, which ranges from gabbroic to granitic with a predominance of granodiorites and tonalites, has a similar major- and trace-element geochemical pattern, irrespective of age (e.g., Pankhurst 1990; Pankhurst et al. 1999; Lucassen et al. 2004). Strontium-Nd-Pb isotope systematics identify the Cretaceous intrusive rocks of our study area as mantle-dominated (90–95%; Lucassen et al. 2004).

In the coastal area between 39°30' S and 40° S, several small, shallow-level, granitoid intrusions are exposed. One of these bodies, the Chaihuín granodiorite close to Valdivia (Fig. 19.2), has been dated at 91.3 ± 4.9 Ma (U/Pb, zircon; Martin et al. 1999). Zircon fission-track data suggest that the entire area of the Coastal Cordillera south of 39°30' S down to central Chiloé experienced transient heating in the late Cretaceous, most likely due to hidden Late Cretaceous intrusions (Glodny et al. in press, Thomson and Hervé 2002).

A possible explanation for mid to Late Cretaceous, fore-arc magmatism is that it is related to the replacement of the Phoenix (or Aluk) plate along the Pacific margin of southern South America by the Farallon (Nazca) Plate. According to most plate reconstructions, the South America-Farallon-Phoenix triple junction migrated southward along the continental margin during the mid to Late Cretaceous. However, to date, the exact chronology of this process is only weakly constrained, but it appears that the triple junction reached the northern Patagonian margin segment in the Late Cretaceous (Beck 1998; Beck et al. 2000). We hypothesize that the fore-arc intrusions near Valdivia relate to basaltic underplating and to the thermal anomaly imposed on the fore-arc by the passage of this triple junction, in the same way as documented by Bourgois et al. (1996) and Guivel et al. (1999) for the Neogene-Quaternary magmatism of the Chile Triple Junction (46–47° S).

In line with this suggestion, the Sr-Nd isotopic signatures of the Chaihuín granodiorite indicate large contributions from mantle sources for magma generation (Lucassen et al. 2004). In addition, other similar, Late Cretaceous fore-arc intrusions younger than 91 Ma are found further south along the margin (e.g., a small 77 ± 2 Ma body at Peninsula Gallegos, 46° S; Pankhurst et al. 1999). Although the Late Cretaceous fore-arc intrusions are clearly a juvenile, magmatic addition to the crust, the net effect of the related processes on continental growth is unclear as ridge-trench interaction might have been associated with considerable subduction erosion, analogous to Pliocene-Holocene subduction erosion near the Chile Triple Junction (Bourgois et al. 1996; Behrmann and Kopf 2001).

Whereas the Jurassic and early Cretaceous saw a predominance of extensional strain in the fore-arc and arc along the south-central Andean margin, the mid Cretaceous has been identified as an episode of shortening and transient mountain-building (Mpodozis and Ramos 1989; Franzese et al. 2003; Rosenau 2004; see also Arancibia 2004; Willner et al. 2005, for the area north of 35° S). During the early Cretaceous (~120 Ma), east of, and even within, the Main Andean Cordillera, marine and continental sediment deposition occurred along a roughly N-S-oriented rift basin that stretched from the present central Andean fore-arc to the intra-arc and back-arc regions of Patagonia (e.g., Ramos and Aleman 2000). Patches of this sediment are preserved in our study area between 41 and 42° S, close to the Chile-Argentina border (SERNAGEOMIN 2003), and further east in Argentina (Chubut group).

Here, mid Cretaceous shortening probably caused sedimentation to stop, and also closed the links between the Neuquén Basin and the Pacific (Cobbold et al. 1999). However, the extra-Andean sedimentary record points to only limited, erosion-driven exhumation in response to this shortening event. The Neuquén back-arc basin was also a major depocenter in the Cretaceous (Legarreta and Uliana 1991) and contains up to 1300 vertical meters of Late Cretaceous, continental, molasse deposits (Cazau and Uliana 1973). However, it was also partly fed from extra-Andean sources and the proportions of sediment supplied by different regions are a matter of debate. Eppinger and Rosenfeld (1996) suggest that, even in the early Cretaceous, Andean-arc sources dominated, but other authors (Legarreta and Gulisano 1989; Uliana and Legarreta 1993) claim that the Andean arc was only a subordinate source until the Paleocene, and only then did the proportion of arc detritus from the west increase.

In the Late Cretaceous, a series of sedimentary basins west of the present-day Coastal Cordillera evolved in the offshore fore-arc region (Fig. 19.1). Related sediment is found at the base of the Itata Basin (36–37° S), at the base of the Arauco Basin (37°–37° 40' S) and also in the Chiloé Basin (42° S; Mordojovich 1981; González 1989). Cretaceous sediment thicknesses vary and reach maxima greater than 1500 m. The oldest sediment in these basins appears to be Maastrichtian in age (~71–65 Ma, Quiriquina formation; Stinnesbeck 1986).

### 19.5.3 Paleocene – Eocene

In the Paleogene, the Andes were again a partly and occasionally submerged magmatic arc. This is evident from the sedimentary history of the retro-arc Nirihuao Basin in Argentina (Fig. 19.1; 41–43° S), which records marine transgressions from the Pacific in the late Eocene (Ramos 1982). However, in the Neuquén Basin north of the region considered here, the effects of the incipient rise of the Andes are visible from the Eocene onwards (Vergani et al. 1995).

Paleocene or Eocene intrusions appear to be rare or absent in south central Chile and adjacent Argentina. Instead, Paleogene, magmatic-arc activity is manifested by calc-alkaline ignimbrite deposits and, in particular, by andesites (e.g. by the Pilcaniyeu Formation and the Serie Andesitica) in the back-arc, eastern flank of the Main Andean Cordillera (e.g., Rapela et al. 1988). Paleocene to early Eocene volcanic and volcano-sedimentary rocks form the basement of the Oligocene-Miocene Nirihuao and Cura-Mallín Basins in the eastern flank of the Andes (Fig. 19.1; Jordan et al. 2001 and references therein) and are also present in the western Neuquén Basin (Cobbold et al. 1999). However, the overall volume of these volcanic products appears to be small. The late Eocene to early Oligocene has been, as in other parts of the southern Andes, a period of little or no volcanic or tectonic activity (Jordan et al. 2001 and references therein).

In the offshore fore-arc, regional subsidence of fore-arc basins continued and also led to the initiation of the Valdivia Basin, the basal sediments of which are Eocene (González 1989). However, the total amount of Paleocene-Eocene sediment in the fore-arc basins is small and appears to decrease from north to south within our study area (Mordojovich 1981; González 1989).

### 19.5.4 Oligocene – Miocene: Magmatism, Basin Formation, and Andean Morphogenesis

A primary event during the Andean evolution in the Oligocene to Miocene was the break-up of the Farallon plate into the Cocos and Nazca Plates, which was completed by the early Miocene (Tebbens and Cande 1997 Lonsdale 2005). This break-up was related to a change from oblique to more orthogonal convergence between the Nazca and South America plates, and to a marked increase in convergence rates (Somoza 1998). The change in convergence angle was linked to a general increase in magmatic activity along the central and southern Andes (Stern 2004).

This process caused crustal shortening and uplift in the central Andes, but extension and crustal thinning in the south-central and southern Andes (Muñoz et al. 2000 and references therein; Jordan et al. 2001). In turn, this led to the evolution of a network of fore-arc, intra-arc and back-arc basins between 35 and 42° S in the late Oligocene to early Miocene (Suárez and Emparán 1995; Kemnitz et al., in press; Jordan et al. 2001; Charrier et al. 2002; Folguera et al. 2002).

A remarkable, extension-related, magmatic event was the generation of a transient, late Oligocene to early Miocene, fore-arc magmatic belt located within the Central Depression at 37° S, in the eastern flank of the Coastal Cordillera at 41° S, and within the Coastal Cordillera of Chiloé Island (Fig. 19.2) (López-Escobar and Vergara 1997). The related, mafic, volcanic rocks, dated at 29 to 19 Ma (Muñoz et al. 2000), partly interfinger with marine sediment that was deposited during the early stages of Central Depression subsidence. Muñoz et al. (2000) ascribed the activity of this volcanic belt, partly based on distinct geochemical attributes, to transient opening of a slab window and asthenospheric upwelling in response to the changes in subduction geometry. This fore-arc magmatism was accompanied by contemporaneous, intra-Andean, crustal-extension-related volcanism (Muñoz et al. 2000; Folguera et al. 2002), parts of which, at least, have arc-indicative trace element patterns (Kemnitz et al. 2005). The late Oligocene extrusion of the voluminous, 'OIB-like' Somuncura Plateau basalts in the back-arc in Argentina, 40–42° S (De Ignacio et al. 2001; Kay et al. 2004) is also correlated with the Farallon-Nazca Plate reorganization.

The widespread extension that occured in the late Oligocene to early Miocene has also been invoked as initiating the Central Depression Basin (Figs. 19.1 and 19.2). This is an asymmetric, graben structure containing the Los Angeles-Temuco, Osorno, and Llanquihue subbasins in which Oligocene-Miocene sediments locally pile up to nearly 2 km (González 1989, Cisternas and Frutos 1994, Muñoz et al. 2000). Further sedimentation into the pre-existing offshore fore-arc basins was similarly favored by extension (Mordojovich 1981). These shelf basins present thick (up to 2 km) Oligocene-Miocene sequences. They evolved intermittently as marine and subaerial basins, and collected detritus mainly derived from the episodically exposed and eroded Coastal Cordillera (e.g., le Roux and Elgueta 2000). The different depocenters were occasionally interconnected as marine basins across the Coastal Cordillera with the evolving Central Depression Basin (González 1989; Le Roux and Elgueta 2000). Miocene extension partly reactivated pre-existing, major fault structures, in particular the Lanalhue Fault Zone (Fig. 19.1), as evident from the creation of small, probably transtensional, basins along the fault at that time (cf. Ardiles 2003; Burón 2003).

Since the mid Miocene (18–15 Ma), extension in the fore-arc and intra-arc basins slowed or ended, and basin fills were partly folded by the late Miocene, as recorded in the intra-arc basin of the Lonquimay area (38° S, Fig. 19.2; Suárez and Emparán 1997, Kemnitz et al. in press). However, the overall amount of mid to late Miocene cross-arc shortening has been small, of the order of 10% or less (Rosenau 2004). Orogenic shortening appears to have continued only locally until the present day (Diraison et al. 1998), whereas in other regions it terminated in the early Pliocene, and was even followed by an episode of strike-slip and transtension (Giacosa and Heredia 2004; Rosenau 2004; Melnick et al. 2006; Rosenau et al., in press), or was negligible (Pankhurst et al. 1992b). A middle Miocene change from horizontal extension to horizontal shortening is also seen in the Nirihuao retro-arc basin of Argentina (41–43° S). It was an active, pull-apart basin in the Oligocene (Dalla Salda and Franzese 1987). It was then inverted and has been an a narrow, fold-and-thrust belt since the mid Miocene (Diraison et al. 1998; Giacosa and Heredia 2004). This change marks the final emergence above sea level, incipient surface uplift, and the onset of significant mountain-building in the south-central Andes.

Contemporaneously, by the mid Miocene, the magmatic arc was re-established as a narrow feature, close to its present-day location, where a chain of voluminous, granitoid bodies was intruded (Muñoz et al. 2000, and references therein; Jordan et al. 2001; Folguera et al. 2002; SERNAGEOMIN 2003; Kemnitz et al. 2005). In the study area, Miocene magmatic activity in the Andean Cordillera appears to have been most intense from (18)15–9 Ma (mostly K-Ar mineral and Rb/Sr whole rock ages; Munizaga et al. 1988; Emparán et al. 1992; Suárez and Emparán 1997; Lara and Moreno 1998; and compilation in Adriasola 2003). The location of the Miocene plutonic chain not only coincides with the trace of the active Liquiñe-Ofqui Fault Zone (LOFZ; Hervé 1976) and the Quaternary, volcanic-arc front (within < 15–30 km) but also with the western edge of the mid Cretaceous intrusions (SERNAGEOMIN 2003, and references therein).

### 19.5.5 Plio-Pleistocene to Holocene Evolution

The Nazca Plate currently converges with South America at a rate of 65 mm yr$^{-1}$ to the ENE (Angermann et al. 1999). The continental margin is oriented NNE-SSW and the resulting convergence obliquity triggers, possibly together with the Chile Ridge subduction, dextral slip along the intra-arc LOFZ (Figs. 19.1 and 19.2). The LOFZ is an active, transcrustal shear zone with Miocene and earlier precursors, and it apparently controls the location of most of the Pleistocene-Holocene volcanic centers (López-Escobar et al. 1995) of the Southern Volcanic Zone of the Andes (cf. Stern 2004 and references therein). This fault zone system is located ~30–70 km west of the continental divide and ~280 km east of the deep-sea trench and it stretches for more than 1 000 km in a mostly trench-parallel, N10° E direction (Forsythe and Nelson 1985; Hervé et al. 1993; Cembrano et al. 1996; Lavenu and Cembrano 1999; Cembrano et al. 2000; Rosenau 2004). Kinematic modeling indicates that in south central Chile the LOFZ has accomodated between ~70 and ~120 km of dextral shear since the Pliocene (Rosenau et al. in press).

Pliocene to Holocene volcanic rocks are typically basaltic andesite, tholeiitic and high-Al basalt, all of predominantly calc-alkaline affinity (Hildreth and Moorbath 1988; Stern 1989; Hickey-Vargas et al. 1989; López-Escobar et al. 1995). Strontium, Pb, and Nd isotopic signatures preclude any significant assimilation of continental crust (Stern 2004 and references therein). The magmas come from the sub-arc mantle and $^{10}$Be and U-Th-Ra isotopic disequilibria (Sigmarsson et al. 2002; Hickey-Vargas et al. 2002) indicate that they have been modified by slab-derived, dehydration fluids of both sedimentary and oceanic-lithospheric origin.

Late Pliocene and early Pleistocene volcanic rocks are also abundant east of the present-day volcanic front, near the crest and in the eastern flank of the Andes, partly as plateau-forming basalts (Fig. 19.2; cf. Lara et al. 2001; Stern 2004, and references therein). Geochemical and geochronological investigations have shown that all Pliocene to Holocene magmatism is very similar. This has been interpreted as evidence for a broader, magmatic arc in the Plio-Pleistocene, the narrowing of which in the late Pleistocene has been correlated with synchronous changes in convergence velocity (Lara et al. 2001). In any case, the position of the western magmatic-arc front remained largely stable during the Pliocene to Holocene (cf. Lara et al. 2001), contrasting with previous suggestions of a slight, westward migration (e.g., Parada 1990, and references therein).

Plio-Pleistocene to Holocene sediment accumulation is focussed into several major fore-arc depocenters, whereas sediment of this age is rare in the back-arc (e.g., Diraison et al. 1998). Firstly, subbasins within the Central Depression contain up to 2 000 m of Pliocene to Quaternary sediment accumulated over Oligo-Miocene strata (González 1989; Jordan et al. 2001). Secondly, the offshore shelf basins generally contain at least several hundred meters of Plio-Pleistocene deposits (Mordojovich 1981; Bangs and Cande 1997). Further, the deep-sea trench off the

Chilean coast is filled with up to 2 000 m of Pleistocene and Holocene sediments. These sediments are, at least near their surface, basaltic-andesitic and probably represent recycled, Mesozoic to Holocene, magmatic-arc material (Lucassen et al. 2005; Völker et al. 2006, Chap. 9 of this volume).

Similar trench sediment most likely constitutes the small frontal accretionary wedge found at the toe of the continental slope in south-central Chile (Bangs and Cande 1997). This wedge is only 20–30 km wide and is less than 1–2 million years old. From its presence, Bangs and Cande (1997) concluded that, at present, the continental plate is frontally growing by accretion. However, this mode of behaviour has not been consistently maintained in the Neogene, so that the margin cannot be classified as accretionary on the long term. Instead, the presence of shelf basins and of pre-Cenozoic basement rocks close to the trench suggest that todays morphology of the margin is predominantly shaped by subduction-erosion related processes.

For the continental margin north of 36° S, northward-increasing, post-Oligocene, tectonic erosion has been inferred from progressive, eastward relocation of the arc front (Kay et al. 2005). South of 36° S, the nearly stable position of the magmatic arc through time (with the exception of transient, early Miocene, fore-arc magmatism) indicates that the current trenchline has remained nearly the same since the Jurassic. This means that the active frontal accretion occurring at the toe of the upper plate is not reflected in arc-front migration. More generally, the almost constant position of the arc front since the Jurassic suggests that the outer fore-arc has, in fact, episodically shifted between accretionary and tectonic-erosional modes in response to the availability of trench sediment (Bangs and Cande 1997). Concomitant uplift and subsidence of the shelf and Coastal Cordillera is seen in the fore-arc sedimentary record, but probably occurred within short intervals, not allowing for major net growth or mass loss through time.

The episodic development of glaciers during the Pliocene-Pleistocene has been an important factor shaping the present-day landscape in the south-central Andes. The oldest glaciation event in Patagonia is dated at between 7 and 4.6 Ma (Mercer and Sutter 1982), and up to 40 further glaciations have been identified between then and the last glacial maximum (e.g., Rabassa and Clapperton 1990; Llibroutry 1999). North of 38° S, glaciation was mostly restricted to high elevations within the Andes, but further south, glaciers covered most of the Central Depression and reached the open ocean at ~42° S (cf. Rabassa and Clapperton 1990; Bangs and Cande 1997). Glacial events have a potentially critical effect on the mass transfer patterns at continental margins; the increased release of detritus can affect trench-sediment thickness and, thus, accretion (cf. von Huene and Scholl 1991; Bangs and Cande 1997).

## 19.6 Surface Mass Transfer

Erosional aspects of regional, surface mass transfer can partly be reconstructed using: available fission-track data; petrologic information on the regional exhumation record (e.g. from fluid inclusion thermometry); estimates of granitoid intrusion levels; and Al-in-hornblende geobarometry. However, specific processes, such as the erosion of transient volcanic covers, may go undetected with the above methods. The redeposition history of eroded material, which also constrains facets of the erosion history, is partly recorded in the regional sedimentary basins. However, not all eroded material is retained within the region. Mobile material from the active margin system is lost through sediment subduction, long-range submarine and subaerial transport, and the transport of solutes into the ocean, all of which are difficult to quantify. Above, we have already summarized the sedimentary record of the area; here, we outline the regional erosion history.

### 19.6.1 Carboniferous to Triassic Erosion

A major constraint on the late Paleozoic (Pennsylvanian and Permian) exhumation and erosion history of the onshore region north of the Lanalhue fault (Fig. 19.1, 38° S), and also, probably, of today's occurrence of the Eastern Series rocks in the Main Cordillera (south of 39° S), comes from the metamorphic imprint of the Eastern Series. This imprint indicates a total cumulative exhumation of nearly 10 km (Martin et al. 1999). In places, most of this exhumation was accomplished by the early Triassic, as indicated by the formation of Triassic basins (e.g., foothills of the Main Cordillera at 40° S; Martin et al. 1999) and by Permian paleotemperatures of ~150 °C, equivalent to crustal depths of 3–4 km, for rocks now exposed in the southern Cordillera Nahuelbuta (Collao et al. 2003).

In contrast, the area now occupied by the Western Series has been an emergent fore-arc high, probably since the Pennsylvanian. Its emergence is correlated with high, long-term erosion rates of 0.6 mm yr$^{-1}$ (cf. Glodny et al. 2005). Such high erosion rates were probably maintained throughout the local activity period of basal-accretionary processes, i.e., from ~305 to ~290 Ma north of 38° S, from ~305 to ~250 Ma in the Coastal Cordillera immediately south of 38° S (Table 19.2), and between ~305 and 200 Ma near Valdivia (40° S) and further south (Glodny et al. 2005, and this work, Table 19.2).

### 19.6.2 Coastal Cordillera since the Jurassic

Zircon and apatite fission-track data indicate that long-term denudation in the Coastal Cordillera was rather uniform after the late Triassic; it does not clearly identify episodes of pronounced erosion. Average erosion rates were in the range of 0.03 to 0.04 mm yr$^{-1}$ in our study area (Glodny et al. in press) and also in adjacent stretches further north and south (Willner et al. 2005; Thomson and Hervé 2002). Episodic emergence of fore-arc highs and mountain-building, such as seen in today's Cordillera Nahuelbuta or near Valdivia in the early Miocene (cf. le Roux and Elgueta 2000), did not result in pronounced long-term denudation. To date, it is only reflected in the available apatite fission-track data by a short-term signal for Plio-Pleistocene denudation in places in the Cordillera Nahuelbuta (Glodny et al. in press). This also indicates that the Andean fore-arc has been remarkably stable since the late Triassic, apart from oscillations between uplift and subsidence that are possibly correlated with episodic changes between fore-arc accretion and subduction erosion (cf. Bangs and Cande 1997). It remains quite unclear how the late Paleozoic to Triassic fore-arc architecture was preserved at a subduction zone over such a long period of time. A possible explanation may be that most of the time there was sufficient sediment entering the subduction zone to prevent the upper plate from being truncated. In any case, it appears that no large-scale, long-term, basal underplating, accretion, trench-parallel tilting, or tectonic erosion processes have affected the Coastal Cordillera during the last 200 million years.

### 19.6.3 Andean Main Cordillera

The subaerial erosion history of the Main Cordillera is complex. Preservation of Triassic sediments in the Lake Region (40° S) in the western flank of the Cordillera (cf. Martin et al. 1999), local preservation of Jurassic volcanic rocks (Kemnitz et al. 2005) and, in particular, preservation of low-grade, Paleozoic metasediments near the crest of the Andes at 40° S (Rosenau 2004) suggest that there has been only minor total erosion (less than 3–12 km) since the late Permian, and less than 10–12 km since the Pennsylvanian. This concurs with the inference above that from the Permian to the Oligocene, the Andean arc was partly submerged and only episodically a major source of detritus.

Using Al-in-hornblende barometry, Seifert et al. (in press) found a southward increase in exhumed crystallization depths for both Cretaceous and Miocene, Andean arc granitoids. These depths were less than 3 km at 37–38° S and increased to =10 km at 41–42° S. This consistent exhumation pattern, together with an absence of significant extension (Rosenau 2004), suggests that substantial erosion was initiated during the late Miocene, with rates of less than 0.3 mm yr$^{-1}$ at 37–38° S and ~1 mm yr$^{-1}$ at 41–42° S. These rates are compatible with those derived from apatite fission-track data that similarly show a prominent, N-S gradient in exhumation rates, from a long-term value of 0.07 mm yr$^{-1}$ since the Eocene-Oligocene near 38° S, to 1–2 mm yr$^{-1}$ since the late Pliocene at 40° S (Glodny et al., in press).

Besides this N-S-gradient of intra-Andean denudation, an E-W gradient of apatite fission-track ages also exists in the southern part of our study area (Adriasola et al. 2002; Adriasola 2003; Glodny et al., in press), indicating that the most pronounced Pliocene to Holocene erosion occurs near the Andean watershed. The overall erosion pattern has been linked to climatic factors, in particular to the intensity of local glaciation (Mpodozis and Ramos 1989; Rosenau 2004, Seifert et al., in press; Glodny et al., in press). Because magnitudes of tectonic shortening are insufficient to account for the erosional mass loss (Rosenau 2004), and because the northern part of our study area is even mildly extending (Potent 2003; Melnick et al. 2006), continuous Andean uplift and erosion is most probably sustained by deep magmatic addition to the crust (cf. Nelson et al. 1999).

## 19.7 Considering Crustal Evolution and Mass Balance

Using the geological record of the south-central Chilean margin as outlined above, we try to semi-quantitatively evaluate the volumetric importance of specific mass transfer processes and their influence on the evolution of the continental crust. In this context, crustal evolution is understood both in terms of mass balance (i.e. changes in the net crustal volume) and in terms of the petrological, chemical, and isotopic evolution of the continental crust in the active margin setting. Thus, we need to distinguish between processes that only redistribute continental material within the margin system, processes which transfer juvenile material from the subducting plate or the mantle to the continental crust, and processes inducing net mass loss. Only the latter two are important for crustal mass balance.

### 19.7.1 Juvenile Additions to the Crust and Crustal Growth

Continental crustal growth is, in general, only achievable by mass transfer from the lower plate and the mantle into the crust. Primary mechanisms are juvenile, mantle-derived, magmatic input, and accretion by scraping off lower-plate igneous material in the fore-arc. Minor contributions to crustal growth may originate from the solute load of subduction fluids that enter the upper plate.

While the proportion of lower-plate lithologies in the Western Series accretionary complex, as well as its geometry, is fairly well known (Kato 1976; Krawczyk and the SPOC Team 2003), the distribution of juvenile magmatic additions within the continental margin system is less clear because there are uncertainties in estimating the magmatic productivity of the arc. Along the Andes, the intensities of plutonic and volcanic activity within the magmatic arc are inversely correlated, corresponding to changes in convergence obliquity and upper-plate deformation characteristics (Hutton et al. 1990; Grocott et al. 1994; McNulty et al. 1998; Folguera et al. 2002). Erosion may, therefore, erase the products of arc activity generated at conditions favoring volcanism much more easily than those formed during 'plutonic' episodes. Estimating arc productivity through time thus requires volume estimates of plutons and their roots (e.g., Jicha et al., in press), as well as considering the arc detritus in arc-adjacent, sedimentary basins.

A second, related reason for the uncertainties regarding the distribution of magmatic additions is the fragmentary constraint on the structure of the deep crust. Hints of the subsurface composition of the present-day, north Patagonian magmatic arc come from:

a  Geochemical studies that indicate the chemically evolved but isotopically 'juvenile', Mesozoic-Cenozoic, arc granitoids formed by intracrustal differentiation under moderate pressure. This, in turn, implies voluminous mafic bodies underneath the arc (Lucassen et al. 2004).
b  The inference that Neogene exhumation may be, at least partially, balanced out by magmatic additions to the crust (Nelson et al. 1999; Glodny et al., in press).
c  Analogy with the Sierra Nevada Batholith of California which has a mafic root at least twice as voluminous as the granitoid lid (Ducea and Saaleby 1998).

The lack of major input of old crustal components into the Southern Volcanic Zone (e.g. McMillan et al. 1989; Stern 2004) suggests that beneath the present-day arc there is minimal old crust left for the juvenile magmas to interact with. Together with the scattered occurrence of pre-Mesozoic rocks in the intra-arc domain, this suggests that the old crust was magmatically inflated and replaced by mostly juvenile, mafic material. Possible replacement mechanisms may have included surface erosion, balanced out by mafic input at depth, and preferential volcanic eruption and subsequent removal of felsic or hybrid magmas rich in old crustal components. Replacement of old crust by mafic arc magma agrees well with results from flexural modeling of the correlation between relief and gravity along the southern Andes, which point to a clearly more mafic crust in the south-central Andes compared to the central Andes (Tassara and Yáñez 2003). Juvenile magmatic additions are also envisaged to have occurred in minor proportions in the Carboniferous-Permian arc, in the Cretaceous and Miocene fore-arc igneous rocks, and in Mesozoic to Holocene back-arc volcanics.

## 19.7.2 Crustal Thickness

Crustal thickness is an important variable for estimating continental crustal growth. Absence of pronounced mountain-building in the south central Andes during the Mesozoic to early Cenozoic indicates that crustal thickness since the Triassic was close to typical (~30–40 km) at the active margin, as it is today (Introcaso et al. 1992; Bohm et al. 2002; Lüth et al. 2003). Further, our study area contains no clear indication of extensive mountain-building with later orogenic collapse during the Pennsylvanian to early Permian, thus, even then, crustal thickness might have been close to typical.

The main implications are that ($a$) net vertical crustal growth by magmatic crustal thickening appears to be unimportant in the south-central Andes, and ($b$) continental mass loss by lower-crust delamination most probably did not occur, because such a process is usually a consequence of transient, crustal shortening and overthickening (e.g., Meissner and Mooney 1998). In addition, the close-to-invariable crustal thickness indicates that internal deformation of the crust is almost irrelevant for calculating the crustal mass budget, and that net continental growth can, in fact, be measured from movements of the magmatic-arc front, given that the arc-trench gap has a nearly constant width (cf. Kay et al. 2005).

## 19.7.3 Crust Destruction and Mass Wasting

At a continental margin of typical crustal thickness that does not allow for delamination, destruction mechanisms for continental crust comprise mainly tectonic erosion (including sediment subduction), mass removal after subaerial weathering, and chemical erosion. Studies of fluvial systems indicate that the total dissolved load could amount to ~20% or more of the suspended load (Sarin et al. 1989; Hu et al. 1982). This means that eroded continental material can only be retained partially as sediment within the continental margin system; a significant proportion is lost to the ocean.

Tectonic erosion is, however, probably the more important mechanism for continental mass loss, in this case into the mantle. Various workers (von Huene and Scholl 1991; Kay et al. 2005) argue that the vast majority of eroded fore-arc material is, in fact, lost into the mantle and that only minor amounts are recycled and returned into the crust as arc-magma components. Even at active margins with frontal or basal accretion, considerable proportions of the trench sediment are likely subducted to depth (von Huene and Scholl 1991). A major episode of tectonic erosion during the early Permian is inferred for the area north of 38° S (see above), and short-term tectonic erosion has also possibly occurred in the Late Cretaceous (see above) and since the late Miocene (Bangs and Cande 1997).

Continental material loss into the mantle is not restricted to these episodes but it certainly also occurred, to some extent, whenever significant trench sediment was available. These periods include the Pennsylvanian-Triassic activity of the Western Series complex and the intermittent, Jurassic to Holocene, fore-arc denudation. However, this did not lead to a significant reduction in the width of the fore-arc, which has been stable since the Jurassic, as evident from the consequent superposition of magmatic belts and the maintenance of the same location of the arc front through time. We speculate that the absence of fore-arc width reduction, a feature clearly contrasting with Andean margin segments north of 36° S (cf. Kay et al. 2005), indicates that, since the Jurassic, most of the time sufficient sediment has been present in the trench to prevent the older, consolidated margin units from being truncated.

## 19.7.4 Mass Flux Pattern through Time

Using the above constraints on composition and geological history of the active-margin continental crust, we try to track the evolution of the south-central Chilean margin segment by estimating duration and flux rates for all geologically documented, mass transfer processes since the initiation of subduction. To do this, we constructed two schematic profiles through the margin crust, using the pre-subduction, passive-margin setting, a 'modern' arc-trench-gap of 280 km, and a 200 km-wide, intra-arc/back-arc zone bordering 'stable' South America as a reference frame (Fig. 19.5a). One of the profiles depicts the margin segment between 36–38° S (Fig. 19.5b), the other at 38–42° S (Fig. 19.5c). The profiles are intended to illustrate today's location and volumes of major Pennsylvanian to Holocene juvenile additions to the crust, as well as the proportion of such additions within specific geological margin units.

Using the available geochronological data together with estimates of total volumes of relocated material

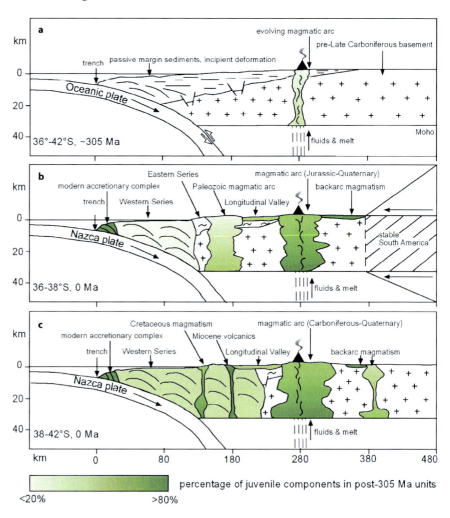

**Fig. 19.5.**
Schematic model of margin geometries, showing the distribution of juvenile additions within the continental crust. **a** Situation immediately after the initiation of subduction, with incipient arc magmatism and progressively deforming, former, passive-margin sequences. **b** Today's architecture north of 38° S. Displacement of both the Paleozoic arc and stable South America towards the trench signifies margin truncation by ~100 km in the early Permian. **c** Today's architecture south of 38° S. Net volume conservation of continental-margin crust is associated with abundant reworking of old crust and replacement by juvenile additions

(based on the hypothetical cross sections through the margin, Fig. 19.5), we are able to constrain mass transfer rates for individual mass transfer processes (Fig. 19.6). Volumes and rates are estimated for all mass transfer processes that are documented in either the erosional, sediment-depositional, accretionary, or magmatic history. These volume (or mass) and rate estimates together constrain the long-range evolution of the margin in terms of continental growth or mass wasting.

For estimating the net effects of mass flux on continental mass at the south-central Chilean margin, a simple mass balancing equation has the following form:

$$\frac{d_{mass}}{dt} = (I_{Mantle-Arc} + I_{MORB-Accret.wedge}) - (L_{Continent-Mantle(SE)} + L_{Continent-Ocean})$$

Terms in this equation are continental growth (or input, $I$) terms considering ($a$) magmatic mass flux from the mantle to the continental plate, mainly to the magmatic arc, and ($b$) addition of lower plate MORB material to a growing accretionary wedge, or by underplating. Mass loss ($L$) terms consider ($a$) mass wasting by transfer from the continent into the mantle by way of subduction erosion (SE) and ($b$) loss of dissolved matter into the ocean, a term which is assumed to be of minor importance in the case of the south-central Chilean margin. Possible mass wasting by delamination is not considered as this probably did not occur at the here considered margin segment. For visualization, in Fig. 19.6 mass transfer processes causing continental growth are color-coded in green, and mass loss processes are coded red. The following constraints, assumptions and procedures have been used to establish the corresponding database (Table 19.3) which gives estimates of mass flux (in km³ of material, per km of margin in NNE-SSW direction and per Myr [km³ km⁻¹ Myr⁻¹]):

- Crustal thickness is assumed at ~35 km, with the exception of the outer fore-arc where today's upper plate crustal thickness is less than 35 km and where the subduction angle of ~30° is used to calculate near-trench crustal thickness and volumes of mobile material.
- For differentiated, magmatic rocks (granitoids) a root zone within the crust is envisaged, with nearly the same

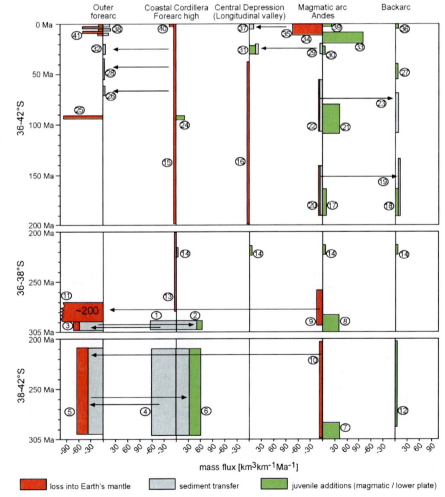

**Fig. 19.6.**
Estimates of mass flux through time since the Pennsylvanian among different morphotectonic units of south-central Chile, and the Earth's mantle. Individual mass flux processes and events are identified by labels corresponding to process labelling in the database of Table 19.3. For process #11 mass flux is out of scale; a value of ~200 km³ km⁻¹ Myr⁻¹ applies here. Arrows indicate mass flux directions; loss of continental crust into Earth's mantle is generally by way of transport through a subduction channel along the plate interface. Flux estimates are based on an average crustal thickness of 35 km, and disregard segment-internal redistribution processes (e.g., formation of locally fed sedimentary basins). Flux rates for magmatic additions only consider the mantle-derived component and ignore crustal melt contributions to total magma production. See text for discussion

**Table 19.3.** Semi-quantitative estimates of mass flux through time (cf. Fig. 19.6), in and among different active margin segments of south-central Chile and the Earth's mantle

| No. in Fig. 19.5 | Time interval (Ma) | Process, location (deg S) | Face of 'mobile' unit[a] or erosion rate × E-W-width of eroded unit | Mass flux (km³ km⁻¹ Ma⁻¹) | Material path: sources – sinks |
|---|---|---|---|---|---|
| 1 | 305 – 290 | Surface erosion, fore-arc high, 36–38 | 0.6 km Ma⁻¹ × 100 km | 60 | Fore-arc high – outer fore-arc |
| 2 | 305 – 290 | Basal accretion, fore-arc high, 36–38 | 0.6 km Ma⁻¹ × 100 km | 60 | Outer fore-arc + 20% oceanic plate – fore-arc high |
| 3 | 305 – 290 | Subduction, outer fore-arc, 36–38 | ? | ~20? | Outer fore-arc – mantle |
| 4 | 305 – 200 | Surface erosion, fore-arc high, 38–42 | 0.6 km Ma⁻¹ × 100 km | 60 | Fore-arc high – outer fore-arc |
| 5 | 305 – 290 | Subduction, outer fore-arc, 38–42 | ? | ~20–30 | Outer fore-arc – mantle |
| 6 | 305 – 200 | Basal accretion, fore-arc high, 38–42 | 0.6 km Ma⁻¹ × 100 km | 60 | Outer fore-arc + 45% oceanic plate – fore-arc high |
| 7 | 305 – 285 | Magmatic addition, arc, 38–42 | (60 × 35 km) × 0.4[b] | 40 | Mantle-arc crust |
| 8 | 305 – 285 | Magmatic addition, arc, 36–38 | (60 × 35 km) × 0.4[b] | 40 | Mantle-arc crust |
| 9 | 300 – 260 | Surface erosion, arc, 36–38 | 7 × 80 km | 14 | Arc – mantle, via subduction |
| 10 | 300 – 200 | Surface erosion, arc, 38–42 | ? | 5–10 ? | Arc – outer fore-arc (mantle) |
| 11 | 290 – 270 | Tectonic erosion, 36–38 | 5 km Ma⁻¹ | ~200 | Outer fore-arc – mantle |
| 12 | 280 – 200 | Magmatism, back-arc, 38–42 | (30 × 35 km) × 0.4[b] | 5 | Mantle – back-arc |
| 13 | 280 – 200 | Surface erosion, fore-arc high, 36–38 | ? | ~2 ? | Fore-arc high – mantle |
| 14 | 225 – 215 | Magmatic additions, 36–38 | (4 × 35 km) × 0.3[b] | ~4? | Mantle – margin cont. crust |
| 15 | 200 – 0 | Surface erosion, fore-arc high | 0.04 km Ma⁻¹ × 100 km | 4 | Fore-arc high – outer fore-arc basins and mantle |
| 16 | 200 – 30 | Surface erosion, Central Depression | ? | ? | Central Depression – mantle |
| 17 | 190 – 160 | Magmatic additions, arc | (10 × 35 km) × 0.7[b] | 8 | Mantle – arc crust |
| 18 | 190 – 160 | Magmatic additions, back-arc | (8 × 35 km) × 0.7[b] | 6 | Mantle – back-arc crust |
| 19 | 190 – 160 | Sediment transfer, arc-back-arc (Neuquen) | 1.5 × 60 km | 3 | Arc – back-arc |
| 20 | 190 – 160 | Surface erosion, arc | ? | ? | Arc – outer fore-arc/mantle (?) |
| 21 | 110 – 80 | Magmatic additions, arc | (40 × 35 km) × 0.9[b] | 40 | Mantle – arc crust |
| 22 | 110 – 60 | Surface erosion, arc | 3 × 70 km | 7 ? | Arc-back-arc and outer fore-arc |
| 23 | 110 – 70 | Sediment transfer, arc – back-arc | 1.3 × 80 km | 5 ? | Arc-back-arc |
| 24 | 93 – 90 | Magmatic additions, fore-arc high | 35 × 2 km | 23 | Mantle-fore-arc high |
| 25 | 93 – 90 | Subduction erosion, outer fore-arc | (5 × 35 km) × 1.5[c] | 80 (?) | Outer fore-arc-mantle |
| 26 | 75 – 65 | Sedimentation, outer fore-arc | 0.8 × 25 km | 2 | Fore-arc high – outer fore-arc |
| 27 | 55 – 40 | Magmatic additions, back-arc | 2 × 35 km × 0.9[b] | 5 | Mantle – back-arc crust |
| 28 | 55 – 35 | Sedimentation, outer fore-arc | 0.5 × 20 km | 0.5 | Fore-arc high – outer fore-arc |
| 29 | 29 – 19 | Sedimentation, Central Depression | 0.5 × 50 km | 2.5 | Arc + fore-arc high – Central Depression |
| 30 | 29 – 19 | Volcanic additions, intra-arc basins | 1.5 × 20 km | 3 | Mantle – arc crust |
| 31 | 29 – 19 | Magmatic additions, Central Depression | 5 × 35 km | 17 ? | Mantle – Central Depression |
| 32 | 29 – 19 | Sedimentation, outer fore-arc | 0.8 × 40 km | 3 | Fore-arc high – outer fore-arc |
| 33 | 18 – 9 | Magmatic additions, arc | (30 × 35 km) × 0.9[b] | 100 | Mantle – arc crust |
| 34 | 12 – 0 | Surface erosion, Main Cordillera 38–42 | 1.5 km Ma⁻¹ × 50 km | 90 | Arc-Central Depression, mantle |
| 35 | 12 – 0 | Surface erosion, Main Cordillera 36–38 | 0.3 km Ma⁻¹ × 60 km | 18 | Arc-Central Depression, mantle |
| 36 | 5 – 0 | Magmatic additions, back-arc | 0.5 × 30 km | 5 | Mantle-back-arc |
| 37 | 4 – 0 | Sedimentation, Central Depression | 0.5 × 30 km | 4 | Arc-Central Depression |
| 38 | 4 – 0 | Sedimentation, outer fore-arc basins | 0.6 × 40 km | 6 | Arc-outer fore-arc (episodic) |
| 39 | 3 – 0 | Magmatic additions, arc | (0.5 × 70 km) × 0.9[b] | 15 | Mantle-arc |
| 40 | 3 – 0 | Increased surface erosion, fore-arc high | ? | ~5 (?) | Fore-arc high – trench |
| 41 | 1.5 – 0 | Sedimentation, trench (alternating with tectonic erosion) | ~70 km² | 50 ? | Arc-trench |

[a] Face of unit in cross section (cf. Fig. 19.5; vertical extent × W-E-width, in km). [b] Proportion of mantle-derived component in magma (net juvenile addition). [c] Geometry factor for tectonic (subduction) erosion.

cross-sectional width as the surface exposure. Allowance is made for some mostly hidden plutons, such as the Late Cretaceous fore-arc intrusions.
- Undifferentiated, mafic, volcanic rocks have no voluminous, deep-crustal counterparts; their volume within the crust is assumed to be only slightly higher than the volume of extruded material.
- For local geological units, or for units with highly variable volumes along strike of the continental margin (e.g., Late Cretaceous fore-arc intrusions), we estimated an average volume per margin kilometer.
- Average compositions with respect to juvenile versus recycled crustal components have been estimated for compositionally heterogeneous units, based on literature and field data (e.g., for the proportion of mafic, originally igneous, juvenile material within the Western Series).
- Local, margin–internal, mass redistribution (e.g., Triassic basin fill) is not considered beause it does not affect the mass balance. Major sediment transfer between different, upper-crustal, morphotectonic units does not affect the mass balance either but is shown in Fig. 19.6 for clarity.
- Magmatic activity is only considered with its proportion of juvenile mantle components; contributions of reworked crust to total magma production are not relevant for mass balance.

It needs to be noted that with such an approach, for many mass-transfer processes, only rough estimates of mass flux can be established. Furthermore, several mass-transfer modes and processes could go undetected. Quantification is particularly difficult for lateral, margin-parallel mass flux such as that of the extensive, along-trench, turbidite migration offshore from south-central Chile (e.g., Thornburg et al. 1990), or for any exogenic, long-range, mass transport by rivers that might have extracted material from the south-central Chilean margin segment.

Another major source of uncertainty is the erosion of volcanic products from the magmatic arc. Even productive episodes of volcanism in the distant past might not be recorded as volcanic rocks are easily eroded, and, in absence of local basins, possibly lost in the subduction system without leaving any signature. However, despite the various difficulties regarding exact quantification of mass-flux volumes, rates, and directions, the mass-flux pattern in Fig. 19.6 enables us to identify those mass-transfer processes that were volumetrically important for active-margin crustal evolution. It also gives an impression of the overall losses and gains of continental crustal material within a 300 Myr period of oceanic-plate subduction.

It appears that important juvenile additions in our crustal segment originate from both magmatic input and the accretion of igneous lower-plate components in the fore-arc (Fig. 19.6). South of 38° S, these two sources are nearly equally important, whereas north of 38° S, juvenile material is mostly magmatic in origin. Magmatic input was apparently episodic, with pluton formation interrupted by either quiescence or undetected volcanic activity. The geological record of the south central Chilean margin indicates that the Miocene magmatic pulse was particularly productive, estimated at ~100 km$^3$ km$^{-1}$ Myr$^{-1}$, whereas magmatic volumes were otherwise, with ~40 km$^3$ km$^{-1}$ Myr$^{-1}$, at the lower edge for the average range of magmatic arc productivities (cf. Jicha et al., in press; Holbrook et al. 1999).

The main point evident from the mass balance (Fig. 19.6) is that the juvenile input into the system is nearly compensated for by loss of continental material into the mantle through subduction. Besides the volumetrically dominant, early Permian episode of tectonic erosion north of 38° S, major amounts of continental material have been tectonically extracted from the system during the activity of the Permo-Triassic, basal-accretionary system, and by nearly permanent subduction of surface-erosion detritus.

## 19.8 Continental Growth at Active Margins? The South-Central Chilean Perspective

The schematic cross sections through the south-central Chilean, active, continental margin (Fig. 19.5) show E-W profiles that include the distribution of pre-subduction (pre-305 Ma) material and juvenile (post-305 Ma) additions to the crust. It appears that north of ~38° S, the margin has been a site of net mass wasting instead of crustal growth (Fig. 19.5b). Mainly owing to the late Permian episode of tectonic erosion, nearly 100 km of fore-arc width has been lost. In addition, a considerable amount of pre-Carboniferous crust has been replaced by subduction-related, juvenile additions of arc magma. South of 38° S, the total volume of continental crust in the margin system appears to have been nearly constant through time (Fig. 19.5c). However, here, the effect of replacing pre-Carboniferous, continental material with input of either mantle-derived magmatic arc or igneous oceanic plate juvenile material appears to be even more pronounced than further north owing to the large volumes of metabasites in the Western Series accretionary complex (cf. Kato 1976) and, also, an apparently broader, Andean magmatic belt.

The distribution patterns of juvenile additions of igneous components versus pre-existing crust (Fig. 19.5) show that an estimated 30 to 50% of the pre-subduction, continental crustal volume is now made up by juvenile input. However, the geological record of the margin shows that nearly equal amounts of material have been lost from the system (Fig. 19.6). Denudational loss from units other than the Western Series amounts to, on average, ~10 km since the Pennsylvanian, which already accounts for nearly 30% of the total crustal volume. Pre-existing continental crust has also been lost from greater depths, namely by surface exposure and erosion of crustal melts within magmatically active domains. In the case of the Western Series, the entire crustal thickness has been involved in a cyclic, mass-transfer system that replaced an average 30–40% of pre-subduction material by MORB-type metabasalts. Therefore, we conclude that the compositional signature of the active margin crust does not signify crustal growth but reflects, over the long term, a nearly balanced replacement of wasted continental crust by juvenile additions.

During the Andean period (the last 200 million years), the margin configuration in our study area remained in a volumetrically 'steady state'. Using narrower time-slices, this steady state appears delicately balanced, probably characterized by alternating periods of net material gain, non-accretion (no gain, no loss) and tectonic erosion (with loss of material), as has been inferred for the Miocene to Holocene (Bangs and Cande 1997).

Altogether, our case study shows that the long-standing south-central Chilean margin (36–42° S) has not been a site of continental crustal growth. Instead, it has been the location of both continental mass loss, and crustal replenishment and replacement by juvenile additions. The net effect of oceanic-crust subduction on continental crustal volume has been clearly negative north of 38° S, and close to zero south of 38° S. Active-margin, mass-transfer processes, such as sediment transport and magmatic assimilation, have a tendency to stir and compositionally homogenize the crust. Furthermore, replenishment of the crust by juvenile additions has led to an overall 'rejuvenation' of the continental margin. Without a systematic analysis of crustal evolution, the occurrence of isotopically juvenile units at continental margins may be misinterpreted in terms of net crustal growth. Oceanic-plate subduction beneath a continent does not, apparently, provide a setting for significant, net, continental crustal growth, not even if the plate interface is sediment-saturated, as has been the case for prolonged periods of time in south-central Chile.

## Acknowledgments

We thank J. Herwig, G. Arnold, V. Kuntz and M. Dziggel for their help with sample preparation and figure drawing, and all individuals from the Concepción geoscience community and the Berlin-Potsdam SFB 267 group who supported this work. Very careful and constructive reviews by F. Hervé and D. Scholl are gratefully acknowledged.

## References

Adriasola A (2003) Low temperature thermal history and denudation along the Liquiñe-Ofqui Fault Zone in the Southern Chilean Andes (41–42°S). PhD thesis, Ruhr–Universität Bochum, pp 1–119

Adriasola A, Stöckhert B, Hervé F (2002) Low temperature thermochronology and tectonics in the Chiloé region, Southern Chilean Andes (41°–43°S: 72°–74°W). In: 5th International Symposium on Andean Geodynamics, Toulouse, Abstract volume, pp 15–18

Aguirre L, Hervé F, Godoy E (1972) Distribution of metamorphic facies in Chile – an outline. Krystalinikum 9:7–19

Albarède F (1998) The growth of continental crust. Tectonophysics 296:1–14

Angermann D, Klotz J, Reigber C (1999) Space-geodetic estimation of the Nazca-South America Euler vector. Earth Planet Sci Lett 171:329–334

Arancibia G (2004) Mid-cretaceous crustal shortening: evidence from a regional-scale ductile shear zone in the Coastal Range of central Chile (32°S). J S Am Earth Sci 17:209–226

Ardiles M (2003) La Serie Occidental del basamento metamórfico, centro sur de la Cordillera de Nahuelbuta, Chile, área Quidico-Capitán Pastene. Petrografía, mesoestructura y análisis microtectónico. Memoria de titulo, Dept. de Ciencias de la Tierra, Universidad de Concepción, Chile, pp 1–132

Augustsson C, Bahlburg H (2003) Active or passive continental margin? Geochemical and Nd isotope constraints of metasediments in the backstop of a pre-Andean accretionary wedge in southernmost Chile (46°30'–48°30'S). In: McCann T, Saintot A (eds) Tracing tectonic deformation using the sedimentary record. Geol Soc London Spec Pub 208, pp 253–268

Bahlburg H, Hervé F (1997) Geodynamic evolution and tectonostratigraphic terranes of northwestern Argentina and northern Chile. Geol Soc Am Bull 109:869–884

Bangs NL, Cande SC (1997) Episodic development of a convergent margin inferred from structures and processes along the southern Chile margin. Tectonics 16:489–503

Beck ME Jr (1998) On the mechanism of crustal block rotations in the central Andes. Tectonophysics 299:75–92

Beck ME Jr, Garcia RA, Burmester RF, Munizaga F, Hervé F, Drake RE (1991) Paleomagnetism and geochronology of late Paleozoic granitic rocks from the Lake District of southern Chile: implications for accretionary tectonics. Geology 19:332–335

Beck ME Jr, Burmester RF, Cembrano J, Drake R, Garcia A, Hervé F, Munizaga F (2000) Paleomagnetism of the North Patagonian batholith, southern Chile. An exercise in shape analysis. Tectonophysics 326:185–202

Behrmann JH, Kopf A (2001) Balance of tectonically accreted and subducted sediment at the Chile Triple Junction. Int J Earth Sci 90(4):753–768

Bohm M, Lüth S, Echtler H, Asch G, Bataille K, Bruhn C, Rietbrock A, Wigger P (2002) The Southern Andes between 36° and 40°S latitude: seismicity and average seismic velocities. Tectonophysics 356:275–289

Bourgois J, Martin H, Lagabrielle Y, Le Moigne J, Frutos Jara J (1996) Subduction erosion related to spreading-ridge subduction: Taitao Peninsula (Chile margin triple junction area). Geology 24(8): 723–726

Branlund J, Regenauer-Lieb K, Yuen DA (2000) Fast ductile failure of passive margins from sediment loading. Geophys Res Lett 25(13): 1989–1992

Burgess PM, Flint S, Johnson S (2000) Sequence stratigraphic interpretation of turbiditic strata: an example from Jurassic strata of the Neuquén Basin, Argentina. Geol Soc Am Bull 112:1650–1666

Burón P (2003) Petrografía, estructuras y microtectónica del área de contacto entre las series metamórficas del basamento Paleozoico entre los 38°08' y 38°21'S, Cordillera de Nahuelbuta, Chile. Memoria de titulo, Dept. de Ciencias de la Tierra, Universidad de Concepción, Chile, pp 1–144

Caminos R, Llambias EJ, Rapela CW, Parica CA (1988) Late Paleozoic-Early Triassic magmatic activity of Argentina and the significance of new Rb-Sr ages from northern Patagonia. J S Am Earth Sci 1:137–145

Cazau L, Uliana M (1973) El Cretácico superior continental de la cuenca Neuquina. V Congreso Geológico Argentino, Buenos Aires, Actas 3, pp 131–164.

Cembrano J, Hervé F, Lavenu A (1996) The Liquiñe-Ofqui fault zone: a long-lived intra-arc fault system in southern Chile. Tectonophysics 259:55–66

Cembrano J, Schermer E, Lavenu A, Sanhueza A (2000) Contrasting nature of deformation along an intra-arc shear zone, the Liquiñe-Ofqui fault zone, southern Chilean Andes. Tectonophysics 319:129–149

Charrier R (1979) El Triásico en Chile y regiones adyacentes de Argentina: una reconstrucción paleogeografica y paleoclimatica. Comunicaciones N°26, Departamento de Geología, Universidad de Chile, Santiago de Chile, pp 1–37

Charrier R, Muñoz N (1994) Jurassic-Cretaceous paleogeographic evolution of the Chilean Andes at 23–24° and 34–35°S latitude: a comparative analysis. In: Reutter K-J, Scheuber E, Wigger P (eds) Tectonics of the Southern Central Andes. Springer-Verlag, Berlin Heidelberg New York, pp 233–242

Charrier R, Baeza O, Elgueta S, Flynn JJ, Gans P, Kay SM, Muñoz N, Wyss AR, Zurita E (2002) Evidence for Cenozoic extensional basin development and tectonic inversion south of the flat-slab segment, southern Central Andes, Chile (33°–36° S.L.). J S Am Earth Sci 15:117–139

Cingolani C, Dalla Salda L, Hervé F, Munizaga F, Pankhurst RJ, Parada MA, Rapela CW (1991) The magmatic evolution of northern Patagonia: new impressions of pre-Andean and Andean tectonics. Geol Soc Am Spec P 265:29–44

Cisternas ME, Frutos J (1984) Evolución tectónica paleogeográfica de la cuenca Terciaria de los Andes del sur de Chile. VII Congreso Geológico Chileno, Concepción, Actas 1, pp 6–12

Clift P, Vannucchi P (2004) Controls on tectonic accretion versus erosion in subduction zones: implications for the origin and recycling of the continental crust. Rev Geophys 42: doi 10.1029/2003RG000127

Cobbold PR, Diraison M, Rossello EA (1999) Bitumen veins and Eocene transpression, Neuquén Basin, Argentina. Tectonophysics 314:423–442

Collao S, Glodny J, Bascuñán S, Esparza E, Pérez S, Aguilar G (2003) Microtermometría y Cronología en cuarzo de estructuras post-Carbonífero en el basamento metamórfico de Chile Centro-Sur. X Congreso Geologico Chileno, Concepción, Abstract Vol. 11

Collot J-Y, Delteil J, Lewis KB, Davy B, Lamarche G, Audru J-C, Barnes P, Chanier F, Chaumillon E, Lallemand S, De Lepinay BM, Orpin A, Pelletier B, Sosson M, Toussaint B, Uruski C (1996) From oblique subduction to intra-continental transpression: structures of the southern Kermadec-Hikurangi margin from multibeam bathymetry, side-scan sonar and seismic reflection. Marine Geophys Res 18:357–381

Dalla Salda L, Franzese J (1987) Las megaestructuras del Macizo y la Cordillera Norpatagónica argentina y la génesis de las cuencas volcano-sedimentarias terciarias. Rev Geol Chile 13:3–14

Dalla Salda LH, Varela R, Cingolani C, Aragón E (1994) The Rio Chico Paleozoic crystalline complex and the evolution of Northern Patagonia. J S Am Earth Sci 7(3-4):377–386

Dalziel IWD, Storey BC, Garrett SW, Grunow AM, Herrod LDB, Pankhurst RJ (1987) Extensional tectonics and the fragmentation of Gondwanaland. In: Dewey JF, Coward MP, Hancock P (eds) Continental Extensional Tectonics. Geol Soc London Spec Pub 28, pp 433–441

Defant MJ, Drummond M (1990) Derivation of some modern arc magmas by melting of young subducted lithosphere. Nature 347:662–665

De Ignacio C, López I, Oyarzun R, Márquez A (2001) The northern Patagonia Somuncura plateau basalts: a product of slab–induced, shallow asthenospheric upwelling? Terra Nova 13:117–121

De Paolo DJ, Wasserburg GJ (1979) Petrogenetic mixing models and Nd-Sr isotopic patterns. Geochim Cosmochim Acta 43(4):615–627

Diraison M, Cobbold PR, Rossello EA, Amos AJ (1998) Neogene dextral transpression due to oblique convergence across the Andes of northwestern Patagonia, Argentina. J S Am Earth Sci 11(6):519–532

Drummond MS, Defant MJ (1990) A model for trondhjemite-tonalite-dacite genesis and crustal growth via slab melting: Archean to modern comparisons. J Geophys Res 95:21503–21521

Ducea MN, Saaleby JB (1998) The age and origin of a thick mafic-ultramafic keel from beneath the Sierra Nevada batholith. Contributions to Mineralogy and Petrology 133:169–185

Duhart P, McDonough M, Muñoz J, Martin M, Villeneuve M (2001) El Complejo Metamórfico Bahía Mansa en la cordillera de la Costa del centro-sur de Chile (39°30'-42°00'S): geocronología K-Ar, $^{40}Ar/^{39}Ar$ y U-Pb e implicancias en la evolución del margen sur-occidental de Gondwana. Rev Geol Chile 28(2):179–208

Emparán C, Suárez M, Muñoz J (1992) Hoja Curacautín, Regiones de la Araucanía y del Biobio. Mapa, escala 1:250.000. Carta Geológica de Chile, No. 71, Sernageomin, Santiago de Chile

Eppinger KJ, Rosenfeld U (1996) Western margin and provenance of sediments of the Neuquén Basin (Argentina) in the Late Jurassic and Early Cretaceous. Tectonophysics 259:229–244

Ernst WG (1975) Systematics of large-scale tectonics and age progressions in Alpine Circum-Pacific blueschist belts. Tectonophysics 26:229–246

Féraud G, Alric V, Fornari M, Bertrand H, Haller M (1999) $^{40}Ar/^{39}Ar$ dating of the Jurassic volcanic province of Patagonia: migrating magmatism related to Gondwana break-up and subduction. EPSL 172(1–2):83–96

Folguera A, Ramos VA, Melnick D (2002) Partición de la deformación en la zona del arco volcánico de los Andes neuquinos (36–39°S) en los últimos 30 millones de años. Rev Geol Chile 29(2):151–165

Forsythe R, Nelson E (1985) Geological manifestations of ridge collision: evidence from the Golfo de Peñas-Taitao basin, southern Chile. Tectonics 4:477–495

Franz G, Lucassen F, Kramer W, Trumbull RB, Romer RL, Wilke H-G, Viramonte JG, Becchio R, Siebel W (2006) Crustal evolution at the central Andean continental margin: a geochemical record of crustal growth, recycling and destruction. In: Oncken O, Chong G, Franz G, Giese P, Götze H-J, Ramos VA, Strecker MR, Wigger P (eds) The Andes – active subduction orogeny. Frontiers in Earth Science Series, Vol 1. Springer-Verlag, Berlin Heidelberg New York, pp 45–64, this volume

Franzese JR, Spalletti LA, (2001) Late Triassic-early Jurassic continental extension in southwestern Gondwana: tectonic segmentation and pre-break-up rifting. J S Am Earth Sci 14:257–270

Franzese J, Spalletti L, Gomez Perez I, Macdonald D (2003) Tectonic and paleoenvironmental evolution of Mesozoic sedimentary basins along the Andean foothills of Argentina (32°S–54°S). J S Am Earth Sci 16:81–90

Giacosa RE, Heredia N (2004) Structure of the North Patagonian thick-skinned fold-and-thrust belt, southern central Andes, Argentina (41–42°S). J S Am Earth Sci 18:61–72

Glodny J, Bingen B, Austrheim H, Molina JF, Rusin A (2002) Precise eclogitization ages deduced from Rb/Sr mineral systematics: The Maksyutov complex, Southern Urals, Russia. Geochim Cosmochim Acta 66:1221–1235

Glodny J, Lohrmann J, Echtler H, Gräfe K, Seifert W, Collao S, Figueroa O (2005) Internal dynamics of a paleoaccretionary wedge: insights from combined isotope tectonochronology and sandbox modeling of the South–Central Chilean forearc. Earth Planet Sci Lett 231:23–39

Glodny J, Gräfe K, Echtler H (in press) Mesozoic to Quaternary continental margin dynamics in South Central Chile (36°–42°S): the apatite and zircon fission track perspective. Int J Earth Sci

Glodny J, Echtler H, Collao S, Ardiles M, Burón P, Figueroa O (submitted) Differential Late Paleozoic active margin evolution in South-Central Chile (37°S–40°S) – the Lanalhue Fault Zone. J S Am Earth Sci

Godoy E, Kato T (1990) Late Paleozoic serpentinites and mafic schists from the Coast Range accretionary complex, central Chile: their relation to aeromagnetic anomalies. Geol Rundsch 79:121–130

González E (1989) Hydrocarbon resources in the Coastal Zone of Chile. In: Ericksen GE, Cañas Pinochet MT, Reinemund JA (eds) Geology of the Andes and its relation to hydrocarbon and mineral resources. Circum–Pacific Council for Energy and Mineral Resources, Earth Science Series Vol. 11, Houston TX, pp 383–404

González Bonorino F, Aguirre L (1970) Metamorphic facies series of the crystalline basement of Chile. Geol Rundsch 59:979–994

Gordon A, Ort MH (1993) Edad y correlación del plutonismo subcordillerano en las provincias de Rio Negro y Chubut (41°–42°30' L.S). XII Congreso Geológico Argentino y II Congreso de Exploración de Hidrocarburos, Mendoza, Actas 4:120–127

Grocott J, Brown M, Dallmeyer RD, Taylor GK, Treloar PJ (1994) Mechanisms of continental growth in extensional arcs: An example from the Andean plate-boundary zone. Geology 22:391–394

Grove M, Jacobson CE, Barth AP, Vucic A (2003) Temporal and spatial trends of Late Cretaceous-early Tertiary underplating of Pelona and related schist beneath southern California and southwestern Arizona. In: Johnson SE, Paterson SR, Fletcher JM, Girty G, Kimbrough DL (eds) Tectonic evolution of northwestern Mexico and the southwestern USA. Geol Soc Am Spec P 374:381–406

Guivel C, Lagabrielle Y, Bourgois J, Maury RC, Fourcade S, Martin H, Arnaud N (1999) New geochemical constraints for the origin of ridge-subduction-related plutonic and volcanic suites from the Chile Triple Junction (Taitao Peninsula and Site 862, LEG ODP141 on the Taitao Ridge). Tectonophysics 311:83–111

Haller MJ, Linares E, Ostera HA (1999) Petrology and geochronology of the subcordilleran plutonic belt of Patagonia. In: South American Symposium on Isotope Geology, No. 2, Actas, Servicio Geológico Minero Argentino, Buenos Aires, Anales 34, pp 210–214

Hartley AJ, Chong G, Turner P, May G, Kape SJ, Jolley EJ (2000) Development of a continental forearc: a Neogene example from the Central Andes, northern Chile. Geology 28:331–334

Hervé M (1976) Estudio geológico de la falla Liquiñe-Reloncaví en la área de Liquiñe: antecendentes de un movimiento transcurrente (Provincia de Valdivia). I Congreso Geológico Chileno, Actas 1(B):39–56

Hervé F (1977) Petrology of the crystalline basement of the Nahuelbuta mountains, southcentral Chile. In: Ishikawa T, Aguirre L (eds) Comparative studies on the geology of the Circum-Pacific Orogenic Belt in Japan and Chile. Japan Soc Prom Sci Tokyo, pp 1–51

Hervé F (1988) Late Paleozoic subduction and accretion in Southern Chile. Episodes 11(3):183–188

Hervé F, Munizaga F (1978) Evidencias de un magmatismo intrusivo Triásico Superior-Jurásico Inferior en la Cordillera de la Costra entre los 35°30' y 36°30'S. Congreso Geológico Argentino, Actas 2:43–52

Hervé F, Thiele R, Parada MA (1976) Observaciones geológicas en el Triásico de Chile central entre las latitudes 35°30' y 40°00' sur. I Congreso Geologico Chileno Santiago, Actas 1A:297–313

Hervé F, Godoy E, Parada MA, Ramos V, Rapela C, Mpodozis C, Davidson J (1987) A general view of the Chilean-Argentinian Andes, with emphasis on their early history. In: Monger JWH, Francheteau J (eds) Circum-Pacific orogenic belts and evolution of the Pacific Ocean basin. AGU, Geodyn Ser 18, pp 97–114

Hervé F, Munizaga F, Parada MA, Brook M, Pankhurst RJ, Snelling NJ, Drake R (1988) Granitoids of the Coast Range of central Chile: Geochronology and geologic setting. J S Am Earth Sci 1:185–194

Hervé F, Pankhurst RJ, Brook M, Alfaro G, Frutos J, Miller H, Schira W, Amstutz GC (1990) Rb-Sr and Sm-Nd data from some massive sulfide occurrences in the metamorphic basement of South-Central Chile. In: Fontboté L, Amstutz GC, Cardozo M, Cedillo E, Frutos J (eds) Stratabound ore deposits in the Andes. Springer-Verlag, Berlin Heidelberg New York

Hervé F, Pankhurst RJ, Drake R, Beck ME Jr, Mpodozis C (1993) Granite generation and rapid unroofing related to strike–slip faulting, Aysén, Chile. Earth Planet Sci Lett 120:375–386

Hervé F, Fanning CM, Pankhurst RJ (2003) Detrital zircon age patterns and provenance of the metamorphic complexes of southern Chile. J S Am Earth Sci 16:107–123

Hickey Vargas R, Moreno H, López-Escobar L, Frey F (1989) Geochemical variations in Andean basaltic and silicic lavas from the Villarrica-Lanín volcanic chain (39.5°S): an evaluation of source heterogeneity, fractional crystallization and crustal assimilation. Contrib Min Petrol 103:361–386

Hickey Vargas R, Sun M, López-Escobar L, Moreno H, Reagan MK, Morris JD, Ryan JG (2002) Multiple subduction components in the mantle wedge: evidence from eruptive centers in the Central Southern volcanic zone, Chile. Geology 30:199–202

Hildreth W, Moorbath S (1988) Crustal contributions to arc magmatism in the Andes of Central Chile. Contrib Min Petrol 98:455–489

Hofmann AW, White WM (1980) The role of subducted oceanic crust in mantle evolution. Carnegie Institution, Washington, Yearbook 79:477–483

Holbrook WS, Lizarralde D, McGeary S, Bangs N, Diebold J (1999) Structure and composition of the Aleutian island arc and implications for continental crustal growth. Geology 27:31–34

Hu M, Stallard RF, Edmond JM (1982) Major ion chemistry of some large Chinese rivers. Nature 298:550–553

Hurley PM, Rand JR (1969) Pre-drift continental nuclei. Science 164: 1229–1242

Hutton DWH, Dempster TJ, Brown PE, Becker SM (1990) A new mechanism of granite emplacement: intrusion in active extensional shear zones. Nature 343:452–455

Introcaso A, Pacino MC, Fraga H (1992) Gravity, isostasy, and Andean crustal shortening between 30 and 35°S. Tectonophysics 205:31–48

Jacobson CE, Oyarzabal FR, Haxel GB (1996) Subduction and exhumation of the Pelona-Orocopia-Rand schists, southern California. Geology 24:547–550

Jicha BR, Scholl DW, Singer BS, Yogodzinski GM, Kay SM (in press) Revised age of Aleutian Island Arc formation implies high rate of magma production. Geology

Jordan TE, Burns WM, Veiga R, Pángaro F, Copeland P, Kelley S, Mpodozis C (2001) Extension and basin formation in the southern Andes caused by increased convergence rate: a mid-Cenozoic trigger for the Andes. Tectonics 20:308–324

Karig DE, Sharman GF (1975) Subduction and accretion in trenches. Earth Planet Sci Lett 21:209–212

Kato TT (1976) The relationship between low-grade metamorphism and tectonics in the Coast Range of Central Chile. PhD thesis, University of California, Los Angeles, pp 1–238

Kato TT (1985) Pre-Andean orogenesis in the Coast Ranges of central Chile. Geol Soc Am Bull 96:918–924

Kay SM, Ramos V, Mpodozis C, Sruoga P (1989) Late Paleozoic to Jurassic silicic magmatism at the Gondwana margin: analogy to the Middle Proterozoic in North America? Geology 17: 324–328

Kay SM, Gorring M, Ramos, VA (2004) Magmatic sources, setting and causes of Eocene to Recent Patagonian plateau magmatism (36°S to 52°S latitude). Rev Asoc Geol Argentina 59(4):556–568

Kay SM, Godoy E, Kurtz A (2005) Episodic arc migration, crustal thickening, subduction erosion, and magmatism in the south-central Andes. Geol Soc Am Bull 117:67–88

Kemnitz H, Kramer W, Rosenau M (2005) Jurassic to Tertiary tectonic, volcanic, and sedimentary evolution of the Southern Andean intra-arc zone, Chile (38°S–39°S) a survey. N Jb Geol Paläont 236:19–42

Kramer W, Siebel W, Romer RL, Haase G, Zimmer M, Ehrlichmann R (2005) Geochemical and isotopic characteristics and evolution of the Jurassic volcanic arc between Arica (18°30'S) and Tocopilla (22°S), North Chilean Coastal Cordillera. Chem Erde 65:47–78

Krawczyk CM, SPOC Team (2003) Amphibious seismic survey images plate interface at 1960 Chile earthquake. EOS 84 (32):301, 304–305

Lara L, Moreno H (1998) Geología preliminar area de Liquiñe-Neltume, Region de Los Lagos. Mapa 13, escala 1:100.000. In: Estudio Geológico-Economico de la X Region Norte, Informe Registrado IR-98-15, Santiago de Chile

Lara L, Rodríguez C, Moreno H, Pérez de Arce C (2001) Geocronología K-Ar y geoquímica del volcanismo plioceno superior-pleistoceno de los Andes del sur (39–42°S). Rev Geol Chile 28(1): 67–90

Latorre CO, Vattuone ME, Linares E, Leal PR (2001) K-Ar ages of rocks from Lago Aluminé, Rucachoroi and Quillen, North Patagonian Andes, Neuquén, Republica Argentina. III South American Symposium on Isotope Geology, Abstract Vol, pp 577–580

Laursen J, Scholl DW, von Huene R (2002) Neotectonic deformation of the central Chile margin: deepwater forearc basin formation in response to hot spot ridge and seamount subduction. Tectonics 21(5): doi 10.1029/2001TC901023

Lavenu A, Cembrano J (1999) Compressional- and transpressional-stress pattern for Pliocene and Quaternary brittle deformation in fore arc and intra-arc zones (Andes of Central and Southern Chile). J Struct Geol 21(12):1669–1691

Legarreta L, Gulisano CA (1989) Análisis estratigráfico secuencial de la Cuenca Neuquina (Triásico superior-Terciario inferior, Argentina). In: Chebli G, Spalletti L (eds) Cuencas Sedimentarias Argentinas. Universidad Nacional de Tucumán, Serie Correlación Geológica 6:221–243

Legarreta L, Uliana MA (1991) Jurassic-Cretaceous marine oscillations and geometry of back arc basin fill, Central Argentine Andes. In: Macdonald DIM (ed) Sea level changes at active plate margins: process and product. Int Assoc Sedimentol Spec Pub 12: 429–450

Legarreta L, Uliana MA (1996) The Jurassic succession in west-central Argentina: stratal patterns, sequences and paleogeographic evolution. Palaeogeog Palaeoclimat Palaeoecol 120:303–330

Le Roux JP, Elgueta S (2000) Sedimentologic development of a Late Oligocene-Miocene forearc embayment, Valdivia Basin Complex, southern Chile. Sediment Geol 130:27–44

Lewis KB, Collot J-Y, Lallemand SE (1998) The dammed Hikurangi Trough: a channel-fed trench blocked by subducting seamounts and their wake avalanches (New Zealand-France GeodyNZ Project). Basin Res 10(4):441–468

Linares E, Cagnoni MC, Do Campo M, Ostera HA (1988) Geochronology of metamorphic and eruptive rocks of southeastern Neuquén and northwestern Río Negro Provinces, Argentine Republic. J S Am Earth Sci 1(1):53–61

Lliboutry L (1999) Glaciers of the Wet Andes. In: Williams, RS, Ferrigno JG (eds) Satellite image atlas of glaciers of the world: South America. US Geol Survey Prof P 1386-I, http://pubs.usgs.gov/prof/p1386i/index.html

Lohrmann J (2002) Identification of parameters controlling the accretive and tectonically erosive mass-transfer mode at the South-Central and North Chilean forearc using scaled 2D sandbox experiments. GeoForschungsZentrum Potsdam Scientific Technical Report STR02/10, http://www.gfz-potsdam.de/bib/zbstr.htm

Lonsdale P (2005) Creation of the Cocos and Nazca plates by fission of the Farallon plate. Tectonophysics 404:237–264

López de Luchi MG, Ostera H, Cerredo ME, Cagnoni MC, Linares E (2000) Permian magmatism in Sierra de Mamil Choique, North Patagonian Massif, Argentina. IX Congreso Geológico Chileno, Actas 2(4):750–754

López-Escobar L, Cembrano J, Moreno H (1995) Geochemistry and tectonics of the Chilean Southern Andes basaltic Quaternatry volcanism (37°–46°S). Rev Geol Chile 22(2):219–234

López-Escobar L, Vergara M (1997) Eocene-Miocene longitudinal depression and Quaternary volcanism in the Southern Andes, Chile (33–42.5°S). Rev Geol Chile 24:227–244

López Gamundí O, Rossello EA (1993) Devonian-Carboniferous unconformity in Argentina and its relation to the Eo-Hercynian orogeny in southern South America. Geol Rundsch 82:136–147

López Gamundí O, Limarino CO, Cesari SN (1992) Late Paleozoic paleoclimatology of central west Argentina. Palaeogeog Palaeoclimat Palaeoecol 91:305–329

Lucassen F, Becchio R, Harmon R, Kasemann S, Franz G, Trumbull R, Wilke H-G, Romer RL, Dulski P (2001) Composition and density model of the continental crust at an active continental margin – the Central Andes between 21° and 27°S. Tectonophysics 341: 195–223

Lucassen F, Trumbull R, Franz G, Creixell C, Vásquez P, Romer RL, Figueroa O (2004) Distinguishing crustal recycling and juvenile additions at active continental margins: the Paleozoic to recent compositional evolution of the Chilean Pacific margin (36°–41°S). J S Am Earth Sci 17:103–119

Lucassen F, Franz G, Wiedicke M (2005) Complete recycling of the magmatic arc in Chile (36°–39°S) – evidence from chemical and isotopic composition of (sub)recent trench sediments. Int Symposium on Andean Geodynamics, Barcelona, Abstract Volume

Lüth S, Wigger P, ISSA Research Group (2003) A crustal model along 39°S from a seismic refraction profile: ISSA (2000). Rev Geol Chile 30(1):83–101

Martin MW, Kato TT, Rodriguez C, Godoy E, Duhart P, McDonough M, Campos A (1999) Evolution of the late Paleozoic accretionary complex and overlying forearc-magmatic arc, south central Chile (38°–41°S): Constraints for the tectonic setting along the southwestern margin of Gondwana. Tectonics 18(4):582–605

McMillan NJ, Harmon RS, Moorbath S, López-Escobar L, Strong DF (1989) Crustal sources involved in continental arc magmatism: A case study of volcano Mocho-Choshuenco, southern Chile. Geology 17:1152–1156

McNulty B, Farber D, Wallace G, López R, Palacios O (1998) Role of plate kinematics and plate slip vector partitioning in continental magmatic arcs: evidence from the Cordillera Blanca, Peru. Geology 26(9):827–830

Meissner R, Mooney W (1998) Weakness of the continental lower crust: a condition for delamination, uplift, and escape. Tectonophysics 296:47–60

Melnick D, Echtler HP (2006) Morphotectonic and geologic digital map compilations of the South-Central Andes (36°–42° S). In: Oncken O, Chong G, Franz G, Giese P, Götze H-J, Ramos VA, Strecker MR, Wigger P (eds) The Andes – active subduction orogeny. Frontiers in Earth Science Series, Vol 1. Springer-Verlag, Berlin Heidelberg New York, pp 565–568, this volume

Melnick D, Rosenau M, Folguera A, Echtler HP (2006) Neogene tectonic evolution of the Neuquén Andes western flank (37–39°S). In: Kay SM, Ramos VA (eds) Evolution of an Andean margin: a tectonic and magmatic view from the Andes to the Neuquén Basin (35°–39°S lat). Geol Soc Am Spec P 407:73–95

Mercer JH, Sutter JF (1982) Late Miocene-earliest Pliocene glaciation in southern Argentina: implications for global ice-sheet history. Paleogeog Paleoclimat Paleoecology 38:185–206

Meschede M, Zweigel P, Kiefer E (1999) Subsidence and extension at a convergent plate margin: evidence for subduction erosion off Costa Rica. Terra Nova 11:112–117

Mordojovich C (1981) Sedimentary basins of Chilean Pacific offshore. In: Halbouty MT (eds) Energy resources of the Pacific region. AAPG Stud Geol 12, pp 63–82

Mpodozis C, Ramos V, (1989) The Andes of Chile and Argentina. In: Ericksen GE, Cañas Pinochet MT, Reinemund JA (eds) Geology of the Andes and its Relation to Hydrocarbon and Mineral Resources, Circum-Pacific Council for Energy and Mineral Resources Earth Science Series, Vol. 11, Houston, Texas, pp 59–90

Munizaga F, Hervé F, Drake R, Pankhurst RJ, Brook M, Snelling N (1988) Geochronology of the Lake Region of south–central Chile (39°–42°S): Preliminary results. J S Am Earth Sci 1:09–316 DOI 10.1016/0895–9811(88)90009–0

Muñoz J, Troncoso R, Duhart P, Crignola P, Farmer L, Stern CR (2000) The relation of the mid–Tertiary coastal magmatic belt in south-central Chile to the late Oligocene increase in plate convergence rate. Rev Geol Chile 27:177 203

Nelson ST, Davidson JP, Heizler MT, Kowallis BJ (1999) Tertiary tectonic history of the southern Andes: the subvolcanic sequence to the Tatara-San Pedro volcanic complex, lat 36°S. Geol Soc Am Bull 111:1387–1404

Nielsen SN (in press) The Triassic Santa Juana Formation at the lower Biobío River, south central Chile. J S Am Earth Sci

Pankhurst RJ (1990) The Paleozoic and Andean magmatic arcs of West Antarctica and southern South America. In: Kay SM, Rapela CW (eds) Plutonism from Antarctica to Alaska. Geol Soc Am Spec P 241:1–7

Pankhurst RJ, Rapela CW, Caminos R, Llambias E, Parica C (1992a) A revised age for the granites of the central Somuncura Batholith, North Patagonian Massif. J S Am Earth Sci 5:321–325

Pankhurst RJ, Hervé F, Rojas L, Cembrano J (1992b) Magmatism and tectonics in continental Chiloé, Chile (42°–42°30'S). Tectonophysics 205:283–294

Pankhurst RJ, Weaver SD, Hervé F, Larrondo P (1999) Mesozoic–Cenozoic evolution of the North Patagonian Batholith in Aysén, southern Chile. J Geol Soc, London, 156:673–694

Pankhurst RJ, Rapela CW, Loske WP, Márquez M, Fanning CM (2003) Chronological study of the pre-Permian basement rocks of southern Patagonia. J S Am Earth Sci 16:27–44

Parada MA (1975) Estudio geológico de los alrededores de los lagos Calafquén, Panguipulli y Riñihue, Provincia de Valdivia. Memoria de título, Departamento de Geologia, Universidad de Chile, Santiago

Parada MA (1990) Granitoid plutonism in central Chile and its geodynamic implications: a review. Geol Soc Am Spec P 241:51–66

Potent S (2003) Kinematik und Dynamik neogener Deformationsprozesse des südzentralchilenischen Subduktionssystems, nördlichste Patagonische Anden (37°–40°S). PhD thesis, Universität Hamburg

Rabassa J, Clapperton CM (1990) Quaternary glaciations of the Southern Andes. Quat Sci Rev 9:153–174

Ramos VA (1982) Las ingresiones pacificas del Terciario en el norte de la Patagonia. III Congreso Geológico Chileno, Actas 1: 262–288

Ramos V (2000) The Southern Central Andes. In: Cordani UG, Milani EJ, Thomaz Filho A, Campos DA (eds) Tectonic evolution of South America, Rio de Janeiro, pp 561–604

Ramos VA, Aleman A (2000) Tectonic evolution of the Andes. In: Cordani UG, Milani EJ, Thomaz Filho A, Campos DA (eds) Tectonic evolution of South America. Rio de Janeiro, pp 635–685

Ramos V, Kay SM (1991) Triassic rift basalts of the Cuyo basin, central Argentina. In: Harmon RS, Rapela CW (eds) Andean magmatism and its tectonic setting. Geol Soc Am Spec Pub 265:79–91

Rapalini AE (1998) Syntectonic magnetization of the Mid–Paleozoic Sierra Grande Formation: further constraints on the tectonic evolution of Patagonia. J Geol Soc 155:105–114

Rapela CW, Kay SM (1988) Late Paleozoic to Recent magmatic evolution of northern Patagonia. Episodes 11:175–82

Rapela CW, Pankhurst RJ (1992) The granites of northern Patagonia and the Gastre Fault System in relation to the break-up of Gondwana. In: Storey BC, Alabaster T, Pankhurst RJ (eds) Magmatism and the causes of continental break-up. Geol Soc Spec Pub 68:209–220

Rapela CW, Pankhurst RJ (1996) Monzonite suites: the innermost Cordilleran plutonism in Patagonia. Trans Royal Soc Edinburgh Earth Sci 87:193–203

Rapela CW, Spalletti LA, Merodio JC, Aragón E (1988) Temporal evolution and spatial variation of early Tertiary volcanism in the Patagonian Andes (40°S–42°30'S). J S Am Earth Sci 1:75–88

Rapela CW, Pankhurst RJ, Harrison SM (1992) Triassic 'Gondwana' granites of the Gastre district, North Patagonian Massif. Trans Royal Soc Edinburgh Earth Sci 83:291–304

Rapela CW, Pankhurst RJ, Fanning CM, Hervé F (2005) Pacific subduction coeval with the Karoo mantle plume: the Early Jurassic Subcordilleran Belt of northwestern Patagonia. Geol Soc Spec Pub 246:217–239

Reymer A, Schubert G (1984) Phanerozoic addition rates to the continental crust and crustal growth. Tectonics 3:63–77

Riley TR, Leat PT, Pankhurst RJ, Harris C (2001) Origins of large volume rhyolitic volcanism in the Antarctic Peninsula and Patagonia by crustal melting. J Petrol 42:1043–1065

Rolando AP, Hartmann LA, Santos JOS, Fernandez RR, Etcheverry RO, Schalamuk IA, McNaughton NJ (2002) SHRIMP zircon U-Pb evidence for extended Mesozoic magmatism in the Patagonian Batholith and assimilation of Archean crustal components. J S Am Earth Sci 15:267–283

Rosenau MR (2004) Tectonics of the Southern Andean Intra-arc Zone (38°–42°S). PhD thesis, Freie Universität Berlin, http://www.diss.fu-berlin.de/2004/280/index.html

Rosenau M, Melnick D, Echtler H (2006) Kinematic constraints on intra-arc shear and strain partitioning in the Southern Andes between 38° S and 42° S latitude. Tectonics 25(4), TC4013

Sarin MM, Krishnaswamy S, Dilli K, Somayajulu BLK, Moore WS (1989) Major ion chemistry of Ganga-Brahmaputra river system: weathering processes and fluxes of the Bay of Bengal. Geochim Cosmochim Acta 53:997–1009

Sato AM, Tickyj H, Llambias EJ, Sato E (2000) The Las Matras tonalitic-trondhjemitic pluton, central Argentina: Grenvillian-age constraints, geochemical characteristics, and regional implications. J S Am Earth Sci 13:587–610

Seifert W, Rosenau M, Echtler H (in press) Crystallization depths of granitoids of South Central Chile estimated by Al-in-hornblende geobarometry: implications for mass transfer processes along the active continental margin. N Jb Geol Paläont

SERNAGEOMIN (2003) Mapa Geológico de Chile: versión digital, No. 4 (CD-ROM, versión 1.0, 2003). Servicio Nacional de Geología y Minería, Santiago de Chile

Sigmarsson O, Chmeleff J, Morris J, López-Escobar L (2002) Origin of $^{226}$Ra-$^{230}$Th disequilibria in arc lavas from southern Chile and implications for magma transfer time. Earth Planet Sci Lett 196: 189–196

Smith WHF, Sandwell DT (1997) Global seafloor topography from satellite altimetry and ship depth soundings. Science 277:1956–1962

Sobolev AV, Hofmann AW, Sobolev SV, Nikogosian IK (2005) An olivine-free mantle source of Hawaiian shield basalts. Nature 434: 590–597

Somoza R (1998) Updated Nazca (Farallon)-South America relative motions during the last 40 My: implications for the mountain building in the central Andean region. J S Am Earth Sci 11: 211–215

Stern CR (1989) Pliocene to Present migration of the volcanic front, Andean Southern Volcanic Zone. Rev Geol Chile 16(2):145–162

Stern CR (2004) Active Andean volcanism: its geologic and tectonic setting. Rev Geol Chile 31(2):161–206

Stinnesbeck W (1986) Zu den faunistischen und palökologischen Verhältnissen in der Quiriquina Formation (Maastrichtium) Zentral-Chiles. Palaeontographica A194(4–6):99–237

Suárez M, Bell CM (1992) Triassic rift related sedimentary basins in northern Chile (24°–29°S). J S Am Earth Sci 6:109–121

Suárez M, Emparán C (1995) The stratigraphy, geochronology and paleophysiography of a Miocene fresh-water interarc basin, southern Chile. J S Am Earth Sci 8:17–31

Suárez M, Emparán C (1997) Hoja Curacautín, regiones de la Araucanía y del BioBío. Carta Geológica de Chile, No. 71. Servicio Nacional de Geología y Minería, Santiago de Chile

Tarney J, Jones CE (1994) Trace element geochemisty of orogenic igneous rocks and crustal growth models. J Geol Soc 151: 855–868

Tassara A, Yáñez G (2003) Relación entre el espesor elástico de la litosfera y la segmentación tectónica del margen andino (15–47°S). Rev Geol Chile 30(2):159–186

Taylor SR (1967) The origin and growth of continents. Tectonophysics 4:17–34

Taylor SR, McLennan SM (1985) The continental crust: its composition and evolution. Blackwell, Oxford

Tebbens SF, Cande S (1997) Southeast Pacific tectonic evolution from Early Oligocene to Present. J Geophys Res 102:12035–12059

Thomson SN, Hervé F (2002) New time constraints for the age of metamorphism at the ancestral Pacific Gondwana margin of southern Chile (42–52°S). Rev Geol Chile 29(2):151–165

Thornburg TM, Kulm LD, Hussong DM (1990) Submarine-fan development in the southern Chile Trench: a dynamic interplay of tectonics and sedimentation. Geol Soc Am Bull 102(12): 1658–1680

Torsvik TH, Cocks LRM (2004) Earth geography from 400 to 250 Ma: a paleomagnetic, faunal and facies review. J Geol Soc, London, 161:555–572

Uliana MA, Legarreta L (1993) Hydrocarbon habitat in a Triassic to Cretaceous sub-Andean setting: Neuquén basin, Argentina. J Petrol Geol 16(4):397–420

Varela R, Teixeira W, Cingolani C, Dalla Salda L (1994) Edad rubidio-estroncio de granitoids de Aluminé-Rahue, Cordillera Norpatagonica, Neuquén, Argentina. VII Congreso Geológico Chileno, Concepción, Actas II, pp 1254–1258

Vergani GD, Tankard AJ, Belotti HJ, Welsink HJ (1995) Tectonic evolution and paleogeography of the Neuquén Basin, Argentina. In: Tankard AJ, Suárez R, Welsink HJ (eds) Petroleum basins of South America. Am Assoc Petr Geol Mem 62, pp 383–402

Völker D, Wiedicke M, Ladage S, Gaedicke C, Reichert C, Rauch K, Kramer W, Heubeck C (2006) Latitudinal variation in sedimentary processes in the Peru-Chile trench off Central Chile. In: Oncken O, Chong G, Franz G, Giese P, Götze H-J, Ramos VA, Strecker MR, Wigger P (eds) The Andes – active subduction orogeny. Frontiers in Earth Science Series, Vol 1. Springer-Verlag, Berlin Heidelberg New York, pp 193–216, this volume

Von Gosen W, Loske W (2004) Tectonic history of the Calcatapul Formation, Chubut province, Argentina, and the "Gastre fault system". J S Am Earth Sci 18:73–88

von Huene R, Lallemand S (1990) Tectonic erosion along the Japan and Peru convergent margins. Geol Soc Am Bull 102(6):704–720

von Huene R, Scholl DW (1991) Observations at convergent margins concerning sediment subduction, subduction erosion, and the growth of continental crust. Rev Geophys 29(3):279–316

von Huene R, Corvalán J, Flueh ER, Hinz K, Korstgard J, Ranero CR, Weinrebe W (1997) Tectonic control of the subducting Juan Fernández Ridge on the Andean margin near Valparaiso, Chile. Tectonics 16(3):474–488

Wells RE, Blakely RJ, Sugiyama Y, Scholl DW, Dinterman PA (2003) Basin-centered asperities in great subduction zone earthquakes – a link between slip, subsidence and subduction erosion? J Geophys Res 108(B10): doi 10.1029/2002JB002072

Willner A, Hervé F, Massonne H-J (2000) Mineral chemistry and pressure-temperature evolution of two contrasting high-pressure-low-temperature belts in the Chonos archipelago, Southern Chile. J Petrol 41:309–330

Willner AP, Pawlig S, Massonne HJ, Hervé F (2001) Metamorphic evolution of spessartine quartzites (Coticules) in the high-pressure, low temperature complex at Bahia Mansa, coastal Cordillera of South-Central Chile. Can Mineral 39:1547–1569

Willner AP, Glodny J, Gerya TV, Godoy E, Massonne H-J (2004) A counterclockwise PTt-path of high pressure-low temperature rocks from the Coastal Cordillera accretionary complex of South Central Chile: constraints for the earliest stage of subduction mass flow. Lithos 75:283–310

Willner AP, Thomson SN, Kröner A, Wartho J-A, Wijbrans JR, Hervé F (2005) Time markers for the evolution and exhumation history of a Late Paleozoic paired metamorphic belt in North-Central Chile (34°–35°30'S). J Petrol doi 10.1093/petrology/egi036

Wynn RB, Weaver PPE, Masson DG, Stow DAV (2002) Turbidite depositional architecture across three interconnected deep-water basins on the north-west African margin. Sedimentology 49(4):669–695

Yáñez GA, Ranero CR, von Huene R, Díaz J (2001) Magnetic anomaly interpretation across the southern central Andes (32°–34°S): The role of the Juan Fernández Ridge in the Late Tertiary evolution of the margin. J Geophys Res 106(B4):6325–6345

# Appendix

## Sample Characterization

### Triassic Granite, Hualpén

*TET1* (36° 46.343' S, 73° 12.659' W; Peninsula Hualpén, 12 km WNW' Concepción) Granitic intrusion, with miarolitic pegmatites. Supracrustal intrusion level evident from local porphyric textures and presence of miarolitic cavities. Miarolitic pegmatite assemblage: K-feldspar, quartz, muscovite, tourmaline.

## Tectonochronology, Quidico/Tirúa

### Main penetrative deformation

*VAL28* (38° 14.337' S, 73° 29.318' W, Playa Quidico). Quartz-rich micaschist, with quartz mobilisates. Stretching lineation 130°. Local presence of small extensional shear bands, formed at ductile conditions. Assemblage: quartz, plagioclase, white mica, chlorite, garnet, biotite (?), ilmenite, tourmaline, zircon, graphite.

*VAL29* (38° 14.337' S, 73° 29.318' W, Playa Quidico). Garnet micaschist, same location as VAL28. Assemblage: quartz, plagioclase, garnet, white mica, biotite (?), chlorite, titanite (rare), tourmaline.

### Tension Gash

*VAL34* (38° 20.250' S, 73° 30.211' W, Cabo Tirúa). System of tension gashes. Partly open cavities, partly filled with quartz-feldspar-white mica mobilisates. Direction of strike: 130°. Assemblage of mobilisate: quartz (predominates), albite, white mica.

## Tectonochronology, Bahía Mansa

### Main Penetrative Deformation

*VAL42* (40° 35.016' S, 73° 44.341' W, harbour of Bahía Mansa). Quartz-rich micaschist, slightly weathered. Assemblage: quartz, white mica, albite, titanite, chlorite, pyrite, garnet (rare) + colorless, paramagnetic phase with density $> 3.3$ g cm$^{-3}$.

*VAL44* (40° 37.024' S, 73° 45.222' W, Playa TrilTril/Bahía Mansa). Micaschist, with 130° stretching lineations. Assemblage: quartz, white mica, albite, chlorite, tourmaline, garnet, titanite + colorless, paramagnetic phase with density $> 3.3$ g cm$^{-3}$.

### Tension Gash

*VAL38* (40° 32.552' S, 73° 43.034' W, Loc. Pucatrihue/Bahía Mansa). Mobilisates, formed in small fissures within micaschist, resulting from semiductile to brittle deformation. Some open cavities. Assemblage: quartz (predominates), white mica, chlorite.

# Chapter 20

# Links between Mountain Uplift, Climate, and Surface Processes in the Southern Patagonian Andes

Peter M. Blisniuk · Libby A. Stern · C. Page Chamberlain · Peter K. Zeitler · Victor A. Ramos · Edward R. Sobel
Michael Haschke · Manfred R. Strecker · Frank Warkus

**Abstract.** Miocene surface uplift of the southern Patagonian Andes, related to an episode of rapid plate convergence prior to the ~14–10 Ma collision of the Chile Ridge with the South American subduction zone, has produced one of the most pronounced orographic rain shadows on Earth. Apatite fission track ages from the western flank of this Andean segment imply that 3–4 km of denudation has occurred in this region since ~17 Ma. The track-length distribution of the studied samples suggests a complex thermal history with initial cooling followed by reheating, presumably owing to the progressive opening and eastward migration of a slab window after the ridge-trench collision, and ultimately more rapid cooling since ~4 Ma. These thermochronological data are in good agreement with constraints on the elevation history of the southern Patagonian Andes. Based on sedimentological and geochronological data from ~23 to ~14 Ma sedimentary rocks in the eastern foreland, and oxygen isotope data from pedogenic carbonate contained in these deposits, we infer that > 1 km of surface uplift of these mountains occurred between ca. 17 and 14 Ma. Carbon isotope data from the pedogenic carbonate samples demonstrate that this led to strong aridification in the eastern foreland and, presumably, strongly increased precipitation rates on the windward western side of the mountains. Because a thicker trench fill promotes weaker coupling along the plate interface, this implies that progressive surface uplift of the southern Patagonian Andes and the increasing sediment flux to the adjacent segment of the South American trench may have contributed significantly to a decrease in compressive deformation and surface uplift.

## 20.1 Introduction

Tectonically controlled surface uplift in mountain belts leads to changes in topography and relief, and can also have important effects on local- and regional-scale climatic and ecological conditions (e.g., Ruddiman et al. 1997). Correspondingly, such surface uplift can greatly affect the rates and spatial distribution of surface processes such as denudation, sediment transport, and deposition and this, in turn, can feed back on tectonic processes responsible for surface uplift (e.g., Lamb and Davis 2003; Ernst 2004; Blisniuk et al. 2005). However, our quantitative understanding of the links between these processes is limited. One of the most limiting factors is the difficulty of distinguishing between the effects of mountain uplift and climate change on surface processes (e.g., Molnar and England 1990).

Commonly, thermochronological data (e.g., $^{40}$Ar/$^{39}$Ar, fission track, or U-Th/He ages) are used to infer mountain uplift, but these methods really constrain cooling and, presumably, denudation histories which may largely reflect climatic conditions. Similarly, sediment transport and depositional histories derived from studies of syn-orogenic intramontane or foreland basin deposits might be controlled by climate rather than mountain uplift. To better understand the interaction between tectonic deformation, mountain uplift, and surface processes, it is important to also consider local and global climatic conditions during the time interval of interest, and to have independent constraints on paleo-topography.

Therefore much attention has recently been focused on the use of stable oxygen, hydrogen, and carbon isotopes in authigenic (*in-situ* formed) minerals in syn-orogenic sediments. Oxygen and hydrogen isotopes can provide reliable constraints on paleo-topography and paleo-climate, and carbon isotopes can be used as a proxy for paleo-ecologic conditions. Authigenic minerals commonly used for such studies include, for example: pedogenic (soil-formed) minerals in paleosols (fossil soils) (e.g., Cerling 1984; Amundson et al. 1989; Quade et al. 1989; Cerling and Quade 1993; Chamberlain et al. 1999; Garzione et al. 2000; Rowley et al. 2001; Poage and Chamberlain 2002; Horton et al. 2004; Blisniuk et al. 2005; Graham et al. 2005; Takeuchi and Larson 2005); chemically precipitated sediments such as lacustrine carbonate (e.g., Norris et al. 1996; Dettman et al. 2003) or freshwater chert (Horton et al. 2004; Abruzzese et al. 2005); and biogenic materials such as fossil shells (Dettman and Lohmann 2000; Morrill and Koch 2002) or teeth (e.g., Kohn et al. 2002; Fricke 2003).

In this paper, we combine the results of thermochronological and sedimentological studies that imply Miocene surface uplift in the southern Patagonian Andes with inferences on paleo-topography and paleo-climate based on stable isotope values of pedogenic carbonate in syn-orogenic foreland basin deposits east of those mountains. Our study area, located approximately between 47° and 48° S, is ideally suited for the use of oxygen isotopes to reconstruct paleo-topography for the following reasons:

1. Here, the Andes are high enough (3 000–4 000 m) to cause substantial orographic uplift of airmasses as the westerly winds impinge upon them, yet not so high as to divert airmasses around them (Boffi 1949; Seluchi et al. 1998).
2. The southern Patagonian Andes are a narrow mountain belt lying immediately east of the Pacific ocean, and in the "roaring forties" (center of the southern hemisphere westerlies) (Prohaska 1976). Therefore, precipitation is derived from a single dominant moisture source (the Pacific Ocean), and topography has a relatively simple influence on the patterns of atmospheric circulation (Boffi 1949). In fact, a comparison of GCM simulations with and without the high topography of the Andes indicates that precipitation in the Southern Andes is largely orographic (Lenters and Cook 1995).
3. Both the westerly winds and the South American Plate have not significantly changed latitude since the Miocene (e.g., Ziegler et al. 1981; Hartnady et al. 1985). Accordingly, the westerlies have been the primary source of precipitation in the area throughout the time interval relevant to our study. This makes interpreting the changes in the isotopic composition of precipitation throughout the geologic record relatively straightforward.

In the following sections we will, after briefly introducing the geology of the study area, present the main results of: (*1*) thermochronological work within the mountains; (*2*) sedimentological and chronostratigraphical work on syn-orogenic deposits in the eastern foreland basin; and (*3*) oxygen and carbon isotope analysis of pedogenic carbonate from foreland basin deposits. In the subsequent discussion section we focus on how the integration of these results helps us to better understand the history of, as well as causes and consequences of, surface uplift in this region.

## 20.2 Geological Setting

The Andean orogen results from the subduction of the Nazca and Antarctic plates beneath the South American Plate, with a triple junction where the Chile Ridge (~46.5° S, Fig. 20.1a) is presently subducting. While some of the elevation increase that occurred in the Andes during the Cenozoic may be the result of increased early Miocene plutonism (Thomson et al. 2001), most of it is probably owing to a strong increase in the convergence rate, and a decrease in convergence obliquity along the South American subduction zone around 28–26 Ma (Pardo-Casas and Molnar 1987; Somoza 1998). In the region south of the Chile triple junction, the tectonic and topographic evolution of the Cordillera has additionally been influenced by the subduction of a series of spreading ridge segments during the progressive northward movement of the triple junction. Due to the similar orientation of the Chile Ridge and the trench, the northward motion of the triple junction has been relatively fast; the spreading ridge first came into contact with the trench around 14 million years ago at 54–55° S, and migrated ~700 km northward to ~48° S within a few million years (Fig. 20.1a) (Cande and Leslie 1986; Ramos and Kay 1992; Gorring et al. 1997; Thomson et al. 2001; Ramos 2005).

Several previous studies have demonstrated that the subduction of this spreading ridge had a significant influence on the tectonic and topographic evolution of the southern Patagonian Andes. In the northern part of this region, between ~47° and 48° S, apatite fission track data imply that accelerated denudation started around 30–23 Ma near the Pacific coast and subsequently migrated ~200 km eastward to the present-day topographic axis of this cordilleran segment until 12–8 Ma, most likely as the result of subduction erosion (Thomson et al. 2001). A similar eastward migration appears to have characterized the evolution of retro-arc deformation in the fold-and-thrust belt east of the mountains (Ramos 1989; Suárez et al. 2000). It thus seems that important contractional deformation, denudation and subduction erosion occurred in the southern Patagonian Andes prior to spreading ridge subduction.

This is consistent with evidence for high convergence rates (Pardo-Casas and Molnar 1987; Somoza 1998), and with the subduction of young and buoyant oceanic lithosphere with relatively high relief and little, or no, sediment cover (Thomson et al. 2001; Lamb and Davis 2003). Following the initiation of spreading ridge subduction, the tectonic evolution of the southern Patagonian Andes was markedly different; both subduction erosion (Thomson et al. 2001) and major deformation in the eastern fold-and-thrust belt (Ramos 1989; Suárez et al. 2000) ceased, presumably because of the strong decrease in the convergence rate along the trench (from ~9 to ~2 cm yr$^{-1}$) (e.g., Gorring et al. 1997; Thomson et al. 2001).

The data implying important early to middle Miocene deformation and surface uplift in the southern Patagonian Andes are in good agreement with the stratigraphic sequence in the eastern foreland basin, which includes thick Oligocene to Recent syn-orogenic clastic (molasse) deposits. The lower part of these molasse deposits is represented by the Oligocene to early Miocene(?) Centinela Formation, comprising near-shore marine conglomerate, sandstone and shale beds deposited after a transgression from the Atlantic (Malumián and Ramos 1984; Ramos 1989; Flynn et al. 2002). These marine beds are conformably overlain by early to middle Miocene fluvial deposits of the Santa Cruz Formation, which have been related to the main phase of Cenozoic deformation and surface uplift in the Cordillera to the west (Ramos 1989; Ramos and

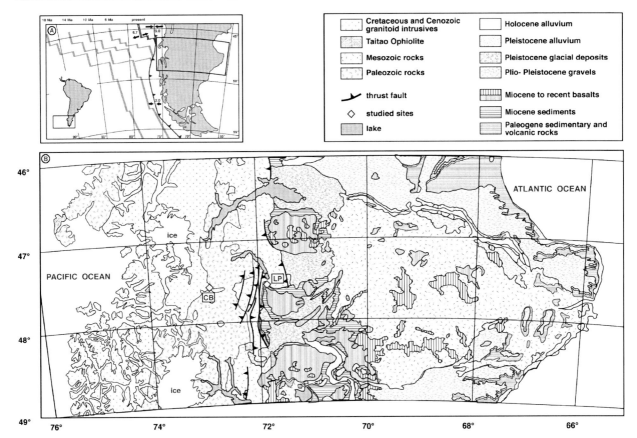

**Fig. 20.1. a** Sketch map showing the geometry of the present-day subduction zone and segments of the active Chile spreading ridge (*black lines*) between the Nazca Plate in the East and the Antarctic Plate in the West. *Gray lines* indicate reconstructed approximate positions of Chile Ridge segments at ~18 Ma, ~14 Ma, ~10 Ma, and ~6 Ma; arrows indicate the present-day spreading rate at the Chile Ridge and convergence rates along the subduction zone (after Cande and Leslie [1986] and Ramos [1989]). Note that the convergence rates are much lower in the south compared to north of the point of ridge collision. **b** Generalized geologic map of the Southern Patagonian Andes (after Thomson et al. [2001] and Lizuaín et al. [1997]). The location of the map is indicated by the *black box* on Fig. 20.1a. The location of sites discussed in the text is indicated by the white diamonds; *CB*, Cerro Barrancos; *LP*, Lago Posadas. Modified after Blisniuk et al. (2005)

Kay 1992; Flynn et al. 2002). In much of the eastern foreland, these deposits are covered by extensive plateau lavas that began erupting about 14–12 Ma. Their formation has been related to slab windows associated with the subduction of segments of the Chile spreading ridge (Ramos and Kay 1992; Gorring et al. 1997). Sediments post-dating the Santa Cruz Fm. are relatively limited in extent and volume, and are dominated by coarse conglomerates related to Pleistocene and older glaciations (Mercer 1976; Mercer and Sutter 1982).

## 20.3 Exhumation History

To better constrain the exhumation history of the southern Patagonian Andes, five samples for apatite fission-track analysis were collected along an elevation transect of Cerro Barrancos (Fig. 20.1b). The samples are granitic rocks from the Cerro Barrancoso Monzogranite, for which early Cretaceous crystallization ages have been reported (Pankhurst et al. 1999; Suárez and De La Cruz 2001). Four samples were collected between 130 and 1 410 m elevation on the western flank, the fifth sample was collected at an elevation of 1 960 m on the eastern side of the mountain. Sample preparation and analysis followed the procedures described in Warkus (2002) and Sobel and Strecker (2003); the full data set is presented in Warkus (2002). For every sample, 26 to 29 grains were dated and ca. 100 confined track-lengths were measured (Table 20.1). The etch pit diameter, Dpar (Donelick et al. 1999), was measured for 20 analysed crystals per sample in order to constrain the kinetic characteristics of the apatites. Extensive track-length modeling was performed with the AFTSolve program (Ketcham et al 2000) using the multicompositional annealing model of Ketcham et al. (1999).

The cooling ages obtained from the samples range from $6.3 \pm 0.6$ Ma to $17.1 \pm 1.2$ Ma (Table 20.1). All samples pass the $\chi^2$ test, indicating that the crystals within each sample

Table 20.1. Summary of apatite fission track data

| Sample ID | Elevation (m) | Lat. S | Long. W | # of xls | $P(\chi^2)$ | Age ±1$\sigma$ (Myr) | Track-length | n |
|---|---|---|---|---|---|---|---|---|
| CB 130 | 130 | 47°34.22' | 72°51.92' | 29 | 99 | 6.3 ±0.6 | 12.10 ±0.33 | 101 |
| CB 550 | 550 | 47°33.85' | 72°50.84' | 26 | 100 | 7.3 ±0.7 | 12.88 ±0.29 | 106 |
| CB 930 | 930 | 47°33.82' | 72°50.04' | 26 | 99 | 11.0 ± 1.1 | 12.74 ±0.30 | 104 |
| CB 1410 | 1410 | 47°33.80' | 72°49.29' | 27 | 92 | 11.3 ±0.7 | 11.94 ±0.29 | 111 |
| CB 1960 | 1960 | 47°33.56' | 72°46.22' | 28 | 99 | 17.1 ±1.2 | 10.30 ±0.24 | 106 |

belong to a single age population. However, the track-length distributions from all of the samples are shortened, indicating significant residence within the partial annealing zone. In addition, the track-length distribution of several samples is bimodal, suggesting that there may have been a reheating event followed by final cooling to surface temperatures. Unfortunately, there are too few geological constraints to allow a unique, best-fitting thermal model. Therefore, we systematically tested a wide range of thermal models in order to delineate possible cooling histories which can explain the track-length data.

Thermal modeling was done using both four and five constraints on the modeled time-temperature paths. The results of the latter are described herein, those of the former yielded similar results. Model runs began at 100 Ma, between 225 and 275 °C, which is consistent with a zircon fission track age of ~100 Ma determined at a nearby locality (Thomson et al. 2001). Models ended at the present, at temperatures between 5 and 15 °C. The three intermediate constraints were shifted systematically such that possible reheating events could be examined. The second constraint was placed at 16 and then at 26 Ma, between 50 and 180 °C. The third constraint was placed at 6, 8, 10, 12 and 16 Ma, at temperatures between 40 and 180 °C. The fourth constraint was positioned between 4 and 1 Myr later than the third constraint, and at temperatures between 60 and 180 °C. Note that the applied constraints do not force reheating; monotonic cooling paths are also possible. Models were run using the constrained random search algorithm, with 20 000 monotonic heating or cooling paths and two halvings between adjacent constraints. This strategy produced 55 models per sample; these were examined to determine the range of constraint values that produced good- or acceptable-fitting results (Table 20.2).

Samples CB130 and CB1410 both yielded good-fitting models (2$\sigma$ fits) for most constraints except when final cooling commenced at 4 Ma. Samples CB550 and CB930 produced acceptable models (1$\sigma$ fits) for most runs (Fig. 20.2). Comparing the best fits from all of the good-fitting models for samples CB130 and CB 1410 shows that the two samples behave quite similarly: The amount of reheating is 65 and 53 °C at average rates of 28 and 27 °C Myr$^{-1}$ while the final cooling is 137 and 119 °C at average rates of 20 and 16 °C Myr$^{-1}$. The standard deviation of the amounts and rates of the reheating are ~24 °C and 29 °C Myr$^{-1}$. The final cooling is better constrained, with standard deviations of about 10 °C and 5 °C Myr$^{-1}$ for the amount and rate, respectively. Using the averaged, best-fit, peak reheating temperatures, the geothermal gradient is 14 °C km$^{-1}$. These model results do not support monotonic cooling for these samples.

Sample CB1960 yielded acceptable fits for this modeling strategy. However, the shape of the length distribution is quite different, suggesting that it was reheated or cooled in a different manner. The differences likely result from the unique location of this sample on the drier, leeward side of the mountains. It has presumably undergone a markedly different exhumation history than the samples from the more humid, windward side.

Several conclusions can be drawn from the fission-track data. First, early Cretaceous pluton emplacement was followed by regional cooling until the late Cenozoic. Then, 60 ± 30 °C of reheating occurred; although poorly constrained, this commenced in the Miocene and must have terminated prior to 4 Ma. Finally, the samples cooled at a higher rate than during the previous regional cooling. The Miocene reheating in the Cerro Barrancos has most likely been caused by thermal processes in the deeper crust; there are no structural data to support burial of the Cerro Barrancos beneath a Miocene thrust fault, nor is there evidence to support burial beneath a several km-thick Miocene sedimentary basin.

The most plausible causes of such a thermal event are progressive opening and eastward migration of a slab window after the collision of the Chile Ridge with the South American subduction zone (e.g., Cande and Leslie 1986; Ramos and Kay 1992; Gorring et al. 1997), or the intrusion of a young pluton in the proximity of the sampling area. The latter scenario cannot be dismissed, but is not supported by the ~6.5 Ma emplacement age of the adjacent San Lorenzo pluton (Welkner, unpub. Thesis, Univ. Chile 1999; Welkner and Suárez 1999; Suárez and De La Cruz 2001), which is the only presently available age constraint on plutonic activity in the study area.

**Table 20.2.**
Summary of AFTSolve model results, sorted by three intermediate constraints (in Ma)

| Constraint | | | Sample | | | | |
|---|---|---|---|---|---|---|---|
| # 2 | # 3 | # 4 | CB 130 | CB 550 | CB 930 | CB 1410 | CB 1960 |
| 16 | 6 | 3 | y | y | y | y | y |
| 16 | 6 | 4 | y | y | y | y | y |
| 16 | 6 | 5 | y | y | y | y | y |
| 16 | 8 | 3 | y | y | y | y | y |
| 16 | 8 | 4 | y | y | y | y | y |
| 16 | 8 | 5 | y | y | y | y | y |
| 16 | 8 | 6 | y | y | y | y | y |
| 16 | 8 | 7 | y | y | y | y | y |
| 16 | 10 | 4 | n | n | y | y | y |
| 16 | 10 | 5 | y | y | y | y | y |
| 16 | 10 | 6 | y | y | y | y | y |
| 16 | 10 | 7 | y | y | y | y | y |
| 16 | 10 | 8 | y | y | y | y | y |
| 16 | 10 | 9 | y | y | y | y | n |
| 16 | 12 | 4 | n | y | y | y | y |
| 16 | 12 | 5 | y | y | y | y | y |
| 16 | 12 | 6 | y | y | y | y | y |
| 16 | 12 | 7 | y | y | y | y | y |
| 16 | 12 | 8 | y | y | y | y | y |
| 16 | 12 | 9 | y | y | y | y | n |
| 16 | 12 | 10 | y | y | y | y | n |
| 16 | 12 | 11 | y | y | y | y | n |
| 26 | 6 | 3 | y | n | y | y | y |
| 26 | 6 | 4 | y | y | y | y | y |
| 26 | 6 | 5 | y | y | y | y | y |
| 26 | 8 | 4 | y | y | y | y | y |
| 26 | 8 | 5 | y | y | y | y | y |
| 26 | 8 | 6 | y | y | y | y | y |
| 26 | 8 | 7 | y | y | y | y | y |
| 26 | 10 | 4 | n | y | y | y | y |
| 26 | 10 | 5 | y | y | y | y | y |
| 26 | 10 | 6 | y | y | y | y | y |
| 26 | 10 | 7 | y | y | y | y | y |
| 26 | 10 | 8 | y | y | y | y | y |
| 26 | 10 | 9 | y | y | y | y | y |
| 26 | 12 | 4 | y | y | n? | y | y |
| 26 | 12 | 5 | y | y | y | y | y |
| 26 | 12 | 6 | y | y | y | y | y |
| 26 | 12 | 7 | y | y | y | y | y |
| 26 | 12 | 8 | y | y | y | y | y |
| 26 | 12 | 9 | y | y | y | y | y |
| 26 | 12 | 10 | y | y | y | y | y |
| 26 | 12 | 11 | y | y | y | y | y |
| 26 | 16 | 4 | y | y | n | n | y |
| 26 | 16 | 5 | y | y | n | n | y |
| 26 | 16 | 6 | y | y | y | n | y |
| 26 | 16 | 7 | y | y | y | y | y |
| 26 | 16 | 8 | y | y | y | y | y |
| 26 | 16 | 9 | y | y | y | y | y |
| 26 | 16 | 10 | y | y | y | y | n |
| 26 | 16 | 11 | y | y | y | y | n |
| 26 | 16 | 12 | y | y | y | y | n |
| 26 | 16 | 13 | y | y | y | y | n |
| 26 | 16 | 14 | y | y | y | y | n |
| 26 | 16 | 15 | y | y | y | y | n |

**y**: Good fit; y: acceptable fit; n: no acceptable fits.

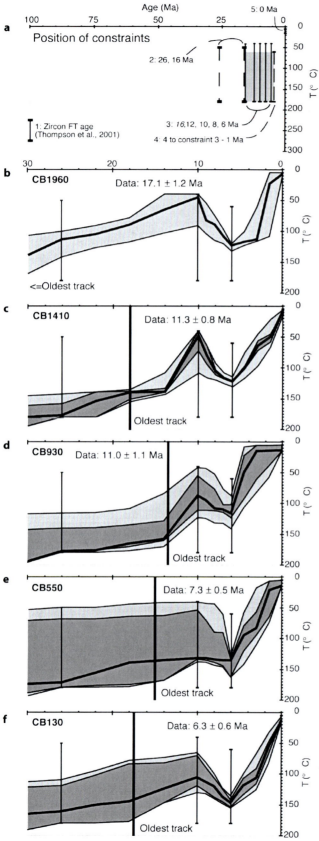

We thus consider it more likely that reheating occurred as the result of slab-window opening, when asthenosphere from beneath the subducted Chile Ridge came into contact with the base of the overriding South American Plate.

## 20.4 Stratigraphy and Geochronology of Miocene Deposits in the Eastern Foreland

To evaluate the effect of Miocene mountain uplift on deposition in the eastern foreland of the southern Patagonian Andes, we studied a 507.5 m thick section of subhorizontally bedded Santa Cruz Fm. deposits, which overlie a ~30 m covered interval above marine deposits of the Oligocene Centinela Fm. (Fig. 20.3b). The top of this section is the surface of a flat-topped mountain which may be a relict land surface, but we cannot exclude the possibility that the top portion of the Santa Cruz Fm. has been eroded. In a locality about 10 km southeast of the studied section, however, the thickness of the (less well exposed) Santa Cruz Fm., between the underlying deposits of the Centinela Fm. and overlying Miocene basalts, is at most 600 m, implying that the studied section is almost complete (Blisniuk et al. 2005).

The Santa Cruz Fm. is a syn-orogenic terrestrial foreland basin sequence dominated by alternating sand-, silt- and claystone as well as minor conglomerate beds. The lowermost part of the section, ~110 m thick, is dominated by silt- and claystone beds, whereas the younger part of the section contains more numerous, thicker and coarser sandstone beds. Important conglomerate beds occur in the uppermost part of the Santa Cruz Formation. These include a 13 m thick debris flow deposit occurring ~100 m below the top of the Santa Cruz Fm. in the Lago Posadas section, and conglomerate beds, up to 5 m thick, observed at or near the top of the formation at many other locations in the study area.

The coarse-grained deposits are compositionally dominated by andesitic clasts and the debris flow also contains a high proportion of intraformational clasts and clasts from Oligocene marine sediments that underlie the Santa Cruz Formation. The sandstones are typically medium to olive gray, moderately well sorted and medium to coarse grained. Sandstone petrography revealed a high proportion of generally sub-rounded andesitic fragments and abundant monomineralic clasts, mostly feldspars and

◀ **Fig. 20.2.** Representative AFTSolve thermal modeling results for all five samples. *Dark and light gray shading* indicate good and acceptable fits, respectively. **a** Schematic diagram illustrating positions of constraints; **b** model results for sample CB 1960; **c** model results for sample CB 1410; **d** model results for sample CB 930; **e** model results for sample CB 550; **f** model results for sample CB 130 amphiboles (Warkus 2002). Individual sandstone beds are laterally continuous for tens to hundreds of meters, and vary in thickness from a few tens of centimeters to several meters thick. The base of the sandstone beds is sometimes a slight erosional unconformity; thick sandstone beds frequently contain basal conglomerates, often with claystone intraclasts. However, commonly observed desiccation cracks and well-preserved soil structures near the tops of the underlying beds attest to the predominantly aggradational nature of these deposits. Trough cross bedding, lateral accretion structures, fining upward cycles and clay-rich sandstones suggest deposition in a meandering stream system with frequent flooding events. The overbank deposits of this stream system are represented by red-brown and gray siltstone and claystone beds.

Numerous paleosol horizons were observed in all lithologies, but are mostly contained in the silt- and claystone units, and are particularly common in the uppermost 100 m of the studied section. In addition to clastic deposits, the section also contains numerous tuff and ash layers that range from a few centimeters to about 2 meters, and are typically a few centimeters to tens of centimeters thick. The tuff layers are generally well sorted and contain idiomorphic feldspars and amphiboles that are up to 1 mm in diameter.

Age control for the studied section is provided by $^{40}Ar/^{39}Ar$ dating of five tuff layers contained in the section, and the age of a tuff from 3 m beneath the top of the Santa Cruz Fm. at a locality ~10 km farther south. The rocks there are overlain by basalts dated at ~12.1 ± 0.7 Ma (Ramos et al. 1991). The ages of the samples from the studied section range from 22.4 ± 0.7 Ma, for a tuff 10 m above the base, to 15.5 ± 0.4 Ma, for a tuff 331.5 m above the base (Fig. 20.3b) (Blisniuk at al. 2005). The tuff from 3 m beneath the top of the Santa Cruz Fm. at the southern locality was dated at 14.2 ± 0.8 Ma (Blisniuk at al. 2005); this provides a minimum age for the top of our section, where some erosion may have occurred.

These age data imply that a dramatic increase in sediment accumulation rates occurred in the lower part of the section (Fig. 20.3a). Between the samples dated at 22.4 ± 0.7 Ma and 18.2 ± 0.3 Ma (taken from 10.0 and 58.8 m above the base of the section, respectively), the mean sediment accumulation rate, not corrected for compaction, was 12 ± 2 m Myr$^{-1}$. This contrasts strongly with rates of 85–270 m Myr$^{-1}$ in higher parts of the section, averaging ~114 ± 25 m Myr$^{-1}$ between the 18.2 ± 0.3 Ma tuff occurring 58.8 m above the base and the top of the section which dates at ~14.2 ± 0.8 Ma. Our data only bracket the main increase in sediment accumulation rates between 22.4 ± 0.7 and 16.7 ± 0.6 Ma (at 10.0 m and 181.2 m above the base of the section, respectively); we note, however, that it may be marked by the change to more frequent and thicker sandstone beds occurring some 110 m above the base of the section.

## 20.5 Stable Isotope Data from Miocene Paleosols in Eastern Foreland Basin Deposits

To provide constraints on Miocene elevation changes within the southern Patagonian Andes, as well as on climatic and ecological changes in the eastern foreland, we studied the oxygen and carbon isotope composition of pedogenic carbonate nodules from paleosols contained in the studied section of the Santa Cruz Formation deposits described above. Isotope ratios are expressed in δ notation, with $\delta^{18}O$ and $\delta^{13}C$ defined as $1000(R_{sample}/R_{standard} - 1)$, where $R$ is $^{18}O/^{16}O$ or $^{13}C/^{12}C$, respectively; in this paper, isotope ratios are reported relative to PDB for carbon, and relative to SMOW for oxygen (Coplen 1995).

The $\delta^{18}O$ value of pedogenic minerals is controlled by the isotopic composition of the soil water present during pedogenesis (and, to a lesser degree, by soil temperature), which in turn largely depends on the $\delta^{18}O$ value of precipitation, and on the amount of soil water evaporation (Hsieh et al. 1998). The $\delta^{18}O$ values of pedogenic minerals are thus potential proxies for the isotopic composition of paleo-precipitation and can, in appropriate settings, be used to estimate paleo-topography (see Poage and Chamberlain 2001, for a review).

The main effect of topography on precipitation is related to adiabatic expansion and cooling of an uplifting airmass, which causes condensation and high amounts of "orographic precipitation". Condensation partitions more of the heavy isotope into the condensed phase, leading to a progressive decrease in the $\delta^{18}O$ of precipitation with increasing local elevation (the "altitude effect") (Dansgaard 1964). The altitude effect is largely controlled by temperature-dependent Rayleigh distillation and is not strictly linear. However, empirical data imply a nearly linear isotopic lapse rate of ca. −0.28‰/100 m net elevation change in many regions of the world, excluding latitudes > 70° and elevations > 5000 m (Chamberlain and Poage 2000; Poage and Chamberlain 2001). Therefore an airmass not only retains less water after passing over a mountain belt, but is also strongly depleted in $^{18}O$; sites with lower $\delta^{18}O$ values of precipitation on the leeward side of mountains are said to have an "isotopic rain shadow" (see Blisniuk and Stern 2005). The magnitude of this depletion should reflect the elevation increase of the airmass during its passage over the mountain belt according to the estimated lapse rate of −0.28‰/100 m.

This is in good agreement with data from present-day precipitation and surface waters sampled along a transect of the southern Patagonian Andes between 47° 20' and 48° S (Fig. 20.4). The transect extended from ~73° 20' W on the western side of the mountains, where maximum windward elevations are some 1 800 m, to ~71° 10' W east of the mountains which reach heights of some 3 700 m. Correspondingly, the data from this transect should re-

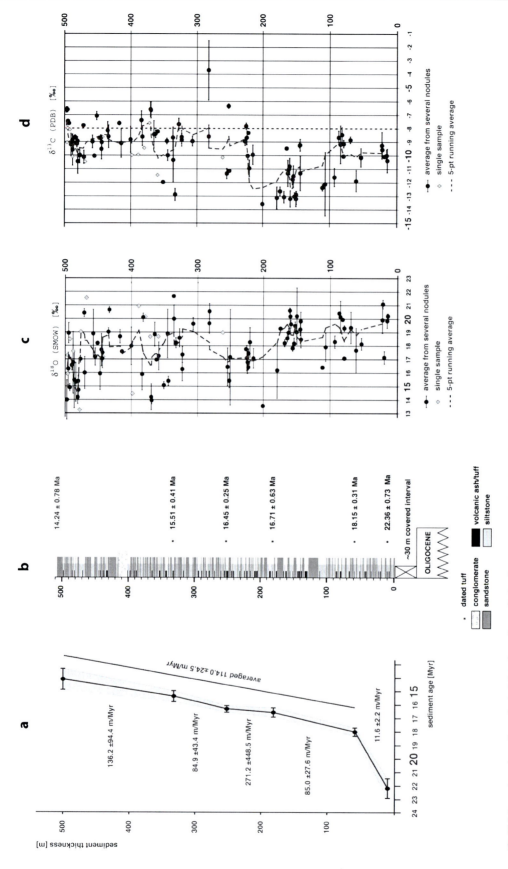

Fig. 20.3. Isotopic and geochronological data from the studied sediment section. **a** Plot of stratigraphic position versus age of dated tuffs from the Santa Cruz Fm. (1σ age uncertainties indicated by error bars). *Numbers* between points of age control are sediment accumulation rates for corresponding time intervals. *Bold line* to the right of the graph indicates average sediment accumulation rate between ~18 and 14 Ma. Sediment accumulation rates are not corrected for compaction. **b** Stratigraphic log of the studied sediment section, with positions and ages of dated tuffs indicated to the right of the section. **c** Oxygen isotope data of pedogenic carbonate nodules in paleosols contained in the studied section. Paleosols from which several individual nodules were analysed are shown by *filled symbols*, with error bars indicating one standard deviation. For symbols without error bars, one standard deviation is smaller than the width of the symbol. Paleosols from which only a single nodule was analysed are indicated by open symbols. **d** Carbon isotope data of pedogenic carbonate nodules in paleosols contained in the studied section. Note the *solid, dashed line* highlighting a δ$^{13}$C value of −8‰, with higher values strongly indicative of the presence of a significant proportion of C4 vegetation. Use of *open* and *filled symbols* as for (**c**). Modified after Blisniuk et al. (2005)

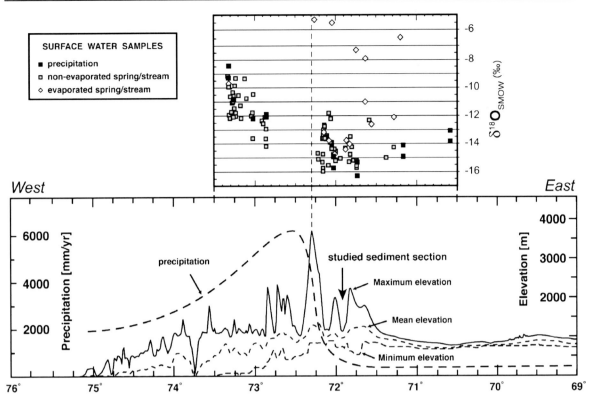

**Fig. 20.4.** Diagram illustrating the present-day rain shadow effect across the southern Patagonian Andes. *Top*: Oxygen isotope values of present-day surface waters collected along a transect of the southern Patagonian Andes. Note that, compared to samples from the windward western side of the mountains, surface waters from the leeward eastern side have significantly lower $\delta^{18}O$ values. Also note that deuterium excess values demonstrate that a relatively high proportion of surface waters east of the mountains has experienced substantial evaporation, which causes an increase in $\delta^{18}O$. As discussed in Stern and Blisniuk (2002), east of the mountains, waters classified as non-evaporated may have experienced some evaporation; in that part of the transect the lower values can be considered to be from the least evaporated waters and thus the best proxy for precipitation. *Bottom*: Maximum, mean, and minimum elevations along an east-west-oriented swath some 70 km wide between 47° 20' S and 48° 00' S, which contains water sampling locations and the studied sediment section. *Stippled bold line* indicates mean annual rate of precipitation along the swath (Hoffman 1975). Modified after Stern and Blisniuk (2002)

flect an elevation increase of ~2000 m for the moisture bearing airmasses carried eastwards over the mountains by the westerly winds.

On the windward, western side of the mountains, the main trend in the data is a steady eastward decrease in the $\delta^{18}O$ values of these waters; Rayleigh distillation modeling and the observed $\delta^{18}O$ values (versus site elevations and maximum elevation of windward topography) imply similar isotopic lapse rates of about −0.3‰ / 100 m (Stern and Blisniuk 2002). This is in good agreement with the isotopic lapse rate estimate of −0.28‰ / 100 m derived from a global compilation (Poage and Chamberlain 2001).

On the leeward, eastern side of the mountains, in contrast, the $\delta^{18}O$ values of precipitation and non-evaporated surface waters are consistently low (4–6‰ lower than at the western end of our transect), and show no significant trend with local elevation (Stern and Blisniuk 2002). This implies that the $\delta^{18}O$ values of precipitation in that region are largely controlled by the progressive condensation and rain-out of moisture on the windward, western side of the southern Patagonian Andes.

We note, though, that many surface water samples from the leeward side of the mountains had deuterium excess values indicating substantial evaporation (Stern and Blisniuk 2002), and that these samples had up to > 5‰ higher $\delta^{18}O$ values than samples of precipitation and apparently non-evaporated waters. The data from the surface water transect thus imply that the $\delta^{18}O$ values of paleosol carbonates on the leeward side of the mountains should, in principle, reflect changes in the surface elevation of this cordilleran segment, but may be affected by evaporation enrichment which, while easy to assess for waters, is difficult to estimate from mineral proxies.

To track potential changes in the isotopic composition of paleo-precipitation, we determined the $\delta^{18}O$ values of pedogenic carbonate nodules from 102 paleosol horizons contained in the ca. 23–14 Ma section of the Santa Cruz Fm. deposits described above (Fig. 20.3c). The main trend in these data is a decrease of ~2‰ in the mean $\delta^{18}O$ values from the lower to the upper part of the section. Most of this decrease occurs in the approximately 17–14 Ma, upper ~300 m of the section, and it is particularly pronounced around 200 m above the base of the studied section, or at ~17 Ma (Blisniuk et al. 2005).

The data set is also characterized by a marked increase in the scatter of the $\delta^{18}O$ values from the lower to the upper part of the section, which is largely directed towards lower $\delta^{18}O$ values. In the upper ~300 m of the section, the range of the $\delta^{18}O$ values has an upper limit similar to that in the lower ~200 m of the section, whereas the lower limit is decreased by 3–4‰. We interpret these changes as the consequence of significant surface uplift in this Andean segment.

As discussed above, influences such as the mixing of isotopically distinct moisture sources, changes in atmospheric circulation patterns and changes in the latitude of the studied site are unimportant at this location. Changes in ocean circulation patterns, sea surface temperatures, and temperatures on land are potentially more important. However, quantitative estimates of those effects imply that they had a small net effect on the $\delta^{18}O$ value of precipitation east of the southern Patagonian Andes during the deposition of the Santa Cruz Fm. (Blisniuk et al. 2005). Correspondingly, the main trend in our data is best explained by surface uplift of the southern Patagonian Andes leading to increased precipitation rates west of the mountains.

This is consistent with a concurrent transition to a more arid climate in the eastern foreland, inferred from the $\delta^{13}C$ values of the carbonate nodules from the Santa Cruz Fm. deposits. The carbon isotope value of pedogenic carbonate is largely controlled by the average carbon isotope composition of vegetation and enables inferences on past plant ecosystems, particularly the relative proportion of low $\delta^{13}C$ C3 plants (typical of relatively humid climates) to high $\delta^{13}C$ C4 plants (typical of more arid climates). Also, greater water stress increases the $\delta^{13}C$ values of C3 plants (Farquhar et al. 1989). The $\delta^{13}C$ values of carbonate nodules from the studied section show a ~3‰ increase at ~17 Ma (Fig. 20.3d), and imply the presence of a significant component of C4 plants, or of great water stress on C3 plants, following this increase. Presumably, significant surface uplift of this Andean segment had established a pronounced orographic rain shadow by that time (Blisniuk et al. 2005).

This provides a plausible explanation for the large scatter of the $\delta^{18}O$ values in the upper ~300 m of the section. The effect of evaporation is hard to quantify exactly, but present-day surface waters with deuterium excess values implying substantial evaporation had $\delta^{18}O$ values up to > 5‰ higher than those of apparently non-evaporated samples (Fig. 20.4) (Stern and Blisniuk 2002). Since surface water composition is an important factor controlling soil water composition, evaporation enrichment of $^{18}O$ is likely to have affected many of our pedogenic carbonate samples, and may have caused the observed increase in the scatter of the data after the rain shadow formed. Correspondingly, the ~2‰ decrease of $\delta^{18}O$ values in the mean soil water composition recorded by our data probably underestimates the shift in the $\delta^{18}O$ values of precipitation. We therefore suggest that the 3–4‰ decrease of the lowest $\delta^{18}O$ values (presumably from the samples least affected by evaporation) is a better measure for the effect of surface uplift in the southern Patagonian Andes on the $\delta^{18}O$ values of precipitation in the eastern foreland. Assuming an isotopic lapse rate of –0.28‰/100 m (Poage and Chamberlain 2001), this would imply that at least 1.3 ± 0.2 km of surface uplift occurred in the southern Patagonian Andes between ~17 and 14 Ma. However, uncertainties related to the effects of climate change, as well as uncertainties related to the isotopic lapse rate will increase the total uncertainty of this estimate (Blisniuk and Stern 2005). It is probably more realistic to infer an isotopic shift of 3.5 ± 1.5‰ and a lapse rate of –0.3 ± 0.1‰/100 m, resulting in an estimate of 1.2 ± 0.5 km for the amount of this surface uplift.

## 20.6 Discussion and Conclusions

The data presented above have several important implications for the tectonic and topographic evolution of the southern Patagonian Andes. From the stable isotope data we conclude that significant surface uplift of this cordilleran segment occurred from ca. 17–14 Ma. This in turn implies that the increase of the sediment accumulation rates in the Lago Posadas section at ~18 Ma is not a consequence of this surface uplift, but reflects either a change to a more erosive climate, or eastward propagation of the eastern fold-and-thrust belt. Because there is no evidence for significant climate change in South America or globally at that time (e.g., Zachos et al. 2001), an eastward propagation of the eastern fold-and-thrust belt appears to be the more likely of these two scenarios.

The observation that sediment accumulation rates were relatively constant from ~18 to ~14 Ma is interesting because this implies that, during that time, any trend towards higher erosion rates resulting from increased relief was roughly balanced by a trend towards decreasing erosion rates resulting from the more arid climate. Owing to the scarcity of eastern foreland deposits post-dating the Santa Cruz Fm., we have no younger paleosol samples and cannot make quantitative estimates of paleo-topography younger than 14 Ma. We emphasize, however, that the observed drastic decrease of the sediment flux and deposition rates east of the mountains is consistent with continued surface uplift and a further intensification of the orographic rain shadow after ~14 Ma; this, in turn, implies greatly increased precipitation and erosion rates on the windward, western side of the mountains, as already discussed by Thomson et al. (2001).

Because plate convergence rates exert a strong influence on the intensity of compressive deformation near subduction zones, major uplift likely continued until the ridge collision in this region (~10(?) Ma), when a strong decrease in plate convergence rates occurred. This scenario is in good agreement with the hypothesis that subduction erosion in the southern Patagonian segment of the upper plate may have ceased at ~10 Ma (Thomson et al. 2001). It is interesting to note, though, that a thicker trench fill is likely to promote weaker coupling along the plate interface (e.g., Lamb and Davis 2003; Ernst 2004). This implies that progressive surface uplift of the southern Patagonian Andes and the increasing sediment flux to the adjacent segment of the South American trench have contributed significantly to a decrease in compressive deformation and surface uplift.

Accordingly, changes in Earth surface processes resulting from mountain uplift may feed back on tectonically controlled surface uplift in subduction orogens. In settings such as the central Andes, characterized by arid conditions on the oceanward side of the mountains and sediment starvation in the trench, such a feed-back would be positive (Lamb and Davis 2003; Ernst 2004). In contrast, in settings like that of the southern Patagonian Andes a negative feed-back would result, because surface uplift leads to increased precipitation, erosion, and sediment flux rates on the oceanward side of the mountains.

## Acknowledgments

We thank George Hilley, Travis Horton and David Völker for discussions, and Susana and Pedro Fortuny for help with local logistics. We also are grateful to the National Park Administration of Argentina and the staff at Parque Nacional Perito Moreno for permitting water collection in the park. Thoughtful comments by Stuart Thomson and an anonymous reviewer helped to improve the original manuscript. This work was supported by the Deutsche Forschungsgemeinschaft (SFB 267).

## References

Abruzzese MJ, Waldbauer RJ, Chamberlain CP (2005) Oxygen and hydrogen isotope ratios in freshwater chert as indicators of ancient climate and hydrologic regime. Geochim Cosmochim Acta 69:1377–1390

Amundson RG, Chadwick OA, Sowers JM, Doner HM (1989) The stable isotope chemistry of pedogenic carbonates at Kyle Canyon, Nevada. Soil Sci Soc Am J 53:201–210

Blisniuk PM, Stern LA (2005) Stable isotope altimetry: a critical review. Am J Sci 305:1033–1074

Blisniuk PM, Stern LA, Chamberlain CP, Idleman B, Zeitler PK (2005) Climatic and ecologic changes during Miocene surface uplift in the southern Patagonian Andes, Earth Planet Sci Lett 230:125–142

Boffi JA (1949) Effect of the Andes Mountains on the general circulation over the southern part of South America. Bull Am Meteor Soc 30:242–247

Cande SC, Leslie RB (1986) Late Cenozoic tectonics of the southern Chile trench. J Geophys Res 91:471–496

Cerling TE (1984) The stable isotopic composition of modern soil carbonate and its relationship to climate. Earth Planet Sci Lett 71:229–240

Cerling TE, Quade J (1993) Stable carbon and oxygen isotopes in soil carbonates. In: Swart PK, Lohmann KC, McKenzie J, Savin S (eds) Climate Change in continental isotopic records. AGU Geophys Monogr, pp 217–231

Chamberlain CP, Poage MA, Craw D, Reynolds RC (1999) Topographic development of the Southern Alps recorded by the isotopic composition of authigenic clay minerals, South Island, New Zealand. Chem Geol 155:279–294

Chamberlain CP, Poage MA (2000) Reconstructing the paleotopography of mountain belts from the isotopic composition of authigenic minerals. Geology, 28:115–118

Coplen TB (1995) Reporting of stable carbon, hydrogen, and oxygen isotopic abundances. In: IAEA (ed) Reference and intercomparison material for stable isotopes of light elements, pp 31–34

Dansgaard W (1964) Stable isotopes in precipitation. Tellus 16:436–468

Dettman DL, Lohmann KC (2000) Oxygen isotope evidence for high-altitude snow in the Laramide Rocky Mountains of North America during the Late Cretaceous and Paleogene. Geology 28:243–246

Dettman DL, Fang X, Garzione CN, Li J (2003) Uplift-driven climate change at 12 Ma: a long $\delta^{18}O$ record from the NE margin of the Tibetan plateau. Earth Planet Sci Lett 214:267–277

Donelick RA, Ketcham RA, Carlson WD (1999) Variability of apatite fission-track annealing kinetics: II. Crystallographic orientation effects. Am Mineralogist 84:1224–1234

Ernst WG (2004) Regional crustal thickness and precipitation in young mountain chains. Proc Nat Acad Sci 101:14998–15001

Farquhar GD, Ehleringer JH, Hubick KT (1989) Carbon isotope discrimination and photosynthesis. Ann Rev Plant Physiol Plant Mol Biol 40:503–537

Flynn JJ, Novacek MJ, Dodson HE, Frassinetti D, McKenna MC, Norell MA, Sears KE, Swisher I, Carl C, Wyss AR (2002) A new fossil mammal assemblage from the southern Chilean Andes: implications for geology, geochronology, and tectonics. J S Am Earth Sci 15:285–302

Fricke HC (2003) Investigation of early Eocene water-vapor transport and paleoelevation using oxygen isotope data from geographically widespread mammal remains. Geol Soc Am Bull 115:1088–1096

Garzione CN, Dettman DL, Quade J, DeCelles PG, Butler RF (2000) High times on the Tibetan Plateau: paleoelevation of the Thakkhola graben, Nepal. Geology 28:439–442

Gorring ML, Kay SM, Zeitler PK, Ramos VA, Rubilio DR, Fernandez MI, Panza JL (1997) Neogene Patagonian plateau lavas: continental magmas associated with ridge collision at the Chile Triple Junction. Tectonics 16:1–17

Graham SA, Chamberlain CP, Yue Y, Ritts BD, Hanson AD, Horton TW, Waldbauer JR, Poage MA, Feng X (2005) Stable isotope records of Cenozoic climate and topography, Tibetan Plateau and Tarim Basin. Am J Sci 305:101–118

Hartnady CJH, Le Roex AP (1985) Southern Ocean hotspot tracks and the Cenozoic absolute motion of the African, Antarctic and South American plates. Earth Planet Sci Lett 75:245–257

Horton TW, Sjostrom DJ, Abruzzese MJ, Poage MA, Waldbauer JR, Hren M, Wooden J, Chamberlain CP (2004) Spatial and temporal variation of Cenozoic surface elevation in the Great Basin and Sierra Nevada. Am J Sci 304:862–888

Hsieh JCC, Chadwick OA, Kelly EF, Savin SM (1998) Oxygen isotopic composition of soil water: quantifying evaporation and transpiration. Geoderma 82:269–293

Ketcham RA, Donelick RA, Carlson WD (1999) Variability of apatite fission-track annealing kinetics: III. Extrapolation to geological time scales. Am Mineralogist 84:1235–1255

Ketcham RA, Donelick RA, Donelick MB (2000) AFTSolve: a program for multi-kinetic modeling of apatite fission-track data. Geol Mater Res 2:1–32

Kohn MJ, Miselis JL, Fremd TJ (2002) Oxygen isotope evidence for progressive uplift of the Cascade Range, Oregon. Earth Planet Sci Lett 204:151–165

Lamb S, Davis P (2003) Cenozoic climate change as a possible cause for the rise of the Andes. Nature 425:792–797

Lenters JD, Cook KH (1995) Simulation and diagnosis of the regional summertime precipitation climatology of South America. J Climatol 8:2988–3005

Lizuaín A, Leanza HA, Panza JL (1997) Mapa Geológica de la República Argentina 1:2 500 000. Ministerio de Economia y Obras y Servicios Públicos

Malumián N, Ramos VA (1984) Magmatic intervals, transgression-regression cycles and oceanic events in the Cretaceous and Tertiary of southern South America. Earth Planet Sci Lett 67:228–237

Mercer JH (1976) Glacial history of southernmost South America. Quat Res 6:125–166

Mercer JH, Sutter JF (1982) Late Miocene-Early Pliocene glaciation in southern Argentina: implications for global ice-sheet history. Palaeogeog Palaeoclimat Palaeoecol 38:185–206

Molnar P, England P (1990) Late Cenozoic uplift of mountain ranges and global climate change: chicken or egg? Nature 346:29–34

Morrill C, Koch PL (2002) Elevation or alteration? Evaluation of isotopic constraints on paleoaltitudes surrounding the Eocene Green River Basin. Geology 30:151–154

Norris RD, Jones LS, Corfield RM, Cartlidge JE (1996) Skiing in the Eocene Uinta Mountains? Isotopic evidence in the Green River Formation for snow melt and large mountains. Geology 24:403–406

Pankhurst RJ, Weaver SD, Hervé F, Larrando P (1999) Mesozoic-Cenozoic evolution of the North Patagonian Batholith in Aysén, southern Chile. J Geol Soc London 156:673–694

Pardo Casas F, Molnar P (1987) Relative motion of the Nazca (Farallon) and South American Plates since late Cretaceous time. Tectonics 6:233

Poage MA, Chamberlain CP (2001) Empirical relationships between elevation and the stable isotope composition of precipitation and surface waters: considerations for studies of paleoelevation change. Am J Sci 301:1–15

Poage MA, Chamberlain CP (2002) Stable isotopic evidence for a Pre-Middle Miocene rain shadow in the western Basin and Range: implications for the paleotopography of the Sierra Nevada. Tectonics 21: doi 10.1029/2001TC001303

Prohaska F (1976) The climate of Argentina, Paraguay and Uruguay. In: Schwerdtfeger W (ed) Climates of Central and South America. World Survey of Climatology, pp 13–73

Quade J, Cerling TE, Bowman JR (1989) Systematic variations in the carbon and oxygen isotopic composition of pedogenic carbonate along elevation transects in the southern Great Basin, United States. Geol Soc Am Bull 101:464–475

Ramos VA (1989) Andean foothills structures in northern Magallanes Basin, Argentina. Am Assoc Petrol Geol Bull 73:887–903

Ramos VA (2005) Seismic ridge subduction and topography: foreland deformation in the Patagonian Andes. Tectonophysics 399:73–86

Ramos VA, Kay SM (1992) Southern Patagonian plateau basalts and deformation: backarc testimony of ridge collisions. Tectonophysics 205:261–282

Ramos VA, Kay SM, Sacomani L (1991) La dacita Puesto Nuevo y otras rocas magmaticas (Cordillera Patagonica Austral): Colisión de un dorsal oceanica Cretacica. VII Congreso Geológico Chileno, Abs Vol 2

Rowley DB, Pierrehumbert RT, Currie BS (2001) A new approach to stable isotope-based paleoaltimetry: implications for paleoaltimetry and paleohypsometry of the High Himalaya since the Late Miocene. Earth Planet Sci Lett 188:253–268

Ruddiman WF, Raymo ME, Prell WL, Kutzbach JE (1997) The uplift-climate connection: a synthesis. In: Ruddiman WF (ed) Tectonic uplift and climate change. Plenum Press, New York London, pp 471–515

Seluchi M, Serafini YV, Le Treut H (1998) The impact of the Andes on transient atmospheric systems: a comparison between observations and GCM results. Monthly Weather Rev 126:895–912

Sobel ER, Strecker MR (2003) Uplift, exhumation, and precipitation: Tectonic and climatic control of Late Cenozoic landscape evolution in the northern Sierras Pampeanas, Argentina. Basin Res 15: doi 10.1046/j.1365–2117.2003.00214.x

Somoza R (1998) Updated Nazca (Farallon)-South America relative motions during the last 40 Myr: implications for mountain building in the central Andean region. J S Am Earth Sci 11:211–215

Stern LA, Blisniuk PM (2002) Stable isotope composition of precipitation across the southern Patagonian Andes. J Geophys Res 107: doi 10.1029/2002JD002509

Suárez M, De La Cruz R (2001) Jurassic to Miocene K-Ar dates from eastern central Patagonian Cordillera plutons, Chile (45°–48° S). Geol Mag 1:53–66

Suárez M, De La Cruz R, Bell CM (2000) Timing and origin of deformation along the Patagonian fold and thrust belt. Geol Mag 137:345–353

Takeuchi A, Larson PB (2005) Oxygen isotope evidence for the late Cenozoic development of an orographic rain shadow in eastern Washington, USA. Geology 33:313–316

Thomson SN, Hervé F, Stöckhert B (2001) Mesozoic-Cenozoic denudation history of the Patagonian Andes (southern Chile) and its correlation to different subduction processes. Tectonics 20:693–711

Warkus F (2002) Die neogene Hebungsgeschichte der Patagonischen Anden im Kontext der Subduktion eines aktiven Spreizungszentrums. PhD thesis, University of Potsdam

Welkner DM, Suárez M (1999) Los plutones del área del Cerro San Lorenzo (47°30′ S): valores K-Ar y Ar-Ar. XIV Congreso Geológico Argentino, Actas, pp 112–113

Zachos J, Pagani M, Sloan L, Thomas E, Billups K (2001) Trends, rhythms, and aberrations in global climate 65 Ma to Present. Science 292:686–693

Ziegler AM, Barret SF, Scotese CR (1981) Palaeoclimate, sedimentation and continental accretion. Phil Trans Royal Soc London A 301:253–264

# Part IV
# The System at Depth: Images and Models

**Chapter 21**
Seismological Studies
of the Central and Southern Andes

**Chapter 22**
Partial Melting
in the Central Andean Crust:
a Review of Geophysical,
Petrophysical, and Petrologic
Evidence

**Chapter 23**
Controls on the Deformation
of the Central and Southern Andes
(10–35° S): Insight from Thin-Sheet
Numerical Modeling

**Chapter 24**
Numerical Study
of Weakening Processes
in the Central Andean Back-Arc

**Chapter 25**
Mechanism of the Andean Orogeny:
Insight from Numerical Modeling

This part summarizes the wealth of geophysical data that provides unprecedented details of the deep architecture of the Central and Southern Andes, and presents a series of numerical models assessing their evolution. The chapter by G. Asch et al. (Chap. 21) sets out by compiling seismicity data from local networks to provide the first high-resolution geometry of the Wadati-Benioff zone in this region. In addition, the Benioff seismicity is used to construct tomographic images of the sub-arc system that reveal an intriguing complexity related to fluid and melt ascent.

F. Schilling et al. (Chap. 22) synthesize and jointly interpret varied geophysical and petrologic datasets and models that relate to the nature and distribution of fluids and melts in the Central Andes crust. Their results complement and corroborate earlier views that major parts of the plateau float on partially molten crust, with significant consequences for the strength of the continental crust.

Using thin sheet modeling, S. Medvedev et al. (Chap. 23) analyze the strength contrast between the Andean lithosphere, the Brazilian shield and the plate interface from the Central to the Southern Andes. Their analysis shows that strength contrasts and their along strike changes by as little as one order of magnitude may be sufficient to explain the north-south variations in distribution and degree of shortening observed in the Andes.

A. Babeyko et al. (Chap. 24) analyze in more detail the rheological properties and their variations with time in the Central Andean back-arc that may contribute to the deformation distribution observed in 2D across the Altiplano Plateau. Their thermo-mechanical modeling shows that thermal weakening due to partial melting, coupled with the localization of weak sedimentary sections in the crust apparently played key roles.

S. Sobolev et al (Chap. 25) conclude this part with comprehensive 2D numerical modeling that attempts to reproduce the spatial distribution and temporal evolution of deformation across the entire Central Andean plate margin. The plateau building in the Central Andes, as well as its absence in the Southern Andes, can be clearly shown as a response to variations in upper plate advance and the degree of mechanical coupling at the plate interface.

# Seismological Studies of the Central and Southern Andes

Günter Asch · Bernd Schurr · Mirjam Bohm · Xiaohui Yuan · Christian Haberland · Benjamin Heit · Rainer Kind
Ingo Woelbern · Klaus Bataille · Diana Comte · Mario Pardo · Jose Viramonte · Andreas Rietbrock · Peter Giese

**Abstract.** The central Andes have formed by the complex interaction of subduction-related and tectonic processes on a lithospheric scale. The deep structure of the entire mountain range and underlying subduction zone has been investigated by passive and active seismological experiments. Detailed tomographic features are interpreted to represent the ascent paths of fluid and melts in the subduction zone and provide new insights about the mechanisms of lithospheric deformation. Receiver functions from teleseismic events have been used to observe the upper-plate continental Moho and subducted oceanic Moho, as well as the interaction of subducted oceanic lithosphere and mantle discontinuities. A second working area was established in the southern Andes to compare two different types of Andean subduction and to identify the principal controlling parameters. Besides the first accurate definition of the Wadati-Benioff zone in south-central Chile, a three-dimensional, tomographic velocity model based on local earthquakes in the southern Andes is presented.

## 21.1 Introduction

In Andean-type margins, oceanic crust is subducted below an upper continental plate, and both plates are influenced and altered, forming the site of complex mass and energy transfer between mantle, crust, hydrosphere and atmosphere. The mechanical properties of the upper plate have a strong influence on its deformation. Thus, detailed information on the distribution of rheological parameters is important for understanding the dynamics of Andean-type orogeny. Reliable information on the lithospheric structure and the distribution of seismic properties is essential for modeling deformation and mountain building (examples of which can be seen in this volume). Integrative seismic models of the lithosphere enable us, through correlation with surface geological observations and petrological data, to infer the state of ongoing processes in the lithosphere.

The Collaborative Research Centre (Sonderforschungsbereich 267), 'Deformation Processes in the Andes', contained two sub-projects (C4 and C4B) that were focused on active and passive seismological investigations of the upper plate in the central and southern Andean subduction zone (Fig. 21.1). Sub-project C4B, 'Structure and rhe-

**Fig. 21.1.**
Survey of the subduction zone of the central and southern Andes. *Left:* Bathymetry (from NOAA 1998) and volcanoes (*triangles*). *Right:* Seafloor age of the Nazca Plate (after Müller et al. 1997), disribution of earthquakes (from NEIC, *black dots*) since 1973 (*M* > 4), and depth of the Wadati-Benioff zone (after Cahill et al. 1992). The *black squares* denote the areas under investigation in the Central and Southern Andes. The Nazca Plate's angle of subduction varies along the western coast between 10 and 30° S and the recent volcanism correlates with steeper subduction angles. In the flat-slab area, no recently active volcanism exists. In general, seismic activity decreases from north to south. In both areas under investigation, subduction angle and subduction velocity are comparable

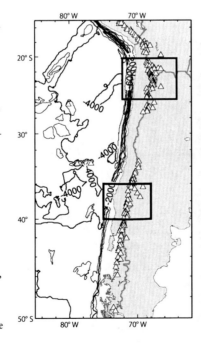
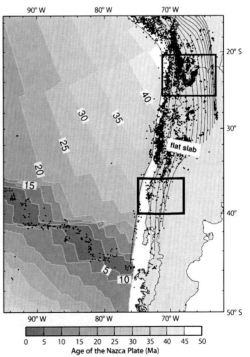

ology of the upper plate from seismological investigations' is a continuation of the 1993–1996, C4 activities headed by P. Giese. Within the project C4, several active and passive seismological data sets were collected during earlier periods of the SFB 267 (Fig. 21.2). These existing data sets form the basis for a wide range of investigations on the structure and physical properties of the upper plate in the central Andes.

An effort was made to merge all the passive seismology data sets in order to derive an integrative, consistent, three-dimensional model for $v_p$, $v_p/v_s$ and $Q_p$ of the entire subduction zone. We used active seismology to interpret seismic reflection and refraction data from the ANCORP'96 transect and earlier SFB 267 experiments and, together with the tomographic results, we built an integrative crustal model for a transect along 21° S. In 2000, the SFB established a new working area in the southern Andes between 36 and 40° S (Fig. 21.3) and acquired new, active and passive, seismological data during project ISSA 2000.

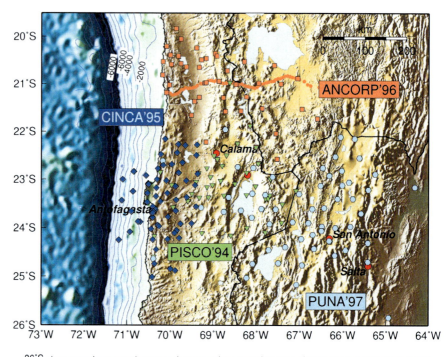

**Fig. 21.2.**
Study area of the seismological experiments within the SFB 267 in the central Andes from 1994 to 1997. Distibution of the seismological stations: PISCO 94 (*green*), CINCA 95 (*blue*), ANCORP (*orange*), and PUNA 97 (*light blue*). Seafloor topography (from NOAA 1998)

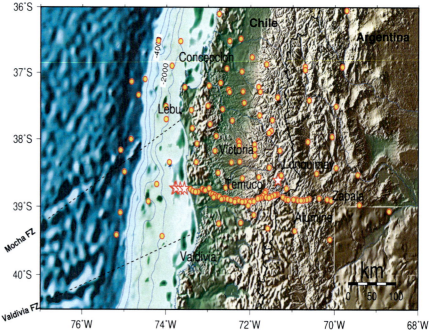

**Fig. 21.3.**
Overview map of the integrated seismological experiment in the southern Andes (ISSA 2000). *Stars* (*) mark shot points, *dashed lines* offshore mark the Mocha and Valdivia fault zones. Seafloor topography from NOAA (1988)

In this part of the Andean subduction zone, nearly identical boundary conditions for subduction (conversion velocity and geometry of the subduction regime) result in very different styles of deformation in the upper plate (no crustal thickening, no plateau formation). Some of the major questions for understanding general and specific Andean-type subduction and deformation processes are:

- the role of fluids and melts;
- the nature of intermediate-depth seismicity;
- the role of upper-plate mantle lithosphere;
- the deformation of the upper-plate crust; and
- the comparison between the central and southern Andes.

These topics are discussed below.

### 21.1.1 Role of Fluids and Melts

Sediments, oceanic crust, and oceanic mantle subducting to greater pressure and temperature undergo various metamorphic transformations that release substantial amounts of water (Peacock 1996). An important water-liberating reaction in oceanic crust is the transformation of blueschist to eclogite. Dehydration rates are highest until the breakdown of amphibole begins. This occurs almost isobarically at pressures of 2.3–2.5 GPa (70–80 km depth). This major fluid pulse at constant depth, together with some lateral transport mechanism in the mantle, has been employed to explain the globally-similar, vertical separation of approximately 120 km between the volcanic front and the Wadati-Benioff zone (Gill 1981). More recent work, however, shows that discrete pulses of water at certain depths are unrealistic. The amphibole-out reaction does not completely dehydrate the slab, and the retained 2 wt% of water is released in continuous reactions down to depths greater than 200 km (Poli and Schmidt 1995). Yet, where dehydration occurs, where and how fluids leave the slab, and how they are transported through the mantle wedge, are still not understood.

### 21.1.2 Nature of Intermediate-Depth Seismicity

Intermediate-depth seismicity is also considered a phenomenon related to dehydration processes in the slab. Water released by metamorphic reactions is thought to reduce the effective strength, thus allowing brittle failure at pressures and temperatures where rocks would otherwise be subject only to ductile deformation (dehydration embrittlement). Kirby et al. (1996) recognized in the global average a peak in intermediate-depth seismicity beneath volcanic fronts. From this coincidence, they suggested that the focus of crustal metamorphism, that is expressed in seismicity, is at the roots of arc volcanism.

### 21.1.3 Role of Upper-Plate Mantle Lithosphere

One of the most intriguing questions in geodynamics is the role of the mantle lithosphere in collisional mountain building. Crustal shortening has been found to be the dominant mechanism for the formation of thick crust and uplift of high continental plateaus (Nur et al. 1982). It seems obvious that the underlying mantle lithosphere must deform and thicken too, if crustal shortening occurs. Surprisingly, unusually thin lithosphere is often found beneath plateaus (e.g., beneath northern Tibet (Kind et al. 2002) and the southern Puna (Schurr et al. 1999)). How is mantle lithosphere removed and how does this effect deformation?

The concept of lithospheric delamination was introduced in the 1980s to explain both the absence of a thick mantle lid beneath elevated plateaus (e.g., Bird 1979; Houseman et al. 1981) and the rapid uplift and mafic magmatism often accompanying the late stage of an orogenic cycle (Kay and Mahlburg Kay 1993). In this model, thickened and cold lithosphere becomes gravitationally unstable, delaminates from the lower crust, and mobilizes the asthenosphere. The missing lithosphere is replaced by an influx of hot asthenosphere that causes uplift of the crust and volcanism.

Although lithospheric delamination has been proposed in many locations based on its symptoms, the process has only been observed geophysically in a few places. (Kay et al. 1993; Seber et al. 1996; Meissner et al. 1998; Kosarev et al. 1999; Kind et al. 2002; Boyd et al. 2004). This may be due to its ephemeral nature: numerical modeling indicates that lithospheric delamination and detachment is a catastrophic event that lasts only a few million years according to some authors (Houseman et al. 1981; Schott and Schmeling 1998). Is it possible to find arguments to support this hypothesis in the unique data set obtained from the central Andes?

### 21.1.4 Deformation of the Upper-Plate Crust

Identification of crustal seismicity is important for understanding present-day deformation. The distribution and frequency of crustal earthquakes help identify active fault systems, and the depth extent of the seismicity helps to estimate the transition from brittle to ductile behavior of the crust. Can inactive fault zones be identified and mapped by seismic methods? Can the mechanism of crustal shortening be derived from such information? Can partial melt be identified?

### 21.1.5 Comparison between the Central and Southern Andes

Subduction in the central Andes has resulted in an extremely thickened continental crust and the formation of a high plateau (Giese et al. 1999; Romanyuk et al. 1999; Schmitz et al. 1999). The crust of the southern Andes (south of 37° S), however, has remained relatively thin (with elevations below 2000 m; Lowrie and Hey 1981) and the back-arc is dominated by basin tectonics. In the central Andes, there have been four, eastward-migrating magmatic arcs built since the Jurassic (Coira et al. 1982; Scheuber and Reutter 1992), whereas in the south, the location of the magmatic arc has remained nearly stationary (Hervé 1994). The continental margin east of the central Andes is erosive, in contrast to the accretive, southern continental edge. Intermediate-depth seismic activity beneath the central Andes is much greater than that beneath the southern Andes and occurs at greater depths (80–100 km) compared to the southern Andes (50–70 m). Because the majority of intermediate-depth earthquakes has magnitudes below $M = 3$. Their depth distribution can only be imaged based on local earthquake observations. Before the ISSA 2000 experiment, knowledge about the distribution of hypocenters in the southern Andes was very poor. Comparison of both subduction regimes has helped to identify the different tectonic processes.

## 21.2 Database

### 21.2.1 Central Andes

The PISCO'94 seismological experiment covered the eastern fore-arc and the volcanic arc of Chile between 22 and 24° S (Fig. 21.2; Graeber and Asch 1999). The PISCO'94 data is complemented by travel times from a temporary deployment south of Calama by the Universidad de Chile, Santiago, and a permanent network around Antofagasta. The CINCA'95 network operated from August to October 1995 in the coastal area west of the PISCO'94 network. For one month, nine Ocean Bottom Hydrophones (OBH) from GEOMAR, Kiel were used to extend the network offshore (Fig. 21.2). The complete aftershock series of the Antofagasta earthquake (30 July 1995, $M = 8.0$) was recorded. For details of the deployment refer to Husen et al. (1999, 2001), Sobiesiak (2000) and Patzig (2000).

The passive part of the ANCORP'96 experiment extended the PISCO'94 coverage northwards to 20° S and also placed eight seismographs in the back-arc of the Bolivian Altiplano. For details of this deployment refer to Haberland (1999) and Haberland and Rietbrock (2001). Additionally, travel times from a deployment by the Universidad de Chile near Iquique are included. The PUNA'97 network (Schurr et al. 1999; Schurr 2001) extended coverage of the PISCO'94 and ANCORP'96 experiments to the east, across the recent magmatic arc and into the back-arc plateau of the Argentine Puna. The Chilean arc and fore-arc were also covered by stations at former PISCO'94 sites and seven, one-component seismographs operated by the Universidad de Chile (Fig. 21.2).

All three experiments were very similar in instrumentation and set-up. Most sites were equipped with PDAS data loggers and MARK L4-3d, short-period (eigenfrequency 1 Hz), three-component seismometers. Average station spacing was about 40 km. The stations ran in continuous mode with sample rates of 100 Hz. The 3-channel raw data (40–50 MByte per station and day) were stored on local hard discs. Each station was synchronized by a GPS-receiver to guarantee a common time base for the whole experiment. All data was stored in the GEOFON archive of the GeoForschungsZentrum Potsdam. In addition, data from two PASSCAL experiments (Beck et al. 1996) have also been used in the receiver function study.

### 21.2.2 Southern Andes

In the first four months of 2000, an integrated active and passive seismological experiment, ISSA 2000, ran in the Southern Andes between 36–40° S and 75–70° W in Chile and Argentina (Asch et al. 2000; Lüth et al. 2000, 2003). The station deployment is shown in Fig. 21.3. During a 100-day period, local seismicity was recorded using 70 receivers. Additionally, broadband instruments were installed for two years to record teleseismic events for a receiver function study. For about one month, fifteen OBHs and one Ocean Bottom Seismograph (OBS), belonging to GEOMAR, Kiel, were installed offshore. This set-up enabled the retrieval of reliable hypocenter and focal mechanism data in the seismic coupling zone between the trench and the coast. The ocean-bottom recorders were deployed and retrieved by a small research vessel (named 'Kay Kay') belonging to the Oceanographic Department of the Universidad de Concepcion. The quality of the recorded data from the southern Andes is not as good as that from the central Andes as cultural noise is much higher in the south. Nevertheless, we achieved a very reasonable database with a comparable threshold magnitude to that of the central Andes.

### 21.2.3 Methods Applied

The results from the seismological SFB 267 experiments document a continuous improvement in understanding subduction-related processes. This work has also resulted in significantly improved methods for obtaining and evaluating seismic data sets.

## 21.2.4 Creation of Base Catalogs

The base for all seismological studies is a catalog that provides reliable source information. For teleseismic and controlled-source experiments this is not a problem, because the time windows of interest are predetermined. In local earthquake studies, this information is not available *a priori* and a large amount of continuous data has to be checked carefully, to identify even very weak earthquakes. This is a very time consuming task, and Asch (1998) developed a method to determine P- and S-onsets automatically. Calculating the coincidence of P- and S-onsets at adjacent stations helps to rule out mis-triggers.

## 21.2.5 Tomography

The basic theory underlying local earthquake tomography was introduced by Thurber (1993). The $v_p$ and $v_p/v_s$ models of the central Andes were computed using SIMULPS, a computer program published by Evans et al. (1994) and further improved by a number of authors (Thurber and Eberhart-Phillips 1999; Eberhart-Phillips and Michael 1998). Rietbrock (1996, 2001) introduced the inversion for $Q_p$, which was applied by Haberland (1999), Haberland and Rietbrock (2001), Schurr (1999), and Schurr (2001). The latter author adjusted the ray-bending algorithm to match the situation in the subduction regime. $Q_p$-tomography based on local earthquake data, in particular, has turned out to be a powerful tool for imaging subduction-related features above the Wadati-Benioff zone. The current knowledge of the area above the slab mainly rests upon the results from these experiments. Data used in this report are P and S-P differential travel times, earthquake locations, and $t^*$ parameters calculated by the inversion of P-wave spectra (Haberland and Rietbrock 2001). For a more detailed description of the inversion process see Schurr (2001).

## 21.2.6 Guided Waves

Based on the papers by Bock et al. (2000) and Martin et al. (2003), we discuss the application of guided waves caused by the slab surface or layering within the subducted Andean lithosphere. Helffrich and Abers (1997), for example, employed converted waves to investigate a low-velocity layer in the eastern Aleutian subduction zone. Likewise, investigations at the northeast Japanese subduction zone revealed undulating velocity contrasts between the accretional wedge and subducted crust at various depths. These velocity contrasts are interpreted as evidence of phase changes in the gabbroic subducted crust (Snoke et al. 1978; Helffrich 1996). Any continuous layered structure that is slow compared to bounding media can act as a wave guide, provided no large heterogeneities are present. This structure causes, for certain source-receiver configurations, internally reflected waves that produce prominent interference patterns called guided waves. The parameters that are investigated are the geometry of the subducting slab, receiver position, the propagation length of signals along the slab, the thickness of subducted crust, and the source position relative to the layer.

## 21.2.7 Receiver Function Method

Receiver function analysis has become a routine method for studying crust and upper-mantle discontinuities (Langston 1977; Vinnik 1977). It detects discontinuities beneath seismic stations by identifying P-to-$S_v$ converted waves. The converted S waves travel more slowly to the stations than the originating P waves do and, therefore, will be recorded after the direct P wave in the P-wave train. The converted energy is usually weak (only a few percent of the P-wave energy) and can be isolated from the P wave with the help of data processing techniques (e.g., Yuan et al. 1997).

We measured the differential time between the converted S and the direct P waves, which is an indicator of the depth of the discontinuity. Along with the primary conversions, multiple phases reverberating between the discontinuity and the Earth's surface. These multiples are useful for improving the constraint of the crustal Poisson's ratio and thickness (Kind et al. 2002), but they also interfere with the Ps conversions at lithospheric and asthenospheric depths. The multiples have different distance dependence (move out) and thus can be distinguished from the Ps. By stacking receiver functions after Ps move-out correction, primary conversions are enhanced while the multiple phases are suppressed. Similarly to reflection seismic processing, depth migration is also performed to produce cross sections (Kosarev et al. 1999).

## 21.3 Results

### 21.3.1 Seismic Observations from the Andean Slab

Conventionally, the slab is defined by its seismicity. Figures 21.4 and 21.5 show the distribution of the seismicity obtained by our experiments in the central and southern Andes, respectively. We identified 250 earthquakes in the ISSA 2000 data set for the southern Andes. This is the first accurate image of the shape of the Wadati-Benioff zone in this area. The depth-frequency distribution of the local earthquakes reaches a maximum at 25 km depth (Fig. 21.5), reflecting the zone of frictional instability. The

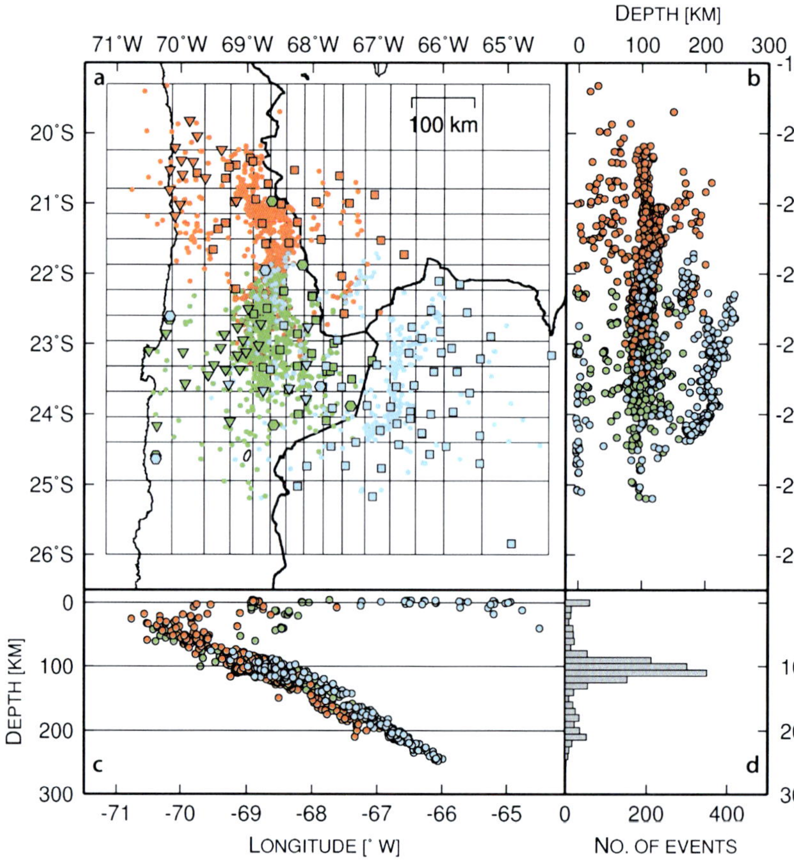

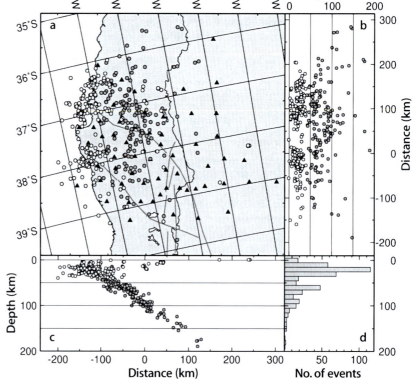

**Fig. 21.4.**
Epicenter and station map (**a**), N-S (**b**) and E-W (**c**) hypocenter projections of events used in the tomographic study. The PISCO (*green*), ANCORP (*orange*) and PUNA (*blue*) stations and events are plotted, as are the broadband (*hexagons*), short-period three-component (*squares*) and one-component (*inverted triangles*) seismograph locations. Most events are of intermediate depth (**d**). Also shown is the inversion grid

**Fig. 21.5.**
Distribution of 440 earthquakes recorded during the ISSA 2000 experiment. The figure is rotated by 11.4° to get a good focus on the Wadati-Benioff zone. **a** *White dots* show crustal events (< 40 km depth) while *gray dots* represent the slab seismicity below this depth. *Black triangles* denote field stations. The main fault systems (Bio-Bio, Gastre, Liquine-Ofqui) are marked as well. **b** N-S and (**c**) E-W depth sections. **d** Frequency depth distribution

**Fig. 21.6.**
Receiver-function cross sections in the central and southern Andes. The IASP91 velocity model is used in both sections for time-depth conversion. Receiver-function amplitudes are color-coded: positive amplitudes (velocity increases with depth) are plotted in *red* and negative (velocity decreases with depth) in *blue*. Significant conversion phases are marked and labeled. *Triangles* denote the seismic stations. Earthquakes from the local catalogs. *ALVZ*: altiplano low velocity zone

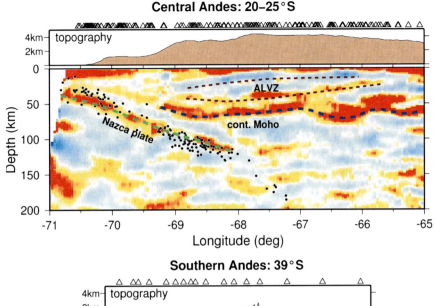

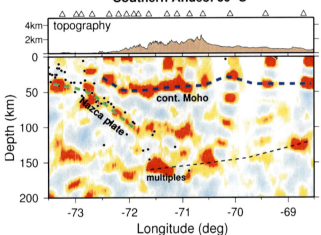

second and deeper peak at 60 km correlates with the depth at which we expect to find intermediate-depth earthquakes (Kirby et al. 1996; Davies 1999). In the central Andes, the majority of the these earthquakes are located at 100–120 km depth (Fig. 21.4). In the younger, and therefore hotter, subducted oceanic crust in the south, the blueschist to eclogite transformation takes place at depths shallower than 100 km. Peacock (1993) determined that this reaction has a lower limit of approximately 500 °C and 1.3 GPa (40–50 km). We think that we have identified the southernmost part of the Chilean continental margin where this phase transformation can be observed.

Receiver function images to 200 km depth in the central and southern Andes are shown in Fig. 21.6. The downgoing Nazca Plate can be observed to a depth of about 120 km as a positive (red, velocity increases with depth) image of the oceanic Moho. The receiver function waveforms have been modeled using a 5–10 km-thick, low-velocity layer with an S-wave velocity contrast of about 15% (Yuan et al. 2000). This layer thickness and velocity contrast to the surrounding mantle corresponds to non-eclogitized, mafic oceanic crust versus garnet peridotite. The likely cause of the breakdown in the observation of the oceanic crust at about 120 km depth is probably the completion of the gabbro to eclogite transformation in most of the oceanic crust. This would make the oceanic crust seismically almost indistinguishable from the peridotitic mantle for the relatively long-period, teleseismic receiver functions. As eclogite is the thermodynamically stable modification of both water-containing and dry mafic rocks at depths greater than 80 km (Peacock 1993), observation of non-eclogitized crust deeper than 80 km provides direct evidence for the kinematically delayed metamorphic reactions in the subducting plate that were previously invoked to explain intermediate-depth earthquakes (Kirby et al. 1996).

Bock et al. (2000) and Martin et al. (2003) used P-to-S converted and guided waves of local events at intermediate depths to show that these earthquakes are located in the subducted oceanic crust or underlying oceanic mantle. Their observations also indicated that the crustal wave guide exists to 160 km depth.

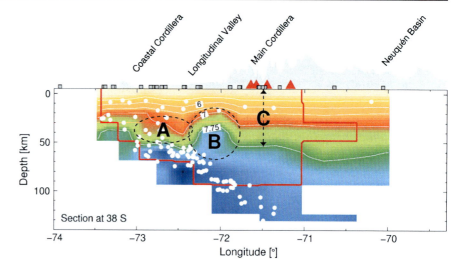

**Fig. 21.7.**
Cross section (30 km N-S extension) of $v_p$ tomography results along 38° S and its most important features. *Dots* mark local earthquakes used in the tomographic inversion. On top topography, seismological stations (*squares*) and active volcanoes (*red triangles*) are shown. **a** Reduced $v_p$ velocities below the Coastal Cordillera; **b** increased $v_p$ velocities below the Longitudinal Valley; and **c** mantle velocities not until 55 km below the magmatic arc

Husen et al. (2001) and Patzig (2000) derived $v_p$ and $v_p/v_s$ velocity models from local earthquake tomography in the central Andes, covering the coastal area around Antofagasta. They found high $v_p$ velocities in the upper crust that are associated with large quantities of basic to intermediate rocks formed by Jurassic magmatic activity. Their models show no great lateral variations and no prominent anomalies, in contrast to those of Graeber et al. (1999), Haberland et al. (2001) and Schurr (2001) for the active, central Andean magmatic arc. Only in front of the seismic coupling zone is the $v_p$ velocity reduced. This occurs between 20 and 30 km depth below the Mejillones Peninsula, which has been uplifted since the Pliocene (Ortlieb et al. 1996). Husen et al. (2001) explain this low velocity zone by accretion of eroded material.

A similar situation can be found in the southern Andes in the Arauco Peninsula. Bohm (2004) presented a $v_p$ velocity model for the southern Andes between 36 and 40° S (Fig. 21.7). The most striking difference to the $v_p$ models of the central Andes is a thin crust below the longitudinal valley associated with high $v_p$ velocities in the lower crust. Between the Coastal Cordillera and the Longitudinal Valley, reduced $v_p$ velocities characterize the lower crust and this is interpreted as being caused by a serpentinized mantle wedge.

## 21.3.2 Paths of Ascending Fluids from the Slab

Based on a 3D tomographic model from local earthquake observations in the Central Andes (Schurr et al. 1999) P, P-S, and $t^*$ parameters were used to calculate $Q_p$ models by the inversion of P-wave spectra (Schurr 2001; Schurr et al. 2003; Schurr et al. 2006). Figure 21.8 shows a suite of $Q_p$ tomographic sections through the central Andes. In the sections north of 22.1° S, the low $Q_p$ anomalies appear to originate at a prominent earthquake cluster occurring at 100 km depth (Haberland and Rietbrock 2001).

From there, the ascent is straight up, or with a considerable lateral westward component, to the base of the arc volcanoes. Further south, the volcanic arc is displaced about 100 km to the east by the Atacama block. The arc volcanoes east of this block (south of 22.1° S) have lost their connection to the 100 km-deep cluster beneath the fore-arc and, instead, appear to be fed from a deeper cluster to the east.

We believe that the low $Q_p$ areas belong to (hot) magmas or other fluids. These images suggest that the often implicitly assumed transport paths of magmas straight up to the base of the volcanic front are not adequate, but that large lateral distances can also be covered. An especially good example is the section at 24.2° S (Fig. 21.8). Here, the arc volcanoes and the back-arc volcano, Cerro Tuzgle, located 200 km east of the volcanic front, appear to have the same source at 200 km depth. For Cerro Tuzgle, the ascent is straight up, whereas magmas feeding the arc volcanoes appear to be transported along the slab to the base of the crust and then up to a crustal magma chamber.

At sub-solidus, intrinsic Q is a linear function of homologous temperature. Water substantially reduces the melting temperature of peridotite (Kawamoto and Holloway 1997) and thus greatly amplifies seismic attenuation (inverse Q). Hence, low $Q_p$ values may indicate water as well as partial melts at temperatures greater than 600–700 °C. The intriguing feature in the cross sections for $Q_p$ (Fig. 21.8) is that the strong, low-$Q_p$ anomalies that cross the mantle wedge and penetrate the crust at the base of the arc volcanoes, appear to originate at the distinct earthquake clusters occurring at 100 and 200 km depth.

It is very suggestive that these anomalies are due to water fluxed into the mantle wedge by earthquakes, reducing the solidus and, probably, subsequently inducing partial melting. Once water enters the mantle wedge, it will ascend and act as a flux to aid melting where the wet solidus of peridotite is exceeded. The low $Q_p$ anomalies

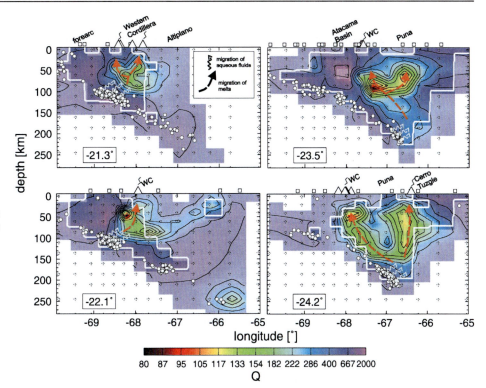

**Fig. 21.8.**
Four cross sections for $Q_p$ with interpretation. Water is released from the slab at the discrete earthquake clusters (maybe through hydro-fracturing triggered by faulting, see text). Water causes flux melting in the overlying, hot mantle wedge. Regions of $Q_p$ higher 200 probably contain significant amounts of partial melt. Melt ascent pathways are not straight up, as is often implicitly assumed, but have a significant horizontal component. The paths of magmas vary, probably because of their source locations, deviatoric tectonic stresses, and asthenospheric flow

probably represent the ascent paths and source regions of fluids and partial melts. In almost all sketches of subduction zones in present-day tectonic textbooks, the sources of lavas erupting at the volcanic arc are assumed to originate atop the subducted slab directly beneath the volcanoes. Transport is either drawn as ascending blobs of magma or as simple, thin, vertical channels.

### 21.3.3 Lithospheric Delamination in the Central Andes

Between 23 and 24° S in the northern Puna of the central Andes, we imaged a fast, high-$Q_p$ body that dips for 100 km from the base of the easternmost Puna crust to the subducted Nazca Plate (Fig. 21.9). South of this body, the mantle wedge has low velocity and Q, and the Nazca Plate thickens by several tens of kilometers above the sharply defined Wadati-Benioff zone. This region of mantle wedge and slab is well resolved, and synthetic tests show that both the slab and structures in the mantle wedge can be retrieved quite accurately (Schurr 2001). Anomalies in velocity and attenuation have, to a first degree, the same physical cause: variations in homologous temperature. Because velocity and $Q_p$ images are based on independent measurements (travel times and spectral shapes, respectively), their similarity provides additional confidence. We interpret the fast, high-$Q_p$ body as delaminating, and probably detaching, South American lithosphere that has been shortened and thickened beneath the Eastern Cordillera and parts of the Puna Plateau. The cold, sub-crustal mantle sinks much faster into the underlying hotter asthenosphere than the time needed for thermal diffusion, causing the strong velocity and $xQ_p$ contrasts. Further south, at 24° S, where the mantle wedge is characterized by low $v_p$ and low $Q_p$, and the slab appears to thicken above the Wadati-Benioff zone, we suggest that detachment has already been completed with cold continental lithosphere now resting atop the subducted Nazca Plate.

We interpret the low $v_p$ and $Q_p$ anomalies engulfing the detached lithosphere as expressions of viscous heating and decompressional melting owing to rapid influx and ascent of hot asthenospheric mantle. Hot, fresh mantle at the base of the crust probably causes partial melting in the lower crust and is responsible for the penetrating "hot" anomalies observed there. We also see very low P-wave velocities in the uppermost crust and large, positive station corrections for $Q_p$ point to a shallow low-$Q_p$ anomaly (Schurr 2001). Extensive, shallow, low-velocity zones in the back-arc crust have also been detected with teleseismic receiver functions (Yuan et al. 2000). They were interpreted as large regions of melt accumulation that possibly feed the large ignimbrite fields that have erupted in the back-arc since the Miocene.

Lithospheric delamination might be the ultimate heat source in the mantle that mobilizes the asthenosphere and triggers large-scale crustal melting and widespread felsic volcanism. Kay and Mahlburg-Kay (1993) proposed loss of the lithospheric lid at around 26° S, just south of our

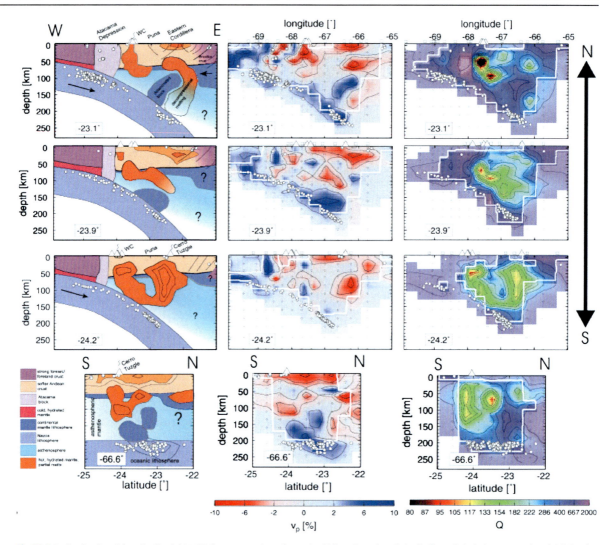

**Fig. 21.9.** Latitudinal and longitudinal (66.6° W) cross sections for $v_p$ (*middle column*) and $Q_p$ (*left*), and their interpretation (*right*). The fast, high-$Q_p$ region is interpreted as detached South American lithosphere. The apparent thickening of the Nazca Plate is caused by detached lithospheric mantle lying atop the oceanic slab

study area, based on low-$Q_p$ mantle (Whitman et al. 1992) and eruptions of ocean-island-type basalts and calc-alkaline lavas one to two million years ago. Shoshonitic lavas, thought to have a lithospheric mantle source, are found to 24° S, the southern limit of our delamination event. At 24° S, above a zone of low velocity and high attenuation in the crust and mantle (Fig. 21.9) stands the Quaternary stratovolcano, Cerro Tuzgle. In its vicinity, shoshonites as well as ocean-island-type basalts and ignimbrites occur. We believe that this volcanism is the result of a recent detachment of lithosphere, of which the remains are now lying on top of the Nazca Plate. The shoshonitic, ocean-island-type, and ignimbrite lavas probably accompanied the delamination. The youngest, andesitic lavas could be subduction related, owing to fluids released from the slab into the now hot, asthenospheric mantle. Because there is no mantle lid left, magmas can reach the surface, as they do 200 km to the west, in the volcanoes of the present volcanic arc.

There are further arguments for the loss of lithosphere beneath the Puna. The Puna Plateau has an average elevation of 4.5 km, almost 1 km higher than the Altiplano to the north. Yet, new estimates for Moho depth from receiver functions (Yuan et al. 2000) indicate significantly thinner crust for the Puna (55–60 km vs. ~70 km for the Altiplano). In addition, shortening estimates from cross-sectional balancing for the Puna are consistently lower than they are for the Altiplano to the north (Kley and Monaldi 1998; Grier et al. 1991). It can explain only 30% of the crustal cross-sectional area owing to tectonic shortening in the Puna. Even if shortening is underestimated, other mechanisms lifting the plateau are certainly required. A missing lithospheric root could explain the discrepancy.

## 21.3.4 Crustal Seismicity

Besides the first accurate definition of the Wadati-Benioff zone in south-central Chile, the pattern of seismicity bears important information on the tectonics of the continental upper plate. The distribution of seismic release in the upper plate (depth < 40 km) shows a clear concentration of active deformation in the fore-arc of the active margin system. In contrast, seismicity is scarce in the Main Cordillera, with the prominent Liquine-Ofqui intra-arc discontinuity, and the Longitudinal Valley (Fig. 21.5).

The oceanic lithosphere north of the Mocha fracture zone (Fig. 21.3) was formed at the Pacific-Farallon spreading center 32 million years ago (Herron 1981; Cifuentes 1989). South of the Valdivia fracture zone, the oceanic lithosphere was generated within 20 million years at the Chile Ridge (Antarctic-Nazca spreading center). The Valdivia and Mocha fracture zones and the Chile Trench (39–40° S) define a triangle, and it is assumed that this area was also generated at the Pacific-Farallon spreading center. The age is unclear, but it is probably younger than the crust north of the Mocha fracture zone. This area delineates the boundary between oceanic crusts of different ages and of different structural grains. The change in seismicity from north to south found in teleseismic observations (NEIC catalog, Fig. 1) as well in the regional data set obtained during the ISSA experiment (Fig. 21.5) reflects this situation, with seismicity in the younger oceanic crust reduced compared to the older crust.

The main area of seismic activity along the coast is identical to the axis of modern major uplift and the postulated center of basal accretion and antiformal stacking at depth (Lohrmann et al. 2000). Regional centers of seismic release south and north of the Arauco Peninsula can be correlated with the prominent Gastre and Bio-Bio fault zones, respectively. Both trend NW-SE and cross the upper plate into the Main Cordillera. Geological field observations show the active character of the faults, whereas their seismically active part seems restricted to the fore-arc zone. It is striking that these two faults limit a zone with apparently no seismicity, corresponding to the Arauco Peninsula and Nahuelbuta Range, which, however, have the highest uplift and topography in the fore-arc (Fig. 21.5).

In the Puna, all historical seismicity and micro-seismicity is located at the El Toro-Olacapato-Calama Lineament (TOCL) and south of it, suggesting that the TOCL marks a structural boundary in the plateau. This suggestion is supported by the fact that young, mafic, back-arc volcanism is also concentrated in the southern Puna and reaches north only up to the TOCL (Allmendinger et al. 1997). We think that the Puna north of the TOCL behaves like a stable block with no internal deformation, whereas to the south, the plateau is weaker and internally deforming. The depths for events occurring inside the plateau crust are very shallow (<10 km), indicating a hot geothermal and a shallow brittle-ductile transition.

## 21.3.5 Mantle Discontinuities

In Fig. 21.10, we compare the receiver-function section of the central Andes with teleseismic tomography (Bijwaard et al. 1998). The 410 km discontinuity is poorly imaged

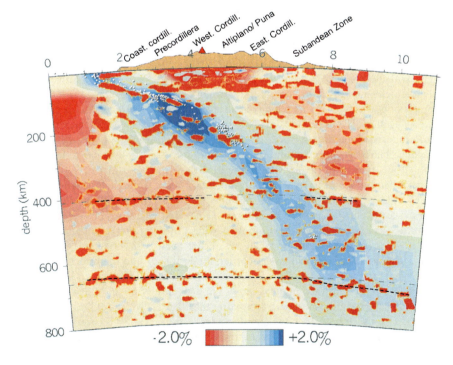

**Fig. 21.10.**
Comparison of the results from receiver functions and teleseismic tomography (after Bijwaard et al. 1998). An E-W section of receiver functions from diverse seismic arrays in the central Andes is plotted in *red*, while the tomography section at 23° S is plotted in *blue* and red as background. Surface topography along with major geological units are plotted at the top. *Thick dashed lines* mark the 410 km and 600 km discontinuities observed by receiver functions

in the slab region, as is indicated by high P-wave velocity in the tomography, while it is well observed in the western part of the profile. The 660 km discontinuity is depressed by 30–40 km within the slab. The 660 km discontinuity is depressed by 30–40 km within the slab. This discontinuity is commonly associated with phase transformation of gamma-spinel to perovskite and magnesio-wüstite. The depth, at which the phase transformation occurs, depends on ambient mantle temperature. In a cold mantle the 660 km discontinuity may be depressed due to a negative Clapeyron slope ($dP/dT$). Taking $-3.0$ MPa K$^{-1}$ as the Clapeyron slope of this transformation (Bina and Helffrich 1994), we estimate the temperature decrease corresponding to the 30–40 km depression of the 660 km boundary to be 300 K at the base of the mantle transition zone.

## 21.4 Discussions

### 21.4.1 Role of Fluids and Melts

In the past, different mechanisms for the transport of aqueous fluids and melts in the mantle wedge have been proposed, such as porous flow along grain boundaries and the currently popular, buoyancy-driven crack propagation theory. Regarding the first mechanism, it has been shown that wetting of mantle-mineral grain boundaries might be complete (dihedral angles = 60°) at the pressures and temperatures of the mantle wedge (Mibe et al. 1999) thus, principally, allowing the porous flow of water in the mantle. Iwamori (1998) developed a numerical model for permeable flow and melt generation in the mantle wedge. In his results, water released from a down-dragged serpentine and chlorite layer (similar to Tatsumi's (1989) model) rises straight up to the melt source. Melts are then transported horizontally trenchwards by corner flow to the base of the volcanic arc. Such a geometry does not agree with our images. Yet, models of permeable flow that include deformation of the matrix may produce more complex flow fields, better matching our results.

Propagation of isolated, fluid-filled fractures is the second proposed mode of transport for aqueous fluids and magmas. The propagation of fluid-filled fractures is controlled by apparent buoyancy forces, the directions of principal stresses, and tectonic stress gradients. Dahm (2000) modeled the propagation of fluid-filled fractures in a subduction environment. The stress field was calculated for a stationary corner-flow model. Owing to the deviatoric stresses, fractures propagate significant horizontal distances away from the wedge corner, similarly to the paths found in the northern sections of our model. For steeper dips (= 45°), ascent paths can be upwards and along the slab, resembling those we observe for the deeper cluster. The real stress field in the mantle wedge is probably more complex than simulated by the corner-flow model, and this is especially true for the compressive environment of the Andes. Additionally, variations in viscosity and the interaction of fluids with mantle material may render things more complicated in the real Earth. It is also debatable if arrays of isolated, fluid-filled cracks can cause the large, continuous anomalies we observe.

### 21.4.2 Role of the Upper-Plate Lithosphere during Deformation

Considering that the magnitude of lithospheric shortening across the Altiplano is much greater than that across the Puna, the Altiplano seems a likelier place for delamination to occur. Nonetheless, under the Altiplano, there still appears to be a lithospheric lid (Reutter et al. 1988; Whitman et al. 1992; Wigger et al. 1994; Myers et al. 1998), whereas beneath the Puna, the lithosphere is, or is in the process of being, completely removed.

These two provinces are also distinguished by different styles of upper-crustal foreland deformation (Whitman et al. 1996). Allmendinger and Gubbels (1996) interpreted the thin-skinned deformation of the Altiplano foreland as indicative of a simple-shear mode of lithospheric shortening, and the thick-skinned deformation of the Puna and its foreland as an expression of pure-shear shortening. This means that beneath the Altiplano, strong Brazilian lithosphere may be underthrust as a whole, intact plate. In contrast, beneath the Puna, shortening of the lithosphere is distributed.

Whitman et al. (1996) suggested that segmentation of the plateau and foreland reflect pre-Andean differences in the structure of crust and mantle lithosphere. The restriction of thick Paleozoic sedimentary sections and tin mineralization to the regions north of 23.5° S might constrain the limit of old cratonic basement (Allmendinger and Gubbels 1996; Whitman et al. 1996). In contrast, crustal xenoliths from the Cretaceous Salta Rift east of the Puna indicate high temperatures in the crust and lithosphere at 90 Ma (Lucassen et al. 1998).

The inherited different states of the Altiplano and Puna lithospheres and their different modes of deformation result in two fundamentally different lithospheric rheologies that might explain why large-scale detachment of mantle lithosphere takes place beneath the Puna but not beneath the Altiplano. Numerical modeling shows that a normal olivine rheology is too strong to allow delamination (Schott and Schmeling 1998). The lithosphere has to be weakened (e.g., by faulting) to deform, detach and sink into the underlying mantle. We suggest that the present difference in lithospheric thickness between the Altiplano and Puna Plateaus are not inherited, but are caused by loss of lithosphere beneath the Puna, which, in turn, is aided by pre-Andean, weak rheology and weakening owing to pure-shear deformation.

## 21.4.3 Comparison between the Central and Southern Andes

While subduction processes with compressive forces formed an extremely thickened (up to 70 km) continental crust in the central Andes (Zandt et al. 1994; Beck et al. 1996; Yuan et al. 2000), the crust is "only" 40–50 km thick in the southern Andes (Lomnitz 1962; Bohm 2004).

The activity of intermediate-depth seismicity in the central Andes is about 10 times higher than that in the Southern Andes. Also, the depth of the maximum release of seismic energy differs remarkably from 50–70 km in the south to 80–110 km in the north, which is explained by the age (i.e. temperature) of the subducted Nazca Plate. In the central Andes, the dip angle of the Wadati-Benioff zone is about 19–20° (Husen et al. 1999) in the uppermost 40 km and 35° at greater depths (Schurr 2001). In the south, these angles are specified by Bohm et al. (2002) for the respective depths to be 10° and 30°. $v_p$ velocities in the subducted oceanic crust are also similar for the upper 80 km ($v_p = 8.2$ km s$^{-1}$). In the model introduced by Graeber et al. (1999), $v_p$ increases to 8.6–9.0 km s$^{-1}$. In the southern Andes, this depth range could not be resolved. In all models, an increased $v_p/v_s$ ratio for the subducted plate is verified and is explained by a low-velocity channel on top of the subducted Nazca Plate (Martin et al. 2003).

A number of wide-angle profiles has resolved the complex crustal structure in the central Andes (Wigger et al. 1994; Schmitz et al. 1999; Patzwahl et al. 1999). In the fore-arc, high $v_p$ velocities are already reached at depths of 10 km, slightly decreasing from west to east (Coastal Cordillera: 6.6 km s$^{-1}$; Precordillera: 6.3 km s$^{-1}$). It is interpreted as Jurassic lower crust that was elevated owing to erosion of the upper crust in the uplifted coastal range. Furthermore, high velocity regions were observed 20–25 km below the coast (7.2 km s$^{-1}$) and 35–40 km below the Precordillera (6.8–7.2 km s$^{-1}$) in the lower continental crust. Beneath this, no typical mantle velocities were observed. At a depth of 65–70 km below the Precordillera, a strong increase in $v_p$ velocity (= 7.5 km s$^{-1}$) is interpreted as the crust/mantle boundary. The subjacent area is characterized by an increased $v_p/v_s$ ratio (Graeber et al. 1999).

The low $v_p$ and increased $v_p/v_s$ ratio possibly originates from hydration (serpentinization) of the continental mantle wedge, probably owing to fluids released from the subducted plate (Graeber et al. 1999; Giese et al. 1999). This serpentinized mantle wedge below the Coastal Cordillera and Precordillera at 40–70 km depth is observed in the southern Andes as well. There, it is confirmed by reduced $v_p$ velocities below the Coastal Cordillera (Bohm 2004) and wide-angle measurements from the ISSA and SPOC experiments. Regrettably, these observations cannot be verified by $v_p/v_s$ observations, which are not available for this region to date. The shallow depth of the serpentinized mantle wedge in the south indicates that metamorphic processes within the subducted plate take place at lesser depth in the southern Andes.

The reason for the extreme crustal thickness in the central Andes is still under discussion. Whereas former publications favor magmatic addition (James 1971), more recent papers prefer a combination of several processes. An essential point is tectonic shortening with a small contribution by magmatic addition and lithospheric thinning, hydration of the lithospheric mantle wedge, and tectonic underplating in the fore-arc (Isacks 1988; Allmendinger et al. 1996; Giese et al. 1999). A fault-and-thrust belt, as seen in the Subandean in the central Andes did not exist in the back-arc of the southern Andes, so tectonic crustal thickening can be ruled out.

## References

Allmendinger R, Gubbels T (1996) Pure and simple shear plateau uplift, Altiplano-Puna, Argentina and Bolivia. Tectonophysics 259:1–13

Allmendinger R, Jordan TE, Kay SM, Isacks BL (1997) The evolution of the Altiplano-Puna plateau of the central Andes. Ann Rev Earth Planet Sci 25:139–174

Asch G (1998) Präzise und schnelle Herdparameterbestimmung – eine Herausforderung an die moderne Seismologie. Habilitationsschrift, Freie Universität Berlin

Asch G, et al. (2000) The ISSA 2000 passive seismological experiment. EOS 81(48):S71A-09

Beck SL, Zandt G, Myers SC, Wallace TC, Silver PG, Drake L (1996) Crustal thickness variations in the Central Andes. Geology 24:407–410

Bijwaard H, Spakman W, Engdahl ER (1998) Closing the gap between regional and global travel time tomography. J Geophys Res 103:30055–30078

Bina CR, Helffrich G (1994) Phase transition Clapeyron slopes and transition zone seismic discontinuity topography. J Geophys Res 99:15853–15860

Bird P (1979) Continental delamination and the Colorado plateau. J Geophys Res 84:7561–7571

Bock G, Schurr B, Asch G (2000) High-resolution receiver function image of the oceanic Moho in the subducting Nazca plate from converted waves. Geophys Res Lett 27:3929–3932

Bohm M (2004) 3-D Lokalbebentomographie der südlichen Anden zwischen 36° und 40° S. GeoForschungZentrum Potsdam, STR 04/15, 141p. http://www.gfz-potsdam.de/bib/pub/str0415/0415.htm

Bohm M, Lüth S, Echtler H, Asch G, Bataille K, Bruhn C, Rietbrock A, Wigger P (2002) The Southern Andes between 36° and 40° S latitude: seismicity and average seismic velocities. Tectonophysics 356(4):275–289

Cahill T, Isacks BL (1992) Seismicity and shape of the subducted Nazca Plate. J Geophys Res 97:17503–17529

Cifuentes IL (1989) The 1960 Chilean Earthquake. J Geophys Res 94:665–680

Coira B, Davidson J, Mpodozis C, Ramos VA (1982) Tectonic and magmatic evolution of the Andes of northern Argentina and Chile. Earth Sci Rev 18:303–332

Dahm T (2000) Numerical simulations of the propagation path and the arrest of fluid-filled fractures in the earth. Geophys J Int 141:623–638

Davis JH (1999) The role of hydraulic fractures and intermediate-depth earthquakes in generating subduction-zone magmatism. Nature 398:142–14

Eberhart-Phillips D, Michael AJ (1998) Seismotectonics of the Loma Prieta, California, region determined from three-dimensional Vp, Vp/Vs, and seismicity. J Geophys Res 103:21099–21120

Evans J, Eberhard-Phillips D, Thurber CH (1994) User's manual for simulps12 for imaging Vp and Vp/Vs: a derivative of the "Thurber" tomographic inversion simul3 for local earthquakes and explosions. USGS Open File Report, pp 94–431

Giese P, Scheuber E, Schilling F, Schmitz M, Wigger P (1999) Crustal thickening processes in the central Andes and the different natures of the Moho-discontinuity. J S Am Earth Sci 12:201–220

Gill JB (1981) Orogenic andesites and plate tectonics. Springer-Verlag, Berlin Heidelberg New York

Graeber F, Asch G (1999) Three-dimensional models of P-wave velocity and P-to-S-velocity ratio in the southern Central Andes by simultaneous inversion of local earthquake data. J Geophys Res 104:20237–20256

Grier ME, Salfity JA, Allmendinger R (1991) Andean reactivation of Cretaceous Salta rift northwestern Argentina. J S Am Earth Sci 4:351–372

Gudmundson O, Sambridge M (1998) A regionalized upper mantle (RUM) seismic model. J Geophys Res 103:7121–7136

Haberland C (1999) Die Verteilung seismischer Absorption in den Zentralen Anden. Ph.D. thesis, Freie Universität Berlin

Haberland C, Rietbrock A (2001) Attenuation tomography in the western Central Andes: a detailed insight into the structure of a magmatic arc. J Geophys Res 106:11151–11167

Helffrich GR (1996) Subducted lithospheric slab velocity structure: observations and mineralogical inferences. In: Subduction top to bottom. Geophys Monogr Ser 96:215–222

Helffrich GR, Abers GA (1997) Slab low-velocity layer in the eastern Aleutian subduction zone. Geophys J I 130:640–648

Herron EM (1981) Chile Margin near 38° S: evidence for a genetic relationship between continental and marine geologic features or a case of curious coincidences? Memoir 154:755–766

Hervé F (1994) The Southern Andes between 39° and 44° S latitude: the geological signature of a transpressive tectonic regime related to a magmatic arc. In: Tectonics of the Southern Central Andes. Springer-Verlag, Berlin Heidelberg New York, pp 243–248

Houseman GA, McKenzie DP, Molnar P (1981) Convective instability of a thickened boundary layer and its relevance for thermal evolution of continental convergent belts. J Geophys Res 86:6115–6132

Husen S, Kissling E, Flueh E, Asch G (1999) Accurate hypocentre determination in the seismogenic zone of the subducting Nazca plate in northern Chile using a combined on-/offshore network. Geophys J Int 138:687–701

Husen S, Kissling E, Flueh E (2001) Local earthquake tomography of shallow subduction in north Chile: a combined onshore and offshore study. J Geophys Res 105:28183–28198

Iwamori H (1998) Transportation of $H_2O$ and melting in subduction zones. Earth Planet Sci Lett 160:65–80

Kawamoto T, Holloway J (1997) Melting temperature and partial melt chemistry of $H_2O$-saturated mantle peridotite to 11 GPa. Science 276:240–243

Kay RW, Mahlburg Kay S (1993) Delamination and delamination magmatism. Tectonophysics 219:177–189

Kind R, et al. (2002) Seismic images of crust and upper mantle beneath Tibet: evidence for Eurasian plate subduction. Science 298:1219–1221

Kirby S, Engdahl ER, Denlinger R (1996) Intermediate depth intraslab earthquakes and arc volcanism as physical expression of crustal and uppermost mantle metamorphism in subducting slabs. In: Bebout G, Scholl D, Kirby S, Platt J (eds) Subduction: Top to Bottom, Vol 96. AGU, pp 195–214

Kley J, Monaldi C (1998) Tectonic shortening and crustal thickness in the Central Andes: how good is correlation? Geology 26:723–726

Kosarev G, Kind R, Sobolev S, Yuan X, Hanka W, Oreshin S (1999) Seismic evidence for detached lithospheric mantle beneath Tibet. Science 283:1306–1309

Langston CA (1977) Oregon, crustal and upper mantle structure from teleseismic p and s waves. Bull Seism Soc Am 67:713–724

Lohrmann J, Kukowski N, Oncken O (2000) Identification of controlling parameters of the accretionary southern Chilean margin (38° S) with analogue models. In: 17. Geowissenschaftliches Lateinamerika-Kolloquium. Stuttgart, Abstract p 50

Lowrie A, Hey R (1981) Geological and geophysical variations along the western margin of Chile near latitude 33° to 36° S and their relation to Nazca plate subduction. Geol Soc Am Mem 154: 741–754

Lucassen F, Leverenz S, Franz G, Viramonte J, Mezger K (1998) Metamorphism, isotopic ages and composition of lower crustal granulite xenoliths from the Cretaceous Salta Rift, Argentina. Contrib Mineral Petrol 134:325–341

Lüth S, Wigger P, Asch G, Bohm M, Bruhn C, Rietbrock A (2000) The crustal structure of the southern central Andes based on 3-component refraction and wide-angle seismic data. EOS 81(48):S71D-11

Lüth S, Wigger P, ISSA Working Group (2003) A crustal model along 39° S from a seismic refraction profile – ISSA (2000). Rev Geol Chile 30:83–101

Martin S, Rietbrock A, Haberland C, Asch G (2003) Guided waves propagating in subducted oceanic crust. J Geophys Res 108(B11):8/1–8/15

Meissner R, Mooney W (1998) Eakness of the lower continetal crust: a condition for delamintaion, uplift, and escape. Tectonophysics 296:47–60

Mibe K, Fuji T, Yasuda A (1999) Control of the location of the volcanic front in island arcs by aqueous fluid connectivity in the mantle wedge. Nature 401:259–262

Müller R, Roest W, Royer J, Gahagan L, Sclater J (1997) Digital isochrons of the world's ocean floor. J Geophys Res 102:3211–3214

Myers SC, Beck S, Zandt G, Wallace T (1998) Lithospheric-scale structure across the Bolivian Andes from tomographic images of velocity and attenuation for P and S waves. J Geophys Res 103: 21233–21252

Nur A, Ben-Avraham Z (1982) Oceanic plateaus, the fragmentation of continents, and mountain building. J Geophys Res 87: 3644–3661

Ortlieb L, Barrientos S, Guzman N (1996) Coseismic coastal uplift and coralline algae record in Northern Chile: the 1995 Antofagasta earthquake case. Quatenary Sci Rev 15:949–960

Patzig R (2000) Lokalbeben-Tomographie der Umgebung von Antofagasta (Nordchile) sowie Betrachtung der Magnituden-Häufigkeits-Parameter in dieser Region. Ph.D. thesis, Freie Universität Berlin

Peacock SM (1993) The importance of blueshist-eclogite dehydration reactions in subducting oceanic crust. Bull Seism Soc Am 105:684–694

Peacock SM (1996) Thermal and petrologic structure of subduction zones. In: Bebout G, Scholl D, Kirby S, Platt J (eds) Subduction: Top to Bottom, Vol 96. AGU, pp 119–133

Poli S, Schmidt MW (1995) $H_2O$ transport and release in subduction zones: experimental constraints on basaltic and andesitic systems. J Geophys Res 100:22299–22314

Reutter K-J, Giese P, Götze H-J, Scheuber E, Schwab K, Schwarz G, Wigger P (1988) Structures and crustal development of the Central Andes between 21° and 25°. In: Bahlburg G (ed) Lecture Notes in Earth Science 17, Springer-Verlag, Berlin Heidelberg New York, pp 231–261

Rietbrock A (1996) Entwicklung eines Programmsystems zur konsistenten Auswertung großer seismologischer Datensätze mit Anwendung auf die Untersuchung der Absorptionsstruktur der Loma-Prieta-Region, Kalifornien. Ph.D. thesis, LMU München

Rietbrock A (2001) P-wave attenuation structure in the fault area of the 1995 Kobe earthquake. J Geophys Res 106:4141–4154

Romanyuk T, Götze H-J, Halvorson P (1999) A density model of the Andean subduction zone. The Leading Edge 18(2):264–268

Rudloff A (1998) Bestimmung von Herdflächenlösungen und detaillierte Spannungsinversion aus ausgewählten Ereignissen des seismologischen Netzes PISCO '94 (Nord-Chile). GeoForschungszentrum Potsdam Scientific Technical Report STR98/01

Scheuber E, Reutter K-J (1992) Magmatic arc tectonics in the Central Andes between 21° and 25° S. Tectonophysics 205:127–140

Schmitz M, et al. (1999) The crustal structure of the central Andean forearc and magmatic arc as derived from seismic studies – the PISCO '94 experiment in northern Chile (21–23° S) J S Am Earth Sci 12:237–260

Schott B, Schmeling H (1998) Delamination and detachment of a lithospheric root. Tectonophysics 296:25–247

Schurr B (2001) Seismic Structure of the Central Andean Subduction Zone from Local Earthquake Data, Scientific Technical Report STR01/01, GeoForschungsZentrum Potsdam

Schurr B, Asch G, Rietbrock A, Kind R, Pardo M, Heit B, Monfret T (1999) Seismicity and average velocity beneath the Argentine Puna. Geophys Res Lett 26:3025–3028

Schurr B, Asch G, Rietbrock A, Trumbull R, Haberland C (2003) Complex patterns of fluid and melt transport in the central Andean subduction zone revealed by attenuation tomography. EPSL 215:105–119

Schurr B, Rietbrock A, Asch G, Kind R, Oncken O (2006) Evidence for lithospheric detachment in the central Andes from local earthquake tomography. Tectonophysics 415:203–223

Seber D, Barazangi M, Ibrahim D, Demnati A (1996) Geophysical evidence for lithospheric delamination beneath the Alboran Sea and Rift-Benthic mountains. Nature 379:785–790

Snoke JA, Sacks IS, Okada H (1978) Determination of the subducting lithosphere boundary by use of converted phases. Bull Seism Soc Am 101:1051–1060

Sobiesiak M (2000) Fault plane structure of the Antofagasta, Chile earthquake of 1995. Geophys Res Lett 27:577–600

Tatsumi Y (1989) Migration of fluid phases and genesis of basalt magmas in subduction zones. J Geophys Res 94:4697–4707

Thurber C (1993) Local earthquake tomography: velocities and $V_p/V_s$-theory In: Iyer H, Hirahara K (eds) Seismic tomography: theory and practice. Chapman and Hall, London, pp 563–583

Thurber C, Eberhart-Phillips D (1999) Local earthquake tomography with flexible gridding. Comput Geosci 25:809–818

Vinnik LP (1977) Detection of waves converted from P to SV in the mantle. Phys Earth Planet Inter 15:39–45

Whitman D, Isacks BL, Chatelain JL, Chiu JM, Perez A (1992) Attenuation of high-frequency seismic waves beneath the central Andean plateau. J Geophys Res 97:19929–19947

Whitman D, Isacks BL, Kay SM (1996) Lithospheric structure and along-strike segmentation of the central Andean plateau: seismic Q, magmatism, flexure, topography and tectonics. Tectonophysics 259:29–40

Wigger P, et al. (1994) Variation of the crustal structure of the southern central Andes deduced from seismic refraction investigations, In: Reutter K-J, Scheuber E, Wigger P (eds) Tectonics of the Southern Central Andes. Springer-Verlag, Berlin Heidelberg New York

Yuan X, Ni J, Kind R, Mechie J, Sandvol E (1997) Lithospheric and upper mantle structure of southern Tibet from a seismological passive source experiment. J Geophys Res 102:27491–27500

Yuan X, Sobolev SV, Kind R, Oncken O, Bock G, Asch G, Schurr B, Graeber F, Rudloff A, Hanka W, Wylegalla K, Tibi R, Haberland C, Rietbrock A, Giese P, Wigger P, Röwer P, Zandt G, Beck S, Wallace T, Pardo M, Comte D (2000) Subduction and collision processes in the Central Andes constrained by converted seismic phases. Nature 408:958–961

# Partial Melting in the Central Andean Crust: a Review of Geophysical, Petrophysical, and Petrologic Evidence

Frank R. Schilling · Robert B. Trumbull · Heinrich Brasse · Christian Haberland · Günter Asch · David Bruhn · Katrin Mai
Volker Haak · Peter Giese · Miguel Muñoz · Juliane Ramelow · Andreas Rietbrock · Edgar Ricaldi · Tim Vietor

**Abstract.** The thickened crust of the Central Andes is characterized by several first-order geophysical anomalies that seem to reflect the presence of partial melts. Magnetotelluric and geomagnetic deep-sounding studies in Northern Chile have revealed a high conductivity zone (HCZ) beneath the Altiplano Plateau and the Western Cordillera, which is extreme both in terms of its size and integrated conductivity of > 20 000 Siemens. Furthermore, this region is characterized by an extremely high seismic attenuation and reduced seismic velocity. The interrelation between the different petrophysical observations, in combination with petrological and heat-flow density studies, strongly indicates a huge area of partially molten rocks that is possibly topped with a thin, saline fluid film. The average melt fraction is deduced to be ~20 vol.%, which agrees with typical values deduced from eroded migmatites. Based on the distribution and geochemical composition of Pliocene to Quaternary silicic ignimbrites in this area, this zone is thought to be dominated by crustally-derived rhyodacite melts with minor andesitic contribution. An interconnected melt distribution – typical for migmatites – would satisfy both the magnetotelluric and seismic observations. The high melt fraction in this mid-crustal zone should lead to strong weakening, which may be a main cause for the development of the flat topography of the Altiplano Plateau.

## 22.1 Introduction

The Andes are located at the convergence zone between the oceanic Nazca Plate and the continental South American Plate. This Cordilleran-type mountain belt is the largest active, subduction-controlled orogen on Earth. It has a length of some 7 500 km and in the central part, between 16° to 25° S (Fig. 22.1), the Andes have extreme dimensions and topographic relief; the mountain belt is 800 km wide and elevations reach more than 6 000 m. It has been appreciated for several decades that the crust is also extremely thick, with seismic and gravimetric data indicating a thickness of up to 70 km below the Altiplano and Western Cordillera (James 1971; Ocola and Meyer 1972; Zandt et al. 1994; Wigger et al. 1994). More recent studies have resolved the problem of imaging the Moho discontinuity beneath the volcanic arc by using seismic tomography (Haberland 1998; Rietbrock 1999) and receiver function methods (Yuan et al. 2000, 2002), so that, now, a fairly complete map of Moho depth exists (Yuan et al. 2002). The Bouguer gravity anomaly of this thickened subduction orogen reaches –450 mgal in the magmatic arc (Götze et al. 1994).

The prominent landscape, active volcanic chain, world-class ore deposits and the type locality of the Andean orogeny have motivated a great number of detailed studies which make the Central Andes one of the best, or even *the* best, studied examples of subduction-related orogeny. Emphasis has been placed on the strong variations in properties that exist across the Central Andes from west to east, as a consequence of the inherent polarity of the subduction system. However, it is equally important to consider the striking variations in properties parallel to the active margin.

The Central Andes crust is not only thick but is characterized by first order, internal geophysical anomalies. One such anomaly is the extreme high conductivity zone (HCZ), revealed by magnetotelluric and geomagnetic deep-sounding studies, beneath the Altiplano Plateau and the Western Cordillera in Northern Chile (Schwarz et al.

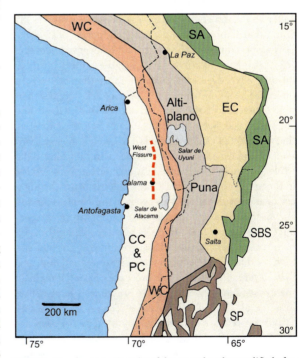

**Fig. 22.1.** Morpho-tectonic units of the central Andes modified after Oncken et al. (2006). *CC*: Coastal Cordillera, *PC*: Precordillera, *WC*: Western Cordillera, *EC*: Eastern Cordillera, *SA*: Subandean Ranges, *SBS*: Santa Barbara System, *SP*: Sierras Pampeanas

1994; Soyer and Brasse 2001, Brasse et al. 2002). Nowhere else has a comparable conductivity anomaly been observed, both in terms of its size and its high, integrated conductivity of > 20 000 S (Siemens).

Other primary geophysical anomalies have been revealed by detailed seismic studies conducted between 20° and 26° S, mainly within the framework of the Berlin-Potsdam Special Research Program 267. These studies detected complex structure within the crust, correlating well with observations by Beck et al. (1996) and Beck and Zandt (2002). Seismic discontinuities and broad, low-velocity zones were deduced from active seismic experiments (e.g., Wigger et al. 1994) and receiver function analysis of passive seismic data gave independent confirmation, and better definition, of the crustal low-velocity zone beneath the Western Cordillera and Plateau (Chmielowski et al. 1999; Yuan et al. 2000; Zandt et al. 2003). Passive seismic experiments enabled identification of areas with high $v_P/v_S$ ratios (compressional and shear wave velocity, respectively) and low $Q$ values (seismic quality factor, describing the attenuation behavior) by three-dimensional (3D) tomographic inversion (e.g., Haberland 1998; Rietbrock 1999; Schurr et al. 2003, 2004).

Furthermore, anomalously high heat-flow values of >100 mW m$^{-2}$ have been found in the Western Cordillera and Altiplano Plateau (Henry and Pollack 1988; Giese 1994; Hamza and Muñoz 1996; Springer and Förster 1998). There is a controversy in how to explain these values (Henry and Pollack 1988; Giese 1994; Hamza and Muñoz 1996; Le Pichon et al. 1997; Springer and Förster 1998). Some thermal models show that conductive heat transport alone cannot account for the high observed heat flow (e.g., Springer and Förster 1998). In these models an additional advective heat transfer of more than 30% of the observed heat flow would be required to explain the thermal structure of the Central Andes (e.g., Giese and Schilling 2000; Babeyko et al. 2002). However, consistent thermal models can be derived (e.g., Le Pichon et al. 1997, Pollack and Chapman 1977; Chapman and Furlong 1992) while assuming an enhanced radiogenic heat production in the upper and lower crust and while ignoring the highest heat flow densities observed in the Central Andes. However, both sets of models allow temperatures exceeding the solidus of felsic rocks at mid-crustal depth.

In many past studies, these geophysical anomalies – high conductivity, seismic low velocity zones and high heat flow – were attributed to the presence of partial melts in the crust (Schilling et al. 1997; Brasse and Soyer 2001; Brasse et al. 2002; Schmitz et al. 1997; Babeyko et al. 2002). This interpretation is strongly supported by the presence of extensive fields of Neogene ignimbrites in the plateau region of the Central Andes (Fig. 22.1), many of which have been shown to be derived from crustal melting (deSilva 1989; Francis et al. 1989; Ort et al. 1996). The largest concentration of ignimbrites regionally is in the so-called Altiplano-Puna Volcanic Complex (APVC) described by deSilva (1989). The APVC covers some 50 000 km$^2$ and contains at least 30 000 km$^3$ of erupted ignimbrites. The youngest of the huge (> 1 000 km$^3$) ignimbrite eruptions in the Central Andes was from Cerro Galán about 2 Ma ago (Francis et al. 1989) but resurgent domes and smaller ignimbrites of one million years age down to less than 0.1 million years suggest that magmatism is still alive.

Thus, the existence of partial melts seems to be well established by a number of independent observations. Indeed, Chmielowski et al. (1999) and Zandt et al. (2003) proposed in that region the existence of a sill-like magma body at 20 km depth which they termed the Altiplano-Puna magma body. However, until now, there has been no attempt to combine all available observations and inferences to derive a self-consistent model for the nature of the APVC. Such a model could then be used for interpreting how the melting zone formed and its consequences for the deformation regime and tectonic evolution of the plateau region. This is the aim of the present review.

Our purpose in this review is to combine geophysical observations, petrological and geological studies, laboratory and numerical experiments to address and answer the following questions: Is the evidence for melts conclusive and consistent with all observations or are there alternatives? Where are the melts located? What proportion of partial melt is needed and how must the melt phase be distributed to explain the observations? And what effect does it have on the thermal structure of the crust and its mechanical behavior?

The increasing quantity and quality of geophysical observation in the Central Andes, now clearly show major variations in the crustal properties and characteristics from north to south. The N-S variation in arc magmatism is well established (e.g., Coira et al. 1993) and so are the variations in the patterns of deformation (e.g., Allmendinger et al. 1998; Riller and Oncken 2003; Vietor and Echtler 2006, Chap. 18 of this volume). The present-day crust and mantle also vary strongly from north to south and while some of the geophysical anomalies described above seem to have a large N-S extent, others are smaller. It has been common in the past to explain the Central Andes orogen using two-dimensional (2D) cross sections, but, as emphasized in this review, considering the Central Andes as a complex 3D system enables better understanding of the processes and mechanisms in an active orogen.

## 22.2 Results

Detailed descriptions of the various geophysical and geological features relevant to this review are deliberately kept to a minimum and the reader is referred to the original literature for detailed information and justification for the interpretations and models described below.

## 22.2.1 3D Variations in Electrical Conductivity

Over the past decade, a large amount of long-period magnetotelluric (MT) data has been collected in the Central Andes. The aim of these investigations was to define the conductivity distribution in the deeper crust and upper mantle by utilizing natural, electromagnetic field variations at periods between 10 s and approx. 20 000 s. Here, we concentrate on results obtained from data in the fore-arc, arc and Altiplano Plateau regions of the Central Andes. Other MT studies of the back-arc (Puna section of the high plateau, Eastern Cordillera, Subandean belt and Chaco) are not included (but see, e.g., Lezaeta et al. 2000, Lezaeta and Brasse 2001).

Figure 22.2 summarizes the results of 2D inverse modeling along various profiles in the Central Andes and places them in a regional context. The inversions were carried out with a non-linear, conjugate gradient algorithm, based on finite differences (developed by Rodi and Mackie 2001), while the forward model studies employed the finite element code of Wannamaker et al. (1986). To deal with the trade-off between model resolution (roughness) and data fit, the L-curve criterion (Hansen 1998) was routinely applied. Although the data are modeled in terms of electrical resistivity, we find it more convenient to discuss the results in terms of conductivity. Further details concerning data analysis, dimensionality, strike determination, and modeling techniques are described in Soyer and Brasse (2001), Brasse et al. (2002), Lezaeta (2002), Schwalenberg et al. (2002) and Lezaeta and Haak (2003).

Profile I (Fig. 22.2) shows part of a resistivity cross section along 22° S (redrawn from Schwarz and Krüger 1997), which traverses the Western Cordillera and Altiplano and reaches the Eastern Cordillera. The data for this profile were obtained by the research group "Mobility of Active Continental Margins" (Schwarz and Krüger 1997) and interpreted by 2D forward modeling only, since reliable and sufficiently fast inversion codes were not widely available at that time (Krüger 1994). The most significant finding is a zone of high conductivity at depths > 20 km below the volcanic arc (G in profile I, Fig. 22.2) and the high plateau, where forward modeling suggested a ramp-like structure shallowing to the east (G in profile I, Fig. 22.2). The inferred maximum conductivity below the arc is about 1 S m$^{-1}$.

In contrast to profile I, the volcanic arc does not appear as a conductive feature along the 21° S ANCORP profile (II in Fig. 22.2) or along the other magnetotelluric profiles further north (III and IV). Instead, all three northern profiles show a low conductivity zone in the arc between the Western Cordillera and Altiplano (D, Fig. 22.2). However, there is a prominent, high-conductivity anomaly (C) in the fore-arc in profiles II and III, which is spatially correlated with the surface trace of the West Fissure (Falla Oeste) (Fig. 22.1). The depth extent of this feature is not well resolved. While forward models for profile III indicate that the conductive zone may reach into the upper mantle (Echternacht et al. 1997), the inversion results shown in Fig. 22.2 suggest that the anomaly is restricted to the brittle part of the crust. This has been corroborated by the 3D modeling studies of Lezaeta (2001). In a more detailed study of the West Fissure, Hoffmann-Rothe et al. (2004) produced a detailed image of the zone to about 2 km depth for a segment of the fault north of the Chuquicamata copper mine.

A common feature of profiles II and III is the very low conductivity of the fore-arc beneath the Coastal Cordillera (A in Fig. 22.2). The inversion image shows the low conductivity extending through into the upper mantle. One might expect a conductive "fluid curtain" to show up in the model above the dehydrating slab, but such a feature is neither required nor contradicted by the magnetotelluric responses.

The most striking conductivity anomaly in this part of the Central Andes is the broad and intense anomaly (E, Fig. 22.2) in the mid-crust beneath the plateau (Brasse et al. 2002). The following discussion mainly concentrates on this feature, which has conductivities around 1 S m$^{-1}$, or below, and an overall conductance exceeding 20 000 S. The response of this conductor is expressed not only in the magnetotelluric transfer functions *sensu strictu*, but also, and exceptionally clearly, in the analysis of magnetovariational functions (Soyer and Brasse 2002). As for all electromagnetic soundings, the lower boundary of the conductive zone is not well resolved, but it seems to be at least 40–50 km deep, as inferred from sensitivity studies by Schwalenberg et al. (2002).

This extremely good, mid-crustal conductor severely attenuates the electromagnetic fields and hinders penetration into deeper crustal and upper mantle levels. However, there is evidence that a conductive root (F) may reach into the mantle wedge. This "root zone" is also detected in section III and, even though it lies at the margin of the profile, its presence seems to be required by the data. Note also, that the color-scaling adopted for visualizing this structure is slightly different from the original presentation in Brasse et al. (2002).

In order to check the three-dimensional extent of the southern Altiplano conductor, additional measurements were carried out in the northern Altiplano at 18–19° S (profile IV in Fig. 22.2). The inversion cross section for this profile is quite different again from those further south. The large, mid-crustal conductor is absent here and electromagnetic (EM) fields can penetrate to much greater depths. The result is the detection of a high-conductivity zone in the upper mantle (F in profile IV, Fig. 22.2), which is tentatively interpreted as representing the mantle wedge. This explanation supports the inference of a conductive zone in the mantle wedge seen in the profiles further south (F in profiles II and III). However, as discussed more fully below, there are also problems concerning (*i*) the fluid or melt fraction required to explain the conductivity, and (*ii*) the position of the anomaly well to the east of the magmatic arc.

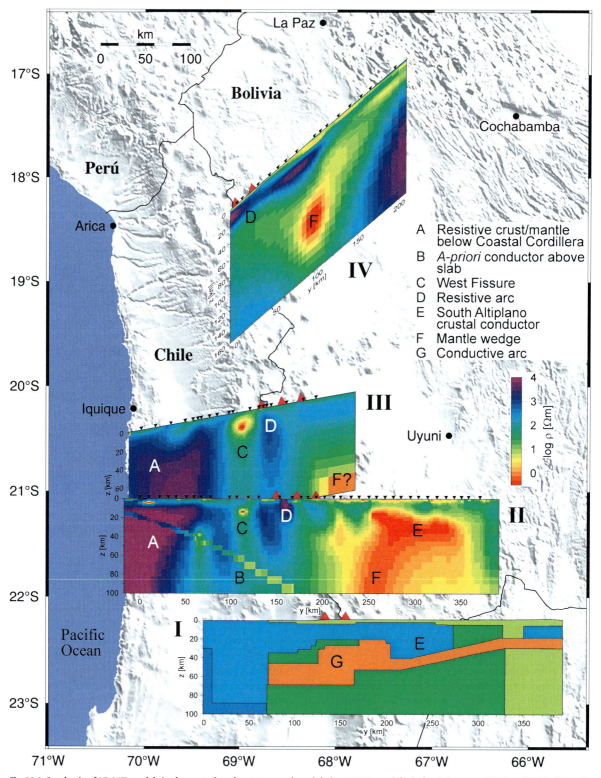

**Fig. 22.2.** Synthesis of 2D MT models in the central Andes. *I*: Forward model along 22° S, modified after Schwarz and Krüger (1997); *II*: result of 2D inversion of the ANCORP profile along 21° S; *III*: 2D inversion for the Pica profile at 19.5° S; *IV*: inversion result of a recent profile across the northern Altiplano (Brasse et al. 2005). For further explanation see text

## 22.2.2 3D Variations in Seismic Properties

A wide variety of seismic and seismological techniques have been applied to constrain the petrophysical properties of the crust and mantle in the Central Andes. Several seismic profiles gained with active techniques that gathered both reflection and wide angle data were accomplished, with the ANCORP line at 21° S being the most comprehensive and with highest spatial resolution (ANCORP working group 2003). Furthermore, a large number of passive seismological arrays where deployed for receiver function and tomographic studies (see Asch et al. 2006, Chap. 21 of this volume). Since the results and newest interpretation of these seismic studies and receiver function analyses are summarized in detail within this volume (Asch et al. 2006, Chap. 21; Sick et al. 2006, Chap. 7), we focus attention here on the tomography results. One of the most striking features, found with receiver functions, is the Altiplano low velocity zone between 20 and 25° S (Yuan et al. 2000). It coincides with the upper limit of the strong conductivity anomaly at 21° S (Fig. 22.2, profile I).

Several studies over the last years have produced a detailed image of seismic attenuation in the crust and uppermost mantle in the Central Andes (Haberland and Rietbrock 2001; Schurr et al. 2003; Haberland et al. 2003; Schurr and Rietbrock 2004). They show a complex attenuation pattern including extremely high attenuation values (low $Q_P$) in diverse regions of the crust and upper mantle. These were attributed mainly to the effects of high

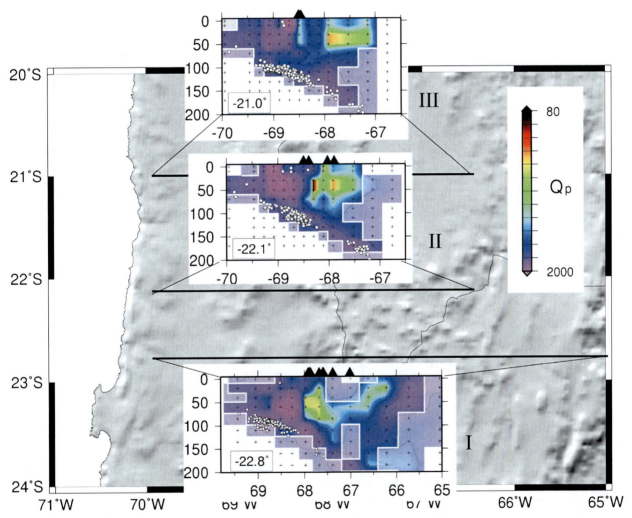

**Fig. 22.3.** W-E depths sections (along 22.8°, 22.1°, and 21.0° S) of the attenuation structure of seismic P waves (synthesis of 3D attenuation tomography models). The seismic quality factor $Q_P$ is color coded, unresolved regions are not shown. The abscissa in the sections is longitude (° W), ordinate is depth (km). The position of the recent volcanoes are indicated by *black triangles*. Profiles II and III are close to the MT profiles I and II (Fig. 22.2). Profile I after Schurr et al. (2003). Profiles II and III are based on the model by Haberland et al. (2003)

temperature and the presence of fluids and/or partial melts. The $Q_p$ tomography results are shown for three profile sections (Fig. 22.3) and the main features of the three profiles are summarized briefly below.

Seismic waves traveling in the subducting slab and the fore-arc crust are nearly unaffected by attenuation, which is consistent with a cold crust containing only minor free fluids. As described above for the magnetotelluric observations, the $Q_p$ tomography shows no indication of a fluid "curtain" from the descending slab in the fore-arc region. In contrast, some regions of the fore-arc, particularly beneath the Salar de Atacama (Fig. 22.1), show extremely low attenuation values, and this probably indicates an especially strong, cold lithospheric block (Schurr and Rietbrock 2004).

Along the southernmost profile I (Fig. 22.3), the $Q_p$ values are slightly reduced beneath the magmatic arc at 68° W compared to the fore-arc, and low $Q_p$ values also extend further east into the back-arc region. Further to the north (profiles II and III, Fig. 22.3), attenuation of seismic waves beneath the magmatic arc diminishes and there is a much greater degree of absorption (low $Q_p$ values) in the mid crust beneath the Altiplano Plateau. This midcrustal anomaly coincides spatially with the electrical conductivity anomaly described above and with the surficial extent of Neogene silicic magmatism (ignimbrites) in the Altiplano-Puna-Volcanic-Complex (APVC), as described in the following section (Fig. 22.4). The zones with low $Q_p$ in this region of the crust also show a number of other seismic properties, including low seismic velocities and elevated $v_p/v_s$ ratio (Asch et al. 2006, Chap. 21 of this volume). These have been interpreted as reflecting the presence of partial melts (e.g., Schilling et al. 2001).

It is, at first, somewhat surprising that the region underlying the volcanic front in profiles II and III shows little or no indication for enhanced attenuation (Fig. 22.3, where the triangles indicate the position of Pleistocene volcanic centers). But this is, at least, consistent with the lack of anomalously high conductivity in the crust beneath the arc front in magnetotelluric lines II and III, noted above. Possible reasons for this are discussed below, but we note here that there is a good correlation of low $v_p$ and low $v_s$ zones in the crust with the position of the young volcanic arc (Sobolev et al. 2006, Chap. 25 of this volume).

Although the focus of this paper is on melt-related features within the crust, it is important to point out that there are also strong contrasts in the $Q_p$ distribution within the mantle wedge under the Central Andes. Whitman et al. (1992) identified zones of high and low attenuation in the upper mantle beneath the Puna Plateau and, from more extensive station arrays, Schurr et al. (2002) were able to generate $Q_p$ tomography images of the mantle wedge in the northern Puna region (23–24° S) and farther north (i.e., profiles I, II, III in Fig. 22.3). Zones with anomalously low $Q_p$ values in the upper mantle were interpreted as reflecting high temperature and/or fluid content and thus revealing the ascent pathways of melts in the asthenosphere. The broad areas with low $Q_p$ values in this part of the upper mantle could indicate inflow of asthenosphere as a response to the detachment and sinking of dense lithospheric mantle (e.g., Kay et al. 1994; Sobolev et al. 2005).

### 22.2.3 Constraints on Crustal Melting from Studies of Neogene Ignimbrites

The plateau region of the Central Andes is host to the world's largest Neogene ignimbrite province (deSilva 1989; Coira et al. 1993; see Fig. 22.4). A detailed description of the andesite-dacite arc volcanoes of the Central Andes is beyond the scope of this contribution and we refer to, e.g., Coira et al. (1993), Davidson et al. (1991), Kay et al. (1999) or Trumbull et al. (1999). Very large-volume (i.e., > 1 000 km³) ignimbrite fields in the Central Andes occur in several locations and erupted at different times (deSilva and Francis 1991), beginning at about 20 Ma in Bolivia and with the youngest example being the 2 Ma Cerro Galán ignimbrite in Argentina (Fig. 22.4). The peak of ignimbrite activity, however, was more focused and occurred at about 10–4 Ma in the APVC region at the Bolivia-Argentina-Chile border. According to Baker and Francis (1978) and deSilva et al. (1989) the volume

**Fig. 22.4.** Sketch map of part of the central Andes showing the distribution of Quaternary arc volcanoes, major salars and the approximate outcrop extent of Neogene ignimbrites with known calderas, figure modified from Lindsay et al. 2001. The *dashed line* labeled APVC is the area of ignimbrite concentration known as the Altiplano-Puna Volcanic Complex (deSilva 1989)

ratio of ignimbrites to andesites in the APVC is about 6:1, a factor of 30 times greater than the typical value for continental arcs. As pointed out by Chmielowski et al. (1999) and Zandt et al. (2003), the APVC coincides spatially with the region of strongest seismic velocity anomalies in the mid-crust. Thus, geochemical-petrologic studies of the ignimbrite magmas are of obvious relevance to interpreting the meaning of the geophysical anomalies and the most pertinent results from these studies are briefly summarized here.

Detailed petrologic and geochemical studies of ignimbrites from the APVC have established the following "family resemblance" among the units which suggests similar sources and magma evolution:

1. Extremely large eruptive volumes, exceeding 1 000 km$^3$ for individual units.
2. A compositional monotony dominated by high-K, calc-alkaline dacite and only about 5% rhyolites. Most large units lack any major compositional variation.
3. High crystallinity (30–50%) of pumice. Major phenocryst phases are plagioclase, quartz, hornblende and orthopyroxene.

Geothermometry, based on mineral equilibria of phenocrysts in several examples of the dacitic APVC ignimbrites (magnetite-ilmenite, 2-feldspar, plagioclase-hornblende), indicates pre-eruptive magma temperatures of 750–850 °C and pre-eruptive water contents of 3–4 wt.%, the latter measured in melt-inclusions (Lindsay et al. 2001; Schmitt et al. 2001; Schmitt 2001). Oxygen fugacities determined by Fe-Ti-oxide equilibria are approximately 2.5 log units above the quartz–fayalite–magnetite buffer. Considering that these samples comprise about 50% anhydrous crystals, the original bulk magma water contents were about 2–3 wt.%, which constrain minimum confining pressures between 110 and 150 MPa. Similar pressure estimates were made by the Al-in-hornblende barometry (110–240 MPa) and this indicates that ignimbrite magma was stored and equilibrated prior to eruption between 5 and 9 km depth, assuming a density of 2.7 g cm$^{-3}$ for the lithostatic column. Very similar values of pre-eruptive magma temperature and oxygen fugacity were estimated for the dacitic ignimbrites of Cerro Galán by Francis et al. (1989).

There is consensus among most workers that the large-volume, felsic ignimbrites in the Central Volcanic Zone (CVZ) mainly reflect crustal melting (e.g., deSilva 1989; Francis et al. 1989; Coira et al. 1993; Ort et al. 1996; Lindsay et al. 2001; Schmitt et al. 2001). However, the presence of mixed pumice in some units, the comparison of ignimbrite isotopic compositions with those of the Andean basement (Lucassen et al. 2001), and the mismatch of ignimbrite major element compositions with experimental melts of natural crustal rocks indicate that the magmas are not pure crustal melts (Ort et al. 1996; Schmitt et al. 2001). Mixing – assimilation – fractional crystallization (AFC) modeling results (Fig. 22.5) yield estimates of 70–80% crustal component in the ignimbrites, based on the average composition of Paleozoic granites as proxy for partial melts of the mid crust, and the composition of isotopically least-contaminated, basaltic andesite for the input of mafic magmas from the arc (Lindsay et al. 2001; Schmitt et al. 2001).

The source depth of magma generation is difficult to estimate directly from the ignimbrite samples because mineral equilibria were established and/or reset in upper crustal storage chambers before eruption. However, a lower limit of source depths can be constrained based on the REE composition of ignimbrites compared with melts produced from garnet-bearing source rocks. As discussed by Schmitt et al. (2001) most of the dacitic APVC ignimbrites have La/Yb ratios of less than 20, suggesting that garnet was not present in the source. Since the ignimbrites are crustally-derived and experimental melting of biotite- or hornblende-bearing gneisses, as are typical for the Andean basement, have been shown to produce garnet at > 1.3 GPa pressure (Patino-Douce and Beard 1995), we thus conclude that the melting zone was shallower than 30–40 km (depth-integrated value for crustal density of 2.8 g cm$^{-3}$, see Lucassen et al. 2001).

## 22.3 Discussion

### 22.3.1 Are There Alternatives to Melts?

The geochemical evidence from the CVZ ignimbrites demonstrates that crustal melts were generated and erupted over a considerable area and time span, with the youngest known examples being the Holocene Chao dacite (< 0.1 Ma, deSilva et al. 1994). The above-described geophysical observations beneath the Altiplano have been interpreted as

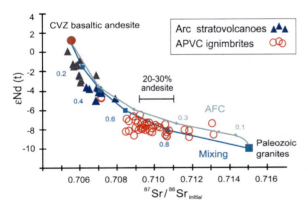

**Fig. 22.5.** Sr and Nd isotope composition of CVZ andesites (data from Trumbull et al. 1999), large-volume APVC ignimbrites (Schmitt et al. 2001) and average Paleozoic granite from the CVZ (Lucassen et al. 2001). The mixing and AFC curves demonstrate that the ignimbrite magmas can be considered to consist of 20–30% mafic andesite melts and 70–80% partial melts from the Paleozoic basement

the result of a present-day extensive area of partially molten crust (e.g., Schilling et al. 1997; Schmitz et al. 1997; Chemielowski et al. 1999; Brasse et al. 2002), but it must be said that this remains an untested interpretation. Before attempting to develop and test concepts and scenarios for generating *and maintaining* a partially molten crust in this region, it is worth restating the key lines of evidence derived from the independent geophysical observations.

The extreme, high electrical conductivity zone (HCZ) beneath the Altiplano (regions E and F in Fig. 22.2) requires an interconnected network of highly conductive phases in the crustal rocks. Fluids, melts, ore minerals, or graphite are possible candidates for effectively enhancing the conductivity of the crust (e.g., Schilling et al. 1997). Graphite is, of course, highly conductive and it can be distributed along grain boundaries to produce a conductive network. However, the petrological evidence of a high oxygen fugacity in the APVC ignimbrites (> 2 log unit above the QFM buffer – see above) would argue against the widespread existence of reduced carbon in the crust. Furthermore, graphite alone, or any other conductive minerals such as ores, would not cause the reduction in seismic velocities and damping of seismic waves that is observed.

A variety of mechanisms can cause the attenuation of seismic waves. The most important mechanisms are grain-boundary sliding (frictional dissipation due to relative movements along grain boundaries, Johnston et al. 1979), viscous relaxation processes, fluid flow in pores and along grain boundaries (Mavko 1980), and scattering of the seismic waves at inhomogeneities (Aki and Chouet 1975). The extremely low $Q$ values ($Q_P < 100$) observed in the Altiplano crust (Fig. 22.3) are far beyond the values usually attributed to grain-boundary sliding. However, near-solidus temperatures could lead to the $Q$ values reported in Fig. 22.3, although this would require the presence of an ~50 km-thick zone beneath the Altiplano that is close to, but does not exceed, its solidus temperature throughout. For any reasonable geothermal profile it is quite unrealistic that this condition would be met in the Central Andes (Figs. 22.6 and 22.7).

Part of the observed $Q$ anomaly might well be the result of near-solidus absorption behavior in the vicinity of a partially molten body; but it is unwarranted to equate the whole extent of the low-$Q$ anomaly with a partially molten zone. However, in zones where low $Q$ values are accompanied by high conductivity, near-solidus temperatures alone are not a viable explanation for the combined observation. The best explanation for the conductivity and seismic anomalies in these zones is therefore a distributed melt phase.

A further point for consideration is the high observed heat flow density. The heat flow database is not sufficient to define a detailed 2D-distribution of heat-flow density in the Central Andes (Hamza and Muñoz 1996). Nevertheless, the available compilations of heat flow density distribution show a strong variation in the observed values, both parallel and perpendicular to the trench (Hamza and Muñoz 1996; Springer and Förster 1998).

Springer and Förster (1998) compiled all available heat-flow-density values of the Central Andes and ordered them into a schematic E-W profile according to the main

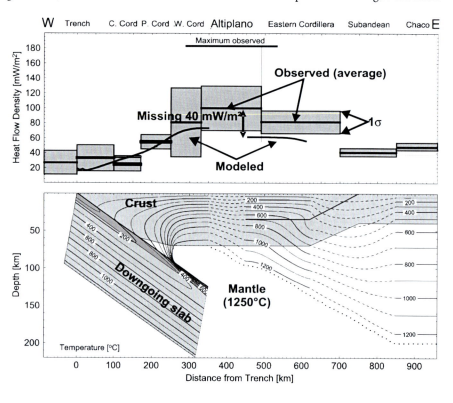

**Fig. 22.6.**
Heat-flow density and conductive 2D thermal model of the central Andes, compiled using tectonostratigraphic units, simplified after Springer and Förster (1998). The *gray bands* represent an errors of 1σ

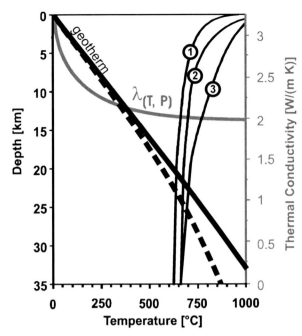

Fig. 22.7. One-dimensional, temperature conductive model, assuming an internal heat production of 0.9 µW m$^{-3}$ (similar to Springer and Förster 1998) and 1.5 µW m$^{-3}$, shown as *solid* and *dotted line* respectively. A surface heat-flow density of 80 mW m$^{-2}$ (reduced average heat flow density of Altiplano-Puna Plateau according to Hamza and Muñoz 1996) is used. Solidus temperatures for wet crustal rocks are taken from Wyllie 1979, for a muscovite granite (①), a tonalite (②), and a gabbro (③). The used and presented pressure and temperature dependent thermal conductivity is derived according to Seipold (1998)

tectono-stratigraphic units. The average value and range for 1 standard deviation are shown from this compilation in Fig. 22.6. Some of the observations exceed 140 mW m$^{-2}$, which seems to be partly the result of near surface effects such as fluid flow (Hamza and Muñoz 1996) or shallow magma chambers (Springer and Förster 1998). If these high values are neglected, an average heat flow density of 78 mW m$^{-2}$ is deduced for the Altiplano-Puna Plateau (Hamza and Muñoz 1996).

Springer and Förster (1998) produced a conductive heat-flow model for the Central Andes whose parameters were chosen to maximize the predicted surface heat-flow density in the Altiplano-Western-Cordillera area (thermal conductivity of 3 W mK$^{-1}$ for the crust, high internal heat production of 0.9 µW m$^{-3}$ and 1 250 °C at the crust-mantle boundary). Even for these conditions, the conductive heat transfer model (~60 mW m$^{-2}$) cannot explain the observed heat-flow-density values for the Altiplano region ~100 mW m$^{-2}$ (Fig. 22.6). This result is unchanged when an uncertainty of 20 mW m$^{-2}$ on the heat-flow density values is assumed or when the reduced average value of 78 mW m$^{-2}$ proposed by Hamza and Muñoz (1996) is used. Even if the reduced average value of 78 mW m$^{-2}$ (Hamza and Muñoz 1996) is used for a simple 1D thermal model, the solidus of usual crustal compositions can be reached between 20 and 25 km depth, if an internal heat production of 0.9 µW m$^{-3}$ (Springer and Förster 1998) or 1.5 µW m$^{-3}$ is assumed (Fig. 22.7, solid and dashed lines, respectively). Even if other thermal models are conceivable, this model shows that the observed heat flow density supports partial melting in the mid-crust.

Crustal melting is therefore the best explanation for the observed surface heat-flow density (Fig. 22.6) in the Altiplano region (Arndt et al. 1997; Springer and Förster 1998), in conjunction with the anomalous density-velocity relationship in the crust beneath the Altiplano (e.g., Schmitz et al. 1997) and melting behavior of crustal rocks (Fig. 22.7). It also best explains the increased $v_P/v_S$ ratios, and, finally, the strong P- to S-wave conversion beneath the APVC area (Chmielowski et al. 1999; Yuan et al. 2000; Zandt et al. 2003). In summary, although there may be alternative explanations to partial melts for some of the individual geophysical anomalies under consideration, the combination of electrical, geothermal, and seismic anomalies in the same crustal zone seem to require a partial-melting hypothesis. This is also the scenario most compatible with the record of recent massive eruption of crustally-derived magmas in this area.

### 22.3.2 Where Are These Crustal Melts Located and How Are They Distributed?

Both the high conductivity and low-Q zones in the Central Andean crust are spatially concentrated east of the arc. To best constrain the region of partial melting beneath the Altiplano, one might invert the conductivity and Q data jointly. However, as pointed out above, neither the entire area of high conductivity nor that of low Q needs to be completely occupied by partially molten rock.

Partial melts are arguably the main cause of the anomalous seismic and conductivity properties of the crustal zone, and most of the areas that appear as anomalous in the magnetotelluric and seismological images (Figs. 22.2 and 22.3, E) do seem to reflect partially molten rocks. However, because of the expected fluid release upon crystallization of individual melt bodies, and the low density and high salinity of magmatic fluids, one should expect that a thin fluid layer (brine) exists, at least transiently, at the top of a partially molten zone. Thus, the upper limit of the observed HCZ at around 20 km depth may equate with such a thin fluid layer directly above a partially molten zone. The lower boundary of the partially molten zone is poorly defined by the geophysical data; it may reach the crust-mantle boundary (~60–70 km depth) or even extend into the mantle (Fig. 22.2 F).

The magnetotelluric observations are restricted to the four available profiles, whereas seismic velocity-tomography and receiver function data exist from a 2D array of seismic stations. Therefore, in plan view, the seismic data best constrain the spatial distribution of the anomalous zone and, as mentioned by Chmielowski et al. (1999) and Zandt et al. (2003), this zone is concentrated in the back-arc region between about 22 and 24° S and below the APVC area of caldera concentration. Regarding the depth of the partially melted zone within the crust, both the seismic and magnetotelluric sections concur with an upper limit of about 20 km, but a lower limit to the zone is less well constrained.

The distribution of $Q_P$ also shows a strong variation from north to south. Schurr and Rietbrock (2004) show an area of extremely low $Q_P$ below the Salar de Atacama, cutting the crust down to the subducted slab. It seems to hinder fluid and melts from propagating to the west and is most probably responsible for the remarkable shift of the volcanic arc, of up to 100 km, towards the east in the Central Andes. The fluids and melts also have a complex pattern of ascent which varies with latitude. Schurr et al. (2003) trace these ascending paths down to areas in which clusters of slab-related seismicity have been recorded. At 23.1° S, the source is at about 100 km depth, and at 24.2° S, they are located at 200 km depth and shifted to the east by more than 100 km.

### 22.3.3 What Is the Proportion of Melts and Their Distribution?

The degree of partial melting and the interconnectivity of melts are key parameters in the evolution of continental crusts, both from a geochemical and tectonic point of view (Brown et al. 1999). Therefore, the processes of partial melting have been intensively studied since the 1970s, both experimentally and in the field (e.g., migmatite observations, e.g., Mehnert et al. 1973; Büsch et al. 1974).

From the observed seismic and conductivity properties of the partially melted zone, it is possible to place constraints on the amount of melt present. The seismic attenuation and velocity values are less amenable to quantification than conductivity because they can be affected by a wider range of rock properties (fracture porosity, temperature, anisotropy effects). In addition, electrical conductivity is a much more sensitive parameter for the evaluation of melt fractions because of the very large difference in electrical conductivity between melts (up to 10 S m$^{-1}$) and solid rock (< 0.01 S m$^{-1}$) compared to the relatively small variations in the elastic properties of the two phases ($v_p$(melt) ≈ 3 km s$^{-1}$; $v_p$(rock) ≈ 7 km s$^{-1}$).

The electrical conductivity of two-phase composites (e.g., melt and rock) can be modeled by the Hashin-Shtrikman upper and lower bounds (HS$^+$ and HS$^-$) (Hashin and Shtrikman 1962). The average DC (direct current) conductivity $\sigma$ in the HS$^+$ and HS$^-$ model is described by:

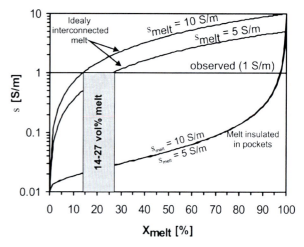

**Fig. 22.8.** Calculated electrical conductivity of a partially molten rock as a function of the melt portion (Eq. 22.1). An interconnected melt along grain boundaries is assumed for modeling (*bold solid line* – Hashin Shtrikman upper boundary). A highly conducting (10 S m$^{-1}$) and an intermediately conducting (5 S m$^{-1}$) melt are used in the calculation. The conductivity of the solid rock is set to 0.01 S m$^{-1}$. Between 14 and 27 vol.% of melt is, at least, necessary to obtain the 1 S m$^{-1}$ observed beneath the Altiplano (ideally interconnected melt). If the melt is concentrated in insulated melt pockets, ~90 vol.% of melt is required to describe the observed conductivity

$$\sigma_{HS^+} = \sigma_I \left( 1 - \frac{3(1-x_I)\Delta\sigma}{3\sigma_I - x_I\Delta\sigma} \right) \quad (22.1)$$

$$\sigma_{HS^-} = \sigma_{II} \left( 1 + \frac{3x_I\Delta\sigma}{3\sigma_{II} + (1-x_I)\Delta\sigma} \right)$$

with

$$\Delta\sigma = \sigma_I - \sigma_{II}$$

where $x_I$ is the volume fraction of the melt and $\sigma_I$ and $\sigma_{II}$ are the conductivities of the melt and solid phases, respectively. For HS$^+$, phase I is assumed to be interconnected within a 3D network (ideal interconnected network of melt), while the HS$^-$ scenario assumes that the melt is ideally insulated in pockets. The modeled conductivities for these two ideal cases are shown in Fig. 22.8.

If the observed electrical conductivity (Fig. 22.2) is assumed to be caused by a highly conductive silicate melt (5 and 10 S m$^{-1}$, Satherly and Smedley 1995; Partzsch et al. 2000; Gaillard 2004) at the expected temperatures of 1 000 °C, we conclude that at least 14–27 vol.% of interconnected melt is needed to explain the observed conductivities in the Central Andes (Fig. 22.8). These values must be considered minimum estimates because they assume that the melts form an ideally interconnected network in the zone. Due to percolation problems (e.g., Bahr 1997), and the likelihood of some melt occurring in melt pockets (e.g., Partzsch et al. 2000), the real amount of melt will probably be higher. Even if all melt is interconnected, part of the melt will contribute only little to the overall conductance, as the conductivity is dominated by the smallest junctions of the highly conducting phase. Therefore, we suggest 20 vol.% as a conservative estimate for the amount of melt in the HCZ beneath the Altiplano.

If partially molten rocks are used to model the elastic behavior in the Central Andes, approximately the same amount of melt as in the electrical conductivity estimations (15–20 vol%) is necessary to explain the seismological observations (Schmitz et al. 1997). In addition, the observed anomalies in parts of the plateau's crust spatially coincide. It is therefore conceivable that the Altplano low velocity zone as identified from receiver functions (typical wavelength > 10 km, Yuan et al. 2000), the reflectivity pattern observed from active seismic methods (typical wavelengths 0.5–5 km, ANCORP Working Group 2003), and the prominent, low-$Q_P$ anomalies in the mid crust beneath the plateau (wavelengths 0.5–10 km, e.g., Haberland et al. 2003) are all manifestations of the same, single structure. The wavelength of the seismic signals, derived by different methods, varies from 1 km to 10 km, implying that the related structures are consistent over at least one order of magnitude. If these anomalies are caused by the presence of partial melts, these melts must also be spatially distributed in this range (i.e. from 1 to > 20 km) over a broad volume.

The concept of dihedral angles (contact angle between a melt and two adjoining crystalline grains, see Waff and Faul 1992) and ideal wetting behavior between mineral and melt phases has traditionally been prominent in studies of the physical properties of partially molten rocks (e.g., Beere 1975a, 1975b; Laporte et al. 1997). This concept is still widely used, even though it only approximates the surface energies of minerals and melts as being isotropic (Holness 1995). In the context of electrical conductivity, melt can only enhance crustal conductivity in a particular zone if it is interconnected to a high degree. Such an interconnected network of melt can develop at relatively low melt fractions of < 1% (Laporte et al. 1997) for dihedral angles less than about 60°. In this case, the melt phase extends along grain edges forming channels and thin films connecting the neighboring melt-filled grain corners (Best and Christiansen 2001). However, because this structure also provides a high degree of permeability, melt can segregate at low melt fractions, and conductivity may be a transient feature.

Recent studies of partially molten rocks, however, have shown strong deviations from an ideal wetting behavior at different scales. Examples are laboratory studies of partially-molten-rock properties (e.g., Lupulescu and Watson 1999; Partzsch et al. 2000) and three-dimensional imaging of migmatites (e.g., Brown et al. 1999) all showing a deviation from equilibrium properties.

Of particular importance for the discussion of melt distribution is the distinction between hydrostatic versus non-hydrostatic stress conditions (e.g., Jin et al. 1994; Daines and Kohlstedt 1997). Non-hydrostatic conditions must be expected for the Central Andes during formation of the Altiplano Plateau. Melts and fluids in deforming rocks may be more concentrated in high-strain shear zones relative to the surrounding rocks and this melt/fluid focusing weakens the shear zone, causing positive feedback by further strain localization. This interaction between deformation and melt convergence into finer-grained regions has been studied in experimentally deformed rock samples (e.g., Bruhn et al. 2005) and can be inferred from field studies of deformed migmatites (Brown et al. 1999).

In general, deformation greatly enhances the percolation of melts and leads to a channelization of melt flow (e.g., Rosenberg and Handy 2000; Holtzman et al. 2003) and to permeability anisotropy as shear zones develop and serve as conduits for melt transport (Rosenberg and Handy 2000). Even melts that are predicted to be immobile based on high interfacial energies, have been found to percolate and localize in channels during deformation (Bruhn et al. 2000).

Since the distribution of melt has a critical influence on the physical properties of rocks, independent information about these properties can potentially help to infer the amount and distribution of melt within the Andean crust. Unfortunately, neither the laboratory experiments or theoretical equilibrium models described above, nor the geophysical and geological data from the Central Andes are sufficiently detailed and complete by themselves to infer the internal structure of the partially melted body in detail. However, the combination of various methods does allow a deeper insight into the geometry and melt distribution in the anomalous zone. In magnetotellurics the commonly best-resolved quantity is conductance, which is the average conductivity integrated over a certain thickness (Edwards et al. 1981). The principle of equivalence dictates that a model assuming discrete melt layers, such as intercalated magmatic sills, will essentially yield – within given error margins – a similar conductance response as a single magmatic body and therefore other, independent information is needed to constrain details of the melt distribution. In the case of the Altiplano conductor attenuation of electromagnetic energy, due to the thick highly conducting layer, makes it difficult to resolve a lower boundary of the partially molten zone. There are indications that the conductive body extends to the mantle at its western margin (region F of model II in Fig. 22.2), but this is not a stable feature in the data inversions and other geometries are equally possible.

There are independent physical observations and arguments that can put constraints on the geometry of melts within the Altiplano crustal conductor. We can rule out one huge highly molten body ("magma chamber") for a number of reasons. First, if the conductive zone were a large magma chamber (> 100 km$^3$), buoyancy forces would lead to the fast ascent of magmas (several km yr$^{-1}$ range using simple Stokes-law velocities and reasonable crustal and melt densities) which would easily overcome the strength of the surrounding rocks. This would cause strong surface expression as volcanism or geothermal fields which are not observed, except very locally, in the region, whereas the main phase of ignimbrite eruptions ended 4 Ma ago. Furthermore, the observation that S-waves can propagate through the HCZ, even though they are strongly attenuated, indicates that a solid network in the zone exists. A horizontally interconnected network of melt channels would satisfy both the magnetotelluric and seismic observations.

Geological observations of migmatite terrains in deeply-eroded orogens throughout the world seem to give a reasonable analog for the partially molten zone in the Altiplano crust. Estimates of the volume of partial melts in migmatite terrains suggest that 20 vol.% of melt is a common value (Mehnert 1981; Nyman et al. 1995; Brown et al. 1999). More importantly, the melt portions of the migmatite tend to be interconnected in stringers or lenses of leucosome within the unmelted mesosome (see the classic descriptions in Mehnert 1981 and Atherton and Gribble 1983). The common observation of around 20 vol.% leucosome material in migmatite zones may represent an upper limit of the amount of melt that can stably exist over a longer time-span in a solid matrix before segregation and movement of melt from the matrix take place. There is no single threshold value for melt segregation in all rocks since this depends on compositional variables of melt and matrix, as well as on the degree of differential stress during partial melting, but values of 10–20 vol.% for granitic melts are commonly cited (e.g., Arzi 1979; Vigneresse et al. 1996; Petford et al. 2000).

We suggest that in the case of the Central Andes, the overall ~20 vol.% of partial melt needed to explain the conductivity zone represents the present, perhaps transient state of a dynamic system where melt may be locally mobile at any given position in the zone as a function of many variables including strain localization, permeability and cooling of individual melt lenses and injection of fresh melt laterally or vertically. The evidence provided from ignimbrite studies in this region supports the concept of a dynamic melt zone undergoing recharge and mixing from more mafic magmas. Relevant observations include the presence of mixed pumice fragments (comingled rhyolitic and andesitic melts, see Lindsay et al. 2001) and geochemical evidence for mafic recharge (Schmitt et al. 2001; Ort et al. 1996). Furthermore, there must be a periodic segregation and upward "drainage" from the zone of melt generation and hybridization to the shallow location of pre-eruptive magma storage (5–8 km estimated for the La Pacana and Purico systems, see Lindsay et al. 2001 and Schmitt et al. 2001).

Compositional studies of the Central Andes basement rocks (Lucassen et al. 2001) and the large-volume dacitic ignimbrites from the APVC have shown that the ignimbrites have isotopic and geochemical compositions that are not compatible with a purely crustal origin, and some contribution from andesitic magma is indicated by the mixed pumices (Lindsay et al. 2001; Schmitt et al. 2001; Ort et al. 1996). As shown in Fig. 22.5, the observed isotopic data can be explained if the erupted magmas contain ~20–30 vol.% andesitic material. Importantly, the large-volume, hybrid magmas erupted in the APVC are remarkably homogeneous in terms of crystal content and chemical composition (e.g., deSilva 1991), and there is a great deal of similarity of ignimbrite compositions erupted at different times and from different calderas within the APVC. The degree of homogeneity observed within and between different large-volume ignimbrite units is unlikely to result from mixing of crustal and andesitic melts in shallow-level magma chambers because cooling at these levels should be too fast to permit efficient convective stirring. We suggest that a better explanation for the compositional homogeneity of hybrid magmas is that the mixing of crustal melts with andesitic magmas took place largely within the source region in the mid crust, and that the processes of segregation and ascent of magmas from this zone produced well-mixed, hybrid magmas represented in the large-volume ignimbrites.

### 22.3.4 Influence on Rheology

Partial melting of rocks can affect their strength by several processes. One is enhanced diffusion along the grain boundaries that are wetted by the melt. This was proposed for partially molten, granitic aggregates by Dell'Angelo et al. (1987) and for olivine aggregates with basaltic melt by Hirth and Kohlstedt (1995), among others. Another possibility at increased melt fraction is that a melt network within solid rock weakens the material because of the much lower viscosity of the melt. In general, sliding of grains past each other is significantly enhanced along melt-containing grain-boundaries (Hirth and Kohlstedt 1995). The dependence of aggregate strength on melt fraction is subdivided into three regions:

A  where aggregate strength is mainly controlled by the solid framework;
B  where it is transitional; and
C  where it is primarily controlled by the melt.

The melt fraction at the transition between regions A and C has been called the "rheologically critical melt percentage" (Rosenberg and Handy 2005). Early experimental examples of such behavior were published by Arzi (1978) for aplite aggregates and by van der Molen and Paterson (1979) for granitic samples. In both of these studies, a strength drop by several orders of magnitude was observed for samples with melt volumes between 12% and 25%, whereas smaller amounts of melt reduced the strength by less than an order of magnitude. In contrast to these studies, Rutter and Neumann (1995) report no sudden drop in strength for partially molten granite samples and, thus, no evidence of a "rheologically critical melt percentage", but rather a continuous weakening with increasing melt fraction.

The usefulness of the concept of a "rheologically critical melt percentage" was further questioned in a recent review of experimental work on deformation of partially molten rocks by Rosenberg and Handy (2005). By looking at the absolute drops in material strength with increasing melt fraction, Rosenberg and Handy (2005) argue that the most significant decrease occurs with relatively small melt volumes (< 7%). The changes observed for higher melt contents span more orders of magnitude but are much smaller in terms of absolute numbers. This drop in strength at low melt volumes is considered to be a far more important mechanism for strain softening and localization than the effect described by the "rheologically critical melt percentage".

The expected melt fractions for the high conductivity zone beneath the Altiplano Plateau exceed 15% and, therefore, regardless of the softening mechanism, the zone can be expected to be much weaker than its surroundings, by at least an order of magnitude. As described above, compressional deformation of the partially molten crust may lead to positive feedback by localization of the melt into channels (Holtzman et al. 2003) and thus causing a partitioning of the crust into subhorizontal high- and low-strain zones. Such a strain partitioning could be a mechanism for explaining the anisotropy in seismic velocity within the partially molten zone which was emphasized in the receiver function results by Zandt et al. (2003).

### 22.3.5 Influence on Plateau Formation

Recent thermal-mechanical modeling of the Andean Plateau formation (Babeyko et al. 2005; Vietor and Echtler 2006, Chap. 18 of this volume) reveals the importance of a weak crust for both the shortening process and for the flat relief of the Altiplano Plateau. Thus, one may suggest a relationship between the establishment of a zone of partial melt with strong shortening and plateau formation.

Numerical and analytical studies have shown that the extent of the flat region in the center of the high plateau requires a similarly sized, weak zone within or at the base of the crust (Beaumont et al. 2001, 2004; Williams et al. 1994).

The large partially molten zone in the Central Andean crust described in this paper would constitute such a weak zone. This weak zone reduces the maximum surface angle and the taper angle of the entire orogenic wedge and hence the transition from the wedge-shaped flanks to the rectangular plateau section can be linked to the extent of the partially molten layer with reduced strength (Wdowinski and Bock 1994b; Wdowinski and Bock 1994a; Beaumont et al. 2001, 2004; Jamieson et al. 2004; Babeyko et al. 2002; Vietor and Oncken 2005).

A close connection between the cross-sectional shape and inferred thermal weakening of the Andean orogen owing to extensive crustal melting is in line with the lateral changes of geology and morphology along strike. In the southern Central Andes, for example, at around 30 to 33° S, upper-plate shortening again reaches magnitudes of more than 150 km (Allmendinger et al. 1990). However, in the absence of crustal melting in this part of the orogen and because of the shallow dip of the subducted Nazca Plate, the mountain belt retains a double-wedge-shaped geometry and lacks a central, flat, plateau section (Ramos et al. 2002). Thus, the presented 3D distribution of melt within the Andean orogen seems to be a requirement for the evolution of the plateau.

## Acknowledgments

This contribution summarizes results of several projects and many individuals within the SFB-267 team, all of whom are gratefully acknowledged. We thank Bernd Schurr in particular for providing his attenuation model. This work was mainly supported by the DFG and by the BMBF (TIPTEQ). The constructive reviews by Claudio Rosenberg and Claire Curie greatly enhanced the quality of the manuscript. Allison Britt significantly improved the English.

## References

Aki K, Chouet B (1975) Origin of coda waves: Source, attenuation, and scattering effects. Phys Earth Planet Inter 80:3322–3342

Allmendinger RW, Figueroa D, Snyder D, Beer J, Mpodozis C, Isacks BL (1990) Foreland shortening and crustal balancing in the Andes at 30° S latitude. Tectonics 9:789–809

Allmendinger RW, Jordan TE, Kay SM, Isacks BL (1997) The evolution of the Altiplano-Puna plateau of the Central Andes. Ann Rev Earth Planet Sci Lett 25:139–174

ANCORP Working Group (2003) Seismic imaging of an active continental margin and plateau in the Central Andes (Andean Continental Research Project (ANCORP) 1996). J Geophys Res 108(B7): doi 10.1029/2002JB001771

Arndt J, Bartel T, Scheuber E, Schilling FR (1997) Thermal and rheological properties of granodioritic rocks from the Central Andes, North Chile. Tectonophysics 271:75–88

Arzi AA (1979) Critical phenomena in the rheology of partially molten rocks. Tectonophysics 44:173–184

Asch G, Schurr B, Bohm M, Yuan X, Haberland C, Heit B, Kind R, Woelbern I, Bataille K, Comte D, Pardo M, Viramonte J, Rietbrock A, Giese P (2006) Seismological studies of the Central and Southern Andes. In: Oncken O, Chong G, Franz G, Giese P, Götze H-J, Ramos VA, Strecker MR, Wigger P (eds) The Andes – active subduction orogeny. Frontiers in Earth Science Series, Vol 1. Springer-Verlag, Berlin Heidelberg New York, pp 443–458, this volume

Atherton MP, Gribble CD (1983) Migmatites, melting and metamorphism. Shiva Geology Series, Shiva Publishing, Nantwich UK

Babeyko AY, Sobolev SV, Trumbull RB, Oncken O, Lavier LL (2002) Numerical models of crustal scale convection and partial melting beneath the Altiplano-Puna Plateau. Earth Planet Sci Lett 199:373–388

Bahr K (1997) Electrical anisotropy and conductivity distributions of fractal random networks and of the crust: the scale effect of connectivity. Geophys J Int 130:649–660

Baker MCW, Francis PW (1978) Upper Cenozoic volcanism in the Central Andes – ages and volumes. Earth Planet Sci Lett 41:175–187

Beaumont C, Jamieson RA Nguyen MH, Lee B (2001) Himalayan tectonics explained by extrusion of a low-viscosity crustal channel coupled to focused surface denudation. Nature 414:738–742

Beaumont C, Jamieson RA, Nguyen MH, Medvedev S (2004) Crustal channel flows: 1. Numerical models with applications to the tectonics of the Himalayan-Tibetan orogen. J Geophys Res 109(B06406): doi 10.1029/2003JB002809

Beck S, Zandt G (2002) The nature of orogenic crust in the Central Andes. J Geophys Res 107: doi 10.1029/2000JB000124

Beck S, Zandt G, Myers SC, Wallace TC, Silver PG, Drake L (1996) Crustal thickness variations in the Central Andes. Geology 24:407–410

Beere W (1975a) A unifying theory of the stability of penetrating liquid phases and sintering pores. Acta Metall 23:131–138

Beere W (1975b) The second stage sintering kinetics of powder compacts. Acta Metall 23:139–145

Bock G, Schurr B, Asch G (2000) High-resolution image of the oceanic Moho in the subducting Nazca plate from P-to-S converted waves. Geophys Res Let 27:3929–3932

Brasse H (2005) The mantle wedge in the Bolivian orocline in the view of deep electromagnetic soundings. Extended Abstract, 6$^{th}$ International Symposium on Andean Geodynamics, Barcelona

Brasse H, Soyer W (2001) A magnetotelluric study in the Southern Chilean Andes. Geophys Res Let 28(19):3757–3760

Brasse H, Lezaeta P, Rath V, Schwalenberg K, Soyer W, Haak V (2002) The Bolivian Altiplano conductivity anomaly. J Geophys Res 107(B5) doi 10.1029/2001JB000391

Brown MA, Brown M, Carlson WD, Denison C (1999) Topology of syntectonic melt-flow networks in deep crust: inferences from three-dimensional images of leucosome geometry in migmatites. Am Mineral 84:1793–1818

Bruhn D, Groebner N, Kohlstedt DL (2000) An interconnected network of core-forming melts produced by shear deformation. Nature 403:883–886

Bruhn D, Kohlstedt DL, Lee KH (2005) The effect of grain size and melt distributions on the rheology of partially molten olivine aggregates. In: Bruhn D, Burlini L (eds) High strain zones: structures and physical properties. Geol Soc Spec Publ 245

Büsch W, Schneider G, Mehnert KR (1974) Initial Melting at grain boundaries. Part II: Melting in rocks of granodioritic, quartz dioritic and tonalitic compositions. N Jb Miner Monatshefte 8:345–370

Chapman DS, Furlong KP (1992) The thermal state of the lower crust. In: Fountain DM, Arculus RJ, Kay RM (eds) Continental Lower Crust. Development in Geotectonics Volume 23, Elsevier, Amsterdam, pp 179–199

Chmielowski J, Zandt G, Haberland C (1999) The Central Andean Altiplano-Puna magma body. Geophys Res Lett 26(6):783–786

Coira B, Kay SM, Viramonte J (1993) Upper Cenozoic magmatic evolution of the Argentine Puna – a model for changing subduction geometry. Int Geol Rev 35:677–720

Daines MJ, Kohlstedt DL (1997) Influence of deformation on melt topology in peridotites. J Geophys Res 107:10257–10271

Davidson J, Harmon R, Wörner G (1991) The source of Central Andean magmas: some considerations. Geol Soc Am Spec P 265:233–243

De Silva SL (1989) Altiplano-Puna volcanic complex of the Central Andes. Geology 17:1102–1106

De Silva SL (1991) Styles of zoning in Central Andean ignimbrites. Insights into magma chamber processes. Geol Soc Am Spec P 265:217–232

De Silva SL Francis P (1991) Volcanoes of the Central Andes. Springer-Verlag, Berlin Heidelberg New York

Echternacht F, Tauber S, Eisel M, Brasse H, Schwarz G, Haak V (1997) Electromagnetic study of the active continental margin in Northern Chile. Phys Earth Planet Int, 102(1–2):69–88

Feeley TC, Davidson J (1994) Petrology of calc-alkaline lavas at Volcan Ollagüe and the origin of compositional diversity at Central Andean stratovolcanoes. J Petrol 35:1295–1340

Francis PW, Sparks RSJ, Hawkesworth CJ, Thorpe RS, Pyle DM, Tait SR, Mantovani MS, McDermott F (1989) Petrology and geochemistry of volcanic rocks of the Cerro Galán caldera, northwest Argentina. Geol Mag 126:515–547

Gardeweg M, Ramírez CF (1987) The La Pacana caldera and the Atana ignimbrite – a major ash-flow and resurgent caldera complex in the Andes of Northern Chile. Bull Volcanology 49:547–566

Giese P (1994) Geothermal structure of the Central Andean crust – implications for heat transport and rheology. In: Reutter K-J, Scheuber E, Wigger P (eds) Tectonics of the Southern Central Andes. Springer-Verlag, Berlin Heidelberg New York, 69–76

Giese P, Schilling FR (2000) Some consequences from the temperature-heat-paradox in the Altiplano (Central Andes). EOS 81(48): F1136

Giese P, Scheuber E, Schilling FR, Schmitz M, Wigger P (1999) Crustal thickening processes in the Central Andes and the different natures of the Moho-discontinuity. J South Am Earth Sci 12:201–220

Götze H-J, Lahmeyer B, Schmidt S, Strunk S (1994) The lithospheric structure of the Central Andes (20–25° S) as inferred from quantitative interpretation of regional gravity. In: Reutter K-J, Scheuber E, Wigger P (eds) Tectonics of the Southern Central Andes. Springer-Verlag, Berlin Heidelberg New York, 23–48

Haberland C (1998) Die Verteilung seismischer Absorption in den Zentralen Anden. Ph.D. thesis, Freie Universität Berlin

Haberland C, Rietbrock A (2001) Attenuation tomography in the western Central Andes: a detailed insight into the structure of a magmatic arc. J Geophys Res 106(B6):11151–11167

Haberland C, Rietbrock A, Schurr B, Brasse H (2003) Coincident anomalies of seismic attenuation and electrical resistivity beneath the southern Bolivian Altiplano Plateau. Geophys Res Lett 30(18): doi 10.1029/2003GL017492

Hamza VM, Muñoz M (1996) Heat flow map of South America. Geothermics 25(6):599–646

Hansen PC (1998) Rank deficient and discrete ill-posed problems. Numerical aspects of linear inversion. Siam, Philadelphia

Hashin Z, Shtrikman A (1962) A variational approach to the theory of effective magnetic permeability of multiphase materials. J Appl Phys 33:3125–3131

Henry SG, Pollack HN (1988) Terrestrial heat flow above the Andean subduction zone in Bolivia and Peru. J Geophys Res 93:15153–15162

Hirth G, Kohlstedt DL (1995) Experimental constraints on the dynamics of the partially molten upper mantle: deformation in the diffusion creep regime. J Geophys Res 100:1981–2001

Hirth G, Kohlstedt DL (1996) Water in the oceanic upper mantle: implications for rheology, melt extraction and the evolution of the lithosphere. Earth Planet Sci Lett 144:93–108

Hoffmann-Rothe A, Ritter O, Janssen C (2004) Correlation of electrical conductivity and structural damage at a major strike-slip fault in Northern Chile. J Geophys Res 109(B10): doi 10.1029/2004JB003030

Holness MB (1995) The effect of feldspar on quartz-$H_2O$-$CO_2$ dihedral angles at 4 kbar, with consequences for the behaviour of aqueous fluids in migmatites. Contrib Mineral Petrol 118:356–364

Holtzman BK, Kohlstedt DL, Zimmerman ME, Heidelbach F, Hiraga T, Hustoft J (2003) Melt segregation and strain partitioning: implications for seismic anisotropy and mantle flow. Science 1227–1230

James DE (1971) Andean crustal and upper mantle structure. J Geophys Res 76:3246–3271

Jamieson RA, Beaumont C, Medvedev S, Nguyen MH (2004) Crustal channel flows: 2. Numerical models with implications for metamorphism in the Himalayan-Tibetan orogen. J Geophys Res 109(B06407) doi 10.1029/2003JB00281

Jin ZM, Green HW, Zhou Y (1994) Melt topology in partially molten mantle peridotite during ductile deformation. Nature 372:164–167

Johnston D, Toksoz M, Timor A (1979) Attenuation of seismic waves in dry and saturated rocks. II. Mechanisms. Geophysics 44(4):691–711

Kay SM, Coira B, Viramonte J (1994) Young mafic back arc volcanic rocks as indicators of continental lithospheric delamination beneath the Argentine Puna plateau, Central Andes. J Geophys Res 99:24323–24339

Kay SM, Mpodozis C, Coira B (1999) Neogene magmatism, tectonism, and mineral deposits of the Central Andes (22° to 33° S latitude). In: Skinner B (ed) Geology and ore deposits of the Central Andes. Soc Econ Geol Spec Pub 7, pp 27–59

Krüger D (1994) Modellierungen zur Struktur elektrisch leitfähiger Zonen in den südlichen zentralen Anden. Berliner Geowiss Abh B21

Laporte D, Rapaille C, Provost A (1997) Wetting angles, equilibrium melt geometry, and the permeability threshold of partially molten crustal protholiths. In: Bouchez JL, Hutton DHW, Stephens WE (eds) Granite: from segregation of melt to emplacement fabrics. Kluwer Academic Publishers, Dordrecht Boston London, pp 31–54

Le Pichon X, Henry P, Goffé B (1997) Uplift of Tibet: from eclogites to granulites – implications for the Andean Plateau and the Variscan belt. Tectonophysics 273:57–76

Lezaeta P (2001) Distortion analysis and 3-modeling of magnetotelluric data in the Southern Central Andes. Ph.D. thesis, Freie Universität Berlin

Lezaeta P (2002) The confidence limit of the magnetotelluric phase sensitive skew. Earth Planet Space 54(5):451–457

Lezaeta P, Brasse H (2001) Electrical conductivity beneath the volcanoes of the NW Argentinian Puna. Geophys Res Lett 28 (24): 4651–4654

Lezaeta P, Haak V (2003) Beyond magnetotelluric decomposition: Induction, current channeling, and magnetotelluric phases over 90°. J Geophys Res 108(B5): doi 10.1029/2001JB000649

Lezaeta P, Muñoz M, Brasse H (2000) Magnetotelluric image of the crust and upper mantle in the backarc of the NW Argentinean Andes. Geophys J Int 142:841–854

Lindsay JM, Schmitt AK, Trumbull RB, De Silva SL, Siebel W, Emmermann R (2001) Magmatic evolution of the La Pacana Caldera system, Central Andes, Chile: compositional variation of two cogenetic, large-volume felsic ignimbrites and implications for contrasting eruption mechanisms. J Petrol 42:459–486

Lucassen F (1992) Geologie, Metamorphosegeschichte und Geochemie neugebildeter basischer Kruste im jurassischen magmatischen Bogen der Küstenkordillere Nordchiles, Region Antofagasta 23°25'–24°20' S. Ph.D. thesis, Technische Universität Berlin

Lucassen F, Becchio R, Harmon R, Kasemann S, Franz G, Trumbull R, Romer RL, Dulski P (2001) Composition and density model of the continental crust in an active continental margin – the Central Andes between 18° and 27° S. Tectonophysics 341:195–223

Lupulescu A, Watson EB (1999) Low melt fraction connectivity of granitic and tonalitic melts in a mafic crustal rock at 800 °C and 1GPa. Contrib Miner Petrol 134(2–3):202–216.

Mavko GM (1980) Velocity and attenuation in partially molten rocks. J Geophys Res 85(B10):5173–5189

Mehnert KR (1981) Migmatites and the origin of granitic rocks. Elsevier, Amsterdam

Mehnert KR, Büsch W, Schneider G (1973) Initial Melting at grain boundaries of quartz and feldspar in gneisses and granulites. N Jb Miner Monatshefte 4:165–183

Nyman MW, Pattison DRM, Ghent ED (1995) Melt extraction during formation of sillimanite and K-feldspar migmatites, west of Revelstoke, British Columbia. J Petrol 36:351–372

Ocola LC, Meyer RP (1972) Crustal low-velocity zones under the Peru-Bolivia Altiplano. Geophys J R Astr Soc 30:199–209

Oncken O, Hindle D, Kley J, Elger K, Victor P, Schemmann K (2006) Deformation of the central Andean upper plate system – facts, fiction, and constraints for plateau models. In: Oncken O, Chong G, Franz G, Giese P, Götze H-J, Ramos VA, Strecker MR, Wigger P (eds) The Andes – active subduction orogeny. Frontiers in Earth Science Series, Vol 1. Springer-Verlag, Berlin Heidelberg New York, pp 3–28, this volume

Ort M, Coira BL, Mazzoni MM (1996) Generation of a crust-mantle magma mixture: magma sources and contamination at Cerro Panizos, Central Andes. Contrib Miner Petrol 123:308–322

Partzsch GM (1998) Elektrische Leitfähigkeit partiell geschmolzener Gesteine: experimentelle Untersuchungen, Modellrechnungen und Interpretation einer elektrisch leitfähigen Zone in den zentralen Anden. Berliner Geowiss Abh B26

Partzsch GM, Arndt J, Schilling FR (2000) Electrical conductivity and melt distribution of a crustal rock during partial melting. Tectonophysics 317:189–203

Patiño Douce AE, Beard JS (1995) Dehydration-melting of biotite gneiss and quartz amphibolite from 3 to 15 kbar. J Petrol 36:707–738

Petford N, Cruden AR, McCaffrey KJW, Vigneresse JL (2000) Granite magma formation, transport and emplacement in the Earth's crust. Nature 408:669–673

Pollack HN, Chapman DS (1977) On the regional variation of heat flow, geotherms, and lithospheric thickness. Tectonophysics 38:279–296

Ramos VA, Cristillini EO, Perez DJ (2002) The Pampean flat slab of the Central Andes, J S Am Earth Sci 15:59–78

Rietbrock A (1999) Velocity structure and seismicity in the Central Andes of Northern Chile and Southern Bolivia. AGU Fall Meeting 1999 Abstracts, EOS supplement

Riller U, Oncken O (2003) Growth of the Central Andean Plateau by tectonic segmentation is controlled by the gradient in crustal shortening. J Geol 111:367–384

Rodi W, Mackie RL (2001) Nonlinear conjugate gradients algorithm for 2-D magnetotelluric inversions. Geophysics 66:174–187

Rosenberg CL, Handy MR (2005) Experimental deformation of partially-melted granite revisited: implications for the continental crust. J Metamorph Geol 23:19–28

Rutter EH, Neumann DH (1995) Experimental deformation of partially molten Westerly granite under fluid-absent conditions, with implications for the extraction of granitic magmas. J Geophys Res 100:15697–15715

Satherley J, Smedley SL (1985) The electrical conductivity of some hydrous and anhydrous molten silicates as a function of temperature and pressure. Geochim Cosmochim Acta 49:769–777

Schilling FR (2003) Does the electrical conductivity of partially molten rocks depend on grain-boundaries? AGU Fall Meeting 2003 Abstracts, EOS supplement

Schilling FR, Partzsch GM, Brasse H, Schwarz G (1997) Partial melting below the magmatic arc in the Cantral Andes deduced from geoelectromagnetic field experiments and laboratory data. Phys Earth Planet Inter 103:17-31

Schmitt AK (2001) Gas-saturated crystallization and degassing in large-volume, crystal-rich dacitic magmas from the Altiplano-Puna, Northern Chile. J Geophys Res 106(B12):30561-305678

Schmitt AK, De Silva SL, Trumbull RB, Emmermann R (2001) Magma evolution in the Purico ignimbrite complex, Northern Chile: evidence for zoning of a dacitic magma by injection of rhyolitic melts following mafic recharge. Contrib Mineral Petrol 140:680-700

Schmitz M, Heinsohn W, Schilling FR (1997) Seismic gravity and petrological evidence for partial melt beneath the thickened Central Andean crust (21-23° S). Tectonophysics 270:313-326

Schurr B, Rietbrock A (2004) Deep seismic structure of the Atacama basin, Northern Chile. Geophys Res Lett 31: doi 10.1029/2004GL019796

Schurr B, Asch G, Rietbrock A, Trumbull R, Haberland C (2003) Complex patterns of fluid and melt transport in the Central Andean subduction zone revealed by attenuation tomography. Earth Planet Sci Lett 215(1-2):105-119

Schwalenberg K, Haak V, Rath V (2002) The application of sensitivity studies on a two-dimensional resistivity model from the Central Andes. Geophys J Int 150:673-686

Schwarz G, Krüger D (1997) Resistivity cross section through the southern Central Andes as inferred from magnetotelluric and geomagnetic deep soundings. J Geophys Res 102 (B6), 11957-11978

Schwarz G, Chong DG, Krüger D, Martinez M, Massow W, Rath V, Viramonte J (1994) Crustal high conductivity zones in the southern Central Andes. In: Reutter K-J, Scheuber E, Wigger P (eds) Tectonics of the Southern Central Andes. Springer-Verlag, Berlin Heidelberg New York

Seipold U (1998) Temperature dependence of thermal transport properties of crystalline rocks – a general law. Tectonophysics 291:161-171

Sick C, Yoon M-K, Rauch K, Buske S, Lüth S, Araneda M, Bataille K, Chong G, Giese P, Krawczyk C, Mechie J, Meyer H, Oncken O, Reichert C, Schmitz M, Shapiro S, Stiller M, Wigger P (2006) Seismic images of accretive and erosive subduction zones from the Chilean margin. In: Oncken O, Chong G, Franz G, Giese P, Götze H-J, Ramos VA, Strecker MR, Wigger P (eds) The Andes – active subduction orogeny. Frontiers in Earth Science Series, Vol 1. Springer-Verlag, Berlin Heidelberg New York, pp 147-170, this volume

Sobolev SV, Babeyko AY (2005) What drives orogeny in the Andes? Geology 33(8):617-620

Sobolev SV, Babeyko AY, Koulakov I, Oncken O (2006) Mechanism of the Andean orogeny: insight from numerical modeling. In: Oncken O, Chong G, Franz G, Giese P, Götze H-J, Ramos VA, Strecker MR, Wigger P (eds) The Andes – active subduction orogeny. Frontiers in Earth Science Series, Vol 1. Springer-Verlag, Berlin Heidelberg New York, pp 513-536, this volume

Soyer W, Brasse H (2001) A magneto-variation array study in the Central Andes of N Chile and SW Bolivia. Geophys Res Lett 28(15):3023-3026

Springer M, Förster A (1998) Heat-flow density across the Central Andean subduction zone. Tectonophysics 291:123-139

Trumbull RB, Wittenbrink R, Hahne K, Emmermann R, Büsch W, Gerstenberger H, Siebel W (1999) Evidence for Late Miocene to Recent contamination of arc andesites by crustal melts in the Chilean Andes (25-26° S) and its geodynamic implications. J S Am Earth Sci 12:135-155

Van der Molen I, Paterson M (1979) Experimental deformation of partially melted granite. Contrib Mineral Petrol 70:299-318

Vietor T, Echtler H (2006) Episodic Neogene southward growth of the Andean subduction orogen between 30° S and 40° S – plate motions, mantle flow, climate, and upper-plate structure. In: Oncken O, Chong G, Franz G, Giese P, Götze H-J, Ramos VA, Strecker MR, Wigger P (eds) The Andes – active subduction orogeny. Frontiers in Earth Science Series, Vol 1. Springer-Verlag, Berlin Heidelberg New York, pp 375-400, this volume

Vietor T, Oncken O (2005) Controls on the shape and kinematics of the Central Andean plateau flanks: insights from numerical modeling. Earth Planet Sci Lett 236(3-4)814-827

Vigneresse JL, Barbey P, Cuney M (1996) Rheological transitions during partial melting and crystallisation with application to felsic magma segretation and transfer. J Petrol 37: 1579-1600

Waff HS, Faul UH (1992) Effects of crystalline anisotropy on fluid distribution in ultramafic partial melts. J Geophys Res 97: 9003-9014

Wannamaker PE, Stodt JA, Rijo L (1986) Two-dimensional topographic responses in magnetotelluric models using finite elements. Geophysics 51(11):2131-2144

Wark DA, Watson BE (2000) Effect of grain size on the distribution and transport of deep-seated fluids and melts. Geophys Res Lett 27:2029-2032

Wdowinski S, Bock Y (1994a) The evolution of deformation and topography of high elevated plateaus 1. Model, numerical analysis, and general results. J Geophys Res 99:7103-7119

Wdowinski S, Bock Y (1994b) The evolution of deformation and topography of high elevated plateaus 2. Application to the Central Andes. J Geophys Res 99:7121-7130

Whitman D, Isacks BL, Chalelain JL, Chiu JM, Perez A (1992) Attenuation of high-frequency seismic waves beneath the Central Andean Plateau. J Geophys Res 97:19929-19947

Wigger P, Schmitz M, Araneda M, Asch G, Baldzuhn S, Giese P, Heinsohn WD, Martinez E, Ricaldi E, Röwer P, Viramonte J (1994) Variation of the crustal structure of the southern Central Andes deduced from seismic refraction investigations. In: Reutter K-J, Scheuber E, Wigger P (eds) Tectonics of the Southern Central Andes. Springer-Verlag, Berlin Heidelberg New York, pp 23-48

Willett SD, Beaumont C (1994) Subduction of Asian lithospheric mantle beneath Tibet inferred from models of continental collision. Nature 369:642-645

Williams CA, Connors C, Dahlen FA, Price EJ, Suppe J (1994) Effect of the brittle-ductile transition on the topography of compressive mountain belts on Earth and Venus. J Geophys Res 99: 19947-19974

Yuan X, Sobolev SV, Kind R (2002) Moho topography in the Central Andes and its geodynamic implications. Earth Planet Sci Lett 199(3-4):389-402

Yuan X, Sobolev SV, Kind R, Oncken O, Bock G, Asch G, Schurr B, Graefer F, Rudloff A, Hanka W, Wylegalla K, Tibi R, Haberland C, Rietbrock A, Giese P, Wigger P, Röwer P, Zandt G, Beck S, Wallace T, Pardo M, Comte D (2000) Subduction and collision processes in the Central Andes constrained by converted seismic phases. Nature 408:958-961

Zandt G, Velasco AA, Beck SL (1994) Composition and thickness of the southern Altiplano crust, Bolivia. Geology 22:1003-1006

Zandt G, Leidig M, Chmielowski J, Baumont D, Yuan X (2003) Seismic detection and characterization of the Altiplano-Puna magma body, Central Andes. Pure Appl Geophys 160:789-807

**Chapter 23**

# Controls on the Deformation of the Central and Southern Andes (10–35° S): Insight from Thin-Sheet Numerical Modeling

Sergei Medvedev · Yuri Podladchikov · Mark R. Handy · Ekkehard Scheuber

**Abstract.** What mechanisms and conditions formed the Central Andean orocline and the neighboring Altiplano Plateau? Why does deformation decrease going from the central to the Southern Andes? To answer these questions, we present a new thin-sheet model that incorporates three key features of subduction orogenesis: (1) significant temporal and spatial changes in the strength of the continental lithosphere in the upper plate; (2) variable interplate coupling along a weak subduction channel with effectively anisotropic mechanical properties; and (3) channeled flow of partially molten lower crust in the thickened upper plate. Application of this model to the present kinematic situation between the Nazca and South American Plates indicates that the deformed Andean lithosphere is significantly weaker than the undeformed South American foreland, and that channel flow of partially melted lower crust smoothes topographic relief. This channel flow is, therefore, inferred to control intra-orogenic topography and is primarily responsible for the development of the Andean Plateau since the Miocene. A parameter study shows that the decrease in shortening rates from the central to the Southern Andes can be attributed to the weakening of the orogenic Andean lithosphere and to along-strike variations in interplate coupling within the subduction zone. The current rates of deformation are reproduced in the model if: the Andean lithosphere is assumed to be 5–15 times weaker than the lithosphere of the Brazilian shield; interplate coupling is assumed to be relatively weak, such that the subduction zone in the vicinity of the Central Andes is some 10–20 times weaker than the Andean lithosphere; and coupling itself decreases laterally by some 2–5 times going from the central to the Southern Andes.

## 23.1 Introduction

The bend of the Andes, known as the Bolivian orocline (Fig. 23.1), is a first-order structure visible from space. Oroclinal bending in the Andes was originally proposed by Carey (1958) to have involved anticlockwise rotation of the Andes north of the Arica bend (Fig. 23.1) superposed on an initially straight Andean chain. Indeed, paleomagnetic studies have revealed a consistent and roughly symmetrical pattern of anticlockwise rotations north of the Arica bend and of clockwise rotations south of it (e.g., MacFadden et al. 1995). Today, however, there is consensus that the Central Andes were never entirely straight (Sheffels 1995) and that the original curvature of the plate margin was enhanced during Andean orogenesis rather than having been imposed later (Isacks 1988).

Here, we refer to the segment of the Central Andes containing the Andean Plateau as the Central Andes (lower case in "central"), reaching from ~16° S to 25° S. Parts of Andes south of this latitude to the southern limit of our model we term the Southern Andes. Our subdivision differs from the traditional division of the Andean chain into Central and Southern Andes at ~30° S (Gansser 1973), and reflects the north-to-south change in morphology and rates of deformation, as discussed throughout this paper.

Several explanations have been given for the development of the Bolivian orocline. Isacks (1988) and Allmendinger et al. (1997) argued that the bending resulted from

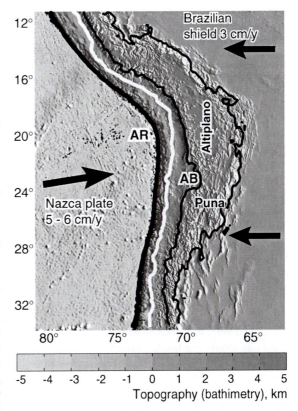

**Fig. 23.1.** Topography and bathymetry of the adjoining Nazca and South American Plates. Plate boundary (*thick black line*), coastal line (*white line*), 3 km elevation line (*thin black line*), Atacama Basin (*AB*), Altiplano and Puna Plateaus, Arica bend (*AR*)

differential shortening along the Andean chain, with shortening increasing towards the Arica bend. Kley et al. (1999) have shown how oroclinal bending could have been achieved by differential shortening. In their model, several fault-bounded blocks underwent various degrees of displacement and differential rotation. Shortening estimates require that rotation of the limbs of the orocline did not exceed 5–10°. Riller and Oncken (2003) also related oroclinal bending to orogen-parallel gradients in horizontal shortening plus block rotations and strike-slip movements. These authors linked these movements to the orogen-normal growth of the Central Andean Plateaus.

Numerical models of Liu et al. (2002) and Yang et al. (2003) set boundary conditions at the western margin of the South America that effectively prescribe the evolution of this margin and preclude modeling the bending of the orocline. Yanez and Cembrano (2004) used a thin sheet approximation model to show that along-strike variations in coupling between Nazca and South American Plates can result in bending of the western boundary of the Central Andes. Sobolev et al. (2006, Chap. 25 of this volume) uses a model that incorporates laws of empirical flow and a coupled thermomechanical approach to compare the evolution of the central and Southern Andes, yet the application of their two-dimensional (2D) modeling to the problem of oroclinal bending, effectively a three-dimensional (3D) process, should be tested using a 3D approach sometime in future. Here, we present a complementary approach to that of Sobolev et al. (2006, Chap. 25 of this volume) by considering a map view of deformational processes in the Andes.

As noted above, the Bolivian orocline is spatially related to another primary structure, the Central Andean Altiplano-Puna Plateau, which, following the Tibetan Plateau, is the second largest plateau on Earth (Fig. 23.1). The thick crust underlying the plateau (> 70 km, James 1971; Wigger et al. 1994; Zandt et al. 1994; Giese et al. 1999; Yuan et al. 2000) is mainly attributed to crustal shortening (Isacks 1988). There are differences between the Altiplano and Puna parts of the plateau: The Puna is slightly higher and has greater relief than the Altiplano. Moreover, the lithosphere is thicker beneath the central and eastern parts of the Altiplano (70–80 km), but less thick (60–70 km) beneath the westernmost Altiplano and Puna (Whitman et al. 1992).

In a synthesis of space-geodetic measurements, Neogene shortening data, and paleomagnetic data, Lamb (2000) has shown that the Altiplano is currently growing eastwards (see also Oncken et al. 2006, Chap. 1 of this volume; Sobolev et al. 2006, Chap. 25 of this volume) without significant uplift of the plateau (in contrast to the results of Yang et al. 2003). Numerical models of orogenesis that incorporate plateau evolution attribute the relatively low relief in the central part of the Himalayan and Andean chains to the lateral flow of weak, possibly partially molten, rock in a middle to lower crustal channel within the thickened orogenic crust (Beaumont et al. 2001, 2004; Vanderhaeghe et al. 2003; Royden 1996; Shen et al. 2001).

Channel flow is driven by differential pressure associated with topographic gradients, such that highly mobile, viscous crust flows from beneath areas of greater to lesser elevation. In flattening topographic gradients, channel flow can form orogenic plateaux as shown in a number of numerical experiments (e.g., Shen et al. 2001; Husson and Sempere 2003; Medvedev and Beaumont, in press). Thermal conditions beneath the Altiplano Plateau favor partial melting (Springer 1999; Springer and Förster 1998; Brasse et al. 2002; Babeyko et al. 2002) and therefore makes channel flow a viable mechanism for the formation of the Andean Plateaux, as already proposed by Husson and Sempere (2003) and Gerbault et al. (2005).

The change in style and in the amount of deformation along the Andean orogenic belt also requires explanation. Deformation of the Central Andes exceeds that in the Southern Andes, as manifest by progressive southward narrowing of the 3 km elevation contour on the map (Fig. 23.1) and by the higher average elevation of the Central Andes. The southward decrease in shortening is accompanied by an increase in crustal thickness to 70 km or more in the Central Andes.

Present shortening rates averaged over last several million years show a similar north-south trend: the Central Andes are currently shortening at rates of 1–1.5 cm yr$^{-1}$ (see Oncken et al. 2006, Chap. 1 of this volume, and discussion within), while in the Southern Andes the rates are so low ($< 0.5$ cm yr$^{-1}$) as to be difficult to estimate (Kley and Monaldi 1998). Several workers attribute this variation in shortening and shortening rates to along-strike differences in the degree of coupling between the Nazca and South American Plates (Lamb and Davis 2003; Yanez and Cembrano 2004; Sobolev et al. 2006, Chap. 25 of this volume; Hoffmann-Rothe et al. 2006, Chap. 6 of this volume), possibly induced by latitudinal, climate-induced variations in erosion rates and the mechanical properties of the trench fill as well as the age and strength of the oceanic plate.

To better understand these first-order deformational features of the Andes, we developed and applied a new numerical model that incorporates some of the salient, physical features of subduction orogenesis. We employ a backward-modeling approach; i.e., we incorporate data on the recent kinematic, gravitational and thermal states of the South America–Nazca Plate boundary system together with knowledge of the current bathymetry, topography and rheology from geophysical studies to back-calculate the local rates of shortening, thickening and rotation in the upper plate.

These calculations are mechanism-specific; they combine lateral shortening, topography and channel flow. The values obtained for primary orogenic features (e.g., orogenic geometry in map view, lithospheric thickness, rotation rates) are then compared with the available data on current rates of shortening (from geological estimations) and rotation (from paleomagnetic studies) to gain insight into the driving mechanisms and parameters controlling Andean orogenesis. Therefore, the criterion for judging the relative importance of different mechanisms is the degree to which predicted, back-calculated values match current, measured values.

We emphasize that our model is not intended to simulate Andean orogenesis or even to provide a detailed picture of Andean orogenic evolution. Rather, it is a vehicle for conducting a series of experiments and parameter studies, each of which is designed to test the plausibility of various mechanisms proposed for the first-order Andean features described above. From a methodological standpoint, our approach is similar to that adopted by Lithgow-Bertelloni and Guynn (2001) to predict stress field variations over the entire Earth. However, our model deals with a much smaller region and is tailored to account for the specific characteristics of recent Andean subduction orogenesis. The numerical basis of our model (thin viscous sheet approximation), of course, limits our analysis precluding analysis of faulting and thrusting processes.

In the next section, we present details of our mechanical model, including the rheologies of the oceanic lower plate, the continental upper plate and the interplate subduction channel, as well as the way in which we have adapted the thin-sheet approach to examine the proposed orogenic processes. We then present a basic model relevant to Andean orogenesis and analyze how changing model parameters affect Andean deformation, particularly deformation related to the formation of the Bolivian orocline and the Andean Plateau.

## 23.2 Thin-Sheet Approximation Applied to Nazca-South American Subduction Orogenesis

The thin-sheet approximation of England and McKenzie (1982) involved determining the balance of stresses in continental lithosphere subjected to tectonic forces exerted by an undeformable, indenting plate (in their case, Asia and India, respectively). The continental lithosphere was assumed to be a uniform sheet that is not subjected to any basal traction, and whose width and breadth far exceed its thickness. England and McKenzie (1982) suggested that the strain rate of the continental lithosphere does not vary vertically, and that a balance of vertically integrated stresses can approximate the balance of stresses in the lithosphere. Thus, simple 2D equations can be used to approximate the 3D deformation of the continental lithosphere. To apply this general approach to the Nazca-South American subduction orogeny, we have had to make some modifications, as outlined below.

Firstly, we consider the deformation of two plates (Nazca and South America) rather than of just a single indented plate, as in England and McKenzie's (1982) original approach. The Nazca and South American Plates have different characteristics and deformational states: Whereas the western part of the South American continent comprises the Andes and is highly deformed, the Nazca Plate is oceanic and virtually undeformed (Fig. 23.1). The lower plate is, therefore, inferred to be stronger than the upper continental plate.

Subduction of the Nazca Plate beneath South America precludes direct application of the thin-sheet approximation in its classical formulation because the lower plate imposes a basal traction on the overlying continental lithosphere in the vicinity of the subduction zone. This poses a problem because simply introducing additional force into the thin-sheet force balance (e.g., Husson and Ricard 2004) is not compatible with the principle assumption of no basal traction in the thin-sheet approximation (England and McKenzie 1982). We therefore treat the plates' interface as a separate mechanical entity, our so-called "subduction zone", which comprises a thin channel of un- or partly consolidated sediments and their metamorphosed equivalents sandwiched between adjacent parts of the upper and lower plates (Fig. 23.2).

The rocks within this channel, known as the subduction channel (Peacock 1987; Cloos and Shreve 1988a,b; Hoffmann-Rothe et al. 2006, Chap. 6 of this volume), are intrinsically weak, the more so if they are subjected to high pore-fluid pressure, as is reasonably expected for subducting oceanic sediments undergoing prograde metamorphic dehydration (e.g., Hacker et al. 2003). These weakening agents are inferred to reduce overall coupling between the upper and lower plates. Thus, the vertically integrated strength of the subduction zone is not only a function of the strengths of the oceanic and continental plates, but also of the dip and strength of the subduction channel. In the Andean case, the subduction channel is assumed to have the same inclination as that of the oceanic slab, dipping 15–30° to the east within the 20–50 km depth interval. This depth interval coincides with the depth interval of seismic coupling (Isacks 1988; Hoffmann-Rothe et al. 2006, Chap. 6 of this volume).

For the general purposes of our model, the upper, South American Plate is assumed to comprise a weak Andean orogenic belt and a stronger orogenic foreland. In mak-

**Fig. 23.2.**
Subduction zone (*above*) and subduction zone element (*below*) described in the text. See Appendix 23.B for integrated rheology of the subduction zone element

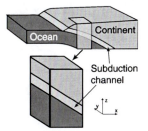

ing this simple assumption, we purposefully ignore evidence that the South American Plate is more complex, with at least two foreland domains (Brazilian shield and basement arches of the Sierras Pampeanas; Kley at al. 1999; Sobolev et al. 2006, Chap. 25 of this volume) and with variations in thickness and thermal properties along the strike of the Andes (e.g., differences between Altiplano and Puna lithospheres, Whitman et al. 1992; Kay et al. 1994). Significant though these differences may seem, they turn out to be small in comparison to the truly large differences between the Andean chain and its foreland.

The assumption that the Andean lithosphere is weaker than its foreland lithosphere is based on several studies indicating that the lithosphere under high-elevated Andes is thinner (e.g., Whitman et al. 1992; Kay et al. 1994) and hotter (Henry and Pollack 1988; Springer and Förster 1998; Springer 1999) than the lithosphere under undeformed parts of South America (east from Andes). This is especially true for the Puna part of the plateau. The modest deformation of the Andean foreland (east from the orogenic chain, Fig. 23.1) is also a qualitative indication that it is stronger than the orogenic lithosphere.

The existence of weak, mid to lower crust is another property of the Andes that cannot be modeled with the classical thin-sheet approximation. England and McKenzie's (1982) formulation assumes uniform deformation throughout the continental lithosphere. However, this

**Table 23.1.** Parameters of the model

| Variable | Description | Value in reference model |
|---|---|---|
| $g$ | Acceleration due to gravity | $9.8$ m c$^{-2}$ |
| $\rho_m$ | Density of mantle lithosphere | $3.3 \times 10^3$ kg m$^{-3}$ |
| $\Phi$ | Isostatic amplification factor | $1 - \rho/\rho_m$ |
| **Oceanic plate** | | |
| $\mu_o$ | Characteristic viscosity of lithosphere | $2 \times 10^{24}$ Pa s |
| $L_o$ | Characteristic thickness of lithosphere | 50 km |
| $\rho$ | Density of oceanic crust | $3 \times 10^3$ kg m$^{-3}$ |
| **Subduction zone** | | |
| $\mu_{xx}$ | Viscosity of the subduction zone in the X-direction (east-west) | $3 \times 10^{20}$ Pa s |
| $\mu_{yy}$ | Viscosity of the subduction zone in the Y-direction (north-south) | $7 \times 10^{22}$ Pa s |
| $\mu_{xy}$ | Viscosity of the subduction zone in a simple shear | $10^{19}$ Pa s |
| $L_s$ | Characteristic thickness of the lithosphere in the vicinity of the subduction zone | 100 km |
| **Continental plate** | | |
| $\mu_{c0}$ | Characteristic viscosity of undeformed foreland lithosphere | $7 \times 10^{22}$ Pa s |
| $\mu_{c1}$ | Characteristic viscosity of orogenic lithosphere | $2.8 \times 10^{21}$ Pa s |
| $L_c$ | Characteristic thickness of lithosphere | 100 km |
| $\rho$ | Density of crust | $2.8 \times 10^3$ kg m$^{-3}$ |
| $H_0$ | Thickness of undeformed crust | 39 km |
| $H^*$ | Critical thickness of crust for channel flow | 62 km |
| **Mid to lower crustal channel** | | |
| $h$ | Thickness of the channel | 10 km |
| $\mu_{ch}$ | Viscosity of the channel rock (changes at critical thickness of the crust, $H^*$) | $10^{20}/10^{22}$ Pa s |
| **Characteristic ratios of the model** | | |
| $\bar{\mu}_{c1}/\bar{\mu}_{c0}$ | Strength of the orogenic lithosphere relative to the strength of the foreland lithosphere | 0.04 |
| $\bar{\mu}_{xx}/\bar{\mu}_{c1}$ | Strength of the subduction zone relative to the strength of the orogenic Andean lithosphere | 0.11 |
| $R_{c/s}$ | Relative strength of the subduction zone at 19° S and at 32° S | 1 |

is unreasonable for the Andes in light of abundant seismic (Swenson et al. 2000; Yuan et al. 2000; Haberland and Rietbrock 2001) and magnetotelluric (Brasse 2002; Haberland et al. 2003) evidence for partial melting (Babeyko et al. 2002; Zandt et al. 2003) in the lower crust beneath the plateau regions. Experimental studies indicate that even minute (= 0.1 vol.%) amounts of partial melt can reduce rock strength by several orders of magnitude (Rosenberg and Handy 2005). Therefore, the partially molten, lower Andean crust is reasonably assumed to be highly mobile, resulting in channel flow (Husson and Sempere 2003; Yang et al. 2003; Gerbault et al. 2005) and non-uniform deformation of the Andean lithosphere.

The parameters listed in Table 23.1 reflect these Andean characteristics and account for the following basic observations mentioned in the introduction:

1. A shortening rate for the Central Andes of 1–1.75 cm yr$^{-1}$ and for the Southern Andes of 0–0.5 cm yr$^{-1}$; this corresponds to a difference in shortening rates between center and south of at least 0.9 cm yr$^{-1}$.
2. Ongoing oroclinal bending of the Andes.
3. Orogen-normal, eastward growth of the Andean Plateau in the absence of increasing altitude.

The plate motion rates in Fig. 23.1 are with respect to the mid-Atlantic Ridge (Silver et al. 1998). If we assume that the undeformed Andean foreland is moving westwards at a rate of 3 cm yr$^{-1}$, then shortening of the Andes accommodates only part of this plate motion, while the other part is taken up by the westward motion of the western margin of the South American continent. Because shortening rates decrease from the central to southern parts of the orogen, the motion of the western boundary of South America must decrease accordingly. The slower westward advancement of the Central Andes results in bending of the plate boundary and development of the Bolivian orocline. Thus, our model automatically fulfills condition 2 if condition 1 is satisfied. The mechanism of bending described above is similar to that proposed by Isaacks (1988).

## 23.3 Model Characteristics

We modified the classical thin-sheet approximation of England and McKenzie (1982) by using the higher order, asymptotic analysis of Medvedev and Podladchikov (1999 a, b). Our analysis shows that although the balance of forces presented in England and McKenzie (1982) are applicable to our problem, the kinematic model must be modified to account for mid to lower crustal flow beneath the Andean Plateau (Appendix 23.A; England and McKenzie 1982; Medvedev and Podladchikov 1999a). This results in the following equations that relate force balance to the thickness and density of the lithosphere:

$$2\frac{\partial \bar{\tau}_{xx}}{\partial x} + \frac{\partial \bar{\tau}_{yy}}{\partial x} + \frac{\partial \bar{\tau}_{xy}}{\partial y} = g\Phi\rho H \frac{\partial H}{\partial x} \quad (23.1)$$

$$\frac{\partial \bar{\tau}_{xx}}{\partial y} + 2\frac{\partial \bar{\tau}_{yy}}{\partial y} + \frac{\partial \bar{\tau}_{xy}}{\partial x} = g\Phi\rho H \frac{\partial H}{\partial y}$$

where $\bar{\tau}$ is a deviatoric stress tensor, $H$ is the thickness of the crust with density $\rho$, $\Phi = (1 - \rho/\rho_m)$ is the buoyancy amplification factor, and $\rho_m$ is the density of the mantle lithosphere. The overbar stands for integration over the depth of the lithosphere. Equation 23.1 assumes no traction at the base of the lithosphere and lithostatic approximation for pressure. The lithosphere is regarded as a viscous fluid, such that the integrated stresses relate to strain rates in the following way:

$$\bar{\tau} = 2\bar{\mu}\dot{\varepsilon} \quad (23.2)$$

$$\dot{\varepsilon}_{ij} = \frac{\left(\frac{\partial V_i}{\partial x_j} + \frac{\partial V_j}{\partial x_i}\right)}{2}$$

where only the horizontal components of stress and strain rate are considered: $\{x_i, y_j\} = \{x, y\}$ are the horizontal coordinates, and $\{V_i, V_j\} = \{V_x, V_y\}$ are the depth-invariant horizontal velocities. In the following subsections, we discuss, in some detail, the parameters of the rheological relation in Eq. 23.2, as well as the deformational processes in the system. The model system is divided into three main types of lithosphere: oceanic, continental, and transitional (the subduction zone).

### 23.3.1 Oceanic Lithosphere

The strength of the oceanic plate depends on its thermal state, which, in turn, is a function of its age (Turcotte and Schubert 1982). Therefore, we adopt the following relation:

$$\bar{\mu}_o = \mu_o L_o \sqrt{\frac{\text{age}}{\text{average\_age}}} \quad (23.3)$$

where $\mu_o$ is the characteristic viscosity of the oceanic plate and $L_o$ is the characteristic thickness of the plate. The average age in this equation is calculated from the 26 to 50 Myr age range of the oceanic crust in this area (Fig. 23.1, Müller et al. 1993). This results in age-dependent variations in the strength of the oceanic plate (Eq. 23.3) of up to a factor of two (Table 23.1).

### 23.3.2 Subduction Zone Elements

The subduction zone elements (Fig. 23.2) comprise layers of oceanic and continental lithosphere, and the subduction channel (Peacock 1987; Cloos and Shreve 1988a,b). This composite structure requires a detailed analysis to complete the integrated rheological relation in Eq. 23.2. We therefore consider stresses associated with each component of the strain-rate tensor in Appendix 23.B and show that the rheological relation in Eq. 23.2, as applied to the subduction zone, can be expressed as:

$$\begin{pmatrix} \bar{\tau}_{xx} \\ \bar{\tau}_{yy} \\ \bar{\tau}_{xy} \end{pmatrix} = 2L_s \begin{bmatrix} \mu_{xx} & 0 & 0 \\ 0 & \mu_{yy} & 0 \\ 0 & 0 & \mu_{xy} \end{bmatrix} \cdot \begin{pmatrix} \dot{\varepsilon}_{xx} \\ \dot{\varepsilon}_{yy} \\ \dot{\varepsilon}_{xy} \end{pmatrix} \quad (23.4)$$

where $L_s$ is the characteristic thickness of the lithosphere in the subduction zone, and $\mu_{yy}$, $\mu_{xx}$, and $\mu_{xy}$ are viscosities of the subduction zone in the directions of consideration. Note that the thin-sheet approximation used in our study relates the stresses integrated over the depth of the lithosphere to the average strain rate of the lithosphere. Thus, Eq. 23.4 considers only horizontal directions, with the $X$ axis oriented east-west and the $Y$ axis north-south.

Appendix 23.B lists the constituent properties of the subduction zone elements. These properties are based on poorly known parameters (e.g., the thickness and viscosity of the subduction channel). Therefore, Eq. 23.4 uses only the most general constraints in Appendix 23.B regarding the anisotropic stiffness of the subduction zone elements. We assume $\mu_{yy} > \mu_{xx} > \mu_{xy}$ in our reference model (Table 23.1).

### 23.3.3 Continental Lithosphere

The integrated strength of continental lithosphere strongly depends on the thermal state and thickness of the lithosphere (England and McKenzie 1982; Ranalli 1995). However, data on the Earth's thermal field is sparse and South America is no exception. We have, therefore, tried to find some empirical dependence for the strength of the lithosphere on a temperature-dependent parameter that can be easily extracted from available data sets.

To understand variations in the strength of the continental lithosphere, we compared the integrated strength of the lithosphere before and after deformation (Fig. 23.3a). Sudden, uniform thickening of the lithosphere (Fig. 23.3b, *dashed line*) results in thickening of the strong crustal and upper-mantle layers and, therefore, in a significant increase of integrated strength. Integrated strength of the lithosphere increases less significant if thickening is gradual. Thermal relaxation and radioactive heating increase average temperature of the lithosphere and make it weaker (Fig. 23.3b, *solid line*). Applied to Andes, however, this mechanism is proved to be insignificant (model 1 of Babeyko et al. 2002).

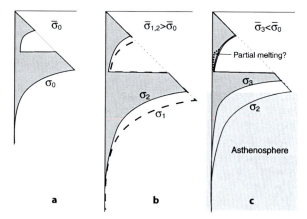

**Fig. 23.3.** Schematic diagram illustrating inferred changes in the strength envelope of the continental lithosphere during orogenesis: **a** Initial strength of the lithosphere $\bar{\sigma}_0$; **b** strength $\bar{\sigma}_1$ after sudden thickening of the lithosphere (*dashed line*). Thermal relaxation weakens the lithosphere to $\bar{\sigma}_2$ (*solid line*); **c** strength $\bar{\sigma}_3$ after weakening due to heat advected from the asthenosphere. Note that partial melting in the lower crust does not significantly decrease the integrated strength of the lithosphere

Geophysical evidence for detachment of the lithospheric mantle beneath parts of the Andean Plateau (e.g., Yuan et al. 2000) lends credence to the idea that lithospheric weakening is triggered by the upwelling of hot, asthenospheric mantle. The weakening associated with this heat advection from below supersedes the initial strengthening associated with lithospheric thickening. Thus, the contraction of the lithosphere can result in its significant weakening (Dewey 1988; Kay et al. 1994; Babeyko et al. 2006, Chap. 24 of this volume; Sobolev et al. 2006, Chap. 25 of this volume).

A reliable measure of the amount of deformation in the lithosphere is the thickness of the crust. Weakening associated with orogenic thickening is assumed to have affected the entire part of South America considered in our study. Despite the fact that the Andean foreland crust is thinner than the Andean orogenic crust ($H_0 < H^*$), it is much stronger and more viscous. Thus, we relate the integrated strength of the continental lithosphere to the thickness of crust with the following empirical relation (Fig. 23.4):

$$\bar{\mu}_c = \begin{cases} \mu_{c0} L_c & (H < H_0) \\ \mu_{c0} L_c \exp[b(H - H_0)^2] & (H_0 < H < H^*) \\ \mu_{c1} L_c & (H > H^*) \end{cases} \quad (23.5)$$

where $\mu_{c0}$ and $L_c$ are the characteristic viscosity and thickness of the continental lithosphere, respectively, and the coefficient $b = \log(\mu_{c1}/\mu_{c0})/(H_0 - H^*)^2$ is chosen to ensure that viscosity varies continuously from $\mu_{c0}$ to $\mu_{c1}$ as a function of crustal thickness, $H$. Thus, the changes of rheology, in Eq. 23.4, not only represent changes owing to increased heat advection during thickening, but also to the direct effects of thickening of the orogenic lithosphere.

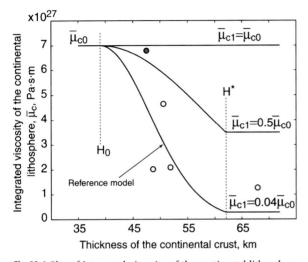

**Fig. 23.4.** Plot of integrated viscosity of the continental lithosphere versus thickness of the continental crust for different ratios (1, 0.5 and 0.04) of the integrated viscosities of the continental foreland ($\bar{\mu}_{c0}$) and orogenic lithosphere ($\bar{\mu}_{c1}$). $H_0$ and $H^*$ are the characteristic thicknesses of the crust in the continental foreland and in the orogen, respectively. Circles indicate values obtained from the reference model of Sobolev et al. (2006, Chap. 25 of this volume): *solid circle* = strength of the western margin of the Brazilian shield, *open circles* = strength of the orogenic crust

The thickness of the thickened continental crust, $H^*$, is chosen to be 62 km for all the models. This is less than the maximum observed value of 70 km and corresponds to the critical thickness of crust required to drive channel flow, as discussed and estimated in the next section. Thus, the most important rheological parameter of the continental crust is the viscous strength ratio of the orogenic and foreland lithospheres, $\bar{\mu}_{c1}/\bar{\mu}_{c0} = \mu_{c1}/\mu_{c0}$. This ratio is 0.04 in the reference model (Fig. 23.4), meaning that the orogenic Andes are 25 times weaker than the Andean foreland. In a separate analysis, we tested different values of this ratio, including situations where the orogenic lithosphere is assumed to be stronger than the foreland ($\bar{\mu}_{c1}/\bar{\mu}_{c0} > 1$).

We note that the viscosity distribution in our model is similar to the integrated effective viscosities in the models of Sobolev et al. (2006, Chap. 25 of this volume; circles and reference model in Fig. 23.4), but that the drop in the integrated viscosity of our reference model is greater than in their model. The 2D models of the Central Andes orogeny in Sobolev et al. (2006, Chap. 25 of this volume) suggest a higher ratio $\bar{\mu}_{c1}/\bar{\mu}_{c0}$ (0.1) than in our reference model (0.04), as discussed below.

Our model integrates the properties of the continental lithosphere with depth (Fig. 23.5), but, in reality, the strength of the continental lithosphere varies unevenly with depth (Fig. 23.3). Strong layers in the upper crust and lithospheric mantle sandwich a weak, mid to lower crustal layer. Appendix 23.A shows that although the contribution of the weak, lower crustal layer to force balance may be insignificant, this layer contributes significantly to the kinematics of continental deformation.

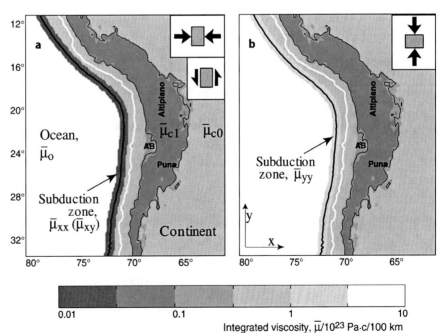

**Fig. 23.5.** Areal distribution of directional integrated viscosity for different shear configurations in the reference model. **a** Pure shear in an east-west direction or simple shear in a north-south direction; **b** pure shear in a north-south direction. The viscosity of the oceanic plate, $\bar{\mu}_{c1}$, is isotropic and depends on the age of the oceanic lithosphere. Likewise, the viscosity of the continental lithosphere, $\bar{\mu}_c$, is isotropic but depends strongly on crustal thickness. The strength of the subduction zone is anisotropic (Eq. 23.4). The values of integrated viscosity in this diagram are the integrated viscosity normalized to the viscosity of a characteristic lithosphere (thickness $L^* = 100$ km, average viscosity $\mu^* = 10^{23}$ Pa s). Map symbols as in Fig. 23.1

## 23.3.4 Channel Flow and the Orogenic Plateau

We distinguish three types of orogenic processes (Fig. 23.6). The Nazca Plate opposes the westward motion of the South American continent, creating lateral compressional forces, $F_t$, that lead to shortening in the Andes (Fig. 23.6b). We term this process "tectonic thickening". The thickened lithosphere is subjected to a second process, gravity spreading, i.e., deformation of the lithosphere under the force of gravity and buoyancy, $F_g$ (Fig. 23.6c). Tectonic and gravitational forces, $F_t$ and $F_g$, are the most important forces acting on the lithosphere, as they are responsible for most deformation in the system. Most of thin-sheet approximations applied to model large orogenic systems took into account these two main mechanisms (e.g., England and McKenzie 1982; Yanez and Cembrano 2004), and the Eq. 23.1 in our model balances the integrated stresses caused by these forces.

In Appendix 23.A we consider a third mechanism when analyzing deformation of the continental crust: channel flow in the mid to lower crust (Fig. 23.6d). The importance of channel flow within orogenic continental crust has been discussed extensively in the literature (e.g., Beaumont et al. 2001, 2004; Royden 1996; Clark and Royden 2000; Shen et al. 2001, Gerbault 2005) as being a requisite for development of orogenic plateaus (Royden 1996; Vanderhaeghe et al. 2003; Husson and Sempere 2003). The classical thin-sheet approach (England and McKenzie 1982) was developed before channel flow was considered, so that thin-sheet models developed, so far, have not reproduced the topographical features of orogenic plateaus (England and Houseman 1988).

The force $F_c$ associated with the channel flow is a result of variations of lithostatic loads at the base of thickened continental crust. Differential load of upper crust results in the flow and corresponding drug by the low viscosity channel material on the more competent parts of the lithosphere. This drug is insufficient to contribute to the integrated balance of forces ($F_c \ll F_g$ and $F_c \ll F_t$, Fig. 23.3c, Appendix 23.A), but potential contribution of the channel flow to the kinematics of orogenesis is significant. The following equation describes the evolution of the crustal thickness accounting for both, bulk deformation of the crust and the channel flow (Appendix 23.A):

$$\frac{\partial H}{\partial t} + \left(\frac{\partial V_x H}{\partial x} + \frac{\partial V_y H}{\partial y}\right) - \frac{\rho g h_c^3}{12\mu_{ch}}\left(\frac{\partial^2 S}{\partial x^2} + \frac{\partial^2 S}{\partial y^2}\right) = 0 \quad (23.6)$$

where $V_x$ and $V_y$ are the horizontal velocities obtained from the solution of Eqs. 23.1–23.5. The second bracketed term corresponds to flow in the lower crustal channel and $S$ is the elevation of topography with respect to sea level (Appendix 23.A, parameters listed in Table 23.1). The influence of the channel increases significantly when the material in the channel becomes very weak during partial melting. Medvedev and Beaumont (in press) show that a decrease of viscosity to $\mu_{ch} = \mu_p \approx 10^{18}$–$10^{20}$ Pa s is sufficient to establish a plateau above the channel. Note, that Eq. 23.6 is used in our model to evaluate the rates of thickness changes ($\partial H/\partial t$), but we do not consider finite-time evolution of the model. Variations of viscosity in the channel are represented in the model by the relation:

$$\mu_{ch} = \begin{cases} \mu_p & (H < H^*) \\ \mu_t & (H > H^*) \end{cases} \quad (23.7)$$

The viscosity in the crustal channel not directly beneath the plateau, $\mu_t = 10^{22}$ Pa s, was chosen to render the influence of the channel in these regions insignificant. The condition for the transition $\mu_{ch} = \mu_p \to \mu_{ch} = \mu_t$ ($H = H^*$) should be consistent with the geometry of the orogen. In particular, the tip of the low-viscosity channel should coincide with the edge of the flat part of the orogen. We term a channel with this configuration a "geometrically consistent channel" (GCC, Fig. 23.7a). If the tip of the channel extends beyond the edge of a plateau, termed an "overdeveloped channel" (OC, Fig. 23.7a), the flux of weak,

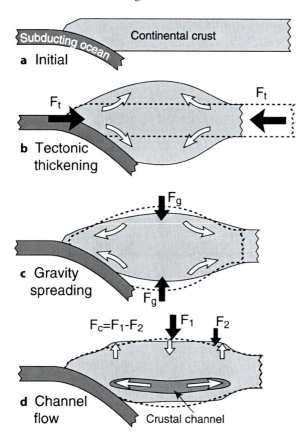

**Fig. 23.6.** Deformation of the upper plate in a subduction orogen. **a** Initial configuration; **b** tectonic thickening; **c** gravity spreading; **d** channel flow creating a plateau. Black arrows illustrate forces acting on the orogenic system, white arrows indicate material flow

partially molten rock is too high to support the overlying plateau, and the plateau collapses. Alternatively, if the channel is too short, an "underdeveloped channel" (UC, Fig. 23.7a), the plateau never becomes flat during shortening, and the margins of the plateau are characterized by high topographic gradients. Of course, the geometry of the Andean orogenic plateau is not as simple as in our experimental model (Fig. 23.7a) and a direct estimation of $H^*$ is impossible for the Andes.

To specify the condition for a geometrically consistent channel, we consider channel flux for different transitional positions in our experimental model (Fig. 23.7a,b). An estimate of the average flux in the channel for all possible transitions (Fig. 23.7c) shows that the change from a transient to a stable flux occurs where the channel has exactly the same lateral extent as the overlying plateau (point GCC, Fig. 23.7c). An absolute value for the flux of low viscosity material in the channel can be calculated for the Central Andes with the following equation (see also Appendix 23.A):

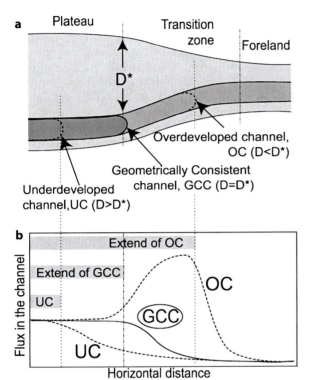

$$|q_c| = \sqrt{q_{cx}^2 + q_{cy}^2} = \frac{\rho g h_c^3}{12\mu_{ch}(H^*)} \sqrt{\left(\frac{\partial S}{\partial x}\right)^2 + \left(\frac{\partial S}{\partial y}\right)^2} \quad (23.8)$$

The absolute value of the flux depends on two parameters: the elevation, $S$, and the critical crustal thickness, $H^*$. Figure 23.8 presents our calculated average fluxes in low-viscosity channel for different critical thicknesses using actual elevation data from the considered domain of South America (Fig. 23.1). The results in Fig. 23.8 compare well with our analytical predictions (Fig. 23.7) and yield a critical thickness value for Andean crust of $H^* = 62$ km. This corresponds to a plateau elevation of 3.5 km. We use these parameters in our numerical models.

Equation 23.8 clearly relates the intensity of the channel flow with variations of topography, because the terms in brackets characterize the absolute value of topographic gradient. Results on Fig. 23.8 show that high-elevated Andes (with crustal thickness > 57 km) average topographic gradient decreases with increase of elevation and reaches stable minimal value at 3.5 km elevation ($H = 62$ km). Thus, the results in Fig. 23.8 indicate that the Andes are typically flatter at elevations exceeding 3.5 km.

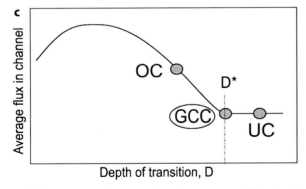

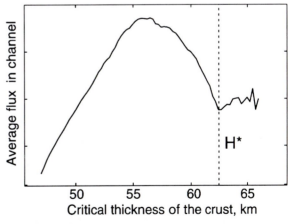

**Fig. 23.7.** General relationship between the lower crustal channel and the orogenic plateau. **a** General view of the plateau margin with different locations for the edge of the low-viscosity channel. GCC = channel with its edge directly below the plateau margin at a critical depth, $D^*$. OC = channel with its edge at a depth less than the critical value needed for sustaining a plateau. UC = channel with its edge beneath the plateau. OC and UC are inconsistent with topography. **b** Variation of flux along the three types of channel. **c** Average flux in the channel versus depth of the channel edge

**Fig. 23.8.** Average flux in channels for different values of critical crustal thickness, $H^*$. Comparison with Fig. 23.7c indicates that the most suitable value of $H^*$ is 62 km, corresponding to a plateau elevation of 3.5 km

### 23.3.5 The Numerical Model

The MATLAB code developed for this study is based on the finite element method (Kwon and Bang 1997). The code uses data regarding the topography and bathymetry (termed S), the age of the ocean floor, and the plate boundary from the SFB267 database (*http://userpage.fu-berlin.de/~data/Welcome.html*). Then, the topographic data is converted into crustal thicknesses by using an Airy isostatic model, $H = H_0 + S/\Phi$ (Turcotte and Schubert 1982). To evaluate unknown initial (undeformed) thickness of the continental crust, $H_0$, and isostatic amplification factor, $\Phi$, we used estimations of crustal thickness from Götze and Kirchner (1997) and Kirchner et al. (1996). Finally, the code separates the region in question into three parts: oceanic plate, continental plate, and a 100 km wide subduction zone. Table 23.1 lists the parameters of the reference model.

The system of Eqs. 23.1–23.7 was solved using the finite-element approximation based on a second-order interpolation inside the elements. The calculations were performed on a serial PC with a grid of up to $120 \times 100$ elements (up to $360 \times 300$ integrating points). The model was designed so that the density of the finite-element grid is a parameter. We used that parameter to test the robustness of the numerical model by comparing results based on the fine and coarse grids.

The boundary conditions for Eq. 23.1 are eastward motion of the Nazca Plate at 5 cm yr$^{-1}$ along its western boundary, westward motion of the Brazilian shield at 3 cm yr$^{-1}$ along its eastern boundary, and free-slip along the northern and southern boundaries of the model. In the reference model, we ignore the oblique motion of the Nazca Plate by assuming that the northward component of oblique motion can cause local effects along the plate margins, for example, coast-parallel motion and faulting in the South American western fore-arc. We assume, however, that these effects do not significantly affect the general pattern of deformation at the scale of the Andes (Hoffmann-Rothe et al. 2006, Chap. 6 of this volume).

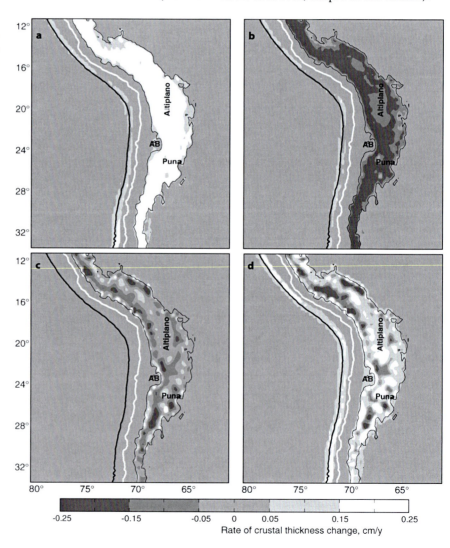

**Fig. 23.9.**
Rates of thickness change in the reference model due to different modes of orogenic deformation (as in Fig. 23.6). **a** tectonic thickening; **b** gravity spreading; **c** channel flow; **d** all modes simultaneously

## 23.4 Results and Discussion

### 23.4.1 The Reference Model

In this section, we present the results of a reference model that employs the parameters listed in Table 23.1. These parameters were chosen to help us understand the development of primary Andean features, such as the modestly deforming Southern Andes, the bending of the Bolivian orocline, and the development of the Andean Plateau. The model is not designed to match the observations completely. Rather, the goal is to demonstrate that the processes considered above are physically possible given the geodynamic and rheological constraints on Andean orogenesis.

The reference model does not present a unique solution. Analyzing the series of numerical experiments based on different sets of parameters, we chose our reference model (Table 23.1) to illustrate typical results. We discuss variations of parameters in the following sections.

Figure 23.9 illustrates the rates of thickness change, particularly how the three processes considered above (Fig. 23.6) act to change the thickness of the Andean crust. The rate of tectonic thickening (Fig. 23.9a) can be evaluated by eliminating gravity (e.g., by setting the density of the crust to 0 in Eqs. 23.1 and 23.7). As expected, the weaker, orogenic parts of the South American Plate accommodate most of the tectonic thickening in this case. The main factor controlling the rate and distribution of tectonic thickening is the relative integrated strengths of the continental plate, the subduction zone, and the oceanic plate (Fig. 23.5, Table 23.1). Weak orogenic lithosphere, therefore, favors faster, more localized and pronounced thickening in the upper plate, whereas stronger orogenic lithosphere engenders a broader zone of thickening.

The amount and rate of gravitational spreading associated with existing topographic gradients can be evaluated by setting the boundary velocities to 0 (Fig. 23.9d). The main parameter characterizing the rate and distribution of gravity spreading is the strength of the continental plate with respect to the magnitude of the gravity forces acting on the plate surface. Accordingly, a narrow, weak and heavy upper plate favors faster, and more widely distributed, gravitational spreading than a broad, strong, and light upper plate.

The rate of thickness change in the upper plate owing to channel flow was evaluated by setting the bulk velocities in Eq. 23.6 to $V_x = 0$ and $V_y = 0$ (in this case the kinematic update, Eq. 23.6, becomes independent from the system Eqs. 23.1–23.5). The influence of the crustal channel is recognizable only in orogenic lithosphere directly beneath the area within 3 km elevation line, where it flattens the topographic relief to a plateau-like morphology. This thickness change rate is heterogeneous, with thinner patches being thickened, and areas of greater thickness becoming somewhat thinner. This smoothing of topography is ascribed to channel flow of partially molten rock. For example, heterogeneous relief of Puna results in uneven distribution rates of thickening, whereas flat topography of the central Altiplano do not result in significant channel flow and in associated thickness changes (cf. Fig. 23.1).

Figure 23.9d presents rates of thickness changes accounting for all three considered mechanisms. The rate of thickening within 3 km elevation line is heterogeneous, tracking the influence of channel flow discussed in the paragraph above. The important feature of the total rate of thickness changes is the active thickening of the crust in the vicinity of the 3 km elevation line. As thickening progresses, the 3 km elevation line expands, tracking the lateral expansion of the Andes. Note, that our numerical model is designed to estimate rates of thickening, not to calculate long-term evolu-

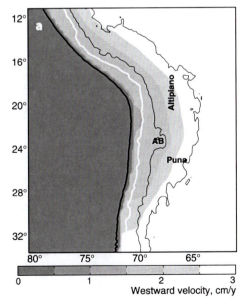
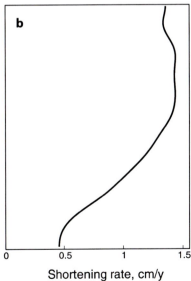

**Fig. 23.10.** Distribution of westward velocity (**a**) and rate of the horizontal shortening along the Andean chain (**b**) in the reference model

tion of Andes. Thus, the conclusion about lateral expansion of Andes illustrated by Fig. 23.9d is valid only for the recent state of the Andean orogeny.

Figure 23.10a indicates that the faster westward advancement of the Southern Andes (2.5 cm yr$^{-1}$ from 30–35° S) than of the Central Andes (1–1.5 cm yr$^{-1}$ from 15–20° S) is consistent with the observed bend of the Bolivian orocline. The shortening rate of the Andes in the model (Fig. 23.10b) is equal to the rate at which the foreland approaches the western boundary of the South American Plate. In the model, all of this convergence is accommodated by the deformation of the orogenic lithosphere. The shortening rate varies strongly along strike of the Andean chain, from up to 1.5 cm yr$^{-1}$ at 15–20° S (in agreement with Oncken et al. 2006, Chap. 1 of this volume) to less than 0.5 cm yr$^{-1}$ in the Southern Andes.

## 23.4.2 3D Aspects of the Thin-Sheet Model

The model used to obtain the results above incorporates an analysis of the horizontal thin sheet with distributed properties, such as integrated viscosity. Similar results within the limits of current accuracy can be obtained by using a series of east-west, cross-sectional models (e.g., Husson and Ricard 2004). The advantage of our model lies in its ability to predict the distribution of some of the 3D features of the Andean orogeny, such as strike-parallel flow of material, and rotations of parts of the Andean crust.

Figure 23.11a depicts the north-south flow of crustal material in the Andes. The flow is presented by an equivalent velocity field for a normalized, 40 km thick crust. Thus, a 3 mm yr$^{-1}$ velocity along the southern border of the Puna Plateau represents flow of the normalized, 40 km thick crust at this rate. This significant mass flow must be accounted for when restoring the deformation of the Andes, as already shown by Gerbault et al. (2005).

Figure 23.11b shows a clockwise rotation of the southern part of the Andes and an anti-clockwise rotation of the northern part, consistent with field and paleomagnetic observations (Isacks 1988; Butler et al. 1995; Hindle and Kley 2002; Riller and Oncken 2003; Rousse et al. 2002, 2003). These vorticity values correspond to rotations of 1.5–2.2 degrees per million years, which is comparable to the estimates of Rousse et al. (2003) and Riller and Oncken (2003). The least rotation occurs in the Central Andes between 20 and 24° S, in agreement with results showing that most rotation in this region occurred earlier, during the Paleogene (Arriagada et al. 2003).

Another 3D aspect of our model is that the properties of the subduction zone can be depth integrated (Appendix 23.B). We performed a sensitivity analysis in order to examine the influence of effective mechanical anisotropy of the subduction elements on Andean-type orogenesis. We did this by varying the viscosity parameters in Eq. 23.5. The reference model (Table 23.1) is based on the relation $\mu_{yy} \gg \mu_{xx} \gg \mu_{xy}$. Changing this relationship to $\mu_{xx} \approx \mu_{xy}$ or $\mu_{xx} \approx \mu_{yy}$ or $\mu_{yy} \approx \mu_{xy}$ and tuning rheology to have reasonable results for central and southern parts of our model results in the domination of gravity spreading over tectonic thickening in the Peruvian and Bolivian Andes. This scenario is unrealistic, suggesting that the assumption made in our reference model of subduction zone elements with strongly directional mechanical properties is quite reasonable. This is confirmed by calculations in Appendix 23.B (where it is shown that $\mu_{yy}$ is of order of $\mu_c$, or even stronger; and that $\mu_{xx} \gg \mu_{xy}$) and depicted by the ratios in the gray-shaded domains of Figs. 23.12 and 23.13 (for which $\mu_c \gg \mu_{xx}$, see next section). Thus, our analytical investigations (Appendix 23.B) and numerical experiments (next chapter) supports the relation of the reference model, $\mu_{yy} \gg \mu_{xx} \gg \mu_{xy}$.

**Fig. 23.11.**
Distribution of south-north flow in the crust (**a**, northward flow is positive) and vorticity (**b**) in the reference model

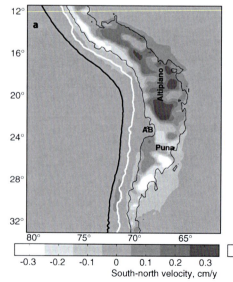
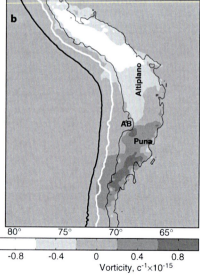

### 23.4.3 Formation of the Bolivian Orocline and Shortening of the Andes

In this section, we search for parameter values that are consistent with the observed amount and rates of shortening and rotation in the Andes. A series of numerical experiments revealed that the main parameters responsible for oroclinal bending and differential shortening of the Andes are the viscosity ratios $\bar{\mu}_{c1}/\bar{\mu}_{c0}$ and $\bar{\mu}_{xx}/\bar{\mu}_{c1}$ (Fig. 23.12). The ratio $\bar{\mu}_{c1}/\bar{\mu}_{c0}$ (Fig. 23.4) determines the degree to which the orogenic lithosphere weakens as the crust thickens, as well as the degree to which the orogenic lithosphere is weaker than the foreland lithosphere. The ratio $\bar{\mu}_{xx}/\bar{\mu}_{c1}$ (Fig. 23.5a) is the strength contrast between the orogenic Andean lithosphere and the subduction zone in the direction of subduction. Figure 23.12 depicts the results of 400 numerical experiments conducted to determine which strength ratios are consistent with observed amounts and rates of deformation, all other parameters being held constant at the values listed in Table 23.1. The results show, for example, that the orogenic Andean lithosphere must be at least 14 times weaker than the foreland lithosphere ($\bar{\mu}_{c1}/\bar{\mu}_{c0} < 0.07$, Fig. 23.12) for the model to match the observed deformation. Otherwise, the difference in shortening rates between the central and Southern Andes is too low, or the deformation of the Southern Andes is too high.

The calculations used to obtain the results in Fig. 23.12 involve no strike-parallel changes in the rheology of the subduction zone. However, several workers (Yanez and Cembrano 2004; Lamb and Davis 2003; Sobolev et al. 2006, Chap. 25 of this volume; Hoffmann-Rothe et al. 2006, Chap. 6 of this volume) emphasize that in the Central Andes the coupling between the upper and lower plates and, therefore, the subduction zone in between are stronger than to the south.

To test the influence of lateral variations in subduction zone rheology, we ran a series of experiments in which the strength of the subduction zone, $\bar{\mu}_{xx}$, decreased linearly away from the Central Andes (19° S), both to the north and the south. The measure of this change, $R_{c/s}$, is defined as the ratio of $\bar{\mu}_{xx}$ at 19° S and at 32° S (in the Southern Andes). Figure 23.13 presents results for $R_{c/s} = 1$, 2, and 5. To fit the observed shortening rates in the Andes, models with $R_{c/s} > 1$ require less weakening of the orogenic lithosphere (i.e. smaller values for ratio $\bar{\mu}_{c1}/\bar{\mu}_{c0}$) than the reference model with $R_{c/s} = 1$. We note that weakening of the orogenic lithosphere, ratio $\bar{\mu}_{c1}/\bar{\mu}_{c0}$, in our reference model is several times greater than that predicted by Sobolev et al. (2006, Chap. 25 of this volume), as shown in Fig. 23.4. However, we obtain comparable values of weakening if we use $R_{c/s} > 1$, as shown in Fig. 23.13.

Models with an along-strike variation in strength contrast of $R_{c/s} > 1$ can be considered as a combination of two end-member models. In one (our reference model), the weakening of the Andean orogenic lithosphere compared to the foreland lithosphere is the prime cause of along-strike variations in Andean shortening and oroclinal bending. This model also faithfully reproduces

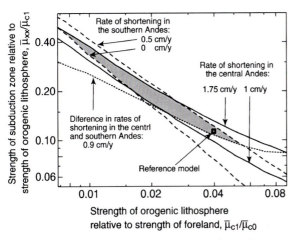

**Fig. 23.12.** Plot of relative strengths showing the range of strength ratios in the model (*area shaded gray*) that yield deformational rates that are consistent with observed values in the Andes. The lines delimiting this domain represent bounds on the observed shortening rates. *Solid lines*: 1–1.75 cm yr$^{-1}$ for the Central Andes (18–20° S), *dashed lines*: 0–0.5 cm yr$^{-1}$ for the Southern Andes (30–32° S), *dotted line*: > 0.9 cm yr$^{-1}$ for the difference in shortening rates between the central and Southern Andes. *Rectangle* refers to ratios in our reference model

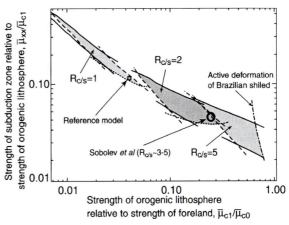

**Fig. 23.13.** Plot of relative strengths with domains of strength ratios (*shaded gray*) that yield realistic Andean deformational rates for three different values of $R_{c/s}$ (1, 2 and 5). $R_{c/s}$ describes the along-strike gradient in the strength of the subduction zone between the Nazca and South American Plates. This strength decreases linearly along strike of the plate margin from $\bar{\mu}_{vx}$ at 19° S to $\bar{\mu}_{xx}/R_{c/s}$ at 32° S. The same along-strike gradient is assumed for the subduction zone to the north of 19° S. The lines delimiting the shaded domains are the same as in Fig. 23.12, except that results for $R_{c/s} = 5$ are limited by the additional constraint that deformation of the Brazilian shield is insignificant. The rectangle refers to ratios in our reference model, the circle refers to ratios in the models of Sobolev et al. (2006, Chap. 25 of this volume)

the observed coincidence of accelerating Andean shortening and decelerating convergence between the Nazca and South American Plates (Hindle et al. 2002) during the most recent 10 Myr. This requires weakening of the orogenic lithosphere by a factor of more than 15, and can only be obtained by assuming unrealistic rheologies or thermal fields. In comparison, the fully numerical cross sectional models in Sobolev et al. (2006, Chap. 25 of this volume) exhibit weakening by no more than a factor of 10.

In the other end-member model, the kinematics of Andean orogenesis are primarily determined by a systematic, along-strike change in the degree of coupling between the Nazca and South American Plates. The importance of that property of the Andean orogeny was outlined in a number of works (Isacks 1988; Lamb and Davis 2003; Yanez and Cembrano 2004; Sobolev et al. 2006, Chap. 25 of this volume). This explains the initiation of oroclinal bending in the absence of orogenic structures along the western side of South America. Without meeting the condition that the orogenic lithosphere is weaker than the foreland, the Andes cannot develop any further and the deformation of South America will be well distributed. Figure 23.13 indicates that if $\bar{\mu}_{c1}/\bar{\mu}_{c0} \sim 1$ or higher, the Brazilian shield will be actively deforming.

Both end-member models have advantages and disadvantages in explaining the salient features of Andean orogenesis, so it is not surprising that reality is best explained by a combination of these end-members. A model in which the strength of the thickened Andean lithosphere decreases by only 5–10 times (Fig. 23.13) and the strength of the subduction zone decreases from the central to the Southern Andes by a factor of 2–5 yields shortening rates that are consistent with observed Recent values.

### 23.4.4 Comparison with Previous Models

Thickening of the lithosphere yields significant weakening according to Eq. 23.5 (Fig. 23.4). As mentioned above, this weakening is responsible for the formation of the Bolivian orocline and the Andean Plateau, and also explains the modest shortening of the Southern Andes in our reference model. This contrasts with the analog model of Marques and Cobbold (2002), in which the thickened continental lithosphere becomes a strong crustal indentor that effects oroclinal bending. In later stages, this indentor is incorporated into the Andean Plateau. The Andean Plateau develops because the proto-plateau is stronger than the foreland, according to conclusion of Marques and Cobbold (2002) model.

Numerical experiments in which lithospheric strength increases with crustal thickness ($\bar{\mu}_{c1}/\bar{\mu}_{c0} > 1$) exhibit deformation of the entire Brazilian shield comparable or even higher than deformation of Andes (Fig. 23.13), which, of course, is entirely unrealistic. The underlying cause of the discrepancy between the results of our reference model (with its requirement that $\bar{\mu}_{c1}/\bar{\mu}_{c0} < 1$) and the Marques and Cobbold (2002) model (2002, $\bar{\mu}_{c1}/\bar{\mu}_{c0} > 1$) is the different conditions at the base of the model: in the analog experiments of Marques and Cobbold (2002), deformation is influenced by the friction along the base of the box, and therefore cannot propagate far from the indentor. In contrast, the base of our numerical model is frictionless (emulating a low-viscosity asthenosphere at the base of the model lithosphere), allowing deformation to propagate far into the upper continental plate, where the weak orogenic lithosphere accommodates most of the shortening.

Channel flow of partially molten rock beneath the high Andes is responsible for flattening of the Andean Plateaux. Without this mechanism, the Andes would thicken continuously in the center instead of growing laterally by thickening at the plateau margins (Lamb 2000; Sobolev et al. 2006, Chap. 25 of this volume). The model of Liu et al. (2002) and Yang et al. (2003) shows continuous and nearly uniform uplift of the Andes because the impenetrable boundary in that model is set just outside the Andes, precluding any flow exterior to the these mountains and, therefore, precluding lateral expansion of the Andes. We should note that the models of Liu et al. (2002) and Yang et al. (2003) demonstrated clearly the importance of low-crustal flow in the formation of flat topography of the Andean Plateaus.

The age-dependent strength of the oceanic plate (Eq. 23.3) does not significantly influence the pattern of deformation in the Andes in our reference model. This is inconsistent with the results of Yanez and Cembrano's (2004) model in which the age of the subducting Nazca Plate was deemed to be one of the major factors responsible for the development of the Bolivian orocline. In assessing this discrepancy, we point out that the oceanic plate plays a very different role in the two models.

In Yanez and Cembrano's (2004) model, the age of the oceanic lithosphere is used only to estimate the degree of coupling between the undeformable, oceanic, lower plate and the deformable, continental, upper plate The older and colder the oceanic plate, the greater the degree of interplate coupling. In our model, however, the oceanic plate is strong but deformable, and the interaction between the oceanic and continental plates is specified along their contact within the subduction zone. If we were to relate the properties of the subduction zone to the age of the subducting oceanic plate (e.g., by a special value of the parameter $R_{c/s}$), the distribution of deformation in our model would probably also depend on the age of the oceanic plate.

## 23.5 Conclusions

We present a new numerical thin-sheet model to analyze the recent state of deformation in the Andean orogen. Our numerical experiments show that several first-order features of the Central and Southern Andes can be explain by three aspects which we incorporate as novel features of our model: the effective mechanical anisotropy of the subduction zone, the distribution of integrated strength of the upper continental plate, and channeled flow of partially molten rock at the base of the thickened crust within this upper plate.

Special analysis shows that the subduction zone behaves anisotropically depicting the influence of weak subduction channel and corresponding weakness in the direction of subduction. The degree of that weakness measures the coupling between the Nazca and South American Plates. We also assume that the strength of continental lithosphere depends on the thickness of the crust. Thus, our model enabled us to assess the relative importance of interplate coupling and strength of the orogenic lithosphere with respect to the foreland lithosphere. The validity of the models was determined by their ability to reproduce observed, Recent shortening rates across the Andes.

For uniform coupling between the Nazca and South American Plates, realistic shortening rates were obtained only when the orogenic lithosphere was 20–100 times weaker than the foreland lithosphere. For coupling between the plates that varies along strike of the trench (i.e., from north to south), we found that southward reduction of interplate coupling eases the requirement for strong weakening of the orogenic lithosphere of Andes. Thus, models with calculated shortening rates that are consistent with currently measured rates involve a decrease in interplate coupling by a factor of 2 to 5 from the central to the Southern Andes, and a decrease in the strength ratio of Andean orogenic lithosphere to Andean foreland by a factor of only 5 to 10.

The incorporation of a channel of weak, probably partially molten, middle crust in the upper plate of the model subduction orogen reduces topographic relief in the overlying crust and provides a simple mechanism for the development of the Andean Plateau. The weak crust forms a cushion which flattens the orogen and allows the orogenic plateau to grow laterally, as observed in the Andes today. We ascertain that channel flow is active beneath parts of Andes with an altitude greater than 3.5 km.

Numerical experiments allow us to constrain the rheological parameters associated with presently observed variations in shortening and vorticity along the Andean chain. These variations are also reflected in the measured rates of bending of the Bolivian orocline.

## Acknowledgments

Tough, but adequate reviews by S. Sobolev, M. Gerbault, and J. Cembrano improved significantly the early version of manuscript. Help from H.-J. Götze (associate editor), R. Hackney (geophysical data), A. Britt (English and style corrections), and A. Kellner (some illustrations) is appreciated. This work was supported by the Deutsche Forschungsgemeinschaft in the form of the Collaborative Research Centre "Deformation Processes in the Andes" Grant (SFB 267, Project G2).

## References

Allmendinger RW, Jordan TE, Kay SM, Isacks BL (1997) The evolution of the Altiplano-Puna Plateau of the Central Andes. Ann Rev Earth Planet Sci 25:139–174

Arriagada C, Roperch P, Mpodozis C, Dupont-Nivet G, Cobbold PR, Chauvin A, Cortés J (2003) Paleogene clockwise tectonic rotations in the forearc of Central Andes, Antofagasta region, Northern Chile. J Geophys Res 108(B1): doi 10.1029/2001JB001598

Babeyko AY, Sobolev SV, Trumbull RB, Oncken O, Lavier LL (2002) Numerical models of crustal scale convection and partial melting beneath the Altiplano-Puna Plateau. Earth Planet Sci Lett 199:373–388

Babeyko AY, Sobolev SV, Vietor T, Oncken O, Trumbull RB (2006) Numerical study of weakening processes in the Central Andean back-arc. In: Oncken O, Chong G, Franz G, Giese P, Götze H-J, Ramos VA, Strecker MR, Wigger P (eds) The Andes – active subduction orogeny. Frontiers in Earth Science Series, Vol 1. Springer-Verlag, Berlin Heidelberg New York, pp 495–512, this volume

Beaumont C, Jamieson RA, Nguyen MH, Lee B (2001) Himalayan tectonics explained by extrusion of a low-viscosity channel coupled to focused surface denudation. Nature 414:738–742

Beaumont C, Jamieson RA, Nguyen MH, Medvedev S (2004) Crustal channel flows: 1. Numerical models with applications to the tectonics of the Himalayan-Tibetan orogen. J Geophys Res 109(B06406): doi 10.1029/2003JB002809

Brasse H, Lezaeta P, Rath V, Schwalenberg K, Soyer W, Haak V (2002) The Bolivian Altiplano conductivity anomaly. J Geophys Res 107(B5): doi 10.1029/2001JB000391

Butler RF, Richards DR, Sempere T, Marshall LG (1995) Paleomagnetic determinations of vertical-axis tectonic rotations from Late Cretaceous and Paleocene strata of Bolivia. Geology 23:799–802

Carey SW (1958) The orocline concept in geotectonics. Proc Royal Soc Tasmania 89:255–288,

Clark MK, Royden LH (2000) Topographic ooze: building the eastern margin of Tibet by lower crustal flow. Geology 28:703–706

Cloos M, Shreve RL (1988a) Subduction-channel model of prism accretion, melange formation, sediment subduction, and subduction erosion at convergent plate margins: 1. Background and description. Pure Appl Geophys 128:455–500

Cloos M, Shreve RL (1988b) Subduction-channel model of prism accretion, melange formation, sediment subduction, and subduction erosion at convergent plate margins: 2. Implications and discussions. Pure Appl Geophys 128:501–545

Dewey JF (1988) Extensional collapse of orogens. Tectonics 7(6):1123–1139

England PC, Houseman GA (1988) The mechanics of the Tibetan Plateau. Phil Trans Royal Soc London A 326:301–320

England P, McKenzie D (1982) A thin viscous sheet model for continental deformation. Geophys J R Astr Soc 70:295–321. [Correction: England P, McKenzie D (1983) Correction to: A thin viscous sheet model for continental deformation Geophys J R Astr Soc 73:523–532]

Gansser A (1973) Facts and theories on the Andes. J Geol Soc London 129:93–131

Gerbault M, Martinod J, Hérail G (2005) Possible orogeny-parallel lower crustal flow and thickening in the Central Andes. Tectonophysics 399:59–72

Giese P, Scheuber E, Schilling FR, Schmitz M, Wigger P (1999) Crustal thickening processes in the Central Andes and the different natures of the Moho-Discontinuity. J S Am Earth Sci 12:201–220

Götze H-J, Kirchner A (1997) Interpretation of gravity and geoid in the Central Andes between 20° and 29° S. J S Am Earth Sci 10:179–188

Haberland C, Rietbrock A (2001) Attenuation tomography in the western sentral Andes: a detailed insight into the structure of a magmatic arc. J Geophys Res 106(B6):11151–11167

Haberland C, Rietbrock A, Schurr B, Brasse H (2003) Coincident anomalies of seismic attenuation and electrical resistivity beneath the southern Bolivian Altiplano plateau. Geophy Res Lett 30(18): doi 10.1029/2003GL017492,2003

Hacker BR, Peacock SM, Abers GA, Holloway SD (2003) Subduction factory 2. Are intermediate-depth earthquakes in subducting slabs linked to metamorphic dehydration reactions? J Geophys Res 108(B1): doi 10.1029/2001JB001129

Henry SG, Pollack HN (1988) Terrestrial heat flow above the Andean subduction zone, Bolivia and Peru. J Geophys Res 93:15153–15162

Hindle D, Kley J (2002) Displacements, strains and rotations in the Central Andean Plate boundary. In: Stein S, Freymuller J (eds) Plate boundary zones. AGU Geodynamic Series 30, pp 135–144

Hindle D, Kley J, Klosko E, Stein S, Dixon T, Norabuena E (2002) Consistency of geologic and geodetic displacements in Andean orogenesis. Geophys Res Lett 29: doi 10.1029/2001GL013757

Hoffmann-Rothe A, Kukowski N, Dresen G, Echtler H, Oncken O, Klotz J, Scheuber E, Kellner A (2006) Oblique convergence along the Chilean margin: partitioning, margin-parallel faulting and force interaction at the plate interface. In: Oncken O, Chong G, Franz G, Giese P, Götze H-J, Ramos VA, Strecker MR, Wigger P (eds) The Andes – active subduction orogeny. Frontiers in Earth Science Series, Vol 1. Springer-Verlag, Berlin Heidelberg New York, pp 125–146, this volume

Husson L, Ricard Y (2004) Stress balance above subduction: application to the Andes. Earth Planet Sci Lett 222:1037–1050

Husson L, Sempere T (2003) Thickening the Altiplano crust by gravity-driven crustal channel flow. Geophys Res Lett 30: doi 10.1029/2002GL016877

Isacks BL (1988) Uplift of the Central Andean plateau and bending of the Bolivian Orocline. J Geophys Res 93:3211–3231

James DE (1971) Andean crust and upper mantle structure. J Geophys Res 76:3246–3271

Kay SM, Coira B, Viramonte J (1994) Young mafic back arc volcanic rocks as indicators of continental lithospheric delamination beneath the Argentine Puna plateau, Central Andes. J Geophys Res 99(B12):24323–24339

Kirchner A, Götze H-J, Schmitz M (1996) 3D-density modelling with seismic constraints in the Central Andes. Phys Chem Earth 21:289–293

Kley J, Monaldi CR (1998) Tectonic shortening and crustal thickening in the Central Andes: how good is the correlation? Geology 26(8):723–726

Kley J, Monaldi CR, Salfity JA (1999) Along-strike segmentation of the Andean foreland: causes and consequences. Tectonophysics 301:75–94

Kwon YW, Bang H (1997) The finite element method using MATLAB. CRC Press, New York

Lamb S (2000) Active deformation in the Bolivian Andes, South America. J Geophys Res 105:25627–25653

Lamb S, Davis P (2003) Cenozoic climate change as a possible cause for the rise of the Andes. Nature 425:792–797

Lithgow-Bertelloni C, Guynn JH (2004) Origin of the lithospheric stress field. J Geophys Res 109(B01408): doi 10.1029/2003JB002467

Liu M, Yang Y, Stein S, Klosko E (2002) Crustal shortening and extension in the Central Andes: insights from a viscoelastic model. In: Stein S, Freymueller J (eds) Plate boundary zones. AGU Geodynamics Series 30, doi 10/1029/030GD19

MacFadden B, Anaya F, Swisher C III (1995) Neogene paleomagnetism and oroclinal bending of the Central Andes of Bolivia. J Geophys Res 100:8153–8167

Marques FO, Cobbold PR (2002) Topography as a major factor in the development of arcuate thrust belts: insights from sandbox experiments. Tectonophysics 348:247–268

Medvedev S, Beaumont C (in press) Growth of continental plateaux by channel injection: constraints and thermo-mechanical consistency. Geol Soc London Spec Pub

Medvedev SE, Podladchikov YY (1999a) New extended thin sheet approximation for geodynamic applications – I. Model formulation. Geophys J Int 136:567–585

Medvedev SE, Podladchikov YY (1999b) New extended thin sheet approximation for geodynamic applications – II. 2D examples. Geophys J Int 136:586–608

Müller RD, Roest WR, Royer JY, Gahagan LM, Sclater JG (1992) A digital age map of the ocean floor. SIO Reference Series 93–30D

Oncken O, Hindle D, Kley J, Elger K, Victor P, Schemmann K (2006) Deformation of the central Andean upper plate system – facts, fiction, and constraints for plateau models. In: Oncken O, Chong G, Franz G, Giese P, Götze H-J, Ramos VA, Strecker MR, Wigger P (eds) The Andes – active subduction orogeny. Frontiers in Earth Science Series, Vol 1. Springer-Verlag, Berlin Heidelberg New York, pp 3–28, this volume

Peacock SM (1987) Thermal effects of metamorphic fluids in subduction zones. Geology 15:1057–1060

Ranalli G (1995) Rheology of the Earth, 2nd Ed. Chapman Hall

Riller U, Oncken O (2003) Growth of the Central Andean Plateau by tectonic segmentation is controlled by the gradient in crustal shortening. J Geol 111:367–384

Rosenberg CL, Handy MR (2005) Experimental deformation of partially melted granite revisited: implications for the continental crust. J Metamorph Geol 23:19–28

Rousse S, Gilder S, Farber D, McNulty B, Torres V (2002) Paleomagnetic evidence of rapid vertical-axis rotation in the Peruvian Cordillera, ca. 8 Ma. Geology 30:75–78

Rousse S, Gilder S, Farber D, McNulty B, Patriat P, Torres V, Sempere T (2003) Paleomagnetic tracking of mountain building in the Peruvian Andes since 10 Ma. Tectonics 22(5): doi 10.1029/2003TC001508

Royden L (1996) Coupling and decoupling of crust and mantle in convergent orogens: implications for strain partitioning in the crust. J Geophys Res 101:17679–17705

Sheffels BM (1995) Mountain building in the Central Andes: an assessment of the contribution of crustal shortening. Inter Geol Rev 37:128–153

Shen F, Royden LH, Burchfiel BC (2001) Large-scale crustal deformation of the Tibetan Plateau. J Geophys Res 106:6793–6816

Silver PG, Russo RM, Lithgow-Bertelloni C (1998) Coupling of South American and African plate motion and plate deformation. Science 279:60–63

Sobolev SV, Babeyko AY, Koulakov I, Oncken O (2006) Mechanism of the Andean orogeny: insight from numerical modeling. In: Oncken O, Chong G, Franz G, Giese P, Götze H-J, Ramos VA, Strecker MR, Wigger P (eds) The Andes – active subduction orogeny. Frontiers in Earth Science Series, Vol 1. Springer-Verlag, Berlin Heidelberg New York, pp 513–536, this volume

Springer M (1999) Interpretation of heat-flow density in the Central Andes. Tectonophysics 306:377–395

Springer M, Förster A (1998) Heatflow density across the Central Andean subduction zone. Tectonophysics 291:123–139

Swenson J, Beck S, Zandt G (2000) Crustal structure of the Altiplano from broadband regional waveform modeling: implications for the composition of thick continental crust. J Geophys Res Vol 105(B1):607–621

Turcotte DL, Schubert G (1982) Geodynamics – applications of continuum physics to geological problems. John Wiley, New York

Vanderhaeghe O, Medvedev S, Beaumont C, Fullsack P, Jamieson RA (2003) Evolution of orogenic wedges and continental plateaux: insights from crustal thermal-mechanical models overlying subducting mantle lithosphere. Geophys J Int 153:27–51

Whitman D, Isacks BL, Chatelain JL, Chiu JM, Perez A (1992) Attenuation of high-frequency seismic waves beneath the Central Andean plateau. J Geophys Res 97(B13):19929–19947

Wigger P, Schmitz M, Araneda M, Asch G, Baldzuhn S, Giese P, Heinsohn WD, Martinez E, Ricaldi E, Röwer P, Viramonte J (1994) Variation in the crustal structure of the southern Central Andes deduced from seismic refraction investigations. In: Reutter K-J, Scheuber E, Wigger P (eds) Tectonics of the Southern Central Andes. Springer-Verlag, Berlin Heidelberg New York, pp 23–48

Yang Y, Liu M, Stein S (2003) A 3-D geodynamic model of lateral crustal flow during Andean mountain building. Geophys Res Lett 30(21): doi 10.1029/2003GL018308

Yañez G, Cembrano J (2004) Role of viscous plate coupling in the late Tertiary Andean tectonics. J Geophys Res 109(B02407)

Yuan X, Sobolev SV, Kind R, Oncken O, Bock G, Asch G, Schurr B, Graeber F, Rudloff A, Hanka W, Wylegalla K, Tibi R, Haberland C, Rietbrock A, Giese P, Wigger P, Röwer P, Zandt G, Beck S, Wallace T, Pardo M, Comte D (2000) Subduction and collision processes in the Central Andes constrained by converted seismic phases. Nature 408:958–961

Zandt G, Velasco A, Beck S (1994) Composition and thickness of the southern Altiplano crust, Bolivia. Geology 22:1003–1006

Zandt G, Leidig M, Chmielowski J, Baumont D (2003) Seismic detection and characterization of the Altiplano-Puna Magma Body, Central Andes. Pure Appl Geophys 160:789–807

## Appendix 23.A  Stress and Mass Balance in the Thin Viscous Sheet Approach

In this section we derive the governing equations of the thin viscous sheet and assess the validity of the simplifications related to this approach. Here, we analyze only a specific 2D case, as the general theory of the thin-sheet approximation is presented in Medvedev and Podladchikov (1999a,b). The force balance in terms of stresses is:

$$\frac{\partial \sigma_{zz}}{\partial x} + 2\frac{\partial \tau_{xx}}{\partial x} + \frac{\partial \tau_{xz}}{\partial z} = 0 \quad (23.A1)$$

$$\frac{\partial \tau_{xz}}{\partial x} + \frac{\partial \sigma_{zz}}{\partial z} = \rho g \quad (23.A2)$$

where $\sigma$ is the full stress tensor and $\tau$ is its deviatoric part. We use an equality which results from mass conservation and viscous rheology ($\tau_{xx} + \tau_{zz} = \mu(\varepsilon_{xx} + \varepsilon_{zz}) = 0$, and thus pressure $p = -\tau_{zz} - \sigma_{zz} = \tau_{xx} - \sigma_{zz}$). Integration of Eq. 23.A1 over the depth of the lithosphere yields (Medvedev and Podladchikov 1999a):

$$\frac{\partial \bar{\sigma}_{zz}}{\partial x} + 2\frac{\partial \bar{\tau}_{xx}}{\partial x} = 0 \quad (23.A3)$$

where we use the condition of zero traction on the top and at the base of the lithosphere. The overbars refer to integration over the thickness of the lithosphere. Double integration of Eq. 23.A2 over the lithospheric depth gives:

$$\bar{\bar{\sigma}}_{zz} = R_z H + \bar{\rho} g - \iint \frac{\partial \tau_{xz}}{\partial x} dz\, dz \quad (23.A4)$$

where $R_z$ is the pressure at the compensation depth and double overbar stands for double integration. The rheology is:

$$\tau_{xz} = \mu\left(\frac{\partial v_x}{\partial z} + \frac{\partial v_z}{\partial x}\right) \quad (23.A5)$$

and therefore the distribution of the horizontal velocity is:

$$v_x(x,z) = V_x(x) + \int_{-L}^{z}\left(\frac{\tau_{xz}}{\mu} - \frac{\partial v_z}{\partial x}\right)dz \quad (23.A6)$$

where $V_x$ is the velocity at the base of the lithosphere, $z = -L$, and $v_z$ is vertical velocity.

Equations 23.A1–23.A6 are exact (not approximated) equations for depth integration in the lithosphere (Medvedev and Podladchikov 1999a). The only simplifying assumption made in that system of equations is that the lithosphere behaves as incompressible viscous fluid.

England and McKenzie (1982) simplified these equations by making two assumptions: (i) they ignored shear stress in Eq. 23.A4, $\tau_{xz} = 0$; and (ii) they ignored velocity variations in Eq. 23.A6 and set $v_x = V_x$. After substitution of the shortened Eqs. 23.A4 and 23.A6, the governing Eq. 23.A3 becomes:

$$4\frac{\partial}{\partial x}\left(\bar{\mu}\frac{\partial V_x}{\partial x}\right) = g\Phi\rho H \frac{\partial H}{\partial x} \quad (23.A7)$$

where $H$ is the thickness of the crust with density $\rho$, $\Phi = (1 - \rho/\rho_m)$ is the buoyancy amplification factor, and $\rho_m$ is the density of the mantle lithosphere.

Simplifications (i) and (ii) in England and McKenzie (1982) are justified for a thin sheet with a nearly even strength distribution and subjected to no friction at the top and bottom boundaries. Oceanic lithosphere fulfills these conditions adequately. However, they are not valid for continental lithosphere characterized by significant strength variations (Fig. 23.3). Whereas the weak part of the mantle lithosphere (bounded at the base by $\tau_{xz} = 0$) should not exert any significant shear stress, the strong variations in strength owing to partial melting in the lower crust can result in finite shear stresses.

There is no way of estimating exactly the influence of this shear stress, so we must make some simplifying assumptions. Here, we ease conditions (*i*) and (*ii*) by assuming that the shear stress $\tau_{xz}$ is small, thus rendering the vertical stress in Eq. 23.A4 close to the lithostatic pressure.

The velocity distribution (Eq. 23.A6) can change in the lower crustal channel, especially if partial melting occurs (Fig. 23.3c). Hence, we assume the total velocity to be:

$$v_x = V_x + v_c \quad (23.\text{A}8)$$

where $v_c$ is the velocity in the lower crustal channel. This velocity has a finite value within the channel and is zero outside of it. Note that this perturbation of velocity in the lower crustal channel does not significantly change the normal stress in Eq. 23.A3, as in the following equation:

$$\bar{\tau}_{xx} = 2\bar{\mu}\frac{\partial V_x}{\partial x} + 2\int_h \mu_{\text{ch}}\frac{\partial v_c}{\partial x}\text{d}z \approx 2\bar{\mu}\frac{\partial V_x}{\partial x} \quad (23.\text{A}9)$$

This is because $\mu_{\text{ch}}$ is much smaller than the average viscosity of the continental lithosphere. The integral in Eq. 23.A9 is taken over the thickness of the low-viscosity channel in the lower crust, $h$. Thus, Eq. 23.A7 describes the stress balance for the lithosphere for conditions that are not as restrictive as in England and McKenzie (1982).

Although the velocity correction in Eq. 23.A8 does not change the balance of integrated stresses in the thin sheet, it is significant enough to change both the balance of masses and the changes in crustal thickness. These changes are described with the equation:

$$\frac{\partial H}{\partial t} + \frac{\partial q_x}{\partial x} = 0; \quad q_x = q_{xV} + q_{xv} \quad (23.\text{A}10)$$

where $q_{xV} = (V_x H)$ is the flux corresponding to the bulk velocity, $V_x$, and $q_{xv}$ is the flux corresponding to the additional flow in the lower crustal channel. We adopted a simplified approach to the estimations of the flux, $q_{xv}$ (Clark and Royden 2000), in which the channel of constant thickness, $h_c$ occupies the entire width of the continental crust. The flux is then described by:

$$q_{xv} = -\frac{\rho g h_c^3}{12\mu_{\text{ch}}}\frac{\partial S}{\partial x} \quad (23.\text{A}11)$$

where $S$ is the elevation of the continent. Medvedev and Beaumont (in press) show that the viscosity distribution, $\mu_c$, should decrease significantly at the edges of the plateau, such that $\mu_c = \mu_p$ beneath the plateau. Otherwise, $\mu_{\text{ch}} = \mu_t \gg \mu_p$, assuming that partial melting in the lower crust is responsible for the development of the plateau. Medvedev and Beaumont (in press) show that orogens with a 10 km thick channel can acquire a plateau-like geometry if $\mu_p \sim 10^{18}\text{--}10^{20}$ Pa s, and $\mu_t \geq 25\mu_p$. Here we associate that rheological transition with a critical thickness of the crust, $H^*$:

$$\mu_{\text{ch}} = \begin{cases} \mu_p & (H < H^*) \\ \mu_t & (H > H^*) \end{cases}$$

An appropriate value for $H^*$ is discussed in the main part of the text.

We analyzed here the classical thin-sheet approach and show that the 3D analog of Eq. 23.A7 (England and McKenzie 1982) can describe the evolution of the two main parts of our model: the Nazca and South American Plates (see Eqs. 23.A1 and 23.A2). The 3D analog of kinematic update (Eqs. 23.A10 and 23.A11) can be easily obtained by adding $y$-related component of flux into Eq. 23.A10:

$$\frac{\partial H}{\partial t} + \frac{\partial q_x}{\partial x} + \frac{\partial q_y}{\partial y} = 0$$

where estimation of $q_y$ can be obtained the same way as for $q_x$, simply replacing subscripts and differentiations from $x$ to $y$. Equation 23.A7 does not apply to the subduction zone and is considered in more detail in the text and in Appendix 23.B.

## Appendix 23.B Rheological Properties of the Subduction Zone Elements

The subduction zone elements (Fig. 23.2) include layers of oceanic and continental lithosphere, and the subduction channel. The complicated geometry of the subduction zone requires a detailed analysis of the depth integration. We therefore consider the stresses associated with each component of strain-rate tensor. Even though the subduction zone consists of several elements in the east-west direction (Fig. 23.2), we analyze the entire zone and assume average values for strains and related stresses across it. The following derivations present an approximate analysis aimed mainly at understanding the general characteristics of the thin-sheet representation of the subduction zone. The approximate analysis of this appendix ignores difference between thickness of the subduction zone lithosphere, $L_s$, its width, $D$, and length of the subduction channel, $l$. We understand

**Fig. 23.B1.**
Estimates of the integrated normal stress due to variations in velocity, $V_x$, in the X-direction within the subduction zone. **a** General view; **b** break down of $V_x$ into velocities parallel and normal to the subduction channel; **c** and **d** flow in the channel at velocity $v_x$ tangential and normal to the channel boundaries, respectively

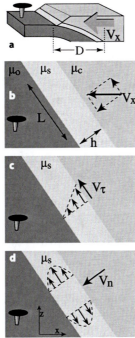

**Fig. 23.B2.**
Estimates of the integrated normal stresses in the subduction zone due to variations in velocity, $V_y$, in the X-direction (**a** general view, **b** associated flow), and in the Y-direction (**c** general view, **d** associated flow)

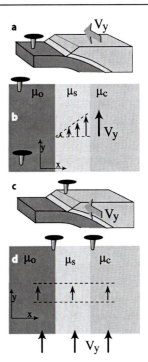

that these parameters can be different by several times, but we limit our analysis to the qualitative analysis distinguishing only parameters that differ by orders of magnitude. We leave the quantitative estimations of the parameters of the subduction zone to the numerical experiments.

In our analysis, we assume the continental and oceanic crust to be much stronger than the rocks in the subduction channel (Sobolev et al. 2006, Chap. 25 of this volume). This assumption allows us to ignore deformation in the continental and oceanic parts of the subduction zone elements, and to concentrate on stresses associated with deformation of the subduction channel only. We also assume that the length of the subduction channel, $l$, is greater than its thickness, $h$, and introduce the channel aspect ratio, $\Delta = l/h \gg 1$ (Fig. 23.B1b).

Consider stresses associated with variations in the east-west velocity, $V_x$, and in the east-west direction, $X$ (Fig. 23.B1a). Those variations represent the $xx$ term of the strain-rate tensor,

$$\dot{\varepsilon}_{xx} = \frac{\partial V_x}{\partial x} \approx \frac{V_x}{D}$$

($D$ is the width of the subduction zone). The total variations of velocity can be broken down into two components: channel-parallel and channel-normal, $V_x = V_\tau + V_n$ (Fig. B1b). The channel-parallel velocity induces Couette flow (Fig. 23.B1c), whereas channel-normal velocity squeezes the channel, inducing Poiseuille flow (Fig. 23.B1d). The average stresses associated with the two components of $V_x$ are:

$$\sigma_\tau = \frac{\mu_s V_\tau}{h}$$

$$\sigma_n = \frac{2\mu_s V_n l^2}{h^3}$$

where $\mu_s$ is the viscosity of the rock in the subduction channel. The first equation is a standard expression of the shear stress associated with Couette flow (e.g., Turcotte and Schubert 1982). The second equation represents the average pressure from the channel response on the channel-perpendicular squeezing. Depth-integration yields the shear and normal stresses on the subduction channel:

$$\bar{\sigma}_\tau = \Delta \mu_s V_x$$

$$\bar{\sigma}_n = 2\Delta^3 \mu_s V_x$$

In these expressions, we assume an undeformable continental lithosphere in the presence of the weak subduction channel, rendering $V_x$ to be invariant with depth. The stress associated with motion parallel to the channel boundaries, $\sigma_\tau$, is much smaller than $\sigma_n$. Thus, the stress-strain-rate relationship for the $xx$ component of deviational stress can be approximated as:

$$\bar{\tau}_{xx} \approx \bar{\sigma}_n = 2\Delta^3 D \mu_s \dot{\varepsilon}_{xx} \qquad (23.B1)$$

The shear-stress component, $\tau_{xy}$, is reduced to influence of Couette flow in the subduction channel. This is initiated by the difference in the north-south motions across the subduction zone, $V_y$ (Fig. 23.B2a,b). The shear stress in the channel is therefore:

$$\tau_{xy} = \frac{\mu_s V_y}{h}$$

In this expression we ignore the stresses associated with variations of velocity $V_x$ in north-south direction, $\partial V_x/\partial y$. In the assumptions of our simplified estimations (preferable deformation of the subduction channel) this type of deformation results in rotation of stronger parts and corresponding Couette flow in the channel, which in turn results in stresses of the same order as already estimated. Integration and simplification ($\dot{\varepsilon}_{xy} \approx V_y/D/2$) yield the expression:

$$\bar{\tau}_{xy} = 2\Delta D \mu_s \dot{\varepsilon}_{xy} \qquad (23.B2)$$

The $yy$ component of the stress tensor is caused by north-south variations in the north-south velocity, $V_y$ (Fig. 23.B2c,d). The corresponding component of the strain-rate tensor is

$$\dot{\varepsilon}_{yy} = \frac{\partial V_y}{\partial y} \approx \frac{V_y}{D_y}$$

where $D_y$ is the size of the subduction zone element in the Y direction. The strength of the subduction zone element in this direction is limited by the strength of the competent layers as the deformation cannot be limited to subduction channel only:

$$\tau_{yy} = \frac{2\mu_c V_y}{D_y} \qquad (23.B3)$$

$$\bar{\tau}_{yy} = 2\mu_a L_s \dot{\varepsilon}_{yy}$$

where $\mu_a$ is an average viscosity in the subduction zone, and should be limited by values of viscosity of the strong parts of the zone, $\mu_c$ and $\mu_o$. Summarizing Eqs. 23.B1–23.B3, we obtain the following relation for the anisotropic rheology of the subduction zone elements:

$$\begin{pmatrix} \bar{\tau}_{xx} \\ \bar{\tau}_{yy} \\ \bar{\tau}_{xy} \end{pmatrix} = 2 \begin{bmatrix} \Delta^3 D \mu_s & 0 & 0 \\ 0 & L_s \mu_a & 0 \\ 0 & 0 & \Delta D \mu_s \end{bmatrix} \cdot \begin{pmatrix} \dot{\varepsilon}_{xx} \\ \dot{\varepsilon}_{yy} \\ \dot{\varepsilon}_{xy} \end{pmatrix}$$

$$\approx 2 L_s \begin{bmatrix} \Delta^3 \mu_s & 0 & 0 \\ 0 & \mu_a & 0 \\ 0 & 0 & \Delta \mu_s \end{bmatrix} \cdot \begin{pmatrix} \dot{\varepsilon}_{xx} \\ \dot{\varepsilon}_{yy} \\ \dot{\varepsilon}_{xy} \end{pmatrix} \qquad (23.B4)$$

The approximate derivations of this section aimed to give the qualitative conclusion that averaged over lithospheric thickness subduction zone behaves anisotropically. Moreover, the parameters of Eq. 23.B4 are difficult to obtain in nature, so we use a simplified version of Eq. 23.B4 (see Eq. 23.4) in our calculations. The analysis presented here, however, can conclude that the viscosity associated with the shear stress component is smaller than one associated with $\tau_{xx}$, because $\Delta \gg 1$. The above analysis cannot estimate the difference in strength of subduction zone in $x$ and $y$ directions, because the components of viscosity in these directions involve $\Delta \gg 1$ and $\mu_a \gg \mu_s$ and we leave this analysis to numerical experiments.

# Chapter 24

# Numerical Study of Weakening Processes in the Central Andean Back-Arc

Andrey Y. Babeyko · Stephan V. Sobolev · Tim Vietor · Onno Oncken · Robert B. Trumbull

**Abstract.** We employed a 2D numerical thermo-mechanical modeling technique to study the dynamics of the plateau-foreland system, with particular application to the Central Andean back-arc. Our model back-arc consists of a weak plateau indented by a relatively stronger foreland. A "normal" shortening scenario implies pure-shear tectonic inflation of the plateau. Forces that drive tectonic shortening encounter major resistance from the growing gravitational potential of the rising plateau. Other resistance factors are the brittle strength of the upper crust and the viscous resistance of ductile lower crust and upper mantle. The overall resistance to shortening is up to $6-8 \times 10^{12}$ N m$^{-1}$ for plateau heights of less than 5 km. Several thermo-mechanical processes within the back-arc have a gross weakening effect on a back-arc scale and, thus, may be responsible for episodes of shortening rate acceleration in the past. These processes include: (*i*) on-going eclogitization of the lower mafic crust beneath the plateau, (*ii*) heating and convection in the plateau felsic crust, (*iii*) mechanical failure of the foreland sediments and (*iv*) high erosion and exhumation rates at the plateau margin (effective in monsoon areas but apparently not relevant for the Central Andes). The largest reduction in the effective lithospheric strength (~$3 \times 10^{12}$ N m$^{-1}$, i.e., 40–50% of the total strength) could result from the mechanical failure of the foreland sediments, leading to the migration of shortening deformation into the Altiplano foreland (Subandean thrust belt). This failure implies a drastic reduction of both sediment cohesion (almost to zero values) and friction angle (to at least a third of the normal value of 30°). The second important factor is on-going eclogitization of the lower crust beneath the plateau, which increases the average density of the crust and thus prevents rising of the plateau. The eclogitization may decrease the effective lithospheric strength by ~$2 \times 10^{12}$ N m$^{-1}$ (25–35% of the total strength). The third important factor is intracrustal convection, which follows eclogitization-driven lithospheric delamination and reduces the brittle strength of the uppermost crust (overall effect ~$1 \times 10^{12}$ N m$^{-1}$, or 10–15% of the total strength). High erosion rate at the plateau margin may efficiently weaken the orogenic lithosphere but, in the case of the low to moderate erosion rates typical of the Central Andes, its effect is negligible. In general, internal weakening of the overriding plate is, in addition to plate tectonic forces, among the major controls on the Andean orogeny.

## 24.1 Introduction

The large-scale temporal and spatial evolution of the Central Andean orogen involves tectonic shortening of the South American plate margin bounded by the Nazca Plate to the west and the indenting Brazilian shield to the east (Fig. 24.1a) (e.g., Isacks 1988; Allmendinger et al. 1997; Lamb et al. 1997). The external far-field forces that drive tectonic shortening in the back-arc encounter major resistance from (*i*) the growing gravitational potential of the rising plateau, (*ii*) the brittle strength of the upper crust, and (*iii*) the viscous resistance of the ductile lower crust and upper mantle. Clearly, in the limiting case of homogeneous, pure-shear plateau inflation ('vise' plateau model in the terminology of Ellis et al. 1988), the overall resistance against shortening steadily grows as the plateau becomes higher with time. Continued shortening in this mode requires a corresponding increase of the far-field driving forces. Alternatively, any thermo-mechanical processes in the back-arc that are able to reduce the

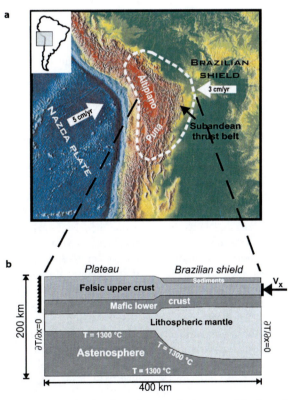

**Fig. 24.1. a** Map showing the major tectonic elements of the Central Andes. *White dashed line* contours the model area. **b** Cartoon illustrating the set-up for the back-arc model. Material parameters are given in Table 24.1. The initial thickness of the lithosphere under the plateau varies between 70 and 90 km

resistance caused by one or more of these factors would have a gross weakening effect and thus contribute to the acceleration of shortening.

Oncken et al. (2006, Chap. 1 of this volume) showed that the bulk shortening rate in the Central Andes was not constant during the Neogene but, instead, significant acceleration occurred at about 10 Ma. In their large-scale thermo-mechanical model of the Andean orogen, Sobolev et al. (2006, Chap. 25 of this volume) inferred that the intensity of the tectonic shortening in the back-arc was controlled by both external (accelerating drift of the South American Plate and shear coupling at the plate interface) as well as internal factors. Among the internal processes they mentioned delamination of the lower crust and mantle lithosphere as well as mechanical failure of the foreland sediments.

The present paper provides a systematic study of the four different thermomechanical processes which operate in the Central Andean back-arc and can be expected to have a gross weakening effect assisting tectonic shortening. They are: (1) lithospheric delamination driven by the gabbro-eclogite transformation in the thickened mafic lower crust under the plateau; (2) intracrustal convection in the felsic crust of the plateau that followed lithospheric delamination; (3) mechanical failure of the foreland sediments; and (4) high erosion and exhumation rates at the plateau margin. All these processes are triggered by ongoing tectonic shortening and, thus, have a positive feedback with respect to it.

Our major goal is a quantitative estimation of the absolute and relative weakening potential of the aforementioned processes. Since our numerical models are kinematically constrained, this estimation can be done by continuous monitoring of the integral force which is required to drive the tectonic shortening at a given rate.

We restrict our numerical modeling to the Central Andean back-arc region. That allows us to run models with at least twice as much resolution as in the large-scale model of the Andean orogen presented by Sobolev et al. (2006, Chap. 25 of this volume). Part of the present work is an extension of our previous studies (Babeyko et al. 2002; Babeyko and Sobolev 2005), whereas models of the lithospheric delamination triggered by eclogitization in the lower crust and, especially, models of erosion and exhumation at the plateau margin are from parallel ongoing projects.

The paper is organized as follows. First, we describe the numerical technique used for thermo-mechanical simulations. After that, we proceed to the numerical models of the aforementioned processes, concentrating on the conditions under which they may operate and on their consequences for the overall dynamics of the plateau-foreland system. The four subsections (Sect. 24.3.1 to 24.3.4) may be considered as individual, independent studies. Each subsection has its own introduction, results and some discussion. The paper is completed by a summarizing discussion which addresses the relative importance of the four weakening processes.

## 24.2 Modeling Approach

### 24.2.1 Governing Equations and Constitutive Laws

Thermo-mechanical processes are modeled by numerical integration of the coupled system of conservation equations for momentum (Eq. 24.1), mass (Eq. 24.2) and energy (Eq. 24.3).

$$-\frac{\partial p}{\partial x_i} + \frac{\partial \tau_{ij}}{\partial x_j} + \rho g_i = 0 \tag{24.1}$$

$$\frac{1}{K}\frac{dp}{dt} - \alpha\frac{dT}{dt} + \frac{\partial v_i}{\partial x_i} = 0 \tag{24.2}$$

$$\rho C_p \frac{dT}{dt} = \frac{\partial}{\partial x_i}\left(\lambda \frac{\partial T}{\partial x_i}\right) + \tau_{ij}\dot{\varepsilon}_{ij} + \rho A \tag{24.3}$$

In the above equations, the Einstein summation rule applies and $x_i$ are coordinates, $t$: time, $v_i$: velocities, $p$: pressure, $\tau_{ij}$ and $\dot{\varepsilon}_{ij}$: stress and strain rate deviators, respectively, $d/dt$: convective time derivative, $K$: bulk modulus, $\rho$: density, $g_i$: gravity vector, $T$: temperature, $C_p$: heat capacity, $\lambda$: heat conductivity, $A$: radioactive heat production.

These equations are solved together with the appropriate constitutive laws that establish the link between deformation and stress. Our numerical code employs composite elasto-visco-plastic rheology. The overall deviatoric strain rate, $\dot{\varepsilon}_{ij}$, is decomposed into elastic, viscous and plastic components:

$$\dot{\varepsilon}_{ij} = \dot{\varepsilon}_{ij}^{(e)} + \dot{\varepsilon}_{ij}^{(v)} + \dot{\varepsilon}_{ij}^{(pl)} = \frac{1}{2G}\frac{\hat{d}\tau_{ij}}{dt} + \frac{\tau_{ij}}{2\eta_{eff}} + \dot{\varepsilon}_{ij}^{(pl)} \tag{24.4}$$

Here $G$ is elastic shear modulus, $\hat{d}\tau_{ij}/dt$ is the Jaumann co-rotational deviatoric stress rate, and $\eta_{eff}$ is effective viscosity. To calculate effective viscosity, we followed the approach of Kameyama et al. (1999) in which they applied three competing creep mechanisms for olivine: diffusion, dislocation and Peierls creep. According to this approach, at a given temperature and stress, the mechanism that produces the highest viscous strain rate becomes the dominant creep mechanism:

$$\dot{\varepsilon}_{eff}^{(v)} = \dot{\varepsilon}_d + \dot{\varepsilon}_n + \dot{\varepsilon}_p \tag{24.5}$$

- diffusion creep:

$$\dot{\varepsilon}_d = B_d \tau_{II} \exp\left(-\frac{E_d}{RT}\right) \tag{24.6}$$

- dislocation, power-law creep:

$$\dot{\varepsilon}_n = B_n \tau_{II}^n \exp\left(-\frac{E_n}{RT}\right) \quad (24.7)$$

- Peierls creep:

$$\dot{\varepsilon}_p = B_p \exp\left[-\frac{E_p}{RT}\left(1 - \frac{\tau_{II}}{\tau_p}\right)^q\right] \quad (24.8)$$

Effective viscosity is then calculated as:

$$\eta_{\text{eff}} = \frac{\tau_{II}}{2\dot{\varepsilon}_{\text{eff}}^{(v)}} \quad (24.9)$$

In these expressions, $\tau_{II}$ is the stress tensor norm defined as the square root of the second invariant of the stress tensor, $R$ is the gas constant, $T$ is temperature, and other parameters are material constants. Kameyama et al. (1999) showed that, at typical geological strain rates ($> 10^{-15}$ s$^{-1}$) and moderate temperatures, deformation in the mantle is mainly accommodated by power-law creep. Diffusion creep becomes dominant at very low strain rates or high temperatures. Peierls creep becomes the main deformation mechanism at high deviatoric stress ($> 500$ MPa). In our models, Peierls creep plays an important role in limiting stress inside the bending subducting slab.

In the low-temperature upper crust, where viscosities and, hence, viscous stresses are extremely high, stress is limited by the rocks' plastic shear failure. In particular, plastic shear failure is responsible for faulting in the upper crust. In our model, we employed the Mohr-Coulomb failure criterion with non-associated shear flow (e.g., Vermeer and de Borst 1984):

$$f_s = \sigma_1 - \sigma_3 \frac{1 + \sin\phi}{1 - \sin\phi} + 2C_h \sqrt{\frac{1 + \sin\phi}{1 - \sin\phi}} \leq 0 \quad (24.10)$$

Here, $\sigma_1$, $\sigma_3$ are maximal and minimal principal stresses, respectively, $\phi$ is the angle of friction, and $C_h$ is cohesion.

Equation 24.10 describes the failure envelope in terms of minimal and maximal principal stresses ($\sigma_1, \sigma_3$). In explicit time-stepping algorithm plastic corrections to stresses are straightforward. At each computational time step we first calculate visco-elastic stress increment corresponding to the strain update. Whenever the resulting stress violates the Mohr-Coulomb failure criterion, shear failure is declared and the stress point is placed back on the envelope, $f_s = 0$, using a non-associated flow rule derived by means of the potential function, $g_s = \sigma_1 - \sigma_3$.

Material parameters, the angle of friction, $\phi$, and cohesion, $C_h$, are generally not constant. Typically, their values decrease as plastic deformation progresses, the process being called "plastic strain softening". We also employed similar strain softening for viscous creep. Both strain softening processes tend to localize deformation in the lithosphere. Note, that faults in our models are not predefined.

### 24.2.2 Modeling Technique

As a numerical tool, we used the 2D, parallel, thermo-mechanical, finite element code LAPEX-2D (LAgrangian Particle EXplicit), which is based on the prototype code PAROVOZ (Poliakov et al. 1993). Our code combines the explicit Lagrangian finite element algorithm FLAC (Poliakov et al. 1993; Cundall and Board 1988) with the particle technique of the particle-in-cell method (Sulsky et al. 1995; Moresi et al. 2003). The solution proceeds on a moving Lagrangian grid using the explicit time integration of Eqs. 24.1–24.3.

For this reason, the inertial term

$$\rho^{\text{inert}} \frac{\partial v_i}{\partial t}$$

is included in the right-hand side of the momentum conservation equation (Eq. 24.1). Here, inertial density, $\rho^{\text{inert}}$, actually plays the role of the dynamic relaxation parameter. The incorporation of the inertial term allows explicit time integration of nodal velocities and displacement increments, thus "driving" the solution towards a quasi-static equilibrium. The magnitude of the inertial term is kept small in comparison to tectonic forces, typically $10^{-3}$–$10^{-5}$, so the solution remains quasi-static.

In comparison to the classical, implicit finite element algorithm, memory requirements for this method are very moderate since no global matrices are formed and inverted. Accordingly, the computational costs for one time-step are very low. However, explicit time integration imposes very strict restrictions on the computational time-step – for typical geodynamic applications, the stable Courant time-step is of the order of one year. Despite this disadvantage, the small computational time-step allows physical non-linearities (such as plastic flow, for example) to be treated in a natural way, without additional iterations or problems with convergence. The algorithm is easy to parallelize. A typical model run of 10 million years of evolution on a 300 × 100 element grid takes 1–2 days on a single IBM SP4 eight-processor node.

Solutions on the moving Lagrangian grid become inaccurate when the grid becomes too distorted. At this point, remeshing should take place. During remeshing, a new grid is built and solution variables are interpolated from the old, distorted grid. Remeshing is inevitably related to the problem of numerical diffusion. In the case of a history-dependent solution, such as the presence of elastic stresses or strong stress gradients (e.g., the subducting slab versus its surroundings), uncontrollable numerical diffusion might strongly affect the solution. In order to sustain it, we have implemented in our code a particle-in-cell, or material point, technique (Sulsky et al. 1995; Moresi et al. 2003).

In this technique, particles, which are distributed throughout the mesh, are more than simple material tracers. In our code, the particles (typically 10 to 60 per element) track not only material properties but also all history-dependent variables, including full strain and stress tensors. Between remeshings, the particles provide no additional computational costs since they are frozen in the moving Lagrangian grid. This grid is composed of constant-stress triangles and there is no need to update the particle properties at this stage; additional computation arises only during remeshing. Firstly, stress and strain increments that have accumulated since the previous remeshing are mapped from the Lagrangian elements to the particles. After the new mesh is constructed, stresses and strains are projected back from the particles onto the new grid. This marker-based remeshing procedure minimizes numerical diffusion and allows us to preserve very high stress gradients through many hundreds of remeshings.

### 24.2.3 Back-Arc Model Set-Up

Our general set-up for the back-arc model is presented in Fig. 24.1b. The model box is 400 km wide and 200 km deep and is meshed with uniform grid of $2 \times 2$ km (in some runs, $1 \times 1$ km). The left-hand side of the model represents the plateau with thermally activated lithosphere beneath; the right-hand side represents the colder Brazilian shield indenting the plateau from the east. As we did not intend to simulate the whole convergence history of the orogen, the plateau crust in our input model is already somewhat pre-thickened. In some models, the uppermost crust of the foreland contains an additional 6–10 km-thick layer of Paleozoic sediments whose frictional strength may differ from that of the 'normal' crust. The whole system is being driven by shortening with a constant convergence velocity applied to the right side of the box. The left side is fixed over the upper 60 km. Beneath that depth, the material is free to flow out of the box. The bottom of the box is also open. These boundary conditions simulate the east-to-west indenting of the cold Brazilian shield into the thermally activated lithosphere of the Central Andean Plateau. Actual model box dimensions may vary for some individual models; thus, for models in Sect. 24.3.3 a 300 km wide box was used.

The initial thermal distribution corresponds to a stable geotherm with a given lithosphere-asthenosphere boundary, and the position of this boundary varies laterally in

**Table 24.1.** Material parameters

| Parameter | Sediments | Felsic crust | Mafic crust | Mantle |
|---|---|---|---|---|
| Density at room conditions, $\rho$ (kg m$^{-3}$) | 2670 | 2800 | 3000 | 3300 |
| Thermal expansion, $\alpha$ (K$^{-1}$) | $3.7 \times 10^{-5}$ | $3.7 \times 10^{-5}$ | $2.7 \times 10^{-5}$ | $3.0 \times 10^{-5}$ |
| Elastic moduli, $K, G$ (GPa) | 55, 36 | 55, 36 | 63, 40 | 122, 74 |
| Heat capacity, $C_p$ (J kg$^{-1}$ K$^{-1}$) | 1200 | 1200 | 1200 | 1200 |
| Heat conductivity, $\lambda$ (W K$^{-1}$ m$^{-1}$) | 2.5 | 2.5 | 2.5 | 3.3 |
| Heat productivity, $A$ (µW m$^{-3}$) | 1.3 | 1.3 | 0.3 | 0 |
| Initial friction angle, $\phi$ (degree) | 3 … 30 | 30 | 30 | 30 |
| Initial cohesion, $C_h$ (MPa) | 1 … 20 | 20 | 40 | 40 |
| Plastic strain softening | 10-times linear decrease of $\phi$ and $C_h$ from 0.5 to 1.5 strain | | | |
| Pre-exponential constant for dislocation creep, $\log(B_n)$ (Pa$^{-n}$ s$^{-1}$) | −28.0 | −28.0 | −21.05 | −16.3 |
| Power law exponent, $n$ | 4.0 | 4.0 | 4.2 | 3.5 |
| Activation energy for dislocation creep, $E_n$ (kJ mol$^{-1}$) | 223 | 223 | 445 | 535 |
| Activation volume (cm$^3$) | – | – | – | 15 |
| Viscous softening | 10-times linear decrease of viscosity from 0.5 to 1.5 strain | | | |

Dislocation creep law parameters – sediments and felsic crust: wet quartzite by Gleason and Tullis (1995); mafic crust: Pikwitonei granulite by Wilks and Carter (1990); mantle: dry peridotite by Hirth and Kohlstedt (1996).

the model. Under the Brazilian shield, the bottom of the lithosphere is 150 km deep. The lithosphere-asthenosphere boundary gradually moves upwards towards the plateau. Beneath the plateau, the initial lithospheric thickness is 70 to 90 km, depending on the model.

Table 42.1 lists material parameters accepted in the present study. Dislocation creep is the dominant viscous creep mechanism at model strain rates. For the mantle material, we also employed Peierls creep, as well as diffusion creep, in accordance with Kameyama et al. (1999). The Peierls mechanism was not considered for the crustal material owing to a lack of corresponding experimental data. Some additional model parameters, used individually for particular model set-ups, are discussed below in the corresponding sections.

One of the main goals of our modeling was to estimate the driving force of the tectonic shortening in the back-arc. To do this, we numerically integrated the deviatoric horizontal stress along the right-hand side of the model box:

$$F_x = \int_{y_{min}}^{y_{max}} (\sigma_{xx} - \sigma_I/3)\,dy \qquad (24.11)$$

Here, $\sigma_I$ is the first invariant of the stress tensor.

## 24.3 Weakening Processes in the Upper Plate

### 24.3.1 Ongoing Gabbro-to-Eclogite Transformation in the Lower Crust

Mantle lithosphere delamination under the Central Andean back-arc has been previously inferred from several petrological and geophysical studies: Kay and Kay (1993) and Kay et al. (1994) argued for late Pliocene lithospheric delamination under the Puna Plateau based on the analysis of young, mafic, back-arc volcanic rocks and geophysical evidence. This delamination scenario under the Puna Plateau is also consistent with the high elevation of the plateau, the change in the regional stress field, as demonstrated by the fault kinematic studies of Allmendinger et al. (1989) and Marrett et al. (1994), and with the regional, seismic-wave attenuation pattern (Whitman et al. 1996). Recently, Beck and Zandt (2002) and Yuan et al. (2002) also argued for localized lithospheric delamination under the northern Altiplano segment of the Central Andean Plateau, mainly based on the seismic tomography study by Myers et al. (1998) and the interpretation of seismic-wave conversions. They also speculated about the possible physical mechanism of the delamination process: delamination might be driven by density instabilities caused by the eclogitization of the mafic lower crust. The absence of a high-velocity lower crust in the seismic cross sections (Wigger et al. 1994; Zandt et al. 1996) implies either an eclogitic mafic lower crust, with velocities similar to those of the mantle, or a mafic crust that has been delaminated.

A plausible delamination scenario begins with pure-shear tectonic shortening that thickens the crust causing the depressed mafic lower crust to enter the eclogite stability field, with a corresponding increase in density. The density of the lower crust becomes higher than that of the underlying mantle and growing Rayleigh-Taylor instability finally leads to the delamination of the eclogitic lowermost crust. The sinking eclogite blobs entrain large portions of the relatively cold and, hence, also unstable sub-Moho mantle. The delaminated crust-mantle material is then replaced by asthenospheric inflow from below.

Previous numerical studies on lower-crustal eclogitization and delamination (Jull and Kelemen 2001; Doin and Henry 2001) primarily addressed the thermal and rheological conditions for lower-crustal instability. In this study, we considered another aspect to the processes of eclogitization and delamination, namely, we studied numerically the influence of the above processes on the overall strength of the back-arc.

A qualitative consideration of the delamination scenarios suggests at least two weakening mechanisms operating in the shortened back-arc. Firstly, progressive eclogitization of the lower crust increases its average density and, hence, works against rising of the tectonically inflated plateau. Secondly, the delamination removes cold and, thus, rheologically strong sub-plateau mantle lithosphere and substitutes it with hot, low-viscosity asthenosphere. To test these possibilities numerically, we incorporated the model of the gabbro-to-eclogite transformation into our thermo-mechanical calculations.

Sobolev and Babeyko (1994) calculated synthetic phase diagrams representing the gabbro-to-eclogite type transformation for the broad range of bulk chemical magmatic compositions. They also computed the corresponding density and seismic velocity variations. At elevated pressures, the density of mafic granulites and eclogites ($SiO_2 < 50\%$) can become higher than the density of the underlying mantle. For strongly mafic compositions, such as olivine gabbro, the density of eclogite can reach more than 3.5 g cm$^{-3}$ (at room conditions).

For this study, we parameterized the synthetic phase diagram for bulk gabbro chemical composition, as computed by Sobolev and Babeyko (1994). According to this parameterization, we can assign to each point in the pressure-temperature space a value representing the degree of the transformation. This value varies from 0 (low-pressure assemblage, gabbro) to 1 (high-pressure assemblage, eclogite). Actual density is then calculated as a weighted average between the two end-members, gabbro (Table 24.1) and eclogite, and then corrected for in-situ conditions. The density of eclogite is a model parameter with two values: 3.45 g cm$^{-3}$ ('normal' gabbro) or 3.55 g cm$^{-3}$ (olivine gabbro).

The effect of the gabbro-to-eclogite transformation on material viscosity is controversial. Jin et al. (2001), based on experimental measurements, argued for strong eclogite rheology that is comparable to that of peridotite. In contrast, Piepenbreier and Stoeckhert (2001) showed from natural observations that the flow strength of omphacite (the main eclogite mineral) is much lower than that of diopside (gabbro) at similar temperatures and strain rates. Moreover, eclogites derived from normal gabbro compositions contain about 10 wt% of weak quartz, which is absent at lower pressures. The bulk viscosity of eclogite could be further decreased owing to textural changes during transformation, such as grain size reduction. Taking into account the above considerations, we assumed that the gabbro-to-eclogite transformation does not influence the creep parameters of the lower crust.

Another model parameter is the initial thickness of the lithosphere under the plateau. Here, we employed values of 90 and 70 km. Thinner lithosphere means higher mantle and crustal temperatures and, hence, lower viscosities. In turn, this means better conditions for delamination and reaction kinetics. At the same time, however, higher temperatures shift the gabbro-to-eclogite transition to greater depths because of the positive slope of the transition isograds ($dP/dT > 0$). Thicker (colder) lithosphere means earlier eclogitization but less favorable conditions for delamination (see Jull and Kelemen 2001 for a discussion on the trade-offs of eclogitization and delamination conditions). Since our models always have a lower-crustal temperature higher than 650–700 °C, and because at least minor water inflow can be expected owing to the presence of the subduction zone, we did not assume any kinetic restrictions on the transformation.

Figure 24.2 presents a time-series of density plots that illustrate progressive eclogitization of the thickened crust under the plateau. The lithosphere is initially 90 km thick and the felsic crust and mafic lower crust are already pre-thickened to 30 and 15 km, respectively. Figure 24.2a shows the reference case without the gabbro-to-eclogite transformation in the mafic crust under the plateau. The

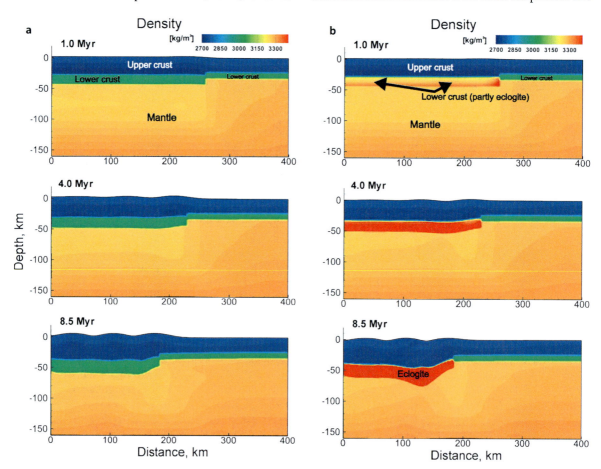

**Fig. 24.2.** Time-series of density plots illustrating progressive eclogitization of the thickened plateau crust. The initial lithospheric thickness under the plateau is 90 km. **a** Reference case without phase transformation in the lower crust. **b** Mafic lower crust transforms into eclogite due to thickening and cooling. The PT-diagram corresponds to parameterization of the synthetic phase diagram for gabbroic bulk composition by Sobolev and Babeyko (1994). Maximum density of eclogite is 3.55 g cm$^{-3}$. Relatively "cold" and viscous lithosphere prevents delamination. Note the lower surface topography in case (**b**) which means less work against gravity and, hence, less resistance to shortening

lithosphere under the plateau is thermally weaker than the lithosphere of the indenting, cold Brazilian shield and, hence, all the shortening is accommodated inside the plateau region with pure shear. This case corresponds to the well-studied 'vise' model presented by Ellis et al. (1998). Tectonic inflation of the plateau results in the isostatic rise of its average topography.

Figure 24.2b shows the case where mafic lower crust undergoes transformation into eclogite due to thickening and cooling. Here, the density of eclogite is 3.55 g cm$^{-3}$. Note that, as qualitatively considered above, the process of progressive eclogitization effectively works against a rise in topography. The dynamic effect of progressive eclogitization is well illustrated by the evolution of the shortening driving force over time in Fig. 24.3a. This figure shows the force that has to be applied to the right side of the model box so as to drive shortening at a constant given rate (here, 1 cm yr$^{-1}$). The short-dashed line corresponds to a reference case without eclogitization (Fig. 24.2a). In this case, pure-shear crustal thickening produces a steadily rising plateau which, in turn, increases resistance to shortening. Solid and long-dashed lines correspond to models with eclogitization (the maximum density of eclogite is 3.55 g cm$^{-3}$ for the solid line and 3.45 g cm$^{-3}$ for the long-dashed line). In the two latter cases, the isostatic uplift caused by the crustal thickening is compensated by isostatic subsidence, which results from the increase in density of the lower crust. Therefore, the driving force remains almost constant, meaning that ongoing eclogitization effectively makes the back-arc system weaker during tectonic shortening.

Note that in the model with an initially 90 km-thick lithosphere, delamination did not take place (Fig. 24.2b). Despite the high eclogite density, relatively low temperatures were responsible for the high viscosities of the lower crust and the surrounding mantle lithosphere ($10^{21}$–$10^{22}$ Pa s), making delamination impossible on a 10 to 20 million-year timescale. In contrast, Figs. 24.3b and 24.4 correspond to the case of successful delamination with an initial lithospheric thickness of 70 km. Higher lower-crustal temperatures shifted the onset of the gabbro-to-eclogite transformation to greater depths (e.g., compare Fig. 24.2 at 4 Myr with Fig. 24.4 at 5 Myr). However, while forming, eclogitic material easily forms large blobs that sink into the mantle (viscosity $10^{19}$–$10^{20}$ Pa s) and entrain large volumes of the subcrustal mantle lithosphere.

After the whole lower crust is delaminated, the response of the back-arc to tectonic shortening becomes similar to the response of our reference (eclogite-free) model. This is well illustrated by Fig. 24.3b. Note the change in slope of the solid and long-dashed lines after ~6 and 8 Myr, respectively, showing the times when the plateau crust finally lost its mafic component. Later on, there are no more mechanisms working against isostatic plateau uplift and, hence, continued shortening is accompanied by a growing resistance, similar to the reference, eclogite-free scenario (dashed line).

After the delamination of the rheologically strong mafic crust and lithospheric mantle, the lithosphere beneath the plateau is composed of relatively weak felsic crust underlain by the hot asthenosphere (Fig. 24.4 at 9.5 Myr). It can be expected that the overall strength of such lithosphere would be considerably lower than prior to delamination, and that this 'post-delaminational weakness' would be manifested by a decrease in the compression force required to drive crustal shortening at the constant rate. However, this is not observed in Fig. 24.3b (solid and long-dashed lines). Instead, forces demonstrate a

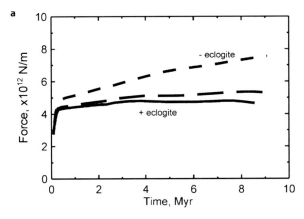

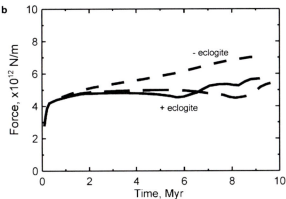

**Fig. 24.3.** Evolution of the driving force, i.e., force, that should be applied to the right side of the model box in order to drive tectonic shortening at a constant rate of 1 cm yr$^{-1}$. **a** Case with an initially 90 km thick lithosphere under the plateau (Fig. 24.2). *Short-dashed line* corresponds to the reference case without eclogitization (Fig. 24.2a). Pure-shear crustal thickening produces steadily growing plateau that, in turn, increases resistance to shortening. *Solid and long-dashed lines* demonstrate the effect of ongoing eclogitization working against increase of topographic elevation (max. density of eclogite is 3.55 g cm$^{-3}$ for the solid line and 3.45 g cm$^{-3}$ for the *long-dashed line*). **b** Case with an initially 70 km-thick lithosphere. After lower-crustal delamination (at about 6 Myr for higher-density eclogite see also Fig. 24.4) or at 8 Myr for lower-density eclogite) further thickening of the crust results in growing resistance similar to the reference case, as demonstrated by the changing slope of the force versus time curve

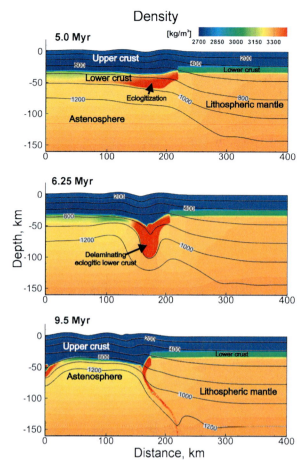

**Fig. 24.4.** Time-series of density plots illustrating progressive eclogitization and delamination of the thickened plateau crust. The initial lithospheric thickness under the plateau is 70 km. Higher crustal temperature shifts eclogitization to greater depths, but, otherwise, makes delamination easier. The driving force for this case is shown in Fig. 24.3b. Note the extremely high temperatures just below the felsic crust of the plateau after delamination

growing resistance to shortening, as noted earlier. The reason for this behavior is that the viscous resistance of ductile lower crust and upper mantle is only minor compared to the gravitational load of the high plateau. The effect of another possible mechanism of shortening resistance, i.e., the strength of the brittle upper crust, is considered in the following section.

### 24.3.2 Intracrustal Convection

Lithospheric delamination under the plateau and associated asthenospheric upwelling (see Fig. 24.4 at 9.5 Myr) might have resulted in the establishment of an extremely high, mantle heat flux at the base of the plateau crust. As the crust had probably lost much of its most refractive mafic component at this time, the high heat flux from the mantle might have triggered extensive melting at the base of the crust followed by melt segregation and transportation to the surface. This thermal impact is probably recorded by the late Miocene and Pliocene peak of ignimbrite activity in the Altiplano-Puna volcanic complex (APVC; de Silva 1989; see also Trumbull et al. 2006, Chap. 2 of this volume). Moreover, an extensive seismic low-velocity zone (Yuan et al. 2000) in the middle crust below the plateau, and other geophysical observations such as anomalous electrical conductivity (Brasse et al. 2002) and high surface heat flow (Springer and Foerster 1998), indicate that partial melting in the mid-crust is still present under the ignimbrite province today (see comprehensive discussion in Schilling et al. 2006, Chap. 22 of this volume).

The available data place tight temporal constraints on the crustal heating rate. Thus, the age of ignimbrite activity in the APVC indicates that the anomalous thermal structure of the crust in this region was established by at least 10 Ma (Coira et al. 1982; Ort et al. 1996; Trumbull et al. 2006, Chap. 2 of this volume).

The timing of tectonic shortening is less definitive. Plate convergence and subduction have been ongoing since the Triassic, but compressional deformation in the Central Andes only began in the Eocene (Lamb et al. 1997; Oncken et al. 2006, Chap. 1 of this volume) with minor deformation at the margins of the future plateau. Deformation in the plateau region accelerated at around 30–25 Ma (Allmendinger et al. 1997; Oncken et al. 2006, Chap. 1 of this volume), probably mostly due to the accelerating westward drift of the South American Plate (Sobolev and Babeyko 2005; Sobolev et al. 2006, Chap. 25 of this volume).

Hence, according to these constraints, heating and partial melting of the middle crust were achieved within some 10 to 20 million years after the initiation of intensive tectonic shortening. It is worth noting that pure-shear thickening is usually accompanied by absolute cooling at a given crustal depth owing to deepening of the isotherms.

Babeyko et al. (2002) employed thermo-mechanical numerical modeling to study all the possible mechanisms that could be responsible for heating the tectonically shortened crust. They considered the following heat sources: (*i*) enhanced radiogenic heating owing to crustal thickening, (*ii*) shear heating caused by active deformation, (*iii*) heating by intrusion and cooling of arc andesites, and (*iv*) heating as a result of increased mantle heat flow caused by crust/mantle delamination and asthenospheric upwelling. Babeyko et al. (2002) found that increased mantle heat flow (> 60 mW m$^{-2}$) coupled with melting of the lower felsic crust and bulk intracrustal convection is the only mechanism capable of explaining the observed heating rates. They also formulated additional conditions for the intracrustal convection: weak, quartz-dominated crustal rheology and active tectonic shortening.

In this paper, we illustrate another aspect of the effects of crustal convection: its influence on the bulk strength of the plateau crust. Preliminary qualitative con-

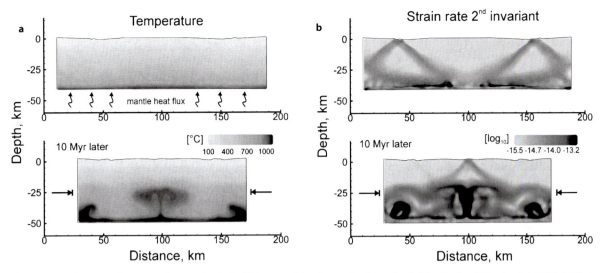

**Fig. 24.5. a** Development of intracrustal convection in the felsic crust of the plateau after lithospheric delamination (Babeyko et al. 2002). **b** Effect of the intracrustal convection on localization of the upper-crustal deformation. The growing plume effectively thins the brittle layer above it

siderations imply that overall crustal heating should decrease crustal viscosity, as well as shift the brittle-ductile transition zone towards the surface. Both these factors decrease the effective strength of the plateau in response to tectonic shortening (see also Schilling et al. 2006, Chap. 22 of this volume).

Figure 24.5a illustrates the development of the intracrustal convection under the plateau resulting from a strong increase in the heat flux after lithospheric delamination. (For details of crustal convective models see Babeyko et al. 2002.) The accompanying strain-rate plot (Fig. 24.5b) shows the effect of intracrustal convection on the localization of upper-crustal deformation. Intracrustal convection produces a laterally inhomogeneous thermal pattern with alternating warm zones of upwelling and cold zones of downwelling.

These warm and cold regions correspond to varying thickness of the brittle crust and these lateral variations in the upper-crustal strength possibly influence the deformation pattern at the surface (Fig. 24.5b). Initially, crustal shortening in the uppermost crust was accommodated by the two fault zones at the sides of the model box. Later, upwelling of a hot diapir decreased the thickness of the brittle crust just above it and the effective strength of the upper crust became laterally highly inhomogeneous. The minimum effective strength occurred above the diapir and a new fault zone was activated.

The overall effect of intracrustal convection on the dynamics of the back-arc system is illustrated in Fig. 24.6. This figure compares two numerical models: with and without intracrustal convection. In both models, lithospheric delamination takes place about six million years after shortening begins (see previous section, Figs. 24.3b and 24.4). Figure 24.6a presents the second invariant of the stress tensor

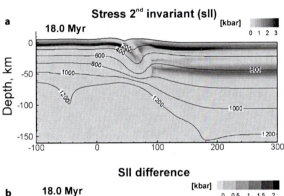

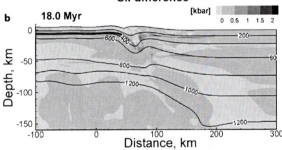

**Fig. 24.6.** Effect of the intracrustal convection on the brittle strength of the plateau upper crust. **a** Second invariant of the stress tensor ($s_{II}$) for the model without convection. Note that mechanical resistance to shortening is restricted to the brittle layer in the uppermost crust of the plateau. *Solid lines* represent isotherms. **b** Difference between $s_{II}$ for the two models: $s_{II}$ without convection minus $s_{II}$ with convection. *Solid lines* are isotherms for the model with convection

$$s_{II} = \sqrt{(s_{xx} - s_{yy})^2 + 4s_{xy}^2}$$

for the model without convection ten million years after the delamination event. In the case of a convective crust, the brittle-ductile transition is shifted a few kilometers

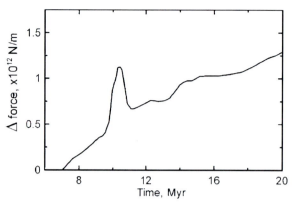

**Fig. 24.7.** Progressive weakening effect of the intracrustal convection as illustrated by the difference in driving forces for the models without and with convection. The total effect of the convection is about $1.2 \times 10^{12}$ N m$^{-1}$, which is about half the effect of eclogitization. Note that the peak at about 10–11 Myr is not directly related to the intracrustal convection, but is explained by the later development of a thrust fault near the plateau margin in one of the models

towards the surface owing to the uplifted isotherms (compare isotherms in Fig. 24.6a,b). The stress difference between these two cases is presented in Fig. 24.6b. As expected, this difference is maximal in the lower part of the brittle upper crust where it reaches 2 kbars and more. The isotherms in Fig. 24.6b correspond to the model with a convective crust. Note the quasi-isothermal lower crust under the plateau that manifests the thermal effect of intracrustal convection. At the same time, convection-accelerated heat transport cools the asthenosphere more effectively (compare the position of the 1200 °C isotherm).

According to Fig. 24.6b, the brittle strength of the convecting felsic crust should be lower than that of the case without convection. Clearly, this difference should increase with time; as convection develops, the brittle-ductile transition progressively moves upwards. This trend is indeed observed in the models, and Fig. 24.7 demonstrates the excessive compressive force in the non-convective model relative to the model with intracrustal convection. As shown in Fig. 24.7, the total effect of the convection is about $1.2 \times 10^{12}$ N m$^{-1}$; this is about half of the effect of the eclogitization discussed in the previous section.

### 24.3.3 Mechanical Failure of the Uppermost Crust in the Altiplano Foreland

The shortening history in the Central Andes (for review see Oncken et al. 2006, Chap. 1 of this volume) illustrates another possible weakening mechanism that may be related to the mechanical failure of the uppermost crust in the foreland. The magnitude and style of tectonic shortening in the Central Andes varies along strike. Allmendinger and Gubbels (1996) subdivided the Central Andean Plateau into two distinct segments, with the boundary between these segments lying at 23–24° S. North of 23° S, in the Altiplano domain, the foreland east of the plateau comprises a well-developed, thin-skinned, Subandean fold-and-thrust belt (Baby et al. 1992; Kley 1993) with total shortening reaching 150 km (compilation by Kley and Monaldi 1998; Oncken et al. 2006, Chap. 1 of this volume). In contrast, south of 24° S, in the Puna, the foreland lacks a thin-skinned thrust belt. Instead, foreland structures demonstrate thick-skinned tectonics with the entire upper crust involved in the deformation (the Santa Barbara System; Allmendinger et al. 1989; Marrett et al. 1994). The total shortening in the Puna foreland is only 50 km (Kley and Monaldi 1998).

Allmendinger and Gubbels (1996) noticed the striking correlation between the shortening style in the Central Andes and the distribution of Paleozoic sedimentary basins: a Cambrian to Carboniferous section of Southern Bolivia, several kilometers thick, pinches out south of 23° S. Their interpretation was that the shallower basement of the Santa Barbara System did not have suitable geometry for forming a thin-skinned thrust belt. Conversely, farther north in the Altiplano, the mechanical failure of thick foreland sediments was responsible for the thin-skinned thrusting.

Despite the presence of Paleozoic sediments in the Altiplano foreland during the entire Andean orogeny, thin-skinned foreland thrusting started only in the late Miocene. The timing of the deformation across 21° S (Allmendinger and Gubbels 1996; Elger et al. 2005; Oncken et al. 2006, Chap. 1 of this volume) demonstrates two contrasting shortening stages that have almost no overlap. The earlier stage, occurring up to the late Miocene, can be characterized as pure-shear shortening. In this shortening mode, deformation in both the upper and lower crust occurs in the same vertical rock column. During this earlier stage, deformation was restricted to the Altiplano and Cordilleras and occurred with a quasi-random, space/time surface pattern.

In the late Miocene (~13–9 Ma), however, tectonic shortening ceased almost completely in the Altiplano and migrated eastwards into the foreland and started to form a broad, thin-skinned, Subandean thrust belt. Simultaneously, the uplift rate of the Altiplano Plateau increased (Gregory-Wodzicki 2000), suggesting intensified lower-crustal thickening under the plateau. The global back-arc shortening mode switched to simple shear, where the loci of upper- and lower-crustal shortening are separated laterally. Based on the tomography images of Myers et al. (1998), Beck and Zandt (2002) interpreted this shortening mode on a regional scale and concluded that the margin of the Brazilian shield shallowly underthrusted beneath the Eastern Cordillera.

Recently, Babeyko and Sobolev (2005) explained this switch between pure- and simple-shear shortening along 21° S as resulting from rapid mechanical failure of the foreland sediments in the late Miocene (drop of cohesion and friction angle). Thermo-mechanical modeling has shown that shortening always occurs in a pure-shear mode unless the uppermost crust of the foreland becomes considerably weaker than the upper crust of the plateau.

Figure 24.8a,b shows the two end-members corresponding to the cases of strong and weak sediments, respectively. In the case of strong sediments, the cohesion and friction angle in the sedimentary layer are the same as in the crust of the plateau (20 MPa and 30°). The model (Fig. 24.8a) demonstrates the pure-shear behavior of a perfect 'vise' model (using the terminology of Ellis et al. 1998) with homogeneous thickening of the thermally weakened crust and mantle under the plateau and with almost no deformation in the cold and stiff foreland. Continued shortening and accompanying plateau uprising do not lead to the lateral growth of the plateau by gravitational spreading: the upper crust of the indenting shield is too cold and, hence, too strong to deform significantly over a ten-million-year period. The corresponding driving force (Fig. 24.9, *dashed line*) demonstrates the growing resistance of the rising plateau to shortening.

In the case of the second end-member (Fig. 24.8b), with very weak sediments in the foreland, the cohesion in the sedimentary layer is set to 1 MPa and the friction angle to 3°, i.e. ten times less than normally accepted values. The shortening mode results in a prominent, simple-shear, thin-skinned pattern. Underthrusting of the shield is accompanied by rapid eastward-propagation of the thin-skinned thrust belt within the sedimentary layer. A shallowly dipping decollement is located between 8 and 12–14 km depth, in good agreement with observations (Allmendinger and Gubbels 1996). It is worth noting that in this model, the deformation front starts to propagate into the Subandean prior to the major uplift of the plateau, in accordance with paleobotanical and geomorphological records (Gregory-Wodzicki 2000).

Also in this model, the shallow underthrusting of the Brazilian shield drives material into the lower crust of the plateau, intensifying its uplift. The domain of the thin-skinned deformation in the foreland is linked to the thickened lower crust beneath the plateau by a low-angle megathrust. The upper crust of the plateau remains virtually undeformed by shortening, and the driving force corresponding to this case is shown in Fig. 24.9 (*solid line*). This force is considerably lower (by about $3 \times 10^{12}$ N m$^{-1}$) than that for the pure-shear plateau inflation (dashed line), despite the fact that plateau uplift continues. Thus, mechanical failure of the uppermost crust of the foreland, switching the system from pure- to simple-shear mode, has an overall weakening effect on a back-arc scale.

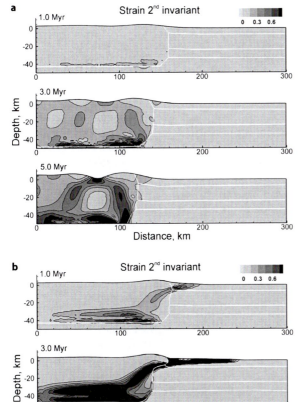

**Fig. 24.8.** Time-series showing finite strain accumulation for (**a**) pure shear and (**b**) simple shear end-member modes of back-arc tectonic shortening (only the uppermost 50 km of the model is shown). *White solid lines* contour the current positions of lithological units (see model set-up in Fig. 24.1b). The weak sedimentary layer is responsible for the formation of the thrust-and-fold foreland belt in the Subandean north of 23° S

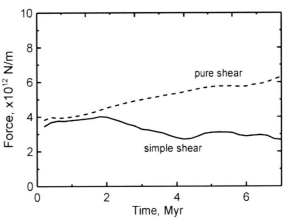

**Fig. 24.9.** Force necessary for driving shortening in pure-shear (*dashed line*) or simple-shear (*solid line*) mode. The mechanical failure of the foreland sediments (drop of cohesion and friction coefficient) switches the system from pure to simple shear and effectively weakens the whole back-arc

We ran a series of additional numerical models in which we varied the cohesion and friction angle of the foreland sediments over the respective ranges of 1 to 20 MPa and 3 to 30°. The aim of these experiments was to study deformation modes that are intermediate between the two end-members (Fig. 24.8a,b).

Figure 24.10 presents a full spectrum of these intermediate modes, showing the finite strain after five million years of shortening at a constant rate of 1 cm yr$^{-1}$. Figure 24.10b demonstrates that even a large drop (from the normal 30° to 3°) in the friction coefficient alone is not enough for the development of a broad, thin-skinned thrust belt; material cohesion (parameter $C_h$ in Eq. 24.10) should also be reduced. The same is true for a drop in the material cohesion alone (Fig. 24.10c). Thus, mechanical failure of the foreland sediments, responsible for the late Miocene 'jump' of the shortening deformation from the plateau region into the foreland, should involve prominent reduction of both material parameters. A series of low-cohesion experiments (Fig. 24.10c–g) provides an upper-boundary estimate of the friction angle that is compatible with the development of a broad thrust-and-fold belt in the foreland. This angle must be less than 10°, i.e., less than a third of its 'normal' value.

The physical nature of the mechanical weakening of the foreland sediments in the late Miocene remains unclear. At least three hypothetical mechanisms can be suggested that attribute the mechanical weakening to an increase in sedimentary pore pressure. The first hypothesis is based on an idea by Cobbold and Castro (1997) who suggest that the increase in pore pressure in the thick and bitumen-rich Paleozoic sediments of the Subandean Ranges might be because of the rapid generation of hydrocarbons. We speculate that the initial uplift of the Eastern Cordillera might have led to the burial of the foreland sediments by eroded and transported material. This, in turn, raised the temperature in the sedimentary layer and at about 10 Ma, maturation conditions were reached at the bottom of the layer.

Another possible mechanism for explaining increased sedimentary pore pressure might be the local increase in the precipitation rate caused by the uplift of the Eastern Cordillera and the establishment of the topographic barrier in the late Miocene. The increase in precipitation on the eastern flank of the plateau combined with the absolute plateau uplift might have increased the lateral gradient of the sedimentary pore pressure and produced an effective eastwardly propagating wave of fluid overpressure in the sediments. The third possibility is that the pore pressure in the sediments was elevated because of compaction associated with high tectonic compression stress.

Without regard for the origin of the mechanical weakening of the sediment, it is clear that this process is consistent with the observed thin-skinned tectonic pattern and it also provides a plausible explanation of the changing deformation styles in the Central Andes at ~10 Ma. In terms of the reduction of effective lithospheric stress, this mechanism is the most efficient from the mechanisms considered in the previous sections.

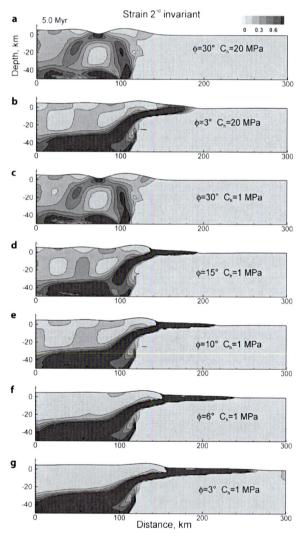

**Fig. 24.10.** From pure to simple shear. Finite strain after five million years of shortening at a constant rate of 1 cm yr$^{-1}$. The models (**a**) to (**g**) differ in the mechanical properties of the foreland sedimentary layer (see model set-up in Fig. 24.1b). The broad, foreland thrust-and-fold belt requires low values for both the material cohesion and friction angle

### 24.3.4 Erosion-Induced Weakening at the Plateau Rim

Strong north-south variation in the annual precipitation on the eastern flank of the Altiplano Plateau resulted in significant differences in erosion rates in the region. Thus, up to 8 km of overburden has been removed along the north-eastern humid flank of the Eastern Cordillera in the La Paz area since about 10–15 Ma (Benjamin et al.

1987, Masek et al. 1994). This implies a denudation rate of about 0.6–0.8 mm yr$^{-1}$. In contrast, in the semi-arid regions south of the Santa Cruz elbow (south of 18° S) erosion rates are two to four times smaller (Masek et al. 1994).

According to Masek et al. (1994), the morphology of the eastern flank of the Altiplano Plateau is also, to a large extent, controlled by the local climate. The steep frontal slope and high peaks of the north-eastern Altiplano largely reflect the high orographic precipitation and erosion rates in this region, whereas the more gentle topography of the semi-arid central Altiplano (wide Subandean thrust belt) reflects the minor effect of erosion on tectonic landform. The strong impact that climate-controlled erosion has on the morphology of the orogen probably means that erosion also has the potential to control the dynamics of orogenesis.

In this section, we test the idea that high erosion rates at the plateau rim can influence the dynamics of the plateau-foreland system. To do this, we have implemented the model of erosion processes into our numerical thermo-mechanical code LAPEX-2D as described below.

Although many erosion processes operate during tectonic uplift, only a few are important. Hovius et al. (1997) showed, using an example from the Southern Alps of New Zealand, that two processes could account for all mass eroded and removed from the system: fluvial incision as long-range transport and landslides as short-range transport. Fluvial incision dissects actively uplifting regions and creates deep mountain valleys. Landslides move material from hill slopes to the valley floors. Later, rivers transport this material out of the system. Here, fluvial incision is the controlling process: as rivers incise valleys, local relief increases, thereby accelerating the rate of hillside erosion.

The erosion rate of bedrock river channels can be described as a simple function of drainage area and local slope (Howard et al. 1994):

$$\frac{dh}{dt} \sim A^m S^n \quad (24.12)$$

where $A$ is the drainage area, $S$ is the slope of the local channel bed, and $m$ and $n$ are constants whose ratio depends on the underlying physics (Whipple and Tucker 1999). Following many authors (e.g., Tucker and Singerland 1994; Willett 1999), we set $m = n = 1$, and, for a one-dimensional case, the above relation takes the form:

$$\frac{dh}{dt} = K_f |x - x_r| \frac{dh}{dx} \quad (24.13)$$

Here, $dh/dx$ is the local slope, $|x - x_r|$ is the absolute distance from the drainage divide, and $K_f$ is the erosion coefficient for fluvial incision. Hillside processes (landslides) can be described by the normal diffusion equation:

$$\frac{dh}{dt} = K_l \frac{d^2h}{dx^2} \quad (24.14)$$

The implementation of the above erosion equations into our numerical thermo-mechanical code allowed us to study the coupled processes of tectonic deformation and erosion at the plateau margin, with particular interest in the dynamics of the system. Similarly to previous sections, we used our general back-arc set-up as a starting model (Fig. 24.1). Fluvial erosion coefficient ($K_f$) was the only variable model parameter. The initial lithospheric thickness under the plateau was 70 km, the shortening rate was fixed at 1 cm yr$^{-1}$, and the foreland sediments had the same frictional strength as the felsic crust of the plateau (Table 24.1). Since fluvial incision is the controlling erosion process, the diffusion coefficient $K_l$ (Eq. 24.14) was varied in accord with variations of $K_f$.

Figure 24.11 shows the distribution of finite strain (second invariant of the strain tensor) after 70 km of shortening in the plateau-foreland system for different erosion rates. Figure 24.11a corresponds to the case without fluvial erosion and shows the same 'vise' model of plateau inflation as discussed in previous sections. As the erosion rate increases (Fig. 24.11b–d), the distribution of deformation in the system changes. For convenience, the coefficients of the fluvial erosion are given relative to a

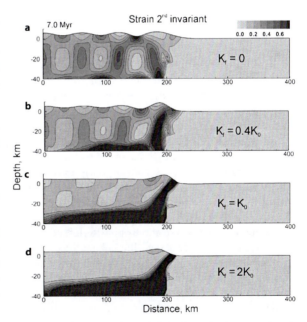

**Fig. 24.11.** Finite strain after 70 km of shortening for the series of models with increasing fluvial erosion. Coefficients of fluvial erosion ($K_f$) are given relative to the selected value of $K_0 = 7.5 \times 10^{-7}$ yr$^{-1}$. **a** Case of no erosion, corresponding to pure-shear shortening with deformation distributed inside the growing plateau. **b** Case of $K_f = 0.4K_0$ corresponding to a maximum denudation depth ($D_{max}$) of 9 km. **c** Case of $K_f = K_0$ and $D_{max} = 13$ km. Deformation is largely accommodated by the exhumation fault at the plateau rim. **d** Case of $K_f = 2K_0$ and $D_{max} = 17$ km. Almost nothing is transmitted into the plateau region

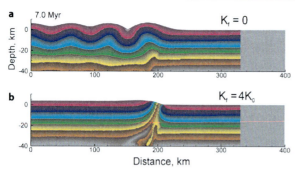
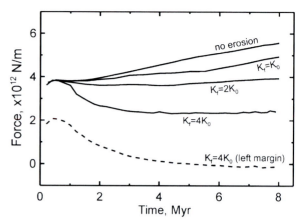

**Fig. 24.12.** Deformation and exhumation pattern after 70 km of shortening as demonstrated by the distortion of initially horizontal layers of particle tracers. Gray area at the right corresponds to markers added to the model during shortening to preserve total width. **a** No fluvial erosion. **b** High fluvial erosion rate ($K_f = 4K_0$). This stimulates effective exhumation at the plateau margin (max. denudation depth, $D_{max} = 35$ km). The plateau material is effectively squeezed along this single fault and later removed from the system by erosion. The plateau does not grow with time

**Fig. 24.13.** Shortening driving forces for progressive increasing values of fluvial erosion ($K_f$ from 0 to $4K_0$). High fluvial erosion and exhumation result in an effective weakening of the lithosphere in response to tectonic shortening. This effect is, however, insignificant for the Bolivian Andes ($K_f < K_0$)

selected value $K_0$ ($K_0 = 7.5 \times 10^{-7}$ yr$^{-1}$). Figure 24.11 should be analyzed together with Fig. 24.12, which shows the distribution of numerical particle tracers initially aligned in horizontal rows. The distortion of the initial tracer rows indicates the deformation and exhumation pattern in the model.

High fluvial erosion rates stimulate very effective exhumation at the plateau margin (Fig. 24.12b). In this end-member case, the exhumation rate along the single super-fault (Fig. 24.11d) is large enough to compensate for the tectonically driven influx of material into the plateau. As a result, the plateau does not grow over time and, consequently, its resistance to shortening also does not increase (Fig. 24.13, lines with $K_f > K_0$). Plateau material is being effectively squeezed along this single fault (Figs. 24.11d and 24.12b) and then removed from the system by erosion. With high erosion rates, the super-fault at the plateau margin accommodates the whole shortening deformation, i.e., no deformation is being transmitted into the upper crust of the plateau (Fig. 24.11d). This is further illustrated by the horizontal force measured at the opposite, left margin of the plateau (Fig. 24.13, dashed line). This force is close to zero or even slightly negative, which means that the plateau itself could even be extending despite the continuous bulk shortening in the back-arc.

With lower fluvial erosion rates, shortening deformation is partly distributed among the plateau and the main exhumation fault (Fig. 24.11b,c).

As one can see in Fig. 24.13, the reduction in effective lithospheric strength owing to erosion and exhumation can be very large if the fluvial erosion rate is high ($K_f = 2K_0$). In our numerical models, such high fluvial erosion rates correspond to maximum denudation depths of more than 15 km and denudation rates higher than 1.5–2 mm yr$^{-1}$.

This is, however, not the case for the Central Andes, where a maximum of 8 km has been eroded since 10–15 Ma (Benjamin et al. 1987, Masek et al. 1994). We speculate that this strong weakening effect might be expected for Himalayan tectonics, characterized by extreme exhumation rates (Pik et al. 2005) and a denudation depth of more than 30 km, as modeled by Beaumont et al. (2000). If, instead, the denudation rate is close to the case of the Bolivian Andes (0.6–0.8 mm yr$^{-1}$, maximum denudation depth: 8 km), the lithospheric strength is reduced insignificantly. The effect is even smaller than shown by the line labeled "$K_f = K_0$" in Fig. 24.13, which corresponds to 13 km of maximum denudation. In the case of low fluvial erosion (typical for the dry Central Andes), the erosion effect on the lithospheric strength is negligible.

From this, we conclude that although the high fluvial erosion focused at the plateau margin has a strong weakening potential for the overall dynamics of the plateau-foreland system, erosion is likely to be relatively unimportant for the Central Andean Plateau, where denudation rates are moderate to low.

## 24.4 Summary and Discussion

We employed 2D numerical thermo-mechanical modeling to study the dynamics of the plateau-foreland system with application to the Central Andean back-arc. Our model back-arc consists of two different regions: a thermally activated and, hence, mechanically weaker plateau; and a cold, stiff foreland that is indenting the plateau from the east at a constant rate (Brazilian shield). In such a bimodal back-arc system, shortening normally occurs in a pure shear, 'vise' type mode (following the terminology of Ellis et al. 1998). The plateau rises as a result of tectonic inflation, and the increasing gravitational potential provides growing resistance to shortening. Horizontal spreading of the plateau is insignificant over a ten-million-year timescale; the foreland is mechanically too strong and indents the plateau without any significant deformation.

The growing potential energy of the rising plateau is the major factor resisting external shortening. Additional resistance is provided by the brittle strength of the upper crust and the rate-dependent, viscous resistance of the system. The overall resistance to shortening is not larger than $6-8 \times 10^{12}$ N m$^{-1}$.

The two groups of processes control shortening rate in the Central Andean back-arc. One group is related to changes of external forces. Processes of that group are considered in the companion paper by Sobolev et al. (2006, Chap. 25 of this volume). They include accelerated westward drift of the South American Plate and changing coupling rate between the South America and Nazca Plates. These processes are likely responsible for the overall increase of the shortening rate during the Cenozoic (acceleration of the westwards drift) and may also contribute to the fluctuations of the shortening rate (Sobolev et al. 2006, Chap. 25 of this volume).

The second group of processes control the strength of the upper-plate lithosphere. We show that several thermomechanical processes in the Central Andean back-arc have a gross weakening effect and may be at least partially responsible for episodes when the shortening rate accelerated in the past. These processes are (ordered by their significance for the Central Andes): (*1*) mechanical failure of the foreland sediments, (*2*) on-going eclogitization of the lower mafic crust beneath the plateau, (*3*) intracrustal convection in the felsic crust of the plateau and (*4*) fluvial erosion at the plateau margin.

Our modeling shows that the late Miocene migration of the shortening deformation from the Altiplano into the foreland can be explained by mechanical failure of the thick Paleozoic sediments in the foreland. This failure implies a drastic reduction of both sediment cohesion (almost to zero values) and the angle of internal friction (to at least a third of the normal value of 30°). The sediment failure has a gross weakening effect on the whole back-arc. The driving force for the shortening could be reduced by $3 \times 10^{12}$ N m$^{-1}$ by this process, possibly explaining the pulse in the bulk shortening rate during the late Miocene.

Ongoing eclogitization of the mafic crust of the plateau increases the average density of the crust and works against an increase in topographic elevation on the tectonically inflated plateau. This results in a reduction of the effective lithospheric strength by ~$2 \times 10^{12}$ N m$^{-1}$. After the eclogitic crust is removed by delamination, the plateau starts to grow, increasing its resistance to shortening.

Lower crustal delamination brings hot asthenosphere directly to the base of the felsic crust, creating the condition that the latter may partially melt and undergo convection, thus providing rapid heating and melting at the mid-crustal level (Babeyko et al. 2002). Intracrustal convection reduces the brittle strength of the crust by another $1 \times 10^{12}$ N m$^{-1}$ and convective thinning of the brittle, uppermost crust may result in the redistribution of the surface deformation pattern and the large-scale eruption of silicic ignimbrites (Altiplano-Puna volcanic complex at 23–24° S).

High fluvial erosion at the plateau margin may compensate for the tectonically driven influx of material into the plateau and, thus, effectively prevent further uplift of the plateau. This dramatically reduces the effective lithospheric strength. If this occurs, the resistance of the back-arc to shortening will not increase with time. Moreover, the whole shortening deformation might be accommodated at the single zone of exhumation, while the converging plateau and foreland remain almost undeformed. However, this effect becomes important only with very high denudation rates (> 1.5–2 mm yr$^{-1}$), typical for monsoonal areas of the Himalayas. With the moderate erosion rates of the northern Altiplano (< 0.8 mm yr$^{-1}$), and the even lower erosion rates typical for the rest of the Central Andes, erosion-induced weakening is insignificant.

For better illustration of the relative hierarchy of the aforementioned weakening processes we summarize our modeling results in Fig. 24.14. Figure 24.14a shows the loci of operation for the four processes considered and Fig. 24.14b presents a sketch of their relative importance expressed as a percentage of effective lithospheric strength (overall resistance to shortening) relative to the reference 'vise' plateau model (pure-shear plateau thickening without gabbro-to-eclogite transformation, intracrustal convection and erosion). Because different processes take place at different stages of the orogen evolution, the time axis of Fig. 24.14b presents the relative time ($t^*$), i.e., duration of the correspondent process after it starts. For example, intracrustal convection starts after the completion of lithospheric delamination, which is driven by the gabbro-to-eclogite transformation. Hence, the corresponding $t^*$ for the two processes lie about 10 Myr apart.

Figure 24.14b presents our estimations of the maximal weakening effects of the four processes for the case of the Central Andes. Please note, that these maximal values cannot be directly combined with each other. Thus, the maximal weakening effect of on-going eclogitization (magenta line, see also Fig. 24.3a) corresponds to the case of initially thicker, 90-km lithosphere, which failed to delaminate in the time scale of our modeling, 10 Myr (Fig. 24.2b), and, therefore, did not give rise to intracrustal convection. In turn, the model with successful lithospheric delamination (Figs. 24.3b and 24.4) demonstrates less weakening followed by effective hardening after the

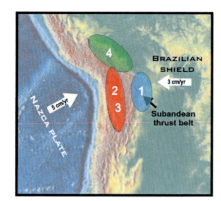

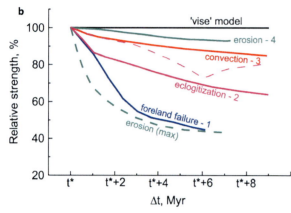

**Fig. 24.14. a** Schematic representation of loci of the four weakening processes operating in the back-arc. The processes are numbered according to their effective weakening potential: *1* – mechanical sediment failure in the foreland, which gave rise to the formation of the Subandean thrust belt; *2* – eclogitization of the thickened lower crust under the Altiplano and Puna; *3* – intracrustal convection after the lithospheric delamination; *4* – high erosion and denudation at the north-east flank of the northern Altiplano. **b** Sketch showing our estimations of the *maximal* weakening effects of the four processes relative to the reference 'vise' plateau model. Also shown in dashed lines are the effect of eclogitization followed by lithospheric delamination and effective hardening (*magenta dashed line*), and the effect of very high erosion (denudation rates > 1.5–2 mm yr$^{-1}$), which is, however, not relevant for the Central Andes (*green dashed line*)

delamination event (magenta dashed line). For reference, we also include the model with monsoon-type erosion rates (dashed green line, see Fig. 24.13, model with $K_f = 4 K_0$); however, such high erosion is not relevant for the Central Andes (compare with solid green line).

As a final remark, we refer to the companion paper by Sobolev et al. (2006, Chap. 25 of this volume) which shows that the fast and accelerating westward drift of the South American Plate during the late Cenozoic, together with high interplate coupling in the Central Andes, established strong compression at the leading edge of the South American Plate. This created the conditions needed for intensive tectonic shortening of the upper plate. However, without the upper-plate weakening processes considered in this paper, there would be less than half of the tectonic shortening that is predicted when these processes are included (Sobolev and Babeyko 2005; Sobolev et al. 2006, Chap. 25 of this volume). This means that mechanical weakening of the overriding plate during tectonic shortening is among the first-order controls on the Andean orogeny.

## Acknowledgments

This work was supported by the collaborative research program SFB-267 "Deformation Processes in the Andes" and by the Alexander von Humboldt Foundation Research Fellowship to A.Y. B. We thank Nina Kukowski and Ron Hackney for useful comments, and Allison Britt for her help in editing the manuscript. Supercomputer facilities were provided by Central Institute for Applied Mathematics (Research Centre Juelich, Germany) and Potsdam Institute for Climate Impact Research.

## References

Allmendinger RW, Gubbels T (1996) Pure and simple shear plateau uplift, Altiplano-Puna, Argentina and Bolivia. Tectonophysics 259:1–13

Allmendinger RW, Strecker MA, Eremchuk JE, Francis PW (1989) Neotectonic deformation of the Southern Puna plateau, northwestern Argentina. J S Am Earth Sci 2:111–130

Allmendinger RW, Jordan TE, Mahlburg Kay S, Isacks BL (1997) The evolution of the Altiplano-Puna Plateau of the Central Andes. Ann Rev Earth Planet Sci 25:139–174

Babeyko AY, Sobolev SV (2005) Quantifying different modes of the Late Cenozoic shortening in the Central Andes. Geology 33:621–624

Babeyko AY, Sobolev SV, Trumbull RB, Oncken O, Lavier LL (2002) Numerical models of crustal scale convection and partial melting beneath the Altiplano-Puna plateau. Earth Planet Sci Lett 199:373–388

Baby P, Hérail G, Salinas R, Sempere T (1992) Geometry and kinematic evolution of passive roof duplexes deduced from cross section balancing: example from the foreland thrust system of the southern Bolivian Subandean zone. Tectonics 11(3):523–536

Beaumont C, Jamieson RA, Nguyen MH, Lee B (2000) Himalayan tectonics explained by extrusion of a low-viscosity crustal channel coupled to focused surface denudation. Nature 414:738–742

Beck SL, Zandt G (2002) The nature of orogenic crust in the Central Andes J Geophys Res 107: doi 10.1029/2000JB000124

Benjamin MT, Johnson NM, Naeser CW (1987) Recent rapid uplift in the Bolivian Andes: evidence from fission-track dating. Geology 15:680–683

Brasse H, Lezaeta P, Rath V, Schwalenberg K, Soyer W, Haak V (2002) The Bolivian Altiplano conductivity anomaly: J Geophys Res B107: doi 10.1029/2001JB000391

Cobbold PR, Castro L (1997) Evidence that hydrocarbon generation leads to fluid overpressures and source-rock detachments in thrust belts. In: VI Simposio Bolivariano, Resumenes, Asoc Colomb Geol Geofis Petrol, Bogota

Coira B, Davidson J, Mpodozis C, Ramos V (1982) Tectonic and magmatic evolution of the Andes of Northern Argentina and Chile. Earth Sci Rev 18:303–322

Cundall PA, Board M (1988) A microcomputer program for modelling large-strain plasticity problems. In: Swoboda G (ed) 6th International conference in numerical methods in geomechanics. AA Balkema, Brookfield VT, pp 2101–2108

De Silva SL (1989) Altiplano-Puna volcanic complex of the Central Andes. Geology 17:1102–1106

Doin MP, Henry P (2001) Subduction initiation and continental crust recycling: the roles of rheology and eclogitization. Tectonophysics 342:163–191

Elger K, Oncken O, Glodny J (2005) Plateau-style accumulation of deformation – the Southern Altiplano. Tectonics 24: doi 10.1029/2004TC001675

Ellis S, Beaumont C, Jamieson R, Quinlan G (1998) Continental collision including a weak zone: the vise model and its application to the Newfoundland Appalachians. Can J Earth Sci 35: 1323–1346

Gleason GC, Tullis J (1995) A flow law for dislocation creep of quartz aggregates determined with the molten salt cell. Tectonophysics 247:1–23

Gregory-Wodzicki KM (2000) Uplift history of the Central and Northern Andes: a review. Geol Soc Am Bull 112:1091–1105

Hirth G, Kohlstedt DL (1996) Water in the oceanic upper mantle: implications for rheology, melt extraction and the evolution of the lithosphere. Earth Planet Sci Lett 144:93–108

Hovius N, Stark CP, Allen PA (1997) Sediment flux from a mountain belt derived by landslide mapping. Geology 25:231–234

Howard AD, Dietrich WE, Seidl MA (1994) Modelling fluvial erosion on regional to continental scales. J Geophys Res B99: 13971–13986

Isacks BL (1988) Uplift of the Central Andean plateau and bending of the Bolivian Orocline. J Geophys Res 93:3211–3231

Jin ZM, Zhang J, Green HW II, Jin S (2001) Eclogite rheology: implications for subducted lithosphere. Geology 29:667–670

Jull M, Kelemen PB (2001) On the conditions for the lower crustal instability. J Geophys Res 106:6423–6446

Kameyama M, Yuen DA, Karato SI (1999) Thermal-mechanical effects of low-temperature plasticity (the Peierls mechanism) on the deformation of a viscoelastic shear zone. Earth Planet Sci Lett 168:159–172

Kay RW, Kay SM (1993) Delamination and delamination magmatism. Tectonophysics 219:177–189

Kay SM, Coira B, Viramonte J (1994) Young mafic back-arc volcanic rocks as indicators of continental lithospheric delamination beneath the Argentine Puna plateau, Central Andes. J Geophys Res B99:24323–24339

Kley J (1993) Der Übergang vom Subandin zur Ostkordillere in Südbolivien (21°15–22° S): Geologische Struktur und Kinematik. Ph.D. thesis, Freie Universität Berlin

Kley J, Monaldi CR (1998) Tectonic shortening and crustal thickness in the Central Andes: how good is the correlation? Geology 26:723–726

Lamb S, Hoke L, Kennan L, Dewey J (1997) Cenozoic evolution of the Central Andes in Bolivia and Northern Chile. In: Burg JP, Ford M (eds) Orogeny through time. Geol Soc Spec Pub 121, pp 237–264

Marrett RA, Allmendinger RW, Alonso RN, Drake RE (1994) Late Cenozoic tectonic evolution of the Puna plateau and adjacent foreland, northwestern Argentine Andes. J S Am Earth Sci 7:179–208

Masek JG, Isacks BL, Gubbels TL, Fielding EJ (1994) Erosion and tectonics at the margins of continental plateaus. J Geophys Res B99:13941–13956

Moresi LN, Dufour F, Muhlhaus HB (2003) A Lagrangian integration point finite element method for large deformation modelling of viscoelastic geomaterials. J Comp Phys 184:476–497

Myers SC, Beck S, Zandt G, Wallace T (1998) Lithospheric-scale structure across the Bolivian Andes from tomographic images of velocity and attenuation for P and S waves. J Geophys Res 103: 21233–21252

Oncken O, Hindle D, Kley J, Elger K, Victor P, Schemmann K (2006) Deformation of the central Andean upper plate system – facts, fiction, and constraints for plateau models. In: Oncken O, Chong G, Franz G, Giese P, Götze H-J, Ramos VA, Strecker MR, Wigger P (eds) The Andes – active subduction orogeny. Frontiers in Earth Science Series, Vol 1. Springer-Verlag, Berlin Heidelberg New York, pp 3–28, this volume

Ort MH, Coira BL, Mazzoni MM (1996) Generation of a crust-mantle magma mixture: magma sources and contamination at Cerro Panizos, Central Andes: Contrib. Mineral Petrol 123:308–322

Piepenbreier D, Stoeckhert B (2001) Plastic flow of omphacite in eclogites at temperatures below 500 °C – implications for interplate coupling in subduction zones. Int J Earth Sci 90:197–210

Pik R, France-Lanord C, Carignan J (2005) Extreme uplift and erosion rates in eastern Himalayas (Siang–Brahmaputra basin) revealed by detrital (U-Th)/He thermochronology. EGU General Assembly 2005, Geophys Res Abs 7

Poliakov AN, Cundall PA, Podladchikov YY, Lyakhovsky VA (1993) An explicit inertial method for the simulation of the viscoelastic flow: an evaluation of elastic effects on diapiric flow in two- and three-layers models. In: Stone DB, Runcorn SK (eds) Flow and creep in the Solar System: observations, modelling and theory. Kluwer Academic Publishers, pp 175–195

Schilling FR, Trumbull RB, Brasse H, Haberland C, Asch G, Bruhn D, Mai K, Haak V, Giese P, Muñoz M, Ramelow J, Rietbrock A, Ricaldi E, Vietor T (2006) Partial melting in the Central Andean crust: a review of geophysical, petrophysical, and petrologic evidence. In: Oncken O, Chong G, Franz G, Giese P, Götze H-J, Ramos VA, Strecker MR, Wigger P (eds) The Andes – active subduction orogeny. Frontiers in Earth Science Series, Vol 1. Springer-Verlag, Berlin Heidelberg New York, pp 459–474, this volume

Sobolev SV, Babeyko AY (1994) Modelling of mineralogical composition, density and elastic wave velocities in anhydrous magmatic rocks. Survey Geophys 15:515–544

Sobolev SV, Babeyko AY (2005) What drives orogeny in Central Andes? Geology 33:617–620

Sobolev SV, Babeyko AY, Koulakov I, Oncken O (2006) Mechanism of the Andean orogeny: insight from numerical modeling. In: Oncken O, Chong G, Franz G, Giese P, Götze H-J, Ramos VA, Strecker MR, Wigger P (eds) The Andes – active subduction orogeny. Frontiers in Earth Science Series, Vol 1. Springer-Verlag, Berlin Heidelberg New York, pp 513–536, this volume

Springer M, Foerster A (1998) Heat-flow density across the Central Andean subduction zone. Tectonophysics 291:123–139

Sulsky D, Zhou SJ, Schreyer HL (1995) Application of a particle-in-cell method to solid mechanics. Comp Phys Commun 87:236–252

Trumbull RB, Riller U, Oncken O, Scheuber E, Munier K, Hongn F (2006) The time-space distribution of Cenozoic volcanism in the South-Central Andes: a new data compilation and some tectonic implications. In: Oncken O, Chong G, Franz G, Giese P, Götze H-J, Ramos VA, Strecker MR, Wigger P (eds) The Andes – active subduction orogeny. Frontiers in Earth Science Series, Vol 1. Springer-Verlag, Berlin Heidelberg New York, pp 29–44, this volume

Tucker GE, Singerland RL (1994) Erosional dynamics, flexural isostasy, and long_lived escarpments: a numerical modelling study. J Geophys Res B99:12229–12243

Vermeer, PA, De Borst R (1984) Non-associated plasticity for soils, concrete and rock. Heron 29:3–64

Whipple KX, Tucker GE (1999) Dynamics of the stream-power river incision model: implications for height limits of mountain ranges, landscape response timescales, research needs. J Geophys Res B104:17661–17674

Wigger P, Baldzuhn S, Giese P, Heinsohn WD, Schmitz M, Araneda M, Martínez E, Ricaldi E, Viramonte A (1994) Variations of the crustal structure of the Southern Central Andes deduced from seismic refraction investigations. In: Reutter K-J, Scheuber E, Wigger P (eds) Tectonics of the Southern Central Andes. Springer Verlag, Berlin Heidelberg New York, pp 23-48

Wilks KR, Carter NL (1990) Rheology of some continental lower crustal rocks. Tectonophysics 182:57-77

Willett SD (1999) Orogeny and orography: the effect of erosion on the structure of mountain belts. J Geophys Res B104:28957-28981

Yuan X, Sobolev SV, Kind R, Oncken O, et al. (2000) Subduction and collision processes in the Central Andes constrained by converted seismic phases. Nature 408:958-961

Yuan X, Sobolev SV, Kind R (2002) New data on Moho topography in the Central Andes and their geodynamic implications. Earth Planet Sci Lett 199:389-402

Zandt G, Beck S, Ruppert S, Ammon C, Rock D, Minaya E, Wallace T, Silver P (1996) Anomalous crust of the Bolivian Altiplano, Central Andes: constraints from broadband regional seismic waveforms. Geophys Res Lett 23:1159-1162

# Mechanism of the Andean Orogeny: Insight from Numerical Modeling

Stephan V. Sobolev · Andrey Y. Babeyko · Ivan Koulakov · Onno Oncken

**Abstract.** The Andes were formed by Cenozoic tectonic shortening of the South American plate margin overriding the subducting Nazca Plate. Using coupled, thermo-mechanical, numerical modeling of the dynamic interaction between subducting and overriding plates, we searched for factors controlling the intensity of the tectonic shortening. From our modeling, constrained by geological and geophysical observations, we infer that the most important factor was fast and accelerating (from 2 to 3 cm yr$^{-1}$) westward drift of the South American Plate, whereas possible changes in the convergence rate were not as important. Other important factors are the crustal structure of the overriding plate and the shear coupling at the plate interface.

The model in which the South American Plate has a thick (40–45 km at 35 Ma) crust and relatively high friction coefficient (0.05) at the Nazca-South American plate interface generates more than 300 km of tectonic shortening over the past 35 million years and replicates well the crustal structure and evolution of the high Central Andes. However, modeling does not confirm that possible climate-controlled changes to the sedimentary trench-fill during the last 30 million years might have significantly influenced the upper-plate shortening rate. The model with initially thinner (less than 40 km) continental crust and a lower friction coefficient (less than 0.015) results in less than 40 km of shortening in the South American Plate, replicating the situation in the Southern Andes.

During upper-plate deformation, the processes that cause a reduction in lithospheric strength and an increase in interplate coupling are particularly important. The most significant of these processes appears to be: (1) delamination of the lower crust and mantle lithosphere, driven by gabbro-eclogite transformation in the thickening lower crust, and (2) mechanical failure of the foreland sediments. The modeling demonstrates that delaminating lithosphere interacts with subduction-zone corner flow, influencing both the rate of tectonic shortening and magmatic-arc productivity, and suggests an anti-correlation between these two parameters. Our model also predicts that the down-dip limit of the frictional coupling domain between the Nazca and South American Plates should be ~15–20 km deeper in the Southern Andes (south of 28° S) compared to the high Central Andes, which is consistent with GPS and seismological observations.

## 25.1 Introduction and Key Questions

The South American Plate is drifting westwards, at a rate that has increased from ~2 to 3 cm yr$^{-1}$ over the last 30 million years (Silver et al. 1998), over the Nazca Plate, which is subducting eastwards at about 5 cm yr$^{-1}$ (Fig. 25.1). The Andean mountain belt stretches along the entire western margin of the South American Plate where it overlies the subducting Nazca Plate. There is a dramatic

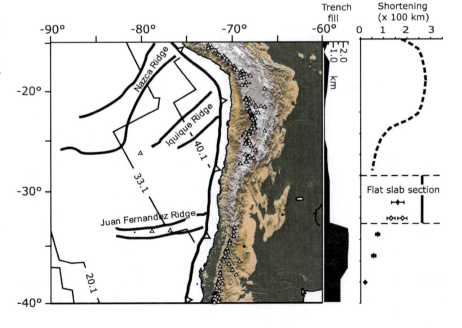

**Fig. 25.1.**
Surface topography of the Andes, thickness of sedimentary trench-fill and magnitude of tectonic shortening. The trench adjacent to the high Central Andes, where the tectonic shortening is the largest, has no sedimentary fill, which might increase friction in the subduction channel (Lamb and Davis 2003)

difference between the central part (~17–27° S) and the rest of the Andes. The Altiplano-Puna Plateau of the Central Andes is the second greatest plateau in the world (after Tibet) and has an average elevation of about 4 km and an area of more than 500 000 km$^2$ (Fig. 25.1). The plateau has resulted from up to 300 km of late Cenozoic, crustal shortening at the western edge of the South American Plate. This shortening generated unusually thick, hot and felsic, continental crust (Isacks 1988; Allmendinger and Gubbels 1996, Allmendinger et al. 1997; Lamb et al. 1997; Kley and Monaldi 1998; Giese et al. 1999; Lucassen et al. 2001; Yuan et al. 2002; Beck and Zandt 2002; Lamb and Davis 2003). In contrast to the Central Andes, no high plateau exists in the Northern and Southern Andes (Fig. 25.1), where only minor (less than 50 km) tectonic shortening has been reported (e.g., Allmendinger et al. 1997; Lamb et al. 1997; Kley and Monaldi 1998). The tectonic shortening was not only much less intensive in the Southern Andes than in Central Andes but also started much later (Vietor and Echtler 2006, Chap. 18 of this volume).

In addition to this first-order difference between the central and Southern Andes, somewhat subordinate, although still very important, is the difference between the structure and deformation style in the Altiplano and Puna segments of the Central Andes (Allmendinger and Gubbels 1996; Kley and Monaldi 1998). North of 23° S, in the Altiplano segment, the tectonic shortening started early (at about 50 Ma) and was most intensive during the last 10 million years, forming a broad, thin-skinned Subandean thrust belt. South of 23° S, in the Puna, tectonic shortening started later, was much less intense and generated a thick-skinned deformation pattern (Allmendinger and Gubbels 1996; Kley and Monaldi 1998).

Perhaps the key question about Andean orogeny is why the high plateau has developed only in the Central Andes and only in Cenozoic times (mostly during the last 30 million years), even though the Nazca Plate has been subducting along the entire western margin of South America for more than the last 200 million years (e.g., Isacks 1988; Allmendinger et al. 1997; see Oncken et al. 2006, Chap. 1 of this volume, for a review of the deformation history of the Central Andes).

Several ideas have been proposed in answer to this question. Isacks (1988) suggested that before about 25–30 Ma, large parts of the Central Andes were underlain by a shallow-dipping slab that became steeper at ~25 Ma, causing thermal weakening and intensive tectonic shortening in the compressed lithosphere of the overriding plate. In another hypothesis (Mégard et al. 1984; Pardo-Casas and Molnar 1987; Somoza 1998), the beginning of intensive tectonic shortening in the Andes was associated with major reorganization of plate motion followed by an increase in the Nazca-South American convergence rate at ~25–30 Ma. Russo and Silver (1994) attributed the Andean orogeny to the Cenozoic increase in the westward drift rate of the South American Plate and associated mantle flow, whose north-south heterogeneity determined the difference in the intensity of deformation between the center and the rest of the Andes. Lamb and Davis (2003) proposed that the high shear stress at the interface between the Nazca and South American Plates, caused by sediment starvation in the Central Andean trench, was crucial for the deformation of the upper plate. Based on the correlation between the shortening rate in the Central Andes and the global temperature of the ocean, these authors have also suggested that this sediment starvation was caused by global climate change beginning at 30 Ma and, therefore, climate was the sole factor responsible for the formation of high Andes.

In relation to the different deformation styles of the Altiplano and Puna segments of the Central Andes, Allmendinger and Gubbels (1996) noticed the striking correlation between the deformation styles and the distribution of thick, Paleozoic, sedimentary basins. These basins are abundant in the foreland of the Altiplano but absent in the foreland of the Puna. Another clear difference in the lithospheric structure beneath the Puna and Altiplano segments are the much higher seismic-wave attenuation and seismic velocities beneath the Puna region (Whitman et al. 1996; Koulakov et al. 2006), indicating significant north-south variations in lithospheric temperature and, hence, in the mechanical strength of the foreland. These observations suggest that the shortening style of the whole system might be controlled by lateral, north-to-south variation in the mechanical properties of the lithosphere.

Another key question related to the Andean orogeny is what happened to the mantle lithosphere under the plateau. A simple argument of mass conservation suggests that the entire lithosphere must be doubled in the places where the crustal thickness has also been tectonically doubled. The consequence would be low crustal temperature and reduced surface heat flow for at least a few tens of millions of years (e.g., Babeyko et al. 2002). However, this contradicts the high values of surface heat flow in the Altiplano (Springer and Förster 1998) and the evidence of partial melting in the Altiplano mid-crust (Yuan et al. 2000). Kay and Kay (1993) suggested that the mantle lithosphere might have been delaminated, but found evidence for this process only for the Puna Plateau and not for the Altiplano. Pope and Willet (1998) postulated the ablative subduction scenario for the Central Andes in which the thickening mantle lithosphere of the upper plate was eroded by the subducting plate. However, this condition was kinematically predefined in their model and, to date, its validity has not been tested by a dynamic model.

Another striking feature of the Andean orogeny is the remarkable correlation between surface topography and the dip of the slab, and their symmetry with the symmetry axis at ~19° S (Gephart 1994), at the so-called Bolivian orocline. Based on these observations, Gephart (1994) stated that the relationship between these parameters suggests that the surface topography of the Andes strongly depends on subduction dynamics, but he did not specify a mechanism.

The diversity of the suggested hypotheses reflects the complexity of the deformation processes responsible for the Andean orogeny, and also indicates the lack of quantitative understanding of these processes. One way to improve such understanding is to study the temporal correlation between tectonic shortening and those processes that possibly contribute to the deformation of the upper plate (Oncken et al. 2006, Chap. 1 of this volume). Such analyses show that the shortening rate in the Central Andes correlates well with the westward drift velocity of the South American Plate, whereas it does not correlate with the convergence rate between the Nazca and South American Plates. As mentioned above, the shortening rate also correlates well with the global ocean temperature (Lamb and Davis 2003; Oncken et al. 2006, Chap. 1 of this volume), apparently suggesting a strong relationship between the Andean orogeny and changing global climate (Lamb and Davis 2003). However, close examination of this issue using geological arguments (Oncken et al. 2006, Chap. 1 of this volume) and our modeling results, presented below, suggest an illusive character to this relationship, reminding us that good correlation between processes does not necessarily mean there is a genetic relationship.

We believe that the best way to understand complicated deformation processes at the active continental margin is by using advanced physical-mathematical modeling that is constrained by robust geological and geophysical observations. In this study, we used finite-element numerical modeling to consider the coupled, thermo-mechanical, dynamic interaction between the subducting and overriding plates over a few tens of millions of years at the spatial scale of the upper mantle. We studied mass and heat fluxes associated with this interaction, particularly focusing on the dependence of tectonic deformation in the overriding plate on: (*1*) convergence rate; (*2*) overriding rate; (*3*) strength of mechanical coupling between subducting and overriding plates; and (*4*) initial lithospheric structure. We derived models consistent with observations of the central and Southern Andes and defined key predictions of these models. Finally we suggest our preferred scenario for the Andean orogeny.

## 25.2 Method and Model Set-Up

According to the aim of this study, the temporal scale for our modeling is a few tens of millions of years and the spatial scale is 1 000 km. Long-term temporal analysis requires that we fully consider coupled, thermo-mechanical processes. Moreover, modeling the dynamic interaction between subducting and overriding plates demands realistic rheological models of both plates, including elasticity, plasticity, as well as temperature- and stress-dependent viscosity. Modeling of this type, and at this scale, is exceptionally difficult and is not yet possible in three dimensions. Therefore, we have exploited the fact that the variation in structure parallel to the strike of the Andes is much smaller than across strike, which enabled the deformation processes to be modeled with practically independent, two-dimensional (2D) cross sections oriented perpendicular to the strike of the Andes.

### 25.2.1 Basic Equations

The deformation process was modeled by numerical integration of the fully coupled system of 2D conservation equations for momentum (Eq. 25.1), mass (Eq. 25.2) and energy (Eq. 25.3). These equations are solved together with rheological relations (Eqs. 24.4 and 25.5) including those for a Maxwell visco-elastic body with temperature- and stress-dependent viscosity (Eq. 25.4), and a Mohr-Coulomb failure criterion with non-associated (zero dilation angle), shear flow potential (Eq 25.5). It was assumed that viscous deformation consists of competing dislocation, diffusion and Peierls creep mechanisms (Kameyama et al. 1999).

$$-\frac{\partial p}{\partial x_i} + \frac{\partial \tau_{ij}}{\partial x_j} + \rho g_i = 0, \quad i = 1, 2 \quad (25.1)$$

$$\frac{1}{K}\frac{dp}{dt} - \alpha\frac{dT}{dt} = -\frac{\partial v_i}{\partial x_i} \quad (25.2)$$

$$\rho C_p \frac{dT}{dt} = \frac{\partial}{\partial x_i}\left(\lambda(x_i, T)\frac{\partial T}{\partial x_i}\right) + \tau_{ij}(\dot{\varepsilon}_{ij}^v + \dot{\varepsilon}_{ij}^p) + \rho A \quad (25.3)$$

$$\frac{1}{2G}\frac{\hat{d}\tau_{ij}}{dt} + \frac{1}{2\eta}\tau_{ij} + \dot{\varepsilon}_{ij}^p = \dot{\varepsilon}_{ij} \quad (25.4)$$

$$\frac{1}{\eta(\tau,T)} = \frac{1}{\eta_{\text{dif}}(T)} + \frac{1}{\eta_{\text{dis}}(\tau,T)} + \frac{1}{\eta_p(\tau,T)}$$

$$\sigma_1 - \sigma_3 \frac{1+\sin\phi}{1-\sin\phi} + 2c\sqrt{\frac{1+\sin\phi}{1-\sin\phi}} = 0 \quad (25.5)$$

$$g_s = \sigma_1 - \sigma_3$$

Here the Einstein summation convention applies and $x_i$ are coordinates, $t$ – time, $v_i$ – velocities, $p$ – pressure, $\tau_{ij}$ and $\dot{\varepsilon}_{ij}$ – stress and strain-rate deviators, $\dot{\varepsilon}_{ij}^v$ and $\dot{\varepsilon}_{ij}^p$ are viscous and plastic strain-rate deviators, respectively, $d/dt$ – convective time derivative, $\hat{d}\tau_{ij}/dt$ – Jaumann co-rotational deviatoric stress rate, $\rho$ – density, $g_i$ – gravity vector, $K$ and $G$ – bulk and shear moduli, $\eta$ – viscosity, $\eta_{dif}$ – diffusion creep viscosity, $\eta_{dis}$ – dislocation creep viscosity, $\eta_P$ – Peierls creep viscosity, $\tau$ – square root of second invariant of stress tensor, $R$ – gas constant, $T$ – temperature, $\sigma_1, \sigma_3$ – maximal and minimal principal stresses, $\phi$ – angle of friction, $c$ – cohesion, $g_s$ – shear plastic flow potential, $C_p$ – heat capacity, $\lambda$ – heat conductivity, and $A$ – radioactive heat production.

### 25.2.2 The Plate Interface and Gabbro-Eclogite Transformation

The interface between the slab and the upper plate was modeled as a 12 km-thick subduction channel with plastic rheology using three finite elements (two elements for the oceanic-slab side and one element for the continental-plate side). The yield stress was defined as the smallest of either the (Mohr-Coulomb) frictional stress:

$$\tau = c + \mu \sigma_n \qquad (25.6)$$

or the temperature-dependent, viscous shear stress (Peacock 1996):

$$\tau = \tau_0 \exp\left(-\frac{T - T_0}{\Delta T}\right) \qquad (25.7)$$

In both equations, $\tau$ is a stress norm defined as the square root of the second invariant of the stress tensor, $c$ is cohesion, $\sigma_n$ is normal stress, $\mu$ is the subduction-channel, effective friction coefficient, which includes the effect of fluid pressure, $T$ and $T_0$ are local and reference temperatures, and $\tau_0$ and $\Delta T$ are parameters. The parameters of Eq. 25.7, $T_0, \Delta T$ and $\tau_0$, are assumed to be 400 °C, 75 °C and 60 MPa, respectively, close to the values of Peacock (1996).

With this approach, the shallow, low-temperature part of the subduction channel has frictional (brittle) rheology with shear stress increasing with depth (Fig. 25.2, *solid* and *dashed curves*). At greater depth and higher temperature, the viscous flow mechanism takes over and shear stress in the channel decreases with depth (Fig. 25.2, *solid curve*). The depth where frictional rheology changes to viscous rheology depends on the friction coefficient and it is larger where friction is lower (compare *dashed* and *solid curves* in Fig. 25.2). We considered the friction coefficient in the subduction channel as a model parameter and changed it from 0 to 0.15, which is in agreement with previous estimates (Bird 1998; Tichelaar and Ruff 1993; Peacock 1996; Hassani et al. 1997).

We also took into account gabbro-eclogite transformation in both oceanic crust and continental lower crust. For simplicity, we used the same gabbro-eclogite phase diagram for both crusts, calculated for the average gabbro composition using the free Gibbs energy minimization technique (Sobolev and Babeyko 1994) and do not consider neither related volume changes nor latent heat effects. The density of the eclogite (at room conditions) was set to 3 450 kg m$^{-3}$ and, in all models, the kinetic blocking temperature for the gabbro-eclogite transforma-

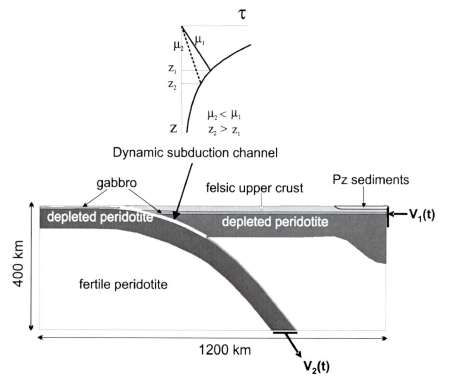

**Fig. 25.2.**
Model set-up and boundary conditions. The subducting plate is 45 million years old. The initial thickness of the continental lithosphere is 100–130 km, with the thickest lithosphere in the eastern (*right*) part of the model corresponding to the Brazilian Shield margin. South America is drifting to the west (*left*) with the velocity increasing from 2 to 3 cm yr$^{-1}$ during the last 35 million years. The lower end of the Nazca Plate is pulled down with the velocity changing from 5 to 13 cm yr$^{-1}$. The inlet shows the shear stress in the subduction channel. Note that with a lower friction coefficient in the channel, the friction (brittle) domain extends deeper

tion was 800 °C for the oceanic crust and 700 °C for the lower continental crust. We do not consider here the details of the processes in the fore-arc. Therefore we omit a number of hydration and dehydration reactions in the subducting lithosphere (see Poli and Schmidt 2002; Hacker et al. 2003 for reviews), which lead to interesting geodynamic consequences at the fore-arc scale (Gerya et al. 2002; Gerya and Yuen 2003).

### 25.2.3 Numerical Method and Rheology

We integrated Eqs 25.1–25.7 using a 2D, parallel, thermo-mechanical, finite-element/finite-difference code called "LAPEX-2D" (Babeyko and Sobolev 2005; Sobolev and Babeyko 2005; see also Babeyko et al. 2002, for a description of the previous version of the code). This code combines the explicit Lagrangian algorithm FLAC (Polyakov et al. 1993; Cundall and Board 1988) with the particle technique that is similar to the particle-in-cell method (Sulsky et al. 1995; Moresi et al. 2001). Particles track material properties and the full stress tensor, thereby minimizing numerical diffusion related to re-meshing. The method allows the employment of realistic temperature- and stress-dependent, visco-elastic rheology combined with Mohr-Coulomb plasticity for layered oceanic and continental lithosphere (Fig. 25.2).

The rheological parameters were taken from published experimental and theoretical studies and are presented in Table 25.1. Owing to the presence of the subduction zone, we have used rheological parameters for "wet" rocks everywhere in the model except for the slab and mantle lithosphere of the shield margin. In the crust, we employed friction and viscosity strain softening, which we assumed to be more intensive in the Paleozoic sediments of the Subandean Zone (Fig. 25.2). The viscous deformation in the mantle was considered to be driven by competing dislocation, diffusion and Peierls creep mechanisms (Kameyama et al. 1999) and the numerical method routinely included shear heating.

At high strains, rocks may mechanically weaken owing to a number of processes (Handy et al. 1999). We considered this in the model by introducing plastic and viscous strain-softening behavior. For felsic and mafic continental rocks, we assumed that cohesion and friction angles

**Table 25.1.** Material parameters

| Parameter | Sediments | Felsic crust | Gabbro continent/ocean | Mantle lithosphere of slab and shield | Mantle lithosphere/asthenosphere |
|---|---|---|---|---|---|
| Density at room conditions, $\rho$ (kg m$^{-3}$)[a] | 2670 | 2800 | 3000 | 3280 | 3280/3300 |
| Thermal expansion, $\alpha$ (K$^{-1}$)[b] | $3.7 \times 10^{-5}$ | $3.7 \times 10^{-5}$ | $2.7 \times 10^{-5}$ | $3.0 \times 10^{-5}$ | $3.0 \times 10^{-5}$ |
| Elastic moduli, $K, G$ (GPa)[b] | 55, 36 | 55, 36 | 63, 40 | 122, 74 | 122, 74 |
| Heat capacity, $C_p$ (J kg$^{-1}$ K$^{-1}$)[b] | 1200 | 1200 | 1200 | 1200 | 1200 |
| Heat conductivity, $\lambda$ (W K$^{-1}$ m$^{-1}$) | 2.5 | 2.5 | 2.5 | 3.3 | 3.3 |
| Heat productivity, $A$ ($\mu$W m$^{-3}$) | 1.3 | 1.3* | 0.2 | 0 | 0 |
| Initial friction angle, $\phi$ (degree) | 30 | 30 | 30 | 30 | 30 |
| Initial cohesion, $C_h$ (MPa) | 2 | 20 | 40 | 40 | 40 |
| Diffusion creep preexp. factor, log($A$) (Pa$^{-n}$ s$^{-1}$) | – | – | – | –10.59 | –10.59 |
| Diffusion creep activation energy (kJ mol$^{-1}$) | – | – | – | 300 | 300 |
| Grain size (mm), (Grain size exponent) | – | – | – | 0.1 (2.5) | 0.1 (2.5) |
| Dislocation creep preexp. factor, log($A$) (Pa$^{-n}$ s$^{-1}$) | –28.0 | –28.0 | –15.4/–25.9 | –16.3 | –14.3 |
| Dislocation creep activation energy (kJ mol$^{-1}$) | 223 | 223 | 356/485 | 535 | 515 |
| Power law exponent, $n$ | 4.0 | 4.0 | 3.0/4.7 | 3.5 | 3.5 |
| Peierls creep preexp. factor, log($A$) (Pa$^{-n}$ s$^{-1}$) | | | | | |
| Peierls stress, $\sigma_P$ (GPa) | – | – | – | 8.5 | 8.5 |
| Peierls creep activation energy (kJ mol$^{-1}$) | 223 | 223 | 445 | 535 | 535 |

[a] (Lukassen et al. 2001). [b] (Sobolev and Babeyko 1994). Sources for dislocation creep laws – sediments and felsic crust: weakened ($A$ increased by 10 times) quartzite by Gleason and Tullis (1995); mafic crust of continent: wet plagioclase by Rybacki and Dresen (2000); mafic crust of ocean: dry Columbian basalt by Mackwell et al. (1998); mantle lithosphere of slab and shield: dry peridotite by Hirth and Kohlstedt (1996); mantle lithosphere of South America (except shield) and asthenosphere: dry peridotite by Hirth and Kohlstedt (1996). Source for diffusion and Peierls creep laws in mantle: Kameyama et al. (1999).

linearly decrease by a factor of three when accumulated plastic strain changes from 1 to 2. For sediment, the softening was assumed to be faster; the friction angle decreases by a factor of ten when the accumulated plastic strain changes from 0 to 0.5. The viscosity of all continental crustal materials was assumed to decrease by a factor of ten (log linearly) when finite strain changes from 0.5 to 1.0.

### 25.2.4 Model Set-Up

We modeled the initial crustal structures expected for the central and Southern Andes at 35 Ma. The initial crustal structure for the Central Andes (Fig. 25.4a) contains thick, felsic, upper crust and thinner, mafic, lower crust, with a total crustal thickness of 40–45 km, assuming that the crust was already significantly shortened by 35 Ma (Allmendinger et al. 1997; Lamb et al. 1997). The initial crust for the Southern Andes (see below Fig. 25.4c) consists of equally thick, upper felsic and lower mafic layers and has a total thickness of 35–40 km.

The geometry and boundary conditions incorporated into all our models are schematically shown in Fig. 25.2. In all models, we explored the interaction of the 45 million-year-old, subducting Nazca Plate with the 100 to 130 km-thick lithosphere of the overriding South American Plate during the last 35 million years. Relatively thin (100 km) lithosphere of the South America plate margin between the magmatic arc and Brazilian Shield margin is consistent with the concept of weak lithosphere of the back-arc mobile belt (Hyndman et al. 2005). Initial lithosphere of Brazilian Shield margin was taken to be 130 km. We assumed also a rather low-angle geometry for the subducting plate, consistent with the present-day structure in the Andes. The model box was 1 200 km long and 400 km high (Fig. 25.2) and moved to the left (west) together with the overriding plate. The drift of the overriding plate and subduction were generated by pushing the overriding plate at its right boundary and by pulling the slab from below, with the velocities taken from plate-tectonics reconstructions (Somoza 1998; Silver et al. 1998). In this approach, velocity boundary conditions move together with the overriding plate (push) and the retreating slab (pull). All other parts of the model box boundary were open for the free motion of material. The surface was treated as a stress free boundary and lower boundary as an open Winkler foundation boundary. Temperature was constant at the surface (0 °C) and at the lower boundary (1 350 °C). At the left and right boundaries horizontal heat flow was assumed to be 0.

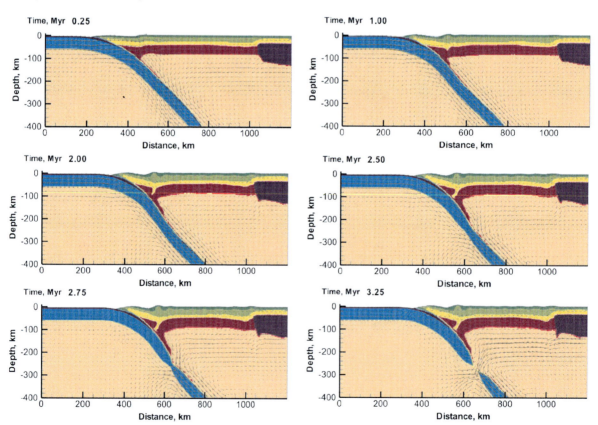

**Fig. 25.3.** Evolution of the slab in the model with high interplate friction coefficient (0.10). Colors represent rock types and vectors show velocity vectors. At $t = 3$ Myr the slab (*blue*) breaks off. The similar picture is observed at interplate friction coefficients higher than 0.10

## 25.3 Modeling Results

### 25.3.1 Model for the Central Andes (Latitude of the Altiplano)

Firstly, we sought the model that best fits the observations from the Central Andes (latitude of the Altiplano) for the time between 35 Ma to the present day. To do that, we assumed thick (40–45 km) crust as a starting model and used kinematic boundary conditions that mimicked subduction and overriding velocities according to the plate-tectonics reconstructions (Somoza 1998; Silver et al. 1998), as shown in Fig. 25.2. With this set-up, we performed several numerical experiments by changing the friction coefficient in the subduction channel "$\mu$". All model-runs with $\mu > 0.10$ resulted in slab break-off and the termination of subduction (Fig. 25.3, see also animation 1 on the Supplementary DVD). When $\mu$ was in the range of 0.05–0.10, subduction survived but large interplate coupling led to shortening in the overriding plate that was too strong.

The model replicates the case of the Central Andes most closely when $\mu$ is about 0.05 (Fig. 25.4b). In this model, 58% of the westward drift of South America during the last 35 million years is accommodated by trench roll-back and the remaining 42% by tectonic shortening (37%) and subduction erosion (5%) of the South American margin. During shortening, the felsic crust almost doubled its thickness, while the thickness of the mafic lower crust and mantle lithosphere actually reduced (compare Fig. 25.4a,b). The reason for this is delamination of the lower crust and mantle lithosphere driven by gabbro-eclogite transformation in the lower crust, first discussed in the Andean context by Kay and Kay (1993).

During the thickening of the crust to more than 45 km, the mineral reactions in the mafic lower crust generated dense garnet and omphacite at the expense of low-density plagioclase, thus increasing the rock density to values higher than that of mantle peridotite (3 300 kg m$^{-3}$). The bodies of the dense lower crust and entrained mantle lithosphere tend to sink into the less dense asthenosphere.

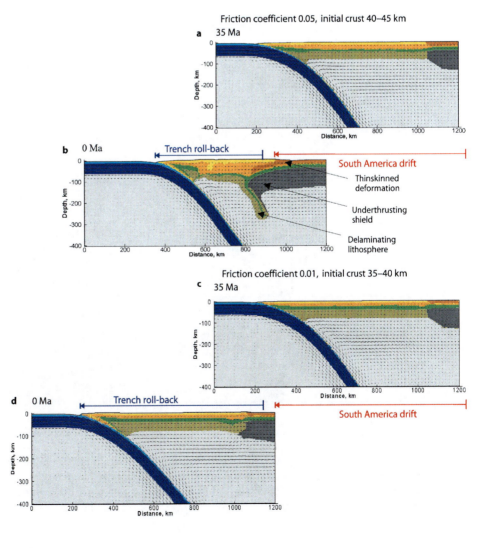

**Fig. 25.4.**
Time snapshots of the evolution of tectonic shortening for the models of the Central Andes (**a, b**) and Southern Andes (**c, d**). Color codes correspond to rock types. In the Central Andes' model, about 60% of the South American westward drift is accommodated by trench roll-back and about 40% by tectonic shortening at the South American margin. Note the intensive thickening of the felsic upper crust (*yellow, orange*) and the loss of mafic lower crust (*green*) in the South American Plate in the Central Andes. Note, also, that mantle lithosphere (*light green*) in the South American Plate in the Central Andes becomes thinner during tectonic shortening. At about 10 Ma, the sedimentary cover of the shield margin (*red*) fails and the shield begins to underthrust the growing plateau. In the Southern Andes' model, 95% of the South American westward drift is accommodated by trench roll-back and the South American Plate remains largely undeformed

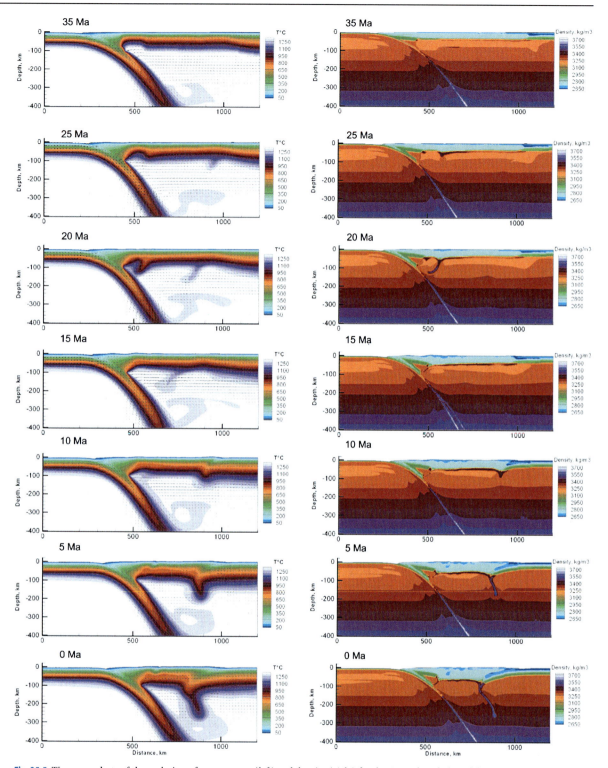

**Fig. 25.5.** Time snapshots of the evolution of temperature (*left*) and density (*right*) for the Central Andes' model

While sinking, most of these bodies are moved by corner flow towards the trench, join the slab and are then subducted into the mantle (Fig. 25.5, see also animations 2, 3 and 4 on the Supplementary DVD).

The model shows at 10 Ma that tectonic shortening generated high topography near the magmatic arc and in the back-arc close to the shield margin (Fig. 25.6, orange curve). These large topographic gradients initiated intensive flow in the lower and middle crust. They also evened out crustal thickness and surface topography and produced a 4 km-high plateau during the last 5–10 million years (Fig. 25.6). At the same time, tectonic shortening reached 350 km, in agreement with the geological estimations of the maximum shortening in the Central Andes (Kley and Monaldi 1998).

The model also predicts failure of the foreland sediments at about 10 Ma followed by underthrusting of the shield margin and a switch from a pure-shear to a simple-shear mode of shortening. This concurs with both the geological model by Allmendinger and Gubbels (1996) and the back-arc-focused numerical study by Babeyko and Sobolev (2005). Figure 25.7 demonstrates the evolution of tectonic deformation, showing calculated, finite-strain distribution at 18 Ma and 0 Ma. At 18 Ma, deformation is moderate and it is dominated by pure shear. In contrast, the final stage of deformation is dominated by simple shear, which is accommodated by underthrusting of the shield margin and lower-crustal flow beneath the plateau.

During tectonic shortening, the temperature of the crust markedly changes (Fig. 25.5). The delamination of the lower crust and mantle lithosphere results in strong heat input into the crust, which leads to partial melting and then convection within the thick, felsic crust (Babeyko et al. 2002). As the resolution of the current model was too low to reproduce small-scale convection in the crust, we emulated such convection by increasing the heat conductivity of the felsic material by ten times if its temperature exceeded 750 °C. The result of this process is a zone of high temperature (higher than 750 °C) rapidly growing in the thickened crust (Fig. 25.5, see also animation 3 at Supplementary DVD) in accordance with present-day high heat flow in the Altiplano (Springer and Förster 1998).

### 25.3.2 Sensitivity of Tectonic Shortening

Here we examine the sensitivity of tectonic shortening to potentially important factors such as the overriding rate, the strength of mechanical coupling between the subducting and overriding plates, the initial lithospheric structure, and the convergence rate. To do that, we modified the Central Andes' model described above by "switching off" different factors and examining the consequences for tectonic shortening over the last 35 million years of evolution.

The shortening curve corresponding to the Central Andes' model is shown in Fig. 25.8b by black solid circles. Firstly, we "switched off" the acceleration of the South American westward drift in the Central Andes' model (e.g., drift velocity remained 2 cm yr$^{-1}$ instead of increasing from 2 to 3 cm yr$^{-1}$), leaving all other parameters unchanged. This modified model (curve denoted by open boxes in Fig. 25.8b) generated only about 60% of the shortening achieved in the original Central Andes' model. Moreover, if the drift velocity was decreased to 1 cm yr$^{-1}$,

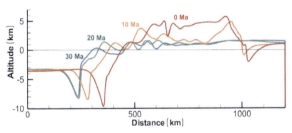

**Fig. 25.6.** Evolution of surface topography in the Central Andes' model. Note the formation of high topography followed by a plateau growth during the last 10 Myr

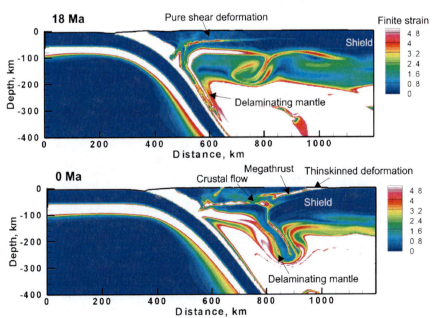

**Fig. 25.7.** Calculated finite strain in the Central Andes' model at 18 Ma and 0 Ma. Note the clear signatures of mega-thrust, thin-skinned deformation in the foreland, lower crustal flow, underthrusting of the shield margin and delamination of mantle lithosphere

there was no tectonic shortening at all. This indicates that in the case of the low-angle subduction and relatively high subduction-channel friction which we consider here, the overriding velocity is among major factors effecting deformation of the overriding plate.

Next, we "switched off" the relatively high, subduction-channel friction in the Central Andes' model by setting $\mu$ to 0.015 instead of 0.05. The resulting model (Fig. 25.8, curve indicated by open circles) still generated a large amount of tectonic shortening, i.e. 74% of the shortening achieved in the original model. This model shows that large changes in the friction coefficient (by three to four times) alone do not lead to dramatic changes in the rate of tectonic shortening. Now, if in addition to the change in the friction coefficient, we adopt a thin-crust (35–40 km) initial model instead of a thick-crust (40–45 km) model, then the tectonic shortening at 0 Ma is reduced to less than 40 km (Fig. 25.8b, filled diamonds). The reason for such behavior is discussed in the next section.

From the shape of the shortening curve for the Central Andes' model (Fig. 25.8b), it can be seen that the high convergence rate of 13–15 cm yr$^{-1}$ achieved between 30 and 25 Ma had little effect on the shortening rate. This is a typical feature in all our models, demonstrating that large changes in the subduction rate (leading to large changes in the convergence rate) do not intesify shortening in the overriding plate.

In several numerical experiments, we changed the geometry of the subducting plate and kept all other parameters fixed. As expected, we observed an increase in the compression stress in the overriding plate when the subduction angle decreased (flatter subduction). Although these effects clearly deserve a more systematic investigation, it is clear that at low interplate friction (coefficient about 0.01), relatively small changes in the subduction angle can result in changes to the stress field of the overriding plate from slight compression to slight tension.

In addition to the factors considered above, some other processes can also influence the rate of tectonic shortening. As previously discussed, the clear increase in the shortening rate in the Central Andes' model at about 10 Ma (Fig. 25.8b) is associated with the failure of the foreland sediments followed by underthrusting of the shield margin and a switch from a pure-shear to a simple-shear mode of shortening. Another important process leading to the weakening of the upper plate, and also to a periodic change in the strength of viscous coupling between the upper and lower plates, is delamination of the lower crust and mantle lithosphere of the upper plate, which we discuss in the following section.

### 25.3.3 Consequences of Mantle Delamination

Before delamination takes place, mineral reactions result in a density increase in the thickening, lower, mafic crust of the overriding plate (Fig. 25.5). The consequence is that little surface uplift occurs despite the increasing thick-

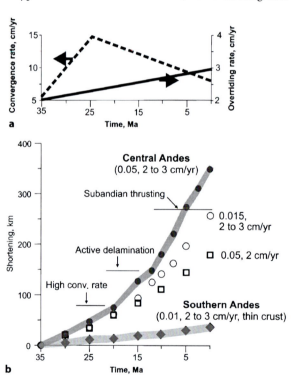

**Fig. 25.8. a** Variation over time in the convergence rate and overriding rate implemented in the model through boundary conditions; **b** Resulting tectonic shortening versus time for different models. *Numbers* near the models indicate the subduction-channel friction coefficient (*first number*) and the South American westward drift velocity (*second group of numbers*). Also shown are the time ranges of some critical processes in the Central Andes' model

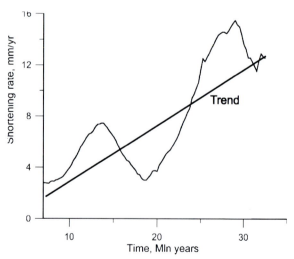

**Fig. 25.9.** Variation in the shortening rate with time in one version of the Central Andes' model. The general tendency for increased shortening is due to the rising rate of the South American westward drift. The reason for periodic variation is explained in Fig. 25.10

ness of the crust (see Fig. 25.6, blue and green curves corresponding to the model time of 30 and 20 Ma respectively). This, in turn, makes tectonic shortening energetically very efficient because no mechanical work is spent against gravitational spreading of the rising surface topography. (For more discussion concerning this process see Babeyko et al. 2006, Chap. 24 of this volume).

To look at the effect of mantle delamination in more detail, we considered a modification of the Central Andes' model, which has a slightly different distribution of initial crustal thickness than the model discussed above. This modified model shows basically the same features as the Central Andes' model presented below, but delamination and related processes are more pronounced in this particular model. Figure 25.9 shows the evolution of the tectonic shortening rate computed for this particular model which has two prominent peaks at ~13 and 29 Ma.

The origin of the first peak is illustrated by Fig. 25.10 (see also animation 7 at Supplementary DVD), which shows a sequence of snapshots of the modeled thermal

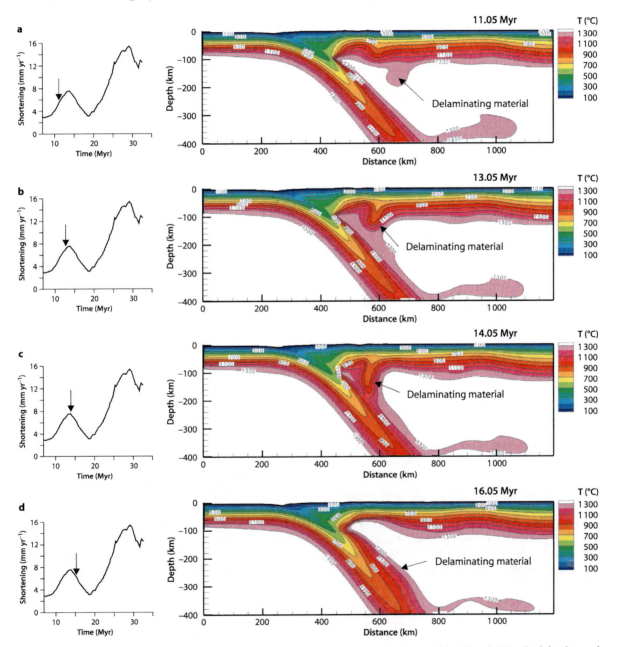

**Fig. 25.10.** Time snapshots of the temperature distribution in a version of the Central Andes' model (*right column*). The *left column* of figures shows the shortening-rate with *arrows* indicating the time of the corresponding temperature snapshots. Note that the maximum shortening rate corresponds to the blocking of the asthenospheric wedge by delaminating mantle material

structure. It is seen that when the tip of the corner flow is occupied by hot asthenospheric material (white), the shortening rate of the upper plate is relatively low. However, when delaminated mantle material, moving by corner flow, blocks its tip, the shortening rate of the upper plate is at a maximum. This behavior is caused by increased viscous coupling between the plates when the asthenospheric wedge is blocked by the cool, delaminated mantle material. Increased mechanical coupling increases the upper plate's resistance to overriding, which intensifies its deformation.

Another important consequence of mantle delamination is heating of the crust. Figure 25.10d shows that after the delaminated material is transported down by the slab, the hot asthenosphere (*white*) rises almost to the level of the Moho. This creates a strong heating impulse that partially melts the crust and may also trigger convection in the felsic crust (Babeyko et al. 2002). The mechanical consequence of this process is the weakening of the crust (Babeyko et al. 2006, Chap. 24 of this volume).

Overall, the consequence of mantle + lower-crust delamination and the preceding mineral reactions in the mafic lower crust of the overriding plate is a significant increase in the susceptibility of the upper plate to tectonic shortening. This result leads to an important conclusion about the possible evolution of tectonic shortening in the upper plate: Because mantle delamination is driven by gabbro-eclogite transformation and the latter occurs only when crustal thickness exceeds some 45 km, the shortening rate may be expected to be relatively low before the crust of the upper plate reaches this critical thickness. After the critical thickness is achieved, the shortening rate increases.

## 25.4 Addressing Key Questions of the Andean Orogeny

### 25.4.1 Why Has the High Plateau Developed Only during the Last 30 Million Years?

There are many arguments for the idea that the most important condition for creating the Altiplano-Puma Plateau was crustal thickening caused by extensive tectonic shortening that occurred in the Central Andes during the Cenozoic (Isacks 1988; Allmendinger and Gubbels 1996, Allmendinger et al. 1997; Lamb et al. 1997; Kley and Monaldi 1998; Giese et al. 1999). Therefore, the question about plateau formation is subdivided: (*1*) Why has the extensive tectonic shortening occurred only during the last 30 million years? and (*2*) How has tectonic shortening created the high plateau?

As shown in the previous section, a modeled increase in convergence rate from 6 to about 15 cm yr$^{-1}$ did not change the shortening rate of the overriding plate significantly (see Fig. 25.8). Moreover, our modeling shows no correlation between the shortening and convergence rates (compare Fig. 25.8a,b). This concurs with data for the Central Andes, which also lacks such a correlation (Oncken et al. 2006, Chap. 1 of this volume), and for subduction zones in general (Heuret and Lallemand 2005). An increase in subduction rate does not lead to upper plate shortening because subduction driven by slab pull induced by slab negative buoyancy always leads to trench roll-back (Christensen 1996). If the subduction rate increases the trench roll-back rate should also increase, resulting in extension, rather than compression of the overriding plate. From this, we infer that it is unlikely that an increased convergence rate caused by the faster subduction of the Nazca Plate could have led to the intensive orogeny in the Central Andes.

It is argued (Lamb and Davis 2003) that another possible cause of the Cenozoic tectonic shortening is late Cenozoic aridization of the global climate, which resulted in reduced sedimentary fill in the trench that, in turn, led to an increase in interplate coupling. This idea looks attractive at first glance because it is consistent with two independent and remarkable correlations: (*1*) the temporal correlation between global ocean temperature and the shortening rate in the Central Andes during the last 25 million years (Lamb and Davis 2003; Oncken et al. 2006, Chap. 1 of this volume) and (*2*) the spatial correlation between the magnitude of tectonic shortening and the thickness of the sedimentary trench-fill off the south-Central Andes (Lamb and Davis 2003; see also Fig. 25.1). Oncken at al. (2006, Chap. 1 of this volume) discuss the Lamb and Davis (2003) hypothesis from a geological viewpoint, whereas we discuss it here based on the results of our modeling.

As seen in the previous section, a three- to fourfold change in the interplate (subduction-channel) friction coefficient at constant other model parameters, may result in a change in the magnitude of tectonic shortening by about 30% (~100 km) in the Central Andes (see Fig. 25.8b). This result shows that if a climate-controlled increase in the subduction-channel friction coefficient were of the order of five times or more, then climate would indeed be an important control on the tectonic shortening rate.

According to the geological data (Oncken et al. 2006, Chap. 1 of this volume), the average thickness of the sedimentary trench-fill was less than 0.5 km off the Central Andes between 30 and 50 Ma, and less than 0.2 km since 30 Ma. Our calibration of the relationship between subduction-channel friction and the thickness of the trench-fill (Fig. 25.12, see Sect. 25.5.1 for explanation) translates the above values into interplate friction coefficients of 0.03–0.05 between 30 and 50 Ma and close to 0.05 since 30 Ma. According to our modeling results, such small changes in the friction coefficient (by 1.6 or less) cannot significantly

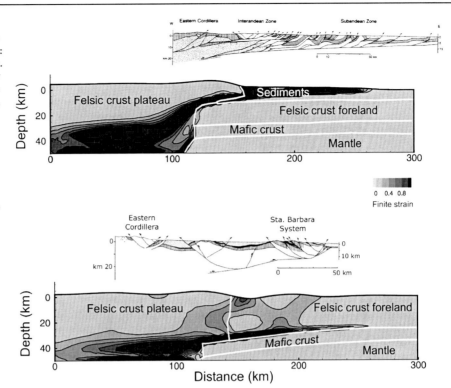

Fig. 25.11.
Models and observations for two cross sections in the Central Andes: (a) the Altiplano and (b) the Puna. Geological sections are from Kley et al. (1999) and modeling results from Babeyko and Sobolev (2005). *Gray colors* in the model cross sections indicate the magnitude of the cumulative, finite-strain norm (square root of the second invariant of the strain tensor). Note the significant difference in the deformation styles between the Altiplano and Puna segments, and the good correspondence between the models and the geological data

change the tectonic shortening rate. From this, we conclude that it is unlikely that climate was the primary reason for the increased rate of tectonic shortening in the Central Andes during the Cenozoic, as suggested by Lamb and Davis (2003). However, although our modeling does not support this part of the Lamb and Davis (2003) hypothesis, we agree with these authors that the large difference in interplate friction between the Central Andes and rest of the Andes might have contributed to the different magnitudes of tectonic shortening in these regions.

Of all the factors considered in our models, tectonic shortening appears to be the most sensitive to the overriding rate of the upper plate (Fig. 25.8). With a low overriding rate of 1 cm yr$^{-1}$ or less, and other parameters fixed, we did not observe any shortening. With a constant overriding rate of 2 cm yr$^{-1}$, the shortening reached 180 km at 0 Ma and when the overriding rate increased from 2 to 3 cm yr$^{-1}$, the tectonic shortening reached 350 km. These results strongly suggest that the late Cenozoic acceleration of the westward drift of the South American Plate might have been a primary cause of the Central Andean orogeny, in agreement with the earlier proposal of Russo and Silver (1994). This also concurs with the observed correlation between the overriding velocity of South America and the shortening rate in the Central Andes (Oncken et al. 2006, Chap. 1 of this volume), and with such correlations for subduction zones in general (Heuret and Lallemand 2005).

Our models show that large amounts (over 300 km) of tectonic shortening that are accompanied by delamination of the mantle lithosphere and mafic lower crust, always generate a high plateau. The exclusion from this rule may be those cases where an exceptionally large erosion rate exists (Babeyko et al. 2006, Chap. 24 of this volume). This does not apply, however, to the Central Andes. One reason that large amounts of tectonic shortening generates high plateau is that the weak, felsic, middle-lower crust cannot maintain the high stresses associated with the large gradients in surface topography created by the tectonic shortening. The consequence is middle-lower crustal flow that tends to flatten the Moho and surface topography, which has also been demonstrated in earlier models (e.g., Royden 1997, Husson and Sempere 2003). Additionally, the basins-and-ranges structure created by the buckling of the shortened lithosphere is smoothed by internal erosion and drainage, which also contributes to the formation of the plateau-like surface topography.

### 25.4.2 Why Are the Central and Southern Andes so Different?

While the high and increasing overriding rate of the South American Plate during the last 30 million years can explain the most extensive tectonic shortening over that time, it does not explain the large amounts of shortening in the Central Andes and the small amounts in the Southern Andes. Russo and Silver (1994), suggested that South America's trenchward motion is driven by deep mantle flow coupled to its base which "collides" with the Nazca Plate; the stagnation point and maximum compression stress occur in the Central Andes. However, in this model it remains unclear how the stresses would differ (if at all) between in the center and periphery of the South American leading edge. Moreover, recent seismic data (Helffrich et al. 2002) argue against the mantle flow pattern predicted by the Russo-Silver model (Russo and Silver 1994). We do not think that these arguments allow a rejection of the mantle-flow model by Russo and Silver (1994), but they indicate the possibility of alternatives.

Leaving this decision for further three-dimensional numerical analysis, we suggest here another explanation for the concentration of deformation in the Central Andes. We note two important differences in the crustal structure of the central and Southern Andes. Firstly, the thickness of the sedimentary fill in the trench off the Southern Andes is more than 1.5 km, whereas it is less than 0.5 km off the Central Andes (see Fig. 25.1), and a similar situation could be expected for the entire Cenozoic (Oncken et al. 2006, Chap. 1 of this volume). According to our model (see Sect. 25.5.1), this corresponds to an interplate friction coefficient that is three to five times lower in the Southern Andes than in the Central Andes. Secondly, in the Southern Andes, the thickness of the crust is, even now, less than 40 km and it has a relatively thick, mafic, lower part (Lüth and Wigger 2003; Tassara 2004). In the Central Andes, however, the crust is likely to have already significantly thickened (probably to more than 40–45 km) by 35 Ma (Lamb et al. 1997), and even at that time it might have been rather felsic (Lucassen et al. 2001; Tassara 2004).

The numerical model that lowers the interplate friction coefficient by five times in the Southern Andes and also uses a thinner crust there, compared to the Central Andes at 35 Ma, generates less than 40 km of tectonic shortening at the present time, even with a high, and increasing, overriding rate for the South American Plate (Fig. 25.8b). In this model, most of the South American Plate's drift is accommodated by roll-back of the Nazca Plate (Figs. 25.4c,d). As a result, the South American Plate overrides the Nazca Plate more than 300 km further in the Southern Andes' model (Fig. 25.4d) than in the Central Andes' model (Fig. 25.4b).

Another explanation for the difference between the central and Southern Andes was suggested by Yanez and Cembrano (2004). These authors proposed that the younger age of the subducting Nazca Plate in the Southern Andes results in regionally lower, shear coupling between the plates, which, in turn, leads to the smaller amount of deformation in the Southern Andes upper plate. Our modeling does not confirm this suggestion. If a subduction channel is introduced in the model, then shear coupling between the plates is controlled by the mechanical (mostly frictional) properties of the rocks within the channel rather than by the age-dependent temperature of the deeper part of the slab. Therefore, it may be expected that the presence or absence of a lubricant (trench sediment) is much more important in determining the magnitude of shear coupling between the plates than the age of the slab. This does not mean, however, that the slab age cannot affect deformation of the overriding plate. Such dependency might in fact exist is because trench roll-back also requires deformation (bending) of the subducting slab, which may well depend on slab age. Therefore, the trench could roll back more easily when the slab is younger, with consequently less deformation of the overriding plate.

### 25.4.3 Why Are the Altiplano and Puna Segments of the Central Andes Different?

We have run several numerical experiments focused on the deformation in the Andean back-arc (Babeyko and Sobolev 2005). In this model set-up, we prescribed the total magnitude of tectonic shortening with an eye to the factors that control deformation styles. We varied the temperature of the lithosphere and the strength of the sediments in the foreland. The finite strain distribution of two models replicates well the deformation patterns in the Altiplano (Fig. 25.11a) and Puna (Fig. 25.11b) segments of the Central Andes. Also shown are corresponding geological cross sections by Kley et al. (1999). The significantly different deformation styles result from the two segments differing in upper crustal strength (weak sediment in the Altiplano foreland and no such sediment in the Puna foreland) and lithospheric temperature (cold lithosphere in the Altiplano foreland and warm in the Puna foreland). For more details see Babeyko and Sobolev (2005).

Therefore, our modeling supports the idea of Allmendinger and Gubbels (1996) that the presence of thick Paleozoic sediment in the Altiplano foreland and the higher lithospheric temperature in the Puna foreland might have been the major cause of the different styles of tectonic shortening in these regions. To accomplish this, the sediments had to become very weak (their effective friction angle must drop below 10°) at about 10 Ma. For discussion of the effects of the weakening lithosphere see Babeyko and Sobolev (2005) and Babeyko et al. (2006, Chap. 24 of this volume).

### 25.4.4 Which Scenario Is Most Plausible for the Andean Orogeny?

From our modeling, we infer the following scenario for the tectonic evolution of the Andes. The South American Plate has a long history of westward drift with its velocity of 1–1.5 cm yr$^{-1}$ increasing to 2–3 cm yr$^{-1}$ during the last 30 million years (Silver et al. 1998). With these conditions, the dipping angle of the relatively young slab (younger than 50 million years in our case) decreases with time (von Hunen et al. 2004) increasing coupling of the slab with the overriding plate. Maximum coupling was achieved in the Central Andes, where the arid climate caused a lack of sediment in the trench (no lubrication), resulting in a relatively high, interplate friction coefficient of about 0.05 during the entire Cenozoic. With this high friction, the stress in the overriding plate became compressional when the overriding velocity exceeded 1–1.5 cm yr$^{-1}$. Therefore, the upper plate in the Central Andes was subjected to significant shortening at ~70 Ma (the Peruvian shortening phase) and in the late Cenozoic, beginning from some 50 Ma (Oncken et al. 2006, Chap. 1 of this volume). At the same time, in the Southern Andes, the upper plate was still under extension because of the lower interplate friction ($\mu \approx 0.01$) caused by the extensive inflow of lubricating sediment into the trench.

Long-term westward drift combined with high interplate friction in the Central Andes has resulted in ~200 km of continental crustal erosion since the Jurassic (Scheuber et al. 1994). Despite the upper plate in the Central Andes being under compression since ~50 Ma, its deformation rate was relatively low before ~30 Ma (Oncken et al. 2006, Chap. 1 of this volume). We suggest that there were two major reasons for this. Firstly, the crust was still relatively thin (< 45 km), which did not allow for eclogitization mineral reactions in the lower crust. In such a case, the mechanical work required to shorten the crust remains very high (Babeyko et al. 2006, Chap. 24 of this volume). Secondly, the overriding velocity was still too low to create the high compression needed for extensive deformation of the strong lithosphere.

At about 30 Ma, after the establishment of a "flat slab" in the Central Andes (Altiplano latitude) and after the associated episode of enhanced tectonic shortening (Isacks 1988), the crustal thickness in a large part of the Central Andes reached the threshold value (of about 45 km) when eclogitization reactions become activated. At this critical point, the lithosphere of the upper plate was effectively weakened (Babeyko et al. 2006, Chap. 24 of this volume) and continued to shorten extensively, even after the slab became steeper again at about 25 Ma (Isacks 1988).

At the same time, westward drift of the South American Plate had begun to accelerate, increasing compression and the rate of tectonic shortening in the upper plate. The tectonic shortening was concentrated in the Eastern and Western Cordilleras, where the lithosphere was relatively weaker and where the crustal thickness first reached the threshold value of 45 km. Continuing tectonic shortening, accompanied by the eclogitization of the thickening lower crust, triggered delamination of the lower crust and mantle lithosphere. The delamination was followed by the thickening, heating, and partial melting of the felsic part of the crust (Babeyko et al. 2002). Finally, the ductile flow in the heated, and weak, felsic crust evened out the Moho and surface topography and created the high plateau in the Central Andes.

The above scenario applies to the Altiplano segment of the Central Andes. Tectonic evolution of the Puna segment was different most likely because of the following factors: (*1*) the flat slab episode in the Puna occurred some 10 to 20 million years later (Kay et al. 1999), so the crust only reached the critical thickness of 45 km at about 10–15 Ma; (*2*) the higher temperature in the foreland lithosphere in the Puna allowed thick-skinned tectonic shortening in the foreland; and (*3*) the absence of thick and weak Paleozoic sediment in the Puna foreland prevented thin-skinned deformation. As a result, tectonic shortening in the Puna and its foreland was not as intensive and occurred later than that at the latitude of the Altiplano. The consequence was a significantly smaller magnitude of the total shortening (Kley and Monaldi 1998) and a more recent episode of lower-crust + mantle-lithosphere delamination (Kay and Kay 1993).

The shortening rate in the Central Andes has generally tended to increase with time since about 50 Ma because of the increasing overriding velocity of the South American Plate. However, the shortening rate has varied owing to several processes that led either to the weakening of the upper plate (such as failure of the foreland sediment) or an increase in interplate coupling caused by, for example, delaminating mantle material blocking corner flow.

In contrast to the Central Andes, in the Southern Andes, the high overriding rate of the South American Plate was almost entirely accommodated by the roll-back of the Nazca Plate. This was possible because of the combined effect of low friction in the well-lubricated (by sediment) subduction channel and the relatively strong lithosphere with a thin (< 40 km) crust. The upper plate was in a low-stress regime during the entire Cenozoic. Changes in the stress field from little extension to small amounts of compression in the Southern Andes (Vietor and Echtler 2006, Chap. 18 of this volume) might have been caused by the increasing overriding rate of the South American Plate and by changes in the geometry of the subducting plate, with the latter probably related to the concurrent processes in the Central Andes.

Finally, the combined effect of the different, late Cenozoic, trench roll-back distances (> 300 km difference) and the amount of Cenozoic, upper-plate erosion (about 200 km difference) in the central and Southern Andes is the spectacular Bolivian orocline. This orocline, according to our model, marks the region of maximum mechanical coupling between the plates during the last 100 million years.

### 25.4.5 Why Do Andean Surface Topography and Slab Geometry Correlate?

Although the Nazca slab dips at rather low angle everywhere beneath the South American Plate, there are significant north-south variations (Isacks 1988; Gephart 1994). Interestingly, the slab is steepest beneath the highest surface topography in the Central Andes, and it is less steep, or even flat, beneath the lower Southern and Northern Andes (Gephart 1994). This correlation is counter-intuitive because the strongest coupling between the plates, and the most deformation in the upper plate, could be expected to occur at a shallow dipping angle and, therefore, has a large coupling interface with the upper plate. Gephard (1994) suggested that the observed correlation indicates that slab geometry controls deformation of the upper plate. However, based on our modeling results, we suggest that the situation may, in fact, be vice versa, i.e. that the deformation of the upper plate may significantly influence the slab geometry.

The results of numerical models for the central and Southern Andes show the importance of upper-plate crustal structure and the interplate friction. The boundary conditions (e.g., slab pull rate and South American-plate push rate) and the initial geometry of the slab are exactly the same in both models. After 35 million years of evolution, the slab in the Central Andes' model became steeper (Fig. 25.4a,b), while in the Southern Andes' model, the slab became less steep (Fig. 25.4c,d). The present-day slab beneath the strongly shortened crust of the high Central Andes (Fig. 25.4b) is dipping at significantly larger angle than beneath the much less deformed crust of the low Southern Andes (Fig. 25.4d).

This is because the geometry of the subducted plate is largely controlled by the rate of trench roll-back (e.g., von Hunen et al. 2004); the higher the roll-back velocity, the less steeper the plate becomes. In the Southern Andes' model, almost the entire rate of upper-plate westward drift was accommodated by trench roll-back, and therefore the plate became less steep with time. In the Central Andes, a large part of the drift rate was accommodated by the shortening of the upper plate and, therefore, the rate of trench roll-back was lower than in the Southern Andes' model, with the consequence of a more steeply dipping slab. Moreover, as suggested by the numerical model of von Hunen et al. (2004) it is much easier to create a "flat slab" regime for a fast trench roll-back, than for a slow trench roll-back. This may explain the appearance of flat slab segments in the Southern and Northern Andes.

## 25.5 Deriving and Testing Model Predictions

A number of testable predictions can be derived from the model of the Andean orogeny described above. We discuss these predictions below.

### 25.5.1 Magnitude and Along-Strike Variation in Interplate Friction

Our modeling suggests that the friction coefficient in the subduction channel must be lower than 0.1 to allow the subduction zone to survive. Moreover, the friction should be even lower to fit the observed magnitudes of tectonic shortening in the Central Andes ($\mu \approx 0.05$) and Southern Andes ($\mu \approx 0.01$). These values are in good agreement with estimates of the subduction zone's friction coefficients based on geothermal constraints (Tichelaar and Ruff 1993, Peacock 1996). However, they are well below the lowest friction coefficients measured for rocks in laboratory experiments (0.1–0.15 for clay minerals according to Kopf and Brown 2003). This means that additional factors decrease the friction coefficient, such as hydrostatic fluid overpressure in the subduction channel rocks.

As discussed above, our modeling suggests that the completely different styles of tectonic evolution during the last 30 million years in the central and Southern Andes could have resulted because the Central Andes had both a ~5–10 km-thicker crust at 35 Ma and a higher (by 3–5 times) friction coefficient in the subduction channel compared to the Southern Andes. The regionally greater thickness of the crust in the Central Andean back-arc at 35 Ma was likely the cumulative result of a long history of tectonic shortening, enhanced during a "flat slab" episode (Isacks 1988) that was absent in the Southern Andes. The difference in subduction-channel friction between the two regions may reflect climate-controlled variations in the supply of sediment fill to the trench (Lamb and Davis 2003). However, the question arises as to whether there is any independent evidence for a variation (by several times) in trench friction between the central and Southern Andes.

One prediction of our model is that the down-dip limit of the frictional (brittle) domain of the subduction channel in the Southern Andes is significantly deeper than in the Central Andes, caused by the difference in the model friction coefficients (see inlet in Fig. 25.2). With a friction coefficient of $\mu = 0.05$ in the Central Andes, our model predicts the brittle-ductile transition in the channel to occur at ~37–40 km depth, whereas at $\mu = 0.015$–0.01, this transition in the Southern Andes occurs at 55–60 km depth. Note, that while higher friction is applied to a narrower brittle zone in the Central Andes than in the Southern Andes, the total coupling (shear) force in the subduction zone still appears to be two to three times higher in the Central Andes.

Assuming that the stick-slip flow mechanism in the seismic coupling zone is related only to the frictional (brittle) rheology, we predict that the down-dip limit of the seismic slip zone (constrained by seismological data) as well as the down-dip limit of the locked zone (constrained by Geographical Positioning System = GPS data) should be significantly deeper in the southern than in the Central Andes. This model prediction agrees remarkably well with both seismological and GPS observations. Tichelaar and Ruff (1993) estimated the down-dip limit for the seismic coupling zone for large earthquakes to be 36–41 km in the Central Andes (north of the 28° S) and 48–53 km in the Southern Andes. At the same time, GPS observations by Khazaradze and Klotz (2003) and Klotz et al. (2006, Chap. 4 of this volume), completely independently of the seismological data of Tichelaar and Ruff (1993), infer maximum depths for mechanical coupling of about 33 km in the region north of 30° S and 50 km in the Southern Andes.

Using the GPS data (Khazaradze and Klotz 2003) and our modeling results, we can also try to estimate the variation in the interplate friction coefficient along the trench. From the definition of the depth to the brittle-ductile transition (see Eqs. 25.6 and 25.7 in Sect. 25.2.2) and assuming that the pressure and temperature change linearly with depth in the subduction channel, we obtain:

$$\mu \cdot H \approx C \exp(-\gamma H) \qquad (25.8)$$

where $\mu$ is the friction coefficient, $H$ is the depth of the brittle-ductile transition in the subduction channel, and $C$ and $\gamma$ are constants. We can rewrite Eq. 25.8 in the form:

$$\frac{\mu}{\mu_0} \approx \frac{H}{H_0} \exp[\gamma(H - H_0)] \qquad (25.9)$$

where index 0 denotes the reference value of the friction coefficient and the corresponding depth to the brittle-ductile transition. From our models, we estimate parameter $\gamma$ to be 0.052. Using Eq. 25.9 and locking depths from the GPS data (Khazaradze and Klotz 2003), we estimate the ratio between friction coefficients in the central and Southern Andes to be about 4.5. This is remarkably close to the ratio of 3–5 which we obtained to explain the differences in the evolution of the central and Southern Andes over the geological timescale.

Now, we can estimate the relation between the subduction-channel friction coefficient and the thickness of the sedimentary fill in the trench (Fig. 25.12). On the ordinate, we show the friction coefficient calculated from Eq. 25.9 and GPS data on locking depths, assuming that the maximum friction coefficient equals 0.05. On the abscissa, we show the thickness of the trench sediment at the same latitudes as the GPS estimates of the locking depths according to Bevis et al. (1999). Despite the limited number of data points, Fig. 25.12 demonstrates systematic and logical changes in the friction coefficient with the sediment thickness saturated at about 1.5 km thickness. We interpret this behavior that the entire subduction channel is filled by sediment (lubricant) and further

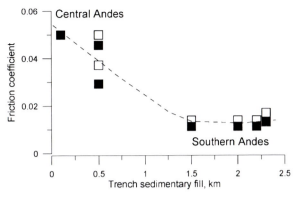

**Fig. 25.12.** Estimated interplate friction coefficient versus observed thickness of the sedimentary trench-fill between 20 and 40° S. *Open boxes*: It is assumed that the depth of the brittle-ductile transition corresponds to the bottom of the locked zone (see Khazaradze and Klotz 2003). *Solid boxes*: It is assumed that the depth of the brittle-ductile transition corresponds to the bottom of the transition zone (see Khazaradze and Klotz 2003). Note that the friction coefficient changes significantly with sedimentary thickness until the thickness reaches some 1–1.5 km; thicker than this, the friction coefficient reaches its asymptotical value of about 0.01

sedimentary addition does not change its friction properties, when the thickness of the trench sediment becomes higher than 1–1.5 km. Interestingly, this conclusion agrees with the observation of a global correlation between the thickness of trench fill and the transition from erosion- to accretion-type subduction (Clift and Vannucchi 2004).

### 25.5.2 Relationship between Arc Magmatism and Tectonic Shortening

Another model prediction follows from the comparison between the evolution of tectonic shortening and the evolution of the mantle temperature beneath the magmatic arc. As we have shown before (Fig. 25.10, see also animation 7 at Supplementary DVD), processing the delaminating material through the asthenospheric wedge by corner flow results in an increasing shortening rate. Simultaneously, the temperature of the asthenospheric wedge beneath the magmatic arc decreases, which must lead to reduced magmatic arc activity over the time period depending on the volume of the delaminating material (typically a few million years). This suggests an anti-correlation between the arc magmatic activity and tectonic shortening. Figure 25.13 shows a histogram of the magmatic activity in the Central Andes (Trumbull et al. 2006, Chap. 2 of this volume; Oncken et al. 2006, Chap. 1 of this volume) together with the rate of tectonic shortening (Oncken et al. 2006, Chap. 1 of this volume). Although the estimation of magmatic activity is still rather controversial (Trumbull et at. 2006, Chap. 2 of this volume), there is, indeed, indication of the anti-correlation between arc magmatic activity and shortening rate.

Our model also predicts strong heating of the overriding-plate crust some 5–10 million years after a large volume of delaminated material was processed through the asthenospheric wedge by corner flow (Fig. 25.5, see also animations 3 and 7 at Supplementary DVD). This may result in the ignimbrite volcanism 5–10 million years after a peak of arc activity, the latter associated with the input of hot asthenospheric material in the wedge just after delamination (Fig. 25.5, see also animations 3 and 7 at Supplementary DVD). The peak of ignimbrite activity in the Central Andes at 8 Ma has indeed followed about 7 million years after the large peak of magmatic arc activity (Trumbull et al. 2006, Chap. 2 of this volume).

### 25.5.3 Seismic Structure of the Crust and Upper Mantle

Here we compare the seismic tomographic images of the lithosphere-asthenosphere system of the Central Andes to expectations based on our modeling results. Figure 25.14 shows depth sections through the recent tomographic model by Koulakov et al. (2006). The tomographic images of P- and S-wave seismic velocities as well as the $Q_p$ factor show similar features that are remarkably different in the crust and upper mantle as well as beneath the Altiplano and Puna Plateau. In the Altiplano (north of 23° S), the dominant low-velocity and high attenuation anomaly is located in the crust (see upper

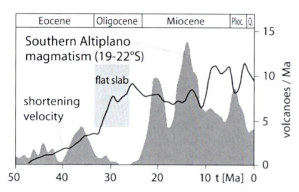

**Fig. 25.13.** Histogram of arc magmatic activity in the Central Andes (Trumbull et al. 2006, Chap. 2 of this volume; Oncken et al. 2006, Chap. 1 of this volume) together with the rate of tectonic shortening (Oncken et al. 2006, Chap. 1 of this volume). Note the tendency for anti-correlation between these two curves

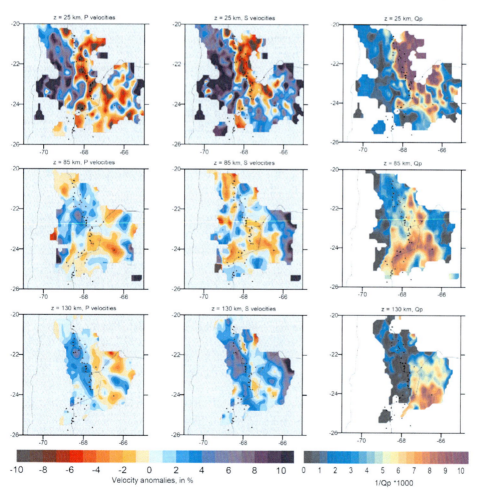

**Fig. 25.14.**
Depth slices through the recent P- and S-wave and $Q_p$ tomographic models of the Central Andes (Koulakov et al., in press). Note the significant change in the pattern from the crust to the upper mantle

row of images in Fig. 25.14), but in the Puna (south of 23° S), the dominant low-velocity anomaly is located in the mantle (see middle and lower row of images in Fig. 25.14).

This is an expected feature based on the tectonic history of these regions and our modeling results. In the Altiplano, tectonic shortening was already active for a few tens of millions of years (Allmendinger and Gubbels 1996, Oncken et al. 2006, Chap. 1 of this volume), which according to our model should have already resulted in delamination of the mantle lithosphere some 15 million years ago. The model predicts that the asthenospheric wedge should now be rather cool while the crust should be very hot (see Fig. 25.5, time snapshot at 0 Ma). In the Puna, most of the tectonic shortening was quite recent (Allmendinger and Gubbels 1996) and mantle delamination also happened recently (Kay and Kay 1993). In this case, the crust had no time to be heated, while the asthenospheric wedge should be occupied by hot, fresh, asthenospheric material that has recently replaced delaminated mantle (e.g., see Fig. 25.10d).

The high velocity and low attenuation anomaly in the mantle east of 67–68° W (see middle and lower sections in Fig. 25.14) probably represents the advancing margin of the Brazilian shield and is also clearly visible in all our geodynamic models. Figure 25.15 shows the results of direct, joint inversion of the P- and S-wave travel times and attenuation seismic data for temperature and water content in olivine of the mantle wedge at 23.5° S (Sobolev et al., in prep.). Estimated temperatures (Fig. 25.15a) are consistent with the advancing margin of the shield and with on-going delamination of the mantle lithosphere. The proxy for the water content in the mantle wedge, shown in Fig. 25.15b, is the pre-exponent factor in the relation for the seismic attenuation (Sobolev et al. 1996).

## 25.6 Looking beyond the Presented Models

The advanced numerical models presented in this paper are still limited and incomplete. Our 2D models ignore strike-parallel variations in structure, stress and velocity fields. One consequence is that the 3D focusing effect of the tectonic shortening (Hindle et al. 2005) is ignored. This means that the magnitude of tectonic shortening required for generating thick crust and high surface topography in the Central Andes is overestimated in our 2D models by up to 40% (Hindle et al. 2005). Our Central Andes' model requires 350 km of tectonic shortening since 35 Ma and some 100 km of shortening before that time. After correcting for the 3D focusing effect, we estimate the magnitude of the total tectonic shortening in the Central Andes to be about 320–350 km. This is close to the estimates by Oncken et al. (2006, Chap. 1 of this volume), but significantly less than estimates by McQuarrie and De Celles (2001).

Another consequence of our 2D approach is that we ignore the obliquity of subduction in the Andes. Because this obliquity was relatively small during the last 30 million years (Somoza 1998), its effect on major compression deformation is also likely to have been small. However, it may control strike-slip deformation in the Andes (Hoffmann-Rothe et al. 2006, Chap. 6 of this volume). A relatively straightforward way to consider the subduction obliquity is to apply an extended 2D modeling technique (Sobolev et al. 2005) that takes into account the 3D vector and tensor fields with no along-strike variation in structure and displacements, and therefore also enables the consideration of strike-slip deformation.

It is more difficult to account for any possible effects on trench roll-back and deformation of the upper plate caused by trench curvature and 3D flow in the mantle.

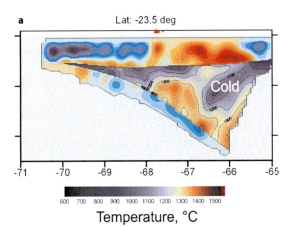 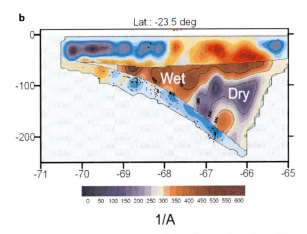

**Fig. 25.15.** Results of the direct inversion of the seismic data for temperature (**a**) and the inverse A factor (**b**) in the mantle wedge of the Central Andes (Sobolev et al., in preparation). The *colors* in the crust and in the slab show P-wave velocity variations in the tomographic model by Koulakov et al. (in press). The inverse A factor is a pre-exponent factor of the mineral physics relation for seismic attenuation as applied to olivine-dominated, ultramafic rocks (e.g., Sobolev et al. 1996), which is believed to be proportional to the water content of olivine

Thin-plate approximations, even extended to incorporate lower-crustal flow and interplate coupling (Medvedev et al. 2006, Chap. 23 of this volume), are not suitable for such a purpose. One possibility is a simplified 3D modeling technique that exploits the relatively slow variation in those parameters parallel to the strike of the Andes. Recently, this kind of technique was developed and employed for thermo-mechanical modeling of a pull-apart basin (Petrunin and Sobolev 2006).

As mentioned earlier in this chapter (Sect. 25.4.2), we cannot reject the possibility that the age of the Nazca slab has influenced trench roll-back mobility and, therefore, deformation in the overriding plate. This possibility can be tested with 2D models similar to the models presented in this paper. The reason we have not yet done this, is that, in addition to the slab age (temperature), the slab rigidity could also be affected by factors such as serpentinization (Ranero et al. 2003). Hydration can also significantly change rheology of the upper plate, not only switching from "dry" to "wet" peridotite rheology in the mantle lithosphere implemented in our model, but also through serpentinization of the fore-arc mantle (Gerya et al. 2002; Gerya and Yuen 2003).

In this work, we did not explicitly model the effect of a flat slab on deformation in the overriding plate, although we recognize its importance for the initial thickening of the crust to the critical thickness of 45 km. Partially, this kind of modeling can also be done using a 2D approach. However, our preliminary modeling results show that it is technically very difficult (if not impossible) to return from the flat-slab to the steep-slab configuration using 2D models (also mentioned by von Hunen et al. 2004), which limits the application of such modeling for the Andes.

Finally, what is still missing, is calculation of seismic velocities from the data on composition, temperature and pressure which all are available in our models, as recently presented by Gerya et al. (2006), but with full consideration of effects of anelasticity, water content and lattice preferred orientation on seismic velocities and seismic anisotropy. This is an interesting task for future studies.

## 25.7 Summary

We developed a coupled thermo-mechanical model of the interaction between subducting and overriding plates that employs realistic temperature- and stress-dependent, visco-elasto-plastic rheology and is focused on the late Cenozoic (last 35 million years) evolution of the Andes.

In the numerical experiments, we studied the conditions under which the overriding plate was tectonically shortened and we tested the sensitivity of shortening to several parameters including convergence rate, overriding rate, interplate friction and crustal structure of the upper plate. From this sensitivity study, we conclude that the major factor that controls tectonic shortening in the Andes is the fast and accelerating, westward drift of the South American Plate. However, we also infer that neither this drift nor any other factor was *alone* responsible for the orogeny in the Central Andes.

Our modeling suggests that the extreme orogeny in the Andes took place, at the time and in the location, when and where at least three major conditions were met: (*1*) a high overriding rate of the South American Plate; (*2*) thick crust (45 km) in the back-arc; and (*3*) a friction coefficient of about 0.05 in the subduction channel. The first condition was achieved over the last 30 million years and especially during the last 10 million years across the entire South American Plate. The second condition occurred only in the Central Andes due to tectonic shortening before 25–30 Ma, partially because shortening was enhanced during a "flat-slab" episode at about 30 Ma. The third condition was fulfilled also only in the Central Andes, where less than 0.5 km of trench sediment was accumulated owing to the arid climate. As a result, intensive tectonic shortening occurred only in the late Cenozoic and only in the Central Andes.

Our numerical models for the central and Southern Andes explain well the major features of Andean evolution during the last 35 million years. This includes the more than 300 km of tectonic shortening in the Central Andes accompanied by the underthrusting of the foreland and the formation of a 4 km-high plateau. Our models also explain the converse situation in the Southern Andes of less than 40 km of shortening and the absence of a high plateau. We show that the main reason for the difference in the late Cenozoic evolution of the central and Southern Andes could be the combined effect of the Southern Andes having a 5–10 km-thinner crust at 35 Ma and about five-times-lower interplate friction compared to the Central Andes. The result of the different magnitude of tectonic shortening and subduction erosion in the central and Southern Andes is the Bolivian orocline and a more steeply dipping subducting plate in the Central Andes than in the rest of Andes.

The models also demonstrate the important role of several processes that cause internal mechanical weakening of the South American Plate during tectonic shortening in the Central Andes. The main processes are gabbro-eclogite mineral reactions in the thickened, continental lower crust followed by lithospheric delamination, as well as the mechanical failure of the sediment covering the shield margin.

The modeling also suggests that the reason for the observed fluctuations in the shortening rate in the Central Andes may be increased viscous coupling between the plates caused when the asthenospheric wedge is blocked by the cool, delaminating, mantle material that is moving to the tip of the asthenospheric wedge by subduction corner flow. Increased mechanical coupling increases resistance in the overriding of the upper plate which intensifies its deformation.

Several model predictions are consistent with observations. We show that the assumption of an interplate friction that is approximately five times smaller in the Southern Andes compared to the Central Andes is consistent with seismological and GPS data from the deeper seismogenic and locking zones of these respective regions. The model prediction of an anti-correlation between tectonic shortening and magmatic arc activity is also consistent with observations. And, finally, we have shown that our model predictions agree well with the known seismic structure of the crust and mantle in the Central Andes.

## Acknowledgments

This paper is a contribution to the Deutsche Forschungsgemeinschaft project SFB-267 "Deformation Processes in the Andes". We thank members of the SFB-267 team for fruitful discussions and Ron Hackney and Allison Britt for useful comments and help in editing the manuscript. Tim Vietor is acknowledged for providing version of Fig. 25.1. Taras Gerya's thorough review helped to improve the paper. The Potsdam Institute of Climate has provided supercomputer facilities.

## References

Allmendinger RW, Gubbels T (1996) Pure and simple shear plateau uplift, Altiplano-Puna, Argentina and Bolivia: Tectonophysics 259:1–13
Allmendinger RW, Jordan TE, Kay SM, Isacks BL (1997) The evolution of the Altiplano-Puna plateau of the Central Andes. Ann Rev Earth Planet Sci 25:139–174
Babeyko AY, Sobolev SV (2005) Quantifying different modes of the Late Cenozoic shortening in the Central Andes, Geology 33:621–624
Babeyko AY, Sobolev SV, Trumbull RB, Oncken O, Lavier LL (2002) Numerical models of crustal-scale convection and partial melting beneath the Altiplano-Puna plateau. Earth Planet Sci Lett 199:373–388
Babeyko AY, Sobolev SV, Vietor T, Oncken O, Trumbull RB (2006) Numerical study of weakening processes in the Central Andean backarc. In: Oncken O, Chong G, Franz G, Giese P, Götze H-J, Ramos VA, Strecker MR, Wigger P (eds) The Andes – active subduction orogeny. Frontiers in Earth Science Series, Vol 1. Springer-Verlag, Berlin Heidelberg New York, pp 495–512, this volume
Beck SL, Zandt G (2002) The nature of orogenic crust in the Central Andes. J Geophys Res 107: doi 10.1029/2000JB000124
Bevis M, Smalley R Jr, Herring T, Godoy J, Galban F (1999) Crustal motion north and south of the Arica deflection: Comparing recent geodetic results from the Central Andes. Geochem Geophys Geosyst 1: doi 1999GC000011
Bird P (1978) Stress and temperature in subduction zones: Tonga and Mariana. Geophys J R Astron Soc 55:411–434
Christensen UR (1996) The influence of trench migration on slab penetration into the lower mantle. Earth Planet Sci Lett 140(1–4): 27–39
Clift PD, Vannucchi P (2004) Controls on tectonic accretion versus erosion in subduction zones: implications for the origin and recycling of the continental crust. Rev Geophys 42: doi 10.1029/2003RG000127
Cundall PA, Board M (1988) A microcomputer program for modelling large-strain plasticity problems. In: Sowboda G (ed) 6th International Conference in Numerical Methods in Geomechanics, AA Balkema, Brookfield VT, pp 2101–2108
Gerya TV, Yuen DA (2003) Rayleigh-Taylor instabilities from hydration and melting propel 'cold plumes' at subduction zones. Earth Planet Sci Lett 212:47–62
Gerya TV, Stoeckhert B, Perchuk AL (2002) Exhumation of high-pressure metamorphic rocks in a subduction channel – a numerical simulation. Tectonics 6(21):1–19
Gerya TV, Connolly JAD, Yuen DA, Gorczyk W, Capel AM (2006) Seismic implications of mantle wedge plumes. Phys Earth Planet Int doi 10.1016/j.pepi.2006.02.005
Giese P, Scheuber E, Schilling F, Schmitz M, Wigger P (1999) Crustal thickening processes in the Central Andes and the different natures of the Moho-discontinuity. J S Am Earth Sci 12:201–220
Gleason GC, Tullis J (1995) A flow law for dislocation creep of quartz aggregates determined with the molten salt cell. Tectonophysics 247:1–23
Gregory-Wodzicki KM (2000) Uplift history of the Central and Northern Andes: a review. Geol Soc Am Bull 112:1091–1105
Hacker BR, Abers GA, Peacock SM (2003) Subduction factory 1. Theoretical mineralogy, density, seismic wave speeds, and $H_2O$ content. J Geophys Res 108: doi 10.1029/2001JB001127
Handy MR, Wissing S, Streit SE (1999) Strength and structure of mylonite with combined frictional-viscous rheology and varied bimineralic composition. Tectonophysics 303:175–192
Hassani R, Jongmans D, Chery J (1997) Study of plate deformation and stress in subduction processes using two-dimensional numerical models. J Geophys Res 102:17951–17965
Heuret A, Lallemand S (2005) Plate motions, slab dynamics and back-arc deformation. Phys Earth Planet Int 149:31–51
Hindle D, Kley J, Oncken O, Sobolev SV (2005) Crustal flux and crustal balance from shortening in the Central Andes. Earth Planet Sci Lett 230:113–124
Hirth G, Kohlstedt DL (1996) Water in the oceanic upper mantle: implications for rheology, melt extraction and the evolution of the lithosphere. Earth Planet Sci Lett 144:93–108
Hoffmann-Rothe A, Kukowski N, Dresen G, Echtler H, Oncken O, Klotz J, Scheuber E, Kellner A (2006) Oblique convergence along the Chilean margin: partitioning, margin-parallel faulting and force interaction at the plate interface. In: Oncken O, Chong G, Franz G, Giese P, Götze H-J, Ramos VA, Strecker MR, Wigger P (eds) The Andes – active subduction orogeny. Frontiers in Earth Science Series, Vol 1. Springer-Verlag, Berlin Heidelberg New York, pp 125–146, this volume
Husson L, Sempere T (2003) Thickening the Altiplano crust by gravity-driven crustal channel flow. Geophys Res Lett 30: doi 10.1029/2002GL016877
Hyndman RD, Currie CA, Mazzotti SP (2005) Subduction zone backarcs, mobile belts, and orogenic heat. GSA Today 15:4–10
Isacks BL (1988) Uplift of the Central Andean plateau and bending of the Bolivian orocline. J Geophys Res 93:3211–3231
Kameyama M, Yuen DA, Karato SI (1999) Thermal-mechanical effects of low-temperature plasticity (the Peierls mechanism) on the deformation of a viscoelastic shear zone. Earth Planet Sci Lett 168:159–172
Kay RW, Kay SM (1993) Delamination and delamination magmatism. Tectonophysics 219:177–189
Kay SM, Mpodozis C, Coira B (1999) Neogene magmatism, tectonism and mineral deposits of the Central Andes (22° to 33° S) In: Skinner BJ (ed) Geology and ore deposits of the Central Andes. Soc Econ Geol Spec pub 7, pp 27–59

Khazaradze G, Klotz J (2002) Short- and long-term effects of GPS measured crustal deformation rates along the south Central Andes. J Geophys Res B108(6)

Kley J, Monaldi C (1998) Tectonic shortening and crustal thickness in the Central Andes: how good is the estimate? Geology 26:723–726

Kley J, Monaldi CR, Salfity JL (1999) Along-strike segmentation of the Andean foreland: causes and consequences. Tectonophysics 301:75–94

Klotz J, Abolghasem A, Khazaradze G, Heinze B, Vietor T, Hackney R, Bataille K, Maturana R, Viramonte J, Perdomo R (2006) Long-term signals in the present-day deformation field of the Central and Southern Andes and constraints on the viscosity of the Earth's upper mantle. In: Oncken O, Chong G, Franz G, Giese P, Götze H-J, Ramos VA, Strecker MR, Wigger P (eds) The Andes – active subduction orogeny. Frontiers in Earth Science Series, Vol 1. Springer-Verlag, Berlin Heidelberg New York, pp 65–90, this volume

Kopf A, Brown KM (2003) Friction experiments on saturated sediments and their implications for the stress state of the Nankai and Barbados subduction thrusts. Marine Geology 202(3–4):193–210

Koulakov I, Sobolev SV, Asch G (in press) P- and S-velocity images of the lithosphere-asthenosphere system in the Central Andes from local-source tomographic inversion. Geophys J Int

Lamb S, Davis P (2003) Cenozoic climate change as a possible cause for the rise of the Andes. Nature 425:792–797

Lamb S, Hoke L (1997) Origin of the high plateau in the Central Andes, Bolivia, South America. Tectonics 16:623–649

Lamb S, Hoke L, Kennan L, Dewey J (1997) Cenozoic evolution of the Central Andes in Bolivia and Northern Chile. In: Burg JP, Ford M (eds) Orogeny through time. Geol Soc Spec Publ 121, pp 237–264

Lucassen F, Becchio R, Harmon R, Kasemann S, Franz G, Trumbull R, Romer RL, Dulski P (2001) Composition and density model of the continental crust in an active continental margin – the Central Andes between 18° and 27° S. Tectonophysics 341:195–223

Lüth S, Wigger P (2003) A crustal model along 39° S from a seismic refraction profile ISSA2000. Rev Geol Chile 30(1):83–101

Mackwell SJ, Zimmerman ME, Kohlstedt DL (1998) High-temperature deformation of dry diabase with application to tectonics on Venus. J Geophys Res 103:975–984

McQuarrie N, De Celles P (2001) Geometry and structural evolution of the Central Andean backthrust belt, Bolivia. Tectonics 20(5):669–692

Medvedev S, Podladchikov Y, Handy MR, Scheuber E (2006) Controls on the deformation of the Central and Southern Andes (10–35° S): insight from thin-sheet numerical modeling. In: Oncken O, Chong G, Franz G, Giese P, Götze H-J, Ramos VA, Strecker MR, Wigger P (eds) The Andes – active subduction orogeny. Frontiers in Earth Science Series, Vol 1. Springer-Verlag, Berlin Heidelberg New York, pp 475–494, this volume

Moresi LN, Dufour F, Muhlhaus HB (2003) A Lagrangian integration point finite element method for large deformation modeling of viscoelastic geomaterials. J Comp Phys 184:476–497

Oncken O, Hindle D, Kley J, Elger K, Victor P, Schemmann K (2006) Deformation of the central Andean upper plate system – facts, fiction, and constraints for plateau models. In: Oncken O, Chong G, Franz G, Giese P, Götze H-J, Ramos VA, Strecker MR, Wigger P (eds) The Andes – active subduction orogeny. Frontiers in Earth Science Series, Vol 1. Springer-Verlag, Berlin Heidelberg New York, pp 3–28, this volume

Pardo Casas F, Molnar P (1987) Relative motion of the Nazca (Farallon) and South American plates since Late Cretaceous time. Tectonics 6:233–248

Peacock S (1996) Thermal and petrologic structure of subduction zones. In: Bebout G, et al. (eds) Subduction, top to bottom. AGU Geophys Monogr Ser 96, Washington DC, pp 119–134

Petrunin A, Sobolev SV (2005) What controls thickness of sediments and lithospheric deformation at a pull-apart basin? Geology 34:389–392

Poli S, Schmidt MW (2002) Petrology of subducted slabs. Ann Rev Earth Planet Sci 30:207–235

Poliakov AN, Cundall PA, Podladchikov YY, Lyakhovsky VA (1993) An explicit inertial method for the simulation of the viscoelastic flow: an evaluation of elastic effects on diapiric flow in two- and three-layers models. In: Stone DB, Runcorn SK (eds) Flow and creep in the Solar System: observations, modelling and theory. Kluwer Academic Publishers, pp 175–195

Pope DC, Willett SD (1998) Thermo-mechanical model for crustal thickening in the Central Andes driven by ablative subduction. Geology 26:511–514

Ranero CR, Phipps Morgan J, McIntosh K, Relchert C (2003) Bending-related faulting and mantle serpentinization at the Middle America trench. Nature 425:367–373

Royden LH, Burchfiel BC, King RW, Wang E, Chen Z, Shen F, Liu Y (1997) Surface deformation and lower crustal flow in eastern Tibet. Science 276:788–790

Russo R, Silver PG (1996) Cordillera formation, mantle dynamics, and the Wilson cycle. Geology 24:511–514

Rybacki E, Dresen G (2000) Dislocation and diffusion creep of synthetic anorthite aggregates. J Geophys Res 105:26017–26036

Scheuber E, Bogdanic T, Jensen A, Reutter K-J (1994) Tectonic development of the north Chilean Andes in relation to plate convergence and magmatism since the Jurassic. In: Reutter K-J, Scheuber E, Wigger P (eds) Tectonics of the Southern Central Andes. Springer-Verlag, Berlin Heidelberg New York, pp 121–139

Silver PG, Russo RM, Lithgow-Bertelloni C (1998) Coupling of South American and African plate motion and plate deformation. Science 279:60–63

Sobolev SV, Babeyko AY (1994) Modeling of mineralogical composition, density and elastic wave velocities in anhydrous magmatic rocks. Survey Geophys 15:515–544

Sobolev SV, Babeyko AY (2005) What drives orogeny in the Andes? Geology 33:617–620

Sobolev SV, Zeyen H, Stoll G, Werling F, Altherr R, Fuchs K (1996) Upper mantle temperatures from teleseismic tomography of French Massif Central including effects of composition, mineral reactions, anharmonicity, anelasticity and partial melt. Earth Plan Sci Lett 139:147–163

Sobolev SV, Petrunin A, Garfunkel Z, Babeyko AY, DESERT Group (2005) Thermo-mechanical model of the Dead Sea transformation. Earth Planet Sci Lett 238:78–95

Somoza R (1998) Updated Nazca (Farallon)-South America relative motions during the last 40 My: implications for mountain building in the Central Andean region. J S Am Earth Sci 11:211–215

Springer M, Förster A (1998) Heat flow density across the Central Andean subduction zone. Tectonophysics 291:123–139

Sulsky D, Zhou SJ, Schreyer HL (1995) Application of a particle-in-cell method to solid mechanics. Comput Phys Commun 87:236–252

Tassara A (2005) Interaction between the Nazca and South American plates and formation of the Altiplano-Puna plateau: Review of a flexural analysis along the Andean margin (15°–34° S). Tectonophysics 399:39–57

Trumbull RB, Riller U, Oncken O, Scheuber E, Munier K, Hongn F (2006) The time-space distribution of Cenozoic volcanism in the South-Central Andes: a new data compilation and some tectonic implications. In: Oncken O, Chong G, Franz G, Giese P, Götze H-J, Ramos VA, Strecker MR, Wigger P (eds) The Andes – active subduction orogeny. Frontiers in Earth Science Series, Vol 1. Springer-Verlag, Berlin Heidelberg New York, pp 29–44, this volume

Victor P, Oncken O, Glodny J (2004) Uplift of the western Altiplano Plateau: evidence from the Precordillera between 20° S and 21° S, Northern Chile. Tectonics 23: doi 10.1029/2003TC001519

Vietor T, Echtler H (2006) Episodic Neogene southward growth of the Andean subduction orogen between 30° S and 40° S – plate motions, mantle flow, climate, and upper-plate structure. In: Oncken O, Chong G, Franz G, Giese P, Götze H-J, Ramos VA, Strecker MR, Wigger P (eds) The Andes – active subduction orogeny. Frontiers in Earth Science Series, Vol 1. Springer-Verlag, Berlin Heidelberg New York, pp 375–400, this volume

von Hunen J, van den Berg AP, Vlaar NJ (2004) Various mechanisms to induce shallow flat subduction: a numerical parameter study. Phys Earth Planet Int 146:179–194

Whitman D, Isacks BL, Kay SM (1996) Lithospheric structure and along-strike segmentation of the Central Andean Plateau: seismic Q, magmatism, flexture, topography and tectonics. Tectonophysics 259:29–40

Yañez G, Cembrano J (2004) Role of viscous plate coupling in the late Tertiary Andean tectonics. J Geophys Res 109(2):1–21

Yuan X, Sobolev SV, Kind R, Oncken O, et al. (2000) Subduction and collision processes in the Central Andes constrained by converted seismic phases. Nature 408:958–961

Yuan X, Sobolev SV, Kind R (2002) New data on Moho topography in the Central Andes and their geodynamic implications. Earth Planet Sci Lett 199:389–402

## Supplementary DVD

**Animation 1 (AVI).** Break-up of the slab due to high interplate friction coefficient (0.10). Colors represent rock types and arrows show velocity vectors. At $t = 3$ Myr the slab (*blue*) breaks off. The similar picture is observed at interplate friction coefficients higher than 0.10.

**Animation 2 (AVI).** Evolution of structure for the Central Andes' model. Colors represent rock types.

**Animation 3 (AVI).** Evolution of density for the Central Andes' model.

**Animation 4 (AVI).** Evolution of temperature for the Central Andes' model. Arrows show velocity vectors.

**Animation 5 (AVI).** Evolution of finite strain (high strain – *red*) for the thin-skin deformation model.

**Animation 6 (AVI).** Evolution of finite strain (high strain – *red*) for the thick-skin deformation model.

**Animation 7 (PDF).** Sequence of snapshots showing simultaneous evolution of upper-plate shortening rate, crustal and mantle temperatures and magnetic activity for the version of the Central Andes' model.

# Part V
# The Andean Information System: Data, Maps and Movies

**Chapter 26**
Data Management of the SFB 267 for the Andes – from Ink and Paper to Digital Databases

**Chapter 27**
Digital Geological Map of the Central Andes between 20°S and 26°S

**Chapter 28**
Bouguer and Isostatic Maps of the Central Andes

**Chapter 29**
Digital Geological Map of the Southern and Central Puna Plateau, NW Argentina

**Chapter 30**
Morphotectonic and Geologic Digital Map Compilations of the South-Central Andes (36°–42°S)

**Chapter 31**
Introduction to the attached DVD

The last part introduces readers to the database generated by the research program underlying this volume, part of which is contained on the accompanying DVD. H. J. Götze et al. (Chap. 26) provide a complete overview of the database available through a geo-service-tool on the SFB-267 website and on the DVD, comprising data, meta-data, and numerical tools for three-dimensional modeling, mapping, and visualization. In addition, the DVD contains animation files from several of the modeling studies. Of particular value are a series of digital maps containing not only basic geological data but also gravity data from both working areas of the program, the Central and the Southern Andes. Introductory descriptions of these maps and their source documentation are included in the chapters of this part.

# Data Management of the SFB 267 for the Andes – from Ink and Paper to Digital Databases

Hans-Jürgen Götze · Michael Alten · Heinz Burger · Patrick Goni · Daniel Melnick · Sabine Mohr · Kerstin Munier
Norbert Ott · Klaus Reutter · Sabine Schmidt

**Abstract.** The Collaborative Research Center 267 (SFB 267), "Deformation Processes in the Andes", focused on interdisciplinary research in the Central and Southern Andes and involved nearly all geoscientific disciplines. Over a period of 12 years, 17 interdisciplinary and international task groups investigated the crustal processes that act at the convergent plate margin of central South America, through both laboratory work and large field campaigns. Since the early 1990s, the ever-growing digital database of the SFB 267 has required extensive data documentation to guarantee its long-term use and avoid redundancies. A meta-data information system has also been developed to facilitate queries by internet and intranet. The Internet module mingles new data sets from laboratory work, field research, and remote sensing with diverse geoscientific meta-data in a way that makes it more useful to both scientists and the general public. Based on an earlier version, the current SFB 267 website has been turned into a geo-service tool that provides data, meta-data, and numerical tools for three-dimensional modeling, mapping, and visualization.

## 26.1 Introduction

Crustal processes near the convergent plate boundary of South America have been studied for 12 years by the Collaborative Research Center 267 (SFB 267), "Deformation Processes in the Andes". The studies focused on interdisciplinary geoscientific research in the Central Andes involving geology, geophysics, geodesy, petrology and remote sensing. This combination of geoscientific surveys employed active and passive seismology, potential field data, and magnetotellurics, together with monitoring by Global Positioning Systems (GPS), geological field work, petrological and geochemical analyses, age dating, and remote sensing to provide new insights into the structure and tectonic evolution of this huge mountain range. Field activities were concentrated on two segments between 20 and 30° S and 36 and 42° S. These extended from the Pacific Ocean to the eastern Andean foreland, and covered parts of Western Argentina, Southern Bolivia and both Northern and Southern Chile.

This long-term project comprised about 30 task groups which were investigating the crustal processes acting at this convergent plate margin through both laboratory work and field campaigns. Emphasis was placed on the interaction between the upper and lower plates (i.e. between the down going Nazca Plate and the continental wedge of the overriding South American Plate), the state of recent deformation in the continental plate, the generation of magmas through time, and plateau building related to shortening of the orogen. It is clear that an interdisciplinary approach is essential for any kind of modeling and interpretation of lithospheric structures and processes, and, further, that a comprehensive information infrastructure for both data and metadata is required (Ott et al. 2002).

The authors of this paper are responsible for the collection and storage of data and laboratory information from the entire SFB 267 working group and our research partners in a user-friendly, on-line environment. In particular, we, the GIS (Geoscientific Information System) group, are focusing on the construction of an easily accessible database system, visualization of different data types, statistical data research, and installation of both a GIS and a MIS (Meta Information System) for the project. The full MIS system is accessible over the internet at the SFB 267 website (URL: *http://www.cms.fu-berlin.de/sfb/sfb267*).

## 26.2 GGT – an Early Attempt at Integrated Interpretation

As early as 1989, members of the Global Geoscience Transect program (GGT), part of the International Lithosphere Program of IUGG (International Union of Geodesy and Geophysics) and IUGS (International Union of Geological Sciences), emphasized the importance of digitized geoinformation and, in the early 1990s, an ad hoc digitization group began preparing a new basis for digital databases. Using small computers and widely available software, the group established guidelines for digitizing transects (Götze and Williams 1993; Götze and Monger 1991). At that time, two university groups were working on the digitization of transect material: Prof. Rick Williams (Knoxville University) was digitizing the Central Appalachian transect, and some of the authors of this paper were working on the Central Andean transect. The Central Andean transect was initially combined in an analog compilation by an international and interdisciplinary working group from the Universidad Nacional de Salta, Argentina, the Universidad del Norte, Antofagasta, Chile and the former SFB 267 task groups B1, B3, D1 at Freie Universität, Berlin, Germany (Omarini and Götze 1991).

The GGT digitization guidelines were tested at several computer-workshops around the world (e.g., Götze et al. 1993). The broad acceptance by the task group and other users encouraged the American Geophysical Union to electronically publish the digitization guidelines along with two of the transects as case studies. This was an important advance in the move from water-colored transect compilation to digital information. The SFB 267 task groups contributed substantially to this development by linking geological, petrological and geophysical data bases.

The guidelines recommended by the former GGT group was based on early SFB 267 compilations. They contain several specifications for line styles and colors to be used in transect displays. They are still valid and we demonstrate below the enormous development of computer applications in geosciences with two examples. The GGT guidelines contain tables which match the tones of color pencils with computer-generated color schemes (Table. 26.1). The proposed computer color codes for age-related geological maps and cross sections was designed to closely match the color scheme recommended by the Commission for the Geological Map of the World (*http://www.cgmw.net*). Codes for potential field maps, geological maps and cross sections, and for the tectonic settings in the GGTs interpretative cross sections were used in an open file application program which contains subroutines of the GKS package (Graphical Kernel System, *http://iffwww.iff.kfa-juelich.de/gli/*, Encarnacao et al. 1987). Use of these color codes ensured an international uniform presentation standard for all GGT graphical applications.

Data sets with common formats enable the application of various user-written programs (e.g., Fast-Fourier-Transformation of potential field data, two-dimensional (2D) modeling of gravity and magnetics by interactive program tools, statistical correlations and much more), including Fortran 77 programs that have GKS subroutines and can operate on a PC 286/386 with a EGA/VGA color screen. We tested an interactive program, based on the well-known algorithm developed by Talwani et al. (1959), with 2D gravity modeling of the uppermost crustal structures of the western continental margin of South America. The open file program used at this time provided interactive functions to modify geometry and density, to display the model on the graphical screen and to plot the modeled results either via plotter or printer. Figure 26.1 shows a typical software layout of that time: The modeled and measured gravity profiles are shown above the model geometry with related densities given in g cm$^{-3}$ (corresponding to the SI units Mg m$^{-3}$). Today, an inter-

**Table 26.1.** Example of color schemes from the GGT guidelines (Götze and Williams 1993), showing pencil color numbers and equivalent digital color codes for geological maps (tectonic setting)

| Tectonic setting | Index | R% | G% | B% | C% | M% | Y% | H° | L% | S% | Pr | Po | R$_{16}$ | G$_{16}$ | B$_{16}$ |
|---|---|---|---|---|---|---|---|---|---|---|---|---|---|---|---|
| Continental platform and shelf | 73 | 37 | 100 | 100 | 63 | 0 | 0 | 300 | 69 | 100 | 904 | 147 | 6 | 16 | 16 |
| Continental slope and rise | 74 | 0 | 100 | 88 | 100 | 0 | 12 | 293 | 50 | 100 | 903 | 152 | 0 | 16 | 14 |
| Magmatic arc | 75 | 100 | 25 | 37 | 0 | 75 | 63 | 110 | 63 | 100 | 925 | 126 | 16 | 4 | 6 |
| Rift/transform related deposits and intrusives within continental crust | 76 | 100 | 100 | 12 | 0 | 0 | 88 | 180 | 56 | 100 | 915 | 105 | 16 | 16 | 2 |
| Oceanic | 77 | 50 | 25 | 100 | 50 | 75 | 0 | 20 | 63 | 100 | 932 | 138 | 8 | 4 | 16 |
| Orogenic and related sedimentary rocks, clastic wedges of foreland and successor basins | 78 | 37 | 100 | 12 | 63 | 0 | 88 | 223 | 56 | 100 | 911 | 167 | 6 | 16 | 2 |
| Anorogenic magmatic rocks | 79 | 100 | 37 | 12 | 0 | 63 | 88 | 137 | 56 | 100 | 922 | 117 | 16 | 6 | 2 |
| Cratonic basement | 80 | 50 | 25 | 25 | 50 | 75 | 75 | 120 | 38 | 33 | 937 | 192 | 8 | 4 | 4 |
| Continent-continent collision | 81 | 63 | 12 | 0 | 37 | 88 | 100 | 132 | 31 | 100 | 925 | 190 | 10 | 2 | 0 |
| Indeterminant | 82 | 63 | 63 | 63 | 37 | 37 | 37 | – | 63 | 0 | 963 | 196 | 10 | 10 | 10 |

| | |
|---|---|
| R% G% B% | Red, green and blue color levels in percent of maximum. |
| C% M% Y% | Cyan, magenta and yellow pigment levels in percent of maximum. |
| H° L% S% | Hue in degrees. Luminosity and saturation in percent of maximum. |
| Pr Po | Prismacolor and Faber-Castell Polychromos pencil sets. |
| R$_{16}$ G$_{16}$ B$_{16}$ | Red, green and blue colors scaled between 0 and 16. |

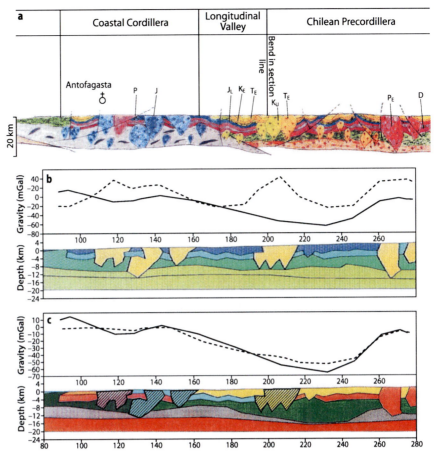

Fig. 26.1.
2D gravity model along the GGT 6 transect (Omarini and Götze 1991). *Top*: Pencil-colored geological cross-section. *Middle*: Initial density model, measured and calculated gravity. *Bottom*: Final density model, measured and calculated gravity. The gravity modeling procedure is based on the method of Talwani et al. (1959). *Dashed areas*: Jurassic and Cretaceous intrusions

active version of this application is available on the internet (*http://www.geophysik.uni-kiel.de/~sabine/ Institut/ModellSchwere2D.htm*). The observed and calculated gravity curves shown in Fig. 26.1 obviously do not fit each other. This is mainly the result of erroneous densities selected for Jurassic and Cretaceous intrusions (see profile between 120–160 km and 200–220 km) and the lack of additional data constraints, such as those from seismic studies.

Digitization of geoscientific data (maps, transects, etc.) does not only facilitate the publication of final products (refer to the accompanying data DVD) and the exchange of databases but, above all, it enables testing and improvement of geological and tectonic hypotheses on the evolution of the Earth, supported by interdisciplinary research. This is an important contribution to a combined interpretation of geological observations and geophysical data. In this context, digitized transect material is the perfect training ground for geoscientists from different disciplines (Schmidt et al. 2003; Götze and Schmidt 2000). However, even today, a vast number of different software applications do not follow any standardization for input and output. This fact, in addition to differences in hardware platforms and operating systems, still hampers free exchange of data.

## 26.3 The Search for Interoperability

The German Collaborative Research Centers were initiated to foster and stimulate interdisciplinary work. However, geoscientific problems have been considered to be *interdisciplinary* by definition for a long time. The heterogeneity of the subprograms of the SFB 267, all focusing on the same object, yet using different methods, data, spatial resolutions and scales convincingly demonstrates this fact. *Interdisciplinarity* shouldn't be confused with *multidisciplinary* interpretation, which is characterized by the fact that different groups are working independently, intending to discuss their individual results to find finally "*a joint interpretation*". In contrast, interdisciplinary interpretation continually tests the research of each working group against the results of other groups during the entire interpretation process (Fig. 26.2). This iterative cycle ensures the ongoing control of interpretations and the early-stage convergence of final models.

From the beginning of the SFB 267, it was realized that substantial scientific progress could be achieved by not restricting the intensive coupling of different geoscientific disciplines to the exchange of data but, rather, there should be a final integration of both data and methods. Indeed, a

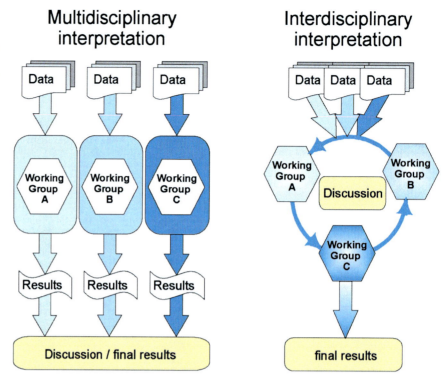

Fig. 26.2.
Interdisciplinary, multidisciplinary and joint interpretation schemes. Independent (parallel) multidisciplinary interpretation (*left*) versus joint (simultaneous) interdisciplinary interpretation, which substantially increases synergy (*right*)

group of some 100 geoscientists, with the ambitious goal of understanding the "Deformation Processes in the Andes", definitely needs a sophisticated working environment that provides the full advantage of interoperability in an easy-to-use "plug and play" manner. Therefore, we initiated the development of an Interoperable Open GIS (IOGIS) in a joint project together with the University of Bonn, financed by the Deutsche Forschungsgemeinschaft from 1996–2000 (Breunig 1996; Schmidt and Götze 1999; Breunig et al. 2000). Data handling and the communication between software components is established in CORBA (Common Object Request Broker Architecture), an object-oriented software architecture for distributed software. This project is currently continued by the computer science group "Informatik III" at the University of Bonn under the name COBIDS (Component-Based Integration of Data and Services) (Bode et al. 2002).

## 26.4 The SFB 267 Web Portal – an Information Gateway

With some 120 scientists from three universities in Germany and some other 10 co-operating partner institutions (universities, geological surveys) in South America, it soon became apparent that good communication within and between all the SFB working groups would be crucial. The World Wide Web is a flexible tool for distributing information at various levels and the SFB 267 website was designed to provide information on two levels: internally and externally. Firstly, we created an intranet module for internal use – administration, exchange of data and ideas, discussion groups, and announcements of colloquia and lectures. Secondly, we developed an internet module that provides information to the global scientific community through the presentation of current research activities, recent results, available maps and data, as well as the contact details for SFB 267 members. The combination of both tasks, and the large amount of data gathered during numerous geological and geophysical field campaigns, required careful design and also needs permanent maintenance of the website.

The creation of the SFB 267 information system started with the presentation of the administrative structure and scientific research proposals of all task groups. The next step was the design of a suitable web-representation of this structure, including traditional items (such as staff lists and details). The tree-like structure of the website is represented in Fig. 26.3, though, in reality, the pages are more intricately linked than is shown. This part of the information system also serves as a noticeboard for the announcement of seminars, meetings, conferences, and new hardware and software available to all SFB 267 members. On-line forms for registration of field work, application forms for travel funds, research grants etc. may all be distributed via the internal mailing lists. The intranet also serves as a forum for discussion of draft papers and preliminary research results.

From the first operational phase of SFB 267, the requirement for remote sensing information was enormous.

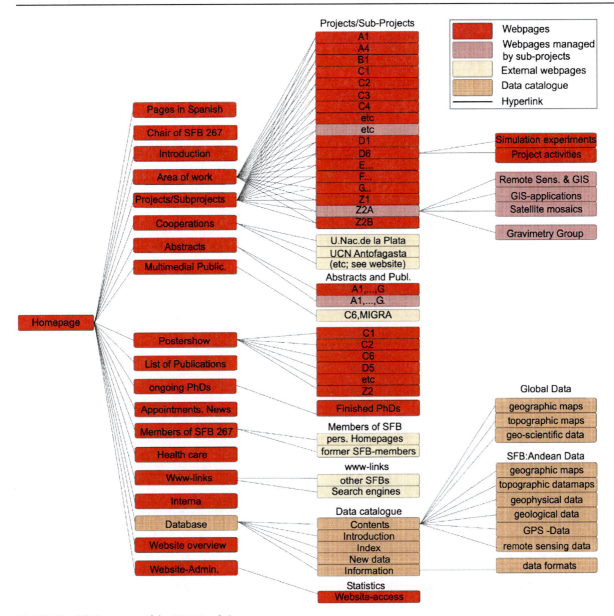

**Fig. 26.3.** Simplified structure of the SFB 267 website

The availability of detailed topographic maps was limited as only topographic maps from the 1960s were available (most of them of scale 1 : 100 000 or 1 : 250 000), and these had few identifiable topographic features shown for orientation in the field. Therefore, each working group could order geo-referenced digital Landsat TM images or working sheets, which contained additional topographic and thematic information for the areas under investigation. These satellite images were specially enhanced and mostly printed at scales of 1 : 250 000. The combination of field work with visual and digital interpretation of the textural and spectral information from satellite images was the main method for refining and correcting existing geological maps and for geological mapping in new areas.

Landsat mosaics were created to cover the whole working area of the SFB 267 (Munier 1997). These mosaics served as GIS layers in the Andean-Information System and they were continuously extended and improved in response to new research outcomes and technical developments in satellite systems, especially for stereoscopic and radar data.

Logistical preparations for field work were supported by lists and previews of available road maps, geological maps and satellite imagery for each working area. A special service provided by the GIS-group was the preparation of enlarged and spectrally-enhanced satellite images. These image maps have proven to be very useful for orientation in remote areas and as a medium for the docu-

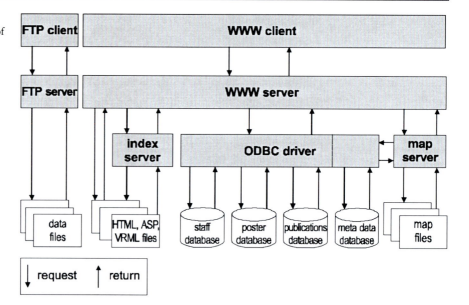

Fig. 26.4.
The client-server architecture of the SFB 267 web server and its simplified internal scheme

mentation of spatial information collected during the field work (e.g., identification of GPS locations for geometric rectification of satellite images).

### 26.4.1 Implementation

The hardware platform of the web server is a rather small PC with 128 MB RAM, 40 GB disk storage, Windows 2000 operating system and Microsoft Internet Information Server 4.0 as the web server. Prominent features are an index server for text retrieval and the connection to several databases via ODBC (open database connector). We use the ASP (active server pages) technology from Microsoft to build pages with dynamically-created contents at runtime. A map server (Autodesk$^©$) is implemented for providing GIS applications via the internet. This approach for developing the website was a very time consuming task, but, at present, it saves substantial maintenance effort and administration (Fig. 26.4).

### 26.4.2 The Current SFB 267 Website

The information system of the SFB 267 has always included various sources of interest to visiting foreign scientists. For a website containing more than 250 pages, navigation tools for finding information are a necessity; navigation bars on each page and a website map provide this support. Access to SFB 267 information is facilitated by full text retrieval and spatial search tools, enabling searches of the existing web pages and also the database. The full text search allows access to individual contact details, participating institutions, the publications database and the data catalog. Additional links connect personal home pages, online publications, posters and other items. Hot spots on maps lead users to a fast overview of research areas, and locations of geophysical transects or seismic networks.

A further improvement was the installation of a map server that provides tools for spatial searches and queries. Various information layers can be imported as ESRI® shapefiles and the user is able to display and overlay, for example, the 1 : 1 000 000 geological map of the Central Andes, the corresponding satellite mosaic, tectonic structure maps and point data. Interactive database queries and spatial search features provide details about faults, volcanoes, the geochemical composition of rock samples, locations with age determination, fission track data and other information (Fig. 26.5).

In 2004, we had about 25 000 visits from outside the host domain (fu-berlin.de) and about fifty percent of the total number of hits were from outside Germany. This popularity has been assisted by a careful selection of keywords (according to the guidelines of GeoGuide; see http://www.geo-guide.de/) in the website's document meta-information, which facilitates contact via web search engines.

### 26.4.3 The SFB Data Catalog

Integrated interpretation of the complex spatial data sets, close cooperation between different research groups, and frequent data requests from groups outside the SFB 267 all require a well structured and easily accessible database, and, at least, basic meta-information. Our database contains all kinds of data types: point data, line data, gridded data and raster images. Most data are available in

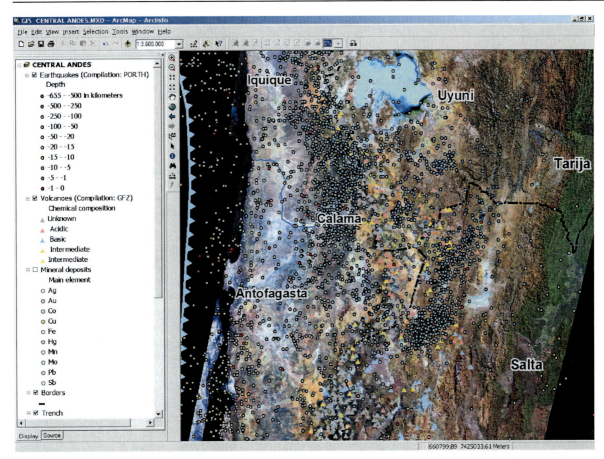

**Fig. 26.5.** Maps on the web: Satellite Mosaic overlain by point layers with classified rock age data and hypocenters of earthquakes. This example illustrates the map-server capability of the SFB 267 website

two different formats: simple ASCII format and the ARC/INFO® format. The underlying motivation for using these formats (Götze and Williams 1993) is maximum flexibility, allowing the storage of any type of geoscience data, as well as simplified data entry for the users who provide digital data files.

An internet database guide provides an overview of available data (URL: *http://userpage.fu-berlin.de/~data/*). This catalog offers information about data types, observation locations, data structure, file size and formats, data sources, and references. Each database entry is accompanied by a short extract from the main data file, content details and the names of those responsible for particular data sets.

The database not only contains data, modeling results, and other information assembled by members of the SFB 267, but also data files and links to data contributed by partners and other institutions throughout the world (e.g., the research group of L. B. Isacks at Cornell University, Ithaca, New York; the United States Geological Survey, Eros Data Center, Sioux Falls, South Dakota; and Scripps Institute of Oceanography, La Jolla, California). Also incorporated are previously published data sets, such as the volcano database of De Silva and Francis (1991). Most of our data can be used by researchers at other universities or governmental agencies for non-commercial purposes, free of charge.

## 26.5 International GIS Activities and Global Scientific Data Archives

The combined use of GIS and satellite images provides prospects for geoscientific research and geological mapping in remote areas which are difficult to access. Apart from the SFB database and information system, there are other information systems such as those at Cornell University (*http://atlas.geo.cornell.edu/geoid/metadata.html*) and the French BRGM (Bureau de recherches géologiques et minières) (*http://www.brgm.fr/Fichiers/LivJaune04/ress.pdf*). The SFB 267 archive provides one of the most complete data sets, particularly for geophysical data.

The internet databases frequently used by our research groups for creating synoptic interpretation maps are listed in Table 26.2.

**Table 26.2.** Internet databases frequently used for creating synoptic interpretation maps

| | | |
|---|---|---|
| Digital elevation: Shuttle Radar Topography Mission (SRTM) | University of Maryland and NASA Global Land Cover Facility, USA | http://srtm.usgs.gov |
| Digital elevation: GTOPO30, a global 30" DEM | USGS's EROS Data Center and others | http://lpdaac.usgs.gov/gtopo30/gtopo30.asp |
| Landsat-MSS, Landsat-TM and Landsat-ETM+ | University of Maryland and NASA Global Land Cover Facility, USA | http://geo.arc.nasa.gov/sge/landsat/landsat.html |
| KMS Global Marine Free-air Gravity Field | Kort- og Matrikelstyrelsen (KMS), Copenhagen, Denmark | http://manicoral.kms.dk/GRAVITY/ (Andersen and Knudsen 1998) |
| Global Gravity data in South America | GETECH, University of Leeds, UK | http://www.getech.com |

## 26.6 Some Results and Products – from Geo-Referenced Images to 3D Modeling

Combined analysis and interpretation of digitized data from various disciplines, such as geophysics, geology, or geochemistry, is essential for both numerical modeling and a synoptic view of the Earth. Integrated analysis, modeling, and visualization can aid in finding relationships between observations from "shallow" geology and "deeper" geophysics, and in defining structures and physical processes within the Earth. It may even help to distinguish contemporary and past tectonic styles.

In the following section we will present some examples of what has been achieved using the SFB 267 data catalog.

### 26.6.1 From Analog to Digital Formats – the Geological Map of the Northern SFB Transect

The fundamental geology of the Central Andes was compiled by K. Reutter to produce the 'Geological Map of the Central Andes 1 : 1 000 000' (Reutter and Götze 1994), and was designed for a print medium. Over many years, this map was the main reference and key publication for those working in the Andes from both the SFB and partner institutions in South America. The compilation was based on many sources (regional geological maps, satellite imagery) and used detailed geological investigations from the field and laboratory.

To enable the overlaying of different geophysical measurements, satellite images and topographic information within a GIS, a digital version of the map was needed. This digital version of the 1 : 1 000 000 map of the Central Andes established the basis for the Andean GIS of the SFB 267 (refer also to accompanying DVD) and provided an important tool for both the interpretation and surface control of deep-sounding geophysics. In its present format, modifications and corrections to the map can be easily made. The GIS can directly compare geological structures and lithological units with geophysical anomalies, and large-scale structures, derived from a combination of the available data, can be extracted.

### 26.6.2 Southern Andes GIS Compilations

With the expansion of SFB 267 activities to the Southern Andes between 36 and 42° S, the existing GIS database also had to be extended. Two compilations at different scales were made: detailed maps for field purposes and local studies, and regional maps for tectonic and geophysical analyses.

For field purposes, we purchased and digitized maps from the state geological surveys of Chile, locally updated by the geological departments of the University of Concepción, other Chilean Universities and from our own field observations (Melnick et al., in press; Melnick et al. 2003). This database is available only for the main SFB 267 area of interest (corresponding to the VIII, IX and X political regions of Chile). For the VIII and IX regions, the compilation was based on 1 : 250 000 scale geological maps produced by the Servicio Nacional de Geología y Minería (SERNAGEOMIN), and unpublished reports from our partners at the University of Concepción. The geological units of each published map from, or report of, these two regions were simplified to obtain uniform maps also at 1 : 250 000 scale. The units of the X Region were digitized from a compilation already made by the Geological Department of the University of Chile, at 1 : 100 000 scale (Cembrano et al. 1993).

As the geological nomenclature used for the VIII, IX and X regions of Chile differ significantly, the maps were not integrated and are presented separately. However, for the main cordillera between 37 and 39° S, the 1 : 250 000 scale maps from the VIII and IX regions, as well as an Argentinean map at 1 : 500 000 scale, compiled by the Neuquén Geological Service (Delpino and Deza 1995), were merged and the geology and structures updated by fieldwork done within the SFB 267 and the Southern Andes projects of the GeoForschungsZentrum in Potsdam.

For regional purposes, tectonic interpretation and correlation between the different geophysical data sets, we compiled the main morpho-tectonic units and regional structures between 36 and 42° S from the literature (e.g., Mpodozis and Ramos 1989) and our own field observations (Fig. 26.6). In order to properly map these units, we used the Landsat satellite mosaics from the SFB 267 data-

**Fig. 26.6.**
Simplified morphotectonic map of the Southern Andes based on a compilation of published data and new data obtained in the field

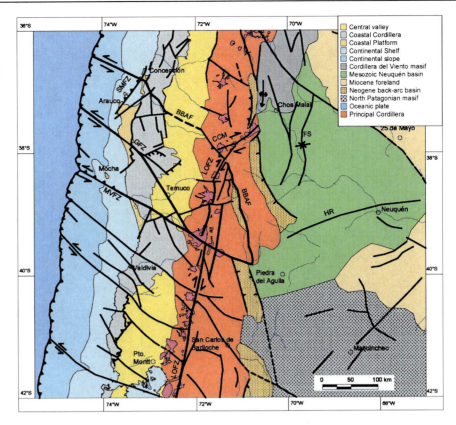

base and a digital elevation model derived from digitized Chilean and Argentinean topographic maps and, later, from the Shuttle Radar Topography Mission (SRTM) data set released in 2002 by NASA (Rabus et al. 2003).

A geological map was also compiled for the area covered by our morpho-tectonic maps by merging the geological maps of Chile (1 : 1 000 000 scale) and Argentina (1 : 2 500 000), compiled by the responsible national geological surveys (SERNAGEOMIN – Servicio Nacional de Geología y Minería 1980; and SEGEMAR – Servicio Geológico Minero Argentino 1996, respectively). The geological units were correlated and simplified to enable the presentation of the map at larger scales. A legend was added describing the lithological characteristics of each unit.

### 26.6.3 Geo-Referenced Images

Visualization techniques are most commonly used to qualitatively compare spatial data from various sources, and enable the superposition of different layers of information (Figs. 26.7 and 26.8). A prerequisite for the discussion of scientific results is a common map base containing geographic and topographic features that can be merged with other geoscientific data. For example, a map of important morpho-structural units (Reutter and Götze 1994) becomes clearer if present-day geographic features are visible for orientation (Fig. 26.7a,b).

An integrated map with age dating results shows that magmatic activity migrated to the east through the later part of Andean history (Fig. 26.7c). Another example of an integrated map is given in Fig. 26.7d, where the location of volcanoes of the Neogene-Holocene magmatic arc (De Silva and Francis 1991) are overlain on geophysical data: Induction vectors, derived from electromagnetic deep soundings (Schwarz et al. 1994), are almost perpendicular to the magmatic belt of the Western Cordillera and the zone of volcanism, which, where wide, is accompanied by a residual gravity low (see also Götze and MIGRA Group 1996; Götze and Krause 2002). This map hints at low density domains that may be related to partially molten material associated with recent volcanoes.

Structures and physical behavior within the Earth may be defined using GIS. This is achieved by analyzing seismic, gravity and magnetic data, and by applying filtering and other numerical data processing techniques. For example, the maps of Fig. 26.8 show how gradient methods (Fig. 26.8a) and shading techniques applied to the residual gravity field (Fig. 26.8b) can be used to distinguish pre-Andean (mainly NW to SE trends in Fig. 26.8b) from younger (N to S) structural trends. Comparisons between gravity lineaments, which were calculated from horizontal derivatives of the residual field, with geological lineaments may help to identify rock boundaries (Fig. 26.8a).

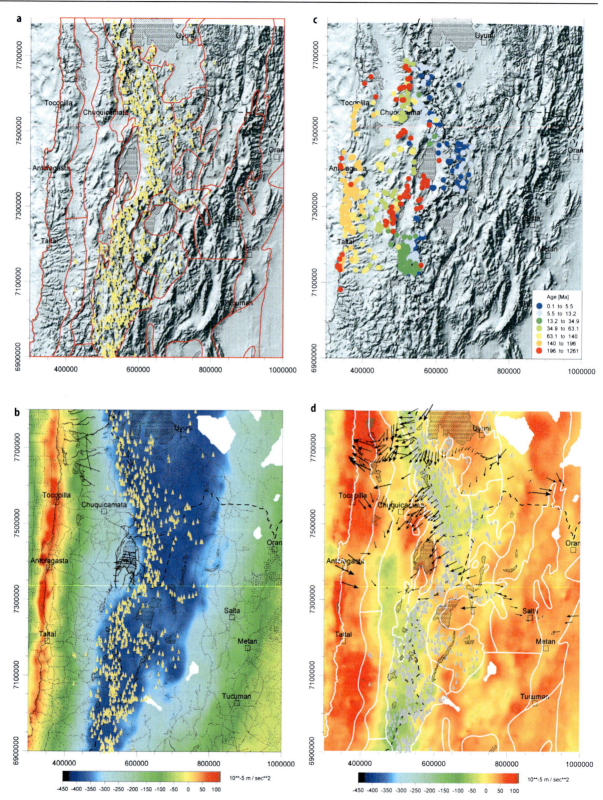

**Fig. 26.7.** Examples from the Central Andean database. **a** Shaded topography from the GTOPO30 global DEM, morphostructural units and volcanoes; **b** Bouguer anomaly (onshore), free air anomaly (offshore), overlain by the entire set of gravity stations; **c** shaded relief of the topography (gtopo30) and radiometric ages (compilation by Scheuber, pers. comm.); **d** residual gravity field, magnetotelluric induction arrows, and volcanoes

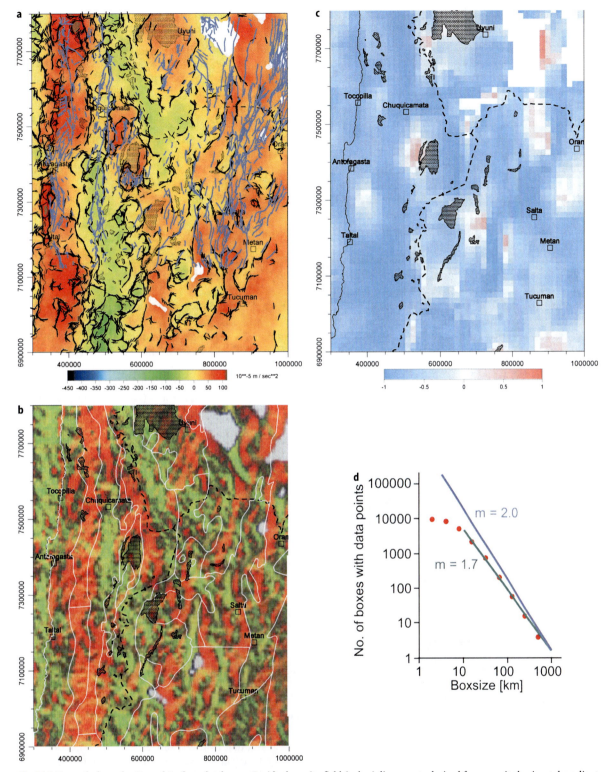

**Fig. 26.8.** Example from the Central Andean database. **a** Residual gravity field (*colors*), lineaments derived from gravity horizontal gradient analysis (*black lines*) and geological lineaments (Reutter and Götze 1994), which are only available north of UTM 7100000 (*blue lines*); **b** example from the Central Andean database: Color-shaded residual gravity field – directional analysis. *Green* colors indicate NW-SE trends, *red* colors indicate NE-SW trends, *white lines* show morphostructural units; **c** example from the Central Andean database: A Pearson correlation between topography and Bouguer anomalies shows a pronounced anti-correlation, which is normal in high mountains; **d** fractal dimension of the gravity station distribution in the Central Andes (the stations are shown in Fig. 26.7b)

In addition to a simple graphical superposition of different data sets, one often desires a quantitative (objective) measure of the correlation. Direct, numerical comparisons, such as the calculation of the spatial variation of correlation coefficient, enable us to quantify the relationship between different data sets. A measure of the size and direction of a linear relationship is the Pearson correlation coefficient. The correlation between the Bouguer anomaly and topography, shown in Fig. 26.8c, has been calculated by using moving-average statistics.

The calculation of this coefficient within moving windows is frequently used to investigate its dependency both on the average value and on the variability of digital spatial data sets; the size of the window depends on the average spacing between data locations. For irregularly spaced data sets, overlapping (with adjacent windows having some data in common) becomes a useful method to show the spatial variation in the relationship. In Fig. 26.8c, a window size of $15 \times 15$ km with a shift of 15 km has been used. As expected, high topographic altitude is anti-correlated with the Bouguer anomaly. However, areas of very strong and weak correlation can also be identified. These variations in the degree of correlation point to density inhomogeneities within the crust or upper mantle.

A further use of the GIS is for determining the quality of the station distribution. This can be derived from fractal dimensions as shown in Fig. 26.8d. Here, a box-counting method shows the spatial distribution of gravity stations (Fig. 26.7b) can be described by a fractal set with a fractal dimension of 1.7 (Mohr 1999; Mohr and Götze 1997).

The benefits of 3D visualization are indisputable. It is used frequently by interdisciplinary groups to aid such discussions as those about geometrical models of crustal structures or the distribution of various physical parameters. Figure 26.9 shows, as an example, the gravimetrically-derived geometry of the Nazca Plate in the Central Andes together with its modeled contribution to the gravity anomaly.

### 26.6.4 Remote Sensing: Results and Products

#### 26.6.4.1 Digital Satellite Mosaics

Remote sensing techniques and products have been used by almost all subprojects of the SFB 267. Geo-referenced and spectrally enhanced Landsat TM/ETM mosaics were created for the two main areas of interest: the Central Andes between 18 and 26° 30' S (Fig. 26.10) and the Southern Andes between 37° 30' and 43° 30' S. The digital mosaic of the Central Andes includes more than 30 Landsat TM/ETM scenes which were resampled on a $50 \times 50$ m grid and spectrally corrected via histogram matching for bands 7,4,1 (SWIR, NIR, VIS BLUE) and coded as RGB colors. For the Southern Andes, the digital mosaic contains about 20 scenes with a spatial resolution of $25 \times 25$ m with the same color composition as the Central Andean mosaic. For both images, detailed color matching was carried out to give a nearly seamless visual impression. These digital mosaics can be used in any GIS application and can be overlain by different information. The mosaics can also be enlarged to provide base maps at more detailed scales (up to 1 : 100 000 for the 25 m resolution data of the Southern mosaic).

#### 26.6.4.2 3D-Modeling Using Satellite Images and A DTM

Remote sensing data have been used for 3D modeling of shallow subsurface structures in order to close the gap between information from the surface (e.g., satellite imagery) and that from the subsurface (e.g., reflection seismic). This is only possible in areas with high relief where the 3D orientation of geometric features can be derived from reference points on a satellite image and corresponding elevation levels from the DTM.

For the area of the Uyuni-Kehnayani Fault Zone (UKFZ) located in the Departamento Lipez of Southern

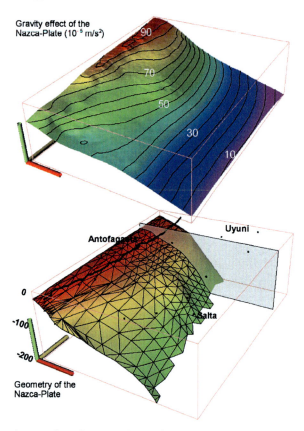

**Fig. 26.9.** Three-dimensional view of the Nazca Plate in the Central Andes, together with its modeled gravity anomaly

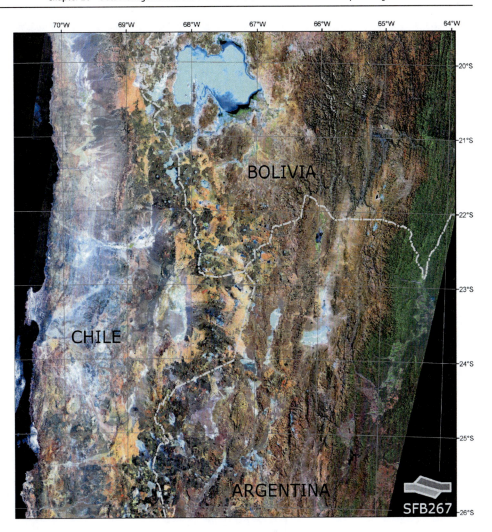

**Fig. 26.10.**
Landsat TM/ETM mosaic of the Central Andes

Bolivia, a 3D model was created from surface geology, strike and dip of bedding planes measured in the field and from aerial photographs, as well as from satellite image interpretation of major thrust and fold structures (Goni 2003). This kind of data extraction requires orthorectified aerial photographs, e.g., orthophotos, which were digitally generated for geometrically-exact measurements of geological features such as strike and dip of faults, and the direction of fold axes.

The central and most difficult task in the modeling process is the extension of surface data into the subsurface domain. The 3D modeling software package gOcad© (Mallet 2002) provides the flexible tool "Discrete Smoothing Interpolation Algorithm" (DSI) for constructing 3D surfaces (e.g., sediment layers, fault planes) and for incorporating uncertain information, such as the blurred seismic data near faults, or constraints like the thickness of a sedimentary sequence.

The resulting geometrical model of the UKFZ clearly shows the geometry of the main tectonic structures associated with this fault zone and their subsurface extension (Fig. 26.11). The non-folded, flat-lying Cenozoic formations east of the UKFZ form a distinct contrast to the upright to overturned orientation of the same formations west of the UKFZ. These overthrusted sediments form the infill of an inverse half-graben basin combined with a large anticline west of the UKFZ. The base of the Cenozoic sediments dips slightly to the north, which is consistent with known regional structures (Elger et al. 2001). Furthermore, the numerical "flattening" of the model gives a value of the finite shortening along the UKFZ that matches the values of previous work (Baby et al. 1997). The model was also validated by comparison with an independent data set from a seismic profile (Elger et al. 2001).

To summarize, the method produces a consistent geological model that can be used as a constraint for other geological methods such as balancing cross sections. The combination of satellite imagery and digital terrain models is an important aid for 3D modeling of structural features in remote areas where only limited field investigations are possible.

**Fig. 26.11.**
3D view of the structural model of the region between UKFZ and the Ines Anticline

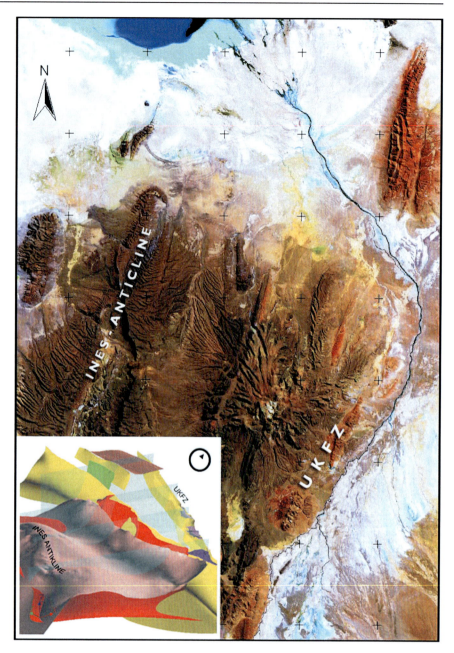

### 26.6.4.3 Integration of Data from Multiple Sources

Digital data from Landsat satellite systems have been successfully used for distinguishing geological units based on their spectral reflectance characteristics. Different mineral combinations can be identified and separated by the presence or absence of spectral features associated with ferric and ferrous iron, hydroxyl and carbonates.

While spectral information from satellite data describes the Earth's surface, geophysical data and potential field anomalies provide information about subsurface geological structures such as magmatic bodies. In combination with geological maps and tectonic features, a synoptic interpretation and digital classification can be achieved. The result is a favorability map indicating potential areas of significant rock alteration and mineralization. The description of typical spectral and geophysical characteristics of deposits are useful for providing the theoretical framework for data interpretation with respect to rock alteration or weathering. This type of GIS-based analysis improves the chances of identifying mineral deposits.

Mineral deposit models are required for data selection and data modeling and, at least, for deciding which spec-

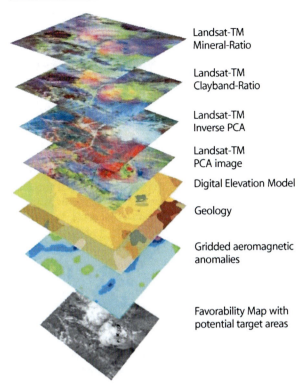

**Fig. 26.12.** Visualization of selected data compilations within the GIS

tral and structural features should be enhanced and analyzed. An example is given in Fig. 26.12, which shows important characteristics of the ore deposit at La Escondida derived from geology, magnetic and remote sensing data, together with the resulting favorability map that indicates potential areas of mineralization.

## 26.7 Learning with the SFB 267 Database: Some General Remarks On GIS

Many geoscientists modeling and using GIS software do not generally deal with geometric entities like 'triangles', 'cubes' or 'lines'. They are accustomed to think in terms of 'fault planes', 'rivers', 'geological formations' or 'increased reflectivity'. Most of us like to have access to the geometry of 'the Atacama Fault' in Chile, rather than to lines 17 through 218. This fact leads directly to the definition of geo-objects, which may be defined as 'existing geoscientific objects, composed by a name, geometry and a thematic description' (Breunig 1996).

For those users involved in complex interpretation of interdisciplinary and heterogeneous geodata, GIS methods and functions, which are based upon a geo-object data structure, become important. It is expected that object-oriented database management will affect most of the tasks related to geosciences (3D/4D models, visualization, validation, geostatistics and more).

Geologists and geophysicists deal with objects that are composed of different geometries and parameters (temperature, pressure, petrological composition, etc.) and, furthermore, they are linked to each other in complicated ways that vary through time. The MOHO, a well known and complicated entity, illustrates the problem. This MOHO geo-object can be defined as:

1. the lower boundary of bodies with crustal densities;
2. a line (2D) or surface (3D) across which the p-wave velocity increases abruptly;
3. the surface at which topography becomes isostatically balanced; or
4. the interface between rigid crustal material and weak lithospheric material.

A tool for object definition should provide the following functions:

1. The ability to define geo-objects with composite geometry, and
2. the ability to link several geometrically independent objects corresponding to the same geo-object.

Some of these features are implemented in the in-house software package IGMAS which has been used for modeling both the Central and Southern Andes (Kirchner et al. 1996; Kösters 1999; Wienecke 2002; Tašárová 2005).

### 26.7.1 Generalization and Simplification

The generalization of geo-data and information using automated computer methods is an ongoing problem in geology and geophysics, and should be the subject for future investigations.

Generalization of geo-information is not only necessary in areas where different modeling methods provide results at different scales, but it is also necessary in the same discipline. For example, a regional geological map represents the results or compilation of one or more detailed mapping projects, drilling data, chemical analyses or other general knowledge of the area. The compilation of this complete (detailed) data set into one map would not necessarily result in a clearly arranged, informative map, and it would be difficult to merge this information with the results of a regional geophysical model. Most of the information must be recompiled, generalized and simplified according to the needs of such varied end users as geologists, geophysicists, industry and administrators.

The same procedure is necessary if data from different sources have to be combined in a joint model that incorporates the results of different methods. Different geophysical disciplines provide modeling results at their own specific scales. For example, a high resolution 3D seismic survey differs in scale from a model of electrical resistivity at logarithmic scale, or from a 3D density model. Information and results at various scales make an interdisciplinary interpretation rather difficult, or at least ineffective. Based on the experience of this interdisciplinary research group, it was agreed that it was essential to use only a few standardized scales and map projections. This was mandatory for all participating scientific teams and greatly assisted with internal communication and a common interpretation.

### 26.7.2 Uncertainties

Models of geo-processes and the Earth's subsurface are based on a limited number of observations and measurements. Interpolation and extrapolation are standard methods used to fill the gaps between observations. As described above, in many cases, modeling and GIS functions enable access to information that is the result of special data processing or modeling routines. Thus, this information – or data, in a broader sense – is a representation of the real world reduced to a limited number of properties and a simple geometry. Therefore, we have to assume a method-dependent uncertainty in our models.

The number of publications on handling uncertainties or quantifying errors in the geosciences is rather limited (Goodchild and Gopal 1989; Myers 1997; Heuvelink 1998; Zhang and Goodchild 2002). Moreover, as far as we know, there is no general method that is able to model error propagation in a modeling process and to quantify the error in the final result. In many cases, however, the data error itself can be quantified (e.g., earthquake locations, analytical errors in geochemistry) and it should be a question of scientific integrity to present and visualize the spatial distribution and quantity of the hard data which are the basis for the model presented.

A more cumbersome way to handle errors is to develop several alternative models which can demonstrate the range of possible model solutions that match the observations within the tolerance allowed by measurement errors (Burger 1997). It is the responsibility of the interpreter(s), to introduce their own ideas or data to select the most probable model from of a sequence of alternatives. An additional problem is that the capabilities of computer techniques used for simultaneous visualization of 3D models and 3D uncertainties are still very limited. These additional data add another parameter to the original model, which causes numerous problems with respect to the clarity of model representation (Houlding 1994).

### 26.8 Conclusions and Call for Cooperation

This paper describes the way in which we have organized a large geoscience project in the direction of free information exchange. Given the mostly unsolved problems discussed in this paper, it is still futuristic to realize easy and automatic exchange of geo-data (geo-objects) between processing, modeling and interpretation software. Our approach does not constitute the state-of-the-art in modern online geo-data management via ODBC as does, for example, the geosciences information portal presently managed by the Australian Government (*http://www.ga.gov.au/map/national*). The website of Geoscience Australia is an excellent example of the benefits provided by freely offering different kinds of geo-data. In our case, the biggest limiting factor is that many scientists and scientific groups are reluctant to allow unlimited distribution of their scientific data and research results, particularly before publication. This reluctance tends to hinder complete and free distribution of information to the scientific community.

As the development of suitable data models and algorithms is an ongoing process, one should be prepared to design these models to be as general as possible. This would enable extension of their use to other problems. The required technology already exists, and it is up to geologists and geophysicists to define their requirements of the modeling software. An important prerequisite for interdisciplinary modeling processes is the development of optimal and general interfaces. This ambitious task can be achieved only by intensive and permanent discussions between computer scientists, geologists and geophysicists. The times in which different groups worked in isolation using only one method, are definitely past. Modeling with the aid of object-oriented GIS will ensure the constant feedback among scientists from various disciplines and across different methods.

### Acknowledgments and Partners

We would like to thank the SERNAGEOMIN and ENAP, both of Santiago, Chile, and YPFB, La Paz, REPSOL-YPF, Buenos Aires, for cooperation and release of data for the SFB 267 map projects. We also would like to thank CODELCO, Santiago, Chile, for fruitful cooperation and providing potential field data for 3D modeling. We also express our thanks to R. Hackney, A. Britt for her editorial and linguistic efforts, R. Hack and two anonymous reviewers for their helpful comments on this paper. This is a publication of the SFB 267, task groups Z2B, C6 and F4 of which the financial support of the Deutsche Forschungsgemeinschaft (DFG) and Freie Universität Berlin is thankfully acknowledged.

## References

Andersen OB, Knudsen P (1998) Global marine gravity field from the ERS-1 and GEOSAT geodetic mission altimetry. J Geophys Res 103(C4):8129–8137

Baby P, Sempere T, Oller J, Hérail G (1997) Evidence for major shortening on the eastern edge of the Bolivian Altiplano: the Calazaya nappe. Tectonophysics 205:155–169

Bode T, Cremers AB, Radetzki U, Shumilov S (2002) COBIDS: A component-based framework for the integration of geo-applications in a distributed spatial data infrastructure. In: Annual Conference of the International Association for Mathematical Geology (IAMG), September 15–20, 2002, Berlin, Germany

Breunig M (1996) Integration of spatial information for geo-information systems. Lecture Notes in Earth Sciences 61, Springer-Verlag, Berlin Heidelberg New York

Breunig M, Cremers AB, Götze H-J, Seidemann R, Schmidt S, Shumilov S, Siehl A (2000) Geologic mapping based on 3D models using an interoperable GIS. GIS J Spatial Inf Decision Making 13(2):12–18

Burger H (1997) 3D-Modeling of multi-layer deposits under uncertainty. Proc 3rd Ann Conf Int Assoc Math Geol Barcelona, Sept. 1997, 1:433–437

Cembrano J, Hervé F, Moreno H, Parada M, Thiele R, Varela J (1993) Compilación Geológica Actualizada de la X Región. Universidad de Chile, Facultad de Ciencias Físicas y Matemáticas, Departamento de Geología, scale 1:100.000

De Silva SL, Francis PW (1991) Volcanoes of the Central Andes. Springer-Verlag, Berlin Heidelberg New York, pp 199–212

Delpino D, Deza M (1995) Mapa geológico y de recursos minerales de la Provincia de Neuquén, 1:500 000. Secretaría de Minería, Servicio Geológico Neuquino, Argentina

Elger K, Oncken O, Kukowski N (2001) History of crustal shortening in the Southern Altiplano in Bolivia and its importance for plateau formation. SFB 267 Deformation processes in the Andes, Report for the research period 1999–2001, SP C1B, Berlin/Potsdam, pp 53–77

Encarnacao J, Bono P, Encarnacao M, Herzner W (1987) PC Graphics with GKS. Hanser Verlag

Goni P (2003) Ein geologisches Modell für den südlichen Altiplano bei 21° S (Bolivien), erstellt mit Fernerkundungs- und GIS-Methoden. Ph.D. thesis, Freie Universität Berlin, http://www.diss.fu-berlin.de/2003/162/

Goodchild M, Gopal S (eds) (1989) Accuracy of spatial databases. Taylor & Francis, London

Götze H-J, Krause S (2002) The Central Andean Gravity High, a relic of an old subduction complex? J S Am Earth Sci 14(8): 799–811

Götze H-J, MIGRA Group (1996) Group updates gravity database for Central Andes. EOS 77(19):183

Götze H-J, Monger JWH (1991) Global Geoscience Transects project: achievements and future goals. Episodes 14(2): 131–138

Götze H-J, Schmidt S (2000) Curso Intensivo sobre Procesamiento y Modelado de Campos Potenciales. Universidad De Buenos Aires, Departamento de Geología, Área de Geofísica, Facultad de Ciencias Exactas y Naturales, http://www.geophysik.uni-kiel.de/~sabine/BsAs2000/

Götze H-J, Williams RT (1993) Digitization of maps and associated geoscience data. AGU ILP Spec P 239:35

Götze H-J, Schmidt S, Barrio L, Kress PR (1993) Aplicaciones de la Computación gráfica en Geología y Geofísica (II). ILP Publication 240, Ottawa, pp 1–51

Heuvelink GBM (1998) Error propagation in environmental modelling with GIS. Taylor & Francis, London

Houlding SW (1994) Geoscience modeling – computer techniques for geological characterization. Springer-Verlag, Berlin Heidelberg New York

Kirchner A, Götze H-J, Schmitz M (1996) 3-D density modelling with seismic constraints in the Central Andes. Phys Chem Earth 21(4)289–293

Kösters M (1999) 3D-Dichtemodellierungen des Kontinentalrandes sowie quantitative Untersuchungen zur Isostasie und Rigidität der zentralen Anden. Berliner Geowiss Abh B32:135

Mallet JL (2002) Geomodeling. Cambridge University Press

Melnick D, Echtler H, Pineda V, Bohm M, Manzanares A, Vietor T (2003) Active faulting and northward growing of the Arauco Peninsula, Southern Chile (37°30' S). X Congreso Geológico Chileno, Concepción, Chile

Melnick D, Rosenau M, Folguera A, Echtler H (2006) Neogene tectonic evolution of the Neuquén Andes western flank. In: Kay SM, Ramos VA (eds) Evolution of an Andean margin: a tectonic and magmatic view from the Andes to the Neuquén Basin (35°–39° S lat). Geol Soc Am Spec P 407:73–95

Mohr S (1999) Strukturierung und Datenanalyse mit Hilfe von GIS-Methoden am Beispiel der Zentralanden. Berliner Geowiss Abh B33

Mohr S, Götze H-J (1997) The "Central Andes GIS", a comprehensive database for studies of deformation processes in the Central Andes. EOS electronic supplement, http://earth.agu.org/eos_elec/96350e.html

Mpodozis C, Ramos V (1989) The Andes of Chile and Argentina. In: Ericksen GE, Canas Pinochet MT, Reinemund JA (eds) Geology of the Andes and its relation to hydrocarbon and mineral resources. Circum-Pacific Council for Energy and Mineral Resources, Earth Science Series 11:59–89.

Munier K (1997) Landsat-TM Satellitenbildmosaik der zentralen Anden (20° S–26° S). Photogramm Fernerk Geoinf 6: 391–392

Myers JC (1997) Geostatistical error management. Van Norstrand Reinhold, New York

Omarini R, Götze H-J (eds) (1991) Global Geoscience Transect 6: Central Andean Transect, Nazca Plate to Chaco Plains, SW Pacific Ocean, N Chile and N Argentina. AGU, Washington DC

Ott N, Götze H-J, Schmidt S, Burger H, Alten M (2002) Meta geoinformation system facilitates use of complex data for study of Central Andes. EOS 83(34)

Rabus B, Eineder M, Roth A, Bamler R (2003) The shuttle radar topography mission – a new class of digital elevation models acquired by spaceborne radar. Photogram Remote Sens 57: 241–262

Reutter K-J, Götze H-J (1994) Comments on the geological and geophysical maps: In: Reutter K-J, Scheuber E, Wigger P (eds) Tectonics of the Southern Central Andes. Springer-Verlag, Berlin Heidelberg New York, pp 329–333

Schmidt S, Götze H-J (1999) Integration of data constraints and potential field modeling – an example from southern Lower Saxony, Germany. Phys Chem Earth A24(3):191–196

Schmidt S, Heubeck C, Götze H-J (2003) Die Erde – der dynamische Planet. Ein Lehrgang über geologische Prozesse. CD, Multimedia Hochschulservice Berlin GmbH

Schwarz G, Chong G, Krüger D, Martínez M, Masow M, Rath V, Viramonte J (1994) Crustal high conductivity zones in the Southern Central Andes. In: Reutter K-J, Scheuber E, Wigger P (eds) Tectonics of the Southern Central Andes. Springer-Verlag, Berlin Heidelberg New York

SEGEMAR (1996) Mapa Geológico de Argentina, digital version, scale 1:2.500.000. Secretaría de Geología y Minería Argentina, Buenos Aires, Argentina

SERNAGEOMIN (1980) Mapa Geológico de Chile, scale 1:1 000 000. Servicio Nacional de Geología y Minería, Santiago, Chile

Talwani M, Worzel, Landisman (1959) Rapid gravity computations for two-dimensional bodies with application to the Mendocino submarine fraction zone. J Geophys Res 64:49–59

Tašárová Z (2005) Gravity data analysis and interdisciplinary 3D modelling of a convergent plate margin (Chile, 36°–42°). Ph.D. thesis, Freie Universität Berlin, http://www.diss.fu-berlin.de/2005/19/

Wienecke S (2002) Homogenisierung und Interpretation des Schwerefeldes entlang der SALT-Traverse zwischen 36°–42° S. Diploma thesis, Freie Universität Berlin

Zhang J, Goodchild M (2002) Uncertainty in geographical information. Taylor & Francis, London

# Digital Geological Map of the Central Andes between 20°S and 26°S

Klaus-J. Reutter · Kerstin Munier

## 27.1 Origin of the Digital Geological Map

Between 1982 and 1993, a relatively small number of geoscientists from the Free University of Berlin and the Technical University of Berlin worked in collaboration with colleagues from Argentina, Bolivia, and Chile on Andean projects. In 1993, the much larger research program, the Collaborative Research Centre (SFB 267), "Deformation processes in the Central Andes", began. It was recognized very early in the life of the SFB that it would be useful to have a geological map covering the Central Andean segment where field research was to be undertaken. Thus, the program's geoscientists would have a starting point from which to base discussions, interpretations and plan further studies.

This was the origin of the "Geological map of the Central Andes between 20° S and 26° S", at a scale of 1 : 1 000 000. It was first completed in 1986 and later published in 1994 (Reutter et al. 1994b), together with Bouguer and isostatic gravity-anomaly maps at the same scale, in a volume dedicated to the tectonics of the south-Central Andes (Reutter et al. 1994a).

The most important feature of this geological map was its coverage across international borders and the compilation of existing national maps, which were at different scales, under a common legend. From its first rough draft until its publication, the map was continually updated whenever new geological data became available. For the presentation of the map, the Lambert Conical Projection was selected with a reference meridian at 67° W and the two standard parallels of 21° S and 25° S. Morphological and geographical data were taken from the Operational Aviation Chart 1 : 1 000 000 (Reutter and Götze 1994).

## 27.2 From Analog to Digital Mapping

As expected, the "Geological map of the Central Andes between 20° S and 26° S" proved to be very useful to the SFB 267, but in order to adapt it to the different requirements of the working groups, it was necessary to digitize it. This was done in two steps: (*1*) the original transparencies of the map-printing process were scanned and automatically vectorized; and (*2*) the lines were corrected manually, if necessary, and converted into polygons. Corresponding geological units were attached to these polygons and a legend was created so that the digital map could be visualized and further modifications could be made. The original Lambert Conformal Conical projection was transformed into UTM projection (zone 19).

There were certain obstacles to the further development of the geological map. The main one was a result of some incongruence with the satellite images of the area. This was most probably caused by the different projections used by the analog topographical base maps. The other was a lack of time and personnel for a thorough revision, although the digitized map made corrections easier. If corrections were introduced they usually concerned small areas.

Nevertheless, the 1994 geological map, in its original and digitized forms, continued to be an important discussion and interpretation tool for the geoscientists collaborating in the SFB 267 until its completion in 2005. In addition, there has been international request from non-SFB-affiliated geoscientists for the map and for permission to obtain and work with the digitized version.

This demand for the map and its possible further development by other groups were the deciding factors for including the digitized map on the DVD that accompanies this final volume of the SFB 267. It is included as an ArcGIS®-Project (version 8), and a digital map version for ArcReader® is also stored on the DVD (together with the necessary software). Thus, the map is now available to all who are interested in the geology of the Central Andes. The authors would be delighted to be occasionally informed by the users of this map about their experiences with it and its further evolution.

## References

Reutter KJ, Götze H-J (1994) Comments on the geological and geophysical maps. In: Reutter K-J, Scheuber E, Wigger P (eds) Tectonics of the Southern Central Andes. Springer-Verlag, Berlin Heidelberg New York

Reutter K-J, Scheuber E, Wigger P (eds) (1994a) Tectonics of the Southern Central Andes. Springer-Verlag, Berlin Heidelberg New York

Reutter KJ, Doebel R, Bogdanic T, Kley J (1994b) Geological map of the Central Andes between 20° S and 26° S, 1:1 000 000. In: Reutter K-J, Scheuber E, Wigger P (eds) Tectonics of the Southern Central Andes. Springer-Verlag, Berlin Heidelberg New York

# Chapter 28
# Bouguer and Isostatic Maps of the Central Andes

Sabine Schmidt · Hans-Jürgen Götze

## 28.1 Introduction

This volume and accompanying DVD contain the results of twenty years of gravity research in the Central Andes (18–44° S). Our research was concentrated on two regions: the central and south-Central Andes, located between 20 and 26° S and 37 and 42° S, respectively. The authors are of the opinion that a set of maps covering these two segments would be a useful accompaniment to this final volume. These gravity maps not only provide a better understanding of those papers dealing with the two regions, but also bridge the gap between surface geology and the lithospheric structures evidenced by these maps.

During the early research activities of the SFB 267, an initial set of gravity maps were compiled at a scale of 1:1 000 000 (Reutter and Götze 1994). These maps played an important role in interdisciplinary interpretation when the first results of geophysical fieldwork became available. Maps and cross sections containing both geological and geophysical data were drawn to common scales so that the results of the different disciplines within the research group could be directly compared.

These early gravity maps were completed and updated for publication in this volume. The resolution of the data sets enables both general overviews and more detailed maps of regional anomalies. It also provides sufficient insight into important local features at a practical map size. The original (non-confidential) and gridded data sets used for the gravity maps are included among the data files of the accompanying DVD.

## 28.2 The Gravity Maps of the Central Andes (18–44° S)

We provide station-complete, Bouguer gravity- and isostatic-anomaly maps (Vening-Meinesz model of isostasy).

### 28.2.1 Database

Prior to the SFB 267, the gravity research group began working in 1982 on a long-term gravity project in close cooperation with the Universidad Católica del Norte (G. Chong, Antofagasta, Chile), the Universidad de Chile (M. Araneda, Santiago, Chile) and the Universidad Nacionál de Salta (J. Viramonte and R. Omarini, Salta, Argentina), with additional support from governmental institutions and the Chilean and Argentinean oil and mining industries: We were aiming to construct a new gravity data base in the Andes for the area covering Northern Chile, north-western Argentina, Southern Chile and Northern Patagonia by collecting and updating already existing measurements and completing them by additional field measurements.

From 1982 to 2002, some 15 000 gravity measurements were made and with all the reprocessed, older industry data included, there is now a database of about 220 000 onshore gravity values available, with satellite altimetry gravity extra to this. The usual spacing of the stations was approximately 5 km along all passable tracks, although there is considerably higher station density in some local areas. The data sources used, besides our own measurements, are:

- Universidad de Chile/Servicios Geofísicos En Ingeniería (SEGMI), M. Araneda, Santiago, Chile
- Instituto Geográfico Militar (IGM), La Paz, (Bolivia)
- Geodetic Institute of the University of Buenos Aires
- REPSOL-YPF, Buenos Aires (National Oil Company of Argentina)
- YPFB, Santa Cruz (National Oil Company of Bolivia)
- Empresa Nacional de Petróleo (ENAP), Santiago, Chile
- Corporación Nacional del Cobre (Codelco), Santiago, Chile
- Shipborne data from offshore Chile (SPOC and CINCA experiments)
- Satellite altimetry (Spacecenter, formerly KMS) Copenhagen, Denmark (Andersen and Knudsen 1998); http://geodesy.spacecenter.dk

All observations are linked to the IGSN71 gravity datum using all available, local, official base stations. In the early years of the research, tidal corrections were neglected in the coastal area because of its small effect compared to the pronounced gravity anomalies observed; later, all stations were usually corrected using the in-house JAVA program DbGrav (Schmidt, pers. comm.).

Logistical problems and the large research area did not always allow us to repeat measurements at each station, which would, thus, determine the drift of the gravimeters. However, even when poor tracks were driven, the drift of the LaCoste & Romberg instruments (models D) rarely exceeded $0.1 \times 10^{-5}$ m s$^{-2}$ per day. It was also impossible to reliably determine geographic coordinates in all places, particularly between 1982 and 1992, as maps of sufficient resolution (e.g., 1:50 000 scale) did not yet exist for the entire area. At worst, the imprecise latitude has caused positioning errors amounting to approximately 0.5 km, which correspond to an error in the gravity anomaly of about $0.25 \times 10^{-5}$ m s$^{-2}$.

Of greater concern was the difficulty of height determination at each station. Only 37% of stations could be directly related to benchmarks such as leveling lines, trigonometric heights, height points of water pipes, railways and spot heights. Until 1992, when we first used a handheld geographical positioning system (GPS) to locate station coordinates directly in the field, stations had to be determined with barometers using available benchmarks as base stations. To improve the quality of the barometric measurements, time-dependent drift corrections were calculated in the manner usually employed for gravity measurements, using as many benchmarks and repeated measurements as possible. Moreover, the profiles of several days were combined in order to eliminate systematic errors. Altimeter types Wallace & Tiernan FA181 and Thommen 384.01.2 were used and the scales of these instruments were calibrated on leveling lines with an altitude difference of about 2 000 m. Error estimations showed that even in the worst case the accuracy was better than 20 m, giving an error in the Bouguer anomaly of about $4 \times 10^{-5}$ m s$^{-2}$, which is less than 1% of the regional anomaly. Later, from 1999 onwards, differential GPS provided both heights and horizontal coordinates more accurately.

### 28.2.2 Correction of Observed Data

The calculation of the gravity anomaly values was based on the following equations:

Bouguer anomaly: $BA = g_{abs} + dg_{top} + \delta g_{BPL} - \gamma_h$

Free-air anomaly: $FA = g_{abs} - \gamma_h$

where:

- $g_{abs}$    absolute gravity at station (measured)
- $dg_{top}$    topographic correction (density: 2.67 Mg m$^{-3}$) up to 167 km
- $\delta g_{BPL}$    Bouguer slab correction (density: 2.67 Mg m$^{-3}$) up to 167 km
- $\gamma_h$    normal gravity at station level $h$ (calculated)

The normal gravity was calculated according to IGSN71 using the International Gravity Formula of 1967. For topographic reduction, a method developed for gravity investigations in the European Alps was used (Ehrismann et al. 1966), after adapting it to the special conditions of the Central Andes. Calculations of topographic correction were based on the USGS digital elevation model *gtopo30* (*http://lpdaac.usgs.gov/gtopo30/gtopo30.asp*).

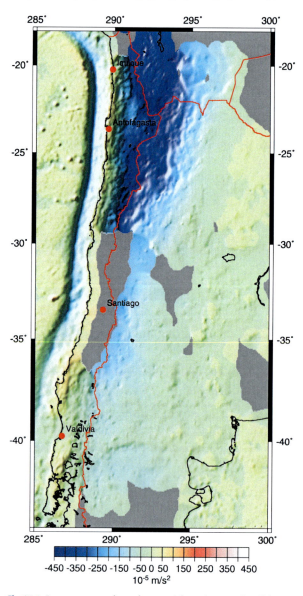

**Fig. 28.1.** Bouguer anomaly onshore and free-air anomaly offshore

## 28.2.3 Bouguer Anomaly

Figure 28.1 shows the calculated Bouguer anomaly map. As usual, in the offshore area, the Bouguer anomaly is replaced by the free-air anomaly. Therefore, a high correlation between trench topography and the gravity field is evident. Onshore, the gravity field decreases to a regional minimum of less than $-400 \times 10^{-5}$ m s$^{-2}$ in the Central Andes, mostly related to crustal thickening caused by isostatic compensation and tectonic processes. Figure 28.2 shows a similar map, but here the free-air anomaly offshore has been replaced by the Bouguer anomaly.

## 28.2.4 Isostatic Anomaly

The effect of isostatic compensation on topography was calculated using the regional compensation model of Vening-Meinesz with the following parameters: crustal density = 2.67 Mg m$^{-3}$, mantle density = 3.2 Mg m$^{-3}$, water density = 1.03 Mg m$^{-3}$, crustal thickness at sea level = 35 km, and crustal rigidity = 1.e23 Nm. The gravity effect of this model was calculated by using fast Fourier techniques and then subtracting it from the Bouguer anomaly (on- and offshore) at station level. The resulting anomaly is the isostatic residual field shown in Fig. 28.3.

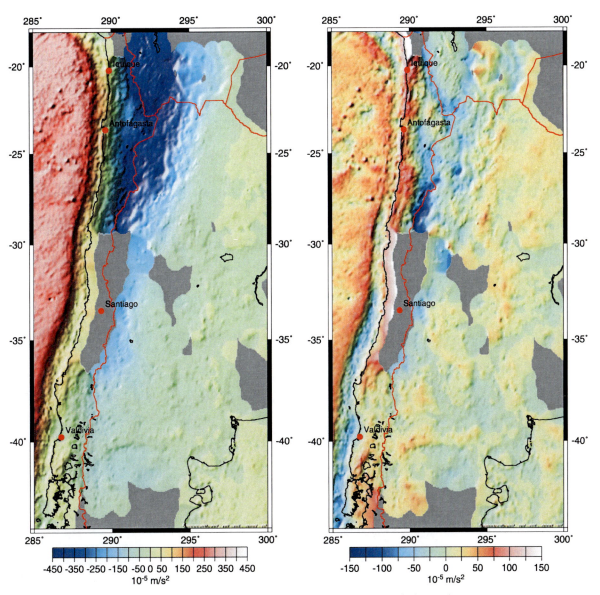

**Fig. 28.2.** Bouguer anomaly on- and offshore

**Fig. 28.3.** Isostatic residual anomaly

## 28.2.5 Data Files and Maps

The data files provided on the DVD consist of only non-confidential data.

File *grav_onshore.dat* contains 27 511 onshore stations with the following columns:

- Station ID
- Longitude (degree)
- Latitude (degree)
- Elevation (m)
- Observed gravity ($10^{-5}$ m s$^{-2}$, mGal)
- Topographic correction, onshore areas ($10^{-5}$ m s$^{-2}$, mGal), density = 2.67 Mg m$^{-3}$
- Topographic correction, offshore areas ($10^{-5}$ m s$^{-2}$, mGal), density = 1.64 Mg m$^{-3}$
- Free-air ($10^{-5}$ m s$^{-2}$, mGal)
- Bouguer ($10^{-5}$ m s$^{-2}$, mGal)

File *grav_offshore.dat* contains 236 718 offshore stations with the following columns:

- Longitude (degree)
- Latitude (degree)
- Bathymetry (m)
- Free-air ($10^{-5}$ m s$^{-2}$, mGal)
- Bouguer ($10^{-5}$ m s$^{-2}$, mGal)

The coordinates are geographical and owing to the high content of old data – the ellipsoid used is mostly unknown – are inconsistent.

Both the Bouguer/free-air gravity and the isostatic residual anomaly were interpolated onto a 0.05×0.05° grid (approx. 5×5 km) using the "Minimum Curvature Method" implemented by Briggs (1974). Please note that the grid values are based on the *complete* data set, which also includes confidential data.

The gridded data are provided in three different files:

1. File *bafa.xyz (bafa.ps)* contains the Bouguer anomaly onshore and the free-air anomaly offshore with the following columns:
   - Longitude (degree)
   - Latitude (degree)
   - Anomaly ($10^{-5}$ m s$^{-2}$, mGal)
2. File *baba.xyz (baba.ps)* contains the Bouguer anomaly on- and offshore with the following columns:
   - Longitude (degree)
   - Latitude (degree)
   - Anomaly ($10^{-5}$ m s$^{-2}$, mGal)
3. File *ia.xyz (ia.ps)* contains the isostatic residual anomaly on- and offshore with the following columns:
   - Longitude (degree)
   - Latitude (degree)
   - Anomaly ($10^{-5}$ m s$^{-2}$, mGal)

## References

Andersen OB, Knudsen P (1998) Global marine gravity field from the ERS-1 and GEOSAT geodetic mission altimetry. J Geophys Res 103(C4):8129–8137

Briggs IC (1974) Machine contouring using minimum curvature. Geophysics 39(1):39–48

Ehrismann W, Müller G, Rosenbach O, Sperlich N (1966) Topographic reduction of gravity measurements by the aid of digital computers. Boll Geofisica teorica ed applicata 8(29)

Reutter K-J, Götze H-J (1994) Comments on the geological and geophysical maps. In: Reutter K-J, Scheuber E, Wigger P (eds) Tectonics of the Southern Central Andes. Springer-Verlag, Berlin Heidelberg New York, pp 329–333

# Digital Geological Map of the Southern and Central Puna Plateau, NW Argentina

Wolfgang B. W. Schnurr · Andreas Risse · Robert B. Trumbull · Kerstin Munier

The digital geological map of the central and Southern Puna presented here is a new compilation of geological data and satellite imagery that covers an area of about 60 000 km² in the provinces of Salta und Catamarca, NW Argentina. The map was compiled at the scale of 1 : 250 000 and is based on published geological maps of the Servicio Geológico y Minero Argentino, SEGEMAR, (Blasco et al. 1996; Hongn et al. 1998; Martínez 1995; Salfity et al. 1998; Seggiaro et al. 2000, 2004). Additional information is incorporated from satellite interpretations (Landsat 5 Thematic Mapper Scenes 19-25 and 19-20) and from field studies undertaken in several projects within the SFB-267 program. Most of the latter work focused on the Salar de Antofalla region and the results were incorporated into an unpublished digital geological map by Erpenstein et al. (1996), from which a reduced version was published in paper form by Kraemer et al. (1999).

Since the publication of Kraemer et al. (1999) other geological projects in the Antofalla area were completed (Voss 2000; Adelmann 2001; Schnurr 2001) and SEGEMAR released two new geological maps at 1 : 250 000 scale (Seggiaro et al. 2000, 2004). The prime motivation for compiling a new digital map of the Puna region was to provide an updated and expanded basis for research projects within the SFB-267 program working in the area. Because of an emphasis of these projects on volcanism, the lithologic attributes and map legend have been modified from the source maps in order to better bring out compositional and age groupings of the Cenozoic volcanic units (Fig. 29.1, next page). The northern limit of the digital map is at 24° S and/or the Chilean border and the map extends to 27° S. The eastern limit is 67° 15' W and the western border is 68° 30' W and/or the Chilean border.

The digital map was produced with the ArcGIS software (version 8.3). The data DVD attached to this volume contains all associated files and full documentation explaining technical details of the compilations and map projections used, lists of files and references incorporated in the compilation, an explanation of how source-map limitations and inconsistencies were resolved, and finally some simple user instructions for viewing the map with ArcGIS.

## Acknowledgments

We wish to thank our colleagues in Argentina, Raul Becchio and Raul Seggiaro, for their help in resolving compilation problems and for providing preliminary versions of the 1 : 250 000 Salar de Antofalla map. Financial support for the first author during work on this map was provided by SFB-267 project G2. Thanks to Peter Pilz, Heinz Burger and Kai Hahne for helpful technical and geological advice.

## References

Adelmann D (2001) Känozoische Beckenentwicklung des zentralandinen Puna-Plateaus (NW-Argentinien). Das Gebiet um den Salar de Antofalla und ein Vergleich zur nördlichen Puna. Berliner Geowiss Abh A210

Blasco G, Zappettini E, Hongn F (1996) Hoja Geológica 2566-I, San Antonio de los Cobres, provincias de Salta y Jujuy, 1:250,000. Servicio Geológico Minero de Argentina, Buenos Aires, Boletín 217

Erpenstein K, Görler K, Heim B, Kraemer B, Prokoph A, Voss R (1996) Sedimentation, Tektonik und Vulkanismus im Gebiet des Salar de Antofalla, südliche Puna (NW-Argentinien). Unpublished report, SFB-267 project D1b, Berlin, pp 755–808

Hongn FD, Monaldi CR, Alonso RN, Gonzalez RE, Seggiaro RE (1998) Hoja Geológica 2566-III-Cachi. Provincias de Salta y Catamarca. Servicio Geológico Minero de Argentina, Buenos Aires, Boletín 248

Kraemer B, Adelmann D, Alten M, Schnurr W, Erpenstein K, Kiefer E, Van den Bogaard P, Görler K (1999) Incorporation of the Paleogene foreland into the Neogene Puna Plateau: the Salar de Antofalla area, NW Argentina. J S Am Earth Sci 12:157–182

Martínez L (1995) Mapa Geológico de la Provincia de Catamarca, Escala 1 : 500 000. Servicio Geológico Minero de Argentina, Buenos Aires

Salfity JA, Monaldi CR (1998) Mapa Geológico de la Provincia de Salta, Escala 1 : 500 000. Servicio Geológico Minero de Argentina, Buenos Aires

Schnurr W (2001) Zur Geochemie und Genese neogener und quartärer felsischer Vulkanite in den südlichen Zentralanden (25°–27° S und 67°–69° W). Berliner Geowiss Abh A211

Seggiaro R, Hongn F, Castillo A, Pereira F, Villegas D, Martinez L (2000) Hoja Geológica 2769-II Paso de San Francisco, Escala 1 : 250 000. Servicio Geológico Minero de Argentina, Buenos Aires, Boletin 294

Seggiaro RE, Becchio R, Schnurr W, Adelmann D, Erpenstein K (2004) Hoja 2569-IV, Antofalla, Escala 1 : 250 000. Servicio Geológico Minero de Argentina, Buenos Aires, Boletin 343

Voss R (2000) Die Geologie der Region um den südlichen Salar de Antofalla (NW-Argentinien). Berliner Geowiss Abh A 208

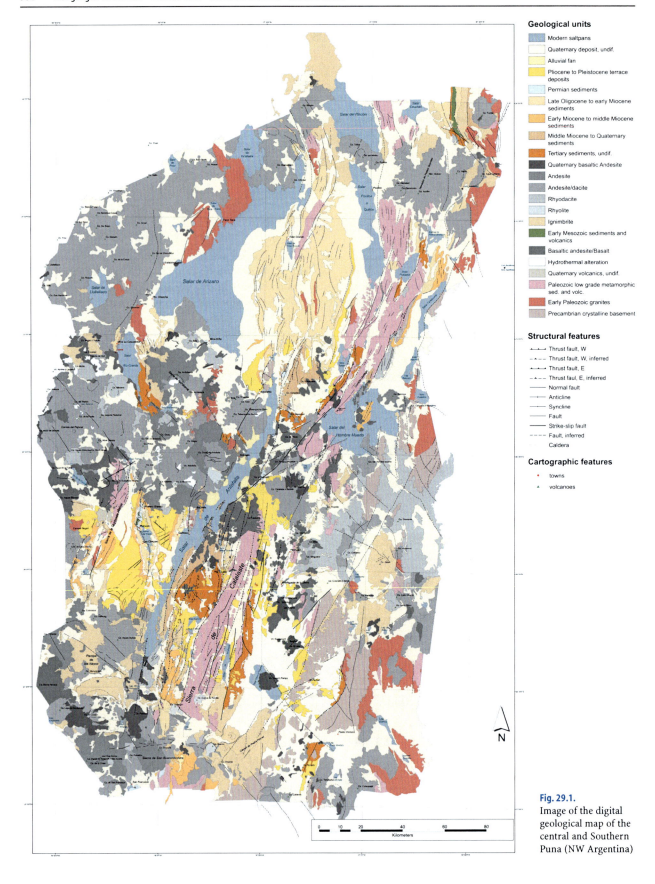

**Fig. 29.1.**
Image of the digital geological map of the central and Southern Puna (NW Argentina)

# Morphotectonic and Geologic Digital Map Compilations of the South-Central Andes (36°–42°S)

Daniel Melnick · Helmut P. Echtler

## 30.1 Introduction

The need of the multidisciplinary SFB-267 project of integrated, large-scale tectonic and geologic base information for the specific and localized subprojects led to produce the two map compilations presented here. Both maps have been conceived at about 1 : 4 000 000 scale and cover the region from 36° to 42° S, between the Pacific Ocean and 67° W in the Argentine pampas (Figs. 30.1 and 30.2).

The first map includes the main morphotectonic units and Cenozoic fault zones as well as major tectonic elements. Off-shore shelf basins and limits of the continental slope and trench plain (Fig. 30.1) are also integrated. These morphotectonic units can be considered as the continental-scale subdivisions of the Andean margin, first described by naturalists like Charles Darwin and Ignacy Domeyko (Darwin 1839; Domeyko 1846). Their tectonic significance was further developed among others by Brüggen (1950), Plafker and Savage (1970), Gansser (1973), Jordan et al. (1983), Mpodozis and Ramos (1989). The off-shore structure of the margin and shelf basins has been integrated from hydrocarbon exploration studies made by Empresa Nacional del Petróleo (ENAP), the Chilean state oil company (Mordojovich 1981; González 1989), interpretation of ENAP seismic reflection profiles and boreholes (Melnick and Echtler 2006) and SPOC multibeam bathymetry (Reichert et al. 2002). The second map integrates the main regional-scale stratigraphic, intrusive, and metamorphic units (Fig. 30.2) comprised within each margin-scale morphotectonic province.

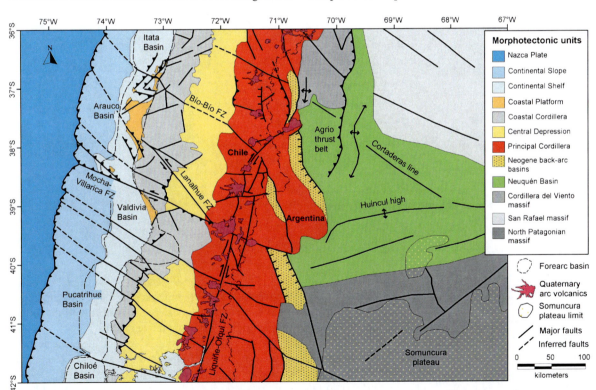

**Fig. 30.1.** Morphotectonic units and major faults of the south-Central Andes

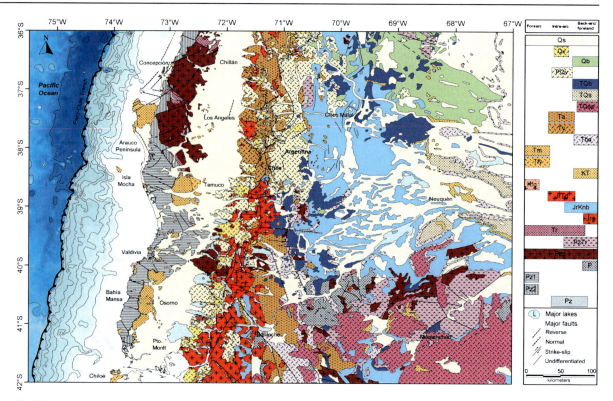

**Fig. 30.2.** Geologic map compilation of the south-Central Andes

**Table 30.1.**
Area and mean and maximum elevations of morphotectonic units derived from SRTM data

| Morphotectonic unit | Area (km²) | Max elevation (m) | Mean elevation (m) |
|---|---|---|---|
| Nazca Plate | 40 890 | −3 081 | −4 355 |
| Continental Slope | 42 690 | −140 | −1 948 |
| Continental Shelf | 35 170 | 0 | −163 |
| Coastal Platform | 3 200 | 382 | 83 |
| Coastal Cordillera | 32 430 | 1 442 | 289 |
| Central Depression | 37 520 | 1 014 | 189 |
| Principal Cordillera | 79 990 | 3 556 | 1 235 |
| Neogene back-arc basin | 11 550 | 1 953 | 1 070 |
| Neuquén Basin | 80 730 | 2 428 | 706 |
| Cordillera del Viento Massif | 11 470 | 4 616 | 2 123 |
| San Rafael Massif | 49 220 | 3 642 | 779 |
| North Patagonian Massif | 65 430 | 1 850 | 988 |

## 30.2 Methods and Data Sources

The morphotectonic units were mapped using the NASA-SRTM digital elevation model, which has a ground resolution of 3 arc sec (Slater et al. 2006). The main criteria used to define the map units were morphology, topographic gradients, geological evolution, and structural boundaries. The precise limit of each unit is based on own field observations, interpretation of the digital elevation model and locally remote sensing data, and published data. Table 30.1 includes area and mean and maximum elevations of the morphotectonic units.

The geologic compilation is based on the available maps of the Chilean and Argentinean Geological Surveys purchased in digital versions at 1 : 1 000 000 and 1 : 2 500 000 scale, respectively (SERNAGEOMIN 1982, 2003; SEGEMAR 1995). Both maps were simplified by merging units based

## Chapter 30 · Morphotectonic and Geologic Digital Map Compilations of the South-Central Andes (36°–42°S)

on chronostratigraphic as well as tectonic considerations; for example, the late Cretaceous to Pliocene sedimentary infill of the Andean fore-arc basins was grouped into 'Marine and continental fore-arc basins'; Jurassic to Miocene intrusive rocks of the Main Cordillera were integrated into 'North Patagonian Batholith magmatic arc root' (e.g., Hervé 1994).

The structural information supplied by both geological surveys has been actualized by studies developed in the frame of the SFB-267, its local partner universities and institutions, and other parallel research groups in the Andes (e.g., Hervé 1988, 1994; Lavenu and Cembrano 1999; Muñoz et al. 2000; Jordan et al. 2001; Radic et al. 2002; Melnick et al. 2002, 2003, 2006a–c; Rosenau 2004; Folguera et al. 2004; Glodny et al. 2005; Ramos and Folguera 2005; Zapata and Folguera 2005; Rosenau et al. 2006). The major fault zones have been revised, updated and mapped using the SRTM digital elevation model. The legend of the geologic map has been organized in order to show the tectonic position of each unit as well as its time span (Fig. 30.3). This facilitates i.e., visualizing shifts in the locus of magmatic activity, sediment transport and distribution, and basin development. Table 30.2 integrates simplified descriptions and age ranges of each unit. All the GIS tables are available in the accompanying DVD in geodetic projection and WGS 1984 datum.

## Acknowledgments

This work benefits from discussions with the SFB-267 community and Berlin-Potsdam geosciences groups as well as colleagues from Chilean and Argentinean institutions.

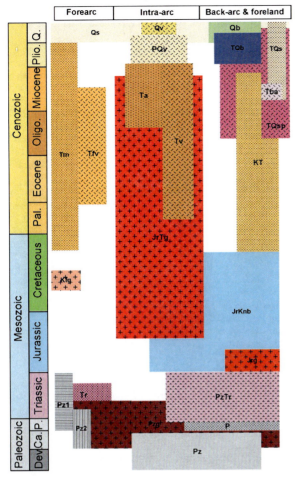

**Fig. 30.3.** Space-time diagram of the geologic units

**Table 30.2.** Description and age range of the geologic units in Fig. 30.2

| Code | Description | Age range |
|---|---|---|
| Qs | Undifferentiated sediments | Quaternary |
| Qv | Volcanic arc edifices | Quaternary |
| Qb | Back-arc plateau volcanics | Quaternary |
| PQv | Volcanic arc edifices and plateaus | Pliocene–Pleistocene |
| TQb | Back-arc volcanics | Tertiary–Quaternary |
| TQs | Foreland synorogenic conglomerates | Tertiary–Quaternary |
| TQsp | Somuncura plateau volcanics | Oligocene–Quaternary |
| Ta | Sedimentary intra-arc basins | Oligocene–Miocene |
| Tv | Volcanic intra-arc basins | Paleocene–Miocene |
| Tba | Back-arc volcanics | Miocene |
| Tm | Marine and continental forearc basins | Late Cretaceous–Pliocene |
| Tfv | Volcanic and continental forearc basins | Oligocene–Miocene |
| KT | Neuquén Basin marine and continental foreland sequences | Late Cretaceous–Tertiary |
| Kfg | Forearc intrusions | Middle Cretaceous |
| JrTg | North Patagonian Batholith magmatic arc root | Jurassic–Miocene |
| JrKnb | Neuquén Basin marine, continental, and volcanic sequences | Jurassic–Cretaceous |
| Jrg | Patagonian intrusions | Jurassic |
| Tr | Rift basins marine, continental, and volcanic sequences | Triassic |
| PzTr | Choyoi group large igenous province | Permian–Triassic |
| Pzg | Coastal Batholith and undifferentiated intrusions | Paleozoic–early Triassic |
| Pz | Metasedimentary Neuquén basement | Paleozoic |
| Pz1 | Western Series high $P$ metasediments and ultramafics | Permian–Triassic |
| Pz2 | Eastern series high $T$ metasediments | Carboniferous–Triassic |
| P | North Patagonian massif metasediments | Permian |

## References

Brüggen J (1950) Fundamentos de la Geología de Chile. Instituto Geográfico Militar, Santiago, Chile

Darwin C (1839) Journal and Remarks 1832-1836. Volume III of Narrative of the surveying voyages of His Majesty's ships Adventure and Beagle between the years 1826 and 1836, describing their examination of the southern shores of South America, and the Beagle's circum navigation. London, pp 370-381

Domeyko I (1846) Memoria sobre la Estructura Geológica de Chile en la Latitud de Concepción, desde la Bahía de Talcahuano hasta la Cumbre de la Cordillera de Pichachén y Descripción del Volcán Antuco. In: Domeyko I (ed) Geología. Imprenta Cervantes, Santiago, pp 123-172

Folguera A, Ramos VA, Hermanns RL, Naranjo J (2004) Neotectonics in the foothills of the southernmost Central Andes (37°-38° S): evidence of strike-slip displacement along the Antiñir-Copahue fault zone. Tectonics 23(5):1-23

Gansser A (1973) Facts and theories on the Andes. J Geol Soc London 129:93-131

Glodny J, Lohrmann J, Echtler H, Gräfe K, Seifert W, Collao S, Figueroa O (2005) Internal dynamics of a paleoaccretionary wedge: insights from combined isotope tectonochronology and sandbox modelling of the South-Central Chilean forearc. Earth Planet Sci Lett 231(1-2):23-39

González E (1989) Hydrocarbon resources in the coastal zone of Chile. In: Ericksen GE, Cañas Pinochet MT, Reinemund JA (eds) Geology of the Andes and its relation to hydrocarbon and mineral resources. Circum-Pacific Council for Energy and Mineral Resources, Texas, pp 383-404

Hervé F (1988) Late Paleozoic subduction and accretion in Southern Chile. Episodes 11(3):183-188

Hervé F (1994) The Southern Andes between 39° and 44° S latitude: the geological signature of a transpressive tectonic regime related to a magmatic arc. In: Reutter K-J, Scheuber E, Wigger P (eds) Tectonics of the Southern Central Andes. Springer-Verlag, Berlin Heidelberg New York, pp 243-248

Jordan TE, Isacks BL, Allmendinger RW, Brewer JA, Ramos VA, Ando CJ (1983) Andean tectonics related to geometry of subducted Nazca plate. Geol Soc Am Bull 94(3):341-361

Jordan TE, Burns WM, Veiga R, Pángaro F, Copeland P, Kelley S, Mpodozis C (2001) Extension and basin formation in the Southern Andes caused by increased convergence rate: a mid-Cenozoic trigger for the Andes. Tectonics 20(3):308-324

Lavenu A, Cembrano J (1999) Compressional- and transpressional-stress pattern for Pliocene and Quaternary brittle deformation in fore arc and intra-arc zones (Andes of Central and Southern Chile). J Struct Geol 21(12):1669-1691

Melnick D, Echtler HP (2006) Inversion of forearc basins in south-central Chile caused by rapid glacial age trench fill. Geology 34(9):709-712

Melnick D, Folguera A, Rosenau M, Echtler H, Potent S (2002) Tectonics from the northern segment of the Liquiñe-Ofqui fault system (37°-39° S), Patagonian Andes. In: 5$^{th}$ International Symposium on Andean Geodynamics, Toulouse, pp 413-417

Melnick D, Sanchez M, Echtler H, Pineda V (2003) Structural geology of Mocha Island, south-central Chile (38°30' S, 74° W): regional tectonic implicances. X Congreso Geológico Chileno, Concepción

Melnick D, Bookhagen B, Echtler H, Strecker M (2006a) Coastal deformation and great subduction earthquakes, Isla Santa María, Chile (37° S). Geol Soc Am Bull 118(9), in press

Melnick D, Folguera A, Ramos VA (2006b) Structural control on arc volcanism: the Caviahue-Copahue complex, Central to Patagonian Andes transition (38° S). J S Am Earth Sci 20(5), in press

Melnick D, Rosenau M, Folguera A, Echtler H (2006c) Neogene tectonic evolution of the Neuquén Andes western flank (37°-39° S), In: Kay SM, Ramos VA (eds) Evolution of an Andean margin: a tectonic and magmatic view from the Andes to the Neuquén Basin (35°-39° S lat). Geol Soc Am Spec P 407:73-95

Mordojovich C (1981) Sedimentary basins of Chilean Pacific Offshore. In: Halbouty MT (ed) Energy resources of the Pacific Region. AAPG Stud Geol, pp 63-82

Mpodozis C, Ramos V (1989) The Andes of Chile and Argentina. In: Ericksen GE, Cañas Pinochet MT, Reinemund JA (eds) Geology of the Andes and its relation to hydrocarbon and mineral resources. Circum-Pacific Council for Energy and Mineral Resources, pp 59-90

Muñoz J, Troncoso R, Duhart P, Crignola P, Farmer L, Stern CR (2000) The relation of the mid-Tertiary coastal magmatic belt in South-Central Chile to the late Oligocene increase in plate convergence rate. Rev Geol Chile 27(2):177-203

Plafker G, Savage JC (1970) Mechanism of the Chilean earthquake of May 21 and 22, 1960. Geol Soc Am Bull 81:1001-1030

Radic J, Rojas L, Carpinelli A, Zurita E (2002) Evolución tectónica de la Cuenca Terciaria de Cura Mallín, región cordillerana chileno-argentina (36°30'-39° S). In: Actas 15° Congreso Geológico Argentino, Calafate, pp 233-237

Ramos VA, Folguera A (2005) Tectonic evolution of the Andes of Neuquén: constraints derived from the magmatic arc and foreland deformation. In: Veiga GD, Spalletti LA, Howell JA, Schwarz E (eds) The Neuquén Basin, Argentina: a case study in sequence stratigraphy and basin dynamics. Geol Soc London Spec Pub, pp 15-35

Reichert C, Schreckenberger B, SPOC team (2002) Fahrtbericht SONNE-Fahrt SO-161, Leg 2&3 SPOC –Subduktionsprozesse vor Chile – BMBF-Forschungsvorhaben 03G0161A. Bundesanstalt für Geowissenschaften und Rohstoffe, Hannover

Rosenau M (2004) Tectonics of the Southern Andean intra-arc zone (38°-42° S). Ph.D. thesis, Freie Universität Berlin

Rosenau M, Melnick D, Echtler H (2006) Kinematic constraints on intra-arc shear and strain partitioning in the Southern Andes between 38° S and 42° S latitude. Tectonics 25(4), TC4013

SEGEMAR (1995) Geologic map of Argentina, digital version 1 : 2 500 000

SERNAGEOMIN (1982) Geologic map of Chile, scale 1 : 1 000 000

SERNAGEOMIN (2003) Geologic map of Chile, digital version, scale 1 : 1 000 000

Slater JA, Garvey G, Johnston C, Haase J, Heady B, Kroenung G, Little J (2006) The SRTM data "finishing" process and products. Photogramm Engin Remote Sens 72(3):237-247

Zapata T, Folguera A (2005) Tectonic evolution of the Andean fold and thrust belt of the Southern Neuquén Basin, Argentina. In: Veiga GD, Spalletti LA, Howell JA, Schwarz E (eds) The Neuquén Basin, Argentina: a case study in sequence stratigraphy and basin dynamics. Geol Soc London Spec Pub, pp 37-56

# Introduction to the Attached DVD

Kerstin Munier · Jörn Levenhagen · Heinz Burger

The attached DVD contains complementary data (e.g., high resolution images and computer animation) related to the contributions in this final volume for the Collaborative Research Centre (SFB 267), "Deformation Processes in the Andes". It also contains additional data, such as GIS projects and movies, which cannot be published using print media. As it was our goal to make the DVD interesting for both scientists and the general public, we have also included virtual flights across the Central Andes and photos from the field work.

All necessary software for viewing the GIS projects and pdf files is also stored on the DVD, in versions for Linux, Windows and Macintosh, in a separate directory. Navigation for the DVD begins automatically when it is inserted into the DVD drive.

The ArcGIS® projects contain the digital geological maps of those parts of the Central Andes (ArcGIS® v. 8) and Southern Andes (ArcGIS® v. 9) that were selected as areas of interest during the life of the SFB 267.

Some digital data sets that may be of special interest to geoscientists working in these areas are:

- A compilation of the ages and locations of Cenozoic volcanic centers in the Central Andes (including a list of references used for the volcano compilation).
- A deformation databank.
- A compilation of 220000 gravity values (location and measurement values).
- Maps of magnetotelluric stations and other geophysical field campaigns.
- A list of published doctoral theses completed during the period of the SFB 267.
- The publication list of the SFB 267.

Last but not least, there is also a list of the SFB 267's collaborative partners. Their contribution to the success of this research program is immeasurable.